Major Geological, Climatic, and Biological Events

There were extensive and repeated periods of glaciation in the Northern Hemisphere, and the Neogene midlatitude savanna faunas became extinct. The genus *Homo* appeared and hominins expanded throughout the Old World. Near the end of the interval, hominins reached the New World. There was an extinction of many large mammals, especially in the New World and in Australia. The remaining carnivorous flightless birds also became extinct.

Cooler and more arid climates persisted, resulting from mountain uplift and the formation of the Isthmus of Panama near the end of the interval. The Arctic ice cap had formed by the end of the interval. The first grasslands spread in the middle latitudes. Modern families of mammals and birds radiated, and marine mammals and birds diversified in the oceans. The first hominins were seen near the end of the interval.

Global climate was warm in the early part of the interval, with forests above the Arctic Circle, but later temperatures fell in the higher latitudes, with the formation of the Antarctic ice cap. Mammals diversified into larger body sizes and a greater variety of adaptive types, including predators and herbivores. Mammal radiations included archaic forms, now extinct, and the earliest members of living orders. Giant carnivorous flightless birds were common as predators.

Further separation of the continents occurred, including the breakup of the southern continent, Gondwana. Teleost fishes radiated, and marine reptiles flourished. Angiosperms first appeared and rapidly diversified to become the dominant land plants by the end of the period. Dinosaurs remained the dominant tetrapods, but small mammals, including the first therians, diversified. Birds and pterosaurs coexisted, and the first snakes appeared. A major mass extinction at the end of the period, defining the end of the Mesozoic, claimed dinosaurs, pterosaurs, and marine reptiles, as well as many marine invertebrates.

The world continent began to break up, with the formation of the Atlantic Ocean. Marine invertebrates began to take on a modern aspect with the diversification of predators, modern sharks and rays appeared, and marine reptiles diversified. Conifers and other gymnosperms were the dominant terrestrial vegetation, and insects diversified. Dinosaurs diversified while mammals remained small and relatively inconspicuous. The first birds, lizards, and salamanders were seen at the end of the period.

The world continent was relatively high, with few shallow seas. No evidence of glaciation existed, and the interior of the continent was arid. Conifers replaced seed fern terrestrial vegetation in the later part of the period. Non-mammalian synapsids declined, while archosaurian reptiles (including dinosaur ancestors) diversified. Remaining large nonamniote tetrapods were all specialized aquatic forms. First appearances by the end of the period included true mammals, dinosaurs, pterosaurs, marine reptiles, crocodiles, lepidosaurs, froglike amphibians, and teleost fishes.

A single world continent, Pangaea, was formed at the end of the period. Glaciation ceased early in the period. The large terrestrial nonamniote tetrapods declined and the amniotes radiated. Amniote diversification included the ancestors of modern reptiles and the ancestors of mammals, the non-mammalian synapsids being the dominant large terrestrial tetrapods. The first herbivorous tetrapods evolved. The largest known mass extinction event occurred on both land and sea at the end of the period. It coincided with low levels of atmospheric O_2 and marked the end of the Paleozoic.

There was a major glaciation in the second half of the period, with low atmospheric levels of CO_2 and high levels of O_2. Coal swamps were prevalent in the then-tropical areas of North America and Europe. Major radiation of insects, including flying forms. Diversification of jawed fishes, including sharklike forms and primitive bony fishes, and first appearance of modern types of jawless fishes. Extensive radiation of non-amniote tetrapods, with the appearance of the first amniotes (including the earliest mammal-like reptiles) by the late part of the period.

There was major mountain building in North America and Europe. Major freshwater basins, containing the first tetrapods, formed in equatorial regions at the end of the period. About the same time there were the first forests with tall trees on land, and terrestrial arthropods diversified. Both jawed and jawless fishes diversified, but both experienced major extinctions toward the end of the period, with the disappearance of the ostracoderms, the armored jawless fishes.

The extensive shallow seas continued, but on dry land there was the first evidence of vascular plants and arthropods. Jawless fishes radiated, and jawed fishes (sharklike forms) were now definitely known.

There were widespread shallow seas over the continents, and the global climate was equable until a sharp glaciation at the end of the period. First evidence of complex plants on land. Major radiation of marine animals, including the first well-known jawless fishes and fragmentary evidence of jawed fishes.

Continental masses of the late Proterozoic now broken up into smaller blocks, covered by shallow seas. Explosive radiation of animals at the beginning of the period, with first appearance of forms with shells or other hard coverings. First appearance of chordates and great diversification of arthropods, including trilobites. First vertebrates appeared early in the period.

Formation of large continental masses. Oxygen first appears in the atmosphere. First eukaryotic organisms appeared around 2 billion years ago. Major diversification of life at 1 billion years ago, with multicellular organisms, including algae. First animals appeared around 600 million years ago, just after a major glaciation.

Formation of the Earth. Major bombardment of the Earth by extraterrestrial bodies, precluding formation of life until 4 billion years ago (first fossils known at 3.8 billion years ago). Small continents. Hydrosphere definite at 3.8 billion years, atmosphere without free oxygen.

F. HARVEY POUGH
Rochester Institute of Technology

CHRISTINE M. JANIS
Brown University

JOHN B. HEISER
Cornell University

VERTEBRATE LIFE

NINTH EDITION

PEARSON

Boston Columbus Indianapolis New York San Francisco Upper Saddle River
Amsterdam Cape Town Dubai London Madrid Milan Munich Paris Montreal Toronto
Delhi Mexico City Sao Paulo Sydney Hong Kong Seoul Singapore Taipei Tokyo

Editor-in-Chief: Beth Wilbur
Executive Director of Development: Deborah Gale
Development and Project Editor: Crystal Clifton, Progressive Publishing Alternatives
Managing Editor: Mike Early
Production Project Manager: Camille Herrera
Production Management: Rose Kernan, Cenveo Publisher Services
Compositor: Cenveo Publisher Services
Interior and Cover Designer: Emily Friel, Integra Software Services, Inc.

Illustrators: Dartmouth Publishing, and Precision Graphics
Art Development Editor: Jennifer Kane
Associate Director of Image Management: Travis Amos
Supervisor Publishing Production, Photos: Donna Kalal
Photo Researcher: Maureen Spuhler
Photo Clearance: Q2A/Bill Smith
Manager of Permissions, Text: Beth Wollar
Permissions Assistant, Text: Tamara Efsen
Manufacturing Buyer: Michael Penne
Marketing Manager: Lauren Harp
Cover Photo Credit: Rod Williams / Nature Picture Library

[Library of Congress Cataloging-in-Publication Data on File]

5 6 7 8 9 10—EBM—17 16 15

www.pearsonhighered.com

ISBN 10:   0-321-77336-5
ISBN 13: 978-0-321-77336-4 (Student edition)

About the Authors

F. Harvey Pough is a Professor of Biology at the Rochester Institute of Technology. He began his biological career at the age of fourteen when he and his sister studied the growth and movements of a population of eastern painted turtles in Rhode Island. His research now focuses on organismal biology, blending physiology, morphology, behavior, and ecology in an evolutionary perspective. Undergraduate students regularly participate in his research and are coauthors of many of his publications. He especially enjoys teaching undergraduates and has taught introductory biology and courses in vertebrate zoology, functional ecology, herpetology, environmental physiology, and animal behavior. He has published more than a hundred papers reporting the results of field and laboratory studies of turtles, snakes, lizards, frogs, and tuatara that have taken him to Australia, New Zealand, Fiji, Mexico, Costa Rica, Panama, and the Caribbean, as well as most parts of the United States. He is a Fellow of the American Association for the Advancement of Science and a past President of the American Society of Ichthyologists and Herpetologists.

Christine M. Janis is a Professor of Biology at Brown University, where she teaches comparative anatomy and vertebrate evolution and is a recipient of the Elizabeth Leduc Prize for Distinguished Teaching in the Life Sciences. A British citizen, she obtained her bachelor's degree at Cambridge University and then crossed the pond to get her Ph.D. at Harvard University. She is a vertebrate paleontologist with a particular interest in mammalian evolution and faunal responses to climatic change. She first became interested in vertebrate evolution after seeing the movie *Fantasia* at the impressionable age of seven. That critical year was also the year that she began riding lessons, and she has owned at least one horse since the age of twelve. Many years later, she is now an expert on ungulate (hoofed mammal) evolution and has recently expanded her interests to the evolution of the Australian mammal fauna, especially the kangaroos. She is a Fellow of the Paleontological Society and is currently President of the Society for the Study of Mammalian Evolution. She attributes her life history to the fact that she has failed to outgrow either the dinosaur phase or the horse phase.

John B. Heiser was born and raised in Indiana and completed his undergraduate degree in biology at Purdue University. He earned his Ph.D. in ichthyology from Cornell University for studies of the behavior, evolution, and ecology of coral reef fishes, research that he continues today with colleagues specializing in molecular biology. For fifteen years, he was Director of the Shoals Marine Laboratory operated by Cornell University and the University of New Hampshire on the Isles of Shoals in the Gulf of Maine. While at the Isles of Shoals, his research interests focused on opposite ends of the vertebrate spectrum—hagfish and baleen whales. He enjoys teaching vertebrate morphology, evolution, and ecology, both in the campus classroom and in the field, and is a recipient of the Clark Distinguished Teaching Award from Cornell University. His hobbies are natural history, travel and nature photography, and videography, especially underwater using scuba. He has pursued his natural history interests on every continent and all the world's major ocean regions. Because of his experience, he is a popular ecotourism leader, having led Cornell Adult University groups to the Caribbean, Sea of Cortez, French Polynesia, Central America, the Amazon, Borneo, Antarctica, and Spitsbergen in the High Arctic.

Brief Contents

Contents

Preface

The theme of *Vertebrate Life* is organismal biology—that is, how the anatomy, physiology, ecology, and behavior of animals interact to produce organisms that function effectively in their environments and how lineages of organisms change through evolutionary time. The ninth edition emphasizes advances in our understanding of vertebrates since the eighth edition was published. Several topics have been greatly expanded.

New to This Editions

- **Molecular biology.** Molecular studies have produced new information about phylogenetic relationships that illuminates events as distant as the origin of jawless vertebrate lineages and as recent as the separation of the clouded leopards on Borneo, Sumatra, and the Asian mainland.

- **Fossil evidence.** Newly described fossils have expanded our understanding of the evolutionary diversity of vertebrate lineages, particularly of dinosaurs (including birds) and our own human lineage.

- **Climate change.** The ever-increasing evidence of global climate change has important implications for the biology and conservation of vertebrates, and new information about atmospheric conditions during the Paleozoic and Mesozoic eras sheds light on vertebrate diversification.

- **Conservation.** As the pace of extinction quickens, specific situations raise acute concerns: the global decline in amphibian populations, part of which can be traced to the worldwide spread of a fungal infection; the threats posed to fisheries by fish farming and transgenic fishes; and the difficulty of preserving large animals that require huge home ranges, and especially the problems associated with large predators, such as tigers, that sometimes eat people.

- **Access to information.** The expansion of electronic databases and the accessibility of online resources give students increased access to the primary literature and authoritative secondary sources, and we have added more citations of printed and online journals and of websites to encourage students to explore these sources.

- **Discussion questions.** We have found that open-ended discussion questions are an effective way to increase active learning in class meetings, and we have added discussion questions in this edition. (And, because we sometimes find ourselves wondering what response the author expected when we use questions from textbooks, we have provided answers to our questions in the Instructor Resource Center at www.pearsonhighered.com.)

- **Lists of derived characters accompanying the cladograms.** The extensive legends accompanying the cladograms provide important information, but they break the flow of text in the chapters. In this edition we have moved the legends to an appendix.

- **Images from the text.** The figures from the text can be downloaded in jpg and PowerPoint formats from the Instructor Resource Center at www.pearsonhighered.com.

Amid all of these changes, the element of *Vertebrate Life* that has always been most important to the authors has remained constant: We are biologists because we care enormously about what we do and the animals we work with. We are deeply committed to passing on the fascination and sheer joy that we have experienced to new generations of biologists and to providing information and perspectives that will help them with the increasingly difficult task of ensuring that the enormous vigor and diversity of vertebrate life do not vanish.

Acknowledgments

We are very grateful to the excellent production team assembled by Pearson for this edition: Editor in Chief, Beth Wilbur; our outstanding Project Editor, Crystal Clifton; Photo Researcher, Maureen Spuhler; and Nesbitt Graphics. Their mastery of every step on the complex path from a manuscript to a bound copy of a book has been enormously comforting to the authors.

We are especially pleased by the return of Jennifer Kane as the artist for this edition. Jennifer first met *Vertebrate Life* when she was a student, and she brings that perspective to her work. Jennifer combines the ability to render anatomical information accurately with an empathy for vertebrates that allows her to produce drawings so lifelike that they appear ready to walk off the page.

Writing a book with a scope as broad as this one requires the assistance of many people. We list below the colleagues who generously provided comments, suggestions, and photographs and who responded to our requests for information.

Mike Barton, *Centre College*
Robin Beck, *University of New South Wales*
Kayla Bell, *Rochester Institute of Technology*
Michael Bell, *Stony Brook University*
Daniel Blackburn, *Trinity College*
Elizabeth Brainerd, *Brown University*
Larry Buckley, *Rochester Institute of Technology*
Gabbie Burns, *Rochester Institute of Technology*
Jennifer Burr, *Nazareth College*
Nathan Cawley, *Rochester Institute of Technology*
Jenny Clack, *University of Cambridge*
Davide Csermely, *University of Parma, Italy*
Andy Derocher, *University of Alberta*
Phillip Donoghue, *University of Bristol*
Robert E. Espinoza, *California State University, Northridge*
Colleen Farmer, *University of Utah*
David Fastovsky, *University of Rhode Island*
Patricia Freeman, *University of Nebraska*
Gene Helfman, *University of Georgia*
Chris Hess, *Butler University*
Morna Hilderbrand, *Rochester Institute of Technology*
Jill Holiday, *University of Florida*
Anne Houtman, *Rochester Institute of Technology*
Rebecca Hull, *Rochester Institute of Technology*
Christopher Irish, *Rochester Institute of Technology*
Tom Kemp, *Oxford University*

Haeja Kessler, *Rochester Institute of Technology*
Jamie Kraus, *Rochester Institute of Technology*
Kyle Mara, *Temple University*
Kaitlyn Moranz, *Rochester Institute of Technology*
Phillip Motta, *University of South Florida*
Henry Mushinksy, *University of South Florida*
Warren Porter, *University of Wisconsin*
Marilyn Renfree, *University of Melbourne*
Tony Russell, *University of Calgary*
Roger Seymour, *University of Adelaide*
Joan Sharp, *Simon Fraser University*
Adrienne Sherman, *Rochester Institute of Technology*
Rick Shine, *University of Sydney*
Susan Smith, *Rochester Institute of Technology*
Harshita Sood, *Rochester Institute of Technology*
Ben Speers-Roesch, *University of British Columbia*
Karen Steudal, *University of Wisconsin*
C. Richard Tracy, *University of Nevada, Reno*
Jason Treberg, *University of Manitoba*
Hugh Tyndale-Biscoe, *Canberra, Australia*
David Wake, *University of California, Berkeley*
John Waud, *Rochester Institute of Technology*
Lisa Whitenack, *Allegheny College*
Amy Yuhas, *Rochester Institute of Technology*

We especially appreciate the work of reviewers of this text: Christine Dahlin, University of Pittsburgh at Johnstown; Margaret B. Fusari, University of California, Santa Cruz; Marion Preest, Keck Science Department of The Claremont Colleges.

We are deeply indebted to Adwoa Boateng and Morna Hilderbrand at the Wallace Center, Rochester Institute of Technology. Their skills and assistance were essential to our extensive review of the enormous literature of vertebrate biology.

Vertebrate Diversity, Function, and Evolution

The more than 63,000 living species of vertebrates inhabit nearly every part of Earth, and other kinds of vertebrates that are now extinct lived in habitats that no longer exist. Increasing knowledge of the diversity of vertebrates was a product of the European exploration and expansion that began in the fifteenth and sixteenth centuries. In the middle of the eighteenth century, Swedish naturalist Carolus Linnaeus developed a binominal classification to catalog the varieties of animals and plants. Despite some problems in reflecting evolutionary relationships, the Linnaean system remains the basis for naming organisms today.

A century later, Charles Darwin and Alfred Russel Wallace explained the diversity of plants and animals as the product of natural selection and evolution. In the early twentieth century, their work was coupled with the burgeoning information about mechanisms of genetic inheritance. This combination of genetics and evolutionary biology, known as the New Synthesis, or neo-Darwinism, continues to be the basis for understanding the mechanics of evolution. Methods of classifying animals also changed during the twentieth century; and classification, which began as a way of trying to organize the diversity of organisms, has become a powerful tool for generating testable hypotheses about evolution.

Vertebrate biology and the fossil record of vertebrates have been at the center of these changes in our view of life. Comparative studies of the anatomy, embryology, behavior, and physiology of living vertebrates have often supplemented the fossil record. These studies reveal that evolution acts by changing existing traits. All vertebrates share basic characteristics that are the products of their common ancestry, and the process of evolution can be analyzed by tracing the modifications of these characters. Thus, an understanding of vertebrate form and function is basic to understanding the evolution of vertebrates and the ecology and behavior of living species.

The Diversity, Classification, and Evolution of Vertebrates

Evolution is central to vertebrate biology because it provides a principle that organizes the diversity we see among living vertebrates and helps to fit extinct forms into the context of living species. Classification, initially a process of attaching names to organisms, has become a method of understanding evolution and planning strategies for conservation. Current views of evolution stress natural selection operating at the level of individuals as the predominant mechanism that produces changes in a population over time. The processes and events of evolution are intimately linked to the changes that have occurred on Earth during the history of vertebrates. These changes have resulted from the movements of continents and the effects of those movements on climates and geography. In this chapter, we present an overview of the scene, the participants, and the events that have shaped the biology of vertebrates.

1.1 The Vertebrate Story

Mention "animal" and most people will think of a vertebrate. Vertebrates are abundant and conspicuous parts of people's experience of the natural world. Vertebrates are also very diverse: The more than 63,000 **extant** (currently living) species of vertebrates range in size from fishes weighing as little as 0.1 gram when fully mature to whales weighing over 100,000 kilograms. Vertebrates live in virtually all the habitats on Earth. Bizarre fishes, some with mouths so large they can swallow prey larger than their own bodies, cruise through the depths of the sea, sometimes luring prey to them with glowing lights. Fifteen kilometers above the fishes, migrating birds fly over the crest of the Himalayas, the highest mountains on Earth.

The behaviors of vertebrates are as diverse and complex as their body forms and habitats. Vertebrate life is energetically expensive, and vertebrates get the energy they need from food they eat. Carnivores eat the flesh of other animals and show a wide range of methods of capturing prey. Some predators search the environment to find prey, whereas others wait in one place for prey to come to them. Some carnivores pursue their prey at high speeds, and others pull prey into their mouths by suction. Many vertebrates swallow their prey intact, sometimes while it is alive and struggling, but other vertebrates have very specific methods of dispatching prey. Venomous snakes inject complex mixtures of toxins, and cats (of all sizes, from house cats to tigers) kill their prey with a distinctive bite on the neck.

Herbivores eat plants. Plants cannot run away when an animal approaches, so they are easy to catch, but they are hard to chew and digest and frequently contain toxic compounds. Herbivorous vertebrates show an array of specializations to deal with the difficulties of eating plants. Elaborately sculptured teeth tear apart tough leaves and expose the surfaces of cells, but the cell walls of plants contain cellulose, which no vertebrate can digest. Herbivorous vertebrates rely on microorganisms living in their digestive tracts to digest cellulose. In addition, these endosymbionts (organisms that live inside another organism) detoxify the chemical substances that plants use to protect themselves.

Reproduction is a critical factor in the evolutionary success of an organism, and vertebrates show an astonishing range of behaviors associated with mating and reproduction. In general, males court females and females care for the young, but these roles are reversed in many species of vertebrates. At the time of birth or hatching, some vertebrates are entirely self-sufficient and never see their parents, whereas other vertebrates (including humans) have extended periods of obligatory parental care. Extensive parental care is found in seemingly unlikely groups of vertebrates—fishes that incubate eggs in their mouths, frogs that carry their tadpoles to water and then return to feed them, and birds that feed their nestlings a fluid called crop milk that is very similar in composition to mammalian milk.

The diversity of living vertebrates is enormous, but the species now living are only a small proportion of the species of vertebrates that have existed. For each living species, there may be more than a hundred extinct species, and some of these have no counterparts among living forms. For example, the dinosaurs that dominated Earth for 180 million years are so entirely different from living animals that it is hard to reconstruct the lives they led. Even mammals were once more diverse than they are now. The Pleistocene epoch saw giants of many kinds—ground sloths as big as modern rhinoceroses and raccoons as large as bears. The number of species of terrestrial vertebrates probably reached its maximum in the middle Miocene epoch, 12 to 14 million years ago, and has been declining since then.

The story of vertebrates is fascinating. Where they originated, how they evolved, what they do, and how they work provide endless intriguing details. In preparing to tell this story, we must introduce some basic information, including what the different kinds of vertebrates are called, how they are classified, and what the world was like as the story of vertebrates unfolded.

Major Extant Groups of Vertebrates

Two major groups of vertebrates are distinguished on the basis of an innovation in embryonic development: the appearance of three membranes formed by tissues that come from the embryo itself. One of these membranes, the amnion, surrounds the embryo, and animals with this structure are called **amniotes**. The division between non-amniotes and amniotes corresponds roughly to aquatic and terrestrial vertebrates, although many amphibians and a few fishes lay non-amniotic eggs in nests on land.

Among the amniotes, we can distinguish two major evolutionary lineages—the sauropsids (reptiles, including birds) and the synapsids (mammals). These lineages separated from each other in the Late Devonian period, before vertebrates had developed many of the characters we see in extant species. As a result, synapsids and sauropsids represent parallel but independent origins of basic characters such as lung ventilation, kidney function, insulation, and temperature regulation.

Figure 1–1 shows the major kinds of vertebrates and the approximate numbers of living species. In the following sections, we briefly describe the different kinds of vertebrates.

Non-Amniotes

The embryos of non-amniotes are enclosed and protected by membranes that are produced by the reproductive tract of the female. This is the condition seen among the invertebrate relatives of vertebrates, and it is retained in the non-amniotes: the fishes and amphibians.

Hagfishes and Lampreys—Myxiniformes and Petromyzontiformes Lampreys and hagfishes are elongate, limbless, scaleless, and slimy and have no internal bony tissues. They are scavengers and parasites and are specialized for those roles. Hagfishes are marine, living on the seabed at depths of 100 meters or more. In contrast, many species of lampreys are migratory, living in oceans and spawning in rivers.

Hagfishes and lampreys are unique among living vertebrates because they lack jaws; this feature makes them important in the study of vertebrate evolution. They have traditionally been grouped as agnathans (Greek *a* = without and *gnath* = jaw) or cyclostomes (Greek *cyclo* = round and *stoma* = mouth), but they are probably not closely related to each other and instead represent two independent (i.e., separate) evolutionary lineages.

Sharks, Rays, and Ratfishes—Chondrichthyes The name Chondrichthyes (Greek *chondro* = cartilage and *ichthyes* =

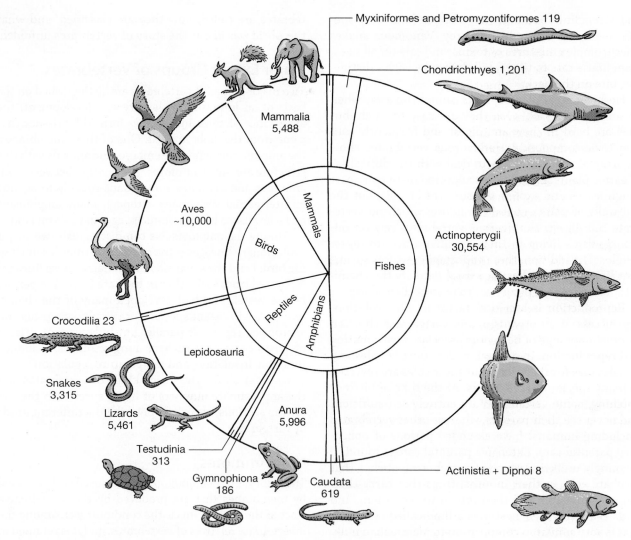

Figure 1–1 Diversity of vertebrates. Areas in the diagram correspond to approximate numbers of living species in each group. (These are estimates, and the numbers change frequently as new species are described.) Common names are in the center circle, and formal names for the groups are on the outer circle. The two major lineages of extant vertebrates are the Actinopterygii (ray-finned fishes) and the Sarcopterygii (lobe-finned fishes), each of which includes more than 30,000 extant species. (The Sarcopterygii includes the lineages Actinistia, Dipnoi, Caudata, Anura, Gymnophiona, Testudinia, Lepidosauria, Crocodilia, Aves, and Mammalia.)

fish) refers to the cartilaginous skeletons of these fishes. Extant sharks and rays form a group called the Neoselachii (Greek *neo* = new and *selach* = shark), but the two kinds of fishes differ in body form and habits. Sharks have a reputation for ferocity that most species would have difficulty living up to. Some sharks are small (15 centimeters or less); and the largest species, the whale shark (which grows to 10 meters), is a filter feeder that subsists on plankton it strains from the water. Rays are mostly bottom feeders; they are dorsoventrally flattened and swim with undulations of their extremely broad pectoral fins.

The second group of chondrichthyans, the ratfishes or chimaerans, gets its name, Holocephalii (Greek *holo* = whole and *cephal* = the head), from the single gill cover that extends over all four gill openings. These are bizarre marine animals with long, slender tails and bucktoothed faces that look rather like rabbits. They live on the seafloor and feed on hard-shelled prey, such as crustaceans and mollusks.

Bony Fishes—Osteichthyes Bony fishes, the Osteichthyes (Greek *osteo* = bone and *ichthyes* = fish), are so diverse that any attempt to characterize them briefly is doomed to failure. Two broad categories can be recognized: the ray-finned fishes (actinopterygians; Greek *actino* = ray and *ptero* = wing or fin) and the lobe-finned or fleshy-finned fishes (sarcopterygians; Greek *sarco* = flesh).

The ray-finned fishes have radiated extensively in fresh and salt water. More than 30,500 species of ray-finned fishes have been named, and several thousand additional species may await discovery. A single project, the Census of Marine Life, is describing 150 to 200 previously unknown species of ray-finned fishes annually. Two major groups can be distinguished among actinopterygians. The Chondrostei (bichirs, sturgeons, and paddlefishes) are survivors of an early radiation of bony fishes. Bichirs are swamp- and river-dwellers from Africa; they are known as African reed fish in the aquarium trade. Sturgeons are large fishes with protrusible, toothless mouths that are used to suck food items from the bottom. Sturgeons are the source of caviar—eggs are taken from the female before they are laid. Of course, this kills the female sturgeon, and many species have been driven close to extinction by overfishing. Paddlefishes (two species, one in the Mississippi drainage of North America and another nearly extinct species in the Yangtze River of China) have a paddlelike snout with organs that locate prey by sensing electrical fields.

The Neopterygii, the modern radiation of ray-finned fishes, can be divided into three lineages. Two of these—the gars and the bowfins—are relicts of earlier radiations. These fishes have cylindrical bodies, thick scales, and jaws armed with sharp teeth. They seize prey in their mouths with a sudden rush or gulp, and they lack the specializations of the jaw apparatus that allow later bony fishes to use more complex feeding modes.

The third lineage of neopterygians, the Teleostei, includes almost 30,500 species of fishes covering every imaginable combination of body size, habitat, and habits. Most of the fishes that people are familiar with are teleosts—the trout, bass, and panfish that anglers seek; the sole (a kind of flounder) and swordfish featured by seafood restaurants; and the salmon and tuna whose by-products find their way into canned catfood. Modifications of the body form and jaw apparatus have allowed many teleosts to be highly specialized in their swimming and feeding habits.

In one sense, only eight species of lobe-finned fishes survive: the six species of lungfishes (Dipnoi) found in South America, Africa, and Australia and the two species of coelacanths (Actinistia), one from deep waters off the east coast of Africa and a second species recently discovered near Indonesia. These are the living fishes most closely related to terrestrial vertebrates, and a more accurate view of sarcopterygian diversity includes their terrestrial descendants—amphibians, mammals, turtles, lepidosaurs (the tuatara, lizards, and snakes), crocodilians, and birds. From this perspective, bony fishes include two major evolutionary radiations—one in the water and the other on land, with each containing more than 30,000 living species.

Salamanders, Frogs, and Caecilians—Caudata, Anura, and Gymnophiona These three groups of vertebrates are popularly known as amphibians (Greek *amphi* = double and *bios* = life) in recognition of their complex life histories, which often include an aquatic larval form (the larva of a salamander or caecilian and the tadpole of a frog) and a terrestrial adult. All amphibians have bare skins (i.e., lacking scales, hair, or feathers) that are important in the exchange of water, ions, and gases with their environment. Salamanders are elongate animals, mostly terrestrial, and usually with four legs; anurans (frogs, toads, and tree frogs) are short-bodied, with large heads and large hind legs used for walking, jumping, and climbing; and gymnophians (caecilians) are legless aquatic or burrowing animals.

Amniotes

An additional set of membranes associated with the embryo appeared during the evolution of vertebrates. They are called fetal membranes because they are derived from the embryo itself rather than from the reproductive tract of the mother. The amnion is one of these membranes, and vertebrates with an amnion are called amniotes. In general, amniotes are more terrestrial than non-amniotes; but there are also secondarily aquatic species of amniotes (such as sea turtles and whales) as well as many species of salamanders and frogs that spend their entire lives on land despite being non-amniotes. However, many features distinguish non-amniotes (fishes and amphibians) from amniotes (mammals and reptiles, including birds), and we will use the terms to identify which of the two groups is being discussed.

By the Permian, amniotes were well established on land. They ranged in size from lizardlike animals a few centimeters long through cat- and dog-size species to the cow-size parieasaurs. Some were herbivores; others were carnivores. In terms of their physiology, however, we can infer that they retained ancestral characters. They had scale-covered skins without an insulating layer of hair or feathers, a simple kidney that could not produce highly concentrated urine, simple lungs, and a heart in which the ventricle was not divided by a septum.

Early in their evolutionary history, terrestrial vertebrates split into two lineages—the synapsids (now represented by mammals) and the sauropsids (the reptiles, including birds). Terrestrial life requires lungs to extract oxygen from air, hearts that can separate oxygen-rich arterial blood from oxygen-poor venous blood, kidneys that can eliminate waste products while retaining water, and insulation and behaviors to keep body temperature stable as the external temperature changes. These features evolved in both lineages; but, because synapsids and sauropsids evolved terrestrial specializations independently, the lungs, hearts, kidneys, and body coverings of synapsids and sauropsids are different.

Sauropsid Amniotes Extant sauropsids are the animals we call reptiles: turtles, the scaly reptiles (tuatara, lizards, and snakes), crocodilians, and birds. Extinct sauropsids include the forms that dominated the world during the Mesozoic era—dinosaurs and pterosaurs (flying reptiles) on land and a variety of marine forms, including ichthyosaurs and plesiosaurs, in the oceans.

Turtles—Testudinia Turtles (Latin *testudo* = a turtle) are probably the most immediately recognizable of all vertebrates. The shell that encloses a turtle has no exact duplicate among other vertebrates, and the morphological modifications associated with the shell make turtles extremely peculiar animals. They are, for example, the only vertebrates with the shoulders (pectoral girdle) and hips (pelvic girdle) inside the ribs.

Tuatara, Lizards, and Snakes—Lepidosauria These three kinds of vertebrates can be recognized by their scale-covered skin (Greek *lepido* = scale and *saur* = lizard) as well as by characteristics of the skull. The tuatara, a stocky-bodied animal found only on some islands near New Zealand, is the sole living remnant of an evolutionary lineage of animals called Sphenodontida, which was more diverse in the Mesozoic. In contrast, lizards and snakes (which are highly specialized lizards) are now at the peak of their diversity.

Alligators and Crocodiles—Crocodilia These impressive animals, which draw their name from *crocodilus*, the Greek word for crocodile, are in the same evolutionary lineage (the Archosauria) as dinosaurs and birds. The extant crocodilians are semiaquatic predators, with long snouts armed with numerous teeth. They range in size from the saltwater crocodile, which has the potential to grow to a length of 7 meters, to dwarf crocodiles and caimans that are less than a meter long. Their skin contains many bones (osteoderms; Greek *osteo* = bone and *derm* = skin) that lie beneath the scales and provide a kind of armor plating. Crocodilians are noted for the parental care they provide for their eggs and young.

Birds—Aves The birds (Latin *avis* = a bird) are a lineage of dinosaurs that evolved flight in the Mesozoic. Feathers are characteristic of extant birds, and feathered wings are the structures that power a bird's flight. Recent discoveries of dinosaur fossils with traces of feathers show that feathers evolved before flight. This offset between the times that feathers and flight appeared illustrates an important principle: the function of a trait in an extant species is not necessarily the same as its function when it first appeared. In other words, current utility is not the same as evolutionary origin. The original feathers were almost certainly structures that were used in courtship displays, and their modification as airfoils, for streamlining, and as insulation in birds is a secondary event.

Synapsid Amniotes The synapsid lineage contains the three kinds of extant mammals: the monotremes (prototheria; the platypus and echidna), marsupials (metatherians), and placentals (eutherians). Extinct synapsids include forms that diversified in the Paleozoic era—pelycosaurs and therapsids—and the rodentlike multituberculates of the late Mesozoic.

Mammals—Mammalia The living mammals (Latin *mamma* = a breast) can be traced to an origin in the late Paleozoic, from some of the earliest fully terrestrial vertebrates. Extant mammals include about 5500 species, most of which are placental mammals. Both placentals and marsupials possess a **placenta**, a structure that transfers nutrients from the mother to the embryo and removes the waste products of the embryo's metabolism. Placentals have an extensive system of placentation and a long gestation period, whereas marsupials have a short gestation period and give birth to very immature young that continue their development attached to a nipple, often in a pouch on the mother's abdomen. Marsupials dominate the mammalian fauna only in Australia. Kangaroos, koalas, and wombats are familiar Australian marsupials. The strange monotremes, the platypuses and the echidnas, are mammals whose young are hatched from eggs. All mammals, including monotremes, feed their young with milk.

New Species

New species of vertebrates are described weekly—this is why we use the words "approximately" and "about" when we cite the numbers of species. More than 300 new species of fishes are described annually. In 2002, a molecular analysis of the relationships of rhacophorid frogs on Madagascar increased the number of species from 18 to more than 100.

Among terrestrial mammals rodents led the way with 174 new species described between 1993 and 2008, and bats were second with 94 species. Most rodents are small and bats are nocturnal, so it is not surprising to find new species in those groups. It is, however, something of a surprise to find that primates, with 55 new species, followed bats in the list because primates are mostly diurnal and large enough to be conspicuous elements of a fauna; other newly identified species are also large.

- In 2005, a new species of mangabey monkey as big as a medium-size dog was discovered in Tanzania. Subsequent study showed that it was so different from related species that a new genus, *Rungwecebus*, was created for it. In addition to being large, this monkey has a loud vocalization and occurs in forest adjacent to cultivated areas. In fact, the type specimen was captured in a trap set by a farmer in a field of maize.
- Three new species of whales have been described since 2000—two rorquals (*Balaenoptera omurai* from the Indo-Pacific region and *B. edeni*, which has a worldwide distribution) and a new right whale (*Eubalaena japonica*) from the North Pacific. These animals are from 11 to 17 meters long.
- In 2001 the largest extant land mammal, the African elephant, was shown to consist of two distinct species: the African savannah elephant (*Loxodonta africana*) and the African forest elephant (*Loxodonta cyclotis*).

Many of these new species occur in isolated populations and some are on the brink of extinction. For example, 221 of the 408 new species of terrestrial mammals that were described between 1993 and 2008 occur in isolated populations, and 34 of those new species are considered to be at risk of extinction. This situation places a biologist who discovers a new species in a paradoxical situation—on one hand, zoological convention calls for killing an individual of the new species and depositing it in a museum to serve as the holotype—that is, the single specimen on which the species is based in the original publication. On the other hand, even one individual may be a significant loss from a small population. The descriptions of two new species of primates and a lizard have been based on photographs, measurements, and tissue samples from living individuals that were released instead of being killed (**Figure 1–2**). (The validity of designating a living holotype rests on interpretation of article 73.1.4 of the 1999 edition of the International Code of Zoological Nomenclature, and this practice has generated a lively controversy.)

(a)

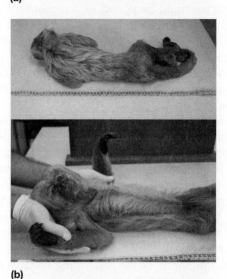

(b)

(c)

Figure 1–2 Three new species of vertebrates that were described on the basis of living holotypes. (a) The highland mangabey, *Rungwecebus kipunji*. The species was initially described on the basis of this photograph. Later, another individual that had been killed by a farmer became available for study. (b) The blonde capuchin monkey, *Cebus queirozi*. These are two of the images from the series of photographs of an anesthetized monkey in the description of the species. (The monkey was released after it recovered consciousness.) (c) The pink Galápagos land iguana, *Conolophus marthae*. An electronic identification tag has been implanted in this individual. When it dies a natural death, its remains will be deposited in a museum.

1.2 Classification of Vertebrates

The diversity of vertebrates (more than 63,000 living species and perhaps a hundred times that number of species now extinct) makes the classification of vertebrates an extraordinarily difficult task. Yet classification has long been at the heart of evolutionary biology. Initially, classification of species was seen as a way of managing the diversity of organisms, much as an office filing system manages the paperwork of the office. Each species could be placed in a pigeonhole marked with its name; when all species were in their pigeonholes, the diversity of vertebrates would have been encompassed. This approach to classification was satisfactory as long as species were regarded as static and immutable: once a species was placed in the filing system, it was there to stay.

Acceptance of the fact that species evolve has made that kind of classification inadequate. Now biologists must express evolutionary relationships among species by incorporating evolutionary information in the system of classification. Ideally, a classification system should not only attach a label to each species but also encode the evolutionary relationships between that species and other species. Modern techniques of systematics (the evolutionary classification of organisms) have become methods for generating testable hypotheses about evolution.

Classification and Names

Our system of naming species is pre-Darwinian. It traces back to methods established by the naturalists of the seventeenth and eighteenth centuries, especially those of Carl von Linné, a Swedish naturalist, better known by his Latin pen name, Carolus Linnaeus. The Linnaean system employs binominal nomenclature to designate species and arranges species into hierarchical categories (**taxa**, singular *taxon*) for classification.

Binominal Nomenclature

The scientific naming of species became standardized when Linnaeus's monumental work, *Systema Naturae* (*The System of Nature*), was published in sections between 1735 and 1758. Linnaeus attempted to give an identifying name to every known species of plant and animal. His method assigns a binominal (two-word) name to each species. Familiar examples include *Homo sapiens* for human beings (Latin *hom* = human and *sapien* = wise), *Passer domesticus* for the house sparrow (Latin *passer* = sparrow and *domesticus* = belonging to the house), and *Canis familiaris* for the domestic dog (Latin *canis* = dog and *familiaris* = of the family).

Why use Latin words? Latin was the early universal language of European scholars and scientists. It has provided a uniform usage that scientists, regardless of their native language, continue to recognize worldwide. The same species may have different colloquial names, even in the same language. For example, *Felis concolor* (Latin for "the uniformly colored cat") is known in various parts of North America as cougar, puma, mountain lion, American panther, painter, and catamount. In Central and South America it is called león colorado, onça-vermelha, poema, guasura, or yaguá-pitá. But biologists of all nationalities recognize the name *Felis concolor* as referring to a specific kind of cat.

Hierarchical Groups

Linnaeus and other naturalists of his time developed what they called a natural system of classification. The **species** is the basic level of biological classification, but the definition of a species has been contentious, partly because criteria that have been used to identify extant species (e.g., reproductive isolation from other species) don't work for fossil species and don't always correspond to genetic differences. Similar species are grouped together in a **genus** (plural *genera*), based on characters that define the genus. The most commonly used characters were anatomical because they can be most easily preserved in museum specimens. Thus all doglike species—various wolves, coyotes, and jackals—were grouped together in the genus *Canis* because they all share certain anatomical features, such as an erectile mane on the neck and a skull with a long, prominent sagittal crest on the top from which massive temporal (jaw-closing) muscles originate. Linnaeus's method of grouping species was functional because it was based on anatomical (and to some extent on physiological and behavioral) similarities and differences. Linnaeus lived before there was any knowledge of genetics and the mechanisms of inheritance, but he used characters that we understand today are genetically determined biological traits that generally express the degree of genetic similarity or difference among groups of organisms. Genera are placed in families, families in orders, orders in classes, and animal classes in phyla (singular *phylum*).

1.3 Phylogenetic Systematics

All methods of classifying organisms, even pre-Linnaean systems, are based on similarities among the included species, but some similarities are more significant than others. For example, nearly all vertebrates have paired limbs, but only a few kinds of vertebrates have mammary glands. Consequently, knowing that the species in question have mammary glands tells you more about the closeness of their relationship than knowing that they have paired limbs. You would thus give more weight to the presence of mammary glands than to paired limbs.

A way to assess the relative importance of different characteristics was developed in the mid-twentieth century by Willi Hennig, who introduced a method of determining evolutionary relationships called **phylogenetic systematics** (Greek *phyla* = tribe and *genesis* = origin). An evolutionary lineage is a **clade** (from *cladus*, the Greek word for a branch), and phylogenetic systematics is also called **cladistics**. Cladistics recognizes only groups of organisms that are related by common descent. The application of cladistic methods has made the study of evolution rigorous. The groups of organisms recognized by cladistics are called natural groups, and they are linked in a nested series of ancestor-descendant relationships that trace the evolutionary history of the group. Hennig's contribution was to insist that these groups can be identified only on the basis of **derived characters**.

"Derived" means "different from the ancestral condition." A derived character is called an **apomorphy** (Greek *apo* = away from [i.e., derived from] and *morph* = form, which is interpreted as "away from the ancestral condition"). For example, the feet of terrestrial vertebrates have distinctive bones—the carpals, tarsals, and digits. This arrangement of foot bones is different from the ancestral pattern seen in lobe-finned fishes, and all lineages of terrestrial vertebrates had that derived pattern of foot bones at some stage in their evolution. (Many groups of terrestrial vertebrates—horses, for example—have subsequently modified the foot bones, and some, such as snakes, have lost the limbs entirely. The significant point is that those evolutionary lineages include species that had the derived terrestrial pattern.) Thus, the terrestrial pattern of foot bones is a **shared derived character** of terrestrial vertebrates. In cladistic terminology, shared derived characters are called **synapomorphies** (Greek *syn* = together, so synapomorphy can be interpreted as "together away from the ancestral condition").

Of course, organisms also share ancestral characters—that is, characters that they have inherited unchanged from their ancestors. These are called **plesiomorphies** (Greek *plesi* = near in the sense of "similar to the ancestor"). Terrestrial vertebrates have a vertebral column, for example, that was inherited from lobe-finned fishes. Hennig called shared ancestral characters **symplesiomorphies** (*sym*, like *syn*, is a Greek root that means "together"). Symplesiomorphies tell us nothing about degrees of relatedness. The principle that only *shared derived* characters can be used to determine evolutionary relationships is the core of cladistics.

The conceptual basis of cladistics is straightforward, although applying cladistic criteria to real organisms can become very complicated. To illustrate cladistic classification, consider the examples presented in **Figure 1–3**. Each of the three **cladograms** (diagrams showing hypothetical sequences of branching during evolution) illustrates a possible evolutionary relationship for the three taxa (plural of *taxon*, which means a species or group of species), identified as 1, 2, and 3. To make the example a bit more concrete, we can consider three characters: the number of toes on the front foot, the skin covering, and the tail. For this example, let's say that in the ancestral character state there are five toes on the front foot, and in the derived state there are four toes. We'll say that the ancestral state is a scaly skin, and the derived state is a lack of scales. As for the tail, it is present in the ancestral state and absent in the derived state.

Figure 1–3 shows the distribution of those three character states in the three taxa. The animals in taxon 1 have five toes on the front feet, lack scales, and have a tail. Animals in taxon 2 have five toes, scaly skins, and no tails. Animals in taxon 3 have four toes, scaly skins, and no tails.

How can we use this information to decipher the evolutionary relationships of the three groups of animals? Notice that the derived number of toes occurs only in taxon 3, and the derived tail condition (absent) is found in taxa 2 and 3. The most **parsimonious** phylogeny (i.e., the evolutionary relationship requiring the fewest number of changes) is represented by Figure 1–3(a). Only three changes are needed to produce the derived character states:

1. In the evolution of taxon 1, scales are lost.
2. In the evolution of the lineage including taxon 2 + taxon 3, the tail is lost.
3. In the evolution of taxon 3, a toe is lost from the front foot.

The other two phylogenies shown in Figure 1–3 are possible, but they would require tail loss to occur independently in taxon 2 and in taxon 3. Any change in a structure is an unlikely event, so the most plausible phylogeny is the one requiring the fewest changes. The second and third phylogenies each require four evolutionary changes, so they are less parsimonious than the first phylogeny we considered.

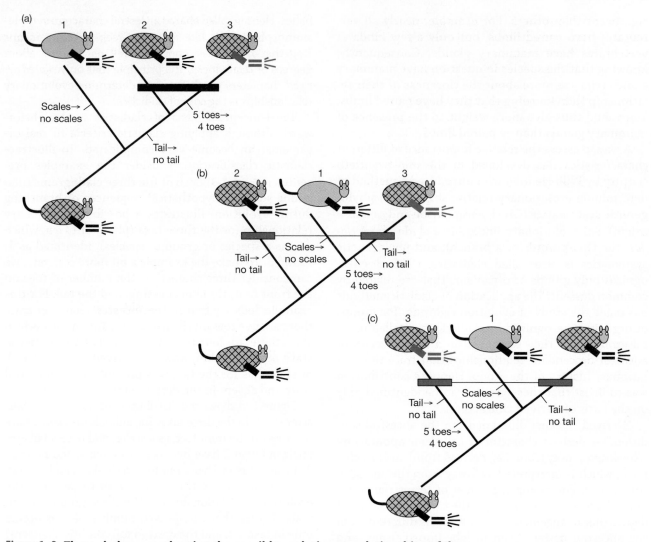

Figure 1–3 Three cladograms showing the possible evolutionary relationships of three taxa. Bars connect derived characters (apomorphies). The black bar shows a shared derived character (a synapomorphy) of the lineage that includes taxa 2 and 3. Colored bars represent two independent origins of the same derived character state that must be assumed to have occurred if there was no apomorphy in the most recent common ancestor of taxa 2 and 3. The labels identify changes from the ancestral character state to the derived condition. Cladogram (a) requires a total of three changes from the ancestral condition to explain the distribution of characters in the extant taxa, whereas cladograms (b) and (c) require four changes. Because cladogram (a) is more parsimonious (i.e., requires the smallest number of changes), it is considered to be the most likely sequence of changes.

A phylogeny is a hypothesis about the evolutionary relationships of the groups included. Like any scientific hypothesis, it can be tested when new data become available. If it fails that test, it is falsified; that is, it is rejected, and a different hypothesis (a different cladogram) takes its place. The process of testing hypotheses and replacing those that are falsified is a continuous one, and changes in the cladograms in successive editions of this book show where new information has generated new hypotheses. The most important contribution of phylogenetic systematics is that it enables us to frame testable hypotheses about the sequence of events during evolution.

So far we have avoided a central issue of phylogenetic systematics: How do scientists know which character state is ancestral (plesiomorphic) and which is derived (apomorphic)? That is, how can we determine the direction (**polarity**) of evolutionary transformation of the characters? For that, we need additional information. Increasing the number of characters we are considering can help, but comparing the characters we are using with an **outgroup** that consists of the closest relatives of the **ingroup** (i.e., the organisms we are studying) is the preferred method. A well-chosen outgroup will possess ancestral character states compared to the ingroup. For example, lobe-finned fishes are an appropriate outgroup for terrestrial vertebrates.

1.4 The Problem with Fossils: Crown and Stem Groups

Evolutionary lineages must have a single evolutionary origin; that is, they must be monophyletic (Greek *mono* = one, single) and include all the descendants of that ancestor. The cladogram depicted in **Figure 1–4** is a hypothesis of the evolutionary relationships of the major living groups of vertebrates. A series of dichotomous branches extends from the origin of vertebrates to the groups of extant vertebrates. Cladistic terminology assigns names to the lineages originating at each branch point. This process produces a nested series of groups, starting with the most inclusive. For example, the Gnathostomata includes all vertebrate animals that have jaws; that is, every taxon above number 2 in Figure 1–4 is included in the Gnathostomata; every taxon above number 3 is included in the Osteichthyes (bony fishes); and so on. Because the lineages are nested, it is correct to say that humans are both gnathostomes and osteichthyans. After number 6, the cladogram divides into Lissamphibia and Amniota, and humans are in the Amniota lineage. The cladogram divides again above number 9 into two lineages, the Sauropsida and Synapsida lineages. Humans are in the Eutheria, which is in the synapsid lineage.

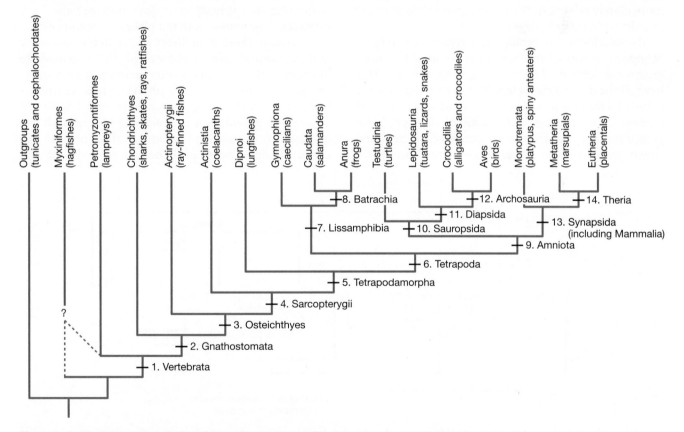

Figure 1–4 Phylogenetic relationships of extant vertebrates. This diagram depicts the probable relationships among the major groups of extant vertebrates. Note that the cladistic groupings are nested progressively; that is, all placental mammals are therians, all therians are synapsids, all synapsids are amniotes, all amniotes are tetrapods, and so on. In a phylogenetic classification lineages can be named at each branching point, although it is not necessary to do so. The dashed line and question mark indicate uncertainty about the branching sequence for hagfishes.

This method of tracing ancestor-descendant relationships allows us to decipher evolutionary pathways that extend from fossils to living groups, but a difficulty arises when we try to find names for groups that include fossils. The derived characters that define the extant groups of vertebrates did not necessarily appear all at the same time. On the contrary, evolution usually acts by gradual and random processes, and derived characters appear in a stepwise fashion. The extant members of a group have all of the derived characters of that group because that is how we define the group today; but, as you move backward through time to fossils that are ancestral to the extant species, you encounter forms that have a mosaic of ancestral and derived characters. The further back in time you go, the fewer derived characters the fossils have.

What can we call the parts of lineages that contain these fossils? They are not included in the extant groups because they lack some of the derived characters of those groups, but the fossils in the lineage are more closely related to the extant group than they are to animals in other lineages.

The solution to this problem lies in naming two types of groups: crown groups and stem groups. The crown groups are defined by the extant species, the ones that have all the derived characters. The stem groups are the extinct forms that preceded the point at which the first member of the crown group branched off. Basically, stem groups contain fossils with some derived characters, and crown groups contain extant species plus those fossils

that have all of the derived characters of the extant species. Stem groups are **paraphyletic** (Greek *para* = beside, beyond); that is, they do not contain all of the descendants of the ancestor of the stem group plus the crown group because the crown group is excluded by definition.

1.5 Evolutionary Hypotheses

Phylogenetic systematics is based on the assumption that organisms in a lineage share a common heritage, which accounts for their similarities. Because of that common heritage, we can use cladograms to ask questions about evolution. By examining the origin and significance of characters of living animals, we can make inferences about the biology of extinct species. For example, the phylogenetic relationship of crocodilians, dinosaurs, and birds is shown in **Figure 1–5**. We know that both crocodilians and birds display extensive parental care of their eggs and young. Some fossilized dinosaur nests contain remains of baby dinosaurs, suggesting that at least some dinosaurs may also have cared for their young. Is that a plausible inference?

Obviously there is no direct way to determine what sort of parental care dinosaurs had. The intermediate lineages in the cladogram (pterosaurs and dinosaurs) are extinct, so we cannot observe their reproductive behavior. However, the phylogenetic diagram in Figure 1–5 provides an indirect way to approach the question by examining the lineage that includes the closest living relatives of dinosaurs, crocodilians and

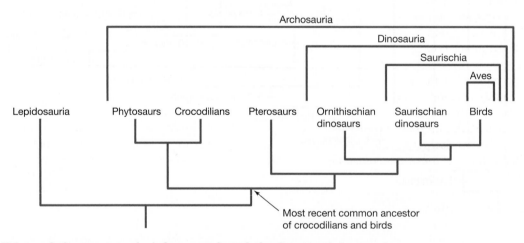

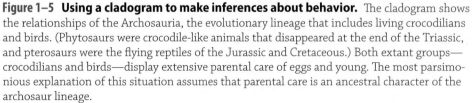

Figure 1–5 Using a cladogram to make inferences about behavior. The cladogram shows the relationships of the Archosauria, the evolutionary lineage that includes living crocodilians and birds. (Phytosaurs were crocodile-like animals that disappeared at the end of the Triassic, and pterosaurs were the flying reptiles of the Jurassic and Cretaceous.) Both extant groups—crocodilians and birds—display extensive parental care of eggs and young. The most parsimonious explanation of this situation assumes that parental care is an ancestral character of the archosaur lineage.

birds. Crocodilians are more basal than pterosaurs and dinosaurs and birds are more derived; together crocodilians and birds form what is called an extant phylogenetic bracket. Both crocodilians and birds, the closest living relatives of the dinosaurs, do have parental care. Looking at living representatives of more distantly related lineages (outgroups), we see that parental care is not universal among fishes, amphibians, or sauropsids other than crocodilians.

The most parsimonious explanation of the occurrence of parental care in both crocodilians and birds is that it had evolved in that lineage *before* the crocodilians separated from dinosaurs + birds. (We cannot prove that parental care did not evolve separately in crocodilians and in birds, but one change to parental care is more likely than two changes.) Thus, the most parsimonious hypothesis is that parental care is a derived character of the evolutionary lineage containing crocodilians + dinosaurs + birds (the Archosauria). That means we are probably correct when we interpret the fossil evidence as showing that dinosaurs did have parental care.

Figure 1–5 also shows how cladistics has made talking about restricted groups of animals more complicated than it used to be. Suppose you wanted to refer to just the two lineages of animals that are popularly known as dinosaurs—ornithischians and saurischians. What could you call them? Well, if you call them dinosaurs, you're not being phylogenetically correct, because the Dinosauria lineage includes birds. So if you say dinosaurs, you are including ornithischians + saurischians + birds, even though any seven-year-old would understand that you are trying to restrict the conversation to extinct Mesozoic animals.

In fact, there is no correct name in cladistic terminology for just the animals popularly known as dinosaurs. That's because cladistics recognizes only monophyletic lineages, and a monophyletic lineage includes an ancestral form *and* all its descendants. The most recent common ancestor of ornithischians, saurischians, and birds in Figure 1–5 lies at the intersection of the lineage of ornithischians with saurischians + birds, so Dinosauria is a monophyletic lineage. If birds are omitted, however, all the descendants of the common ancestor are no longer included; and ornithischians + saurischians minus birds does not fit the definition of a monophyletic lineage. It would be called a paraphyletic group. The stem groups discussed in the previous section are paraphyletic because they do not include all of the descendants of the fossil forms.

Biologists who are interested in how organisms live often want to talk about paraphyletic groups. After all, the dinosaurs (in the popular sense of the word) differed from birds in many ways. The only correct way of referring to the animals popularly known as dinosaurs is to call them nonavian dinosaurs, and you will find that and other examples of paraphyletic groups later in the book. Sometimes even this construction does not work because there is no appropriate name for the part of the lineage you want to distinguish. In this situation we will use quotation marks (e.g., "ostracoderms") to indicate that the group is paraphyletic.

Another important term is **sister group**. The sister group is the monophyletic lineage most closely related to the monophyletic lineage being discussed. In Figure 1–5, for example, the lineage that includes crocodilians + phytosaurs is the sister group of the lineage that includes pterosaurs + ornithischians + saurischians + birds. Similarly, pterosaurs are the sister group of ornithischians + saurischians + birds, ornithischians are the sister group of saurischians + birds, and saurischians are the sister group of birds.

Determining Phylogenetic Relationships

We've established that the derived characters systematists use to group species into higher taxa must be inherited through common ancestry. That is, they are **homologous** (Greek *homo* = same) similarities. In principle, that notion is straightforward; but in practice, the determination of common ancestry can be complex. For example, birds and bats have wings that are modified forelimbs, but the wings were not inherited from a common ancestor with wings. The evolutionary lineages of birds (Sauropsida) and bats (Synapsida) diverged long ago, and wings evolved independently in the two groups. This process is called **convergent evolution**.

Parallel evolution describes the situation in which species that have diverged relatively recently develop similar specializations. The long hind legs that allow the North American kangaroo rats and the African jerboa to jump are an example of parallel evolution in these two lineages of rodents.

A third mechanism, **reversal,** can produce similar structures in distantly related organisms. Sharks and cetaceans (porpoises and whales) have very similar body forms, but they arrived at that similarity from different directions. Sharks retained an ancestral aquatic body form, whereas cetaceans arose from a lineage of terrestrial mammals with well-developed limbs that returned to an aquatic environment and reverted to the aquatic body form.

Convergence, parallelism, and reversal are forms of **homoplasy** (Greek *homo* = same and *plas* = form, shape). Homoplastic similarities do not indicate common ancestry. Indeed, they complicate the process of deciphering evolutionary relationships. Convergence

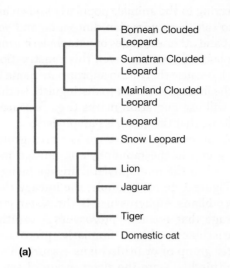

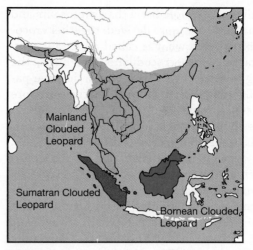

(a)

(b)

(c)

(d)

Figure 1–6 An example of the value of phylogenetic analyses in conservation. (a) A cladogram of clouded leopards and (b) a map showing the geographic locations of surviving populations. The clouded leopard found on the Asian mainland, *Neofelis nebulosa* (c), is as distant genetically from the species of clouded leopard found on Sumatra and Borneo, *Neofelis diardi* (d), as lions are from tigers. The island forms of *Neofelis diardi* are more closely related to each other than either is to the mainland form. Nonetheless, the genetic differences that distinguish the leopards on Sumatra from those on Borneo are large enough to be separated in captive breeding programs for the two forms.

and parallelism give an appearance of similarity (as in the wings of birds and bats) that is not the result of common evolutionary origin. Reversal, in contrast, conceals similarity (e.g., between cetaceans and their four-legged terrestrial ancestors) that is the result of common evolutionary origin.

Phylogeny and Conservation

Combining genetic analysis with cladistic analyses can provide an important tool for biologists concerned with conservation (**Figure 1–6**). For example, some of the new species of mammals described in section 1.1

were identified by comparing their DNA with the DNA of related species. When a genetic difference is large, it means that the two forms have been reproductively isolated from each other and have followed different evolutionary pathways. From a conservationist's perspective, lineages that have evolved substantial genetic differences are Evolutionarily Significant Units (ESUs), and management plans should protect the genetic diversity of ESUs.

For example, a genetic study published in 2007 revealed that the clouded leopards on the islands of Borneo and Sumatra (*Neofelis diardi*) and those on the

Asian mainland (*Neofelis nebulosa*) separated between 1.4 and 2.8 million years ago, and the three forms have been following independent evolutionary pathways since then. The genetic mainland form and the island forms are genetically different. Furthermore, the island populations are reproductively isolated from each other, and the clouded leopards on Borneo and Sumatra are genetically distinct. Thus, the three forms represent three ESUs, and conservation plans should treat the mainland species and the two island species separately. Before this study the three forms were grouped together, and a portion of the genetic diversity of clouded leopards was lost through interbreeding in captivity.

1.6 Earth History and Vertebrate Evolution

Since their origin in the early Paleozoic, vertebrates have been evolving in a world that has changed enormously and repeatedly. These changes have affected vertebrate evolution both directly and indirectly. Understanding the sequence of changes in the positions of continents, and the significance of those positions regarding climates and interchange of faunas, is central to understanding the vertebrate story. These events are summarized inside the front cover of the book, and Chapters 7, 15, and 19 give details.

The history of Earth has occupied three geological eons: the Archean, Proterozoic, and Phanerozoic. Only the Phanerozoic, which began about 542 million years ago, contains vertebrate life, and it is divided into three geological eras: the Paleozoic (Greek *paleo* = ancient and *zoo* = animal), Mesozoic (Greek *meso* = middle), and Cenozoic (Greek *cen* = recent). These eras are divided into periods, which can be further subdivided in a variety of ways, such as the subdivisions called epochs within the Cenozoic era from the Paleocene to the Recent.

Movement of landmasses, called continental drift, has been a feature of Earth's history at least since the Proterozoic, and the course of vertebrate evolution has been shaped extensively by continental movements. By the early Paleozoic, roughly 540 million years ago, a recognizable scene had appeared. Seas covered most of Earth as they do today, large continents floated on Earth's mantle, life had become complex, and an atmosphere of oxygen had formed, signifying that the photosynthetic production of food resources had become a central phenomenon of life.

The continents still drift today—North America is moving westward and Australia northward at the speed of approximately 4 centimeters per year (about the rate at which fingernails grow). Because the movements are so complex, their sequence, their varied directions, and the precise timing of the changes are difficult to summarize. When the movements are viewed broadly, however, a simple pattern unfolds during vertebrate history: fragmentation, coalescence, fragmentation.

Continents existed as separate entities over 2 billion years ago. Some 300 million years ago, all of these separate continents combined to form a single landmass known as Pangaea, which was the birthplace of terrestrial vertebrates. Persisting and drifting northward as an entity, this huge continent began to break apart about 150 million years ago. Its separation occurred in two stages: first into Laurasia in the north and Gondwana in the south, and then into a series of units that have drifted and become the continents we know today.

The complex movements of the continents through time have had major effects on the evolution of vertebrates. Most obvious is the relationship between the location of landmasses and their climates. At the end of the Paleozoic, much of Pangaea was located on the equator, and this situation persisted through the middle of the Mesozoic. Solar radiation is most intense at the equator, and climates at the equator are correspondingly warm. During the late Paleozoic and much of the Mesozoic, large areas of land enjoyed tropical conditions. Terrestrial vertebrates evolved and spread in these tropical regions. By the end of the Mesozoic, much of Earth's landmass had moved out of equatorial regions; and, by the mid-Cenozoic, most terrestrial climates in the higher latitudes of the Northern and Southern Hemispheres were temperate instead of tropical.

A less obvious effect of the position of continents on terrestrial climates comes from changes in patterns of oceanic circulation. For example, the Arctic Ocean is now largely isolated from the other oceans, and it does not receive warm water via currents flowing from more equatorial regions. High latitudes are cold because they receive less solar radiation than do areas closer to the equator, and the Arctic Basin does not receive enough warm water to offset the lack of solar radiation. As a result, the Arctic Ocean is permanently frozen, and cold climates extend well southward across the continents. The cooling of climates in the Northern Hemisphere at the end of the Eocene epoch, around 34 million years ago, may have been a factor leading to the extinction of archaic mammals, and it is partly the result of changes in oceanic circulation at that time.

Another factor that influences climates is the relative levels of the continents and the seas. At some periods in Earth's history, most recently in the late Mesozoic and again in the first part of the Cenozoic, shallow seas flooded large parts of the continents. These **epicontinental seas** extended across the middle of North America and the middle of Eurasia during the Cretaceous period and early Cenozoic. Water absorbs heat as air temperature rises, and then releases that heat as air temperature falls. Thus, areas of land near large bodies of water have maritime climates—they do not get very hot in summer or very cold in winter, and they are usually moist because water that evaporates from the sea falls as rain on the land. Continental climates, which characterize areas far from the sea, are usually dry with cold winters and hot summers. The draining of the epicontinental seas at the end of the Cretaceous probably contributed to the demise of the dinosaurs by making climates in the Northern Hemisphere more continental.

In addition to changing climates, continental drift has formed and broken land connections between the continents. Isolation of different lineages of vertebrates on different landmasses has produced dramatic examples of the independent evolution of similar types of organisms, such as the diversification of mammals in the mid-Cenozoic, a time when Earth's continents reached their greatest separation during the history of vertebrates.

Much of evolutionary history appears to depend on whether a particular lineage of animals was in the right place at the right time. This random element of evolution is assuming increasing prominence as more detailed information about the times of extinction of old groups and radiation of new groups suggests that competitive replacement of one group by another is not the usual mechanism of large-scale evolutionary change. The movements of continents and their effects on climates and the isolation or dispersal of animals are taking an increasingly central role in our understanding of vertebrate evolution. On a continental scale, the advance and retreat of glaciers in the Pleistocene caused homogeneous habitats to split and merge repeatedly, isolating populations of widespread species and leading to the evolution of new species.

Summary

The more than 63,000 species of living vertebrates span a size range from less than a gram to more than 100,000 kilograms. They live in habitats extending from the bottom of the sea to the tops of mountains. This extraordinary diversity is the product of more than 500 million years of evolution, and the vast majority of species fall into one of the two major divisions of bony fishes (Osteichthyes)—the aquatic ray-finned fishes (Actinopterygii) and the primarily terrestrial lobe-finned fishes and tetrapods (Sarcopterygii), each of which contains more than 25,000 extant species.

Phylogenetic systematics, usually called cladistics, classifies animals on the basis of shared derived character states. Natural evolutionary groups can be defined only by these derived characters; retention of ancestral characters does not provide information about evolutionary lineages. Application of this principle produces groupings of animals that reflect evolutionary history as accurately as we can discern it and forms a basis for making hypotheses about evolution and for designing management plans that conserve the genetic diversity of evolutionary lineages.

Earth has changed dramatically during the half-billion years of vertebrate history. Continents were fragmented when vertebrates first appeared; coalesced into one enormous continent, Pangaea, about 300 million years ago; and began to fragment again about 150 million years ago. This pattern of fragmentation, coalescence, and fragmentation has resulted in isolation and renewed contact of major groups of vertebrates on a worldwide scale.

Discussion Questions

1. Why don't phylogenetic (cladistic) classifications have a fixed number of hierarchical categories like those in a Linnaean classification?

2. What aspect of evolution does a phylogenetic classification represent more clearly than a Linnaean classification does?

3. What is the meaning of an Evolutionarily Significant Unit (ESU) in conservation biology?
4. Tetrapoda (node 6 in Figure 1–4) is a crown group, whereas Tetrapodomorpha is the corresponding stem group. What organisms are included in Tetrapoda? In Tetrapodomorpha?
5. What is the difference between parallel and convergent evolution?
6. What is the significance of an extant phylogenetic bracket?

Additional Information

Anderson, R. P., and C. O. Handley, Jr. 2001. A new species of three-toed sloth (Mammalia: Xenarthra) from Panamá, with a review of the genus *Bradypus*. *Proceedings of the Biological Society of Washington* 114:1–33.

Budd, G. 2001. Climbing life's tree. *Nature* 412:48.

Ceballos, G., and P. R. Ehrlich. 2009. Discoveries of new mammal species and their implications for conservation and ecosystem services. *Proceedings of the National Academy of Sciences* 106(10):3841–3846.

Davenport, T. R. B., et al. 2006. A new genus of African monkey, *Rungwecebus*: Morphology, ecology, and molecular phylogenetics. *Science* 312:1378–1381.

de Queiroz, K., and J. Gauthier. 1992. Phylogenetic taxonomy. *Annual Review of Ecology and Systematics* 23:449–480.

Donegan, T. 2009. Type specimens, samples of live individuals and the Galápagos pink land iguana. *Zootaxa* 2201:12–20.

Douglas, M. E., et al. 2011. Conservation phylogenetics of helodermatid lizards using multiple molecular markers and a supertree approach. *Molecular Phylogenetics and Evolution* 55:153–167.

Edwards et al., Ancient Hybridization and an Irish Origin for the Modern Polar Bear Matriline, *Current Biology* (2011), doi:10.1016/j.cub.2011.05.058.

Gentile, G., and H. Snell. 2009. *Conolophus marthae* sp. nov. (Squamata, Iguanidae), a new species of land iguana from the Galápagos archipelago. *Zootaxa* 2201:1–10.

Hennig, W. 1966. *Phylogenetic Systematics*. Urbana, IL: University of Illinois Press.

Jones, T., et al. 2005. The highland mangabey *Lopjocebus kipunji*: A new species of African monkey. *Science* 308:1161–1164.

Kemp. T. S. 1999. *Fossils and Evolution*. Oxford, UK: Oxford University Press.

Kuntner, M. et al. 2011. Phylogeny and conservation priorities of afrotherian mammals (Afrotheria, Mammalia) *Zoologica Scripta* 40:1–15.

May-Collado, L. J. and I. Agnarsson. 2011. Phylogenetic analysis of conservation priorities for aquatic mammals and their terrestrial relatives, with a comparison of methods. *PLoS ONE* 6(7): e22562. doi:10.1371/journal.pone.0022562.

Mendes Pontes, A. R., et al. 2006. A new species of capuchin monkey, genus *Cebus* Erxleben (Cebidae, Primates): Found at the very brink of extinction in the Pernambuco Endemism Centre. *Zootaxa* 1200:1–12.

Nemesio, A. 2009. On the live holotype of the Galápagos pink land iguana, *Conolophus marthae* Gentile & Snell, 2009 (Squamata, Iguanidae): Is it an acceptable exception? *Zootaxa* 2201:21–25.

Olson, L. E., et al. 2008. Additional molecular evidence strongly supports the distinction between the recently described African primate *Rungwecebus kipunji* (Cercopithecidae, Papionini) and *Lophocebus*. *Molecular Phylogenetics and Evolution* 48:789–794.

Pennisi, E. 2001. Linnaeus's last stand? *Science* 291:2304–2307.

Pio, D. V., et al. 2011. Spatial predictions of phylogenetic diversity in conservation decision making. *Conservation Biology* 25:1229–1239.

Roberts, T. E., et al. 2010. The biogeography of introgression in the critically endangered African monkey *Rungwecebus kipunji*. *Biology Letters* 6:233–237.

Roca, A. L., et al. 2001. Genetic evidence for two species of elephant in Africa. *Science* 293:1473–1477.

Rosenbaum, H. C., et al. 2000. World-wide genetic differentiation of *Eubalaena*: Questioning the number of right whale species. *Molecular Ecology* 9:1793–1802.

Rovero, F., et al. 2008. A new species of giant sengi or elephant-shrew (genus *Rhynchocyon*) highlights the exceptional biodiversity of the Udzungwa Mountains of Tanzania. *Journal of Zoology* 274:126–133.

Sangster, G. 2009. Increasing numbers of bird species result from taxonomic progress, not taxonomic inflation. *Proceedings of the Royal Society B* 276:3185–3191.

Singleton, M. 2009. The phenetic affinities of *Rungwecebus kipunji*. *Journal of Human Evolution* 56:25–42.

Swiderski, D. L. (ed.). 2001. Beyond reconstruction: Using phylogenies to test hypotheses about vertebrate evolution. *American Zoologist* 41:485–607.

Templeton, A. R. 2010. The diverse applications of cladistic analysis of molecular evolution, with special reference to nested clade analysis. *International Journal of Molecular Sciences* 11:124–139.

Timm, R. M., et al. 2005. What constitutes a proper description? *Science* 309:2163–2164.

Wado, S., et al. 2003. A newly discovered species of living baleen whale. *Nature* 426:278–281.

Wilting, A., et al. 2007. Clouded leopard phylogeny revisited: Support for species recognition and population division between Borneo and Sumatra. *Frontiers in Zoology* 4:15. http://www.frontiersinzoology.com/content/4/1/15

Zinner, D., et al. 2009. Is the new primate genus *Rungwecebus* a baboon? PLoS ONE 4(3): e4859. doi:10.1371/journal.pone.0004859.

Websites

Census of Marine Life

The Census of Marine Life is a growing global network of researchers in more than 80 nations engaged in a 10-year initiative to assess and explain the diversity, distribution, and abundance of marine life in the oceans—past, present, and future. www.coml.org

Discover Life

Species identification and images. http://www.discoverlife.org/

International Code of Zoological Nomenclature

The International Commission on Zoological Nomenclature sets the procedures for naming species of animals; its goal is to promote "standards, sense, and stability for animal names in science". http://www.nhm.ac.uk/hosted-sites/iczn/code/

Journey into Phylogenetic Systematics

Explanation of the methods, uses, and implications of cladistic classification. http://www.ucmp.berkeley.edu/clad/clad1.html

Tree of Life web project

When complete, the Tree of Life web project will have a page for every species of organism, both living and extinct; web pages follow the evolutionary branching patterns of the lineages. http://tolweb.org/tree/phylogeny.html

CHAPTER 2

Vertebrate Relationships and Basic Structure

In this chapter, we explain the structures that are characteristic of vertebrates, discuss the relationship of vertebrates to other members of the animal kingdom, and describe the systems that make vertebrates functional animals. We need an understanding of the fundamentals of vertebrate anatomy, physiology, and development to appreciate the changes that have occurred during their evolution and to trace homologies between basal vertebrates and more derived ones.

2.1 Vertebrates in Relation to Other Animals

Vertebrates are a diverse and fascinating group of animals. Because we are vertebrates ourselves, that statement may seem chauvinistic, but vertebrates are remarkable in comparison with most other animal groups. Vertebrates are the subphylum Vertebrata of the phylum Chordata. At least 30 other animal phyla have been named, but only the phylum Arthropoda (insects, crustaceans, spiders, etc.) rivals the vertebrates in diversity of forms and habitat. And it is only in the phylum Mollusca (snails, clams, and squid) that we find animals (such as octopus and squid) that approach the very large size of some vertebrates and also have a capacity for complex learning.

The tunicates (subphylum Urochordata) and cephalochordates (subphylum Cephalochordata) are placed with vertebrates in the phylum Chordata. Within the chordates, cephalochordates and vertebrates resemble each other anatomically, but molecular characteristics show that tunicates are more closely related to vertebrates than are cephalochordates.

Characteristics of Chordates

Chordates are united by several shared derived features, which are seen in all members of the phylum at some point in their lives:

- A notochord (a dorsal stiffening rod that gives the phylum Chordata its name)
- A dorsal hollow nerve cord
- A segmented, muscular postanal tail (i.e., extending beyond the gut region)
- An endostyle (a ciliated, glandular groove on the floor of the pharynx that secretes mucus for trapping food particles during filter feeding; generally homologous with the thyroid gland of vertebrates, an endocrine gland involved with regulating metabolism)

Chordates are also characterized by a pharynx (throat region) containing pharyngeal slits. Nonvertebrate

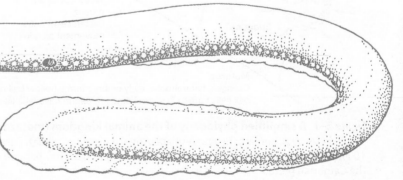

chordates use the pharyngeal slits for filter feeding, and aquatic vertebrates (fishes) use them for respiration. Some other deuterostomes (the larger grouping to which chordates belong) also have pharyngeal slits, however, and they may be a primitive feature for chordates as a group. Chordates also share some nervous system features: they all have structures in the brain corresponding to a pineal eye and hormone-regulating pituitary.

Chordate Relationships

Although chordates are all basically bilaterally symmetrical animals (i.e., one side is the mirror image of the other), they have a left-to-right asymmetry within the body—for example, the position of the heart on the left side and most of the liver on the right side. (Rare human individuals have a condition termed "situs inversus," in which the positions of the major body organs are reversed.)

The relationship of chordates to other kinds of animals is revealed by anatomical, biochemical, and embryonic characters as well as by the fossil record. **Figure 2–1** shows the relationships of animal phyla. Vertebrates superficially resemble other active animals, such as insects, in having a distinct head end, jointed legs, and bilateral symmetry. However, and perhaps surprisingly, developmental and molecular data show that the phylum Chordata is closely related to the phylum Echinodermata (starfishes, sea urchins, and the like), which are marine forms without distinct heads and with pentaradial (fivefold and circular) symmetry as adults.

Characteristics of Deuterostomes

The chordates, echinoderms, and two other phyla (hemichordates and xenoturbellids) are linked as **deuterostomes** (Greek *deutero* = second and *stoma* = mouth) by several unique embryonic features, such as the way in which their eggs cleave into daughter cells after fertilization, their larval form, and some other features discussed later. Hemichordates are a small,

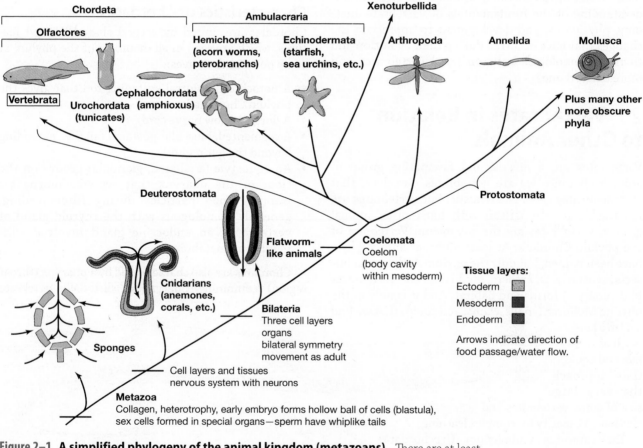

Figure 2–1 A simplified phylogeny of the animal kingdom (metazoans). There are at least 30 phyla today (Chordata, Echinodermata, Annelida, etc. are phyla). Possibly as many as 15 additional phyla are known from the early Paleozoic era; they became extinct at the end of the Cambrian period.

poorly known phylum of marine animals containing the earthwormlike acorn worms and the fernlike pterobranchs. Xenoturbellids are small marine wormlike forms (only two species are known), which have recently been identified as deuterostomes by molecular analysis.

Hemichordates were long considered the sister group of chordates because both groups have pharyngeal slits, and hemichordates also have features of the pharynx that can be interpreted as the precursor to an endostyle. However, we now consider these to be primitive features of the larger group (deuterostomes) in which hemichordates and chordates are placed.

Although modern echinoderms lack pharyngeal slits, some extinct echinoderms appear to have had them. (The diversity of extinct echinoderms is much greater than that of the living forms.) Furthermore, early echinoderms were bilaterally symmetrical, meaning that the fivefold symmetry of modern echinoderms is probably a derived character of that lineage. Molecular characteristics currently unite echinoderms and hemichordates as the Ambulacraria, and xenoturbellids are more closely related to these phyla than to the chordates.

Relationships of Deuterostomes

To consider how deuterostomes are related to other animals, we will start at the bottom of the tree and work upward (see Figure 2–1). All animals (metazoans) are multicellular and share some features of reproduction and embryonic development: they have motile sperm with whiplike tails, the embryo initially forms as a hollow ball of cells (the blastula), and sex cells form in special organs. All metazoans also have the structural protein collagen, which in humans forms the matrix of many of our tissues and organs, including the nose and ears.

Animals more derived than sponges have a nervous system, and their bodies are made of distinct layers of cells, or germ layers, that are laid down early in development at a stage called gastrulation. Gastrulation occurs when the hollow ball of cells forming the blastula folds in on itself, producing two distinct layers of cells and a gut with an opening to the outside at one end. The outer layer of cells is the **ectoderm** (Greek *ecto* = outside and *derm* = skin), and the inner layer forms the **endoderm** (Greek *endo* = within).

Jellyfishes and related animals have only these two layers of body tissue, making them diploblastic (Greek *diplo* = two and *blast* = a bud or sprout). Animals more derived than jellyfishes and their kin add an additional, middle cell layer of **mesoderm** (Greek *mesos* = middle), making them triploblastic (Greek *triplo* = three).

Triploblasts have a gut that opens at both ends (i.e., with a mouth and an anus) and are bilaterally symmetrical with a distinct anterior (head) end at some point in their life. The mesoderm forms the body's muscles, and only animals with a mesoderm are able to be motile as adults. (Larval forms do not need muscles because they are small enough to be powered by hairlike cilia on the body surface.) The **coelom**, an inner body cavity that forms as a split within the mesoderm, is another derived character of most, but not all, triploblastic animals.

Coelomate animals (i.e., animals with a coelom) are split into two groups on the basis of how the mouth and anus form. When the blastula folds in on itself to form a gastrula, it leaves an opening to the outside called the blastopore (Latin *porus* = a small opening). During the embryonic development of coelomates, a second opening develops. In the lineage called protostomes (Greek *proto* = first and *stome* = mouth), the blastopore (which was the first opening in the embryo) becomes the mouth, whereas in deuterostomes the second opening becomes the mouth and the blastopore becomes the anus. Chordates, hemichordates, and echinoderms are deuterostomes, whereas mollusks (snails, clams, and squid), arthropods (insects, crustaceans, and spiders), annelids (earthworms), and many other phyla are protostomes (see Figure 2–1).

Extant Nonvertebrate Chordates

The two groups of extant nonvertebrate chordates are small marine animals. More types of nonvertebrate chordates may have existed in the past, but such soft-bodied animals are rarely preserved as fossils. Some possible Early Cambrian chordates are described at the end of this section.

Urochordates Present-day **tunicates** (subphylum Urochordata) are marine animals that filter particles of food from the water with a basketlike perforated pharynx. There are about 3000 living species, and all but 100 or so are sedentary as adults, attaching themselves to the substrate either singly or in colonies.

Adult tunicates (also known as sea squirts, or ascideans) bear little apparent similarity to cephalochordates and vertebrates. Most tunicate species have a brief free-swimming larval period lasting a few minutes to a few days, after which the larvae metamorphose into sedentary adults attached to the substrate, although some species remain motile as adults. Tunicate larvae have a notochord, a dorsal hollow nerve cord, and a muscular postanal tail that moves in a fishlike swimming pattern (**Figure 2–2**). It was long believed that the earliest chordates would have been sessile as adults (like most

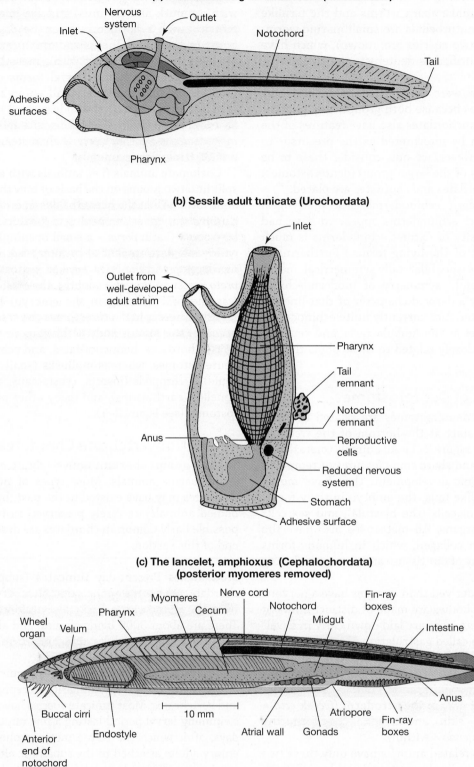

(a) Free-swimming larval tunicate (Urochordata)

Nervous system

Inlet

Outlet

Notochord

Tail

Adhesive surfaces

Pharynx

(b) Sessile adult tunicate (Urochordata)

Inlet

Outlet from well-developed adult atrium

Pharynx

Tail remnant

Notochord remnant

Reproductive cells

Anus

Reduced nervous system

Stomach

Adhesive surface

(c) The lancelet, amphioxus (Cephalochordata) (posterior myomeres removed)

Myomeres

Nerve cord

Pharynx

Cecum

Notochord

Fin-ray boxes

Wheel organ

Velum

Midgut

Intestine

Buccal cirri

Endostyle

Atrial wall

Gonads

Anus

Anterior end of notochord

10 mm

Atriopore

Fin-ray boxes

Figure 2–2 Nonvertebrate chordates. Tunicates have a free-swimming larva (a) that metamorphoses into a sessile adult (b), whereas amphioxus (c) is free-swimming throughout its life.

other deuterostomes) and that cephalochordates and vertebrates evolved from an ancestor that resembled a tunicate larva. However, it now seems more likely that a sessile adult stage (Figure 2–2b) is a derived character for tunicates and that the living species that remain free-swimming as adults most resemble the ancestral chordate. The ancestral chordate (and indeed, the ancestral deuterostome) was probably a free-swimming wormlike creature that used pharyngeal slits for filter feeding.

Cephalochordates The subphylum Cephalochordata contains about 27 species, all of which are small, superficially fishlike marine animals usually less than 5 centimeters long. The best-known cephalochordate is the lancelet (*Branchiostoma lanceolatum*), more commonly known as **amphioxus** (Greek *amphi* = both and *oxy* = sharp). Amphioxus means "sharp at both ends," an appropriate term for an animal in which the front and rear ends are nearly the same shape because it lacks a distinct head. Lancelets are widely distributed in marine waters of the continental shelves and are usually burrowing, sedentary animals as adults, although the adults of a few species retain an active, free-swimming behavior.

A notable characteristic of amphioxus is its fishlike locomotion produced by **myomeres**—blocks of striated muscle fibers arranged along both sides of the body and separated by sheets of connective tissue. Sequential contraction of the myomeres bends the body from side to side, resulting in forward or backward propulsion. The notochord acts as an incompressible elastic rod, extending the full length of the body and preventing the body from shortening when the myomeres contract. While the notochord of vertebrates ends midway through the head region, the notochord of amphioxus extends from the tip of the snout to the end of the tail, projecting well beyond the region of the myomeres at both ends. This anterior elongation of the notochord apparently is a specialization that aids in burrowing.

Figure 2–2c shows some details of the internal structure of amphioxus. Amphioxus and vertebrates differ in the use of the pharyngeal slits. Amphioxus has no gill tissue associated with these slits; its body is small enough for oxygen uptake and carbon dioxide loss to occur by diffusion over the body surface, and the gill slits are used for filter feeding. Water is moved over the gill slits by cilia on the gill bars between the slits, aided by the features of the buccal (mouth region, Latin *bucc* = cheek) cirri and the wheel organ, while the velum is a flap helping to control the one-way flow of water.

In addition to the internal body cavity, or coelom, amphioxus has an external body cavity called the atrium, which is also seen in tunicates and hemichordates—and thus is probably an ancestral deuterostome feature—but is absent from vertebrates. (This atrium is not the same as the atrium of the vertebrate heart; the word *atrium* [plural *atria*] comes from the Latin term for an open space.) The atrium of amphioxus is formed by outgrowths of the body wall (metapleural folds), which enclose the body ventrally. (Imagine yourself wearing a cape and then extending your arms until there is a space between the cape and your body—that space would represent the position of the atrium in amphioxus.) The atrium opens to the outside world via the atriopore, an opening in front of the anus. The atrium appears to work in combination with the beating of the cilia on the pharyngeal bars and the wheel organ in the head to control passage of substances through the pharynx and is probably functionally associated with the ancestral chordate character of using the pharyngeal slits for filter feeding.

Cephalochordates have several anatomical features that are shared with vertebrates but absent from tunicates. In addition to the myomeres, amphioxus has a vertebrate-like tail fin, a circulatory system similar to that of vertebrates, with a dorsal aorta and a ventral heartlike structure that forces blood through the pharyngeal slits, and specialized excretory cells called podocytes.

Olfactores—Tunicates Plus Vertebrates Cephalochordates were long considered to be the sister group of vertebrates on the basis of the shared anatomical characters described above. However, some (but not all) molecular analyses have placed tunicates as the sister group of vertebrates. (The group that includes vertebrates plus tunicates is the Olfactores.) If tunicates are the sister group of vertebrates, the sessile nature of most adult tunicates must be a derived character of the lineage, and both ancestral tunicates and ancestral chordates likely were mobile.

A sister-group relationship of vertebrates and tunicates is now generally accepted, as we show in Figure 2–1. Morphology still strongly supports a vertebrate/cephalochordate association, however, and the features that cephalochordates share with vertebrates may simply be ancestral chordate features.

Cambrian Chordates

The best-known early chordatelike animal is *Pikaia*, which looks a little like an amphioxus. About 100 specimens of this animal are known from the Middle Cambrian Burgess Shale in British Columbia. *Pikaia* is probably not a true cephalochordate: although it has an obvious notochord, the myomeres are straight rather than V-shaped and there is no evidence of gills.

Recently spectacular fossils of soft-bodied animals have been found in the Early Cambrian Chengjiang

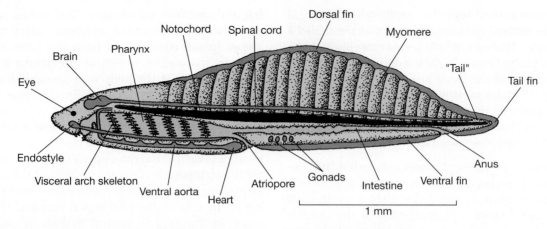

Figure 2–3 **The yunnanozoan *Haikouella*, from the Early Cambrian Chenjiang Fauna of southern China, an early chordate or chordate relative.**

formation in southern China. This 522-million-year-old deposit is at least 17 million years older than the Burgess Shale. The Chengjiang deposit includes the earliest-known true vertebrates (described in Chapter 3) and some intriguing fossils that may be early chordates. (The flattened "road kill-like" nature of the Chengjiang specimens makes it difficult to interpret their structure.)

The most vertebrate-like member of the Chengjiang Fauna is *Haikouella*, which is known from more than 300 individuals (**Figure 2–3**). *Haikouella* has the chordate features of myomeres, a notochord, and a pharynx apparently enclosed in an atrium, as well as derived features that suggest that it is the sister group to vertebrates. These features include a large brain, clearly defined eyes, thickened branchial bars (that appear to be made of a type of cartilage similar to that of lamprey larvae), and an upper lip like that of larval lampreys. The endostyle and tentacles surrounding the mouth suggest that this animal was a suspension feeder, like amphioxus.

Haikouella appears to have developed a vertebrate-like muscular pharynx. The thickened branchial bars appear stout enough to support both gill tissue and pharyngeal muscles, suggesting that the gills were used for respiration as well as for feeding.

2.2 Definition of a Vertebrate

The term *vertebrate* is derived from the vertebrae that are serially arranged to make up the spinal column, or backbone, of vertebrate animals. In ourselves, as in other land vertebrates, the vertebrae form around the notochord during development and also encircle the nerve cord. The bony vertebral column replaces the original notochord after the embryonic period. In many

fishes the vertebrae are made of cartilage rather than bone.

All vertebrates have the uniquely derived feature of a **cranium**, or skull, which is a bony, cartilaginous, or fibrous structure surrounding the brain. Vertebrates also have a prominent head containing complex sense organs. Although many of the genes that determine head development in vertebrates are present in amphioxus, the anterior portion of the vertebrate head does seem to be a new feature of vertebrates.

Hagfish have been considered to lack vertebrae, but new developmental evidence suggests homologs to the ventral portion of the vertebrae in the tail of jawed fishes. The structures corresponding to vertebrae in lampreys are segmental cartilaginous rudiments (arcualia) flanking the nerve cord. Fully formed vertebrae, with a **centrum** (plural *centra*) surrounding the notochord, are found only in **gnathostomes** (jawed vertebrates—see Chapter 3), and many jawed fishes retain a functional notochord as adults.

Because of the apparent lack of vertebrae in hagfishes, some people have preferred the term **Craniata** to Vertebrata for the subphylum. However, we continue to use the familiar term *vertebrate* in this book, and the term *Craniata* is redundant if hagfishes are indeed the sister group of lampreys (see Chapter 3).

Unique Embryonic Features of Vertebrates

Several embryonic features may account for many of the differences between vertebrates and other chordates. The two primary ones are duplication of the *Hox* **gene complex** (homeobox genes) and the appearance of a new embryonic tissue, the **neural crest**.

***Hox* Genes** *Hox* genes regulate the expression of a hierarchical network of other genes that control the process of development from front to back along the body. Vertebrates have more *Hox* genes than other groups of animals: Jellyfishes (and possibly also sponges) have one or two *Hox* genes, the common ancestor of protostomes and deuterostomes probably had seven, and more derived metazoans have up to thirteen. However, vertebrates are unique in having undergone duplications of the entire *Hox* complex.

The first duplication event seems to have occurred at the start of vertebrate evolution because amphioxus and tunicates have a single *Hox* cluster, whereas the living jawless vertebrates have two. A second duplication event had taken place by the evolution of gnathostomes, because all jawed vertebrates have at least four clusters. Finally, an additional duplication event occurred in both teleost fishes (derived bony fishes) and frogs.

Interactions among genes modify the effects of those genes, and more genes allow more interactions that probably produce more complex structures. The doubling and redoubling of the *Hox* gene sequence during vertebrate evolution is believed to have made the structural complexity of vertebrates possible.

Neural Crest Neural crest is a new tissue in embryological development that forms many novel structures in vertebrates, especially in the head region (a more detailed description is given in section 2.3). The evolution of the neural crest is the most important innovation in the origin of the vertebrate body plan. Neural-crest tissue is a fourth germ layer that is unique to vertebrates and is on a par with ectoderm, endoderm, and mesoderm. Neural-crest cells originate at the lateral boundary of the neural plate, the embryonic structure that makes the nerve cord, and migrate throughout the body to form a variety of structures, including pigment cells.

A similar population of cells, with a similar genetic expression, can be found in amphioxus, but here the cells do not migrate and do not change into different cell types. Recently cells resembling migratory neural-crest cells have been identified in the larval stage of one tunicate species, where they differentiate into pigment cells. These cells in tunicates may represent a precursor to the vertebrate neural crest. (This appears to be one morphological feature that tunicates share with vertebrates, but it is not seen in cephalochordates.) If the Cambrian chordate *Haikouella* has been correctly interpreted as having eyes and a muscular pharynx, these features would imply the presence of neural crest in this animal.

Placodes Another new type of embryonic tissue in vertebrates, which is similar to neural crest but probably has a different origin, forms the epidermal thickenings (placodes) that give rise to the complex sensory organs of vertebrates, including the nose, eyes, and inner ear. Some placode cells migrate caudally to contribute, along with the neural-crest cells, to the lateral line system and to the cranial nerves that innervate it.

MicroRNAs The appearance of many microRNAs is a genetic innovation in vertebrates that may contribute to their anatomical complexity. MicroRNAs are noncoding RNA sequences 22 bases long that have been added to the genomes of metazoans throughout their evolutionary history. MicroRNAs regulate the synthesis of proteins by binding to complementary base sequences of messenger RNAs. The phylum Chordata is characterized by the addition of two new microRNAs, another three are shared by vertebrates and tunicates, and all vertebrates possess an additional 41 unique microRNAs. (All vertebrate lineages have independently acquired yet more microRNAs of their own, and mammals in particular have a great number of novel microRNAs.) MicroRNAs are involved in regulating the development of some derived vertebrate structures, including the liver and the kidney.

Brains

The brains of vertebrates are larger than the brains of primitive chordates, and they have three parts—the forebrain, midbrain, and hindbrain. The brain of amphioxus appears simple, but genetic studies show that amphioxus has all the genes that code for the vertebrate brain with the exception of those directing formation of the front part of the vertebrate forebrain, the **telencephalon** (the portion of the brain that contains the cerebral cortex, the area of higher processing in vertebrates). The presence of the genes in amphioxus combined with the absence of a complex brain reinforces the growing belief that the differences in how genes are expressed in different animals are as important as differences in what genes are present. Other unique vertebrate features include a multilayered epidermis and blood vessels that are lined by endothelium.

2.3 Basic Vertebrate Structure

This section serves as an introduction to vertebrate anatomical structure and function. The heart of this section is in **Table 2–1** and **Figure 2–4**, which contrast the basic vertebrate condition with that of a nonvertebrate chordate such as amphioxus. These same systems will

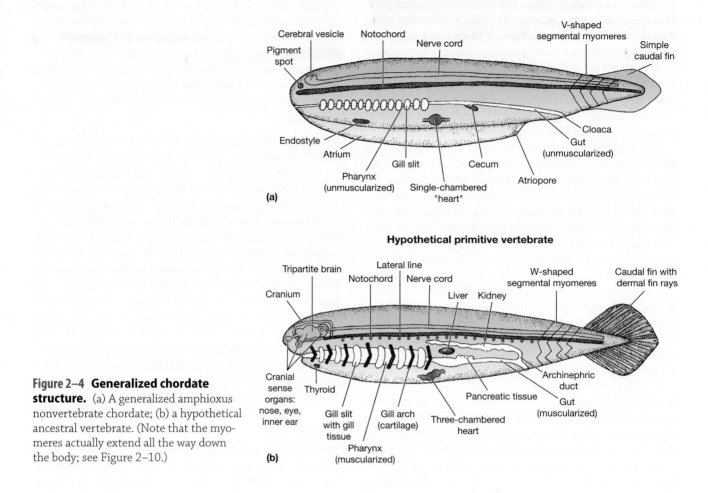

Amphioxus-like nonvertebrate chordate

Cerebral vesicle — Notochord — Nerve cord — V-shaped segmental myomeres — Simple caudal fin

Pigment spot

Endostyle — Atrium — Pharynx (unmuscularized) — Gill slit — Single-chambered "heart" — Cecum — Atriopore — Gut (unmuscularized) — Cloaca

(a)

Hypothetical primitive vertebrate

Tripartite brain — Notochord — Lateral line — Nerve cord — Liver — Kidney — W-shaped segmental myomeres — Caudal fin with dermal fin rays

Cranium

Cranial sense organs: nose, eye, inner ear — Thyroid — Gill slit with gill tissue — Gill arch (cartilage) — Three-chambered heart — Pancreatic tissue — Archinephric duct — Gut (muscularized)

Pharynx (muscularized)

(b)

Figure 2–4 Generalized chordate structure. (a) A generalized amphioxus nonvertebrate chordate; (b) a hypothetical ancestral vertebrate. (Note that the myomeres actually extend all the way down the body; see Figure 2–10.)

be further discussed for more derived vertebrates in later chapters; our aim here is to provide a general introduction to the basics of vertebrate anatomy. More detail can be found in books listed at the end of the chapter.

At the whole-animal level, an increase in body size and increased activity distinguish vertebrates from nonvertebrate chordates. Early vertebrates generally had body lengths of 10 centimeters or more, which is about an order of magnitude larger than the bodies of nonvertebrate chordates. Because of their relatively large size, vertebrates need specialized systems to carry out processes that are accomplished by diffusion or ciliary action in smaller animals. Vertebrates are also more active animals than other chordates, so they need organ systems that can carry out physiological processes at a greater rate. The transition from nonvertebrate chordate to vertebrate was probably related to the adoption of a more actively predaceous mode of life, as evidenced by the features of the vertebrate head

(largely derived from neural-crest tissue) that would enable suction feeding with a muscular pharynx, and a bigger brain and more complex sensory organs for perceiving the environment.

Vertebrates are characterized by mobility, and the ability to move requires muscles and a skeleton. Mobility brings vertebrates into contact with a wide range of environments and objects in those environments, and a vertebrate's external protective covering must be tough but flexible. Bone and other mineralized tissues that we consider characteristic of vertebrates had their origins in this protective integument.

Embryonic Development

Studying embryos can show how systems develop and how the form of the adult is related to functional and historical constraints during development. Scientists no longer adhere to the biogenetic law that "ontogeny recapitulates phylogeny" (i.e., the idea that the embryo

Table 2–1 Comparison of features in nonvertebrate chordates and ancestral vertebrates

Generalized Nonvertebrate Chordate (based on features of the living cephalochordate amphioxus)	Ancestral Vertebrate (based on features of the living jawless vertebrates—hagfishes and lampreys)
Brain and Head End	
Notochord extends to tip of head (may be derived condition).	Head extends beyond tip of notochord.
No cranium (skull).	Cranium—skeletal supports around brain, consisting of capsules surrounding the main parts of the brain and their sensory components plus underlying supports.
Simple brain (= cerebral vesicle), no specialized sense organs (except photoreceptive frontal organ, probably homologous with the vertebrate eye).	Tripartite brain and multicellular sense organs (eye, nose, inner ear).
Poor distance sensation (although the skin is sensitive).	Improved distance sensation: in addition to the eyes and nose, also have a lateral line system along the head and body that can detect water movements (poorly developed lateral line system on the head is found only in hagfishes).
No electroreception.	Electroreception may be an ancestral vertebrate feature (but absent in hagfishes, possibly lost).
Pharynx and Respiration	
Gill arches used for filter feeding (respiration is by diffusion over the body surface).	Gill arches (= pharyngeal arches) support gills that are used primarily for respiration.
Numerous gill slits (up to 100 on each side).	Fewer gill slits (6 to 10 on each side), individual gills with highly complex internal structure (gill filaments).
Pharynx not muscularized (except in wall of atrium, or external body cavity).	Pharynx with specialized (branchiomeric) musculature.
Water moved through pharynx and over gills by ciliary action.	Water moved through pharynx and over gills by active muscular pumping.
Gill arches made of collagen-like material	Gill arches made of cartilage (allows for elastic recoil—aids in pumping).
Feeding and Digestion	
Gut not muscularized: food passage by means of ciliary action.	Gut muscularized: food passage by means of muscular peristalsis.
Digestion of food is intracellular: individual food particles taken into cells lining gut.	Digestion of food is extracellular: enzymes poured onto food in gut lumen, then breakdown products absorbed by cells lining gut.
No discrete liver and pancreas: structure called the midgut cecum or diverticulum is probably homologous to both.	Discrete liver and pancreatic tissue.
Heart and Circulation	
Ventral pumping structure (no true heart, just contracting regions of vessels; = sinus venosus of vertebrates). Also accessory pumping regions elsewhere in the system.	Ventral pumping heart only (but accessory pumping regions retained in hagfishes). Three-chambered heart (listed in order of blood flow): sinus venosus, atrium, and ventricle.
No neural control of the heart to regulate pumping.	Neural control of the heart (except in hagfishes).
Circulatory system open: large blood sinuses; capillary system not extensive.	Circulatory system closed: without blood sinuses (some remain in hagfishes and lampreys); extensive capillary system.
Blood not specifically involved in the transport of respiratory gases (O_2 and CO_2 mainly transported via diffusion). No red blood cells or respiratory pigment.	Blood specifically involved in the transport of respiratory gases. Red blood cells containing the respiratory pigment hemoglobin (binds with O_2 and CO_2 and aids in their transport).

(continued)

Table 2–1 Comparison of features in nonvertebrate chordates and ancestral vertebrates (*Continued*)

Generalized Nonvertebrate Chordate	Ancestral Vertebrate
Excretion and Osmoregulation	
No specialized kidney. Coelom filtered by solenocytes (flame cells) that work by creating negative pressure within cell. Cells empty into the atrium (false body cavity) and then to the outside via the atriopore.	Specialized glomerular kidneys: segmental structures along dorsal body wall; works by ultrafiltration of blood. Empty to the outside via the archinephric ducts leading to the cloaca.
Body fluids same concentration and ionic composition as seawater. No need for volume control or ionic regulation.	Body fluids more dilute than seawater (except for hagfishes). Kidney important in volume regulation, especially in freshwater environment. Monovalent ions regulated by the gills (also the site of nitrogen excretion), divalent ions regulated by the kidney.
Support and Locomotion	
Notochord provides main support for body muscles.	Notochord provides main support for body muscles, vertebral elements around nerve cord at least in all vertebrates except hagfishes.
Myomeres with simple V shape.	Myomeres with more complex W shape.
No lateral fins; no median fins besides tail fin.	Initially no lateral fins. Caudal (tail) fin has dermal fin rays. Dorsal fins present in all except hagfishes.

faithfully passes through its ancestral evolutionary stages in the course of its development) proposed by the nineteenth-century embryologist Haeckel. Nevertheless, embryology can provide clues about the ancestral condition and about homologies between structures in different animals.

The development of vertebrates from a single fertilized cell (the zygote) to the adult condition will be summarized only briefly. This is important background information for many studies, but a detailed treatment is beyond the scope of this book. Note, however, that there is an important distinction in development between vertebrates and invertebrates: invertebrates develop from cell lineages whose fate is predetermined, but vertebrates are much more flexible in their development and use inductive interactions between developing structures to determine the formation of different cell types and tissues.

We saw earlier that all animals with the exception of sponges are formed of distinct tissue layers, or germ layers. The fates of germ layers have been very conservative throughout vertebrate evolution. The outermost germ layer, the ectoderm, forms the adult superficial layers of skin (the epidermis); the linings of the most anterior and most posterior parts of the digestive tract; and the nervous system, including most of the sense organs (such as the eye and the ear). The innermost layer, the endoderm, forms the rest of the digestive tract's lining as well as the lining of glands associated with the gut—including the liver and the pancreas—and most respiratory surfaces of vertebrate gills and lungs. Endoderm also forms the taste buds and the thyroid, parathyroid, and thymus glands.

The middle layer, the mesoderm, is the last of the three layers to appear in development, perhaps reflecting the fact that it is the last layer to appear in animal evolution. It forms everything else: muscles, skeleton (including the notochord), connective tissues, and circulatory and urogenital systems. A little later in development, there is a split within the originally solid mesoderm layer, forming a coelom or body cavity. The coelom is the cavity containing the internal organs. In mammals it is divided into the **pleural cavity** (around the lungs), the **peritoneal cavity** (around the viscera), and the **pericardial cavity** (around the heart). In other animals, which either lack lungs entirely or lack a diaphragm separating the pleural cavity from the peritoneal cavity, these two cavities are united into the **pleuroperitoneal cavity**. These coelomic cavities are lined by thin sheets of mesoderm—the **peritoneum** (around the pleural or peritoneal cavity) and the **pericardium** (around the heart). The gut is suspended in the peritoneal cavity by sheets of peritoneum called **mesenteries**.

Neural crest forms many of the structures in the anterior head region, including some bones and muscles that were previously thought to be formed by mesoderm. Neural crest also forms almost all of the peripheral nervous system (i.e., that part of the nervous

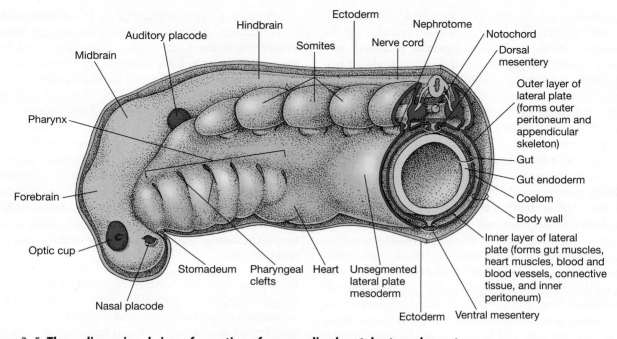

Figure 2–5 Three-dimensional view of a portion of a generalized vertebrate embryo at the developmental stage (called the pharyngula) when the developing gill pouches appear. The ectoderm is stripped off the left side, showing segmentation of the mesoderm in the trunk region and pharyngeal development. The stomadeum is the developing mouth.

system outside of the brain and the spinal cord) and contributes to portions of the brain. Some structures in the body that are new features of vertebrates are also formed from neural crest. These include the adrenal glands, pigment cells in the skin, secretory cells of the gut, and smooth muscle tissue lining the aorta.

Figure 2–5 shows a stage in early embryonic development in which the ancestral chordate feature of pharyngeal pouches in the head region makes at least a fleeting appearance in all vertebrate embryos. In fishes the grooves between the pouches (the pharyngeal clefts) perforate to become the gill slits, whereas in land vertebrates these clefts disappear in later development. The linings of the pharyngeal pouches give rise to half a dozen or more glandular structures often associated with the lymphatic system, including the thymus gland, parathyroid glands, carotid bodies, and tonsils.

The dorsal hollow nerve cord typical of vertebrates and other chordates is formed by the infolding and subsequent pinching off and isolation of a long ridge of ectoderm running dorsal to the developing notochord. The notochord itself appears to contain the developmental instructions for this critical embryonic event, which is probably why the notochord is retained in the embryos of vertebrates (such as us) that no longer have the complete structure in the adult. The cells that

will form the neural crest arise next to the developing nerve cord (the neural tube) at this stage. Slightly later in development, these neural-crest cells disperse laterally and ventrally, ultimately settling and differentiating throughout the embryo.

Embryonic mesoderm becomes divided into three distinct portions, as shown in Figure 2–5, with the result that adult vertebrates are a strange mixture of segmented and unsegmented components. The dorsal (upper) part of the mesoderm, lying above the gut and next to the nerve cord, forms an epimere, a series of thick-walled segmental buds (**somites**), which extends from the head end to the tail end. The ventral (lower) part of the mesoderm, surrounding the gut and containing the coelom, is thin-walled and unsegmented and is called the **lateral plate mesoderm** (or hypomere). Small segmental buds linking the somites and the lateral plate are called nephrotomes (the mesomere or the intermediate mesoderm). The nervous system also follows this segmented versus unsegmented pattern, as will be discussed later.

The segmental somites will eventually form the dermis of the skin, the striated muscles of the body that are used in locomotion, and portions of the skeleton (the vertebral column, ribs, and portions of the back of the skull). Some of these segmental muscles later migrate ventrally from their originally dorsal (**epaxial**) position to form the layer of striated muscles on the underside of the body (the **hypaxial** muscles), and from there they form the muscles of the limbs in **tetrapods** (four-footed land vertebrates). The lateral plate forms all the internal, nonsegmented portions of the body, such as the connective tissue, the blood vascular system, the mesenteries, the peritoneal and pericardial linings of the coelomic cavities, and the reproductive system. It also forms the smooth muscle of the gut and the **cardiac** (heart) muscle. The nephrotomes form the kidneys (which are elongated segmental structures in the ancestral vertebrate condition), the kidney drainage ducts (the **archinephric ducts**), and the gonads.

Some exceptions exist to this segmented versus nonsegmented division of the vertebrate body. The locomotory muscles, both **axial** (within the trunk region) and **appendicular** (within the limbs), and the axial skeleton are derived from the somites. Curiously, however, the limb bones are mostly derived from the lateral plate, as are the tendons and ligaments of the appendicular muscles, even though they essentially form part of the segmented portion of the animal. The explanation for this apparent anomaly may lie in the fact that limbs are add-ons to the basic limbless vertebrate body plan, as seen in the living jawless vertebrates. The boundary of the complex interaction between the somite-derived muscles and structures derived from the outer layer of the lateral plate in the embryo is known as the lateral somitic frontier. This area is involved in the switching on and off of regulatory genes and is thus of prime importance in evolutionary change.

Other peculiarities are found in the expanded front end of the head of vertebrates, which has a complex pattern of development and does not follow the simple segmentation of the body. The head mesoderm contains only somites (no lateral plate), which give rise to the striated eye muscles and **branchiomeric muscles** powering the pharyngeal arches (gills and jaws). Within the brain, the anteriormost part of the forebrain (the front of the telencephalon) and the midbrain are not segmented, but the hindbrain shows segmental divisions during development (rhombomeres).

Adult Tissue Types

There are several kinds of tissue in vertebrates: epithelial, connective, vascular (i.e., blood), muscular, and nervous. These tissues are combined to form larger units called organs, which often contain most or all of the five basic tissue types.

Connective Tissue A fundamental component of most animal tissues is the fibrous protein **collagen**. Collagen is primarily a mesodermal tissue: in addition to the softer tissues of organs, it forms the organic matrix of bone and the tough tissue of tendons and ligaments. Vertebrates have a unique type of fibrillar collagen that may be responsible for their ability to form an internal skeleton. Collagen is stiff and does not stretch easily. In some tissues, collagen is combined with the protein **elastin**, which can stretch and recoil. Another important fibrous protein, seen only in vertebrates, is **keratin**. While collagen forms structures within the mesoderm, keratin is primarily an ectodermal tissue. Keratin is mainly found in the epidermis (outer skin) of tetrapods, making structures such as hair, scales, feathers, claws, horns, and beaks; it also forms the horny toothlike structures of the living jawless vertebrates.

The Integument The external covering of vertebrates, the integument, is a single organ, making up 15 to 20 percent of the body weight of many vertebrates and much more in armored forms. It includes the skin and its derivatives, such as glands, scales, dermal armor, and hair. The skin protects the body and receives information from the outside world. The major divisions of the vertebrate skin are the **epidermis** (the superficial cell layer derived from embryonic ectoderm) and the unique vertebrate **dermis** (the deeper cell layer of mesodermal and neural-crest origin). The dermis extends deeper into a subcutaneous tissue (**hypodermis**) that is derived from mesoderm and overlies the muscles and bones.

The epidermis forms the boundary between a vertebrate and its environment and is of paramount importance in protection, exchange, and sensation. It often contains secretory glands and may play a significant role in osmotic and volume regulation. The dermis, the main structural layer of the skin, includes many collagen fibers that help to maintain its strength and shape. The dermis contains blood vessels, and blood flow within these vessels is under neural and hormonal control (e.g., as in human blushing, when the vessels are dilated and blood rushes to the skin).

The dermis also houses **melanocytes** (melanin-containing pigment cells that are derived from the neural crest) and smooth muscle fibers, such as the ones in mammals that produce skin wrinkling around the nipples. In tetrapods, the dermis houses most of the sensory structures and nerves associated with sensations of temperature, pressure, and pain.

The hypodermis, or subcutaneous tissue layer, lies between the dermis and the fascia overlying the muscles. This region contains collagenous and elastic fibers and is the area in which subcutaneous fat is stored by birds and mammals. The subcutaneous striated muscles of mammals, such as those that enable them to flick the skin to get rid of a fly, are found in this area.

Mineralized Tissues Vertebrates have a unique type of mineral called hydroxyapatite, a complex compound of calcium and phosphorus. Hydroxyapatite is more resistant to acid than is calcite (calcium carbonate), which forms the shells of mollusks. The evolution of this unique calcium compound in vertebrates may be related to the fact that vertebrates rely on anaerobic metabolism during activity, producing lactic acid that lowers blood pH. A skeleton made of hydroxyapatite may be more resistant to acidification of the blood during anaerobic metabolism than is the calcite that forms the shells of mollusks.

Vertebrate mineralized tissues are composed of a complex matrix of collagenous fibers, cells that secrete a proteinaceous tissue matrix, and crystals of hydroxyapatite. The hydroxyapatite crystals are aligned on the matrix of collagenous fibers in layers with alternating directions, much like the structure of plywood. This combination of cells, fibers, and minerals gives bone its complex latticework appearance that combines strength with relative lightness and helps to prevent cracks from spreading.

Six types of tissues can become mineralized in vertebrates, and each is formed from a different cell lineage in development.

- **Mineralized cartilage.** Cartilage is an important structural tissue in vertebrates and many invertebrates but is not usually mineralized. Mineralized cartilage occurs naturally only in jawed vertebrates, where it forms the main mineralized internal skeletal tissue of sharks. (Sharks and other cartilaginous fishes appear to have secondarily lost true bone.) Some fossil jawless vertebrates also had internal calcified cartilage, probably evolved independently from the condition in sharks.
- **Bone.** The internal skeleton of bony fishes and tetrapods is formed by bone. Bone may replace cartilage in development, as it does in our own skeletons, but bone is not simply cartilage to which minerals have been added. Rather, it is composed of different types of cells—osteocytes (Greek *osteo* = bone and *cyte* = cell), which are called osteoblasts (Greek *blasto* = a bud) while they are actually making the bone; in contrast, chondrocytes form cartilage. The cells that form bone and cartilage are derived from the mesoderm, except in the region in the front of the head, where they are derived from neural-crest tissue. Bone and mineralized cartilage are both about 70 percent mineralized.

- **Enamel and dentine.** The other types of mineralized tissues are found in the teeth and in the mineralized exoskeleton of ancestral vertebrates. The enamel and dentine that form our teeth are the most mineralized of the tissues—enamel is about 96 percent mineralized, and dentine is about 90 percent mineralized. This high degree of mineralization explains why teeth are more likely to be found as fossils than are bones. The cells that form dentine (odontoblasts) are derived from neural-crest tissue, and those that form enamel (amyloblasts) are derived from the ectoderm.
- **Enameloid.** Enameloid resembles enamel in its degree of hardness and its position on the outer layer of teeth or dermal scales, but it is produced by mesodermal cells. Enameloid is the enamel-like tissue that was present in ancestral vertebrates and is found today in cartilaginous fishes. Both enamel and enameloid may have evolved independently on a number of occasions.
- **Cementum.** Cementum is a bonelike substance that fastens the teeth in their sockets in some vertebrates, including mammals, and may grow to become part of the tooth structure itself.

Bone

Bone remains highly vascularized even when it is mineralized (ossified). This vascularization allows bone to remodel itself. Old bone is eaten away by specialized blood cells (osteoclasts, from the Greek *clast* = broken), which are derived from the same cell lines as the macrophage white blood cells that engulf foreign bacteria in the body. Osteoblasts enter behind the osteoclasts and deposit new bone. In this way, a broken bone can mend itself and bones can change their shape to suit the mechanical stresses imposed on an animal. This is why exercise builds up bone and why astronauts lose bone in the zero gravity of space. Mineralized cartilage is unable to remodel itself because it does not contain blood vessels.

There are two main types of bone in vertebrates: **dermal bone**, which, as its name suggests, is formed in the skin without a cartilaginous precursor; and **endochondral bone**, which is formed in cartilage. Dermal bone (**Figure 2–6**) is the earliest type of vertebrate bone first seen in the fossil jawless vertebrates called **ostracoderms**, which are described in Chapter 3. Only in the bony fishes and tetrapods is the endoskeleton composed primarily of bone. In these vertebrates, the endoskeleton is initially laid down in cartilage and is replaced by bone later in development.

Dermal bone originally was formed around the outside of the body, like a suit of armor (*ostracoderm*

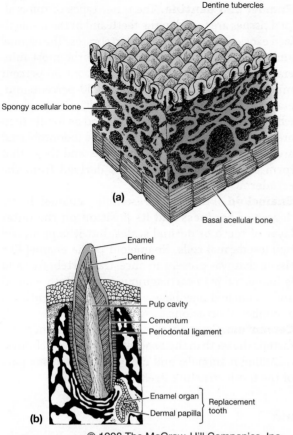

(a)

Dentine tubercles

Spongy acellular bone

Basal acellular bone

(b)

Enamel

Dentine

Pulp cavity

Cementum

Periodontal ligament

Enamel organ
Dermal papilla
} Replacement tooth

© 1998 The McGraw-Hill Companies, Inc.

Figure 2–6 Organization of vertebrate mineralized tissues.
(a) Three-dimensional block diagram of dermal bone from
an extinct jawless vertebrate (heterostracan ostracoderm).
(b) Section through a developing tooth (shark scales are similar).

means "shell-skinned"), forming a type of exoskeleton. We think of vertebrates as possessing only an
endoskeleton, but most of our skull bones are dermal
bones, and they form a shell around our brains. The endoskeletal structure of vertebrates initially consisted
of only the braincase and was originally formed from
cartilage. Thus, the condition in many early vertebrates was a bony exoskeleton and a cartilaginous endoskeleton (**Figure 2–7**).

Teeth

Teeth form from a type of structure called a dermal papilla, so they form only in the skin, usually over dermal bones. When the tooth is fully formed, it erupts
through the gum line. Replacement teeth may start to
develop to one side of the main tooth even before its
eruption. The basic structure of the teeth of jawed vertebrates is like the structure of **odontodes**, which were
the original toothlike components of the original

vertebrate dermal armor, and odontodes are homologous with teeth and the dermal denticles of cartilaginous fishes.

Teeth are composed of an inner layer of dentine and
an outer layer of enamel or enameloid around a central
pulp cavity (Figure 2–6). Shark scales (dermal denticles)
have a similar structure. There has been considerable
controversy about whether dental tissues are always
derived from the ectoderm, or whether they can form
from the endoderm, as seen in pharyngeal teeth in
some fishes. Recent experimental studies have shown
that the critical issue in tooth development is the
neural-crest precursor that forms the dentine, and
that either ectoderm or endoderm may be co-opted to
form the outer layers.

The Skeletomuscular System

The basic endoskeletal structural features of chordates
are the notochord, acting as a dorsal stiffening rod
running along the length of the body, and some sort of
gill skeleton that keeps the gill slits open. The cranium
surrounding the brain was the first part of the vertebrate skeleton to evolve. Next the dermal skeleton of
external plates and the axial skeleton (vertebrae, ribs,
and median fin supports) were added, and still later
the appendicular skeleton (bones of the limb skeleton
and limb girdles) evolved.

The Cranial Skeleton The skull, or cranium, is formed by
three basic components: the **chondrocranium** (Greek
chondr = cartilage [literally "gristle"] and *cran* = skull)
surrounding the brain; the **splanchnocranium** (Greek
splanchn = viscera) forming the gill supports; and the
dermatocranium (Greek *derm* = skin) forming in the
skin as an outer cover that was not present in the earliest vertebrates.

The splanchnocranial components of the vertebrate
skeleton are known by a confusing variety of names.
In general they can be called *gill arches* because they
support the gill tissue and muscles. The anterior elements of the splanchnocranium are specialized into
nongill-bearing structures in all extant vertebrates,
such as the jaws of gnathostomes. Other names for
these structures are *pharyngeal arches* (because they
form in the pharynx region) and *branchial arches*
(which is just a fancy way of saying *gill arches* because the Greek word *branchi* means gill). Yet another
name for these structures is *visceral arches*, because
the splanchnocranium is also known as the visceral
skeleton. (Still another name associated with the
pharyngeal region, *aortic arches*, refers not to the gill
skeleton but to the segmental arteries that supply the
gill arches.)

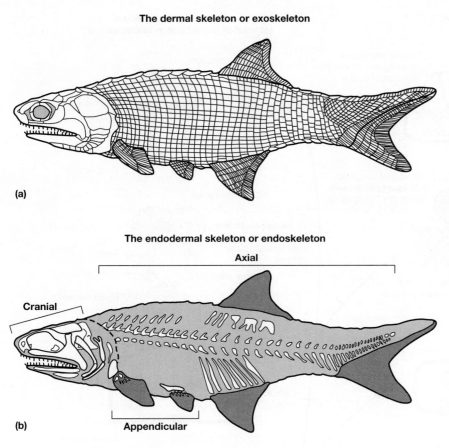

The dermal skeleton or exoskeleton

(a)

The endodermal skeleton or endoskeleton

Axial

Cranial

(b)

Appendicular

Figure 2–7 Vertebrate skeletons.
(a) the originally dermal bone exoskeleton
and (b) the originally cartilaginous bone
endoskeleton. (The animal depicted is an
extinct bony fish.)

We will call these structures *pharyngeal arches* when we are discussing the embryonic elements of their development, and *gill arches* in adults, especially for those arches that actually do bear gill tissue (i.e., arches 3–7).

The vertebrate chondrocranium and splanchnocranium are formed primarily from neural-crest tissue, although a splanchnocranium-equivalent formed by endodermal tissue is present in cephalochordates and hemichordates. Thus, a structure with the same function as the vertebrate splanchnocranium preceded the origin of vertebrates and of neural-crest tissue, although only vertebrates have a true splanchnocranium (i.e., one that is derived from neural-crest tissue).

The chondrocranium and splanchnocranium are formed from cartilage in the ancestral vertebrate condition, but they are made of endochondral bone in the adults of some bony fishes and most tetrapods. The dermatocranium is made from dermal bone, which is formed in a membrane rather than in a cartilaginous precursor. (Because it forms in a membrane it is sometimes called membrane bone.) The dermatocranium is cartilaginous only as a secondary condition in some fishes, such as sturgeons, where ossification of the dermatocranium has been lost. **Figure 2–8** shows a diagrammatic representation

of the structure and early evolution of the vertebrate cranium, and **Figure 2–9** (see page 36) illustrates three vertebrate crania in more detail.

The Cranial Muscles There are two main types of striated muscles in the head of vertebrates: the extrinsic eye muscles and the branchiomeric muscles. Six muscles in each eye rotate the eyeball in all vertebrates except hagfishes, in which their absence may represent secondary loss. Like the striated muscles of the body, these muscles are innervated by somatic motor nerves.

The branchiomeric muscles are associated with the splanchnocranium and are used to suck water into the mouth during feeding and respiration. Branchiomeric muscles are innervated by cranial nerves that exit from the dorsal part of the spinal cord (unlike striated muscles, which are innervated by motor nerves that exit from the ventral part of the spinal cord). The reason for this difference is not clear, but it emphasizes the extent to which the vertebrate head differs in its structure and development from the rest of the body.

The Axial Skeleton and Musculature The notochord is the original "backbone" of all chordates, although it is never actually made of bone. The notochord has a core

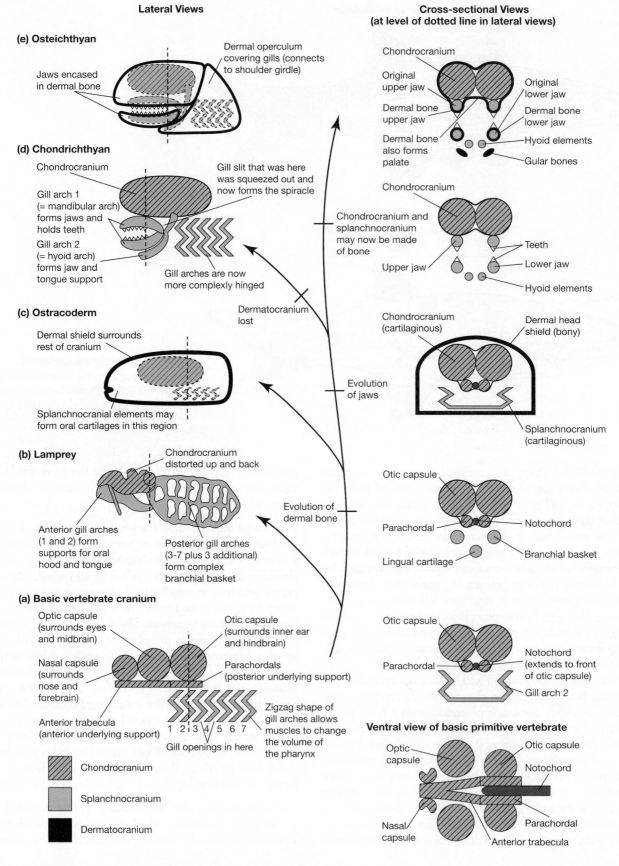

Lateral Views

Cross-sectional Views
(at level of dotted line in lateral views)

(e) Osteichthyan

Jaws encased in dermal bone

Dermal operculum covering gills (connects to shoulder girdle)

Chondrocranium

Original upper jaw

Dermal bone upper jaw

Dermal bone also forms palate

Original lower jaw

Dermal bone lower jaw

Hyoid elements

Gular bones

(d) Chondrichthyan

Chondrocranium

Gill arch 1 (= mandibular arch) forms jaws and holds teeth

Gill arch 2 (= hyoid arch) forms jaw and tongue support

Gill slit that was here was squeezed out and now forms the spiracle

Gill arches are now more complexly hinged

Dermatocranium lost

Chondrocranium and splanchnocranium may now be made of bone

Chondrocranium

Upper jaw

Teeth

Lower jaw

Hyoid elements

(c) Ostracoderm

Dermal shield surrounds rest of cranium

Splanchnocranial elements may form oral cartilages in this region

Evolution of jaws

Chondrocranium (cartilaginous)

Dermal head shield (bony)

Splanchnocranium (cartilaginous)

(b) Lamprey

Chondrocranium distorted up and back

Anterior gill arches (1 and 2) form supports for oral hood and tongue

Posterior gill arches (3-7 plus 3 additional) form complex branchial basket

Evolution of dermal bone

Otic capsule

Parachordal

Lingual cartilage

Notochord

Branchial basket

(a) Basic vertebrate cranium

Optic capsule (surrounds eyes and midbrain)

Nasal capsule (surrounds nose and forebrain)

Anterior trabecula (anterior underlying support)

Otic capsule (surrounds inner ear and hindbrain)

Parachordals (posterior underlying support)

1 2 3 4 5 6 7

Gill openings in here

Zigzag shape of gill arches allows muscles to change the volume of the pharynx

Otic capsule

Parachordal

Notochord (extends to front of otic capsule)

Gill arch 2

Chondrocranium

Splanchnocranium

Dermatocranium

Ventral view of basic primitive vertebrate

Optic capsule

Nasal capsule

Otic capsule

Notochord

Parachordal

Anterior trabecula

Figure 2–8 Diagrammatic view of the form and early evolution of the cranium of vertebrates. The ancestral condition (a) was to have a chondrocranium formed from the paired sensory capsules, one pair for each part of the tripartite brain, with the underlying support provided by paired anterior trabeculae (at least in jawed vertebrates) and parachordals flanking the notochord posteriorly. The splanchnocranium was probably ancestrally made up of seven pairs of pharyngeal arches supporting six gill openings, without any anterior specializations. In the lamprey (b), the mandibular (second segment arch, but termed *arch 1* because there is no arch in the first segment) pharyngeal arch becomes the velum and other supporting structures in the head, and the remainder of the splanchnocranium forms a complex branchial basket on the outside of the gills (possibly in association with the unique mode of tidal gill ventilation). Above the level of the lamprey, the chondrocranium and splanchnocranium are surrounded with a dermatocranium of dermal bone, as first seen in ostracoderms (c). In gnathostomes (d,e) the pharyngeal arches of the mandibular (second) and hyoid (third) head segments become modified to form the jaws and jaw supports. The dermatocranium is lost in chondrichthyans (d). In osteichthyans (e) the dermatocranium forms in a characteristic pattern, including a bony operculum covering the gills and aiding in ventilation in bony fishes. The basics of this pattern are still seen in us.

of large, closely spaced cells packed with incompressible fluid-filled vacuoles wrapped in a complex fibrous sheath that is the site of attachment for segmental muscles and connective tissues. The notochord ends anteriorly just posterior to the pituitary gland and continues posteriorly to the tip of the fleshy portion of the tail. The original form of the notochord is lost in adult tetrapods, but portions remain as components of the intervertebral discs between the vertebrae.

The axial muscles are composed of myomeres that are complexly folded in three dimensions so that each one extends anteriorly and posteriorly over several body segments (**Figure 2–10** on page 37). Sequential muscle blocks overlap and produce undulation of the body when they contract. In amphioxus, myomeres have a simple V shape, whereas in vertebrates they have a W shape. The myomeres of jawed vertebrates are divided into epaxial (dorsal) and hypaxial (ventral) portions by a sheet of fibrous tissue called the horizontal septum.

The segmental pattern of the axial muscles is clearly visible in fishes. It is easily seen in a piece of raw or cooked fish where the flesh flakes apart in zigzag blocks, each block representing a myomere. (This pattern is similar to the fabric pattern of interlocking V shapes known as herringbone, although "herring muscle" would be a more accurate description.) In tetrapods, the pattern is less obvious, but the segmental pattern can be observed on the six-pack stomach of body builders, where each ridge represents a segment of the rectus abdominis muscle (a hypaxial muscle of tetrapods).

Locomotion Many small aquatic animals, especially larvae, move by using cilia to beat against the water. However, ciliary propulsion works only at very small body sizes. Adult chordates use the serial contraction of segmental muscle bands in the trunk and tail for locomotion, a feature that possibly first appeared as a startle response in larvae. The notochord stiffens the body so it bends from side to side as the muscles contract. (Without the notochord, contraction of these muscles would merely compress the body like an accordion.)

Most fishes still use this basic type of locomotion. The paired fins of jawed fishes are generally used for steering, braking, and providing lift—not for propulsion except in some specialized fishes such as skates and rays that have winglike pectoral fins and in some derived bony fishes (teleosts) such as seahorses and coral reef fishes.

Energy Acquisition and Support of Metabolism

Food must be processed by the digestive system into molecules small enough to pass through the walls of the intestine, then transported by the circulatory system to the body tissues. Oxygen is required for this process; the respiratory and the circulatory systems are closely intertwined with those of the digestive system.

Feeding and Digestion Feeding includes getting food into the mouth, mechanical processing ("chewing" in the broad sense—although today only mammals truly chew their food), and swallowing. Digestion includes the breakdown of complex compounds into small molecules that are absorbed across the wall of the gut and transported to the tissues.

Vertebrate ancestors probably filtered small particles of food from the water, as amphioxus and larval lampreys still do. Most vertebrates are particulate feeders; that is, they take in their food as bite-sized pieces rather than as tiny particles. Vertebrates move the food through the gut by rhythmical muscular contractions (peristalsis), and digest it by secreting digestive enzymes produced by the liver and the pancreas

(a) Chondrocranium and splanchnocranium of a lamprey

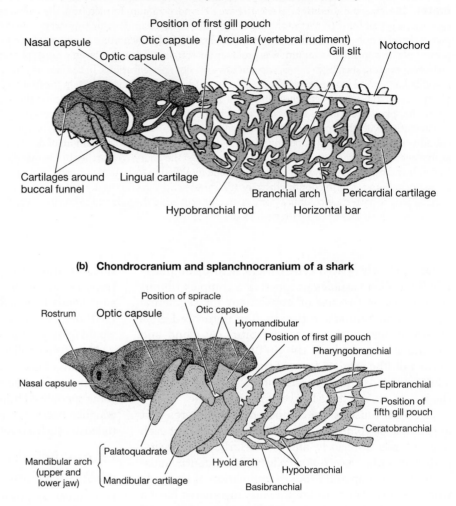

Nasal capsule
Optic capsule
Otic capsule
Position of first gill pouch
Arcualia (vertebral rudiment)
Gill slit
Notochord
Cartilages around buccal funnel
Lingual cartilage
Hypobranchial rod
Branchial arch
Horizontal bar
Pericardial cartilage

(b) Chondrocranium and splanchnocranium of a shark

Rostrum
Optic capsule
Position of spiracle
Otic capsule
Hyomandibular
Position of first gill pouch
Pharyngobranchial
Nasal capsule
Epibranchial
Position of fifth gill pouch
Ceratobranchial
Palatoquadrate
Mandibular arch (upper and lower jaw)
Mandibular cartilage
Hyoid arch
Hypobranchial
Basibranchial

(c) Dermatocranium of a primitive generalized (basal) bony fish

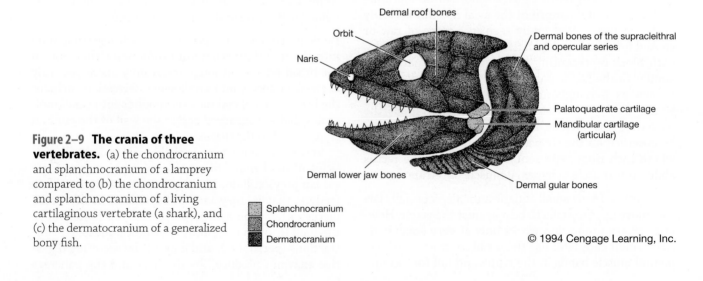

Dermal roof bones
Orbit
Naris
Dermal bones of the supracleithral and opercular series
Palatoquadrate cartilage
Mandibular cartilage (articular)
Dermal lower jaw bones
Dermal gular bones

Splanchnocranium
Chondrocranium
Dermatocranium

Figure 2–9 The crania of three vertebrates. (a) the chondrocranium and splanchnocranium of a lamprey compared to (b) the chondrocranium and splanchnocranium of a living cartilaginous vertebrate (a shark), and (c) the dermatocranium of a generalized bony fish.

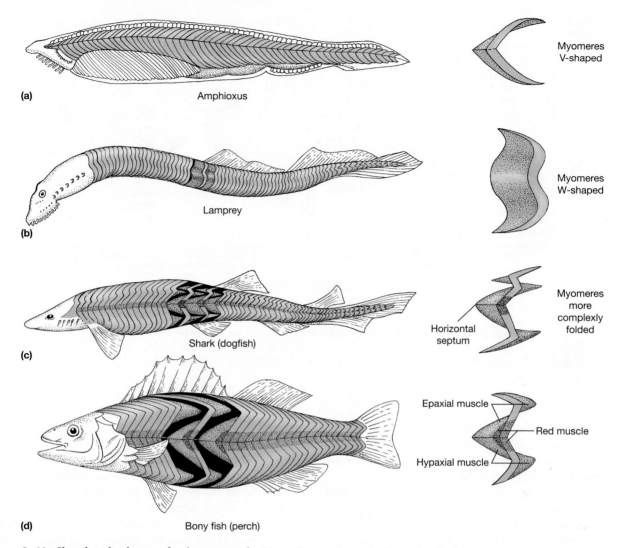

Figure 2–10 Chordate body muscles (myomeres). (a) amphioxus (nonvertebrate chordate), (b) lamprey (jawless vertebrate), and jawed vertebrates, (c) shark, and (d) bony fish.

into the gut. The pancreas also secretes the hormones insulin and glucagon, which are involved in the regulation of glucose metabolism and blood-sugar levels.

In the primitive vertebrate condition, there is no stomach, no division of the intestine into small and large portions, and no distinct rectum. The intestine empties to the **cloaca,** which is the shared exit for the urinary, reproductive, and digestive systems in all vertebrates except therian mammals.

Respiration and Ventilation Ancestral chordates probably relied on oxygen absorption and carbon dioxide loss by diffusion across a thin skin (cutaneous respiration). This is the mode of respiration of amphioxus, which is small and sluggish.

Cutaneous respiration is important for many vertebrates (especially modern amphibians), but the combination of large body size and high levels of activity make specialized gas-exchange structures essential for most vertebrates. Gills are effective in water, whereas lungs work better in air. Both gills and lungs have large surface areas that allow oxygen to diffuse from the surrounding medium (water or air) into the blood.

Cardiovascular System Blood carries oxygen and nutrients through the vessels to the cells of the body, removes carbon dioxide and other metabolic waste products, and stabilizes the internal environment. Blood also carries hormones from their sites of release to their target tissues.

Blood is a fluid tissue composed of liquid plasma, red blood cells (erythrocytes) that contain the iron-rich protein hemoglobin, and several different types of white blood cells (leukocytes) that are part of the

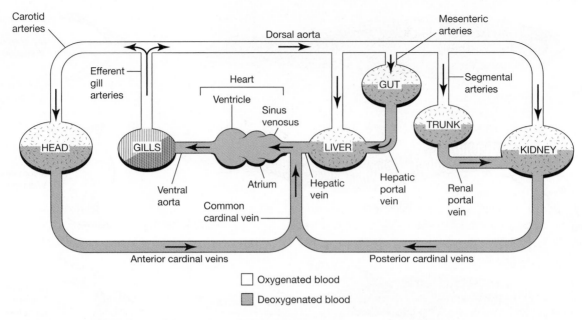

Figure 2–11 Diagrammatic plan of vertebrate cardiovascular circuit. All vessels are paired on the left and right sides of the body except for the midline ventral aorta and dorsal aorta. Note that the cardinal veins actually run dorsally in the real animal, flanking the carotid arteries (anterior cardinals) or the dorsal aorta (posterior cardinals).

immune system. Cells specialized to promote clotting of blood (called platelets or thrombocytes) are present in all vertebrates except mammals, in which they are replaced by noncellular platelets.

Vertebrates have closed circulatory systems; that is, the arteries and veins are connected by capillaries. Arteries carry blood away from the heart, and veins return blood to the heart (**Figure 2–11**). Blood pressure is higher in the arterial system than in the venous system, and the walls of arteries have a layer of smooth muscle that is absent from veins. The following features are typical of vertebrate circulatory systems:

- **Capillary beds.** Interposed between the smallest arteries (arterioles) and the smallest veins (venules) are the capillaries, which are the sites of exchange between blood and tissues. Their walls are only one cell layer thick; so diffusion is rapid, and capillaries pass close to every cell. Collectively the capillaries provide an enormous surface area for the exchange of gases, nutrients, and waste products. Arteriovenous **anastomoses** connect some arterioles directly to venules, allowing blood to bypass a capillary bed, and normally only a fraction of the capillaries in a tissue have blood flowing through them.
- **Portal vessels.** Blood vessels that lie between two capillary beds are called **portal vessels**. The hepatic

portal vein, seen in all vertebrates, lies between the capillary beds of the gut and the liver (see Figure 2–11). Substances absorbed from the gut are transported directly to the liver, where toxins are rendered harmless and some nutrients are processed or removed for storage. Most vertebrates also have a renal portal vein between the veins returning from the tail and posterior trunk and the kidneys (see Figure 2–11). The renal portal system is not well developed in jawless vertebrates and has been lost in mammals.

- **The heart.** The vertebrate heart is a muscular tube folded on itself and is constricted into three sequential chambers: the **sinus venosus**, the **atrium**, and the **ventricle**. Our so-called four-chambered heart lacks a distinct sinus venosus, and the original atrium and ventricle have been divided into left and right chambers.

  The sinus venosus is a thin-walled sac with few cardiac muscle fibers. Suction produced by muscular contraction draws blood anteriorly into the atrium, which has valves at each end that prevent backflow. The ventricle is thick-walled, and the muscular walls have an intrinsic pulsatile rhythm, which can be speeded up or slowed down by the nervous system. Contraction of the ventricle forces the blood into the ventral aorta. Mammals no longer have a distinct structure identifiable as the sinus venosus;

rather, it is incorporated into the wall of the right atrium as the sinoatrial node, which controls the basic pulse of the heartbeat.

- **The aorta.** The basic vertebrate circulatory plan consists of a heart that pumps blood into the single midline ventral aorta. Paired sets of aortic arches (originally six pairs) branch from the ventral aorta (**Figure 2–12**). One member of each pair supplies the left side and the other the right side. In the original vertebrate circulatory pattern, which is retained in fishes, the aortic arches lead to the gills, where the blood is oxygenated and returns to the dorsal aorta. The dorsal aorta is paired above the gills, and the vessels from the most anterior arch run forward to the head as the carotid arteries. Behind the gill region, the two vessels unite into a single dorsal aorta that carries blood posteriorly.

The dorsal aorta is flanked by paired cardinal veins that return blood to the heart (see Figure 2–11). Anterior cardinal veins (the jugular veins) draining the head and posterior cardinal veins draining the body unite on each side in a common cardinal vein that enters the atrium of the heart. In lungfishes and tetrapods, the posterior cardinal veins are essentially replaced by a single midline vessel, the posterior vena cava. Blood is also returned separately to the heart from the gut and liver via the hepatic portal system.

Excretory and Reproductive Systems Although the functions of the excretory and reproductive systems are entirely different, both systems are formed from the nephrotome or intermediate mesoderm, which forms the embryonic nephric ridge (**Figure 2–13** on page 41). The kidneys are segmental, whereas the gonads (**ovaries** in females and **testes** in males) are unsegmented.

The Kidneys The kidneys dispose of waste products, primarily nitrogenous waste from protein metabolism, and regulate the body's water and minerals—especially sodium, chloride, calcium, magnesium, potassium, bicarbonate, and phosphate. In tetrapods the kidneys are responsible for almost all these functions, but in fishes and amphibians the gills and skin also play important roles (see Chapter 4).

The kidney of fishes is a long, segmental structure extending the entire length of the dorsal body wall. In all vertebrate embryos, the kidney is composed of three portions: pronephros, mesonephros, and metanephros (see Figure 2–13). The pronephros is functional only in the embryos of living vertebrates and possibly in adult hagfishes. The kidney of adult fishes and amphibians includes the mesonephric and metanephric portions and is known as an **opisthonephric kidney**. The compact

bean-shaped kidney seen in adult amniotes (the **metanephric kidney**) includes only the metanephros, drained by a new tube, the **ureter**, derived from the basal portion of the archinephric duct.

The basic units of the kidney are microscopic structures called **nephrons**. Vertebrate kidneys work by ultrafiltration: high blood pressure forces water, ions, and small molecules through tiny gaps in the capillary walls. Nonvertebrate chordates lack true kidneys. Amphioxus has excretory cells called solenocytes associated with the pharyngeal blood vessels that empty individually into the false body cavity (the atrium). The effluent is discharged to the outside via the atriopore. The solenocytes of amphioxus are thought to be homologous with the podocytes of the vertebrate nephron, which are the cells that form the wall of the renal capsule.

The Gonads—Ovaries and Testes Although the gonads are derived from the mesoderm, the primordial germ cells that will form the **gametes** (eggs and sperm) are formed in the endoderm and then migrate up through the dorsal mesentery (see Figure 2–5) to enter the gonads. The archinephric duct drains urine from the kidney to the cloaca and from there to the outside world. In jawed vertebrates, this duct is also used for the release of sperm by the testes.

Reproduction is the means by which gametes are produced, released, and combined with gametes from a member of the opposite sex to produce a fertilized zygote. Vertebrates usually have two sexes, and sexual reproduction is the norm—although unisexual species occur among fishes, amphibians, and lizards.

The gonads are paired in jawed vertebrates but are single in the jawless ones: it is not clear which represents the ancestral vertebrate condition. The gonads usually lie on the posterior body wall behind the peritoneum (the lining of the body cavity); it is only among mammals that the testes are found outside the body in a scrotum. The gonads (ovaries in females, testes in males) also produce hormones, such as estrogen and testosterone.

In living jawless vertebrates, which probably represent the ancestral vertebrate condition, there is no special tube or duct for the passage of the gametes. Rather, the sperm or eggs erupt from the gonad and move through the coelom to pores that open to the base of the archinephric ducts. In jawed vertebrates, however, the gametes are always transported to the cloaca via specialized paired ducts (one for each gonad). In males, sperm are released directly into the archinephric ducts that drain the kidneys in non-amniotes and embryonic amniotes. In females, the egg is still released into the coelom but is then transported via a new structure, the **oviduct**. The oviducts produce

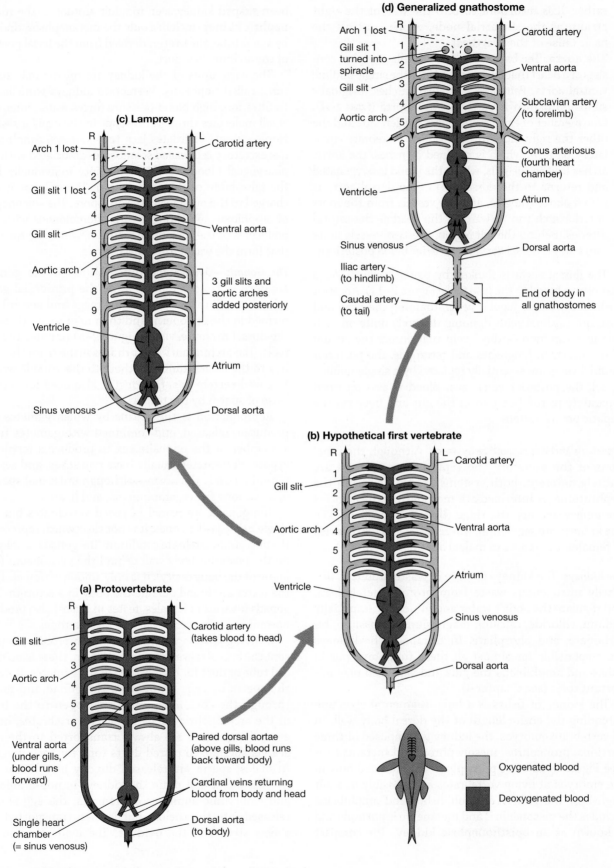

(d) Generalized gnathostome

Arch 1 lost
Gill slit 1 turned into spiracle
Gill slit
Aortic arch

Carotid artery
Ventral aorta
Subclavian artery (to forelimb)
Conus arteriosus (fourth heart chamber)
Atrium

Ventricle

Sinus venosus
Iliac artery (to hindlimb)
Caudal artery (to tail)

Dorsal aorta
End of body in all gnathostomes

(c) Lamprey

Arch 1 lost
Gill slit 1 lost
Gill slit
Aortic arch

Carotid artery

Ventral aorta

3 gill slits and aortic arches added posteriorly

Ventricle

Atrium

Sinus venosus

Dorsal aorta

(b) Hypothetical first vertebrate

Gill slit
Aortic arch

Carotid artery

Ventral aorta

Ventricle

Atrium
Sinus venosus

Dorsal aorta

(a) Protovertebrate

Gill slit
Aortic arch

Carotid artery (takes blood to head)

Ventral aorta (under gills, blood runs forward)

Paired dorsal aortae (above gills, blood runs back toward body)

Cardinal veins returning blood from body and head

Single heart chamber (= sinus venosus)

Dorsal aorta (to body)

Oxygenated blood
Deoxygenated blood

40   CHAPTER 2   Vertebrate Relationships and Basic Structure

Figure 2–12 Diagrammatic view of the form and early evolution of the heart and aortic arches of vertebrates. The view is from the ventral side of the animal. In the protovertebrate (a) and the earliest vertebrate condition (b), there were probably six pairs of aortic arches, just as are seen in the embryos of all living vertebrates, although arch 1 is never seen in the adults. Protovertebrates used these arches for feeding, not for respiration, so the blood is not shown as picking up oxygen in this illustration. In lampreys (c), additional aortic arches are added posteriorly to accommodate more gill openings. In gnathostomes (d), the subclavian and iliac arteries are added to the main circulatory system, supplying the forelimbs and hindlimbs, respectively. A fourth chamber is also added to the heart, the conus arteriosus, which damps out the pulsatile component of the blood flow.

substances associated with the egg, such as the yolk or the shell. The oviducts can become enlarged and fused in various ways to form a single uterus or paired uteri in which eggs are stored and young develop.

Vertebrates may deposit eggs that develop outside the body or retain the eggs within the mother's body until embryonic development is complete. Shelled eggs must be fertilized in the oviduct before the shell and albumen are deposited. Many viviparous vertebrates and vertebrates that lay shelled eggs have some sort of intromittent organ—such as the pelvic claspers of sharks and the penis of amniotes—by which sperm are inserted into the female's reproductive tract.

Coordination and Integration

The nervous and endocrine systems respond to conditions inside and outside an animal, and together they control the actions of organs and muscles, coordinating them so they work in concert.

General Features of the Nervous System Individual cells called neurons are the basic units of the nervous system.

In jawed vertebrates, elongated portions of the neurons, the axons, are encased in a fatty insulating coat, the myelin sheath, that increases the conduction velocity of the nerve impulse. The axons are generally collected like wires in a cable, forming a nerve. Information enters the neuron via short processes called dendrites. The brain and spinal cord are known as the central nervous system (CNS), and the nerves running between the CNS and the body are known as the peripheral nervous system (PNS).

The Spinal Cord The nerves of the PNS are segmentally arranged, exiting from either side of the spinal cord between the vertebrae. The spinal cord receives sensory inputs, integrates them with other portions of the CNS, and sends impulses that cause muscles to contract and glands to alter their secretion. The spinal cord has considerable autonomy in many vertebrates. Even complex movements such as swimming are controlled by the spinal cord rather than the brain, and fishes continue coordinated swimming movements when the brain is severed from the spinal cord. Our familiar knee-jerk

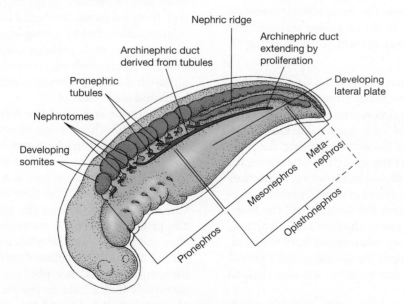

Figure 2–13 Kidney development in a generalized vertebrate embryo, showing the nephrotome regions.

reaction is produced by the spinal cord as a reflex arc. Development of more complex connections within the spinal cord and between the spinal cord and the brain has been a trend in vertebrate evolution.

The Somatic and Visceral Nervous Systems Vertebrates are unique in having a dual type of nervous system: the **somatic nervous system** (known as the voluntary nervous system) and the **visceral nervous system** (called the involuntary nervous system). This dual nervous system mimics the dual pattern of development of the mesoderm. The somatic nervous system innervates the structures derived from the segmented portion (the somites), including the striated muscles that we can move consciously (e.g., the limb muscles), and relays information from sensation that we are usually aware of (e.g., from temperature and pain receptors in the skin). The visceral nervous system innervates the smooth and cardiac muscles that we usually cannot move consciously (e.g., the gut and heart muscles) and relays information from sensations that we are not usually aware of, such as the receptors monitoring the levels of carbon dioxide in the blood.

Each spinal nerve complex is made up of four types of fibers: somatic sensory fibers coming from the body wall, somatic motor fibers running to the body, visceral sensory fibers coming from the gut wall and blood vessels, and visceral motor fibers running to the muscles and glands of the gut and to the blood vessels of both the gut and peripheral structures like the skin. The motor portion of the visceral nervous system is known as the **autonomic nervous system**. In more derived vertebrates, such as mammals, this system becomes divided into two portions: the **sympathetic nervous system** (usually acting to speed things up) and the **parasympathetic nervous system** (usually acting to slow things down).

Cranial Nerves Vertebrates also have nerves that emerge directly from the brain; these **cranial nerves** (10 pairs in the ancestral vertebrate condition, 12 in amniotes) are identified by Roman numerals. Some of these nerves, such as the ones supplying the nose (the olfactory nerve, I) or the eyes (the optic nerve, II), are not really nerves at all, but tracts—that is, parts of the CNS. Somatic motor fibers in cranial nerves innervate the muscles that move the eyeballs and the branchiomeric muscles that power the jaws and gill arches, and sensory nerves convey information from the head, including the sense of taste. The special sensory nerves that supply the lateral line in fishes are also derived from the cranial nerves.

The vagus nerve (cranial nerve X) ramifies through all but the most posterior part of the trunk, carrying the visceral motor nerve supply to various organs. People who break their necks may be paralyzed from the neck down (i.e., lose the function of their skeletal muscles) but still may retain their visceral functions (workings of the gut, heart, etc.) because the vagus nerve is independent of the spinal cord and exits above the break.

Brain Anatomy and Evolution All chordates have some form of a brain, as a thickening of the front end of the nerve cord. The brain of all vertebrates is a tripartite (three-part) structure (**Figure 2–14**), and the telencephalon (front part of the forebrain) and the olfactory receptors are probably true new features in vertebrates. In the ancestral condition, the forebrain is associated with the sense of smell, the midbrain with vision, and the hindbrain with balance and detection of vibrations (hearing, in the broad sense). These portions of the brain are associated with the nasal, optic, and otic capsules of the chondrocranium, respectively (see Figure 2–8).

The vertebrate brain has three parts: the forebrain, midbrain, and hindbrain.

The Forebrain The forebrain has two parts:

• The anterior region of the adult forebrain, the telencephalon, develops in association with the olfactory capsules and coordinates inputs from other sensory modalities. The telencephalon becomes enlarged in different vertebrate groups and is known as the **cerebrum** or cerebral hemispheres.

  Tetrapods developed an area in the cerebrum called the **neocortex** or neopallium, which is the primary seat of sensory integration and nervous control. Bony fishes also evolved a larger, more complex telencephalon, but by a completely different mechanism. Sharks and, perhaps surprisingly, hagfishes independently evolved relatively large forebrains, although a large cerebrum is primarily a feature of birds and mammals.

• The posterior region of the forebrain is the **diencephalon**, which contains structures that act as a major relay station between sensory areas and the higher brain centers. The **pituitary gland**, an important endocrine organ, is a ventral outgrowth of the diencephalon. The floor of the diencephalon (the **hypothalamus**) and the pituitary gland form the primary center for neural-hormonal coordination and integration. Another endocrine gland, the **pineal organ**, is a median dorsal outgrowth of the diencephalon that is a photoreceptor. Many early tetrapods had a hole in the skull over the pineal gland to admit light, and this condition persists in some reptiles (e.g., the tuatara and many lizards).

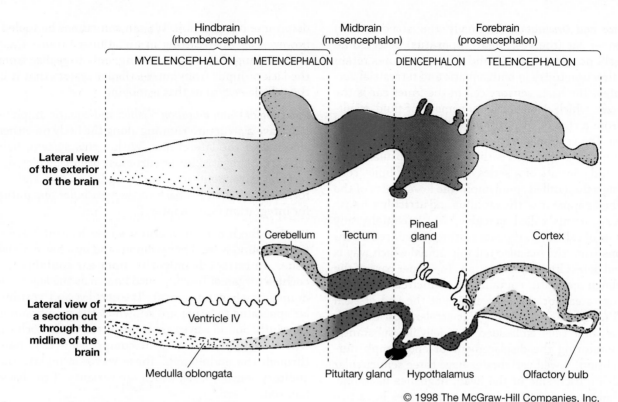

Figure 2–14 **The generalized vertebrate brain.**

© 1998 The McGraw-Hill Companies, Inc.

The Midbrain The midbrain develops in conjunction with the eyes and receives input from the optic nerve, although in mammals the forebrain has taken over much of the task of vision.

The Hindbrain The hindbrain has two portions:

- The posterior portion, the **myelencephalon**, or medulla oblongata, controls functions such as respiration and acts as a relay station for receptor cells from the inner ear.
- The anterior portion of the hindbrain, the **metencephalon**, develops an important dorsal outgrowth, the **cerebellum**—present as a distinct structure only in jawed vertebrates among living forms. The cerebellum coordinates and regulates motor activities whether they are reflexive (such as maintenance of posture) or directed (such as escape movements).

The Sense Organs We think of vertebrates as having five senses—taste, touch, sight, smell, and hearing—but this list does not reflect the ancestral condition, nor does it include all the senses of living vertebrates. Complex, multicellular sense organs that are formed from epidermal placodes and tuned to the sensory worlds of the species that possess them are derived features of many vertebrate lineages.

Chemosensation: Smell and Taste The senses of smell and taste both involve the detection of dissolved molecules by specialized receptors. We think of these two senses as being closely interlinked; for example, our sense of taste is poorer if our sense of smell is blocked because we have a cold. However, the two senses are actually very different in their innervation. Smell is a somatic sensory system—sensing items at a distance, with the sensations being received in the forebrain. Taste is a visceral sensory system—sensing items on direct contact, with the sensations being received initially in the hindbrain.

Vision The receptor field of the vertebrate eye is arrayed in a hemispherical sheet, the **retina**, which originates as an outgrowth of the brain. The retina contains two types of light-sensitive cells, **cones** and **rods**, which are distinguished from each other by morphology, photochemistry, and neural connections.

Electroreception The capacity to perceive electrical impulses generated by the muscles of other organisms is also a form of distance reception, but one that works only in water. Electroreception was probably an important feature of early vertebrates and is seen today primarily in fishes and monotreme mammals. Many extant fishes produce electrical discharge for communication with other individuals or for protection from predators.

Balance and Orientation Originally the structures in the inner ear (the **vestibular apparatus**) detected an animal's position in space, and these structures retain that function today in both aquatic and terrestrial vertebrates. The basic sensory cell in the inner ear is the hair cell, which detects the movement of fluid resulting from a change of position or the impact of sound waves. The vestibular apparatus (one on either side of the animal) is enclosed within the otic capsule of the skull and consists of a series of sacs and tubules containing a fluid called endolymph. The lower parts of the vestibular apparatus, the **sacculus** and **utriculus**, house sensory organs called **maculae**, which contain tiny crystals of calcium carbonate resting on hair cells. Sensations from the maculae tell the animal which way is up and detect linear acceleration. The upper part of the vestibular apparatus contains the **semicircular canals**. Sensory areas located in swellings at the end of each canal (**ampullae**) detect angular acceleration through cristae, hair cells embedded in a jellylike substance, by monitoring the displacement of endolymph during motion. Jawed vertebrates have three semicircular canals on each side of the head, hagfishes have one, and lampreys and fossil jawless vertebrates have two (**Figure 2–15**).

We often fail to realize the importance of our own vestibular senses because we usually depend on vision to determine our position. We can sometimes be fooled, however, as when sitting in a stationary train or car and thinking that we are moving, only to realize from the lack of input from our vestibular system that it is the vehicle *next* to us that is moving.

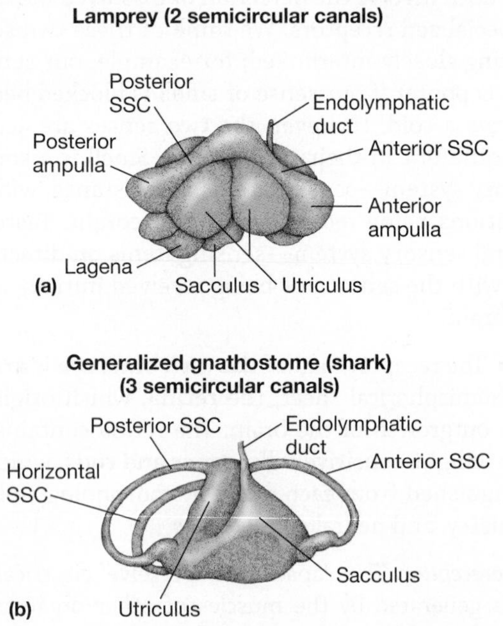

Detection of Water Vibration Fishes and aquatic amphibians have a structure running along the body on either side, called the lateral line. Within this system, hair cells are aggregated into **neuromast organs**, which detect the movement of water around the body, and information is then fed back to the vestibular apparatus for integration (see Chapter 4).

Hearing The inner ear is also used for hearing (reception of sound waves) by tetrapods and by a few derived fishes. In tetrapods only, the inner ear contains the **cochlea** (organ of hearing, also known as the lagena in nonmammalian tetrapods). The cochlea and vestibular apparatus together are known as the membranous labyrinth. Sound waves are transmitted to the cochlea, where they create waves of compression that pass through the endolymph. These waves stimulate the auditory sensory cells, which are variants of the basic hair cell.

The Endocrine System The endocrine system transfers information from one part of the body to another via the release of a chemical messenger (**hormone**) that produces a response in the target cells. The time required for an endocrine response ranges from seconds to hours. Hormones are produced in discrete endocrine glands, whose primary function is hormone production and excretion (e.g., the pituitary, thyroid, thymus, and adrenals), and by organs with other major bodily functions—such as the gonads, kidneys, and gastrointestinal tract. Endocrine secretions are predominantly involved in controlling and regulating energy use, storage, and release, as well as in allocating energy to special functions at critical times. The trend in the evolution of vertebrate endocrine glands has been consolidation from scattered clusters of cells or small organs in fishes to larger, better-defined organs in amniotes.

Lamprey (2 semicircular canals)

Posterior SSC
Endolymphatic duct
Posterior ampulla
Anterior SSC
Anterior ampulla
Lagena
(a)
Sacculus
Utriculus

**Generalized gnathostome (shark)
(3 semicircular canals)**

Posterior SSC
Endolymphatic duct
Horizontal SSC
Anterior SSC
Sacculus
(b)
Utriculus

© 1998 The McGraw-Hill Companies, Inc.

Figure 2–15 Anatomy of the vestibular apparatus in fishes. The lamprey (a) has two semicircular canals (SSC), whereas gnathostomes (represented by a shark, b) have three semicircular canals.

The Immune System

Vertebrates have adaptive immunity, a type of immune system different from that of invertebrates. While all animals have innate, specifically genetically encoded responses to pathogens that are fixed and unchanging, vertebrates additionally have evolved lymphocytes (a type of white blood cell), which provide a system of adjustable antigen recognition. The adaptive immune systems of jawless and jawed vertebrates are somewhat different. While gnathostomes generate lymphocyte receptors via immunoglobulin gene segments, lampreys and hagfishes employ leucine-rich repeat molecules. In addition, lampreys and hagfishes lack a thymus gland, which produces lymphocytes in gnathostomes. The agnathan condition is probably the ancestral condition for vertebrates, as it more closely resembles the mode of antipathogen responses in invertebrates.

Summary

Vertebrates are large, active members of the phylum Chordata, a group of animals whose other members, tunicates and cephalochordates (amphioxus), are small and sluggish or entirely sessile as adults. Chordates share with many other derived animal phyla the features of being bilaterally symmetrical, with a distinct head and tail end. Both embryological and molecular evidence show that chordates are related to other sessile marine animals, such as echinoderms. Chordates are distinguished from other animals by the presence of a notochord, a dorsal hollow nerve chord, a muscular postanal tail, and an endostyle (which is homologous to the vertebrate thyroid gland). Vertebrates appear to be most closely related to tunicates among nonvertebrate chordates, and most of the differences in structure and physiology between vertebrates and other chordates reflect an evolutionary change to larger body sizes, greater levels of activity, and predation rather than filter feeding.

Vertebrates have the unique features of an expanded head with multicellular sense organs and a cranium housing an enlarged, tripartite brain. The features that distinguish vertebrates from other chordates appear to be related to two critical embryonic innovations: a doubling of the *Hox* gene complex and the development of neural-crest tissue. The diverse activities of vertebrates are supported by a complex morphology. Study of embryology can throw light on the genetic basis and developmental pathways underlying these structures. In particular, neural-crest cells, which are unique to vertebrates, are responsible for many of their derived characters, especially those of the new anterior portion of the head.

An adult vertebrate can be viewed as a group of interacting anatomical and physiological systems involved in protection, support and movement, acquisition of energy, excretion, reproduction, coordination, and integration. These systems underwent profound changes in function and structure at several key points in vertebrate evolution. The most important transition was from the prevertebrate condition—as represented today by the cephalochordate amphioxus—to the vertebrate condition, shown by hagfishes and lampreys. Other important transitions, to be considered in later chapters, include the shift from jawless to jawed vertebrates and from fish to tetrapod.

Discussion Questions

1. Suppose new molecular data showed that tunicates and vertebrates are sister taxa. What difference would this make to our assumptions about the form of the original chordate animal? What additional features might this animal have possessed?

2. We noted that many features typical of vertebrates, such as a distinct head and limbs of some sort for locomotion, are seen in invertebrates such as insects and crustaceans, and even in cephalopods (squid and octopuses). What characteristic do these animals share with vertebrates that might have led to the independent evolution of such structures?

3. Why would evidence of sense organs in the head of a fossil animal (such as *Haikouella*) suggest a close relationship with vertebrates—that is, which critical vertebrate feature would have to be present?

4. Vertebrates have been thought of as "dual animals," consisting of both segmented and unsegmented portions. How is this duality reflected in their embryonic development and the structure of the nervous system?

5. Among the extant vertebrates, only the bony fishes (Osteichthyes) possess bone. (a) Why, then, do we make the assumption that the ancestors of cartilaginous fishes must have had bone that was subsequently lost? (b) Why don't we think this is true for the lampreys and hagfishes?

6. Amphioxus obtains its oxygen by diffusion over the body surface. Why aren't vertebrates generally able to do this—that is, why do most aquatic vertebrates rely on gills for gas exchange?

Additional Information

Donoghue, P. C. J., and M. A. Purnell. 2009. The evolutionary emergence of vertebrates from among their spineless relatives. *Evolution Education Outreach* 2:204–212.

Fraser, G. J., et al. 2010. The odontode explosion: The origin of tooth-like structures in vertebrates. *Bioessays* 32:808–817.

Gai, Z. et al. 2011. Fossil jawless fish from China foreshadows early jawed vertebrate anatomy. *Nature* 476:324–327.

Garcia-Fernàndez, J., and E. Benito-Gutiérrez. 2009. It's a long way from amphioxus: Descendants of the earliest chordate. *BioEssays* 31:665–675.

Guo, P., et al. 2009. Dual nature of the adaptive immune system in lampreys. *Nature* 459:796–801.

Huysseune, A., et al. 2009. Evolutionary and developmental origins of the vertebrate dentition. *Journal of Anatomy* 214:465–476.

Janvier, J. 2011. Comparative anatomy: All vertebrates do have vertebrae. *Current Biology* 21:R661–R663.

Kuraku, S. 2011. Hox Gene clusters of early vertebrates: Do they serve as reliable markers for genome evolution? *Genomics, Proteomics & Bioinformatics* 9:97–103.

Lowe, C. B. et al. 2011. Three periods of regulatory innovation during vertebrate evolution. *Science* 333:1019–1024.

Manning, G., and E. Scheeff. 2010. How the vertebrates were made: Selective pruning of a double-duplicated genome. *BMC Biology* 8:144.

Ouilon, S. et al. 2011. Evolution of repeated structures along the body axis of jawed vertebrates, insights from the *Scyliorhinus canicula* Hox code. *Evolution & Development* 13:247–259.

Sansom R. S., et al. 2011. Decay of vertebrate characters in hagfish and lamprey (Cyclostomata) and the implications for the vertebrate fossil record. *Proceedings of the Royal Society* B 278:1150–1157.

Shubin, N. 2009. *Your Inner Fish: A Journey into the 3.5-Billion-Year History of the Human Body*. New York: Vintage Books.

Sire, H.-Y., et al. 2009. Origin and evolution of the integumentary skeleton in non-tetrapod vertebrates. *Journal of Anatomy* 214:409–440.

Websites

Metazoan diversity: The Cambrian explosion
http://palaeo.gly.bris.ac.uk/Palaeofiles/Cambrian/Index.html

Introduction to the Urochordata
http://www.ucmp.berkeley.edu/chordata/urochordata.html

Introduction to the Cephalochordata
http://www.ucmp.berkeley.edu/chordata/cephalo.html

The amphioxus song (annotated)
http://www.flounder.com/amphioxus.htm

***Haikouella* and other Chengjiang animals**
http://www.fossilmuseum.net/Fossil_Sites/Chengjiang/Xidazoon-Haikouella/XidazoonHaikouella.htm

***Hox* genes and metazoan evolution**
http://scienceblogs.com/pharyngula/2006/05/hox_genesis.php

Evolution of the circulatory system
http://www.brown.edu/cis/sta/dev/vascular_evolution/2.html

CHAPTER 3

Early Vertebrates: Jawless Vertebrates and the Origin of Jawed Vertebrates

The earliest vertebrates represented an important advance over the nonvertebrate chordate filter feeders. The most conspicuous new feature of these early vertebrates was a distinct head end, containing a tripartite brain enclosed by a cartilaginous cranium (skull) and complex sense organs. They used the newly acquired pharyngeal musculature that powered the gill skeleton to draw water into the mouth and over the gills, which were now used for respiration rather than for filter feeding. Early vertebrates were active predators rather than sessile filter feeders. Many of them also had external armor made of bone and other mineralized tissues. We know a remarkable amount about the anatomy of some of these early vertebrates because the internal structure of their bony armor reveals much of their soft anatomy.

Gnathostomes represent an advance in the vertebrate body plan for high levels of activity and predation. Jaws themselves are homologous with the structures that form the gill arches and probably first evolved as devices to improve the strength and effectiveness of gill ventilation. Later the jaws were modified for seizing and holding prey. In this chapter we trace the earliest steps in the radiation of vertebrates, beginning more than 500 million years ago, and discuss the biology of both the Paleozoic agnathans (the ostracoderms) and the extant forms (hagfishes and lampreys). We also consider the transition from the jawless condition to the jawed one and the biology of some of the early types of jawed fishes that did not survive the Paleozoic era (placoderms and acanthodians).

3.1 Reconstructing the Biology of the Earliest Vertebrates

The Earliest Evidence of Vertebrates

Until very recently our oldest evidence of vertebrates consisted of fragments of the dermal armor of the jawless vertebrates colloquially known as **ostracoderms**. These animals were very different from any vertebrate alive today. They were essentially fishes encased in bony armor (Greek *ostrac* = shell and *derm* = skin), quite unlike the living jawless vertebrates that lack bone completely. Bone fragments that can be assigned definitively to vertebrates are known from the Ordovician period, some 480 million years ago, although there are some pieces of mineralized material that are tentatively believed to represent vertebrates from the Late Cambrian period, around 500 million years ago. This was about 80 million years before whole-body vertebrate fossils became abundant, in the Late Silurian period.

Early Cambrian Vertebrates

Recent finds of early soft-bodied vertebrates from the Early Cambrian of China extend the fossil record back by another 40 million years or so, to around 520 million years ago. Two Early Cambrian vertebrates, *Myllokunmingia* and *Haikouichthys* (although they may be the same species), are found in the Chengjiang Fauna, the fossil deposit that also yielded the possible vertebrate relative *Haikouella* (see Chapter 2). These specimens are small, fish-shaped, and about 3 centimeters long (**Figure 3–1a**), and *Haikouichthys* is now known from hundreds of well-preserved individuals. A third type of early vertebrate, *Zhongjianichthys*, from the same deposit has a more eel-like body form. Evidence of a notochord and myomeres marks these animals as chordates, and the presence of a cranium (nasal capsules and possibly otic capsules) and paired sensory structures (probably representing eyes) at the head end marks them as vertebrates because these structures are formed from neural-crest tissue (see Chapter 2). These animals also have additional vertebrate features: a dorsal fin and a ribbonlike ventral fin (but without the fin rays seen in other vertebrates); six to seven gill pouches with evidence of filamentous gills and a branchial skeleton; myomeres that are probably W-shaped rather than V-shaped; and segmental structures flanking the notochord that may represent lampreylike arcualia. However, unlike ostracoderms, the Early Cambrian vertebrates lack any evidence of bone or mineralized scales. Additionally, evidence for serial segmental gonads, as in amphioxus, emphasizes their ancestral condition—in derived vertebrates the gonads are nonsegmental.

Ordovician Vertebrates

The next good evidence of early vertebrates is from the Early Ordovician. Several sites have yielded bone fragments, suggesting that by this time vertebrates

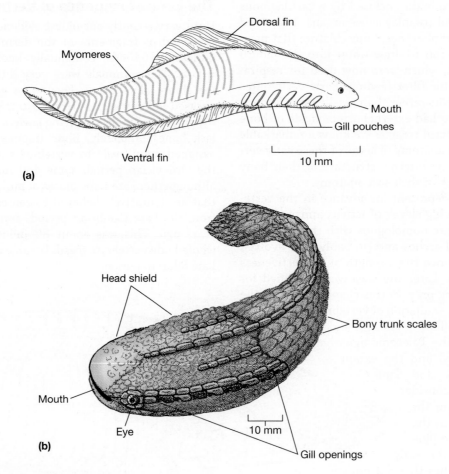

(a)

(b)

Figure 3–1 Some of the earliest vertebrates. (a) The Early Cambrian *Myllokunmingia* from China. (b) The Ordovician pteraspid ostracoderm *Astraspis* from North America.

had diversified and had a worldwide distribution. The earliest complete fossils of vertebrates are from the Late Ordovician of Bolivia, Australia, North America, and Arabia. These were torpedo-shaped jawless fishes, ranging from 12 to 35 centimeters in length, with a bony external armor. The head and gill region were encased by many small, close-fitting, polygonal bony plates, and the body was covered by overlapping bony scales. The bony head shield shows the presence of sensory canals, special protection around the eye, and—in the reconstruction of the North American *Astraspis* (**Figure 3–1b**)—as many as eight gill openings on each side of the head.

The Ordovician was also a time for great radiation and diversification among marine invertebrates, following the extinctions at the end of the Cambrian. The early radiation of vertebrates involved both jawed and jawless groups, although the first evidence of jawed fishes is in the Middle Ordovician, a little later than the first ostracoderms. By the Late Silurian (about 400 million years ago) both complete fossils of ostracoderms and fragmentary fossils of gnathostomes are known from diverse fossil assemblages worldwide.

The Origin of Bone and Other Mineralized Tissues

Mineralized tissues composed of hydroxyapatite are a major new feature of vertebrates. Enamel (or enameloid) and dentine, which occur primarily in the teeth among living vertebrates, are at least as old as bone and were originally found in intimate association with bone in the dermal armor of ostracoderms. However, mineralized tissues did not appear at the start of vertebrate evolution and are lacking in the extant jawless vertebrates.

The Earliest Mineralized Tissues The origin of vertebrate mineralized tissues remains a puzzle: the earliest-known tissues were no less complex in structure than the mineralized tissues of living vertebrates. The basic units of mineralized tissue appear to be odontodes, little toothlike elements formed in the skin. They consist of projections of dentine, covered in some cases with an outer layer of enameloid, with a base of bone. Our own teeth are very similar to these structures (see Figure 2–6 in Chapter 2).

Odontode-like structures are seen today as the tiny sharp denticles in the skin of sharks—and the larger scales, plates, and shields on the heads of many ostracoderms and early bony fishes are interpreted as aggregations of odontodes. Note that these bony elements would not have been external to the skin like a snail's shell. Rather, they were formed within the dermis of the skin and overlain by a layer of epidermis, as with our own skull bones. The original condition for vertebrate bone is to lack cells in the adult form; this type of acellular bone is also known as aspidin. Cellular bone is found only in gnathostomes and in some derived ostracoderms (osteostracans).

What could have been the original selective advantage of mineralized tissues in vertebrates? The detailed structure of the tissues forming the bony armor points to a function more complex than mere protection. Suggestions about the function of mineralized tissues include storage and/or regulation of minerals such as calcium and phosphorus, and insulation around electroreceptive organs (like those of living sharks) that enhanced detection of prey. A rigid bony head shield might also have served a hydrodynamic function in the earlier, finless vertebrates, providing stability while swimming in the open water. Although we think of a skeleton as primarily supportive or protective, our own bone also serves as a store of calcium and phosphorus. These hypotheses about the initial advantage of bone and other mineralized tissues—protection, electroreception, and mineral storage—are not mutually exclusive. All of them may have been involved in the evolutionary origin of these complex tissues.

The Problem of Conodonts

Curious microfossils known as **conodont** elements are widespread and abundant in marine deposits from the Late Cambrian to the Late Triassic periods. Conodont elements are small (generally less than 1 millimeter long) spinelike or comblike structures composed of apatite (the particular mineralized calcium compound that is characteristic of vertebrate hard tissues). They originally were considered to be the partial remains of marine invertebrates, but recent studies of conodont mineralized tissues have shown that they are similar in their microstructure to dentine, which is a uniquely vertebrate tissue. Thus, conodont elements are now considered to be the toothlike elements of true vertebrates. This interpretation has been confirmed by the discovery of impressions of complete conodont animals with vertebrate features (such as a notochord, a cranium, myomeres, and large eyes) and with conodont elements arranged within the pharynx in a complex apparatus (**Figure 3–2**).

The paleontological community is still debating how conodont teeth fit into the evolutionary picture, although the general consensus is that they represent true vertebrates. However, some workers argue that various features of conodonts are not vertebrate-like;

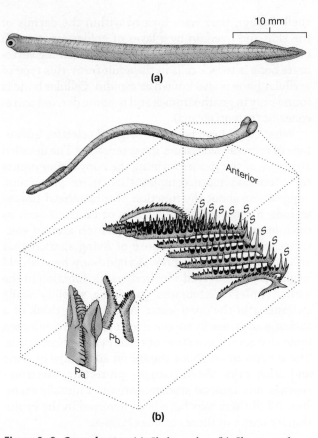

(a)

Anterior

Pb

Pa

(b)

Figure 3–2 Conodonts. (a) *Clydagnathus*. (b) Close-up of the feeding apparatus (conodont elements) inside the head of *Idiognathodus*. The anterior S elements appear to be modified for grasping, and the posterior P elements for crushing.

for example, conodont myomeres have the V shape of nonvertebrate chordates. The ability to form mineralized tissues in the skin is a feature of all vertebrates from ostracoderms onward, but the production of teeth, or toothlike structures, as seen in conodonts, may have evolved independently several different times among vertebrates, as discussed later in the chapter.

Accepting conodonts as vertebrates has changed our ideas about early vertebrate interrelationships, particularly the importance—in the vertebrate phylogeny—of having mineralized tissues. The toothlike structures of conodonts and the bony dermal skeletons of ostracoderms are now thought to make these animals more derived vertebrates than the soft-bodied, jawless fishes living today (**Figure 3–3**).

The Environment of Early Vertebrate Evolution

By the Late Silurian, ostracoderms and early jawed fishes were abundant in both freshwater and marine environments. Under what conditions did the first vertebrates evolve? The vertebrate kidney is very good at excreting excess water while retaining biologically important molecules and ions. That is what a freshwater fish must do, because its body fluids are continuously being diluted by the osmotic inflow of water and it must excrete that water to regulate its internal concentration (see Chapter 4). Thus, the properties of the vertebrate kidney suggest that vertebrates evolved in freshwater.

Despite the logic of that inference, however, a marine origin of vertebrates is now widely accepted. Osmoregulation is complex and fishes use cells in the gills as well as the kidney to control their internal fluid concentration. Probably the structure of the kidney is merely fortuitously suited to freshwater.

Two lines of evidence support the hypothesis of a marine origin of vertebrates:

- The earliest vertebrate fossils are found in marine sediments.
- All nonvertebrate chordates and deuterostome invertebrate phyla are exclusively marine forms, and they have body fluids with approximately the same osmolal concentration as their surroundings. Hagfishes also have concentrated body fluids, and these high body-fluid concentrations probably represent the original vertebrate condition.

3.2 Extant Jawless Fishes

The extant jawless vertebrates—hagfishes and lampreys—once were placed with the ostracoderms in the class "Agnatha" because they lack the gnathostome features of jaws and two sets of paired fins. They also have other ancestral features, such as lack of specialized reproductive ducts, and neither has mineralized tissues.

However, it is now clear that the "Agnatha" is a paraphyletic assemblage, and that ostracoderms are actually more closely related to gnathostomes than are living jawless vertebrates. Hagfishes and lampreys have often been linked as cyclostomes because they have round, jawless mouths, but this grouping may also be paraphyletic, since lampreys and gnathostomes share derived features that hagfishes lack (see Figure 3–3 and **Figure 3–4** on page 52).

Because both hagfishes and lampreys appear to be more distantly related to gnathostomes than were the armored ostracoderms of the Paleozoic, we will look at them before considering the extinct agnathans. The fossil record of the extant jawless vertebrates is sparse. Lampreys are known from the Late Devonian period— *Priscomyzon*, a short-bodied form, from South Africa;

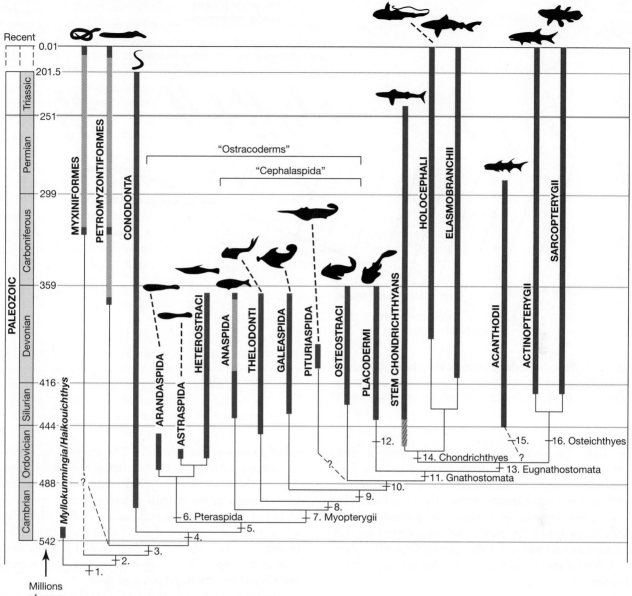

Figure 3–3 Phylogenetic relationships of early vertebrates. This diagram depicts the probable relationships among the major groups of "fishes," including living and extinct jawless vertebrates and the earliest jawed vertebrates. The black lines show relationships only; they do not indicate times of divergence or the unrecorded presence of taxa in the fossil record. Lightly shaded bars indicate ranges of time when the taxon is believed to be present, because it is known from earlier and later times, but is not recorded in the fossil record during this interval. The hatched bar shows probable occurrence based on limited evidence. Only the best-corroborated relationships are shown and question marks indicate uncertainty about relationships. The numbers at the branch points indicate derived characters that distinguish the lineages—see the Appendix for a list of these characters.

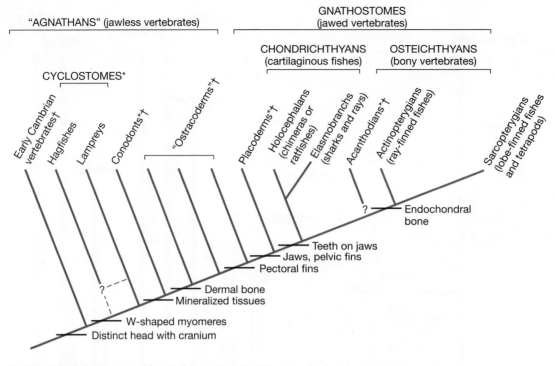

Figure 3–4 Simplified cladogram of vertebrates. Only living taxa and major extinct groups are shown. Quotation marks indicate paraphyletic groups. An asterisk indicates possibly paraphyletic groups. A dagger indicates extinct groups.

the Late Carboniferous period—*Hardistiella* from Montana and *Mayomyzon* from Illinois; and the Early Cretaceous period—*Mesomyzon* from southern China, the first known freshwater form. All these fossil lampreys appear to have been specialized parasites similar to the living forms. *Myxinikela* (an undisputed hagfish) and a second possible hagfish relative, *Gilpichthys*, have been found in the same Carboniferous deposits as *Mayomyzon*.

Hagfishes—Myxiniformes

There are about 75 species of hagfishes in two major genera (*Eptatretus* and *Myxine*). Adult hagfishes (**Figure 3–5**) are elongated, scaleless, pinkish to purple in color, and about half a meter in length. Hagfishes are entirely marine, with a nearly worldwide distribution except for the polar regions. They are primarily deep-sea, cold-water inhabitants. They are the major scavengers of the deep-sea floor, drawn in large numbers by their sense of smell to carcasses.

Structural Characteristics Large mucous glands that open through the body wall to the outside are a unique feature of hagfishes. These so-called slime glands secrete enormous quantities of mucus and tightly

coiled proteinaceous threads. The threads straighten on contact with seawater to entrap the slimy mucus close to the hagfish's body. An adult hagfish can produce enough slime within a few minutes to turn a bucket of water into a gelatinous mess. This obnoxious behavior is apparently a deterrent to predators. When danger has passed, the hagfish makes a knot in its body and scrapes off the mass of mucus, then sneezes sharply to blow its nasal passage clear.

Hagfishes lack any trace of vertebrae, and their internal anatomy shows many additional unspecialized features. For example, the kidneys are simple, and there is only one semicircular canal on each side of the head. Hagfishes have a single terminal nasal opening that connects with the pharynx via a broad tube, and the number of gill openings on each side (1 to 15) varies with the species. The eyes are degenerate or rudimentary and covered with a thick skin.

The mouth is surrounded by six tentacles that can be spread and swept to and fro by movements of the head when the hagfish is searching for food. Two horny plates in the mouth bear sharp toothlike structures made of keratin rather than mineralized tissue. These tooth plates lie to each side of a protrusible tongue and spread apart when the tongue is protruded. When the

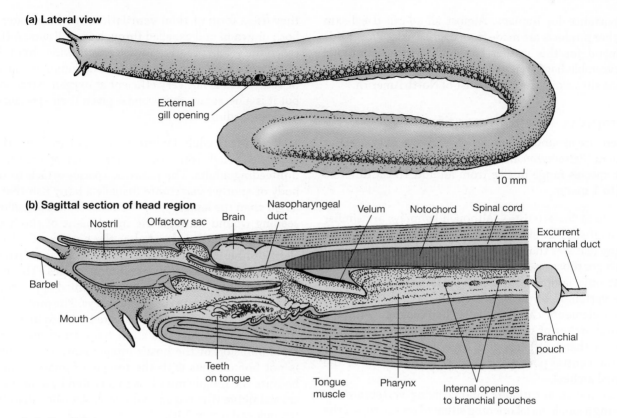

(a) Lateral view

External
gill opening

10 mm

(b) Sagittal section of head region

Nostril — Olfactory sac — Brain — Nasopharyngeal duct — Velum — Notochord — Spinal cord

Excurrent
branchial duct

Barbel

Mouth

Teeth
on tongue

Tongue
muscle

Pharynx

Internal openings
to branchial pouches

Branchial
pouch

Figure 3–5 Hagfishes.

tongue is retracted, the plates fold together and the teeth interdigitate in a pincerlike action.

Hagfishes have large blood sinuses and very low blood pressure. In contrast to all other vertebrates, hagfishes have accessory hearts in the liver and tail regions in addition to the true heart near the gills. These hearts are aneural, meaning that their pumping rhythm is intrinsic to the hearts themselves rather than coordinated via the central nervous system. In all these features, hagfishes resemble the condition seen in amphioxus—although, like other vertebrates, their blood does have red blood cells containing hemoglobin, and the true heart has three chambers.

Feeding Hagfishes attack dead or dying vertebrate prey. Once attached to the flesh, they can tie a knot in their tail and pass it forward along their body until they are braced against their prey and can tear off the flesh in their pinching grasp. They often begin by eating only enough outer flesh to enter the prey's coelomic cavity, where they dine on soft parts. Some recent, but controversial, research suggests that hagfishes can actually absorb dissolved organic nutrients through their skin and gill tissues.

Reproduction In most species, female hagfishes outnumber males by a hundred to one; the reason for this strange sex ratio is unknown. Examination of the gonads suggests that at least some species are hermaphroditic, but nothing is known of mating. The yolky eggs, which are oval and more than a centimeter long, are encased in a tough, clear covering that is secured to the sea bottom by hooks. The eggs are believed to hatch into small, completely formed hagfishes, bypassing a larval stage. Unfortunately, almost nothing is known of the embryology or early life history of any hagfish because few eggs have been available for study. However, some hagfish embryos have recently been examined, and it has been determined that hagfishes possess neural crests like other vertebrates.

Hagfishes and Humans We still know very little about hagfish ecology: we do not know how long hagfishes live; how old they are when they first begin to reproduce; exactly how, when, or where they breed; where the youngest juveniles live; what the diets and energy requirements of free-living hagfishes are; or virtually any of the other information needed for good management of commercially exploited populations of hagfishes. And, strangely enough, hagfishes do have an economic importance for humans. Almost all so-called eel-skin leather products are made from hagfish skin. Worldwide demand for this leather has eradicated economically harvestable hagfish populations in Asian waters and in some sites along the West Coast of North America.

Lampreys—Petromyzontiformes

There are around 40 species of lampreys in two major genera (*Petromyzon* and *Lampetra*); the adults of different species range in size from around 10 centimeters up to 1 meter.

Structural Characteristics Although lampreys are similar to hagfishes in size and shape (**Figure 3–6**), they have many features that are lacking in hagfishes but shared with gnathostomes. Traditionally it has been assumed that only lampreys had structures homologous with the vertebrae of jawed vertebrates: minute cartilaginous elements called arcualia, homologous with the neural arches of vertebrae. However, recent work has shown that hagfishes also have vertebral rudiments in the ventral portion of their tails (homologs of the hemal arches).

Lampreys are unique among living vertebrates in having a single nasal opening situated on the top of the head, combined with a duct leading to the hypophysis (pituitary) and known as a nasohypophysial opening. Development of this structure involves distortion of the front of the head, and its function is not known. Several groups of ostracoderms had an apparently similar structure, which evidently evolved convergently in those groups. The eyes of lampreys are large and well developed, as is the pineal body, which lies under a pale spot just posterior to the nasal opening. In contrast to hagfishes, lampreys have two semicircular canals on each side of the head—a condition shared with the extinct ostracoderms as well as in gnathostomes. In addition, the heart is innervated by the parasympathetic nervous system (the vagus nerve, X), as in gnathostomes, but not in hagfishes. Chloride-transporting cells in the gills and well-developed kidneys regulate ions, water, and nitrogenous wastes, as well as overall concentration of body fluids, allowing lampreys to exist in a variety of salinities.

Lampreys have seven pairs of gill pouches that open to the outside just behind the head. In most other fishes and in larval lampreys, water is drawn into the mouth and then pumped out over the gills in continuous or **flow-through ventilation**. Adult lampreys spend much of their time with their suckerlike mouths affixed to the bodies of other fishes, and during this time they cannot ventilate the gills in a flow-through fashion. Instead,

they use a form of **tidal ventilation** by which water is both drawn in and expelled through the gill slits. A flap called the velum prevents water from flowing out of the respiratory tube into the mouth. The lampreys' mode of ventilation is not very efficient at oxygen extraction, but it is a necessary compromise given their specialized mode of feeding.

Feeding Most adult lampreys are parasitic on other fishes, although some small, freshwater species have nonfeeding adults. The parasitic species attach to the body of another vertebrate (usually a bony fish that is larger than the lamprey) by suction and rasp a shallow, seeping wound through the integument of the host. The round mouth is located at the bottom of a large fleshy funnel (the oral hood), the inner surface of which is studded with keratinized conical spines. The oral hood, which appears to be a hypertrophied upper lip, is a unique derived structure in lampreys. The protrusible tonguelike structure is also covered with spines, and together these structures allow tight attachment and rapid abrasion of the host's integument. This tongue is not homologous with the tongue of gnathostomes because the tongue muscle is innervated by a different cranial nerve (the trigeminal nerve, V, rather than the hypoglossal nerve, XII).

An oral gland secretes an anticoagulant that prevents the victim's blood from clotting. Feeding is probably continuous when a lamprey is attached to its host. The bulk of an adult lamprey's diet consists of body fluids of fishes. The digestive tract is straight and simple, as one would expect for an animal with a diet as rich and easily digested as blood and tissue fluids. Lampreys generally do not kill their hosts, but they do leave a weakened animal with an open wound. At sea, lampreys feed on several species of whales and porpoises in addition to fishes. Swimmers in the Great Lakes, after having been in the water long enough for their skin temperature to drop, have reported initial attempts by lampreys to attach to their bodies.

Reproduction Lampreys are primarily found in northern latitude temperate regions, although a few species are known from southern temperate latitudes. Nearly all lampreys are **anadromous**; that is, they live as adults in oceans or big lakes and ascend rivers and streams to breed. Some of the most specialized species live only in freshwater and do all of their feeding as larvae, with the adults acting solely as a reproductive stage in the life history of the species. Little is known of the habits of adult lampreys because they are generally observed only during reproductive activities or when captured with their host.

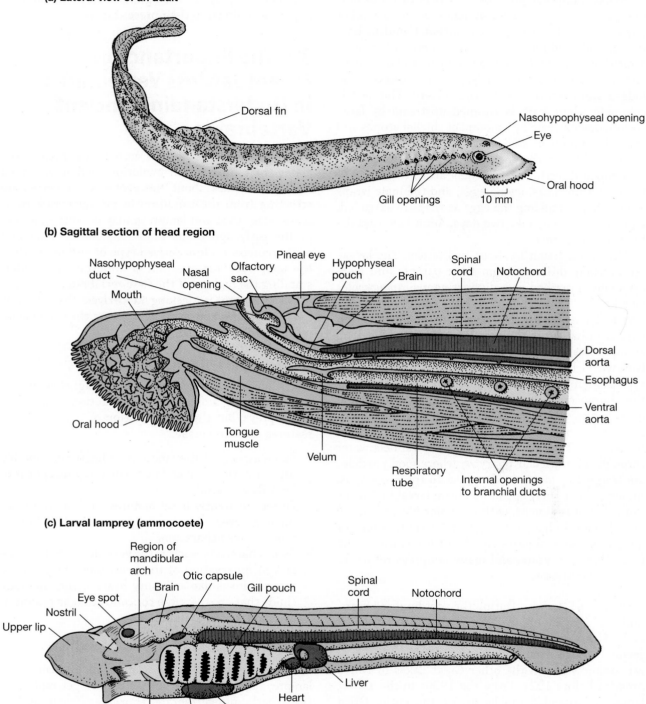

(a) Lateral view of an adult

Dorsal fin

Nasohypophyseal opening

Eye

Oral hood

Gill openings

10 mm

(b) Sagittal section of head region

Nasohypophyseal duct

Nasal opening

Olfactory sac

Pineal eye

Hypophyseal pouch

Brain

Spinal cord

Notochord

Mouth

Oral hood

Tongue muscle

Velum

Respiratory tube

Internal openings to branchial ducts

Dorsal aorta

Esophagus

Ventral aorta

(c) Larval lamprey (ammocoete)

Region of mandibular arch

Otic capsule

Gill pouch

Spinal cord

Notochord

Brain

Eye spot

Nostril

Upper lip

Velum

Gill opening

Endostyle

Heart

Liver

Figure 3–6 **Lampreys.**

Female lampreys produce hundreds to thousands of eggs, about a millimeter in diameter and devoid of any specialized covering such as that found in hagfishes. Male and female lampreys construct a nest by attaching themselves by their mouths to large rocks and thrashing about violently. Smaller rocks are dislodged and carried away by the current. The nest is complete when a pit is rimmed upstream by large stones, downstream by a mound of smaller stones and sand that produces eddies. Water in the nest is oxygenated by this turbulence but does not flow strongly in a single direction. The female attaches to one of the upstream rocks, laying eggs, and the male wraps around her, fertilizing the eggs as they are extruded, a process that may take two days. Adult lampreys die after breeding once.

The larvae hatch in about two weeks. The larvae are radically different from their parents and were originally described as a distinct genus, *Ammocoetes* (see Figure 3–6c). This name has been retained as a vernacular name for the larval form. A week to ten days after hatching, the tiny 6- to 10-millimeter-long ammocoetes leave the nest. They are wormlike organisms with a large, fleshy oral hood and nonfunctional eyes hidden deep beneath the skin. Currents carry the ammocoetes downstream to backwaters and quiet banks, where they burrow into the soft mud or sand and spend three to seven years as sedentary filter feeders. The protruding oral hood funnels water through the muscular pharynx, where food particles are trapped in mucus and carried to the esophagus. An ammocoete may spend its entire larval life in the same bed of sediment, with no major morphological or behavioral change until it is 10 centimeters or longer and several years old. Adult life is usually no more than two years, and many lampreys return to spawn after one year.

Lampreys and Humans During the past hundred years, humans and lampreys have increasingly been at odds. Although the sea lamprey, *Petromyzon marinus*, seems to have been indigenous to Lake Ontario, it was unknown from the other Great Lakes of North America before 1921. From the 1920s to the 1950s, lampreys expanded rapidly across the entire Great Lakes basin, and by 1946 they inhabited all the Great Lakes. Lampreys were able to expand unchecked until sporting and commercial interests became alarmed at the reduction of economically important fish species, such as lake trout, burbot, and lake whitefish. Chemical lampricides as well as electrical barriers and mechanical weirs at the mouths of spawning streams have been employed to bring the Great Lakes lamprey populations down to their present level.

3.3 The Importance of Extant Jawless Vertebrates in Understanding Ancient Vertebrates

The fossil record of the first vertebrates reveals little about their pre-Silurian evolution, and it yields no undisputed clues about the evolution of vertebrate structure from the condition in nonvertebrate chordates. Hagfishes and lampreys may provide examples of the early agnathous radiation, but do hagfishes really represent a less derived type of vertebrate than the lamprey? This question is important for understanding the biology of the first vertebrates: Were the first vertebrates as lacking in derived characters as living hagfishes appear to be, or were they somewhat more complex animals?

Anatomical Evidence

On balance, anatomical features of hagfishes appear to represent a more ancestral condition than those in lampreys, but the interpretation of many of these features is controversial.

- Some aspects of the anatomy of hagfishes, notably aspects of the brain and neuroanatomy, do appear to be truly ancestral.
- Other apparently basal features, such as the rudimentary eyes, may represent a secondary loss of more derived characters.
- Some characters, such as the very simple kidney, the lack of innervation of the heart (and the presence of amphioxus-like accessory hearts), and the body fluids that are the same concentration as seawater, are simply hard to interpret.

Molecular Evidence

The majority of molecular studies link hagfishes and lampreys as sister taxa, but we cannot get molecules from early nonvertebrate chordates (such as the Cambrian *Haikouella*) or from early vertebrates (such as the Cambrian vertebrate *Haikouichthys* and the huge diversity of ostracoderms) for comparison. Missing data of this sort can bias the computer programs that create phylogenies by introducing a statistical artifact known as "long branch attraction" that does not represent the true phylogenetic relationship. We do not claim here that the molecular findings in the case of lampreys and hagfishes are artifacts, but the difference in the results obtained from morphological data versus molecular data is troubling.

Recent molecular studies have focused on small portions of the genome called microRNAs, which are particularly prominent in vertebrates, and which appear to have been accumulating in vertebrate genomes over time (see Chapter 2). Studies of hagfishes and lampreys show that these two types of cyclostomes have four unique families of microRNAs, suggesting that they are in fact closely related. However, microRNAs apparently may be lost over time, so the debate is likely to continue.

3.4 The Radiation of Paleozoic Jawless Vertebrates— "Ostracoderms"

"Ostracoderms" is a paraphyletic assemblage because some more derived types of ostracoderms are clearly more closely related to the gnathostomes (jawed vertebrates) than others (see Figure 3–3). Our interpretation of exactly how different lineages of ostracoderms are related to one another and to living vertebrates has changed considerably since the latter part of the twentieth century.

Ostracoderms are clearly more derived than extant agnathans: Ostracoderms had dermal bone, and impressions on the underside of the dorsal head shield suggest that they had derived (i.e., gnathostome-like) features—a cerebellum in the hindbrain and an olfactory tract connecting the olfactory bulb with the forebrain. (Living agnathans lack a distinct cerebellum, and their olfactory bulbs are incorporated within the rest of the forebrain rather than placed more anteriorly and linked to the head via the olfactory tract [cranial nerve I].)

Characters of Ostracoderms

Most ostracoderms are characterized by the presence of a covering of dermal bone, usually in the form of an extensive armored shell, or **carapace**, but sometimes in the form of smaller plates or scales (e.g., anaspids), and some are relatively naked (e.g., thelodonts). Ostracoderms ranged in length from about 10 centimeters to more than 50 centimeters. Although they lacked jaws, some apparently had various types of movable mouth plates that have no analogues in any living vertebrates. These plates

were arranged around a small, circular mouth that appears to have been located farther forward in the head than the larger, more gaping mouth of jawed vertebrates. Most species of ostracoderms probably ate small, soft-bodied prey.

Most ostracoderms had some sort of midline dorsal fin, and although many heterostracans and anaspids had some sort of anterior, paired, finlike projections, only the more derived osteostracans had true pectoral fins, with an accompanying pectoral girdle and endoskeletal fin supports. As in living jawless vertebrates, the notochord must have been the main axial support throughout adult life. **Figure 3–7** depicts some typical ostracoderms.

Agnathans and Gnathostomes

During the Late Silurian and the Devonian, most major known groups of extinct agnathans coexisted with early gnathostomes, and it is highly unlikely that ostracoderms were pushed into extinction by the radiation of gnathostomes after 50 million years of coexistence. Jawless and jawed vertebrates appear to represent two different basic types of animals that probably exploited different types of resources. The initial reduction of ostracoderm diversity at the end of the Early Devonian may be related to a lowering of global sea levels, with the resulting loss of coastal marine habitats. The extinction of the ostracoderms in the Late Devonian occurred at the same time as mass extinctions among many marine invertebrates, which suggests that their demise was not due to gnathostome competition. Gnathostomes also suffered in the Late Devonian mass extinctions, and the placoderm lineage that dominated the Devonian period became extinct at its end.

3.5 The Basic Gnathostome Body Plan

Gnathostomes are considerably more derived than agnathans, not only in their possession of jaws but also in many other ways. Jaws allow a variety of new feeding behaviors, including the ability to grasp objects firmly, and along with teeth enable the animal to cut food to pieces small enough to swallow or to grind hard foods. New food resources became available when vertebrates evolved jaws: Herbivory was now possible, as was taking bite-sized pieces from large prey items, and many gnathostomes became larger than contemporary jawless vertebrates. A grasping, movable jaw also permits manipulation of objects:

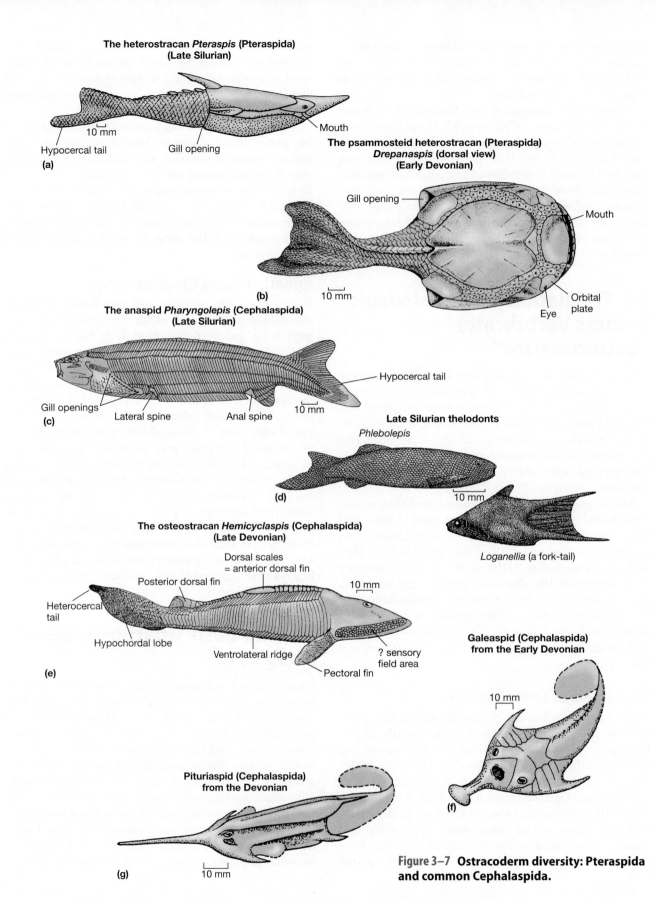

The heterostracan *Pteraspis* (Pteraspida)
(Late Silurian)

10 mm

Mouth

Hypocercal tail   Gill opening

(a)

The psammosteid heterostracan (Pteraspida)
***Drepanaspis* (dorsal view)**
(Early Devonian)

Gill opening

Mouth

10 mm

Orbital
plate

Eye

(b)

The anaspid *Pharyngolepis* (Cephalaspida)
(Late Silurian)

Hypocercal tail

Gill openings   Lateral spine   Anal spine   10 mm

(c)

Late Silurian thelodonts

Phlebolepis

10 mm

Loganellia (a fork-tail)

The osteostracan *Hemicyclaspis* (Cephalaspida)
(Late Devonian)

Dorsal scales
= anterior dorsal fin

Posterior dorsal fin

10 mm

Heterocercal
tail

Hypochordal lobe

Ventrolateral ridge   ? sensory
field area

Pectoral fin

(e)

Galeaspid (Cephalaspida)
from the Early Devonian

10 mm

(f)

Pituriaspid (Cephalaspida)
from the Devonian

(g)   10 mm

**Figure 3–7 Ostracoderm diversity: Pteraspida
and common Cephalaspida.**

jaws are used to dig holes, to carry pebbles and vegetation to build nests, and to grasp mates during courtship and juveniles during parental care. However, it seems that the likely origin of jaws was for more efficient gill ventilation rather than predation, as will be discussed later.

Gnathostome Biology

Gnathostomes are first known with certainty from the Early Silurian, but isolated sharklike scales suggest that they date back to the Middle Ordovician. The difference between gnathostomes and agnathans is traditionally described as the presence of jaws that bear teeth and two sets of paired fins or limbs (pectoral and pelvic). However, the gnathostome body plan (**Figure 3–8**) reveals that they are characterized by many other features, implying that gnathostomes represent a basic step up in level of activity and complexity from the jawless vertebrates. Interestingly, the form of the gnathostome lower jaw seems to be most variable in

the Silurian, when jawed fishes were not very common. By the Devonian, when the major radiation of jawed vertebrates commenced, the variety of jaw forms had stabilized.

These derived features include improvements in locomotor and predatory abilities and in the sensory and circulatory systems. Just as jawless vertebrates show a duplication of the *Hox* gene complex in comparison to nonvertebrate chordates (see Chapter 2), living jawed vertebrates show evidence of a second *Hox* duplication event. Gene duplication would have resulted in a greater amount of genetic information, perhaps necessary for building a more complex type of animal. However, as can be seen in Figure 3–3, a number of extinct ostracoderm taxa lie between living agnathans and gnathostomes, and many features seen today only in gnathostomes might have been acquired in a steplike fashion throughout ostracoderm evolution. Paired pectoral fins, for example, are usually considered to be a gnathostome character, but some osteostracans had evolved paired fins.

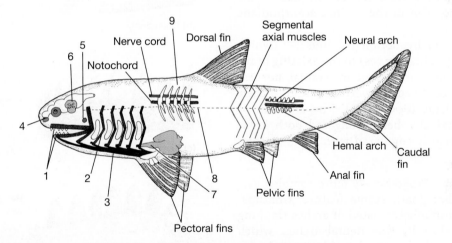

Figure 3–8 Generalized jawed vertebrate (gnathostome) showing derived features compared to the jawless vertebrate (agnathan) condition. Legend: 1. Jaws (containing teeth) formed from the mandibular gill arch. **2.** Gill skeleton consists of jointed branchial arches and contains internal gill rakers that stop particulate food from entering the gills. Gill musculature is also more robust. **3.** Hypobranchial musculature allows strong suction in inhalation and suction feeding. **4.** Two distinct olfactory tracts, leading to two distinct nostrils. **5.** Original first gill slit squeezed to form the spiracle, situated between the mandibular and hyoid arches. **6.** Three semicircular canals in the inner ear (addition of horizontal canal). **7.** Addition of a conus arteriosus to the heart, between the ventricle and the ventral aorta. (Note that the position of the heart is actually more anterior than shown here, right behind the most posterior gill arch.) **8.** Horizontal septum divides the trunk muscle into epaxial (dorsal) and hypaxial (ventral) portions. It also marks the position of the lateral line canal, containing the neuromast sensory organs. **9.** Vertebrae now have centra (elements surrounding the notochord) and ribs, but note that the earliest gnathostomes have only neural and hemal arches, as shown in the posterior trunk.

Jaws and Teeth Extant gnathostomes have teeth on their jaws, but teeth must have evolved after the jaws were in place because early members of the most basal group of jawed fishes—the placoderms—appear to have lacked true teeth (see section 3.6). Another reason to dissociate the evolution of jaws from the evolution of teeth is the presence of pharyngeal toothlike structures in various jawless vertebrates. These are most notable in the conodonts previously discussed (see Figure 3–2) but also have been reported in thelodonts, which are poorly known ostracoderms that lack a well-mineralized skeleton. Developmental studies support the notion that the first teeth were not oral structures but were located in the pharynx.

Bony fishes and tetrapods have teeth embedded in the jawbones (**Figure 3–9**). However, because teeth form from a dermal papilla, they can be embedded only in dermal bones, and cartilaginous fishes such as sharks and rays lack dermal bone. The teeth of cartilaginous fishes form within the skin, resulting in a tooth whorl that rests on the jawbone but is not actually embedded in it. This condition is probably the ancestral one for all gnathostomes more derived than placoderms because it is also seen in the extinct acanthodians (see section 3.6).

Adding jaws and hypobranchial muscles (which are innervated by spinal nerves) to the existing branchiomeric muscles (innervated by cranial nerves) allowed vertebrates to add powerful suction to their feeding mechanisms (**Figure 3–10a**). The cranium of gnathostomes has also been elongated both anteriorly and posteriorly compared with the agnathan condition.

Vertebrate and Ribs Progressively more complex vertebrae are another gnathostome feature. Gnathostome vertebrae initially consisted of arches flanking the nerve cord dorsally (the neural arches, which are homologous with the arcualia of lampreys) with matching arches below the notochord (the hemal arches, which may be present in the tail only—see the posterior portion of the trunk in Figure 3–8). More derived gnathostomes had a vertebral centrum or central elements with attached ribs (Figure 3–10b). Still more complete vertebrae support the notochord and eventually replace it as a supporting rod for the axial muscles used for locomotion (mainly in tetrapods). Well-developed centra were not a feature of the earliest jawed fishes and are unknown in the two extinct groups of fishes—placoderms and acanthodians.

Ribs are another new feature in gnathostomes. They lie in the connective tissue between successive segmental muscles, providing increased anchorage for axial muscles. There is now a clear distinction between the epaxial and hypaxial blocks of the axial muscles, which are divided by a horizontal septum made of thin fibrous tissue that runs the length of the animal. The lateral line canal—containing the organs that sense vibrations in the surrounding water—lies in the plane of this septum, perhaps

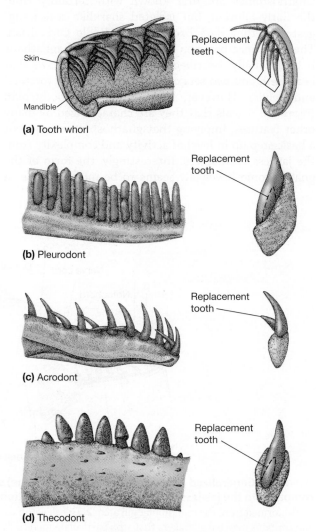

© 1998 The McGraw-Hill Companies, Inc.

Figure 3–9 Gnathostome teeth. (a) Tooth whorl of a chondrichthyan. (b–d) Teeth embedded in the dermal bones of the jaw as seen in osteichthyan fishes and tetrapods. (b) Pleurodont, the basal condition: teeth set in a shelf on the inner side of the jawbone, seen in some bony fishes and in modern amphibians and some lizards. (c) Acrodont condition: teeth fused to the jawbone, seen in most bony fishes and in some reptiles (derived independently). (d) Thecodont condition: teeth set in sockets and held in place by peridontal ligaments, seen in archosaurian reptiles and mammals (derived independently).

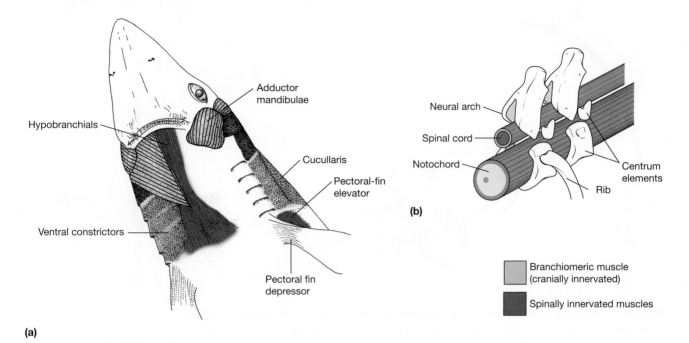

Figure 3–10 Some gnathostome specializations. (a) Ventral view of a dogfish, showing both branchiomeric muscles and the new, spinally innervated, hypobranchial muscles. (b) View of generalized gnathostome vertebral form, showing central elements and ribs.

reflecting improved integration between locomotion and sensory feedback. In the inner ear there is a third (horizontal) semicircular canal, which may reflect an improved ability to navigate and orient in three dimensions.

What About Soft Tissues? Features such as jaws and ribs can be observed in fossils, so we know that they are unique to gnathostomes. However, we cannot know for sure whether other new features within the soft anatomy characterize gnathostomes alone or whether they were adopted somewhere within the ostracoderm lineage.

- We can surmise that some features of the nervous system that are seen only in gnathostomes among living vertebrates were acquired by the earliest ostracoderms. For example, impressions on the inner surface of the dermal head shield reveal the presence of a cerebellum in the brain and olfactory tracts.
- We can see that no ostracoderm possessed a third semicircular canal.
- A new feature of the nervous system in gnathostomes is the insulating sheaths of myelin on the nerve fibers, which increase the speed of nerve impulses.
- The heart of gnathostomes has an additional small chamber in front of the pumping ventricle, the **conus arteriosus**, which acts as an elastic

reservoir that smoothes out the pulsatile nature of the flow of blood produced by contractions of a more powerful heart. (Some workers consider that lampreys, but not hagfishes, have a small conus arteriosus despite their weak hearts and low blood pressures.)
- Living agnathans lack a stomach, but some derived ostracoderms may have had a stomach. Gnathostomes also have a type of cartilage different from the cartilage seen in the hagfishes and lampreys. (See Chapter 2 and the legend to Figure 3–3 in the Appendix for additional gnathostome characters.)
- One particularly important gnathostome feature is in the reproductive system, where the gonads now have their own specialized ducts leading to the cloaca (see Chapter 2).

The Origin of Fins

Guidance of a body in three-dimensional space is complicated. Fins act as hydrofoils, applying pressure to the surrounding water. Because water is practically incompressible, force applied by a fin in one direction against the water produces a thrust in the opposite direction. A tail fin increases the area of the tail, giving more thrust during propulsion, and allows the fin to exert the force needed for rapid acceleration. Rapid adjustments of the body position in the water may be

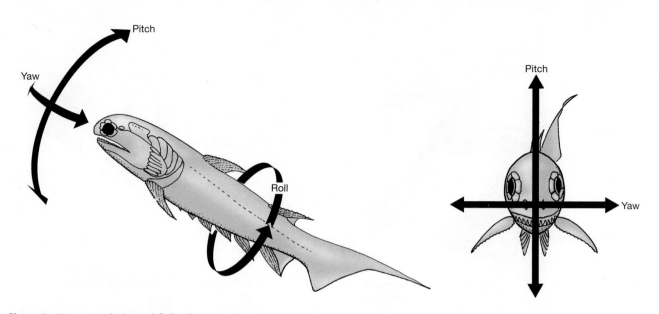

Figure 3–11 An early jawed fish (the acanthodian, *Climatius*). Views from the side and front illustrate pitch, yaw, and roll and the fins that counteract these movements.

especially important for active, predatory fishes such as the early gnathostomes, and the unpaired fins in the midline of the body (the dorsal and anal fins) control the tendency of a fish to roll (rotate around the body axis) or yaw (swing to the right or left) (**Figure 3–11**). The paired fins (pectoral and pelvic fins) can control the pitch (tilt the fish up or down) and act as brakes, and they are occasionally specialized to provide thrust during swimming, like the enlarged pectoral fins of skates and rays.

The basic form of the pectoral and pelvic fins, as seen today in chondrichthyans and basal bony fishes, is for a tribasal condition. This means that three main elements within the fin articulate with the limb girdle within the body, and molecular studies back up the anatomical ones in showing that this type of fin represents the ancestral gnathostome condition. The paired fins of gnathostomes also share patterns of development, involving particular *Hox* genes, with the median fins of lampreys.

Fins have non-locomotor functions as well. Spiny fins are used in defense, and they may become systems to inject poison when combined with glandular secretions. Colorful fins are used to send visual signals to potential mates, rivals, and predators. Even before the gnathostomes appeared, fishes had structures that served the same purpose as fins. Many ostracoderms had spines or enlarged scales derived from dermal armor that acted like immobile fins. Some anaspids had long finlike sheets of tissue running along the flanks, and osteostracans had pectoral fins.

3.6 The Transition from Jawless to Jawed Vertebrates

In Chapter 2 we saw that the branchial arches were a fundamental feature of the vertebrate cranium, providing support for the gills. It has long been known that vertebrate jaws are made of the same material as the skeletal elements that support the gills (cartilage derived from the neural crest), and they clearly develop from the first (mandibular) arch of this series in vertebrates. (Note that while the word "mandible" commonly refers to the lower jaw only, the term "mandibular arch" includes both upper and lower jaws.)

It is helpful at this stage to envisage the vertebrate head as a segmented structure, with each branchial arch corresponding to a segment (**Figure 3–12**). The mandibular arch that forms the gnathostome jaw is formed within the second head segment; jaw supports are formed from the hyoid arch of the third head segment; and the more posterior branchial arches that form the gill supports (arches 3 through 7) are formed in head segments 4 through 8. The branchial arch numbering does not match the numbering of the head segments because no evidence exists for a branchial arch functioning as a gill support structure in the first (premandibular) segment at any point in vertebrate history.

No living vertebrate has a pair of gill-supporting branchial arches in such an anterior position as the

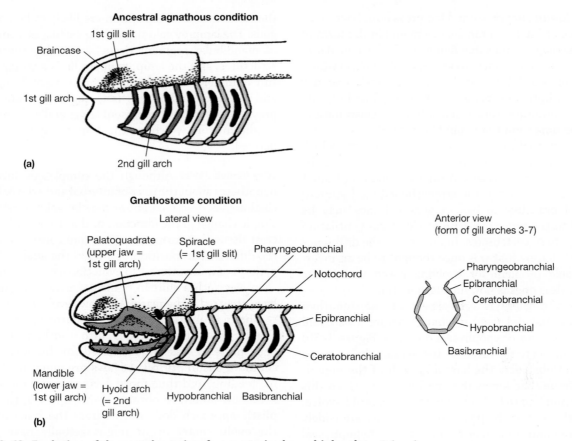

Figure 3–12 Evolution of the vertebrate jaw from anterior branchial arches. Colored shading indicates splanchnocranium elements (branchial arches and their derivatives).

jaws as shown in Figure 3-12(a). However, all vertebrates have some structure in this position (i.e., in the second head segment) that appears to represent the modification of an anterior pair of pharyngeal arches; these are the jaws in gnathostomes and velar cartilages (i.e., the structures that support the velum) in lampreys and hagfishes. Thus it has been proposed that the common ancestor of living vertebrates had an unmodified pair of branchial arches in this position, with a fully functional gill slit lying between the first and second arches, and that living jawless and jawed vertebrates are both divergently specialized from this condition. There is no trace of a gill slit between arches 1 and 2 in living jawless vertebrates; but, in many living cartilaginous fishes, and in a few bony fishes, there is a small hole called the spiracle in this position, which is now used for water intake. Figure 3–12 summarizes the differences in the gill arches between jawed vertebrates and their presumed jawless ancestor and illustrates the major components of the hinged gnathostome gill arches (as seen in arches 3 through 7).

Evolution of Gills of Early Vertebrates

Because ostracoderms are now viewed as "stem gnathostomes," any understanding of the origin of gnathostomes must encompass the view that at some point a jawless vertebrate was transformed into a jawed one. However, for much of the last century, researchers considered jawless and jawed vertebrates as two separate evolutionary radiations. This was because of the apparent nonhomology of their branchial arches: at least in living jawless vertebrates (the situation is less clear for the fossil ones), the gill arches lie lateral to (that is, external to) the gill structures, whereas in jawed vertebrates they lie medially (internal to the gills). Additionally, extant jawless vertebrates have pouched gills with small, circular openings that are different from the flatter, more lens-shaped openings between the gills of gnathostomes. However, in the early twenty-first century a veritable explosion of studies on vertebrate head development (especially studies of lampreys) showed that this difference in gill arch position may be produced simply by a switch in developmental timing.

In addition, hagfishes and lampreys both have some evidence of internal branchial elements in the form of velar cartilages, and some living sharks have evidence of small external branchial cartilages. That observation raises the possibility that the earliest vertebrates may have had both internal and external gill arches, and that is a condition from which both the cyclostome and gnathostome conditions could be derived.

The external branchial basket of lampreys, which squeezes the pharyngeal area from the outside, may be essential for the specialized tidal ventilation of adult lampreys. In contrast, the strengthened (and jointed) internal branchial arches in gnathostomes may be related to the more powerful mode of gill ventilation used by these vertebrates. In any event, the difference in gill skeletons that was once thought to be an insurmountable problem for evolving jawed vertebrates from jawless ones is now inconsequential.

However, there is one aspect of the agnathan anatomy that did need to change before jaws could develop: the nasohypophyseal duct (see Figure 3–6b) that connects to the pituitary in the brain. In lampreys this duct obstructs the lateral growth of the neural-crest tissue that forms the gnathostome jaw, so this connection had to be broken before jaws could evolve. In gnathostomes the paired olfactory sacs are widely separated and no longer connect to the hypophyseal pouch. This separation has now been identified in a derived ostracoderm, one of the galeaspids. Thus, the rearrangement of cranial anatomy necessary for jaw evolution was acquired in jawless fishes closely related to gnathostomes.

Stages in the Origin of Jaws

In recent years, molecular developmental biology has provided fresh insights into the issue of the origin of jaws. It is worth noting that some of the controversies described above relate to a couple of misconceptions about the probable processes of evolution:

- First, we should consider that evolutionary transformation does not involve changing one adult structure into another adult structure; instead, morphology is altered via modifications of gene expression and shifts in developmental timing.
- Second, the fact that the lamprey lies below the position of gnathostomes on the cladogram does not mean that it represents a basal vertebrate condition. On the contrary, lampreys, with their bizarrely hypertrophied upper lip, are highly derived in their own right.

The same genes are expressed in the mandibular segments of lampreys and gnathostomes, indicating that the structures in this position are likely to be homologous. The lamprey velum and velar cartilages, composed of mandibular segment tissue, are highly specialized structures, and the lamprey upper lip is a strange mixture of material from the mandibular (second) segment and the premandibular (first) segment. Thus, lampreys clearly do not represent a generalized ancestral condition from which gnathostomes might have been derived.

Why Evolve Jaws? Although the complex evolutionary hypotheses about the roles of internal and external branchial arches described above may be valid, a relatively simple change in the direction of the streams of neural-crest tissue that form the arches could also account for the difference in position of jaws in the adult. What is perhaps of more interest to evolutionary morphologists than *how* such a shift in branchial arch position could have occurred is the reason *why* jaws evolved at all.

The notion that derived predatory vertebrates should convert gill arches into toothed jaws has been more or less unquestioned for decades. It is a common assumption that jaws are superior devices for feeding, and thus more derived vertebrates were somehow bound to obtain them. However, this simplistic approach does not address the issue of how the evolutionary event might actually have taken place: What use would a protojaw be prior to its full transformation?

And even if early vertebrates needed some sort of superior mouth anatomy, why modify a branchial gill-supporting arch, which initially was located some distance behind the mouth opening? Why not just modify the existing cartilages and plates surrounding the mouth? There is nothing about those structures that prevents them from being modified. Quite the contrary, in fact—the living agnathans have specialized oral cartilages, and various ostracoderms apparently had oral plates. Additionally, as we noted earlier, gnathostome teeth must have evolved after jaws evolved, so the first jaws were toothless; and of what use could a toothless jaw be? (A movie entitled *Gums* wouldn't sell out a theater.)

Jon Mallatt has proposed a novel explanation of the origin of jaws based on the hypothesis that jaws were initially important for gill ventilation rather than predation. Living gnathostomes are more active than jawless vertebrates and have greater metabolic demands, and features of the earliest known ones suggest that this was the condition from the start of their evolutionary history. One derived gnathostome feature associated with such high activity is the powerful mechanism for pumping water over the gills. Gnathostomes have a characteristic series

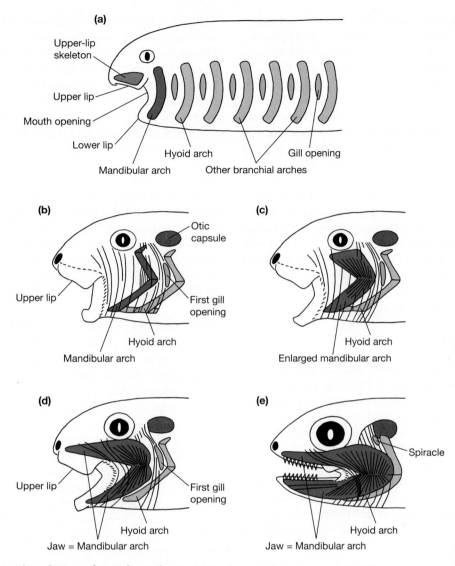

Figure 3–13 Proposed evolution of vertebrate jaws. (a) Jawless gnathostome ancestor. The 'upper lip skeleton' is derived from the mandibular arch and is considered to be the precursor of the upper jaw. (b) Early pre-gnathostome with jointed branchial arches. (c) Late pre-gnathostome with enlarged mandibular arch now employed to firmly close mouth during ventilation. (d) Early gnathostome with mandibular arch now used as a feeding jaw. (e) More derived gnathostome (e.g., a shark) with teeth added to jaw.

of internal branchial muscles (with cranial nerve innovation) as well as the new, external hypobranchials (with spinal nerve innervation). These muscles not only push water through the pharynx in exhalation but also suck water into the pharynx during inhalation. Gnathostome fishes can generate much stronger suction than agnathans, and powerful suction is also a way to draw food into the mouth. Living agnathans derive a certain amount of suction from their pumping velum, but this pump mostly pushes (rather than sucks) water, and its action is weak.

Mallatt proposed that the mandibular branchial arch enlarged into protojaws because it played an essential role in forceful ventilation—rapidly closing and opening the entrance to the mouth (**Figure 3–13**). During strong exhalation, as the pharynx squeezed water back across the gills, water was kept from exiting via the mouth by bending the mandibular arch sharply shut. Next, during forceful inspiration, the mandibular arch was rapidly straightened to reopen the mouth and allow water to enter. To accommodate the forces of the powerful muscles that bent the arch (the adductor mandibulae, modified branchial muscles) and straightened it (the hypobranchial muscles), the mandibular arch enlarged and became more robust.

The advantage of using the mandibular arch would be that the muscles controlling it are of the same functional series as the other ventilatory muscles, and their common origin would ensure that all of the muscles were controlled by the same nerve circuits. In contrast, the muscles of the more anterior oral cartilages would not have been coupled with the musculature of the pharyngeal arches. Probably that would not have mattered if feeding had been the original function of jaws, whereas it would matter if the jaws had to coordinate their movements with the pharyngeal arches that were responsible for ventilation. This line of reasoning may explain why the mandibular arch, rather than the more anterior oral cartilages, became the jaws of gnathostomes.

3.7 Extinct Paleozoic Jawed Fishes

With jaws that can grasp prey, muscles that produce powerful suction, and other features indicative of higher levels of activity, gnathostomes were able to enter ecological niches unavailable to agnathan vertebrates. We have numerous fossils of the entire bodies of gnathostomes (rather than fragments such as teeth and scales) from Devonian sediments. At this point, gnathostomes can be divided into four distinctive clades: two extinct groups—placoderms and acanthodians, and two groups that survive today—chondrichthyans (cartilaginous fishes) and osteichthyans (bony vertebrates):

- Placoderms were highly specialized, armored fishes that appear to be basal to other gnathostomes.
- The acanthodians, or "spiny sharks," were small, more generalized fishes. Although acanthodians have been traditionally grouped with the bony fishes, new studies suggest that the different species occupied a diversity of positions in the vertebrate phylogeny.
- The cartilaginous fishes, which include sharks, rays, and ratfishes, evolved distinctive specializations of small dermal scales, internal calcification, jaw and fin mobility, and reproduction.
- Bony fishes evolved endochondral bone in their internal skeleton, a distinctive dermal head skeleton that included an operculum covering the gills, and an internal air sac forming a lung or a swim bladder.

The bony fishes include the ray-finned fishes (actinopterygians), which comprise the majority of living fishes, and the lobe-finned fishes (sarcopterygians).

Only a few lobe-finned fishes survive today (lungfishes and coelacanths), but they were more diverse in the Paleozoic. (Furthermore, sarcopterygian fishes are the group that gave rise to the tetrapods, and from this perspective there are as many extant species of sarcopterygians as of actinopterygians.) If we take another step back, Osteichthyes includes tetrapods, and we are ourselves highly modified fishes. Bony fishes by themselves constitute a paraphyletic group because their common ancestor is also the ancestor of tetrapods, and the same is true of the lobe-finned fishes.

Before studying the extant groups of jawed fishes, we turn to the placoderms and acanthodians to examine the variety of early gnathostomes. Figures 3–3 and 3–4 show the interrelationships of gnathostome fishes. Living and extinct groups of chondrichthyans are discussed in Chapter 5, and osteichthyans are discussed in Chapter 6.

Placoderms—The Armored Fishes

As the name placoderm (Greek *placo* = plate and *derm* = skin) implies, placoderms were covered with a thick bony shield over the anterior one-third to one-half of their bodies. Unlike the ostracoderms, the bony shield of placoderms was divided into separate head and trunk portions, linked by a mobile joint that allowed the head to be lifted up during feeding (**Figure 3–14a**). The endoskeleton was mineralized by perichondral bone ossification around the rim of the bone (perichondral bone). (Ossification throughout the entire bone [endochondral bone] appears to be limited to osteichthyans.) Many (though not all) researchers consider that placoderms are not a distinctive clade, but rather represent a paraphyletic assemblage at the stem of gnathostome evolution. This would mean that some placoderm species were more closely related to the living groups of gnathostomes than were others.

History of Placoderms Placoderms are known from the Early Silurian to the end of the Devonian and were the most diverse vertebrates of their time in terms of number of species, types of morphological specializations, and body sizes, with the largest reaching a length of 8 meters, the size of a great white shark. Like the ostracoderms, they suffered massive losses in the Late Devonian extinctions; but, unlike any ostracoderm group, a few placoderm lineages continued for another 5 million years, until the very end of the period. Ancestral placoderms were primarily marine, but a great many lineages became adapted to freshwater and estuarine habitats.

Biology of Placoderms Placoderms have no modern analogues, and their massive external armor makes

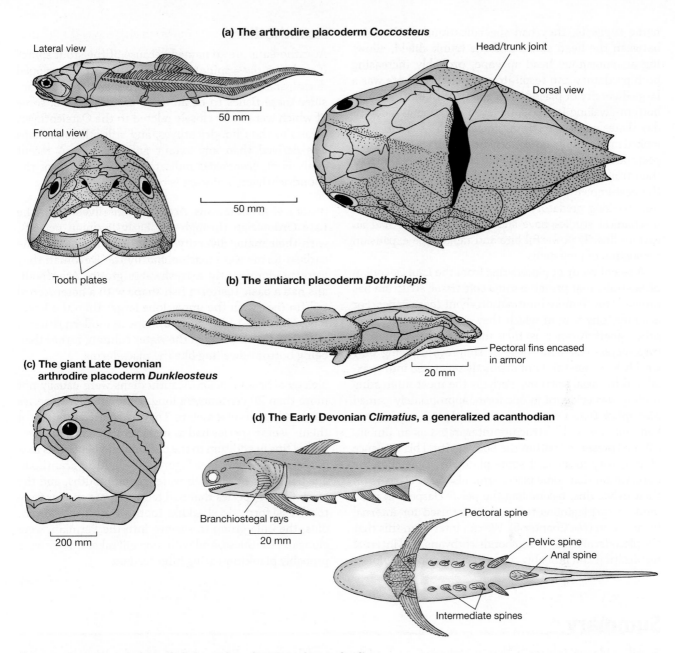

(a) The arthrodire placoderm *Coccosteus*

Lateral view

Head/trunk joint

Dorsal view

50 mm

Frontal view

50 mm

Tooth plates

(b) The antiarch placoderm *Bothriolepis*

Pectoral fins encased in armor

20 mm

(c) The giant Late Devonian arthrodire placoderm *Dunkleosteus*

200 mm

(d) The Early Devonian *Climatius*, a generalized acanthodian

Branchiostegal rays

20 mm

Pectoral spine

Pelvic spine
Anal spine

Intermediate spines

Figure 3–14 Extinct Paleozoic fishes: placoderms and acanthodians.

interpreting their lifestyle difficult, although they were apparently primarily inhabitants of deep water. Most placoderms lacked true teeth; their toothlike structures (tooth plates, see Figure 3–14a) were actually projections of the dermal jawbones that were subject to wear and breakage without replacement. However, some later placoderms did possess true teeth, possibly evolved convergently with teeth of other gnathostomes. It is impossible to know whether placoderms had many of the soft-anatomy characters of living gnathostomes, but it has been determined that, like their extant relatives, they had myelinated nerve sheaths. Structures preserved in one placoderm were initially interpreted as being lungs, but this notion has now been disproved: these supposed "lungs" appear to be portions of the digestive tract that were filled with sediment after death.

More than half of the known placoderms, nearly 200 genera, belonged to the predatory arthrodires (Greek *arthros* = a joint and *dira* = the neck). As their name suggests, they had specializations of the joint between the head shield and the trunk shield, allowing an enormous head-up gape, probably increasing both predatory and respiratory efficiency. There was a large diversity of placoderms, many of them flattened, bottom-dwelling forms, some even resembling modern-day skates and rays. The antiarchs, such as *Bothriolepis*, looked rather like armored catfishes (Figure 3–14b). Their pectoral fins were also encased in the bony shield, so that their front fins looked more like those of a crab. Perhaps the best-known placoderm is *Dunkleosteus*, a voracious, 8-meter-long predatory arthrodire (Figure 3–14c); biomechanical studies have shown that this fish had an extraordinarily powerful bite and rapid gape expansion for suction of prey items.

A recent study of placoderms from the Late Devonian of Australia that preserve some soft-tissue structure has provided much more information about these fishes. For example, one way in which they are less derived than other gnathostomes is that their segmental muscles (myomeres) resemble those of lampreys in being only weakly W-shaped and not distinctly separated into epaxial and hypaxial portions. Perhaps the most interesting finding was evidence in one form, appropriately named *Materpiscis* (Latin *mater* = mother and *piscis* = fish), that had embryos and a structure interpreted as an umbilical cord preserved within the body cavity. This evidence of viviparity in at least some placoderms matches the observation that some placoderms also have claspers on their pelvic fins, resembling the pelvic claspers in male modern cartilaginous fishes that are used for internal fertilization (see Chapter 5). We can infer from this that the placoderms, like living chondrichthyans, had internal fertilization and probably complex courtship behaviors.

Acanthodians

Acanthodians are so named because of the stout spines (Greek *acantha* = spine) anterior to their well-developed dorsal, anal, and paired fins. Most researchers now consider these fishes to be an array of early forms, some of which were more closely related to the Osteichthyes, others to the Chondrichthyes, and still others perhaps less derived than any extant gnathostome. A recent analysis of *Acanthodes* indicates that the form of the chondrichthyan braincase is basal for gnathostomes.

History of Acanthodians Acanthodians lived from the Late Ordovician through the Early Permian periods, with their major diversity in the Early Devonian. The earliest forms were marine, but by the Devonian they were predominantly a freshwater group. Acanthodians had a basic fusiform fish shape with a heterocercal tail fin (i.e., with the upper lobe larger than the lower lobe; Figure 3–14d). This tail shape in modern fishes is associated with living in the water column, rather than being bottom-dwelling like the placoderms.

Biology of Acanthodians Acanthodians were usually not more than 20 centimeters long, although some species reached lengths of 2 meters. They had paired fins, lots of them—some species had as many as six pairs of ventrolateral fins in addition to the pectoral and pelvic fins that gnathostomes have (see Figure 3–14d). Most acanthodians had large heads with wide-gaping mouths, and the teeth (in the species that had teeth—some species were toothless) formed a sharklike tooth whorl. The acanthodids, the only group to survive into the Permian, were elongate, toothless, and with long gill rakers. They were probably plankton-eating filter feeders.

Summary

Fossil evidence indicates that vertebrates evolved in a marine environment. Jawless vertebrates are first known from the Early Cambrian, and there is evidence that the first jawed vertebrates (gnathostomes) evolved as long ago as the Middle Ordovician. The first vertebrates would have been more active than their ancestors, with a switch from filter feeding to more active predation and with a muscular pharyngeal pump for gill ventilation. The first mineralized tissues were seen in the teeth of conodonts, enigmatic animals that have only recently been considered true vertebrates. Bone is a feature of many early vertebrates, although it was absent from the Early Cambrian forms and is also absent in the living jawless vertebrates—the hagfishes and lampreys. Bone is first found with accompanying external layers of dentine and enamel-like tissue in the dermal armor of early jawless fishes called the ostracoderms. Current explanations for the original evolutionary use of bone include protection, a store for calcium and phosphorus, and housing for electroreceptive sense organs.

Among the living jawless fishes, hagfishes are apparently less derived in their anatomy than the lampreys and all other known vertebrates, but molecular data now suggest that they form a clade with lampreys, which would mean that the apparently ancestral anatomical features represent secondary loss of derived characters.

Lampreys, in turn, are less derived than the extinct armored ostracoderms. Ostracoderms are widely known from the Silurian and Early Devonian, and none survived past the end of the Devonian. Ostracoderms were not a unified evolutionary group: some forms (Cephalaspida) were more closely related to gnathostomes than were other forms, sharing with gnathostomes the feature of a pectoral fin. Ostracoderms and gnathostomes flourished together for 50 million years; thus there is little evidence to support the idea that jawed vertebrates outcompeted and replaced jawless ones.

Just as the evolution of vertebrates from nonvertebrate chordates represented an advance in basic anatomy and physiology, so did the evolution of jawed vertebrates from jawless ones. Jaws may have evolved initially to improve gill ventilation rather than to bite prey. In addition to jaws, gnathostomes have a number of derived anatomical features (such as true vertebrae, ribs, and a complete lateral line sensory system), suggesting a sophisticated and powerful mode of locomotion and sensory feedback.

The early radiation of jawed fishes, first known in detail from the fossil record of the Late Silurian, included four major groups. Two groups, the chondrichthyans (cartilaginous fishes) and osteichthyans (bony fishes), survive today. Osteichthyans were the forms that gave rise to tetrapods in the Late Devonian. The other two groups, placoderms and acanthodians, are now extinct. Placoderms did not survive past the Devonian, but acanthodians survived almost until the end of the Paleozoic. Placoderms were armored fishes, superficially like the ostracoderms in their appearance. They were the most diverse fishes of the Devonian and included large predatory forms. Placoderms are considered to be basal to all other gnathostomes, perhaps forming a paraphyletic stem group. Acanthodians are more generalized fishes than the placoderms were. Acanthodians were originally considered to be the sister group to the bony fishes, but are now considered to be a generalized basal gnathostome assemblage.

Discussion Questions

1. Did jawed vertebrates outcompete the jawless ones? Does the fossil record provide any evidence?
2. How did the realization that conodont animals were vertebrates change our ideas about the pattern of vertebrate evolution?
3. How is the body-fluid composition of hagfishes different from that of other vertebrates? How does this difference figure in the debate about whether vertebrates arose in fresh or salt water?
4. Ventilating the gills by taking water in through the gill openings, as well as using the gill openings for water ejection, is not a very efficient way of doing things. Why, then, do adult lampreys ventilate their gills this way?
5. We usually think of jaws as structures that evolved for biting. What might have been a different original use, and what is the evidence for this?
6. Recent fossil evidence shows a placoderm with a developing embryo inside her body. How might we have been able to speculate that placoderms were viviparous in the absence of such evidence?

Additional Information

Ahlberg, P. E. 2009. Birth of the jawed vertebrates. *Nature* 457:1094–1095.

Ahlberg, P., et al. 2009. Pelvic claspers confirm chondrichthyan-like internal fertilization in arthrodires. *Nature* 460:888–889.

Anderson, P. S. L., and M. W. Westneat. 2009. A biomechanical model of feeding kinematics for *Dunkleosteus terrelli* (Arthrodira, Placodermi). *Paleobiology* 35: 251–269.

Anderson, P. S. L., et al. 2011. Initial radiation of jaws demonstrated stability despite faunal and environmental changes. *Nature* 476:206–209.

Brazeau, M. D. 2009. The braincase and jaws of a Devonian 'acanthodian' and modern gnathostome origins. *Nature* 457:305–308.

Cerny, R., et al. 2010. Evidence for the prepattern/cooption model of vertebrate jaw evolution.

Proceedings of the National Academy of Sciences 107:17262–17267.

Coates, M. I. 2009. Beyond the age of fishes. *Nature* 458:413–414.

Davis, S. P. et al. 2012. *Acanthodes* and shark-like conditions in the last common ancestor of modern gnathostomes. *Nature* 486:247-251.

Donoghue, P. C. J., and M. A. Purnell. 2009. The evolutionary emergence of vertebrates from among their spineless relatives. *Evolution Education Outreach* 2:204–212.

Friedman, M., and M. D. Brazeau. 2010. A reappraisal of the origin and basal radiation of the Osteichthyes. *Journal of Vertebrate Paleontology* 30:36–56.

Gai, Z., et al. 2011. Fossil jawless fish from China foreshadows early jawed vertebrate anatomy. *Nature* 476:324–327.

Gillis, J. A., and N. H. Shubin. 2009. The evolution of gnathostome development: Insights from chondrichthyan embryology. *Genesis* 47:825–841.

Glover, C. N., et al. 2011. Adaptations to *in situ* feeding: Novel nutrient acquisition pathways in an ancient vertebrate. *Proceedings of the Royal Society of London B* 278:3096–3101.

Goujet, D. 2011. "Lungs" in Placoderms, a persistent palaeobiological myth related to environmental preconceived interpretations. *Comptes Rendus Palevol* 10: 323–329.

Janvier, J. 2011. Comparative anatomy: All vertebrates do have vertebrea. *Current Biology* 21:R661-R663.

Janvier, P. 2008. Early jawless vertebrates and cyclostome origins. *Zoological Science* 25:1045–1056.

Long, J. A. 2011. *The Rise of Fishes*, 2nd ed. Baltimore, MD: Johns Hopkins University Press.

Long, J. A., et al. 2008. Live birth in the Devonian Period. *Nature* 453:650–653.

Mallatt, J. 2008. Origin of the vertebrate jaw: Neoclassical ideas versus newer, development-based ideas. *Zoological Science* 25:990-998.

Near, T. J. 2009. Conflict and resolution between phylogenies inferred from molecular and phenotypic data sets for hagfish, lampreys, and gnathostomes. *Journal of Experimental Zoology (Mol Dev Evol)* 312B:749–761.

Richardson, M. K., et al. 2010. Developmental anatomy of lampreys. *Biological Reviews* 85:1–33.

Shubin, N. 2009. *Your Inner Fish: A Journey into the 3.5-Billion-Year History of the Human Body.* New York: Random House.

Turner, S., et al. 2010. False teeth: Conodont-vertebrate phylogenetic relationships revisited. *Geodiversitas* 32:545–594.

Young, G. V. 2010. Placoderms (armored fish): Dominant vertebrates of the Devonian Period. *Annual Review of Earth and Planetary Sciences* 38:523–550.

Zhu, M., et al. 2009. The oldest articulated osteichthyan reveals mosaic gnathostome characters. *Nature* 458:469–474.

Websites
Information about living agnathans
http://www.ucmp.berkeley.edu/vertebrates/basalfish/petro.html
Information about basal gnathostomes
http://www.tolweb.org/Gnathostomata/14843
Information about placoderms
http://hoopermuseum.earthsci.carleton.ca//placoderms/fifthb.html
http://www.ucmp.berkeley.edu/vertebrates/basalfish/placodermi.html
Information about acanthodians
http://www.devoniantimes.org/who/pages/acanthodians.html

PART II

Non-Amniotic Vertebrates: Fishes and Amphibians

Vertebrates originated in the sea, and more than half of the species of living vertebrates are the products of evolutionary lineages that have never left an aquatic environment. Water now covers 73 percent of Earth's surface (the percentage has been both higher and lower in the past) and provides habitats extending from deep oceans, lakes, and mighty rivers to fast-flowing streams and tiny pools in deserts. Fishes have adapted to all these habitats, and there are more than 37,000 species of extant cartilaginous and bony fishes.

Life in water poses challenges for vertebrates but offers many opportunities. Aquatic habitats are some of the most ecologically productive on Earth, and energy is plentifully available in many of them. Some aquatic habitats (coral reefs are an example) have enormous structural complexity, whereas others (like the open ocean) have virtually none. The diversity of fishes reflects specializations for this variety of habitats.

The diversity of fishes and the habitats in which they live has offered unparalleled scope for variations in life history. Some fishes produce millions of eggs that are released into the water to drift and develop on their own, other species of fishes produce a few eggs and guard both the eggs and the young, and numerous fishes give birth to young that require no parental care. Males of some species of fishes are larger than females; in others the reverse is true. Some species have no males at all, and a few species of fishes change sex partway through life. Feeding mechanisms have been a central element in the evolution of fishes, and the specializations of modern fishes range from species that swallow prey longer than their own bodies to species that rapidly extend their jaws like a tube to suck up minute invertebrates from tiny crevices. In the Devonian period, vertebrates entered a new world as fishlike forms emerged onto the land and occupied terrestrial environments. In this part of the book, we consider the evolution of this extraordinary array of vertebrates.

Living in Water

Although life evolved in water and the earliest vertebrates were aquatic, the physical properties of water create some difficulties for aquatic animals. To live successfully in open water, a vertebrate must adjust its buoyancy to remain at a selected depth and force its way through a dense medium to pursue prey or to escape its own predators. Heat flows rapidly between an animal and the water around it, and it is difficult for an aquatic vertebrate to maintain a body temperature that is different from the water's temperature. (That phenomenon was dramatically illustrated when the *Titanic* sank—in the cold water of the North Atlantic, most of the victims died from hypothermia rather than by drowning.)

Ions and water molecules move readily between the external environment and an animal's internal body fluids, so maintaining a stable internal environment can be difficult. On the plus side, ammonia is extremely soluble in water so disposal of nitrogenous waste products is easier in aquatic environments than on land. The concentration of oxygen in water is lower than it is in air, however, and the density of water imposes limits on the kinds of gas-exchange structures that can be effective.

Despite these challenges, many vertebrates are entirely aquatic. Fishes, and especially the bony fishes, have diversified into an enormous array of sizes and ways of life. In this chapter we will examine some of the challenges of living in water and the ways aquatic vertebrates have responded to them.

4.1 The Aquatic Environment

Seventy-three percent of the surface of Earth is covered by freshwater or salt water. Most of this water is held in the ocean basins, which are populated everywhere by vertebrates. Freshwater lakes and rivers hold a negligible amount of the water on Earth—about 0.01 percent. This is much less than the amount of water tied up in the atmosphere, ice, and groundwater, but freshwater habitats are exceedingly rich biologically, and nearly 40 percent of all bony fishes live in freshwater.

Water and air are both fluids at biologically relevant temperatures and pressures, but they have different physical properties that make them drastically different environments for vertebrates to live in (Table 4–1). In air, for example, gravity is an important force acting on an animal, but fluid resistance to movement (air resistance) is trivial for all but the fastest birds. In water, the opposite relationship holds—gravity is negligible, but fluid resistance to movement is a major factor with which aquatic vertebrates must contend and most fishes are streamlined. Although each major clade of aquatic vertebrates solved environmental challenges in somewhat different ways, the basic specializations needed by all aquatic vertebrates are the same.

Table 4–1 The physical properties of freshwater and air at 20°C. Most of these properties change with temperature and atmospheric pressure, and some are affected by the presence of solutes as well.

Property	Freshwater	Air	Comparison
Density	1 kg · liter^{-1}	0.0012 kg · liter^{-1}	Water is about 800 times as dense as air.
Dynamic viscosity	1 mPa · s	0.018 mPa · s	Water is about 55 times as viscous as air.
Oxygen content	6.8 ml · liter^{-1}	209 ml · liter^{-1}	The oxygen content of freshwater decreases from about 15 ml · liter^{-1} at 0°C to about 5 ml · liter^{-1} at 40°C. Seawater contains less oxygen than freshwater—5.2 ml · liter^{-1} at 20°C.
Heat capacity	4.18 kJ · kg^{-1} · °K^{-1}	0.0012 kJ · kg^{-1} · °K^{-1}	The heat capacity of water is about 3500 times that of air.
Heat conductivity	0.58 W · m^{-1} · °K^{-1}	0.024 W · m^{-1} · °K^{-1}	Water conducts heat about 24 times as fast as air.

Obtaining Oxygen in Water—Gills

Most aquatic vertebrates have gills, which are specialized structures where oxygen and carbon dioxide are exchanged. Teleosts are derived ray-finned fishes, and this group includes the majority of species of extant freshwater and marine fishes. The gills of teleosts are enclosed in pharyngeal pockets called the opercular cavities (**Figure 4–1**). The flow of water is usually unidirectional—in through the mouth and out through the gills. Flaps just inside the mouth and flaps at the margins of the gill covers (**opercula**, singular *operculum*) of bony fishes act as valves to prevent backflow. The respiratory surfaces of the gills are delicate projections from the lateral side of each gill arch. Two columns of gill filaments extend from each gill arch. The tips of the filaments from adjacent arches meet when the filaments are extended. As water leaves the buccal cavity, it passes over the filaments. Gas exchange takes place at the numerous microscopic projections from the filaments called **secondary lamellae**.

The pumping action of the mouth and opercular cavities (**buccal pumping**) creates a positive pressure across the gills so that the respiratory current is only slightly interrupted during each pumping cycle. Some filter-feeding fishes and many pelagic fishes—such as mackerel, certain sharks, tunas, and swordfishes—have reduced or even lost the ability to pump water across the gills. These fishes create a respiratory current by swimming with their mouths open, a method known as **ram ventilation**, and they must swim continuously. Many other fishes rely on buccal pumping when they are at rest and switch to ram ventilation when they are swimming.

The arrangement of blood vessels in the gills maximizes oxygen exchange. Each gill filament has two arteries, an afferent vessel running from the gill arch to the filament tip and an efferent vessel returning blood to the arch. Each secondary lamella is a blood space connecting the afferent and efferent vessels (**Figure 4–2** on page 75). The direction of blood flow through the lamellae is opposite to the direction of water flow across the gill. This arrangement, known as **countercurrent exchange**, assures that as much oxygen as possible diffuses into the blood. Pelagic fishes such as tunas, which sustain high levels of activity for long periods, have skeletal tissue reinforcing the gill filaments, large gill exchange areas, and a high oxygen-carrying capacity per milliliter of blood compared with sluggish bottom-dwelling fishes, such as toadfishes and flat fishes (**Table 4–2** on page 76).

Obtaining Oxygen from Air—Lungs and Other Respiratory Structures

Although the vast majority of fishes depend on gills to extract dissolved oxygen from water, fishes that live in water with low oxygen levels cannot obtain enough oxygen via gills alone. These fishes supplement the oxygen they get from their gills with additional oxygen obtained from the air via lungs or accessory air respiratory structures.

The accessory surfaces used to take up oxygen from air include enlarged lips that are extended just above the water surface and a variety of internal structures into which air is gulped. The anabantid fishes of tropical Asia (including the bettas and gouramies seen in pet stores) have vascularized chambers in the rear of the head, called labyrinths. Air is sucked into the mouth and transferred to the labyrinth, where gas exchange takes place. Many of these fishes are facultative air breathers; that is, they switch oxygen uptake from their gills to accessory respiratory structures when the

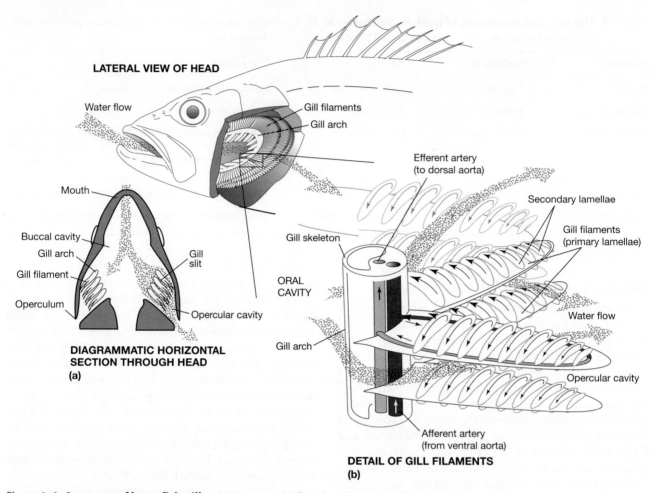

LATERAL VIEW OF HEAD

Water flow

Gill filaments

Gill arch

Mouth

Buccal cavity

Gill arch

Gill filament

Operculum

Gill slit

Opercular cavity

DIAGRAMMATIC HORIZONTAL SECTION THROUGH HEAD
(a)

Efferent artery (to dorsal aorta)

Secondary lamellae

Gill filaments (primary lamellae)

Gill skeleton

ORAL CAVITY

Gill arch

Water flow

Opercular cavity

Afferent artery (from ventral aorta)

DETAIL OF GILL FILAMENTS
(b)

Figure 4–1 Anatomy of bony fish gills. (a) Position of gills in head and general flow of water; (b) countercurrent flow of water (colored arrows) and blood (black arrows) through the gills.

level of oxygen in the water becomes low. Others, like the electric eel and some of the snakeheads, are obligatory air breathers. The gills alone cannot meet the respiratory needs of these fishes, even if the water is saturated with oxygen, and they drown if they cannot reach the surface to breathe air.

We think of lungs as being the respiratory structures used by terrestrial vertebrates, as indeed they are, but lungs first appeared in fishes and preceded the evolution of tetrapods by millions of years. Lungs develop embryonically as outpocketings (evaginations) of the pharyngeal region of the digestive tract, originating from its ventral or dorsal surface. The lungs of bichirs (a group of air-breathing fishes from Africa), lungfishes, and tetrapods originate from the ventral surface of the gut, whereas the lungs of gars (a group of primitive bony fishes) and the lungs of the derived bony fishes known as teleosts originate embryonically from its dorsal surface.

Lungs used for gas exchange need a large surface area, which is provided by ridges or pockets in the wall. This structure is known as an alveolar lung, and it is found in gars, lungfishes, and tetrapods. Increasing the volume of the lung by adding a second lobe is another way to increase the surface area, and the lungs of lungfishes and tetrapods consist of two symmetrical lobes. (Bichirs have non-alveolar lungs with two lobes, but one lobe is much smaller than the other; gars have single-lobed alveolar lungs.)

Adjusting Buoyancy

Holding a bubble of air inside the body changes the buoyancy of an aquatic vertebrate, and bichirs and teleost fishes use the lungs and swim bladders to regulate their position in the water. Air-breathing aquatic vertebrates (whales, dolphins, seals, and penguins, for example) can adjust their buoyancy by altering the volume of air in their lungs when they dive.

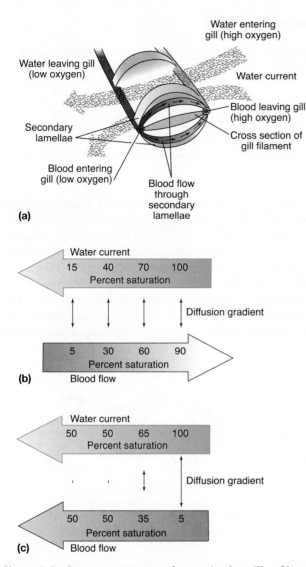

(a)

(b)

(c)

Figure 4–2 Countercurrent exchange in the gills of bony fishes. (a) The direction of water flow across the gill opposes the flow of blood through the secondary lamellae. Blood cells are separated from oxygen-rich water only by the thin epithelial cells of the capillary wall, as shown in the cross section of a secondary lamella. (b) Countercurrent flow maintains a difference in oxygen concentration (a diffusion gradient) between blood and water for the full length of the lamella and results in a high oxygen concentration in the blood leaving the gills. (c) If water and blood flowed in the same direction, the difference in oxygen concentration and the diffusion gradient would be high initially, but would drop to zero as the concentration of oxygen equalized. No further exchange of oxygen would occur, and the blood leaving the gills would have a low oxygen concentration.

Bony Fishes Many bony fishes are neutrally buoyant (i.e., have the same density as water). These fishes do not have to swim to maintain their vertical position in the water column. The only movement they make when at rest is backpedaling of the pectoral fins to counteract the forward thrust produced by water as it is ejected from the gills and a gentle undulation of the tail fin to keep them level in the water. Fishes capable of hovering in the water like this usually have well-developed swim bladders.

The swim bladder is located between the peritoneal cavity and the vertebral column (**Figure 4–3**). The bladder occupies about 5 percent of the body volume of marine teleosts and 7 percent of the volume of freshwater teleosts. The difference in volume corresponds to the difference in density of salt water and freshwater—salt water is denser, so a smaller swim bladder is sufficient. The swim bladder wall, which has smooth walls composed of interwoven collagen fibers without blood vessels, is virtually impermeable to gas.

Neutral buoyancy produced by a swim bladder works as long as a fish remains at one depth, but if a fish swims vertically up or down, the hydrostatic pressure that the surrounding water exerts on the bladder changes, which in turn changes the volume of the bladder. For example, when a fish swims deeper, the additional pressure of the water column above it compresses the gas in its swim bladder, making the bladder smaller and reducing the buoyancy of the fish. When the fish swims toward the surface, water pressure decreases, the swim bladder expands, and the fish becomes more buoyant. To maintain neutral buoyancy, a fish must adjust the volume of gas in its swim bladder as it changes depth.

A bony fish regulates the volume of its swim bladder by secreting gas into the bladder to counteract the increased external water pressure when it swims down and removing gas when it swims up. Primitive teleosts—such as bony tongues, eels, herrings, anchovies, salmon, and minnows—retain a connection, the pneumatic duct, between the gut and swim bladder (see Figure 4–3a). These fishes are called **physostomous** (Greek *phys* = bladder and *stom* = mouth), and goldfish are a familiar example of this group. Because they have a connection between the gut and the swim bladder, they can gulp air at the surface to fill the bladder and can burp gas out to reduce its volume.

The pneumatic duct is absent in adult teleosts from more derived clades, a condition termed **physoclistous** (Greek *clist* = closed). Physoclists regulate the volume of the swim bladder by secreting gas from the blood into the bladder. Both physostomes and physoclists have a gas gland, which is located in the anterior ventral floor of the swim bladder (see Figure 4–3b). Underlying the gas gland is an area with many capillaries arranged to give countercurrent flow of blood entering

Table 4–2 Anatomical and physiological characteristics of three types of fishes

Species of Fishes	Activity	Oxygen Consumption (ml $O_2 \cdot g^{-1} \cdot h^{-1}$)	Gill Area (mm² · g body mass⁻¹)	Oxygen Capacity (ml $O_2 \cdot$ 100 ml blood⁻¹)
Mackerel (*Scomber*)	High, swims continuously	0.73	1160	14.8
Porgy (*Stenotomus*)	Intermediate	0.17	506	7.3
Toadfish (*Opsanus*)	Sluggish, bottom dweller	0.11	197	6.2

and leaving the area. This structure, which is known as a **rete mirabile** ("wonderful net," plural *retia mirabilia*), moves gas (especially oxygen) from the blood to the gas bladder. It is remarkably effective at extracting oxygen from the blood and releasing it into the swim bladder, even when the pressure of oxygen in the bladder is many times higher than its pressure in blood. Gas secretion occurs in many deep-sea fishes despite the hundreds of atmospheres of gas pressure within the bladder.

The gas gland secretes oxygen by releasing lactic acid and carbon dioxide, which acidify the blood in the rete mirabile. Acidification causes hemoglobin to release oxygen into solution (the Bohr and Root effects). Because of the anatomical relations of the rete mirabile, which folds back upon itself in a countercurrent multiplier arrangement, oxygen released from the hemoglobin accumulates and is retained within the rete until its pressure exceeds the oxygen pressure in

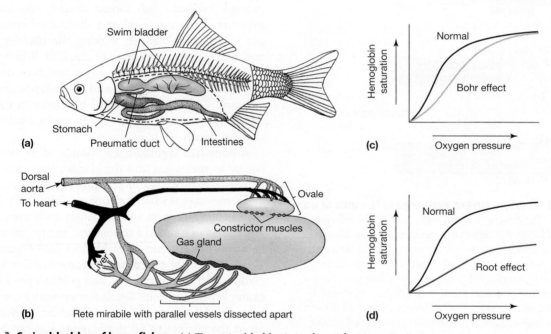

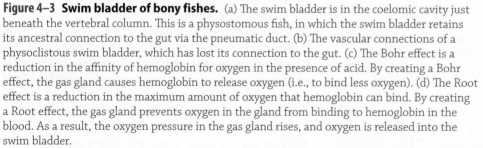

Figure 4–3 Swim bladder of bony fishes. (a) The swim bladder is in the coelomic cavity just beneath the vertebral column. This is a physostomous fish, in which the swim bladder retains its ancestral connection to the gut via the pneumatic duct. (b) The vascular connections of a physoclistous swim bladder, which has lost its connection to the gut. (c) The Bohr effect is a reduction in the affinity of hemoglobin for oxygen in the presence of acid. By creating a Bohr effect, the gas gland causes hemoglobin to release oxygen (i.e., to bind less oxygen). (d) The Root effect is a reduction in the maximum amount of oxygen that hemoglobin can bind. By creating a Root effect, the gas gland prevents oxygen in the gland from binding to hemoglobin in the blood. As a result, the oxygen pressure in the gas gland rises, and oxygen is released into the swim bladder.

the swim bladder. At this point oxygen diffuses into the bladder, increasing its volume. The maximum multiplication of gas pressure that can be achieved is proportional to the length of the capillaries of the rete mirabile, and deep-sea fishes have very long retia. A large Root effect is characteristic only of the blood of ray-finned fishes, and it is essential for the function of the gas gland.

Physoclistous fishes have no connection between the swim bladder and the gut, so they cannot burp to release excess gas from the bladder. Instead, physoclists open a muscular valve, called the ovale, located in the posterior dorsal region of the bladder adjacent to a capillary bed. The high internal pressure of oxygen in the bladder causes it to diffuse into the blood of this capillary bed when the ovale sphincter is opened.

Cartilaginous Fishes Sharks, rays, and ratfishes do not have swim bladders. Instead, these fishes use the liver to create neutral buoyancy. The average tissue densities of sharks with their livers removed are heavier than water—1.06 to 1.09 grams per milliliter compared to about 1.025 grams per milliliter for seawater. The liver of a shark, however, is well known for its high oil content (shark-liver oil). Shark-liver tissue has a density of only 0.95 gram per milliliter, which is lighter than water, and the liver may contribute as much as 25 percent of the body mass. A 4-meter tiger shark (*Galeocerdo cuvieri*) weighing 460 kilograms on land may weigh as little as 3.5 kilograms in the sea. Not surprisingly, bottom-dwelling sharks, such as nurse sharks, have livers with fewer and smaller oil vacuoles in their cells, and these sharks are negatively buoyant.

Nitrogen-containing compounds in the blood of cartilaginous fishes also contribute to their buoyancy. Urea and trimethylamine oxide in the blood and muscle tissue provide positive buoyancy because they are less dense than an equal volume of water. Chloride ions, too, are lighter than water and provide positive buoyancy, whereas sodium ions and protein molecules are denser than water and are negatively buoyant. Overall these solutes provide positive buoyancy.

Deep-Sea Fishes Many deep-sea fishes have deposits of light oil or fat in the gas bladder, and others have reduced or lost the gas bladder entirely and have lipids distributed throughout the body. These lipids provide static lift, just like the oil in shark livers. Because a smaller volume of the bladder contains gas, the amount of secretion required for a given vertical descent is less. Nevertheless, a long rete mirabile is needed to secrete oxygen at high pressures, and the gas gland in deep-sea fishes is very large. Fishes that migrate over large vertical distances depend more on lipids such as wax esters than on gas for buoyancy, whereas their close relatives that do not undertake such extensive vertical movements depend more on gas for buoyancy.

Air-Breathing Divers Air in the lungs of air-breathing aquatic vertebrates reduces their density. Unlike most fishes, air-breathing vertebrates must return to the surface at intervals, so they do not hover at one depth in the water column. Deep-diving animals, such as elephant seals and some whales and porpoises, face a different problem, however. These animals dive to depths of 1000 meters or more and are subjected to pressures more than 100 times higher than at the surface. Under those conditions, nitrogen would be forced from the air in the lungs into solution in the blood and carried to the tissues at high pressure. When the animal rose toward the surface, the nitrogen would be released from solution. If the animal moved upward too fast, the nitrogen would form bubbles in the tissues—this is what happens when human deep-sea divers get "the bends" (decompression sickness). Specialized diving mammals avoid the problem by allowing the thoracic cavity to collapse as external pressure rises. Air is forced out of the lungs as they collapse, reducing the amount of nitrogen that diffuses into the blood. Even these specialized divers would have problems if they made repeated deep dives, however; a deep dive is normally followed by a period during which the animal remains near the surface and makes only shallow dives until the nitrogen level in its blood has equilibrated with the atmosphere. Despite these specializations, the bones of sperm whales, which dive to depths of 2000 m, contain areas of dead tissue caused by the bends.

4.2 Water and the Sensory World of Fishes

Water has properties that influence the behaviors of fishes and other aquatic vertebrates. Light is absorbed by water molecules and scattered by suspended particles. Objects become invisible at a distance of a few hundred meters even in the very clearest water, whereas distance vision is virtually unlimited in clear air. Fishes supplement vision with other senses, some of which can operate only in water. The most important of these aquatic senses is mechanical and consists of detecting water movement via the lateral line system. Small currents of water can stimulate the sensory organs of the lateral line because water is dense and viscous. Electrical sensitivity is another sensory mode that depends on the properties of water and does not operate in air. In this case it is the electrical conductivity of water that is the key. Even vision is different in water and air because of the different refractive properties of the two media.

Vision

Vertebrates generally have well-developed eyes, but the way an image is focused on the retina is different in terrestrial and aquatic animals. Air has an index of refraction of 1.00, and light rays bend as they pass through a boundary between air and a medium with a different refractive index. The amount of bending is proportional to the difference in indices of refraction. Water has a refractive index of 1.33, and the bending of light as it passes between air and water causes underwater objects to appear closer to an observer in air than they really are. The corneas of the eyes of terrestrial and aquatic vertebrates have an index of refraction of about 1.37, so light is bent as it passes through the air-cornea interface. As a result, the cornea of a terrestrial vertebrate plays a substantial role in focusing an image on the retina. This relationship does not hold in water, however, because the refractive index of the cornea is too close to that of water for the cornea to have much effect in bending light. The lens plays the major role in focusing light on the retina of an aquatic vertebrate, and fishes have spherical lenses with high refractive indices. The entire lens is moved toward or away from the retina to focus images of objects at different distances from the fish. Terrestrial vertebrates have flatter lenses, and muscles in the eye change the shape of the lens to focus images. Aquatic mammals such as whales and porpoises have spherical lenses like those of fishes.

Chemosensation: Taste and Odor

Fishes have taste-bud organs in the mouth and around the head and anterior fins. In addition, olfactory organs on the snout detect soluble substances. Sharks and salmon can detect odors at concentrations of less than 1 part per billion. Sharks, and perhaps bony fishes, compare the time of arrival of an odor stimulus on the left and right sides of the head to locate the source of the odor. Homeward-migrating salmon are directed to their stream of origin from astonishing distances by a chemical signature from the home stream that was permanently imprinted when they were juveniles. Plugging the nasal olfactory organs of salmon destroys their ability to home.

Touch

Mechanical receptors detect touch, sound, pressure, and motion. Like all vertebrates, fishes have an internal ear (the labyrinth organ, not to be confused with the organ of the same name that assists in respiration in anabantid fishes) that detects changes in speed and direction of motion. Fishes also have gravity detectors at the base of the semicircular canals that allow them to distinguish up from down. Most terrestrial vertebrates also have an auditory region of the inner ear that is sensitive to sound-pressure waves. These diverse functions of the labyrinth depend on basically similar types of sense cells, the hair cells (**Figure 4–4**). In fishes and aquatic amphibians, clusters of hair cells and associated support cells form **neuromast organs** that are dispersed over the surface of the head and body. In jawed fishes, neuromast organs are often located in a series of canals on the head, and one or more canals pass along the sides of the body onto the tail. This surface receptor system of fishes and aquatic amphibians is referred to as the **lateral line system**. Lateral line systems are found only in aquatic vertebrates because air is not dense enough to stimulate the neuromast organs. Amphibian larvae have lateral line systems, and permanently aquatic species of amphibians, such as African clawed frogs and mudpuppies, retain lateral lines throughout their lives. Terrestrial species of amphibians lose their lateral lines when they metamorphose into adults, however, and terrestrial vertebrates that have secondarily returned to the water, such as whales and porpoises, do not have lateral line systems.

Detecting Water Displacement

Neuromasts of the lateral line system are distributed in two configurations—within tubular canals or exposed in epidermal depressions. Many kinds of fishes have both arrangements. Hair cells have a **kinocilium** placed asymmetrically in a cluster of **kinocilia**. Hair cells are arranged in pairs with the kinocilia positioned on opposite sides of adjacent cells. A neuromast contains many such hair-cell pairs. Each neuromast has two afferent nerves: one transmits impulses from hair cells with kinocilia in one orientation, and the other carries impulses from cells with kinocilia positions reversed by 180 degrees. This arrangement allows a fish to determine the direction of displacement of the kinocilia.

All kinocilia and microvilli are embedded in a gelatinous secretion, the **cupula** (Latin *cupula* = a small tub). Displacement of the cupula causes the kinocilia to bend. The resultant deformation either excites or

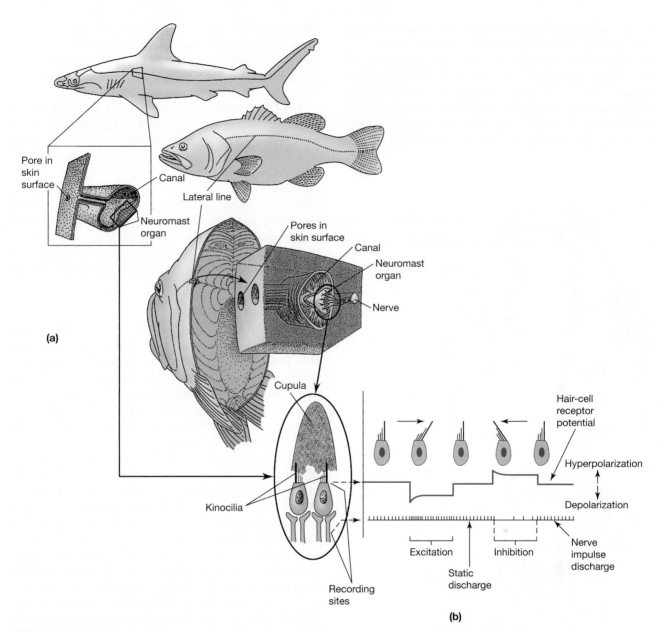

Figure 4–4 Lateral line systems. (a) Semidiagrammatic representations of the two configurations of lateral line organs in fishes. (b) Hair-cell deformations and their effect on hair-cell transmembrane potential (receptor potential) and afferent nerve-cell discharge rates. Deflection of the kinocilium (dark line) in one direction (the right in this diagram) depolarizes the cell and increases the discharge rate (excitation). Deflection of the kinocilium in the opposite direction (to the left in the diagram) hyperpolarizes the cell and reduces the discharge rate (inhibition).

inhibits the neuromast's nerve discharge. Each hair-cell pair, therefore, signals the direction of cupula displacement. The excitatory output of each pair has a maximum sensitivity to displacement along the line joining the kinocilia, and falling off in other directions. The net effect of cupula displacement is to increase the

firing rate in one afferent nerve and to decrease it in the other nerve. These changes in lateral line nerve firing rates thus inform a fish of the direction of water currents on different surfaces of its body.

Several surface-feeding fishes and African clawed frogs provide vivid examples of how the lateral line organs act under natural conditions. These animals find insects on the water surface by detecting surface waves created by their prey's movements. Each neuromast group on the head of the killifish, *Aplocheilus lineatus*, provides information about surface waves coming from a different direction (**Figure 4–5**). The groups of neuromasts have overlapping stimulus fields, allowing the fish to determine the precise

location of the insect. Removing a neuromast group from one side of the head disturbs the directional response to stimuli, showing that a fish combines information from groups on both sides of the head to interpret water movements.

The large numbers of neuromasts on the heads of some fishes might be important for sensing vortex trails in the wakes of adjacent fishes in a school. Many of the fishes that form extremely dense schools (herrings, atherinids, mullets) lack lateral line organs along the flanks and retain canal organs only on the head. These well-developed cephalic canal organs concentrate sensitivity to water motion in the head region, where it is needed to sense the turbulence into which the fish is swimming, and the reduction of flank lateral line elements would reduce noise from turbulence beside the fish.

Electrical Discharge

Unlike air, water conducts electricity, and the torpedo ray of the Mediterranean, the electric catfish from the Nile River, and the electric eel of South America can discharge enough electricity to stun prey animals and deter predators. The weakly electric knifefishes (Gymnotidae) of South America and the elephant fishes (Mormyridae) of Africa use electrical signals for courtship and territorial defense.

All of these electric fishes use modified muscle tissue to produce the electrical discharge. The cells of such modified muscles, called **electrocytes**, are muscle cells that have lost the capacity to contract and are specialized for generating an ion current flow (**Figure 4–6**). When at rest, the membranes of muscle cells and nerve cells are electrically charged, with the intracellular fluids about 84 millivolts more negative than the extracellular fluids. The imbalance is primarily due to sodium ion exclusion. When the cell is stimulated, sodium ions flow rapidly across the smooth surface, sending its potential to a positive 67 millivolts. Only the smooth surface depolarizes; the rough surface remains at –84 millivolts, so the potential difference across the cell is 151 millivolts (from –84 to +67 millivolts). Because electrocytes are arranged in stacks like the batteries in a flashlight, the potentials of many layers of cells combine to produce high voltages. The South American electric eel has up to 10,000 layers of cells and can generate potentials in excess of 600 volts.

Most electric fishes are found in tropical freshwaters of Africa and South America. Few marine forms can generate specialized electrical discharges—among marine cartilaginous fishes, only the torpedo ray (*Torpedo*), the ray genus *Narcine*, and some skates are electric; and

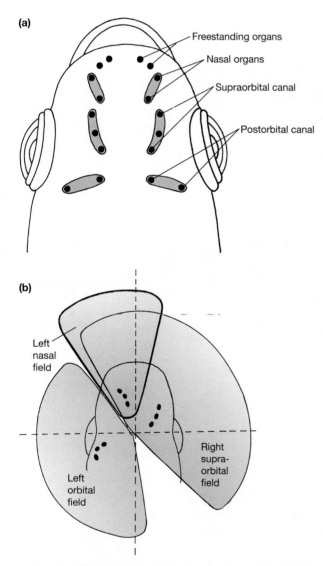

Figure 4–5 Distribution of the lateral line canal organs.
(a) The dorsal surface of the head of the killifish *Fundulus notatus*. (b) The sensory fields of the head canal organs in a different species of killifish, *Aplocheilus lineatus*. The wedge-shaped areas indicate the fields of view for each group of canal organs. Note that fields overlap on opposite sides as well as on the same side of the body, allowing the lateral line system to localize the source of a water movement.

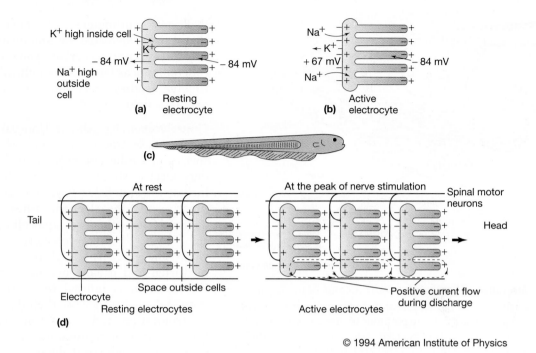

Figure 4–6 Weakly electric fishes. Some fishes use transmembrane potentials of modified muscle cells to produce a discharge. In this diagram the smooth surface is on the left and the rough surface on the right. Only the smooth surface is innervated. (a) At rest, K^+ (potassium ion) is maintained at a high internal concentration and Na^+ (sodium ion) at a low internal concentration by the action of a Na^+/K^+ cell-membrane pump. Permeability of the membrane to K^+ exceeds the permeability to Na^+. As a result, K^+ diffuses outward faster than Na^+ diffuses inward (arrow) and sets up the –84-mV resting potential. (b) When the smooth surface of the cell is stimulated by the discharge of the nerve, Na^+ diffuses into the cell and K^+ diffuses out of the smooth surface, changing the net potential to +67 mV. The rough surface does not depolarize and retains a –84-mV potential, creating a potential difference of 151 mV across the cell. (c) A weakly electric South American gymnotid, showing the location of electrocytes along the sides of the body. (d) By arranging electrocytes in series so that the potentials of individual cells are summed, some electric fishes can generate very high voltages. Electric eels, for example, have 10,000 electrocytes in series and produce potentials in excess of 600 volts.

among marine teleosts, only the stargazers (family Uranoscopidae) produce specialized discharges.

Electroreception by Sharks and Rays The high conductivity of seawater makes it possible for sharks to detect the electrical activity that accompanies muscle contractions of their prey. Sharks have structures known as the **ampullae of Lorenzini** on their heads, and rays have them on the pectoral fins as well. The ampullae are sensitive electroreceptors (**Figure 4–7**). The canal connecting the receptor to the surface pore is filled with an electrically conductive gel, and the wall of the canal is nonconductive. Because the canal runs for some distance beneath the epidermis, the sensory cell can detect a difference in electrical potential between the tissue in which it lies (which reflects the adjacent epidermis and environ-

ment) and the distant pore opening. Thus, it can detect electric fields, which are changes in electrical potential in space.

Electroreceptors of sharks respond to minute changes in the electric field surrounding an animal. They act like voltmeters, measuring differences in electrical potentials across the body surface. Ampullary organs are remarkably sensitive, with thresholds lower than 0.01 microvolt per centimeter, a level of detection achieved by only the best voltmeters.

Sharks use their electrical sensitivity to detect prey. All muscle activity generates electrical potential: motor nerve cells produce extremely brief changes in electrical potential, and muscular contraction generates changes of longer duration. In addition, a steady potential issues from an aquatic organism as a result of

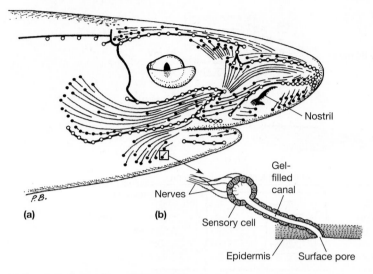

Nostril

Gel-filled canal

Nerves

Sensory cell

Epidermis — Surface pore

P.B.

(a)

(b)

Figure 4–7 Ampullae of Lorenzini. (a) Distribution of the ampullae on the head of a spiny dogfish, *Squalus acanthias*. Open circles represent the surface pores; the black dots are positions of the sensory cells. (b) A single ampullary organ consists of a sensory cell connected to the surface by a pore filled with a substance that conducts electricity.

livolt per centimeter—well above the level that can be detected by ampullary organs. In addition, ocean currents generate electrical gradients as large as 0.5 millivolt per centimeter as they carry ions through Earth's magnetic field.

Electrolocation by Teleosts Unusual arrangements of electrocytes are present in several species of freshwater fishes that do not produce electric shocks. In these fishes—which include the knifefishes (Gymnotidae) of South America and the elephant fishes (Mormyridae) of Africa—the discharge voltages are too small to be of direct defensive or offensive value. These weakly electric fishes are mostly nocturnal and usually live in turbid waters where vision is limited to short distances even in daylight; they use their discharges for electrolocation and social communications. When a fish discharges its electric organ, it creates an electric field in its immediate

the chemical imbalance between the organism and its surroundings. A shark can locate and attack a hidden fish by relying only on this electrical activity (**Figure 4–8**).

Sharks may use electroreception for navigation as well as for locating prey. The electromagnetic field at Earth's surface produces tiny voltage gradients, and a swimming shark could encounter gradients as large as 0.4 mil-

Figure 4–8 Electrolocation capacity of sharks.
(a) A shark can locate a live fish concealed from sight beneath the sand. (b) The shark can still detect the fish when it is covered by an agar shield that blocks olfactory cues but allows the electrical signal to pass. (c) The shark follows the olfactory cues (displaced by the agar shield) when the live fish is replaced by chopped bait that produces no electrical signal. (d) The shark is unable to detect a live fish when it is covered by a shield that blocks both olfactory cues and the electrical signal. (e) The shark attacks electrodes that give off an electrical signal duplicating a live fish without producing olfactory cues. These experiments indicate that when the shark was able to detect an electrical signal, it used that to locate the fish—and it was also capable of homing in on a chemical signal when no electrical signal was present. This dual system allows sharks to find both living and dead food items.

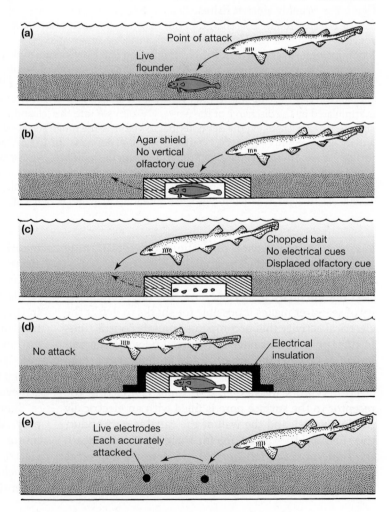

(a) Point of attack
Live flounder

(b) Agar shield
No vertical olfactory cue

(c) Chopped bait
No electrical cues
Displaced olfactory cue

(d) No attack
Electrical insulation

(e) Live electrodes
Each accurately attacked

vicinity (**Figure 4–9**). Because of the high energy costs of maintaining a continuous discharge, electric fishes produce a pulsating discharge. Most weakly electric teleost fishes pulse at rates between 50 and 300 cycles per second, but the knifefishes of South America reach 1700 cycles per second, which is the most rapid continuous firing rate known for any vertebrate muscle or nerve.

The electric field from even weak discharges extends outward for a considerable distance in freshwater because fresh water does not conduct electricity as well as seawater does. The electric field the fish creates will be distorted by the presence of electrically conductive and resistant objects. Rocks are highly resistive, whereas

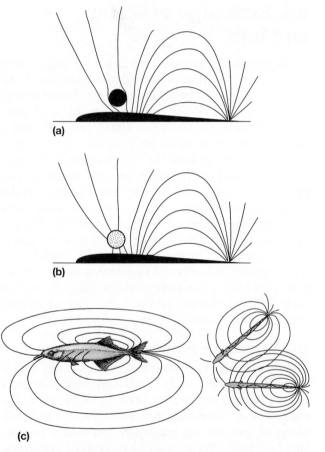

(a)

(b)

(c)

Figure 4–9 Weakly electric fishes. An electric field surrounds a weakly electric fish. Electroreceptors in the skin allow a fish to detect the presence of nearby objects by sensing distortion of the lines of electrical force. (a) Nonconductive objects, such as rocks, spread the field and diffuse potential differences along the body surface. (b) Conductive objects, such as another fish, concentrate the field on the skin of the fish. (c) When two electric fish swim close enough to each other to create interference between their electric fields, they change the frequencies of their discharges.

other fishes, invertebrates, and plants are conductive. Distortions of the field cause a change in the distribution of electrical potential across the fish's body surface. An electric fish detects the presence, position, and movement of objects by sensing where on its body maximum distortion of its electric field occurs.

The skin of weakly electric teleosts contains special sensory receptors: ampullary organs and tuberous organs. These organs detect tonic (steady) and phasic (rapidly changing) discharges, respectively. Electroreceptors of teleosts are modified lateral line neuromast receptors. Like lateral line receptors, they have double innervation—an afferent channel that sends impulses to the brain and an efferent channel that causes inhibition of the receptors. During each electric organ discharge, an inhibitory command is sent to the electroreceptors, and the fish is rendered insensitive to its own discharge. Between pulses, electroreceptors report distortion in the electric field or the presence of a foreign electric field to the brain.

Electric organ discharges vary with habits and habitat. Species that form groups or live in shallow, narrow streams generally have discharges with high frequency and short duration. These characteristics reduce the chances of interference from the discharges of neighbors. Territorial species, in contrast, have long electric organ discharges. Electric organ discharges vary from species to species. In fact, some species of electric fishes were first identified by their electric organ discharges, which were recorded by placing electrodes in water that was too murky for any fishes to be visible. During the breeding season, electric organ discharges distinguish immature individuals, females with eggs, and sexually active males.

Other Electroreceptive Vertebrates Electrogenesis and electroreception are not restricted to a single group of aquatic vertebrates, and monotremes (the platypus and the echidna, early offshoots off the main mammalian lineages that still lay eggs) use electroreception to detect prey. Electrosensitivity was probably an early feature of vertebrate evolution. The brain of the lamprey responds to electric fields, and it seems likely that the earliest vertebrates had electroreceptive capacity. All fishlike vertebrates of lineages that evolved before the neopterygians have electroreceptor cells. These cells, which have a prominent kinocilium, fire when the environment around the kinocilium is negative relative to the cell. Their impulses pass to the midline region of the posterior third of the brain.

Electrosensitivity was apparently lost in neopterygians, and teleosts have at least two separate new evolutions of electroreceptors. Electrosensitivity of teleosts is distinct from that of other vertebrates: teleost electro-receptors *lack* a kinocilium and fire when the environment is *positive* relative to the cell, and nerve impulses are sent to the lateral portions of the brain rather than to the midline.

4.3 The Internal Environment of Vertebrates

Seventy to eighty percent of the body mass of most vertebrates is water, and the chemical reactions that release energy or synthesize new chemical compounds take place in an aqueous environment. The body fluids of vertebrates contain a complex mixture of ions and other solutes. Some ions are cofactors that control the rates of metabolic processes; others are involved in the regulation of pH, the stability of cell membranes, or the electrical activity of nerves. Metabolic substrates and products must diffuse from sites of synthesis to the sites of utilization. Almost everything that happens in the body tissues of vertebrates involves water, and maintaining the concentrations of water and solutes within narrow limits is a vital activity. Water sounds like an ideal place to live for an animal that itself consists mostly of water, but in some ways an aquatic environment can be too much of a good thing. Freshwater vertebrates—especially fishes and amphibians—face the threat of being flooded with water that flows into them from their environment, and saltwater vertebrates must prevent the water in their bodies from being sucked out into the sea.

Temperature, too, is a critical factor for living organisms because chemical reactions are temperature sensitive. In general, the rates of chemical reactions increase as temperature increases, but not all reactions have the same sensitivity to temperature. A metabolic pathway is a series of chemical reactions in which the product of one reaction is the substrate for the next, yet each of these reactions may have a different sensitivity to temperature, so a change in temperature can mean that too much or too little substrate is produced to sustain the next reaction in the series. To complicate the process of regulation of substrates and products even more, the chemical reactions take place in a cellular milieu that itself is changed by temperature because the viscosity of plasma membranes is also temperature sensitive. Clearly, the smooth functioning of metabolic pathways is greatly simplified if an organism can limit the range of temperatures its tissues experience.

Water temperature is more stable than air temperature because water has a much higher heat capacity than air. The stability of water temperature simplifies the task of maintaining a constant body temperature, as long as the body temperature the animal needs to maintain is the same as the temperature of the water around it. An aquatic animal has a hard time maintaining a body temperature different from water temperature, however, because water conducts heat so well. Heat flows out of the body if an animal is warmer than the surrounding water and into the body if the animal is cooler than the water.

In the following sections, we discuss in more detail how and why vertebrates regulate their internal environments and the special problems faced by aquatic animals.

4.4 Exchange of Water and Ions

An organism can be described as a leaky bag of dirty water. That is not an elegant description, but it accurately identifies the two important characteristics of a living animal—it contains organic and inorganic substances dissolved in water, and this fluid is enclosed by a permeable body surface. Exchange of matter and energy with the environment is essential to the survival of the organism, and much of that exchange is regulated by the body surface. Water molecules and ions pass through the skin quite freely, whereas larger molecules move less readily. The significance of this differential permeability is particularly conspicuous in the case of aquatic vertebrates, but it applies to terrestrial vertebrates as well. Vertebrates use both active and passive exchange to regulate their internal concentrations in the face of varying external conditions.

The Vertebrate Kidney

An organism can tolerate only a narrow range of concentrations of the body fluids and must eliminate waste products before they reach harmful levels. The molecules of ammonia that result from the breakdown of protein are especially important because they are toxic. Vertebrates have evolved superb capacities for controlling water balance and excreting wastes, and the kidney plays a crucial role in these processes.

The adult vertebrate kidney consists of hundreds to millions of tubular **nephrons**, each of which produces urine. The primary function of a nephron is removing excess water, salts, waste metabolites, and foreign substances from the blood. In this process, the blood is first filtered through the **glomerulus**, a structure unique to

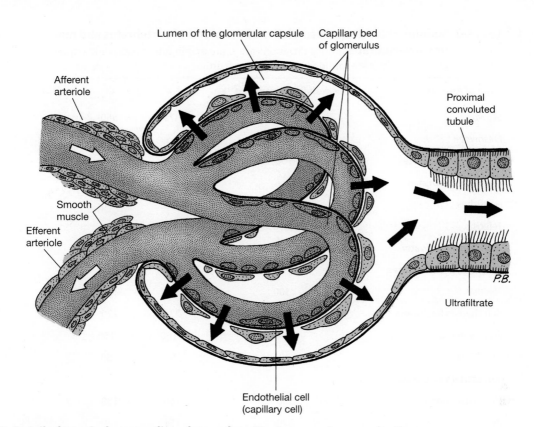

Afferent
arteriole

Lumen of the glomerular capsule

Capillary bed
of glomerulus

Proximal
convoluted
tubule

Smooth
muscle

Efferent
arteriole

P.B.

Ultrafiltrate

Endothelial cell
(capillary cell)

Figure 4–10 Detail of a typical mammalian glomerulus. Blood pressure forces an ultrafiltrate of the blood through the walls of the capillary into the lumen of the glomerular capsule. The blood flow to each glomerulus is regulated by smooth muscles that can close off the afferent and efferent arterioles to adjust the glomerular filtration rate of the kidney as a whole. The ultrafiltrate, which consists of water, ions, and small molecules, passes from the glomerular capsule into the proximal convoluted tubule, where the process of adding and removing specific substances begins.

vertebrates (Figure 4–10). Each glomerulus is composed of a leaky arterial capillary tuft encapsulated within a sieve-like filter. Arterial blood pressure forces fluid into the nephron to form an **ultrafiltrate**, composed of blood minus blood cells and larger molecules. The ultrafiltrate is then processed to return water and essential metabolites (glucose, amino acids, and so on) to the general circulation. The fluid that remains after this processing is urine.

Regulation of Ions and Body Fluids

The salt concentrations in the body fluids of many marine invertebrates are similar to those in seawater, as are those of hagfishes (Table 4–3). It is likely that the first vertebrates also had ion levels similar to those in seawater. In contrast, salt levels are greatly reduced in the blood of all other vertebrates.

In the context of body fluids, a **solute** is a small molecule that is dissolved in water or blood plasma. Salt ions, urea, and some small carbohydrate molecules are the solutes primarily involved in the regulation of

body fluid concentrations. The presence of solutes lowers the potential activity of water. Water moves from areas of high potential to areas of lower potential; therefore, water flows from a dilute solution (one with a high water potential) to a more concentrated solution (with a lower water potential). This process is called **osmosis**.

Seawater has a solute concentration of approximately 1000 millimoles per kilogram of water (mmol · kg^{-1}). Most marine invertebrates and hagfishes have body fluids that are in osmotic equilibrium with seawater; that is, they are **isosmolal** to seawater. Body fluid concentrations in marine teleosts and lampreys are between 300 and 350 mmol · kg^{-1}. Therefore, water flows outward from their blood to the sea (i.e., from a region of high water potential to a region of lower water potential). Cartilaginous fishes retain urea and other nitrogen-containing compounds, raising the osmolality of their blood slightly above that of seawater so water flows from the sea into their bodies. These osmolal

Table 4–3 **Sodium, chloride and osmolality of the blood of vertebrates and marine invertebrates.** Concentrations are expressed in millimoles per kilogram of water; all values are rounded to the nearest 5 units.

Type of Animal	Osmolality (mmol · kg^{-1})	Na$^+$	Cl$^-$	Urea
Seawater	~1000	475	550	
Freshwater	<10	~5	~5	
Marine invertebrates	~1000	~465	~54	
Marine vertebrates				
Hagfishes	~1000	535	540	
Teleosts	<350	180	150	
Bull shark in seawater	1050	290	290	360
Freshwater vertebrates				
Teleosts	<300	140	120	
Bull shark in freshwater	~630	235	220	180
Freshwater sawfish	~340	~160	~150	<1
Amphibians	~200	~100	~80	
Terrestrial vertebrates				
Non-avian reptiles	350	160	130	
Birds	320	150	120	
Mammals	300	145	105	

differences are specified by the terms **hyposmolal** (lower solute concentrations than the surrounding water, as seen in marine teleosts and lampreys) and **hyperosmolal** (higher solute concentrations than the surrounding water, as seen in coelacanths and cartilaginous fishes).

Salt ions, such as sodium and chloride, can also diffuse through the surface membranes of an animal, so the water and salt balance of an aquatic vertebrate in seawater is constantly threatened by outflow of water and inflow of salt and in freshwater by inflow of water and outflow of salt.

Most fishes are **stenohaline** (Greek *steno* = narrow and *haline* = salt); that means they inhabit either freshwater or seawater and tolerate only modest changes in salinity. Because they remain in one environment, the magnitude and direction of the osmotic gradient to which they are exposed are stable. Some fishes, however, move between freshwater and seawater and tolerate large changes in salinity. These fishes are called **euryhaline** (Greek *eury* = wide); water and salt gradients are reversed in euryhaline species as they move from one medium to the other.

Freshwater Vertebrates—Teleosts and Amphibians Several mechanisms are involved in the salt and water regulation of vertebrates that live in freshwater. The body surface of fishes has low permeability to water and to ions. However, fishes cannot entirely prevent osmotic exchange. Gills are permeable to oxygen and carbon dioxide, and they are also permeable to water. As a result, most water and ion movements take place across the gill surfaces. Water is gained by osmosis, and ions are lost by diffusion. A freshwater teleost does not drink water because osmotic water movement is already providing more water than it needs—drinking would only increase the amount of water it had to excrete via the kidneys. To compensate for the influx of water, the kidney of a freshwater fish or amphibian produces a large volume of urine. Salts are actively reabsorbed to reduce salt loss. Indeed, urine processing in a freshwater teleost provides a simple model of vertebrate kidney function.

The large glomeruli of freshwater teleosts produce a copious flow of urine, but the glomerular ultrafiltrate is isosmolal to the blood and contains essential blood salts (**Figure 4–11**). To conserve salt, ions are reabsorbed

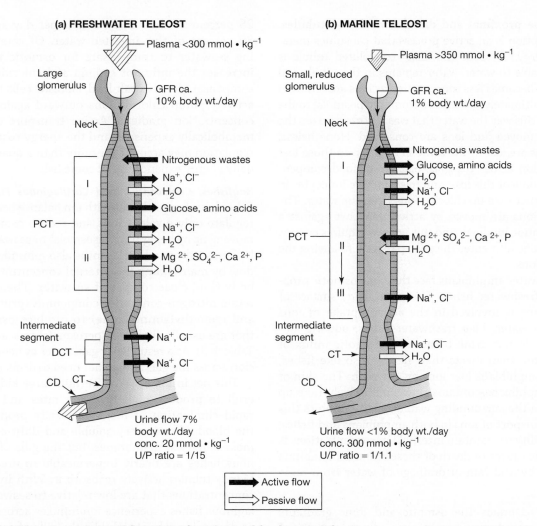

(a) FRESHWATER TELEOST

Plasma <300 mmol • kg^{-1}

Large glomerulus

GFR ca. 10% body wt./day

Neck

Nitrogenous wastes

I

Na$^+$, Cl$^-$

H$_2$O

Glucose, amino acids

PCT

Na$^+$, Cl$^-$

H$_2$O

II

Mg^{2+}, SO$_4^{2-}$, Ca^{2+}, P

H$_2$O

Intermediate segment

Na$^+$, Cl$^-$

DCT

Na$^+$, Cl$^-$

CD   CT

Urine flow 7% body wt./day
conc. 20 mmol • kg^{-1}
U/P ratio = 1/15

(b) MARINE TELEOST

Plasma >350 mmol • kg^{-1}

Small, reduced glomerulus

GFR ca. 1% body wt./day

Neck

Nitrogenous wastes

Glucose, amino acids

I

H$_2$O

Na$^+$, Cl$^-$

H$_2$O

PCT

II

Mg^{2+}, SO$_4^{2-}$, Ca^{2+}, P

H$_2$O

III

Na$^+$, Cl$^-$

Intermediate segment

Na$^+$, Cl$^-$

CT

H$_2$O

CD

Urine flow <1% body wt./day
conc. 300 mmol • kg^{-1}
U/P ratio = 1/1.1

➡ Active flow
⇨ Passive flow

Figure 4–11   Kidney structure and function of marine and freshwater teleosts. Arrows pointing into the lumen of the kidney tubules show movement of substances into the forming urine, and arrows pointing outward show movement from the urine back into the body fluids. Dark arrows show active movements (i.e., those that involve a transport system) and light arrows show passive flow. GFR is the glomerular filtration rate—that is, the rate at which the ultrafiltrate is formed. It is expressed as percentage of body weight per day. Freshwater teleosts are flooded by water that they must excrete; consequently they have high GFRs and low U/P (urine to plasma) concentration ratios. Marine teleosts have the opposite problem; they lose water by osmosis to their surroundings and must conserve water in the kidney—consequently they have low GFRs and high U/P ratios. PCT is the proximal convoluted tubule. (Two segments [I and II] of the PCT are recognized in both freshwater and marine teleosts. Segment III of the PCT of marine fishes is sometimes equated with the DCT [distal convoluted tubule] of freshwater fishes.) Substances a fish needs to conserve (primarily glucose and amino acids) are actively removed from the ultrafiltrate in the PCT, and nitrogenous waste products (ammonia and urea) are actively added to the forming urine. Freshwater fishes actively remove divalent ions (magnesium, sulfate, calcium, and phosphorus) from the forming urine, whereas marine fishes actively excrete those ions into the urine. Sodium and chloride are also removed from the forming urine. That makes sense for freshwater fishes because they are trying to conserve those ions, but it is surprising for marine fishes because they are battling an influx of excess sodium and chloride from seawater. The explanation of that paradox for marine fishes is that movement of sodium and chloride ions is needed to produce a passive flow of water back into the body. Freshwater fishes continue to reabsorb sodium and chloride in the CT (collecting tubule), but the walls of their CTs are not permeable to water so there is no passive uptake at this stage. In contrast, the CTs of marine fishes are permeable to water, allowing further recovery of water at this point. The net effect of the differences in GFR, inward and outward movements, and permeabilities allows freshwater fishes to produce copious amounts of dilute urine that rids them of excess water and allows marine fishes to produce scanty amounts of more concentrated urine that conserves water. Finally the urine passes into the collecting duct (CD) and from there to the outside of the body via the archinephric duct (not shown).

across the **proximal** and **distal convoluted tubules**. (Reabsorption is an active process that consumes metabolic energy.) Because the distal convoluted tubule is impermeable to water, water remains in the tubule and the urine becomes less concentrated as ions are removed from it. Ultimately, the urine becomes hyposmolal to the blood. In this way the water that was absorbed across the gills is removed and ions are conserved. Nonetheless, some ions are lost in the urine in addition to those lost by diffusion across the gills. Salts from food compensate for some of this loss, and teleosts have ionocytes in the gills that take up chloride ions from the water. The chloride ions are moved by active transport against a concentration gradient, and this process requires energy. Sodium ions also enter the gills, passively following the chloride ions.

Freshwater amphibians face the same osmotic problems as freshwater fishes. The entire body surface of amphibians is involved in the active uptake of ions from the water. Like freshwater fishes, aquatic amphibians do not drink because the osmotic influx of water more than meets their needs. Also like fishes, aquatic amphibians lose ions by diffusion. The skin of these amphibians contains cells that actively take up ions from the surrounding water. Acidity inhibits this active transport of ions in both amphibians and fishes, and inability to maintain internal ion concentrations is one of the causes of death of these animals in habitats acidified by acid rain or drainage of water from mine wastes.

Marine Vertebrates The osmotic and ionic gradients of vertebrates in seawater are basically the reverse of those experienced by freshwater vertebrates. Seawater is more concentrated than the body fluids of vertebrates, so there is a net outflow of water by osmosis and a net inward diffusion of ions.

Teleosts The integument of marine fishes, like that of freshwater teleosts, is highly impermeable so that most osmotic and ion movements occur across the gills (**Figure 4–12**). The kidney glomeruli are small, and the glomerular filtration rate is low. Little urine is formed, so the amount of water lost by this route is reduced. Marine teleosts lack a water-impermeable distal convoluted tubule. As a result, urine leaving the nephron is less copious but more concentrated than that of freshwater teleosts, although it is always hyposmolal to blood. To compensate for osmotic dehydration, marine teleosts do something unusual—they drink seawater. Sodium and chloride ions are actively absorbed across the lining of the gut, and water flows by osmosis into the blood. Estimates of seawater consumption vary, but many species drink in excess of

25 percent of their body weight per day and absorb 80 percent of this ingested water. Of course, drinking seawater to compensate for osmotic water loss increases the influx of sodium and chloride ions. To compensate for this salt load, chloride cells in the gills actively pump chloride ions outward against a large concentration gradient. Active transport of ions is metabolically expensive, and the energy cost of osmoregulation may account for more than a quarter of the daily energy expenditure of some fish.

Hagfishes, Coelacanths, and Cartilaginous Fishes Hagfishes have few problems with ion balance because they regulate only divalent ions and reduce osmotic water movement by being nearly isosmolal to seawater. Cartilaginous fishes and coelacanths also minimize osmotic flow by maintaining the internal concentration of the body fluid close to that of seawater. These animals retain nitrogen-containing compounds (primarily urea and trimethylamine oxide) to produce osmolalities that are usually slightly hyperosmolal to seawater (see Table 4–3). As a result, they gain water by osmotic diffusion across the gills and do not need to drink seawater.

This net influx of water permits large kidney glomeruli to produce high filtration rates and therefore rapid elimination of metabolic waste products from the blood. Urea is very soluble and diffuses through most biological membranes, but the gills of cartilaginous fishes are nearly impermeable to urea and the kidney tubules actively reabsorb it. With internal ion concentrations that are low relative to seawater, cartilaginous fishes experience ion influxes across the gills, as do marine teleosts. Unlike the gills of marine teleosts, those of cartilaginous fishes have low ion permeabilities (less than 1 percent those of teleosts).

Cartilaginous fishes generally do not have highly developed salt-excreting cells in the gills. Rather, they achieve ion balance by secreting from the rectal gland a fluid that is approximately isosmolal to body fluids and seawater but contains higher concentrations of sodium and chloride ions than do the body fluids.

Sharks in Freshwater About 5 percent of cartilaginous fishes are euryhaline. Bull sharks (*Carcharhinus leucas*) have a worldwide distribution. This species readily enters rivers and may venture thousands of kilometers from the sea. Bull sharks have been recorded from Indiana (in the Ohio River) and from Illinois (in the Mississippi), and there is a landlocked population of bull sharks in Lake Nicaragua. In seawater bull sharks retain high levels of urea, but in freshwater their blood urea levels decline.

The true freshwater sharks (five poorly known species in the genus *Glyphis*) and stingrays in the family

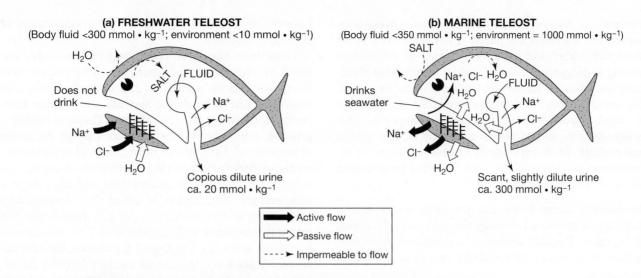

Figure 4–12 Water and salt regulation by freshwater and marine teleosts. The body fluids of freshwater fishes are more concentrated than the water surrounding them; consequently they gain water by osmosis and lose sodium and chloride by diffusion. They do not drink water; they actively absorb sodium and chloride through the gills, and they have kidneys with large glomeruli that produce a large volume of dilute urine. Marine fishes are less concentrated than the water they live in; consequently they lose water by osmosis and gain sodium and chloride by diffusion. Marine fishes drink water and actively excrete the sodium and chloride through the gills. They have kidneys with small glomeruli that produce small volumes of more concentrated urine.

Potamotrygonidae spend their entire lives in freshwater. Potamotrygonids have blood sodium and chloride ion concentrations that are lower than those in bull sharks and only slightly above levels typical of freshwater teleosts (see Table 4–3). The potamotrygonids may have lived in the Amazon basin for tens of millions of years, and their reduced salt and water gradients may reflect long adaptation to freshwater. When exposed to increased salinity, potamotrygonids do not increase the concentration of urea in the blood as euryhaline species do, even though the enzymes required to produce urea are present. Apparently their long evolution in freshwater has led to an increase in the permeability of their gills to urea and reduced the ability of their kidney tubules to reabsorb it.

Amphibians in Seawater Most amphibians are found in freshwater or terrestrial habitats. One of the few species that occurs in salt water is the crab-eating frog, *Fejervarya* (formerly *Rana*) *cancrivora*. This frog inhabits intertidal mudflats in Southeast Asia and is exposed to 80 percent seawater at each high tide. During seawater exposure, the frog allows its blood ion concentrations to rise and thus reduces the ionic gradient. In addition, ammonia is removed from proteins and rapidly converted to urea, which is released into the blood. The blood urea concentration rises from 20 to 30 mmol · kg^{-1}, and the frogs become hyperosmolal to the surrounding

water. In this sense, the crab-eating frog acts like a shark and absorbs water osmotically. Frog skin is permeable to urea, however, so the urea that crab-eating frogs synthesize is rapidly lost. To compensate for this loss, the activity of the urea-synthesizing enzymes is very high.

Tadpoles of the crab-eating frog, like most tadpoles, lack urea-synthesizing enzymes until late in their development. Thus, tadpoles of crab-eating frogs must use a method of osmoregulation different from that of adults. The tadpoles have salt-excreting cells in the gills; and, by pumping ions outward as they diffuse inward, the tadpoles maintain their blood hyposmolal to seawater in the same manner as do marine teleosts.

Nitrogen Excretion by Aquatic Vertebrates

Carbohydrates and fats are composed of carbon, hydrogen, and oxygen, and the waste products from their metabolism are carbon dioxide and water molecules that are easily voided. Proteins and nucleic acids are another matter, for they contain nitrogen. When protein is metabolized, the nitrogen is enzymatically reduced to ammonia through a process called deamination. Ammonia is very soluble in water and diffuses readily, but it is also extremely toxic. Rapid excretion of ammonia is therefore crucial.

Nitrogenous Wastes Differences in how nitrogenous wastes are excreted are partly a matter of the availability of water and partly the result of differences among phylogenetic lineages. Most vertebrates eliminate nitrogen as ammonia, as urea, or as uric acid. Excreting nitrogenous wastes primarily as ammonia is called **ammonotelism**, excretion primarily as urea is **ureotelism**, and excretion primarily as uric acid is **uricotelism**. Most vertebrates excrete a mixture of nitrogenous waste products, with different proportions of the three compounds.

Ammonotely Bony fishes are primarily ammonotelic and excrete ammonia through the skin and gills as well as in urine. Because ammonia is produced by deamination of proteins, no metabolic energy is needed to produce it.

Ureotely Mammals are primarily ureotelic, although they excrete some nitrogenous wastes as ammonia and uric acid. Normal values for humans, for example, are 82 percent urea and 2 percent each ammonia and uric acid. The remaining 14 percent of nitrogenous waste is composed of other nitrogen-containing compounds, primarily amino acids and creatine. Urea is synthesized from ammonia in a cellular enzymatic process called the **urea cycle**. Urea synthesis requires more energy than does ammonia production, but urea is less toxic than ammonia. Because urea is not very toxic, it can be concentrated in urine, thus conserving water.

Uricotely Reptiles, including birds, are primarily uricotelic, but here again all three of the major nitrogenous compounds are present. The pathway for synthesis of uric acid is complex and requires more energy than synthesis of urea. The advantage of uric acid lies in its low solubility: it precipitates from the urine and is excreted as a semisolid paste. The water that was released when the uric acid precipitated is reabsorbed, so uricotely is an excellent method of excreting nitrogenous wastes while conserving water. Some species of reptiles change the proportions of the three compounds depending on the water balance of the animal, excreting more ammonia and urea when water is plentiful and shifting toward uric acid when it is necessary to conserve water.

4.5 Responses to Temperature

Vertebrates occupy habitats that extend from cold polar latitudes to hot deserts. To appreciate their adaptability, we must consider how temperature affects an aquatic vertebrate such as a fish or amphibian that has little capacity to maintain a difference between its body temperature and the temperature of the water around it.

Organisms have been called bags of chemicals catalyzed by enzymes. This description emphasizes that living systems are subject to the laws of physics and chemistry just as nonliving systems are. Because temperature influences the rates at which chemical reactions proceed, temperature vitally affects the life processes of organisms. The rates of most chemical reactions increase or decrease when the temperature changes. The ratio of the rate at one temperature and the rate at a temperature 10°C higher is called Q_{10}. Because Q_{10} is the *ratio* of the rates, a Q_{10} of 1.0 means that the rate stays the same, a Q_{10} greater than 1 means the rate increases, and a Q_{10} less than 1.0 indicates the rate decreases (**Figure 4–13**). Q_{10} can be used to describe the effect of temperature on biological processes at all levels of biological organization from whole animals down to molecules.

The **standard metabolic rate** (SMR) of an organism is the minimum rate of oxygen consumption needed to sustain life. That is, the SMR includes the costs of ventilating the lungs or gills, of pumping blood through the circulatory system, of transporting ions across membranes, and of all the other activities necessary to maintain the integrity of an organism. The SMR does not include the costs of activities like locomotion or growth. The SMR is temperature sensitive, and that means the energy cost of living is affected by changes in body temperature. If the SMR of a fish is 2 joules per minute at 10°C and the Q_{10} is 2, the fish will use 4 joules per minute at 20°C and 8 joules per minute at 30°C. That increase in energy use translates to a corresponding increase in the amount of food the fish must eat.

Controlling Body Temperature

Because the rates of many biological processes are affected by temperature, it would be advantageous for any animal to be able to control its body temperature. However, the high heat capacity and heat conductivity of water make it difficult for most fishes or aquatic amphibians to maintain a temperature difference between their bodies and the surrounding water. Air has both a lower heat capacity and a lower heat conductivity than water, however, and the body temperatures of most terrestrial vertebrates are at least partly independent of the air temperature.

Many terrestrial vertebrates and some aquatic vertebrates do have body temperatures substantially above the temperature of the air or water around them. Maintaining those temperature differences requires

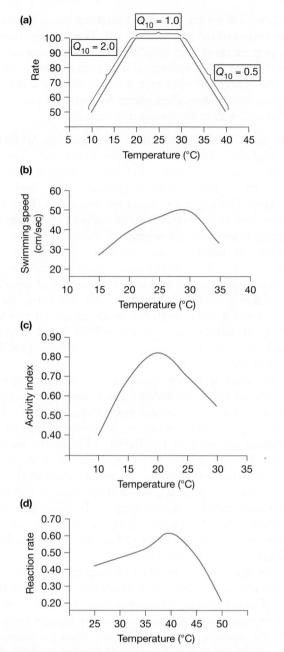

thermoregulatory mechanisms, and these are well developed among vertebrates.

Poikilothermy and Homeothermy Vertebrates were commonly described as poikilotherms (Greek *poikilo* = variable and *therm* = heat) and homeotherms (Greek *homeo* = the same) through the middle of the twentieth century. Poikilotherms were animals with variable body temperatures, and homeotherms were animals with stable body temperatures. Fishes, amphibians, and reptiles were called poikilotherms, and birds and mammals were homeotherms. This terminology has become less appropriate as our knowledge of the temperature-regulating capacities of animals has become more sophisticated. Poikilothermy and homeothermy describe the variability of body temperature, and these terms cannot readily be applied to groups of animals.

For example, some mammals allow their body temperatures to drop 20°C or more from their normal levels at night and in the winter, whereas many fishes live in water that changes temperature less than 2°C in an entire year. That example presents the contradictory situation of a homeotherm that experiences 10 times as much variation in body temperature as a poikilotherm.

Ectothermy and Endothermy Complications like these make it very hard to use the words *homeotherm* and *poikilotherm* rigorously. Most biologists concerned with temperature regulation prefer the terms *ectotherm* and *endotherm*. These terms are not synonymous with the earlier words because, instead of referring to the variability of body temperature, they refer to the sources of energy used in thermoregulation.

Ectotherms (Greek *ecto* = outside) gain their heat largely from external sources—by basking in the sun, for example, or by resting on a warm rock. **Endotherms** (Greek *endo* = inside) largely depend on metabolic production of heat to raise their body temperatures. The source of heat used to maintain body temperature is the major difference between ectotherms and endotherms because their body temperatures are quite similar. Terrestrial ectotherms (like lizards and turtles) and endotherms (like birds and mammals) have activity temperatures between 30°C and 40°C.

Endothermy and ectothermy are not mutually exclusive mechanisms of temperature regulation, and many animals use them in combination. In general, birds and mammals are primarily endothermal, but some species make extensive use of external sources of heat. For example, roadrunners are predatory birds living in the deserts of the southwestern United States and adjacent Mexico. On cold nights, roadrunners allow their body temperatures to fall from the normal level of 38°C or 39°C down to 35°C or lower. In the mornings they

Figure 4–13 Examples of the effect of temperature on organisms. (a) A hypothetical reaction for which the rate initially increases and then falls as temperature rises. Between 10°C and 20°C the rate doubles from 50 to 100, which is a Q_{10} of 2.0. The rate does not change between 20°C and 30°C, so the Q_{10} for this temperature range is 1.0. The rate drops from 100 units at 30°C to 50 units at 40°C, so the Q_{10} is 0.5. (b) The maximum swimming speed of a goldfish increases up to about 30°C and then falls. (c) Spontaneous activity by a goldfish peaks at around 20°C and falls as the temperature increases. (d) Activity of the enzyme lactic dehydrogenase from a lungfish increases slowly from 25°C to 35°C (Q_{10} about 1.2), more steeply between 35°C and its maximum at 40°C (Q_{10} about 1.4), and declines rapidly between 40°C and 50°C (Q_{10} about 0.4).

bask in the sun, raising the feathers on their backs to expose an area of black skin. Calculations indicate that a roadrunner can save 132 joules per hour by using solar energy instead of metabolism to raise its body temperature.

Deviations from general patterns of temperature regulation go the other way as well. Snakes are normally ectothermal, but the females of several species of pythons coil around their eggs and produce heat by rhythmic contraction of their trunk muscles. The rate of contraction increases as air temperature falls, and a female Indian python is able to maintain her eggs close to 30°C at air temperatures as low as 23°C. This heat production entails a substantial increase in the python's metabolic rate—at 23°C, a female python uses about 20 times as much energy when she is brooding as she does normally. Thus, generalizations about the body temperatures and thermoregulatory capacities of vertebrates must be made cautiously, and the actual mechanisms used to regulate body temperature must be studied carefully.

Regional Heterothermy Regulation of body temperature is not an all-or-nothing phenomenon for vertebrates. **Regional heterothermy** is a general term used to refer to different temperatures in different parts of an animal's body. Dramatic examples of regional heterothermy are found in several fishes that maintain some parts of their bodies at temperatures 15°C warmer than the water in which they are swimming. That's a remarkable accomplishment for a fish because each time the blood passes through the gills it comes into temperature equilibrium with the water. Thus, to raise its body temperature by using endothermal heat production, a fish must limit the loss of heat to the water via the gills.

Warm Muscles The mechanism used to retain heat is a countercurrent system of blood flow in retia mirabilia. As cold arterial blood from the gills enters the warm part of the body, it flows through a rete and is warmed by heat from the warm venous blood that is leaving the tissue. This arrangement is found in some sharks, especially species in the family Lamnidae (including the mako, great white shark, and porbeagle), which have retia mirabilia in the trunk. These retia retain the heat produced by activity of the swimming muscles, with the result that those muscles are kept 5°C to 10°C warmer than water temperature.

Scombroid fishes, a group of teleosts that includes the mackerels, tunas, and billfishes (i.e., the swordfish and spearfish as well as the sailfishes and marlins), have also evolved endothermal heat production. Tunas have an arrangement of retia that retains the heat produced by myoglobin-rich swimming muscles located close to the vertebral column (**Figure 4–14**). The temperature of these muscles is held near 30°C at water temperatures from 7°C to 23°C. Additional heat exchangers are found in the brains and eyes of tunas and sharks, and these organs are warmer than water temperature but somewhat cooler than the swimming muscles.

Hot Eyes The billfishes have a somewhat different arrangement in which only the brain and eyes are warmed, and the source of heat is a muscle that has changed its function from contraction to heat production. The superior rectus eye muscle of these billfishes has been extensively modified. Mitochondria occupy more than 60 percent of the cell volume, and changes in cell structure and biochemistry result in the release of heat by the calcium-cycling mechanism that is usually associated with contraction of muscles. A related scombroid, the butterfly mackerel, has a thermogenic organ with the same structural and biochemical characteristics found in billfishes, but in the mackerel it is the lateral rectus eye muscle that has been modified.

An analysis of the phylogenetic relationships of scombroid fishes by Barbara Block and her colleagues suggests that endothermal heat production has arisen independently three times in the lineage—once in the common ancestor of the living billfishes (by modification of the superior rectus eye muscle), once in the butterfly mackerel lineage (by modification of the lateral rectus eye muscle), and a third time in the common ancestor of tunas and bonitos (by development of countercurrent heat exchangers in muscle, viscera, and brain, and development of red muscle along the horizontal septum of the body).

The ability of these fishes to keep parts of the body warm may allow them to venture into cold water that would otherwise interfere with body functions. Block has pointed out that modification of the eye muscles and the capacity for heat production among scombroids is related to the temperature of the water in which they swim and capture prey. The metabolic capacity of the heater cells of the butterfly mackerel, which is the species that occurs in the coldest water, is the highest of all vertebrates. Swordfishes, which dive to great depths and spend several hours in water temperatures of 10°C or lower, have better-developed heater organs than do marlins, sailfishes, and spearfishes, which spend less time in cold water.

Marine Mammals The temperature equilibration of blood and water that occurs in the gills is the primary obstacle to whole-body endothermy for fish. Countercurrent systems allow them to keep critical parts of

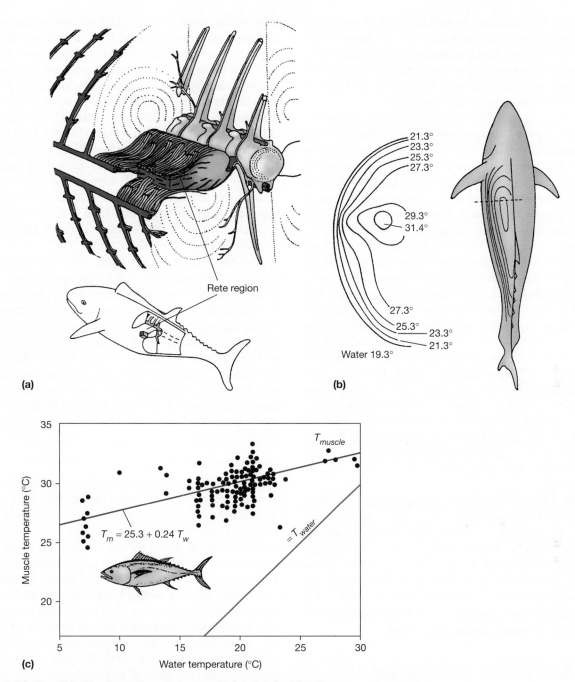

(a)

Rete region

(b)

Water 19.3°

21.3°
23.3°
25.3°
27.3°

29.3°
31.4°

27.3°
25.3°
23.3°
21.3°

(c)

T_{muscle}

$T_m = 25.3 + 0.24\,T_w$

$= T_{water}$

Muscle temperature (°C)

Water temperature (°C)

Figure 4–14 Details of body temperature regulation by the bluefin tuna. (a) The red muscle and retia are located adjacent to the vertebral column. (b) Cross-sectional views showing the temperature gradient between the core (at 31.4°C) and water temperature (19.3°C). (c) Core muscle temperatures of bluefins compared to water temperature.

their bodies warm, but other parts are at water temperature. Air-breathing aquatic tetrapods avoid that problem because they have lungs instead of gills.

Mammals are endotherms, and their high metabolic rates combined with the muscular metabolism that accompanies activity release heat that warms the body.

For terrestrial mammals, the furry body covering (the **pelage**) traps metabolic heat in the dead air spaces between hairs and reduces the movement of heat out of the body, just as the air trapped between strands of fiberglass insulations reduces the movement of heat through the wall of a building. A furry body covering works well in air but not as well in water. If water displaces the air between the hairs, the coat loses its insulative properties. Many semiaquatic mammals, such as otters and beavers, have water-repellent coats, but they must emerge onto land frequently and groom their coats to keep them water-repellent.

The most specialized marine mammals, cetaceans (whales and porpoises) and pinnipeds (seals, sea lions, and walruses), use **blubber** (a layer of fat beneath the skin) rather than fur for insulation. Blubber is an extremely effective insulator, so good that some seals risk death by overheating if they undertake prolonged strenuous activity on land. Even in water, strenuous activity can lead to overheating.

Aquatic mammals have countercurrent exchange systems in their flippers that allow them to retain heat in the body or release it to the ocean. The venous blood returning from the flipper is cold because it has passed through a flat structure with a large surface area that is in contact with the ocean water. Those are ideal conditions for heat exchange, and the blood that leaves the flippers is very close to water temperature. When a marine mammal needs to retain heat in the body, blood returning from the flippers flows through veins that are closely associated with the arteries that carry blood from the body to the flippers. Cold venous blood is heated by warm arterial blood flowing out from the core of the body. By the time the venous blood reaches the body, it is nearly back to body temperature.

When a bout of rapid swimming produces enough heat to increase body temperature above normal, the animal changes the route that venous blood takes as it returns from the flipper. Instead of flowing through veins that are pressed closely against the walls of the arteries, the returning blood is shunted through vessels distant from the arteries. The arterial blood is still hot when it reaches the flipper, and that heat is dissipated into the water, cooling the animal.

4.6 Body Size and Surface/Volume Ratio

Body size is an extremely important element in the exchange occurring between an organism and its environment. For objects of the same shape, volume increases as the cube of linear dimensions, whereas surface area increases only as the square of linear dimensions. Consider a cube that is 1 centimeter (cm) on each side. Each face of the cube is 1 cm^2 (that is, 1 cm · 1 cm), and a cube has six faces (numbered 1 through 6 if the cube comes from a set of dice), so the total surface area is 6 cm^2. The volume of that cube is one cubic centimeter (1 cm · 1 cm · 1 cm), and the ratio of surface to volume is 6 cm^2/1 cm^3 **(Figure 4–15a)**.

If we increase the linear dimensions of the cube so that a side becomes 5 centimeters long, each face has a surface area of 25 cm^2 (that is, 5 cm · 5 cm), and the total surface area is 150 cm^2 (6 · 25 cm^2). The volume of this larger cube is 125 cm^3 (that is, 5 cm · 5 cm · 5 cm), and the ratio of surface to volume is 150 cm^2/125 cm^3, which reduces to 1.2 cm^2/1 cm^3. Thus the larger cube has only one-fifth as much surface area per unit volume as the smaller cube. A cube 10 centimeters on a side has a surface/volume ratio of 600 cm^2/1000 cm^3, which is 0.6 cm^2/1 cm^3—only one-tenth the surface/volume ratio of the smallest cube.

The pattern emerging from these calculations is shown in Figure 4–15b: as an object gets larger, it has progressively less surface area in relation to its volume. Exchange between an animal and its environment occurs through its body surface, and large animals have proportionally less area for exchange in relation to the volume (or mass) of their bodies than small animals do. The biological significance of that relationship lies in the conclusion that bigger species exchange energy with the environment less rapidly than smaller species, merely because of the difference in surface/volume ratios.

Gigantothermy

Simply being big gives an animal some independence of external temperature because heat cannot flow rapidly into or out of a large body through its relatively small surface. If an animal is large enough, its body temperature will be stable simply because it's so big; this is a form of thermoregulation called **gigantothermy**. The enormous dinosaurs that lived in the Jurassic and Cretaceous periods would have had very stable body temperatures just because they were so large. It would take many days for a huge dinosaur to warm or cool as its environment changed temperature. Even elephants (which are only one-twentieth the size of the largest dinosaurs) are big enough to feel the consequences of surface/volume ratio in body-temperature regulation. Elephants can easily overheat when they are active. When that happens, they dump heat by sending large volumes of blood flowing through their ears and waving them to promote cooling.

Being big makes temperature regulation in water easier, as leatherback sea turtles dramatically illustrate. Leatherbacks (*Dermochelys coriacea*) are the largest living turtles, reaching adult body masses of 850 kilograms or more. They are also the most specialized sea turtles, having lost the bony shell that covers most

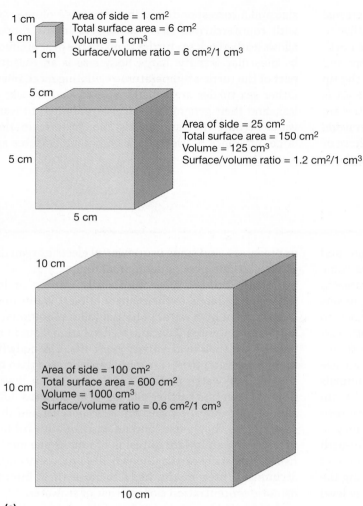

1 cm
1 cm
1 cm
Area of side = 1 cm^2
Total surface area = 6 cm^2
Volume = 1 cm^3
Surface/volume ratio = 6 cm^2/1 cm^3

5 cm
5 cm
5 cm
Area of side = 25 cm^2
Total surface area = 150 cm^2
Volume = 125 cm^3
Surface/volume ratio = 1.2 cm^2/1 cm^3

10 cm
10 cm
10 cm
Area of side = 100 cm^2
Total surface area = 600 cm^2
Volume = 1000 cm^3
Surface/volume ratio = 0.6 cm^2/1 cm^3

(a)

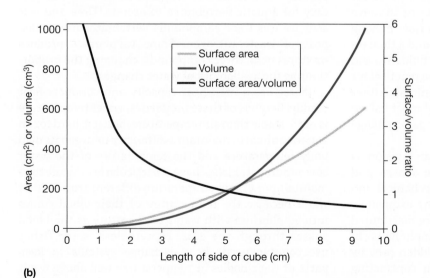

(b)

Figure 4–15 Linear dimensions, surface area, and volume. A cube illustrates how the surface/volume ratio changes with size. (a) As the length of the side of the cube is increased from 1 cm to 10 cm, the total surface area of the cube increases as the square of that length, whereas the volume of the cube increases as the cube of the length of a side. (b) Because volume increases more rapidly than surface area, the surface/volume ratio of the cube decreases as the size of the cube increases. Functionally this means that as an object becomes larger, it has less surface area relative to its volume. Thus the rate of exchange with the environment decreases. For example, if you take two cubes—one that is 1 cm on a side and the other 10 cm on a side—and then heat them to the same temperature and put them side by side on a table, the small cube will cool to room temperature faster than the large cube.

turtles and replaced it with a thick, leatherlike external body covering. Leatherback turtles are **pelagic** (live in the open ocean); their geographic range extends north to Alaska and halfway up the coast of Norway and south past the southern tip of Africa almost to the tip of South America. Water temperatures in these areas are frigid, and the body temperatures of the turtles are as much as 18°C higher than the water temperature. Turtles have metabolic rates much lower than those of mammals. Nonetheless, the combination of large body size and a correspondingly small surface/volume ratio with countercurrent heat exchangers in the flippers allows leatherback sea turtles to retain the heat produced by muscular activity. Large body size is an essential part of the turtle's temperature-regulating mechanism. Other sea turtles are half the size of leatherbacks or less, and their geographic ranges are limited to warm water because they are not big enough to maintain a large difference between their body temperatures and water temperature.

Summary

The properties of water offer both advantages and disadvantages for aquatic vertebrates. Water is some 800 times more dense than air, and vertebrates in water maintain nearly neutral buoyancy. That means the skeletons of aquatic vertebrates do not have to resist the force of gravity, and the largest aquatic vertebrates are substantially larger than the largest terrestrial forms. Animal body tissues (especially muscle and bone) are denser than water, and aquatic animals offset that weight with lighter tissues (air-filled swim bladders and lungs, or oil-filled livers) to achieve neutral buoyancy. A teleost fish can adjust its buoyancy so precisely that it can hang stationary in the water with only a backpedaling of its pectoral fins to counteract the forward propulsion generated by water leaving the gills and a gentle undulation of the tail fin to stay level in the water.

The density and viscosity of water create problems for animals trying to move through water or to move water across gas-exchange surfaces. Even slow-moving aquatic vertebrates must be streamlined, and a tidal respiratory system (moving the respiratory fluid into and out of a lung) takes too much energy to be practical for vertebrates that breathe water. These animals either have flow-through ventilation (the gills of most fishes) or use the entire body surface for aquatic gas exchange (amphibians).

Water is less transparent than air and vision is often limited to short distances, but the density and electrical conductivity of water make mechanical and electrical senses effective. Sensors that are exquisitely responsive to the movement of water are distributed over the bodies of fishes and aquatic amphibians. In addition, cartilaginous fishes can find hidden prey by detecting the electrical discharges from contracting muscles as the prey breathes. Other fishes create electric fields in the water to detect the presence of prey or predators, and some use powerful electric organ discharges to stun prey or deter predators.

Aquatic animals are continuously gaining or losing water and ions to their surroundings. Water flows from areas of high water potential (dilute solutions) to low water potential (concentrated solutions), and ions move down their own activity gradients. The body fluids of freshwater fishes are more concentrated than the surrounding water, so they are flooded by an inward flow of water and further diluted by an outward diffusion of ions. Marine fishes are less concentrated than seawater, so they must contend with an outward flow of water and an inward diffusion of ions. Some marine fishes (especially cartilaginous fishes and coelacanths) accumulate urea in the body to raise their internal osmotic concentration close to that of seawater.

Ammonia is a waste product from deamination of proteins. It is toxic but very soluble in water, so it is easy for aquatic vertebrates to excrete. Urea and uric acid are less toxic compounds vertebrates use to dispose of waste nitrogen. Some vertebrates produce mixtures of all three compounds, changing the proportions as the availability of water changes.

Water has a high heat capacity and conducts heat readily. Because of these properties, water temperature is more stable than air temperature, and it is hard for an aquatic animal to maintain a difference between its own body temperature and the temperature of the water surrounding it. Fishes have a particularly difficult time maintaining a body temperature different from that of water because the temperature of their blood comes into equilibrium with water temperature as the blood passes through the gills. Nonetheless, some fishes use countercurrent heat exchange systems to keep parts of their bodies at temperatures well above water temperature. The largest extant sea turtle, the leatherback, is able to maintain a body temperature 18°C above water temperature. Part of the temperature difference can be traced to countercurrent heat exchangers, which minimize loss of heat from the flippers; the enormous body size of these turtles is also a critical factor. Body surface area increases as the square of linear dimensions, whereas body volume increases as the cube of linear dimensions. Consequently, large animals exchange heat energy with the environment more slowly than do smaller animals. Merely being big confers a degree of stability to the internal environment of an animal.

Discussion Questions

1. What is the difference between the effects of the density and the viscosity of water from the perspective of an aquatic vertebrate?

2. When you walk past the tanks in an aquarium store, you see both physostomous and physoclistous fishes. Goldfish, for example, are physostomous, and cichlids (such as the popular convict cichlid) are physoclistous. Conduct a thought experiment comparing the responses of the two kinds of fish to changes in air pressure. (Don't actually do this because it would be painful for the fishes.)
   a. Imagine that you put a goldfish and a cichlid into a jar half filled with water and then screwed on a lid to make an airtight seal. Protruding from the lid is a tube to which you can attach a vacuum pump and a valve that allows you to close off the tube.
   b. Pump some air out of the jar. What effect will doing that have on the air pressure inside the jar? How will each of the fishes respond to that change in pressure? How are those responses related to being physostomous or physoclistous?
   c. Now open the valve and allow air to enter the jar, returning the pressure quickly to its starting level. How will each of the fishes respond to this change in pressure? How are these responses related to being physostomous or physoclistous?

3. When deep-sea fishes are pulled quickly to the surface, they often emerge with their swim bladders protruding from their mouths.
   a. Why does this happen and what does it tell you about whether these deep-sea fishes are physoclistous or physostomous?
   b. Why could you have figured out whether deep-sea fishes are physostomous or physoclistous simply by considering how air enters the swim bladder of a physostomous fish?

4. The text points out that the swim bladders of deep-sea fishes that migrate over large vertical distances are nearly filled with lipids, leaving space for only a small volume of gas, whereas deep-sea fishes that do not make large vertical migrations have more gas and less lipid in their swim bladders. What is the functional significance of this difference?

5. If you look at pictures of frogs you will see that a few species have conspicuous lateral lines. What does the presence of lateral lines tell you about the ecology of those species of frogs?

Additional Information

Ballantyne, J. S., and J. W. Robinson. 2010. Freshwater elasmobranchs: A review of their physiology and biochemistry. *Journal of Comparative Physiology B* 180:475–493.

Berenbrink, M., et al. 2005. Evolution of oxygen secretion in fishes and the emergence of a complex physiological system. *Science* 307:1752–1757.

Berenbrink, M. 2006. Evolution of vertebrate haemoglobins: Histidine side chains, specific buffer value and Bohr effect. *Respiratory Physiology and Neurobiology* 154:165–184.

Bleckmann, H., and R. Zelick. 2009. Lateral line system of fish. *Integrative Zoology* 4:13–25.

Block, B. A., et al. 1993. Evolution of endothermy in fish: Mapping physiological traits on a molecular phylogeny. *Science* 260:210–214.

Bostrom, B. L., et al. 2010. Behaviour and physiology: The thermal strategy of leatherback turtles. PLOS One 5(11):e13925.

Carleton, K. 2009. Cichlid fish visual systems: Mechanisms of spectral tuning. *Integrative Zoology* 4:75–86.

da Costa, D. C. F., and A. M. Landeira-Fernandez. 2009. Thermogenic activity of the Ca^{2+}-ATPase from blue marlin heater organ: Regulation by KCl and temperature. *American Journal of Physiology—Regulatory, Integrative, and Comparative Physiology* 297:R1460–R1468.

Dangles, O., et al. 2009. Variability in sensory ecology: Expanding the bridge between physiology and evolutionary biology. *Quarterly Review of Biology* 84:51–74.

Evans, D. H. 2008. Teleost fish osmoregulation: What have we learned since August Krogh, Homer Smith, and Ancel

Keys? *American Journal of Physiology: Regulative, and Comparative Physiology* 295:R704–R713.

Evans, D. H., et al. 2005. The multifunctional fish gill: Dominant site of gas exchange, osmoregulation, acid-base regulation, and excretion of nitrogenous waste. *Physiological Reviews* 85:97–177.

Faucher, K., et al. 2010. Fish lateral system is required for accurate control of shoaling behaviour. *Animal Behaviour* 79:679–687.

Feulner, P. G. D., et al. 2008. Electrifying love: Electric fish use species-specific discharge for mate recognition. *Biology Letters* 5:225–228.

Gardiner, J. M., and J. Atema. 2010. The function of bilateral odor time differences in olfactory orientation of sharks. *Current Biology* 20:1187–1191.

Glass, M. L., and S. C Wood (eds.). 2009. *Cardio-Respiratory Control in Vertebrates: Comparative and Evolutionary Aspects*. New York: Springer-Verlag.

Kimber, J. A., et al. 2011. The ability of a benthic elasmobranch to discriminate between biological and artificial electric fields. *Marine Biology* 158:1–8.

Little, A. G., et al. 2010. Evolutionary affinity of billfishes (Xiphiidae and Istiophoridae) and flatfishes (Plueronectiformes): Independent and trans-subordinal origins of endothermy in teleost fishes. *Molecular Phylogenetics and Evolution* 56:897–904.

Martin, R. A. 2005. Conservation of freshwater and euryhaline elasmobranchs: A review. *Journal of the Marine Biology Association U.K.* 85:1049–1073.

Pillans, R. D., et al. 2009. Observations on the distribution, biology, short-term movements and habitat requirements of river sharks *Glyphis* spp. in Northern Australia. *Endangered Species Research* 10:321–332.

Runcie, R. M., et al. 2009. Evidence for cranial endothermy in the opah (*Lampris guttatus*). *Journal of Experimental Biology* 212:461–470.

Sabbah, S., et al. 2010. Functional diversity in the color vision of cichlid fishes. *BMC Biology* 8:133. http://www.biomedcentral.com/1741-7007/8/133

Sepulveda, C. A., et al. 2008. Elevated red myotomal muscle temperatures in the most basal tuna species, *Allothunnus fallai*. *Journal of Fish Biology* 73:241–249.

Shiels, H. A., et al. 2011. Warm fish with cold hearts: Thermal plasticity of excitation–contraction coupling in bluefin tuna. *Proceedings of the Royal Society B* 278:18–27.

Siebeck, U. E., et al. 2010. A species of reef fish that uses ultraviolet patterns for covert face recognition. *Current Biology* 20:407–410.

Stoddard, P. K. 1999. Predation enhances complexity in the evolution of electric fish signals. *Nature* 400:254–256.

Tada, T., et al. 2009. Evolutionary replacement of UV vision by violet vision in fish. *Proceedings of the National Academy of Science USA* 106:17457–17462.

von der Emde, G. 1998. Electroreception. In D. H. Evans (ed.), *The Physiology of Fishes*. Boca Raton, FL: CRC Press, 313–314.

Weihrauch D., M. P. Wilkie, and P. J. Walsh. 2009. Ammonia and urea transporters in gills of fish and aquatic crustaceans. *Journal of Experimental Biology* 212:1716–1730.

Westby, G. W. M. 1988. The ecology, discharge diversity, and predatory behavior of gymnotiform electric fish in the coastal streams of French Guiana. *Behavioral Ecology and Sociobiology* 22:341–354.

Wilson, S. G., and B. A. Block. 2009. Habitat use in Atlantic bluefin tuna *Thunnus thynnus* inferred from diving behavior. *Endangered Species Research* 10:355–367.

Withers, P. C., et al. 1994. Buoyancy role of urea and TMAO in an elasmobranch fish, the Port Jackson shark, *Heterodontus portjacksoni*. *Physiological Zoology* 67:693–705.

Websites

Physical properties of air

http://www.engineeringtoolbox.com/air-properties-d_156.html

Physical properties of water

http://www.engineeringtoolbox.com/water-properties-d_1258.html

Anatomy and physiology of fish gills

http://physrev.physiology.org/content/85/1/97.full

Anatomy and physiology of the swim bladder

http://www.scienceisart.com/B_SwimBladder/SwimBladder.html

Allometry and surface/volume ratio

http://www.nature.com/scitable/knowledge/library/allometry-the-study-of-biological-scaling-13228439

Temperature effects in biological systems

http://www.nature.com/scitable/knowledge/library/body-size-and-temperature-why-they-matter-15157011

Radiation of the Chondrichthyes

The appearance of jaws and internally supported, paired appendages (described in Chapter 3) was the basis for a new radiation of vertebrates with diverse predatory and locomotor specializations. The extant cartilaginous fishes—sharks, rays, and chimaerans—are the descendants of one clade of this radiation, and we consider their origins and radiations in this chapter.

5.1 Chondrichthyes—The Cartilaginous Fishes

Chondrichthyans make a definite first appearance in the fossil record in the Early Devonian period, although isolated scales and possibly teeth appear in Ordovician sediments, and undoubted chondrichthyan scales are known from Early Silurian sediments. *Doliodus problematicus* from the Early Devonian of Canada is the oldest articulated fossil of a chondrichthyan. (*Doliodus* was sharklike in appearance but not a true shark; true sharks are unknown until the Mesozoic.)

Extant forms can be divided into two groups: those with a single gill opening on each side of the head and those with multiple gill openings on each side (**Figures 5–1** and **5–2** and **Table 5–1**).

- The Neoselachi (Greek *neo* = new and *selach* = a shark) have multiple gill openings on each side of the head. This group includes the extant sharks (mostly cylindrical forms with five to seven gill openings on each side of the head) and skates and rays (flattened forms with five pairs of gill openings on the ventral surface of the head). The sharklike form was probably the ancestral chondrichthyan condition.

- The Holocephali (Greek *holo* = whole and *cephalo* = head), so named for the undivided appearance of the head that results from having a single external gill opening, have individual gill slits covered by an operculum formed by soft tissue that is supported on the hyoid arch by an opercular cartilage. The common names of the extant members of this group—ratfishes and chimaerans—come from their bizarre form: a fishlike body, a long flexible tail, and a head that resembles a caricature of a rabbit with big eyes and broad tooth plates.

Distinctive Characters of Chondrichthyans

Chondrichthyans differ from most other vertebrates in having a cartilaginous skeleton, a lipid-filled liver, high blood urea concentrations, and an unusual pattern of energy metabolism. These characters may be functionally related to each other.

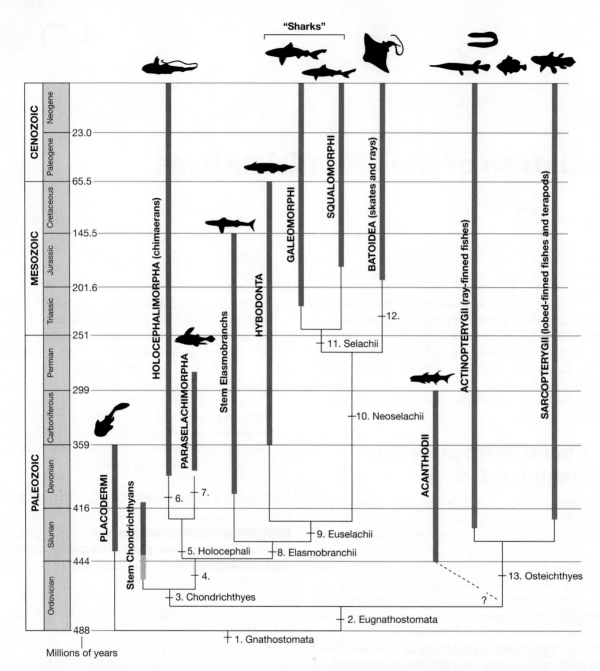

Figure 5–1 Phylogenetic relationships of cartilaginous fishes. This diagram shows the general diversification of jawed vertebrates, focusing on the probable relationships among the major groups of chondrichthyans. The black lines show interrelationships only; they do not indicate times of divergence or the unrecorded presence of taxa in the fossil record. Only the best-corroborated relationships are shown. The numbers at the branch points indicate derived characters that distinguish the lineages—see the Appendix for a list of these characters.

Cartilaginous Skeleton The absence of bone in the endoskeleton of chondrichthyans is the character that gives the cartilaginous fishes their name. A cartilaginous endoskeleton is the ancestral condition in vertebrates, seen today in the hagfishes and lampreys, but all vertebrates more derived than the living agnathans (including the Paleozoic radiation of jawless fishes, the ostracoderms) have a bony exoskeleton. Although

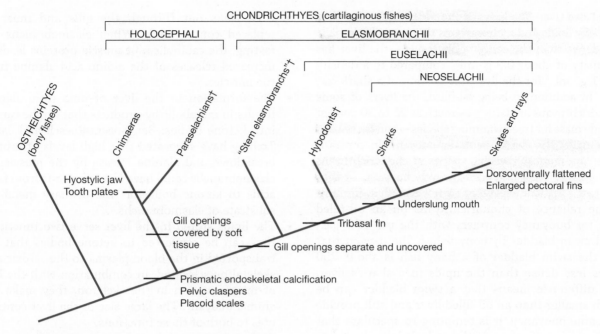

Figure 5–2 Simplified cladogram of cartilaginous fishes. Quotation marks indicate para-phyletic groups. A dagger (†) indicates an extinct taxon.

Table 5–1 Classification of Chondrichthyes, the cartilaginous fishes

Neoselachii (sharks, skates, and rays)

Galeomorphi: About 394 species of sharklike fishes with an anal fin, including ragged toothed sharks, mackerel sharks, megamouth sharks, thresher sharks, requiem sharks, hornsharks, wobbegongs, nurse sharks, and whale sharks. Galeomorphs range in length from less than 1 m to 12 m and possibly to 18 m.

Squalomorphi: About 162 species of sharklike fishes without an anal fin, including most deep-sea sharks, dogfish sharks, angel sharks, and saw sharks. Squalomorphs range in length from 15 cm to more than 7 m. (The "Squalomorphi" lineage is not monophyletic, but the rela-tionships of the six lineages it contains are not yet understood. Some systematists combine squalomorphs with skates and rays, which also lack an anal fin.)

Batoidea (skates and rays): At least 638 species of electric rays, stingrays, manta rays, a large number of skates, and a few entirely freshwater species. They range from less than 1 m to more than 6 m long and more than 6 m wide.

Holocephali (ratfishes)

Chimaeriformes: Three families with about 49 species of mostly deepwater fishes: the plownose chimaerans (Callorhynchidae), shortnose chimaerans (Chimaeridae), and longnose chimaerans (Rhinochimeridae). Chimaerans range in length from 60 cm to nearly 1.5 m.

chondrichthyans have evidently lost the extensive plates of dermal bones seen in other jawed fishes, they retain dentine, enameloid, and even traces of bone in their teeth and their scales. These unique scales are called **placoid scales**, or dermal denticles. Some degree of perichondral (= around the edge) ossification of the endoskeleton is seen in some derived ostracoderms as well, and occasionally among chondrichthyans; it may be a general gnathostome feature.

Whether the ancestors of the chondrichthyans ever possessed the fully ossified bony endoskeleton of the osteichthyans is unclear. Mineralization does occur in the axial and appendicular skeletons of chondrich-thyans as globular or stellate deposits of crystalline calcium in the superficial layers of cartilage matrix. This condition, known as tesserate or prismatic endo-skeletal calcification, is a unique derived character of Chondrichthyes. Cartilage is lighter than bone, and the increased buoyancy that cartilaginous fishes gained from the absence of a fully mineralized skeleton is believed to be the functional significance of this character. Cartilage is not necessarily weaker than bone; calcified cartilage can be made extremely strong when there are multiple lay-ers and internal struts, as in the jaws of rays that crunch mollusks.

Lipid-Filled Liver The livers of chondrichthyans are filled with oily lipids and hydrocarbons that are substantially less dense than the other body tissues—the liver has a density of about 0.9 g · ml^{-1} compared to a density of 1.1 g · ml^{-1} for the liver-free portion of a shark carcass. In addition to being oil-filled, the livers of some chondrichthyans are large—as much as 25 to 30 percent of body mass for free-swimming species—and these large livers make the sharks neutrally buoyant. In contrast, skates and bottom-dwelling species of chondrichthyans have smaller livers and are negatively buoyant, so they sink to the seafloor when they are not actually swimming.

The reliance of chondrichthyans on an oil-filled liver for buoyancy contrasts with the use of a gas-filled swim bladder by bony fishes. The oxygen that fills the swim bladder of a bony fish is about 750 times less dense than the lipids in a shark's liver. This difference means that a swim bladder can be much smaller than an oil-filled liver and still provide the same buoyancy. It is tempting to speculate that the loss of ossification of the endoskeleton by chondrichthyans and their reliance on an oil-filled liver for buoyancy may be related specializations. That is, the reduction in density produced by having a cartilaginous skeleton may allow an oil-filled liver to make a shark neutrally buoyant.

High Blood Urea Concentrations Chondrichthyans accumulate urea, making their body tissues slightly hyperosmolal to seawater (see Chapter 4), but urea molecules diffuse outward through the gills and must be replaced continuously. When an elasmobranch is feeding it breaks down protein from its prey to replenish urea, but when an elasmobranch is fasting it must break down its own proteins. Catabolism of protein produces amino acids, such as alanine, that are transported to the liver and converted to urea.

Unusual Energy Metabolism The skeletal and cardiac muscles of most vertebrates, including teleost fishes and mammals, oxidize fatty acids, but in chondrichthyans those tissues have little capacity for fatty-acid oxidation. Instead, the skeletal and cardiac muscles of elasmobranchs use ketone bodies—high-energy molecules such as beta-hydroxybutyrate that are synthesized in the liver—as metabolic substrates.

Synthesis A speculative, but plausible, hypothesis links these four characteristics of elasmobranchs:

- Retention of urea may be energetically efficient because it reduces the concentration gradient between the body fluids and seawater, thereby reducing the metabolic cost of osmoregulation.

- Urea leaks out through the gills and must be replaced continuously. When elasmobranchs are fasting, the catabolism of muscle proteins leads to increased releases of the amino acid alanine from the muscles.

- The formation in the liver of urea from alanine results in metabolic by-products that can be turned into ketone bodies. Ben Speers-Roesch and Jason Treberg have suggested that high levels of protein breakdown and alanine release by the muscles of elasmobranchs could have favored a shift from fatty acids to ketone bodies as the primary metabolic substrate of elasmobranchs.

- The lipids stored in the liver serve two functions: they can be converted to ketone bodies that are transported in the blood plasma to the cardiac and skeletal muscles and, in combination with the loss of mineralization in the skeleton, they make the animal buoyant. The large size of the liver contributes to both of these functions.

5.2 Evolutionary Diversification of Chondrichthyes

Early chondrichthyans, like the extant species, were diverse in form and habitats. In the Late Devonian, many of them were found in freshwater habitats, in contrast to their primarily marine distribution today. Through time, different lineages of Chondrichthyes developed similar but not identical modifications in feeding and locomotor structures. In the following sections, we will trace the radiations of chondrichthyans through the Paleozoic, Mesozoic, and Cenozoic.

5.3 The Paleozoic Chondrichthyan Radiation

Recent molecular studies suggest that diversification of the modern lineages occurred somewhat earlier than the first appearance of fossils. The Holocephali appear to have split from the lineage that included the ancestors of sharks and rays during the Silurian (although the earliest known fossil is from the Middle Devonian), and the split between sharks and rays appears to have occurred in Early Permian (although the earliest fossil of the modern elasmobranchs is Early Triassic).

Diversity of Stem Chondrichthyans

The stem chondrichthyans, identified mostly from scales and teeth, are known from the Silurian to the Early Devonian, but their origins may extend back into the

Ordovician. By the Middle Devonian all chondrichthyans could be assigned to either the Elasmobranchii or the Holocephali.

Diversity of Stem Elasmobranchs

The first elasmobranchs appeared in the Early Devonian. Many were generally sharklike in form, although they lacked several derived characters of modern sharks. A conspicuous difference between the stem elasmobranchs and modern sharks is the position of the mouth; stem elasmobranchs had a "terminal mouth" located at the front of the head, rather than a "subterminal mouth" beneath a protruding snout (or rostrum) as in modern sharks. *Cladoselache* is a representative example of a stem elasmobranch (Figure 5–3). It was about 2 meters long when fully grown, with large fins and mouth and five separate external gill openings on each side of the head. The mouth opened terminally, forming a large gape, and the upper jaw (palatoquadrate) was attached to the skull by several points of articulation. The lower jaw (mandible) also obtained some support from the upper portion of the second branchial arch, the hyomandibula. The term **amphistylic** (Greek *amphi* = both and *styl* = pillar or support) is used to describe this mode of multiple sites of upper-jaw suspension, which may have been the ancestral condition for gnathostomes, although the earliest gnathostomes probably lacked involvement of the hyoid arch in jaw suspension.

The stem elasmobranchs are identified by the form of the teeth common to most members—basically multicusped with little root development. Although there is evidence of bone around their bases, the teeth are primarily composed of dentine capped with an enameloid coat. As the teeth are used, they become worn; cusps break off and cutting edges grow dull. In sharks, each tooth on the functional edge of the jaw is but one member of a tooth whorl, attached to a ligamentous band that courses down the inside of the jaw cartilage deep below the fleshy lining of the mouth (Figure 5–4 on page 105). A series of developing teeth is aligned in each whorl in a row directly behind the functional tooth.

Essentially the same dental apparatus is present in all elasmobranchs, living and extinct; tooth whorls are also seen in the paraselachians (extinct forms related to chimaerans) and in the extinct acanthodians, part of the early gnathostome radiation (see Chapter 3). Tooth whorls thus may be a general ancestral feature of eugnathostomes (i.e., excluding the placoderms, which appear to be less derived than other gnathostomes on several grounds, one of which is the absence

of teeth in most forms). While not all extinct elasmobranchs shed their teeth, tooth replacement in extant sharks can be rapid: young modern sharks replace each lower-jaw tooth as often as every 8.2 days and each upper-jaw tooth every 7.8 days, although this varies considerably with species, age, and general health of the shark, as well as environmental factors such as water temperature.

Ecology of Paleozoic Elasmobranchs

We can piece together the lives of many of the early elasmobranchs from their morphology and fossil locations. Most of these forms, and *Cladoselache* in particular, were probably predators that swam in a sinuous manner, engulfing their prey whole or slashing it with daggerlike teeth. The lack of internal skeletal mineralization suggests a tendency to reduce weight and thereby increase buoyancy.

Xenacanthus, a stem elasmobranch, had a braincase, jaws, and jaw suspension very similar to those of *Cladoselache*. But there the resemblance ends. The xenacanths were freshwater bottom-dwellers, with very robust fins and heavily calcified skeletons that would have decreased their buoyancy, and some had elongated, eel-like bodies. The xenacanths appeared in the Devonian and survived until the Triassic period, when they died out without leaving direct descendants.

Stethacanthid sharks from the Early Carboniferous showed remarkable sexual dimorphism. The five species of stethacanthids that have been described ranged in length from about 30 centimeters to nearly 3 meters. The first dorsal fin and spine of males in this family were elaborated into a structure that was probably used in courtship. In two genera of stethacanthids, *Orestiacanthus* and *Stethacanthus*, this structure projected more or less upward and ended in a blunt surface that was covered with spines. Males of two other genera of stethacanthids, *Damocles* and *Falcatus*, had swordlike spines that projected forward parallel to the top of the head. One of the fossils of *Falcatus* may be a pair that died in a precopulatory courtship position, with the female grasping the male's dorsal spine in her jaws. These spines are odd-looking structures, but they are not unique—males of some extant chimaerans have a modification of the anterior spine called a cephalic clasper that is used during courtship.

Paleozoic Holocephalans

The earliest-known holocephalan comes from the Middle Devonian, and from their first appearance holocephalans included two distinct groups (sometimes

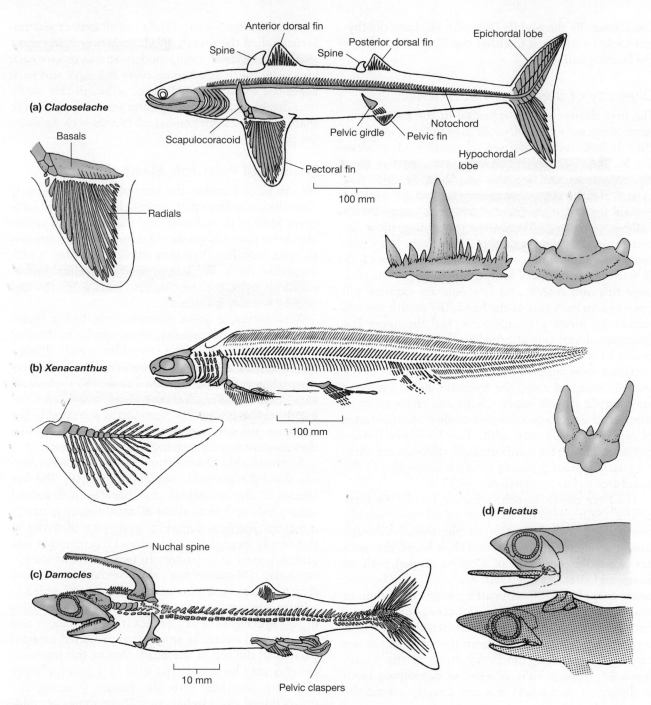

Figure 5–3 Paleozoic elasmobranchs. (a) *Cladoselache* and (b) *Xenacanthus* showing details of their fin structure and teeth. (c) A male *Damocles serratus*, a 15-centimeter-long sharklike form from the Late Carboniferous period showing the sexually dimorphic nuchal spine and pelvic claspers. Females had neither of those structures. (d) A pair of *Falcatus falcatus*, a species closely related to *Damocles*, shown as they were fossilized, possibly in courtship position. The female is clasping the male's nuchal spine in her mouth. *Falcatus* was 25 to 30 centimeters long.

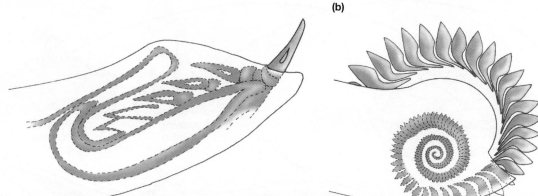

Figure 5–4 Tooth replacement by chondrichthyans. (a) Cross section of the jaw of an extant shark, showing a single functional tooth backed by a band of replacement teeth in various stages of development. Together, these form a tooth whorl. (b) Lateral view of the symphysial (middle of the lower jaw) tooth whorl of the early chondrichthyan edestoid *Helicoprion*, showing the chamber into which the lifelong production of teeth spiraled.

grouped together as the Euchondrocephali). The Holocephalimorpha includes the extant order Chimaeriformes (the chimaerans, first known from the Late Carboniferous) plus half a dozen other orders that were restricted to the late Paleozoic. Paraselachimorpha, which may be paraphyletic, includes half a dozen late Paleozoic orders, none of which persisted past the Early Permian. As their name suggests, paraselachians looked like sharks; they had sharklike teeth, and their jaw suspension was not holostylic (i.e., their upper jaw was not fused to the cranium). Paraselachians included forms called iniopterygians that had pectoral fins set up high on their sides above the gill openings, rather than ventrally as in most other fishes.

5.4 The Early Mesozoic Elasmobranch Radiation

Further elasmobranch evolution involved changes in feeding and locomotor systems. Species exhibiting these modifications appear in the Carboniferous, and this radiation of stem chondrichthyans (the Hybodonta) flourished until the end of the Cretaceous. Hybodont "sharks" and extant elasmobranchs can be grouped together in the Euselachii. *Hybodus* is a well-known genus of the Late Triassic through the Cretaceous. Complete skeletons 2 meters in length have been found that look very much like modern sharks except that the mouth is terminal (**Figure 5–5**), although many hybodonts were much smaller than this. Hybodonts have been found in both marine and freshwater settings.

Dentition

Hybodonts are distinguished by a heterodont dentition (that is, different-shaped teeth along different regions of the jaw). The anterior teeth had sharp cusps and may have been used for piercing, holding, and slashing softer foods. The posterior teeth were stout, blunt versions of the anterior teeth that may have crushed hard-bodied prey, such as crabs.

Fins

The hybodonts also showed advances in the structure of the pectoral and pelvic fins that made them more mobile than the broad-based fins of earlier elasmobranchs. Both pairs of fins were supported on narrow stalks formed by three narrow, platelike basal cartilages that replaced the long series of basals seen in earlier sharks. The narrow base allowed the fin to be rotated to different angles as the shark swam up or down, making the fins more mobile and flexible than those of earlier elasmobranchs and allowing for more sophisticated types of locomotion. The blade of the fin also changed: the cartilaginous radials were segmented and did not extend to the fin margin. Flexible fin rays called **ceratotrichia**, made from a keratin-like protein, extended from the outer radials to the margin of the fin. (Similar fin rays are seen in the fins of bony fishes, where they are called leptotrichia. Leptotrichia differ from ceratotrichia in being made of stiffer dermal bone and being segmented.)

Tail

Along with changes in the paired fins, the caudal fin assumed new functions, and an anal fin appeared.

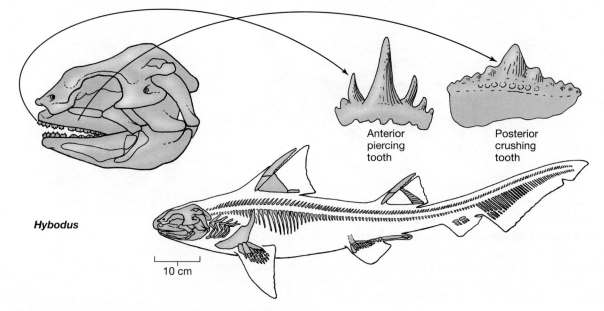

Figure 5–5 The Late Cretaceous elasmobranch *Hybodus*. The dentition consisted of pointed teeth at the front of the mouth and blunt teeth in the rear.

Reduction of the lower (hypochordal) lobe, division of its radials, and addition of flexible ceratotrichia altered the caudal fin shape so that the upper lobe was distinctively larger than the lower lobe. This tail-fin arrangement is known as **heterocercal** (Greek *hetero* = different and *kerkos* = tail).

The value of the euselachian heterocercal tail lies in its flexibility (because of the more numerous radial skeletal elements) and the control of shape made possible by the intrinsic musculature. When it was undulated from side to side, the fin twisted so that the flexible lower lobe trailed behind the stiff upper one. This distribution of force produced forward and upward thrust that could lift a shark from a resting position or counteract its tendency to sink as it swam horizontally.

Other New Internal and External Features

Other morphologic changes in euselachians include the appearance of a complete set of hemal arches in the vertebral column that protected the arteries and veins running below the notochord; well-developed ribs; and narrow, more pointed dorsal-fin spines closely associated with the leading edges of the dorsal fins. These spines were ornamented with ridges and grooves and studded with barbs on the posterior surface, suggesting that they were used in defense. Despite their variety and success during the Mesozoic, members of this second radiation of elasmobranchs became increasingly rare and disappeared at the end of the Mesozoic.

5.5 Extant Lineages of Elasmobranchs

The first representatives of the extant radiation of elasmobranchs, the Neoselachii, appeared at least as early as the Triassic. By the Jurassic period, species of modern appearance had evolved, and a surprising number of Jurassic and Cretaceous genera are still extant. The most conspicuous difference between most members of the earlier radiations and neoselachians is that most extant forms have subterminal, "underslung" mouths (**Figure 5–6**). The characters distinguishing the neoselachian clades are subtle, and the interrelationships among the extant lineages are not yet fully understood. Molecular studies unite all of the sharks in one clade (the Selachii) and the skates and rays in another (Batoidea), whereas anatomical characters indicate that the skates and rays are related to the squalomorph sharks.

Galeomorph sharks are the dominant carnivores of shallow, warm, species-rich regions of the oceans. This lineage includes many of the sharks most often featured on television—the large carnivores such as the white shark and the hammerhead. The galeomorphs also include the largest living fishes: the enormous whale shark, which grows to a length of 15 meters, and the

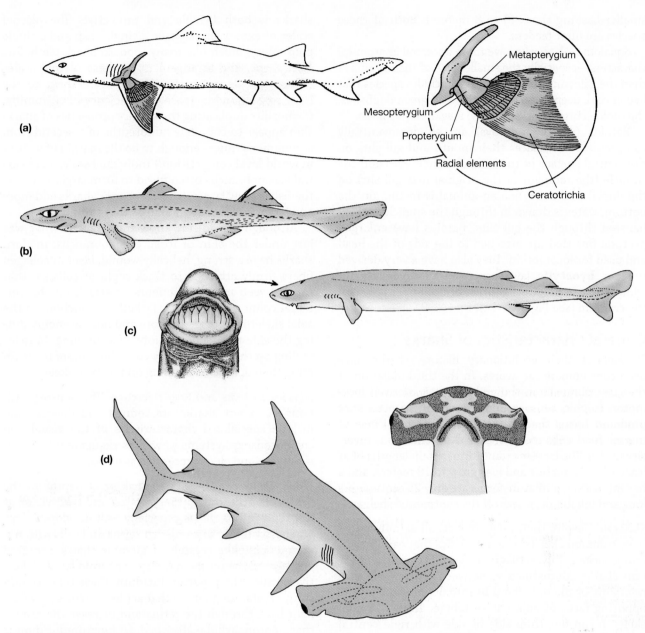

Figure 5–6 Representative extant sharks. Compare the rostrum that extends forward of the mouth in extant sharks to the terminal position of the mouth in the Late Cretaceous shark *Hybodus* shown in Figure 5–5. (a) *Negaprion brevirostris*, the lemon shark, grows to a length of 3.4 meters. The internal anatomy of the pectoral girdle and fin is shown. The mesopterygium, protperygium, and metapterygium are the basal elements. (b) *Etmopterus virens*, the green lantern shark, is a bioluminescent miniature shark only 25 centimeters long, yet it feeds on much larger prey items. (c) *Isistius brasiliensis*, the cookie-cutter shark, is a small shark whose curious mouth (*left*) is able to take chunks from fishes and cetaceans much larger than itself. (d) The great hammerhead shark (*Sphyrna mokarran*) grows to a length of more than 6 meters. The peculiar shape of the head may increase the directional sensitivity of the mechanoreceptors in the lateral lines and the electroreceptors in the ampullae.

smaller basking shark (6 to 8 meters). Both of these species are filter feeders.

Squalomorph sharks have more ancestral anatomical characters—especially the smaller size of their brains—than the galeomorphs. This lineage includes species that live in cold, deep water—the spiny and green dogfishes, the cookie-cutter sharks, and the megamouth sharks.

Batoids (skates and rays) are all dorsoventrally compressed. This places their mouth and gill slits on the ventral surface of the head and their eyes and the spiracle (the remnant of the original first gill slit) on the dorsal surface. When an animal is resting on the bottom, water is drawn in through the spiracle and exhausted through the gill slits. Batoids have enlarged pectoral fins that are attached to the side of the head and used for locomotion. They also have a very derived type of **hyostylic** jaw suspension, **euhyostyly**, in which there is no direct contact of the upper jaw with the cranium (see Figure 5–7e).

General Characteristics of Sharks

Throughout their evolutionary history, sharks have been consummate carnivores. In the third radiation of the elasmobranchs in the mid-Mesozoic, derived locomotor, trophic, sensory, and behavioral characteristics produced forms that still dominate the top levels of marine food webs today. Sharks show enormous diversity in size. The largest extant forms attain lengths of at least 12 to 15 meters and may grow to 18 meters, and a few interesting miniature forms are only 25 centimeters long and inhabit deep seas off the continental shelves.

Skeleton Despite their range in size, all extant sharks have common skeletal characteristics that earlier radiations lacked. The cartilaginous vertebral centra of extant sharks are distinctive. Between centra, spherical remnants of the notochord fit into depressions on the opposing faces of adjacent vertebrae. Thus, the axial skeleton can flex from side to side with rigid central elements of calcified cartilage swiveling on ball-bearing joints of notochordal remnants. In addition to the neural and hemal arches, extra elements that are not found in the axial skeletons of other vertebrates (the intercalary plates) protect the spinal cord above and the major arteries and veins below the centra.

Skin The placoid scales of sharks also changed in the third elasmobranch radiation. Although scales of the same general type are known from earlier chondrichthyans, they are often found in clusters or fused into larger plates rather than being distributed evenly across the body surface as they are in modern sharks. The shagreen (sharkskin) body covering of extant

sharks is both flexible and protective. The placoid scales of extant sharks have a single cusp and a single pulp cavity, much like many modern shark teeth. The size, shape, and arrangement of these placoid scales reduce turbulence in the flow of water next to the body surface and increase the efficiency of swimming. (Swimsuits duplicating the surface properties of sharkskin appear to cut a few hundredths of a second from a swimmer's time—enough to be the margin of victory in world-level competition.) Individual scales in ancestral elasmobranchs often fused to form larger scales as the fishes grew larger, whereas extant sharks add more and larger scales to their skin as they grow.

Although we do not know much about what was just under the skin of fossil elasmobranchs, modern sharks have a strong, helically wound, layer of collagen fibers firmly attached to thick septa of collagen that run between muscle segments to attach on the vertebral column. This elastic jacket with anchors to the axial skeleton appears to store and release energy during the side-to-side oscillations of swimming. Muscles pulling on the skin probably contribute more to swimming than do muscles pulling on the axial skeleton.

Sensory Systems and Prey Detection The sensory systems of extant sharks are refined and diverse, and include specialized characteristics of the visual and chemosensory systems as well as exquisitely sensitive mechano- and electroreceptive systems.

Vision Many sharks feed at dusk or at depths in the sea where little sunlight penetrates, and their vision at low light intensities is especially well developed. This sensitivity is due to a rod-rich retina and cells with numerous platelike crystals of guanine that are located just behind the retina in the eye's choroid layer. Collectively called the **tapetum lucidum**, these cells contain shiny crystals of guanine that act like mirrors to reflect light back through the retina and increase the chance that photons will be absorbed. (A tapetum lucidum is found in many nocturnal animals and accounts for the eyeshine of animals seen in the headlights of a car.)

Chemoreception Sharks have been described as swimming noses, so acute is their sense of smell. Experiments have shown that some sharks respond to chemicals in concentrations as low as 1 part in 10 billion, turning in the direction of the nostril that first receives an odor stimulus to locate its source.

Electroreception In Chapter 4 we described the extraordinary sensitivity of the neuromast organs and ampullae of Lorenzini. The ampullae of Lorenzini, which are composed of soft tissues, do not normally fossilize, but their concentration on the rostrum of modern sharks and the near-universal appearance of the projecting rostrum as a new character of the extant radiation of sharks make a strong case for the role of electrosensitivity in the success of the modern forms. The strange heads of hammerhead sharks of the genus *Sphyrna* (see Figure 5–6d) may spread the ampullae of Lorenzini over a larger area than does the standard shark rostrum. This arrangement might increase the sensitivity to electrical impulses from buried prey and to minute geomagnetic gradients used for navigation.

Integration of Sensory Information for Predation

Sharks use their sensory modalities in an ordered sequence to locate, identify, and attack prey. Olfaction is often the first sense that alerts a shark to potential prey, especially when the prey is wounded or otherwise releasing body fluids. Because of its exquisite sensitivity, a shark can use smell as a long-distance sense.

Not as useful as olfaction over great distances, but more directional over a wide range of environmental conditions, is another distance sense—the vibration sensitivity of mechanoreception. The lateral line system and the sensory areas of the inner ear detect vibrations such as those produced by a struggling fish. The effectiveness of mechanoreception in drawing sharks from considerable distances to a sound source has been demonstrated by using underwater speakers to broadcast vibrations like those produced by a struggling fish.

Once a shark is close to the stimulus, vision takes over as the primary mode of prey detection. If the prey is easily recognized, a shark may proceed directly to an attack. Unfamiliar prey is treated differently, as studies aimed at developing shark deterrents have discovered. A circling shark may suddenly turn and rush toward unknown prey. Instead of opening its jaws to attack, however, the shark bumps and scrapes the surface of the object with its rostrum. Opinions differ on whether this is an attempt to determine texture through mechanoreception, make a quick electrosensory appraisal, or use the rough placoid scales to abrade the surface and release fresh olfactory cues.

Following further circling and apparent evaluation of all sensory cues from the potential prey, the shark may either wander off or attack. In the latter case, the rostrum is raised and the jaws protrude. In the last moments before contact, some sharks draw an opaque eyelid (the nictitating membrane) across each eye to protect it. At this point, it appears that sharks shift entirely to electroreception to track prey. This hypothesis was developed while studying the attacks by large sharks on bait suspended from boats. Divers in submerged protective cages watched the attacks. After occluding their eyes with the nictitating membranes or approaching so close that the rostrum blocks visual contact with the bait, attacking sharks frequently veered away from the bait and bit some inanimate object near the bait (including the observer's cage, much to the dismay of the divers). Apparently, the metal of the cages distorted the local electric fields, and the sharks mistakenly attacked these artificial sources of electrical activity.

Jaws, Teeth, and Feeding Most living elasmobranchs have a derived type of **hyostylic** jaw suspension (Figure 5–7), which allows them to move the upper jaw independently from the skull, effecting greater suction and precision in feeding. In contrast to the amphistylic jaw suspension described for *Cladoselache*, the original points of bony articulation of the upper jaw (palatoquadrate) with the skull are replaced by ligaments. (Note that, technically, any type of jaw suspension involving the hyoid arch can be termed "hyostylic," but "amphistylic" is the traditional term retained for this type of jaw suspension with an attached palatoquadrate.) Protrusion of the upper jaw is aided by a new muscle, the preorbitalis, running forward from the upper jaw to the front of the skull. The hyomandibula can indirectly influence the movement of the lower jaw via a chain of cartilages, as it is attached to the ceratohyal in the lower part of the hyoid arch, which in turn attaches by ligaments to the lower jaw. There are various different types of upper-jaw mobility in chondrichthyans, indicating a degree of parallel evolution of this condition. Jaw opening in sharks, as in all gnathostomes (see Chapter 3), is via the hypobranchial muscles, running from the ventral portion of the jaws and branchial arches to the coracoid bar that forms the ventral part of the fused left and right halves of the pectoral girdle, the scapulocoracoid cartilage.

The advantages of the jaws of extant sharks are displayed when the upper jaw is protruded. In sharks that use suction to capture prey, such as the nurse shark, muscles swing the hyomandibula laterally and anteriorly to increase the distance between the right and left jaw articulations and thereby increase the volume of the orobranchial chamber. This expansion, which sucks water and food forcefully into the mouth, is not possible with the more ancestral amphistylic jaw suspension because the palatoquadrate is tightly attached to the chondrocranium. With hyomandibular extension, the palatoquadrate is protruded to the limits of the elastic ligaments on its orbital processes. In ram feeding or sharks that bite their prey, such as bull sharks, protrusion drops the mouth away from the head to allow a shark to bite an organism much larger than itself (see Figure 5–7g). The dentition of the

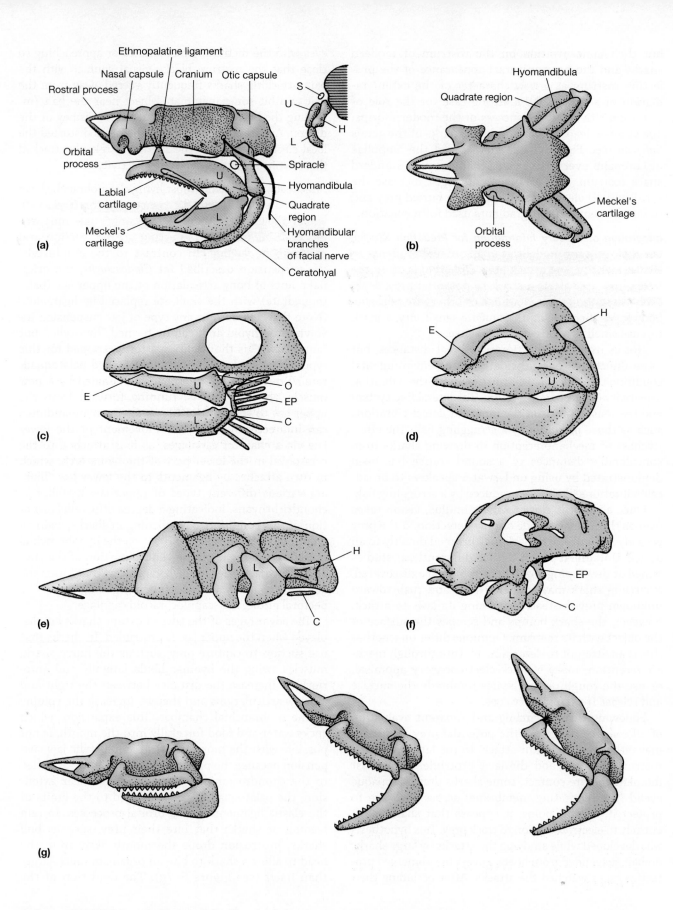

Figure 5–7 Anatomical relationships of the jaws and chondrocranium of chondrichthyans.
(a) Lateral and cross-sectional (half of skull shown) views of the head skeleton of dogfish (*Scyllium*)
with the jaws closed, showing the hyostylic jaw suspension typical of modern elasmobranchs, where
the upper jaw is attached to the cranium only by ligaments and the hyoid apparatus suspends the
lower jaw. (S = spiracle; Q = quadrate region of the palatoquadrate; H = hyomandibula; M = mandible, U = upper jaw, L = lower jaw) (b) Dorsal view of the skull of a blacktip shark (*Carcharhinus
limbatus*). (c) Head of a hypothetical ancestral gnathostome, where the hyoid arch is still unmodified (and bearing gill structures) and the upper jaw articulates with the cranium at several points:
this is the unmodified type of amphistylic jaw suspension (= autodiastyly) and can be observed in a
very ancestral chimaera. (d) Head of the Mesozoic elasmobranch *Hybodus*, showing amphistylic jaw
suspension (upper jaw articulates with cranium in several places, modified hyoid arch supports the
lower jaw). (e) Head of the ray *Rhinobatus*, showing euhyostylic jaw suspension (upper jaw has no
direct contact with the cranium). (f) Head of the extant chimaeran *Callorhinchus*, showing autostylic
(or holostylic) jaw suspension (upper jaw is fused to the cranium, hyoid apparatus has no part in jaw
suspension). (g) During jaw opening and upper-jaw protrusion in a hyostylic shark, the hyomandibula rotates from a position at a 45° angle to the long axis of the cranium to a 90° angle to that axis.
(c–f: E= ethmoidal process, O = orbital process, U = upper jaw, L = lower jaw, Ep = epihyal [dorsal
portion of unmodified hyoid arch], H = hyomandibula, C = ceratohyal)

palatoquadrate is specialized for attacking prey too
large to be swallowed whole—the teeth on the palatoquadrate are stouter than those in the mandible and
are often recurved or strongly serrated.

When feeding on large prey, a shark opens its mouth,
sinks its lower and upper teeth deeply into the prey, and
then protrudes its upper jaw ever more deeply into the
slash initiated by the teeth. As the jaws reach their maximum initial penetration, some sharks throw their bodies into exaggerated lateral undulations, which results
in a violent side-to-side shaking of the head. In these
sharks, the head movements bring the serrated upper
teeth into action to sever a piece of flesh from the prey.

Modern sharks have a wide variety of tooth shapes—
ranging from sharp, narrow points to broad, flat surfaces—and many sharks have teeth of different shapes
in different locations. The hornshark gets its generic
name, *Heterodontus* (Greek *hetero* = different and *dont*
= tooth), from this characteristic; it has sharp teeth at
the front of the mouth and broad, flat, molarlike teeth
at the rear. The mako shark has narrow, pointed teeth
at the anterior end of the jaw and broader, more bladelike teeth farther back in the tooth row, and the knife-
tooth dogfish has pointed teeth in the upper jaw and
broader, bladelike teeth in the lower jaw.

The shape of a shark's teeth has been assumed to
tell us something about their function and, by extension, what kind of prey that shark eats. Three major
functions have been described, each with a characteristic tooth shape:

- **Tearing.** Teeth with sharply pointed tips and a
  distinct cutting edge, like those of the mako shark,
  have been considered good for penetrating the body
  surface of prey.

- **Cutting.** Teeth that are flattened into a blade (i.e.,
  in the labial-lingual/front-to-back direction) with a
  relatively blunt tip and serrations along the margins
  of the tooth, like those of the great white shark,
  appear to be suitable for slicing through skin and
  muscle.

- **Crushing.** Broad flat teeth, like those in the rear
  of the jaws of the bonnethead and hornsharks, have
  been considered necessary for crushing hard-shelled
  prey, such as crabs.

A biomechanical analysis of the functional characteristics of shark teeth by Lisa Whitenack and her
colleagues has shown that the situation is more complicated than this, however (**Figure 5–8**). In the first place,
examples of parallel or convergent evolution are widespread and the shapes of sharks' teeth show no clear
relationship to their phylogeny. In addition, the functional properties of teeth are less distinct than previous studies had assumed. Most of the teeth tested
were able to carry out at least two of the three functions, and almost half of the teeth tested could do all
three of them.

The way that shark teeth are fastened to the jaw appears to be important in allowing a single tooth type to
perform multiple functions. Unlike the teeth of most
vertebrates, sharks' teeth are not set rigidly in bony
sockets or fused to the jaw bone. Instead, a collagen attachment holds them in place within the tooth whorl.
This arrangement provides flexibility that allows limited movement and rotation of the teeth about the jaw
as force is applied during a bite. In addition, each tooth
acts in concert with adjacent teeth because the bases
of teeth overlap and forces exerted on one tooth may
be transmitted to adjacent teeth.

In addition, the characteristics of prey vary—other elasmobranchs, bony fishes, and marine mammals are all soft-bodied prey, but the structures of the skin and the underlying tissues are quite different. When the complexity of the interactions among teeth is combined with the differences in prey, it is not surprising that the tooth structure of sharks provides limited information about their diets.

Predatory Behavior Rare observations of sharks feeding under natural conditions indicate that these fishes are versatile and effective predators. Great white sharks (*Carcharodon carcharias*) swim back and forth parallel to the shoreline of islands with seal and sea lion breeding sites, attempting to intercept the adults as they come and go from the rookeries. The sharks kill mammalian prey, such as seals, by exsanguination— bleeding them to death. A shark holds a seal tightly in its jaws until it is no longer bleeding and then bites down, removing an enormous chunk of flesh.

Attacks by great white sharks on sea lions are quite different from those on seals because sea lions—unlike seals—have powerful front flippers that can be used effectively in defense. Sharks often seize and release sea lions repeatedly until they die from blood loss. These behavioral observations support a hypothesis that emerged from the biomechanical analysis of shark teeth described earlier: perhaps the main selective factor that acts on shark dentition is not efficiency, but rather the severity of damage that a bite inflicts.

Some deep-sea sharks lure prey to them with bioluminescent organs. The megamouth shark has luminous organs called photophores on the upper jaw that may attract the shrimp and other small prey items it eats. Photophores on the belly of the cookiecutter shark emit a greenish glow that may attract predators, such as tunas and cetaceans. The cookie-cutter shark turns the tables on its predators, and instead of being eaten it gouges its trademark circular bites from their flesh (**Figure 5–9**).

Suction feeding appears to be the primary feeding mode of the carpet sharks, but it is also used by species of sharks in other

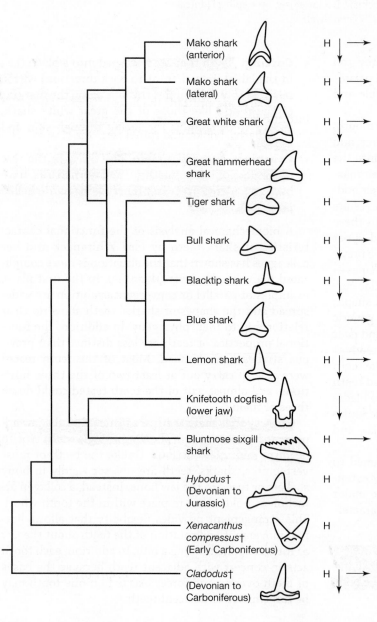

Figure 5–8 Shapes and functional characteristics of the teeth of modern and fossil sharks. The shapes of the teeth of sharks vary widely, and parallel or convergent evolution is seen in the similarity of the teeth of distantly related species. Even within a single species the shape of the teeth in different positions may be different, as is the case for the mako shark. The functional characteristics of each tooth are shown on the right. (H = able to penetrate hard-shelled prey, such as crabs; ↓ = able to penetrate soft-bodied prey; → = able to slice through the flesh of prey). The bottom three teeth are from extinct species (indicated by daggers), and the others are from extant species.

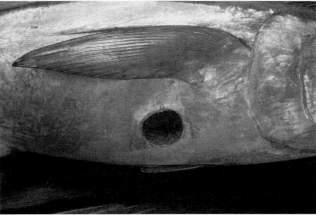

(a) **(b)**

Figure 5–9 A bite by a cookie-cutter shark leaves a distinctive wound. The shark uses its mouth to create suction that holds it to the side of its prey while the small teeth in the upper jaw and the large, triangular teeth in the lower jaw bite into the prey. Then the shark spins, cutting a cookie-shaped piece of skin and flesh from the prey. (a) A yellowfin tuna with a wound from a cookie-cutter shark; (b) a close-up of the wound.

evolutionary lineages. Leopard sharks can suck worms out of their burrows, and the epaulette shark and white-spotted bamboo shark sometimes thrust their heads into the substrate, apparently using buccal expansion to suck up buried prey.

The largest sharks—the whale sharks, basking sharks, and megamouth sharks—are filter feeders. Their strange morphology and feeding habitats appear to have been evolved convergently, at least between the whale sharks on one hand, and the basking and megamouth sharks on the other. Whale and basking sharks feed at or near the surface, whereas the megamouth spends the day at depths of 100 meters or more and moves upward to 20 meters below the surface at night. Phil Motta and his colleagues studied the feeding behavior of whale sharks off the Yucatán Peninsula of Mexico. The sharks spend most of the daylight hours engaged in ram feeding, swimming with the mouth open at a speed of about 1 meter per second. A whale shark with a length of 6.22 meters would filter just over 4600 cubic meters of water per day and harvest about 21 kilograms of plankton.

Reproduction Much of the success of the extant grade of neoselachians may be attributed to their sophisticated breeding mechanisms. Internal fertilization is universal, as was probably true for all chondrichthyans. The pelvic claspers of males have a solid skeletal structure that may increase their effectiveness. During copulation (**Figure 5–10a**), a single clasper is bent at 90° to the long axis of the body, so the dorsal groove on the clasper lies directly under the cloacal papilla from

which sperm are emitted. The flexed clasper is inserted into the female's cloaca and locked there by an assortment of barbs, hooks, and spines near the clasper's tip. Sperm from the genital tract are ejaculated into the clasper groove. Simultaneously, a muscular subcutaneous sac extending anteriorly beneath the skin of the male's pelvic fins contracts. This siphon sac has a secretory lining and is filled with seawater by the pumping of the male's pelvic fins before copulation. Seminal fluid from the siphon sac washes sperm down the groove into the female's cloaca, from which point the sperm swim up the female's reproductive tract.

Male sharks of small species secure themselves *in copulo* by wrapping around the female's body. Large sharks swim side by side, their bodies touching, or enter copulation in a sedentary position with their heads on the substrate and their bodies angled upward. Many male sharks and skates bite the female's flanks or hold onto one of her pectoral fins with their jaws, and the skin on the back and flanks of a female of these species may be twice as thick as the skin of a male the same size.

Fetal Nutrition With the evolution of internal fertilization, sharks adopted a reproductive strategy favoring the production of a small number of offspring that are retained, protected, and nourished for varying periods within the female's body. A female can invest energy in nourishing the embryo in the form of an egg yolk (as in birds) or deliver energy to the embryo from the reproductive tract of the female. **Lecithotrophy** (Greek *lecith* = egg and *troph* = nourishment) refers to the

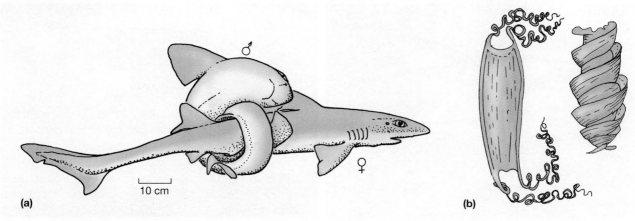

Figure 5–10 Reproduction by sharks. (a) Copulation by the European spotted catshark *Scyliorhinus*. Only a few other species of sharks and rays have been observed *in copulo*, but all assume postures so that one of the male's claspers can be inserted into the female's cloaca. (b) The egg cases of two oviparous sharks: *Scyliorhinus* (*left*) and *Heterodontus* (*right*). (Not to same scale as part a.)

situation in which yolk supplies most of the nourishment for the embryo, whereas **matrotrophy** (Greek *matro* = mother) means that the reproductive tract of the female supplies most of the energy. (Lecithotrophy and matrotrophy are at the ends of a continuum, and many vertebrates use a combination of lecithotrophy and matrotrophy in which the embryo receives some nourishment from a yolk and some from the reproductive tract of the mother.)

Eggs or Babies? Another pair of terms describes the way in which a baby enters the world: in **oviparity** the baby hatches from an egg that is deposited outside the body of the mother (as in birds, for example), and in **viviparity** a fully developed baby is born (as in placental mammals). The embryo of an oviparous species is lecithotrophic, but viviparous species include a range of modes of fetal nourishment extending from lecithotrophy to matrotrophy. There also are species of vertebrates in which the young hatch and begin feeding while they are still within the oviduct, consuming cells from the walls of the oviduct or eating unhatched eggs containing what would have been their siblings.

- **Oviparity.** Most oviparous elasmobranchs, such as hornsharks and skates, produce large eggs with large yolks (the size of a chicken yolk or larger). A specialized structure at the anterior end of the oviduct, the nidimental gland, secretes a proteinaceous case around the fertilized egg. Protuberances on the cases become tangled with vegetation or wedged into protected sites on the substrate (see **Figure 5–10b**). The embryo obtains nutrition exclusively from the yolk during the six- to ten-month developmental period. Movements of the embryo produce a flow of water through openings in the shell that flushes out organic wastes and brings in dissolved oxygen. The young are generally miniature replicas of the adults when they hatch and seem to live much as they do when mature.

- **Viviparity.** Prolonged retention of the fertilized eggs in the reproductive tract was a significant step in the evolution of shark reproduction. The eggs hatch within the oviducts, and the young may spend as long in their mother after hatching as they did within the shell, eventually emerging as miniatures of the adults. Most sharks with this reproductive mode have about a dozen young at a time.

Social Behavior Sharks have long been considered solitary and asocial, but this view is changing. Accumulating field observations indicate that sharks of many species sometimes aggregate in great numbers:

- More than 60 giant basking sharks have been observed milling together and occasionally circling in head-to-tail formations in an area off Cape Cod in the summer. Forty more individuals have been seen nearby.

- More than 200 hammerhead sharks have been seen near the surface off the eastern shore of Virginia in successive summers.

- Divers on seamounts that reach to within 30 meters of the surface in the Gulf of California have observed enormous aggregations of hammerheads schooling in an organized manner around the seamount tip. Some of these hammerhead observations include behavior thought to be related to courtship. The schools are initially formed by large numbers of females, with the largest females aggressively securing a central position within the schools. Eventually males arrive, dash into the center of the schools, and mate with the largest females.

- More than 1000 individuals of the blue shark have been observed near the surface over canyons on the edge of the continental shelf off Ocean City, Maryland.
- Fishermen are all too familiar with the large schools of spiny dogfishes that seasonally move through shelf regions, ruining fishing by destroying gear, consuming bottom fishes and invertebrates, and displacing commercially more valuable species. These dogfish schools are usually made up of individuals that are all the same size and the same sex. The distribution of schools is also peculiar: female schools may be inshore and males offshore, or male schools may all be north of some point and females south.

Our understanding of these phenomena is slim, but it is clear that not all sharks are solitary all the time and aggregations are often related to reproduction. A new field of research, which has been called social oceanography, focuses on the behavior of top oceanic predators, such as the large sharks.

Conservation of Sharks

The shark way of life has been successful, since sharks evolved 350 to 400 million years ago, but alterations of their habitat and predation by humans now threaten the survival of many species. Although young sharks are relatively large compared with other fishes, they are subject to predation, especially by other sharks. Many species depend on protected nursery grounds—usually shallow inshore waters, which are the areas most subject to human disturbance and alteration. In addition, adult sharks are increasingly falling prey to humans. A rapid expansion in recreational and commercial shark fishing worldwide threatens numerous species of these long-lived, slowly reproducing top predators.

Shark-Eating Humans It's headline news around the world when a shark attacks a swimmer, but the tens of thousands of sharks that are captured by fishing boats every year are largely unnoticed. People like to eat sharks, whether they know what they are eating or not. Spiny dogfish (*Squalus acanthias*), marketed as "grayfish," has long been served up as "fish and chips" in Europe and began to make its unheralded way into the American prepared-food market in the 1980s. "Mako" shark steaks (sliced from sharks of two

genera, *Isurus* and *Lamna*) have become an alternative to swordfish in the fresh-seafood cases of nearly every American supermarket.

Shark Finning At about the same time as domestic consumption of sharks increased in the United States, an even more powerful economic force exploded on the scene—export of shark fins to Asian markets. Shark-fin soup, which is reputed to have medicinal properties, may fetch $90 or more a bowl in restaurants in Asia. Dried shark fins sell at wholesale in Hong Kong for more than $500 a kilogram; in the United States, fresh wet fins sell for $200 or more per kilogram. A large fin from a whale shark or basking shark brings $10,000.

The growth of Asian economies during the 1980s and 1990s created an almost unlimited market for fins. Because the rest of the carcass is worthless by comparison, shark finning—the barbaric practice of catching sharks, cutting off the fins, and throwing the rest of the animal, often still alive, back into the sea—became a worldwide phenomenon. In late 2003, the United Nations General Assembly passed a resolution asking nations to ban shark finning, and many countries have outlawed the practice. Unfortunately, the resolution includes no provisions for enforcement, and there is no effective way to control what happens in international waters.

Big, Fierce Animals Are Rare Large predatory animals, such as great white sharks, are at the top of the food web—they eat species smaller than themselves and there are few, if any, predators large enough to attack them once they are adults. That sounds like an ideal situation, but there is a downside—big, fierce animals are rare because the top of the food web cannot support large populations of these **apex predators**.

Only a few individual sharks occur in any area, except perhaps during breeding or in other social aggregations. For example, the entire population of great white sharks off the coast of central California is estimated to be only 219 individuals, and they are widely dispersed. Only nine to fourteen great white sharks were observed over a period of 5 years around the South Farallon Islands, which are near San Francisco. The same individuals returned each fall when prey numbers were high. When only four great white sharks were killed near the sea-lion colony that attracted them, the number of attacks on seals and sea lions in the area dropped by half for the next 2 years.

Losing Apex Predators The predators at the top of the food web often exert a top-down control over populations of species beneath them in the web, and the loss of apex predators can have ramifications throughout the system. Populations of large sharks along the eastern coast of North America have declined by 87 to 99 percent since 1970. Smaller species of sharks and rays are the normal prey of the large sharks, and populations of the smaller species increased as much as 20-fold during the same period. Many of the smaller species of elasmobranchs feed on clams and scallops (bivalves), and populations of these shellfish have also plunged. In 2003 the commercial harvest of bivalves in the Chesapeake Bay was 300 metric tons, compared to an estimated consumption of 840,000 tons of bivalves by cownose rays.

Similar patterns of drastic declines in the populations of apex marine predators and cascading effects on the ecosystems in which they occurred can be cited for almost anywhere on Earth. Indeed, the effect is so widespread and so pervasive that Jeremy Jackson and Enric Sala have suggested that the old fishing lament should have a new twist—"You should have seen the ocean that got away."

The Outlook Is Bleak Because of their biologic characteristics, sharks are particularly susceptible to near extirpation by fishing. They grow slowly, they mature late in their lives, they have few young at a time, and, because of the great amount of energy invested in those young, the females of some species reproduce only every other year (Figure 5–11). Fisheries management specialists say there is little chance that populations of many species of overfished sharks will recover significantly in less than half a century, even with strict fishing limitations.

There is, however, one small but brighter note in the story of elasmobranch conservation: Viewing large sharks and rays in the wild has become an underwater tourist attraction in several parts of the world. Rigorous protection of these local populations of sharks has begun to afford a limited measure of protection for some species.

5.6 Batoidea: Skates and Rays

The skates and rays are more diverse than the sharks (Figure 5–12). Their characteristic suite of specializations relates to their early assumption of a bottom-dwelling, durophagous (Latin *duro* = hard and Greek *phagus* = to eat) habit. The teeth are almost all flat-crowned plates that form a pavementlike dentition. The mouth is often highly and rapidly protrusible to provide powerful suction used to dislodge shelled invertebrates from the substrate. Skates and rays have undergone considerable diversification in the adaptations of their jaws and teeth for feeding, and many similar-appearing morphologies in living forms have been evolved convergently. The dorsoventral flattening of their bodies and lateral extension of their pectoral fins provide a large surface over which ampullae of Lorenzini are distributed.

Body Form

Skates and rays are similar in being dorsoventrally flattened, with radial cartilages extending to the tips of the greatly enlarged pectoral fins. The anteriormost basal elements fuse with the chondrocranium in front of the

	White shark *Carcharodon carcharias*	Sandbar *Carcharhinus plumbeus*	Scalloped hammerhead *Sphyrna lewini*	Spiny dogfish *Squalus acanthius*	Atlantic cod *Gadus morhua*
Age to maturity (years)	m 9–10, f 12–14	m 13–16	m 4–10, f 4–15	m 6–14, f 10–12	m 2–4
Size at maturity (centimeters)	m 350–410, f 400–430	m 170, f >180	m 140–280, f 150–300	m 60, f 70	m 32–41
Life span (years)	m 15(?)	m 25–35	m 35	m 35, f 40–50	m 20+
Litter size	2–10 pups	8–13 pups	12–40 pups	2–14 pups	2 million–11 million eggs
Reproductive frequency	Biennial(?)	Biennial	(?)	Biennial	Annual
Gestation period (months)	>12	9–12	9–12	18–24	n/a

Figure 5–11 Life-history parameters of sharks. The life-history characteristics of sharks (late maturing, small litters, and biennial reproduction) make them more vulnerable to overexploitation than fishes (such as the cod) that mature quickly, have large litters, and reproduce annually. (m = male; f = female)

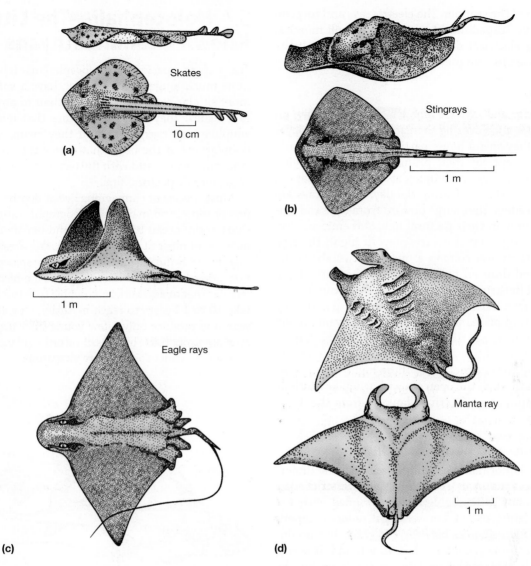

Figure 5–12 Representative extant skates and rays. (a, b) Benthic forms; (c, d) forms that frequently swim high in the water column or are truly pelagic. (a) *Raja* is a typical skate. (b) *Dasyatis* is a typical ray. (c) *Aetobatus* is representative of the eagle rays, wide-ranging predators of shelled invertebrates. (d) *Manta* is representative of its family of gigantic fishes that feed exclusively on zooplankton. Extensions of the pectoral fins anterior to the eyes (the horns of these devil rays) help funnel water into the mouth during filter feeding.

eye and with one another in front of the rest of the head. Skates and rays swim by undulating these massively enlarged pectoral fins. The placoid scales so characteristic of the integument of a shark are absent from large areas of the bodies and pectoral fins of skates and rays. The few remaining denticles are often greatly enlarged to form scales called bucklers along the dorsal midline.

Skates are distinguished from rays by their tails: skates have a long, thick tail stalk supporting two dorsal fins and a terminal caudal fin, whereas a typical ray has a whiplike tail stalk with fins replaced by one or more enlarged, serrated, and venomous dorsal barbs. Derived forms, such as stingrays (family Dasyatidae), have a few greatly elongated and venomous spines derived from modified placoid scales at the base of the tail.

The electric skates (family Rajidae) have specialized tissues in their long tails that are capable of emitting a weak electrical discharge. Each species appears to have a unique pattern of discharge, and the discharges may identify conspecifics (members of the same species) in the gloom of the seafloor. The electric rays and torpedo rays (family Torpedinidae) have modified gill muscles, producing electrical discharges as high as 200 volts that are used to stun prey.

Ecology

Many skates and rays are ambush hunters, resting on the seafloor and covering themselves with a thin layer of sand. They spend hours partially buried and nearly invisible except for their prominent eyes, surveying their surroundings and lunging at a prey item that approaches too closely. During the day, electric rays are ambush feeders: they surge upward from the sand, enclose the prey in their pectoral fins, and emit an electrical discharge that stuns the prey. At night, the rays hover in the water column a meter or two above the bottom and drop on prey that passes beneath them, cupping it in their pectoral fins while they stun it.

Other rays hunt in the water column, positioning themselves in places where tidal water movement will carry prey to them. The largest rays, like the largest sharks, are plankton strainers. Devilfishes or manta rays are up to 6 meters wide. These highly specialized rays swim through the open sea with winglike motions of the pectoral fins, filtering plankton from the water as they go.

Courtship and Reproduction

The dentition of many benthic rays is sexually dimorphic; males have sharper teeth in the front of the jaw than females. Different dentitions coupled with the generally larger size of females might reduce competition for food resources between the sexes, but no difference in stomach contents has been found. It seems more likely that sexual selection is the basis for the difference in dentition of males and females because a male uses its teeth to hold a female before and during copulation. Males of the stingray *Dasyatis sabina* have blunt teeth like those of females for most of the year; but, during the breeding season, males grow sharp-cusped teeth. The precopulatory bites of males of this species, as well as other species of rays, can leave lasting scars on females that are used for courtship.

Skates are oviparous, laying eggs enclosed in horny shells popularly called "mermaid's purses." Fetal nutrition is entirely lecithotrophic, and embryonic development can extend over many months. Rays, in contrast, are viviparous. In the early stages of development the egg yolk nourishes the embryos, and when the yolk is exhausted the embryos consume a nutrient secretion produced by the walls of the oviduct; thus, they shift from lecithotrophy to matrotrophy during their development.

5.7 Holocephali—The Little Known Chondrichthyans

The extant species of holocephalans (chimaerans), none much longer than a meter, have a soft anatomy more similar to sharks and rays than to any other extant fishes. They have long bodies that may end in a whiplike tail (**Figure 5–13**), and they swim with lateral undulations of the body that throw the long tail into sinusoidal waves and with fluttering movements of the large, mobile pectoral fins.

Most species of chimaerans live at depths of 500 meters or more, and many have geographic ranges that include entire ocean basins. Male and female chimaerans may spend most of the year in separate areas and come together only during an annual inshore spawning migration. Some chimaerans have the same life-history characteristics that make sharks vulnerable to extinction—they take 10 to 12 years to reach maturity, reproduce once a year, and produce only a few young. All extant chimaerans are oviparous, but fossil-record evidence suggests that some extinct species were viviparous.

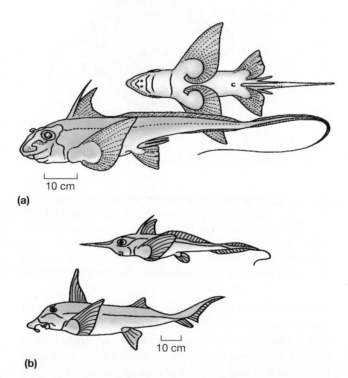

Figure 5–13 Holocephalans. (a) A chimaera, the spotted ratfish (*Hydrolagus colliei*), is an extant holocephalan. (b) Representatives of two other extant groups: (*upper*) Pacific longnose chimaeran (*Harriotta raleighana*) and (*lower*) plownose chimaeran (*Callorhinchus callorynchus*).

Chimaerans have long been grouped with elasmobranchs as Chondrichthyes because of the shared specializations described in section 5.1, but they have a bizarre suite of unique features as well. They have only four gill openings, which are covered by a soft tissue flap so that only one opening is visible to the outside; they have no spiracle; the branchial skeleton is below the cranium (rather than behind it, as in sharks); and their teeth have been reduced to tooth plates.

Several species of chimaerans have elaborate rostral extensions that are densely studded with lateral line mechanoreceptors and ampullary electroreceptors. These receptors are especially dense around the mouth, which is relatively small and faces downward. Holocephalans prefer soft substrates, and they feed on invertebrates and small fishes that live on the seafloor. They consume soft-bodied organisms, such as anemones and jellyfishes, but their tooth plates can crush hard-bodied prey, such as crabs. Like other fishes that have crushing tooth plates (e.g., lungfishes) chimaerans

have a derived type of jaw suspension, where the upper jaw is completely fused to the cranium (see Figure 5–7f). This type of jaw suspension is called **holostyly** in chimaerans, and they retain an unmodified hyoid arch. (The similar condition in lungfishes and tetrapods evolved independently in a different fashion from a condition where the hyoid arch was incorporated into the jaw suspension; this is called **autostyly**.)

Some species of holocephalans have a poison gland that is associated with a mobile dorsal spine. The spine can be erected when the chimaera is attacked, and a predator that was stabbed in the mouth as it tried to swallow a chimaeran might well decide to release the fish and seek less noxious prey. Males of some species of chimaeran have a spine-encrusted cephalic clasper that is used in courtship. Oddly, although the clasper is on the top of the head, it is controlled by muscles that also move the lower jaw. A male uses the cephalic clasper to pin the female's pectoral fin against his forehead while he is attempting copulation.

Summary

Fossil remains of chondrichthyans first appear in the fossil record in the Ordovician period, although definitive dental remains are not known until the Early Silurian. Chondrichthyans are distinguished by an entirely cartilaginous endoskeleton with a unique form of prismatic mineralization, and by the presence of pelvic claspers in the males, indicating reproduction via internal fertilization. Chondrichthyans are subdivided into the holocephalans (chimaerans and fossil relatives) and the elasmobranchs (sharks, rays, and fossil relatives).

Three radiations of elasmobranchs—Paleozoic, Mesozoic, and Cenozoic—can be traced through increasingly derived characters of the jaws and fins. Modern sharks and rays have modifications of the skull that allow the jaws to be protruded, enabling the fishes to bite chunks of flesh from prey too large to be swallowed whole. Sharks are mainly midwater predators.

Skates and rays are adapted for life on the sea bottom. They are dorsoventrally flattened, with eyes and spiracles on the tops of their heads. Many skates and rays lie buried in sand and ambush their prey, but the largest species—the manta rays—are plankton feeders.

The holocephalans were much more diverse in the Paleozoic than they are today. The paraselachians are a separate clade from the one that gave rise to the extant chimaerans, which are a small group of bizarre fishes (also known as ratfishes). Chimaerans generally occur in deep water, and little is known of their natural history and behavior.

The life histories of most species of elasmobranchs are based on producing a few relatively large young at a time. This reproductive strategy depends on a high survival rate for the young and long life expectancies for adults. It worked well for more than 350 million years, but loss of coastal habitat and outrageous overfishing within the past 50 years have now brought many species of sharks to the edge of extinction.

Discussion Questions

1. "Chondrichthyes" means "cartilaginous fishes." Biologists used to assume that bone was not evolved until the Osteichthyes, or bony fishes. What clues do we have that shark ancestors, at some level, had bone? How can we infer this from (a) evidence from the fossil record and (b) evidence from comparative anatomy?

2. Why do we call Paleozoic chondrichthyans such as *Cladoselache* "sharks," and why isn't that terminology taxonomically correct?

3. The peculiar heads of hammerhead sharks are called "cephalofoils," a term that reflects an interpretation of the head as a hydrodynamic structure that helps to direct the shark's head upward while it is swimming or to make the shark more maneuverable. An alternative hypothesis is that the head increases the shark's chemosensitivity or electrosensitivity. Proponents of the two interpretations have used anatomical, behavioral, and evolutionary information to support their views and to discredit the alternative hypothesis.

Hydrodynamic Function
- The cross section of the cephalofoil is shaped like a wing, and the angle of its trailing edge can be altered in a manner reminiscent of changing the angle of the flaps on the trailing surface of an airplane's wing to control the amount of lift.
- The sizes of the two anterior planing surfaces, the cephalofoil and the pectoral fins, are inversely related. That is, species with wide cephalofoils have relatively small pectoral fins, and vice versa. As a result of that relationship, the total surface area of those surfaces is fairly constant.
- Hammerheads are among the least buoyant species of sharks.

Sensory Function
- Hammerheads swing the head from side to side when they are searching for prey buried in the sediment. This is the same way a person searches for buried objects with a metal detector.
- The ampullae of Lorenzini are distributed over the entire ventral surface of the hammer.

Additional Observations
- The least-derived extant species of hammerhead, the winghead shark (*Eusphyra blochii*), has the broadest hammer; its width is nearly 50 percent of the total length of the shark.
- The size of the hammer in more derived species of hammerheads does not change unidirectionally according to current phylogenetic interpretations of the lineage (that is, there appears to be no tendency for the size of the hammer to have increased or decreased during the evolution of the extant hammerheads).
- The nostrils of the winghead shark are near the midline of the hammer, but they lie toward the ends of the hammer in more derived species.
- The nostrils of the scalloped hammerhead (*Sphyrna lewini*), which is a well-studied species, lie near the ends of the hammer and collect water via a prenarial groove that probably increases the volume of water flowing across the olfactory epithelium.
- The surface area of the olfactory epithelium of the scalloped hammerhead is no larger than that of non-hammerhead sharks.

It appears that observations have taken us as far as they can and it's time to try a different approach to testing hypotheses about the functional significance of the hammerhead morphology.

a. What experimental tests can you propose to evaluate the two hypotheses? (Don't worry too much about how you would carry out a manipulation—if a test involves putting sharks into a flow tank, for example, assume that you have access to a tank and sharks of the appropriate species and sizes.)

b. Is there another possibility that you miss if you consider the hydrodynamic and sensory functions to be mutually exclusive hypotheses?

4. John Musick and Julia Ellis have proposed that the ancestral reproductive mode for elasmobranchs was "yolk-sac viviparity." That is, embryos were retained throughout their development in the oviducts of the female and emerged as miniatures of the adults (i.e., viviparity), but nutrition was provided by yolk that was deposited at the time the egg was formed, not from the mother during development (i.e., lecithotrophy, not matrotrophy). They suggest that oviparity (depositing eggs that develop outside the body of the mother) was associated with the evolution of small body size because it increased the fecundity of small species of elasmobranchs.

What is their reasoning? That is, why would oviparity provide greater fecundity than viviparity for small species of elasmobranchs? What other factors might make one mode superior to the other?

5. A hyostylic jaw suspension in chondrichthyans allows protrusion of the upper jaw independent of the lower jaw. Evolutionary biologists have assumed that the ability to protrude the upper jaw is advantageous for feeding—in other words, that it is an adaptive derived character. An alternative hypothesis is that a hyostylic jaw suspension is a neutral ancestral character (i.e., one that is neither advantageous nor disadvantageous).

These two hypotheses (i.e., hyostylic jaw is advantageous/hyostylic jaw is neutral) generate different predictions about the phylogenetic distribution of hyostyly. What are those predictions, and which one is supported by the phylogenetic distribution of hyostyly?

Additional Information

Abel, R. A., et al. 2010. Functional morphology of the nasal region of a hammerhead shark. *Comparative Biochemistry and Physiology, Part A* 155:464–475.

Barnett, L. A. K., et al. 2009. Maturity, fecundity, and reproductive cycle of the spotted ratfish, *Hydrolagus colliei*. *Marine Biology* 156:301–316.

Carrier, J. R., and J. A. Musick. 2012. *Biology of Sharks and Their Relatives*, 2nd ed. Boca Raton, FL: CRC Press.

Carrier, J. C., J. A. Musick, and M. R. Heithaus (eds.). 2010. *Sharks and Their Relatives II: Biodiversity, Adaptive Physiology, and Conservation*. Boca Raton, FL: CRC Press.

Chappie, T. K., et al. 2011. A first estimate of white shark, *Carcharodon carcharias*, abundance off Central California. *Biology Letters* 7:581–583.

Estes, J. A., et al. 2010. Some effects of apex predators in higher-latitude coastal oceans. In J. Terborgh and J. A. Estes (eds.), *Trophic Cascades: Predators, Prey, and the Changing Dynamics of Nature*. Washington, DC: Island Press.

Ferretti, F., et al. 2010. Patterns and ecosystem consequences of shark declines in the ocean. *Ecology Letters* 13:1055–1071.

Gardiner, J. M., and J. Atema. 2010. The function of bilateral odor time differences in olfactory orientation of sharks. *Current Biology* 20:1187–1191.

Heinicke, M. P., et al. 2009. Cartilaginous fishes (Chondrichthyes). In S. B. Hedges and S. Kumar (eds.), *The Timetree of Life*. Oxford, UK: Oxford University Press, 320–327.

Helfman, G. S., et al. 2009. *The Diversity of Fishes*, 2nd ed. Malden, MA: Blackwell.

Inoue, J. G., et al. 2010 Evolutionary origins and phylogeny of the modern holocephalans (Chondrichthyes: Chimaeriformes): A mitogenomic perspective. *Molecular Biology and Evolution* 27:2576–2586.

Jacques, P. J. 2010. The social oceanography of top oceanic predators and the decline of sharks: A call for a new field. *Progress in Oceanography* 86:192–203.

Jorgansen. S. J., et al. 2010. Philopatry and migration of great white sharks. *Proceedings of the Royal Society B* 277:679–688.

Lisney, T. J. 2010. A review of the sensory biology of chimaeroid fishes (Chondrichthyes: Holocephali). *Reviews in Fish Biology and Fisheries* 20:571–590.

Long, J. A. 2011. *The Rise of Fishes: 500 Million Years of Evolution*, 2nd ed. Baltimore, MD: The Johns Hopkins University Press.

Lucifora, L.O., et al. 2011. Global diversity hotspots and conservation priorities for sharks. PLoS ONE 6(5):e19356.

Motta, P. J., and D. R. Huber. 2012. Prey capture behavior and feeding mechanics of elasmobranchs. In J. C. Carrier and J. A. Musk (eds.), *Biology of Sharks and Their Relatives*, 2nd ed. Boca Raton, FL: CRC Press.

Motta, P. J., et al. 2010. Feeding anatomy, filter-feeding rate, and diet of whale sharks *Rhincodon typus* during surface ram filter feeding off the Yucatan Peninsula, Mexico. *Zoology* 113:199–212.

Musick, J. A. 2010. Chondrichthyan reproduction. In K. S. Cole (ed.), *Reproduction and Sexuality in Marine Fishes: Patterns and Processes*. Berkeley, CA: University of California Press, 3–20.

Rivera-Vicente, A. C., et al. 2011. Electrosensitive spatial vectors in elasmobranch fishes: Implications for source localization. PLoS ONE 6:e16008. doi:10.1371/journal.pone.0016008.

Simpfendorfer, C. A., and P. M. Kyne. 2009. Limited potential to recover from overfishing raises concerns for deep-sea sharks, rays, and chimaeras. *Environmental Conservation* 36:97–103.

Speers-Roesch, B., and J. R. Treberg. 2010. The unusual energy metabolism of elasmobranch fishes. *Comparative Biochemistry and Physiology A* 155:417–434.

Taylor, R. 2010. When predator becomes prey. ECOS, CSIRO Publishing. http://www.ecosmagazine.com/?paper=EC156p14/

Taylor, S., et al. 2011. Fine-scale spatial and seasonal partitioning among large sharks and other elasmobranchs in south-eastern Queensland, Australia. *Marine and Freshwater Research* 62:638–647.

Whitenack, L. B., and P. J. Motta. 2010. Performance of shark teeth during puncture and draw: Implications for the mechanics of cutting. *Biological Journal of the Linnean Society* 100:271–286.

Whitenack, L. B., et al. 2011. Biology meets engineering: The structural mechanics of fossil and extant shark teeth. *Journal of Morphology* 272:169–179.

Websites

General information about cartilaginous fishes

American Elasmobranch Society http://elasmo.org

Florida Museum of Natural History, Sharks http://www.flmnh.ufl.edu/fish/Sharks/ISAF/ISAF.htm

ReefQuest Centre for Shark Research http://elasmo-research.org/index.html

Shark Research Institute http://www.sharks.org/

Fossil chondrichthyans

Elasmo.com—The Life and Times of Long Dead Sharks http://www.elasmo.com/

Fossil Fishes of Bear Gulch: information about stem chondrichthyans, including the photograph of the fossilized pair of *Falcatus falcatus*, with the male's spine inserted through the gills of the female, that is shown in Figure 5–3d http://www.sju.edu/research/bear_gulch/pages_fish_species/Falcatus_falcatus.php

Conservation of chondrichthyans

European Elasmobranch Association: a coalition of national organizations that promote sustainable management of shark and ray fisheries on a regional basis http://www.sharkalliance.org/

Shark Alliance: a coalition of nongovernmental organizations dedicated to the conservation of sharks http://www.sharkalliance.org/

CHAPTER 6

Dominating Life in Water: The Major Radiation of Fishes

By the end of the Silurian period, the agnathous fishes had diversified and the cartilaginous gnathostomes were starting to radiate. The stage was set for the appearance of the largest extant group of vertebrates, the bony fishes. The first fossils of bony fishes (Osteichthyes) occur in the Late Silurian. The osteichthyan radiation was in full bloom by the middle of the Devonian, with two major groups diverging: the ray-finned fishes (Actinopterygii) and the lobe-finned fishes (Sarcopterygii). Specialization of feeding mechanisms is a key feature of the evolution of these major groups of vertebrates. Increasing mobility among the bones of the skull and jaws allowed the ray-finned fishes to exploit a wide range of prey types and predatory modes. Specializations in locomotion, habitat, behavior, and life histories accompanied the specializations of feeding mechanisms.

6.1 The Origin of Bony Fishes

The Devonian is known as the Age of Fishes because all major lineages of fishes, extant and extinct, coexisted in the fresh and marine waters of the planet for its 57-million-year duration (Figures 6–1 and 6–2 on pages 123–124). Most groups of gnathostome fishes significantly diversified during this period. Among them was the most species-rich and morphologically diverse lineage of vertebrates, the Osteichthyes or bony fishes (Greek oste = bone and ichthy = fish). Most authorities believe that Osteichthyes is a monophyletic lineage. Bony fishes, as their name suggests, possess bone, but as we have seen in earlier chapters, dermal bone (or its remnants, such as placoid scales) is seen somewhere in all other types of jawed fishes and also in the jawless ostracoderms. Most osteichthyans have a fully ossified endoskeleton with endochondral bone in addition to the dermal and perichondral bone seen in other fishes.

Osteichthyans have several additional important features that distinguish them from other fishes:

- Osteichthyes have a unique pattern of dermal bones surrounding their jaws and braincase, which we can see retained in ourselves.
- The original dermal jaws are covered with marginal mouth bones that bear rooted teeth: the **maxilla** and **premaxilla** in the upper jaw, and the **dentary** in the lower jaw.

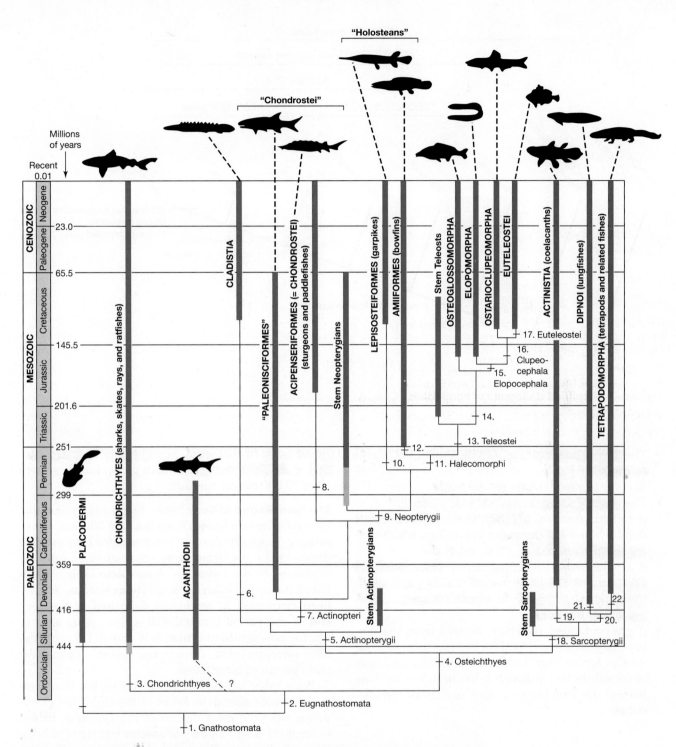

Figure 6–1 Phylogenetic relationships of bony fishes. This diagram depicts the probable relationships among the major groups of basal gnathostomes. The black lines show interrelationships only; they do not indicate times of divergence or the unrecorded presence of taxa in the fossil record. Lightly shaded bars indicate ranges of time when the taxon is believed to be present, because it is known from earlier times, but is not recorded in the fossil record during this interval. Only the best-corroborated relationships are shown, and uncertainties are indicated by dashed lines. The numbers at the branch points indicate derived characters that distinguish the lineages—see the Appendix for a list of these characters.

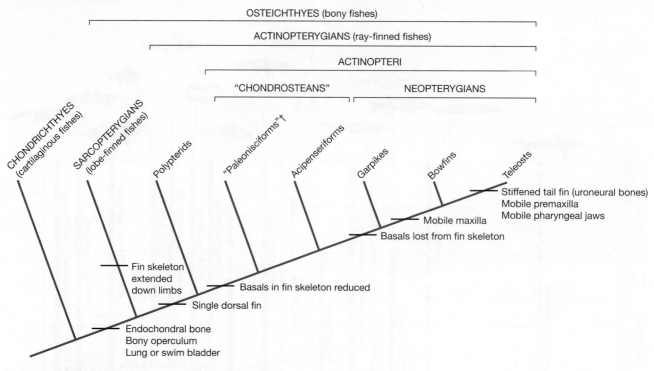

Figure 6–2 Simplified cladogram of bony fishes. Quotation marks indicate paraphyletic groups.

- The dermal bones extend into the roof of the mouth to cover the palate, and many osteichthyes (as well as some tetrapods) have palatal teeth.
- The dermal head bones extend posteriorly (the postcleithral series) to attach to the pectoral girdle; dermal bones also form the operculum, which covers the gills and aids in gill ventilation.
- A fanlike series of dermal bones, **branchiostegal rays**, form the floor of the gill chamber and aid in the rapid expansion of the mouth for suction feeding and respiration.
- A gas-containing structure, derived from the embryonic gut tube and used for buoyancy, for respiration, or for both functions, is an ancestral condition of osteichthyans, although it has been lost in some bottom-dwelling lineages, such as darters and flatfishes.

Earliest Osteichthyes and the Major Groups of Bony Fishes

The Osteichthyes include two major groups: the ray-finned fishes (the Actinopterygii, Greek *actin* = a ray and *pter* = a fin or wing) and the lobe-finned fishes (the Sarcopterygii, Greek *sarco* = flesh). The Actinopterygii account for the largest number of *fishes* in the conven-tional sense of that word, but all the terrestrial vertebrates are Sarcopterygii, so each group includes more than 30,000 extant species.

The Fossil Record of Bony Fishes Fragmentary remains of bony fishes are known from the Late Silurian, representing "Lophosteiformes"—stem osteichthyans less derived than any living form—that survived into the beginning of the Devonian. A complete skeleton of a Late Silurian fish (*Guiyu oneiros*) (**Figure 6–3**) has recently been discovered in sediments deposited at the very end of the period (around 418 million years ago) in what is now southern China. It is a small (33 centimeters) sarcopterygian, close to the ancestor of the two main groups of bony fishes.

Bony fishes radiated in the Devonian, but they were not the most abundant fishes of the time. The placoderms were the most diverse of the Devonian fishes, in terms of both numbers of species and types of body forms, and acanthodians were also prominent. The sarcopterygians were the most diverse of the bony fishes during the Devonian, in contrast to their diminished role after this period. It is not until the extinctions at the end of the Devonian that the actinopterygians (and the chondrichthyans) became the predominant aquatic vertebrates, a condition that persists to this day. Figures 6–1 and 6–2 show how bony fishes are

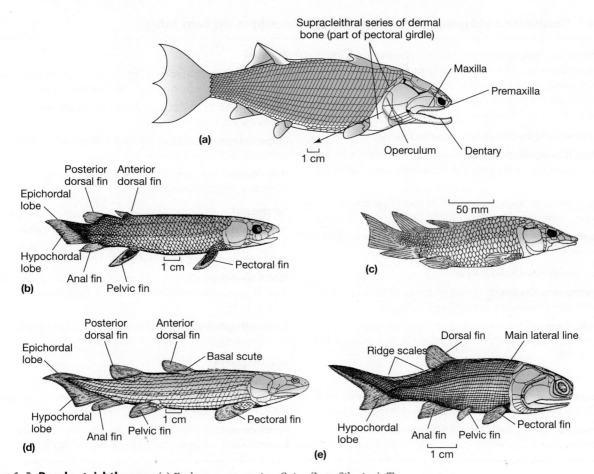

Figure 6–3 Basal osteichthyans. (a) Early sarcopterygian *Guiyu* (Late Silurian). The arrow shows where the water exits from the gills. (b) Relatively unspecialized sarcopterygian (dipnoan) *Dipterus*, Middle Devonian. (c) Long-snouted sarcopterygian (dipnoan) *Griphognathus*, Late Devonian. (d) Osteolepiform sarcopterygian (tetrapodomorph) *Osteolepis*, Middle Devonian. (e) Typical early actinopterygian *Moythomasia*, Late Devonian. Note the differences in the scale bars.

interrelated, and **Table 6–1** lists the modern groups of bony fishes. Some representative early osteichthyans are shown in Figure 6–3.

Fins As their names suggest, actinopterygians and sarcopterygians differ in the form of their fins. The ancestral condition for the ostheichthyan fin (**Figure 6–4** on page 127) is rather like that of the chondrichthyans—a row of basals that articulate with the limb girdles (scapulocoracoid for the pectoral girdle, ischiopubic plate for the pelvic girdle), then a row of radials, elements that articulate with the basals, and finally fin rays that branch out from the radials and support the web of the fin. The basals and radials are endochondral bone, whereas the fin rays (lepidotrichia) have a dermal origin. This ancestral form of osteichthyan fin is seen in some living actinopterygians, such as the polypterids

and Acipenseriformes. The tendency in actinopterygian evolution is to reduce the endoskeletal fin elements. In teleosts (the most derived forms of actinopterygians) the basals have been lost, the radials attach directly to the limb girdles, and the visible external part of the fin consists of only the fin rays.

In contrast, all but the earliest sarcopterygians have a monobasal fin in which only a single basal articulates with the limb girdles. The metapterygial bony axis extends from the most posterior of the basals down into the fin. Bony elements called mesomeres form a straight line, with radials branching off to one or both sides, and the fin muscles follow the bones to form the fleshy lobe of the fin.

In the tetrapodomorph sarcopterygian fishes (the precursors of tetrapods), the fin skeleton is asymmetrical; the radials branch from only the anterior (preaxial)

Table 6.1 Classification and geographic distribution of Osteichthyes, the bony fishes

Only the evolutionarily or numerically most important groups are listed. The subdivision of the Neopterygii varies greatly from author to author. Groups in quotation marks are not monophyletic, but relationships are not yet understood. Names in square brackets are alternative names for the groups.

Sarcopterygii (fleshy-finned fishes and tetrapods)

Actinistia [Coelacanthiformes]: 2 species of coelacanths from the western Indian Ocean and central Indonesia, deep-water marine, 1 to 1.5 m

Dipnoi: 6 species of lungfishes from the Southern Hemisphere, freshwater, shorter than 1 to 1.8 m

Tetrapoda: More than 40,000 species of terrestrial and secondarily aquatic vertebrates

Actinopterygii (ray-finned fishes)

Polypteriformes [Cladistia]: At least 16 species of bichirs and the reedfishes, Africa, freshwater, shorter than 30 to 90 cm

Acipenseriformes [Chondrostei]: 27 species of sturgeons and paddlefishes, Northern Hemisphere, coastal and freshwater, about 2 m to at least 5 m

Neopterygii

Lepisosteiformes [Ginglymodi]: 7 species of gars, North and Central America, freshwater and brackish water, less than 1 m to about 3 m

Amiiformes: 1 species, the bowfin, North America, freshwater, up to 90 cm

Teleostei

Osteoglossomorpha: At least 219 species of bonytongues, worldwide, mostly tropical freshwater, 10 cm to at least 2.5 m

Elopomorpha: At least 856 species of tarpons and eels, worldwide, mostly marine, 1 to 4 m

Clupeomorpha: About 364 species of herrings and anchovies, worldwide, especially marine, 6 cm to 1 m

Ostariophysi: More than 7931 species of catfishes and minnows, worldwide, mostly freshwater, 1 cm to 5 m

Euteleostei

"Protacanthopterygii": About 366 species of trout, salmon, and their relatives, temperate Northern and Southern Hemisphere, freshwater, from about 7 cm to at least 1.4 m

"Stem Neoteleosts": About 916 species of lanternfishes and their relatives, worldwide, mostly mesopelagic or bathypelagic marine, 7 cm to 1.8 m

Paracanthopterygii: About 1340 species of cod and anglerfishes, Northern Hemisphere, marine and freshwater, 6 cm to 2 m

Acanthopterygii

Atherinomorpha: About 1624 species of silversides, killifishes, and their relatives, worldwide, surface-dwelling, freshwater and marine, less than 4 cm to about 2 m

Perciformes: More than 13,173 species of perches and their relatives, worldwide, primarily marine, 8 mm to 3 m

---

edge of the skeleton so that the mesomeres and radials form a "one bone, two bones" pattern, foreshadowing the pattern of tetrapod limbs. It has long been assumed that the symmetrical arrangement seen in the extant lungfishes and coelacanths is the ancestral sarcopterygian condition, and that the alternating pattern in tetrapodomorphs is derived. However, new skeletal material of a primitive Devonian coelacanth, *Shoshonia arctopteryx*, shows an asymmetrical pattern, suggesting that asymmetry may be the ancestral sarcopterygian condition.

Skulls Other ways in which actinopterygians and sarcopterygians differ include their feeding mechanisms and skull anatomy. Although early actinopterygians, such as polypterids, retain a complete dermal skull roof and have no mobility of their upper jaw, more derived actinopterygians have reduced the dermal bone in their skull and obtained greater mobility of the marginal mouth bones of the upper jaw, the maxilla and premaxilla. The sarcopterygians have not reduced the dermal bone of the skull, and if they have jaw mobility, it is derived from movement within the skull itself, between the anterior and posterior portions of the chondrocranium.

Brains Many sarcopterygians (such as ourselves) may develop large brains, but they retain the ancestral form of brain development seen in other vertebrates, where the cerebral hemispheres fold inward on themselves during growth. In contrast actinopterygians have brains that develop by folding the cerebral hemispheres outward.

Scales Although most living bony fishes have reduced, thin scales (elasmoid scales), early forms had thick scales, which differed between the two groups: actinopterygian scales had a thick layer of an enamel-like material (ganoine), while sarcopterygian scales had a thick layer of a dentinelike material (cosmine).

6.2 Evolution of the Actinopterygii

Stem actinopterygians include a variety of taxa that were formerly placed in a group of extinct fishes, the "paleonisciformes," that is no longer considered to be

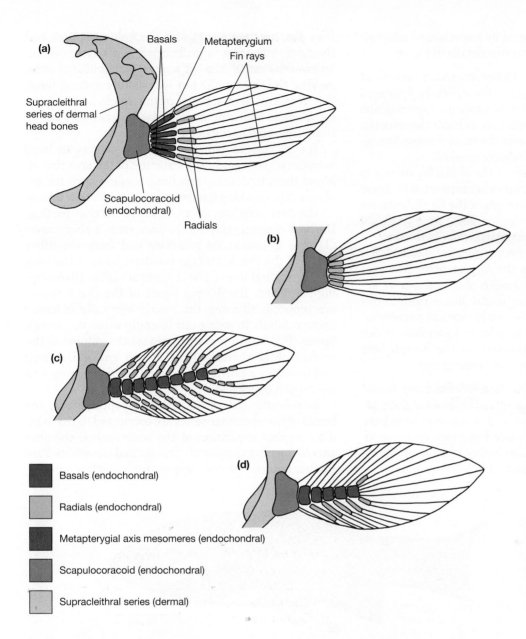

(a)

Basals · Metapterygium · Fin rays

Supracleithral series of dermal head bones

Scapulocoracoid (endochondral)

Radials

(b)

(c)

(d)

Basals (endochondral)

Radials (endochondral)

Metapterygial axis mesomeres (endochondral)

Scapulocoracoid (endochondral)

Supracleithral series (dermal)

Figure 6–4 Fin structure in ostheichthyans (diagrammatic). All drawings show a left pectoral fin with the animal facing left and the preaxial (leading) edge of the fin downward: (a) hypothetical ancestral type of osteichthyan fin, with similarities to those seen today in actinopterygians such as *Polyodon* and *Polypterus*; (b) derived type of actinopterygian fin, typical of teleosts; (c) derived type of sarcopterygian fin, resembling that seen in the living lungfish *Neoceratodus*; (d) ancestral type of sarcopterygian fin, reconstructed from the condition in the Devonian coelacanth *Styloichthyes*: this fin foreshadows the type of fin seen in the tetrapodomorph fishes.

monophyletic. Although fragmentary fossils of Late Silurian actinopterygians exist, complete fossil skeletons are not known before the late Middle Devonian.

Early actinopterygians were small fishes (usually 5 to 25 centimeters long, although some were longer than a meter) with a single dorsal fin and a strongly heterocercal, forked caudal fin with little fin web. The paired fins still retained the ancestral condition of multiple basals (see Figure 6–4a). The interlocking scales were thick, like those of sarcopterygians, and covered in ganoine (derived from enamel). Parallel arrays of closely packed radial bones supported the bases of the fins, and the number of bony rays supporting the fin membrane was greater than the number of supporting radials.

Biology of Early Actinopterygians

The earliest actinopterygians did not possess the specialized fins that give the group its name, and these more generalized types of fins are retained in basal living actinopterygians (polypterids like the bichir, and acipensiforms like the sturgeon). So how can we identify an early member of this lineage? One characteristic feature of early actinopterygians is the presence of a single dorsal fin (see Figure 6–3e)—early sarcopterygians had two dorsal fins. Other actinopterygian features include teeth capped by a specialized mineralized tissue, acrodin, and various details of the skull.

Fins Near the end of the Paleozoic, modifications of the ancestral actinopterygian fin structure appeared in a lineage of ray-finned fishes called neopterygians (= new fins). This is the clade that includes the teleosts, and neopterygians have been the most diverse lineage of bony fishes from the Mesozoic onward.

The upper and lower lobes of the caudal fin are nearly symmetrical, the fin membranes are supported by fewer bony rays, and the basal elements of the fin skeleton are reduced to a small metapterygium. These morphological changes probably increased the flexibility of the fins.

One of these stem neopterygians, *Leedsichthys* of Jurassic England, was the largest bony fish ever known, with a maximum length of 16 meters. It appears to have been a filter feeder like a present-day whale shark. Relicts of these early types of neopterygians remain today as the bowfin and garpikes, which are sometimes grouped together as the paraphyletic "holosteans" (i.e., nonteleost neopterygians).

Scales The dermal scales of late Paleozoic ray-finned fishes were also reduced compared to those of their ancestors. The changes in fins and armor may have been complementary—more mobile fins mean more versatile locomotion, and a greater ability to avoid predators may have permitted a reduction in heavy armor.

Jaws Early actinopterygians had amphistylic jaws, and the lower jaw was snapped closed in a scissors action by contraction of the adductor mandibulae muscle, moving the jaw rapidly but with little crushing force. The close-knit dermal bones of the cheeks left no space for a larger adductor mandibulae muscle; this condition is retained in the extant polypterids.

The success of later neopterygians rests in large measure on changes in the structure of the jaws that allowed them to develop functions ranging from the application of crushing force to a precise pincerlike action.

The first step was more effective suction feeding. Derived neopterygians have jaws with a short maxilla that is free at its posterior end from the other bones of the cheek, and the quadrate bone at the back of the original upper jaw is vertical rather than sloping backward. The dermal bones of the cheek region are reduced, allowing the nearly vertically oriented hyomandibula to swing out laterally when the mouth opens. This action rapidly increases the volume of the **orobranchial chamber** (the mouth and gill region) and produces a powerful suction that draws prey into the mouth (**Figure 6–5**).

In teleosts, the most derived neopterygians, the bones of the opercular series are connected to the mandible so that expansion of the orobranchial chamber aids in opening the mouth. The dermal cheekbones are further reduced, providing space for larger jaw muscles.

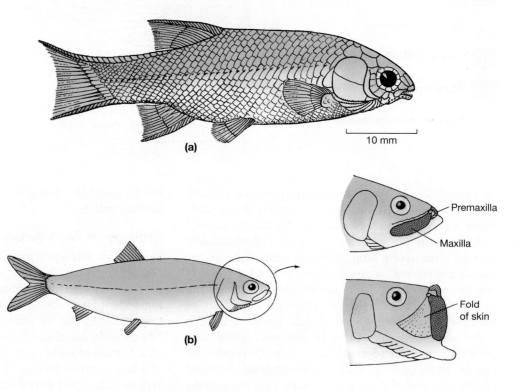

Figure 6–5 Neopterygians. (a) *Acentrophorus* of the Permian illustrates an early member of the late Paleozoic neopterygian radiation. (b) *Leptolepis*, an Early Jurassic teleost with enlarged mobile maxillae that form a nearly circular mouth when the jaws are fully opened. Membranes of skin close the gaps behind the protruded bony elements. Modern herrings have a similar jaw structure.

10 mm

(a)

(b)

Premaxilla

Maxilla

Fold of skin

The anterior articulated end of the maxilla developed a ball-and-socket joint with the neurocranium. Because of its ligamentous connection to the mandible, the free posterior end of the maxilla now rotates forward as the mouth opens. This points the maxilla's marginal teeth forward and helps to grasp prey. The folds of skin covering the maxilla change the shape of the gape from a semicircle to a circular opening. These changes increase the suction produced during opening of the mouth. The result is greater directionality of suction and elimination of a possible side-door escape route for prey.

Specializations of the Teleosts

Fossil teleosts are first known from the Late Triassic. Molecular divergence times suggest a much earlier origin for the group, but conservative evaluation of the molecular data and the fossil record places the origin of teleosts in the Late Permian. By the Jurassic, teleosts were well established in both marine and freshwater habitats, and most of the more than 400 families of modern teleosts had evolved by the start of the Cenozoic.

Teleosts continued and expanded the changes in fins and jaws that contributed to the diversity of more basal neopterygians. Teleosts have radiated into nearly 32,000 species occupying almost every ecological niche imaginable for an aquatic vertebrate—including some kinds of fishes that spend substantial amounts of time on land.

Jaw Protrusion In addition to rapid and forceful suction, many teleosts have great mobility in the skeletal elements that rim the mouth. This mobility allows the grasping margins of the jaws to be extended forward from the head, often at remarkable speed. The functional result, called **protrusible** jaws, has evolved three or four times in different teleost clades, as shown by differences in the details of jaw anatomy. (Sturgeons and some sharks also generate suction with completely different mechanisms of jaw protrusion.)

All teleost jaw-protrusion mechanisms involve complex ligamentous attachments that allow the ascending processes of the premaxilla to slide forward on top of the cranium without dislocation. In addition, since no muscles are in position to pull the premaxillae forward, they must be pushed by leverage from behind. Two sources provide the necessary leverage:

- Opening the lower jaw may protrude the premaxillae through ligamentous ties between the mandible and the posterior tip of the premaxillae (**Figure 6–6**).
- Leverage may be provided by complex movements of the maxillae, which become isolated from the rim

of the mouth by long, posterior projections of the premaxillae that often bear teeth. The independent movement of the protrusible upper jaw also permits the mouth to be closed by extension of the premaxillae while the orobranchial cavity is still expanded, trapping prey in the mouth.

Pharyngeal Jaws Ancestrally, early neopterygian fishes had numerous dermal tooth plates in the pharynx. These plates were aligned with (but not fused to) both dorsal and ventral skeletal elements of the gill arches, and there was a general trend of fusion of these tooth plates to one another and to a few gill arch elements above and below the esophagus. Movements of the gill arches for feeding, as well as for gill ventilation, are possible in bony fishes because the operculum is providing most of the ventilatory power.

These consolidated tooth plates and gill arches were originally not very mobile, a condition retained today in bowfins and garpikes, where the jaws are used primarily to hold and manipulate prey in preparation for swallowing it whole. In teleosts, however, these gill arches attained great mobility and are now termed *pharyngeal jaws*. This second set of jaws has been modified convergently among different teleost lineages.

In the minnows and their relatives the suckers, the primary jaws are toothless but protrusible. The pharyngeal jaws are greatly enlarged and close against a horny pad on the base of the skull. These specializations allow the fishes to extract nutrients from thick-walled plant cells, and the minnows and suckers represent one of the largest radiations of herbivores among vertebrates.

In the most derived teleosts, the muscles associated with the branchial skeletal elements supporting the pharyngeal jaws have undergone radical evolution, resulting in a variety of powerful movements of the pharyngeal jaw tooth plates. Not only are the movements of these second jaws completely unrelated to the movements and functions of the primary jaws, but in a variety of derived teleosts the upper and lower tooth plates of the pharyngeal jaws move quite independently of each other. Some moray eels can extend their pharyngeal jaws out of their throats and into their oral cavity to grasp struggling prey and pull it back into the throat and esophagus. With so many separate systems to work with, it is little wonder that some of the most extensive adaptive radiations among teleosts have been in fishes endowed with protrusible primary jaws and specialized mobile pharyngeal jaws.

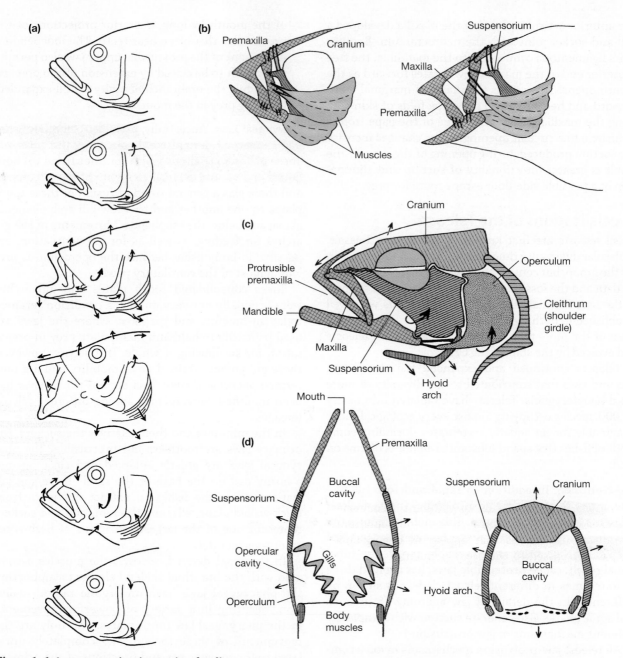

Figure 6–6 Jaw protrusion in suction feeding. (a) *Top to bottom:* Sequence of jaw movements in an African cichlid fish, *Serranochromis*. (b) Muscles, ligaments, and bones involved in movements during premaxillary protrusion. (c) Skeletal movements and ligament actions during jaw protrusion. (d) Frontal section (*left*) and cross section (*right*) of buccal expansion during suction feeding.

Specializations of the Fins The caudal fin of adult actinopterygians is supported by a few enlarged and modified hemal spines, called hypural bones, that articulate with the tip of the abruptly upturned vertebral column. In general, the number of hypural bones decreases during the transition from the earliest actinopterygians to the more derived teleosts. Modified posterior neural arches—the uroneurals—add further support to the dorsal side of the tail. These uroneurals are a derived character of teleosts. Thus supported, the caudal fin of teleosts is symmetrical and flexible. This type of caudal structure is known as homocercal.

In conjunction with a swim bladder that adjusts buoyancy, a homocercal tail allows a teleost to swim horizontally without using its paired fins for lift, as sharks must. In burst and sprint swimming, the tail produces a symmetrical force, but during steady-speed swimming, intrinsic muscles in the tail may produce an asymmetrical action that increases maneuverability without requiring the use of the lateral fins. Relieved of responsibility for controlling lift, the paired fins of teleosts are flexible, mobile, and diverse in shape, size, and position. They have become specialized for activities that include food gathering, courtship, sound production, walking, and gliding over the surface of the water as well as turning and braking during swimming.

6.3 Extant Actinopterygii—Ray-Finned Fishes

With nearly 32,000 extant species the actinopterygians are enormously diverse in body form, ecology, and behavior. A phenomenon called **ecological speciation** appears to be important in the remarkable ability of fishes to form new species. The classic hypothesis for species formation, **geographic (allopatric) speciation**, requires a physical barrier that prevents interbreeding between two or more populations of the parent species. Because these populations are in different places they are exposed to different selective forces, and natural selection leads to different frequencies of alleles in the geographically isolated populations. In addition, mutations that appear in one population are not passed to the other populations because the geographic barrier prevents gene flow. Eventually the populations differ enough so that individuals from one population will not interbreed with individuals of another population if the geographic barrier is removed.

Geographic speciation certainly accounts for some instances of species formation by actinopterygians, but other examples do not appear to fit that model. Populations of fishes may live and breed in bays or inlets, but these habitats are not really isolated from other bays and inlets because they all open to a larger body of water that provides a route for gene exchange.

Ecological speciation is a mechanism that can lead to the formation of new species even when gene flow is possible, because the critical factor is not isolation but rather selective mating between individuals that are especially well adapted to local conditions. This hypothesis proposes that secondary sexual characteristics—ornaments such as colors and morphological structures—identify those individuals and that matings between these individuals produce more surviving offspring than matings with less adapted (and less ornamented) individuals. Thus, sexual selection could produce **assortative mating** (mating with like individuals), increasing the effect of limited geographic isolation and promoting ecological speciation. Both mathematical models and molecular analyses have provided support for this hypothesis.

Polypteriformes and Acipenseriformes—Bichirs, Reedfishes, Sturgeons, and Paddlefishes

Although stem actinopterygians were largely replaced during the early Mesozoic by neopterygians, a few specialized forms of these early ray-finned fishes have survived (**Figure 6–7**). The Polypteriformes (bichirs and reedfishes) and the Acipenseriformes (sturgeons and paddlefishes) have retained many ancestral characters, including basals in the fins, a spiral intestinal valve, and a gas bladder or lung that lies beneath the esophagus rather than above it like the lungs or gas bladders of neopterygians. Although there is some controversy about the systematic position of the polypteriforms, they are now usually regarded as less derived than any of the known fossil actinopterygians. Acipenseriformes are the paraphyletic sister taxon to the neopterygians, but they are more closely related to the neopterygians than the paraphyletic assemblage of Paleozoic "palaeoniscoids" (see Figures 6–1 and 6–2).

Bichirs and Reedfishes Polypteriformes are elongate, heavily armored fishes from Africa. The name Polypteriformes (Greek *poly* = many, *ptery* = fin, and *form* = shape) refers to the peculiar flaglike dorsal finlets. Like the Paleozoic actinopterygians, bichirs and reedfishes retain heavy scales with a layer of ganoine. These scales deform when the fish expires air from its lungs, storing elastic energy that helps to draw in fresh air—a mechanism called recoil aspiration.

Polypteriformes are slow-moving fishes, less than a meter long, with modified heterocercal tails. Bichirs (*Polypterus*) are heavy bodied and feed primarily on other fishes. The reedfish (*Erpetoichthys*), also known as the rope fish or snake fish, is slimmer and more eel-like than the bichir. It moves through thick vegetation with sinuous locomotion and feeds primarily on snails and invertebrates.

Sturgeons and Paddlefishes Acipenseriformes secondarily lack endochondral bone and have lost much of the dermal skeleton of more ancestral actinopterygians;

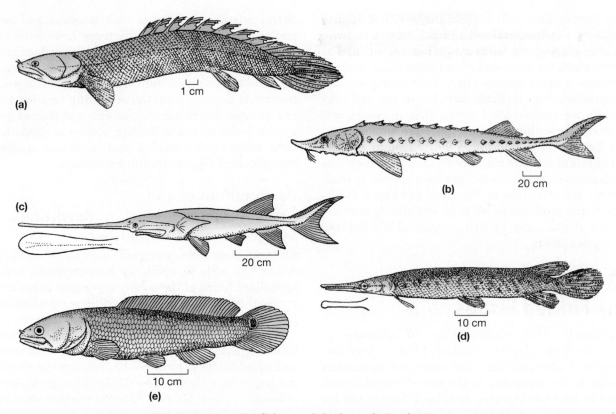

Figure 6–7 Extant nonteleostean actinopterygian fishes and the least derived extant neopterygians. Actinopterygians: (a) *Polypterus*, a bichir; (b) *Acipenser*, a sturgeon; and (c) *Polyodon spathula*, one of two extant species of paddlefishes. The outlines show the shape of the jaws as seen from above. Neopterygians: (d) *Lepisosteus*, a gar; and (e) *Amia calva*, the bowfin. Note the differences in the scale bars.

in fact, they are unique among vertebrates in having dermal head bones that are formed of cartilage. They resemble teleosts in having an upper jaw that is not fused to the cranium and in having transformed their lung into a swim bladder, although these features have evolved independently of the teleost condition.

Sturgeons are large (1 to 6 meters), active, benthic fishes. Sturgeons have a strongly heterocercal tail armored with a specialized series of scales extending from the dorsal margin of the caudal peduncle (the elongated base of the tail where it connects to the trunk) and continuing along the upper edge of the caudal fin. Most sturgeons have five rows of enlarged armorlike scales (called "scutes") along the body that represent the remnants of ganoid scales.

The protrusible jaws of sturgeons make them effective suction feeders, and they suck small crustaceans and insect larvae from the substrate. Sturgeons are found only in the Northern Hemisphere; some live in freshwater, and others are marine forms that ascend into freshwater to breed.

The paddlefishes, Polyodontidae (Greek *poly* = many and *odon* = tooth), are closely related to the sturgeons (although the lineages have been separate since the Late Cretaceous) but have a greater reduction of dermal ossification. Their most outstanding feature is a greatly elongate and flattened rostrum, which extends nearly one third of their 2-meter length. The rostrum is richly innervated with ampullary organs that detect minute electric fields.

The two extant species of paddlefishes have a disjunct zoogeographic distribution, with one species in China and the other in North America. The Chinese paddlefish occurs in the Changjiang (Yangtze) River valley of China. It has a protrusible mouth and feeds on small fishes and crustaceans that it sucks from the river bottom. It may be extinct—the last adult specimens were recorded in 2002, and an intensive search of the upper Yangtze River from 2006 to 2008 failed to find any individuals.

The geographic range of the North American paddlefish includes the length and breadth of the Mississippi River drainage of the United States, from western

New York to central Montana and from Canada to Louisiana. Unlike the bottom-feeding Chinese paddlefish, the American paddlefish is a filter feeder that captures copepods (small crustaceans) in its gaping mouth as it swims. The enormous geographic range of the species and the tendency of adult paddlefish to travel long distances and to establish local subpopulations provide a buffer against extinction, but because of the loss of spawning habitats (clear-flowing stretches of large rivers with gravel substrates) this species is classified as Vulnerable on the IUCN Red List.

Gars and the Bowfin

Gars (Lepisosteiformes) and the bowfin (Amiiformes) retain many ancestral neopterygian characters. Both retain heavy scales and have a dorsally placed gas bladder that is well vascularized and serves as a respiratory organ as well as a hydrostatic device. They are widely divergent types of fishes, and their relationships have long been controversial, although most researchers consider the bowfin to be more closely related to teleosts than are the gars.

Gars Also called garpikes, gars grow to lengths of 1 to 4 meters and live in warm, temperate fresh and brackish (estuarine) waters in North and Central America and Cuba. Gars feed on other fishes taken unaware when the seemingly lethargic and excellently camouflaged gar eases alongside them and, with a sideways flip of the body, grasps prey with its needlelike teeth. Their interlocking multilayered ganoid scales are similar to those of many Paleozoic and Mesozoic actinopterygians, and alligators are the only natural predators able to cope with the thick armor of an adult gar.

Bowfin The bowfin lives in the same geographic areas and habitats as gars, but it is a very different kind of predator. Bowfins, which are 0.5 to 1 meter long, eat almost any organism smaller than themselves, using suction to draw prey into the mouth. The scales of the bowfin are comparatively thin and made up of a single layer of bone as in teleost fishes; however, the asymmetrical caudal fin is very similar to the heterocercal caudal fin of many Paleozoic and Mesozoic actinopterygians.

Teleosteans

Most extant fishes are teleosts. They share many characters of caudal and cranial structure and are grouped into four clades of varying size and diversity: Osteoglossomorpha, Elopomorpha, Ostarioclupeomorpha, and Euteleostei. Teleosts differ from all other vertebrates in having undergone a further round of duplication of the *Hox* genes, now having seven sets of *Hox* genes rather than four sets like other gnathostomes.

Although teleosts are generally thought to have a swim bladder rather than a lung, many less-derived teleosts use the swim bladder for gas exchange as well as for buoyancy control. A posterior location of the pelvic fins is another ancestral feature retained by some teleosts—in many derived forms the pelvic fins have moved to an anterior position where they are attached to the base of the pectoral girdle, and some teleosts have lost the pelvic fins altogether.

Osteoglossomorpha The extant osteoglossomorphs (Greek *osteo* = bone, *gloss* = tongue, and *morph* = form) live mostly in tropical freshwater. *Osteoglossum* (**Figure 6–8**) is a 1-meter-long predator from the Amazon, familiar to tropical fish enthusiasts as the arawana. *Arapaima* is an even larger Amazonian predator, perhaps the largest strictly freshwater fish. Before intense fishing reduced the populations, they were known to reach a length of at least 3 meters and perhaps as much as 4.5 meters. *Mormyrus*, one of the African elephant fishes, is representative of the small African bottom feeders that use weak electric discharges to communicate with other members of their species. As dissimilar as they seem, the osteoglossomorph fishes are united by unique bony characters of the mouth (including teeth on the lower portions of the hyoid arch that resembles a bony tongue and gives the group its name) and by the mechanics of their jaws.

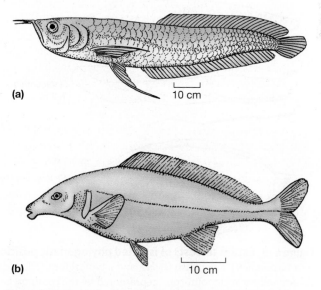

(a)

10 cm

(b)

10 cm

Figure 6–8 Extant osteoglossomorphs. (a) *Osteoglossum*, the arawana, from South America. (b) *Mormyrus*, an elephant-nose from Africa.

Elopomorpha The Elopomorpha (Greek *elop* = a kind of fish) had appeared by the Late Jurassic period. A specialized leptocephalus (Greek *lepto* = small and *cephal* = head) larva is a unique character of elopomorphs (**Figure 6–9**). These larvae spend a long time adrift, usually at the ocean surface, and are widely dispersed by currents.

Most elopomorphs are eel-like and marine, but some species live in freshwater. The common American eel, *Anguilla rostrata*, has one of the most unusual life histories of any fishes. After growing to sexual maturity (which takes as long as 25 years) in rivers, lakes, and ponds, the **catadromous** (migrating from freshwater to seawater to breed) eels enter the sea.

American eels migrate to the Sargasso Sea, an area in the central North Atlantic between the Azores and the West Indies. Here they are thought to spawn and die, presumably at great depth. The eggs and newly hatched leptocephalus larvae float to the surface and drift in the currents. Larval life continues until the larvae reach continental margins, where they transform into miniature eels and ascend rivers to feed and mature.

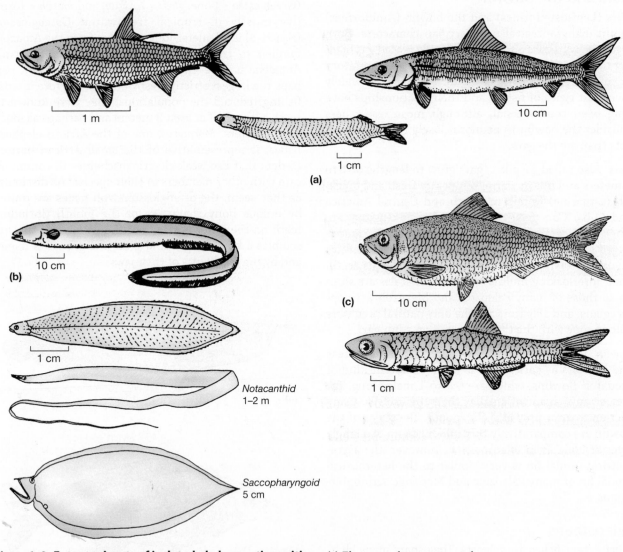

Figure 6–9 Extant teleosts of isolated phylogenetic position. (a) Elopomorpha, represented by a tarpon (*left*), a bonefish (*right*), and a typical fork-tailed leptocephalus larva (*below*). (b) Anguilliform elopomorphs, represented by the common eel, *Anguilla rostrata* (*above*), its leptocephalus larva (*immediately below*), and two other very different eel leptocephali. (c) Clupeomorpha, represented by a herring (*above*) and an anchovy (*below*). Note the differences in the scale bars.

European eels (*Anguilla anguilla*) spawn in the same region of the Atlantic as their American kin but may choose a somewhat more northeasterly part of the Sargasso Sea and perhaps a different—but also deep—depth at which to spawn. Their leptocephalus larvae remain in the clockwise currents of the North Atlantic (principally the Gulf Stream) and ride to northern Europe as well as North Africa, the Mediterranean, and even the Black Sea before entering river mouths to migrate upstream and mature.

In contrast to the eels, the two other major lineages of elopomorphs—bonefishes and tarpons—are prized by sport fishers. Both are fast-swimming predatory fishes that inhabit shallow coastal waters in the tropics and subtropics. Bonefishes reach maximum lengths of about 40 centimeters, but female tarpons (which are larger than the males) can be as long as 2.5 meters.

Ostarioclupeomorpha This group includes the herring-like fishes (the cluepeiforms) and the minnows (ostariophysans), two of the most abundant, widespread, and commercially important groups of fishes.

Most clupeiforms are specialized for feeding on minute plankton gathered by a specialized mouth and gill-straining apparatus. They are silvery, mostly marine schooling fishes of great commercial importance. Common examples are herrings, shad, sardines, and anchovies (see Figure 6–9). Several species are **anadromous** (i.e., migrating from seawater to freshwater to spawn), and the springtime migrations of American shad (*Alosa sapidissima*) from the North Atlantic into rivers in eastern North America once involved millions of individuals. The enormous shad runs of the recent past have been greatly depleted by dams and pollution of aquatic environments.

Ostariophysans are the predominant fishes of the world's freshwaters, representing about 80 percent of the fish species in freshwater and 25 to 30 percent of all extant fishes. Many ostariophysans have protrusible jaws that are specialized to capture prey and pharyngeal teeth in the throat that process a range of food items. Many species have fin spines or armor for protection, and the skin typically contains glands that produce substances used in olfactory communication.

Ostariophysans have two distinctive derived characters. Their name refers to small bones that connect the swim bladder with the inner ear (**Figure 6–10**). Using the swim bladder as an amplifier and the chain of bones as conductors, this **Weberian apparatus** greatly enhances the hearing sensitivity of these fishes. Ostariophysans are more sensitive to sounds and have a broader frequency range of detection than other fishes.

The second derived character uniting the ostariophysans is the presence of a fright or alarm substance in the skin. Chemical signals (pheromones) are released into the water when the skin is damaged, and they produce a fright reaction in nearby members of their own species and other ostariophysan species. The fright reaction may cause the fishes to rush for cover or form a tighter school.

Although all ostariophysans have an alarm substance in the skin and a Weberian apparatus (or a rudimentary precursor of it), they are a diverse group in other respects, with some 8000 species. Ostariophysans include the characins (piranhas, neon tetras, and other familiar aquarium fishes) of tropical America and Africa, the carps and minnows (nearly worldwide in freshwater), the catfishes (nearly worldwide in freshwater and many shallow marine areas), and the highly derived electric knifefishes of Central and South America.

Euteleostei The vast majority of extant teleosts belong to the fourth clade, the Euteleostei (Greek *eu* = true), which evolved before the Late Cretaceous. With so many thousands of species, it is impossible to give more than scant information about them.

We will consider the stem euteleostian stock to be represented today by the generalized salmoniforms, although this interpretation is a matter of dispute. The esocid and salmonid fishes (**Figure 6–11a** on page 137) include important commercial and game fishes.

The esocids (pickerels, pikes, muskellunges, and their relatives) are at the base of the radiation of more derived teleosts. They are ambush predators that eat other fishes, relying on cryptic colors and patterns to conceal them until a prey fish has approached close enough to be captured by a sudden lunge.

The salmonids include the anadromous salmon, which usually spend their adult lives at sea and make epic journeys to inland waters to breed, as well as the closely related trouts, which usually live entirely in freshwater.

The Acanthopterygii (true spiny-rayed fishes) that dominate the open ocean surface and shallow marine waters of the world appear to form a monophyletic lineage within the euteleosts. Among the acanthopterygians, the atherinomorphs have protrusible jaws and specializations of form and behavior that suit them to shallow marine and freshwater habitats. This group includes the silversides, grunions, flying fishes, and halfbeaks as well as the egg-laying and live-bearing cyprinodonts (Figure 6–11b on page 137). Killifishes are examples of oviparous cyprinodonts, and viviparous forms include the guppies, mollies, and swordtails commonly kept in home aquaria.

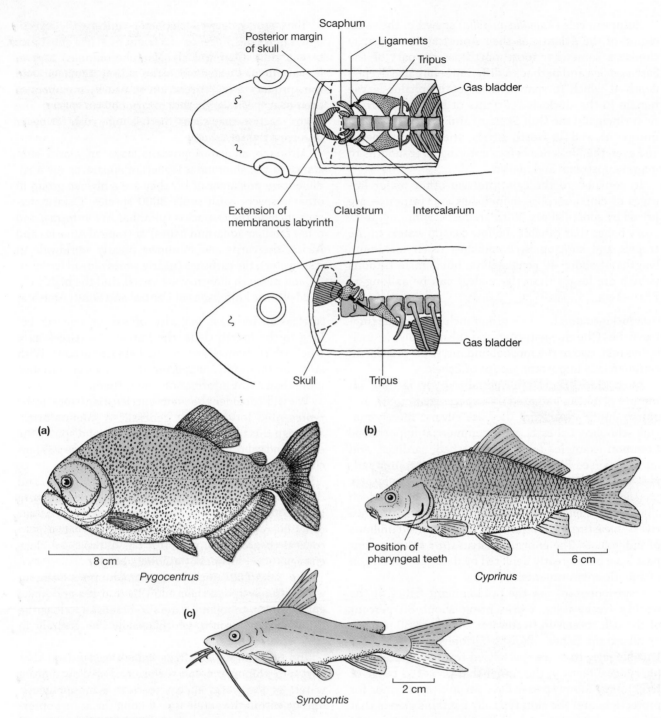

Figure 6–10 The Weberian apparatus. Ostariophysan fishes have a sound-detection system, the Weberian apparatus, which is a modification of the swim bladder and the first few vertebrae and their processes. Typical ostariophysans include (a) characins (here a piranha), (b) minnows, and (c) catfishes.

Most species of acanthopterygians—and the largest order of fishes—are Perciformes, with about 10,000 extant species. Snooks, sea basses, sunfishes, perches, darters, dolphinfish (mahi mahi), snappers, grunts, porgies, drums, cichlids, barracudas, tunas, billfishes, and most of the fishes found on coral reefs are some of the well-known Perciformes.

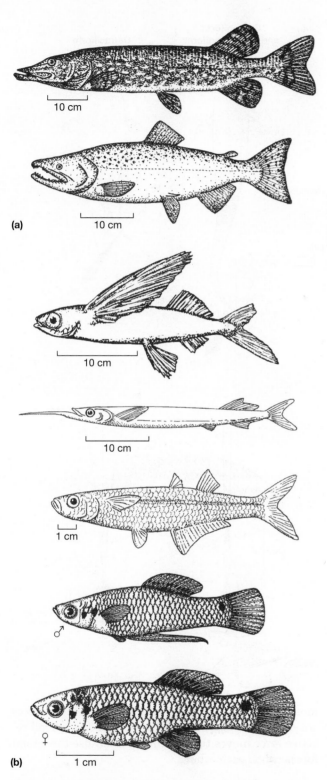

6.4 **Locomotion in Water**

Perhaps the single most recognizable characteristic of the enormous diversity of fishes is their mode of locomotion. Fish swimming is immediately recognizable, aesthetically pleasing, and—when first considered—rather mysterious, at least when compared to the locomotion of most land animals. Fishes swim with anterior-to-posterior sequential contractions of the muscle segments along one side of the body and simultaneous relaxation of the corresponding muscle segments on the opposite side. Thus, a portion of the body momentarily bends, the bend moves posteriorly, and a fish oscillates from side to side as it swims. These lateral undulations are most visible in elongate fishes, such as lampreys and eels (**Figure 6–12**). Most of the power for swimming comes from muscles in the posterior region of the fish.

A variety of swimming movements have been described (**Figure 6–13** on page 139):

- Anguilliform—This movement is typical of highly flexible fishes capable of bending into more than half a sinusoidal wavelength. It is named for the locomotion seen in the true eels, the Anguilliformes.
- Carangiform—Undulations are limited mostly to the caudal region, with the body bending into less than half a wavelength. The movement is named for *Caranx*, a genus of fast-swimming marine fishes known as jacks.
- Ostraciiform—The body is inflexible, and undulation is limited to the caudal fin. The movement is named for the boxfishes, trunkfishes, and cowfishes (family Ostraciidae) whose fused scales form a rigid box around the body, preventing undulations.
- Labriform—Most of the force used for locomotion is provided by the pectoral fins, and the caudal fin plays a relatively small role. This movement is characteristic of wrasses and other fishes that maneuver through complex three-dimensional habitats, such as coral reefs.
- Rajiform, amiiform, gymnotiform, and balistiform swimming—Sine waves are passed along the elongated pectoral fins (rajiform), the dorsal fin (amiiform), the anal fin (gymnotiform), or both the dorsal and anal fins (balistiform). Usually, several complete waves are observed along the fin, and very fine adjustments can be made in the direction of motion, including swimming backward. Because only the fins are moving, these types of locomotion can be stealthy and maneuvering can be precise.

Although these forms of locomotion were named for the groups of fishes that exemplify them, many other types of fishes use these swimming modes. For

Figure 6–11 Euteleosts. (a) Some of the least derived euteleosts: the pike (*above*) and the salmon (*below*). (b) Atherinomorph fishes represented by (*top to bottom*) a flying fish, a halfbeak, an Atlantic silverside and a live-bearing killifish, the male of which has a modified anal fin that is used for internal fertilization.

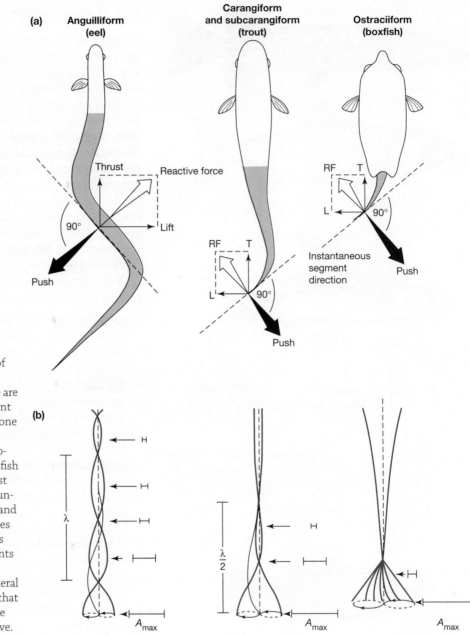

Figure 6–12 Basic movements of swimming fishes. Outlines of some major swimming types. (a) The body regions that undulate are shaded. The lift (lateral) component of the reactive force produced by one undulation's push on the water is canceled by that of the next, oppositely directed undulation, so the fish swims in a straight line. The thrust (forward) component from each undulation is in the same direction and thus is additive, and the fish moves through the water. (b) Waveforms are created by undulations of points along the body and tail (*bottom*). A_{max} represents the maximum lateral displacement of any point. Note that A_{max} increases posteriorly; λ is the wavelength of the undulatory wave.

example, hagfishes, lampreys, most sharks, sturgeons, arawanas, many catfishes, and countless elongate spiny-rayed fishes are not true eels, yet they use an anguilliform swimming mode.

Generating Thrust

In general, fishes swim forward by pushing backward on the water. For every active force, there is an opposite reactive force (Newton's third law of motion). Undulations of the body or fins produce an active force

directed backward and also lateral forces. (These forces are the vectors shown in Figure 6-12.) The overall reactive force moves the fish forward because the lateral forces cancel each other.

Minimizing Drag

A swimming fish experiences two forms of drag: viscous or frictional drag, which is caused by friction between the fish's body and the water, and inertial drag, which is caused by pressure differences created by the

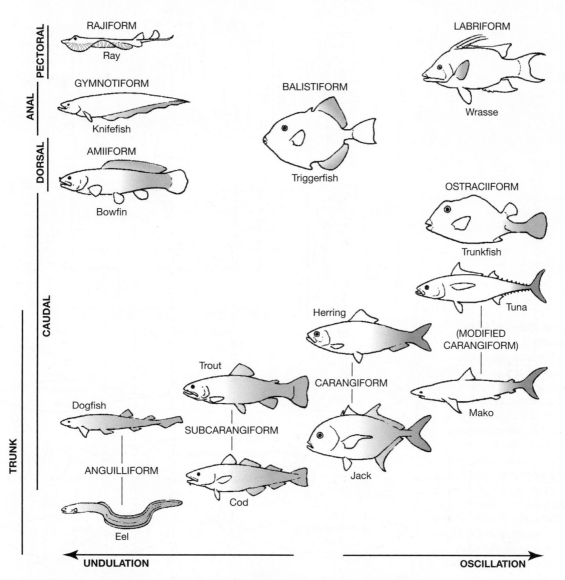

Figure 6–13 Location of swimming movements in various fishes. Shaded areas of the body undulate or oscillate from side to side. Names such as carangiform describe the major types of locomotion found in fishes; they do not imply phylogenetic relationships, nor are they comprehensive.

fish's displacement of water. The two types of drag respond differently to speed and body form.

- Viscous drag is relatively constant over a range of speeds and is sensitive to the surface area of the body and the smoothness of the surface. A thin body has high viscous drag because it has a large surface area relative to its muscle mass; a scaleless skin (like an eel's) has low viscous drag because it is smooth.
- Inertial drag increases as speed increases, and it is sensitive to the shape and cross-sectional area of the body. A thick body induces high inertial drag because it displaces a large volume of water as it

moves forward. Streamlined (teardrop) shapes produce minimum inertial drag when their maximum width is about one-fourth their length and is situated about one-third body length from the leading tip (**Figure 6–14**).

The fastest swimmers, such as mackerels and tunas, use modified carangiform swimming, undulating only the caudal peduncle and caudal fin. These fishes have a caudal peduncle that is flattened from top to bottom and edged by sharp scales, so it creates minimum inertial drag as it sweeps from side to side.

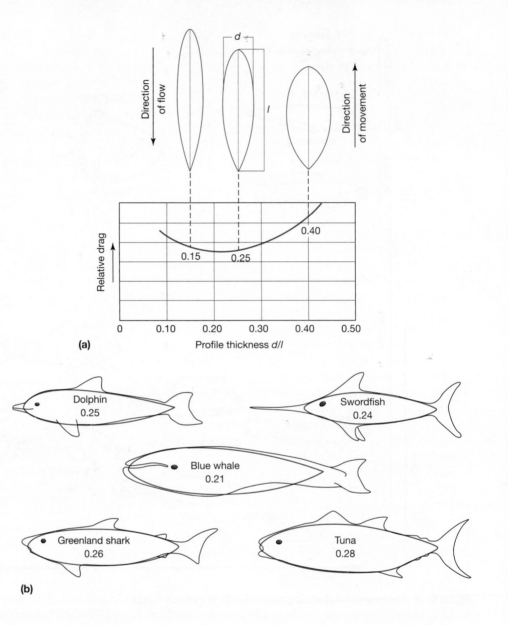

Figure 6–14 Effect of body shape on drag. (a) Streamlined profiles with maximum width (*d*) equal to approximately one-fourth length (*l*) minimize drag. The examples are for solid, smooth test objects with the thickest section about two fifths of the distance from the tip. (b) Width-to-length ratios (*d/l*) for several swimming vertebrates. Like the test objects, these vertebrates tend to be circular in cross section. Note that the ratio is near 0.25, and the general body shape is approximately fusiform (as shown by the colored outlines).

The caudal fin creates turbulent vortexes (whirl-pools of swirling water) in a fish's wake that are part of the inertial drag. The total drag created by the caudal fin depends on its shape. When the aspect ratio of the fin (the dorsal-to-ventral length divided by the ante-rior-to-posterior width) is high, the amount of thrust produced relative to drag is high. The stiff sickle-shaped fin of mackerels, tunas, swordfishes, and cer-tain sharks results in a high aspect ratio and efficient forward motion. Even the cross section of the forks of these caudal fins assumes a streamlined teardrop shape, further reducing drag. Many species with these specializations swim continuously.

The caudal fins of trouts, minnows, and perches are not stiff and seldom have high aspect ratios. These fishes bend the body more than carangiform swimmers, in a swimming mode called subcarangiform. Subcaran-giform swimmers spread or compress the caudal fin to modify thrust and stiffen or relax portions of the fin to produce vertical movements of the posterior part of the body. The caudal peduncle of these subcarangiform swimmers is compressed from side to side and deep from top to bottom, and it contributes a substantial part of the total force of propulsion. These fishes often swim in bursts, usually accelerating rapidly from a standstill.

Stopping, Steering, and Staying in Place

Fishes use all of their fins for maneuvering through their three-dimensional world; observing fishes in an aquarium will illustrate the behaviors described here.

Stopping A fish uses its fins for braking. The pectoral and pelvic fins rotate outward, and the posterior portions of the dorsal, anal, and caudal fins are flexed sideward to create drag.

Steering A fish turns by extending the fins on one side of the body to create drag, just as a rower allows an oar to drag in the water to turn a boat. Any combination of fins can be used to turn—extending just a pectoral fin produces a gradual turn, and extending additional fins increases the drag and sharpens the turn.

Staying in Place Watching a fish hovering in the water is fascinating. Teleosts use the swim bladder to make themselves neutrally buoyant, so you would think that a fish could float in the water just as you can, with no movement of its appendages. A moment's observation will disprove that hypothesis, however—a motionless fish is actually moving several fins continuously.

The pectoral fins maintain a sculling motion, and careful observation will show that the power stroke is directed forward. Looking a bit farther forward, you will see that the opercula are opening and closing as the fish breathes. Each time a fish exhales, water jets backward from beneath the opercula, creating a thrust that would drive the fish forward. The backward sculling movements of the pectoral fins counteract the forward thrust of water leaving the gill chambers.

Acute observation will reveal that a sine wave travels down the posterior margin of the caudal fin and sometimes the margins of the dorsal and anal fins as well. A downward sine wave creates an upward thrust, which should rotate the posterior end of the fish upward, yet the fish does not somersault—it remains horizontal in the water. The position of the swim bladder explains this observation. The volume of gas in the swim bladder can be adjusted to make the fish neutrally buoyant, but the swim bladder is located anterior to the center of gravity of the fish. As a result a fish is tail-heavy and it would float tail-down without the upward thrust produced by the sine wave in its caudal fin.

6.5 Actinopterygian Reproduction

Actinopterygians show more diversity in their reproductive biology than is found in any other vertebrate taxon.

- Most species lay eggs, which may be buried in the substrate, deposited in nests, or released to float in the water, and a few species lay eggs out of water, but viviparity has evolved in several lineages and nutrition of the embryos ranges from lecithotrophy to matrotrophy.
- Parental care ranges from none at all, through attending eggs in nests and protecting the young after they hatch, to carrying eggs in the mouth during development and allowing the young to flee back into the parent's mouth when danger threatens.
- Even sex determination shows diversity—the sex of most species of teleosts is genetically determined and fixed for life, but some species change sex partway through their lives, others are hermaphroditic, and a few species consist entirely of females.

The diversity seen in reproduction by fishes means that there are exceptions to every generalization. The following discussion focuses on the typical characteristics of the groups of fishes described.

Egg Layers

The vast majority of ray-finned fishes are oviparous, and freshwater and marine species show contrasting specializations of oviparous reproduction.

Freshwater Habitats Teleosts that live in freshwater generally produce and care for a relatively small number of large, yolk-rich **demersal** eggs (i.e., eggs that are buried in gravel, placed in a nest, or attached to the surface of a rock or plant). Attachment is important in flowing water that might carry a floating egg away from habitats suitable for development. Because the eggs remain in one place, parental care is possible; males often guard the nests and sometimes the young.

The eggs of freshwater teleosts hatch into young that often have large yolk sacs containing a reserve of yolk that supports their growth for some time after hatching. When the yolk reserve has been consumed, the young fishes (called fry) generally resemble their parents.

Marine Habitats Teleosts that live in the sea generally release large numbers of small, buoyant, transparent eggs into the water. These eggs are left to develop and hatch while drifting in the open sea. The larvae are also small and usually have little yolk reserve remaining after they hatch. They begin feeding on microplankton soon after hatching.

Marine larvae are generally very different in appearance from their parents, and many larvae have been described for which the adult forms are unknown. Such larvae are often specialized for life in the oceanic plankton, feeding and growing while adrift at sea for weeks or months, depending on the species. The larvae eventually settle into the juvenile or adult habitats appropriate for their species. It is not yet generally understood whether arrival at the appropriate adult habitat (deep-sea floor, coral reef, or river mouth) is an active or passive process on the part of larvae, but increasing evidence suggests that many larvae actively choose their settling habitats.

The strategy of producing planktonic eggs and larvae that are exposed to a prolonged and risky pelagic existence appears to be wasteful of gametes. Nevertheless, complex life cycles of this sort are the principal mode of reproduction of marine fishes. Pelagic spawning may offer the following advantages:

- Predators on eggs may be abundant in the parental habitat but scarce in the pelagic realm.
- Microplankton (bacteria, algae, protozoans, rotifers, and minute crustaceans) are abundant where sufficient nutrients reach sunlit waters. If energy is limiting to the parent fishes, it could be advantageous to invest the minimum possible amount of energy by producing eggs that hatch into specialized larvae that feed on pelagic food items too small for the adults to eat.
- Producing floating, current-borne eggs and larvae increases the chances of colonizing all patches of appropriate adult habitat in a large area. A widely dispersed species is not vulnerable to local environmental changes that could extinguish a species with a restricted geographic distribution. Perhaps the predominance of pelagic spawning species in the marine environment reflects millions of years of extinctions of species with reproductive behaviors that did not disperse their young as effectively.

Terrestrial Habitats If you have lived near the shore of California you have probably heard of the grunion runs. California grunions (*Leuresthes*) are small marine fishes that lay their eggs at the top of the tide mark on beaches from March through August. (A different species of grunion is found in the Gulf of California.) Tides reach their maximum for about two nights on each side of the full moon and the new moon, and this is when grunions spawn, riding a wave ashore and squirming tail-first into the wet sand as the wave recedes. A male wraps around a female and releases sperm as she releases eggs, and then both fishes ride a receding wave back into the sea.

The eggs require about 10 days to hatch, and during this period of lower tides the nests are above the reach of waves. At the next high tide cycle, about 2 weeks after the eggs were laid, they have hatched and the young grunions are swept back into the sea.

Even more unfishlike are the reproductive behaviors of at least one species of mudskipper (*Periophthalmodon schlosseri*) and the rockhopper blenny (*Andamia tetradactyla*). Mudskippers are marine fishes of the Indo-Pacific region that live on mudflats, where they construct burrows. During high tide the mudskippers hide in their flooded burrows, and they emerge at low tide to forage on the mud. The burrows contain an upward bend that creates a chamber that is filled with air, even when the burrow entrance is covered by the tide, and the adhesive eggs are deposited on the roof of this chamber, well above the water and exposed only to air. Other species of mudskippers also deposit eggs in their burrows, and it is likely that some of those eggs also develop in air.

Rockhopper blennies forage during low tide on wave-splashed mats of algae on rocky coastlines and shelter during high tide in crevices in the rocks. Females deposit eggs in the crevices and males attend the nests, which are entirely out of the water except during the height of the tide or when they are splashed by waves during storms.

Live Bearers

The widespread occurrence of viviparity among familiar aquarium fishes in the family Poeciliidae (guppies, mollies, platys, swordtails, and related species) gives the impression that it is a common mode of reproduction for teleosts. In fact, that is not true—although viviparity is believed to have evolved in at least 12 lineages of teleosts, only 3 percent of teleosts are viviparous and more than a third of the viviparous teleosts are poeciliids. This scarcity of viviparous lineages of teleosts is a bit surprising considering the diversity that teleosts display in other aspects of their reproductive biology.

The family Syngnathidae (seahorses, sea dragons, and pipefishes) provides an especially interesting example of viviparity—male gestation. Males of all the species in this family carry the fertilized eggs until they hatch, but the details vary: In some species the eggs are carried externally (adhering to the male's abdomen or his tail), in other species the eggs are partially enclosed in a pouch that remains open to the external environment, and in the most derived species the eggs are held within a closed pouch on the male's abdomen.

The method by which eggs in a closed pouch are fertilized has been described for the yellow seahorse (**Figure 6–15**). During mating the male and female are in close contact for only a few seconds. Initially the male and female swim side by side, then turn to face each other, pressing their abdomens together. During this stage, which lasts for only about 6 seconds, the eggs are apparently released and pass across the opening of

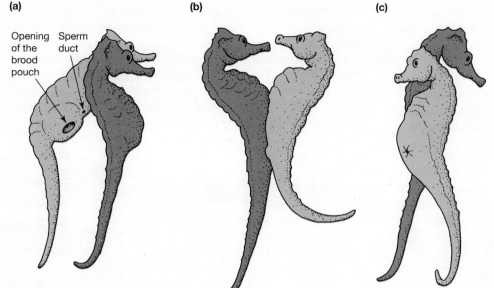

Figure 6–15 Mating behavior of the yellow seahorse. (a) The male and female (the darker individual in these drawings) swim side by side, posturing and changing color. During this phase the male may open its brood pouch and inflate it with seawater. The male's external sperm duct lies about 4.5 mm anterior to the opening of the brood pouch. (b) Egg transfer and fertilization occur during a 6-second period when the female has turned to press her abdomen against the male, aligning the opening of her oviducts with the opening of the brood pouch. (c) The brood pouch closes immediately after the eggs have been transferred.

the sperm duct while sperm are released, and then the eggs and sperm are carried through the opening of the brood pouch, which closes tightly.

The Sex Lives of Teleosts

As mammals, humans are accustomed to thinking that the sex of an individual is determined genetically at the moment of conception—female mammals are XX and males are XY. (If we were birds, we would be equally comfortable with the assumption that females are always ZW and males are always ZZ.)

In fishes, however, the environment plays a role in sex determination, and some species begin life as one sex and change to the other after they are adults. The principle that underlies these differences in life history is Darwinian fitness—that is, what allocation of time and energy to reproduction yields the highest representation of the alleles of one individual in succeeding generations relative to the alleles of other individuals in the population?

Protandry In some species of teleosts, an individual starts life as a male and changes to a female (Greek *prot* = first and *andro* = male). Anemone fishes (about 30 species in the genus *Amphiprion*) are protandrous (**Figure 6–16**).

Many people are familiar with anemone fish from the movie *Finding Nemo*. Nemo and his family are clown anemone fish, *Amphiprion ocellaris*. This species lives in pairs or small groups that defend a territory within the tentacles of a sea anemone. The largest individual is always a female and the second largest is a male—both of these individuals are reproductively mature. Smaller individuals, if they are present, remain immature. If the female dies, the male changes sex and the largest immature individual in the group becomes a male.

The size-advantage hypothesis proposes that the benefit of protandry lies in the relationship between body size and the number of eggs an individual produces—large individuals produce more eggs than small ones, but even a small male can produce all the sperm needed to fertilize the eggs of a large female. Because of this relationship between body size and sperm and egg production, a clownfish can maximize its lifetime reproductive success by starting life as a male and fertilizing the eggs of a large female, and then changing to a female when it has grown to a large size.

Protogyny Protogyny is the reverse of protandry (Greek *gyno* = female). In this case juveniles mature initially as females and subsequently change sex to

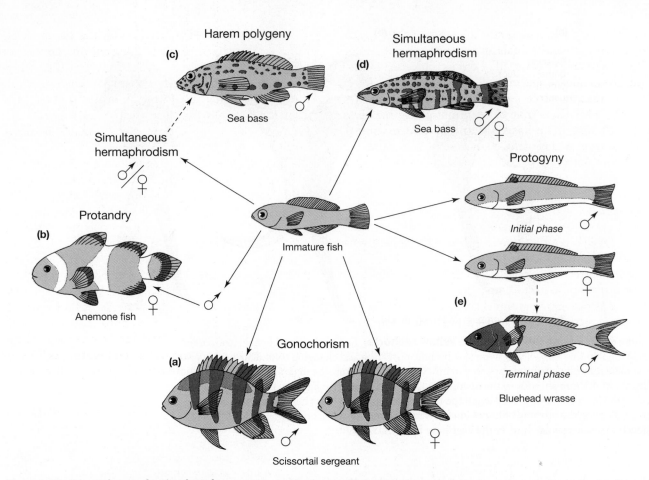

Figure 6–16 Sex and reproduction by teleosts. Teleosts display a variety of sex-determining mechanisms and life-history patterns. (a) In gonochorism sex is determined by genes at the time of conception and does not change during the lifetime of an individual. About 88 percent of teleosts follow this pattern, illustrated here by the scissortail sergeant major, *Abudefduf sexfasciatus*. (b) Protandrous fishes, illustrated by the anemone fish (*Amphiprion ocellaris*), begin life as males and change to females. (c) Some species of sea basses, including the lantern bass (*Serranus baldwini*), start life as simultaneous hermaphrodites and change to males when they are large enough to defend a harem of smaller hermaphrodites. (d) Fishes that are simultaneous hermaphrodites. such as the harlequin bass (*Serranus tigrinus*), alternate the male and female roles during spawning. (e) Some protogynous fishes, such as the bluehead wrasse (*Thalossoma bifasciatum*), begin life as either males or females and grow into an initial phase that is identified by a pattern that is the same for both sexes. Males and females can reproduce in the initial phase, but the harem-based social system of this species greatly limits the reproductive success of initial-phase males. Terminal-phase males defend territories containing harems of females and enjoy far greater reproductive success. When the terminal-phase male in a territory dies, the largest female changes sex, taking the male role in spawning within hours.

become males. Protogyny is characteristic of species such as the bluehead wrasse (*Thalassoma bifasciatum*), in which territorial males defend mating territories. In this situation, large body size is advantageous to a male. Both male and female bluehead wrasses start adult life in a mostly yellow color pattern called the initial phase. Some initial-phase individuals (both males and females) subsequently change into terminal-phase

males, which have blue heads, a black saddle, and a green posterior region.

In small populations of bluehead wrasse, spawning takes place over coral heads that protrude above the general height of the reef, and a terminal-phase male bluehead wrasse defends a territory centered on a coral head, mating with females that come to the coral head to spawn. Initial-phase males also breed, but instead of defending a breeding site they swim in groups of a dozen or more individuals and try to intercept females. When a female spawns, all of the initial-phase males in the group release sperm.

The mating success of initial-phase males is very much lower than that of terminal-phase males. An initial-phase male has only one or two mating opportunities a day, and even then his sperm must compete with sperm from all the other males in his group. In contrast, a terminal-phase male mates 40 to 100 times a day without sperm competition from other males.

Clearly the potential reproductive success of a terminal-phase male is greater than that of even the largest female because the male mates so often. In this situation it is advantageous for a large individual to be a male, and when a terminal-phase male disappears from its territory the largest female in the group changes to a male. The transition is astonishingly rapid—a field study found that the largest female began to display male behavior within minutes of the removal of the terminal-phase male from its territory, and individuals that had spawned as females one day mated as males on the following day.

Hermaphroditism Simultaneous hermaphroditism means that an individual has functional ovaries and functional testes at the same time and can mate either as a male or as a female. This life-history pattern is uncommon, but it does occur in about 20 families of teleosts, including the sea basses (*Hypoplectrus*), which have three mating patterns:

- *Egg traders* (or *parcelers*) alternate male and female roles during a single mating session. The member of the pair that is acting as a female releases a portion of its eggs (the parcel that gives this mating pattern its name), and the individual that is acting as the male releases sperm. Then the two individuals switch roles and continue alternating through the remainder of the spawning session, which may last for more than an hour.
- *Reciprocating monogamous pairs* alternate roles between spawning sessions; that is, the individual that releases eggs in one spawning session releases sperm in the next spawning session, which may occur some days later.
- *Harem polygynous species* combine an initial period of simultaneous hermaphroditism with sex change later in life. They live in harems consisting of as many as seven small hermaphrodites and a single large individual. These large fishes have lost their ovarian tissue, becoming functional males and mating with the hermaphrodites, which act as females.

Although the hermaphrodites in the harem could mate with each other, they almost never do so. The largest hermaphrodites have more testicular tissue than small ones, but it is not known whether these individuals, called "transitional hermaphrodites," will leave the harem to establish their own harems in new territories, or if they remain in the transitional phase until the male in their harem dies.

Only one species of self-fertilizing hermaphrodite has been documented among fishes, *Kryptolebias* (formerly *Rivulus*) *marmoratus*. This small fish (7.5 centimeters) lives in shallow brackish waters from Florida through the Caribbean. About 95 percent of the individuals are born as self-fertilizing hermaphrodites, which produce homozygous clones of the parent when they reproduce, and the remaining 5 percent are males—females have never been described in this species. In areas with temperatures below 20°C, about 60 percent of the hermaphrodites have transformed to secondary males after 3 or 4 years, but in warmer areas where the water temperature is 25°C or higher, the fishes remain hermaphrodites throughout their lives.

Gynogenesis and Hybridogenesis Some species of teleosts consist entirely of females. Although this phenomenon is rare among fishes, four families of freshwater fishes include species that are gynogenetic or hybridogenetic. Both these forms of reproduction begin with interspecific hybridization of two bisexual species, giving rise to a diploid unisexual (all-female) species.

Gynogenesis The Amazon molly (*Poecilia formosa*), a unisexual hybrid of the bisexual sailfin molly (*Poecilia latipinna*) and the bisexual Mexican molly (*Poecilia mexicana*), is gynogenetic (**Figure 6–17a**). Meiosis in Amazon mollies omits the final reduction division, so it produces diploid eggs. These eggs must be activated by sperm from a male Mexican molly in order to begin embryonic development, but the genetic material from the male never enters the egg, which develops as a clone of its mother.

Hybridogenesis The Rio Fuerte topminnow (*Poeciliopsis monacha-lucida*) is a unisexual hybrid that originated from the insemination of a female headwater topminnow (*Poeciliopsis monacha*) by a male clearfin topminnow (*Poeciliopsis lucida*) (Figure 6–17b). All Rio Fuerte topminnows are diploid females that carry a haploid genome from each of the parental species. During meiosis the paternal (clearfin) genome is eliminated, producing eggs with only the maternal (headwater) genome. Female Rio Fuerte topminnows mate with male

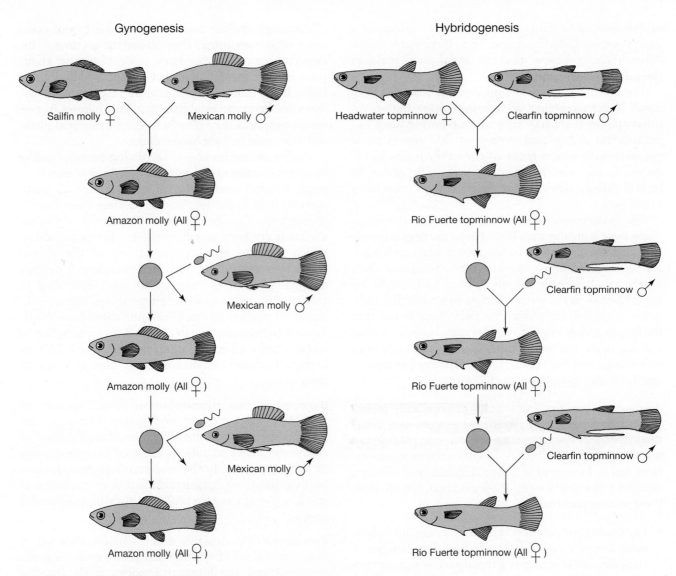

Gynogenesis

Hybridogenesis

Sailfin molly ♀   Mexican molly ♂

Headwater topminnow ♀   Clearfin topminnow ♂

Amazon molly (All ♀)

Rio Fuerte topminnow (All ♀)

Mexican molly ♂

Clearfin topminnow ♂

Amazon molly (All ♀)

Rio Fuerte topminnow (All ♀)

Mexican molly ♂

Clearfin topminnow ♂

Amazon molly (All ♀)

Rio Fuerte topminnow (All ♀)

Figure 6–17 Gynogenesis and hybridogensis. (a) In gynogenesis a diploid egg must be activated by a sperm from a male of a bisexual species before it begins embryonic development. The DNA from the sperm never enters the egg, which develops into a clone of its mother. Amazon mollies are an all-female species that originated from hybridization between two bisexual species, the sailfin molly and the Mexican molly. Amazon mollies produce diploid eggs that are activated by sperm from a male Mexican molly. (b) In hybridogenesis the genome of the male parental species is eliminated during meiosis, producing a haploid egg carrying only the maternal DNA. This egg is fertilized by sperm from a male of a bisexual species, and in this case the DNA from the sperm does enter the egg, producing a diploid female. The DNA from the male is eliminated at meiosis, again producing an egg with only the DNA of the female parental species, and that egg is fertilized by sperm from a male of a bisexual species.

clearfin topminnows, and in this case the male's DNA does enter the egg, producing a new generation of diploid female Rio Fuerte topminnows. These females again eliminate the clearfin genome at meiosis. Thus,

only the genetic contribution of the maternal (headwater) species persists from generation to generation; the chromosomes from the paternal (clearfin) species are lost and replaced at each generation.

6.6 The Adaptable Fishes—Teleosts in Contrasting Environments

Given the great numbers of both species and individuals of extant fishes, especially the teleosts, it is little wonder that they inhabit an enormous diversity of watery environments. Examining two vastly different habitats, the deep sea and coral reefs, allows us to better understand the amazing adaptability of teleost fishes.

Deep-Sea Fishes

Considered as a place to live, the deep sea presents two problems:

- There's no light: Light does not penetrate deeper than 1000 meters, even in the clearest water, so the depths of the sea are perpetually dark.

- There's very little food: Photosynthetic organisms are largely confined to the upper 100 meters of the sea (the epipelagic region, **Figure 6–18**). The food chain of the ocean depths is largely dependent on falling detritus that can range in size from microscopic remains of plankton to carcasses of large fishes and whales. The plankton biomass can reach 500 milligrams per cubic meter of water at the surface, but by the time it has fallen to 10,000 meters only 0.5 milligram of plankton per cubic meter of water remains.

Despite these challenges, a distinctive and bizarre array of deep-sea fishes lives in the mesopelagic and bathypelagic zones. Because there is not much food, deep-sea fishes are small (the average length is about 5 centimeters) and their populations are sparse. Both of those characteristics present problems: small species are vulnerable when they venture into regions with larger predators, and it is difficult to locate a mate

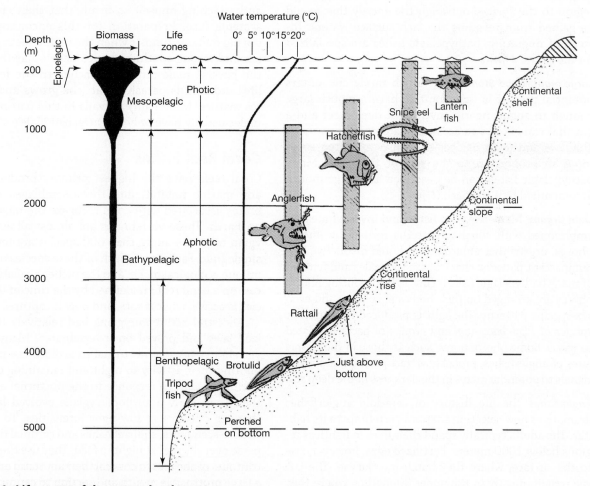

Figure 6–18 Life zones of the ocean depths. Schematic cross section of the life zones within the deep sea. Temperature and biomass are shown, as are the vertical ranges of several fish species, some of which migrate daily.

when it's pitch dark and there is only one female of your species in each cubic kilometer of water. Many of the behavioral, anatomical, and life-history characteristics of deep-sea fishes are related to those challenges.

Mesopelagic Fishes One way of coping with the scarcity of food in the deep sea is to move to shallower water to feed, and fishes and invertebrates that live in the mesopelagic zone can do this. They move toward the surface at dusk to enter areas with more food and descend again near dawn. Vertical migration has both benefits and costs. By rising at dusk, mesopelagic fishes enter a region of higher productivity, where food is more concentrated so they can gather more energy. But the shallower water is also warmer, so their metabolic rates increase and that means that they are using energy faster. The daytime descent into cooler waters lowers their metabolism and conserves energy.

Bathypelagic Fishes The costs in energy and time to migrate several thousand meters from the bathypelagic region to the surface outweigh the energy that would be gained from invading the rich surface waters. Instead of migrating, bathypelagic fishes are specialized for life in the depths.

Large Mouths and Stomachs If a fish rarely encounters potential prey, it is important to have a mouth large enough to engulf nearly anything it does meet and a gut that can expand to accommodate nearly any meal. The jaws and teeth of deep-sea fishes are often enormous in proportion to the rest of the body. Many bathypelagic fishes can be described as a large mouth accompanied by a stomach (**Figure 6–19**).

Light Organs Many deep-sea fishes and invertebrates are emblazoned with startling patterns formed by **photophores**, organs that emit blue light, and distinctive bioluminescent patterns characterize the males and females of many bathypelagic fishes. Tiny, light-producing photophores are arranged on their bodies in species- and even sex-specific patterns. The light is produced by symbiotic species of *Photobacterium* and groups of bacteria related to *Vibrio*. Some photophores, such as those in the fin-ray lures of anglerfishes, probably attract prey. Others act as signals to potential mates in the darkness of the deep sea.

Anglerfishes The life history of ceratioid anglerfishes dramatizes how selection adapts a vertebrate to its habitat. The adults typically spend their lives in lightless regions below 1000 meters. Fertilized eggs, however, rise to the surface, where they hatch into larvae. The larvae remain mostly in the upper 30 meters, where they grow, and later descend to the lightless region. Descent is accompanied by metamorphic changes that differen-

tiate females and males. During metamorphosis, young females descend to great depths, where they feed and grow slowly, reaching maturity after several years.

Female anglerfishes feed throughout their lives, whereas males feed only during the larval stage. Males have a different future—reproduction, literally by lifelong matrimony. During metamorphosis males cease eating and begin an extended period of swimming. The olfactory organs of males enlarge at metamorphosis, and the eyes continue to grow. These changes suggest that adolescent males spend a brief free-swimming period finding a female. The journey is precarious, for males must search vast, dark regions for a single female while running a gauntlet of other deep-sea predators. The sex ratio of young adults is unbalanced—often more than 30 males for every female. Apparently, at least 29 of those males will not locate a virgin female.

Having found a female, a male does not want to lose her. He ensures a permanent union by attaching himself as a parasite to the female, biting into her flesh and attaching himself so firmly that their circulatory systems fuse. Preparation for this encounter begins during metamorphosis when the male's teeth degenerate and strong toothlike bones develop at the tips of the jaws. A male remains attached to the female for life, and in this parasitic state, he grows and his testes mature. Monogamy prevails in this pairing, for females usually have only one attached male.

Coral Reef Fishes

Coral reefs have the highest primary productivity of any marine habitat, and the assemblages of vertebrates associated with coral reefs are the most diverse on Earth. These vertebrates are almost all actinopterygian teleosts—more than 600 species are found on a single Indo-Pacific reef. Each of these species occupies a unique ecological niche, and the niches available to species on a coral reef are defined by the time of day a species is active, what it eats, and how it captures its prey.

Ancestral actinopterygians had relatively unspecialized jaws and preyed on invertebrates. Many reef invertebrates became nocturnal to avoid these predators, limiting their activity to night and remaining concealed during the day. In response to the nocturnal activity of their prey, early acanthopterygians evolved large eyes that are effective at low light intensities. To this day, their descendants, squirrel fishes and cardinal fishes, disperse over the reef at night to feed. They use the irregular contours of the reef to conceal their approach and rely on a large protrusible mouth and suction to capture prey.

The evolution of jaws specialized to take food items hidden in the complex reef surface was a major

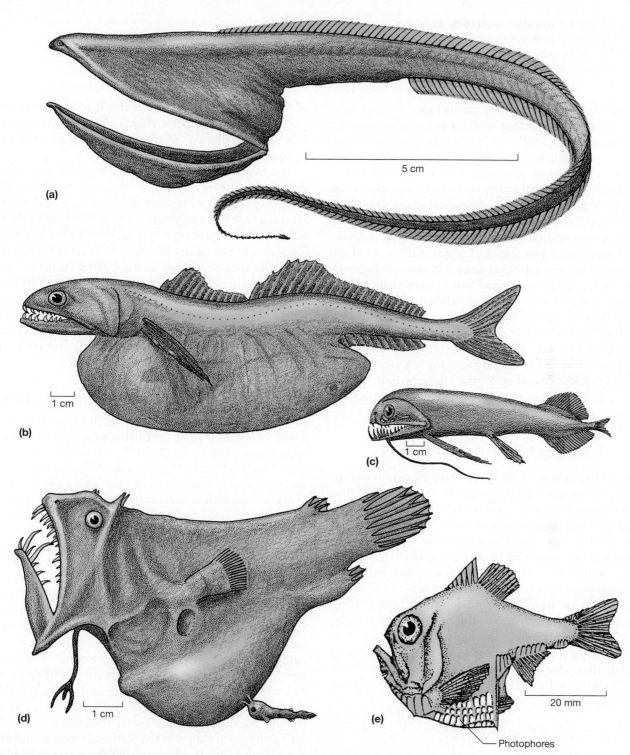

Figure 6–19 Deep-sea fishes. Prey is scarce in the deep sea, and fishes with large mouths and distensible guts are able to eat most prey items they encounter. Some deep-sea fishes lure prey with luminescent organs. (a) Pelican eel, *Eurypharynx pelecanoides*. (b) Deep-sea perch, *Chiasmodus niger,* its belly distended by a fish bigger than itself. (c) Stomiatoid, *Aristostomias grimaldii*. (d) Female anglerfish, *Linophryne argyresca*, with a parasitic male attached to her belly. (e) Hatchetfish, *Sternoptyx diaphana*, with photophores on the ventrolateral region.

Figure 6–20 A Caribbean coral reef. Fish activity and diversity change from day to night as diurnal species retreat into the reef and nocturnal species emerge. Activity on the reef (a) at midnight and (b) at midday. Most reef fishes are Perciformes; of the fishes shown here, only the herring, halfbeak, and squirrel fish are not Perciformes.

advance. Some species rely on suction, whereas others use a forceps action of their protrusible jaws to extract small invertebrates from their daytime hiding places or to snip off coral polyps, bits of sponges, and other exposed reef organisms. This mode of predation demands high visual acuity that can be achieved only in the bright light of day. These selection pressures produced fishes capable of maneuvering through a complex three-dimensional habitat in search of food.

A coral reef has day and night shifts—at dusk the colorful diurnal fishes seek nighttime refuges in the reef, and the nocturnal fishes leave their hiding places and replace the diurnal fishes in the water column (**Figure 6–20**). The timing of the shift is controlled by light intensity, and the precision with which each species leaves or enters the protective cover of the reef day after day indicates that this is a strongly selected behavioral and ecological characteristic. Space, time, and the food resources available on a reef are partitioned through this activity pattern.

6.7 Conservation of Fishes

Fishes face a host of problems. Like all extant organisms, fishes suffer from changes in their habitats that stem directly or indirectly from human activities. In addition, the larval stages of many species occupy distinctly different habitats at different stages of their lives and are vulnerable to changes in all of these habitats. Unlike most other vertebrates, many fishes are enormously important commercially as food and major industries are based on capturing or culturing them. Other species of fishes are captured for the pet trade, sometimes using collection methods that destroy their habitats. As a result of these factors, many species of fishes that were once common have been brought to the verge of extinction, just as we are discovering important new uses for some species.

Conservation Concerns for Freshwater Fishes

Nearly 40 percent of fish species live in the world's freshwaters, and all of them are threatened by the alteration and pollution of lakes, rivers, and streams. Draining, damming, canalization, and diversion of rivers create habitats that no longer sustain indigenous fishes. In addition to the loss and physical degradation of fish habitat, freshwaters in much of the world are polluted by silt and toxic chemicals of human origin. This is especially true of the Western nations and urbanized regions elsewhere.

The United States has had, in recent years, more than 2400 instances *annually* of beaches and flowing waterways closed to human use because of pollution. Sites that are too dangerous for people to play in them are often lethal to the organisms trying to live in them! Sixty-one of the nearly 800 species of native freshwater fishes that once lived in the United States are extinct, and almost 40 percent of the remaining species are imperiled. The situation is no better in Europe, where 38 percent of the species of freshwater fishes are threatened with extinction and 12 are already extinct. Assessments of this sort for Asian countries are only now being initiated.

Conservation Concerns for Marine Fishes

Commercial overfishing is a sadly familiar problem. Many of the world's richest fisheries are on the verge of collapse. The Georges Bank, which lies between Cape Cod, Massachusetts and Cape Sable island, Nova Scotia, is an example of what overfishing can do. For years, conservation organizations called for a reduction in catches of cod, yellowtail flounders, and haddock, but their concerns were not heeded.

Not surprisingly, the populations of many commercial fish species crashed dramatically in the 1990s. By October 1994, the situation was so bad that a government and industry group, the New England Fishery Management Council, directed its staff to devise measures that would reduce the catch of those species essentially to zero. Nearly 20 years later a few encouraging signs indicate that these draconian measures have allowed some species to increase. Populations of cod and haddock are on the rebound, although the average size of the fishes is less than half what it was in the pre-crash years, and increased populations of yellowtail flounders allowed the National Oceanic and Atmospheric Administration to raise the quota in 2011.

These gains, however, must be balanced against losses. Also in 2011 the European Fisheries Commission determined that the fisheries policy that the European Union had been following since 1983 has been a failure and called for far more stringent regulations beginning in 2013.

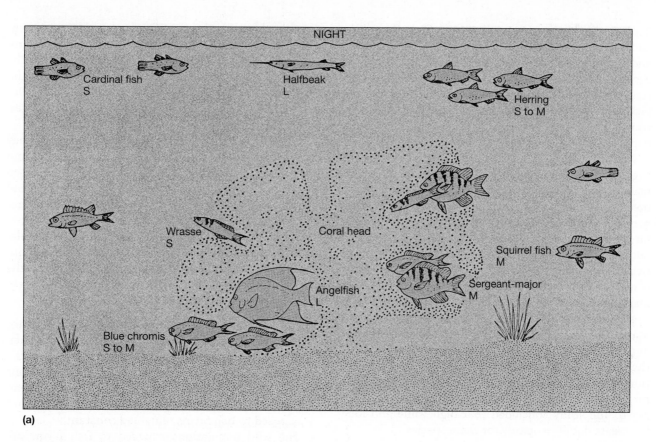

(a)

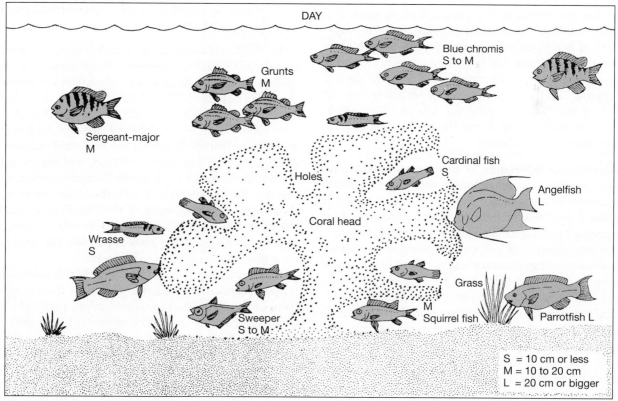

(b)

Reproductive Biology and Overfishing Cod, flounders, and haddock mature relatively fast (in as little as 2 years) and produce vast numbers of eggs—up to 11 million eggs per year for a large cod. Early maturity and high fecundity are characteristics that give a species high reproductive potential. Species of this sort are relatively resistant to overharvesting, and they are able to rebuild populations when excessive harvesting is stopped.

Maturing slowly and producing a small number of eggs relative to adult body size are characteristics that make a species vulnerable to overfishing, and large predatory fishes, tunas and billfishes, fall into this category. On a unit-weight basis, the egg production of bluefin tunas is less than 3 percent that of cod, and bluefins take more than twice as long as cod to reach reproductive size. To make the situation even worse, tuna flesh is considered a delicacy and bluefin tuna fetches high market prices, which rise even higher as the supply falls.

A recent analysis by Bruce Collette of the National Marine Fisheries Service and colleagues from similar organizations around the world found that eight of the ten large predatory fish species that have generation times of 6 years or longer meet the IUCN criteria for being Vulnerable, Endangered, or Critically Endangered. The study concluded that high value and long life create double jeopardy for these species.

Impact of Overfishing on Ecosystems A study published in 2009 reported declines in populations of mesopelagic and bathypelagic fishes at depths from 800 to 2500 meters. These are not commercially desirable species, and in any case commercial trawlers do not fish below 1600 meters. Apparently we are seeing the effect of trawlers capturing the fish at night, when they have moved upward, and the impact is spreading to the deepest ocean basins.

Human-Caused Difficulties in Conservation of Marine Fishes Overfishing is, of course, a problem that is directly attributable to humans, but a different kind of threat to marine fishes results from an activity that is often portrayed as ecologically responsible—fish farming.

As populations of commercially important, overfished species have plummeted, the role of fish farming has increased. Many fish farms are huge netted enclosures that are anchored in rivers or in coastal waters. Young fishes from a hatchery are placed in the enclosures and fed until they have grown to harvestable size. Although fish farming is promoted as an ecologically sound alternative to catching wild fishes, it creates both short-term and long-term problems.

Pollution, Parasites, and Disease In the short term, concentrating thousands of fishes in a small volume of water and feeding them artificial food create a tremendous accumulation of feces and uneaten food that promotes blooms of bacteria and algae. In addition, the fishes crowded into pens are susceptible to disease and parasites that can be transferred to wild fishes. Sea lice are small copepod crustaceans that parasitize fishes, and farmed salmon are heavily infected with them.

When wild salmon swim past the farm pens, they become infected with sea lice—a recent study found that a single fish farm in British Columbia increased the infection of juvenile wild salmon by 73-fold and the effect extended for 30 kilometers along the migration paths the wild fishes followed.

Thirty-five years of records of wild salmon populations were analyzed, comparing 7 rivers that flow into channels with fish farms with 64 rivers in which the wild salmon were not exposed to fish farms. These findings were reached:

- Until 2001, when sea lice appeared in the fish farms, there were no differences in the year-to-year population changes in salmon in the two groups of rivers.
- From 2001 onward, the wild salmon populations that were exposed to fish farms shrank every year, whereas the salmon populations that were not exposed to fish farms remained constant.
- The wild populations exposed to fish farms are shrinking so fast that, if the current trend continues, they will be 99 percent gone in four generations.

In 2011, researchers at Simon Fraser University in British Columbia reported the appearance in wild salmon of a viral disease, infectious salmon anemia. The virus has killed 70 percent or more of the salmon in farms in Norway, Scotland, and Chile, but it had been unknown in North America. The salmon farming industry in British Columbia has imported millions of Atlantic salmon eggs from Scotland, and the virus infecting the Canadian fish is the European strain, suggesting that it was carried with the imported eggs. If those eggs were the source of the Canadian infection, the virus could easily have spread to wild salmon because only a net would have separated the wild fish from the infected salmon in the pens.

Some evidence suggests that wild Pacific salmon are more resistant than Atlantic salmon to the European form of the virus. That resistance might lessen the impact of the disease on native salmon, at least initially. However, the virus that causes infectious salmon anemia is an orthomyxovirus, which is related to the virus that causes human influenza. These are RNA viruses that mutate readily, so strains of the virus that overcome the resistance mechanisms of Pacific salmon could well evolve.

Depletion of Commercial Fish Stocks The rationale for fish farming is that the availability of farmed fishes will reduce the need to catch wild fishes. This reasoning is only partly correct, however, because farmed salmon and sea trouts are raised on fish meal that is produced by catching small fishes and invertebrates, and depleting populations of prey species is likely to deprive wild species of fishes of their food base.

Genetic Change and Evolution The conspicuous problems associated with fish farming, such as pollution, disease, parasites, and depletion of prey species, are not the only concerns. Recent work has focused attention on the interaction of evolutionary processes with ecological and commercial elements of fishery biology.

For example, the reproductive physiologies of farmed and wild salmon are different. One of the general features of the life history of animals is a trade-off between the number of eggs produced and the size of each egg—many small eggs versus fewer large eggs.

The large eggs laid by wild salmon produce large hatchlings that can survive the risks that wild hatchlings face. In contrast, hatchlings of farm-raised salmon are protected from most of the hazards that confront wild hatchlings. As a result, farmed salmon have shifted toward producing many small eggs—in just four generations, the egg size of farmed salmon can decrease by 25 percent.

That change in egg size would not be a problem if farmed fishes remained in farms, but not all of them do—escapes are a regular occurrence and sometimes involve large numbers of fishes. Storms have allowed millions of farmed salmon to escape from their pens and enter the wild salmon population, bringing with them genotypes that produce small eggs. The influence of this sudden genetic input on wild populations of salmon is unknown, but it is potentially harmful to the wild genotype that has been shaped by millennia of natural selection.

6.8 Sarcopterygii—The Lobe-Finned Fishes

With at least three distinct lineages, sarcopterygians are considered a polyphyletic group: Two lineages of sarcopterygian fishes that first appeared in the Devonian have survived as aquatic forms to the present day: the dipnoans (lungfishes) and the actinistians (coelacanths). A third lineage, the tetrapodomorph fishes (discussed in more detail in Chapter 9), contains the ancestors of tetrapods.

Evolution of the Sarcopterygii

Early sarcopterygians were 20 to 70 centimeters long and cylindrical. They had two dorsal fins, an epichordal lobe on the heterocercal caudal fin (a fin area supported by the dorsal side of the vertebral column), and paired fins that were fleshy and scaled and had a bony central axis.

Extant Sarcopterygian Fishes—Lungfishes and Coelacanths

Although they were abundant in the Devonian, the number of sarcopterygian fishes dwindled in the late Paleozoic and Mesozoic eras. Today only four non-tetrapod genera remain (**Figure 6–21**): the freshwater dipnoans or lungfishes (*Neoceratodus* in Australia, *Lepidosiren* in South America, and *Protopterus* in Africa) and the marine actinistian *Latimeria* (the coelacanths) in waters 100 to 300 meters deep off East Africa and central Indonesia.

Dipnoans—Lungfishes Extant Dipnoi are distinguished by the derived features of the loss of the tooth-bearing dermal bones (the premaxilla, maxilla, and dentary) and the fusion of the palatoquadrate to the cranium. Teeth are scattered over the dermal bones of the palate and fused into tooth ridges (tooth plates) along the lateral palatal margins to form a dentition capable of crushing hard foods. This **durophagous** apparatus has persisted throughout lungfish evolution.

The earliest dipnoans were marine. The group was diverse during the Paleozoic, but by the Mesozoic most species belonged to the ceratodonts, the lineage that includes the living forms.

By the Carboniferous, lungfishes had evolved a body form quite distinct from that of the other Osteichthyes. The dorsal, caudal, and anal fins are fused into one continuous fin that extends around the entire posterior third of the body. In addition, the caudal fin has changed from heterocercal to symmetrical, and the mosaic of small dermal bones of the earliest dipnoan skulls has changed to a pattern of several large elements. Most of this transformation can be explained as a result of **paedomorphosis** (Greek *paedo* = child and *morph* = form; the retention of juvenile characters in an adult).

The Australian lungfish *Neoceratodus forsteri*, which is native to parts of southeastern Queensland, is similar in morphology to Paleozoic and Mesozoic dipnoans, retaining lobelike fins and heavier scales than those in the other genera (see Figure 6–21a). Having comparatively well-developed gills and simple lung structure, *Neoceratodus* inhabits large lakes and rivers. These are

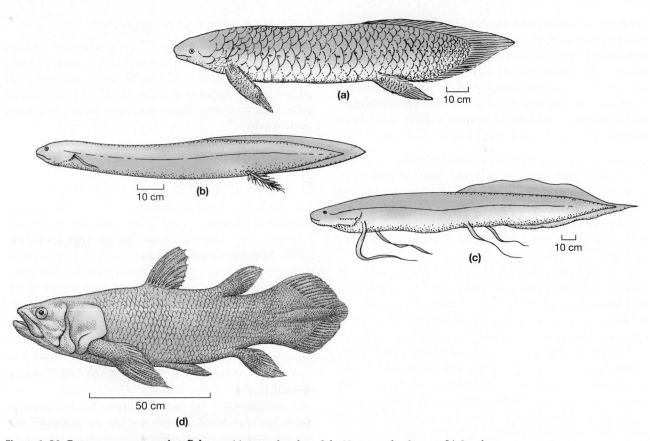

Figure 6–21 Extant sarcopterygian fishes. (a) Australian lungfish, *Neoceratodus forsteri*; (b) South American lungfish, *Lepidosiren paradoxa*, male; (c) African lungfish, *Protopterus*; (d) *Latimeria chalumnae*, the best known of the extant coelacanths.

permanent water bodies, and the Australian lungfish does not estivate.

Neoceratodus attains a length of 1.5 meters and a reported weight of 45 kilograms. It swims by body undulations or slowly walks across the bottom of a pond using its pectoral and pelvic appendages. Chemical senses seem important to lungfishes, and their mouths contain numerous taste buds.

Australian lungfishes have a complex courtship that may include male territoriality, and they are selective about the vegetation upon which they lay their adhesive eggs; however, no parental care has been observed after spawning. The jelly-coated eggs, 3 millimeters in diameter, hatch in 3 to 4 weeks, but the young are elusive and nothing is known of their juvenile life.

Surprisingly little is known about the South American lungfish, *Lepidosiren paradoxa*, but the closely related African lungfishes, *Protopterus*, with four recognized species, are better known. All are thin-scaled, heavy-bodied, elongate fishes with mobile,

filamentous paired appendages. Both genera have weakly developed gills and rely on their lungs to obtain oxygen; they drown if they are prevented from breathing air. The gills are important in eliminating carbon dioxide. Male *Lepidosiren* develop vascularized extensions on their pelvic fins during the breeding season that probably deliver oxygen from the male's blood to the young that he is guarding in the nest cavity.

Unlike Australian lungfishes, African lungfishes live in areas that flood during the wet season and bake during the dry season—habitats that are not available to actinopterygians except by immigration during floods. The lungfishes enjoy the flood periods, feeding heavily and growing rapidly. When the floodwaters recede, a lungfish digs a vertical burrow in the mud with an enlarged chamber at the end. As drying proceeds, the lungfish becomes more lethargic and breathes air from the burrow opening. Eventually the water in the burrow dries up, and the lungfish enters the final stages of estivation—folded into a U shape with its tail over its eyes. Heavy secretions of mucus that the fish has produced since entering the burrow dry and form a protective envelope around its body. Only an opening at its mouth remains to permit breathing.

Estivation is an ancient trait of dipnoans. Fossil burrows containing lungfish tooth plates have been found in Carboniferous- and Permian-era deposits in North America and Europe. Without the unwitting assistance of the lungfishes such fossils might not exist.

Coelacanths Coelacanths are first known from the Early Devonian, around 408 million years ago. They have two derived characters: a first dorsal fin that is supported by a plate of bone but lacks an internal lobe, and a symmetrical three-lobed tail with a central fleshy lobe that ends in a fringe of rays. Coelacanths lack a maxilla and branchiostegal rays, and they have a unique pattern of bones in their paired fins.

Extant coelacanths retain the ancestral fusiform fish shape, but during the Carboniferous there was an explosive radiation resulting in a greater diversity of body forms, including some deep-bodied species with reduced fin lobes that were originally mistaken for actinopterygians. By the Late Devonian coelacanths differed only slightly from the extant species, but during the Cretaceous there was a huge predatory coelacanth, *Mawsonia*, in South America that reached lengths in excess of 4 meters. Some early coelacanths lived in shallow freshwaters, but the fossil remains of coelacanths during the Mesozoic are largely marine.

The fossil record for coelacanths was not believed to extend beyond the Cretaceous period, and until about 70 years ago coelacanths were thought to be extinct. But in 1938 an African fisherman bent over an unfamiliar catch from the Indian Ocean and nearly lost his hand to its ferocious snap. Imagine the astonishment of the scientific community when J. L. B. Smith of Rhodes University announced that the catch was an actinistian. This large fish was so similar to Mesozoic fossil coelacanths that its systematic position was unquestionable. Smith named this living fossil *Latimeria chalumnae* in honor of his former student Marjorie Courtenay Latimer, who saw the strange catch, recognized it as unusual, and brought the specimen to his attention.

Since then more than 150 specimens, ranging in length from 75 centimeters to slightly more than 2 meters, have been caught in the Comoro Archipelago or in nearby Madagascar or Mozambique. Coelacanths are hooked near the bottom, usually in 260 to 300 meters of water about 1.5 kilometers offshore. Strong and aggressive, *Latimeria chalumnae* is steely blue-gray with irregular white spots and golden eyes.

Rather than having a ventral lung that can be used for respiration, as in lungfishes and basal actinopteryg-ians, *Latimeria* has a "lung" (sometimes called a swim bladder) that is filled with fat and has ossified walls. Fat is less dense than water and the "lung" is probably a hydrostatic organ, akin to the lipid-filled livers of sharks. *Latimeria* also has a unique rostral organ, a large cavity in the midline of the snout with three gel-filled tubes that open to the surface through a series of six pores: this organ is almost certainly an electro-receptor.

Latimeria resembles chondrichthyans in retaining urea to maintain a blood osmolality close to that of seawater. This osmoregulatory physiology may be an ancestral gnathostome condition. Like chondrichthy-ans, *Latimeria* has a rectal gland that excretes salt. A fascinating glimpse of the life of the coelacanth was reported by Hans Fricke and his colleagues, who used a small submarine to observe fishes and divers using special deep-water diving techniques. Fricke saw six coelacanths at depths between 117 and 198 meters off a short stretch of the shoreline of one of the Comoro Islands. Coelacanths were seen only in the middle of the night and only on or near the bottom. Divers have found coelacanths in waters 100 to 130 meters deep off the southeastern coast of Africa and the western coast of Madagascar. There they hide in underwater caves during the day and venture out at night to dine on small fishes, squid, and octopuses. Unlike extant lungfishes, the coelacanths did not use their paired fins as props or to walk across the bottom. However, when they swam, the pectoral and pelvic appendages were moved in the same sequence as tetrapods move their limbs.

In 1927, D. M. S. Watson described two small skeletons from inside the body cavity of *Undina*, a Jurassic coelacanth, and suggested that coelacanths were viviparous. In 1975, dissection of a 1.6-meter *Latimeria* confirmed this prediction, revealing five young, each 30 centimeters long and at an advanced stage of embryonic development. Several other females with young in advanced stages of development have now been collected. Internal fertilization must occur, but how copulation is achieved is unknown because males show no specialized copulatory organs.

In 1998 ichthyologists were again surprised when a second species of coelacanth was discovered 10,000 kilometers to the east of the Comoro Islands by Mark Erdmann, a graduate student studying coral reef invertebrates, and his wife Arnaz, who spotted the fish in a market and brought it to Mark's attention. Other specimens have subsequently been caught in shark-fishing nets set 100 to 150 meters deep off the northeast tip of Sulawesi, a large Indonesian island near Borneo.

Subsequently named *Latimeria menadoensis* after the nearest major city, the new coelacanth appears from DNA data to have separated from the East African species 1.8 to 11.0 million years ago.

Extinct Tetrapodomorph Fishes

This group (at one time known as "rhipidisteans") was a predominant part of the Devonian fauna and included many large carnivorous forms. Both morphological and molecular data place lungfishes as the sister group to the tetrapodomorphs, and the Early Devonian fish *Styloichthyes* displays anatomical features shared with lungfishes and tetrapodomorphs, placing it close to the ancestry of these two clades. The origin of tetrapods, the terrestrial descendants of the tetrapodomorph fishes, is described in Chapter 9.

Summary

At their first appearance in the fossil record, osteichthyans, the largest vertebrate taxon, are separable into two lineages, both first known from the Late Silurian period: the Sarcopterygii (lobe-finned fishes, including coelacanths, lungfishes, and tetrapods) and the Actinopterygii (ray-finned fishes). Extant sarcopterygian fishes, predominant in the Devonian but reduced to four genera today, provide information about the fishes immediately ancestral to tetrapods. Actinopterygians inhabit the 73 percent of the Earth's surface that is covered by water, and they are a numerous and species-rich lineage of vertebrates. Several trends in the development of food-gathering and locomotor structures characterize actinopterygian evolution. The radiations of these levels are represented today by small numbers of species that are relicts of groups that once had many more species: polypteriforms (bichirs and reedfishes), acipenseriforms (sturgeons and paddlefishes), and the most ancestral of the surviving lineages of neopterygians (gars and the bowfin). The most derived level of bony fishes, teleosteans, display a wealth of adaptations in their morphology, ecology, behavior, and physiology. Examination of the way teleosts have adapted to habitats like the deep sea and coral reefs helps us to understand how evolution has acted upon the basic teleost body plan, but looking at how humans have overexploited commercial fisheries and changed fish habitats explains why so many fishes are now in danger of extinction.

Discussion Questions

1. How do actinopterygian fishes and sarcopterygian fishes use their pectoral fins? Evaluate the structure of the pectoral fins in those two lineages in the context of the differences in how they are used.

2. Explain briefly why the modifications of the jaws of early neopterygians (see Figure 6–5) represent an important advance in feeding, and how the further modifications seen in the jaws of teleosts (see Figure 6–6) allow still more specialization.

3. What do the pharyngeal jaws of bony fishes do? How is this different from the actions of the anterior jaws (the premaxilla and maxilla)?

4. In general, freshwater teleosts lay adhesive eggs that remain in one place until they hatch, whereas marine teleosts release free-floating eggs that are carried away by water currents. (There are exceptions to that generalization in both freshwater and marine forms.) What can you infer about differences in the biology of newly hatched freshwater and marine teleosts on the basis of that difference in their reproduction?

5. Teleosts have a bewildering variety of patterns of sex change, but protogyny and protandry are the most common. Describe the evolutionary principle that underlies those two patterns.

6. Nowadays we consider lungfishes to be more closely related to tetrapods than the coelacanths are, but in earlier times coelacanths were thought to be the more closely related forms. How have changes in our way of thinking about phylogenetic relationships changed our opinion in this case?

Additional Information

Ahnesjö, I., and J. F. Craig (eds.). 2011. Special issue: The biology of Syngnathidae: pipefishes, seadragons, and seahorses. *Journal of Fish Biology* 78 (issue 6):1597–1869.

Barton, M. 2007. *Bond's Biology of Fishes*, 3rd ed. Belmont, CA: Thomson Brooks/Cole.

Brito, P. M., et al. 2010. The histological structure of the calcified lung of the fossil coelacanth, *Axelrodichthyes araripensis* (Actinistia: Mawsoniidae). *Palaeontology* 53:1281–1290.

Coates, M. I. 2009. Beyond the age of fishes. *Nature* 458:413–414.

Collette, B. B., et al. 2011. High value and long life—double jeopardy for tunas and billfishes. *Science* 333:291–292.

Crews, D. (ed.). 2009. Environmental regulation of sex determination in vertebrates. *Seminars in Cell & Developmental Biology* 20:251–376.

Erisman, B. E., et al. 2009. A phylogenetic test of the size-advantage model: Evolutionary changes in mating behavior influence the loss of sex change in a fish lineage. *American Naturalist* 174:E83–E99.

Friedman, M., and M. D. Brazeau. 2010. A reappraisal of the origin and basal radiation of the Osteichthyes. *Journal of Vertebrate Paleontology* 30:36–56.

Godwin, J. 2009. Social determination of sex in reef fishes. *Seminars in Cell & Developmental Biology* 20:264–270.

Helfman, G. S., et al. 2009. *The Diversity of Fishes*, 3rd ed. Malden, MA: Blackwell Science.

Jarial, M. S., and J. H. Wilkins. 2010. Structure of the kidney in the coelacanth *Latimeria chalumnae* with reference to osmoregulation. *Journal of Fish Biology* 76:655–668.

Kuwamura, T., et al. 2011. Reversed sex change by widowed males in polygynous and protogynous fishes: Female removal experiments in the field. *Naturwissenschaften* 98:1041–1048.

Long, J. A. 2011. *The Rise of Fishes: 500 Million Years of Evolution,* 2nd ed. Baltimore, MD: The Johns Hopkins University Press.

Long, J. H., Jr., et al. 2010. Go reconfigure: How fish change shape as they swim and evolve. *Integrative and Comparative Biology* 50:1120–1139.

Mehta, R. S., and P. C. Wainwright. 2007. Raptorial jaws in the throat help moray eels swallow large prey. *Nature* 449:79–82.

Meunier, F. J. 2011. The Osteichthyes, from the Paleozoic to the extant time, through histology and palaeohistology of bony tissues. *Comptes Rendus Palevol* 10:347–355.

Nelson, J. S. 2006. *Fishes of the World*, 4th ed. New York: Wiley.

Norris, D., and K. Lopez. 2010. *Hormones and Reproduction of Vertebrates*. Volume 1: *Fishes*. London: Academic Press.

Pietsch, T. W. 2009. *Oceanic Anglerfishes: Extraordinary Diversity in the Deep Sea*. Berkeley, CA: University of California Press.

Sallan, L. C., and M. I. Coates. 2010. End-Devonian extinction and a bottleneck in the early evolution of modern jawed vertebrates. *Proceedings of the National Academy of Sciences* 107:10131–10135.

Tierney, K. B. 2011. Swimming performance assessment in fishes. (Video Article) *Journal of Visualized Experiments* 51. http://www.jove.com/index/Details.stp?ID=2572, doi: 10.3791/2572.

Van Look, K. J. W., et al. 2007. Dimorphic sperm and the unlikely route to fertilization in the yellow seahorse. *The Journal of Experimental Biology* 210:432–437.

Videler, J. J., and P. He. 2010. Swimming in marine fish. In P. He (ed.), *Behavior of Marine Fishes: Capture Processes and Conservation Challenges*. Oxford, UK: Wiley-Blackwell.

Zhu, M., et al. 2009. The oldest articulated osteichthyan reveals mosaic gnathostome characters. *Nature* 458:469–474.

Websites
General Information

FishBase provides information about the biology of fishes as well as photographs and a searchable database of fish names. http://www.fishbase.org/home.htm

Fish Online is a companion site for FishBase that focuses on ichthyology as a subject. http://www.fishbase.org/FishOnLine/English/index.htm

The Online Catalog of Fishes of the California Academy of Sciences provides an up-to-date list of the extant species of fishes. http://research.calacademy.org/redirect?url=http://researcharchive.calacademy.org/research/Ichthyology/catalog/fishcatmain.asp

The ichthyology program at the University of Florida provides information about the biology of fishes. http://www.flmnh.ufl.edu/fish/default.htm

Geography and Ecology of the Paleozoic Era

The Paleozoic world was very different from the one we know—the continents were in different places, climates were different, and initially there was little structurally complex life on land. By the Early Devonian period, terrestrial environments supported a substantial diversity of plants and invertebrates, setting the stage for the first terrestrial vertebrates (tetrapods) in the Late Devonian. Early plants were represented by basal groups such as horsetails, mosses, and ferns. The evolution of plants resulted in the production of soils that trapped carbon dioxide. Removal of carbon dioxide from the atmosphere had a profound effect on Earth's climate, resulting in a reverse greenhouse effect—global cooling. Extensive glaciation occurred during the Late Carboniferous and Early Permian periods, when the tetrapods lived in equatorial regions. Terrestrial ecosystems became more complex in the Carboniferous and

Permian; some modern types of plants, such as conifers, appeared; tetrapods diversified; and the first flying insects took to the air. Fluctuations of atmospheric oxygen and carbon dioxide during this time may underlie many events and patterns in vertebrate evolution. Extinctions occurred in terrestrial ecosystems at the end of the Early Permian, and a major extinction (the largest in Earth's history) occurred both on land and in the oceans at the end of the Paleozoic era.

7.1 Earth History, Changing Environments, and Vertebrate Evolution

It is important to realize that the world of today is very different from the world of times past. Our particular

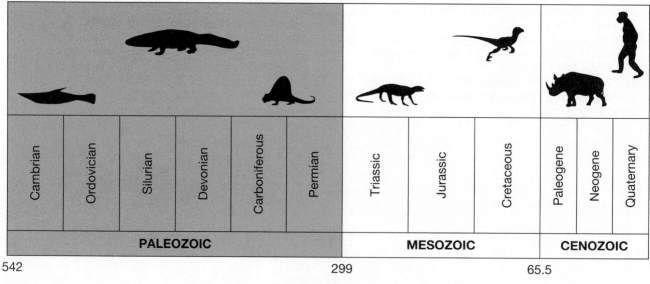

Cambrian	Ordovician	Silurian	Devonian	Carboniferous	Permian	Triassic	Jurassic	Cretaceous	Paleogene	Neogene	Quaternary
PALEOZOIC						**MESOZOIC**			**CENOZOIC**		

542   299   65.5

Million years ago

pattern of global climates, including such features as ice at the poles and the directions of major winds and water currents, results from the present-day positions of the continents. The world today is, in general, cold and dry in comparison with many past times. It is also unusual because the continents are widely separated from one another, and the main continental landmass is in the Northern Hemisphere. Neither of these conditions existed for most of vertebrate evolution.

The Earth's Time Scale and the Early History of the Continents

Vertebrates are known from the portion of Earth's history called the **Phanerozoic eon** (Greek *phanero* = visible and *zoo* = animal). The Phanerozoic began 542 million years ago and contains the Paleozoic (Greek *paleo* = ancient), Mesozoic (Greek *meso* = middle), and Cenozoic (Greek *ceno* = recent) eras. Our own portion of time, the Recent epoch, lies within the Cenozoic era. Each era contains a number of periods, and each period in turn contains a number of subdivisions (see the inside front cover of this textbook). At least 99 percent of described fossil species occur in the Phanerozoic, although the oldest known fossils are from around 3.5 billion years ago, and the origin of life is estimated to be around 4 billion years ago.

The time before the Phanerozoic is often loosely referred to as the Precambrian, because the Cambrian is the first period in the Paleozoic era and thus marks the beginning of the Phanerozoic. However, the Precambrian actually represents seven eighths of the entire history of Earth! Precambrian time is better perceived as two eons, comparable to the Phanerozoic eon: the Archean (Greek *archeo* = first or beginning), commencing with Earth's formation around 4.5 billion years ago (the earlier part of this time period is sometimes referred to as the Hadean), and the Proterozoic (Greek *protero* = former), which began around 2.5 billion years ago. The start of the Proterozoic is marked by the appearance of the large continental blocks seen in today's world. (The entire pre-Proterozoic world would have looked rather like the South Pacific does now—many little volcanic islands separated by large tracts of ocean.)

Although life dates from the Archean, organisms more complex than bacteria are not known until the Proterozoic. The evolution of eukaryotic organisms, which depend on oxygen for respiration, followed shortly after the first appearance of atmospheric oxygen in the middle Proterozoic (around 2.2 billion years ago). Multicellular organisms are first known near the end of the Proterozoic, about 1 billion years ago, and metazoans (animals) started to radiate in earnest by the start of the Phanerozoic, around 540 million years ago.

Continental Drift—History of Ideas and Effects on Global Climate

Our understanding of the dynamic nature of Earth and the variable nature of Earth's climate over time is fairly recent. The notion of mobile continents, or **continental drift**, was established in the late 1960s as the theory of **plate tectonics**. Oceanographic research has demonstrated the spreading of the seafloor as a plausible mechanism for movements of the tectonic plates that underlie the continents. These movements are responsible for the sequence of fragmentation, coalescence, and refragmentation of the continents that has occurred during Earth's history. Plants and animals were carried along as continents slowly drifted, collided, and separated. When continents moved toward the poles, they carried organisms into cooler climates.

As continents collided, terrestrial floras and faunas that had evolved in isolation were mixed together, and populations of marine organisms were separated. A recent (in geological terms) example of this phenomenon is the joining of North and South America around 2.5 million years ago. The faunas and floras of the two continents mingled, which is why we now have armadillos (of South American origin) in Texas and deer (of North American origin) in Argentina. In contrast, the marine organisms originally found in the sea between North and South America were separated, and the populations on the Atlantic and Pacific sides of the land bridge became increasin`gly different from each other with the passage of time.

The positions of continents affect the flow of ocean currents; and, because ocean currents transport enormous quantities of heat, changes in their flow affect climates worldwide. For example, the breakup and northward movements of the continents during the late Mesozoic and Cenozoic led first to the isolation of Antarctica and the formation of the Antarctic ice cap (around 45 million years ago) and eventually to the isolation of the Arctic Ocean, with the formation of Arctic ice by around 5 million years ago. The presence of this ice cap influences global climatic conditions in a variety of ways, and the world today is colder and drier than it was earlier in the Cenozoic.

7.2 Continental Geography of the Paleozoic

The world of the early Paleozoic contained at least six major continental blocks (**Figure 7–1**). **Laurentia** included most of present-day North America plus Greenland, Scotland, and part of northwestern Asia. **Gondwana**

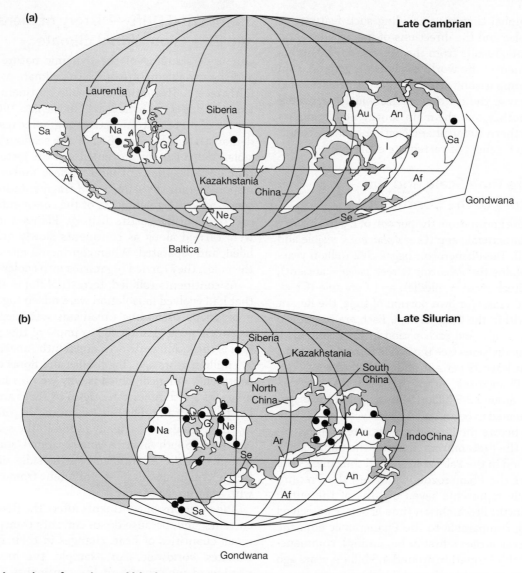

Figure 7–1 Location of continental blocks. (a) In the Late Cambrian (~500 Ma [millions of years ago]) and (b) in the Late Silurian (~420 Ma) periods. The black dots indicate fossil localities where Ordovician vertebrates have been found. The positions of the modern continents are indicated. (Af = Africa, An = Antarctica, Ar = Arabia, Au = Australia, G = Greenland, I = India, Ne = northern Europe, Sa = South America, Se = southern Europe, Na = N. America)

(also known as Gondwanaland) included most of what is now the Southern Hemisphere (South America, Africa, Antarctica, and Australia), and additionally at this time it included portions of Asia and Europe. Four smaller blocks contained other parts of what is now the Northern Hemisphere.

In the Late Cambrian (around 500 million years ago), Gondwana and Laurentia straddled the equator, and the positions of the modern continents within

Gondwana were different from their positions today—Africa and South America appear to be upside down (see Figure 7–1a). By the Late Silurian (around 420 million years ago), the eastern portion of Gondwana was over the South Pole, and Africa and South America were in positions similar to those they occupy today (see Figure 7–1b). Laurentia was still in approximately the same position and had collided with a northward-moving Baltica.

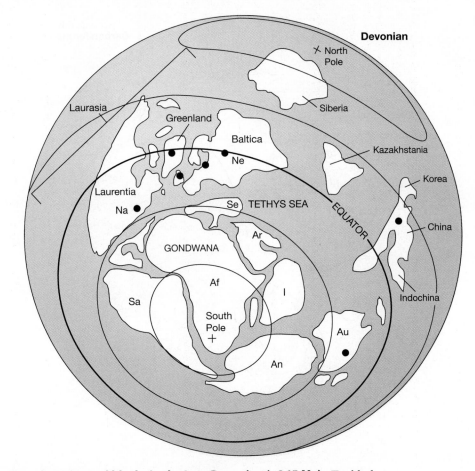

Figure 7–2 Location of continental blocks in the Late Devonian (~365 Ma). The black dots indicate fossil localities where Devonian tetrapods have been found. Laurentia, Greenland, and Baltica lie on the equator. An arm of the Tethys Sea extends westward between Gondwana and the northern continents.

The most dramatic radiation of animal life had occurred by the start of the Cambrian, with the so-called Cambrian explosion of animals with hard (preservable) parts, many lineages of which were extinct by the end of the period. A further radiation of marine animals occurred in the Ordovician period, and many of the groups that were to dominate the ecosystem of the rest of the Paleozoic appeared and radiated at this time. Although vertebrates date from the Early Cambrian, their major early diversification apparently took place during the Ordovician.

From the Devonian through the Permian, the continents were drifting together (**Figure 7–2**). The continental blocks that correspond to parts of modern North America, Greenland, and Western Europe had come into proximity along the equator. With the later addition of Siberia, these blocks formed a northern supercontinent known as **Laurasia**. Most of Gondwana was in the far south, overlying the South Pole. The Tethys Sea, which separated Gondwana from Laurentia, did not close completely until the Late Carboniferous, when Africa moved northward to collide with the east coast of North America (**Figure 7–3**).

During the Carboniferous, the process of coalescence continued, and by the Permian most of the continental surface was united in a single continent, **Pangaea** (sometimes spelled Pangea). At its maximum extent, the land area of Pangaea covered 36 percent of Earth's surface, compared with 31 percent for the present arrangement of continents. This supercontinent persisted for 160 million years, from the mid-Carboniferous to the mid-Jurassic period, and profoundly influenced the evolution of terrestrial plants and animals.

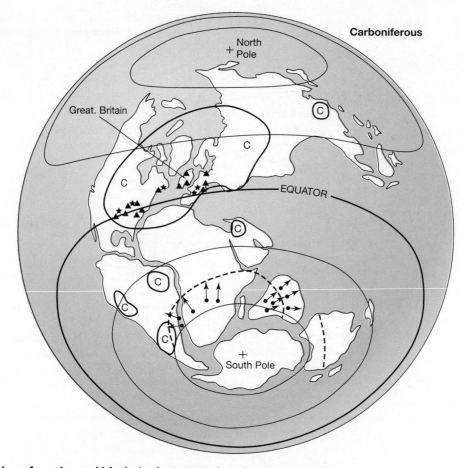

Figure 7–3 Location of continental blocks in the Late Carboniferous (~310 Ma). This map illustrates an early stage of the supercontinent Pangaea. The location and extent of continental glaciation are shown by the dashed lines and arrows radiating outward from the South Pole. The extent of the forests that formed today's coal beds is marked by the heavy lines enclosing each letter "C." The triangles mark the locations of Carboniferous tetrapods.

7.3 **Paleozoic Climates**

During the early Paleozoic, sea levels were at or near an all-time high for the Phanerozoic, and atmospheric carbon dioxide levels were also apparently very high. The oxygen level fluctuated, but in general was lower than the current level of 21 percent. The high levels of carbon dioxide would have resulted in a greenhouse effect, with the land experiencing very hot and dry climates during the Cambrian and much of the Ordovician. These conditions would have been unsuitable for the invasion of the terrestrial realm by early plants. However, a major glaciation in the Late Ordovician—combined with falling levels of atmospheric carbon dioxide—created cooler and moister conditions in at least some places, setting the scene for the development of the Silurian terrestrial ecosystems.

Atmospheric oxygen levels rose during the Silurian, and this may be the reason that the gnathostomes, which had higher energy needs than the agnathans (see Chapter 3), radiated rapidly at this time. Oxygen levels reached a maximum of about 25 percent in the Early Devonian, but by the mid-Devonian they had plummeted to around 12 percent. The increased use of lungs by the sarcopterygian fish precursors of tetrapods may be related to the low level of atmospheric oxygen at this time. A liter of water contains far less oxygen than a liter of air, and when atmospheric oxygen concentrations are low, aquatic oxygen concentrations are lower still. Thus, low atmospheric oxygen concentrations could have spurred not only greater air breathing by fishes but also the emergence of tetrapods onto land in the Late Devonian.

A relatively equable climate continued through the mid-Devonian, but glaciation recurred in the Late Devonian, and ice sheets covered much of Gondwana from the mid-Carboniferous until the mid-Permian. Glaciers regressed in the mid-Permian and an essentially ice-free world persisted until the mid-Cenozoic.

The waxing and waning of the glaciers created oscillations in sea level, which resulted in the cyclic formation of coal deposits, especially in eastern North America and Western Europe (see Figure 7–3). The climate over Pangaea was fairly uniform in the Early Carboniferous, but in the Late Carboniferous and Early Permian it was highly differentiated as the result of glaciation, with significant regional differences in the flora. Most vertebrates were found in equatorial regions during this time.

The spread of land plants in the Devonian, and their resultant modification of soils, profoundly affected Earth's atmosphere and climate, leading to climatic cooling. The formation of soils would speed the breakdown of the underlying rocks as plant roots penetrated them and organic secretions and the decomposition of dead plant material dissolved minerals in the rock. The evidence points to a sharp decrease (by around 90 percent) in atmospheric carbon dioxide levels during the Late Devonian, possibly related to the spread of land plants. Atmospheric levels of carbon dioxide reached an extreme low during the Late Carboniferous and Early Permian, resembling the levels of today's world. (Even with the increase in carbon dioxide concentrations that has occurred during the past century, atmospheric concentrations are much lower now than they were in most of the Phanerozoic.) In contrast, oxygen levels were high during this time, reaching present-day levels by the start of the Carboniferous and peaking at around 30 percent in the later Permian. The reverse greenhouse effect of this low level of atmospheric carbon dioxide probably caused the extensive Permo-Carboniferous glaciations.

7.4 Paleozoic Terrestrial Ecosystems

Photosynthesizing bacteria (cyanobacteria) probably existed in wet terrestrial habitats from the time of their origin in the Archean, and algae, lichens, and fungi have probably occurred on land since the late Proterozoic, but the earliest known complex terrestrial ecosystems, with evidence of associated plants, fungi, and animals (arthropods), appeared in the Early Silurian. There is no firm evidence for any animal life on land until the mid-Ordovician, although some fossil fragments and trackways suggest that arthropods may have come onto land as early as the Cambrian. Fragments of cuticle from the mid-Ordovician suggest the presence of very early plants. Land plants appear to represent a single terrestrial invasion from a particular group of green algae (the charophytes). These earliest plants were actually the first vascular plants, and the plants we consider basal today (bryophytes, represented now by mosses, liverworts, and hornworts) are not known as fossils until the Devonian. The landscape would have looked bleak by our standards—mostly barren, with a few kinds of low-growing vegetation limited to moist areas.

Silurian ecosystems consisted of a minimal terrestrial food web of primary producers (plants, still small forms), decomposers (fungi), secondary consumers (fungus-eating arthropods), and predators (larger arthropods such as millipedes and scorpions). Terrestrial ecosystems increased in complexity through the Early and Middle Devonian, with the appearance of more derived vascular plants at the start of the period, but food webs remained simple. Vascular plants have internal channels for the conduction of water through the tissues, and are thus able to grow taller than nonvascular plants such as mosses. Today plants form the base of the terrestrial food chain, but there is no evidence that Devonian invertebrates fed on living plants. Instead, arthropods such as springtails, millipedes, and mites were probably detritivores, consuming dead plant material and fungi. This process would recycle the plant nutrients to the soil. Spiders (initially burrowing forms rather than web builders) and scorpions were abundant and probably preyed on the smaller invertebrates.

In the Early Devonian the land would still have appeared fairly barren. However, the diversity of plant species was greater than it had been in the Silurian. By the Middle Devonian, the vascular plants attained heights of around 2 meters, and the canopy they created would have modified microclimatic conditions on the ground. Treelike forms evolved independently among several ancient plant lineages, and by the Middle Devonian, there were stratified forest communities consisting of plants of different heights. However, these plants were not closely related to modern trees and were not really like modern trees in their structure. It would not be possible to make furniture out of Devonian trees; they had narrow trunks and would not have provided enough woody tissue.

Although the terrestrial environment of the mid-Devonian was quite complex, it differed in many respects from the modern ecosystems to which we are accustomed. In the first place, there were no terrestrial vertebrates—the earliest of those appeared in the Late

Devonian. Furthermore, plant life was limited to wet places—low-lying areas and the margins of streams, rivers, and lakes. Flying and plant-eating insects were absent.

Terrestrial ecosystems became increasingly complex during the remainder of the Paleozoic. Plants with large leaves first appeared in the Late Devonian. The evolution of large leaves probably coincided with the drop in levels of atmospheric carbon dioxide because leaves with more stomata (pores) admit carbon dioxide more readily. The Late Devonian saw the spread of forests of the progymnosperm *Archaeopteris*, an early seed plant with a trunk up to a meter in diameter and a height of at least 10 meters. Giant horsetails reached heights of several meters, and there were also many species of giant clubmosses (lycophytes), a few of which survive today as small ground plants. Other areas were apparently covered by bushlike plants, vines, and low-growing ground cover. The diversity and habitat specificity of Late Devonian floras continued to expand in the Carboniferous. Most of the preserved habitats represent swamp environments, and vegetation buried in these swamps formed today's coal beds. (The word *carboniferous* means "coal-bearing.")

Most of the major groups of plants evolved during the Early Carboniferous, although the flowering seed plants (**angiosperms**) that are abundant today were as yet unknown. Seed ferns (the extinct pteridosperms) and ferns (which survive today) lived in well-drained areas, and swamps were dominated by clubmosses, with horsetails, ferns, and seed ferns also present. There were forests consisting of trees of varying heights, giving stratification to the canopy, and vinelike plants hung from their branches. The terrestrial vegetation would have looked superficially as it does today, although the actual types of plants present were completely different.

A fundamental transition in global vegetation types occurred near the end of the Late Carboniferous (around 305 million years ago). The coal deposits disappeared, signifying global drying and the collapse of the earlier types of rain-forest biomes. Seed plants such as conifers became an important component of the flora, replacing spore-bearing plants, and by the end of the Paleozoic they were the major group of plants worldwide. This vegetational change had a profound effect on the fauna.

Terrestrial invertebrates diversified during the Carboniferous. Flying insects are known from late in the Early Carboniferous, and fossils of damaged leaves show that herbivorous insects had become well established by the end of the Carboniferous. Higher levels of atmospheric oxygen at this time would have made flight easier to evolve, in part because flight requires high levels of activity and thus large quantities of food and oxygen, and also because an atmosphere with higher oxygen levels is somewhat denser. Higher oxygen levels also enabled insects to attain much larger body sizes than those seen today. Huge dragonflies (one species with a wingspan of 63 centimeters) flew through the air, and the extinct predatory arthropleurids were nearly 2 meters long. New arthropod types appearing in the Permian include hemipterans (bugs), beetles, and forms resembling mosquitoes but not closely related to true mosquitoes.

Terrestrial vertebrates appeared in the Late Devonian and diversified during the Carboniferous. The first amniotes were mid-Carboniferous in age; and, by the Late Carboniferous, amniotes had split into two major lineages—one leading to the mammals (synapsids) and the other to modern reptiles and birds (sauropsids). The diversity of tetrapods in the fossil record correlates with the number of ecological niches occupied. The earliest tetrapods would have radiated into new ecological niches, and their diversity increased as they evolved new and different ways of life.

The Carboniferous was characterized by a diversity of semiaquatic early tetrapods (**Figure 7–4**), but toward the end of the period there was a dramatic extinction of many types of aquatic tetrapods and an increase in the diversity of terrestrial forms, including both amniotes and some derived non-amniotes, such as the herbivorous genus *Diadectes*.

This is the time when the first herbivorous tetrapods appeared and, with them, the capacity to exploit the primary productivity of plants directly. By the Early Permian several vertebrate lineages had given rise to small insectivorous predators, rather like modern salamanders and lizards. Larger vertebrates (up to 1.5 meters long) were probably predators of these small species, and still larger predators topped the food web.

The Late Carboniferous and Early Permian terrestrial faunas showed more regional differentiation than had been the case, probably reflecting habitat fragmentation following the demise of the rain forests. By the Late Permian, terrestrial communities were assuming the pattern we know today, with a multitude of herbivorous vertebrates supporting a smaller variety of carnivorous forms, although the kinds of plants and animals in those ecosystems were still almost entirely different from the ones we know today.

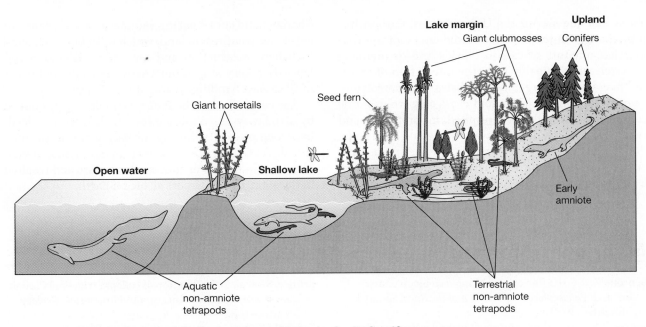

Figure 7–4 Aquatic and terrestrial plants and vertebrates in the Carboniferous. A reconstruction of a scene from a Late Carboniferous lake in Europe and its surroundings. The plants and animals are not shown exactly to scale. The largest animals would have been 1 to 1.5 m long.

7.5 **Paleozoic Extinctions**

The Paleozoic era saw a large number of extinctions, beginning with a major extinction among marine invertebrates in the Ordovician. (The record of vertebrates from that time is too poor to know whether this event affected them as well.) The next two major extinctions, both during the Late Devonian, had severe effects on aquatic organisms. The first affected mainly marine invertebrates, but the second, which occurred at the end of the Devonian, affected both marine and freshwater vertebrates. Thirty-five families of fishes (76 percent of the existing families) became extinct, including the ostracoderms, the placoderms, and many of the lobe-finned fishes.

A smaller, though significant, extinction occurred among land vertebrates at the end of the Early Permian. Fifteen families of tetrapods became extinct, including many non-amniote tetrapods ("amphibians" in the broad sense) and pelycosaurs (early "mammal-like reptiles"). These extinctions may have been related to climatic changes associated with the end of the Carboniferous-Permian glaciations and perhaps also with the accompanying changes in the atmosphere—a decrease in the level of oxygen and an increase in the level of carbon dioxide.

The background extinction rate increased in the Late Permian, suggesting that the environment was becoming less suitable for vertebrates, and the most severe extinctions in the history of life on Earth occurred in both terrestrial and marine environments at the end of the Paleozoic. Approximately 57 percent of marine invertebrate families and 95 percent of all marine species, including 12 families of fishes, disappeared. Twenty-seven families of tetrapods (49 percent) became extinct, with especially heavy losses among the synapsids ("mammal-like reptiles").

Levels of atmospheric oxygen were falling during this time, which may have caused hypoxic stress for the amniotic vertebrates that had evolved and radiated under conditions of oxygen abundance during the Early Permian. Furthermore, lower atmospheric oxygen levels may have limited the maximum altitude at which vertebrates could live. Restricting vertebrates to low altitudes would have reduced the habitat diversity available to them and limited their ability to migrate.

Hypotheses abound for the reason for the massive end-Permian extinction, including the possibility of an asteroid impact. The Permo-Triassic boundary coincides with massive volcanic eruptions in Siberia concentrated in a time period of less than a million years. Enough molten lava was released to cover an area about half the size of the United States. Carbon isotope changes in the boundary sediments indicate that massive amounts of carbon dioxide and/or methane were released, probably from the interaction of volcanic magma with organic-rich shales and petroleum-bearing evaporates. This rapid change in atmospheric composition would have resulted in profound and rapid global warming. Such warming would have had a direct effect not only on organisms but also on oceanic circulation, resulting in stagnation and low oxygen levels, with profound effects on marine organisms.

Charcoal and soot-bearing sediments are evidence of extensive wildfires on land, which would have resulted in habitat destruction and terrestrial ecosystem collapse. Recovery of rainforest ecosystems did not occur until at least 5 million years into the Triassic.

The Permo-Triassic boundary was apparently marked by a runaway greenhouse effect, with a positive feedback loop of events causing increasing global warming that disrupted the normal global environmental mechanisms for hundreds of thousands of years and resulted in the extinction of almost all life on Earth.

Additional Information

Benton, M. J. 2010. The origins of modern biodiversity on land. *Philosophical Transactions of the Royal Society B* 365:3667–3679.

Boyce, C. K. 2010. The evolution of plant development in a paleontological context. *Current Opinion in Plant Biology* 13:102–107.

Brodribb, T. J., et al. 2010. Viewing leaf structure and evolution from a hydraulic perspective. *Functional Plant Biology* 47:488–498.

Dahl, T. W., et al. 2011. Devonian rise in atmospheric oxygen correlated to radiation of terrestrial plants and large predatory fish. *Proceedings of the National Academy of Sciences* 107:17911–17915.

Harrison, J. F., et al. 2010. Atmospheric oxygen level and the evolution of insect body size. *Proceedings of the Royal Society B* 277:1937–1946.

Hazen, R. M. 2010. How old is the earth, and how do we know? *Evolution Education Outreach* 3:198–205.

Labandeira, C. C. 2005. Invasion of the continents: Cyanobacterial crusts to tree-inhabiting arthropods. *Trends in Ecology and Evolution* 20:253–262.

Qua, Q.-W., et al. 2010. Silurian atmospheric O_2 changes and the early radiation of gnathostomes. *Palaeoworld* 19:146–159.

Sahney, S., et al. 2010. Links between global taxonomic diversity, ecological diversity and the expansion of vertebrates on land. *Biology Letters* 6:544–547.

Sahney, S., et al. 2010. Rainforest collapse triggered Carboniferous tetrapod diversification in Euramerica. *Geology* 38:1079–1082.

Sallan, L. C., and M. I. Coates. 2010. End-Devonian extinction and a bottleneck in the early evolution of jawed vertebrates. *Proceedings of the National Academy of Sciences USA* 107:10131–10135.

Shen, S.-z., et al. 2011. Calibrating the end-Permian mass extinction. *Sciencexpress*: doi 10.1126/science.1213454.

Weng, J-K., and C. Chappie. 2010. The origin and evolution of lignin. *New Phytologist* 187:273–285.

Wodniok S., et al. 2011. Origin of land plants: Do conjugating green algae hold the key? *BMC Evolutionary Biology* 11:104.

Websites
Plate tectonics
Animations http://www.scotese.com
Mechanisms responsible for continental drift
http://www.ucmp.berkeley.edu/geology/tectonics.html
Paleozoic world
http://www.ucmp.berkeley.edu/paleozoic/paleozoic.html
http://www.fossilmuseum.net/PaleobiologyVFM.htm
Permo-Triassic extinctions
http://science.nationalgeographic.com/science/
prehistoric-world/permian-extinction.html

Living on Land

The spread of plants and then invertebrates across the land in the late Paleozoic era provided a new habitat for vertebrates and imposed new selective forces. The demands of terrestrial life are quite different from those of an aquatic environment because water and air have different physical properties. Air is less viscous and less dense than water, so streamlining is a minor factor for tetrapods, whereas a skeleton that supports the body against the pull of gravity is essential. Respiration, too, is different because of the properties of air and water. Gills don't work in air because the gill filaments collapse on each other without the support of water. When that happens, the surface area available for gas exchange is greatly reduced. And because water is both dense and viscous, pumping water into and out of a close-ended gas-exchange structure is prohibitively expensive; aquatic animals have flow-through gas-exchange structures. Terrestrial animals need a gas-exchange organ that won't collapse; given the low density and viscosity of air, tetrapods can use a tidal flow of air into and out of a saclike lung. Heat capacity and heat conductivity are additional differences between water and air that are important to terrestrial animals. Terrestrial habitats can have large temperature differences over a very short distance, and even small tetrapods can maintain body temperatures that are substantially

different from air temperatures. As a result, terrestrial animals have far more opportunity than aquatic organisms for regulation of their body temperatures.

8.1 Support and Locomotion on Land

Perhaps the most important difference between water and land is the effect of gravity on support and locomotion. Gravity has little significance for a fish living in water because the bodies of vertebrates are approximately the same density as water, and hence fishes are essentially weightless in water. Gravity is a very important factor on land, however, and the skeleton of a tetrapod must be able to support the body.

Water and air also require different forms of locomotion. Because water is dense, a fish swims merely by passing a sine wave along its body—the sides of the body and the fins push backward against the water, and the fish moves forward. Pushing backward against air doesn't move a tetrapod forward unless it has wings. Most tetrapods (including fliers when they are on the ground) use their legs and feet to transmit a backward force to the substrate. Thus, both the skeletons and modes of locomotion of tetrapods are different from those of fishes.

Bone

The skeleton is composed of bone, which must be rigid enough to resist the force of gravity and the forces exerted as an animal starts, stops, and turns. The remodeling capacity of bone is of great importance for a terrestrial animal as the internal structure of bone adjusts continuously to the changing demands of an animal's life. In humans, for example, intense physical activity results in an increase in bone mass, whereas inactivity (prolonged bed rest or being in space) results in loss of bone mass. In addition, remodeling allows broken bones to mend, and the skeletons of terrestrial animals experience greater stress than those of aquatic animals and are more likely to break. (Chapter 2 provides a background on bone in vertebrates, including a contrast between the endoskeleton of endochondral bone and the "exoskeleton" of dermal bone.)

Amniotes have bone that is arranged in concentric layers around blood vessels forming cylindrical units called Haversian systems (**Figure 8–1a**). The structure of a bone is not uniform. If it were, animals would be very heavy. The external layers of a bone are formed of dense, compact or lamellar bone, but the internal layers are lighter, spongy (cancellous) bone. The joints at the ends of bones are covered by a smooth layer of articular cartilage that reduces friction as the joint moves. (Arthritis occurs when this cartilage is damaged or worn.) The bone within the joint is composed of cancellous bone rather than dense bone, and the entire joint is enclosed in a joint capsule, containing synovial fluid for lubrication (Figure 8–1b).

The Axial System: Vertebrae and Ribs

The vertebrae and ribs of fishes stiffen the body so it will bend when muscles contract, rather than shortening. In tetrapods, the axial skeleton is modified for support on land. Processes called zygapophyses (singular *zygapophysis*) on the vertebrae of tetrapods (**Figure 8–2** on page 170) interlock and resist twisting (torsion) and bending (compression), allowing the spine to act like a suspension bridge to support the weight of the viscera on land (**Figure 8–3** on page 170). (Tetrapods that have permanently returned to the water, such as whales and many of the extinct Mesozoic marine reptiles, have lost the zygapophyses.)

Bony fishes use the opercular bones, formed from the dermal bone, to protect the gills, and these bones play a vital role in gill ventilation. A series of dermal bones behind and above the operculum (the supracleithral series) connect the head to the pectoral girdle. A fish cannot turn its head—instead it must pivot its entire body. Tetrapods have lost the bony connection between the head and pectoral girdle, and, as a result,

tetrapods have a flexible neck region and can move the head separately from the rest of the body. This is also the condition in *Titktaalik*, the fish most closely related to tetrapods (see Chapter 9). The cervical (neck) vertebrae (see Figure 8–3) allow the head to turn from side to side and up and down relative to the trunk, and the muscles that support and move the head are attached to processes on the cervical vertebrae. The two most anterior cervical vertebrae are the atlas and axis, and they are highly differentiated in mammals.

The trunk vertebrae are in the middle region of the body and bear the ribs. In mammals, the trunk vertebrae are differentiated into two regions: the thoracic vertebrae (those that bear ribs) and the lumbar vertebrae (those that have lost ribs).

The sacral vertebrae, which are derived from the trunk vertebrae, fuse with the pelvic girdle and allow the hind limbs to transfer force to the axial skeleton. Early tetrapods and extant amphibians have a single sacral vertebra, extant reptiles usually have two sacral vertebrae, mammals usually have two to five, and some dinosaurs had a dozen or more. The caudal vertebrae, found in the tail, are usually simpler in structure than the trunk vertebrae.

The ribs of early tetrapods were fairly stout and more prominently developed than those of fishes. They may have stiffened the trunk in animals that had not yet developed much postural support from the axial musculature (**Figure 8–4** [number 7] on page 171). The trunk ribs are the most prominent ones in tetrapods in general, and many basal tetrapods had small ribs extending throughout the entire vertebral column. Modern amphibians have almost entirely lost their ribs; in mammals, ribs are confined to the thoracic vertebrae.

Axial Muscles

The axial muscles assume two new roles in tetrapods: postural support of the body and ventilation of the lungs. These functions are more complex than the side-to-side bending produced by the axial muscles of fishes, and the axial muscles of tetrapods are highly differentiated in structure and function. Muscles are important for maintaining posture on land because the body is not supported by water; without muscular action, the skeleton would buckle and collapse. Likewise, the method of ventilating the lungs is different if the chest is surrounded by air rather than by water.

The axial muscles still participate in locomotion in basal tetrapods, producing the lateral bending of the backbone also seen during movement by many amphibians and reptiles. However, in birds and mammals, limb movements have largely replaced the lateral flexion of the trunk by axial muscles. The trunks of birds are rigid,

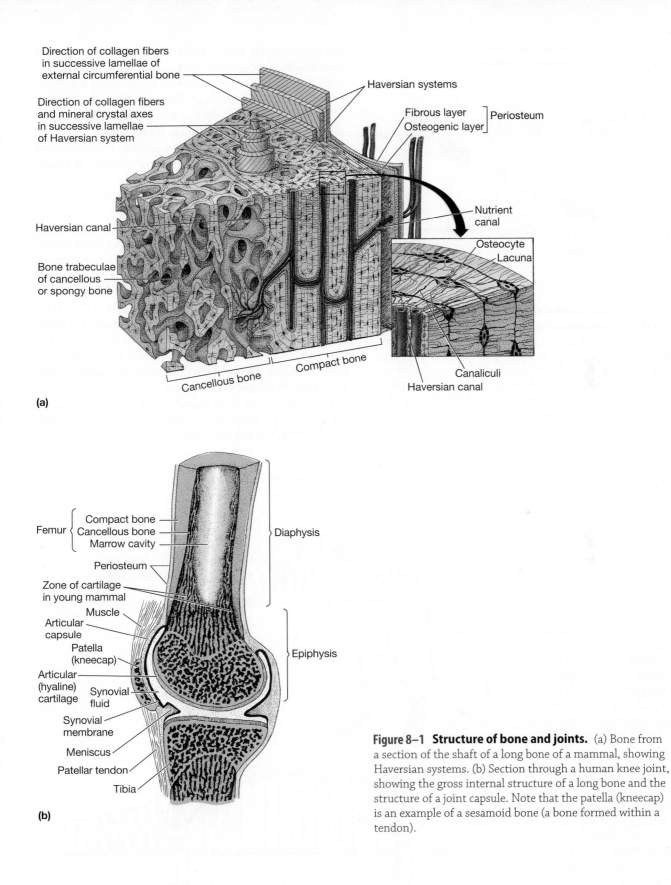

(a)

Direction of collagen fibers in successive lamellae of external circumferential bone

Direction of collagen fibers and mineral crystal axes in successive lamellae of Haversian system

Haversian canal

Bone trabeculae of cancellous or spongy bone

Cancellous bone

Compact bone

Haversian systems

Fibrous layer
Osteogenic layer } Periosteum

Nutrient canal

Osteocyte
Lacuna

Canaliculi

Haversian canal

(b)

Femur { Compact bone
Cancellous bone
Marrow cavity

Periosteum

Zone of cartilage in young mammal

Muscle

Articular capsule

Patella (kneecap)

Articular (hyaline) cartilage

Synovial fluid

Synovial membrane

Meniscus

Patellar tendon

Tibia

Diaphysis

Epiphysis

Figure 8–1 Structure of bone and joints. (a) Bone from a section of the shaft of a long bone of a mammal, showing Haversian systems. (b) Section through a human knee joint, showing the gross internal structure of a long bone and the structure of a joint capsule. Note that the patella (kneecap) is an example of a sesamoid bone (a bone formed within a tendon).

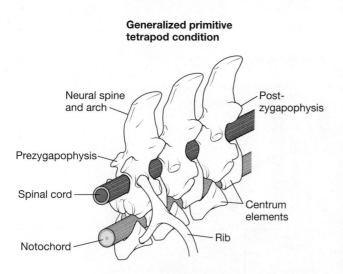

Generalized primitive tetrapod condition

Neural spine and arch

Post-zygapophysis

Prezygapophysis

Spinal cord

Centrum elements

Notochord

Rib

Figure 8–2 Gnathostome vertebrae and ribs. In aquatic vertebrates, the vertebral column primarily stiffens the body so that it bends instead of telescoping when muscles on one side of the body contract. Terrestrial vertebrates require more rigidity. Articulating surfaces (zygapophyses) on adjacent vertebrae of tetrapods allow the vertebral column to resist gravity (additional articulations are seen in some derived tetrapods). In derived tetrapods, the centrum is a single solid structure, obliterating the notochord in the adult form.

but dorsoventral flexion is an important component of mammalian locomotion. In secondarily aquatic mammals (e.g., whales), the axial muscles again assume a major role in locomotion in powering the tail.

In fishes and modern amphibians, the epaxial muscles form an undifferentiated single mass. This was probably their condition in the earliest tetrapods (see Figure 8–4 [number 11]). The epaxial muscles of extant amniotes are distinctly differentiated into three major components, and their primary role is now postural rather than locomotory.

The hypaxial muscles form two layers in bony fishes (the external and internal oblique muscles), but in tetrapods a third inner layer is added, the transversus abdominis (see Figure 8–4 [number 12]). This muscle is responsible for exhalation of air from the lungs of modern amphibians (which, unlike amniotes, do not use their ribs to breathe), and it may have been essential for respiration on land by early tetrapods. Air-breathing fishes use the pressure of the water column on the body to force air from the lungs, but land-dwelling tetrapods need muscular action. In unspecialized amphibians, such as salamanders, both epaxial and hypaxial muscles contribute to the bending motions of the trunk while walking on land, much as they do in fishes swimming in water.

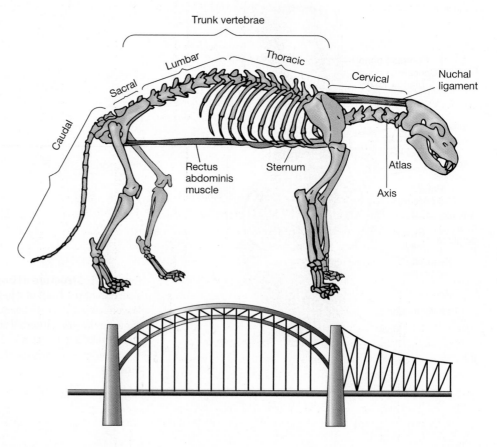

Trunk vertebrae

Lumbar

Thoracic

Cervical

Sacral

Nuchal ligament

Caudal

Rectus abdominis muscle

Sternum

Atlas

Axis

Figure 8–3 Skeleton of a cat. The muscles and ligaments of postural support are shown in comparison with the support elements of a suspension bridge. The front and back legs of the cat are like the pillars of the bridge: the ligaments that run between the vertebrae in the trunk act like the arch of the bridge to support the weight of the body. The nuchal ligament acts in a similar fashion to support the head, while the rectus abdominis muscle acts like the string of a bow to maintain the arch in the vertebral column.

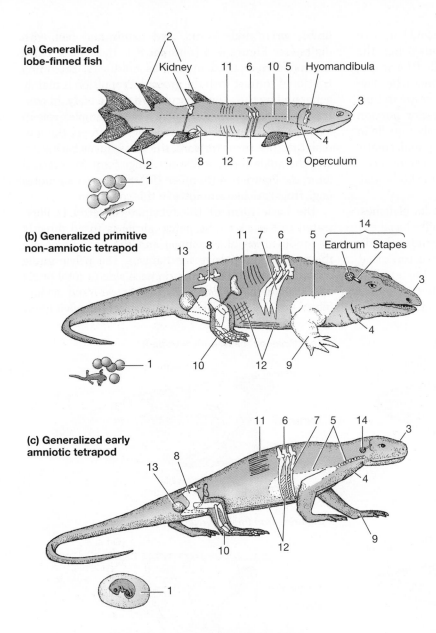

(a) Generalized lobe-finned fish

Kidney 11 6 10 5 Hyomandibula

2 2

3

4 Operculum

8 12 7 9

1

(b) Generalized primitive non-amniotic tetrapod

13 8 11 7 6 5 14 Eardrum Stapes

3

4

10 12 9

1

(c) Generalized early amniotic tetrapod

11 6 7 5 14

13 8 3

4

10 12 9

1

Figure 8–4 Comparison of morphological and physiological characteristics of fishes, basal tetrapods, and amniotes. Numbers indicate the various systems that are referred to in the text, and comparing the three drawings shows where the change occurred—for example, the length of the snout (number 3) changed between fishes and primitive tetrapods. **1.** Mode of reproduction. **2.** Presence of midline and tail fins. **3.** Length of snout. **4.** Length of neck. **5.** Form of lungs and trachea. **6.** Interlocking of vertebral column. **7.** Form of ribs. **8.** Attachment of pelvic girdle to vertebral column. **9.** Form of the limbs. **10.** Form of the ankle joint. **11.** Differentiation of epaxial muscles. **12.** Differentiation of hypaxial muscles. **13.** Presence of urinary bladder. (Note that the kidneys of fishes and non-amniotic tetrapods are in fact elongated structures lying along the dorsal body wall. The kidneys have been portrayed in all the animals as a mammalian bean-shaped form, for familiarity and convenience.) **14.** Form of the acousticolateralis system and middle ear.

The rectus abdominis is another new hypaxial muscle in tetrapods. It runs along the ventral surface from the pectoral girdle to the pelvic girdle, and its role appears to be primarily postural. (This is the muscle responsible for the six-pack stomach of human bodybuilders.) The costal muscles in the rib cage of amniotes are formed by all three layers of the hypaxial muscles and are responsible for inhalation as well as for exhalation. The use of the ribs and their associated musculature as devices to ventilate the lungs was probably an amniote innovation.

Thus, in tetrapods, the axial skeleton and its muscles assume very different roles from their original functions in aquatic vertebrates. The skeleton now participates in postural support and ventilation of the lungs, as well as in locomotion, and some of these functions are incompatible. For example, the side-to-side bending of the trunk that occurs when a lizard runs means that it has difficulty using its ribs for lung ventilation, creating a conflict between locomotion and respiration. More derived tetrapods such as mammals and birds have addressed this conflict by a change in posture from sprawling limbs to limbs that are held more directly underneath the body. These tetrapods are propelled by limb movements rather than by trunk bending.

The Appendicular Skeleton

The appendicular skeleton includes the limbs and limb girdles. In the ancestral gnathostome condition, illustrated by sharks, the pectoral girdle (supporting the front fins) is formed solely from the endoskeleton; it is a simple cartilaginous rod, called the coracoid bar, with a small ascending scapular process. In bony fishes, the endoskeletal portion of the pectoral girdle (the scapulocoracoid) is joined with some dermal bones (the clavicle and cleithrum), and these dermal bones are in turn connected to bones that form the posterior portion of the dermal skull roof (**Figure 8–5**). The pelvic girdle in both kinds of fishes is represented by the puboischiatic plate, which has no connection with the vertebral column but merely anchors the hind fins in the body wall. Neither arrangement works well on land.

The tetrapod limb is derived from the fin of fishes. The basic structure of a fin consists of fanlike basal elements supporting one or more ranks of cylindrical radials, which usually articulate with raylike structures that support most of the surface of the fin web. The tetrapod limb is made up of the limb girdle and five segments that articulate end to end. All tetrapods have jointed limbs, wrist/ankle joints, and hands and feet with digits (see Figure 8–4 [number 9]). The feet of basal tetrapods were used mainly as holdfasts, providing frictional contact with the ground. Propulsion is mainly generated by the axial musculature of the body. In contrast, the feet of amniotes play a more complex role in locomotion. They are used as levers to propel the animal: the knee turns forward, the elbow turns backward, and the ankle forms a distinct hinge joint (mesotarsal joint; see Figure 8–4 [number 10]). Some non-amniotes (e.g., frogs) are like amniotes in this condition.

The basic form of the tetrapod skeleton is illustrated in **Figure 8–6**. The pelvic girdle is fused directly to modified sacral vertebrae, and the hind limbs are the primary propulsive mechanism. The pelvic girdle contains three paired bones on each side (a total of six bones): ilium (plural *ilia*), pubis (plural *pubes*), and ischium (plural *ischia*). The ilia on each side connect the pelvic limbs

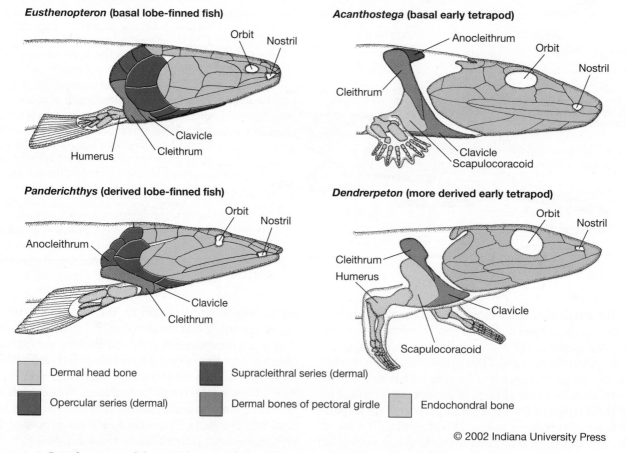

© 2002 Indiana University Press

Figure 8–5 Development of the neck in tetrapods. The bones of the skull are connected to the pectoral girdle by the supracleithral bones in the basal lobe-finned fish *Eusthenopteron* and the derived lobe-finned fish *Panderichthys*. Note that the posterior edge of the operculum is not joined to the bones behind it: the operculum is mobile, and this is where water exits from the gills. The opercular bones and supracleithral series have been lost in the basal tetrapod *Acanthostega* and the more derived tetrapod *Dendrerpeton*.

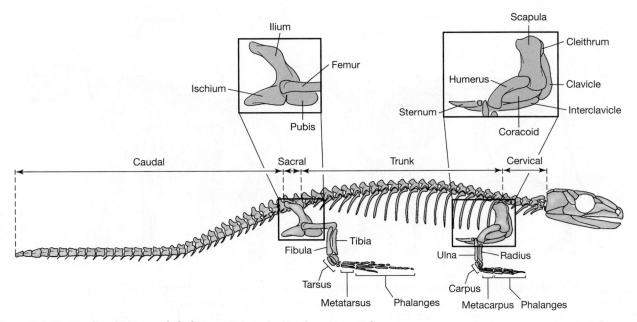

Figure 8–6 Generalized tetrapod skeleton. This is the basal amniote *Hylonomus*.

to the vertebral column, forming an attachment at the sacrum via one or more modified ribs (see Figure 8–4 [number 8]).

Support of the forelimb is only a minor role of the pectoral girdle of fishes, which mainly anchors the muscles that move the gills and lower jaw. In bony fishes, the pectoral girdle and forelimb are attached to the back of the head via a series of dermal bones (see Figure 8–5). In tetrapods these bones are lost, and the pectoral girdle is freed from the dermal skull roof. The main endochondral bones are the scapula and the coracoid; the humerus (upper arm bone) articulates with the pectoral girdle where these two bones meet. However, some dermal bones (the anocleithrum, the cleithrum, the clavicle, and the interclavicle) become incorporated into the pectoral girdle along the anterior border of the scapula. The cleithrum and anocleithrum are seen only in extinct tetrapods. The clavicle (the collar bone of humans) connects the scapula to the sternum or to the interclavicle. The interclavicle is a single midline element lying ventral to the sternum, seen today in lizards and crocodiles. It has been lost in birds and in most mammals but is still present in the monotremes.

Unlike the pelvic girdle, the pectoral girdle does not articulate directly with the vertebral column: this is why you can shrug your shoulders but you can't shrug your hips. (Only in some pterosaurs [extinct flying reptiles] is there an equivalent of a sacrum in the anterior vertebral column, a structure called the notarium.) In all other vertebrates the connection between the pectoral girdle and the vertebral column consists of muscles and connective tissue that hold the pectoral girdle to the sternum and the ribs. The sternum is a midventral structure, formed from endochondral bone and usually segmented, that links the lower ends of right and left thoracic ribs in amniotes. The sternum is extensively ossified only in birds and mammals. Bones in the shoulder girdle of frogs and salamanders (called sternal elements) may not be homologous with the amniote sternum.

The midline fins of fishes—the dorsal and anal fins—help to reduce roll, but they have no function on land and are not present in terrestrial animals (although a dorsal "fin," without any internal fin structure, may be present in secondarily aquatic tetrapods such as dolphins and ichthyosaurs [extinct marine reptiles]). The pectoral and pelvic fins are used for hydrodynamic lift, steering, and braking but not usually for propulsion except in rays and some coral reef fishes. The pectoral and pelvic fins of fishes become the limbs of tetrapods. The appendicular muscles of tetrapods are more complicated and differentiated than those of fishes. The ancestral fish pattern of a major levator and a major depressor is retained, but derived fishes (humans, for example) have many additional muscles in the shoulder region alone—not to mention the ones that move the other joints, including the fingers and toes.

Size and Scaling

Body size is one of the most important things to know about an organism. Because all structures are subject to the laws of physics, the absolute size of an animal profoundly affects its anatomy and physiology. The

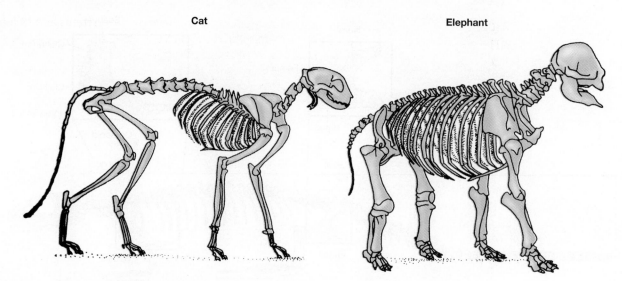

Cat

Elephant

Figure 8–7 Which is bigger? Even though they are drawn to the same size, it is instantly apparent that the animal on the right (an elephant) is larger than the animal on the left (a cat) because of the proportions of the bones, especially the limbs.

structure of the skeletomuscular system is especially sensitive to absolute body size in tetrapods because of the effects of gravity on land. The skeleton must be built not only to support the actual weight of the animal, but must additionally be strong enough to withstand forces generated during locomotion. Vertebrate skeletons are actually "overbuilt" to withstand forces 5-10 times the amount of the simple support of the body weight while standing still. This is known as the "safety factor."

The study of scaling, or how shape changes with size, is also known as **allometry** (Greek *allo* = different and *metric* = measure). If the features of an animal showed no relative changes with increasing body size (i.e., if a larger animal appeared just like a photo enlargement of a smaller one), then all of its component parts would be scaled with isometry (Greek *iso* = the same). However, animals are not built this way, and very few body components scale isometrically.

Underlying all scaling relationships is the issue of how the linear dimensions of an object relate to its surface area and volume: when linear dimensions double (i.e., a twofold change), the surface area increases as the square of the change in linear dimensions (a fourfold change) and the volume increases as the cube of the change in linear dimensions (an eightfold change). Thus, an animal that is twice as tall as another is eight times as heavy.

It is the cross-sectional area of a limb bone that actually supports an animal's weight on land. If an animal increased in size isometrically, its weight would rise as

the cube of its linear dimensions, but the cross section of its bones would increase only as the square of its linear dimensions. As a result, its limbs would be unable to support the increased weight. Thus, the limb bones must have greater cross-sectional area to keep up with increases in weight, and bone diameter scales with positive allometry; that is, bigger animals have proportionally thicker limb bones than smaller ones The skeleton of a bigger animal can easily be distinguished from that of a smaller animal, even when they are drawn to the same size, by the proportionally thicker bones of the large animal (**Figure 8–7**).

Although the cross-sectional area of the limb bones of terrestrial vertebrates does scale with positive allometry, the area does not increase in proportion to the stress the bones experience. Consequently, the limbs of large animals are relatively more fragile than the limbs of small animals, and large animals have a different posture than small animals. Small animals stand with flexed joints, but large animals stand erect on straight legs. (Compare the posture of the limbs of the cat and the elephant in Figure 8–7.) The reason for this difference lies in the mechanical aspects of bone: A bone withstands compressive forces (forces exerted parallel to the long axis of the bone) much better than shearing forces (those exerted at an angle to the long axis). The larger the animal, the less flexed are its limb joints. The pillarlike weight-bearing stance of very large animals, such as elephants and the huge sauropod dinosaurs, reduces the effect of shearing on their limb bones.

Locomotion on Land

The basic form of tetrapod locomotion, still seen today in salamanders, combines lateral axial movements with diagonal pairs of legs moving together, but with the limbs used more as holdfasts than for propulsion. The right front and left hind move as one unit and the left front and right hind move as another in a type of gait known as the walking-trot. Lizards retain a modified version of this mode of locomotion, although their limbs are more important for propulsion. Even though humans are bipedal, relying entirely on the hind legs for locomotion, we retain this ancestral coupling of the limbs in walking, swinging the right arm forward when striding with the left leg, and vice versa. This type of coupled, diagonally paired limb movement is probably an ancestral feature for gnathostomes because sharks also move their fins in this fashion when bottom-walking over the substrate. **Figure 8–8** shows the modes of locomotion used by tetrapods in a phylogenetic perspective.

Amniotes employ the "walk" gait, in which each leg moves independently in succession, usually with three feet on the ground at any one time. The "amble" is a speeded-up walk in which only one foot or two feet are on the ground at any time. The "trot" is used by all amniotes for faster movement. In this gait diagonal pairs of limbs move together (e.g., right front and left rear) as in the ancestral tetrapod condition, but often with a period of suspension when all four legs are off the ground.

Derived amniotes, such as mammals and archosaurs (birds, dinosaurs, and crocodiles), have an upright posture and hold their limbs more nearly underneath the body. While archosaurs tended toward bipedalism, mammals devised some new modes of locomotion with the evolution of the dorsoventral flexion of the vertebral column. The new fast gait that is characteristic of mammals is the "bound," which involves jumping off the hind legs and landing on the forelegs, with the flexion of the back contributing to the length of the stride. In larger mammals the bound is modified into the "gallop"—as seen, for example, in horses—and there is less flexion of the backbone (**Figure 8–9** on page 177).

8.2 Eating on Land

The difference between water and air profoundly affects feeding by tetrapods. In water, most food items are nearly weightless, and aquatic animals can suck the food into their mouth and move it within the mouth by creating currents of water. Aquatic vertebrates, from the tiniest tadpoles to the largest whales, use suction feeding to capture food items suspended in the water column. In contrast, terrestrial animals use their jaws, tongues, and teeth to seize food items and to manipulate items in the mouth.

The skull of early tetrapods is much like that of early bony fishes, with extensive dermal skull bones. (These bones are retained in most extant tetrapods—most of the bones of the human skull represent the legacy of this bony fish dermatocranium.) However, the gill skeleton and the bones connecting the pectoral girdle to the head have been lost in all but the very earliest known tetrapods, and no tetrapod has retained the operculum. The skull of bony fishes has a short snout; movements of the jaws, hyoid apparatus (the lower part of the hyoid arch), and operculum cause water to be sucked into the mouth for both gill ventilation and feeding.

Early tetrapods had wide, flat skulls and longer snouts than their fish ancestors so that most of the tooth row was now in front of the eye. Their flat heads and long snouts combined the functions of feeding and breathing, as do the heads of extant amphibians, which use movements of the hyoid apparatus to ventilate the lungs. This method of lung ventilation is called a positive-pressure mechanism, or buccal pumping. The same expansion of the buccal cavity is used for suction feeding in water. Suction feeding is not an option on land, however, because air is much less dense than the food particles. (You can suck up the noodles in soup along with the liquid, but you cannot suck up the same noodles if you put them on a plate.)

The tongue of jawed fishes is small and bony, whereas the tongue of tetrapods is large and muscular. (The muscular tongues of lampreys and hagfishes are not homologous with the tongues of tetrapods, and they are innervated by different cranial nerves.) The tetrapod tongue works in concert with the hyoid apparatus and is probably a key innovation for feeding on land. Most tetrapods use the tongue to manipulate food in the mouth and transport it to the pharynx. Most terrestrial salamanders and lizards have sticky tongues that help to capture prey and transport it into the mouth—a phenomenon called "prehension." In addition, some tetrapods—such as frogs, salamanders, and the true chameleon lizards—can project their tongue to capture prey. (The mechanism of tongue projection is different in each group; tongue projection is an example of convergent evolution.)

Salivary glands are known only in terrestrial vertebrates, probably because lubrication is not required to swallow food in water. Saliva also contains enzymes that begin the chemical digestion of food while it is still in the mouth. Some insectivorous mammals, two species of lizards, and several lineages of snakes have elaborated salivary secretions into venoms that kill prey.

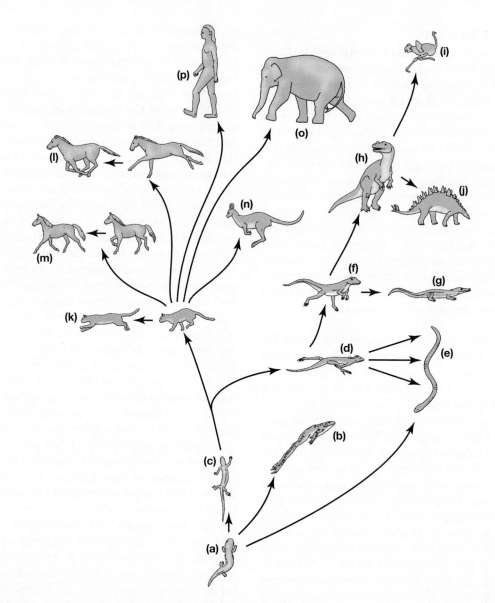

Figure 8–8 Phylogenetic view of tetrapod terrestrial stance and locomotion. (a) Ancestral tetrapod condition, retained today in salamanders: movement mainly via axial movements of the body, limbs moved in diagonal pairs (basic walking-trot gait). (b) Derived jumping form of locomotion in the frog, with highly specialized hind limbs. (c) Ancestral amniote condition, seen in many extant lizards: limbs used more for propulsion, with development of the walk gait (limbs moved one at a time/independently). (d) Diapsid amniote condition with hind limbs longer than forelimbs; tendency for bipedal running, seen in some extant lizards. (e) Derived limbless condition with snakelike locomotion; evolved convergently several times among early tetrapods (e.g., several types of lepospondyls), lissamphibians (caecilians and salamanders), and lepidosaurs (many lizards, snakes, amphisbaenians). (f) Ancestral archosaur condition, with upright posture and tendency to bipedalism. (g) Secondary return to sprawling posture and quadrupedalism in crocodilians. (h) Obligate bipedality in early dinosaurs and (i) birds. (j) Return to quadrupedality several times within dinosaurs. (k) Ancestral mammalian condition: upright posture and the use of the bound as a fast gait with dorsoventral flexion of the vertebral column (all mammals use the walk as a slow gait). In the bound the animal jumps off the two hind limbs, flies through the air with limbs outstretched, and lands on the two forelimbs (or on one forelimb and then the other, as in this half-bounding cat). (l) Condition in larger mammals where the bound is turned into the gallop: the limbs move one at a time, and the period of suspension when all four feet are in the air occurs when the legs are bunched up, as shown. (m) The trot is used at intermediate speeds between the walk and the gallop. The canter is essentially a slower version of the gallop. (n) The ricochet, a derived hopping gait of kangaroos and some rodents. (o) The amble, a speeded-up walk gait seen in the fast gait of elephants and in some horses (e.g., Paso Finos and Icelandic ponies). (p) The unique human condition of upright bipedal striding. Penguins can also walk with an upright trunk, but they waddle rather than stride.

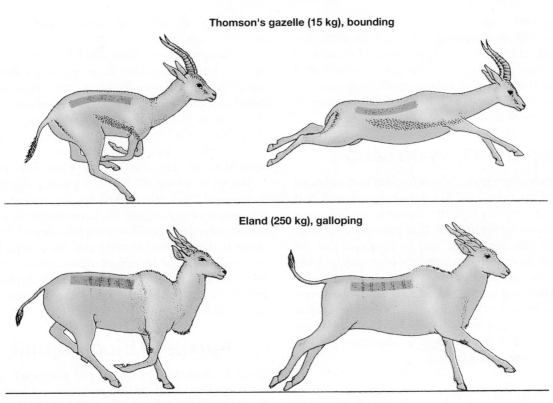

Thomson's gazelle (15 kg), bounding

Eland (250 kg), galloping

Figure 8–9 Bounding and galloping. Small species of antelopes, such as the Thomson's gazelle, bound, whereas larger species, represented by the eland, gallop. The dark bars represent the vertebral column.

With the loss of gills in tetrapods, much of the associated branchiomeric musculature was also lost, but the gill levators are a prominent exception. In fishes, these muscles are combined into a single unit, the cucullaris, and this muscle in tetrapods becomes the trapezius, which runs from the top of the neck and shoulders to the shoulder girdle. In mammals, this muscle helps to rotate and stabilize the scapulae (shoulder blades) in locomotion, and we use it when we shrug our shoulders.

Understanding the original homologies of the trapezius muscle explains an interesting fact about human spinal injuries. Because the trapezius is an old branchiomeric muscle, it is innervated directly from the brain by cranial nerves (cranial nerve XI, which is actually part of nerve X), not by the nerves exiting from the spinal cord in the neck. Thus people who are paralyzed from the neck down by a spinal injury can still shrug their shoulders. Small muscles in the throat—for example, those powering the larynx and the vocal cords—are other remnants of the branchiomeric muscles associated with the gill arches. Ingenious biomedical engineering allows quadriplegic individuals to use this remaining muscle function to control prosthetic devices.

The major branchiomeric muscles that are retained in tetrapods are associated with the mandibular and hyoid arches and are solely involved in feeding (Figure 8–10). The adductor mandibulae remains the major jaw-closing muscle, and it becomes increasingly complex in more derived tetrapods. The hyoid musculature forms two

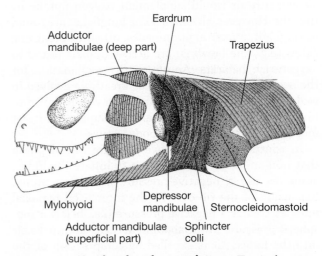

Figure 8–10 Head and neck musculature. This is the generalized tetrapod condition as seen in a tuatara (*Sphenodon*).

new important muscles in tetrapods. One is the depressor mandibulae, running from the back of the jaw to the skull and helping the hypobranchial muscles to open the mouth. The other is the sphincter colli that surrounds the neck and aids in swallowing food. In mammals the sphincter colli has become the muscles of facial expression.

8.3 Reproduction on Land

Sauropsids and synapsids, the tetrapods that are most specialized for terrestrial life, are amniotes, and that observation suggests that the amniotic egg has some special advantage for animals that reproduce on land. More than one mechanism may be involved, and the presence of an egg shell in amniotic eggs may be a key element. An egg shell provides support for the egg, and the shell may be the reason that all the large tetrapods are amniotes. In addition, a shell restricts water movement into or out of the egg, and the shell may allow amniotes to deposit eggs in sites that are not suitable for non-amniotes.

8.4 Breathing Air

In some respects air is an easier medium for respiration than is water. The low density and viscosity of air make tidal ventilation of a saclike lung energetically feasible, and the high oxygen content of air reduces the volume of fluid that must be pumped to meet an animal's metabolic requirements.

The lung is an ancestral feature of bony fishes, and lungs were not evolved for breathing on land. For many years it was assumed that lungs evolved in fishes living in stagnant, oxygen-depleted water where gulping oxygen-rich air would supplement oxygen uptake by the gills. However, although some lungfishes are found in stagnant, anoxic environments, other air-breathing fishes (e.g., the bowfins) are active animals found in oxygen-rich habitats. An alternative explanation for the evolution of lungs is that air breathing evolved in well-aerated waters in active fishes in which the additional oxygen is needed primarily to supply the heart muscle itself rather than the body tissues.

In contrast to the positive-pressure buccal pump that non-amniotic tetrapods use to inflate the lungs, amniotes use a negative-pressure aspiration pump. Expansion of the rib cage by the intercostal hypaxial muscles creates a negative pressure (i.e., below atmospheric pressure) in the abdominal cavity and sucks air into the lungs. Air is expelled by compression of the abdominal cavity, primarily through elastic return of the rib cage to a resting position and contraction of the elastic lungs, as well as by contraction of the transversus abdominis muscle.

The lungs of many extant amphibians are simple sacs with few internal divisions. They have only a short chamber leading directly into the lungs. In contrast, amniotes have lungs that are subdivided, sometimes in very complex ways, to increase the surface area for gas exchange. They also have a long trachea (windpipe), strengthened by cartilaginous rings, that branches into a series of bronchi in each lung (see Figure 8–4 [number 5]).

The form of lung subdivisions is somewhat different in mammals and other amniotes, suggesting independent evolutionary origins. The combination of a trachea and negative-pressure aspiration has allowed many amniotes to develop longer necks than those seen in modern amphibians or in extinct non-amniotic tetrapods. Amniotes also possess a larynx (derived from pharyngeal arch elements) at the junction of the pharynx and the trachea that may be used for sound production.

8.5 Pumping Blood Uphill

Blood is weightless in water, and the heart needs to overcome only fluid resistance to move blood around the body. Circulation is more difficult for a terrestrial animal because blood tends to pool in low spots, such as the limbs, and it must be forced through the veins and back up to the heart by more blood pumped into the arteries. Thus, tetrapods have blood pressures high enough to push blood upward through the veins against the pull of gravity and valves in the limb veins to resist backflow.

The walls of capillaries are somewhat leaky, and high blood pressure forces some of the blood plasma (the liquid part of blood) out of the vessels and into the intercellular spaces of the body tissues. This fluid is recovered and returned to the circulatory system by the lymphatic system of tetrapods. The lymphatic system is a one-way system of blind-ended, veinlike vessels that parallel the veins and allow fluid in the tissues to drain into the venous system at the base of the neck. (A lymphatic system is also well developed in teleost fishes, but it is of critical importance on land, where the cardiovascular system is subject to the forces of gravity.) In tetrapods, valves in the lymph vessels prevent backflow, and contraction of muscles and tissues keeps lymph flowing toward the heart.

Lymph nodes, concentrations of lymphatic tissues, are found in mammals and some birds at intervals along the lymphatic channels. Lymphatic tissue is also involved in the immune system; white blood cells (macrophages) travel through lymph vessels, and the lymph nodes can intercept foreign or unwanted material, such as migrating cancer cells.

With the loss of the gills and the evolution of a distinct neck in tetrapods, the heart has moved posteriorly. In fishes the heart lies in the gill region in front of the shoulder girdle, whereas in tetrapods it lies behind the shoulder girdle in the thorax. The sinus venosus and conus arteriosus are reduced or absent in the hearts of tetrapods.

Tetrapods have evolved a double circulation in which the pulmonary circuit supplies the lungs with deoxygenated blood and the systemic circuit supplies oxygenated blood to the body. The atrium of the heart is divided into left and right chambers (atria) in lungfishes and tetrapods, and the ventricle is divided either by a fixed barrier or by the formation of transiently separate chambers as the heart contracts. The right side of the heart receives deoxygenated blood returning from the body via the systemic veins, and the left side of the heart receives oxygenated blood returning from the lungs via the pulmonary veins. The double circulation of tetrapods can be pictured as a figure eight, with the heart at the intersection of the loops (**Figure 8–11**).

The aortic arches have undergone considerable change in association with the loss of the gills in tetrapods. Arches 2 and 5 are lost in most adult tetrapods (although arch 5 is retained in salamanders). Three major arches are retained: arch 3 (carotid arch) going to the head, arch 4 (systemic arch) going to the body, and arch 6 (pulmonary arch) going to the lungs (**Figure 8–12** on pages 180 and 181).

Modern amphibians retain a fishlike condition, in which the aortic arches do not arise directly from the heart. This condition, with retention of a prominent conus arteriosus and a ventral arterial trunk (the ventral aorta of fishes and the truncus arteriosus of amphibians), was probably found among the earliest tetrapods. In amniotes, the pulmonary artery receives blood from the right ventricle, and the right systemic and carotid arches receive blood from the left ventricle, although details of the heart anatomy suggest that this condition may have evolved independently in mammals and in other amniotes.

In modern amphibians, the skin is of prime importance in the exchange of oxygen and carbon dioxide. In frogs, the pulmonary arch is actually a pulmocutaneous arch, with a major artery branching off the pulmonary artery to supply the skin. The cutaneous vein, now carrying oxygenated blood, feeds back into the systemic system via the subclavian vein and hence into the right atrium. A similar, but less well-differentiated, system exists in other amphibians. Thus, oxygenated blood enters the amphibian ventricle from both the left atrium (supplied by the pulmonary vein) and the right atrium (supplied by veins that return blood from the skin). This type of heart in modern amphibians, with the absence of any ventricular division (in contrast to the lungfish condition with a partial ventricular septum), may be a derived condition associated with using the skin as well as the lungs for respiration.

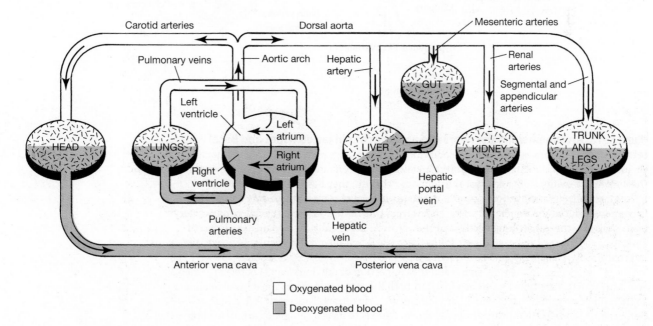

Figure 8–11 Double-circuit cardiovascular system in a tetrapod.

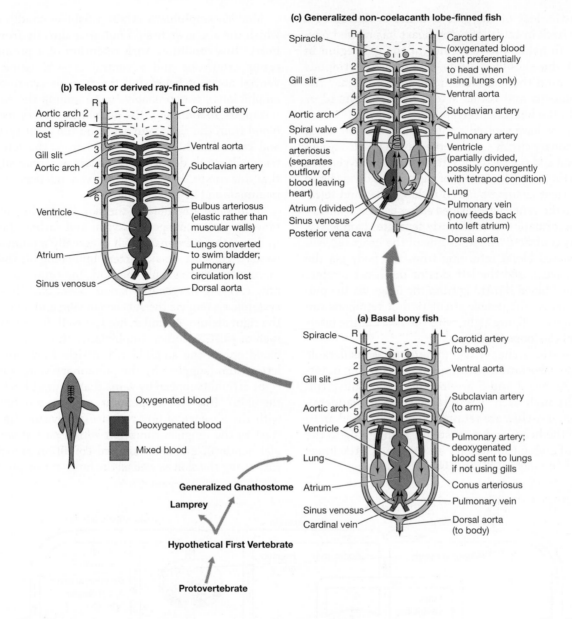

Figure 8–12 Arrangement of the heart and aortic arches in jawed fishes and basal tetrapods. (a) The ancestral bony fish condition is to have lungs and some sort of pulmonary circuit, as shown in the extant actinopterygian with lungs, *Polypterus*. (b) Teleosts have converted the lung into a swim bladder and lost the pulmonary circuit; they have also lost aortic arch 2. (c) The generalized sarcopterygian condition above the level of coelacanths. Some living lungfishes have reduced the number of aortic arches from this condition. Here the pulmonary artery feeds back into the left atrium directly, and the ventricle is partially divided. (d) The inferred early tetrapod (above the *Acanthostega/Ichthyostega* level) is similar to lungfishes; however, the internal gills have been lost, and aortic arch 2 may also have been lost by this stage (as seen in all extant tetrapods). (e) In the frog, arch 5 and the connection of the dorsal aorta between arches 3 and 4 (the ductus carotidus) have been lost, but both these features are retained in salamanders, so this condition cannot have been inherited from an early tetrapod ancestor. A derived condition in some modern amphibians (frogs, as shown here) is to have a cutaneous artery leading

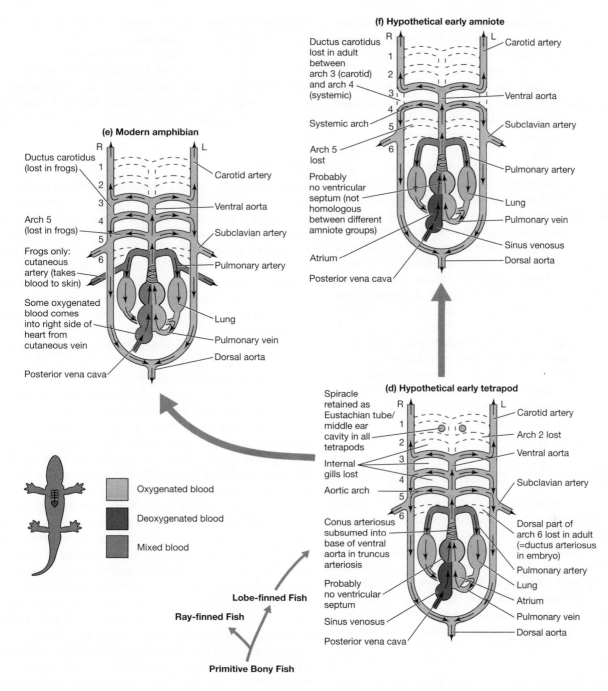

(f) Hypothetical early amniote

Ductus carotidus lost in adult between arch 3 (carotid) and arch 4 (systemic)

Carotid artery

Ventral aorta

Systemic arch

Subclavian artery

Arch 5 lost

Pulmonary artery

Probably no ventricular septum (not homologous between different amniote groups)

Lung

Pulmonary vein

Sinus venosus

Atrium

Dorsal aorta

Posterior vena cava

(e) Modern amphibian

Ductus carotidus (lost in frogs)

Carotid artery

Ventral aorta

Arch 5 (lost in frogs)

Subclavian artery

Frogs only: cutaneous artery (takes blood to skin)

Pulmonary artery

Some oxygenated blood comes into right side of heart from cutaneous vein

Lung

Pulmonary vein

Dorsal aorta

Posterior vena cava

Oxygenated blood

Deoxygenated blood

Mixed blood

(d) Hypothetical early tetrapod

Spiracle retained as Eustachian tube/ middle ear cavity in all tetrapods

Carotid artery

Arch 2 lost

Internal gills lost

Ventral aorta

Aortic arch

Subclavian artery

Conus arteriosus subsumed into base of ventral aorta in truncus arteriosis

Dorsal part of arch 6 lost in adult (=ductus arteriosus in embryo)

Pulmonary artery

Probably no ventricular septum

Lung

Atrium

Sinus venosus

Pulmonary vein

Posterior vena cava

Dorsal aorta

Lobe-finned Fish

Ray-finned Fish

Primitive Bony Fish

Figure 8–12 (*continued*)
from the pulmonary artery taking blood to the skin to be oxygenated; the blood returns via the cutaneous vein that feeds into the subclavian vein. The ventricular septum is absent in modern amphibians: it is not clear if this is a primary or secondary condition. (f) The proposed early amniote condition is similar to that of the frog with the exception of the absence of a cutaneous circuit. The ventricular septum would have been simple, if it was present at all, and there would have been no division of the vessels leaving the heart because the more derived conditions in extant sauropsid and synapsid amniotes are not homologous (i.e., one cannot be derived from the other). (R = right, L = left).

A ventricular septum of some sort is present in all amniotes, but the form is different in the various lineages of amniotes. A transient ventricular septum is formed during ventricular contraction in turtles and lizards, whereas a permanent septum is present in mammals, crocodiles, and birds. This pattern of occurrence suggests that a permanent ventricular septum evolved independently in the sauropsid and synapsid lineages.

The heavy workload and divided ventricle of birds and mammals introduce another complication: how to supply oxygen to the heart muscle. Modern amphibians and nonavian reptiles have lower blood pressures than mammals and birds, their hearts don't work as hard, and their ventricles allow some mixing of oxygenated and deoxygenated blood. The hearts of these animals never evolved coronary arteries, presumably because enough oxygen diffuses into the cardiac muscle from the blood in the lumen of the ventricle. In contrast, the ventricular muscles of mammals and birds are thicker and must work harder than those of amphibians and non-avian reptiles to generate higher blood pressures. In addition, these animals have a permanent ventricular septum, so the right ventricle contains only deoxygenated blood. Both birds and mammals have coronary arteries that supply oxygenated blood to the muscles of both ventricles, and these arteries apparently evolved separately in the two lineages.

8.6 Sensory Systems in Air

Some of the sensory modes that are exquisitely sensitive in water are useless in air, whereas other modes work better in air than in water. Air is not dense enough to stimulate the mechanical receptors of the lateral line system, for example, and does not conduct electricity well enough to support electrosensation. Chemical systems work well on land, however, at least for molecules small enough to be suspended in air, and air offers advantages for both vision and hearing.

Vision

The sense of vision is easier to use in air than in water because light is transferred through air with less disturbance than through water. Air is rarely murky in the way that water can frequently be, and, as a result, vision is more useful as a distance sense in air than in water.

In air, the cornea (the transparent covering of the front of the eye) participates in focusing light on the retina, and tetrapods have flatter lenses than do fishes and focus an image on the retina differently. Fishes focus light by moving the position of the lens within the eye, whereas tetrapods focus by changing the shape of the lens. (Snakes are an odd exception here; they move the lens to focus the eye.)

In air, the eye's surface must be protected and kept moist and free of particles. New features in tetrapods include eyelids, glands that lubricate the eye and keep it moist (including tear-producing lacrimal glands), and a nasolacrimal duct to drain the tears from the eyes into the nose.

Hearing

Sound perception is very different in air than in water. The density of animal tissue is nearly the same as the density of water, and sound waves pass freely from water into animal tissue. Because water is dense, movement of water molecules directly stimulates the hair cells of the lateral line system. Air is not dense enough to move the cilia of hair cells, and the lateral line system has been lost in all tetrapods except for larval and permanently aquatic amphibians. The inner ear assumes the function of hearing airborne sounds, with sound waves transmitted through a bone (or a chain of bones) in a middle ear.

Considerably more energy is needed to set the fluids of the inner ear in motion than most airborne sounds impart, and the middle ear is a sound amplifier. It receives the relatively low energy of sound waves on its outer membrane, the tympanum (eardrum), and these vibrations are transmitted by the bone (or bones) of the middle ear to the oval window of the otic capsule in the skull. The area of the tympanum is much larger than that of the oval window, and the difference in area plus the lever system of the bones that connect the tympanum to the oval window amplifies the sound waves. In-and-out movement of the oval window produces waves of compression in the fluids of the inner ear, and these waves stimulate the hair cells in the organ of Corti. This organ discriminates the frequency and intensity of vibrations it receives and transmits this information to the central nervous system. The organ of Corti lies within a flask-shaped structure, the lagena (Greek *lagenos* = flask) (Figure 8–13). The lagena is larger in derived tetrapods, and in mammals it is called the cochlea.

The middle ear is not an airtight cavity—the Eustachian (auditory) tube, derived from the spiracle of fishes, connects the mouth with the middle ear. Air flows into or out of the middle ear as air pressure changes. (These tubes sometimes become blocked. When that happens, changes in external air pressure can produce a painful sensation in addition to reduced auditory sensitivity. Anyone who has had a bad cold while traveling in an airplane knows about this.)

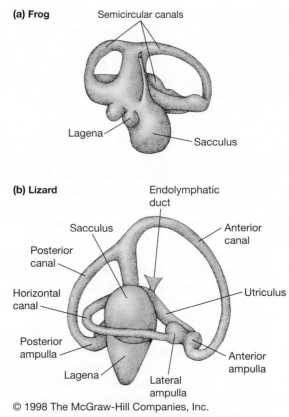

(a) Frog

Semicircular canals

Lagena

Sacculus

(b) Lizard

Endolymphatic duct

Sacculus

Posterior canal

Horizontal canal

Posterior ampulla

Lagena

Anterior canal

Utriculus

Anterior ampulla

Lateral ampulla

© 1998 The McGraw-Hill Companies, Inc.

Figure 8–13 The vestibular apparatus of tetrapods.
(a) Generalized amphibian condition. Note the small lagena for hearing airborne sounds. (b) Generalized nonmammalian amniote condition with a larger lagena.

The middle ear of tetrapods has evolved convergently several times, although in each case the stapes (the old fish hyomandibula, often called the columella in nonmammalian tetrapods) transmits vibrations between the tympanum and the oval window. Modern amphibians have an organization of the inner ear that is different from that of amniotes, indicating an independent evolution of hearing, and a middle ear has evolved independently in mammals and other amniotes. Even in nonmammalian amniotes, differences in anatomy suggest that evolution of the middle ear occurred independently in turtles, lizards, and archosaurs.

Olfaction

The olfactory receptors responsible for the sense of smell are located in the olfactory epithelium in the nasal passages of tetrapods, and air passes over the olfactory epithelium with each breath. The receptors can be extraordinarily sensitive, and some chemicals can be detected at concentrations below 1 part in 1 million trillion (10^{15}) parts of air. Among tetrapods, mammals probably have the greatest olfactory sensitivity; the area of the olfactory epithelium in mammals

is increased by the presence of scrolls of thin bone called turbinates (**Figure 8–14**). Primates, including humans, have a relatively poor sense of smell because our snouts are too short to accommodate large turbinates and an extensive olfactory epithelium.

Tetrapods have an additional chemosensory system located in a unique organ of olfaction in the anterior roof of the mouth—the vomeronasal organ or Jacobson's organ. When snakes flick their tongues in and out of their mouth, they are capturing molecules in the air and transferring them to this organ. Many male ungulates (hoofed mammals) sniff or taste the urine of a female, a behavior that permits them to determine the stage of her reproductive cycle. This sniffing is usually followed by "flehmen," a behavior in which the male curls his upper lip and often holds his head high, probably inhaling

Generalized mammalian condition

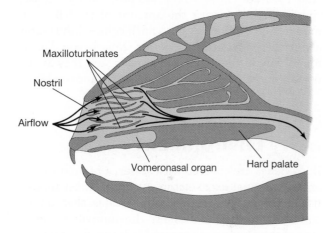

Maxilloturbinates

Nostril

Airflow

Vomeronasal organ

Hard palate

Ungulate condition

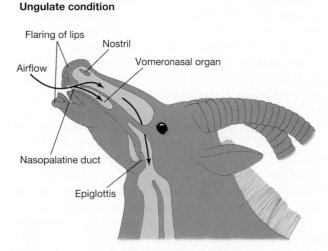

Flaring of lips

Nostril

Airflow

Vomeronasal organ

Nasopalatine duct

Epiglottis

Figure 8–14 Olfactory system of a mammal. (a) Generalized mammal, showing the positions of the turbinates and vomeronasal organ. (b) The flehmen behavior of an ungulate.

molecules of pheromones into the vomeronasal organ. Primates, with their relatively flat faces, were thought to have lost their vomeronasal organs, but some recent work suggests the presence of a remnant of this structure in humans that is used for pheromone detection.

Proprioception—Where Are Your Parts?

Aquatic vertebrates don't have long appendages, and their appendages have relatively little range of movement in relation to the body. Their heads are attached to their pectoral girdles, and their fins move either from side to side or forward and back. That is not true of terrestrial animals, which have necks and limbs that can move in three dimensions with respect to the body. It is important for a terrestrial vertebrate to know where all the parts of the body are, and proprioception provides that information. (It is the proprioceptors in your arm that enable you to touch your finger to your nose when your eyes are closed.) Proprioceptors include muscle spindles, which detect the amount of stretch in the muscle, and tendon organs, which convey information about the position of the joints. Muscle spindles are found only in the limbs of tetrapods, and they are important for determining posture and balance on land.

8.7 Conserving Water in a Dry Environment

Bony fishes are covered with scales, a remnant of the ancient dermal exoskeleton. Although most modern fishes have thin proteinaceous scales that do not contain mineralized tissues, the immediate ancestors of tetrapods were covered in heavy dermal scales containing layers of enamel, dentine, and bone like the scales of gars. These scales were mostly lost in the earliest tetrapods: only scales on the belly remained, and these had lost the enamel and dentine layers.

In terrestrial environments, water is evaporated from the body surface and respiratory system as water vapor and lost through the kidneys and in feces as liquid water. The permeability of the skin of terrestrial vertebrates depends on its structure and varies from very high in most extant amphibians to very low in most amniotes. It is difficult to say what the skin of the earliest tetrapods would have been like. The very thin, glandular skin of modern amphibians appears to be a recent specialization that is associated with using the skin for gas exchange.

The epidermal cells of vertebrates synthesize keratin (Greek *keratin* = horn), which is an insoluble protein that ultimately fills those cells. The outer layer of the skin of vertebrates is composed of layers of keratinized epidermal cells, forming the stratum corneum (Latin *cornu* = horn). The stratum corneum is only a few cell layers deep in fishes and amphibians, but it is many layers deep in the skins of amniotes. These keratinized cells resist physical wear; the presence of an insoluble protein has some waterproofing effect, but lipids in the skin are the main agents that limit evaporative water loss.

The sauropsid and synapsid lineages developed different solutions to the problem of minimizing water loss through the kidney; the urinary systems of the two groups are compared in Chapter 11. What they have in common, however, is a urinary bladder—a saclike structure that receives urine from the kidney and voids it to the outside (see Figure 8–4 [number 13]). A bladder is a new feature of tetrapods, although some bony fishes have a bladderlike extension of the kidney duct, and a bladder was probably an ancestral character of tetrapods. It may have served as a water-recovery device in early tetrapods, as it still does in extant lissamphibians and some sauropsids.

Amniotes have a new duct draining the kidney—the ureter—derived from the base of the archinephric duct. In most vertebrates the urinary, reproductive, and digestive systems reach the outside through a single common opening, the cloaca (**Figure 8–15**). Only in therian mammals (marsupials and placentals) is the cloaca replaced by separate openings for the urogenital and digestive systems.

The penis is a conduit for urine only in therian mammals. In all other amniotes it is purely an intromittent organ, used to introduce sperm into the reproductive tract of the female so the egg can be fertilized before it is encased in a shell.

8.8 Controlling Body Temperature in a Changing Environment

An animal on land is in a physical environment that varies over small distances and can change rapidly. Temperature, in particular, varies dramatically in time and space in terrestrial environments, and changes in environmental temperature have a direct impact on terrestrial animals, especially small ones, because they gain and lose heat rapidly.

This difference in the aquatic and terrestrial environments results from differences in the physical properties of water and air. The heat capacity of water is high, as is its ability to conduct heat. As a result, water temperatures are relatively stable and do not vary much from one place to another in a pond or stream. An aquatic animal has little capacity to change

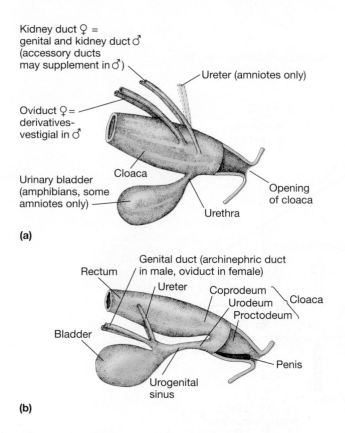

(a)

Kidney duct ♀ =
genital and kidney duct ♂
(accessory ducts
may supplement in ♂)

Ureter (amniotes only)

Oviduct ♀ =
derivatives-
vestigial in ♂

Cloaca

Opening
of cloaca

Urinary bladder
(amphibians, some
amniotes only)

Urethra

(b)

Rectum

Genital duct (archinephric duct
in male, oviduct in female)

Ureter

Coprodeum
Urodeum — Cloaca
Proctodeum

Bladder

Penis

Urogenital
sinus

Figure 8–15 Anatomy of urogenital ducts in tetrapods.
(a) Generalized condition in jawed vertebrates, including non-amniotic tetrapods and some basal amniotes. (b) More derived amniote condition (except for therian mammals), as illustrated by a male monotreme. The urogenital sinus is the name given to the urethra past the point where the genital duct has joined the system. In this example, the entire structure between the bladder and the cloaca is the urogenital sinus, but this is variable in different species of amniotes.

its body temperature by moving or to maintain a body temperature different from the water temperature.

As soon as an animal moves out onto the shore, however, it encounters a patchwork of warm and cool spots with temperatures that may differ by several degrees within a few centimeters. Terrestrial animals can select favorable temperatures within this thermal mosaic, and the low heat conductivity of air means that they can maintain body temperatures that are different from the air temperature.

In fact, thermoregulation (regulating body temperature) is essential for most tetrapods because they encounter temperatures that are hot enough to kill them or cold enough to incapacitate them. In general, tetrapods maintain body temperatures that are higher than the air temperature (some exceptions to this generalization are discussed in subsequent chapters), and

to do this they need a source of heat. The heat used to raise the body temperature to levels permitting normal activity can come from the chemical reactions of metabolism (endothermy) or from basking in the sun or being in contact with a warm object such as a rock (ectothermy). The convergent evolution of endothermy in the synapsid and sauropsid lineages is discussed in Chapter 11; here we focus on ectothermy, which is the ancestral form of thermoregulation by tetrapods.

Ectothermal Thermoregulation

Ectothermy is the method of thermoregulation used by nearly all non-amniotes and by turtles, lepidosaurs, and crocodilians. Despite the evolutionary antiquity of ectothermy, it is a complex and effective way to control body temperature. Ectothermal thermoregulation is based on balancing the movement of heat between an organism and its environment. Many lizards can maintain stable body temperatures with considerable precision and have body temperatures very similar to those of birds and mammals.

A brief discussion of the pathways by which thermal energy moves between a living organism and its environment is necessary to understand the thermoregulatory mechanisms employed by terrestrial ectotherms. Ectotherms and endotherms gain or lose energy by several pathways: solar radiation, thermal (infrared) radiation, convection, conduction, evaporation, and metabolic heat production. Adjusting the flow through various pathways allows an animal to warm up, cool down, or maintain a stable body temperature.

Figure 8–16 illustrates pathways of thermal energy exchange. Heat comes from both internal and external sources, and the flow in some of the pathways can be either into or out of an organism.

- For most organisms, the sun is the primary source of heat, and solar energy always results in heat gain. Solar radiation reaches an animal directly when it is in a sunny spot. In addition, solar energy is reflected from clouds and dust particles in the atmosphere and from other objects in the environment, and reflected solar radiation reaches the animal by these pathways. The wavelength distribution of solar radiation is the portion of the solar spectrum that penetrates Earth's atmosphere. About half this energy is in the visible wavelengths (roughly 390 to 750 nanometers), and most of the rest is in the infrared region of the spectrum (i.e., wavelengths longer than 750 nanometers).

  Energy exchange in the infrared is an important part of the radiative heat balance. All objects, animate or inanimate, radiate energy at wavelengths determined by their absolute temperatures. Objects

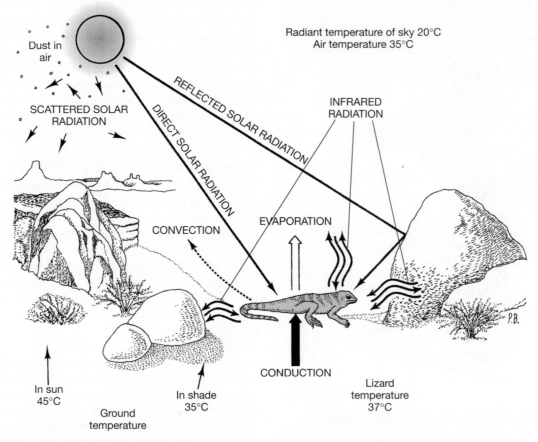

Dust in
air

SCATTERED SOLAR
RADIATION

REFLECTED SOLAR RADIATION

DIRECT SOLAR RADIATION

Radiant temperature of sky 20°C
Air temperature 35°C

INFRARED
RADIATION

CONVECTION

EVAPORATION

CONDUCTION

In sun
45°C

In shade
35°C

Ground
temperature

Lizard
temperature
37°C

P.B.

Figure 8–16 Energy exchange. Energy is exchanged between a terrestrial organism and its environment via several pathways. These are illustrated in simplified form by a lizard resting on the floor of a desert arroyo. Small adjustments of posture or position can change the magnitude and even the direction of energy exchange in the various pathways, giving a lizard considerable control over its body temperature.

in the temperature range of animals and Earth's surface (roughly –20°C to +50°C) radiate in the infrared portion of the spectrum. Animals continuously radiate energy to the environment and receive infrared radiation from the environment. Thus, infrared radiation can lead to either gain or loss of heat, depending on the relative temperature of the animal's body surface and the environmental surfaces, as well as on the radiation characteristics of the surfaces themselves. In Figure 8–16, the lizard is cooler than the sunlit rock in front of it and receives more energy from the rock than it loses to the rock. (The direction of energy flow is indicated by the arrowheads.) However, the lizard is warmer than the shaded rock behind it and has a net loss of energy in that exchange. The radiative temperature of the clear sky is about 20°C, so the lizard loses energy by radiation to the sky.

• Heat is exchanged between objects in the environment and the air via convection—the transfer of heat between an animal and a fluid. Convection can result in either gain or loss of heat. If the air temperature is lower than an animal's surface temperature, convection leads to heat loss—in other words, it's a cooling breeze. If the air is warmer than the animal, however, convection results in heat gain. In still air, convective currents formed by local heating produce convective heat exchange. When the air is moving—that is, when there is a breeze—forced convection replaces natural convection, and the rate of heat exchange is greatly increased. In Figure 8–16, the lizard is warmer than the air and loses heat by convection.

• Heat exchange by conduction occurs where the body and the substrate are in contact—the transfer of energy between an animal and a solid material. Conductive heat exchange resembles convection in that its direction depends on the relative temperatures of the animal and the environment. It can be modified by changing the surface area of the animal in contact with the substrate and by changing the rate of blood flow in the parts of the animal's body that are in contact with the substrate. In Figure 8–16, the lizard gains heat by conduction from the warm ground.

- Evaporation of water occurs from the body surface and from the pulmonary system. Each gram of water evaporated represents a loss of about 2450 joules (the exact value changes slightly with temperature). Evaporation of water transfers heat from the animal to the environment and thus represents a loss of heat. (The inverse situation—condensation of water vapor on an animal—would produce heat gain, but it rarely occurs under natural conditions.)
- Metabolic heat production is the final pathway by which an animal can gain heat. Among ectotherms, metabolic heat gain is usually trivial in relation to the heat gained directly or indirectly from solar energy.

Endothermal Thermoregulation

Endotherms (birds and mammals for the purpose of this discussion) exchange energy with the environment by the same routes as ectotherms. Everyone has had the experience of getting hot in the sun (direct solar radiation) and starting to sweat (evaporation). When you were in that situation you probably moved into the shade, and that is the same behavioral thermoregulatory response that a lizard exhibits.

What is different about endotherms is the magnitude of their metabolic heat production. Endotherms have metabolic rates that are seven to ten times higher than those of ectotherms of the same body mass. During cellular metabolism, chemical bonds are broken, and some of the energy in those bonds is captured in the bonds of other molecules, such as adenosine triphosphate (ATP). Metabolism is an energetically inefficient process, however, and only a portion of the energy released when a bond is broken is captured—the rest is released as heat. This wasted energy from metabolism is the heat that endotherms use to maintain their body temperatures at stable levels.

Ectothermy Versus Endothermy

Neither ectothermy nor endothermy can be considered the better mode of thermoregulation because each one has advantages and disadvantages:

- By producing heat internally, endotherms gain considerable independence from environmental temperatures. Endotherms can live successfully in cold climates and can be nocturnal in situations that would not be possible for ectotherms. These benefits come at the cost of high energy (food) requirements, however.
- Ectotherms save energy by relying on solar heating, and an ectotherm eats less than an endotherm with the same body mass. Because of that difference, ectotherms can live in places that do not provide enough energy to sustain an endotherm.

The examples of ectothermy and endothermy described here represent the extreme ends of a spectrum of thermoregulatory patterns that includes intermediate conditions. Fishes that maintain their body core at a temperature higher than that of the surrounding sea are examples of ectotherms with significant metabolic heat production, and some birds and mammals allow their body temperatures to fall during the night and then bask in the sun to warm up in the morning.

Summary

Life on land differs from life in water in a host of ways because the physical properties of water and air are so different. Aquatic vertebrates are nearly neutrally buoyant in water, so gravity is only a small factor in their lives. The vertebral column of a fish need only resist lengthwise compression as trunk muscles contract. In contrast, the skeleton of a tetrapod has to support the animal's weight. Zygapophyses in the vertebral column of tetrapods transmit forces from one vertebra to the next, resisting the pull of gravity.

The heads of tetrapods are not connected to their pectoral girdles like the heads of fishes, and the vertebral column shows regional modification in the shapes of vertebrae that are associated with functional specializations. The limbs of a tetrapod lift the body off the ground and push against the substrate as the animal moves. Locating the parts of the body in space is the function of the proprioceptive system.

Air is easier to breathe than water because it is less dense and has a higher concentration of oxygen. Tidal respiration is feasible for air-breathing vertebrates, and movement of the ribs in amniotes creates a negative pressure in the abdominal cavity that can draw air into the lungs through a long neck. Suction feeding is ineffective in air, however, and tetrapods use their mobile heads to seize prey and muscular tongues to manipulate it in the mouth.

Even the cardiovascular system feels the effect of gravity because venous blood must be forced upward as it returns from regions of the body that are lower than the heart. Tetrapods have high blood pressures that are created by thick-walled muscular hearts. When a septum divides the ventricle into left and right sides, the right side of the heart receives only deoxygenated blood, and coronary arteries are needed to bring oxygen to the heart muscle itself.

The difference in the function of sensory systems in air and water is profound. The cornea of the eye participates in focusing light on the retina of terrestrial vertebrates, and an image is focused by changing the shape of the lens rather than by moving the lens, as is the case in fishes.

Air is not dense enough to activate the hair cells of a lateral line system. The hair cells of terrestrial vertebrates are found in the ear, and a lever system amplifies sound-pressure waves as they are transmitted from air to the fluid in the inner ear. Somewhat surprisingly, the structural details of the ears of tetrapods show that hearing evolved independently in different lineages of terrestrial vertebrates.

Chemosensation is as important to terrestrial vertebrates as it is to aquatic forms, but the receptor cells are internal. The vomeronasal system is a chemosensory system unique to tetrapods that is intimately involved with reproductive behaviors.

Physical abrasion and evaporation of water through the skin are potential problems for terrestrial vertebrates, and the skin of the earliest tetrapods was probably covered by a stratum corneum containing keratinized epidermal cells that resisted abrasion and lipids that reduced water loss.

Temperature varies far more on land than in water, both from one spot to another and from hour to hour. Aquatic ectotherms have few options for selecting favorable temperatures, and most have little ability to maintain body temperatures that are different from water temperatures. In contrast, terrestrial ectotherms can exploit a mosaic of temperatures created by patches of sunlight and shade, and they can have body temperatures that are very different from air temperatures.

Discussion Questions

1. Why is bone such a useful structural material for tetrapods on land? Calcified cartilage is lighter than bone. Isn't that an advantage for a terrestrial vertebrate?
2. Why would we expect secondarily marine tetrapods (like whales) to have lost the zygapophyses that interconnect their vertebrae?
3. How would you determine whether an isolated vertebra of a mammal came from the thoracic region or the lumbar region?
4. We usually think of our ribs as being used for breathing. How is this different from their original function in fishes, or even their original function in early tetrapods?
5. How can we be almost certain that the earliest tetrapods had a transversus abdominis component of their hypaxial musculature, as in living tetrapods?
6. Why do you think it might be primarily mammals that evolved gaits (such as the bound) that involved dorsoventral (versus lateral) flexion of the backbone?

Additional Information

Carroll, R. L., et al. 2005. Thermal physiology and the origin of terrestriality in vertebrates. *Zoological Journal of the Linnean Society* 143:345–358.

Biewener, A. A. 2005. Biomechanical consequences of scaling. *Journal of Experimental Biology* 208:1665-1676.

Clack, J. A. 2002. *Gaining Ground: The Origin and Evolution of Tetrapods*. Bloomington, IN: Indiana University Press.

Domenici, P., et al. 2007. Environmental constraints upon locomotion and predator–prey interactions in aquatic organisms: An introduction. *Philosophical Transactions of the Royal Society B: Biological Sciences* 362:1929–1936.

Hillenius, W. J. 1992. The evolution of nasal turbinates and mammalian endothermy. *Paleobiology* 18:17–29.

Reilly, S. M., et al. 2006. Tuataras and salamanders show that walking and running mechanics are ancient features of tetrapod locomotion. *Proceedings of the Royal Society B: Biological Sciences* 273:1563–1568.

Websites
"The Biology of B-Movie Monsters"
(An entertaining, and erudite, look at why various types of proposed shrunken or giant animals just don't work in terms of size and scaling.) http://fathom.lib.uchicago.edu/2/21701757/
Animal locomotion
(This site presents some of the work of the Victorian photographer Eadweard Muybridge, who was the first person to photograph running mammals.) http://bowlingsite.mcf.com/movement/locoindex.html

CHAPTER 9

Origin and Radiation of Tetrapods

By the Late Devonian period, the stage was set for the appearance of terrestrial vertebrates, whose origin can be found among the lobe-finned fishes. New fossil evidence shows that the earliest tetrapods were actually aquatic animals and that many of the anatomical changes that were later useful for life on the land were first evolved primarily in the water. The early tetrapods underwent a rapid radiation in the Carboniferous period: many were probably amphibious, but some lineages became secondarily fully aquatic, while others became increasingly specialized for terrestrial life. However, only one of the terrestrial lineages of Paleozoic tetrapods made the next major transition in vertebrate history, developing the embryonic membranes that define the amniotic vertebrates. Amniote diversification shows an initial early split between the synapsids, the lineage that includes mammals, and the sauropsids, the lineage that includes reptiles and birds.

9.1 Tetrapod Origins

Our understanding of the origin of tetrapods is advancing rapidly. The earliest known tetrapod fossils are from the Late Devonian, some 360 million years ago. Until fairly recently, the genus *Ichthyostega* (originally found in East Greenland in 1932) was the only well-known representative of the earliest tetrapods. In the past couple of decades, however, we have discovered new material from this fossil site, including both skulls and skeletons of a different genus, *Acanthostega*, which was a more fishlike animal, and a new undescribed taxon has also been identified. Fragmentary fossil material of other Late Devonian tetrapods has been found in Latvia, Scotland, Australia, China, Russia, and North America (**Table 9–1** on page 198).

Analysis of the new specimens has focused on derived characters, and this perspective has emphasized the sequence in which the characteristics of tetrapods were acquired. The gap between fishes and tetrapods has narrowed, and the earliest tetrapods now appear to have been much more fishlike than we had previously realized. That information provides a basis for hypotheses about the ecology of animals at the transition between aquatic and terrestrial life.

The next stage in the history of tetrapods was their radiation into different lineages and different ecological types during the late Paleozoic and Mesozoic eras (**Figure 9–1** on page 191). By the Early Carboniferous, tetrapods had split into two lineages. One of these lineages is the stem amphibians (or batrachomorphs), which gave rise to some, if not all, of the modern amphibians, and the other lineage is the stem amniotes, from which the amniotes (reptiles, birds, and mammals) were derived. The heyday of both of these groups was in the Late Carboniferous to Early Permian, and most were extinct by the end of the Permian.

By the start of the Cenozoic era, the only remaining non-amniotic tetrapods were the lineages of amphibians that we see today: frogs, salamanders, and caecilians. Amniotes have been the most abundant tetrapods since the late Paleozoic. They have radiated into many of the terrestrial life zones that were

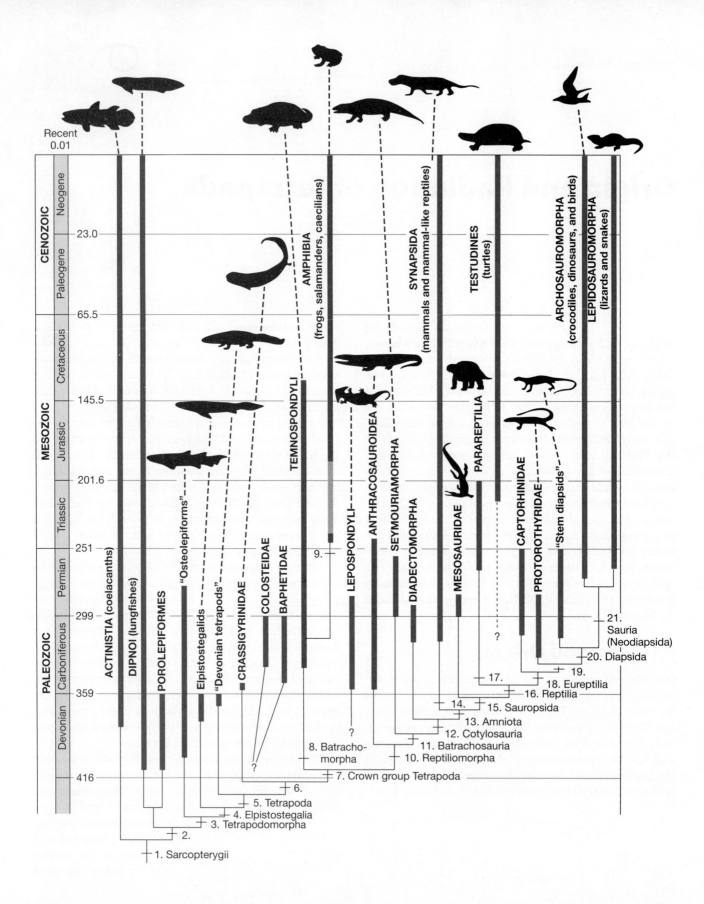

previously occupied by non-amniotes, and they have developed feeding and locomotor specializations that had not previously been seen among tetrapods. Figure 9–1 shows a detailed phylogeny of early tetrapods, and **Figure 9–2** on page 192 shows a simplified cladogram.

Fish-Tetrapod Relationships

Tetrapods are clearly related to the sarcopterygian (lobe-finned) fishes, which survive today as lungfishes, and the coelacanths. The discovery of lungfishes seemed to provide an ideal model of a prototetrapod—what more could one ask for than an air-breathing fish? However, lungfishes are very specialized animals, and many of their apparent similarities to tetrapods (such as the internal nostril, or choana) appear to have been evolved independently.

Coelacanths lack the specializations of lungfishes and, for a while after its discovery in 1938, the coelacanth *Latimeria chalumnae* was hailed as a surviving member of the group ancestral to tetrapods. However, most scientists now consider that lungfishes are more closely related to tetrapods than are coelacanths; coelacanths have a mixture of ancestral characters and unique specializations of their own. Chapter 6 covers the living lobe-finned fishes in more detail. Note that all lobe-finned fishes have the derived condition of the fin in which only a single basal element, rather than several elements, articulates with the limb girdle (i.e., the monobasal condition). This single basal is homologous with the humerus in our arm or the femur in our leg.

Both lungfishes and coelacanths have an extensive Paleozoic fossil record, along with a third group, called the tetrapodomorph fishes. (The formal term *Tetrapodomorpha* includes these fishes plus all tetrapods.) These fishes were originally called "rhipidistians," and rhipidistians and coelacanths were grouped as "crossopterygians." Both terms have now fallen into disuse, as they represent paraphyletic groupings.

Tetrapodomorph fishes include several families, such as the osteolepidids, the rhizodontids, and the tristichopterids (**Figure 9–3**). Basal tetrapodomorph fishes, such as the osteolepidids, were cylindrical-bodied, large-headed fishes with thick scales. They were probably shallow-water predators. Tristichopterids, more derived forms (*Eusthenopteron* is the best known of the tristichopterids), resembled early tetrapods in having paired crescentic vertebrae and teeth with labyrinthine infoldings of enamel. They had also reduced their scaly cover to scales of bone only, lacking the cosmine (dentine) and enamel layers typical of other sarcopterygians.

Several genera, including *Panderichthys*, *Livoniana*, *Elpistostege*, and *Tiktaalik*, are more closely related to tetrapods than the tristichopterids, although they form a paraphyletic stem rather than a distinct clade. These derived fishes, sometimes referred to as elpistostegalians, had eyes on the top of their heads (like crocodilians) and probably lived in shallow water. Tetrapod-like features associated with this lifestyle include the loss of the dorsal and anal fins, a greatly reduced tail fin, dorsoventrally flattened bodies, and flat heads with long snouts. These fishes also shared with early tetrapods a derived form of the humerus (the upper arm bone), indicating powerful forelimbs that would be capable of propping the front end of the animal out of the water, and many details of the skull, including the anatomy of the ear region.

Tiktaalik (pronounced with the accent on the second syllable; the name means "a large freshwater fish seen in the shallows" in the local Inuktitut language) was a spectacular new find of an elpistostegalian fish, on Ellesmere Island in the Canadian Arctic. This fossil has filled a morphological gap between the previously known most-derived sarcopterygian fishes and the first tetrapods (**Figure 9–4**). The fact that it retained fin rays means that it was definitely on the fish side of the transition, but it also possessed many derived, tetrapod-like features.

Tiktaalik lived in the early Late Devonian, around 385 million years ago. That places it approximately

Figure 9–1 Phylogenetic relationships of sarcopterygian fishes and early tetrapods. This diagram depicts the probable relationships among the major groups of sarcopterygian fishes and early tetrapods. The black lines show interrelationships only; they do not indicate times of divergence or the unrecorded presence of taxa in the fossil record. Lightly shaded bars indicate ranges of time when the taxon is believed to be present, because it is known from earlier times, but is not recorded in the fossil record during this interval. Only the best-corroborated relationships are shown. The numbers at the branch points indicate derived characters that distinguish the lineages—see the Appendix for a list of these characters. (An alternative view of lissamphibian relationships is discussed in the text.) Note that some authors prefer to restrict the term *tetrapod* to the crown group, which encompasses only extant taxa and extinct taxa that fall within the range of characters seen in the extant taxa. Thus, under this scheme, the taxonomic term *Tetrapoda* would be shifted to node 7.

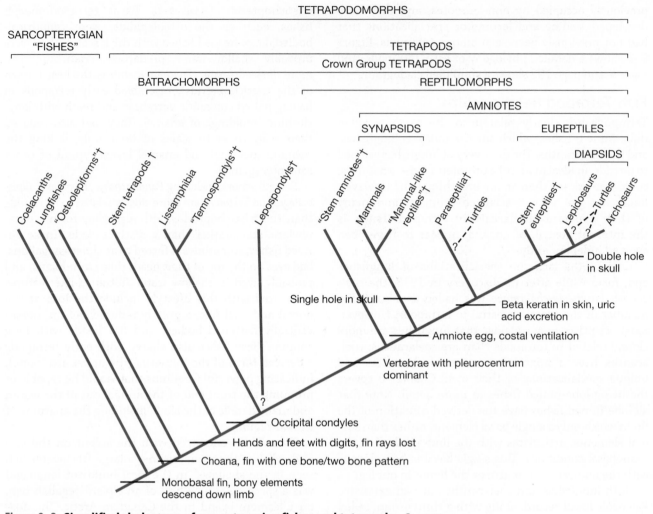

Figure 9–2 Simplified cladogram of sarcopterygian fishes and tetrapods. Quotation marks indicate paraphyletic groups. A dagger (†) indicates an extinct taxon. Note the alternative phylogenetic positions of turtles.

20 million years before *Acanthostega* and *Ichthyostega* and 10 million years before the oldest tetrapod fragments from Scotland and Latvia, but 2 to 3 million years later than the elpistostegalid *Panderichthys*. Perhaps its most tetrapod-like feature is the loss of the bony operculum; this change, which means that *Tiktaalik* would be able to raise its head above its body, was probably related to feeding on prey items outside of the water, snapping at them with its long snout rather than sucking them in with water as other fishes do. However, despite the fact that no adult tetrapod possesses an operculum, traces of an operculum can be seen in chicken embryos, and the genes that code for its formation are the same as those in teleost fishes.

Tiktaalik had large, overlapping ribs, like those of many early tetrapods, suggesting that it could support

its body at least partway out of the water. This behavior is also suggested by the structure of its pectoral fin, which could bend in the middle to act as a prop to raise the anterior part of the body. The pectoral fin also had an elaboration of the distal elements, not precisely homologous to those of the tetrapod wrist and hand, but certainly providing an example of the kind of structures that might later be elaborated into a tetrapod limb. (Fingerlike bones were not unique to *Tiktaalik*; other tetrapodomorph fishes such as the rhizodontid *Sauripterus* and the elpistostegalid *Panderichthys* were also apparently "experimenting" at this time with the development of fingerlike bones at the end of the fin.)

Features of *Tiktaalik* that remained definitely fishlike include well-developed gills, poorly ossified vertebrae, and a long body. The structure of the pelvic region is

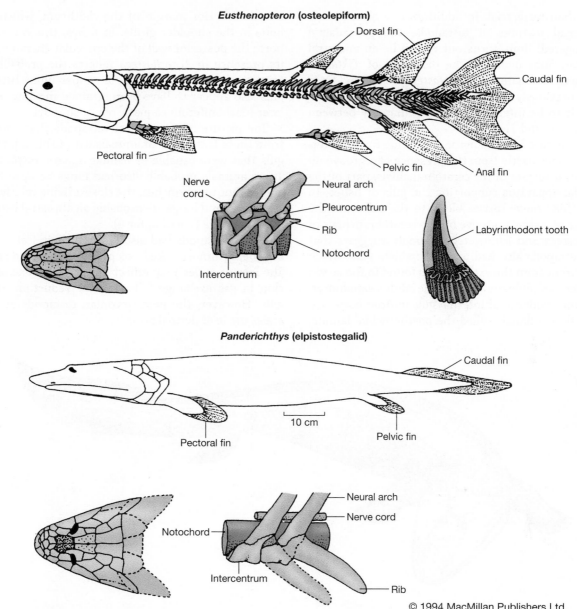

Eusthenopteron (osteolepiform)

Dorsal fin

Caudal fin

Pectoral fin

Pelvic fin

Anal fin

Nerve cord

Neural arch

Pleurocentrum

Rib

Notochord

Intercentrum

Labyrinthodont tooth

Panderichthys (elpistostegalid)

Caudal fin

10 cm

Pectoral fin

Pelvic fin

Neural arch

Nerve cord

Notochord

Intercentrum

Rib

© 1994 MacMillan Publishers Ltd.

Figure 9–3 Devonian tetrapodomorph fishes. The tristichopterid *Eusthenopteron* has a cylindrical body, a short snout, and four unpaired fins in addition to the paired pectoral and pelvic appendages. The elpistostegalid *Panderichthys* has a dorsoventrally flattened body with a long, broad snout and eyes on top of the head. The dorsal and anal fins have been lost, and the caudal fin has been reduced in size. In the vertebral column of *Eusthenopteron* the ribs are short and probably extended dorsally. The ribs are larger in *Panderichthys* and project laterally and ventrally. In the skull roof of *Eusthenopteron* the area anterior to the parietals (sparsely stippled) is occupied by a single, median element (densely stippled). In the skull roof of *Panderichthys* there is a single pair of large frontal bones (densely stippled) immediately anterior to the parietals, as in tetrapods.

still unknown, although related fishes had hindlimbs apparently used to anchor themselves in the substrate. Despite their fishlike elements, these derived tetrapodomorph fishes already had most of the skeletal features that we find in ourselves today (**Figure 9–5**).

The Earliest Tetrapods

The new specimens of the Late Devonian *Acanthostega* and *Ichthyostega* from East Greenland have changed our views about early tetrapod biology. These fossils show that early tetrapods were primarily aquatic rather than terrestrial. In addition, one of the most widespread features of tetrapods, the pentadactyl (five-fingered) limb, turns out not to be an ancestral character. Subsequent to the discovery of *Tiktaalik*, more complete material of *Ventastega*, a tetrapod from Latvia previously known only from fragments, has shown it to be intermediate in its anatomy between *Acanthostega* and *Tiktaalik*.

The evidence for an aquatic way of life for early tetrapods comes partly from the presence of a groove on the ventral surface of the ceratobranchials, part of the branchial apparatus supporting the gills of fishes. In fishes, this groove carries blood to the gills, and the presence of a similar groove on the ceratobranchials of *Acanthostega* and *Ichthyostega* strongly suggests that these tetrapods also had internal fishlike gills, which are different from the external gills found in the larvae of modern amphibians and in some adult salamanders. Additional evidence of internal gills in *Acanthostega* is provided by a flange, called the postbranchial lamina,

on the anterior margin of the cleithrum, which is a bone in the shoulder girdle. In fishes, this ridge supports the posterior wall of the opercular chamber, and its presence in *Acanthostega* reflects the probable retention of some sort of soft-tissue operculum. Internal gills also appear to have been retained in some of the later (Carboniferous) aquatic temnospondyls.

The picture of the earliest tetrapods that emerges from these features is of animals with fishlike internal gills that were capable of fishlike aquatic respiration. These animals probably also had lungs because lungs are present in lungfishes, the closest living relatives of tetrapods, and lungs are probably an ancestral osteichthyian feature (see Chapter 4).

These tetrapods had also reduced their scale cover, now being almost naked except for scales on the belly. The loss of scales may reflect an increased use of the skin in gas exchange following the reduction of the gills. However, the post-Devonian tetrapods evolved scales made of dermal bone.

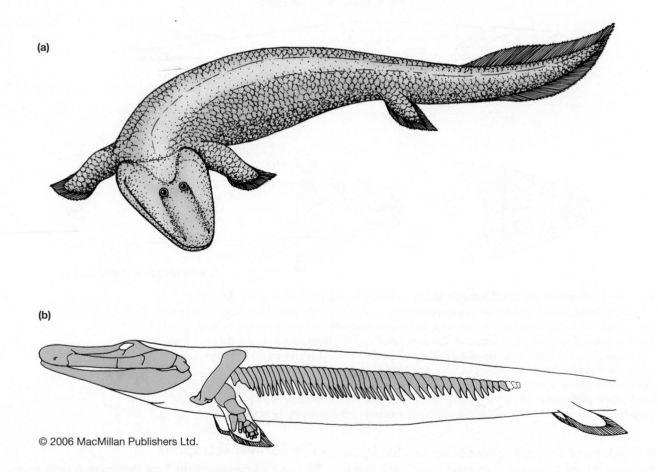

(a)

(b)

Figure 9–4 The Late Devonian "fishapod" *Tiktaalik*. (a) A reconstruction of the probable appearance of *Tiktaalik*. (b) The portions of the skeleton of *Tiktaalik* that have been discovered. Note that the form of hindlimb is conjectural; it has not yet been found.

Another unexpected feature of Devonian tetrapods is polydactyly—that is, having more than five toes. *Acanthostega* had eight toes on its front and hind feet, and *Ichthyostega* had seven toes on its hind foot (its fore foot is unknown). Additionally, *Tulerpeton* (known from Russia) had six toes. These discoveries confound long-standing explanations of the supposed homologies of bones in the fins of sarcopterygians with those in tetrapod hands and feet, but they correspond beautifully with predictions based on the embryology of limb development. An interesting new discovery is that the development of teleost fins involves the same genes as those involved in tetrapod limbs: the evolution of digits represents a tinkering with the original genetic information for making a fin, not a completely new system.

In all extant tetrapods the formation of the digits commences with digit 4 (the ring finger in our hands) and proceeds in an arc across the base of the limb toward the thumb, with the 5th digit (the "pinky") being added in the opposite direction. In living tetrapods the branching of digits from this arc ends with digit 1 (the thumb) or sooner in animals such as horses that have lost digit 1.

If the process of developmental branching continues, however, a polydactylous condition results with additional digits beyond the thumb (**Figure 9–6**). This situation sometimes occurs as an abnormal condition in extant vertebrates, and it was apparently the normal condition in the Late Devonian tetrapods: their additional digits are clearly beyond the thumb.

Although the fossil site where early tetrapods were first discovered, Greenland, was a freshwater environment, information from other Devonian sites indicates that tetrapods evolved in brackish or saline lagoons or even in marine habitats. An unexpected discovery of apparent tetrapod footprints in a marine lagoon setting predates the earliest known tetrapods by about 18 million years and is earlier than the appearance of most of the tetrapodomorph fishes in the fossil record.

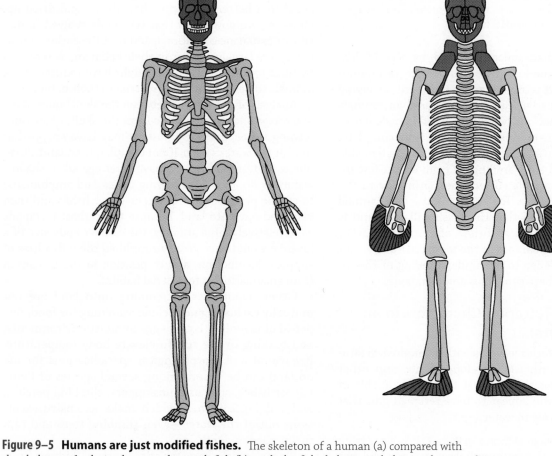

(a) (b)

Figure 9–5 Humans are just modified fishes. The skeleton of a human (a) compared with the skeleton of a derived tetrapodomorph fish (b), with the fish skeleton scaled up to the size of the human's and oriented in a humanlike pose. Darker shading indicates dermal bone or other dermal elements (fin rays in the fishes).

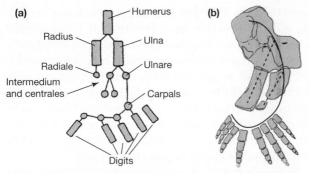

(a) Humerus, Radius, Ulna, Radiale, Ulnare, Intermedium and centrales, Carpals, Digits

© 1994 MacMillan Publishers Ltd.   © MacMillan Publishers Ltd.

Figure 9–6 Developmental hypothesis of the origin of tetrapod limbs. In both cases, the head of the animal is to the left. Thus, the preaxial direction (i.e., anterior to the axis of the limb) is to the left. (a) A diagram of the forelimb skeleton of a mouse during early development. The proximal (nearest to the shoulder) parts of the limb skeleton consist of an axis with preaxial branches (radials), while the digits are formed as postaxial branches. (b) The forelimb of *Acanthostega* showing the polydactylous condition of eight digits, with the inferred position of the preaxial branch (dotted line) and the postaxial branch (solid line). Note that the radius and the carpal bone are formed by preaxial radials, whereas the ulna and the digits originate as postaxial branches.

If tetrapods had indeed evolved by the Middle Devonian, then the known fossils of both tetrapodomorph fishes and tetrapods must represent surviving forms from an earlier radiation rather than animals "caught in the act" of transition. Our knowledge of the fossil record from this period is far from complete. All of these "transitional" forms, on both the fish side and the tetrapod side, are known from only a few individual fossils in isolated localities, primarily in the Northern Hemisphere. The chance that a dead animal will be fossilized and that the fossil will be found is vanishingly small, and an origin of tetrapods earlier than their first appearance in the fossil record does not represent a challenge to our understanding of the interrelationships between fishes and tetrapods.

Evolution of Tetrapod Characters in an Aquatic Habitat

These new discoveries leave us with a paradoxical situation: animals with well-developed limbs and other structural features that suggest they were capable of locomotion on land appear to have retained gills that would function only in water.

How Does a Land Animal Evolve in Water? It is clear from the fossil record that tetrapod characteristics did not evolve because they would someday be useful to animals that would live on land; rather, they evolved because they were advantageous for animals that were still living in water.

Fishes like *Panderichthys* and *Tiktaalik* were large—2 meters or longer—with heavy bodies, long snouts, and large teeth. They presumably could breathe air by swimming to the surface and gulping or by propping themselves on their pectoral fins in shallow water to lift their heads to the surface. Their elongated snouts and the loss of the operculum in *Tiktaalik* would have aided them in this behavior. Their lobed fins may have been useful in slow, careful stalking behavior in dense plants on the bottom of a lagoon or other body of water, and the development of digits, wrists, and ankles would have promoted maneuverability. An African lungfish, *Protopterus annectens*, moves its pelvic fins in tetrapod-like gaits, including walking and bounding, and lifts its body clear of the substrate.

What Were the Advantages of Terrestrial Activity? The classic story of the evolution of tetrapods proposed that the Devonian was a time of seasonal droughts, and that fishes living in lakes and ponds evolved limbs to crawl to a better situation when their pond dried up. However, we now know that tetrapods evolved in marine or estuarine conditions, and that the earliest tetrapods were largely aquatic in their behavior, as reflected by their retained gills. What might have induced these animals to become more terrestrial in their habits?

Analysis of the sutures between the skull bones of *Acanthostega* shows differences from the skulls of suction-feeding fishes, and *Acanthostega* may have snapped at food at the water's edge, if not completely on land. Juvenile early tetrapods might have congregated in shallow water (as do juveniles of living fishes and amphibians) to escape predation from deeper-water fishes and then ventured out onto land. Because the earliest tetrapods were relatively large animals, the smaller body size of a juvenile would have greatly simplified the difficulties of support, locomotion, and respiration in the transition from an aquatic to a terrestrial habitat.

Other hypotheses for coming onto land are not mutually exclusive and include searching for food, dispersal of juveniles, laying eggs in moist environments, and basking in the sun to elevate body temperature. Behavioral and morphological specializations for life on land can be found among several species of living teleost fishes, such as mudskippers, climbing perches, and walking catfishes, which make extensive excursions out of the water—even climbing trees and capturing food on land.

Physiological characteristics, such as kidney function, are important in the transition from water to land. The gills of fishes have functions besides gas exchange—they are sites of sodium and chloride regulation and of nitrogen excretion. With the loss of gills in post-Devonian tetrapods, the kidney now assumed these functions.

9.2 Radiation and Diversity of Non-Amniotic Paleozoic Tetrapods

For more than 200 million years, from the Late Devonian to the Early Cretaceous, non-amniotic tetrapods (excluding the groups of modern amphibians) radiated into a great variety of terrestrial and aquatic forms. Non-amniotic tetrapods are often called amphibians, but that term is now reserved for the extant non-amniotic tetrapods: frogs, salamanders, and caecilians. It is misleading to think of Paleozoic tetrapods as being amphibians, even though they may have had amphibious habits. First, many of them were much larger than any living amphibians and would have been more crocodile-like in appearance and habits. Second, they lacked the specializations of modern amphibians. For example, many forms had dermal scales, making it

unlikely that they relied on their skin for respiration, as do modern amphibians. Finally, and most important, many of them were actually more closely related to amniotes than to amphibians. Table 9–1 lists the different types of Paleozoic non-amniotic tetrapods, and Figure 9–1 illustrates the current consensus about their interrelationships.

Devonian Tetrapods

Ichthyostega and *Acanthostega* have been known since the 1930s, but only in the past couple of decades have other Devonian tetrapods been described, and these taxa are known only from incomplete material (see Table 9–1). They ranged from about 0.5 to 1.2 meters in length, and they differ enough from each other to show that by the end of the Devonian, some 7 million years after their first appearance, tetrapods had diversified into several niches. *Acanthostega* was more fishlike in its anatomy than *Ichthyostega* (**Figure 9–7**), with short ribs and wrists and ankles that would have been incapable of bearing weight on land. However, *Ichthyostega* was specialized in its own unique fashion and has modifications (such as in the vertebral column and in the inner ear) that appear to be for both aquatic and terrestrial life.

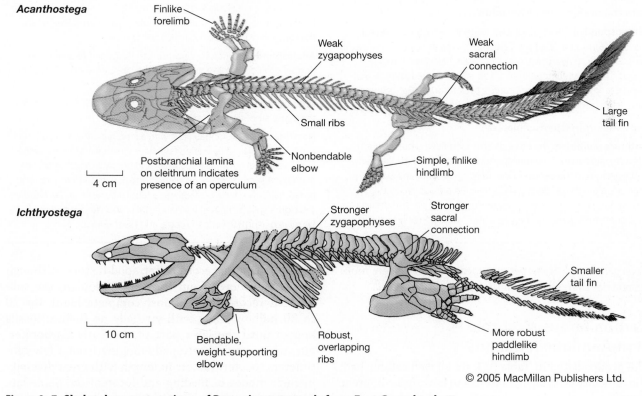

© 2005 MacMillan Publishers Ltd.

Figure 9–7 Skeletal reconstructions of Devonian tetrapods from East Greenland. The notations illustrate the more aquatic nature of *Acanthostega*. (The fore foot of *Ichthyostega* is unknown.)

Table 9–1 Major groups of Paleozoic non-amniotic tetrapods

Stem Tetrapods

Late Devonian taxa: e.g., *Acanthostega* and *Ichthyostega* from Greenland (see Figure 9–8a, b); *Metaxygnathus* (Australia, 1977—lower jaw); *Tulerpeton* (Russia, 1984—skeleton only); *Ventastega* (Latvia, 1994—fragments); *Hynerpeton* (Pennsylvania, North America, 1994—shoulder girdle and partial lower jaw); *Obruchevichthys* (Latvia, 1995—lower jaw); *Elginerpeton* (Scotland, 1995—skeletal fragments); *Denisgnathus* (Pennsylvania, North America, 2000—lower jaw); *Sinostega* (China, 2002—lower jaw); unnamed form (Pennsylvania, North America, 2004—humerus); unnamed form (Belgium, 2004—lower jaw); *Jakubsonia* (Russia, 2004—fragments of the skull and skeleton); *Ymeria* (Greenland, 2012—portions of the skull and shoulder girdle).

Enigmatic late Early Carboniferous taxa: e.g., *Pederpes* and *Crassigyrinus* (see Figure 9–8c) from Europe and *Whatcheeria* from North America.

Colosteidae: Aquatic late Early Carboniferous forms, possibly secondarily so, known from North America and Europe, with elongate, flattened bodies, small limbs, and lateral line grooves (e.g., *Greererpeton, Pholidogaster, Colosteus*).

Baphetidae (formerly Loxomattidae): Late Early and early Late Carboniferous forms from Europe (possibly also North America), with crocodile-like skulls and distinctive keyhole-shaped orbits (e.g., *Eucritta, Megalocephalus*).

Batrachomorphs (stem lissamphibians)

Temnospondyli: The most diverse, longest-lived group, ranging worldwide from the late Early Carboniferous to the Early Cretaceous; possessed large heads with akinetic skulls. Paleozoic forms (e.g., *Eryops, Cacops*) were terrestrial or semiaquatic (see Figure 9–8d); Mesozoic forms (e.g., *Trematosaurus, Cyclotosaurus, Gerrothorax*) were all secondarily fully aquatic (see Figure 9–8e-g).

Reptiliomorphs (stem amniotes)

Anthracosauroidea: The other diverse, long-lived group, though to a lesser extent than the temnospondyls; known from the late Early Carboniferous to the earliest Triassic of North America and Europe. Anthracosauroids had deeper skulls than temnospondyls, with prominent tabular horns, and they retained cranial kinesis. Some forms (e.g., gephyrostegans and chronosuchians) were primarily terrestrial

(see Figure 9–9b). Others, grouped together as embolomeres, were secondarily aquatic (e.g., *Pholiderpeton, Archeria;* see Figure 9–9a).

Seymouriamorpha: Known from the Permian only. Early Permian forms known from North America (e.g., *Seymouria;* see Figure 9–9c) were large and fully terrestrial. Later Permian forms known from Europe and China (discosauriscids and kotlassiids) were secondarily more aquatic.

Diadectomorpha: Known from the Late Carboniferous and the Early Permian of North America and Europe; large, fully terrestrial forms now considered to be the sister group of amniotes. Diadectidae (e.g., *Diadectes;* see Figure 9–9d) had laterally expanded cheek teeth suggestive of a herbivorous diet. Limnoscelidae and Tseajaiidae had sharper, pointed teeth and were probably carnivorous.

Lepospondyls

Microsauria: Distinguished by a single bone in the temporal series termed the tabular. Many (e.g., the tuditanomorphs) were terrestrial and rather lizardlike, with deep skulls and elongate bodies, but had only four toes (*microsaur* = small reptile; e.g., *Pantylus;* see Figure 9–10a). Some forms (microbranchomorphs) were probably aquatic. Known from the Late Carboniferous and the Early Permian of North America and Europe.

Aïstopoda: Limbless forms, lacking limb girdles, with elongate bodies (up to 200 trunk vertebrae) and rather snakelike skulls that may have allowed them to swallow large prey items (e.g., *Lethiscus, Ophiderpeton;* see Figure 9–10c). They may have been aquatic or have lived in leaf litter. Known from the Middle to Late Carboniferous and the Early Permian of North America and Europe.

Adelospondyli: Limbless, long-trunked forms, but retaining the dermal shoulder girdle; known from the late Early Carboniferous of Scotland.

Lysorophia: Elongate forms with greatly reduced limbs; known from the Late Carboniferous and the Early Permian of North America.

Nectridea: Rather elongate, but with a long tail rather than a long trunk; distinguished by having fan-shaped neural and hemal arches in the vertebral column; limbs small and poorly ossified, indicative of an aquatic mode of life. Some nectridians (keraterpetontids) had broad, flattened skulls with enlarged tabular bones (e.g., *Diplocaulus;* see Figure 9–10b). Known from the Late Carboniferous and the Early Permian of North America, Europe, and North Africa.

Tulerpeton and *Hynerpeton* appear to have been more terrestrial than either *Acanthostega* or *Ichthyostega*.

Carboniferous-Mesozoic Non-Amniotic Tetrapods

The groups listed in Table 9–1 are all well established; the difficulty is trying to understand how these different groups are related to one another and to the modern groups of tetrapods: the amphibians and the amniotes. A major problem is that we are missing a critical piece

of the geological record of tetrapod history. Although fossils are known from the Late Devonian, the subsequent record is an almost complete blank for 20 to 30 million years, with virtually no further fossils known until the later part of the Early Carboniferous, when we find tetrapods ranging from a few centimeters to a few meters in length with great diversity in their modes of feeding and locomotion. Some late Early Carboniferous taxa appear to have been quite terrestrial. Included in this group are *Pederpes* (which had five fully formed digits and a remnant of a sixth

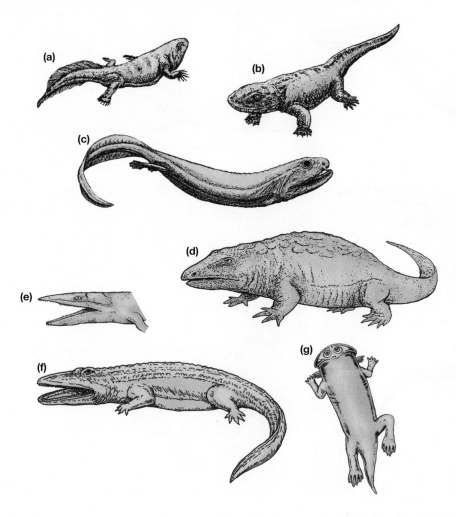

Figure 9–8 Stem tetrapods and temnospondyls. (a) *Acanthostega*, an aquatic Late Devonian stem tetrapod. (b) *Ichthyostega*, an aquatic Late Devonian stem tetrapod. (c) *Crassigyrinus*, an aquatic Early Carboniferous stem tetrapod. (d) *Eryops*, a semiterrestrial Early Permian eryopoid temnospondyl. (e) *Trematosaurus*, an aquatic (marine) Early Triassic trematosaurid temnospondyl. (f) *Cyclotosaurus*, an aquatic Middle Triassic capitosaurid temnospondyl. (g) *Gerrothorax*, an aquatic Late Triassic plagiosaurid temnospondyl. *Eryops* was about 2 m long (the size of a medium-size crocodile). The other animals are drawn approximately to scale.

finger); *Casineria*, a mouse-sized animal with limb anatomy showing that it was fully terrestrial; and the long-bodied, rather snakelike lepospondyl *Lethiscus*, which had lost its limbs completely.

This gap in the fossil record in the earliest part of the Carboniferous may bias our understanding of how the groups known from the later Carboniferous are related to one another. The interrelationship of early tetrapods shown in Figure 9–1 broadly reflects the current consensus, although a plethora of phylogenies have been suggested, some offering dramatically different opinions. The discovery of more Early Carboniferous tetrapod fossils is crucially needed to help confirm or refute the various competing hypotheses. For the purposes of this volume, we will treat the phylogeny in Figure 9–1 as a working hypothesis. The important points are that the earliest known tetrapods were no more closely related to modern amphibians than to modern amniotes, and that some Paleozoic tetrapods were on the stem of modern amphibians, some were

on the stem of modern amniotes, and some were related to neither.

The Paleozoic tetrapods were originally divided into groups called "labyrinthodonts" and "lepospondyls" (see Table 9–1). "Labyrinthodonts," which would include all non-amniotic tetrapods apart from lepospondyls, were mainly larger forms (large lizard to crocodile size, with a skull at least 5 centimeters long; **Figures 9–8** and **9–9**), with a multipartite vertebral centrum, and teeth with complexly infolded enamel (labyrinthodont teeth). Lepospondyls were small forms (salamander or small lizard size, with a skull less than 5 centimeters long; **Figure 9–10** on page 201) with a single, spool-shaped vertebral centrum and without the labyrinthine form of enamel.

The various groups of lepospondyls are probably related to each other, despite the fact that the distinguishing features just mentioned may relate merely to smaller body size, but how they are related to other early tetrapods remains a point of controversy. The

Figure 9–9 Non-amniotic reptiliomorph tetrapods. (a) *Pholiderpeton*, an aquatic Late
Carboniferous embolomere. (b) *Gephyrostegus*, a terrestrial Late Carboniferous anthracosauroid.
(c) *Seymouria*, a terrestrial Early Permian seymouriamorph. (d) *Diadectes*, a terrestrial Early
Permian diadectomorph. (e) *Westlothiana*, an Early Carboniferous form, probably closely related
to amniotes. *Seymouria* is about 1 m long (the size of a golden retriever). The other animals are
drawn approximately to scale (*Diadectes* should be a little larger and *Westlothiana* a little smaller).

large early tetrapods were a diverse taxonomic grouping; some were stem tetrapods, some were closely related to modern amphibians, and others were more
closely related to amniotes. For this reason the term
labyrinthodont is no longer formally employed even
though it is still in common use in some popular texts.

Current consensus recognizes two main groups of
large non-amniotic Paleozoic tetrapods: the stem amphibians and the stem amniotes. Both are paraphyletic
groupings because they contain the origins of the living
amphibians and amniotes, respectively. The stem amphibians (sometimes called batrachomorphs) include
the temnospondyls. The stem amniotes (sometimes
called reptiliomorphs) include the anthracosauroids,

seymouriamorphs, and diadectomorphs. The batrachomorphs were in general more aquatic than the reptiliomorphs. Batrachomorphs were characterized by flat,
non-kinetic skulls and a reduction of the hand to four
fingers. The reptiliomorphs had domed skulls that retained some kinetic ability, and had a five-fingered
hand. The affinities of the lepospondyls are open to
question, although most researchers consider them to
be a monophyletic group.

Non-amniotic tetrapods reached their peak of diversity in the Late Carboniferous and Early Permian,
when they included fully aquatic, semiaquatic, and
terrestrial forms that have been found in central and
western Europe and in eastern North America. Most

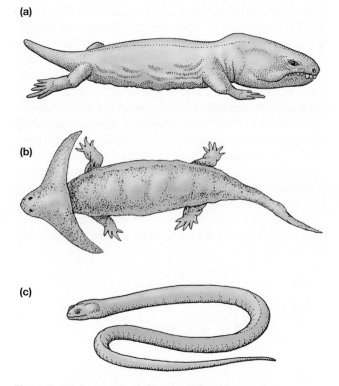

Figure 9–10 Lepospondyl tetrapods. (a) *Pantylus*, a terrestrial Early Permian microsaur. (b) *Diplocaulus*, an aquatic Early Permian nectridian. (c) *Ophiderpeton*, an aquatic (or possibly terrestrial burrowing form) Late Carboniferous aïstopod. *Pantylus* is about 20 cm long (the size of a hamster). The other animals are drawn approximately to scale.

flattened skulls with tabular bones elongated into horns (see Figure 9–10). These tabular horns were up to five times the width of the anterior part of the skull, and skin imprints show that they were covered by a flap of skin extending back to the shoulder (not shown in the figure). These horns may have acted like a hydrofoil to help in underwater locomotion, and they may have supported highly vascularized skin to aid in underwater respiration.

The temnospondyls (see Figure 9–8) were the only group of non-amniotic tetrapods (aside from the lissamphibians) to survive the Paleozoic, and all of the Mesozoic forms were large, flattened, fully aquatic predators. Some trematosaurids evolved the elongated snout characteristic of specialized fish eaters and are found in marine beds, making them the only known fully marine non-amniotic tetrapods. How did these animals osmoregulate in the marine environment? Even if the adults had evolved a reptilelike impermeable skin, the larvae still would have had gills. Perhaps trematosaurids retained high levels of urea to raise their internal osmotic pressure, as do some brackish water frogs today.

In contrast to the batrachomorphs, the reptiliomorphs appear to have been predominantly terrestrial as adults, and many have been mistaken for early reptiles (especially animals such as *Seymouria* and *Diadectes*). Terrestriality also evolved convergently among other early tetrapods, predominately in the microsaurs and the dissorophid temnospondyls that may have been ancestral to frogs. These animals acquired skeletal adaptations such as longer, more slender limb bones.

Modern Amphibians The origins of the modern amphibians (Lissamphibia) are in debate. Some workers propose that lissamphibians are a monophyletic group derived from within the temnospondyls (as shown in Figure 9–1). Other researchers think that frogs and salamanders are derived from the temnospondyls (but from different, albeit related, families), while caecilians are derived from the microsaur lepospondyls. A possible temnospondyl ancestor of frogs and salamanders, *Gerobatrachus*, has been described from the Early Permian of Texas. Still another hypothesis proposes that all modern amphibians are derived from the lepospondyls.

lineages were extinct by the mid-Permian period, although some specialized aquatic lineages of temnospondyls were moderately common in the Triassic, and one type survived into the Early Cretaceous in Australia. Although all living amphibian groups had their origins in the Mesozoic, the generic diversity of non-amniotic tetrapods did not return to Permian levels until the mid-Cenozoic era.

Ecological and Adaptive Trends One of the most striking aspects of early tetrapods is the number and diversity of forms that returned to a fully aquatic mode of life. Among living amphibians, many salamanders and some frogs are fully aquatic as adults, and this was also true of a diversity of Paleozoic forms (see Table 9–1). Many of these forms independently acquired an elongated body with the reduction or loss of the limbs, a morphology that may also be associated with burrowing, as seen in the living caecelian amphibians.

Some of the most bizarre aquatic forms were found among the lepospondyls. Some nectridians had broad,

9.3 Amniotes

Amniotes include most of the tetrapods alive today. Their name refers to the amniotic egg, which is one of the most obvious features distinguishing living amniotes from living amphibians. Amniotes appeared somewhat later in the fossil record than the earliest tetrapods of the Devonian, but they seem to have been well established by the time of the later radiation of non-amniotic tetrapods, although they were a minor part of the Carboniferous fauna. Their initial major radiation occurred in the Permian. Early amniotes are assumed to be a primarily terrestrial radiation on the basis of their anatomy, the depositional environments of the sites where they are found, and the discovery of amniote footprints in Carboniferous deposits from dry environments. However, the microanatomy of the bones suggests that at least one Early Permian form, *Captorhinus*, was amphibious.

The first known candidates for the status of amniote—or near amniote—are from the Early Carboniferous of Scotland, and they are only 20 million years younger than the earliest known tetrapods. These include the mouse-size *Casineria* and the slightly younger salamander-size *Westlothiana*, both discovered in the 1990s. They appear to have been small, agile animals, most likely with insectivorous diets, rather resembling present-day lizards. Whether or not these animals turn out to be true early amniotes, they are certainly representative of what the first amniotes would have been like—small and more terrestrial than other early tetrapods. In Chapter 8, Figure 8–4 compares various features of the biology of amniotes with those of fishes and non-amniote tetrapods.

A key event in the radiation of early tetrapods may have been the great diversification of insects in the Late

Figure 9–11 Diversity of Paleozoic amniotes. Early amniotes varied in size from a few centimeters to a couple of meters long, and their ecological roles were equally diverse. (a) *Hylonomus*, a protorothyridid (lizard size), represents the typical lizardlike body form of many early amniotes. (b) *Haptodus*, a synapsid (dog size). (c) *Mesosaurus*, a mesosaur (cat size). (d) *Captorhinus*, a captorhinid (lizard size). (e) *Petrolacosaurus kansensis*, a stem diapsid (araeoscelid; lizard size). (f) *Procolophon*, a procolophonid (dog size). (g) *Pareiasaurus*, a pareiasaur (cow size). (See Table 9–2 for more details.)

Carboniferous, probably in response to the increasing quantity and diversity of terrestrial vegetation. Probably for the first time in evolutionary history, the terrestrial food supply was adequate to support a diverse fauna of fully terrestrial vertebrate predators. The initial radiation of non-amniotes was carnivorous forms (including fish- and invertebrate-eaters). No living adult amphibian is herbivorous, and there is little evidence in the fossil record to suggest that Paleozoic non-amniotic tetrapods were herbivores.

The Permian *Diadectes*, the probable sister taxon to amniotes, is the only non-amniote that may have been herbivorous: it had flat, crushing teeth like those of some extant herbivorous lizards. Carboniferous amniotes, such as *Hylonomus* (**Figure 9–11**), appear to have been insectivores but, from the start of the Permian onward, terrestrial habitats were dominated by a series of radiations of amniote tetrapods that now included herbivores, with large forms such as pareiasaurs (**Table 9–2**). Other body forms seen among Permian amniotes include the bipedal *Eudibamus* and the arboreal *Suminia*.

Derived Features of Amniotes

Traditionally, living amniotes have been distinguished by the amniotic egg (sometimes called the *cleidoic egg*;

Greek *cleido* = closed or locked) and a waterproof skin. The amniotic egg is characteristic of turtles, lepidosaurs (lizards and their relatives), crocodilians, birds, and monotremes (egg-laying mammals). Furthermore, embryonic membranes that contribute to the placenta of therian mammals (marsupials and placentals) are homologous to certain membranes in the egg. The amniotic egg is assumed to have been the reproductive mode of Mesozoic diapsids, and fossil dinosaur eggs are relatively common. In many other ways, however, amniotes represent a more derived kind of tetrapod than either the living amphibians or the Paleozoic non-amniotic tetrapods.

Skin permeability varies widely among living amphibians and amniotes. Although amniotes have a thicker skin than amphibians and a keratinized epidermis, it is the presence of lipids in the skin that makes the skin relatively impermeable to water. Compared to amphibians, amniotes have a greater variety of skin elaborations—scales, hair, and feathers—all formed from keratin. The lack of similar structures in living amphibians may be related to their use of the skin in respiration. Another important derived amniote feature is costal (rib) ventilation of the lungs. Because amniotes rely on the lungs for gas exchange, the skin does not have to be moist, and cutaneous water loss is reduced.

Table 9–2 Major groups of Paleozoic amniotes

Synapsida

Synapsids: "Mammal-like reptiles" are discussed in Chapter 18. Early synapsids (see Figure 9–11b) were somewhat larger than early eureptiles, and their larger heads and teeth suggest a more specialized carnivorous habit. Small forms first appeared in the Late Carboniferous and then diversified into larger forms in the Permian.

Sauropsida

Mesosaurs: The first secondarily aquatic amniotes (see Figure 9–11c). Known from freshwater deposits in the Early Permian of South Africa and South America, they provide one of the classic pieces of evidence for continental drift because these continents were united in Gondwana at that time. Swimming adaptations include large, probably webbed, hind feet, a laterally flattened tail, and heavily ossified ribs that may have acted as ballast in diving. The long jaws and slender teeth may have been used to strain small crustaceans from the surrounding water.

Parareptilia

Millerettids: Rather like the eureptile protorothyridid shown in Figure 9–11a; known from the Late Permian of South Africa.

Procolophonids: Medium size, with peglike teeth that were laterally expanded in later members of the group, apparently specialized for crushing or grinding, suggestive of herbivory (see Figure 9–11f);

known worldwide (except Australia) from the Late Permian to the Late Triassic.

Pareiasaurs: Large, approaching 3 m long (see Figure 9–11g); known from the Late Permian of Europe, Asia, and Africa. Their teeth were laterally compressed and leaf-shaped, like the teeth of herbivorous lizards. Pareiasaurs were evidently the dominant terrestrial herbivores of the later Permian.

Eureptilia

Protorothyridids: Small, relatively short-legged, rather lizardlike forms, probably insectivorous in habits (see Figure 9–11a); known from the mid-Carboniferous to the Early Permian in North America and Europe.

Captorhinids: Tetrapods with more robust skulls and flatter teeth than protorothyridids and early diapsids and may have had a more omnivorous diet that required crushing (see Figure 9–11d); known from the Late Carboniferous of North America, the Early and mid-Permian of North America and Europe, and the Late Permian of East Africa.

Araeoscelids (stem diapsids): Early diapsids with shorter bodies and longer legs than protorothyridids and probably also insectivorous (see Figure 9–11e); known from the Late Carboniferous and the Early Permian of North America and Europe.

Costal ventilation has other consequences. It allows an animal to have a long neck because movement of the ribs can produce a pressure differential large enough to draw air down a long, thin tube, such as the trachea in the necks of amniotes. In addition, some of the muscles involved in buccal pumping insert into the shoulder girdle, and this arrangement also may limit the development of a long neck in non-amniote tetrapods.

A longer neck provides space for elaboration of the nerves that supply the forelimb. Nerves supplying the forelimb leave the spinal cord in the neck and join together in a nerve complex called the brachial plexus. (There is a similar sacral plexus for nerves supplying the hindlimb.) The brachial plexus in living amphibians is simple; only two nerves are involved, in contrast to the plexus of amniotes, which has at least five nerves. Thus, amniotes have more complex innervation of the forelimb, which improves control of the limb and the ability for manipulation. This example shows how anatomical features may be linked together in evolution in unexpected ways because the animals evolve as an integrated whole. Who would suspect that our distant ancestors' using their ribs to ventilate the lungs could be linked to our ability to use our hands for tasks such as writing?

Structure of the Amniotic Egg An amniotic egg is a remarkable example of biological complexity (**Figure 9–12**). A flexible eggshell is probably the ancestral amniote condition, and it persists in many lizards, snakes, and turtles as well as in monotreme mammals. In other groups of lizards and in turtles the shell is rigid due to the inclusion of calcium deposits, as it is in crocodilians and birds.

The shell is the first line of defense against mechanical damage, while pores in the shell permit the movement of water vapor, oxygen, and carbon dioxide. Albumen (egg white) gives further protection against mechanical damage and provides a reservoir of water and protein. The large yolk is the energy supply for the developing embryo. At the beginning of embryonic development, the embryo is represented by a few cells resting on top of the yolk. As development proceeds, these cells multiply, and endodermal and mesodermal tissue surrounds the yolk, enclosing it in a **yolk sac** that is part of the developing gut. Blood vessels differentiate rapidly in the mesodermal tissue surrounding the yolk sac and transport food and gases to the embryo. By the end of development, only a small amount of yolk remains, and this is absorbed before or shortly after hatching.

While all vertebrates have an extraembryonic membrane (or membranes) enclosing the yolk, amniotes have three additional extraembryonic membranes—the **chorion**, **amnion**, and **allantois**. The chorion and amnion develop from outgrowths of the body wall at the edges of the developing embryo and spread outward and around the embryo until they meet. At their junction, the membranes merge and leave an outer membrane (the chorion), which surrounds the entire contents of the egg, and an inner membrane (the amnion), which surrounds just the embryo itself. The allantoic membrane develops as an outgrowth of the hindgut posterior to the yolk sac and lies within the chorion. The allantois appears to have evolved as a storage place for nitrogenous wastes produced by the metabolism of the embryo, and the urinary bladder of the adult grows out from its base. The allantois also serves as a respiratory organ during later development because it is vascularized and can transport oxygen from the surface of the egg to the embryo and carbon dioxide from the embryo to the surface. The allantois is left behind when the embryo emerges, so the nitrogenous wastes

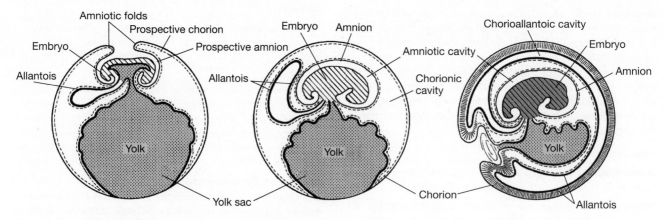

Figure 9–12 Distinctive features of the amniotic egg. Progressive stages in development are illustrated from left to right.

stored in it do not have to be reprocessed. The embryo in an amniotic egg bypasses the larval stage typical of amphibian embryos and does not form external gills at any stage in development. All traces of the lateral line are also lost. This loss of the larval form relates to one significant disadvantage of the amniotic egg—it can no longer be laid in water because the gill-less embryo would drown. Marine amniotes must either come ashore to lay eggs, such as sea turtles and penguins, or be viviparous (bearing live young), such as sea snakes and marine mammals such as whales and sea cows.

Origin of the Amniotic Egg How and why might the amniotic egg have evolved? It is not essential for development on land because many species of extant amphibians, some fishes, and many invertebrates lay non-amniotic eggs that develop quite successfully on land. Both amniotic and non-amniotic eggs must be laid in relatively moist conditions to avoid desiccation, and both types of eggs are usually buried in the soil or deposited under objects such as rocks and logs. (Birds, which represent a highly derived condition, are an exception to this generalization.)

Various plausible explanations for the development of the amniote egg have been proposed. For example, the extraembryonic membranes may improve within-egg respiratory capacities, and the shell may provide mechanical support on land; together these features would allow the evolution of a large egg that produced a large hatchling that in turn grew into a large adult because egg size is related to adult size. However, the truth is that we do not really understand what evolutionary forces would have led to the first amniote eggs, even though this kind of egg was doubtless important in the later evolution of amniotes.

An amniotic egg requires internal fertilization because sperm cannot penetrate the eggshell after the egg has left the female's body. Most male amniotes possess an intromittent (copulatory) organ (a penis). The anatomy of the penis differs among amniote lineages, however, indicating that a penislike copulatory structure evolved independently on four separate occasions: in the ancestors of mammals, in squamates (lizards and snakes; the tuatara lacks a penis and probably represents the ancestral amniote condition), in turtles, and in archosaurs (crocodiles and birds, although most living birds have secondarily lost this organ). Additionally, a copulatory organ has been evolved in a completely independent fashion in the caecilian lissamphibians.

How could we tell if a fossil animal laid an amniotic egg when features such as extraembryonic membranes are not preserved in the fossil record? We can estimate the latest point of origin of the amniote egg from the tetrapod phylogeny (see Figure 9–1). The synapsids (mammals and their extinct relatives) branched off from other amniotes very early, and all other fossil animals that we consider to be amniotes are more closely related to sauropsids (living reptiles and birds) than to mammals. Because the egg membranes of mammals are homologous with those of other living amniotes, all tetrapods higher than node 13 in the cladogram must have inherited an amniotic egg from the common ancestor of mammals and other amniotes.

A more difficult question is whether any fossil tetrapods lower down in the phylogeny might have laid an amniotic egg. We know that this was not true of seymouriamorphs because larval forms with external gills and lateral lines are known. That leaves the diadectomorphs as the only possible candidates for the amniotic condition among the stem amniotes.

Patterns of Amniote Temporal Fenestration Amniotes have traditionally been subdivided by the number of holes in their head—that is, on the basis of **temporal fenestration** (Latin *fenestra* = window). The major configurations that give names to different lineages of amniotes are **anapsid** (Greek *an* = without and *apsid* = junction), seen in early amniotes and in turtles; **synapsid** (single arch; Greek *syn* = joined), seen in mammals and their ancestors; and **diapsid** (double arch; Greek *di* = two), seen in birds and other reptiles. The term *arch* refers to the temporal bars lying below and between the holes. **Figure 9–13** shows the different patterns in the different groups. Note that the phylogenetic pattern of fenestrae suggests that temporal openings arose independently in the synapsid and diapsid lineages because early sauropsids lack holes entirely. Turtles have traditionally been classified with the other anapsids on the basis of morphological characters. More recently, however, molecular studies have suggested that their origins may lie within the diapsid radiation (see Figure 9–2), and some morphological characters support this view. If turtles are diapsids, they have secondarily roofed in their skull from an originally fenestrated condition. If turtles are diapsids, are they more closely related to archosaurs (crocodilians, dinosaurs, and birds) or to lepidosaurs (lizards and snakes)? That question is still unanswered, as is the possibility that turtles are the sister lineage of the archosaurs + lepidosaurs.

Even though the skull of living mammals is highly modified from the ancestral synapsid condition, you can still feel these skull features in yourselves. If you put your hands on either side of your eyes, you can feel your cheekbone (the zygomatic arch)—that is the temporal bar that lies below your synapsid skull

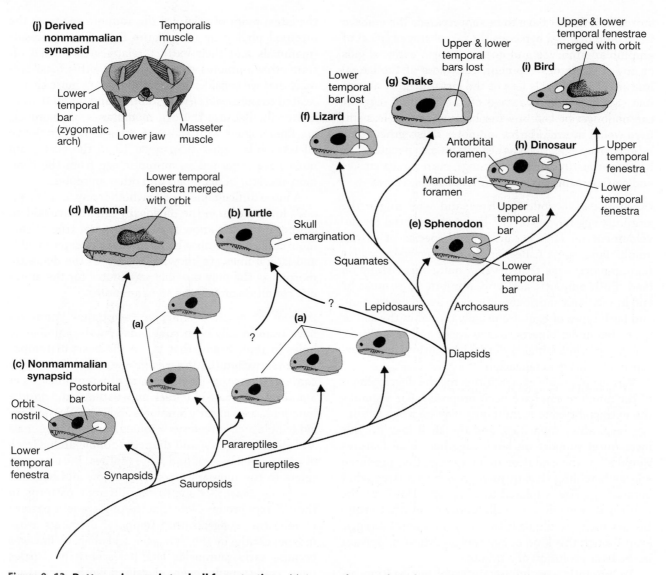

Figure 9–13 Patterns in amniote skull fenestration. (a) Ancestral anapsid condition, as seen in the common ancestor of all amniotes and in basal members of the parareptiles and eureptiles. (b) Modified anapsid condition, with emargination of the posterior portion of the skull, as seen in turtles. (Turtles may be secondarily anapsid; see Chapter 12.) (c) Ancestral synapsid condition, with lower temporal fenestra only. (d) Derived mammalian synapsid condition, in which the orbit has become merged with the temporal opening and dermal bone has grown down from the skull roof to surround the braincase. (e) Ancestral diapsid condition, as seen today in the reptile *Sphenodon*, although the condition here is apparently a secondary one from a more lizardlike anatomy. Both upper and lower temporal fenestrae are present. (f) Lizardlike condition, typical of most squamates, where the lower temporal bar has been lost. (Note that some fossil diapsids apparently regained a slightly different version of this bar.) (g) Snake condition, upper temporal bar lost in addition to the lower bar. (h) Ancestral archosaur diapsid condition, as seen in thecodonts and most dinosaurs. An antorbital foramen and a mandibular foramen have been added to the basic diapsid pattern. (Note that the antorbital foramen is secondarily reduced or lost in crocodiles.) (i) Derived avian archosaur condition, convergent with the condition in mammals. The orbit has become merged with the temporal openings, and the braincase is enclosed in dermal bone. (j) Posterior view through the skull of a synapsid (a cynodont mammal-like reptile) showing how the temporal fenestra allows muscles to insert on the outside of the skull roof. The temporalis and masseter muscles are divisions of the original amniote adductor muscle complex.

opening. Then, if you clench your jaw, you can feel the muscles bulging above the arch. The muscles are passing through the temporal opening, running from their origin from the top of the skull to the insertion on your lower jaw.

What is the function of these holes? As you just discovered, they provide room for muscles to bulge. Amniotes have larger and more complex jaw muscles than non-amniotes, and the notion of room for bulging was originally the preferred evolutionary explanation. But only a large hole will allow enough room for a bulging muscle, so what could be the evolutionary advantage of the initial, small hole? And why doesn't any non-amniote ever develop temporal fenestration?

The key to the evolution of the temporal fenestrae may lie in changes in the complexity and orientation of the jaw-closing (adductor) muscles. The large, flat skull of non-amniotes, which may be related to their buccal-pumping mode of respiration (in which the skull is acting as a pair of bellows), does not permit a change in the orientation of the jaw muscles from the basic fish condition (**Figure 9–14**). A muscle originating from the skull roof, like the amniote adductor mandibularis, would be too short to allow the jaw to open wide because muscles can be stretched for only one third of their resting length. With the evolution of costal ventilation, the head size and shape were no longer important: amniotes were now able to evolve smaller, more domed skulls, allowing differentiation of the simple fishlike jaw adductors into the adductor mandibularis and the pterygoideus. The pterygoideus originates from a distinct pterygoid flange on the base of the skull, which is a characteristic feature of amniotes, revealing that a change in jaw musculature has occurred.

The advantage of this change in musculature would be a change in the feeding abilities. Fishes and non-amniotic tetrapods can only close their jaws with a single snap (inertial feeding), whereas amniotes can snap the jaws closed and also apply pressure with the teeth when the jaws are closed (static pressure feeding). This difference may have allowed more complex types of feeding in amniotes, such as the ability of herbivores to nip off vegetation with their front teeth. Occlusion (i.e., the teeth meeting each other rather than just contacting the food) between the upper and lower teeth is seen for the first time in the probable herbivore *Diadectes* and in the amniotes. Dental occlusion is probably also related to the ability to be herbivorous because herbivores require more oral processing to break down tough food.

This reorientation of the adductor mandibularis may be the underlying reason for skull fenestration in amniotes. The development of a hole in the skull

(a) Non-amniotic tetrapod

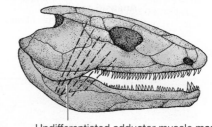

Lateral view

Undifferentiated adductor muscle mass

Palatal view

(b) Amniotic tetrapod (captorhinid)

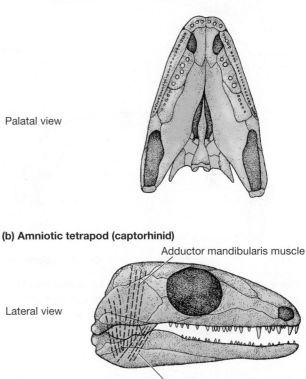

Adductor mandibularis muscle

Lateral view

Pterygoideus muscle

Palatal view

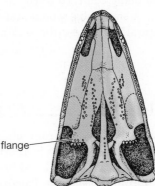

Pterygoid flange

Figure 9–14 Jaw muscles. The skulls show the differences in jaw muscles of (a) non-amniotic and (b) amniotic tetrapods.

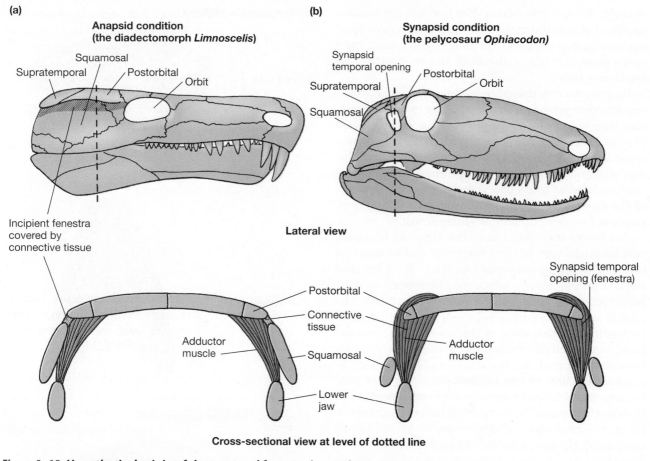

(a)

**Anapsid condition
(the diadectomorph *Limnoscelis*)**

Squamosal

Supratemporal   Postorbital

Orbit

Incipient fenestra
covered by
connective tissue

(b)

**Synapsid condition
(the pelycosaur *Ophiacodon*)**

Synapsid
temporal opening   Postorbital

Supratemporal   Orbit

Squamosal

Lateral view

Postorbital

Connective
tissue

Adductor
muscle

Adductor
muscle

Squamosal

Lower
jaw

Synapsid temporal
opening (fenestra)

Cross-sectional view at level of dotted line

Figure 9–15 Hypothetical origin of the temporal fenestra in amniotes. Even the small hole in the skull produced by a failure of ossification where three bones meet during development as in the skull of *Limnoscelis* (a) could be the starting point for the evolution of the synapsid condition seen in *Ophiacodon* (b).

roof (actually an area where three bones fail to completely meet in development) would allow the muscle to develop an area of origin on the outside of the skull, rather than being confined to the interior of the skull (**Figure 9–15**). A small hole would still reflect an important structural change for the action of the jaw muscles, and a larger hole could then be selected for, which would allow for a larger volume of bulging muscle. Perhaps differences in muscle actions, relating to different feeding styles, encouraged temporal fenestration to take a different form in the synapsid and diapsid groups.

Summary

The origin of tetrapods from elpistostegalid lobe-finned fishes in the Devonian is inferred from similarities in the bones of the skull and braincase, vertebral structure, and limb skeleton. Paleozoic non-amniotic tetrapods comprise about a dozen distinct lineages with relationships that have not yet been determined. One current view divides them into three major groups: batrachomorphs (or stem amphibians), including the predominantly aquatic temnospondyls; reptiliomorphs (or stem amniotes), including the predominantly terrestrial anthracosaurs; and lepospondyls. The batrachomorphs radiated extensively in the Late Carboniferous and Permian, and several aquatic lineages extended into the Mesozoic. At least some of the modern amphibians—salamanders and frogs—may be derived from this lineage. The reptiliomorphs radiated during the Carboniferous and became extinct by the earliest Triassic. The lepospondyls were small forms of uncertain phylogenetic affiliation and may contain the ancestry of the modern caecilians.

The amniotic egg, with its distinctive extraembryonic membranes, is a shared derived character that distinguishes the amniotes (turtles, lepidosaurs, crocodilians, birds, and mammals) from the non-amniotes (fishes and amphibians). The earliest amniotes were small animals, and their appearance coincided with a major radiation of terrestrial insects in the Carboniferous. By the end of the Carboniferous, amniotes had begun to radiate into most of the terrestrial life zones that had been occupied by non-amniotes, and additionally evolved herbivorous primary consumers: only the relatively aquatic groups of non-amniotic tetrapods maintained much diversity throughout the Triassic.

The major groups of amniotes can be distinguished by different patterns of temporal fenestration—holes in the dermal skull roof that reflect the increasing complexity of jaw musculature. The major division of amniotes is into synapsids (mammals and their relatives) and sauropsids (reptiles and birds).

Discussion Questions

1. What information about the Devonian tetrapods leads to the conclusion that tetrapods evolved in the water rather than on land?
2. What is the definition (in phylogenetic terms) of a tetrapodomorph fish?
3. What features of the elpistostegalid fishes lead us to infer that they were shallow-water forms?
4. What is the main new piece of information about early tetrapods that contradicts the old "drying pond" hypothesis of the origins of terrestriality?
5. How can we be sure that fossil amniotes, such as the captorhinomorphs, laid amniotic eggs? How can we be sure that at least most of the non-amniote reptiliomorphs (below the level of diadectomorphs) did not?
6. How can we determine whether an animal was herbivorous? Propose a hypothesis to explain why herbivory is limited to amniotes and their closest relatives.

Additional Information

Ahlberg, P. E., and J. A. Clack. 2006. A firm step from water to land. *Nature* 440:747–749.

Anderson, J. S. 2008. The origin(s) of modern amphibians. *Evolutionary Biology* 35:231–247.

Anderson, J. S., et al. 2008. A stem batrachian from the Early Permian of Texas and the origin of frogs and salamanders. *Nature* 453:515–518.

Boisvert, C. A., et al. 2008. The pectoral fin of *Panderichthys* and the origin of digits. *Nature* 456:636–638.

Callier, V., et al. 2009. Contrasting developmental trajectories in the earliest known tetrapod forelimbs. *Science* 17:364–367.

Clack, J. A. 2006. The emergence of early tetrapods. *Palaeogeography, Palaeoclimatology, Palaeoecology* 232:167–189.

Clack, J. A. 2009. The fish–tetrapod transition: New fossils and interpretations. *Evolution, Education, and Outreach* 2:213–223.

Clack, J. A. 2009. The fin to limb transition: New data, interpretations, and hypotheses from paleontology and developmental biology. *Annual Review of Earth and Planetary Sciences* 37:163–179.

Clack, J. A. 2011. *Gaining Ground: The Origin and Evolution of Tetrapods*, 2nd ed. Bloomington, IN: Indiana University Press.

Clack, J. A., et al. 2012. A new genus of Devonian tetrapod from north-east Greenland, with new information on the lower jaw of Ichthyostega. *Palaeontology* 55: 73–86.

Coates, M. I., et al. 2008. Ever since Owen: Changing perspectives on the early evolution of tetrapods. *Annual Review of Earth and Planetary Sciences* 39:571–592.

Daeschler, E. B., et al. 2006. A Devonian tetrapod-like fish and the evolution of the tetrapod body plan. *Nature* 440:757–763.

Daeschler, E. B., et al. 2009. Late Devonian tetrapod remains from Red Hill, Pennsylvania, USA: How much diversity? *Acta Zoologica* 90:306–317.

Davis, M. C., et al. 2007. An autopodiallike pattern of *Hox* expression in the fins of a basal actinopterygian fish. *Nature* 447:473–476.

Fröbisch, J., and Reisz, R. R. 2009. The Late Permian herbivore *Suminia* and the early evolution arboreality in terrestrial vertebrate ecosystems. *Proceedings of the Royal Society, B* 276:3611–3618.

Janis, C. M., and C. Farmer. 1999. Proposed habits of early tetrapods. Gills, kidneys, and the water-land transition. *Zoological Journal of the Linnean Society* 126:117–126.

Janis, C. M., and J. C. Keller. 2001. Modes of ventilation in early tetrapods: Costal aspiration as a key feature of amniotes. *Acta Palaeontologica Polonica* 46:137–170.

Janvier, P., and G. Clément. 2010. Muddy tetrapod origins. *Nature* 463:40–41.

King, H. M., et al. 2012. Behavioral evidence for the evolution of walking and bounding before terrestriality in sarcopterygian fishes. *Proceedings of the National Academy of Sciences*, in Press.

Laurin, M. 2010. *How Vertebrates Left the Water*. Berkeley, CA: University of California Press.

Long, J. A. 2011. *The Rise of Fishes*. Baltimore, MD: The Johns Hopkins University Press.

Lyson, T. R., et al. 2012. MicroRNAs support a turtle + lizard clade. *Biology Letters* 8:104–107.

Markey, M. J., and C. R. Marshall. 2007. Terrestrial-style feeding in a very early aquatic tetrapod is supported by evidence from experimental analysis of suture morphology. *Proceedings of the National Academy of Sciences* 104:7134–7138.

Pierce, S. E. et al. 2012. Three-dimensional limb joint mobility in the early tetrapod. *Ichthoystega Nature* 486:523–526.

Reilly, S. M. 2001. Prey processing in amniotes: Biomechanical and behavioral patterns of food reduction. *Comparative Biochemistry and Physiology A* 128:397–415.

Reisz, R. R. 2006. Origin of dental occlusion in tetrapods: Signal for vertebrate evolution? *Journal of Experimental Zoology (Mol Dev Evol)* 306B:261–277.

Richardson, J., et al. 2012. The presence of an embryonic opercular flap in amniotes. *Proceedings of the Royal Society, B* 279:224–229.

Schoch, R. R., and Witzmann, F. 2011. Bystrow's paradox—Gills, fossils, and the fish-to-tetrapod transition. *Acta Zoologica* 92:251–265.

Shen, X-X., et al. 2011. Multiple genome alignments facilitate development of NPCL markers: A case study of tetrapod phylogeny focusing on the position of turtles. *Molecular Biology and Evolution* 28:3237–3252.

Shubin, N. H., et al. 2006. The pectoral fin of *Tiktaalik roseae* and the origin of the tetrapod limb. *Nature* 440:764–771.

Woltering, J. M., and D. Duboule. 2010. The origin of digits: Expression patterns versus regulatory mechanisms. *Developmental Cell* 18:526–532.

Zeller, R. 2010. The temporal dynamics of vertebrate limb development, teratogenesis and evolution. *Current Opinion in Genetics & Development* 2010:20:1–7.

Websites

Fish/tetrapod transition and Devonian tetrapods
http://tiktaalik.uchicago.edu/index.html
http://evolution.berkeley.edu/evolibrary/article/0_0_0/evograms_04
Non-amniote tetrapods
http://palaeo.gly.bris.ac.uk/Palaeofiles/Fossilgroups/Temnospondyli/index.html
Lepospondyls
http://palaeo.gly.bris.ac.uk/Palaeofiles/Fossilgroups/Nectridea/index.html
Amniotes
Parareptiles
http://palaeo.gly.bris.ac.uk/Palaeofiles/Fossilgroups/anapsida/subgroup.html
Diapsids
http://palaeo.gly.bris.ac.uk/Palaeofiles/Fossilgroups/Diapsida/index.html

CHAPTER 10

Salamanders, Anurans, and Caecilians

The three lineages of extant amphibians (salamanders, frogs, and caecilians) have very different body forms, but they are identified as a monophyletic evolutionary lineage by several shared derived characters. Some of these characters—especially the moist, permeable skin—have channeled the evolution of the three lineages in similar directions. Frogs are the most successful amphibians, and it is tempting to think that the variety of locomotor modes permitted by their specialized morphology may be related to their success: frogs can jump with simultaneous movements of the hind legs, swim with either simultaneous or alternating leg movements, and walk or climb with alternating leg movements. In contrast, salamanders retain the ancestral tetrapod locomotor pattern of lateral undulations combined with alternating limb movements in a walking-trot.

The range of reproductive specializations of amphibians is nearly as great as that of fishes—a remarkable fact considering that there are more than five times as many species of fishes as amphibians. The ancestral reproductive mode of amphibians probably consisted of laying large numbers of eggs that hatched into aquatic larvae, and many amphibians still reproduce this way. An aquatic larva gives a terrestrial species access to resources that would not otherwise be available to it. Modifications of the ancestral reproductive mode include bypassing the larval stage, viviparity, and parental care of eggs and young, including females that feed their tadpoles.

The permeable skin of amphibians is central to many aspects of their lives. The skin is a major site of respiratory gas exchange and must be kept moist. Evaporation of water from the skin limits the activity of most amphibians to relatively moist microenvironments. The skin contains glands that produce substances used in courtship as well as other glands that produce toxic substances that deter predators. Many amphibians advertise their toxicity with bright warning colors, and some nontoxic species deceive predators by mimicking the warning colors of toxic forms.

10.1 Amphibians

The extant amphibians, or Amphibia, are tetrapods with moist, scaleless skins. The group includes three distinct lineages: anurans (frogs), urodeles (salamanders), and gymnophionans (caecilians or apodans). Most amphibians have four well-developed limbs, although a few salamanders and all caecilians are limbless. Frogs lack tails (hence the name *anura*, which means "without a tail"), whereas most salamanders have long tails. The tails of caecilians are short, as are those of other groups of elongate, burrowing animals.

Table 10–1 Shared derived characters of amphibians

Structure of the skin and the importance of cutaneous gas exchange: All amphibians have mucous glands that keep the skin moist. A substantial part of an amphibian's exchange of oxygen and carbon dioxide with the environment takes place through the skin. All amphibians also have poison (granular) glands in the skin.

Papilla amphibiorum: All amphibians have a special sensory area, the papilla amphibiorum, in the wall of the sacculus of the inner ear. The papilla amphibiorum is sensitive to frequencies below 1000 hertz (Hz; cycles per second), and a second sensory area, the papilla basilaris, detects sound frequencies above 1000 Hz.

Operculum-columella complex: Most amphibians have two bones that are involved in transmitting sounds to the inner ear. The columella, which is derived from the hyoid arch, is present in salamanders, caecilians, and most frogs. The operculum develops in association with the fenestra ovalis of the inner ear. The columella and operculum are fused in anurans, caecilians, and some salamanders.

Green rods: Salamanders and frogs have a distinct type of retinal cell, green rods which have maximum sensitivity to blue light, in addition to the red rods that are found in all vertebrates and are sensitive to red light. Caecilians apparently lack green rods; however, the eyes of caecilians are extremely reduced, and the green rods may have been lost.

Pedicellate teeth: Nearly all modern amphibians have teeth in which the crown and base (pedicel) are composed of dentine and are separated by a narrow zone of uncalcified dentine or fibrous connective tissue. A few amphibians lack pedicellate teeth, and the boundary between the crown and base is obscured in some other genera. Pedicellate teeth also occur in some actinopterygian fishes, which are not thought to be related to amphibians.

Structure of the levator bulbi muscle: This muscle is a thin sheet in the floor of the orbit that is innervated by the fifth cranial nerve. It causes the eyes to bulge outward, thereby enlarging the buccal cavity. This muscle is present in salamanders and anurans and in modified form in caecilians.

At first glance, the three lineages of amphibians appear to be very different kinds of animals: frogs have long hindlimbs and short, stiff bodies that don't bend when they walk; salamanders have forelimbs and hindlimbs of equal size and move with lateral undulations; and caecilians are limbless and burrow with alternating extensions and contractions known as concertina locomotion. These obvious differences are all related to locomotor specializations, however, and closer examination shows that amphibians have many derived characters in common, indicating that they form a monophyletic evolutionary lineage derived from the temnospondyls (Table 10–1).

We will see that many of these shared characters play important roles in the functional biology of amphibians. Perhaps the most important derived character of extant amphibians is a moist, permeable skin. The name applied to the lineage, Lissamphibia, refers to the texture of the skin (Greek *liss* = smooth). Many of the Paleozoic non-amniote tetrapods had dermal armor in the form of bony scutes in the skin; a permeable, unadorned skin is a derived character shared by amphibians.

All living adult amphibians are carnivorous, and relatively little morphological specialization is associated with different dietary habits within each group. Amphibians eat almost anything they are able to catch and swallow. The tongue of aquatic forms is broad, flat, and relatively immobile, but some terrestrial amphibians can protrude the tongue from the mouth to capture prey. The size of the head is an important determinant of the maximum size of prey that can be taken, and

species of salamanders that live in the same habitat frequently have markedly different head sizes, suggesting that this may be a feature that reduces competition for food. Frogs in the tropical American genera *Lepidobatrachus* and *Ceratophrys*, which feed largely on other frogs, have such large heads that they are practically walking mouths.

The anuran body form probably evolved from a more salamander-like starting point. Both jumping and swimming have been suggested as the mode of locomotion that made the change advantageous. Salamanders and caecilians swim as fishes do—by passing a sine wave down the body. Anurans have inflexible bodies and swim with simultaneous thrusts of the hind legs.

Some paleontologists have proposed that the anuran body form evolved because of the advantages of that mode of swimming. An alternative hypothesis traces the anuran body form to the advantage gained by an animal that could rest near the edge of a body of water and escape aquatic or terrestrial predators with a rapid leap followed by locomotion on either land or water. The stem anuran *Triadobatrachus* may be an example of that body form.

The oldest fossils that may represent modern amphibians are isolated vertebrae of Permian age that appear to include both salamander and anuran types. The oldest froglike fossils are from the Early Jurassic period, but fossils of salamanders and caecilians are not known before the Jurassic. Clearly, the modern orders of amphibians have had separate evolutionary histories for a long time. The continued presence

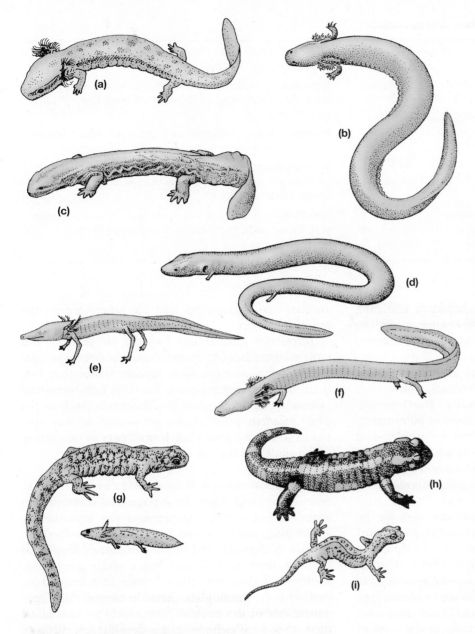

Figure 10–1 Diversity of salamanders. The body forms of salamanders reflect differences in their life histories and habitats. Aquatic salamanders may retain gills as adults, as in (a) the North American mudpuppy (*Necturus maculosus*) and (b) the North American siren (*Siren lacertina*). Others have folds of skin that are used for gas exchange or rely on lungs and the body surface, as in (c) the North American hellbender (*Cryptobranchus alleganiensis*) and (d) the North American Congo eel (*Amphiuma means*). Specialized cave-dwelling salamanders, such as (e) the Texas blind salamander (*Eurycea* [*Typhlomolge*] *rathbuni*) and (f) the European olm (*Proteus anguinus*), are white and lack eyes. Terrestrial salamanders usually have sturdy legs, like (g) the North American tiger salamander (*Ambystoma tigrinum*) and its aquatic larva, (h) the European fire salamander (*Salamandra salamandra*), and (i) the North American slimy salamander (*Plethodon glutinosus*).

of such common characters as a permeable skin, after at least 190 million years of independent evolution, suggests that the shared characters are central to the lives of modern amphibians. In other characters, such as reproduction, locomotion, and defense, amphibians show tremendous diversity.

Salamanders—Caudata or Urodela

The salamanders have the most generalized body form and locomotion of the living amphibians. Salamanders are elongate, and all but a very few species of completely aquatic salamanders have four functional limbs (**Figure 10–1**). Their walking-trot gait is probably similar to that employed by the earliest tetrapods. It combines the lateral bending characteristic of fish locomotion with leg movements.

The nine families, containing approximately 619 species, are almost entirely limited to the Northern Hemisphere; their southernmost occurrence is in northern South America (**Table 10–2**). North and Central America have the greatest diversity of salamanders—more species of salamanders are found in Tennessee than in all of Europe and Asia combined. Paedomorphosis is widespread among salamanders, and several families of aquatic salamanders consist solely of paedomorphic

Table 10–2 Families of salamanders

Ambystomatidae: Small to large terrestrial salamanders with aquatic larvae (37 species in North America; up to 30 cm)

Amphiumidae: Elongate aquatic salamanders lacking gills (three species in North America; about 1 m)

Cryptobranchidae: Very large paedomorphic aquatic salamanders with external fertilization of the eggs (one species in North America and two species in Asia; 1–1.5 m)

Hynobiidae: Terrestrial or aquatic salamanders with external fertilization of the eggs and aquatic larvae (55 species in Asia; up to 30 cm)

Plethodontidae: Aquatic or terrestrial salamanders, some with aquatic larvae and others that lay eggs on land and bypass the larval stage (about 409 species in North, Central, and northern South

America, plus eight species in Mediterranean Europe and one on the Korean peninsula; 3–30 cm)

Proteidae: Paedomorphic aquatic salamanders with external gills (six species in North America [*Necturus*] and one in Europe [*Proteus*]; up to 30 cm)

Rhyacotritonidae: Semiaquatic salamanders with aquatic larvae (four species in North America; shorter than 10 cm)

Salamandridae: Terrestrial and aquatic salamanders with aquatic larvae (89 species in Europe, Asia, North America, and extreme northwestern Africa; up to 20 cm)

Sirenidae: Elongate aquatic salamanders with external gills and lacking the pelvic girdle and hindlimbs (four species in North America; 15–75 cm)

---

forms. Paedomorphs retain larval characters, including larval tooth and bone patterns, the absence of eyelids, a functional lateral line system, and (in some cases) external gills.

The largest living salamanders are the Japanese and Chinese giant salamanders (*Andrias*), which reach lengths of 1 meter or more. The related North American hellbenders (*Cryptobranchus*) grow to 60 centimeters. All are members of the Cryptobranchidae and are paedomorphic and permanently aquatic. As their name indicates (Greek *crypto* = hidden and *branchus* = a gill), they do not retain external gills, although they do have other larval characters.

Another group of large aquatic salamanders are the mudpuppies (*Necturus*), which consist of paedomorphic species that retain external gills. Mudpuppies occur in lakes and streams in eastern North America. The congo eels (three species of aquatic salamanders in the genus *Amphiuma*) live in the lower Mississippi Valley and coastal plain of the United States. They have well-developed lungs and can estivate in the mud of dried ponds for up to 2 years.

Several lineages of salamanders have adapted to life in caves. The constant temperature and moisture of caves make them good salamander habitats, and cave-dwelling invertebrates supply food. The brook salamanders (*Eurycea*, Plethodontidae) include species that form a continuum from those with fully metamorphosed adults inhabiting the twilight zone near cave mouths to fully paedomorphic forms in the depths of caves or sinkholes. The Texas blind salamander, *Eurycea rathbuni*, is a highly specialized cave dweller—blind, white, with external gills, extremely long legs, and a flattened snout used to probe underneath pebbles for food. The unrelated European olm (*Proteus*) is

another cave salamander that has converged on the same body form.

Terrestrial salamanders like the North American mole salamanders (*Ambystoma*) and the European salamanders (*Salamandra*) have aquatic larvae that lose their gills at metamorphosis. The most fully terrestrial salamanders, the lungless plethodontids (such as the slimy salamander, *Plethodon glutinosus*), include species in which the young hatch from eggs as miniatures of the adult and there is no aquatic larval stage.

Feeding Specializations of Plethodontid Salamanders Lungs seem an unlikely organ for a terrestrial vertebrate to abandon, but among salamanders the evolutionary loss of lungs has been a successful tactic. The Plethodontidae is characterized by the absence of lungs and contains more species and has a wider geographic distribution than any other lineage of salamanders. Furthermore, many plethodontids have evolved specializations of the hyobranchial apparatus that allow them to protrude the tongue a considerable distance from the mouth to capture prey. This ability has not evolved in salamanders that have lungs, probably because the hyobranchial apparatus in these forms is an essential part of the respiratory system.

Bolitoglossine plethodontids (Greek *bola* = dart and *glossa* = tongue) can project the tongue a distance equivalent to their head plus trunk length and can pick off moving prey (**Figure 10–2**). This ability requires fine visual discrimination of distance and direction, and the eyes of bolitoglossines are placed more frontally on the head than the eyes of less specialized plethodontids. Furthermore, the eyes of bolitoglossines have a large number of nerves that travel to the ipsilateral (same side) visual centers of the brain as well as the

Figure 10–2 A European bolitoglossine salamander, *Hydromantes*. This species captures prey by trapping the prey on the sticky tip of its tongue, which can be projected from the mouth.

strong contralateral (opposite side) visual projection that is typical of salamanders. Because of this neuroanatomy, bolitoglossines have a complete dual projection of the binocular visual fields to both hemispheres of the brain. They can estimate their distance from a prey object very exactly and rapidly.

The ability to project the tongue is intertwined with the life histories of plethodontid salamanders, including their reproductive modes. Aquatic larval salamanders employ suction feeding, opening the mouth and expanding the buccal cavity to create a current of water that carries a prey item with it. The hyobranchial apparatus is an essential part of this feeding mechanism, and the first ceratobranchial becomes well developed during the larval period. In contrast, enlargement of the second ceratobranchial is associated with the tongue-projection mechanism of adult plethodontids. Furthermore, larval salamanders have laterally placed eyes, and the optic nerves project mostly to the contralateral side of the brain. Thus the morphological specializations that make aquatic plethodontid larvae successful are different from the specializations of adults that allow tongue projection, and this situation creates a conflict between the selective forces that act on juveniles and adults.

The bolitoglossines do not have aquatic larvae, and the morphological specializations of adult bolitoglossines appear during embryonic development. In contrast, hemidactyline plethodontids do have aquatic larvae that use suction feeding. As adults, hemidactylines have considerable ability to project the tongues, but they retain the large first ceratobranchial that appears in the larvae. This is a mechanically less efficient arrangement than the large second ceratobranchial of bolitoglossines, and the ability of hemidactylines to project their tongues is correspondingly less than that of bolitoglossines. Thus, the development of a specialized feeding mechanism by plethodontid salamanders has gone hand in hand with such diverse aspects of their biology as respiratory physiology and life history and demonstrates that organisms evolve as whole functioning units, not as collections of independent characters.

Social Behavior of Plethodontid Salamanders Plethodontid salamanders can be recognized externally by the nasolabial groove that extends ventrally from each external naris (nostril opening) to the lip of the upper jaw (**Figure 10–3**). These grooves are an important part of the chemosensory system of plethodontids. As a plethodontid salamander moves about, it repeatedly presses its snout against the substrate. Fluid is drawn into the grooves and moves upward to the external

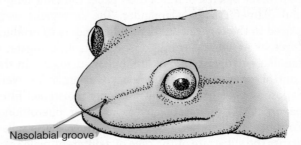

Nasolabial groove

Figure 10–3 Nasolabial grooves of a plethodontid salamander.

nares, into the nasal chambers, and over the chemoreceptors of the vomeronasal organ.

Studies of plethodontid salamanders have contributed greatly to our understanding of the roles of competition and predation in shaping the structure of ecological communities. Much work in this area has focused on experimental manipulations of animals in the field or laboratory. Because plethodontid salamanders have small home ranges and often remain in a restricted area for their entire lives, they are excellent species to use for these studies.

Males of many plethodontid salamanders defend all-purpose territories that are used for feeding and reproduction. Studies of these salamanders have revealed patterns of social behavior and foraging that seem remarkably complex for animals with skulls the size of a match head and brains little larger than the head of a pin. Robert Jaeger and his colleagues have studied the territorial behavior of the red-backed salamander, *Plethodon cinereus*, a common species in woodlands of eastern North America. Male red-backed salamanders readily establish territories in cages in the laboratory. A resident male salamander marks the substrate of its cage with pheromones. A salamander can distinguish between substrates it has marked and those marked by another male salamander or by a female salamander. Male salamanders can also distinguish between the familiar scent of a neighboring male salamander and the scent of a male they have not previously encountered, and they react differently to those scents.

In laboratory experiments, red-backed salamanders select their prey in a way that maximizes their energy intake: when equal numbers of large and small fruit flies are released in the cages, the salamanders first capture the large flies. This is the most profitable foraging behavior for the salamanders because it provides the maximum energy intake per capture. In a series of experiments, Jaeger and his colleagues showed that territorial behavior and fighting interfere with the ability of salamanders to select the most profitable prey. These experiments used surrogate salamanders that were made of a roll of moist filter paper the same length and diameter as a salamander. The surrogates were placed in the cages of resident salamanders to produce three experimental conditions: a control surrogate, a familiar surrogate, and an unfamiliar surrogate. In the control experiment, male red-backed salamanders were exposed to a surrogate that was only moistened filter paper; it did not carry any salamander pheromone. For both of the other groups, the surrogate was rolled across the substrate of the cage of a different male salamander to absorb the scent of that salamander before being placed in the cage of a resident male.

The experiments lasted seven days; the first six days were conditioning periods, and the test itself occurred on the seventh day. For the first six days, the resident salamanders in both experimental groups were given surrogates bearing the scent of another male salamander. The residents thus had the opportunity to become familiar with the scent of that male. On the seventh day, however, the familiar and unfamiliar surrogate groups were treated differently. Resident salamanders in the familiar surrogate group once again received a surrogate salamander bearing the scent of the same individual it had been exposed to for the previous six days, whereas the residents in the unfamiliar surrogate group received a surrogate bearing the scent of a different salamander, one to which they had never been exposed before. After a 5-minute pause, a mixture of large and small fruit flies was placed in each cage, and the behavior of the resident salamander was recorded.

The salamanders in the familiar surrogate group showed little response to the now-familiar scent of the other male salamander. They fed as usual, capturing large fruit flies. In contrast, the salamanders that were exposed to the scent of an unfamiliar surrogate began to show threatening and submissive displays, and their rate of prey capture decreased as a result of the time they spent displaying. In addition, salamanders exposed to unfamiliar surrogates did not concentrate on catching large fruit flies, so the average energy intake per capture also declined. The combined effects of the reduced time spent feeding and the failure to concentrate on the most profitable prey items caused an overall 50 percent decrease in the rate of energy intake for the salamanders exposed to the scent of an unfamiliar male.

The ability of male salamanders to recognize the scent of another male after a week of habituation in the laboratory cages suggests that they would show the same behavior in the woods. That is, a male salamander could learn to recognize and ignore the scent of a male in the adjacent territory, while still being able to recognize and attack a strange intruder. Learning not to respond to the presence of a neighbor may allow a salamander to forage more effectively, and it may also help to avoid injuries that can occur during territorial encounters.

Resident male red-backed salamanders challenge strange intruders, and the encounters involve aggressive and submissive displays and biting (**Figure 10–4**). Bites on the body can drive another male away but are not likely to do permanent damage. A bite to the tail may cause the bitten salamander to autotomize (break off) its tail. Salamanders store fat in their tails, and this injury may delay reproduction for a year while the tail is regenerated. Most bites are directed at the snout of

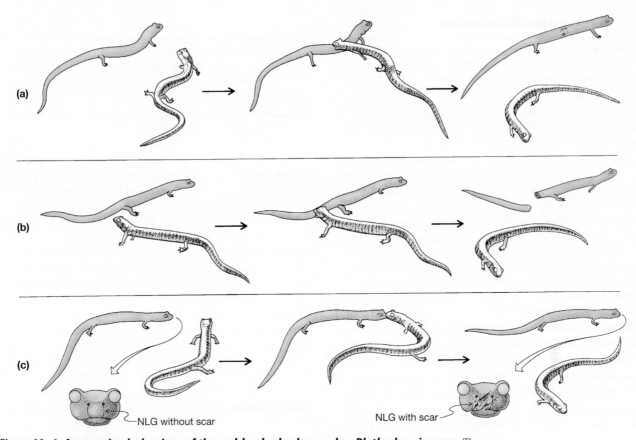

NLG without scar

NLG with scar

Figure 10–4 Aggressive behaviors of the red-backed salamander, *Plethodon cinereus*. The intruder is colored in these drawings. (a) Resident bites the intruder on the body. (b) Bitten on the tail by the resident, the intruder autotomizes its tail to escape. (c) Resident bites intruder on the snout, injuring the nasolabial grooves (NLG).

an opponent and may damage the nasolabial grooves. Because the nasolabial grooves are used for olfaction, these injuries can reduce a salamander's success in finding prey. Twelve salamanders that had been bitten on the snout were able to capture an average of only 5.8 fruit flies in a 2-hour period compared with an average of 18.6 flies for 12 salamanders that had not been bitten. In a sample of 144 red-backed salamanders from the Shenandoah National Forest, 11.8 percent had been bitten on the nasolabial grooves, and these animals weighed less than the unbitten animals, presumably because their foraging success had been reduced.

The possibility of serious damage to an important sensory system during territorial defense provides an additional advantage for a red-backed salamander in being able to distinguish neighbors (which are always there and are not worth attacking) from intruders (which represent a threat and should be attacked). The phenomenon of being able to recognize territorial neighbors has been called "dear enemy recognition"

and may be generally advantageous because it minimizes the time and energy that territorial individuals expend on territorial defense and also minimizes the risk of injury during territorial encounters. Similar dear enemy recognition has been described among territorial birds that show more aggressive behavior on hearing the songs of strangers than they do when hearing the songs of neighbors.

Frogs and Toads—Anura

In contrast to the limited number of species of salamanders and their restricted geographic distribution, the anurans (Greek *an* = without and *uro* = tail) include 45 families with nearly 5400 species and occur on all the continents except Antarctica (**Table 10–3**). Specialization of the body for jumping is the most conspicuous skeletal feature of anurans (**Figure 10–5** on page 219). The hindlimbs and muscles form a lever system that can catapult an anuran into the air, and numerous morphological specializations are associated with this type of locomotion;

Table 10–3 Major families of anurans

Bufonidae: Mainly terrestrial frogs; most have aquatic larvae, but some species of *Nectophrynoides* are viviparous (558 species worldwide except for Australia, Madagascar, and the Oceanic islands; 2–25 cm)

Centrolenidae: Small, arboreal frogs commonly called glass frogs because the internal organs are visible through the skin of the ventral surface of the body (146 species from Central America and northern South America; most about 3 cm)

Dendrobatidae: Small terrestrial frogs, many brightly colored and extremely toxic; terrestrial eggs hatch into tadpoles that are transported to water by an adult (179 species in Central and northern South America; 2–6 cm)

Eleutherodactylidae: Small arboreal frogs; many eleutherodactylids have direct development and one species, *Eleutherodactylus jasperi* from Puerto Rico, retains the eggs within the oviducts until the young have passed through metamorphosis and emerge as froglets; this species has not been seen since 1981 and is believed to be extinct (201 species in Central and northern South America and the Greater and Lesser Antilles; 1–9 cm)

Hylidae: Mostly tree frogs, but a few species are aquatic or terrestrial (901 species in North, Central, and South America, the West Indies, Europe, Asia, and the Australo-Papuan region; 2–15 cm)

Hyperoliidae: Small, brightly colored arboreal frogs (209 species from sub-Saharan Africa, Madagascar and adjacent islands; 2–8 cm)

Leptodactylidae: Aquatic and terrestrial frogs with aquatic larvae (100 species in southern North America, Central, and northern South America and the West Indies; 5–25 cm)

Mantellidae: The major group of frogs on Madagascar; most are terrestrial, but the family includes arboreal, aquatic, and fossorial species; species in the genus *Mantella* resemble tropical American dendrobatids in being brightly colored terrestrial frogs with alkaloid toxins in the skin (191 species restricted to Madagascar and the Comoros Islands; 3–10 cm)

Megophryidae: Terrestrial frogs, some remarkably camouflaged to resemble dead leaves (156 species from Pakistan to western China, and to the Indonesian archipelago, west of Wallace's line and in the Philippines; 2–12 cm)

Microhylidae: Terrestrial or arboreal frogs; many have aquatic larvae, but some species have nonfeeding tadpoles and others lay eggs on land and bypass the tadpole stage (487 species in North, Central, and South America, sub-Saharan Africa, India, and Korea to northern Australia; 5–10 cm)

Pipidae: Specialized aquatic frogs; *Xenopus, Hymenochirus, Pseudohymenochirus*, and some species of *Pipa* have aquatic larvae; other species of *Pipa* have eggs that develop directly into juvenile frogs (33 species in South America and sub-Saharan Africa; 2–15 cm)

Ranidae: Aquatic or terrestrial frogs; most have aquatic tadpoles (347 species worldwide except for southern South America and most of Australia; the genus *Rana* contains 48 species from temperate Eurasia to Indochina and western North America; the genus *Lithobates* includes 49 species from eastern North America through Central America to southern Brazil; 10–30 cm)

Rhacophoridae: Asian tree frogs that have converged on the body form of hylids and have large heads, large eyes, and enlarged toe disks (321 species in sub-Saharan Africa, southern Asia, Japan, and the Philippines; 1.5–12 cm)

---

the hind legs are elongate, and the tibia and fibula are fused. A powerful pelvis strongly fastened to the vertebral column is clearly necessary, as is stiffening of the vertebral column. The ilium is elongate and reaches far anteriorly, and the posterior vertebrae are fused into a solid rod, the **urostyle**. The pelvis and urostyle render the posterior half of the trunk rigid. The vertebral column is short, with only five to nine presacral vertebrae, and these are strongly braced by zygapophyses that restrict lateral bending. The strong forelimbs and flexible pectoral girdle absorb the impact of landing. The eyes are large and are placed well forward on the head, giving binocular vision. Specializations of the locomotor system can be used to distinguish many different kinds of anurans (**Figure 10–6** on page 220).

The hindlimbs generate the power to propel the frog into the air, and this high level of power production results from structural and biochemical features of the limb muscles. The internal architecture of the semimembranosus muscle and its origin on the ischium and insertion below the knee allow it to operate at the length that produces maximum force during the entire period of contraction. The muscle shortens faster and generates more power than muscles from most other animals. Furthermore, the intracellular physiological processes of muscle contraction continue at the maximum level throughout contraction rather than declining, as is the case in muscles of most vertebrates.

The diversity of anurans exceeds the number of common names that can be used to distinguish various ecological specialties (**Figure 10–7** on page 221). Animals called frogs usually have long legs and move by jumping, and this body form is found in many lineages. Semiaquatic frogs are moderately streamlined and have webbed feet. Stout-bodied terrestrial anurans that make short hops instead of long leaps are often called toads. They usually have blunt heads, heavy bodies, relatively short legs, and little webbing between the toes. True toads (the family Bufonidae) have this body form, and very similar body forms are found in other families, including the spadefoot toads of western North America and the horned frogs of South America.

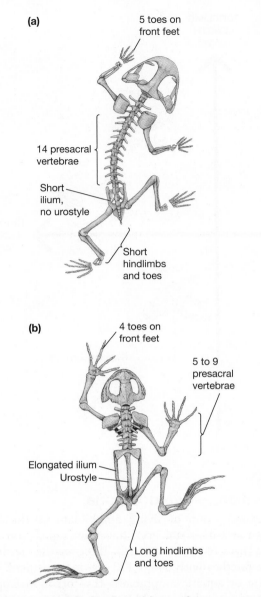

(a)

5 toes on front feet

14 presacral vertebrae

Short ilium, no urostyle

Short hindlimbs and toes

(b)

4 toes on front feet

5 to 9 presacral vertebrae

Elongated ilium
Urostyle

Long hindlimbs and toes

Figure 10–5 *Triadobatrachus* and a modern anuran. The Triassic fossil *Triadobatrachus* (a) is considered the sister group of the modern anuran (b). The derived characters of anurans visible in this comparison are shortening of the body, elongation of the ilia, and fusion of the posterior vertebrae to form a urostyle.

Arboreal frogs, such as the tree frogs in the families Hylidae and Rhacophoridae, usually have large heads and eyes and often slim waists and long legs. Arboreal frogs in many different families move by quadrupedal walking and climbing as much as by leaping. Many arboreal species of hylids and rhacophorids have enlarged toe discs and are called tree frogs. The surfaces of the toe pads consist of an epidermal layer with peglike projections separated by spaces or canals

(**Figure 10–8** on page 222). Mucous glands distributed over the discs secrete a viscous solution of long-chain, high-molecular-weight polymers in water. Arboreal species of frogs use a mechanism known as wet adhesion to stick to smooth surfaces. (This is the same mechanism by which a wet scrap of paper sticks to glass.) The watery mucus secreted by the glands on the toe discs forms a layer of fluid between the disc and the surface and establishes a meniscus at the interface between air and fluid at the edges of the toes. As long as no air bubble enters the fluid layer, a combination of surface tension (capillarity) and viscosity holds the toe pad and surface together.

Frogs can adhere to vertical surfaces and even to the undersides of leaves. Cuban tree frogs (*Osteopilus septentrionalis*) can cling to a sheet of smooth plastic as it is rotated past the vertical; the frogs do not begin to slip until the rotation reaches an average of 151 degrees—that is, 61 degrees past vertical. Adhesion and detachment of the pads alternate as a frog walks across a leaf. As a frog moves forward, its pads are peeled loose, starting at the rear, as the toes are lifted.

Tree frogs are not able to rest facing downward because in that orientation the frog's weight causes the toe pads to peel off the surface. Frogs invariably orient their bodies facing upward or across a slope, and they rotate their feet if necessary to keep the toes pointed upward. When a frog must descend a vertical surface, it moves backward. This orientation keeps the toes facing upward. During backward locomotion, toes are peeled loose from the tip backward by a pair of tendons that insert on the dorsal surface of the terminal bone of the toe.

Toe discs have evolved independently in several lineages of frogs and show substantial convergence in structure. Expanded toe discs are not limited exclusively to arboreal frogs; many terrestrial species that move across fallen leaves on the forest floor also have toe discs.

Several aspects of the natural history of anurans appear to be related to their different modes of locomotion. In particular, short-legged species that move by hopping are frequently wide-ranging predators that cover large areas as they search for food. This behavior makes them conspicuous to their own predators, and their short legs prevent them from fleeing rapidly enough to escape. Many of these anurans have potent defensive chemicals that are released from glands in the skin when they are attacked. Species of frogs that move by jumping, in contrast to those that hop, are usually sedentary predators that wait in ambush for prey to pass their hiding places. These species are usually cryptically colored, and they often lack chemical defenses. If they are discovered,

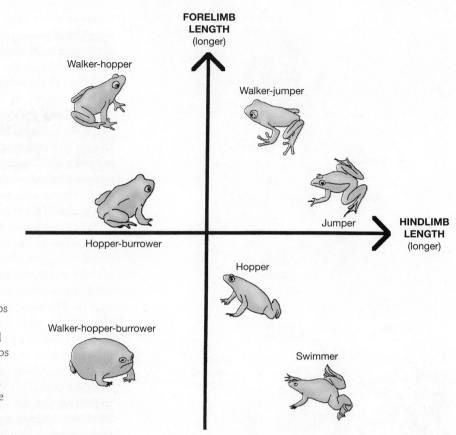

Figure 10–6 The relation of body form and locomotor mode among anurans. Short forelimbs and hindlimbs are associated with burrowing, whereas long forelimbs and hindlimbs are found in species that climb and leap. Hindlimbs that are distinctly longer than the fore-limbs usually indicate that a species is a hopper if the amount of webbing on the hind feet is limited or a swimmer if the hind feet are fully webbed.

they rely on a series of rapid leaps to escape. Anurans that forage widely encounter different kinds of prey from those that wait in one spot, and differences in dietary habits may be associated with differences in locomotor mode.

Aquatic anurans use suction to engulf food in water, but most species of semiaquatic and terrestrial anurans have sticky tongues that can be flipped out to trap prey and carry it back to the mouth (**Figure 10–9** on page 223). Most terrestrial anurans use a cata-pult-like mechanism to project the tongue. As the mouth is opened, contraction of the genioglossus muscles causes the front of the tongue to stiffen. Simultaneously, contraction of a short muscle at the front of the jaws (the submentalis) provides a ful-crum, and the stiffened tongue rotates forward over the submentalis and flips out of the mouth. Inertia causes the rear portion of the tongue to elongate as it emerges, and, because the tongue has rotated, its dorsal surface slams down on the prey. The tongue is drawn back into the mouth by the hyoglossus muscle, which originates on the hyoid apparatus and inserts within the tongue.

Caecilians—Gymnophiona

The third group of living amphibians is the least known and does not even have an English common name (**Figure 10–10** on page 224). These are the caecilians, three families with about 173 species of legless, bur-rowing or aquatic amphibians that occur in tropical habitats around the world (**Table 10–4**). The eyes of cae-cilians are greatly reduced and covered by skin or even by bone. Some species lack eyes entirely, but the reti-nas of other species have the layered organization that

Table 10–4 Families of caecilians

Caeciliidae: Terrestrial and aquatic caecilians with both oviparous and viviparous species; no aquatic larval stage (125 species in Central and South America, Africa, India, and the Seychelles Islands; 10 cm to 1.5 m)

Ichthyophiidae: Terrestrial caecilians with aquatic larvae (50 species in the Philippines, India, Thailand, southern China, and the Malayan Archipelago; up to 50 cm)

Rhinatrematidae: Terrestrial caecilians believed to have aquatic larvae (11 species in northern South America; up to 30 cm)

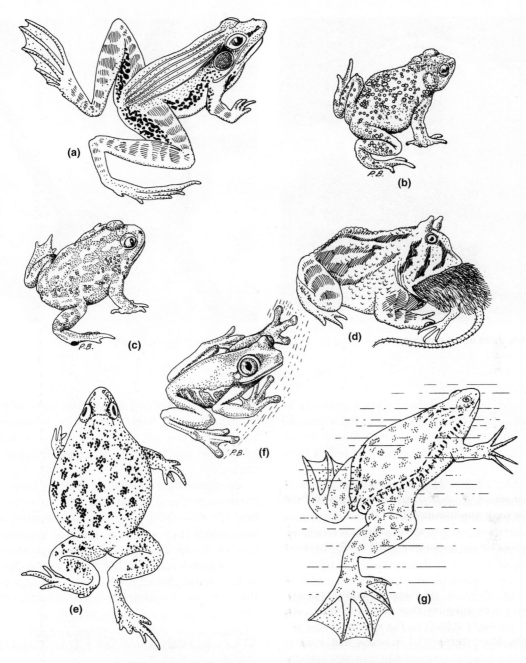

Figure 10–7 Anuran body forms. Body shape and limb length reflect specializations for different habitats and different methods of locomotion. (a) Semiaquatic frogs that both jump and swim, such as the African ridged frog (*Ptychadena oxyrhynchus*), have streamlined heads and bodies and long limbs with webbed hind feet, whereas terrestrial walkers and hoppers like (b) the spotted toad (*Anaxyrus punctatus*), (c) the western spadefoot toad (*Spea hammondii*), and (d) the Argentine horned frog (*Ceratophrys ornata*) have blunt heads, stout bodies, short limbs, and hind feet with little or no webbing. Frogs that burrow into the ground head first, like (e) the African shovel-nosed frog (*Hemisus marmoratus*), have short limbs and pointed snouts, whereas arboreal frogs, like (f) the red-eyed leaf frog (*Agalychnis callidryas*), have long limbs and broad snouts. Specialized aquatic frogs, like (g) the African clawed frog (*Xenopus laevis*), have a smooth surface contour and a well-developed system of lateral lines.

Figure 10–8 Toe discs of a hylid frog. *Left:* A single toe pad. *Right:* Detail of the polygonal plates.

is typical of vertebrates, and these species appear to be able to detect light. Conspicuous dermal folds (annuli) encircle the bodies of caecilians. The primary annuli overlie vertebrae and myotomal septa and reflect body segmentation.

Many species of caecilians have dermal scales in pockets in the annuli; scales are not known in the other groups of living amphibians. A second unique feature of caecilians is a pair of protrusible tentacles, one on each side of the snout between the eye and the nostril. Some structures that are associated with the eyes of other vertebrates have become associated with the tentacles of caecilians. One of the eye muscles, the retractor bulbi, has become the retractor muscle for the tentacle; the levator bulbi moves the tentacle sheath; and the Harderian gland (which moistens the eye in other tetrapods) lubricates the channel of the tentacle of caecilians. The tentacle is probably a sensory organ that allows chemical substances to be transported from the animal's surroundings to the vomeronasal organ on the roof of the mouth. The eye of caecilians in the African family Scolecomorphidae is attached to the side of the tentacle near its base. When the tentacle is protruded, the eye is carried along with it, moving out of the tentacular aperture beyond the roofing bones of the skull.

The earliest caecilian known is *Eocaecilia,* an Early Jurassic fossil from the Kayenta formation of western North America. It has a combination of ancestral and derived characters and is the sister taxon of extant caecilians. *Eocaecilia* has a fossa for a chemosensory tentacle, which is a unique derived character of caecilians, but it also has four legs, whereas all living caecilians are legless.

Caecilians feed on small or elongate prey—termites, earthworms, and larval and adult insects—and the tentacle may allow them to detect the presence of prey when they are underground. Females of some species of caecilians brood their eggs, whereas other species give birth to fully formed young. The embryos of terrestrial species have long, filamentous gills, and the embryos of aquatic species have saclike gills.

10.2 Diversity of Life Histories of Amphibians

Of all the characters of amphibians, none is more remarkable than their range of reproductive modes. Most species of amphibians lay eggs. The eggs may be deposited in water or on land, and they may hatch into aquatic larvae or into miniatures of the terrestrial adults. The adults of some species of frogs carry eggs attached to the surface of their bodies. Others carry their eggs in pockets in the skin of the back or flanks, in the vocal sacs, or even in the stomach. In still other species, the females retain the eggs in the oviducts and give birth to metamorphosed young. Many amphibians have no parental

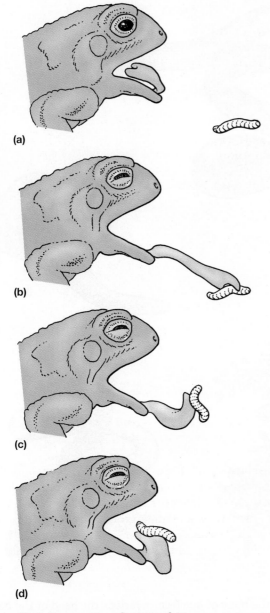

(a)

(b)

(c)

(d)

Figure 10–9 Prey capture by a toad. The tongue is attached at the front of the lower jaw and pivots around a stiffened muscle as it is flipped out. The tip of the tongue (the portion that is at the rear of the mouth when the tongue is retracted) has glands that excrete sticky mucus that adheres to the prey as the tongue is retracted into the mouth.

care of their eggs or young, but in many other species, a parent remains with the eggs and sometimes with the hatchlings or transports tadpoles from the nest to water. In a few species, an adult even feeds the tadpoles.

Amphibians have two characters that make their population ecology hard to study. First, fluctuation in size appears to be a normal feature of amphibian populations. Many species of amphibians lay hundreds of eggs, and the vast majority of these eggs never reach maturity. In a good year, however, the survival rate may be unusually high and a large number of individuals may be added to the population. Conversely, in a year of drought, the entire reproductive output of a population may die and no individuals will be added to the population in that year. Thus, year-to-year variation in recruitment creates natural fluctuations in populations that can obscure long-term trends in population size.

In addition, many species of amphibians live in metapopulations in which individual animals move among local populations that are often centered on breeding sites. In the shifting existence of a metapopulation, breeding populations may disappear from some sites while a healthy metapopulation continues to exist and breed at other sites. A limited study might conclude that a species was vanishing, whereas a broader analysis would show that the total population of the species had not changed. W. Ronald Heyer compiled *Measuring and Monitoring Biological Diversity: Standard Methods for Amphibians* to provide standard methods for studies of amphibian populations so that data from different studies can be compared and combined.

Caecilians

The reproductive adaptations of caecilians are as specialized as their body form and ecology. A male intromittent organ that is protruded from the cloaca accomplishes internal fertilization. Some species of caecilians lay eggs, and the female may coil around the eggs, remaining with them until they hatch. Viviparity is widespread, however, and about 75 percent of the species are viviparous and matrotrophic. At birth young caecilians are 30 to 60 percent of their mother's body length.

A female *Typhlonectes* 500 millimeters long may give birth to nine babies, each 200 millimeters long. The initial growth of the fetuses is supported by yolk contained in the egg at the time of fertilization, but this yolk is exhausted long before embryonic development is complete. *Typhlonectes* fetuses have absorbed all of the yolk in the eggs by the time they are 30 millimeters long. Thus, the energy they need to grow to 200 millimeters (a 6.6-fold increase in length) must be supplied by the mother. The energetic demands of producing nine babies, each one increasing its length 6.6 times and reaching 40 percent of the mother's length at birth, must be considerable.

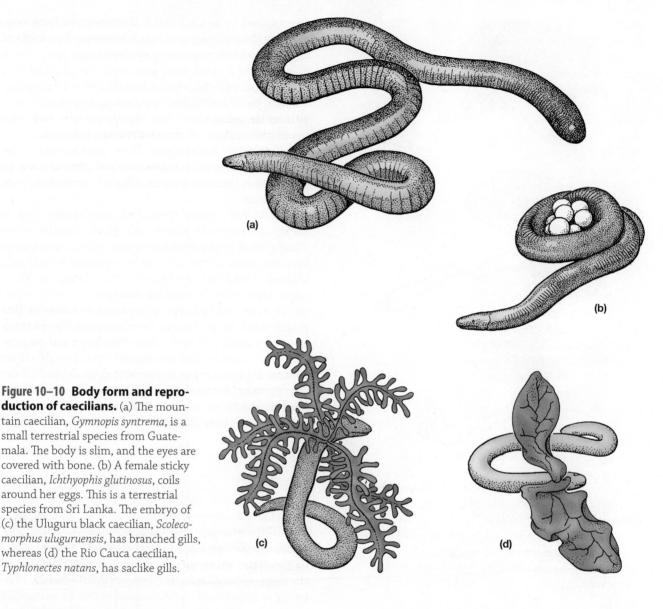

Figure 10–10 Body form and reproduction of caecilians. (a) The mountain caecilian, *Gymnopis syntrema*, is a small terrestrial species from Guatemala. The body is slim, and the eyes are covered with bone. (b) A female sticky caecilian, *Ichthyophis glutinosus*, coils around her eggs. This is a terrestrial species from Sri Lanka. The embryo of (c) the Uluguru black caecilian, *Scolecomorphus uluguruensis*, has branched gills, whereas (d) the Rio Cauca caecilian, *Typhlonectes natans*, has saclike gills.

The fetuses obtain this energy by scraping material from the walls of the oviducts with specialized embryonic teeth. The epithelium of the oviduct proliferates and forms thick beds surrounded by ramifications of connective tissue and capillaries. As the fetuses exhaust their yolk supply, these beds begin to secrete a thick, white, creamy substance that has been called uterine milk. When their yolk supply has been exhausted, the fetuses emerge from their egg membranes, uncurl, and align themselves lengthwise in the oviducts. The fetuses apparently bite the walls of the oviduct, stimulating secretion and stripping some epithelial cells and muscle fibers that they swallow with the uterine milk. Small fetuses are regularly spaced along the oviducts. Large fetuses have their heads spaced at intervals,

and the body of one fetus overlaps the head of the next. This spacing probably gives all the fetuses access to the secretory areas on the walls of the oviducts.

Gas exchange appears to be achieved by close contact between the fetal gills and the walls of the oviducts. All the terrestrial species of caecilians have fetuses with a pair of triple-branched filamentous gills. In preserved specimens, the fetuses frequently have one gill extending forward beyond the head and the other stretched along the body. In the aquatic genus *Typhlonectes*, the gills are saclike but are usually positioned in the same way. Both the gills and the walls of the oviducts are highly vascularized, and it seems likely that exchange of gases, and possibly of small molecules such as metabolic substrates and waste products, takes place across the adjacent gill and oviduct. The gills are absorbed before birth, and cutaneous gas exchange may be important for fetuses late in development.

A remarkable method of feeding the young has been described for a caecilian from Kenya, *Boulengerula taitana*. Females give birth to young that are in a relatively undeveloped (altricial) stage, and the young remain with their mother in subterranean nests. The cells in the outer layer of the skin (the stratum corneum) of brooding females are thickened and contain vesicles filled with lipids. The young have a specialized fetal dentition that allows them to peel off single layers of skin cells from the surface of their mother's body.

Salamanders

Most groups of salamanders use internal fertilization, but the Cryptobranchidae, Hynobiidae, and probably the Sirenidae retain external fertilization. Internal fertilization in salamanders is accomplished not by an intromittent organ but by the transfer of a packet of sperm (the **spermatophore**) from the male to the female (Figure 10–11). The form of the spermatophore differs in various species of salamanders, but all consist of a sperm cap on a gelatinous base. The base is a cast of the interior of the male's cloaca, and in some species it reproduces the ridges and furrows in accurate detail. Males of the Mediterranean salamandrid *Euproctus* deposit a spermatophore on the body of a female and then, holding her with their tail or jaws, use their feet to insert the spermatophore into her cloaca. Females of the hynobiid salamander *Ranodon sibiricus* deposit egg sacs on top of a spermatophore. In derived species of salamanders, the male deposits a spermatophore on the substrate, and the female picks up the cap with her cloaca. The sperm are released as the cap dissolves, and fertilization occurs in the oviducts.

Courtship Courtship patterns are important for species recognition, and they show great interspecific variation. Males of some species have elaborate secondary sexual characters that are used during courtship. Pheromones are released primarily by males and play a large role in the courtship of salamanders; they probably contribute to species recognition and may stimulate endocrine activity that increases the receptivity of females.

Pheromone delivery by most salamanders that breed on land involves physical contact between a male and female, during which the male applies secretions of specialized courtship glands (hedonic glands) to the nostrils or body of the female (Figure 10–12). Several modes of pheromone delivery have been described. Males of many plethodontids (e.g., *Plethodon jordani*) have a large gland beneath the chin (the mental gland), and secretions of the gland are applied to the nostrils of the female with a slapping motion.

The anterior teeth of males of many species of *Desmognathus* and *Eurycea* (both members of the Plethodontidae) hypertrophy during the breeding season. A male of these species spreads secretion from his mental gland onto the female's skin and then abrades the skin with his teeth, inoculating the female with the pheromone. Males of two small species of *Desmognathus* use specialized mandibular teeth to bite and stimulate the female. Male salamandrids (Salamandridae) rub the female's snout with hedonic glands on their cheeks (the red-spotted newt, *Notophthalmus viridescens*), chin (the rough-skinned newt, *Taricha granulosa*), or cloaca (the Spanish newt, *Pleurodeles waltl*).

The males of many species of newts perform elaborate courtship displays in which the male vibrates its tail to create a stream of water that wafts pheromones secreted by a gland in his cloaca toward the female.

Two trends are apparent within evolutionary lineages of newts: an increase in diversity of the sexual displays

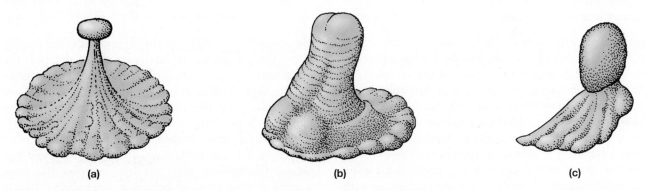

(a)   (b)   (c)

Figure 10–11 Spermatophores. Male salamanders deposit a spermatophore that contains a capsule of sperm supported on a gelatinous base: (a) red-spotted newt, *Notophthalmus viridescens*; (b) dusky salamander, *Desmognathus fuscus*; (c) two-lined salamander, *Eurycea bislineata*.

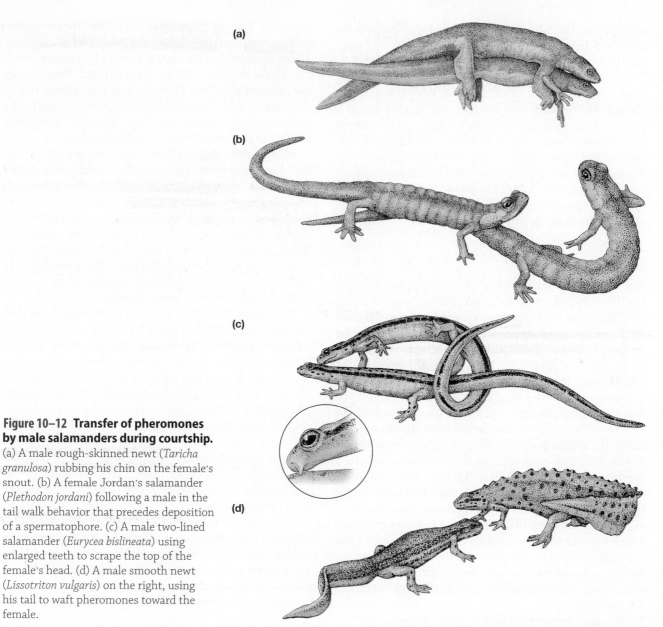

Figure 10–12 Transfer of pheromones by male salamanders during courtship. (a) A male rough-skinned newt (*Taricha granulosa*) rubbing his chin on the female's snout. (b) A female Jordan's salamander (*Plethodon jordani*) following a male in the tail walk behavior that precedes deposition of a spermatophore. (c) A male two-lined salamander (*Eurycea bislineata*) using enlarged teeth to scrape the top of the female's head. (d) A male smooth newt (*Lissotriton vulgaris*) on the right, using his tail to waft pheromones toward the female.

performed by the male and an increase in the importance of positive feedback from the female. The behaviors seen in *Ichthyosaura alpestris* may represent the ancestral condition. This species shows little sexual dimorphism, and the male's display consists of only fanning (a display in which the tail is folded back against the flank nearest the female and the tail tip is vibrated rapidly). The male's behavior is nearly independent of response by the female—a male *I. alpestris* may perform his entire courtship sequence and deposit a spermatophore without active response by the female he is courting.

A group of large newts, including *Triturus cristatus* and *Ommatotriton vittatus*, are highly sexually dimor-

phic (**Figure 10–13**). Males of these species defend display sites. Their displays are relatively static and lack the rapid fanning movements of the tail that characterize the displays of other groups of newts. A male of these species does not deposit a spermatophore unless the female he is courting touches his tail with her snout.

A group of small-bodied newts includes *Lissotriton vulgaris* and *Lissotriton boscai*. These species show less sexual dimorphism than the large species and have a more diverse array of behaviors, including a nearly static lateral display, whipping the tail violently against the female's body, fanning with the tail tip, and other displays (with names like wiggle and flamenco) that

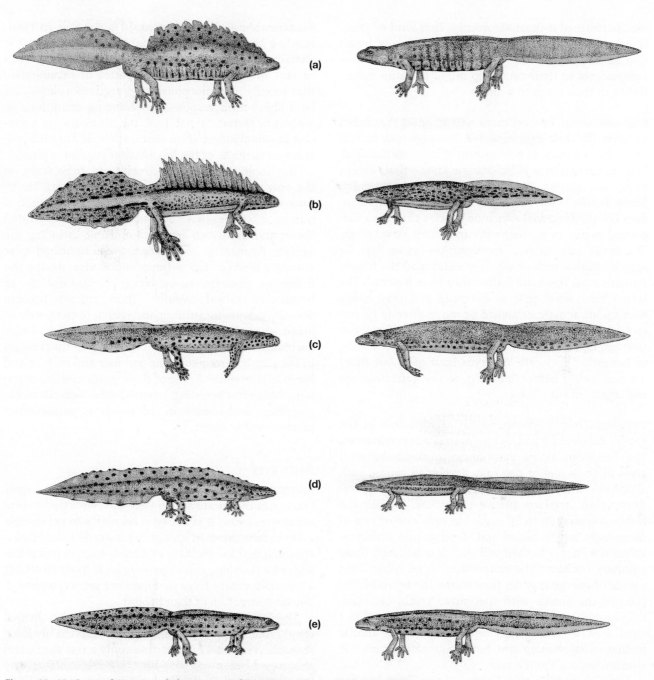

Figure 10–13 Secondary sexual characters of European newts (Salamandridae). The male is on the left; the female is on the right. (a) The great crested newt, *Triturus cristatus*. (b) The banded newt, *Ommatotriton vittatus*. (c) The alpine newt, *Mesotriton alpestris*. (d) The smooth newt, *Lissotriton vulgaris*. (e) Bosca's newt, *Lissotriton boscai*.

occur in some species in the group. Response by the female is an essential component of courtship for these species—a male will not move on from the static display that begins courtship to the next phase unless the female approaches him repeatedly, and he will not deposit a spermatophore until the female touches his tail.

These trends toward greater sexual dimorphism, more diverse displays, and more active involvement of the female in courtship may reflect sexual selection by females within the derived groups. Tim Halliday has suggested that in the ancestral condition there was a single male display, and females mated with the males that performed it most vigorously. That kind of selection by females would produce a population of males that all display vigorously, and males that added new components to their courtship might be more attractive than their rivals to females.

Eggs and Larvae In most cases, salamanders that breed in water lay their eggs in water. The eggs may be laid singly or in a mass of transparent gelatinous material. The eggs hatch into gilled aquatic larvae that, except in aquatic species, transform into terrestrial adults. Some families—for example, the lungless salamanders (Plethodontidae)—include species that have dispensed in part or entirely with an aquatic larval stage. The dusky salamander, *Desmognathus fuscus*, lays its eggs beneath a rock or log near water, and the female remains with them until after they have hatched. The larvae have small gills at hatching and may either take up an aquatic existence or move directly to terrestrial life. The red-backed salamander, *Plethodon cinereus*, lays its eggs in a hollow space in a rotten log or beneath a rock. The embryos have gills, but these are reabsorbed before hatching, so the hatchlings are miniatures of the adults.

Viviparity Only a few species of salamanders in the genera *Salamandra* and *Lyciasalamandra* are viviparous. The European alpine salamander (*Salamandra atra*) gives birth to one or two fully developed young, each about one-third the adult body length, after a gestation period that lasts from 2 to 4 years. Initially the clutch contains 20 to 30 eggs, but only one or two of these eggs are fertilized and develop into embryos. When the energy in their yolk sacs is exhausted, these embryos consume the unfertilized eggs; when that source of energy is gone, they scrape the reproductive tract of the female with specialized teeth. *Lyciasalamandra antalyana*, a species of salamandrid found in a small area of Turkey, has a reproductive mode similar to that of *Salamandra atra*, but the gestation period is shorter, lasting about a year.

Females in some populations of the European fire salamander (*Salamandra salamandra*) produce 20 or more small larvae, each about one-twentieth the length of an adult. The embryos probably get all the energy needed for growth and development from egg yolk. The larvae are released in water and have an aquatic stage that lasts about 3 months. In other populations of this species, the eggs are retained in the oviducts and, when all of the unfertilized eggs have been consumed, some of the embryos cannibalize other embryos. The surviving embryos pass through metamorphosis in the oviducts of the female.

Paedomorphosis Paedomorphosis is the rule in families like the Cryptobranchidae and Proteidae, and it characterizes most cave dwellers. It also appears as a variant in the life history of species of salamanders that usually metamorphose, and paedomorphosis can be a short-term response to changing conditions in aquatic or terrestrial habitats. The life history of a species of salamanders from eastern North America provides an example of the flexibility of paedomorphosis.

The mole salamander, *Ambystoma talpoideum*, is the only species of mole salamander in eastern North America that displays paedomorphosis, although a number of species of *Ambystoma* in the western United States and in Mexico are paedomorphic including the axolotl, *Ambystoma mexicanum*. Small-mouthed salamanders breed in the autumn and winter; during the following summer, some larvae metamorphose to become terrestrial juveniles. These animals become sexually mature by autumn and return to the ponds to breed when they are about a year old. Ponds in South Carolina also contain paedomorphic larvae that remain in the ponds throughout the summer and mature and breed in the winter. Some of these paedomorphs metamorphose after breeding, whereas others do not metamorphose and remain in the ponds as permanently paedomorphic adults.

Anurans

Anurans are the most familiar amphibians, largely because of the vocalizations associated with their reproductive behavior. It is not even necessary to get outside a city to hear them. In springtime, a weed-choked drainage ditch beside a highway or a trash-filled marsh at the edge of a shopping center parking lot is likely to attract a few toads or tree frogs that have not yet succumbed to human usurpation of their habitat.

The mating systems of anurans can be divided roughly into **explosive breeding**, in which the breeding season is very short (sometimes only a few days), and **prolonged breeding**, with breeding seasons that may last for several months. Explosive breeders include many species of toads and other anurans that breed in temporary aquatic habitats, such as vernal ponds or pools that form after rainstorms in the desert. Because these bodies of water do not last very long, breeding congregations of anurans usually form as soon as the site is available. Males and females arrive at the breeding sites nearly simultaneously and often in very large numbers. The numbers of males and females present are approximately equal because the entire population breeds in a short time. Time is the main constraint on how many females a male is able to court, and mating success is usually approximately the same for all the males in a chorus.

In species with prolonged breeding seasons, the males usually arrive at the breeding sites first. Males of some species, such as green frogs (*Lithobates clamitans*), establish territories in which they spend several months, defending the spot against the approach of other males. The males of other species move between daytime retreats and nocturnal calling sites on a daily basis. Females come to the breeding site to breed and leave when they have finished. Only a few females arrive every day, and the number of males at the breeding site is greater than the number of females every night. Mating success may be skewed, with many of the males not mating at all and a few males mating several times. Males of anuran species with prolonged breeding seasons compete to attract females, usually by vocalizing. The characteristics of a male frog's vocalization (pitch, length, and repetition rate) may provide information that a female frog can use to evaluate his quality as a potential mate. This is an active area of study in anuran behavior.

Vocalizations Anuran calls are diverse; they vary from species to species, and most species have two or three different sorts of calls used in different situations. The most familiar calls are the ones usually referred to as mating calls, although a less specific term such as **advertisement calls** is preferable. These calls range from the high-pitched *peep* of a spring peeper to the nasal *waaah* of a spadefoot toad or the bass *jug-o-rum* of a bullfrog. The characteristics of a call identify the species and sex of the calling individual. Many species of anurans are territorial, and males of at least one species, the North American bullfrog (*Lithobates catesbeianus*) recognize one another individually by voice.

An advertisement call is a conservative evolutionary character, and among related species there is often considerable similarity in advertisement calls. Superimposed on the basic similarity are the effects of morphological factors, such as body size, as well as ecological factors that stem from characteristics of the habitat. Most toads have an advertisement call that consists of a train of repeated pulses, and the pitch of the call varies with body size, extending downward from 5200 hertz for the oak toad (*Anaxyrus quercicus*), which is only 2 or 3 centimeters long; to 1800 hertz for the American toad (*Anaxyrus americanus*), about 6 centimeters long; and down to 600 hertz for the giant toad (*Rhinella marina*), with a length of nearly 20 centimeters. A Bornean frog (*Metaphrynella sundana*) calls from cavities in trees, and males of this species adjust the frequency of their calls to match the resonant frequency of the hole from which they are calling, thereby increasing the amplitude

(loudness) of the call so it carries farther and can be heard by more potential mates.

Female frogs are responsive to the advertisement calls of males of their species for a brief period when their eggs are ready to be laid. The hormones associated with ovulation are thought to sensitize specific cells in the auditory pathway that respond to the species-specific characteristics of the male's call. Mixed choruses of anurans are common in the mating season; a dozen species may breed simultaneously in one pond. A female's response to her own species' mating call is a mechanism for species recognition in that situation.

Costs and Benefits of Vocalization The vocalizations of male frogs are costly in two senses. The actual amount of energy that goes into call production can be very large, and the variations in calling pattern that accompany social interactions among male frogs in a breeding chorus can increase the energy cost per call. Another cost of vocalization for a male frog is the risk of predation. A critical function of vocalization is to permit a female frog to locate a male, but female frogs are not the only animals that can use vocalizations as a cue to find male frogs; predators of frogs also find that calling males are easy to locate.

The túngara frog (*Engystomops pustulosus*) is a small terrestrial anuran that lives in Central America (**Figure 10–14**). Túngara frogs breed in small pools, and breeding assemblies range from a single male to choruses of several hundred males. The advertisement call of a male túngara frog is a strange noise, a whine that sounds as if it would be more at home in an arcade of video games than in the tropical night. The whine starts at a fundamental frequency of 1000 hertz and sweeps downward to 500 hertz in about 400 milliseconds (**Figure 10–15** on page 231).

The whine may be produced by itself, or it may be followed by one or several *chucks*. When a male túngara frog is calling alone in a pond, it usually gives only the whine portion of the call; however, as additional males join a chorus, more and more of the frogs produce calls that include chucks. Male túngara frogs calling in a breeding pond added chucks to their calls when they heard playbacks of calls of other males. That observation suggested that it was the presence of other calling males that stimulated frogs to make their calls more complex by adding chucks to the end of the whine.

What advantage would a male frog in a chorus gain from using a whine-chuck call instead of a whine? Perhaps the complex call is more attractive to female frogs than the simple call. Michael Ryan and Stanley Rand tested that hypothesis by placing female túngara frogs

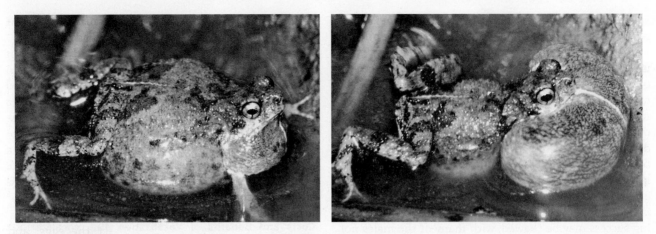

Figure 10–14 Male túngara frog, _Engystomops pustulosus._ As the frog calls, air is forced from the lungs (_left_) into the vocal sacs (_right_).

in a test arena with a speaker at each side. One speaker broadcast a whine call, and the second speaker broadcast a whine-chuck. When female frogs were released individually in the center of the arena, 14 of the 15 frogs tested moved toward the speaker broadcasting the whine-chuck call.

If female frogs are attracted to whine-chuck calls in preference to whine calls, why do male frogs give whine-chuck calls only when other males are present? Why not always give the most attractive call possible? One possibility is that whine-chuck calls require more energy than whines, and males save energy by using whine-chucks only when competition with other males makes the energy expenditure necessary. However, measurements of the energy expenditure of calling male túngara frogs showed that the energy cost was not related to the number of chucks.

Another possibility is that male frogs that give whine-chuck calls are more vulnerable to predators than frogs that give only whine calls. Túngara frogs in breeding choruses are preyed upon by frog-eating bats, _Trachops cirrhosus_, and the bats locate the frogs by homing on their vocalizations. In a series of playback experiments, Ryan and Merlin Tuttle placed pairs of speakers in the forest and broadcast vocalizations of túngara frogs. One speaker played a recording of a whine and the other a recording of a whine-chuck. The bats responded as if the speakers were frogs: they flew toward the speakers and even landed on them. In five experiments at different sites, the bats approached speakers broadcasting whine-chuck calls twice as frequently as those playing simple whines (168 approaches versus 81). Thus, female frogs are not alone in finding whine-chuck calls more attractive than simple whines—frog-eating bats also locate male frogs

that are producing chucks more readily than males that emit only whines.

Predation can be a serious risk for male túngara frogs. Ryan and his colleagues measured the rates of predation in choruses of different sizes. The major predators were frog-eating bats, a species of opossum (_Philander opossum_), and a larger species of frog (_Leptodactylus pentadactylus_); the bats were the most important predators of the túngara frogs. Large choruses of frogs did not attract more bats than small choruses, and consequently the risk of predation for an individual frog was less in a large chorus than in a small one. Predation was an astonishing 19 percent of the frogs per night in the smallest chorus and a substantial 1.5 percent per night even in the largest chorus.

When a male frog shifts from a simple whine to a whine-chuck call, it increases its chances of attracting a female, but it simultaneously increases its risk of attracting a predator. In small choruses, the competition from other males for females is relatively small, and the risk of predation is relatively large. Under those conditions it is apparently advantageous for a male túngara frog to give simple whines. However, as chorus size increases, competition with other males also increases while the risk of predation falls. In that situation, the advantage of giving a complex call apparently outweighs the risks.

When a frog detects a predator, it falls silent rather than reverting to a simple whine. Adjacent frogs also fall silent and the cessation of calling sweeps through the chorus. Playback experiments showed that a single frog's silence is sufficient to begin the process, and three males falling silent simultaneously are more effective than a single male. Furthermore, cessation of complex calls is more effective than cessation of simple

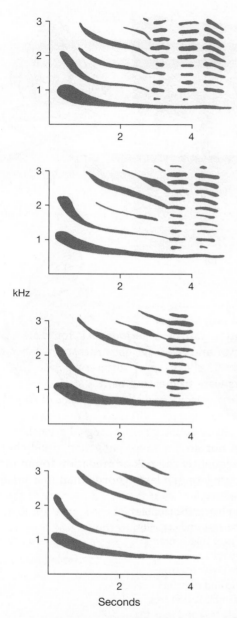

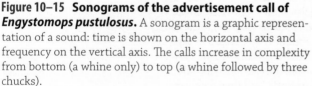

Figure 10–15 Sonograms of the advertisement call of
Engystomops pustulosus. A sonogram is a graphic representation of a sound: time is shown on the horizontal axis and frequency on the vertical axis. The calls increase in complexity from bottom (a whine only) to top (a whine followed by three chucks).

calls in silencing a chorus. It does not matter whether the final call of a frog is completed or ends in mid-call.

Modes of Reproduction

Fertilization is external in most anurans; the male uses his fore legs to clasp the female in the pectoral region (axillary amplexus) or pelvic region (inguinal amplexus). Amplexus may be maintained for several hours or even days before the female lays eggs. Males of the tailed frog of the Pacific Northwest (*Ascaphus truei*) have an extension of the cloaca (the "tail" that gives them their name) that is used to introduce sperm into the cloaca of the female. Internal fertilization has been demonstrated for the Puerto Rican coquí (*Eleutherodactylus coqui*) and may be widespread among frogs that lay eggs on land. Fertilization must also be internal for the few species of viviparous anurans.

Anurans show even greater diversity in their modes of reproduction than do salamanders (**Figure 10–16**). Similar reproductive habits have clearly evolved independently in different lineages of anurans, and these often include behaviors that protect the eggs, and sometimes the tadpoles as well, from predation. For example, the land may be a safer place than water for a frog egg to develop, and a study of Amazon rain forest frogs revealed a positive relationship between the intensity of predation on frogs' eggs in a pond and the proportion of frog species in the area that laid eggs in terrestrial situations.

Many arboreal frogs lay their eggs on the leaves of trees overhanging water. The eggs undergo embryonic development out of the reach of aquatic egg predators, and when the tadpoles hatch they drop into the water and take up an aquatic existence. Other frogs, such as *Engystomops pustulosus*, achieve the same result by constructing foam nests that float on the water surface. The female emits mucus during amplexus, which the pair of frogs beat into foam with their hind legs. The eggs are laid in the foam mass, and, when the tadpoles hatch, they drop through the foam into the water.

Although these methods reduce egg mortality, the tadpoles are subjected to predation and competition. Some anurans avoid both problems by finding or constructing breeding sites free from competitors and predators. Some frogs, for example, lay their eggs in the water that accumulates in bromeliads—epiphytic tropical plants that grow in trees and are morphologically specialized to collect rainwater. A large bromeliad may hold several liters of water, and the frogs pass through egg and larval stages in that protected microhabitat.

Many tropical frogs lay eggs on land near water. The eggs or tadpoles may be released from the nest sites when pond levels rise after a rainstorm. Other frogs construct pools in the mud banks beside streams. These volcano-shaped structures are filled with water by rain or seepage and provide a favorable environment for the eggs and tadpoles. Some frogs have eliminated the tadpole stage entirely. These frogs lay large eggs on land that develop directly into little frogs. This reproductive mode, called direct development, is characteristic of about 20 percent of all anuran species.

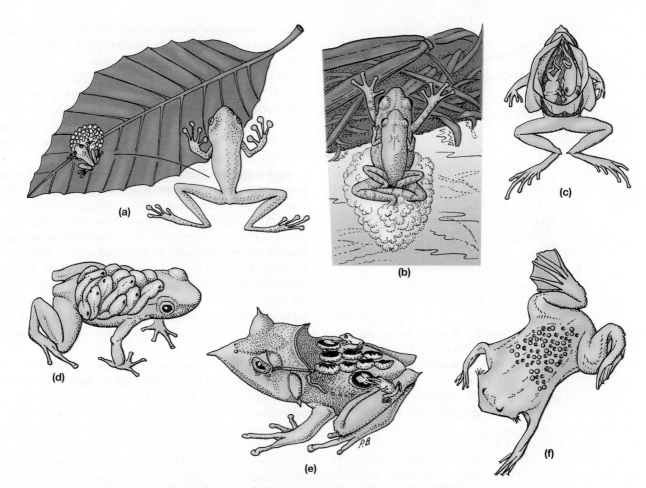

Figure 10–16 Reproductive modes and parental care among anurans. (a) The female
northern glass frog (*Hyalinobatrachium fleischmanni*) attaches the eggs to the underside of a leaf
overhanging a stream, and the male frog remains with the eggs. When the eggs hatch, the tadpoles
drop into the water. (b) A male túngara frog (*Engystomops pustulosus*) takes the eggs and egg jelly
with his hind legs as the female releases them and beats the jelly into a foam nest that contains
the eggs and floats on the surface of the water. When the tadpoles hatch, they release an enzyme
that dissolves the foam and allows them to drop into the water. (c) The female Darwin's frog
(*Rhinoderma darwinii*) deposits eggs on the ground; the male frog picks them up and transfers
them to his vocal sacs, where they develop through metamorphosis. (d) The female rocket frog
(*Colostethus inguinalis*) deposits eggs in a nest site within the territory of a male frog and remains
with the eggs until they hatch into tadpoles. The newly hatched tadpoles squirm onto the female's
back and are carried to water. (e) As a female Spix's horned frog (*Hemiphractus scutatus*) releases
her eggs, the male pushes them onto her back and into the thickened skin. They develop in capsules that form around them, passing through metamorphosis and hatching as tiny frogs. (f) A
female Surinam toad (*Pipa pipa*) carries eggs on her back. The eggs of this species hatch as tadpoles, whereas other species of *Pipa* retain the eggs through metamorphosis, and they hatch as
tiny frogs.

Parental Care Adults of many species of frogs guard
their eggs. In some cases it is the male that protects the
eggs; in others it is the female. In most cases it is not
clearly known which sex is involved because external
sex identification is difficult with many anurans. Many
of the frogs that lay their eggs over water remain with
them. Some species sit beside the eggs; others rest on
top of them. Most of the terrestrial frogs that lay direct-
developing eggs remain with the eggs and will attack an
animal that approaches the nest. Removing the guarding frog frequently results in the eggs being eaten by
predators or desiccating and dying before hatching.

Male African bullfrogs (*Pyxicephalus adspersus*) guard
their eggs and then continue to guard the tadpoles after

they hatch. The male frog moves with the school of tadpoles and will even dig a channel to allow the tadpoles to swim from one pool in a marsh to an adjacent one. Tadpoles of several species in the tropical American frog genus *Leptodactylus* follow their mother around the pond. These species of *Leptodactylus* are large and aggressive, and the adult frogs are able to deter many potential predators.

Some of the dart-poison frogs of the American tropics deposit their eggs on the ground, and one of the parents remains with the eggs until they hatch into tadpoles. The tadpoles adhere to the adult and are transported to water. Females of the Panamanian frog *Colostethus inguinalis* carry their tadpoles for more than a week, and the tadpoles increase in size during this period. The largest tadpoles being carried by females had small amounts of plant material in their stomachs, suggesting that they had begun to feed while they were still being transported by their mother. Females of another Central American dart-poison frog, *Oophaga pumilio*, release their tadpoles in small pools of water in the leaf axils of plants and then return at intervals to the pools to deposit unfertilized eggs that the tadpoles eat.

Other anurans, instead of remaining with their eggs, carry the eggs with them. The male of the European midwife toad (*Alytes obstetricians*) gathers the egg strings about his hind legs as the female lays them. He carries them with him until they are ready to hatch, at which time he releases the tadpoles into water. The male of the terrestrial Darwin's frog (*Rhinoderma darwinii*) of Chile snaps up the eggs the female lays and carries them in his vocal pouches, which extend back to the pelvic region. The embryos pass through metamorphosis in the vocal sacs and emerge as fully developed froglets.

Male frogs are not alone in caring for eggs. The females of a group of tree frogs from the American tropics carry the eggs on their back in an open oval depression, a closed pouch, or individual pockets. The eggs develop into miniature frogs before they leave their mother's back. A similar specialization is seen in the completely aquatic Surinam toad, *Pipa pipa*. In the breeding season, the skin of the female's back thickens and softens. During egg laying, the male and female in amplexus swim in vertical loops in the water. On the upward part of the loop, the female is above the male and releases a few eggs, which fall onto his ventral surface. He fertilizes them and, on the downward loop, presses them against the female's back. They sink into the soft skin, and a cover forms over each egg, enclosing it in a small capsule.

Tadpoles of the two species of the Australian frog genus *Rheobatrachus* are carried in the stomach of the female frog. The female swallows eggs or newly hatched larvae and retains them in her stomach through metamorphosis. This behavior was first described in *Rheobatrachus silus* and is accompanied by extensive morphological and physiological modifications of the stomach. These changes include distension of the proximal portion of the stomach, separation of individual muscle cells from the surrounding connective tissue, and inhibition of hydrochloric acid secretion, perhaps by prostaglandin released by the tadpoles. In January 1984, a second species of gastric-brooding frog, *R. vitellinus*, was discovered in Queensland. Strangely, this species lacks the extensive structural changes in the stomach that characterize the gastric brooding of *R. silus*. The striking differences between the two species suggest the surprising possibility that this bizarre reproductive mode might have evolved independently. Both species of *Rheobatrachus* disappeared within a few years of their discovery and are thought to be extinct, victims of a worldwide decline of amphibian populations.

Females of the Jamaican frog *Eleutherodacytlus cundalli* and males of two microhylid frogs from New Guinea (*Liophryne schlaginhaufeni* and *Sphenophryne cornuta*) carry froglets on their backs. *Liophryne* males carried the froglets for periods extending from three to nine nights, and a few froglets jumped off each night.

Viviparity Species in the African bufonid genus *Nectophrynoides* show a spectrum of reproductive modes. One species deposits eggs that are fertilized externally and hatch into aquatic tadpoles, two species produce young that are nourished by yolk, and other species have embryos that feed on secretions from the walls of the oviduct. The golden coquí of Puerto Rico (*Eleutherodactylus jasperi*) also gives birth to fully formed young, but in this case the energy and nutrients come from the yolk of the egg. (The golden coquí has not been seen since 1981 and is presumed to be extinct.)

The Ecology of Tadpoles

Although many species of frogs have evolved reproductive modes that bypass an aquatic larval stage, a life history that includes a tadpole has certain advantages. A tadpole is a completely different animal from an adult anuran, both morphologically and ecologically.

Tadpoles are as diverse in their morphological and ecological specializations as adult frogs, and they occupy nearly as great a range of habitats (**Figure 10–17**).

- Tadpoles that live in still water usually have ovoid bodies and tails with fins that are as large as the muscular part of the tail, whereas tadpoles that live in fast-flowing water have more streamlined bodies and smaller tail fins.

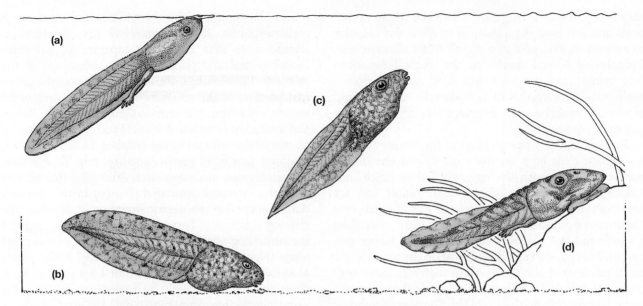

Figure 10–17 Body forms of tadpoles. (a) The tadpole of the Kwangshien spadefoot toad, *Xenophrys minor*, is a surface feeder. The mouthparts unfold into a platter over which water and particles of food on the surface are drawn into the mouth. (b) The tadpole of the red-legged frog, *Rana aurora*, scrapes food from rocks and other submerged objects. (c) The tadpole of the red-eyed leaf frog, *Agalychnis callidryas*, is a midwater suspension feeder that filters particles of food from the water column. This species shows the large fins and protruding eyes that are typical of midwater tadpoles. It maintains its position in the water column with rapid undulations of the end of its tail, which is thin and nearly transparent. (d) A stream-dwelling tadpole (an unidentified species of Australasian tree frog, *Litoria*) that adheres to rocks in swiftly moving water with a suckerlike mouth while scraping algae and bacteria from the rocks. The low fins and powerful tail are characteristic of tadpoles living in swift water.

- Semiterrestrial tadpoles wiggle through mud and leaves and climb on damp rock faces; they are often dorsoventrally flattened and have little or no tail fin, and many tadpoles that live in bromeliads have a similar body form.
- Direct-developing tadpoles have large yolk supplies and reduced mouthparts and tail fins.
- The mouthparts of tadpoles also show variation that is related to diet (**Figure 10–18**). Filter-feeding tadpoles that hover in midwater lack keratinized mouthparts, whereas species that graze from surfaces have small beaks that are often surrounded by rows of denticles.
- Predatory tadpoles have larger beaks that can bite pieces from other tadpoles. Funnel-mouthed, surface-feeding tadpoles have greatly expanded mouthparts that skim material from the surface of the water.

Tadpoles of most species of anurans are filter-feeding herbivores, whereas all adult anurans are carnivores that catch prey individually. Because of these differences, tadpoles can exploit resources that are not available to adult anurans. This advantage may be a factor that has led many species of frogs to retain the ancestral pattern of life history in which an aquatic larva matures into a terrestrial adult. Many aquatic habitats experience annual flushes of primary production, when nutrients washed into a pool by rain or melting snow stimulate the rapid growth of algae. The energy and nutrients in this algal bloom are transient resources that are available for a brief time to the organisms able to exploit them.

Tadpoles are excellent eating machines. All tadpoles extract suspended food particles from water, and feeding and ventilation of the gills are related activities. The stream of water that moves through the mouth and nares to ventilate the gills also carries particles of food. As the stream of water passes through the branchial basket, small food particles are trapped in mucus secreted by epithelial cells. The mucus, along with the particles, is moved from the gill filters to the ciliary grooves on the margins of the roof of the pharynx and then transported posteriorly to the esophagus.

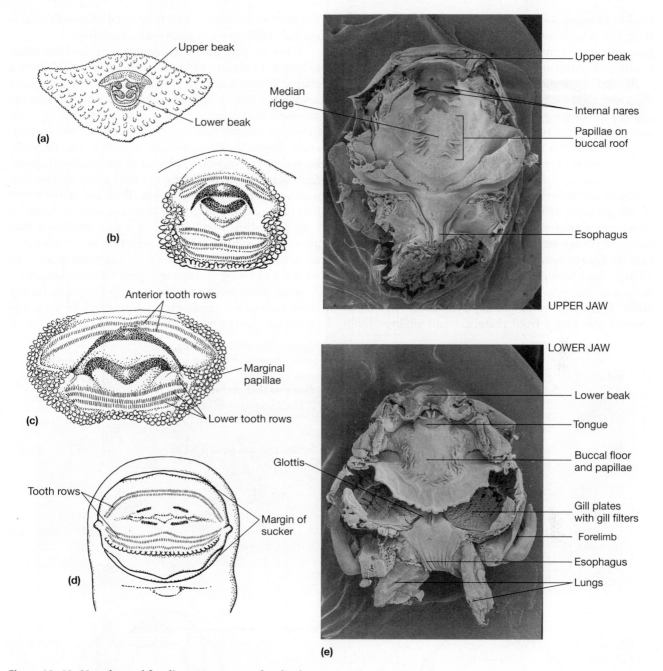

Figure 10–18 Mouths and feeding structures of tadpoles. (a) Surface feeder, *Xenophrys minor*. (b) Surface scraper, *Rana aurora*. (c) Midwater feeder, *Agalychnis callidryas*. (d) Stream-dweller, *Litoria*. (e) Scanning electron micrograph of the inside of the mouth and buccal region of a tadpole of the island spiny-chested frog, *Alsodes monticola*.

Although all tadpoles filter food particles from a stream of water that passes across the gills,  the method used to put the food particles into suspension differs among species. Some tadpoles filter floating plankton from the water. Tadpoles of this type are represented in several families of anurans, especially the Pipidae

and Hylidae, and usually hover in the water column. Midwater-feeding tadpoles are out in the open, where they are vulnerable to predators, and they show various characteristics that may reduce the risk of predation. Tadpoles of the African clawed frog (*Xenopus*), for example, are nearly transparent and may be hard for predators to see. Some midwater tadpoles form schools that, like schools of fishes, may confuse a predator by presenting so many potential prey items that it has difficulty concentrating its attack on one individual.

Many tadpoles are bottom feeders that scrape bacteria and algae off the surfaces of rocks or the leaves of plants. The rasping action of their keratinized mouthparts frees the material and allows it to be whirled into suspension in the water stream entering the mouth, and then filtered out by the branchial apparatus. Some bottom-feeding tadpoles, such as toads and spadefoot toads, form dense aggregations that create currents to lift particles of food into suspension in the water. These aggregations may be groups of siblings. Tadpoles of American toads (*Anaxyrus americanus*) and cascade frogs (*Rana cascadae*) are able to distinguish siblings from nonsiblings, and they associate preferentially with siblings. They probably recognize siblings by scent. Toad tadpoles can distinguish full siblings (both parents the same) from maternal half-siblings (only

the mother the same), and they can distinguish maternal half-siblings from paternal half-siblings.

Some tadpoles are carnivorous and feed on other tadpoles. Predatory tadpoles have large mouths with a sharp, keratinized beak. Predatory individuals appear among the tadpoles of some species of anurans that are normally herbivorous. Some species of spadefoot toads in western North America are famous for this phenomenon (**Figure 10–19**). Spadefoot tadpoles are normally herbivorous, but when tadpoles of the southern spadefoot toad (*Spea multiplicata*) eat freshwater shrimp that occur in some breeding ponds, they are transformed into the carnivorous morph. These carnivorous tadpoles have large heads and jaws and a powerful beak that allow them to bite off bits of flesh that are whirled into suspension and then filtered from the water stream. In addition to eating shrimp, they prey on other tadpoles.

In an Amazonian rain forest, tadpoles are by far the most important predators of frog eggs. In fact, egg predation decreases as the density of fish increases,

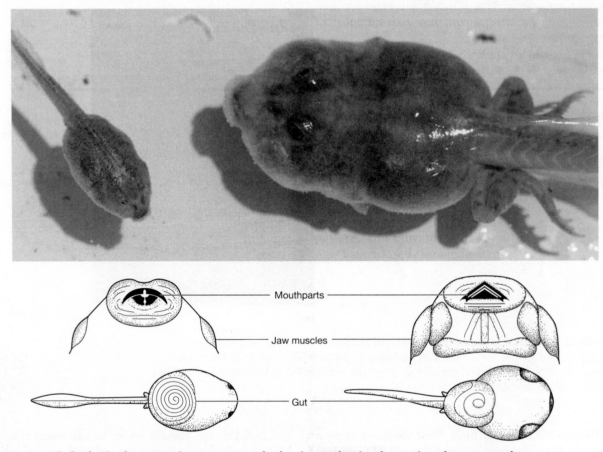

Figure 10–19 Tadpoles in the genus *Spea* express polyphenism—that is, alternative phenotypes that are dependent on the environment. Some individuals (*left*) develop as typical omnivorous anuran larvae that possess small jaw muscles (shown dissected to allow size comparison), smooth and unserrated mouthparts, and a long, coiled intestine. In contrast, other individuals (*right*) develop as a carnivorous morph, possessing large jaw muscles, notched and serrated mouthparts, and a short, relatively uncoiled gut.

apparently because the fish eat tadpoles that would otherwise eat frog eggs. Carnivorous tadpoles are also found among some species of frogs that deposit their eggs or larvae in bromeliads. These relatively small reservoirs of water may have little food for tadpoles. It seems possible that the first tadpole to be placed in a bromeliad pool may feed largely on other frog eggs—either unfertilized eggs deliberately deposited by the mother of the tadpole, as is the case for the dart-poison frog *Oophaga pumilio,* or fertilized eggs subsequently deposited by unsuspecting female frogs.

The feeding mechanisms that make tadpoles such effective collectors of food particles suspended in the water allow them to grow rapidly, but that growth contains the seeds of its own termination. As tadpoles grow bigger, they become less effective at gathering food because of the changing relationship between the size of food-gathering surfaces and the size of their bodies. The branchial surfaces that trap food particles are two-dimensional. Consequently, the food-collecting apparatus of a tadpole increases in size approximately as the square of the linear dimensions of the tadpole. However, the food the tadpole collects must nourish its entire body, and the volume of the body increases in proportion to the cube of the linear dimensions of the tadpole. The result of that relationship is a decreasing effectiveness of food collection as a tadpole grows; the body it must nourish increases in size faster than does its food-collecting apparatus.

10.3 Amphibian Metamorphosis

The transition from tadpole to frog involves a complete metamorphosis in which tadpole structures are broken down and their chemical constituents are rebuilt into the structures of adult frogs. In the early twentieth century, Friedrich Gudersnatch discovered the importance of thyroid hormones for amphibian metamorphosis quite by accident when he induced rapid precocious metamorphosis in tadpoles by feeding them extracts of beef thyroid glands. Some details of the interaction of neurosecretions and endocrine gland hormones have been worked out, but no fully integrated explanation of the mechanisms of hormonal control of amphibian metamorphosis is yet possible.

Anuran larval development is generally divided into three periods: (1) during premetamorphosis, tadpoles increase in size with little change in form; (2) in prometamorphosis, the hind legs appear and growth of the body continues at a slower rate; and (3) during metamorphic climax, the fore legs emerge and the tail regresses. These changes are stimulated by the actions of thyroxine, and the production and release of thyroxine are controlled by thyroid-stimulating hormone (TSH), which is a product of the pituitary gland.

The metamorphosis of a tadpole to a frog involves readily visible changes in almost every part of the body. The tail is absorbed and recycled into the production of adult structures. The small tadpole mouth that was sufficient for eating algae broadens into the huge mouth of an adult frog. The long tadpole gut, characteristic of herbivorous vertebrates, changes to the short gut of a carnivorous animal. The action of thyroxine on larval tissues is both specific and local. In other words, it has a different effect in different tissues, and that effect is produced by the presence of thyroxine in the tissue; it does not depend on induction by neighboring tissues. The particular effect of thyroxine in a given tissue is genetically determined, and virtually every tissue of the body is involved (Table 10–5).

Metamorphic climax begins with the appearance of the forelimbs and ends with the disappearance of

Table 10–5 Some morphological and physiological changes induced by thyroid hormones during amphibian metamorphosis

Body form and structure

Formation of dermal glands

Restructuring of mouth and head

Intestinal regression and reorganization

Calcification of skeleton

Appendages

Degeneration of skin and muscle of tail

Growth of skin and muscle of limbs

Nervous system and sense organs

Increase in rhodopsin in retina

Growth of extrinsic eye muscles

Formation of nictitating membrane of the eye

Growth of cerebellum

Growth of preoptic nucleus of the hypothalamus

Respiratory and circulatory systems

Degeneration of the gill arches and gills

Degeneration of the operculum that covers the gills

Development of lungs

Shift from larval to adult hemoglobin

Organs

Pronephric resorption in the kidney

Induction of urea-cycle enzymes in the liver

Reduction and restructuring of the pancreas

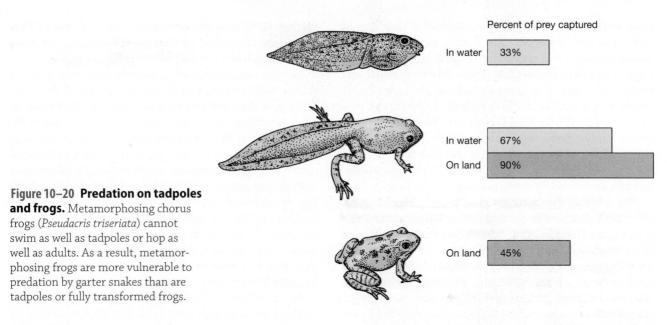

Percent of prey captured

In water 33%

In water 67%
On land 90%

On land 45%

Figure 10–20 Predation on tadpoles and frogs. Metamorphosing chorus frogs (*Pseudacris triseriata*) cannot swim as well as tadpoles or hop as well as adults. As a result, metamorphosing frogs are more vulnerable to predation by garter snakes than are tadpoles or fully transformed frogs.

the tail. This is the most rapid part of metamorphosis, taking only a few days after a larval period that lasts for weeks or months. One reason for the rapidity of metamorphic climax may be the vulnerability of larvae to predators during this period. A larva with legs and a tail is neither a good tadpole nor a good frog: the legs inhibit swimming, and the tail interferes with jumping. As a result, predators are more successful at catching anurans during metamorphic climax than they are in prometamorphosis or following the completion of metamorphosis. Metamorphosing chorus frogs (*Pseudacris triseriata*) are most vulnerable to garter snakes when they have developed legs but still retain a tail. Both tadpoles (with a tail and no legs) and metamorphosed frogs (with legs and no tail) were more successful than the metamorphosing individuals at escaping from snakes (**Figure 10–20**). In water, the snakes captured 33 percent of the tadpoles that were offered, compared with 67 percent of the transforming frogs. On land, the snakes captured 45 percent of the fully transformed frogs that were offered and 90 percent of the transforming frogs. Life-history theory predicts that selection will act to shorten the periods in the lifetime of a species when it is most vulnerable to predation, and the speed of metamorphic climax may be a manifestation of that phenomenon.

10.4 Exchange of Water and Gases

Amphibians have a glandular skin that lacks external scales and is highly permeable to gases and water. Both the permeability and glandularity of the skin have been of major importance in shaping the ecology and evolution of amphibians. Mucous glands are distributed over the entire body surface and secrete glycopeptides and probably glycerol that keep the skin moist. For an amphibian, a dry skin means reduced permeability. That, in turn, reduces oxygen uptake and the ability of the animal to use evaporative cooling to maintain its body temperature within equable limits. Experimentally produced interference with mucous gland secretion leads to lethal overheating of frogs undergoing normal basking activity.

Both water and gases pass readily through amphibian skin. In biological systems, permeability to water is inseparable from permeability to gases, and amphibians depend on cutaneous respiration for a significant part of their gas exchange. Although the skin permits the passive movement of water and gases, it controls the movement of other compounds. Sodium is actively transported from the outer surface to the inner, and urea is retained by the skin. These characteristics are important in the regulation of osmolal concentration and in facilitating the uptake of water by terrestrial species.

Blood Flow in Larvae and Adults

Larval amphibians rely on their gills and skin for gas exchange, whereas adults of species that complete full metamorphosis lose their gills and develop lungs. Lungs develop at different larval stages in different lineages of amphibians, and as the lungs develop, they are increasingly used for respiration. Late in their development, tadpoles and partly metamorphosed froglets can be seen swimming to the surface to gulp air.

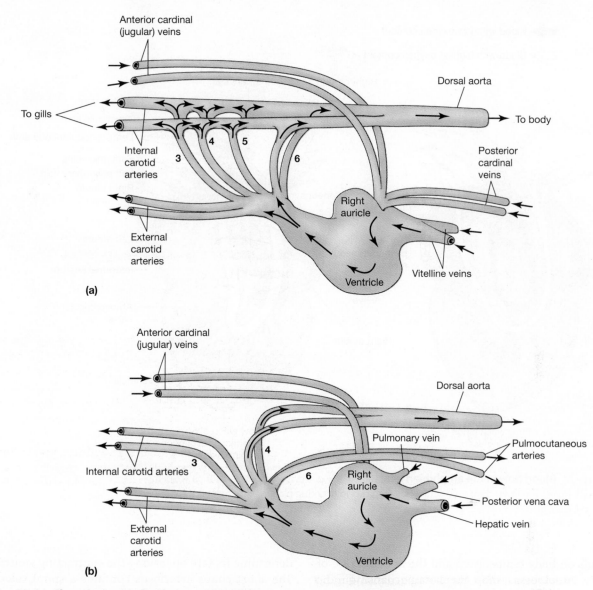

Figure 10–21 Changes in circulation at metamorphosis. Blood flow through the aortic arches of (a) a larval amphibian and (b) an adult without gills. The head is to the left. Arches 3, 4, and 5 carry blood to the gills of larvae, and arch 6 flows into the dorsal aorta. In adults, arch 3 carries blood to the brain via the internal carotid arteries, and arch 6 sends blood to the pulmocutaneous arteries. The posterior cardinal veins return blood from the posterior body, the vitelline veins carry blood from the intestine, and the hepatic vein brings blood from the liver.

As the gills lose their respiratory function, the carotid arches also change their roles (**Figure 10–21**). Arches 1 and 2 are lost early in embryonic development. In tadpoles, arches 3 through 5 supply blood to the gills and thence to the internal carotid arteries that carry the blood to the head. Arch 6 carries blood to the dorsal aorta via a connection called the ductus arteriosus. At metamorphosis, arch 3 becomes the supply vessel for the internal carotid arteries. Initially, arches 4 and 5 supply blood to the dorsal aorta; however,

arch 5 is usually lost in anurans, so arch 4 becomes the main route by which blood from the heart enters the aorta. Arch 6 primarily supplies blood to the lungs and skin via the pulmocutaneous arteries.

Cutaneous Respiration

All amphibians rely on their skin surface for gas exchange, especially for the release of carbon dioxide. The balance between cutaneous and pulmonary uptake of oxygen varies among species, and within a species it

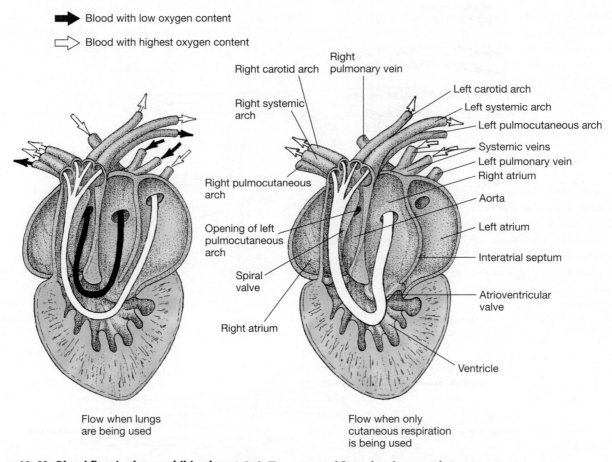

Blood with low oxygen content

Blood with highest oxygen content

Left diagram labels:
Right carotid arch
Right systemic arch
Right pulmocutaneous arch
Opening of left pulmocutaneous arch
Spiral valve
Right atrium

Flow when lungs are being used

Right diagram labels:
Right pulmonary vein
Left carotid arch
Left systemic arch
Left pulmocutaneous arch
Systemic veins
Left pulmonary vein
Right atrium
Aorta
Left atrium
Interatrial septum
Atrioventricular valve
Ventricle

Flow when only cutaneous respiration is being used

Figure 10–22 Blood flow in the amphibian heart. *Left:* The pattern of flow when lungs are being ventilated. *Right:* The flow when only cutaneous respiration is taking place. (Light arrows = oxygen-rich blood; dark arrows = oxygen-poor blood.)

depends on body temperature and the animal's rate of activity. Amphibians show increasing reliance on the lungs for oxygen uptake as temperature and activity increase.

The patterns of blood flow within the hearts of adult amphibians reflect the use of two respiratory surfaces. The following description is based on the anuran heart (**Figure 10–22**). The atrium of the heart is divided anatomically into left and right chambers by a septum. Oxygen-rich blood from the lungs flows into the left side of the heart, and oxygen-poor blood from the systemic circulation flows into the right side.

The ventricle shows greater anatomical division into left and right sides in species that depend heavily on pulmonary respiration than in species that rely mostly on cutaneous respiration. The spongy muscular lumen of the ventricle minimizes the mixing of blood with high and low oxygen content, and the position within the ventricle of a particular parcel of blood appears to determine its fate on leaving the contracting ventricle. The short conus arteriosus contains a spiral valve of tissue that differentially guides the blood from the left and right sides of the ventricle to the aortic arches.

The anatomical relationships within the heart are such that oxygen-rich blood returning to the heart from the pulmonary veins enters the left atrium, which injects it on the left side of the common ventricle. Contraction of the ventricle tends to eject blood in laminar streams that spiral out of the pumping chamber, carrying the left-side blood into the ventral portion of the spirally divided conus. This half of the conus is the one from which the carotid (head-supplying) and systemic aortic arches arise.

Thus, when the lungs are actively ventilated, oxygen-rich blood returning from them to the heart is selectively distributed to the tissues of the head and body. Oxygen-poor venous blood entering the right atrium is directed into the dorsal half of the spiral valve in the conus. It goes to the pulmocutaneous arch, destined for oxygenation in the lungs.

However, when the skin is the primary site of gaseous exchange, as it is when a frog is underwater, the highest oxygen content is in the systemic veins that drain the skin. In this situation the lungs may actually be net users of oxygen, and, because of vascular constriction, little blood passes through the pulmonary circuit.

Because the ventricle is undivided and the majority of the blood is arriving from the systemic circuit, the ventral section of the conus receives blood from an overflow of the right side of the ventricle. The scant left atrial supply to the ventricle also flows through the ventral conus. Thus the most oxygenated blood coming from the heart flows to the tissues of the head and body during this shift in primary respiratory surface, a phenomenon possible only because of the undivided ventricle. Variability of the cardiovascular output in amphibians is an essential part of their ability to use alternative respiratory surfaces effectively.

Permeability to Water

The internal osmolal pressure of amphibians is approximately two-thirds that of most other vertebrates. The primary cause of the dilute body fluids of amphibians is low sodium content—approximately $100 \text{ mM} \cdot \text{kg}^{-1}$ compared with $150 \text{ mM} \cdot \text{kg}^{-1}$ in other vertebrates. Amphibians can tolerate a doubling of the normal sodium concentration, whereas an increase from $150 \text{ mM} \cdot \text{kg}^{-1}$ to $170 \text{ mM} \cdot \text{kg}^{-1}$ is the maximum humans can tolerate.

Amphibians are most abundant in moist habitats, especially temperate and tropical forests, but a surprisingly large number of species live in dry regions. Anurans have been by far the most successful amphibian invaders of arid habitats. All but the harshest deserts have substantial anuran populations, and different families have converged on similar specializations. Avoiding the harsh conditions of the ground surface is the most common mechanism by which amphibians have managed to invade deserts and other arid habitats. Anurans and salamanders in deserts may spend 9 or 10 months of the year in moist retreat sites, sometimes more than a meter underground, emerging only during the rainy season and compressing feeding, growth, and reproduction into just a few weeks.

Many species of arboreal frogs have skins that are less permeable to water than the skin of terrestrial frogs, and a remarkable specialization is seen in a few tree frogs. The African rhacophorid *Chiromantis xerampelina* and the South American hylid *Phyllomedusa sauvagii* lose water through the skin at a rate only one-tenth

that of most frogs. *Phyllomedusa* has been shown to achieve this low rate of evaporative water loss by using its legs to spread the lipid-containing secretions of dermal glands over its body surface in a complex sequence of wiping movements, but the basis for the impermeability of *Chiromantis* is not yet understood. These two frogs are also unusual because they excrete nitrogenous wastes as precipitated salts of uric acid (as do lizards and birds) rather than as urea. This method of disposing of nitrogen provides still more water conservation.

Behavioral Control of Evaporative Water Loss

For animals with skins as permeable as those of most amphibians, the main difference between rain forests and deserts may be how frequently they encounter a water shortage. The Puerto Rican coquí (*Eleutherodactylus coqui*) lives in wet tropical forests; nonetheless, it has elaborate behaviors that reduce evaporative water loss during its periods of activity. Each male coquí has a calling site on a leaf in the understory vegetation from which it calls to attract a female. Male coquís emerge from their daytime retreat sites at dusk and move to their calling sites, remaining there until shortly before dawn, when they return to their daytime retreats.

The activities of the frogs vary from night to night, depending on rainfall. On nights after a rainstorm, when the forest is wet, the coquís begin to vocalize soon after dusk and continue until about midnight, when they fall silent for several hours. They resume calling briefly just before dawn. When they are calling, coquís extend their legs and raise themselves off the surface of the leaf (**Figure 10–23**). In this position they lose water by evaporation from the entire body surface.

On dry nights, the behavior of the frogs is quite different. The males move from their retreat sites to their calling stations, but they call only sporadically. Most of the time they rest in a water-conserving posture in which the body and chin are flattened against the leaf surface, the eyes are closed, and the limbs are pressed against the body. A frog in this posture exposes only half its body surface to the air, thereby reducing its rate of evaporative water loss. The effectiveness of the postural adjustments is illustrated by the water losses of frogs in the forest at El Verde, Puerto Rico, on dry nights. Frogs in one test group were placed individually in small wire-mesh cages on leaf surfaces. A second group was composed of unrestrained frogs sitting on leaves. The caged frogs spent most of the night climbing around the cages trying to get out. This activity, like vocalization, exposed the entire body surface to the air, and the

(a)

(b)

(c)

Figure 10–23 A male Puerto Rican coquí, *Eleutherodactylus coquí.* (a) During vocalization, nearly the entire body surface is exposed to evaporation. (b) In the alert posture in which frogs wait to catch prey, most of the body surface is exposed. (c) In the water-conserving posture adopted on dry nights, half the body surface is protected from exposure.

caged frogs had an evaporative water loss that averaged 27.5 percent of their initial body mass. In contrast, the unrestrained frogs adopted water-conserving postures and lost an average of only 8 percent of their initial body mass by evaporation.

Experiments showed that the jumping ability of coquís was not affected by an evaporative loss of as much as 10 percent of the initial body mass, but a loss of 20 percent or more substantially decreased the distance frogs could jump. Thus, coquís can use behavior to limit their evaporative water losses on dry nights to levels that do not affect their ability to escape from predators or to capture prey. Without those behaviors, however, they would probably lose enough water by evaporation to affect their survival.

Uptake and Storage of Water

The mechanisms that amphibians use for obtaining water in terrestrial environments have received less attention than those used for retaining it. Amphibians do not drink water. Because of the permeability of their skins, species that live in aquatic habitats face a continuous osmotic influx of water that they must balance by producing urine. The impressive adaptations of terrestrial amphibians are ones that facilitate rehydration from limited sources of water. One such specialization is the **pelvic patch.** This is an area of highly vascularized skin in the pelvic region that is responsible for a very large portion of an anuran's cutaneous water absorption. Toads that are dehydrated and completely immersed in water rehydrate only slightly faster than those placed in water just deep enough to wet the pelvic area. In arid regions, water may be available only as a thin layer of moisture on a rock or as wet soil. The pelvic patch allows an anuran to absorb this water.

Aquaporins, tubular proteins that are inserted into the plasma membranes of cells, are the channels for the movement of water through the pelvic patch. The aquaporin family of genes appears to have undergone an extensive radiation in anurans, producing genes that code for aquaporins with different sites of activity and different functions. The h3-like AQPa2 aquaporins provide channels for water uptake through the pelvic patch, and the AQP3 and AQP5 aquaporins may control the movement of water, glycerol, and mucus to keep the skin moist.

The urinary bladder plays an important role in the water relations of terrestrial amphibians, especially anurans. Amphibian kidneys produce urine that is hyposmolal to the blood, so the urine in the bladder is dilute. Amphibians can reabsorb water from urine to replace water they lose by evaporation, and terrestrial amphibians have larger bladders than aquatic species.

Storage capacities of 20 to 30 percent of the body mass of the animal are common for terrestrial anurans, and some species have still larger bladders: the Australian desert frogs *Notaden nichollsi* and *Neobatrachus wilsmorei* can store urine equivalent to about 50 percent of their body mass, and a bladder volume of 78.9 percent of body mass has been reported for the Australian frog *Heleioporus eyrei*.

Behavior is as important in facilitating water uptake as it is in reducing water loss. Leopard frogs, *Lithobates pipiens*, spend the summer activity season in grassy meadows where they have no access to ponds. The frogs spend the day in retreats they create by pushing vegetation aside to expose moist soil. In the retreats, the frogs rest with the pelvic patch in contact with the ground, and tests have shown that the frogs are able to absorb water from the soil. On nights when dew forms, many frogs move from their retreats and spend some hours in the early morning sitting on dew-covered grass before returning to their retreats.

Leopard frogs show a daily pattern of water gain and loss during a period of several days when no rain falls: in the morning the frogs are sleek and glistening with moisture, and they have urine in their bladders. That observation indicates that in the morning the frogs have enough water to form urine. By evening the frogs have dry skins and little urine in the bladder, suggesting that as they lost water by evaporation during the day, they had reabsorbed water from the urine to maintain the water content of their tissues.

By the following morning the frogs have absorbed more water from dew and are sleek and well hydrated again. Net gains and losses of water are shown by daily fluctuations in the body masses of the frogs; in the mornings they are as much as 4 or 5 percent heavier than their overall average mass, and in the evenings they are lighter than the average by a similar amount. Thus, these terrestrial frogs are able to balance their water budgets by absorbing water from moist soil and from dew to replace the water they lose by evaporation and in urine. As a result, they are independent of sources of water like ponds or streams and are able to colonize meadows and woods far from any permanent sources of water.

10.5 Poison Glands and Other Defense Mechanisms

The mucus that covers the skin of an amphibian has a variety of properties. In at least some species, it has antibacterial activity, and a potent antibiotic that may have medical applications has been isolated from the skin of the African clawed frog (*Xenopus laevis*). It is mucus that makes some amphibians slippery and hard for a predator to hold.

Other species have mucus that is extremely adhesive. Many salamanders, for example, have a concentration of mucous glands on the dorsal surface of the tail. When one of these salamanders is attacked by a predator, it bends its tail forward and slaps its attacker. The sticky mucus causes debris to adhere to the predator's snout or beak, and with luck the attacker soon concentrates on cleaning itself, losing interest in the salamander. When the California slender salamander (*Batrachoseps attenuatus*) is seized by a garter snake, the salamander curls its tail around the snake's head and neck. This behavior makes the salamander hard for the snake to swallow and also spreads sticky secretions on the snake's body. A small snake can find its body glued into a coil from which it is unable to escape.

Although secretions of the mucous glands of some species of amphibians are irritating or toxic to predators, an amphibian's primary chemical defense system is located in the poison glands (**Figure 10–24**). These

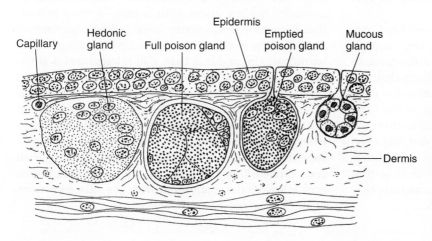

Figure 10–24 Amphibian skin. This cross section of skin is from the base of the tail of a red-backed salamander, *Plethodon cinereus*. Three types of glands can be seen: hedonic glands produce pheromones used in social interactions with other individuals of the species, poison glands produce toxins that deter predators, and mucous glands secrete a mucopolysaccharide that helps to keep the skin moist and may have antibacterial properties.

glands are concentrated on the dorsal surfaces of the animal, and defense postures of both anurans and salamanders present the glandular areas to potential predators.

A great diversity of pharmacologically active substances has been found in the skins of amphibians. Some of these substances are extremely toxic; others are less toxic but capable of producing unpleasant sensations when a predator bites an amphibian. Biogenic amines such as serotonin and histamine, peptides such as bradykinin, and hemolytic proteins have been found in frogs and salamanders belonging to many families. Many of these substances are named for the animals in which they were discovered—bufotoxin, epibatidine, leptodactyline, and physalaemin are examples.

Cutaneous alkaloids are abundant and diverse among the dart-poison frogs, the family Dendrobatidae, of the New World tropics. Most of these frogs are brightly colored and move about on the ground surface in daylight, making no attempt at concealment. More than 200 new alkaloids have been described from dendrobatids. Most of the alkaloids found in the skins of dart-poison frogs are similar to those found in ants, beetles, and millipedes that live in the leaf litter with the frogs, suggesting that frogs obtain alkaloids from their prey.

That hypothesis was tested by John Daly and his colleagues by raising juvenile Panamanian dart-poison frogs, *Dendrobates auratus*, under three dietary conditions. Dart-poison frogs live in the leaf litter on the forest floor, and leaf litter was used as the cage bedding for all experiments. Frogs with the most restricted diet were kept in a glass cage with carefully sealed joints and were fed only fruit flies. (Even these measures did not prevent a few tiny myrmicine ants from entering the cage during the experiment.) To ensure that the frogs in the glass cage had no natural prey, the leaf litter in their cage was frozen for 2 weeks before the experiment to kill the ants and other arthropods that live in it.

Frogs receiving the other two diets were kept outdoors in screened cages with a mesh size large enough to allow fruit flies, ants, and other small arthropods to enter. The frogs in one cage lived in leaf litter that had been frozen, and they ate primarily fruit flies that were attracted by bananas and fruit fly medium in the cage. Fruit flies were present in such abundance in this cage that they made up the bulk of the frogs' diet, but myrmicine ants also got into the cage. Frogs in the other screened cage lived in freshly gathered leaf litter that was replaced weekly; these frogs ate the ants and other arthropods that came in with the leaf litter.

At the end of the experiment, analyses of the skins of the frogs showed that one frog from the glass cage had no detectable alkaloids and a second had a trace amount of an alkaloid found in myrmicine ants— probably this frog had eaten the ants that had entered the cage despite the sealed joints. Frogs raised in the screened cage that was baited with bananas to attract fruit flies contained four alkaloids, all characteristic of myrmicine ants. Frogs in the cage with fresh leaf litter had at least 16 different alkaloids in their skins that could be traced to millipedes, beetles, and ants.

Wild-caught frogs from the area of the experiment had still more alkaloids in their skins—more than 40 different compounds. This experiment shows clearly that the frogs obtain toxic alkaloids from the prey that they eat, and that the more varied a frog's diet is, the more different alkaloids it contains. Furthermore, some species of *Dendrobates* can modify alkaloids they obtain from their prey, converting them to a more toxic form by adding a hydroxyl group to the molecule.

The name "dart-poison frog" refers to the use by South American Indians of the toxins of some of these frogs to poison the tips of the blowgun darts used for hunting. The use of frogs in this manner appears to be limited to three species of *Phyllobates* that live in western Colombia; plant poisons like curare are used to poison blowgun darts in other parts of South America. A unique alkaloid, batrachotoxin, is found in the genus *Phyllobates*. Batrachotoxin is a potent neurotoxin that prevents the closing of sodium channels in nerve and muscle cells, leading to irreversible depolarization and producing cardiac arrhythmias, fibrillation, and cardiac failure.

The bright yellow *Phyllobates terribilis* is the largest and most toxic species in the genus. The Emberá Choco Indians of Colombia use *Phyllobates terribilis* as a source of poison for their blowgun darts. The dart points are rubbed several times across the back of a frog and set aside to dry. The Indians handle the frogs carefully, holding them with leaves—a wise precaution because batrachotoxin is exceedingly poisonous. A single frog may contain up to 1900 micrograms of batrachotoxin, and less than 200 micrograms is probably a lethal dose for a human if it enters the body through a cut. Batrachotoxin is also toxic when it is eaten. In fact, the investigators inadvertently caused the death of a dog and a chicken in the Indian village in which they were living when the animals got into garbage that included plastic bags in which the frogs had been carried. Cooking destroys the poison and makes prey killed by darts anointed with the skin secretions of *Phyllobates terribilis* safe to eat.

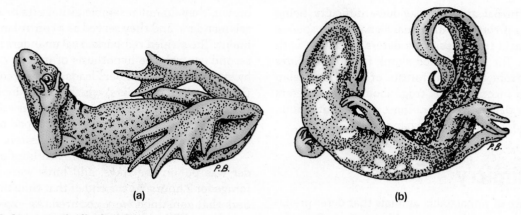

(a)   (b)

Figure 10–25 Aposematic displays by amphibians. Brightly colored species of amphibians
have displays that present colors as warnings that predators can learn to associate with the animals'
toxic properties. (a) The European fire-bellied toad (*Bombina bombina*) has a cryptically colored dorsal
surface and a brightly colored underside that is displayed in the *unken* reflex when the animal is
attacked. (b) The Hong Kong newt (*Paramesotriton hongkongensis*) has a brownish dorsal surface and
a mottled red and black ventral surface that is revealed by its aposematic display.

Many amphibians advertise their distasteful properties with conspicuous **aposematic** (warning) colors
and behaviors. A predator that makes the mistake of
seizing one is likely to spit it out because it is distasteful. The toxins in the skin may also induce vomiting
that reinforces the unpleasant taste for the predator.
Subsequently, the predator will remember its unpleasant experience and avoid the distinctly marked animal
that produced it. Some toxic amphibians combine a
cryptic dorsal color with an aposematic ventral pattern. Normally, the cryptic color conceals them from
predators, but if they are attacked, they adopt a posture that displays the brightly colored ventral surface
(**Figure 10–25**).

Three species of salamanders have a morphological
specialization that enhances the defensive effects of their
chemical secretions. The European salamander *Pleurodeles waltl* and two genera of Asian salamanders (*Echi-
notriton* and *Tylotriton*) have ribs that pierce the body
wall when a predator seizes the salamander. You can
imagine the shock for a predator that bites a salamander and finds its tongue and palate impaled by a dozen or
more bony spikes! Even worse, the ribs penetrate poison
glands as they emerge through the body wall, and each
rib carries a drop of poison into the wound.

Many anurans make long leaps to escape a predator,
and others feign death. Some cryptically colored frogs
extend their legs stiffly when they play dead. In this
posture they look so much like the leaf litter on the
ground that they may be hard for a visually oriented
predator such as a bird to see. Very large frogs attack
potential predators. They increase their apparent size

by inflating the lungs and hop toward the predator,
often croaking loudly. That alone can be an unnerving
experience, and some of the carnivorous species such
as the horned frogs of South America (*Ceratophrys*),
which have recurved teeth on the maxillae and toothlike serrations on the mandibles, can inflict a painful
bite.

Red efts (the terrestrial life stage of the red-spotted
newt, *Notophthalmus viridescens*) are classic examples
of aposematic animals (see the color insert). They are
bright orange and are active during the day, making
no effort to conceal themselves. Efts contain tetrodo-
toxin, which is a potent neurotoxin. Touching an eft
to your lips produces an immediate unpleasant numb-
ness and tingling sensation, and the behavior of animals that normally prey on salamanders indicates that
it affects them the same way. As a result, an eft that is
attacked by a predator is likely to be rejected before it
is injured. After one or two such experiences, a predator will no longer attack efts. Support for the belief
that this protection may operate in nature is provided
by the observation that four of eleven wild-caught blue
jays (*Cyanocitta cristata*) refused to attack the first red
eft they were offered in a laboratory feeding trial. Their
behavior suggests that these four birds had learned to
avoid red efts before they were captured. The remaining seven birds attacked at least one eft but dropped it
immediately. After one or two experiences of this sort,
the birds made retching movements at the sight of an
eft and refused to attack.

Of course, aposematic colors and patterns work to
deter predation only if a predator can see the aposematic

signal. Nocturnal animals may have difficulty being conspicuous if they rely on visual signals. One species of dendrobatid frog apparently deters predators with a foul odor. The aptly named skunk frog (*Aromobates nocturnus*) from the cloud forests of the Venezuelan Andes is an inconspicuous frog, about 5 centimeters long with a dark olive color. These frogs emit a foul, skunklike odor when they are handled.

10.6 Mimicry

The existence of unpalatable animals that deter predators with aposematic colors and behaviors offers the opportunity for other species that lack noxious qualities to take advantage of predators that have learned by experience to avoid the aposematic species. In this phenomenon, known as **mimicry**, the mimic (a species that lacks noxious properties) resembles a noxious model, and that resemblance causes a third species—the dupe—to mistake the mimic for the model. Some of the best-known cases of mimicry among vertebrates involve salamanders. One that has been investigated involves two color morphs of the common red-backed salamander, *Plethodon cinereus*.

Red-backed salamanders normally have dark pigment on the sides of the body, but in some regions an erythristic (Greek *erythr* = red) color morph is found that lacks the dark pigmentation and has red-orange on the sides as well as on the back. These erythristic morphs resemble red efts and could be mimics of efts. Red-backed salamanders are palatable to many predators, and mimicry of the noxious red efts might confer some degree of protection on individuals of the erythristic morph. That hypothesis was tested in a series of experiments. Salamanders were put in leaf-filled trays from which they could not escape, and the trays were placed in a forest where birds were foraging. The birds learned to search through the leaves in the trays to find the salamanders. This is a very lifelike situation for a test of mimicry because some species of birds are important predators of salamanders. For example, red-backed salamanders and dusky salamanders (*Desmognathus ochrophaeus*) made up 25 percent of the prey items fed to their young by hermit thrushes in western New York.

Three species of salamanders were used in the experiments, and the number of each species was adjusted to represent a hypothetical community of salamanders containing 40 percent dusky salamanders, 30 percent red efts, 24 percent striped red-backed salamanders, and 6 percent erythristic red-backed salamanders. The dusky salamanders are palatable to birds and are light brown; they do not resemble either efts or red-backed salamanders, and they served as a control in the experiment. The striped red-backed salamanders represent a second control: the hypothesis of mimicry of red efts by erythristic salamanders leads to the prediction that the striped salamanders, which do not look like efts, will be eaten by birds, whereas the erythristic salamanders, which are as palatable as the striped ones but do look like the noxious efts, will not be eaten.

A predetermined number of each kind of salamander was put in the trays, and birds were allowed to forage for 2 hours. At the end of that time the salamanders that remained were counted. As expected, only 1 percent of the efts had been taken by birds, whereas 44 percent to 60 percent of the palatable salamanders had disappeared. As predicted, the birds ate fewer of the erythristic form of the red-backed salamanders than they ate of the striped form.

These results show that the erythristic morph of the red-backed salamander does obtain some protection from avian predators as a result of its resemblance to the red eft. In this case the resemblance is visual, but mimicry can exist in any sensory mode to which a dupe is sensitive. Olfactory mimicry by amphibians might be effective against predators such as shrews and snakes, which rely on scent to find and identify prey. This possibility has scarcely been considered, but careful investigations may yield fascinating new examples.

10.7 Why Are Amphibians Vanishing?

Biologists from many countries met in England in 1989 at the First World Congress of Herpetology. In a week of formal presentations of scientific studies and in casual conversations at meals and in pubs, the participants discovered that an alarmingly large proportion of them knew of populations of amphibians that had once been abundant and now were rare or even entirely gone. Events that had appeared to be isolated instances began to look like parts of a global pattern. As a result of that discovery, David Wake, of the University of California at Berkeley, persuaded the National Academy of Sciences to convene a meeting of biologists concerned about vanishing amphibians. Biologists from all over the world met at the West Coast center of the Academy in February 1990. All reported that populations of amphibians in their countries were disappearing, and often there was no apparent reason.

Following that meeting, an international effort to identify the causes of amphibian declines was initiated by the Declining Amphibian Populations Task Force of the Species Survival Commission of the World Conservation Union (IUCN). In 1998 another meeting was convened, this time in Washington, DC, that brought together authorities in disciplines ranging from herpetology and population biology through toxicology and infectious diseases to climate change and science policy. The conference concluded that "there is compelling evidence that, over the last 15 years, unusual and substantial declines have occurred in abundance and numbers of populations in globally distributed geographic regions." Rapid declines occurred during the 1950s and 1960s and are continuing; at least 9 species, and perhaps more than 100, have become extinct since 1980.

The current high rate of declines and extinctions of amphibians contrasts dramatically with the fossil record of the Pleistocene, in which amphibian species persisted over millions of years while birds and mammals disappeared. For example, the amphibian fauna of a Middle Pleistocene fossil site in Italy consists exclusively of extant species that lived with species of mammals that are either extinct (saber-toothed tigers) or no longer found in Italy (apes, elephants, rhinoceroses, and leopards). Clearly, something new is happening that affects amphibians more severely than other terrestrial vertebrates.

Habitat Loss

Habitat loss is probably the most widespread cause of amphibian declines. Paving land for highways and shopping centers completely destroys the habitat for amphibians. Logging and construction of residential subdivisions are almost equally destructive because frogs and salamanders depend on cool, moist microhabitats on the forest floor. When the forest canopy is removed, sunlight reaches the ground and conditions become too hot and dry for amphibians, and when forest remnants are separated from streams by cultivated fields, stream-breeding species of frogs cannot reach their breeding sites.

Diseases

In the past 50 years, disease-causing organisms have emerged as a major cause of the extinction of amphibians. This phenomenon is something new because most amphibian species long ago established stable relationships with their local pathogens. But when animals carrying these pathogens were moved to regions where the pathogens are unknown, lethal epidemics spread, just as they did when Europeans carried previously unknown pathogens to the New World. The World Organization for Animal Health has identified two epidemic diseases of amphibians—ranavirus disease and chitridiomycosis.

Ranavirus Disease Ranaviruses infect frogs and salamanders and can lead to the extinction of local populations. The occurrence of ranavirus appears to have shaped the biology of some populations of tiger salamanders (*Ambystoma tigrinum*). In Arizona, tiger salamanders have two larval forms, the normal morph and a cannibalistic morph with an enlarged head and jaws. Some populations consist entirely of the normal morph, whereas in other populations some individuals develop into cannibalistic larvae that eat individuals of the normal morph. A survey of the occurrence of normal and cannibalistic forms revealed that cannibals are absent from populations that are infected by ranaviruses but occur regularly in populations without the viruses. Apparently in virus-free populations it is safe to eat other larvae, but, when viruses are present, a cannibal risks infection when it eats other larval salamanders.

The correspondence between the presence of ranaviruses and the life-history structure of salamander populations suggests that this host-pathogen association is ancient and that the salamander hosts and their ranavirus pathogens have evolved a stable relationship. This generalization probably applies to other amphibians that serve as hosts for ranaviruses, and it appears that iridoviruses cause fluctuations in amphibian populations but are not usually responsible for extinctions.

Chytridiomycosis A chytrid fungus, *Batrachochytrium dendrobatidis*, is responsible for the disappearances of amphibians in the Americas, Europe, Australia, and New Zealand. The chytrid, which is often referred to as Bd because its full name is so cumbersome to write and read, has motile reproductive zoospores that live in water and can penetrate the skin of an amphibian, causing a disease called chytridiomycosis. When a zoospore enters an amphibian, it matures to form a spherical reproductive body, the zoosporangium, which has branching structures that extend through the skin. These structures apparently interfere with respiration and control of water movement, and adult frogs die.

On a worldwide basis, about 40 percent all species of frogs and salamanders are in decline, and chytridiomycosis has played a large role in the loss of amphibians. The epidemic has been studied especially well in Central and South America. In the late 1980s, two charismatic species of frogs, the golden toad (*Incilius periglenes*) and the harlequin frog (*Atelopus varius*), abruptly vanished from the amphibian community of the Monteverde Reserve in the mountains of northern Costa Rica. The disappearance of those two species was followed by others, and 20 of the 49 species that once lived in the reserve have vanished. The reason frogs were vanishing remained a mystery for several years, until Bd was detected and its pathogenic effect was recognized.

From Monteverde, the epidemic moved southward through Costa Rica in the 1990s, entering Panama around 2000 (**Figure 10–26**). The impact of the epidemic on communities of amphibians has been appalling. Thirty species of anurans were lost when the epidemic reached El Copé, Panama, and those species represented 41 percent of the total diversity of evolutionary lineages at this location.

Additional waves of infection spread along mountain chains in South America: one wave started in Venezuela in the late 1970s and spread southward, meeting the northward-moving branch of an epidemic that had originated in Ecuador. The southward-moving branch of this epidemic has reached Chile.

The source of the infection at Monteverde was puzzling because the disease appeared suddenly and seemingly from nowhere. An ingenious application of the polymerase chain reaction (PCR) has solved that mystery by detecting the presence of Bd in specimens of frogs and salamanders collected in southern Mexico in the 1970s, at the time of a hitherto-unexplained die-off of amphibians in that region. Careful tests have demonstrated a precise association between the dates of collection, the dates of die-offs, and the presence of Bd: the fungus is absent from museum specimens that were collected before the die-offs began, and its appearance at each location corresponds to the first reports of mass mortality at that site.

The PCR analysis pushed the origin of the Central American epidemic northward to southern Mexico and the earliest occurrences to the 1970s, but how did Bd appear in Mexico? Apparently African clawed frogs (*Xenopus laevis* and other species in the genus *Xenopus*) brought the fungus to the Americas. These species are resistant to the infection, but studies of preserved specimens of *Xenopus* in museums in South Africa showed that the fungus was present in that country as early as 1938. The worldwide spread of chytridiomycosis can be traced to the discovery in 1934 that female *Xenopus* provide a convenient assay for human pregnancy.

World War II made access to clawed frogs difficult, but after the war enormous numbers of clawed frogs were exported from South Africa. Some of these frogs were released or escaped from laboratories and came into contact with native frogs. Soon chytrid infections began to appear in native frogs. Although chytridiomycosis was not recognized at the time, retrospective studies have revealed chytrid infections in frogs in North America in 1961, in South America in 1977, in Australia in 1978, and in Europe in 1997. North American bullfrogs (*Lithobates catesbeianus*) have been implicated as a source of the epidemic of Bd in South America.

A variety of interacting factors make definitive statements about chytridiomycosis difficult. Climate change was initially suspected as the cause of the decline of amphibians at Monteverde, and temperature and humidity do play a role in the susceptibility of populations to infection by Bd. The fungus grows best in cool, moist conditions, and tropical amphibians that live in mountains are at greater risk of infection than are lowland species. In both the Americas and Australia, the epidemic has spread along mountain chains rather than through the lowlands. Climate change is no longer regarded as a primary cause of the amphibian declines in these regions, but it might contribute to the probability of infection.

Habitat destruction is a second factor that is responsible for the decline of amphibians, and habitat destruction and Bd display an unexpected interaction: human-altered habitats are generally warmer and drier than undisturbed sites. Amphibians in these altered habits are less likely to be infected by Bd than those in undisturbed habitats because the fungus survives well in cool, moist conditions.

Both innate and acquired immune defenses play a role in resistance to Bd. The antimicrobial skin peptides of amphibians provide some protection, and a symbiotic bacterium, *Janthinobacterium lividum*, that is normally present on the skin can inhibit the growth of Bd. Furthermore, both the antimicrobial skin peptides and *J. lividum* can be transmitted from a female frog to her embryos.

Species of amphibians vary in their susceptibility to Bd. In laboratory experiments, southern toads (*Anaxyrus terrestris*) and wood frogs (*Lithobates sylvaticus*) were at least 20 times more likely to contract a lethal Bd infection than were upland chorus frogs (*Pseudacris feriarum*) and gray tree frogs (*Hyla versicolor*). Furthermore, there is individual variation in susceptibility to Bd within a species, and some of that variation is associated with individual differences in the major histocompatibility complex genes that initiate an immune response to the presence of a pathogen, such as Bd.

The Additive Effects of Multiple Threats

The challenges that amphibians face do not act alone—they come in combinations. An ambitious analysis by Christian Hof and his colleagues evaluated the exposure

Figure 10–26 Vanishing amphibians.
Waves of amphibian population disappearances, indicated by the dark ribbons, have flowed along the mountain ranges of Central and South America as a result of infections by *Batrachochytrium dendrobatidis*. (a) Disappearances began in southern Mexico and Guatemala in the 1970s and early 1980s. In 1987 disappearances were noted at Monteverde, Costa Rica, and by 2007 the wave had swept through most of Panama and was approaching the border of Colombia. (b) In South America the first population decline occurred near Caracas, Venezuela, in 1977, and a wave of infection spread westward into Colombia along the Cordillera Oriental. The oldest record of *Batrachochytrium dendrobatidis* in South America is from 1980 in Cañar, Ecuador, and that infection spread northward into Colombia and Venezuela along both the Cordillera Occidental and the Cordillera Oriental. A third wave spread from Cañar southward into Peru along the Cordillera Occidental and the Cordillera Oriental and has now reached southern Chile. An additional introduction of *Batrachochytrium dendrobatidis* into the Atlantic coastal forest of Brazil in 1981 is not shown.

of 5527 different species of amphibians to the three major threats: habitat loss, climate change, and Bd. The effects of habitat loss and climate change are most severe in parts of Africa, northern South America, and the Andes Mountains. The Andes are also severely affected by Bd. Furthermore, areas that are rich in amphibian species are disproportionately affected by one or more threats. The geographic distribution of the threats and their overlaps are shown on the fourth page of the color insert.

Summary

Locomotor adaptations distinguish the lineages of amphibians. Salamanders (Urodela) usually have short, sturdy legs that are used with lateral undulation of the body in walking. Aquatic salamanders use lateral undulations of the body and tail to swim, and some specialized aquatic species are elongate and have very small legs. Frogs and toads (Anura) are characterized by specializations of the pelvis and hindlimbs that permit both legs to be used simultaneously to deliver a powerful thrust used both for jumping and for swimming. Many anurans walk quadrupedally when they move slowly, and some are agile climbers. The caecilians (Gymnophiona) are legless tropical amphibians; some are burrowers, and others are aquatic.

The diversity of amphibian reproductive modes exceeds that of any other group of vertebrates except the fishes. Fertilization is internal in derived salamanders, but most frogs rely on external fertilization. All caecilians have internal fertilization. Many species of amphibians have aquatic larvae. Tadpoles, the aquatic larvae of anurans, are specialized for life in still or flowing water, and some species of frogs deposit their tadpoles in very specific sites, such as the pools of water that accumulate in the leaf axils of bromeliads or other plants. Specializations of tadpoles are entirely different from specializations of frogs, and metamorphosis causes changes in all parts of the body. Direct development that bypasses the larval stage is also widespread among anurans and is often combined with parental care of the eggs. Viviparity occurs in all three orders.

In many respects the biology of amphibians is determined by properties of their skin. Hedonic glands are key elements in reproductive behaviors, poison glands protect the animals against predators, and mucous glands keep the skin moist, facilitating gas exchange. Above all, the permeability of the skin to water limits most amphibians to microhabitats in which they can control water gain and loss. That sounds like a severe restriction, but in the proper microhabitat, amphibians can utilize the permeability of their skin to achieve a remarkable degree of independence of standing water. Thus the picture that is sometimes presented of amphibians as animals barely hanging on as a sort of evolutionary oversight is misleading. Only a detailed examination of all facets of their biology can produce an accurate picture of amphibians as organisms.

An examination of that sort reinforces the view that the skin is a dominant structural characteristic of amphibians. This is true not only for the limitations and opportunities presented by the skin's permeability to water and gases but also as a result of the intertwined functions of the skin glands in defensive and reproductive behaviors. The structure and function of the skin may be primary characters that have shaped the evolution and ecology of amphibians, and it may also be responsible for some aspects of their susceptibility to pollution. All over the world, populations of amphibians are disappearing at an alarming rate—about 40 percent of all species are in decline. Habitat loss is responsible for the loss of many populations of amphibians, but two epidemic diseases—ranavirus disease and chytridiomycosis—are wiping out populations and even causing the extinction of entire species.

Discussion Questions

1. Deserts appear to be the least likely habitats in which to find animals with skins that are very permeable to water, yet amphibians, especially anurans, inhabit deserts all over the world. How is this possible?

2. Four species of salamanders in the genus *Desmognathus* form a streamside salamander guild in the mountains of North Carolina. The four species are found at different distances from the stream, as listed in the table. Suggest one or more hypotheses

to explain the relationship between the body sizes of these species and the distance each is found from the water. What prediction can you make for each hypothesis, and how could you test that prediction?

Species	D. quadramaculatus	D. monticola	D. ochrophaeus	D. aeneus
Adult body length	100–175 mm	83–125 mm	70–100 mm	44–57 mm
Distance from stream	Less than 10 cm	Less than 1 m	1–4 m	4–7 m

3. Why do you think most of the male coquís in the population emerge from their daytime retreats on dry nights? Only female coquís that are seeking a mate are active on these nights; the remaining females remain in their shelters. Why don't males do that as well?

4. An aquatic egg and larva are ancestral characters of anurans, but terrestrial eggs and even direct development that bypasses the larval stage entirely have evolved independently in many lineages of anurans. Why didn't all lineages evolve that way? That is, why might it be advantageous for a species of frog to have retained an aquatic egg and a larval stage?

5. There are about 10 times as many species of anurans as there are species of salamanders. Can you think of a reason why anurans might be more diverse than salamanders are?

6. *Ranavirus* and *Batrachochytrium dendrobatidis* both cause lethal infections, but *Ranavirus* decimates local populations of amphibians, whereas *Batrachochytrium dendrobatidis* is wiping out amphibians on a continental scale in North and South America. What difference between the histories of these pathogens in the Americas might account for the difference in the severity of their effects?

Additional Information

Bourke, J., et al. 2011. New records of *Batrachochytrium dendrobatidis* in Chilean frogs. *Diseases of Aquatic Organisms* 95:259–261.

Brodie, E. D., Jr., and E. D. Brodie III. 1980. Differential avoidance of mimetic salamanders by free-ranging birds. *Science* 208:181–182.

Cannatella, D. C., et al. 2009. Amphibians (Lissamphibia). In S. B. Hedges and S. Kumar (eds.), *The Timetree of Life*. Oxford, UK: Oxford University Press.

Cheng, T. L., et al. 2011. Coincident mass extirpation of neotropical amphibians with the emergence of the infectious fungal pathogen *Batrachochytrium dendrobatidis*. *Proceedings of the National Academy of Sciences* 108:9502–9507.

Conlon, J. M. 2011. The contribution of skin antimicrobial peptides to the system of innate immunity in anurans. *Cell and Tissue Research* 343:201–212.

Crawford, A. J., et al. 2010. Epidemic disease decimates amphibian abundance, species diversity, and evolutionary history in the highlands of central Panama. *Proceedings of the National Academy of Sciences* 107:13777–13782.

Daly, J. W., et al. 2000. Arthropod-frog connection: Decahydroquinoline and pyrrolizidine alkaloids common to microsympatric myrmicine ants and dendrobatid frogs. *Journal of Chemical Ecology* 26:73–85.

Dapper, A. L., et al. 2011. The sounds of silence as an alarm cue in túngara frogs, *Physalaemus pustulosus*. *Biotropica* 43:380–385.

Herrel, A. and G. J. Measey. 2010. The kinematics of locomotion in caecilians. *Journal of Experimental Zoology* A 315:301–309.

Heyer, W. R., et al. (ed.). 1993. *Measuring and Monitoring Biological Diversity: Standard Methods for Amphibians*. Washington, DC: Smithsonian Institution Press.

Hof, C., et al. 2011. Additive threats from pathogens, climate and land-use change for global amphibian diversity. *Nature* 480:516–519.

Jaeger, R. G. 1981. Dear enemy recognition and the costs of aggression between salamanders. *American Naturalist* 117:962–974.

Jaeger, R. G., et al. 1983. Foraging tactics of a terrestrial salamander: Costs of territorial defense. *Animal Behaviour* 31:191–198.

Lips, K. R., et al. 2008. Riding the wave: Reconciling the roles of disease and climate change in amphibian declines. *PLoS Biology* 6(3):e72. doi:10.1371/journal.pbio.0060072.

Rosenblum, E. B., et al. 2010. The deadly chytrid fungus: A story of an emerging pathogen. *PLoS Pathogens* 6(1):e1000550. doi:10.1371/journal.ppat.1000550.

Rovito, S. M., et al. 2009. Dramatic declines in neotropical salamander populations are an important part of the global amphibian crisis. *Proceedings of the National Academy of Sciences* 106:3231–3236.

Searle, C. L., et al. 2011. Differential host susceptibility to *Batrachochytrium dendrobatidis*, an emerging amphibian pathogen. *Conservation Biology* 25:965–974.

Searle, C. L., et al. 2011. A dilution effect in the emerging amphibian pathogen *Batrachochytrium dendrobatidis*. *Proceedings of the National Academy of Sciences* 108:16322–16326.

Sigurdsen, T., and D. M. Green. 2011. The origin of modern amphibians: A re-evaluation. *Zoological Journal of the Linnean Society* 162:457–469.

Suzuki, M., and S. Tanaka. 2009. Molecular and cellular regulation of water homeostasis in anuran amphibians by aquaporins. *Comparative Biochemistry and Physiology, Part A* 153:231–241.

Takahashi, Y., et al. 2001. Distribution of blue-sensitive photoreceptors in amphibian retinas. *FEBS Letters* 501:151–155.

Walke, J. B., et al. 2011. Social immunity in amphibians: Evidence for vertical transmission of innate defenses. *Biotropica* 43:396–400.

Weldon, C., et al. 2004. Origin of the amphibian chytrid fungus. *Emerging Infectious Diseases* 10:2100–2105 www.cdc.gov/eid.

Wells, K. D. 2007. *The Ecology and Behavior of Amphibians*. Chicago: University of Chicago Press.

Websites

General

Amphibian Species of the World 5.5, an Online Reference, American Museum of Natural History
http://research.amnh.org/vz/herpetology/amphibia/
AmphibiaWeb
http://amphibiaweb.org/index.html
Center for North American Herpetology
http://www.cnah.org/index.asp

Amphibian Conservation

Amphibian Ark
http://www.amphibianark.org/
Global Amphibian Forum
http://www.amphibians.org/redlist/
Global Bd-Mapping Project
http://www.bd-maps.net/maps/
Save the Frogs
http://www.savethefrogs.org

PART III

Sauropsida: Turtles, Lepidosaurs, and Archosaurs

An early division of amniotes produced the two evolutionary lineages that include the vast majority of extant terrestrial vertebrates, the Synapsida and Sauropsida. Mammals are synapsids, and turtles, tuatara, lizards, snakes, crocodilians, and birds are sauropsids, as are all extinct reptiles except for the so-called mammal-like reptiles (more correctly called nonmammalian synapsids).

The lineages can be distinguished in the fossil record by the mid-Carboniferous period, and they show remarkable similarities and differences in the solutions they found to the challenges of life on land. Both approaches were successful. We tend to think of mammals as the preeminent terrestrial vertebrates, but that opinion reflects our own position in the synapsid lineage. The extant species of sauropsids greatly outnumber mammals, and sauropsids have exploited virtually all of the terrestrial adaptive zones occupied by mammals plus many that mammals have never penetrated, such as the gigantic body size achieved by some dinosaurs and the elongate body form of snakes.

CHAPTER 11

Synapsids and Sauropsids:
Two Approaches to Terrestrial Life

The terrestrial environment provided opportunities for new ways of life that amniotes have exploited. The amniotic egg may be a critical element of the success of synapsids and sauropsids because amniotic eggs are larger than non-amniotic eggs and produce larger hatchlings that grow into larger adults. Early in their evolutionary history, amniotes split into the two evolutionary lineages that dominate terrestrial habitats today, the Sauropsida and the Synapsida. Extant sauropsids include turtles, the scaly reptiles (tuatara, lizards, and snakes), crocodilians, and birds, whereas mammals are the only extant synapsids. Both lineages underwent great radiations in the late Paleozoic and Mesozoic eras that include animals that are now extinct and have no modern equivalents—the dinosaurs and pterosaurs were sauropsids, and the pelycosaurs and therapsids were synapsids.

At the time the sauropsid and synapsid lineages separated, amniotes had evolved few derived characters associated with terrestrial life. As a result, the sauropsid and synapsid lineages independently developed most of the derived characters that are necessary for terrestrial life, such as respiratory and excretory systems that conserve water and locomotor systems that are compatible with high rates of lung ventilation.

Both lineages developed fast-moving predators that could pursue fleeing prey (as well as fleet-footed prey that could run away from predators), and both lineages included species capable of powered flight. Both lineages had members that became endothermal, evolving high metabolic rates and insulation to retain metabolic heat in the body, and both lineages evolved extensive parental care and complex social behavior.

Despite the parallel evolutionary trends in synapsids and sauropsids, differences in the way they carry out basic functions show that they evolved those derived characters independently:

- A terrestrial animal that runs for long distances must eliminate the conflict between respiratory movements and locomotion that is characteristic of early amniotes. Derived sauropsids (pterosaurs, dinosaurs, and birds) and derived synapsids (therapsids and mammals) both reduced the side-to-side bending of the rib cage by placing their legs more under the body (upright posture) and relying more on movement of the legs than of the trunk. The derived sauropsids became bipedal and retained expansion and contraction of the rib cage as the primary method of creating the pressure differences that move air into and out of the lungs. In contrast, the derived synapsids remained quadrupedal, and more derived therapsids (including mammals) also added a diaphragm to aid with lung ventilation.

- The high rates of oxygen consumption that are needed to sustain rapid locomotion require respiratory systems that can take up oxygen and release carbon dioxide rapidly and still conserve water. In derived sauropsids (birds and crocodilians), these functions are accomplished with a one-way flow of air through the lung (a through-flow lung), whereas synapsids retain the basic tetrapod condition of in-and-out airflow (a tidal-flow lung).

- High rates of oxygen consumption during activity produce large amounts of heat, and a layer of insulation allows an animal to retain that heat to maintain a high body temperature. Derived sauropsids have feathers for insulation and derived synapsids have hair, and both have developed a subcutaneous layer of fat.

- A terrestrial animal requires an excretory system that eliminates nitrogenous wastes while conserving water. Sauropsids do this by having an insoluble waste product (uric acid), kidneys that do not produce concentrated urine (some marine birds are an exception to that generalization), and glands that secrete salt, whereas synapsids excrete a highly soluble waste product (urea) through kidneys that can produce very concentrated urine, and they lack salt-secreting glands.

These differences in structural and functional characters of sauropsids and synapsids show that there is more than one way to succeed as a terrestrial amniotic vertebrate.

11.1 Taking Advantage of the Opportunity for Sustained Locomotion

Running involves much more than just moving the legs rapidly. If an animal expects to run very far, the muscles used to move the limbs require a steady supply of oxygen, and that is where the ancestral form of vertebrate locomotion encounters a problem. Early tetrapods moved with lateral undulations of the trunk, as salamanders and lizards do today. The axial muscles provide the power in this form of locomotion, bending the body from side to side. The limbs and feet are used in alternate pairs (i.e., left front and right rear, left rear and right front) to provide purchase on the substrate as the trunk muscles move the animal.

Figure 11–1 Lung ventilation and locomotion. The effect of axial bending on lung volume of a running lizard (top view) and a galloping dog (side view). The bending axis of the lizard's thorax is between the right and left lungs. As the lizard bends laterally, the lung on the concave (*left*) side is compressed and air pressure in that lung increases (shown by +), while air pressure on the convex side decreases (shown by –). Air moves between the lungs (*arrow*), but little or no air moves into or out of the animal. In contrast, the bending axis of a galloping mammal's thorax is dorsal to the lungs. As the vertebral column bends, the volume of the thoracic cavity decreases; pressure in both lungs rises (shown by +), pushing air out of the lungs (*arrow*). When the vertebral column straightens, the volume of the thoracic cavity increases, pressure in the lungs falls (shown by –), and air is pulled into the lungs (*arrow*).

Tetrapodal locomotion based on the trunk muscles works for only short dashes. The problem with this ancestral locomotor mode is that the axial muscles are responsible for two essential functions—bending the trunk unilaterally for locomotion and compressing the rib cage bilaterally to ventilate the lungs—and these activities cannot happen simultaneously. **Figure 11–1** illustrates the problem: the side-to-side bending of the lizard's rib cage compresses one lung as it expands the other, so air flows from one lung into the other, interfering with airflow in and out via the trachea.

Short sprints are feasible for animals that use lateral bending of the trunk for locomotion because the energy for a sprint is supplied initially by a reservoir of high-energy phosphate compounds (such as adenosine triphosphate [ATP] and creatine phosphate [CP]) that are present in the muscle cells. When those compounds are used up, the muscles switch to anaerobic metabolism, which draws on glycogen stored in the cells and does not require oxygen. The problem arises when rapid locomotion must be sustained beyond a minute or two because, at that point, the supply of glycogen and high-energy phosphate compounds in the muscles has been consumed. Because of this conflict between running and breathing, lizards that retain the ancestral modes of locomotion and ventilation are limited to short bursts of activity.

Sustained locomotion requires a way to separate respiration from locomotion. Synapsids and sauropsids both developed modes of locomotion that allow them to hold the trunk rigid and use limbs as a major source of propulsion, but the ways they did this are quite different.

Locomotion and Respiration of Synapsids

Early nonmammalian synapsids retained the short limbs, sprawling posture, and long tail that are ancestral characters of amniotes and are still seen in extant lizards and crocodilians. Later nonmammalian synapsids, a group called therapsids, adopted a more upright posture with limbs held more underneath the trunk (although not as fully as extant mammals) (**Figure 11–2**). Limbs in this position can move fore and aft with less bending of the trunk.

A second innovation in the synapsid lineage also contributed to resolving the conflict between locomotion and respiration. Ancestrally, contraction of the trunk muscles produced the reduced pressure within the trunk that draws air into the lungs for inspiration, but this situation was modified with the development of a diaphragm in mammals, possibly as early as some derived therapsids (cynodonts).

The diaphragm is a sheet of muscle that separates the body cavity into an anterior portion (the pulmonary cavity) and a posterior portion (the abdominal cavity). The diaphragm is convex anteriorly (i.e., it bulges toward the head) when it is relaxed and flattens when it contracts. This flattening increases the volume of the pulmonary cavity, creating a negative pressure that draws air into the lungs. Simultaneous contraction of the hypaxial muscles pulls the ribs forward and outward, expanding the rib cage—you can feel this change when you take a deep breath. Relaxation of the diaphragm permits it to resume its domed shape, and relaxation of the hypaxial muscles allows elastic recoil of the rib cage. These changes raise the pressure in the pulmonary cavity, causing air to be exhaled from the lungs.

Movements of the diaphragm do not conflict with locomotion, and in fact the bounding gait of therian mammals further resolves the conflicting demands of locomotion and respiration. The inertial backward and forward movements of the viscera (especially the liver) with each bounding stride work with the diaphragm to force air into and out of the lungs. Thus, in derived mammals, respiration and locomotion work together in a synergistic fashion rather than conflicting.

Humans have little direct experience of this basic mammalian condition because our bipedal locomotion has separated locomotion and ventilation, but locomotion and respiration interact in many quadrupedal mammals, with a coupling of gait and breathing; that is, an animal inhales and exhales in synchrony with limb movements.

Locomotion and Respiration of Sauropsids

Early sauropsids were quadrupedal animals that moved with lateral undulations of the trunk, just as early synapsids did and as nearly all lizards do today. Derived sauropsids (birds and many dinosaurs) found a different solution to the problem of decoupling locomotion and respiration, however. They rely on bipedal locomotion, using only the hindlimbs without movements of the trunk.

Instead of developing a diaphragm, sauropsids appear to have incorporated pelvic movements and the ventral ribs (**gastralia**, bones in the ventral abdominal wall of some reptiles) into lung ventilation. Extant crocodilians provide a model for understanding the respiratory mechanism used by stem archosaurs. Crocodilians move with lateral undulations of the trunk, but lung ventilation is not limited by locomotor movements—quite the contrary; in fact, alligators hyperventilate during locomotion. Examination of alligators shows that they use three methods of changing the volume of the trunk to move air into and out of the lungs: movement of the ribs, movement of the liver, and rotation of the pubic bones. These movements are shown in **Figure 11–3** on page 258.

Gastralia are an ancestral character of amniotes that consist of V-shaped bony rods located in the ventral body wall with the apex of the V pointing forward. Today they are seen in only crocodilians and the tuatara (*Sphenodon*), but they were present in many dinosaurs (theropods—including the ancestors of birds—and prosauropods) as well as in pterosaurs (Mesozoic flying reptiles), plesiosaurs (Mesozoic marine reptiles), and early synapsids. Dinosaurs may have used the gastralia in combination with the ribs and with extended pelvic bones to change the volume of the thoracic cavity by a mechanism called cuirassal breathing, a term that is derived from *cuirass*—a type of armor that protects the chest (**Figure 11–4** on page 259).

Evolution of Lung Ventilation and Locomotor Stamina

The conflicting demands placed on the hypaxial trunk muscles by their dual roles in locomotion and respiration probably limited the ability of early amniotes to

(a) Pelycosaur *(Mycterosaurus)*

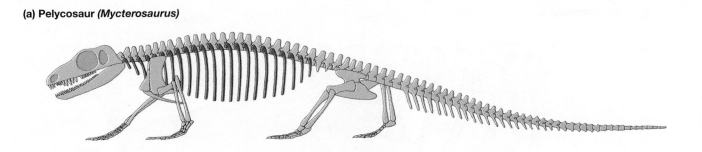

(b) Noncynodont therapsid *(Titanophoneus)*

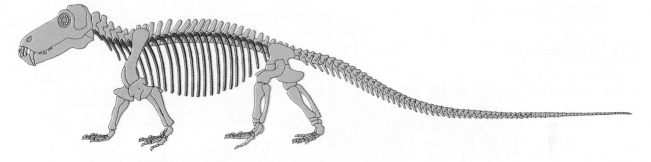

(c) Primitive cynodont therapsid *(Thrinaxodon)*

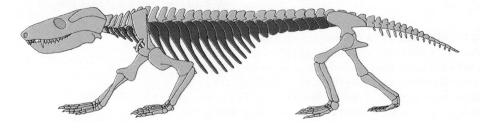

(d) Derived cynodont therapsid *(Massetognathus)*

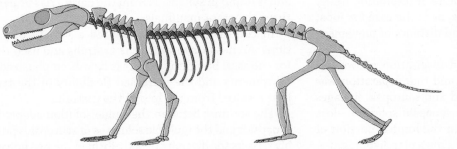

Figure 11–2 Changes in the anatomy of synapsids. Early synapsids such as *Mycterosaurus* (*top*) retained the ancestral conditions of ribs on all thoracic vertebrae, short legs, and long tails, whereas later synapsids like *Massetognathus* (*bottom*) had lost ribs from the posterior vertebrae and had longer legs and shorter tails. These changes probably coincided with the development of a diaphragm for respiration and fore-and-aft movement of the legs during locomotion.

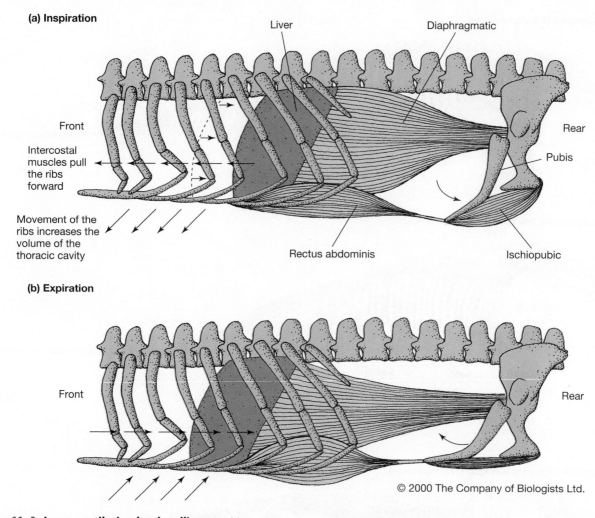

(a) Inspiration

Liver

Diaphragmatic

Front

Intercostal
muscles pull
the ribs
forward

Movement of the
ribs increases the
volume of the
thoracic cavity

Rear

Pubis

Rectus abdominis

Ischiopubic

(b) Expiration

Front

Rear

© 2000 The Company of Biologists Ltd.

Figure 11–3 **Lung ventilation by the alligator.** (a) During inspiration, contraction of the intercostal muscles (not shown) moves the ribs anteriorly, contraction of the diaphragmatic muscle pulls the liver posteriorly, and contraction of the ischiopubic muscle rotates the pubic bones ventrally, increasing the volume of the thoracic cavity. (b) During expiration, the rectus abdominis and transversus abdominis muscles rotate the pubic bones dorsally. This movement forces the viscera anteriorly as the diaphragmatic and intercostal muscles relax, reducing the volume of the thorax and forcing air out of the lungs.

occupy many of the adaptive zones that are potentially available to a terrestrial vertebrate. If respiration nearly ceases when an animal moves, as is the case for most modern lizards, both speed and distance of movement are limited.

Separating locomotion and respiration allows tetrapods to move far and fast, and that separation was achieved in both the synapsid and sauropsid lineages (**Figure 11–5** on page 260). The synapsid solution—loss of the gastralia and the ribs in the lumbar portion of the trunk and development of a muscular diaphragm—

appeared fairly early in the development of the lineage and is found in synapsids from the Early Triassic period through extant mammals.

In contrast, sauropsids devised a variety of solutions. Archosaurs retained the gastralia and used them for cuirassal breathing, whereas lizards emphasized rib movements and the increased flexibility of the trunk that resulted from the loss of the gastralia.

The contrast between the single solution adopted by synapsids and the multiple solutions of sauropsids probably reflects the diversity of body form in the two lineages.

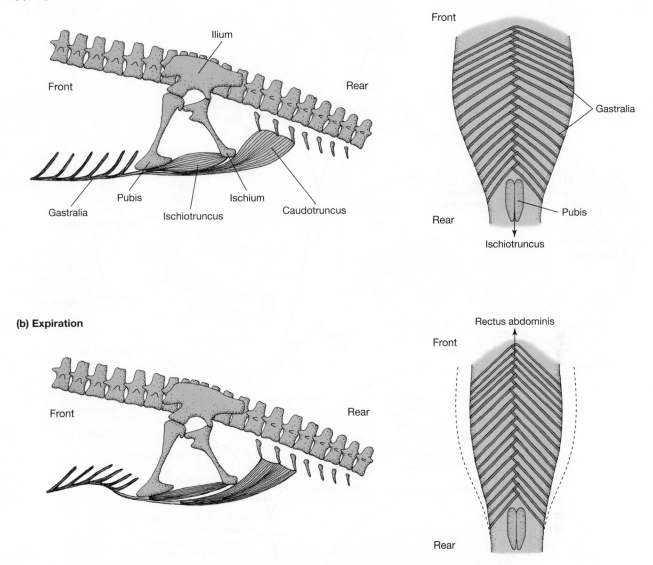

(a) Inspiration

Front

Ilium

Rear

Front

Gastralia

Pubis   Ischium

Gastralia   Ischiotruncus   Caudotruncus

Front

Gastralia

Pubis

Rear

Ischiotruncus

(b) Expiration

Front

Rear

Rectus abdominis

Front

Rear

Figure 11–4 Proposed mechanism of cuirassal breathing by a nonavian sauropsid. This reconstruction is based on the carnivorous theropod dinosaur *Allosaurus*. (a) As the dinosaur inhales, the ischiotruncus and caudotruncus muscles pull the gastralia posteriorly, pushing the body wall laterally and ventrally. The expanded area on the ventral end of the elongate pubis may have been a guide that oriented the pull of the muscles. The pubis extends anteriorly and the ischium posteriorly, and the distal ends of those bones are widely separated. As a result, the ischiotruncus muscle is long, which is mechanically significant because muscles can contract by about one-third of their resting length. (b) Expiration is accomplished by contracting the rectus abdominis muscle, which pulls the gastralia anteriorly, narrowing the V's and pulling the body wall inward and upward.

Synapsids remained quadrupedal, and terrestrial through the Mesozoic, while sauropsids became enormously diverse, with species ranging from the size of modern lizards to dinosaurs 30 meters or longer, as well as quadrupedal, bipedal, flying, and secondarily aquatic species.

11.2 Increasing Gas Exchange: The Trachea and Lungs

Ventilating the lungs by moving the ribs is an ancestral feature of amniotes, and a trachea is probably ancestral

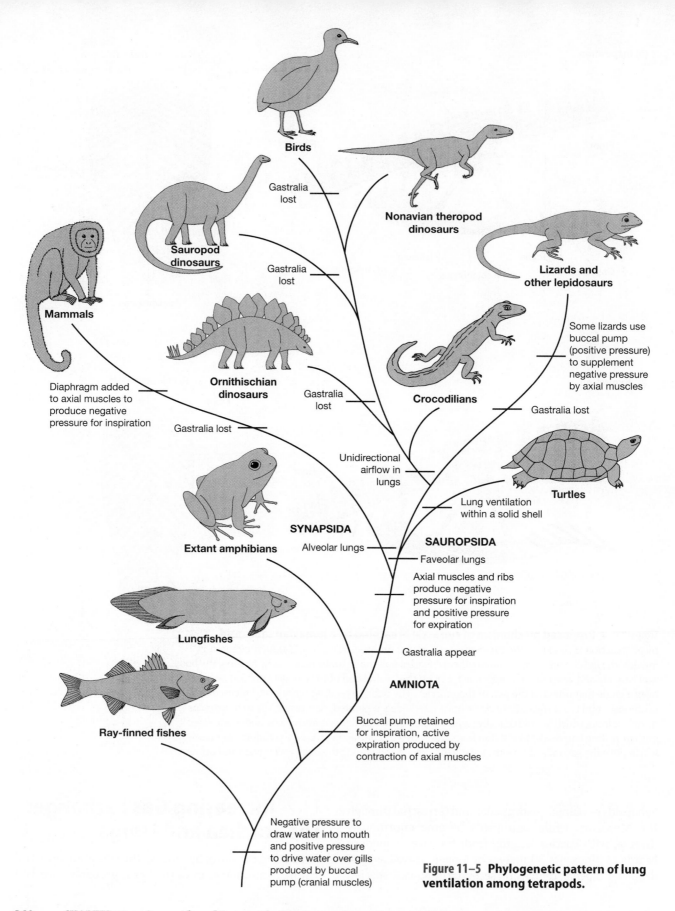

Birds

Nonavian theropod dinosaurs

Gastralia lost

Sauropod dinosaurs

Gastralia lost

Lizards and other lepidosaurs

Mammals

Ornithischian dinosaurs

Some lizards use buccal pump (positive pressure) to supplement negative pressure by axial muscles

Gastralia lost

Crocodilians

Gastralia lost

Diaphragm added to axial muscles to produce negative pressure for inspiration

Gastralia lost

Unidirectional airflow in lungs

Lung ventilation within a solid shell

Turtles

SYNAPSIDA

SAUROPSIDA

Extant amphibians

Alveolar lungs

Faveolar lungs

Axial muscles and ribs produce negative pressure for inspiration and positive pressure for expiration

Lungfishes

Gastralia appear

AMNIOTA

Buccal pump retained for inspiration, active expiration produced by contraction of axial muscles

Ray-finned fishes

Negative pressure to draw water into mouth and positive pressure to drive water over gills produced by buccal pump (cranial muscles)

Figure 11–5 Phylogenetic pattern of lung ventilation among tetrapods.

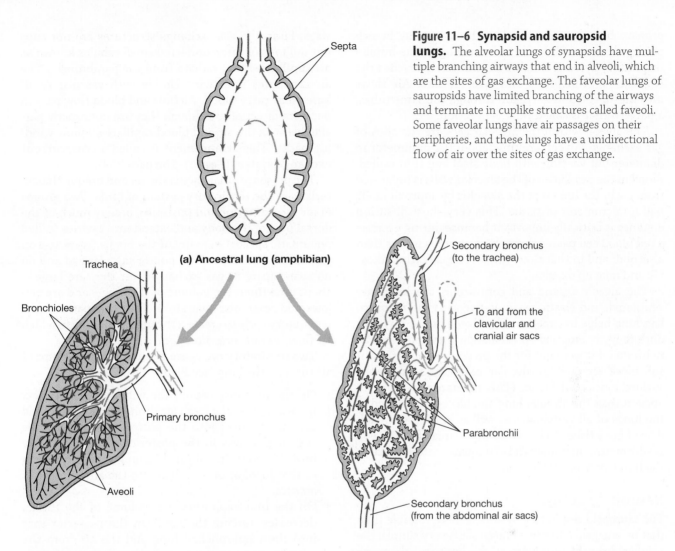

Figure 11–6 Synapsid and sauropsid lungs.
The alveolar lungs of synapsids have multiple branching airways that end in alveoli, which are the sites of gas exchange. The faveolar lungs of sauropsids have limited branching of the airways and terminate in cuplike structures called faveoli. Some faveolar lungs have air passages on their peripheries, and these lungs have a unidirectional flow of air over the sites of gas exchange.

Septa

(a) Ancestral lung (amphibian)

Trachea

Bronchioles

Primary bronchus

Aveoli

(b) Alveolar lung (mammal)

Secondary bronchus (to the trachea)

To and from the clavicular and cranial air sacs

Parabronchii

Secondary bronchus (from the abdominal air sacs)

(c) Faveolar lung (bird)

as well. The limited speed and endurance of early tetrapods minimized the importance of high rates of gas exchange. Simple lungs—basically internal sacs in which inhaled air could exchange oxygen and carbon dioxide with blood in capillaries of the lung wall—were probably sufficient for these animals.

Rates of oxygen consumption would have increased as sustained locomotion appeared, and a larger surface area would have been needed in the lungs for gas exchange. Complex lungs appeared in both the synapsid and sauropsid lineages, but the additional surface area for gas exchange and the network of air-conducting tubes are organized in very different ways (**Figure 11–6**). In synapsids the conducting airways have a treelike dichotomous pattern of branching, where the walls of the last generations of the branches contain cuplike chambers, the alveoli, that are densely populated by blood capillaries. This type of lung is called an **alveolar lung**.

Sauropsids developed a multitude of different branching patterns in the lungs, and the size and distribution of the gas-exchange units vary enormously depending on lineage. These units are cuplike chambers that line the walls of regions of the airways. They are known by different names—ediculi, faveoli, or air capillaries—depending on the diameter and depth of the cups. This type of lung is called a **faveolar lung**. The location and distribution of the ediculi or faveoli vary, and in many lineages, such as snakes, the posterior region of the lung does not contain gas-exchange tissues but acts as a storage space and a bellows.

Synapsid Lungs

The structure of the synapsid respiratory system is an elaboration of the saclike lungs of the earliest tetrapods. Air passes from the trachea through a series of progressively smaller passages—beginning with the primary bronchi and extending through many branch points; there are 23 levels of branching in the human respiratory system. Inhaled air ultimately reaches the respiratory bronchioles and alveolar sacs. Air flows into the alveoli and out again through the same tubes, a pattern that is called tidal ventilation.

The alveoli within the alveolar sacs are the sites of gas exchange. Alveoli are tiny (about 0.2 millimeter in diameter but varying with body size) and thin walled. Blood in the capillaries of the alveolar walls is separated from air in the lumen of the alveolus by approximately half a micrometer of tissue. This very short diffusion distance is critically important because during exercise a red blood cell passes through an alveolus in less than a second, and in that time it must release carbon dioxide and take up oxygen.

The alveoli expand and contract as the lungs are ventilated, and elastic recoil of alveoli in the mammalian lung helps to expel air. The alveoli are so tiny that once they are emptied on exhalation they could not be reinflated if it were not for the presence of a surfactant substance secreted by alveolar cells that reduces the surface tension of water. (This substance is far more ancient than the alveolar lung and has been detected in the lungs of all vertebrates as well as in the swim bladders of bony fishes.) The total surface area of the alveoli is enormous—in humans it is 70 square meters, equal to the floor space of a large room.

Sauropsid Lungs

The sauropsid respiratory system is more variable than that of synapsids. In some lineages it is very simple: the primary bronchi terminate at the lung and there are no air-conducting structures within the lung itself; the gas-exchange cups simply line the wall of the lungs. This morphology is characteristic of many kinds of lizards.

In other lineages of lizards, such as monitor lizards, and in some turtles, the primary bronchi continue into the lungs, where they give rise to secondary bronchi. These secondary bronchi can branch to give rise to tertiary bronchi. In birds and crocodilians, the secondary bronchi are connected to one another through this third level of tubes called the **parabronchi** (**Figure 11–7**).

Thus, the lungs of birds and crocodilians have only three levels of branching rather than the 23 levels seen in humans and other synapsids. Furthermore, air flows in the same direction during inhalation and exhalation in most of the secondary and tertiary bronchi. This unidirectional flow is produced by the alignment of the primary and secondary bronchi.

The Respiratory System of Birds The respiratory system of birds is unique among extant vertebrates in two ways. First, the gas-exchange structures are not cups but millions of interconnected small tubules known as **air capillaries** that radiate from the parabronchi. The air capillaries intertwine closely with vascular capillaries that carry blood. Airflow and blood flow pass in opposite directions, although they are not exactly parallel because the air and blood capillaries follow winding paths. This arrangement is called a **crosscurrent exchange** system (**Figure 11–8** on page 264).

The presence of air sacs is the second unique characteristic of the respiratory system of birds. Two groups of air sacs, anterior and posterior, occupy much of the dorsal part of the body and extend into cavities (called pneumatic spaces) in many of the bones (**Figure 11–9** on page 265). The air sacs are poorly vascularized and do not participate in gas exchange, but they are large—about nine times the volume of the lung—and are bellows and reservoirs that store air during parts of the respiratory cycle to create a through-flow lung in which air flows in only one direction.

Two respiratory cycles are required to move a unit of air through the lung (see Figure 11–7).

- On the first inspiration, the volume of the thorax increases, drawing fresh air through the trachea and primary bronchi into the posterior air sacs, while the air that was in the posterior set of secondary bronchi and the parabronchial lung at the beginning of that inspiration is pulled into the anterior set of air sacs.
- On the first expiration, the volume of the thorax decreases, forcing the air from the posterior sacs into the parabronchial lung and the air from the anterior sacs into the primary bronchus and then to the exterior via the trachea.
- The second inspiration draws that unit of air into the anterior air sacs, and the second expiration sends it out through the trachea.

The one-way passage of air in the bird lung may facilitate the extraction of oxygen from lung gases. The crosscurrent flow of air in the air capillaries and blood in the blood capillaries allows an efficient exchange of gases, like the countercurrent flow of blood and water in the gills of fishes. In addition, the volumes of the secondary and tertiary bronchi and the air capillaries change very little during ventilation, so the blood vessels are not stretched at each respiratory cycle. As a result, the walls of the air capillaries and blood capillaries of birds can be thinner than the walls of mammalian alveoli, reducing the distance that carbon dioxide and oxygen must diffuse. Rapid diffusion of gas between blood and air is probably one of the mechanisms that allows birds to breathe at very high altitudes.

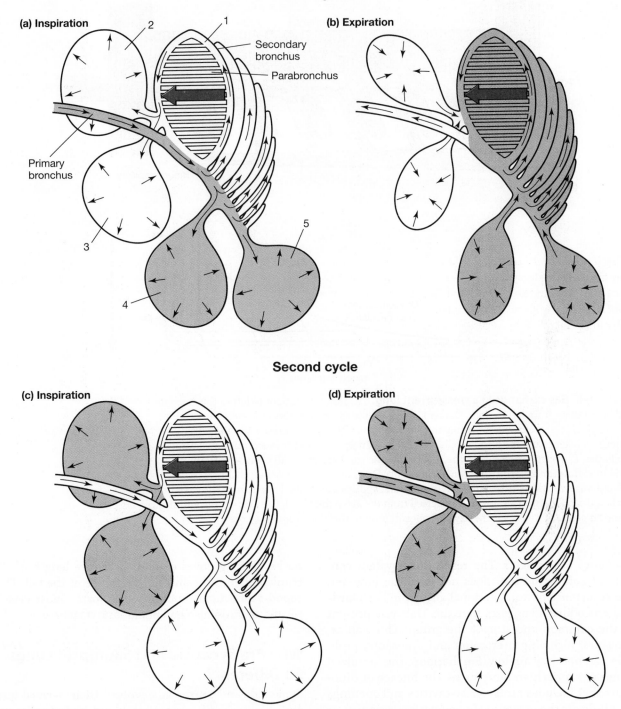

First cycle

(a) Inspiration

2

1

Secondary
bronchus

Parabronchus

Primary
bronchus

3

4

5

(b) Expiration

Second cycle

(c) Inspiration

(d) Expiration

Figure 11–7 Pattern of airflow during inspiration and expiration by a bird. Note that air
flows in only one direction through the parabronchial lung: **1.** parabronchial lung; **2.** clavicular air
sac; **3.** cranial thoracic air sac; **4.** caudal thoracic air sac; **5.** abdominal air sacs.

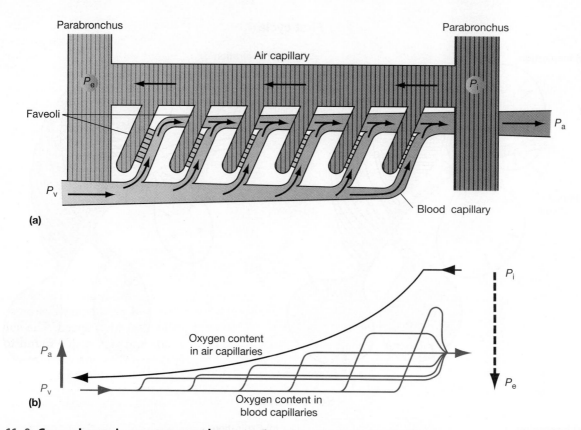

Figure 11–8 Gas exchange in a crosscurrent lung. Air flows from right to left in this diagram, and blood flows from left to right. Darker shading indicates a higher concentration of oxygen. (P_e = oxygen pressure in the air exiting the parabronchus; P_v = oxygen pressure in the mixed venous blood entering the blood capillaries; P_a = oxygen pressure in the blood leaving the blood capillaries; P_i = oxygen pressure in the air entering the parabronchus.) (a) General pattern of air and blood flow through the parabronchial lung. (b) Diagrammatic representation of crosscurrent gas exchange. Air flows from right to left in the air capillaries and blood flows from left to right in the blood capillaries. As oxygen is transferred from the air to the blood, the oxygen concentration in the air capillaries falls and the oxygen concentration in the blood capillaries rises.

The Lungs of Dinosaurs The respiratory system consists of soft tissue that does not fossilize; however, the occurrence of unidirectional airflow in both birds and crocodilians suggests that this trait was present in the common ancestor of the group. Thus, all archosaurs, including pterosaurs and dinosaurs, probably had one-way airflow. Furthermore the fossilized bones of saurischian dinosaurs—the lineage of dinosaurs that includes birds—have cavities and openings that indicate the presence of pneumatic spaces and air sacs in life.

This condition is called pneumaticity, and the most spectacular examples of pneumaticity are found among the huge, secondarily quadrupedal, long-necked sauropod dinosaurs, in which the vertebrae have grooves showing the presence of four large air sacs. In *Diplodocus* and related species, which had exceptionally long

necks, these grooves extend the entire length of the trunk and onto the anterior vertebrae of the tail. Theropod saurischian dinosaurs, the forms most closely related to birds, also had pneumatic vertebrae.

Why Are Synapsid and Sauropsid Lungs So Different?

Synapsids and sauropsids evolved their derived lung structures at different times, and Colleen Farmer suggests that differences in the atmospheric oxygen concentration at those times may account for the differences in their lungs. Synapsids were the predominant large forms in the late Paleozoic when oxygen levels were high. Mammalian lungs combine a large internal surface area with relatively long diffusion distances between blood and air in the alveoli. High

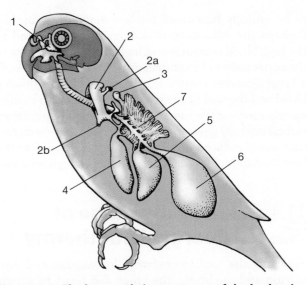

Figure 11–9 **The lung and air sac system of the budgerigar.** **1.** Infraorbital sinus; **2.** clavicular air sac; **2a.** axillary diverticulum to the humerus; **2b.** sternal diverticulum; **3.** cervical air sac; **4.** cranial thoracic air sac; **5.** caudal thoracic air sac; **6.** abdominal air sac; **7.** parabronchial lung. (Only the left side is shown.)

atmospheric oxygen levels provide the large pressure gradients that are needed when the diffusion distance is long.

By the early Mesozoic, when derived archosaurs rose to prominence, oxygen levels had fallen. Ancestral archosaurs probably had lungs like those of crocodiles, where the pumping of the heart promotes unidirectional airflow through the lung when the animal is not actively breathing. This unidirectional airflow may have facilitated partitioning of the respiratory system into air sacs and gas-exchange regions. That separation of functions permits the lungs to have a thin barrier between air and blood that did not require high atmospheric oxygen levels to drive diffusion. The lung morphology of archosaurs may have given them an advantage over synapsids as the world was repopulated following the end-Permian extinctions.

11.3 Transporting Oxygen to the Muscles: Structure of the Heart

Changes in the mechanics of lung ventilation resolved the conflict between locomotion and breathing, and internal divisions of the lungs increased the capacity for gas exchange. These features were essential steps toward occupying adaptive zones that require sustained locomotion, but another element is necessary—oxygen must be transported rapidly from the lungs to the muscles and carbon dioxide from the muscles to the lungs to sustain high levels of cellular metabolism.

A powerful heart can produce enough pressure to move blood rapidly, but there is a complication: although high blood pressure is needed in the systemic circulation to drive blood from the heart to the limbs, high blood pressure would be bad for the lungs. Lungs are delicate structures because of the very short diffusion distances between blood and air that are needed for rapid gas exchange, and high blood pressure in the lungs forces plasma out of the thin-walled capillaries into the air spaces. When these spaces are partly filled with fluid instead of air—as in pneumonia, for example—gas exchange is reduced. Thus, amniotes must maintain different blood pressures in the systemic and pulmonary systems while they are pumping blood at high speed. The solution that derived synapsids and sauropsids found to that problem was separation of the ventricle into systemic and pulmonary sides with a permanent septum. Differences in the hearts of the two lineages indicate that this solution was reached independently in each lineage and more than once in the sauropsids.

The ancestral amniote heart probably lacked a ventricular septum. The flow of blood through the ventricle was probably directed by a spongelike internal structure and perhaps by a spiral valve in a conus arteriosus, as in extant lungfishes and lissamphibians. Turtles and lizards do not have a permanent complete septum in the ventricle formed by tissue. Instead, during ventricular contraction the wall of the ventricle presses against a muscular ridge in the interior of the ventricle, keeping oxygenated and deoxygenated blood separated. This anatomy, which is undoubtedly derived and more complex than the ancestral amniote condition, plays an important functional role in the lives of turtles and lizards because it allows them to shunt blood between the systemic and pulmonary circuits in response to changing conditions.

Differences in resistance to flow in the pulmonary and systemic circuits are important in controlling the movement of blood through the hearts of turtles and lizards, and their blood pressures and rates of blood flow are low compared to those of birds and mammals. It may be that higher blood pressures and higher rates of blood flow made a permanent division necessary for derived synapsids (mammals) and derived sauropsids (crocodilians and birds) **(Figure 11–10** on page 268). However, a corollary of this apparent "improvement" in heart structure is that each heartbeat must now send the same volume of blood to the lungs of mammals and birds as it does to the body.

Because the volume of blood in the pulmonary circuit is much smaller than the volume in the systemic circuit, this restriction may limit blood flow to the body. Additionally, blood can no longer be shunted from the (oxygenated) left ventricle to the (deoxygenated) right ventricle, and ventricular coronary vessels must be developed to oxygenate the heart muscle. The muscles in the right ventricle receive oxygenated blood via the **coronary arteries**, which branch off from the aorta. (These are the vessels in which blockage causes a heart attack.) All amniotes have some blood supply to the heart, but an extensive ventricular coronary system has been evolved independently in mammals and derived sauropsids (archosaurs).

When the ventricle is permanently divided, the relative resistance to blood flow in the systemic and pulmonary circuits no longer determines where blood goes when it leaves the ventricle; instead, blood can flow only into the vessels that exit from each side of the ventricle: the right ventricle leads to the pulmonary circuit and the left ventricle to the systemic circuit. Synapsids and sauropsids have both reached this stage, but they must have done it independently because the relationship of the systemic arches to the left ventricle differs in the two lineages. Mammals retain the left systemic arch as the primary route for blood flow from the left ventricle, whereas birds retain the right arch. Portions of the old right systemic arch remain in adult mammals as the right brachiocephalic artery that gives rise to the right carotid artery (or to both carotids in some mammals) and the right subclavian artery. This situation contrasts with the usual sauropsid condition, in which the carotids branch from the right systemic arch.

While most sauropsids retain both aortic arches, in birds the left arch is lost entirely. In some nonavian sauropsids, notably turtles, crocodilians, and varanid lizards, the left systemic arch assumes a new function. In these animals the celiac artery, which is the major artery supplying the stomach and anterior intestines, branches off from the base of the left systemic arch, rather than from the dorsal aorta as in other vertebrates. This arrangement routes acidic, deoxygenated blood into the gut, perhaps aiding digestion.

Why have birds and mammals each lost one of the systemic arches (or, in the case of mammals, the bottom part of the right systemic arch)? Developmental studies show that mammalian and avian embryos both start off with two arches that are subsequently reduced to a single one. The independent reduction to a single arch in each lineage suggests that one arch is somehow better than two arches, although two arches appear to be entirely functional for less derived sauropsids. Perhaps the advantage of a single arch is related to the high blood pressures and high rates of blood flow in the aortic arches of mammals and birds. One vessel with a large diameter creates less friction between flowing blood and the wall of the vessel than would two smaller vessels carrying the same volume of blood. In addition, turbulence may develop where the two arches meet, and that would reduce flow. Thus, a single arch may be the best conduit for blood leaving the heart at high pressure.

11.4 Taking Advantage of Wasted Energy: Endothermy

Resolving the conflicts between locomotion and ventilation and modifying the lungs and heart to supply oxygen to muscles did more than just increase the endurance of synapsids and sauropsids—it produced a lot of heat. The synthesis and consumption of chemical energy in compounds like ATP are not very efficient, and a substantial amount of energy is lost as heat. This is why you get hot when you exercise vigorously, and the increase in body temperature during exercise can be substantial; it is overheating rather than exhaustion that forces a cheetah to end its pursuit of a gazelle within a minute of starting its sprint.

The body forms of derived synapsids and derived theropod dinosaurs clearly indicate that increasing activity was developing in both lineages. Synapsids appear to have reorganized their hindlimb musculature in association with the new limb position more extensively than birds and dinosaurs did. The increasing locomotor activity and endurance suggested by the changes in body forms would have been accompanied by increased metabolic rates that would have generated substantial amounts of heat during activity, and that heat could have been a critical step toward endothermy.

A Paradox in the Evolution of Endothermy

Endothermy and ectothermy are both effective methods of temperature regulation, but there is a barrier to the evolutionary transition from ectothermy (which is the ancestral condition for amniotes) to endothermy. Endothermy requires two characteristics that ectotherms lack—a high metabolic rate and insulation that retains heat in the body. The difficulty in moving from ectothermy to endothermy is that neither a high metabolic rate nor insulation *by itself* is sufficient for endothermy. *Both* characters must be present for an animal to maintain a high and stable body temperature.

Figure 11–10 (pages 268–269) Diagrammatic view of the heart and aortic arches in synapsids and sauropsids. (a) Early amniote condition. A conus arteriosus with a spiral valve and a truncus arteriosus are retained, and a ventricular septum is lacking. This condition, basically like that of living amphibians, is proposed here to account for the differences in these structures between synapsids and sauropsids. (b) Hypothetical early synapsid condition. Mammal ancestors cannot have had the sauropsid pattern of dual systemic arches, as it would be impossible with the sauropsid condition to retain the left arch only, as seen in mammals. Here a separation of the truncus arteriosus into separate pulmonary and (single) systemic trunks is proposed. Some degree of shunting between pulmonary and systemic circuits and within the heart may have been possible with an incomplete ventricular septum. The sinus venosus is shown as retained because a small sinus venosus is still present in monotremes. (c) Mammal (therian). The ventricular septum is complete, and the lower portion of the right systemic arch has been lost. (d) Generalized sauropsid condition, similar to that seen in turtles and lepidosaurs. The truncus arteriosus is divided into three parts: a pulmonary arch and two separate systemic arches (the left arch exits from the right side of the heart, and the right arch exits from the left side). The ventricular septum is incomplete, although the ventricle may be complexly subdivided. This system allows blood to be shunted within the heart and between pulmonary and systemic circuits. (e) Crocodile. The ventricular septum (possibly a separate development in archosaurs) is now complete, but shunting between the left and right systemic arches is still possible via the foramen of Panizza. (f) Bird. The entire left systemic arch is now lost, and the sinus venosus has been subsumed into the right atrium. (R = right; L = left.)

A small endotherm, such as a mouse or a sparrow, has a metabolic rate that is about 10 times that of an ectotherm of the same body size, such as a lizard. The heat produced by metabolism in the endotherm is retained by insulation (hair for the mouse, feathers for the bird, and subcutaneous fat in both), and metabolic heat raises body temperature. A lizard lacks insulation, so metabolic heat would be lost, but heat from sunlight is rapidly absorbed. Adding a layer of insulation to the lizard does not allow it to be endothermal because it still lacks a high metabolic rate, but it does prevent the lizard from absorbing heat from sunlight. Raymond Cowles demonstrated that fact in 1958 when he made small fur coats for lizards and measured their rates of warming and cooling. The potential benefit of a fur coat for a lizard is, of course, that it will keep the lizard warm as the environment cools off. However, the well-dressed lizards in Cowles's experiments never realized that benefit because when they were wearing fur coats they were unable to get warm in the first place.

So the evolution of endothermy involves a catch-22—insulation provides no advantage to an animal without a high metabolic rate, and the heat produced by a high metabolic rate is lost unless the animal has insulation. This paradox confounded discussions of the evolution of endothermy for decades as authors offered scenarios in which incipient insulation could initially have functioned in ectothermal thermoregulation before it became effective in retaining metabolic heat.

These proposals were not very convincing and were not generally accepted.

The relationship of locomotor activity to the evolution of endothermy was not appreciated until the late twentieth century. Now it seems clear that endothermy evolved in a stepwise process in which the appearance of one new feature created conditions in which another new feature could be advantageous; the complication is deciphering the sequence in which these changes occurred.

Endothermy in Synapsids

The evolution of a high metabolic rate via increased capacity for locomotor activity involved changes in many parts of the body of synapsids. Changes in the skeleton can be seen in fossils, but changes in the soft tissues and in physiological characteristics are not fossilized. These changes must be inferred from the parts of an animal that do fossilize. **Table 11–1** on page 270 lists several characteristics of derived amniotes that are associated with the evolution of endothermy.

The changes in locomotion and respiration that we have described set the stage for additional demands on the physiology of synapsids, and one of the most important is associated with high rates of respiration. The air that an endothermal animal inhales is normally cooler than the temperature deep inside the animal's body, and it is not fully saturated with water vapor. Lungs are delicate tissues that dry out if they are ventilated with air that is not saturated with water vapor at the core

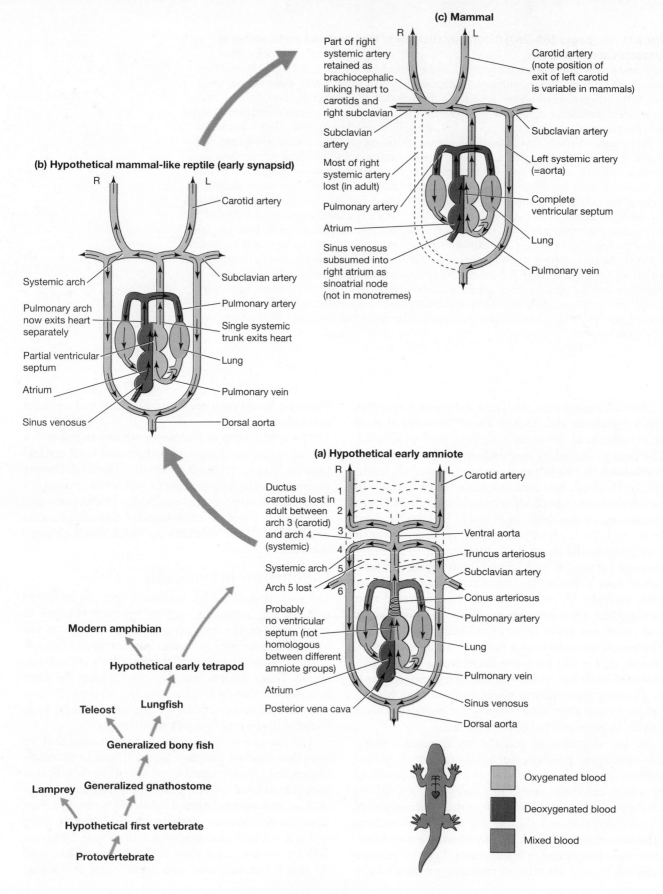

(c) Mammal

R L

Part of right systemic artery retained as brachiocephalic linking heart to carotids and right subclavian

Subclavian artery

Most of right systemic artery lost (in adult)

Pulmonary artery

Atrium

Sinus venosus subsumed into right atrium as sinoatrial node (not in monotremes)

Carotid artery (note position of exit of left carotid is variable in mammals)

Subclavian artery

Left systemic artery (=aorta)

Complete ventricular septum

Lung

Pulmonary vein

(b) Hypothetical mammal-like reptile (early synapsid)

R L

Carotid artery

Systemic arch

Subclavian artery

Pulmonary arch now exits heart separately

Pulmonary artery

Single systemic trunk exits heart

Partial ventricular septum

Lung

Atrium

Pulmonary vein

Sinus venosus

Dorsal aorta

(a) Hypothetical early amniote

R L

Carotid artery

Ductus carotidus lost in adult between arch 3 (carotid) and arch 4 (systemic)

1
2
3
4
5
6

Ventral aorta

Truncus arteriosus

Subclavian artery

Conus arteriosus

Pulmonary artery

Lung

Pulmonary vein

Sinus venosus

Dorsal aorta

Systemic arch

Arch 5 lost

Probably no ventricular septum (not homologous between different amniote groups)

Atrium

Posterior vena cava

Modern amphibian

Hypothetical early tetrapod

Teleost

Lungfish

Generalized bony fish

Lamprey

Generalized gnathostome

Hypothetical first vertebrate

Protovertebrate

Oxygenated blood

Deoxygenated blood

Mixed blood

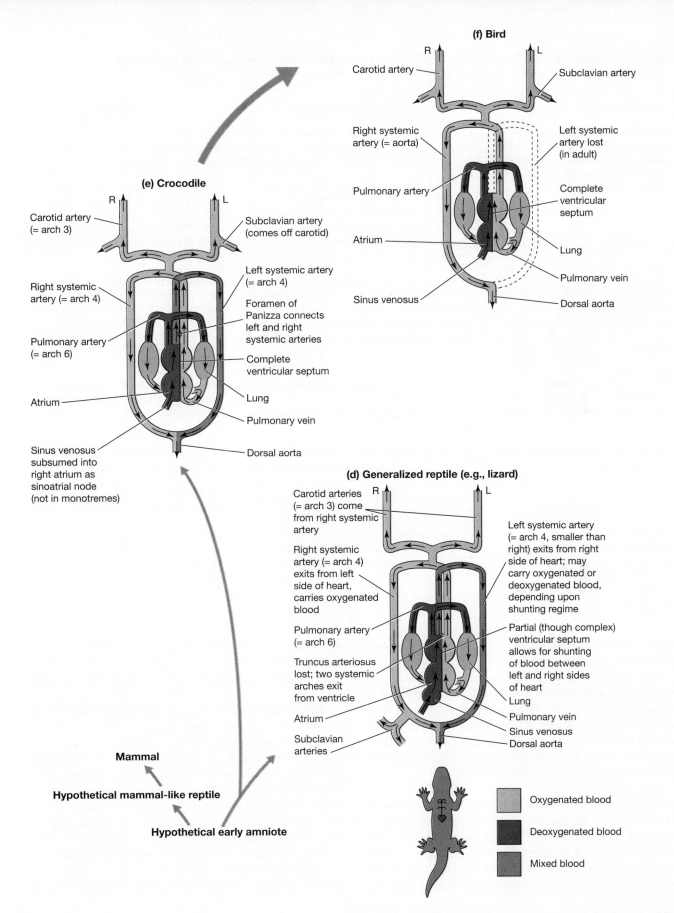

(f) Bird

Carotid artery
Subclavian artery
Right systemic artery (= aorta)
Left systemic artery lost (in adult)
Pulmonary artery
Complete ventricular septum
Atrium
Lung
Pulmonary vein
Sinus venosus
Dorsal aorta

(e) Crocodile

Carotid artery (= arch 3)
Subclavian artery (comes off carotid)
Right systemic artery (= arch 4)
Left systemic artery (= arch 4)
Foramen of Panizza connects left and right systemic arteries
Pulmonary artery (= arch 6)
Complete ventricular septum
Atrium
Lung
Pulmonary vein
Sinus venosus subsumed into right atrium as sinoatrial node (not in monotremes)
Dorsal aorta

(d) Generalized reptile (e.g., lizard)

Carotid arteries (= arch 3) come from right systemic artery
Left systemic artery (= arch 4, smaller than right) exits from right side of heart; may carry oxygenated or deoxygenated blood, depending upon shunting regime
Right systemic artery (= arch 4) exits from left side of heart, carries oxygenated blood
Pulmonary artery (= arch 6)
Partial (though complex) ventricular septum allows for shunting of blood between left and right sides of heart
Truncus arteriosus lost; two systemic arches exit from ventricle
Lung
Atrium
Pulmonary vein
Subclavian arteries
Sinus venosus
Dorsal aorta

Mammal

Hypothetical mammal-like reptile

Hypothetical early amniote

Oxygenated blood
Deoxygenated blood
Mixed blood

Table 11–1 Changes associated with the development of endothermy in synapsids and sauropsids

Physiological Issue	Anatomical Correlates	
	Synapsids (mammals)	Sauropsids (birds)
Need to resolve conflict between locomotion and lung ventilation that results from the ancestral tetrapod method of locomotion with axial flexion of the trunk	Derived members of both lineages adopted upright posture so the trunk does not bend laterally as the limbs move.	
	Ancestral quadrupedal posture was retained with derived changes in the hindlimb muscles (e.g., gluteal muscles rather than caudofemoral muscles are used to retract the hindlimb).	Derived upright posture developed in correlation with bipedality, and the ancestral pattern of hindlimb muscles was retained.
	Changes in the limbs are visible in the fossil record.	Changes in the limbs are visible in the fossil record.
Need for more oxygen to support the high metabolic rates associated with sustained locomotion	Development of the diaphragm to aid in lung ventilation can be inferred in fossils from loss of the lumbar ribs.	Development of the flow-through lung with one-way passage of air can be inferred in fossils from cavities and openings in bones that reveal the presence of air sacs in some dinosaurs and perhaps from their very long necks.
	A secondary palate developed for eating and breathing at the same time is preserved in fossils.	
Need to warm and humidify large volumes of air on inspiration and recover water and heat on expiration	Turbinate bones in the nasal passages provide a large, moist surface.	Narrow nasal passages suggest that cartilaginous turbinates were a late development.
	Turbinates are bony, and traces can be seen in fossils.	
Need for more food to fuel high rates of metabolism	Complex teeth with precise occlusion to reduce the particle size of food are preserved in fossils.	A muscular gizzard used to reduce the particle size of food is not preserved in fossils, but gizzard stones have been found in association with the fossils of dinosaurs.
	The volume of the jaw musculature increased to masticate food. The presence of large muscles can be inferred from skull features of fossils.	
Need to retain heat produced by metabolism within the body	Hair and perhaps subcutaneous fat deposits (blubber) developed. No fossils yet from pre-Cretaceous sediments show such fine detail.	Feathers are visible in fossils preserved in very fine-grained sediments.

body temperature, so inhaled air must be warmed and humidified before it reaches the lungs.

Furthermore, when the air has been saturated with water vapor and warmed to the core body temperature, exhaling the air would entail a loss of water and heat that the animal cannot afford. Thus, animals need a way to warm and humidify the air they inhale and then to recover water and heat when they exhale—that is, a recycling mechanism for water and heat.

Recycling Heat and Water Recycling heat and water is not difficult for animals with low rates of metabolism and lung ventilation. Extant lizards do this, and the moist walls of their tubular nasal passages provide enough surface area to meet their needs. As metabolic rate increases, however, the rate of ventilation increases

and places greater demands on the recycling system. An animal that inhales and exhales air rapidly needs a larger surface area than just the walls of tubular nasal passages.

Additional surface area for recycling heat and water is provided in extant mammals by an array of thin sheets of bone or cartilage in the nasal passages that are covered in life with moist tissue (**Figure 11–11**). These are the turbinates, which are also called conchae because they form spiral curves that look like the interiors of some shells. Mammals have two kinds of turbinates: olfactory and respiratory. The olfactory turbinates support the olfactory epithelium that contains the sensory cells used for olfaction. The olfactory turbinates are located above and behind the nasal passages, out of the direct flow of air. (This is why you sniff when you are trying

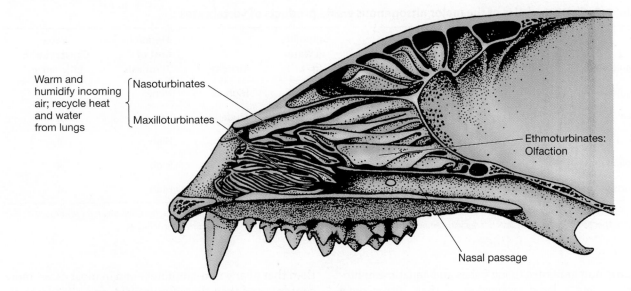

Warm and humidify incoming air; recycle heat and water from lungs { Nasoturbinates

Maxilloturbinates

Ethmoturbinates: Olfaction

Nasal passage

Figure 11–11 Longitudinal section through the snout of a raccoon.

to smell something—the abrupt inhalation draws air over the olfactory turbinates.)

The respiratory turbinates protrude directly into the main pathway of respiratory airflow, and air passes over them with each inspiration and expiration. Inhaled air is warmed and humidified, and the turbinates are cooled by the combination of cool outside air and the cooling effect of water evaporating from their surface. Then the warm, moist air leaving the lungs is cooled as it passes back over the turbinates, and some of the water vapor in the air condenses. The warm air and condensation of water rewarm the turbinates, preparing them for the next inhalation. Thus, the turbinates recycle both water and heat as air flows into and out of the lungs. A recycling system of this sort is probably essential because without it the loss of heat and water would be too large for an endotherm to sustain. Respiratory turbinates are found in all extant mammals, and traces of the ridges that support them can be seen in the nasal passages of derived therapsids and early mammals.

Endothermy in Sauropsids

Sauropsids faced the same challenges to the evolution of endothermy as synapsids, and they found similar solutions, though with some differences in detail. The capacity to sustain locomotor activity was probably the starting point for the evolution of endothermy by sauropsids, just as it was for synapsids. The body forms of dromeosaurs (derived theropod dinosaurs that are close to the transition to birds) strongly suggest that they were fleet-footed predators that pursued their prey. The faveolar lung is effective in gas exchange, and derived dromeosaurs probably had a system of air sacs and flow-through ventilation of the lungs.

Extant birds have turbinates that are at least as effective as those of mammals in recycling water and heat, but they often remain cartilaginous and do not have bony contacts with the nasal passages that could be seen in fossils. Turbinates have not been identified in dinosaurs, and computed tomography (CT) scans of the fossilized skulls of theropod dinosaurs and basal birds suggest that their nasal passages were narrow. Thus, pre-avian sauropsids and early birds probably did not have turbinates, and we cannot yet identify the time at which the turbinates seen in extant birds appeared.

11.5 Getting Rid of Wastes: The Kidneys and Bladder

High metabolic rates are beneficial for locomotor endurance and thermoregulation, but they require high rates of food intake and that means high rates of nitrogenous waste production. Nitrogenous wastes are excreted as urine, but urine is mostly water and for a terrestrial animal water is too valuable to waste. The challenge is to excrete nitrogen while retaining water, and synapsids and sauropsids found different ways to do this.

Metabolism of protein produces ammonia, NH_3. Ammonia is quite toxic, but it is very soluble in water and diffuses rapidly because it is a small molecule (**Table 11–2**).

Table 11–2 Characteristics of the major nitrogenous waste products of vertebrates

Compound	Chemical Formula	Molecular Weight	Solubility in Water $(g \cdot L^{-1})$	Toxicity	Metabolic Cost of Synthesis	Water Conservation Efficiency*
Ammonia	NH_3	17	890	High	None	1
Urea	$CO(NH_2)_2$	60	1190	Moderate	Low	2
Uric acid	$C_5H_4O_3N_4$	168	0.065	Low	High	4
Sodium urate	$C_5H_4O_3N_4Na_2$	212	0.88	Low	High	4
Potassium urate	$C_5H_4O_3N_4K_2$	244	2.32	Low	High	4

*The efficiency of water conservation is expressed as the number of nitrogen (N) atoms per osmotically active particle; higher ratios mean more nitrogen is excreted per osmotic unit. Solubility changes with temperature; these values refer to the normal body temperatures of vertebrates.

Aquatic non-amniotes (bony fishes and aquatic amphibians) excrete a large proportion of their nitrogenous waste as ammonia, and ammonia is a nitrogenous waste product of terrestrial amniotes as well—human urine and sweat contain small amounts of ammonia.

Ammonia can be converted to urea, $CO(NH_2)_2$, which is less toxic and even more soluble than ammonia. Because it is both soluble and relatively nontoxic, urea can be accumulated within the body and released in a concentrated solution in urine, thereby conserving water. Urea synthesis is an ancestral character of amniotes and probably of all gnathostomes. Synapsids retained the ancestral pattern of excreting urea and developed a kidney that is extraordinarily effective in producing concentrated urine.

A complex metabolic pathway converts several nitrogen-containing compounds into uric acid, $C_5H_4O_3N_4$. Unlike ammonia and urea, uric acid is insoluble, and it readily combines with sodium ions and potassium ions to precipitate as a salt of sodium or potassium urate. Sauropsids developed the capacity to synthesize and excrete uric acid and recover the water that is released when it precipitates.

Nitrogen Excretion by Synapsids: The Mammalian Kidney

The mammalian kidney is a highly derived organ composed of millions of nephrons, the basic units of kidney structure that are recognizable in nearly all vertebrates (Figure 11–12 on page 274). Each nephron consists of a glomerulus that filters the blood and a long tube in which the chemical composition of the filtrate is altered. A portion of this tube, the loop of Henle, is a derived character of mammals that is largely responsible for their ability to produce concentrated urine. The mammalian kidney can produce urine more concentrated

than that of any non-amniote—and in most cases, more concentrated than that of sauropsids as well (Table 11–3).

The Formation of Urine Understanding how the mammalian kidney works is important for understanding how mammals can thrive in places that are seasonally or chronically short of water. Urine is concentrated by removing water from the ultrafiltrate that is produced in the glomerulus when water and small molecules are forced out of the capillaries. Because cells are unable to transport water directly, they manipulate the movement of water molecules by transporting ions to create osmotic gradients. In addition, the cells lining the nephron actively reabsorb substances important to the body's economy from the ultrafiltrate and secrete toxic substances into it.

The nephron's activity is a six-step process, with each step localized in a region that has special cell characteristics and distinctive variations in the osmotic environment. The first step is production of an ultrafiltrate at the glomerulus. The ultrafiltrate is isosmolal with blood plasma and resembles whole blood after the removal of (1) cellular elements, (2) substances with a molecular weight of 70,000 or higher (primarily proteins), and (3) substances with molecular weights between 15,000 and 70,000, depending on the shapes of the molecules. Humans produce about 120 milliliters of ultrafiltrate per minute—that is 170 liters (45 gallons) of glomerular filtrate per day! We excrete only about 1.5 liters of urine because the kidney reabsorbs more than 99 percent of the ultrafiltrate produced by the glomerulus.

The second step in the production of the urine is the action of the proximal convoluted tubule (PCT) in decreasing the volume of the ultrafiltrate. The PCT cells have greatly enlarged surface areas that actively transport sodium ions, glucose, and amino acids from

Table 11–3 Maximum urine concentrations of some synapsids and sauropsids

Species	Maximum Observed Urine Concentration (mosm · kg⁻¹)	Approximate Urine: Plasma Concentration Ratio
Synapsids		
Human (*Homo sapiens*)	1430	4
Bottlenose porpoise (*Tursiops truncatus*)	2658	7.5
Hill kangaroo (*Macropus robustus*)	2730	7.5
Camel (*Camelus dromedarius*)	2800	8
Laboratory rat (*Rattus norvegicus*)	2900	8.9
Marsupial mouse (*Dasycercus cristicauda*)	3231	10
Cat (*Felis domesticus*)	3250	9.9
Desert woodrat (*Neotoma lepida*)	4250	12
Vampire bat (*Desmodus rotundus*)	6250	20
Kangaroo rat (*Dipodomys merriami*)	6382	18
Australian hopping mouse (*Notomys alexis*)	9370	22
Sauropsids		
American alligator (*Alligator mississippiensis*)	312	0.95
Desert iguana (*Dipsosaurus dorsalis*)	300	0.95
Desert tortoise (*Gopherus agassizii*)	622	1.8
Pelican (*Pelecanus erythrorhynchos*)	700	2
House sparrow (*Passer domesticus*)	826	2.4
House finch (*Carpodacus mexicanus*)	850	2.4
Savannah sparrow (*Passerculus sandwichensis*)	2000	5.8

the lumen of the tubules to the exterior of the nephron. Chloride and water move passively through the PCT cells in response to the removal of sodium ions. By this process, about two-thirds of the salt is reabsorbed in the PCT, and the volume of the ultrafiltrate is reduced by the same amount. Although the urine is still very nearly isosmolal with blood at this stage, the substances contributing to the osmolality of the urine after it has passed through the PCT are at different concentrations than in the blood.

The next alteration occurs in the descending limb of the loop of Henle. The thin, smooth-surfaced cells freely permit diffusion of water. Because the descending limb of the loop passes through tissues of increasing osmolality as it plunges deeper into the kidney, water is lost from the urine and it becomes more concentrated. In humans the osmolality of the fluid in the descending limb may reach 1200 mmol · kg⁻¹, and other mammals can achieve considerably higher concentrations.

The fourth step takes place in the thick ascending limb of the loop of Henle, which has cells with numerous large, densely packed mitochondria. The ATP produced by these organelles is used to remove sodium ions from the forming urine. Because these cells are impermeable to water, the volume of urine does not decrease as sodium ions are removed, and because the sodium ions were removed without loss of water, the urine is hyposmolal to the body fluids as it enters the next segment of the nephron. Although this sodium ion-pumping, water-impermeable, ascending limb does not concentrate or reduce the volume of the forming urine, it contributes to setting the stage for these important processes.

The very last portion of the nephron changes in physiological character, but the cells closely resemble those of the ascending loop of Henle. This region, the terminal portion of the distal convoluted tubule (DCT), is permeable to water. The osmolality surrounding the DCT is the same as that of the body fluids, and water in the entering hyposmolal fluid flows outward and equilibrates osmotically. This process reduces the fluid volume to as little as 5 percent of the original ultrafiltrate.

The final touch in the formation of a small volume of highly concentrated mammalian urine occurs in the collecting ducts. Like the descending limb of the loop of Henle, the collecting ducts course through tissues of increasing osmolality, which withdraw water from the urine. The significant phenomenon associated with the collecting duct and with the terminal portion of the DCT is variable permeability to water. During excess fluid intake, the collecting duct demonstrates low water permeability: only half of the water entering it may be reabsorbed and the remainder excreted. In this way, copious dilute urine can be produced.

Antidiuretic hormone (ADH, also known as vasopressin) is released into the circulation in response to an increase in blood osmolality or when blood volume

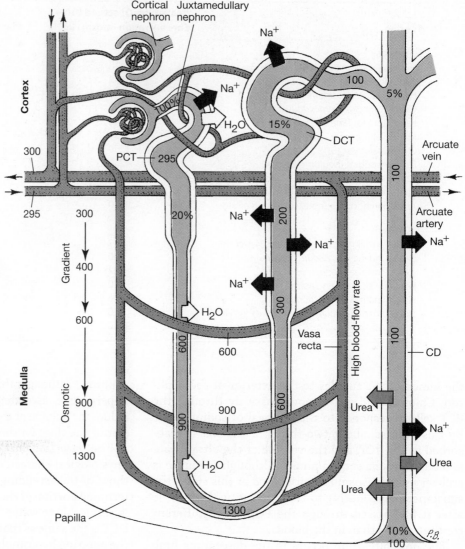

(a) BODY HYDRATED—ADH ABSENT—COPIOUS DILUTE URINE

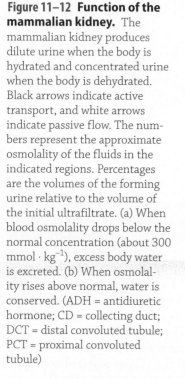

Figure 11–12 Function of the mammalian kidney. The mammalian kidney produces dilute urine when the body is hydrated and concentrated urine when the body is dehydrated. Black arrows indicate active transport, and white arrows indicate passive flow. The numbers represent the approximate osmolality of the fluids in the indicated regions. Percentages are the volumes of the forming urine relative to the volume of the initial ultrafiltrate. (a) When blood osmolality drops below the normal concentration (about 300 mmol · kg^{-1}), excess body water is excreted. (b) When osmolality rises above normal, water is conserved. (ADH = antidiuretic hormone; CD = collecting duct; DCT = distal convoluted tubule; PCT = proximal convoluted tubule)

drops. ADH is a polypeptide that is produced in the hypothalamus and stored in the posterior pituitary. When ADH is carried by the circulation to the cells of the terminal portion of the DCT and the collecting duct, it causes them to insert proteins called aquaporins into the plasma membranes of the cells lining the collecting duct.

Aquaporins are tubular proteins with non-polar (hydrophobic) amino acid residues on the outside and polar (hydrophilic) residues on the inside. When they are inserted into the plasma membranes of the cells, the hydrophobic residues on the outer surface anchor them in the phospholipid bilayer of the membrane and the hydrophilic residues on the inner surface form a channel through which water molecules can flow. When aqua-

porins are inserted into plasma membrane, the cells of the terminal portion of the DCT and the collecting duct become permeable to water, which flows outward following its osmotic gradient.

The water that is reabsorbed from the DCT and collecting duct enters the blood, reducing its osmolality and increasing its volume. As water is removed, the volume of the urine in the collecting duct decreases and its concentration increases. The final urine volume leaving the collecting duct may be less than 1 percent of the original ultrafiltrate volume. In some desert rodents, so little water remains that the urine crystallizes almost immediately upon urination. Alcohol inhibits the release of ADH, inducing a copious urine flow, and this can result in dehydrated misery the morning after a drinking binge.

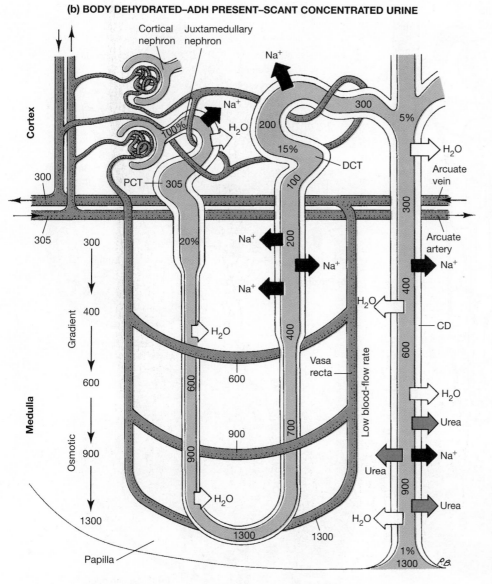

(b) BODY DEHYDRATED–ADH PRESENT–SCANT CONCENTRATED URINE

Figure 11–12 *(Continued)*

By the time the filtrate enters the collecting duct, its volume has been reduced to about 5 percent of the initial volume of the filtrate from Bowman's capsule, but up to this point the walls of the nephron have been impermeable to urea. As a result, the concentration of urea in the fluid that enters the collecting duct is very high. The walls of the terminal portion of the collecting duct are very permeable to urea, and urea diffuses outward from the collecting duct into the extracellular fluid deep in the kidney. This urea contributes to the high osmotic concentration of the tissues at the lower end of the loop of Henle.

The Structure of the Nephron The structural characteristics of the cells that form the walls of the nephron are directly related to the processes that take place in that portion of the nephron (**Figure 11–13**). The cells of the PCT contain many mitochondria and have an enormous surface area produced by long, closely spaced microvilli. These structural features reflect the function of the PCT in actively moving sodium ions from the lumen of the tubule to the peritubular space and capillaries. Chloride follows the electric charge gradient created by the movement of sodium ions, and water moves in response to the concentration gradient produced by the movement of those two ions.

Farther along the nephron, the cells of the thin segment of the loop of Henle are waferlike and contain fewer mitochondria. The descending limb of the loop of Henle permits passive flow of water. Cells in the thick

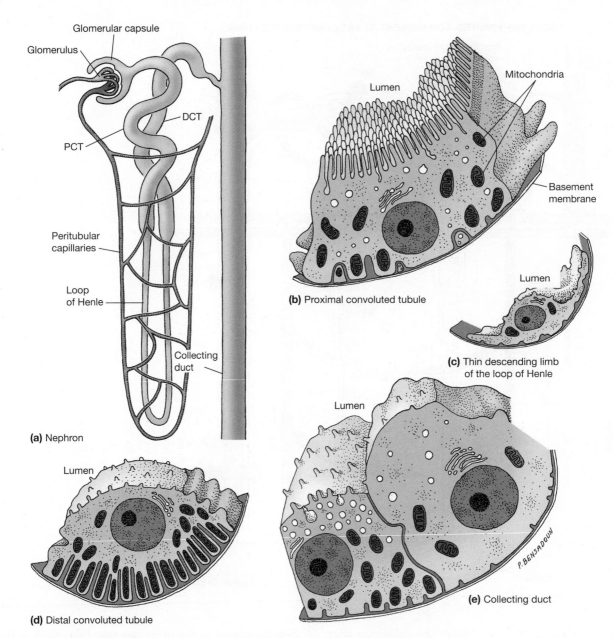

Figure 11–13 The mammalian nephron. (a) Structure of the nephron. (b–e) Fine structure of the cells lining the walls of the nephron. (PCT = proximal convoluted tubule; DCT = distal convoluted tubule.)

ascending limb are similar to those in the PCT and actively remove sodium ions from the ultrafiltrate.

Finally, cells of the collecting duct appear to be of two kinds. Most seem to be suited to the relatively impermeable state characteristic of periods of sufficient body water. Other cells are rich in mitochondria and have a larger surface area. They are probably the cells that respond to the presence of ADH from the pituitary gland, triggered by insufficient body fluid. Under the influence of ADH, the collecting duct becomes permeable to water, which flows from the lumen of the duct into the concentrated peritubular fluids.

The Loop of Henle The key to concentrated urine production is the passage of the loops of Henle and collecting ducts through tissues with increasing osmolality. These osmotic gradients are formed and maintained within the mammalian kidney as a result of its structure (**Figure 11–14**).

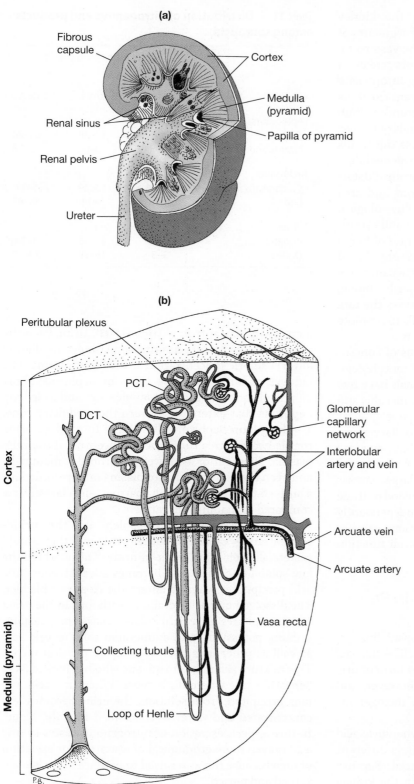

(a)

Fibrous capsule

Cortex

Medulla (pyramid)

Papilla of pyramid

Renal sinus

Renal pelvis

Ureter

(b)

Peritubular plexus

PCT

DCT

Glomerular capillary network

Interlobular artery and vein

Cortex

Arcuate vein

Arcuate artery

Medulla (pyramid)

Vasa recta

Collecting tubule

Loop of Henle

P.B.

Figure 11–14 Gross morphology of the mammalian kidney. (a) Structural divisions of the kidney and proximal end of the ureter. (b) Enlarged diagram of a section extending from the outer cortical surface of the apex of a renal pyramid, the renal papilla. (DCT = distal convoluted tubule; PCT = proximal convoluted tubule.)

The structural arrangements within the kidney medulla of the descending and ascending segments of the loop of Henle and its blood supply, the vasa recta, are especially important. These elements create a series of parallel tubes with flow passing in opposite directions in adjacent vessels (countercurrent flow). As a result, sodium ions pumped from the ascending limb of the loop of Henle diffuse into the medullary tissues to increase their osmolality, and this excess salt forms part of the steep osmotic gradient within the medulla.

The final concentration of a mammal's urine is determined by the concentration of sodium ions and urea accumulated in the fluids of the medulla. Physiological alterations in the concentration in the medulla result primarily from the effect of ADH on the rate of blood flushing the medulla. When ADH is present, blood flow into the medulla is retarded and salt accumulates to create a steep osmotic gradient. Another hormone, aldosterone, from the adrenal gland increases the rate of sodium-ion reabsorption into the medulla to promote an increase in medullary salt concentration.

In addition to these physiological means of concentrating urine, a variety of mammals have morphological alterations of the medulla. Most mammals have two types of nephrons: those with a cortical glomerulus and abbreviated loops of Henle that do not penetrate far into the medulla, and those with juxtamedullary glomeruli, deep within the cortex, with loops that penetrate as far as the papilla of the renal pyramid. Obviously, the longer, deeper loops of Henle experience large osmotic gradients along their lengths. The flow of blood to these two populations of nephrons seems to be independently controlled. Juxtamedullary glomeruli are more active in regulating water excretion; cortical glomeruli function in ion regulation.

Nitrogen Excretion by Sauropsids: Renal and Extrarenal Routes

All extant representatives of the sauropsid lineage, including turtles and birds, are uricotelic—that is, they can excrete nitrogenous wastes in the form of uric acid. They are not limited to uricotely, however, and many sauropsids can facultatively excrete nitrogenous wastes as ammonia or urea (Table 11–4).

The strategy for water conservation when uric acid is the primary nitrogenous waste is entirely different from that required when urea is produced. Because urea is so soluble, concentrating urine in the kidney can conserve water, but concentrating uric acid would cause it to precipitate and block the nephrons. The kidneys of lepidosaurs lack the long loops of Henle that allow mammals to reduce the volume of urine and

Table 11–4 Distribution of nitrogenous end products among sauropsids

Group	Total Urinary Nitrogen (%)		
	Ammonia	Urea	Salts of Uric Acid
Lepidosaurs			
Tuatara	3–4	10–28	65–80
Lizards and snakes	Small	0–8	90–98
Archosaurs			
Crocodilians	49–86	12–39	1–11
Birds	1–28	2–14	53–85
Turtles			
Aquatic	4–44	45–95	1–24
Desert	3–8	15–50	20–50

increase its concentration (**Figure 11–15**). Urine from the kidneys of lepidosaurs has the same (or even a slightly lower) osmotic concentration as the blood plasma. The kidneys of birds have two types of nephrons: short loop nephrons like those of lepidosaurs and long loop nephrons that extend down into the medullary cone. The long loop nephrons allow birds to produce urine that is two to three times more concentrated than the plasma. These ratios are lower than those of mammals, and even the highest urine to plasma ratio recorded for a bird—5.8 for the savannah sparrow—is relatively low compared to mammalian ratios.

If sauropsids depended solely on the urine-concentrating capacity of their kidneys, they would excrete all their body water in urine. This is where the low solubility of uric acid becomes advantageous: uric acid precipitates when it enters the cloaca or bladder. The dissolved uric acid combines with ions in the urine and precipitates as a light-colored mass that includes sodium, potassium, and ammonium salts of uric acid as well as other ions held by complex physical forces. This mixture is familiar to anyone who has parked a car beneath a tree where birds roost. When uric acid and ions precipitate from solution, the urine becomes less concentrated and water is reabsorbed into the blood. In this respect, excretion of nitrogenous wastes as uric acid is even more economical of water than is excretion of urea because the water used to produce urine is reabsorbed and reused.

Water is not the only substance that is reabsorbed from the cloaca, however. Many sauropsids also reabsorb sodium and potassium ions and return them to the bloodstream. At first glance, that seems a remarkably

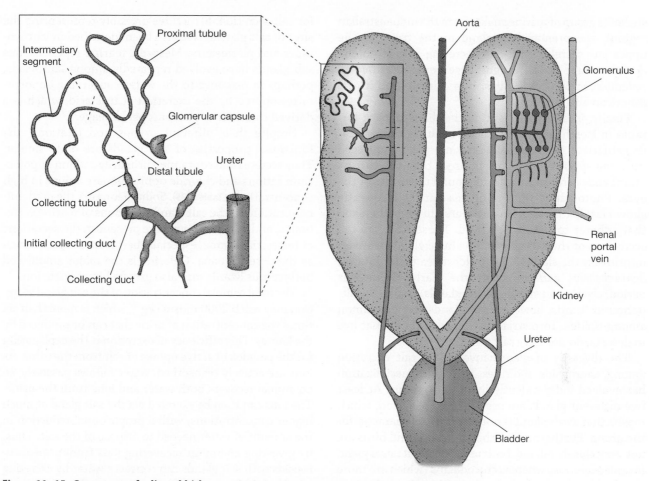

Figure 11–15 Structure of a lizard kidney. The left side shows a nephron in detail (*inset*) and its position within the kidney. The right side shows the relationship of glomeruli and nephrons within a segment of the kidney.

inefficient thing to do. After all, energy was used to create the blood pressure that forced the ions through the walls of the glomerulus into the urine in the first place, and now more energy is being used in the cloaca or bladder to operate the active transport system that returns the ions to the blood. The animal has used two energy-consuming processes, and it is back where it started—with an excess of sodium and potassium ions in the blood. Why do that?

The solution to the paradox lies in a water-conserving mechanism that is present in many sauropsids: salt-secreting glands that provide an extrarenal (i.e., in addition to the kidney) pathway that disposes of salt with less water than urine. Nasal salt glands are found in many families of lizards, although they are not necessarily found in all species within a family.

When salt glands are present in lizard, they are the lateral nasal glands. The secretions of the glands empty into the nasal passages, and a lizard expels them by sneezing or by shaking its head. In birds, also, the lateral nasal gland has become specialized for salt excretion. The glands are situated in or around the orbit, usually above the eye. Marine birds (pelicans, albatrosses, penguins) have well-developed salt glands, as do many freshwater birds (ducks, loons, grebes), shorebirds (plovers, sandpipers), storks, flamingos, carnivorous birds (hawks, eagles, vultures), upland game birds (pheasants, quail, grouse), parrots, the ostrich, and the roadrunner. Depressions in the supraorbital region of the skull of the extinct aquatic birds *Hesperornis* and *Ichthyornis* suggest that salt glands were present in these forms as well.

In sea snakes and the marine elephant trunk snakes, the posterior sublingual gland secretes a salty fluid into the tongue sheath, from which it is expelled when the tongue is extended. In some species of homalopsine snakes (a group of marine snakes from the Indoaustralian region), the premaxillary gland in the front of the upper jaw secretes salt. Salt-secreting glands on the dorsal surface of the tongue have been identified in several species of crocodiles, in a caiman, and in the American alligator.

Finally, in sea turtles and in the diamondback terrapin (a North American species of turtle that inhabits estuaries), the lacrimal glands are greatly enlarged (in some species each gland is larger than the turtle's brain) and secrete a salty fluid around the orbits of the eyes. Photographs of nesting sea turtles frequently show clear paths streaked by tears through the sand that adheres to the turtle's head. Those tears are the secretions of the salt glands. The huge glands leave an imprint on the skull, and the oldest sea turtle known, *Santanachelys gaffneyi* from the Early Cretaceous period, clearly had salt-secreting glands. Unlike the situation in lizards, however, salt glands are uncommon among turtles. Terrestrial turtles, even those that live in deserts, do not have salt glands.

The diversity of glands involved in salt excretion among sauropsids indicates that this specialization has evolved independently in various groups. At least five different glands are used for salt secretion, which means that a salt gland is not an ancestral character for the group. Furthermore, although lizards and birds are not very closely related, both use the lateral nasal gland for salt secretion, whereas crocodilians (which are more closely related to birds than to lizards) use lingual glands for salt secretion. Structures probably representing the sites of salt glands have also been identified in ichthyosaurs and plesiosaurs, Mesozoic marine reptiles. Thus salt glands have evolved repeatedly among sauropsids, perhaps in response to the water-conserving opportunities offered by the excretion of uric acid, which is a derived character of the sauropsid lineage.

Despite their different origins and locations, the functional properties of salt glands are quite similar. They secrete fluid containing primarily sodium or potassium cations and chloride or bicarbonate anions in high concentrations (Table 11–5). Sodium ions are the predominant cations in the salt gland secretions of marine vertebrates, and potassium ions are present in the secretions of terrestrial lizards, especially herbivorous species such as the desert iguana. Chloride is the major anion, and herbivorous lizards may also excrete bicarbonate ions.

The total osmolal concentration of the salt gland secretion may reach 2000 mosm · kg^{-1}, which is more than six times the concentration of urine that can be produced by the kidney. This efficiency of excretion is the explanation for the paradox of active uptake of salt from the urine. As ions are actively reabsorbed, water follows passively, so an animal recovers both water and ions from the urine. The ions can then be excreted via the salt gland at much higher concentrations, with a proportional reduction in the amount of water needed to dispose of the salt. Thus, by investing energy in recovering ions from urine, sauropsids with salt glands can conserve water by excreting ions through the more efficient extrarenal route.

Table 11–5 **Salt gland secretions from sauropsids**

Species and Condition	Ion Concentration (mmol · kg^{-1})		
	Na$^+$	K$^+$	Cl$^-$
Turtles			
Loggerhead sea turtle (*Caretta caretta*), seawater	732–878	18–31	810–992
Diamondback terrapin (*Malaclemys terrapin*), seawater	322–908	26–40	N/R
Lizards			
Desert iguana (*Dipsosaurus dorsalis*), estimated field conditions	180	1700	1000
Fringe-toed lizard (*Uma scoparia*), estimated field conditions	639	734	465
Snakes			
Sea snake (*Pelamis platurus*), salt loaded	620	28	635
Estuarine snake (*Cerberus rhynchops*), salt loaded	414	56	N/R
Crocodilian			
Saltwater crocodile (*Crocodylus porosus*), natural diet	663	21	632
Birds			
Black-footed albatross (*Diomedea nigripes*), salt loaded	800–900	N/R	N/R
Herring gull (*Larus argentatus*), salt loaded	718	24	N/R

N/R = not reported

11.6 Sensing and Making Sense of the World: Eyes, Ears, Tongues, Noses, and Brains

In many respects synapsids and sauropsids perceive the world around them very differently because the lineages are quite different in some of their sensory capacities. Most synapsids are exquisitely sensitive to odors but have relatively poor vision. (Primates in general—and humans in particular—are an exception to that generalization; indeed, primates and especially humans have lost many of the genes associated with olfactory reception.)

Sauropsids have the reverse combination of characters—most have good vision, including good color vision, and a rather poor sense of smell. These sensory capacities are reflected in the social behaviors of the groups: scent marking is a common element of the territorial behavior of mammals but not of birds, whereas territorial displays that emphasize color and pattern are common among lizards and birds but rare among mammals.

Vision

The vertebrate retina contains two types of cells that respond to light—rods and cones. Rod cells are sensitive to a wide range of wavelengths, and the electrical responses of many rod cells are transmitted to a single bipolar cell that sums their inputs. Because of these characteristics, rod cells are sensitive to low light levels but do not produce high visual acuity.

The population of cone cells in the retina of a vertebrate includes subgroups of cells that contain pigments that are sensitive to different wavelengths of light. Blue light stimulates one set of cells, red light another, and so on. The variety of cone cells found in vertebrate eyes is extensive and includes cells with pigments that are sensitive to wavelengths from deep red to ultraviolet light. Only a few cone cells transmit their responses to a bipolar cell, so cone cells require higher levels of illumination than rod cells; however, they are capable of producing sharper images than do rod cells.

Fishes (teleosts, at least), amphibians, and sauropsids use retinal cone cells to perceive colors. Thus the capacity for good color vision is probably ancestral for amniotes. At some point in the evolution of the synapsid lineage, the ability to perceive color was evidently reduced, possibly in connection with the adoption of a nocturnal lifestyle by early mammals. Most mammals have dichromatic (two pigments) vision—a long-wavelength pigment with maximum absorption in the green portion of the spectrum and a short-wavelength pigment with maximum sensitivity at blue wavelengths.

Our ability to perceive a richly colored world with subtle variations in hues is the result of a type of trichromatic vision that we share with only the other anthropoid primates (apes and monkeys). We anthropoids perceive three primary colors—short wavelengths (blue), medium wavelengths (green), and long wavelengths (red). The spectral sensitivities of the three types of cones overlap, and intermediate hues are produced by graded responses from the cones in response to mixtures of wavelengths in the light entering the eye. The excellent color vision of anthropoid primates is probably related to the increased reliance of anthropoids on vision and the importance of perceiving ripe fruit (usually red or yellow) against a background of green foliage.

The full ancestral array of cones and color vision was retained in sauropsids, and many species have colored oil droplets in the cone cells. The oils range from reddish through orange to yellow, and they may improve visual acuity by filtering out the very short wavelengths of light that are not focused effectively by the vertebrate eye. (You can easily demonstrate the difficulty of focusing blue light by comparing the sharpness of the image produced by a blue Christmas tree bulb compared to a red one.)

Many birds have tetrachromatic (four pigments) color vision, with red, green, and blue pigments like primates and an ultraviolet-sensitive pigment as well. Surprisingly, the ultraviolet-sensitive pigment in the retinas of birds is not the same as the ancestral short-wavelength pigment of vertebrates; instead it is a new pigment that evolved when birds "re-invented" ultraviolet sensitivity.

Chemosensation: Gustation and Olfaction

Tasting and smelling are forms of chemosensation mediated by receptor cells that respond to the presence of chemicals with specific characteristics. Tasting and smelling appear distinct from our perspective as terrestrial animals, but the distinction is blurred among aquatic non-amniotes where taste buds may be widely distinguished across the body surface.

Taste buds are clusters of cells derived from the embryonic endoderm that open to the exterior through pores. Chemicals must be in contact with the sensory cells in a taste bud to elicit a response, and they produce a relatively narrow range of responses. Amniotes have taste buds in the oral cavity (especially on the tongue) and in the pharynx. The taste buds of mammals are broadly distributed over the tongue and oral cavity, whereas in sauropsids they are on the back of the tongue and palate. We perceive chemicals as combinations of salt, sweet, sour, and bitter sensations.

Olfactory cells are derived from neural crest tissue and are distributed in the epithelium of the olfactory turbinates that are adjacent to the nasal passages. Chemicals that stimulate olfactory cells are called odorants, and some mammals are exquisitely sensitive to olfactory stimuli. Even humans, who have poor olfactory sensitivity by mammalian standards, may be able to detect 10,000 different odors. The chemical structures of odorant molecules allow them to bind to only specific receptor proteins in the membranes of olfactory cells. Genetically defined families of receptor proteins respond to different categories of odorants, but it is not clear how we distinguish so many odors. Olfaction and gustation interact to produce the sensation we call taste, and the process can be highly stylized, as in the sniffing, sipping, and swirling ritual of wine tasting. In general, olfaction is better developed among synapsids than sauropsids.

Hearing

A hearing ear was not a feature of the earliest tetrapods—it evolved independently in amniotes and non-amniotes, and separately in different lineages of aminotes. The lagena (called the cochlea in mammals) is a structure in the inner ear that is devoted to hearing, and it seems to be an ancestral feature for all amniotes. In contrast, the middle ear appears to have evolved independently several times among amniotes. Mammals have a completely enclosed middle ear involving the stapes (the old hyomandibula) and two additional bones, whereas all sauropsids have a single-boned middle ear, consisting of only the stapes. And beyond those differences, details of the ear structure suggest that among sauropsids the final condition of the middle ear evolved independently in turtles, lepidosaurs, and archosaurs, and within synapsids a fully enclosed middle ear evolved independently in monotremes and therian mammals.

Brains

All amniotes have a relatively enlarged forebrain in comparison with amphibians, especially with respect to the telencephalon. Furthermore, the brains of mammals and birds are larger in proportion to their body size than are the brains of nonavian reptiles (**Figure 11–16**). For a 1-kilogram animal, an average-size mammalian brain would weigh 9.9 grams, a bird's brain 6.7 grams, and a nonavian reptile's brain only

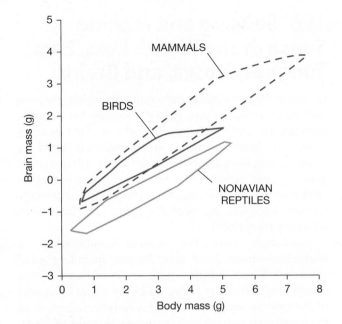

Figure 11–16 Body size and brain size. The relationship between body size and brain size, both expressed as 10^n grams. (The numbers on the axes are values of n.) The polygons enclose the values for 301 genera of mammals, 174 species of birds, and 62 species of nonavian reptiles.

0.7 gram. Thus both birds and mammals are "brainy" compared to nonavian reptiles, but their cerebrums are not strictly homologous.

Both birds and mammals have larger brains in proportion to their body size than do basal sauropsids and synapsids, but the increase in size has been achieved differently in the two groups (**Figure 11–17**). In both there is a distinct cerebrum, formed by the expansion of the dorsal portion of the telencephalon, the so-called dorsal pallium. The dorsal pallium has two components in the original amniote condition, which have different fates in the sauropsid and synapsid lineages.

Sauropsids enlarged one of these portions into what is called the dorsal ventricular ridge, while the other portion (the lemnopallium) remained small. In synapsids, the reverse occurred, with the cerebrum formed from the lemnopallium. In addition, the mammalian dorsal pallium forms a distinctive six-layered structure, now known as the **neocortex**. This structure is highly convoluted in larger-brained mammals, resulting in the type of wrinkled surface that is so apparent in the human cerebrum.

We often tend to think of birds as being less intelligent than mammals (the epithet "bird-brained" is a common one, after all), but new studies show that at least some birds are capable of highly complex behavior and that birds might attain conscious experience.

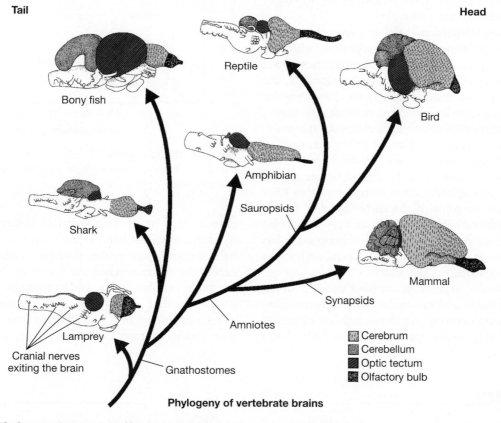

Figure 11–17 Phylogenetic pattern of brain enlargement among vertebrates.

Differences also exist in the way visual information is processed. Birds and other sauropsids retain the ancestral vertebrate condition of relying mainly on the optic lobes of the midbrain for visual processing. In mammals, the optic lobes are small and are used mainly for optical reflexes such as tracking motion, while it is a new portion of the cerebrum (the visual cortex) that actually interprets the information.

Evaluating how brainy an animal is can be problematic. For a start, the size of brains scales with negative allometry; that is, larger animals have proportionally smaller brains for their size than do small animals. The reason for this negative scaling is not known for certain but may be related to the fact that bigger bodies don't need absolutely more brain tissue to operate. For example, if 200 nerve cells in the brain were needed to control the right hind leg of a mouse, there is no reason to suppose that more cells would be needed to control the right hind leg of an elephant. Of course, larger animals have brains that are absolutely larger than smaller ones; they are just not quite as large as one would predict for their size.

The size of brains is usually expressed in relation to body size. One such estimate is the **encephalization quotient (EQ)**, a measure of the actual brain size in relation to the expected brain size for an animal of that body mass. In other words, the actual mass of the brain is divided by the body mass. An EQ of 1.0 means that the brain is exactly the size expected, whereas values less than 1 indicate brains that are smaller than expected and values greater than 1 are brains larger than expected.

Many mammals (e.g., rodents, shrews, marsupials) have EQ values less than 1, whereas others (including dogs, cats, horses, elephants, primates, and whales) have values greater than 1. Great apes have values of about 3, and humans have a value of about 8. Comparisons of brain size in the popular media often neglect differences in body size. For example, television programs often state that the brains of dolphins are just as big as human brains to emphasize how intelligent dolphins are, but most dolphins are much bigger than humans so they would be expected to have *larger* brains. In fact, the EQ of the bottlenose dolphin

(*Tursiops truncatus*) is 4.5, which is only slightly more than half of the human value.

Although we think of the tendency to evolve a large brain as a natural outcome of mammalian evolution, the situation is not that simple. Pack-hunting carnivores have larger brains than solitary ones, for example, but many small-brained animals display complex behavior and substantial capacity for learning. **Figure 11–18** shows the changes in the EQs of hooved herbivorous mammals (ungulates) over time—graphs for other mammalian orders would have similar shapes. Although the average brain size has increased and the largest brains known belong to extant species, some living ungulates have brains as small as some of the early Cenozoic forms. In other words, rather than seeing an *overall* increase in brain size, we find that the *range* of brain sizes has increased and small-brained species coexist with much larger-brained forms.

Nonetheless, a large brain contains more neurons than a small brain and can make more different connections among neurons. A greater number of neural connections may confer greater behavioral flexibility

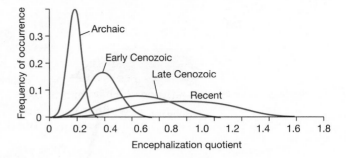

Figure 11–18 Brain gain. Changes in the encephalization quotient of Northern Hemisphere ungulates during the Cenozoic era.

than a small brain. One way this behavioral flexibility might manifest itself is the ability to adjust to new habitats, and some evidence supports that hypothesis: Across the entire spectrum of tetrapods—amphibians, nonavian reptiles, birds, and mammals—species that have succeeded in establishing themselves when they were transported to a new area do have larger brains than species that were unsuccessful invaders.

Summary

Synapsids (represented now by mammals) and sauropsids (represented by turtles, tuatara, lizards and snakes, crocodilians, and birds) dominate the terrestrial vertebrate fauna, and both include lineages specialized for flight, burrowing, and secondarily aquatic life. The two lineages have been separate at least since the late Paleozoic, and they have faced the same evolutionary challenges. The solutions they developed to the problems of life on land are similar in some respects and different in others. In some cases ancestral characters were retained, and in others derived characters appeared. For example, synapsids retained the ancestral pattern of tidal-flow lung ventilation and excretion of urea, while some sauropsids evolved through-flow lung ventilation and excretion of uric acid. Both lineages developed new forms of locomotion, endothermal thermoregulation, and complex hearts and brains, achieving functionally equivalent conditions by different routes.

More important than the differences themselves are the consequences of some of the differences in terms of the adaptive zones occupied by synapsids and sauropsids. Sauropsids have specialized in diurnal activity, with excellent color vision and extensive use of color, pattern, and movement in social displays. Synapsids may have passed through a stage of being largely nocturnal, when color vision was limited and scent and hearing were the primary senses.

Most important is that the different solutions that synapsids and sauropsids have found to the challenges of terrestrial life emphasize that there is more than one way to succeed as a terrestrial amniote.

Discussion Questions

1. How does the ancestral locomotor system of tetrapods interfere with respiration?

2. Why is reabsorption of water from the bladder or cloaca essential for sauropsids?

3. By the time you feel thirsty, you are already somewhat dehydrated. What is happening to the ADH in the pituitary gland at that point? How does that response contribute to regulating the water content of the body?

4. Birds regularly reach altitudes higher than human mountain climbers can ascend without using auxiliary breathing apparatus, both as residents and during migration. For example, radar tracking of migrating birds shows that they sometimes fly as high as 6500 meters, the alpine chough lives at altitudes of around 8200 meters on Mount Everest, and bar-headed geese pass directly over the summit of the Himalayas at altitudes of 9200 meters during their migrations. Explain the features of the sauropsid respiratory system that allow birds to breathe at such high altitudes.

5. Understanding the evolution of endothermy requires the resolution of a paradox: Increasing the resting metabolic rate can provide heat to raise the body temperature, but without insulation (hair or feathers) that heat is rapidly lost to the environment. But a layer of insulation does not benefit a vertebrate unless the animal has a high metabolic rate. Discuss the evidence suggesting that Mesozoic synapsids and sauropsids had developed both high metabolic rates and insulation, and identify the puzzling gaps in the information currently available.

6. The nasal salt glands of insectivorous lizards respond to injections of chloride ions by increasing secretions of sodium or potassium chloride, whereas injections of sodium or potassium ions do not increase the activity of the nasal salt glands. What difference in the routes of excretion of sodium and potassium ions compared to chloride ions accounts for the different responses of the lizards?

Additional Information

Amiel, J. J., et al. 2011. Smart moves: Effects of relative brain size on establishment success of invasive amphibians and reptiles. *PLoS ONE* 6(4):e18277. doi:10.1371/journal.pone.0018277.

Butler, A. B., et al. 2005. Evolution of the neural basis of consciousness: A bird-mammal comparison. *BioEssays* 27:923–936.

Carvalho, O., and C. Goncalves. 2011. Comparative physiology of the respiratory system in the animal kingdom. *The Open Biology Journal* 4:35–46.

Clack, J. A. 2011. *Gaining Ground: The Origin and Evolution of Tetrapods*, 2nd ed. Bloomington, IN: Indiana University Press.

Cowles, R. B. 1958. Possible origin of dermal temperature regulation. *Evolution* 12:347–357.

Farmer, C. G. 1999. Evolution of the vertebrate cardiopulmonary system. *Annual Review of Physiology* 61:573–592.

Farmer, C. G. 2010. The provenance of alveolar and parabronchial lungs: Insights from paleoecology and the discovery of cardiogenic, unidirectional airflow in the American alligator (*Alligator mississipiensis*). *Physiological and Biochemical Zoology* 83:561–575.

Farmer, C. G. 2011. On the evolution of arterial vascular patterns of tetrapods. *Journal of Morphology* 272:1325–1341.

Farmer, C. G., and K. Sanders. 2010. Unidirectional airflow in the lungs of alligators. *Science* 327:338–340.

Geist, N. R. 2000. Nasal respiratory turbinate function in birds. *Physiological Biochemistry and Zoology* 73:581–589.

Hazard, L. C., et al. 2010. Secretion by the nasal salt glands of two insectivorous lizard species is initiated by an ecologically relevant dietary ion, chloride. *Journal of Experimental Zoology (Ecol Genet Physiol)* 313A:442–451.

Hunt, D. M., et al. 2009. Evolution and spectral tuning of visual pigments in birds and mammals. *Philosophical Transactions of the Royal Society B* 364:2941–2955.

Kemp, T. A. 2006. The origin of mammalian endothermy: A paradigm for the evolution of complex biological structure. *Zoological Journal of the Linnean Society* 147:473–488.

Makovicky, P. J., and L. E. Zanno. 2011. Theropod diversity and the refinement of avian characteristics. In G. Dyke and G. Kaiser (eds.), *Living Dinosaurs: The Evolutionary History of Modern Birds*. Hoboken, NJ: John Wiley & Sons, 9–29.

Nespolo, R. F., et al. 2011 Using new tools to solve an old problem—The evolution of endothermy in vertebrates. *Trends in Ecology and Evolution* 26:414–423.

O'Connor, P. M. 2009. Evolution of archosaurian body plans: Skeletal adaptations of an air-sac-based breathing apparatus in birds and other archosaurs. *Journal of Experimental Zoology (Ecol Genet Physiol)* 311A:629–646.

Perry, S. F., et al. 2009. Implications of an avian-style respiratory system for gigantism in sauropod dinosaurs. *Journal of Experimental Zoology (Ecol Genet Physiol)* 311A:600–610.

Sabat, P. 2000. Birds in marine and saline environments: Living in dry habitats. *Revista Chilena de Historia Natural* 73:401–410.

Schachner, E. R., et al. 2011. Evolution of the dinosauriform respiratory apparatus: New evidence from the postcranial axial skeleton. *The Anatomical Record* 294:1532–1547.

Walsh, S., and A. Milner. 2011. Evolution of the avian brain and senses. In G. Dyke and G. Kaiser (eds.), *Living Dinosaurs: The Evolutionary History of Modern Birds.* Hoboken, NJ: John Wiley & Sons, 282–305.

Wedel, M. J. 2009. Evidence for bird-like air sacs in saurischian dinosaurs. *Journal of Experimental Zoology (Ecol Genet Physiol)* 311A:611–628.

Websites

Respiration

Avian respiratory system http://people.eku.edu/ritchisong/birdrespiration.html

Renal system

Avian renal system http://people.eku.edu/ritchisong/bird_excretion.htm

Mammalian renal system http://www.vetmed.vt.edu/education/Curriculum/VM8054/Labs/Lab23/lab23.htm

Thermoregulation

Avian thermoregulation http://people.eku.edu/ritchisong/birdmetabolism.html

Mammalian thermoregulation http://openlearn.open.ac.uk/mod/oucontent/view.php?id=398668&section=6

Turtles

Turtles provide a contrast to amphibians in the relative lack of diversity in their life histories. All turtles lay eggs, and none exhibits parental care of the hatchlings. Turtles show morphological specializations associated with terrestrial, freshwater, and marine habitats, and marine turtles make long-distance migrations rivaling those of birds. Probably turtles and birds use many of the same navigation mechanisms to find their way. Most turtles are long-lived animals with relatively poor capacities for rapid population growth. Many turtles, especially sea turtles and large tortoises, are endangered by human activities. Some efforts to conserve turtles have apparently been frustrated by a feature of the embryonic development of many species of turtles—the sex of an individual is determined by the temperature to which it is exposed during embryonic development.

12.1 Everyone Recognizes a Turtle

Turtles evolved a successful approach to life in the Triassic period and have scarcely changed since. The shell, which is the key to their success, has also limited the group's diversity. For obvious reasons, flying or gliding turtles have never existed, and arboreality is only slightly developed. Perhaps the most distinctive turtles are an extinct group of very large terrestrial turtles with horns and frills on the head and a clublike expansion on the end of the tail, the Meiolaniidae (**Figure 12–1**). Meiolaniids lived in South America in the Cretaceous period and Eocene epoch, and in Australia and New Caledonia from the Miocene through the Pleistocene epochs.

Shell morphology reflects the ecology of turtle species: The most terrestrial forms, the tortoises of the family Testudinidae, have high, domed shells and elephant-like feet. Many smaller species of tortoises have flat, spadelike front feet that they use for burrowing (**Figure 12–2** on page 289). The gopher tortoises of North America are an example of this body form—their front legs are flattened into scoops, and the dome of the shell is reduced. The Bolson tortoise of northern Mexico constructs burrows a meter or more deep and several meters long in the hard desert soil. These tortoises bask at the mouths of their burrows, and when a predator appears, they throw themselves down the steep entrance tunnels of the burrows to escape, just as an aquatic turtle dives off a log.

Figure 12–1 An extinct horned turtle. *Meiolania platyceps* is from Lord Howe Island, New South Wales, Australia. This late Pleistocene species grew to a shell length of 2.5 m, and its head was about 60 cm wide.

The pancake tortoise of Africa is a radical departure from the usual tortoise morphology. Its shell is flat and flexible because its ossification is much reduced. This species lives in rocky foothill regions and scrambles over the rocks with nearly as much agility as a lizard. When threatened by a predator, the pancake tortoise crawls into a rock crevice and uses its legs to wedge itself in place. The flexible shell presses against the overhanging rock and creates so much friction that it is almost impossible to pull the tortoise out.

Other terrestrial turtles have a moderately domed **carapace** (upper shells), like the box turtles of the family Emydidae. This is one of several kinds of turtles that have evolved flexible regions in the **plastron** (lower shell), which allow the front and rear lobes to be pulled upward to close the openings of the shell. Aquatic turtles have low carapaces that offer little resistance to movement through water. The Emydidae and Bataguridae contain a large number of pond turtles, including the painted turtles and the red-eared turtles often seen in pet stores and anatomy and physiology laboratory courses.

The snapping turtles (family Chelydridae) and the mud and musk turtles (family Kinosternidae) prowl along the bottoms of ponds and slow rivers and are not particularly streamlined. The mud turtle has a hinged plastron, but the musk and snapping turtles have very reduced plastrons. They rely on strong jaws for protection. A reduction in the size of the plastron

makes these species more agile than most turtles, and musk turtles may climb several feet into trees, probably to bask. If a turtle falls on your head while you are canoeing, it is probably a musk turtle.

The soft-shelled turtles (family Trionychidae) are fast swimmers. The ossification of the shell is greatly reduced, which lightens the animal, and the feet are large with extensive webbing. Soft-shelled turtles lie in ambush partly buried in debris on the bottom of the pond. Their long necks allow them to reach considerable distances to seize the invertebrates and small fishes on which they feed.

Extant turtles can be placed into 13 families with about 313 species (**Table 12–1** on page 290). The two lineages of living turtles can be traced through fossils to the Mesozoic. The **cryptodires** (Greek *crypto* = hidden and *dire* = neck) retract the head into the shell by bending the neck in a vertical S shape. The **pleurodires** (Greek *pleuro* = side) retract the head by bending the neck horizontally. All the turtles discussed so far have been cryptodires, and this is the largest group. Cryptodires are the only turtles now found in most of the Northern Hemisphere, and there are aquatic and terrestrial cryptodires in South America and terrestrial ones in Africa. Only Australia has no cryptodires. Pleurodires are now found only in the Southern Hemisphere, although they had worldwide distribution in the late Mesozoic and early Cenozoic eras. *Stupendemys*, a pleurodire from the Pliocene epoch of Venezuela, had a carapace more than 2 meters long. All the living pleurodires are at least semiaquatic, but some fossil pleurodires had high, domed shells that suggest they may have been terrestrial. The most terrestrial of the living pleurodires are probably the African pond turtles, which readily move overland from one pond to another.

The snake-necked pleurodire turtles (family Chelidae) are found in South America, Australia, and New Guinea. As their name implies, they have long, slender necks. In some species, the length of the neck is considerably greater than that of the body. These forms feed on fishes that they catch with a sudden dart of the head. Other snake-necked turtles have much shorter necks. Some of these feed on mollusks and have enlarged palatal surfaces used to crush shells. The same specialization for feeding on mollusks is seen in certain cryptodire turtles.

An unusual feeding method among turtles is found in a pleurodire, the matamata of South America. Large matamatas reach shell lengths of 40 centimeters. They are bizarre looking animals, with broad heads and flat shells and flaps of skin projecting from the sides of the

Figure 12–2 Body forms of turtles. (a) Tortoise, *Testudo*, Testudinidae. (b) Pancake tortoise, *Malacochersus*, Testudinidae. (c) Terrestrial box turtle, *Terrapene*, Emydidae. (d) Pond turtle, *Trachemys*, Emydidae. (e) Soft-shelled turtle, *Apalone*, Trionychidae. (f) Mud turtle, *Kinosternon*, Kinosternidae. (g) Alligator snapping turtle, *Macrochelys*, Chelydridae. (h) African pond turtle, *Pelusios*, Pelomedusidae. (i) Australian snake-necked turtle, *Chelodina*, Chelidae. (j) South American matamata, *Chelus*, Chelidae. (k) Loggerhead sea turtle, *Caretta*, Cheloniidae. (l) Leatherback sea turtle, *Dermochelys*, Dermochelyidae. (The turtles are not drawn to scale.)

head and the broad neck. To these are added trailing bits of adhering algae. The effect is exceedingly cryptic. It is hard to recognize a matamata as a turtle even in an aquarium, and it is practically impossible to see one against the mud and debris of a river bottom. In addition to obscuring the shape of the turtle, the flaps of skin on the head are sensitive to minute vibrations in water caused by the passage of a fish. When it senses the presence of prey, the matamata abruptly opens its mouth and expands its throat. Water rushes in, carrying the prey with it, and the matamata closes its mouth, expels the water, and swallows the prey. The matamata lacks the horny beak that other turtles use for seizing prey or biting off pieces of plants.

Table 12–1 **The major families of turtles**

Cryptodira

Testudinidae: About 58 species of small (15 cm) to very large (1 m) terrestrial turtles with a worldwide distribution in temperate and tropical regions. Although tortoises are clumsy swimmers, they float well and withstand long periods without food or water. These characteristics have allowed tortoises swept to sea by flooding rivers to populate oceanic islands, such as the Galápagos Islands.

Emydidae: About 50 species of small (12 cm) to large (60 cm) freshwater, semiaquatic, and terrestrial turtles, mostly in North America; 1 genus in Central and South America; and 1 in Europe, Asia, and North Africa.

Geoemydidae: About 70 species of small (12 cm) to large (75 cm) freshwater, semiaquatic, and terrestrial turtles, primarily Asian.

Trionychidae: About 31 species of small (25 cm) to very large (130 cm) freshwater turtles with flattened bodies and reduced ossification of the shell from North America, Africa, and Asia.

Kinosternidae: About 25 species of small (11 cm) to medium (40 cm) bottom-dwelling freshwater turtles from North America and South America.

Cheloniidae: 6 species of large (70 cm) to very large (150 cm) sea turtles with bony shells covered with epidermal scutes and paddlelike forelimbs, found worldwide in tropical and temperate oceans.

Dermochelyidae: A single species, the leatherback, it is the largest extant turtle (up to 240 cm). This is a marine turtle in which the shell is reduced to thousands of small bones embedded in a leathery skin. It occurs in oceans worldwide, and its range extends north and south into seas too cold for other sea turtles.

Chelydridae: 4 species of large (*Chelydra*, 50 cm) to very large (*Macrochelys*, 70 cm and 80 kg) freshwater turtles in North and Central America.

Pleurodira

Chelidae: About 52 species of small (15 cm) to large (50 cm) aquatic turtles from South America, Australia, and New Guinea.

Pelomedusidae: 19 species of small aquatic turtles from Africa, Madagascar, and the Seychelles Islands. All extant pelomedusids inhabit freshwater, but some extinct species may have been marine.

Podocnemididae: 8 species of aquatic turtles found in northern South America and Madagascar. *Podocnemis expansa*, which occurs in the Amazon and Orinoco Rivers, is the largest extant pleurodire; females reach shell lengths of 90 cm. The extinct *Stupendemys* (from the Late Tertiary of Venezuela) was more than 2 m long.

Marine turtles are cryptodires. The families Cheloniidae and Dermochelyidae show more extensive specialization for aquatic life than any freshwater turtle; for example, the forelimbs are modified as flippers in both families. The largest of the sea turtles of the family Cheloniidae is the loggerhead, which once reached weights exceeding 400 kilograms. The largest marine turtle, the leatherback, reaches shell lengths of more than 2 meters and weights in excess of 600 kilograms. In this species the dermal ossification has been reduced to bony platelets embedded in dense connective tissue, which gives the leatherback its name. This is a pelagic turtle that ranges far from land, and it has a wider geographic distribution than any other ectothermal amniote. Leatherback turtles penetrate far into cool temperate seas and have been recorded in the Atlantic from Newfoundland to Argentina and in the Pacific from Japan to Tasmania. Leatherback turtles dive to depths of more than 1000 meters. One dive that drove the depth recorder off the scale is estimated to have reached 1200 meters, which exceeds the deepest dive recorded for a sperm whale (1140 meters). Leatherback turtles feed largely on jellyfish, whereas the smaller hawksbill sea turtles eat sponges that are defended by spicules of silica (glass) as well as a variety of chemicals (including alkaloids and terpenes) that are toxic to most vertebrates. Green turtles (*Chelonia mydas*) are the only herbivorous marine turtles.

12.2 But What Is a Turtle? Phylogenetic Relationships of Turtles

Turtles show a combination of ancestral features and highly specialized characters that are not shared with any other group of vertebrates, and their phylogenetic affinities are not fully understood. The turtle lineage probably originated among the early amniotes of the Late Carboniferous period. Like those animals, turtles have anapsid skulls, but the shells and postcranial skeletons of turtles are unique. One view emphasizes the importance of the anapsid skull and places the affinities of turtles among the parareptiles. A radically different opinion regards the anapsid skull of turtles as being a secondarily derived feature and places turtles among the diapsids. The case for a diapsid origin is stronger, but that hypothesis leads to the further question of whether turtles are the sister group of lepidosaurs (tuatara, snakes, and lizards) or archosaurs (crocodilians and birds). A lively debate is in progress and the issue is far from resolved (**Figure 12–3**).

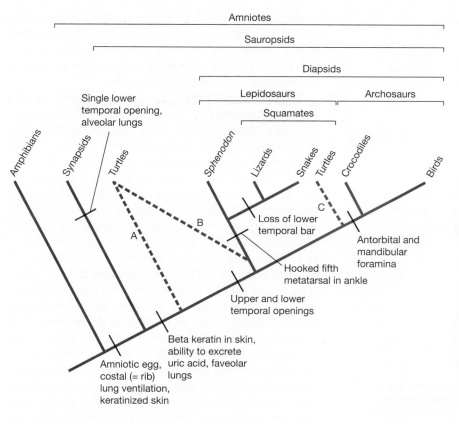

Figure 12–3 Simplified cladogram of tetrapods. The three competing hypotheses of the phylogenetic position of turtles are shown with broken lines. A, the most extreme, places the origin of turtles among the parareptiles and considers turtles to be the sister group of the other reptile lineages. B places turtles as the sister group of the lepidosaurs, and C considers turtles to be the sister group of archosaurs. Positions B and C for turtles imply the secondary loss of temporal foramina.

The earliest turtles are found in Late Triassic deposits in Germany, Thailand, and Argentina. These animals had nearly all the specialized characteristics of derived turtles and shed no light on the phylogenetic affinities of the group. *Proganochelys*, from Triassic deposits in Germany, was nearly a meter long (larger than most living turtles) and had a high, arched shell. The marginal teeth had been lost, and the maxilla, premaxilla, and dentary bones were probably covered with a horny beak, just as they are in derived turtles. The skull of *Proganochelys* retained the supratemporal and lacrimal bones and the lacrimal duct, and the palate had rows of denticles; derived turtles have lost all these structures. The plastron of *Proganochelys* also contained some bones that have been lost by derived turtles, and the vertebrae of the neck lack specializations that would have allowed the head to be retracted into the shell.

Turtles with neck vertebrae specialized for retraction are not known before the Cretaceous, but differences in the skulls and shells allow the pleurodire and cryptodire lineages to be traced back to the Late Triassic (shells of pleurodires) and Late Jurassic (skulls of cryptodires). The otic capsules of all turtles beyond the proganochelids are enlarged, and the jaw adductor muscles bend posteriorly over the otic capsule (**Figure 12–4**). The muscles pass over a pulleylike structure, called the trochlear process. In cryptodires the trochlear process is formed by the anterior surface of the otic capsule itself, whereas in pleurodires it is formed by a lateral process of the pterygoid. Fusion of the pelvic girdle to the carapace and plastron further distinguishes pleurodires from cryptodires, which have a bony connection attaching the shell to the girdle. The beginnings of these changes are seen in *Australochelys*.

12.3 Turtle Structure and Function

Turtles are among the most derived vertebrates. They are encased in bone, with the limbs inside the ribs and with horny beaks instead of teeth—if turtles had become extinct at the end of the Mesozoic, they would rival dinosaurs in their novelty. However, because they survived, they are regarded as commonplace and are often used in comparative anatomy courses to represent basal amniotes (inappropriately, because they are so specialized).

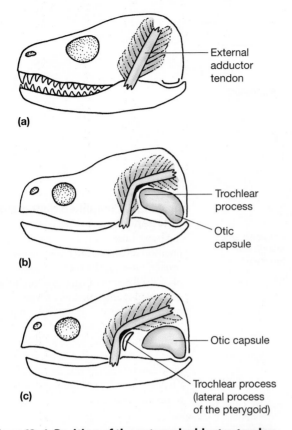

Figure 12–4 Position of the external adductor tendon.
(a) In the ancestral (parareptilian) condition; (b) in cryptodire turtles; (c) in pleurodire turtles.

Shell and Skeleton

The shell is the most distinctive feature of a turtle (**Figure 12–5**). The carapace is composed of dermal bone that typically grows from 59 separate centers of ossification. Eight plates along the dorsal midline form the neural series and are fused to the neural arches of the vertebrae. Lateral to the neural bones are eight paired costal bones, which are fused to the broadened ribs. The ribs of turtles are unique among tetrapods in being external to the girdles. Eleven pairs of peripheral bones, plus two unpaired bones in the dorsal midline, form the margin of the carapace. The plastron is formed largely from dermal ossifications, but the entoplastron is derived from the interclavicle; the paired epiplastra anterior to it are derived from the clavicles. Processes from the hyoplastron and hypoplastron fuse with the first and fifth pleurals, forming a rigid connection between the plastron and carapace.

The bones of the carapace are covered by horny scutes of epidermal origin that do not coincide in number or position with the underlying bones. The carapace has a row of five central scutes, bordered on each side by four lateral scutes. Ten to twelve marginal scutes on each side turn under the edge of the carapace. The plastron is covered by a series of six paired scutes.

Flexible areas, called hinges, are present in the shells of many turtles. The most familiar examples are the North American and Asian box turtles (*Terrapene* and *Cuora*), in which a hinge between the hyoplastral and hypoplastral bones allows the anterior and posterior lobes of the plastron to be raised to close off the front and rear openings of the shell. Mud turtles (*Kinosternon*) have two hinges in the plastron; the anterior hinge runs between the epiplastra and the entoplastron (which is triangular in kinosternid turtles rather than diamond shaped), and the posterior hinge is between the hypoplastron and the xiphiplastron. Some species of tortoises have plastral hinges; in *Testudo* the hinge lies between the hypoplastron and the xiphiplastra, as it does in *Kinosternon*, but in another genus of tortoise (*Pyxis*), the hinge is anterior and involves a break across the entoplastron. The African forest tortoises (*Kinixys*) have a hinge on the posterior part of the carapace. The margins of the epidermal shields and the dermal bones of the carapace are aligned, and the hinge runs between the second and third pleural scutes and the fourth and fifth costals. The presence of hinges is sexually dimorphic in some species of tortoises. The erratic phylogenetic occurrence of kinetic shells and differences among related species indicate that shell kinesis has evolved many times in turtles.

Modifications of the bony structure of the shell are seen in some families. Soft-shelled turtles lack peripheral ossifications and epidermal scutes. The distal ends of the broadened ribs are embedded in flexible connective tissue, and the carapace and plastron are covered with skin. The New Guinea river turtle (*Carrettochelys*) is also covered by skin instead of scutes, but in this species the peripheral bones are present and the edge of the shell is stiff. The leatherback sea turtle (*Dermochelys*) has a carapace formed of cartilage with thousands of small polygonal bones embedded in it, and the plastral bones are reduced to a thin rim around the edge of the plastron. The neural and costal ossifications of the pancake tortoise (*Malacochersus*) are greatly reduced, but the epidermal plates are well developed.

Extant turtles have only 10 vertebrae in the trunk and 8 in the neck. The centra of the trunk vertebrae are elongated and lie beneath the dermal bones in the dorsal midline of the shell. The centra are constricted in their centers and fused to each other. The neural arches in the

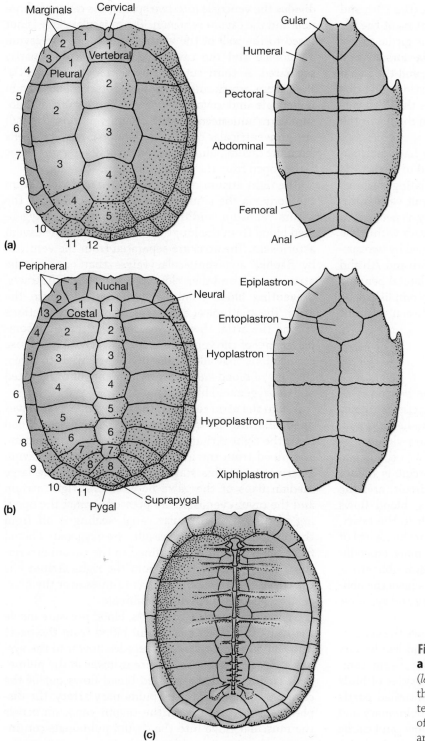

Figure 12–5 Shell and vertebral column of a turtle. (a) Epidermal scutes of the carapace (*left*) and plastron (*right*). (b) Dermal bones of the carapace (*left*) and plastron (*right*). (c) Vertebral column of a turtle, seen from the inside of the carapace. Note that anteriorly the ribs articulate with two vertebral centra.

anterior two-thirds of the trunk lie between the centra as a result of anterior displacement, and the spinal nerves exit near the middle of the preceding centrum. The ribs are also shifted anteriorly; they articulate with the anterior part of the neurocentral boundary; and, in the anterior part of the trunk, where the shift is most pronounced, the ribs extend onto the preceding vertebra.

Cryptodires have two sacral vertebrae (the 19th and 20th vertebrae) with broadened ribs that meet the ilia of the pelvis. Pleurodires have the pelvic girdle firmly fused to the dermal carapace by the ilia dorsally and by the pubic and ischial bones ventrally, and the sacral region of the vertebral column is less distinct. The ribs on the 17th, 18th, 19th, and sometimes the 20th vertebrae are fused to the centra and end on the ilia or the ilio-carapacial junction.

The cervical vertebrae of cryptodires have articulations that permit the S-shaped bend used to retract the head into the shell. Specialized articulating surfaces between vertebrae, called ginglymi, permit vertical rotation. This type of rotation, ginglymoidy, is peculiar to cryptodires, but the anatomical details vary within the group. In most families, the hinge is formed by two successive ginglymoidal joints between the 6th and 7th and the 7th and 8th cervical vertebrae. The lateral bending of the necks of pleurodire turtles is accomplished by ball-and-socket or cylindrical joints between adjacent cervical vertebrae.

The Heart

The circulatory systems of tetrapods can be viewed as consisting of two circuits: The systemic circuit carries oxygenated blood from the heart to the head, trunk, and appendages, whereas the pulmonary circuit carries deoxygenated blood from the heart to the lungs. The blood pressure in the systemic circuit is higher than the pressure in the pulmonary circuit, and the two circuits operate in series. That is, blood flows from the heart through the lungs, back to the heart, and then to the body. The morphology of the hearts of derived synapsids and sauropsids (mammals, crocodilians, and birds) makes this sequential flow obligatory, but the hearts of turtles and lepidosaurs have the ability to shift blood between the pulmonary and systemic circuits.

The route of blood flow is flexible because the ventricular chambers in the hearts of turtles and lepidosaurs are in anatomical continuity, instead of being completely divided by a septum like the ventricles of birds and mammals. The flow of blood is controlled partly by the relative resistance to flow in the pulmonary and systemic circuits. The pattern of blood flow can best be explained by considering the morphology of the heart and how intracardiac pressure changes during a heartbeat. **Figure 12–6** shows a schematic view of the heart of a turtle. The left and right atria are completely separate, and three subcompartments can be distinguished in the ventricle. A muscular ridge in the core of the heart divides the ventricle into two spaces, the cavum pulmonale and the cavum venosum. The muscular ridge is not fused to the wall of the ventricle, and thus the cavum pulmonale and the cavum venosum are only partly separated. A third subcompartment of the ventricle, the cavum arteriosum, is located dorsal to the cavum pulmonale and cavum venosum. The cavum arteriosum communicates with the cavum venosum through an intraventricular canal. The pulmonary artery opens from the cavum pulmonale, and the left and right aortic arches open from the cavum venosum.

The right atrium receives deoxygenated blood from the body via the sinus venosus and empties into the cavum venosum, and the left atrium receives oxygenated blood from the lungs and empties into the cavum arteriosum. The atria are separated from the ventricle by flaplike atrioventricular valves that open as the atria contract and then close as the ventricle contracts, preventing blood from being forced back into the atria. The anatomical arrangement of the connections among the atria, their valves, and the three subcompartments of the ventricle is crucial because it is those connections that allow pressure differentials to direct the flow of blood and to prevent mixing of oxygenated and deoxygenated blood.

When the atria contract, the atrioventricular valves open, allowing blood to flow into the ventricle. Blood from the right atrium flows into the cavum venosum, and blood from the left atrium flows into the cavum arteriosum. At this stage in the heartbeat, the large median flaps of the valve between the right atrium and the cavum venosum are pressed against the opening of the intraventricular canal, sealing it off from the cavum venosum. As a result, the oxygenated blood from the left atrium is confined to the cavum arteriosum. Deoxygenated blood from the right atrium fills the cavum venosum and then continues over the muscular ridge into the cavum pulmonale.

When the ventricle contracts, blood pressure inside the heart increases. Ejection of blood from the heart into the pulmonary circuit precedes flow into the systemic circuit because resistance is lower in the pulmonary circuit. As deoxygenated blood flows out of the cavum pulmonale into the pulmonary artery, the displacement of blood from the cavum venosum across the muscular ridge into the cavum pulmonale continues. As the ventricle shortens during contraction, the muscular ridge comes into contact with the wall of the ventricle and closes off the passage for blood between the cavum venosum and cavum pulmonale.

Simultaneously, blood pressure inside the ventricle increases, and the flaps of the right atrioventricular valve

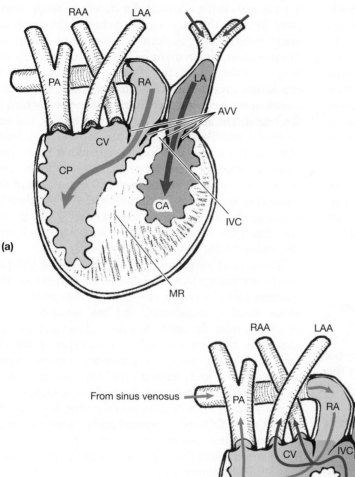

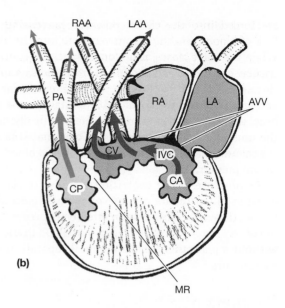

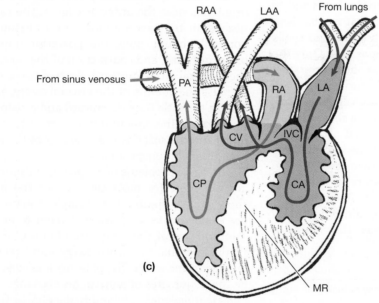

Figure 12–6 Blood flow in the heart of a turtle. (a) As the atria contract, oxygenated blood (*colored arrows*) from the left atrium (LA) enters the cavum arteriosum (CA), while deoxygenated blood (*gray arrow*) from the right atrium (RA) first enters the cavum venosum (CV) and then crosses the muscular ridge (MR) and enters the cavum pulmonale (CP). The atrioventricular valve (AVV) blocks the intraventricular canal (IVC) and prevents mixing of oxygenated and deoxygenated blood. (b) As the ventricle contracts, the deoxygenated blood in the cavum pulmonale is expelled through the pulmonary arteries (PA); the AVV closes, no longer obstructing the IVC; and the oxygenated blood in the cavum arteriosum is forced into the cavum venosum and expelled through the right and left aortic arches (RAA and LAA). The adpression of the wall of the ventricle to the muscular ridge prevents mixing of deoxygenated and oxygenated blood. (c) Summary of the pattern of blood flow through the heart of a turtle.

are forced into the closed position, preventing backflow of blood from the cavum venosum into the atrium. When the valve closes, it no longer blocks the intraventricular canal. Oxygenated blood from the cavum arteriosum can now flow through the intraventricular canal and into the cavum venosum. At this stage in the heartbeat, the wall of the ventricle is pressed firmly against the muscular ridge, separating the oxygenated blood in the cavum venosum from the deoxygenated blood in the cavum pulmonale.

As pressure in the ventricle continues to rise, oxygenated blood in the cavum venosum is ejected into the aortic arches. This system effectively prevents mixing of oxygenated and deoxygenated blood in the heart without a permanent morphological separation of the two circuits.

Respiration

As we have seen, basal amniotes probably used movements of the rib cage to draw air into the lungs and to force it out, and lizards still employ that mechanism. The fusion of the ribs of turtles with their rigid shells makes that method of breathing impossible. Only the openings at the anterior and posterior ends of the shell contain flexible tissues. The lungs of a turtle, which are large, are attached to the carapace dorsally and laterally. Ventrally, the lungs are attached to a sheet of nonmuscular connective tissue that is itself attached to the viscera (**Figure 12–7**). The weight of the viscera keeps this diaphragmatic sheet stretched downward.

Turtles produce changes in pressure in the lungs by contracting muscles that force the viscera upward, compressing the lungs and expelling air, followed by contracting other muscles that increase the volume of the visceral cavity, allowing the viscera to settle downward. Because the viscera are attached to the diaphragmatic sheet, which in turn is attached to the lungs, the downward movement of the viscera expands the lungs, drawing in air. In turtles, both inhalation and exhalation require muscular activity. The viscera are forced upward against the lungs by the contraction of the transverse abdominis muscle posteriorly and the pectoralis muscle anteriorly. The transverse abdominus inserts on the cup-shaped connective tissue (the posterior limiting membrane) that closes off the posterior opening of the visceral cavity. Contraction of the transverse abdominus flattens the cup inward, thereby reducing the volume of the visceral cavity. The pectoralis draws the shoulder girdle back into the shell, further reducing the volume of the visceral cavity.

The inspiratory muscles are the abdominal oblique, which originates near the posterior margin of the plastron and inserts on the external side of the posterior limiting membrane, and the serratus, which originates near the anterior edge of the carapace and inserts on the pectoral girdle. Contraction of the abdominal oblique pulls the posterior limiting membrane outward, and contraction of the serratus rotates the pectoral girdle outward. Both of these movements increase the volume of the visceral cavity, allowing the viscera to settle back downward and causing the lungs to expand. The in-and-out movements of the forelimbs and the soft tissue at the rear of the shell during breathing are conspicuous.

The basic problems of respiring within a rigid shell are the same for most turtles, but the mechanisms show some variation. For example, aquatic turtles can use the hydrostatic pressure of water to help move air into and out of the lungs. In addition, many aquatic turtles are able to absorb oxygen and release carbon dioxide to the water. The pharynx and cloaca appear to be the major sites of aquatic gas exchange. In 1860, in *Contributions to the Natural History of the United States of America*, Louis Agassiz pointed out that the pharynx of soft-shelled turtles contains fringelike processes and suggested that these structures are used for underwater respiration. Subsequent study has shown that soft-shelled turtles use movements of the hyoid apparatus to draw water into and out of the mouth and pharynx (oropharynx) when they are confined underwater, and that pharyngeal respiration accounts for most of the oxygen absorbed from the water. Musk turtles are also capable of aquatic gas exchange, and histological examination shows that the oropharyngeal region is lined by flat-topped papillae. These structures are highly vascularized and are probably the site of gas exchange.

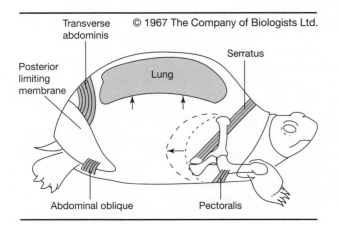

© 1967 The Company of Biologists Ltd.

Transverse abdominis

Serratus

Posterior limiting membrane

Lung

Abdominal oblique

Pectoralis

Figure 12–7 Schematic view of the lungs and respiratory movements of a tortoise.

The Australian turtle *Rheodytes leukops* uses cloacal respiration. Its cloacal orifice may be as large as 30 millimeters in diameter, and the turtle holds it open. Large bursae (sacs) open from the wall of the cloaca, and the bursae have a well-vascularized lining with numerous projections (villi). The turtle pumps water into and out of the bursae at rates of 15 to 60 times per minute. Captive turtles rarely surface to breathe, and experiments have shown that the rate of oxygen uptake through the cloacal bursae is very high.

Patterns of Circulation and Respiration

The morphological complexity of the hearts of turtles and of squamates allows them to adjust blood flow through the pulmonary and systemic circuits to meet short-term changes in respiratory requirements. The key to these adjustments is changing pressures in the systemic and pulmonary circuits.

Recall that in the turtle heart, deoxygenated blood from the right atrium normally flows from the cavum venosum across the muscular ridge and into the cavum pulmonale. The blood pressure inside the heart increases as the ventricle contracts, and blood is first ejected into the pulmonary artery because the resistance to flow in the pulmonary circuit is normally less than the resistance in the systemic circuit. However, the resistance to blood flow in the pulmonary circuit can be increased by muscles that narrow the diameter of blood vessels. When this happens, the delicate balance of pressure in the heart that maintained the separation of oxygenated and deoxygenated blood is changed. When the resistance of the pulmonary circuit is essentially the same as that of the systemic circuit, blood flows out of the cavum pulmonale and cavum venosum at the same time, and some deoxygenated blood bypasses the lungs and flows into the systemic circuit (**Figure 12–8**). This process is called a right-to-left intracardiac shunt. *Right-to-left* refers to the shift of deoxygenated blood from the pulmonary circuit into the systemic circuit, and *intracardiac* means that it occurs in the heart rather than by flow between the major blood vessels.

Why would it be useful to divert deoxygenated blood from the lungs into the systemic circulation? The ability to make this shunt is not unique to turtles—it occurs also among lizards and snakes and in crocodilians. The heart morphology of lizards is much like that of turtles, and the same mechanism of changing pressures in the pulmonary and systemic circuits is used to achieve an intracardiac shunt. Crocodilians have hearts in which the ventricle is permanently divided into right and left

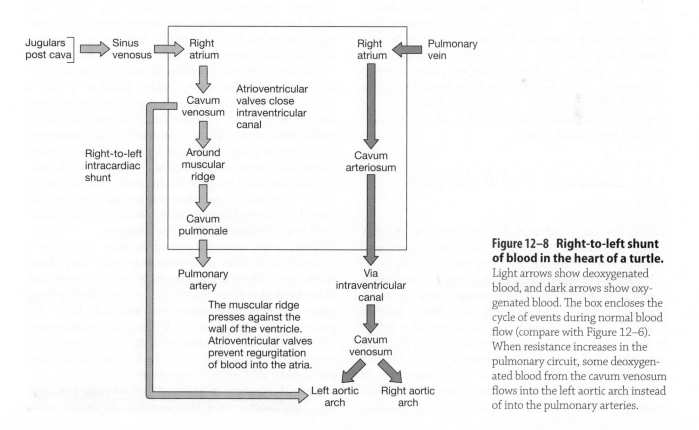

Figure 12–8 Right-to-left shunt of blood in the heart of a turtle. Light arrows show deoxygenated blood, and dark arrows show oxygenated blood. The box encloses the cycle of events during normal blood flow (compare with Figure 12–6). When resistance increases in the pulmonary circuit, some deoxygenated blood from the cavum venosum flows into the left aortic arch instead of into the pulmonary arteries.

halves by a septum, and they employ a different mechanism to achieve a right-to-left shunt.

The most general function for blood shunts may be the ability they provide to match the patterns of lung ventilation and pulmonary gas flow. Lizards, snakes, crocodilians, and turtles normally breathe intermittently, and periods of lung ventilation alternate with periods of **apnea** (no breathing). A mathematical model indicates that a combination of right-to-left and left-to-right shunts could stabilize oxygen concentration in blood during alternating periods of apnea and breathing.

Another function of intracardiac shunts may be to reduce blood flow to the lungs during breath-holding to permit more effective use of the oxygen stored in the lungs. Diving is one situation in which reptiles hold their breath. Many reptiles are excellent divers, and even terrestrial and arboreal forms such as green iguanas may dive into water to escape predators, but that is not the only situation in which defensive behaviors interfere with breathing. Turtles are particularly prone to have periods of breath-holding because their method of lung ventilation means they cannot breathe when they withdraw their heads and legs into their shells.

Yet another function of blood shunts may be to enhance digestion: a right-to-left shunt that retains carbon dioxide-rich blood in the body could enhance secretion of gastric acid in the digestive system. That hypothesis was tested by blocking the shunt in one group of American alligators (*Alligator mississippiensis*) and performing a sham operation on a control group of alligators. As predicted by the hypothesis, the group in which the shunt was blocked had lower rates of gastric-acid secretion and digested bone more slowly than did the control group.

12.4 Ecology and Behavior of Turtles

Turtles are long-lived animals. Even small species like the painted turtle (*Chrysemys picta*) do not mature until they are 7 or 8 years old, and they may live to be 14 or older. Larger species of turtles live longer. Estimates of centuries for the life spans of tortoises are exaggerated, but large tortoises and sea turtles may live at least as long as humans, and even box turtles may live longer than 50 years. These longevities make the life histories of turtles hard to study. Furthermore, a long lifetime is usually associated with a low replacement rate of individuals in the population, and species with those characteristics are at risk of extinction when hunting

or habitat destruction reduces their numbers. Conservation efforts for sea turtles and tortoises are especially important areas of concern.

Temperature Regulation and Body Size

Turtles can achieve considerable stability in body temperature by regulating their exchange of heat energy with the environment. Turtles basking on a log in a pond are a familiar sight in many parts of the world because few pond turtles are large enough to maintain body temperatures higher than the temperature of the water surrounding them. Emerging from the water to bask is the only way most pond turtles can raise their body temperatures to speed digestion, growth, and the production of eggs. In addition, basking may help aquatic turtles to rid themselves of algae and leeches. Exposure to ultraviolet light may activate vitamin D, which is involved in controlling calcium deposition in their bones and shell. A few turtles are quite arboreal; these turtles have small plastrons that allow considerable freedom of movement for the limbs. The big-headed turtle (*Platysternon megacephalum*) from Southeast Asia lives in fast-flowing streams at high altitudes and is said to climb on rocks and trees to bask. In North America, musk turtles (*Sternotherus*) bask on overhanging branches and drop into the water when they are disturbed.

Small terrestrial turtles, such as box turtles and small species of tortoises, can thermoregulate by moving between sunlight and shade. Small tortoises warm and cool quite rapidly, and they appear to behave very much like other small reptiles in selecting suitable microclimates for thermoregulation. Familiarity with a home range may assist this type of thermoregulation, and turtles can locate suitable microhabitats within an unfavorable area. A study conducted in Italy compared the thermoregulation of resident Hermann's tortoises (animals living in their own home ranges) with individuals that were brought to the study site and tested before they had learned their way around. The resident tortoises warmed faster and maintained more stable shell temperatures than did the strangers.

Turtles are unusual among reptiles in having a substantial number of species that reach large body sizes. The bulk of a large tortoise provides considerable thermal inertia, and large species like the Galápagos and Aldabra tortoises heat and cool slowly. The giant tortoises of Aldabra Atoll (*Dipsochelys dussumieri*), which weigh 60 kilograms or more, allow their body temperatures to rise to 32°C to 33°C on sunny days and cool to 28°C to 30°C overnight.

Large body size slows the rate of heating and cooling, but it can make temperature regulation more difficult. A small turtle can find shade beside a bush or even a clump of grass, but a giant tortoise needs a bigger object—a tree, for example. In open, sunny habitats, overheating can be a problem for giant tortoises. The difficulty is particularly acute for some tortoises on Grande Terre, an island in the Indian Ocean. During the rainy season each year, some of the turtles on the island move from the center of the island to the coast. This movement has direct benefits because the migrant turtles gain access to a seasonal flush of plant growth on the coast. As a result of the extra food, migrant females are able to lay more eggs than females that remain inland. There are risks to migrating, however, because shade is limited on the coast and the rainy season is the hottest time of the year. Tortoises on the coast must limit their activity to the vicinity of patches of shade, which may be no more than a single tree in the midst of a grassy plain. During the morning tortoises forage on the plain, but as their body temperatures rise, they move back toward the shade of the tree. As the day grows hotter, tortoises try to get into the deepest shade, and the biggest individuals do this most successfully. As the big tortoises (which are mostly males) push their way into the shade, they force smaller individuals (most of which are females) out into the sunlight, and some of these tortoises die of overheating.

Marine turtles are large enough to achieve a considerable degree of endothermy. A body temperature of 37°C was recorded by telemetry from a green turtle swimming in water that was 20°C. The leatherback turtle is the largest living turtle; adults may weigh up to 1000 kilograms. Leatherbacks range far from warm equatorial regions and in the summer can be found off the coasts of New England and Nova Scotia in water as cool as 8°C to 15°C. Body temperatures of these turtles appear to be at least 18°C higher than water temperatures, and a countercurrent arrangement of blood vessels in the flippers may contribute to retaining heat produced by muscular activity.

Social Behavior and Courtship

Turtles use tactile, visual, and olfactory signals during social interactions. Many pond turtles have distinctive stripes of color on their heads, necks, and forelimbs and on their hindlimbs and tail. These patterns are used by herpetologists to distinguish the species, and they may be species-isolating mechanisms for the turtles as well. During the mating season, male pond turtles swim in pursuit of other turtles, and the color and pattern on the posterior limbs may enable males to identify females of their own species. At a later stage of courtship, when the male turtle swims backward in front of the female and vibrates his claws against the sides of her head, both sexes can see the patterns on their partner's head, neck, and forelimbs (Figure 12–9).

Among terrestrial turtles, the behavior of tortoises is best known. Many tortoises vocalize while they are mating; the sounds they produce have been described as grunts, moans, and bellows. The frequencies of the calls that have been measured range from 500 hertz to 2500 hertz. Some tortoises have glands that become enlarged during the breeding season and appear to produce pheromones. The secretion of the subdentary gland found on the underside of the jaw of tortoises in the North American genus Gopherus appears to identify both the species and the sex of an individual. During courtship, males and females of the Florida gopher tortoise rub their subdentary gland across one or both forelimbs, and then extend the limbs toward the other individual, which may sniff at them. Males also sniff the cloacal region of other tortoises, and male tortoises of some species trail females for days during the breeding season. Fecal pellets may be territorial markers; fresh fecal pellets from a dominant male tortoise have been reported to cause dispersal of conspecifics.

Tactile signals used by tortoises include biting, ramming, and hooking. These behaviors are used primarily by males, and they are employed against other males and also against females. Bites are usually directed at the head or limbs, whereas ramming and hooking are directed against the shell. The epiplastral region is used for ramming, and in some species the epiplastral bones of males are elongated and project forward beneath the neck. A tortoise about to ram another individual raises itself on its legs, rocks backward, and then plunges forward, hitting the shell of the other individual with a thump that can be heard from a distance of 100 meters in large species. During hooking, the epiplastral projections are placed under the shell of an adversary, and the aggressor lifts the front end of its shell and walks forward. The combination of lifting and pushing hustles the adversary along and may even overturn it.

Movements of the head appear to act as social signals for tortoises, and elevating the head is a signal of dominance in some species. Herds of tortoises have social hierarchies that are determined largely by aggressive encounters. Ramming, biting, and hooking are employed in these encounters, and the larger individual is usually the winner—although experience may play some role. These social hierarchies are expressed in the priority of different individuals in access to food or forage areas, mates, and resting sites. Dominance relationships also appear to be involved in determining the sequence in

Figure 12–9 Social behavior of turtles. (a) A male painted turtle (*Chrysemys picta*) courting a female by vibrating the elongated claws of his fore feet against the sides of her head. (b) The head-raising dominance posture of a Galápagos tortoise, *Chelonoidis*. (This behavior can sometimes be elicited by crouching in front of a male tortoise and raising your arm.)

which individual tortoises move from one place to another. The social structure of a herd of tortoises can be a nuisance for zookeepers trying to move the animals from an outdoor pen into an enclosure for the night because the tortoises resist moving out of their proper rank sequence.

12.5 **Reproductive Biology of Turtles**

All turtles are oviparous. Female turtles use their hindlimbs to excavate a nest in sand or soil and deposit a clutch that ranges from four or five eggs for small species to more than a hundred eggs for the largest sea turtles. Turtles in the families Cheloniidae, Dermochelyidae, and Chelydridae lay eggs that have soft, flexible shells, as do most species in the families Bataguridae, Emydidae, and Pelomedusidae. The eggs of turtles in the families Carettochelyidae, Chelidae, Kinosternidae, Testudinidae, and Trionychidae have rigid shells.

Embryonic development typically requires 40 to 60 days and, in general, soft-shelled eggs develop more rapidly than hard-shelled eggs. Some species of turtles lay their eggs in late summer or fall, and the eggs have a diapause (a period of arrested embryonic development) during the winter and resume development when temperatures rise in the spring.

Moisture and Egg Development

The amount of moisture in the soil surrounding a turtle nest is an important variable during embryonic development of the eggs. Moist incubation conditions produce larger hatchlings than do dry conditions, apparently because water is needed for metabolism of the yolk. When water is limited, turtles hatch early and at smaller body sizes, and their guts contain a quantity of yolk that was not used during embryonic development. Hatchlings from nests under wetter conditions are larger and contain less unmetabolized yolk. The large hatchlings that emerge from moist nests are able to run and swim faster than hatchlings from drier nests and, as a result, they may be more successful at escaping predators and catching food.

Temperature-Dependent Sex Determination

The discovery that the sex of some reptiles is determined by the temperature they experienced during embryonic development has important implications for understanding patterns of life history as well as conservation of these species. Temperature-dependent sex determination (TSD) is widespread among turtles, is apparently universal among crocodilians, and is known for the tuatara and a few species of lizards. The switch from one sex to the other occurs within a span of 3°C or 4°C (**Figure 12–10**). Three patterns of TSD have been described:

- Type 1 produces males at high temperatures and females at low temperatures.
- Type 2 produces females at high temperatures and males at low temperatures.
- Type 3 produces females at both low and high temperatures and males at intermediate temperatures.

Allowing environmental factors to determine the sex of one's offspring sounds like a risky proposition, and several hypotheses have sought to find a benefit to TSD. An early suggestion proposed that a female turtle could select a nest site that would produce male or female young, depending on the ratio of the sexes in the adult population—presumably it would be beneficial to produce young of the rarer sex. A more recent hypothesis holds that TSD may be correlated with sexual size dimorphism of adults—that is, for each species high incubation temperatures produce the sex that benefits from being larger as an adult.

A third possibility is that selection acts on a different effect of incubation temperature and sex is merely a by-product of that selection. Incubation temperature affects the body size, growth rate, swimming and running speeds, and the mode of escape behavior employed by hatchling turtles, and selection could act on any one of these traits or on a combination of them. For example, hatchling snapping turtles from eggs incubated at 28°C attempted to flee to escape predators, whereas hatchlings produced at 26°C or 30°C remained motionless and avoided detection. In a field enclosure, hatchlings from 26°C and 30°C had significantly higher survival rates after one year than hatchlings from 28°C.

Temperatures of natural nests are not completely stable, of course. There is some daily temperature variation superimposed on a seasonal cycle of changing

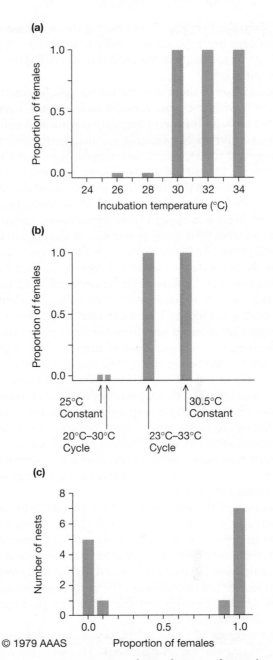

© 1979 AAAS

Figure 12–10 Temperature-dependent sex determination. (a) Eggs of the European pond turtle *Emys orbicularis* hatch into males when they are incubated at 26°C and 28°C and into females at 30°C and warmer. (b) The North American map turtle *Graptemys ouachitensis* shows the same pattern. A temperature that cycles between 20°C and 30°C produces males, whereas a temperature cycle of 23°C to 33°C produces females. (c) Natural nests of map turtles produce predominantly males or females, depending on the nest temperature.

environmental temperatures. The middle third of embryonic development is the critical period for sex determination; the sex of the embryos depends on the temperatures they experience during those few weeks. When eggs are exposed to a daily temperature cycle, the high temperature in the cycle is most critical for sex determination.

In addition to these external sources of temperature variation within a nest, a turtle embryo may have a limited capacity for thermoregulation while it is still in the egg. Embryos of a Chinese softshell turtle (*Pelodiscus sinensis*) rotate within their eggshells so that their backs point toward the source of heat. Two sets of eggs were incubated in the laboratory. Initially all of the eggs were buried in the substrate with their backs facing upward. Heat was applied from above to one group of eggs and from 90 degrees to one side of the eggs in the second group. Embryos in the eggs that received heat from above rotated about 10 degrees in one direction during the first 15 days of the experiment, and then rotated back to approximately their original orientation during the second 15 days. In contrast, embryos in the eggs that received heat from the side rotated about 30 degrees, which is one-third of the initial 90-degree offset between the top of the shell and the direction of the source of heat. Furthermore, when the source of heat was shifted to the opposite side of the container on day 15, the embryos responded by rotating nearly 50 degrees to bring their dorsal surfaces closer into alignment with the new direction of the heat source. Temperature measurements revealed a 1°C difference between the sides of the embryo facing toward and away from the source of heat.

Because of the narrowness of the thermal windows involved in sex determination and the variation that exists in environmental temperatures, both sexes are produced under field conditions but not necessarily in the same nests. A study of painted turtles (*Chrysemys picta*) found that two-thirds of the nests produced only males or only females and the remaining nests produced hatchlings of both sexes.

The wetness of a nest interacts with temperature in determining the sex of hatchlings. Dry substrates induced the development of some female painted turtles at low temperatures (26.5°C and 27°C) that would normally have produced only males. The wetness of the substrate did not affect the sex of turtles from eggs incubated at 30.5°C and 32°C: all the hatchlings from these eggs were females, as would be expected on the basis of temperature-dependent sex determination alone.

Some conservation efforts were initially confounded by temperature-dependent sex determination in sea turtles. Many programs collect hundreds of eggs from natural nests and incubate them under controlled conditions. Unfortunately, the unnaturally uniform conditions of incubation used by some programs have produced hatchlings of only one sex.

12.6 Hatching and the Behavior of Baby Turtles

Internal and external cues appear to be important in synchronizing the hatching of turtle embryos. Temperature variation within the nest causes some embryos to develop more rapidly than others, but the slow embryos increase their rates of development in a brief catch-up period shortly before hatching. Dryness inhibits hatching by some species of turtles that wait until the nest is moistened or even flooded before emerging from the eggs. Once emergence starts, the vibrations produced by the first embryos to emerge from their shells appear to stimulate other embryos to initiate emergence, This positive feedback cycle promotes simultaneous hatching and emergence of all the eggs in a nest, perhaps saturating predators and increasing the chances that some of the hatchlings will survive.

Nest Emergence

The first challenge that a turtle faces after hatching is escaping from the nest, and in some instances interactions among all of the hatchlings in a nest may be essential. Sea-turtle nests are quite deep; the eggs may be buried 50 centimeters beneath the sand, and the hatchling turtles must struggle upward through the sand to the surface. After several weeks of incubation, the eggs all hatch within a period of a few hours, and a hundred or so baby turtles find themselves in a small chamber at the bottom of the nest hole. Spontaneous activity by a few individuals sets the whole group into motion, crawling over and under one another. The turtles at the top of the pile loosen sand from the roof of the chamber as they scramble about, and the sand filters down through the mass of baby turtles to the bottom of the chamber.

Periods of a few minutes of frantic activity are interspersed with periods of rest, possibly because the turtles' exertions reduce the concentration of oxygen in the nest and they must wait for more oxygen to diffuse into the nest from the surrounding sand. Gradually, the entire group of turtles moves upward through the sand as a unit until it reaches the surface. As the baby turtles approach the surface, high sand temperatures probably inhibit further activity, and the turtles wait a few centimeters below the surface until night falls, when a decrease in temperature triggers emergence. All the babies emerge from a nest in a very brief time, and all the babies in different nests that are ready to emerge on a given night leave their nests at almost the same time, probably because their behavior is cued by temperature. The result is the sudden appearance of hundreds or even thousands of baby turtles on the beach, each one crawling toward the ocean as fast as it can.

Reaching the Water

Simultaneous emergence is an important feature of the hatching of sea turtles because the babies suffer a high mortality crossing the few meters of beach and surf. Terrestrial predators—crabs, foxes, raccoons, and other predators—gather at the turtles' breeding beaches at hatching time and await their appearance. Some of the predators come from distant places to prey on the baby turtles. In the surf, sharks and bony fishes patrol the beach. Few, if any, baby turtles would get past that gauntlet were it not for the simultaneous emergence that brings all the babies out at once and temporarily swamps the predators.

The Early Years

Turtles are self-sufficient at hatching, and the adults provide no parental care to hatchlings. Baby turtles are secretive and are rarely encountered in the field. Probably small turtles spend most of their time in concealment because they are vulnerable to predators. A hard shell is a turtle's defense, and the adults of most species of turtles have shells that predators cannot crush. Baby turtles are in a very different position, however: they are bite-sized and their shells are not rigid enough to resist crushing—baby turtles are Oreo cookies from a predator's perspective.

We know even less about the biology of baby sea turtles than we do about terrestrial and freshwater species. Where the turtles go in the period following hatching has been a long-standing puzzle in the life cycle of sea turtles. For example, green turtles hatch in the late summer at Tortuguero on the Caribbean coast of Costa Rica. The turtles disappear from sight as soon as they are at sea, and they are not seen again until they weigh 4 or 5 kilograms. Apparently, they spend the intervening period floating in ocean currents. Material drifting on the surface of the sea accumulates in areas where currents converge, forming drift lines of flotsam that include sargassum (brown algae) and the vertebrate and invertebrate fauna associated with it. These drift lines are probably important resources for juvenile sea turtles.

Navigation and Migration

Pond turtles and terrestrial turtles usually lay their eggs in nests that they construct within their home ranges. The mechanisms of orientation that they use to find nesting areas are probably the same ones they use to find their way among foraging and resting areas. Familiarity with local landmarks is an effective method of navigation for these turtles, and they may also use the sun for orientation. Sea turtles have a more difficult time, partly because the open ocean lacks conspicuous landmarks but also because feeding and nesting areas are often separated by hundreds or thousands of kilometers. Most sea turtles are carnivorous. The leatherback turtle feeds on jellyfishes, loggerhead and Ridley sea turtles eat crabs and other benthic invertebrates, and the hawksbill turtle uses its beak to scrape encrusting organisms (sponges, tunicates, bryozoans, mollusks, and algae) from reefs. Juvenile green turtles are carnivorous, but the adults feed on vegetation, particularly turtle grass (*Thalassia testudinum*), which grows in shallow water on protected shorelines in the tropics. The areas that provide food for sea turtles often lack the characteristics needed for successful nesting, and many sea turtles move long distances between their feeding grounds and their breeding areas.

The ability of sea turtles to navigate over thousands of kilometers of ocean and find their way to nesting beaches that may be no more than tiny coves on a small island is astonishing. The migrations of sea turtles, especially the green turtle, have been studied for decades. Turtles captured at breeding sites in the Caribbean and Atlantic Oceans have been individually marked with metal tags since 1956, and tag returns from turtle catchers and fishing boats have allowed the major patterns of population movements to be established (**Figure 12–11**).

Four major nesting sites of green turtles have been identified in the Caribbean and South Atlantic: one at Tortuguero on the coast of Costa Rica, one on Aves Island in the eastern Caribbean, one on the coast of Suriname, and one on Ascension Island between South America and Africa. Male and female green turtles congregate at these nesting grounds during the nesting season. The male turtles remain offshore, where they court and mate with females, and the female turtles come ashore to lay eggs on the beaches. A typical female green turtle at Tortuguero produces three clutches of eggs about 12 days apart. About a third of the female turtles in the Tortuguero population nest in alternate years, and the remaining two-thirds of the turtles follow a 3-year breeding cycle. The coast at Tortuguero lacks the beds of turtle grass on which green turtles feed, and the turtles come to Tortuguero only for nesting. In the intervals between breeding periods, the turtles disperse around the

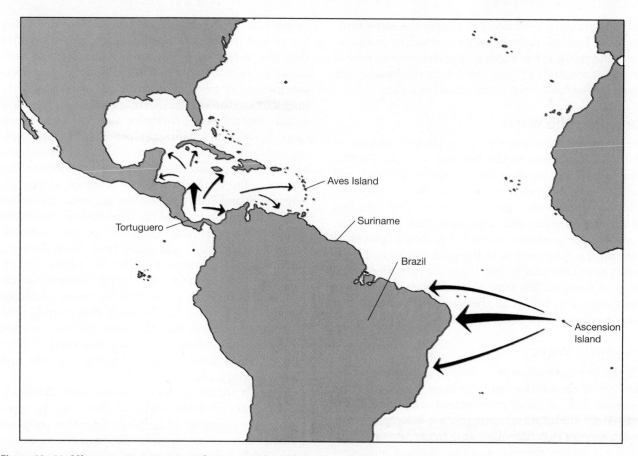

Figure 12–11 Migratory movements of green turtles (*Chelonia mydas*). The population that nests on beaches in the Caribbean is drawn from feeding grounds in the Caribbean and Gulf of Mexico. The turtles that nest on Ascension Island feed along the coast of northern South America.

Caribbean. The largest part of the Tortuguero population spreads northward along the coast of Central America. The Miskito Bank off the northern coast of Nicaragua appears to be the main feeding ground for the Tortuguero colony. A smaller number of turtles from Tortuguero swim south along the coast of Panama, Colombia, and Venezuela. Female green turtles return to their natal beaches to nest, and the precision with which they home is astonishing. Female green turtles at Tortuguero return to the same kilometer of beach to deposit each of the three clutches of eggs they lay in a breeding season.

Probably the most striking example of the ability of sea turtles to home to their nesting beaches is provided by the green turtle colony that has its feeding grounds on the coast of Brazil and nests on Ascension Island, a small volcanic peak that emerges from the ocean. The island is 2200 kilometers east of Brazil and less than 20 kilometers in diameter—a tiny target in the vastness of the South Atlantic.

Navigation by Adult Turtles How do adults turtles migrating to or from breeding sites find their way across thousands of kilometers of ocean? Other migratory animals use a variety of cues for navigation, and sea turtles probably do so as well. Chemosensory information may be one important component of their navigation. For example, the South Atlantic equatorial current flows westward, washing past Ascension Island and continuing toward Brazil, and the odor plume of the island may help to guide female turtles back to the island to nest. That is, a female turtle leaving the coast of Brazil may swim upstream in the South Atlantic equatorial current (i.e., up the odor gradient) to locate Ascension Island.

It is impractical to locate female turtles off the coast of Brazil as they are about to begin their journey to the island, but it is easy to find turtles that have completed nesting at Ascension Island and are ready to start back to Brazil. Five female green turtles were tracked on their return trip using the Argos satellite system. The

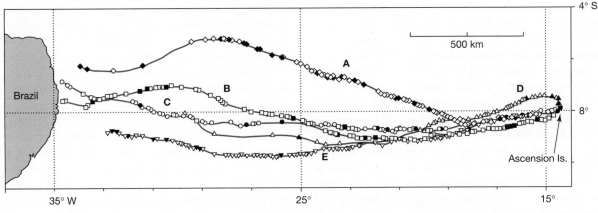

Figure 12–12 Turtle tracks. The paths followed by five green turtles (*Chelonia mydas*) were tracked by satellite as they returned from Ascension Island to Brazil. The dark symbols are the most accurate positions (within 1 km).

turtles traveled 1777 to 2342 kilometers and reached Brazil in 33 to 74 days.

For the first 500 kilometers of the journey, they followed a west-southwest heading that carried them slightly south of a direct route toward the bulge of Brazil (**Figure 12–12**). At this stage they were following the route of the South Atlantic equatorial current. Perhaps the turtles were simply being carried off course by the current, but they may have been using the same guidance system that they rely on for their outward journey—that is, staying within the plume of the island's scent. Even though the current carries them slightly south of a direct route to Brazil, they may save energy and move faster by initially staying in the current.

If they remained in the current for the entire trip, they would be carried too far south, so the turtles make a midcourse correction. The new course heads west-northwest on a nearly direct route to the bulge on the coast of Brazil. The shift in direction might be triggered by the waning strength of the scent plume. The turtles spend more than 90 percent of the journey underwater, suggesting that they may be sampling the plume in three dimensions.

Studies of green sea turtles homing across the Mozambique Channel to feeding grounds on the east coast of Africa suggest that they initially used an unidentified cue, possibly scent or the prevailing direction of wave movement, for navigation. As they approach their goal they shifted to magnetic orientation at distances of tens of kilometers, and as they approached the shore they probably relied on landmarks.

Navigation by Hatchling and Juvenile Sea Turtles Several studies of navigation by hatchling loggerhead turtles have shown that they also use at least three cues for orientation: light, wave direction, and magnetism. These stimuli played sequential roles in the turtles' behavior. When they emerged from their nests, the hatchlings crawled toward the brightest light they saw. The sky at night is lighter over the ocean than over land, so this behavior brought them to the water's edge. (Lights at shopping centers, streetlights, and even porch lights on beachfront houses can confuse these and other species of sea turtles and lead them inland, where they are crushed on roads or die of overheating the next day.)

In the ocean, the loggerhead hatchlings swam into the waves. This response moved them away from shore and ultimately into the Gulf Stream and then into the Atlantic Gyre, the current that sweeps around the Atlantic Ocean in a clockwise direction, northward along the coast of the United States and then eastward across the Atlantic. Off the coast of Portugal, the gyre divides into two branches. One turns north toward England, and the other swings south past the bulge of Africa and eventually back westward across the Atlantic. It is essential for the baby turtles to turn right at Portugal; if they fail to make that turn, they are swept past England into the chilly North Atlantic, where they perish. If they do turn southward off the coast of Portugal, they are eventually carried back to the coast of tropical America—a round-trip that takes 5 to 7 years.

Magnetic orientation appears to keep the turtles in the Atlantic Gyre and to tell them when to turn right to catch the current that will carry them to the South Atlantic. We usually think of Earth's magnetic field as providing two-dimensional information—north-south and east-west—but it's more complicated than that. The field loops out of Earth's north and south magnetic poles. At the equator, the field is essentially parallel to Earth's surface (in other words, it forms an angle of 0 degrees), and at the poles it intersects the surface at an angle of 90 degrees. Thus, the three-dimensional orientation of Earth's magnetic field provides both directional information (which way is magnetic north?) and information about latitude (what is the angle at which the magnetic field intersects Earth's surface?).

When loggerheads in a pool on land that had no waves were exposed to an artificial magnetic field at the 57-degree angle of intersection with Earth's surface that is characteristic of Florida, they swam toward artificial east—even when the magnetic field was changed 180 degrees, so that the direction they thought was east was actually west. That is, they were able to use a compass sense to determine direction. But that wasn't all they could do: subsequent studies showed that as the three-dimensional magnetic field was changed to match points around the North Atlantic Gyre, the hatchlings turned in the appropriate direction at each location to remain in the North Atlantic Gyre and ultimately to return to the coast of Florida (**Figure 12–13**). Thus, it appears that young loggerhead turtles can use magnetic sensitivity to recognize both direction and latitude.

12.7 **Conservation of Turtles**

Slow rates of growth and delayed maturity are characteristics that make a species vulnerable to the risk of extinction when changing conditions increase the mortality of adults or drastically reduce recruitment of juveniles into the population. The plight of large tortoises and sea turtles is particularly severe, not only because these species are among the largest and slowest growing turtles but also because other aspects of their biology expose them to additional risk.

The largest living tortoises are found on the Galápagos and Aldabra Islands. The relative isolation of these small and (for humans) inhospitable landmasses has probably been an important factor in the survival of tortoises. Human colonization of the islands has brought with it domestic animals such as goats and donkeys, which compete with tortoises for the limited quantities of vegetation found in these arid habitats as well as dogs, cats, and rats that prey on tortoise eggs and on baby tortoises.

The limited geographic range of a tortoise that occurs on only a single island makes it vulnerable to extinction. In 1985 and again in 1994 brush fires on the island of Isabela in the Galápagos Archipelago threatened the 20 surviving individuals of *Chelonoidis vicina* and emphasized the advantage of moving some or all of the turtles to the breeding facility operated by the Charles Darwin Research Station on Santa Cruz Island. This station has a successful record of breeding and releasing another species of Galápagos tortoise, *Chelonoidis hoodensis*, which is native to Española Island. In the early 1960s, only 14 individuals of this form could be located. All were adults and apparently had not bred successfully for many years. All of the tortoises were moved to the research station, and the first babies were produced in 1971. On March 24, 2000, the one-thousandth captive-bred tortoise was released on Española. This success story shows that carefully controlled captive breeding and release programs can be an effective method of conservation for endangered species of turtles.

Entire turtle faunas are threatened in some areas. Nearly all species of turtles in Southeast Asia are now at risk because of economic and political changes in the region. Turtles have traditionally been used in China for food and for their supposed medicinal benefits. A 2-day survey in just two Chinese food markets found an estimated 10,000 turtles for sale. If the turnover time is estimated conservatively to be a week, those two markets would consume more than a quarter of a million turtles annually. When that rate is extrapolated to all of the markets in China, the estimate rises to 12 million turtles sold annually in China alone.

As long-lived animals with low reproductive rates, turtles have exactly the wrong characteristics to withstand heavy predation. Very little is known about the natural history of Chinese turtles. In fact, some species such as *Cuora mccordi* are known scientifically only from specimens purchased in markets—wild populations have never been described. These species may never be known in the wild; specimens have not been seen in markets for several years, and the species may be extinct.

Although China is considered the biggest black hole for turtles, it is by no means the only one. Madagascar is home to endangered species of many kinds of animals, including tortoises. Although these species are protected by international treaty, they are smuggled out of the country by the score for sale as pets in Japan, Europe, and North America, where they fetch high prices. In the United States, our own protected species are collected and sold illegally as pets, and the most endangered species command the highest prices.

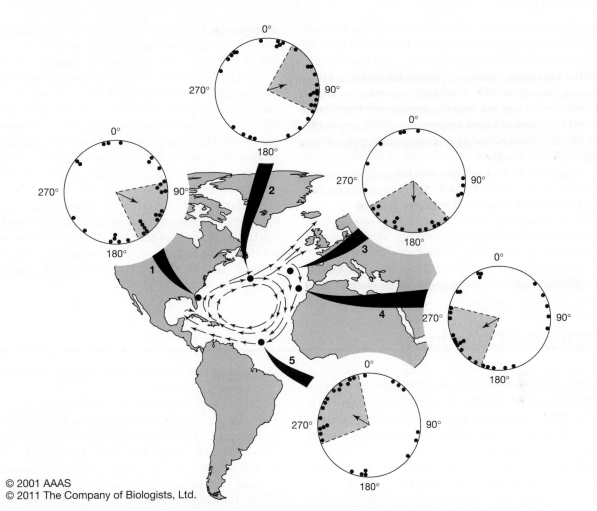

Figure 12–13 Orientation of hatchling loggerheads in magnetic fields duplicating locations along their migratory route. The direction of flow of the North Atlantic Gyre is shown by arrows. In the orientation circles, each dot represents a single hatchling, the arrow in the center of each circle indicates the mean angle of the group, and the pie-shaped segment shows the 95 percent confidence interval of the mean angle. (1) Turtles tested in a magnetic field like that found off the coast of Florida oriented to the southeast, the direction of swimming that would bring them into the gyre. (2) In a magnetic field matching the mid-Atlantic at the latitude of New York City, the turtles oriented northeast, following the path of the gyre. (3) A field like that found off the westward bulge of Spain caused turtles to begin their right turn by orienting to the southeast, and (4) when they were tested with a field like that off the southern coast of Portugal, they had completed the turn and were oriented to the southwest. (5) In a field like that at the southern boundary of the gyre, the turtles were oriented northwest, the direction that would keep them in the gyre and bring them back to Florida.

Summary

The earliest turtles known, fossils from the Triassic, have nearly all the features of derived turtles. The first Triassic forms were not able to withdraw their heads into the shell, but this ability appeared in the two major lineages of living turtles, which were established by the Late Triassic. The cryptodire turtles retract the head with a vertical flexion of the neck vertebrae, whereas the pleurodires use a sideward bend.

Turtles are among the most morphologically specialized vertebrates. The shell is formed of dermal bone that is fused to the vertebral column and ribs. In most turtles, the dermal shell is overlain by a horny layer of epidermal scales. The limb girdles are inside the rib cage. Breathing presents special difficulties for an animal that is encased in a rigid shell: exhalation is accomplished by muscles that squeeze the viscera against the lungs, and inhalation is accomplished by muscles that increase the volume of the visceral cavity, thereby allowing the lungs to expand. The heart of turtles (and of lepidosaurs as well) is able to shift blood between the pulmonary and systemic circuits in response to the changing requirements of gas exchange and thermoregulation.

The social behavior of turtles includes visual, tactile, and olfactory signals used in courtship. Dominance hierarchies shape the feeding, resting, and mating behaviors of some of the large species of tortoises. All species of turtles lay eggs, and none provides parental care to their young. Coordinated activity by hatchling sea turtles may be necessary to enable them to dig themselves out of the nest, and simultaneous emergence of baby sea turtles from their nests helps them to evade predators as they rush down the beach into the ocean. Sea turtles migrate tens, hundreds, and even thousands of kilometers between their feeding areas and their nesting beaches and use a variety of cues for navigation.

The life history of many turtles makes them vulnerable to extinction. Slow rates of growth and long times required to reach maturity are characteristic of turtles in general and of large species of turtles in particular. Turtles cannot withstand commercial exploitation of the sort that has occurred historically on many oceanic islands and is currently in progress in parts of Asia, or the worldwide loss and degradation of habitat.

Discussion Questions

1. Considering the life-history characteristics of most turtles, which of these approaches to conservation of a species of turtle is likely to be more effective, and why: protecting eggs and nests so that the hatchling turtles can reach the water or protecting adult female turtles?

2. Why is it so difficult to determine the phylogenetic relationships of turtles to other sauropsids?

3. Does temperature-dependent sex determination offer any benefits to turtles, or is it better regarded as a possibly ancestral system of sex determination that has persisted in most lineages of turtles because it works well enough?

4. What simple experiment could you conduct to test the hypothesis that hatchling sea turtles use orientation to Earth's magnetic field to swim in the correct direction when they reach the ocean?

5. The absence of a complete ventricular septum allows turtles to reduce the blood flow to the lungs by shunting blood from the right (pulmonary) circuit to the left (systemic) circuit. When and why would a right-to-left blood shunt be advantageous?

Additional Information

Anquetin, J., et al. 2009. A new stem turtle from the Middle Jurassic of Scotland: New insights into the evolution and palaeoecology of basal turtles. *Proceedings of the Royal Society B* 276:879–886.

Benhamou, S., et al. 2011. The role of geomagnetic cues in green turtle open sea navigation. *PLoS ONE* 6(10):e26672. doi:10.1371/journal.pone.0026672.

Colbert, P. L., et al. 2010. Mechanism and cost of synchronous hatching. *Functional Ecology* 24:112–121.

Doody, J. S. 2011. Environmentally cued hatching in reptiles. *Integrative and Comparative Biology* 51:49–61.

Du, W.-G., et al. 2011. Behavioral thermoregulation by turtle embryos. *Proceedings of the National Academy of Sciences* 108:9513–9515.

Fuentes, M. M. P. B., et al. 2011. Vulnerability of sea turtle nesting grounds to climate change. *Global Change Biology* 17:140–153.

Gaffney, E. S., et al. 2011. Evolution of the side-necked turtles: The Family Podocnemidae. *Bulletin of the American Museum of Natural History* 350:1–237.

Heiss, E., et al. 2010. The fish in the turtle: On the functionality of the oropharynx in the common musk turtles, *Sternotherus odoratus* (Chelonia, Kinosternidae) concerning feeding and underwater respiration. *The Anatomical Record* 293:1416–1424.

Jaffee, A. L., et al. 2011. The evolution of island gigantism and body size variation in tortoises and turtles. *Biology Letters* 7:558–561.

Luschi, P. et al. 1998. The navigational feats of green sea turtles migrating from Ascension Island investigated by satellite telemetry. *Proceedings of the Royal Society B* 265:2279-2284.

Lyson, T. R., et al. 2012. MicroRNAs support a turtle + lizard clade. *Biology Letters* 8:104-107.

Marcovaldi, M. A., et al. 2010. Satellite-tracking of female loggerhead turtles highlights fidelity behavior in northeastern Brazil. *Endangered Species Research* 12:263–272.

McGlashan, J. K., et al. 2012. Embryonic communication in the nest: Metabolic responses of reptilian embryos to developmental rates of siblings. *Proceedings of the Royal Society B*, in press.

Merrill, M. W., and M. Salmon. 2011. Magnetic orientation by hatchling loggerhead sea turtles (*Caretta caretta*) from the Gulf of Mexico. *Marine Biology* 158:101–112.

Mitchell-Campbell, M. A., et al. 2011. Staying cool, keeping strong: Incubation temperature affects performance of a freshwater turtle. *Journal of Zoology* 285:266–273.

Olmo, E. (ed.). 2010. *Sex Determination and Differentiation in Reptiles*. Basel, Switzerland: Karger.

Putnam, N. F., et al. 2010. Sea turtle nesting distributions and oceanographic constraints on hatchling migration. *Proceedings of the Royal Society B* 277:3631–3637.

Shoemaker, C. M., and D. Crews. 2009. Analyzing the coordinated gene network underlying temperature-dependent sex determination in reptiles. *Seminars in Cell & Developmental Biology* 20:293–303.

Southwood, A., and L. Avens. 2010. Physiological, behavioral, and ecological aspects of migration in reptiles. *Journal of Comparative Physiology B* 180:1–23.

Spencer, R.-J., and F. J. Janzen. 2011. Hatching behavior in turtles. *Integrative and Comparative Biology* 51:100–110.

Tyson, T. R., et al. 201. Transitional fossils and the origin of turtles. *Biology Letters* 6:830–833.

Wallace, B. P., et al. 2011. Global conservation priorities for marine turtles. *PLoS ONE* 6(9):e24510. doi:10.1371/journal.pone.0024510.

Websites

Biology and conservation of turtles

Desert Tortoise Information & Collaboration http://www.deserttortoise.org/

Sea Turtle Conservancy http://www.conserveturtles.org/

The Leatherback Trust http://www.leatherback.org/

Tortoise Trust http://www.tortoisetrust.org/

Turtle Survival Alliance http://turtlesurvival.org/

Satellite tracking of sea turtles

http://www.conserveturtles.org/satelliteturtles.php

http://www.seaturtle.org/tracking/

CHAPTER 13

The Lepidosaurs: Tuatara, Lizards, and Snakes

Some aspects of the biology of lepidosaurs may give us an impression of the ancestral way of life of amniotes, although many of the structural characters of lizards and snakes are derived. One derived characteristic of lizards is determinate growth; that is, increase in body size stops when the growth centers of the long bones ossify. This mechanism sets an upper limit to the size of individuals of a species and may be related to the specialization of most lizards as predators of small animals, such as insects. The predatory behavior of lizards ranges from sitting in one place and ambushing prey to seeking food by traversing a home range in an active, purposeful way. Broad aspects of the biology of lizards are correlated with these foraging modes, including morphology, exercise physiology, reproductive mode, defense against predators, and social behavior. The anatomical specializations of snakes are associated with their elongate body form and include modifications of the jaws and skull that allow them to subdue and swallow large prey.

13.1 The Lepidosaurs

Lepidosaurs are the largest group of nonavian reptiles, containing nearly 5500 species of lizards and more than 3300 species of snakes in addition to the tuatara (Table 13–1). Lepidosaurs are predominantly terrestrial tetrapods with some secondarily aquatic species, especially among snakes.

The skin of lepidosaurs is covered by scales and is relatively impermeable to water. The outer layer of the epidermis is shed at intervals. Tuatara and most lizards have four limbs; however, reduction or complete loss of limbs is widespread among some groups of lizards, and all snakes are limbless. Lepidosaurs have a transverse cloacal slit rather than the longitudinal slit that characterizes other tetrapods.

Lepidosaurs are the sister lineage of archosaurs (crocodilians and birds). Within the Lepidosauria, the Sphenodontidae (tuatara) is the sister group of Squamata (lizards and snakes). Within the squamates, lizards can be distinguished from snakes in colloquial terms but not phylogenetically because snakes are derived from lizards. Thus "lizards" is a paraphyletic group because it does not include all the descendants of a common ancestor. Nonetheless, lizards and snakes are distinct in many

Table 13–1 Major lineages of lepidosaurs

Rhynchocephalia (tuatara)

Sphenodontidae: The sister group of squamates. One extant species, *Sphenodon punctatus*, which now occurs only on islands off the coast of New Zealand.

Squamata (lizards and snakes)

Amphisbaenidae: About 164 species of legless, burrowing lizards found in the West Indies, South America, sub-Saharan Africa, and around the Mediterranean Sea. From 10 to 80 cm in length.

Lacertidae: About 306 species of small- to medium-size terrestrial lizards found in Europe, Africa, and Asia. From less than 10 cm to about 80 cm in length.

Teiidae: About 135 species of active terrestrial lizards ranging from near the Canadian border in North America to central Argentina, and including the West Indies. From 20 cm to about 1 m in length.

Gymnophthalmidae: About 231 species that live in the leaf litter of neotropical forests. Limb reduction is widespread in this lineage. Less than 6 cm in length.

Anguidae: About 118 species of anguids in North and South America, Europe, the Middle East, and southern China. About 20 cm to more than 1 m in length.

Helodermatidae: Two species of heavy-bodied lizards found in the southwestern United States and Mexico. Helodermatids are venomous. From 50 cm to about 1 m in length.

Varanidae: About 73 species in Africa, Asia, and the East Indies; about half the species are limited to Australia. Varanids have the greatest size range of any extant genus of vertebrate, extending from 20 cm to 3 m in length, and at least one species is venomous.

Iguanidae: About 38 species of herbivorous lizards—iguanas—found primarily in the New World but with representatives on the Galapagos Islands and Fiji. From about 25 cm to more than 1 m in length.

Phrynosomatidae: About 137 species of terrestrial, rock-dwelling, and arboreal lizards, mostly from North America. This family includes the wide-ranging genus *Sceloporus* and the horned lizards (*Phrynosoma*). From 10 to 30 cm in length.

Dactyloidae: About 377 species of arboreal lizards, mostly in the genus *Anolis*. From the Americas and the West Indies. From 10 to 50 cm in length.

Chamaeleonidae: About 187 species of primarily arboreal lizards but including a few grassland and terrestrial species; found in Africa and Madagascar and extending into southern Spain and along the west coast of the Mediterranean. From about 3 cm to 0.5 m in length.

Agamidae: About 416 species of lizards from the Middle East and parts of Asia, Africa, the Indo-Australian archipelago, and Australia. From about 10 cm to 1 m in length.

Scincidae: With 1503 species, this is the most species-rich lineage of lepidosaurs. Skinks occur on all continents except Antarctica. From about 10 to 40 cm in length.

Gekkonidae: About 881 species, some with modified scales (setae) on the bottoms of the toes that allow them to climb vertical surfaces and even to hang by a single toe. The geographic distribution of gekkonids includes every continent except Antarctica. From about 30 mm to 30 cm in length.

These are total lengths–i.e., nose to tail tip. Our understanding of the phylogenetic relationships of lizards is changing rapidly. This list is based largely on The Reptile Database (http://www.reptile-database.org/).

aspects of their ecology and behavior, and a colloquial separation is useful in discussing them.

13.2 Radiation of Sphenodontids and the Biology of Tuatara

The sphenodontids were a diverse group in the Mesozoic era, including terrestrial, arboreal, and marine forms and both insectivores and herbivores. Triassic forms were small, with body lengths of only 15 to 35 centimeters, but during the Jurassic and Cretaceous periods, some sphenodontids reached lengths of 1.5 meters.

The tuatara is the only extant sphenodontid (**Figure 13–1**). (*Tuatara* is a Māori word meaning "spines on the back." No -*s* is added to form the plural.) Tuatara

Figure 13–1 Tuatara. *Sphenodon punctatus* has natural populations on about 30 small islands off the coast of New Zealand and introduced populations on an additional three islands.

formerly inhabited the north and south islands of New Zealand, but the advent of humans and their associates (cats, dogs, rats, sheep, and goats) exterminated tuatara on the mainland about 800 years ago. Now wild populations are found on only 32 small islands off the coast.

The number of species of tuatara recognized by biologists has changed several times. Nineteenth-century biologists described some of the island populations as separate species, but by the twentieth century opinions had changed, and only a single species, *Sphenodon punctatus*, was recognized. A study published in 1990 concluded that the population of tuatara on North Brother Island was distinct enough from the other populations to be called by its nineteenth-century name, *Sphenodon guntheri*, but a study of variation in mitochondrial DNA published in 2010 concluded that *Sphenodon punctatus* should be recognized as a single species with a substantial amount of geographic variation among the island populations.

Adult tuatara are about 60 centimeters long. They are nocturnal, and in the cool, foggy nights that characterize their island habitats, they cannot raise their body temperatures during activity by basking in sunlight as lizards do. Body temperatures ranging from 6°C to 16°C have been reported for active tuatara, and these are low compared with the temperatures of most lizards. During the day, tuatara do bask, and they raise their body temperatures to 28°C or higher. Tuatara feed largely on invertebrates, with an occasional frog, lizard, or seabird for variety. The jaws and teeth of tuatara produce a shearing effect during chewing: The upper jaw contains two rows of teeth, one on the maxilla and the other on the palatine bones. The teeth of the lower jaw fit between the two rows of upper teeth, and the lower jaw closes with an initial vertical movement, followed by an anterior sliding movement. As the lower jaw slides, the food item is bent or sheared between the triangular cusps on the teeth of the upper and lower jaws.

Tuatara live in burrows that they may share with nesting seabirds. The burrows are spaced at intervals of 2 to 3 meters in dense colonies, and both male and female tuatara are territorial. They use vocalizations, behavioral displays, and color change in their social interactions.

The ecology of tuatara rests to a large extent on exploitation of the resources provided by colonies of seabirds. Tuatara feed on the birds, which are most vulnerable to predation at night. In addition, the quantities of guano produced by the birds, the scraps of food they bring to their nestlings, and the bodies of dead nestlings attract huge numbers of arthropods, which are eaten by tuatara. These arthropods are largely nocturnal and must be hunted when they are active.

Thus the nocturnal activity of tuatara and the low body temperatures resulting from being active at night are probably specializations that stem from the association of tuatara with colonies of nesting seabirds. This pattern of behavior and thermoregulation probably does not represent the ancestral condition even for sphenodontids, and there is no reason to interpret it as being ancestral for lepidosaurs or diapsids.

13.3 Radiation of Squamates

Determinate growth may be the most significant derived character of squamates. Growth occurs as cells proliferate in the cartilaginous epiphyseal plates at the ends of long bones, continues while the epiphyseal plates are composed of cartilage, and stops completely when the epiphyses fuse to the shafts of the bones, obliterating the cartilaginous plates. Determinate growth of this sort is characteristic of squamates (and also of birds and mammals). Crocodilians and turtles continue to grow throughout their lives, although the growth rates of adults are much slower than those of juveniles. Development of determinate growth in squamates may initially have been associated with the insectivorous diet that researchers believe was characteristic of early lepidosaurs. Generalized lizard-size animals can readily capture insects, whereas large insect-eating vertebrates, such as mammalian anteaters, require morphological or ecological specializations to capture tiny prey.

The fossil record of lizards is largely incomplete through the middle of the Mesozoic, but Late Jurassic deposits in China and Europe include members of most lineages of extant lizards. The major groups of lizards had probably diverged by the end of the Jurassic.

Lizards

Lizards range in size from diminutive geckos and chameleons less than 3 centimeters long to the Komodo monitor lizard, which is 3 meters long at maturity and weighs about 75 kilograms. A reconstruction of the skeleton of a fossil monitor lizard, *Megalania prisca*, from the Pleistocene epoch of Australia, is 5.5 meters long, and in life the lizard may have weighed more than 1000 kilograms.

Lizards display an enormous variety of body forms (**Figure 13–2** on page 314). Among agamid lizards, terrestrial and semiarboreal species display more variation in body shape than arboreal and rock-dwelling forms do, and that generalization probably applies to phrynosomatid lizards as well. About 80 percent of extant lizards weigh less than 20 grams as adults and are insectivorous. Spiny swifts and japalures are examples of these small, generalized insectivores. Other small lizards have specialized diets: the North American horned lizards and the Australian spiny devil feed on ants. Most geckos are nocturnal, and many species are closely associated with human habitations.

Lizards are adaptable animals that occupy habitats ranging from swamp to desert and even above the timberline in some mountains. Many species are arboreal. The most specialized of these are frequently laterally flattened, and they often have peculiar projections from the skull and back that help to obscure their outline. The Old World chameleons (Chamaeleonidae) are the most specialized arboreal lizards. The toes on their **zygodactylous** (Greek *zygo* = joined and *dactyl* = digit) feet are arranged in two opposable groups that grasp branches firmly, and additional security is provided by a prehensile tail. The tongue and hyoid apparatus are specialized, and the tongue can be projected forward more than a body's length to capture insects that adhere to its sticky tip. This feeding mechanism requires good eyesight, especially the ability to gauge distances accurately so that the correct trajectory can be employed. The chameleon's eyes are elevated in small cones and can move independently. When the lizard is at rest, the eyes swivel back and forth, viewing its surroundings. When it spots an insect, the lizard fixes both eyes on it and cautiously stalks to within shooting range.

Most large lizards are herbivores. Many iguanas are arboreal inhabitants of the tropics of Central and South America. Large terrestrial iguanas live on islands in the West Indies and the Galápagos Islands, probably because the absence of predators has allowed them to spend a large part of their time on the ground. Smaller terrestrial herbivores like the black iguanas live on the mainland of Mexico and Central America, and still smaller relatives such as the chuckwallas and desert iguanas range as far north as the southwestern United States. Many species of lizards live on beaches, but few extant species actually enter the water. The marine iguana of the Galápagos Islands is an exception. The feeding habits of the marine iguana are unique. It feeds on seaweed, diving 10 meters or deeper to browse on algae growing below the tide mark.

An exception to the rule of herbivorous diets for large lizards is found in the monitor lizards (family Varanidae). Varanids are active predators that feed on a variety of vertebrate and invertebrate animals, including birds and mammals. They circumvent the conflict between locomotion and lung ventilation that constrains the activity of other lizards by using a positive-pressure gular pump to assist the axial muscles, and they are able to sustain high levels of activity. The Komodo monitor lizard is capable of killing adult water buffalo, but deer

and feral goats are the usual prey. Large monitor lizards were widely distributed on the islands between Australia and Indonesia during the Pleistocene and may have preyed on pygmy elephants that also lived on the islands.

The hunting methods of the Komodo monitor are very similar to those employed by mammalian carnivores, showing that a non-mammalian brain is capable of complex behavior and learning. In the late morning, a Komodo monitor waits in ambush beside the trails deer use to move down from the hilltops, where they rest during the morning, to the valleys, where they sleep during the afternoon. The lizards, familiar with trails used by the deer, often wait where several deer trails converge. If no deer pass the lizard's ambush, it moves into the valleys, systematically stalking the thickets where deer are likely to be found. This purposeful hunting behavior, which demonstrates familiarity with the behavior of prey and with local geography, is in strong contrast to the opportunistic seizure of prey that characterizes the behavior of many lizards, but it is very similar to the hunting behavior of some snakes.

The skull of the Komodo monitor lizard is remarkably lightly built for a predator that routinely attacks and kills large prey. The calculated bite force of a Komodo monitor is only 15 percent of the force exerted by a crocodile with the same skull size. Rather than seizing and holding prey as crocodilians do, Komodo monitors deliver a slashing bite and allow the prey to flee. Venom from a gland in the mandible contains a mixture of enzymes that produces an abrupt drop in blood pressure and persistent hemorrhage that rapidly send the prey into shock, allowing the lizard to follow the path of flight to the carcass. The extinct giant monitor lizard, *Megalania prisca,* may have had a venom-delivery system like that of the Komodo monitor.

The effectiveness of monitor lizards as predators is reflected in reciprocal geographic distributions of monitors and small mammalian carnivores in the Indo-Australian Archipelago, Australia, and New Guinea. Wallace's Line is a zoogeographic boundary that extends between the islands of Borneo (on the west) and Sulawesi (to the east). It traces the location of a deep-water trench, and areas to the west of the line have never had a land connection with areas to the east. Because of this ancient separation, Wallace's Line marks the easternmost boundary of the occurrence of many species from the Asian mainland and the westernmost boundary of species from Australia and New Guinea. In particular, small carnivorous placental mammals are found west of the line and small marsupial carnivores to the east. Large species of monitors occur on both sides of the line, but small species are found on only the eastern side. West of the line, the ecological niche for small carnivores is filled by mammals (such as small cats, civets, mongooses, and weasels), but east of the line—especially in Australia and New Guinea—most of the small carnivores are monitor lizards.

Limb reduction has evolved more than 60 times among lizards, and every continent has one or more families with legless, or nearly legless, species. Leglessness in lizards is usually associated with life in dense grass or shrubbery, in which a slim, elongate body can maneuver more easily than a short one with functional legs. Some legless lizards crawl into small openings among rocks and under logs, and a few are subterranean.

The amphisbaenians are extremely **fossorial** (Latin *fossor* = a digger) lizards with specializations that are different from those of other squamates. The earliest amphisbaenian known is a fossil from the Late Cretaceous. Most amphisbaenians are legless, but the three species in the Mexican genus *Bipes* have well-developed fore legs that they use to assist entry into the soil but not for burrowing underground.

The skin of amphisbaenians is distinctive. The **annuli** (rings; singular *annulus*) that pass around the

circumference of the body are readily apparent from external examination, and dissection shows that the integument is nearly free of connections to the trunk. Thus, it forms a tube within which the amphisbaenian's body can slide forward or backward. The separation of trunk and skin is employed during locomotion through tunnels. Integumentary muscles run longitudinally from annulus to annulus. The skin over this area of muscular contraction is then telescoped and buckles outward, anchoring that part of the amphisbaenian against the walls of its tunnel. Next, contraction of muscles that pass anteriorly from the vertebrae and ribs to the skin slide the trunk forward within the tube of integument. Amphisbaenians can move backward along their tunnels with the same mechanism by contracting muscles that pass posteriorly from the ribs to the skin. (The name amphisbaenian is derived from Greek roots [*amphi* = double and *baen* = walk] that refer to the ability of amphisbaenians to move forward and backward with equal facility.)

The dental structure of amphisbaenians is also distinctive: They possess a single median tooth in the

Figure 13–2 Variation in body form among lizards.
Small, generalized insectivores: (a) spiny swift, *Sceloporus* (Phrynosomatidae); (b) japalure, *Calotes* (Agamidae). Herbivores: (c) black iguana, *Ctenosaura* (Iguanidae); (d) mastigure, *Uromastyx* (Agamidae). Ant specialists: (e) horned lizard, *Phrynosoma* (Phrynosomatidae); (f) spiny devil, *Moloch* (Agamidae). Nocturnal lizard: (g) Tokay gecko, *Gekko* (Gekkonidae). Arboreal lizard: (h) African chameleon, *Chamaeleo* (Chamaeleonidae). Legless lizard: (i) North American glass lizard, *Ophisaurus* (Anguidae). Large predator: (j) monitor lizard, *Varanus* (Varanidae).

upper jaw—a feature unique to this group of vertebrates. The median tooth is part of a specialized dental battery that makes amphisbaenians formidable predators, capable of subduing a wide variety of invertebrates and small vertebrates. The upper tooth fits into the space between two teeth in the lower jaw and forms a set of nippers that can bite out a piece of tissue from a prey item too large for the mouth to engulf as a whole.

The skulls of amphisbaenians are used for tunneling, and they are rigidly constructed. Some species have blunt heads; the rest have either vertically keeled or horizontally spade-shaped snouts. The burrowing habits of amphisbaenians make them difficult to study, but three major functional categories can be recognized (**Figure 13–3**):

- Blunt-snouted forms burrow by ramming their heads into the soil to compact it. Sometimes an oscillatory rotation of the head with its heavily keratinized scales is used to shave material from the face of the tunnel.
- Shovel-snouted amphisbaenians ram the end of the tunnel and then lift the head to compact soil into the roof.
- Wedge-snouted forms ram the snout into the end of the tunnel and then use the snout or the side of the neck to compress the material into the walls of the tunnel.

In parts of Africa, representatives of the three types occur together and share the subsoil habitat.

Figure 13–2 (*Continued*)

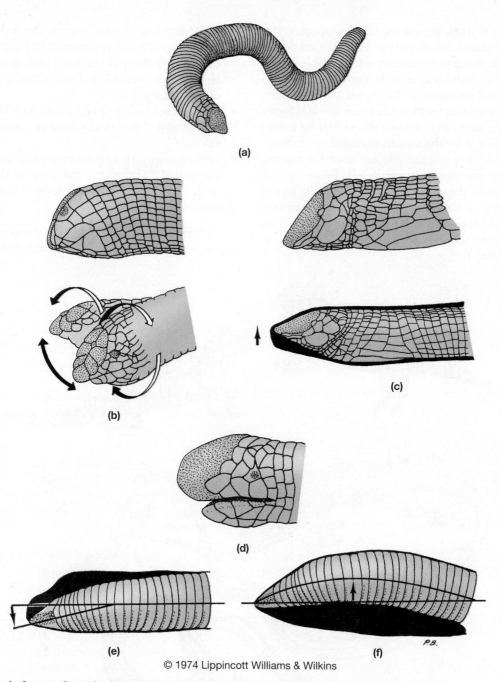

Figure 13–3 Body forms of amphisbaenians. The dark areas show how the tunnel is widened. (a) *Monopeltis* from Africa. Variation in snout shape: (b) blunt snouts (represented by *Agamodon*, from Africa); (c) shovel snouts (represented by *Rhineura* from Florida); (d) wedge snouts (represented by *Anops*, Brazil). Widening the tunnel (e) with movements of the head in loose soil or (f) with the anterior part of the body in dense soil.

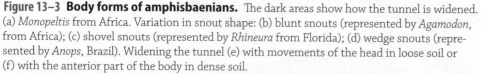

The unspecialized blunt-headed forms live near the surface where the soil is relatively easy to tunnel through, and the specialized forms live in deeper, more compact soil. The geographic range of unspecialized forms is greater than that of specialized ones, and in areas where only a single species of amphisbaenian occurs, it is a blunt-headed species.

The ecological relationship between unspecialized and specialized burrowers is complex. One might predict that the specialized forms with their more elaborate

Table 13–2 Major families of snakes

Leptotyphlopidae: About 117 species of small fossorial snakes found in Africa, southwestern Asia, South and Central America, and southwestern North America. From 10 to 20 cm in length.

Typhlopidae: About 369 species of small- to medium-size fossorial snakes with reduced eyes. Typhlopids are found on all continents except North America and Antarctica. From 20 to 75 cm in length.

Boidae: About 52 species of terrestrial, arboreal, and semiaquatic snakes from western North America through subtropical South America and the West Indies. From about 75 cm to almost 9 m in length.

Pythonidae: About 41 species of terrestrial and arboreal snakes found in Africa, Asia, and Australia. From about 1 m to almost 10 m in length.

Viperidae: About 305 species of venomous snakes in which the maxillae are rotated about their attachment to the prefrontals, allowing the fangs to rest horizontally when the mouth is closed. True vipers are found in Eurasia and Africa; pit vipers are found in the New World and in Asia. Viperids are absent from Australia and Antarctica. From about 75 cm to 2 m in length.

Elapidae: About 347 species of venomous snakes with hollow fangs near the front of relatively immobile maxillae. Elapids occur on all continents except Antarctica and are most diverse in Australia. The sea snakes are elapids. From about 25 cm to 5 m in length.

Lamprophiidae: About 299 species of African ground-dwelling or burrowing snakes. From 25 to 75 cm in length.

Colubridae: About 1755 species of snakes. Colubrids are found on all continents except Antarctica. Many colubrids have glands that secrete venom that kills prey, but they lack hollow teeth specialized for injecting venom. From 20 cm to about 3 m in length.

Our understanding of the phylogenetic relationships of snakes is changing rapidly. This list is based partly on the publications by Zahler et al. (2009) and Pryon et al. (2011) listed at the end of the chapter.

methods of burrowing would replace the unspecialized ones, but this has not happened. The explanation may lie in the conflicting selective forces on the snout. On one hand, it is important to have a snout that will burrow through soil, but, on the other hand, it is important to have a mouth capable of tackling a wide variety of prey. The specializations of the snout that make it an effective structure for burrowing appear to reduce its effectiveness for feeding. The blunt-headed amphisbaenians may be able to eat a wider variety of prey than the specialized forms can. Thus, in loose soil where it is easy to burrow, the blunt-headed forms may have an advantage. Only in soil too compact for a blunt-headed form to penetrate might the specialized forms find the balance of selective forces shifted in their favor.

Snakes

Snakes range in size from diminutive burrowing species, which feed on termites and grow to only 10 centimeters, to the large constrictors, which approach 10 meters in length (**Table 13–2**). The phylogenetic affinities of snakes are hotly debated. Clearly they are nested within the squamates, and they appear to have evolved in a terrestrial environment. The leptotyphlopids and typhlopids probably represent the ancestral condition for snakes. These small burrowing snakes (worm snakes) have shiny scales and reduced eyes. Traces of the pelvic girdle remain in most species, but the braincase is snakelike. Their anatomical and ecological characters appear consistent with a long-standing hypothesis that

snakes evolved from a subterranean lineage of lizards with greatly reduced eyes. The hypothesis that the eyes of extant surface-dwelling snakes were redeveloped after nearly disappearing during a fossorial stage in their evolution could explain some puzzling differences in the eyes of snakes and lizards.

Burrowing snakes in the families Aniliidae and Uropeltidae use their heads to dig through soil, and the bones of their skulls are solidly united. The sole xenopeltid, the sunbeam snake of Southeast Asia, is a ground-dwelling species that takes its common name from its highly iridescent scales. Boa constrictors (Boidae) are mostly New World snakes, whereas pythons (Pythonidae) are found in the Old World. The anaconda, a semiaquatic species of boa from South America, is considered the largest extant species of snake—it probably approaches a length of 10 meters—and the reticulated python of Southeast Asia is nearly as large. Not all boas and pythons are large, however; some secretive and fossorial species are considerably less than 1 meter long as adults.

The family Colubridae contains half of the extant species of snakes. The diversity of the group makes characterization difficult. Colubrids have lost all traces of the pelvic girdle, they have only a single carotid artery, and the skull is very kinetic. Many colubrid snakes are venomous, and snakes in the families Elapidae and Viperidae have hollow fangs at the front of the mouth that inject extremely toxic venom into their prey.

The body form of even a generalized snake such as the milk snake (**Figure 13–4**) is so specialized that little

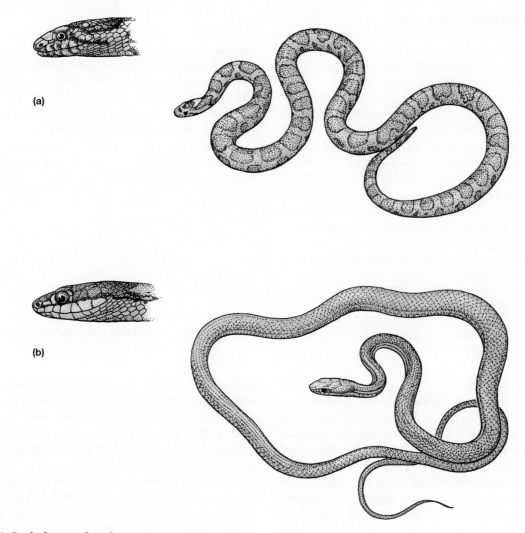

(a)

(b)

Figure 13–4 Body forms of snakes. Slow-moving constrictors such as the milk snake, (a) *Lampropeltis*, are relatively short and stout. Active, visually oriented snakes such as racers, (b) *Masticophis*, are longer and faster moving. Arboreal snakes such as the parrot snake, (c) *Leptophis*, are still more elongate and can follow their prey out among the twigs at the ends of branches. Burrowing snakes such as the blind snakes, (d) *Typhlops*, have small rounded or pointed heads with little distinction between head and neck, short tails, and smooth, often shiny scales; their eyes are greatly reduced in size. Vipers, especially the African vipers like the puff adder, (e) *Bitis arietans*, have large heads and stout bodies that accommodate large prey. Sea snakes, such as (f) *Laticauda*, have a tail that is flattened from side to side and valves that close the nostrils when they dive.

further morphological specialization is associated with different habits or habitats. King snakes and milk snakes are constrictors, and they crawl slowly, poking their heads under leaf litter and into holes that might shelter prey. Chemosensation is an important means of detecting prey for these snakes. Snakes have forked tongues, with widely separated tips that can move independently. When the tongue is projected, the tips are waved in the air or touched to the ground. The tongue then is retracted, and chemical stimuli are transferred

to the paired vomeronasal organs. The forked shape of the tongue of snakes (which is seen also among the Amphisbaenia, Lacertiformes, and Varanoidea) may allow them to detect gradients of chemical stimuli and localize objects.

Nonconstrictors such as the whip snakes and racers move quickly and are visually oriented. They forage by crawling rapidly, frequently raising the head to look around. Many arboreal snakes are extremely elongated and frequently have large eyes. Their length distributes

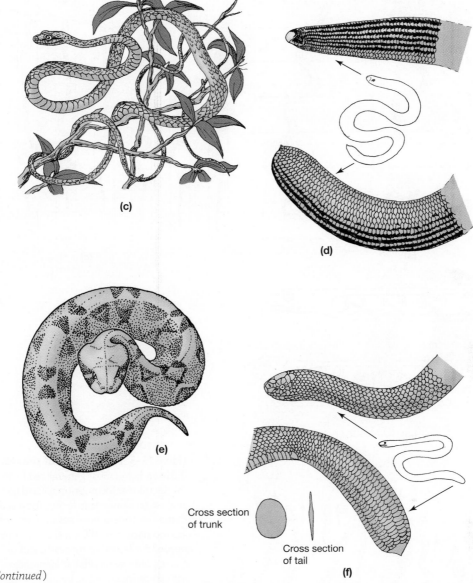

(c)

(d)

(e)

Cross section
of trunk

Cross section
of tail

(f)

Figure 13–4 (*Continued*)

their weight and allows them to crawl over even small twigs without breaking them. Burrowing snakes, at the opposite extreme of snake body form, are short and have blunt heads and very small eyes. The head shape assists in penetrating soil, and a short body and tail create less friction in a burrow than would the same mass in an elongate body. Vipers, especially forms like the African puff adder, are heavy bodied with broad heads.

The sea snakes are derived from terrestrial elapids. Sea snakes are characterized by extreme morphological specialization for aquatic life: The tail is laterally flattened into an oar, the large ventral scales are reduced or absent in most species, and the nostrils are located dorsally on the snout and have valves that

exclude water. The lung extends back to the cloaca and apparently has a hydrostatic role in adjusting buoyancy during diving as well as having a respiratory function. Oxygen uptake through the skin during diving has been demonstrated in sea snakes. *Laticauda* are less specialized than other sea snakes and may represent a separate radiation into the marine habitat. They retain enlarged ventral scales and emerge onto land to bask and to lay eggs. The other sea snakes are so specialized for marine life that they are helpless on land, and these species are viviparous.

Snakes use different methods of locomotion, depending on the body form of the species being considered and the substrate over which it is moving (**Figure 13–5**). In

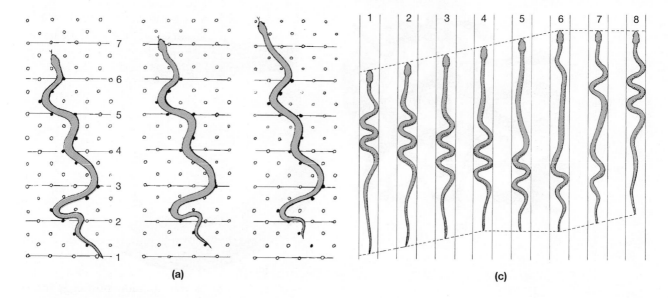

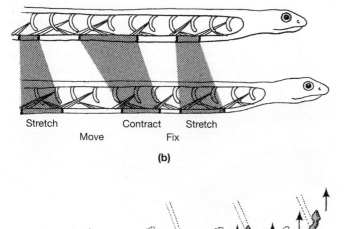

Figure 13–5 Locomotion of snakes. (a) Lateral undulation is the most generalized form of locomotion. (b) Rectilinear locomotion is used by heavy-bodied snakes. (c) Concertina locomotion is used by snakes that enter rodent burrows in search of prey. Burrows are too narrow to allow a snake to form the curves needed for lateral undulation, and only heavy-bodied snakes can use rectilinear locomotion. (d) Sidewinding can be imagined as bending a stiff wire into one and a half turns and rolling it across the ground. On a soft substrate the wire would leave a track like the track behind the snake in the drawing. Sidewinding is effective on loose sand because most of the force is directed downward instead of to the side. Many small snakes use sidewinding locomotion when they cross hot surfaces to minimize contact between the body and the substrate.

lateral undulation (also called serpentine locomotion), the body is thrown into a series of curves. The curves may be irregular, as shown in the illustration of a snake crawling across a board dotted with fixed pegs. Each curve presses backward; the pegs against which the snake is exerting force are shown in solid color. The lines numbered 1 to 7 are at 3-inch intervals, and the positions of the snake at intervals of 1 second are shown.

Rectilinear locomotion is used primarily by heavy-bodied snakes, such as large vipers, boas, and pythons. Alternate sections of the ventral integument are lifted off the ground and pulled forward by muscles that originate on the ribs and insert on the ventral scales. The intervening sections of the body rest on the ground and support the snake's body. Waves of contraction pass from anterior to posterior, and the snake moves in a straight line. Rectilinear locomotion is slow, but it is effective even when there are no surface irregularities strong enough to resist the sideward force exerted by serpentine locomotion. Because the snake moves slowly and in a straight line, it is inconspicuous, and rectilinear locomotion is used by some snakes when stalking prey.

Concertina locomotion is used in narrow passages such as rodent burrows that do not provide space for the broad curves of serpentine locomotion. A snake anchors the posterior part of its body by pressing several loops against the walls of the burrow and extends the front part of its body. When the snake is fully extended, it forms new loops anteriorly and anchors itself with these while it draws the rear end of its body forward.

Sidewinding locomotion is used primarily by snakes that live in deserts where windblown sand provides a substrate that slips away during serpentine locomotion. A sidewinding snake raises its body in loops, resting its weight on two or three points that are the only body parts in contact with the ground. The loops are swung forward through the air and placed on the ground, with the points of contact moving smoothly along the body. Force is exerted downward; the lateral component of the force is so small that the snake does not slip sideward. This downward force is shown by imprints of the ventral scales in the tracks. Because the snake's body is extended nearly perpendicular to its line of travel, sidewinding is an effective means of locomotion only for small snakes that live in habitats with few plants or other obstacles.

Snake skeletons are delicate structures that do not fossilize readily. In most cases, we have only vertebrae, and little information has been gained from the fossil record about the origin of snakes. The earliest fossils known are from Cretaceous deposits. A 3.5-meter-long snake from the Late Cretaceous of India, *Sanajeh indicus*, was apparently fossilized in the act of raiding a nest of dinosaur eggs. The snake's body is coiled around a crushed eggshell, and its head is facing an embryo that might have been inside the egg (**Figure 13–6**).

13.4 **Ecology and Behavior of Squamates**

The past quarter-century has seen an enormous increase in the number and quality of field studies of the ecology and behavior of snakes and lizards. Studies of lizards—especially studies of *Anolis* lizards on islands in the Caribbean—have been particularly fruitful, in large measure because many species of lizards are conspicuous and active during the day. These conspicuous, diurnal species dominate the literature—much less is known about species with cryptic habits. The discussions in this section rely on studies of particular species, and it is important to remember that no single species or family is representative of lizards or snakes as a group.

Foraging and Feeding

The methods that snakes and lizards use to find, capture, subdue, and swallow prey are diverse, and they are important in determining the interactions among species in a community. Astonishing specializations have evolved: blunt-headed snakes with long lower jaws that can reach into a shell to winkle out a snail, nearly toothless snakes that swallow bird eggs intact and then slice them open with sharp ventral processes (hypapophyses) on the neck vertebrae, and chameleons that project their tongues to capture insects or small vertebrates on the sticky tips are only a sample of the diversity of feeding specializations of squamates.

Many of the feeding specializations of squamates are related to changes in the structure of the skull and jaws. The most conspicuous of these is the loss of the lower temporal bar and the quadratojugal bone that formed part of that bar (**Figure 13–7** on page 323). This modification is part of a suite of structural changes in the skull that contribute to the development of a considerable degree of kinesis. The living tuatara show the ancestral condition for squamates, with the quadratojugal linking the jugal and the quadrate bones to form a complete lower temporal arch. (This fully diapsid condition is not characteristic of all sphenodontids, however; some of the Mesozoic forms did not have a complete lower temporal arch.)

Early lizards are not well known. The fossil genera *Paliguana* and *Palaeagama* from the Late Permian and Early Triassic periods of South Africa are probably not true lizards, but they do show changes in the structure of the skull that probably parallel the changes that occurred in early squamates. The gap between the quadrate and jugal widened, and the complexly interdigitating suture between the frontal and parietal bones on the roof of the skull became straighter and more like a hinge. Additional areas of flexion evolved at the front and rear of the skull and in the lower jaw of some lizards. These changes were accompanied by the development of a flexible connection at the articulation of the quadrate bone with the squamosal, which provided some mobility to the quadrate. This condition, known as streptostyly, increases the force the pterygoideus muscle can exert when the jaws are nearly closed.

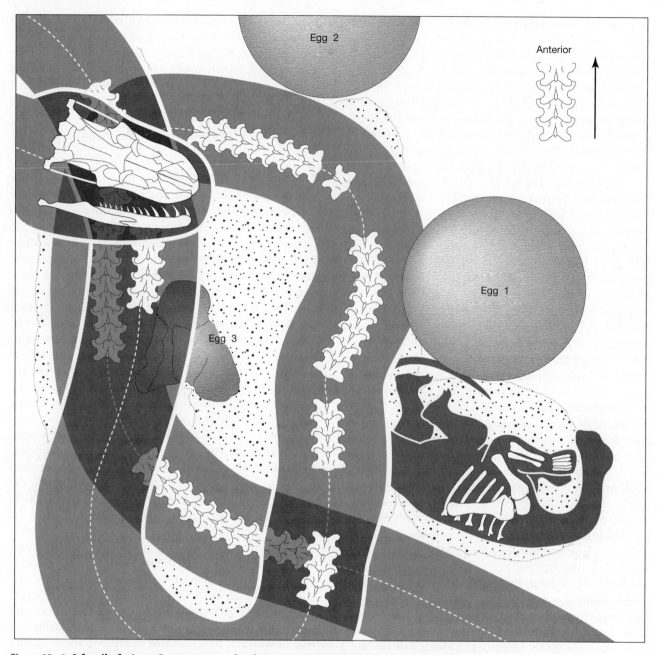

Figure 13–6 A fossil of a Late Cretaceous snake that was apparently buried by a sandstorm while it was feeding in a nest of dinosaur eggs. Two intact eggs (eggs 1 and 2) are shown, and a broken eggshell can be seen beneath the snake's coils (egg 3). The dinosaur embryo on the right may have been expelled from the broken egg. The sketch in the upper right corner shows the anterior-posterior orientation of vertebrae.

In snakes, the flexibility of the skull was increased still further by loss of the second temporal bar, which was formed by a connection between the postorbital and squamosal bones. A further increase in the flexibility of the joints between other bones in the palate and the roof of the skull produced the extreme flexibility of snake skulls. The third group of squamates has a completely different sort of skull specialization. The amphisbaenians are small, legless, burrowing animals. They use their heads as rams to construct tunnels in the soil, and the skull is heavy with rigid joints between the bones. Their specialized dentition allows them to bite small pieces out of large prey.

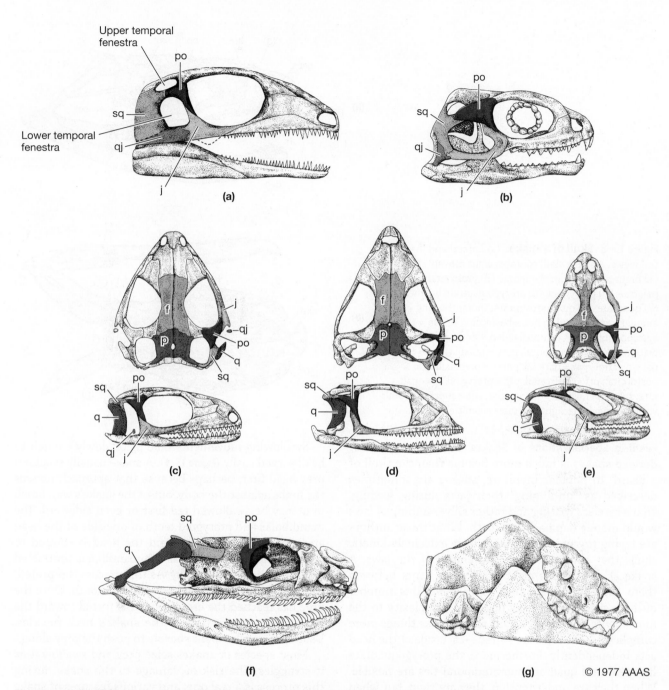

Figure 13–7 Modifications of the diapsid skull among lepidosauromorphs. Fully diapsid forms like the Permian *Petrolacosaurus* (a) retain two complete arches of bone that define the upper and lower temporal fenestrae. This condition is seen in living tuatara, *Sphenodon* (b). Lizards have achieved a kinetic skull by developing a gap between the quadrate and quadratojugal and by simplifying the suture between the frontal and parietal bones, as shown by the modern collared lizard *Crotaphytus* (e). Probable transitional stages allowing increasing skull kinesis that occurred in a nonsquamate lineage are illustrated by *Paliguana* (c) and *Palaeagama* (d). In snakes (f), skull kinesis is further increased by loss of the upper temporal arch. Amphisbaenians (g), which use their heads for burrowing through soil, have specialized akinetic skulls. (f = frontal; j = jugal; p = parietal; po = postorbital; q = quadrate; qj = quadratojugal; sq = squamosal.)

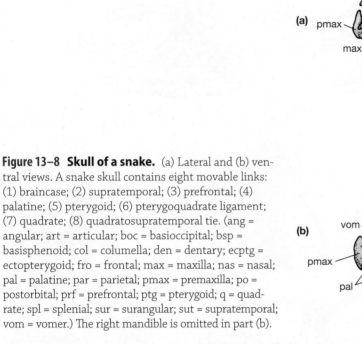

Figure 13–8 Skull of a snake. (a) Lateral and (b) ventral views. A snake skull contains eight movable links: (1) braincase; (2) supratemporal; (3) prefrontal; (4) palatine; (5) pterygoid; (6) pterygoquadrate ligament; (7) quadrate; (8) quadratosupratemporal tie. (ang = angular; art = articular; boc = basioccipital; bsp = basisphenoid; col = columella; den = dentary; ecptg = ectopterygoid; fro = frontal; max = maxilla; nas = nasal; pal = palatine; par = parietal; pmax = premaxilla; po = postorbital; prf = prefrontal; ptg = pterygoid; q = quadrate; spl = splenial; sur = surangular; sut = supratemporal; vom = vomer.) The right mandible is omitted in part (b).

Feeding Specializations of Snakes The entire skull of derived snakes is much more flexible than the skull of a lizard. In popular literature, snakes are sometimes described as "unhinging" their jaws during feeding. That's careless writing and rather silly—unhinged jaws would merely flap back and forth. What those authors are trying to say is that snakes have extremely kinetic skulls that allow extensive movement of the jaws. A snake skull contains eight links, with joints between them that permit rotation (**Figure 13–8**). This number of links gives a staggering degree of complexity to the movements of the snake skull, and to make things more complicated, the links are paired—each side of the head acts independently. Furthermore, the pterygoquadrate ligament and quadrato-supratemporal ties are flexible. When they are under tension, they are rigid, but when they are relaxed, they permit sideward movement as well as rotation. All of this results in a considerable degree of three-dimensional movement in a snake's skull.

The mandibles of lizards are joined at the front of the mouth by a rigid bony connection, but in snakes the mandibles are attached by only muscles and skin so they can spread sideward and move forward or backward independently. Loosely connected mandibles and flexible skin in the chin and throat allow the jaw tips to spread, so that the widest part of the prey passes ventral to the articulation of the jaw with the skull.

Swallowing movements take place slowly enough to be observed easily (**Figure 13–9**). A snake usually swallows prey head first, perhaps because that approach presses the limbs against the body, out of the snake's way. Small prey may be swallowed tail first or even sideward. The mandibular and pterygoid teeth of one side of the head are anchored in the prey, and the head is rotated to advance the opposite jaw as the mandible is protracted and grips the prey ventrally. As this process is repeated, the snake draws the prey item into its mouth. Once the prey has reached the esophagus, it is forced toward the stomach by contraction of the snake's neck muscles. Usually the neck is bent sharply to push the prey along.

Most species of snakes seize prey and swallow it as it struggles. The risk of damage to the snake during this process is a real one, and various features of snake anatomy seem to give some protection from struggling prey. The frontal and parietal bones of a snake's skull extend downward, entirely enclosing the brain and shielding it from the protesting kicks of prey being swallowed. Possibly the kinds of prey that can be attacked by snakes without a specialized feeding mechanism are limited by the snake's ability to swallow the prey without being injured in the process.

Constriction and venom are predatory specializations that permit a snake to tackle large prey with little risk of injury to itself. Constriction is characteristic of the boas

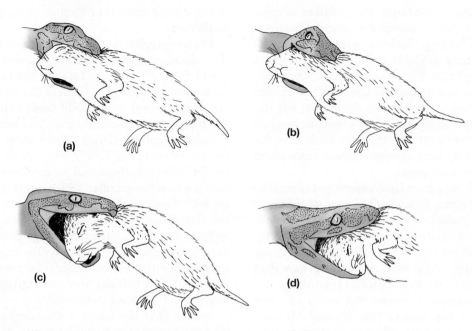

Figure 13–9 Jaw movements of a snake during feeding. Snakes use a combination of head movements and protraction and retraction of the jaws to swallow prey. (a) Prey grasped by left and right jaws at the beginning of the swallowing process. (b) The upper and lower jaws on the right side have been protracted, disengaging the teeth from the prey. (c) The head is rotated counterclockwise, moving the right upper and lower jaws over the prey. The recurved teeth slide over the prey like the runners of a sled. (d) The upper and lower jaw on the right side are retracted, embedding the teeth in the prey and drawing it into the mouth. Notice that the entire head of the prey has been engulfed by this movement. Next the left upper and lower jaws will be advanced over the prey by clockwise rotation of the head. The swallowing process continues with alternating left and right movements until the entire body of the prey has passed through the snake's jaws.

and pythons as well as some colubrid snakes. Despite travelers' tales of animals crushed to jelly by a python's coils, the process of constriction involves very little pressure. A constrictor seizes prey with its jaws and throws one or more coils of its body around the prey. The loops of the snake's body press against adjacent loops, and friction prevents the prey from forcing the loops open. Each time the prey exhales, the snake takes up the slack by tightening the loops slightly. Two hypotheses have been proposed to explain the cause of death from constriction. The traditional view holds that prey suffocates because it cannot expand its thorax to inhale. In addition, the increased internal pressure may interfere with, and eventually stop, the heart.

Interactions of Feeding and Locomotion Snakes that constrict their prey must be able to throw the body into several loops of small diameter to wrap around the prey. Constrictors achieve these small loops by having short vertebrae and short trunk muscles that span only a few vertebrae from the point of origin to the point of insertion. Contraction of these muscles produces sharp bends in the trunk that allow constrictors to press tightly against their prey. However, the trunk muscles of snakes are also used for locomotion, and the short muscles of constrictors produce several small-radius curves along the length of the snake's body. That morphology limits the speed with which constrictors can move because rapid locomotion by snakes is accomplished by throwing the body into two or three broad loops. This is the pattern seen in fast-moving species such as whip snakes, racers, and mambas. The muscles that produce these broad loops are long, spanning many vertebrae, and the vertebrae are longer than those of constrictors.

In North America, fast-moving snakes (colubrids and related lineages) first appear in the fossil record during the Miocene, a time when grasslands were expanding. Constrictors predominated in the snake fauna of the early Miocene, but by the end of that epoch the snake fauna was composed primarily of colubrids. Fast-moving snakes may have had an advantage over slow-moving constrictors in the more open habitats that developed during the Miocene.

The Evolution of Venom and Fangs The radiation of colubrids may have involved a complex interaction between locomotion and feeding. Rodents were probably the most abundant prey available to snakes, and rodents are dangerous animals for a snake to swallow while they are alive and able to bite and scratch. Constriction provided a relatively safe way to kill rodents, but the long vertebrae and long trunk muscles that allowed colubrids to move rapidly through the open habitats of the later Miocene would have prevented them from using constriction to kill their prey.

Early colubrids may have used venom to immobilize prey. Duvernoy's gland, found in the upper jaw of many extant colubrid snakes, is homologous to the venom glands of viperids and elapids and produces a toxic secretion that immobilizes prey. Some extant colubrids have venom that is dangerously toxic, and humans have died as a result of being bitten by the African boomslang.

Thus, the evolution of venom that could kill prey may have been a key feature that allowed Miocene colubrid snakes to dispense with constriction and become morphologically specialized for rapid locomotion in open habitats. The presence of Duvernoy's gland appears to be an ancestral character for colubrid snakes, as this hypothesis predicts. Some colubrids, including the rat snakes (*Elaphe*), gopher snakes (*Pituophis*), and king snakes (*Lampropeltis*), have lost the venom-producing capacity of the Duvernoy's gland, and these are the groups in which constriction has been secondarily developed as a method of killing prey.

In this context, the front-fanged venomous snakes (Elapidae and Viperidae) are not a new development, but instead represent alternative specializations of an ancestral venom-delivery system. Given the ancestral nature of venom for snakes, you might expect that different specializations for venom delivery would be represented in the extant snake fauna, as indeed they are. A variety of snakes have enlarged teeth (fangs) on the maxillae. Three categories of venomous snakes are recognized (**Figure 13–10**): opisthoglyphous, proteroglyphous, and solenoglyphous. A fourth category, the aglyphous snakes, is reserved for snakes with no fangs. This classification is descriptive and represents convergent evolution by different phylogenetic lineages.

Opisthoglyphous (Greek *opistho* = behind and *glyph* = knife) snakes have one or more enlarged teeth near the rear of the maxilla, with smaller teeth in front. In some forms the fangs are solid; in others there is a groove on the surface of the fang that may help to conduct venom into the wound. Several African and Asian opisthoglyphs can deliver a dangerous or even lethal bite to large animals, including humans, but their primary prey are lizards and birds, which are often held in the mouth until they stop struggling and are then swallowed.

Proteroglyphous snakes (Greek *prot* = first) include the cobras, mambas, coral snakes, and sea snakes in the Elapidae. The hollow fangs of the proteroglyphous snakes are located at the front of the maxilla, and there are often several small, solid teeth behind the fangs. The fangs are permanently erect and relatively short.

Solenoglyphous (Greek *solen* = pipe) snakes include the pit vipers of the New World and the true vipers of the Old World. In these snakes, the hollow fangs are the only teeth on the maxillae, which rotate so that the fangs are folded against the roof of the mouth when the jaws are closed. This folding mechanism permits solenoglyphous snakes to have long fangs that inject venom deep into the tissues of the prey. The venom, a complex mixture of enzymes and other substances (**Table 13–3** on page 328), first kills the prey and then speeds its digestion after it has been swallowed.

Snakes that can inject a disabling dose of venom into their prey have evolved a very safe prey-catching method. A constrictor is in contact with its prey while it is dying and runs some risk of injury from the prey's struggles, whereas a solenoglyphous snake needs only to inject venom and allow the prey to run off to die. Later the snake can follow the scent trail of the prey to find its corpse. This is the prey-capture pattern of most vipers, and experiments have shown that a viper can distinguish the scent trail of a mouse it has bitten from the scent trails left by uninjured mice.

Several features of the body form of vipers allow them to eat larger prey in relation to their own body size than can most nonvenomous snakes. Many vipers, including rattlesnakes, the jumping viper, the African puff adder, and the Gaboon viper, are very stout snakes. The triangular head shape that is usually associated with vipers is a result of the outward extension of the rear of the skull, especially the quadrate bones. The wide-spreading quadrates allow bulky objects to pass through the mouth, and even a large meal makes little bulge in the stout body and thus does not interfere with locomotion. Vipers have specialized as relatively sedentary predators that wait in ambush and can prey even on quite large animals. The other family of terrestrial venomous snakes, the elapids—cobras, mambas, and their relatives—are primarily slim-bodied snakes that actively search for prey.

Hearts and Stomachs Kinetic skulls, stout bodies, and safe ways to kill large prey items enable snakes to consume large meals, but having swallowed the prey they must digest it. Binge-feeding snakes and lizards—species that eat large meals at infrequent intervals—conserve energy by allowing the intestine to shrink

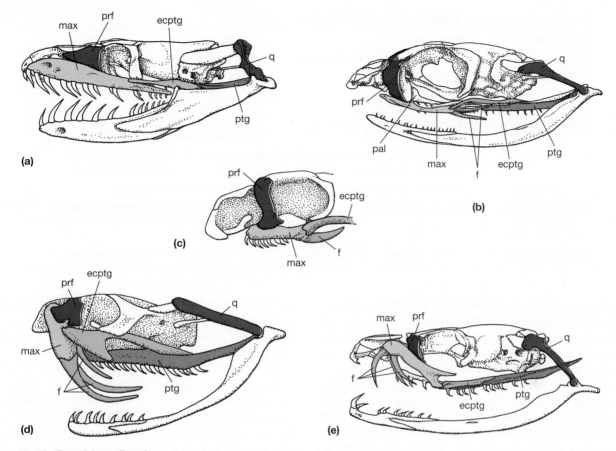

Figure 13–10 Dentition of snakes. (a) Aglyphous (without fangs), African python, *Python sebae*. (b, c) Opisthoglyphous (fangs in the rear of the maxilla), African boomslang, *Dispholidus typus*, and Central American false viper, *Xenodon rabdocephalus*. (d) Solenoglyphous (fangs on a rotating maxilla), African puff adder, *Bitis arietans*. (e) Proteroglyphous (permanently erect fangs at the front of the maxilla), African green mamba, *Dendroaspis jamesoni*. The fangs of solenoglyphs (d) are erected by an anterior movement of the pterygoid that is transmitted through the ectopterygoid and palatine to the maxilla, causing it to rotate about its articulation with the prefrontal, thereby erecting the fang. Some opisthoglyphs, especially *Xenodon* (c), have the same mechanism of fang erection. (ecptg = ectopterygoid; f = fang; max = maxilla; pal = palatine; prf = prefrontal; ptg = pterygoid; q = quadrate.)

between meals and to hypertrophy rapidly following a meal. The small intestine of a Burmese python doubles in mass while a meal is being digested and shrinks again when digestion is completed. Digestion requires a large increase in blood flow to the intestine, and the mass of the ventricle of the heart increases by 40 percent within 48 hours after a python swallows a meal. The masses of the liver, pancreas, and kidney also increase, as do the activities of several digestive enzymes. The bacteria in the gut also show a dramatic post-feeding increase in overall species diversity and a shift in the dominant species of bacteria.

Foraging Behavior and Energetics of Lizards The activity patterns of lizards span a range from extremely sedentary species that spend hours in one place to species that are in nearly constant motion. Field observations of the tropidurid lizard *Leiocephalus schreibersi* and the teiid *Ameiva chrysolaema* in the Dominican Republic revealed two extremes of behavior. *Leiocephalus* rested on an elevated perch from sunrise to sunset and was motionless for more than 99 percent of the day. Its only movements were short, rapid dashes to capture insects or to chase away other lizards. These periods of activity never lasted longer than 2 seconds, and the frequency of movements averaged 9.6 per hour. In contrast, *Ameiva* were active for only 4 or 5 hours in the middle of the day, but they were moving more than 70 percent of that time, and their velocity averaged one body length every 2 to 5 seconds.

Table 13–3 Components of the venoms of squamates

Compound	Occurrence	Effect
Proteinases	All venomous squamates, especially vipers	Destroys tissue
Hyaluronidase	All venomous squamates	Increases tissue permeability, hastening the spread of other constituents of venom through the tissues
L-amino acid oxidase	All venomous squamates	Attacks a wide variety of substrates and causes great tissue destruction
Cholinesterase	High in terrestrial elapids; may be present in sea snakes; low in vipers	Unknown; is not responsible for the neurotoxic effects of elapid venom
Phospholipases	All venomous squamates	Destroys cell membranes
Phosphatases	All venomous squamates	Breaks down high-energy compounds such as ATP, preventing cells from repairing damage
Basic polypeptides	Terrestrial elapids and sea snakes	Blocks neuromuscular transmission

The same difference in behavior was seen in a laboratory test of spontaneous activity: *Ameiva* was more than 20 times as active as *Leiocephalus*. In fact, the teiids were as active in exploring their surroundings as small mammals tested in the same apparatus. The island night lizard, a xantusiid, tested in the laboratory apparatus had a pattern of spontaneous activity that fell approximately midway between those of the teiid and the tropidurid. Thus, a spectrum of spontaneous locomotor activity is apparent in lizards, extending from species that are nearly motionless through species that move at intermediate rates to species that are as active as mammals.

For convenience, the extremes of the spectrum are frequently called sit-and-wait predators and widely foraging predators, and the intermediate condition has been called a cruising forager. Other field studies have shown that this spectrum of locomotor behaviors is widespread in lizard faunas. In North America, for example, spiny swifts (*Sceloporus*) are sit-and-wait predators, many skinks (*Eumeces*) appear to be cruising foragers, and whiptail lizards (*Aspidoscelis*) are widely foraging predators. The ancestral locomotor pattern for lizards may have been that of a cruising forager, and both sit-and-wait predation and active foraging may represent derived conditions. (A spectrum of foraging modes is not unique to lizards; it probably applies to nearly all kinds of mobile animals, including fishes, mammals, birds, frogs, insects, and zooplankton.)

The ecological, morphological, and behavioral characters that are correlated with the foraging modes of different species of lizards appear to define many aspects of the biology of these animals. For example, sit-and-wait predators and widely foraging predators consume different kinds of prey and fall victim to different kinds of predators. They have different social systems, probably emphasize different sensory modes, and differ in some aspects of their reproduction and life history.

These generalizations are summarized in **Table 13–4** and are discussed in the following sections. However, a weakness of this analysis must be emphasized: sit-and-wait species of lizards (at least the ones that have been studied most) are primarily phrynosomatids, whereas widely foraging species are mostly teiids. That phylogenetic split raises the question of whether the differences we see between sit-and-wait and widely foraging lizards are really the consequences of the differences in foraging behavior or ancestral characters of the two lineages of lizards. If the latter is true, the association with different foraging modes may be misleading. In either case, however, the model presented in Table 13–4 provides a useful integration of a large quantity of information about the biology of lizards; it represents a hypothesis that may be modified as more information becomes available.

Lizards with different foraging modes use different methods to detect prey. Sit-and-wait lizards normally remain in one spot from which they can survey a broad area. These motionless lizards detect the movement of an insect visually and capture it with a quick dash from their observation site. Sit-and-wait lizards may be most successful in detecting and capturing relatively large insects like beetles and grasshoppers. Active foragers spend most of their time on the ground surface, moving steadily and poking their snouts under fallen leaves and into crevices in the ground. These lizards apparently rely largely on chemical cues to detect insects, and they probably seek out local concentrations of patchily distributed prey such as termites. Widely foraging species of lizards

Table 13–4 Ecological and behavioral characters associated with the foraging modes of lizards. Foraging modes are presented as a continuum from sit-and-wait predators to widely foraging predators. In most cases, data are available only for species at the extremes of the continuum. (See the text for details.)

Character	Sit-and-Wait	Cruising Forager	Widely Foraging
Foraging behavior			
Movements/hour	Few	Intermediate	Many
Distance moved/hour	Low	Intermediate	High
Sensory modes	Vision	Vision and olfaction	Vision and olfaction
Exploratory behavior	Low	Intermediate	High
Types of prey	Mobile, large	Intermediate	Sedentary, often small
Predators			
Risk of predation	Low	?	Higher
Types of predators	Widely foraging	?	Sit-and-wait and widely foraging
Body form			
Trunk	Stocky	Intermediate?	Elongate
Tail	Often short	?	Often long
Physiological characters			
Endurance	Limited	?	High
Sprint speed	High	?	Intermediate to low
Aerobic metabolic capacity	Low	?	High
Anaerobic metabolic capacity	High	?	Low
Heart mass	Small	?	Large
Hematocrit	Low	?	High
Energetics			
Daily energy expenditure	Low	?	Higher
Daily energy intake	Low	?	Higher
Social behavior			
Size of home range	Small	Intermediate	Large
Social system	Territorial	?	Not territorial
Reproduction			
Mass of clutch (eggs or embryos) relative to mass of adult	High	?	Low

appear to eat more small insects than do lizards that are sit-and-wait predators. Thus, the different foraging behaviors of lizards lead to differences in their diets, even when the two kinds of lizards occur in the same habitat.

The different foraging modes also have different consequences for the exposure of lizards to their own predators. A lizard that spends 99 percent of its time resting motionless is relatively inconspicuous, whereas a lizard that spends most of its time moving is more easily seen. Sit-and-wait lizards are probably most likely to be discovered and captured by predators that are active searchers, whereas widely foraging lizards are likely to be caught by sit-and-wait predators. Because of this difference, foraging modes may alternate at successive levels in the food chain: Insects that move about may be captured by lizards that are sit-and-wait predators, and

those lizards may be eaten by widely foraging predators. Insects that are sedentary are more likely to be discovered by a widely foraging lizard, and that lizard may be picked off by a sit-and-wait predator.

The body forms of sit-and-wait lizard predators may reflect selective pressures different from those that act on widely foraging species. Sit-and-wait lizards are often stout bodied, short tailed, and cryptically colored. Many of these species have dorsal patterns formed by blotches of different colors that probably obscure the outlines of the body as the lizard rests motionless on a rock or tree trunk. Widely foraging species of lizards are usually slim and elongate with long tails, and they often have patterns of stripes that may produce optical illusions as they move. However, one predator-avoidance mechanism, the ability to break

(a)

(b)

(c)

Figure 13–11 Autotomy as a defensive mechanism. The freshly autotomized tail of a skink writhes and jerks, engaging the attention of a predatory king snake while the lizard escapes. In this sequence of photographs, (a) a king snake seizes a skink by the tail. (b) The skink autotomizes its tail and runs off, (c) leaving the snake struggling to swallow the wriggling tail.

off the tail when it is seized by a predator (**autotomy**), does not differ among lizards with different foraging modes (**Figure 13–11**).

What physiological characters are necessary to support different foraging modes? The energy requirements of a dash that lasts for only a second or two are quite different from those of locomotion that is sustained nearly continuously for several hours. Sit-and-wait and widely foraging species of lizards differ in their relative emphasis on the two metabolic pathways that provide adenosine triphosphate (ATP) for activity and in how long that activity can be sustained. Sit-and-wait lizards move in brief spurts, and they rely largely on anaerobic metabolism to sustain their movements. Anaerobic metabolism uses glycogen stored in the muscles as a metabolic substrate and produces lactic acid as its end product. It is a way to synthesize ATP

quickly (because the glycogen is already in the muscles and no oxygen is required), but it is not good for sustained activity because the glycogen is soon exhausted and lactic acid inhibits cellular metabolism. Lizards that rely on anaerobic metabolism can make brief sprints but become exhausted when they are forced to run continuously. In contrast, aerobic metabolism uses glucose that is carried to the muscles by the circulatory system as a metabolic substrate, and it produces carbon dioxide and water as end products. Aerobic exercise can continue for long periods because the circulatory system brings more oxygen and glucose and carries carbon dioxide away. As a result, widely foraging species can sustain activity for long periods without exhaustion.

The differences in exercise physiology are associated with differences in the oxygen transport systems of the lizards: Widely foraging species of lizards have larger hearts and more red blood cells in their blood than do sit-and-wait species. As a result, each beat of the heart pumps more blood, and that blood carries more oxygen to the tissues of a widely foraging species of lizard than a sit-and-wait species.

Sustained locomotion is probably not important to a sit-and-wait lizard that makes short dashes to capture prey or to escape from predators, but sprint speed might be vitally important in both these activities. Speed may be relatively unimportant to a widely foraging lizard that moves slowly, methodically looks for prey under leaves and in cavities, and can hide under a bush to confuse a predator. As you might predict from these considerations, sit-and-wait lizards generally have high sprint speeds and low endurance, whereas widely foraging species usually have lower sprint speeds and greater endurance.

The continuous locomotion of widely foraging species of lizards is energetically expensive. Measurements of energy expenditure of lizards in the Kalahari showed that the daily energy expenditure of a widely foraging species averaged 1.5 times that of a sit-and-wait species. However, the energy that the widely foraging species invested in foraging was more than repaid by its greater success in finding prey. The daily food intake of the widely foraging species was twice that of the sit-and-wait predator. As a result, the widely foraging species had more energy available to use for growth and reproduction than did the sit-and-wait species, despite the additional cost of its mode of foraging.

Social Behavior

Squamates employ a variety of visual, auditory, chemical, and tactile signals in the behaviors they use to maintain territories and to choose mates. The various sensory modalities employed by animals have biased the amount of information we have about the behaviors of different species. Because humans are primarily visually oriented, we perceive the visual displays of other animals quite readily. The auditory sensitivity of humans is also acute, and we can detect and recognize vocal signals that are used by other species. However, the olfactory sensitivity of humans is low and we lack a well-developed vomeronasal system, so we are unable to perceive most chemical signals used by squamates. One result of our sensory biases has been a concentration of behavioral studies on organisms that use visual signals. Because of this focus, the extensive repertoires of visual displays of Anolis lizards figure largely in the literature of behavioral ecology but much less is known about the chemical and tactile

signals that are probably important for other lizards and for snakes.

The social behaviors of squamates appear to be limited in comparison with those of crocodilians, but many species show dominance hierarchies or territoriality. The signals used in agonistic encounters between individuals are often similar to those used for species and sex recognition during courtship. Parental attendance at a nest during the incubation period of eggs occurs among squamates, but extended parental care of the young is rare.

Agamid, dactyloid, and phrynosomatid lizards employ primarily visual displays during social interactions. The dactyloid genus Anolis includes some 400 species of small- to medium-size lizards that occur primarily in tropical America. Male Anolis have gular fans, areas of skin beneath the chin that can be distended by the hyoid apparatus during visual displays. The brightly colored scales and skin of the gular fans of many Anolis species are conspicuous signaling devices, and they are used in conjunction with movements of the head and body.

Figure 13–12 shows the colors of the gular fans of eight species of Anolis that occur in Costa Rica. (Images of some Anolis gular fans are shown on the third page of the color insert.) Since no two species of Anolis in the habitat have the same combination of colors on their gular fans, it is possible to identify a species solely by seeing the colors it displays. Each species also has a behavioral display that consists of raising the body by straightening the fore legs (called a push-up), bobbing the head, and extending and contracting the gular fan. The combination of these three sorts of movements allows a complex display. The three movements can be represented graphically by an upper line that shows the movements of the body and head and a lower line that shows the expansion and contraction of the gular fan. This representation is called a display action pattern. No two display action patterns are the same, so it is possible to identify any of the eight species of Anolis by seeing its display action pattern.

The behaviors that territorial lizards use for species and sex recognition during courtship are very much like those employed in territorial defense—push-ups, head bobs, and displays of the gular fan. A territorial male lizard is likely to challenge any lizard in its territory, and the progress of the interaction depends on the response it receives. An aggressive response indicates that the intruder is a male and stimulates the territorial male to defend its territory, whereas a passive response from the intruder identifies a female and stimulates the territorial male to initiate courtship. These behaviors are illustrated by the displays of a male Anolis carolinensis shown in **Figure 13–13**. The first

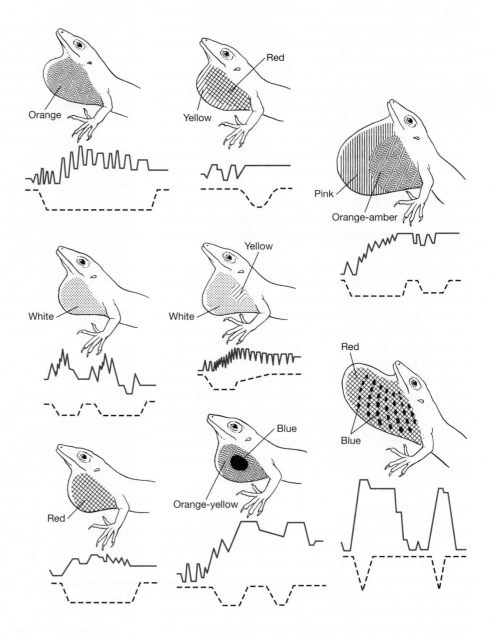

(a) SIMPLE   **(b)** COMPOUND   **(c)** COMPLEX

Figure 13–12 Species-typical displays of *Anolis* lizards.
Eight species of *Anolis* from Costa Rica can be separated into three groups based on the size and color pattern of their gular fans. *Simple* fans are unicolored (a), *compound* fans are bicolored (b), and *complex* fans are bicolored and very large (c). Display action patterns for each species are graphed beneath the lizard drawings. The horizontal axis is time (the duration of these displays is about 10 seconds), and the vertical axis is vertical height. Solid line shows movements of the head; dashed line indicates extension of the gular fan.

response of a territorial lizard to an intruder is the assertion-challenge display: The gular fan is extended, and the lizard bobs at the intruder. The nuchal (neck) and dorsal crests are also slightly raised, and a black spot appears behind the eye.

The next stage in the interaction depends on the intruder's sex and its response to the territorial male's challenge (**Figure 13–14** on page 334). If the intruder is a receptive female, the territorial male initiates courtship. If the intruder is a male and does not retreat from the initial challenge, both males become more aggressive.

During aggressive posturing the males orient laterally to each other, the nuchal and dorsal crests are fully erected, the body is compressed laterally, the black spot behind the eye darkens, and the throat is swelled. All these postural changes make the lizards appear larger and presumably more formidable to the opponent.

The differences in color and movement that characterize the gular fans and display action patterns of *Anolis* are conspicuous to human observers, but do the lizards also rely on them for species identification? Indirect evidence suggests that the lizards probably do use

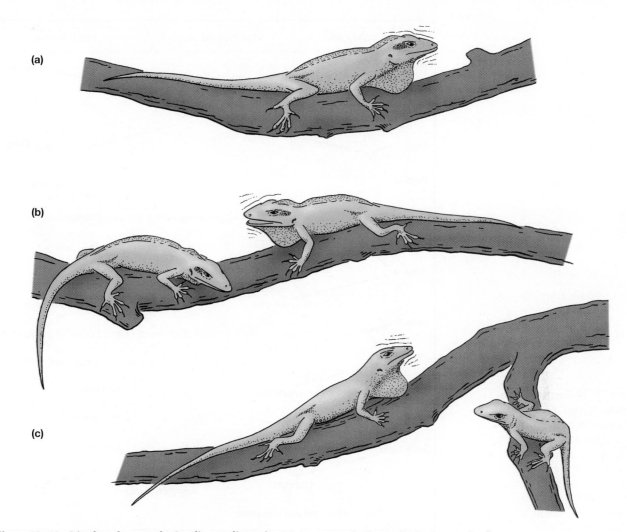

Figure 13–13 Displays by a male *Anolis carolinensis*. (a) Assertion-challenge display by a male. (b) Aggressive posturing between two males. (c) Courtship: male on the left, female on the right. Note the extension of the dewlap, the species-typical head bob, and the absence of the dorsal and nuchal crests and the eyespot.

gular fan color and display action patterns for species identification. For example, examination of communities of *Anolis* containing many species shows that the differences in colors of gular fans and in display action patterns are greatest for those species that encounter one another most frequently. Movement of windblown vegetation is visual "noise" for *Anolis* lizards, and on Puerto Rico male *Anolis* display most frequently during periods of low noise when their signals are readily perceived by neighboring males and females.

Studies of a common species of lizard in western North America, the side-blotched lizard *Uta stansburiana*, have revealed a complex association between gular color and territorial and reproductive behavior and sexual size dimorphism. In some populations male lizards have one of three colors in the gular region—blue, orange, or yellow. Males with blue throats are territorial, maintaining territories that overlap with the home ranges of one or more females and mating with those females. Males with orange throats are more aggressive than blue males; they do not maintain territories themselves but displace blue-throated males from their territories and mate with the females. Yellow-throated males try to sneak into the territories of blue-throated males and steal a mating before they are chased away by the territorial male. The throat color of a male is determined by the level of testosterone in its blood and is fixed during early development.

The most effective mating tactic depends on the number of lizards using each tactic. Yellow sneakers beat aggressive orange males, who beat territorial blue males, who beat yellow sneakers, like a game of

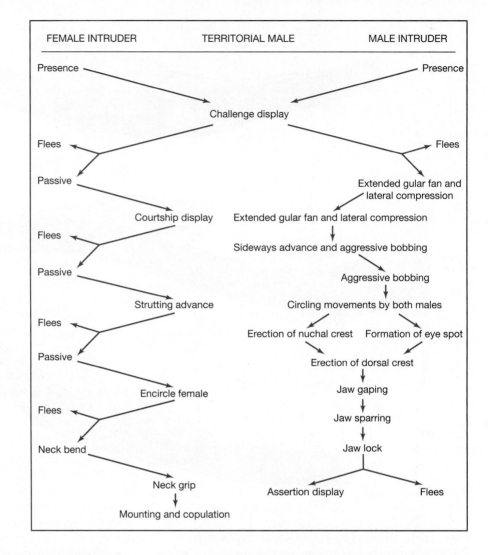

Figure 13–14 Territorial behavior. Normal sequence of behaviors for a territorial male *Anolis carolinensis* confronting an intruding male or female anole. A territorial male challenges any intruder, and the response of the intruder determines the subsequent behavior of the territorial male.

rock, paper, scissors. The fitness of any one of the color forms relative to the others depends on the proportions of the three forms in the population and also on the spatial distribution of the forms within a habitat. Blue-throated males that are neighbors of other blue-throated males have three times the fitness of blue-throated males that have orange- or yellow-throated neighbors, and orange-throated males do best when they are not near another orange-throated male.

Not all populations of *Uta stansburiana* contain males with different throat colors—in some populations the males are monomorphic for throat color; that is, all of the males have the same throat color. In some monomorphic populations all the males have orange throats, and in other populations all the males have blue throats. Selection operates differently in populations with monomorphic versus polymorphic throat colors. In monomorphic populations, males maximize their reproductive success by mating with the largest females available because large females produce more eggs than do smaller females. In these populations selection has acted on female body size, leading to the evolution of large females. In contrast, selection has acted to produce large males in the polymorphic populations where competition between males is the key to their reproductive success.

Territoriality, the relative importance of vision compared with olfaction, and foraging behavior appear to be broadly correlated among lizards. The elevated perches from which sit-and-wait predators survey their home ranges allow them to see both intruders and prey, and they dash from the perch to repel an intruder or to catch an insect. In contrast, widely foraging lizards are almost entirely nonterritorial, and olfaction is as important as vision in their foraging behavior. These lizards spend most of their time on the ground, where their field of vision is limited and they probably have little opportunity to detect intruders.

Sociality In general lizards and snakes are not considered to be social animals; that is, they generally do not appear to live in groups. The *Egernia* lineage of Australian skinks is a well-studied exception to that generalization, however. The social behavior of species within this group ranges from solitary species with only transient contacts between individuals to species that form stable, long-term family groups. *Egernia stokesii* is a highly social species that lives in groups of up to 17 individuals, consisting of a monogamous pair of adults and their offspring. Group members share a common shelter site, bask close to one another, and defecate on a common scat pile.

Sociality based on long-term associations among parents or siblings may be more common than we currently realize. For example, a cross-fostering experiment with desert night lizards (*Xantusia vigilis*) found that juvenile lizards released in the field with their biological siblings were twice as likely to form aggregations as juveniles that were released with unrelated individuals, and the aggregations formed by juveniles released with their siblings were more than three times larger than the aggregations formed by unrelated juveniles.

Reproduction

Squamates show a range of reproductive modes from **oviparity** (development occurs outside the female's body and is supported entirely by the yolk—i.e., lecithotrophy) to **viviparity** (eggs are retained in the oviducts, and development is supported by transfer of nutrients from the mother to the fetuses—matrotrophy). Intermediate conditions include retention of the eggs for a time after they have been fertilized and the production of **precocial** young that were nourished primarily by material in the yolk.

Oviparity is assumed to be the ancestral condition, and viviparity has evolved more than 100 times among squamates. A fossil of a pregnant lizard containing at least 15 embryos has been described from the Early Cretaceous of China. Viviparous squamates have specialized chorioallantoic placentae; in the Brazilian skink, *Mabuya heathi*, more than 99 percent of the mass of the fetus results from transport of nutrients across the placenta.

Viviparity is usually a high-investment reproductive strategy. Females of viviparous squamates generally produce relatively small numbers of large young, although there are exceptions to that generalization. Viviparity is not evenly distributed among lineages of squamates: nearly half the origins of viviparity in the group have occurred in the family Scincidae, whereas it is unknown in teiid lizards and occurs in only two genera of lacertids. Viviparity has advantages and disadvantages as a mode of reproduction. The most commonly cited benefit is the opportunity it provides for a female snake or lizard to use her own thermoregulatory behavior to control the temperature of the embryos during development. This hypothesis is appealing in an ecological context because a relatively short period of retention of the eggs by the female might substantially reduce the total amount of time required for development, especially in a cold climate.

Viviparity potentially lowers reproductive output because a female who is retaining one clutch of eggs cannot produce another. Lizards in warm habitats may produce more than one clutch of eggs in a season, but that is not possible for a viviparous species because development takes too long. In a cold climate, lizards are not able to produce more than one clutch of eggs in a breeding season anyway, and viviparity would not reduce the annual reproductive output of a female lizard. Phylogenetic analyses of the origins of viviparity suggest that it has evolved most often in cold climates, as this hypothesis predicts, but other origins appear to have taken place in warm climates; more than one situation favoring viviparity among squamates appears likely.

Viviparity has other costs. The agility of a female lizard is substantially reduced when her embryos are large. Experiments have shown that pregnant female lizards cannot run as fast as nonpregnant females and that snakes find it easier to capture pregnant lizards than nonpregnant ones. Females of some species of lizards become secretive when they are pregnant, perhaps in response to their vulnerability to predation. They reduce their activity and spend more time in hiding places. This behavioral adjustment may contribute to the reduction in body temperature seen in pregnant females of some species of lizards, and it probably reduces their rate of prey capture as well.

In general, large species of squamates produce more eggs or fetuses than do small species, and within a single species, large individuals often have more offspring in a clutch than do small individuals. Both phylogenetic and ecological constraints play a role in determining the number of young produced, however. All geckos have a clutch size of either one or two eggs, and all *Anolis* produce only one egg at a time. Lizards with stout bodies usually have clutches that are a greater percentage of the mother's body mass than do lizards with slim bodies. The division between stout and slim bodies approximately parallels the division between sit-and-wait predators and widely foraging predators. It is tempting to infer that a lizard that moves about in search of prey finds a bulky clutch of eggs more hindrance than a lizard that spends 99 percent of its time resting motionless. However, because some of the divisions among modes of predatory behavior, body form, and relative clutch mass also correspond to the phylogenetic division between phrynosomatid and teiid lizards, it is not possible to decide which characters are ancestral and which may be derived.

Parthenogenesis

All-female (**parthenogenetic**) species of squamates have been identified in six families of lizards and one snake. The phenomenon is particularly widespread in the teiids and lacertids and occurs in several species of geckos. Parthenogenetic species are known or suspected to occur among chameleons, agamids, xantusiids, and typhlopids. However, parthenogenesis is probably more widespread among squamates than this list indicates because parthenogenetic species are not conspicuously different from bisexual species. Parthenogenetic species are usually detected when a study undertaken for an entirely different purpose reveals that a species contains no males. Confirmation of parthenogenesis can be obtained by getting fertile eggs from females raised in isolation or by making reciprocal skin grafts between individuals. Individuals of bisexual species usually reject tissues transplanted from another individual because genetic differences between them lead to immune reactions. Parthenogenetic species, however, produce progeny that are genetically identical to the mother, so no immune reaction occurs and grafted tissue is retained.

The chromosomes of lizards have allowed the events that produced some parthenogenetic species to be deciphered. Many parthenogenetic species appear to have had their origin as interspecific hybrids. These hybrids are diploid ($2n$), with one set of chromosomes from each parental species. For example, the diploid parthenogenetic whiptail lizard, *Aspidoscelis tesselatus*, is the product of hybridization between the bisexual diploid species *A. tigris* and *A. septemvittatus* (**Figure 13–15**). Some parthenogenetic species are triploids ($3n$). These forms are usually the result of a backcross of a diploid parthenogenetic individual to a male of one of its bisexual parental species or, less commonly, the result of hybridization of a diploid parthenogenetic species with a male of a bisexual species different from its parental species. A parthenogenetic triploid form of *A. tesselatus* is apparently the result of a cross between the parthenogenetic diploid *A. tesselatus* and the bisexual diploid species *A. sexlineatus*.

It is common to find the two bisexual parental species and a parthenogenetic species living in overlapping habitats. Parthenogenetic species of *Aspidoscelis* often occur in habitats like the floodplains of rivers that are subject to frequent disruption. Disturbance of the habitat may bring together closely related bisexual species, fostering the hybridization that is the first step in establishing a parthenogenetic species. Once a parthenogenetic species has become established, its reproductive potential is twice that of a bisexual species because every individual of a parthenogenetic species is capable of producing young. Thus, when a flood or other disaster wipes out most of the lizards, a parthenogenetic species can repopulate a habitat faster than a bisexual species.

Temperature-Dependent Sex Determination

The phylogenetic distribution of temperature-dependent sex determination (TSD) is more limited in lepidosaurs than it is in turtles or crocodilians: it has been demonstrated in the tuatara and in some species of skinks, agamids, and gekkonids, and it may be present in other families as well.

TSD among lepidosaurs has a feature that is not present in turtles—some viviparous species of lizards have TSD, and a pregnant female lizard can determine the sex of her embryos by adjusting her thermoregulatory behavior. Free-ranging female Australian snow skinks, *Niveoscincus ocellatus*, that maintained high body temperatures for many hours a day gave birth to litters that were about 65 percent female, whereas females that were warm for shorter periods of time had litters that were about 75 percent male.

The warmer females gave birth earlier in the season than the cooler females, and that observation suggests an advantage for producing females at high incubation temperatures: birth date influences the adult body size of snow skinks—individuals that are born early in the year are larger when they reproduce than individuals that are born later. Large body size is probably more advantageous for female snake skinks than for males because large females produce more young than do small females. A small body size may not be a disadvantage for a male snow skink because males of this species do not compete with other males to hold territories, and even a small male can fertilize the eggs of a large female.

Sex determination is still more complicated in the oviparous Australian three-lined skink, *Bassiana duperreyi*. In this species genetic sex determination (GSD) and TSD interact with egg size to determine the sex of hatchlings. This species has heteromorphic sex

(a)

(b)

(c)

(d)

(e)

Figure 13–15 Unisexual lizards. The photographs show the species of whiptail lizards involved in a sequence of crosses leading to the formation of diploid and triploid unisexual species of *Aspidoscelis*. Hybridization of the bisexual diploid species (a) *A. tigris* and (b) *A. septemvittatus* produced a unisexual diploid form with half of its genetic complement derived from each of the parental species (an allodiploid). This parthenogenetic form is called *A. tesselatus* (c). Hybridization between a diploid *A. tesselatus* and a male of the bisexual species *A. sexlineatus* (d) produced a unisexual triploid form with its genetic complement derived from three different parental species (an allotriploid). This parthenogenetic triploid form is also called *A. tesselatus* (e). Thus *A. tesselatus* consists of clones of both diploid and triploid lineages, although taxonomists soon will probably treat these two forms as separate species.

chromosomes—normally females are XX and males are XY, but low incubation temperatures override GSD to produce some XX males. In addition, egg size varies within clutches: larger eggs produce females and smaller eggs produce males.

The occurrence of TSD among lepidosaurs has implications for their conservation. The management strategy for tuatara includes establishing new populations in protected sites on the North and South Islands of New Zealand. The climate at the south end of the South Island is colder than it is in the current range of the species, and laboratory studies have shown that only females are produced at the soil temperatures found at some of these sites. At the Orokonui Ecosanctuary (at about 45 degrees south latitude), for example, only a few artificial nest sites had temperatures that were potentially warm enough to produce males. Researchers concluded that establishing a population of tuatara at Orokonui now would require artificial incubation of some eggs to ensure production of male hatchlings, but they noted that the 3°C increase in soil temperature that is expected to result from global climate change will improve conditions for tuatara at Orokonui.

An optimistic view of the consequences of climate change does not apply to most species of lepidosaurs with TSD, however, and species in which the shift from one sex to the other occurs over a narrow range of temperatures are considered to be especially vulnerable.

Parental Care

Parental care has been recorded for more than a hundred species of squamates. A few species of snakes and a larger number of lizards remain with the eggs or nest site. Some female skinks remove dead eggs from the clutch. Some species of pythons brood their eggs: the female coils tightly around the eggs, and, in some species, muscular contractions of the female's body produce sufficient heat to raise the temperature of the eggs to about 30°C, which is substantially warmer than air temperature. One unconfirmed report exists of baby pythons returning at night to their empty eggshells, where their mother coiled around them and kept them warm. Little interaction between adult and juvenile squamates has been documented. In captivity, female prehensile-tailed skinks (*Corucia zebrata*) have been reported to nudge their young toward the food dish, as if teaching them to eat. Prehensile-tailed skinks, which occur only on the Solomon Islands, are herbivorous and viviparous.

Free-ranging baby green iguanas have a tenuous social cohesion that persists for several months after they hatch. The small iguanas move away from the nesting area in groups that may include individuals from several different nests. One lizard may lead the way, looking back as if to see that others are following. The same individual may return later and recruit another group of juveniles. During the first 3 weeks after they hatch, juvenile iguanas move up into the forest canopy and are seen in close association with adults. During this time, the hatchlings probably ingest feces from the adults, thereby inoculating their guts with the symbiotic microbes that facilitate digestion of plant material. After their fourth week of life, the hatchlings move down from the forest canopy into low vegetation, where they continue to be found in loosely knit groups of two to six or more individuals that move, feed, and sleep together. This association might provide some protection from predators. If the hatchlings continue to eat fecal material, it is another opportunity to ensure that each lizard has received its full complement of gut microorganisms.

13.5 Behavioral Control of Body Temperatures by Ectotherms

The behavioral mechanisms involved in ectothermal temperature regulation are quite straightforward and are employed by insects, birds, and mammals (including humans) as well as by ectothermal vertebrates. Lizards, especially desert species, are particularly good at behavioral thermoregulation. Movement back and forth between sunlight and shade is the most obvious thermoregulatory mechanism they use. Early in the morning or on a cool day, lizards bask in the sun, whereas in the middle of a hot day, they retreat to shade and make only brief excursions into the sun. Sheltered or exposed sites may be sought out. In the morning, when a lizard is attempting to raise its body temperature, it is likely to be in a spot protected from the wind. Later in the day, when it is getting too hot, the lizard may climb into a bush or onto a rock outcrop where it is exposed to the breeze and its convective heat loss is increased.

An animal can alter the amount of solar radiation it absorbs by changing its orientation to the sun, its body contour, and its skin color. Lizards use all of these mechanisms. An animal oriented perpendicular to the sun's rays intercepts the maximum amount of solar radiation, and one oriented parallel to the sun's rays intercepts minimum radiation. Lizards adjust their orientation to control heat gained by direct solar radiation. Many lizards can spread or fold their ribs to change the shape of the trunk. When the body is oriented perpendicular to the sun's rays and the ribs are spread, the surface area exposed to the sun is maximized and heat gain is increased. Compressing the ribs decreases the surface area exposed to the sun and can be combined with orientation parallel to the rays to minimize heat gain. Horned lizards (*Phrynosoma*) provide a good example of this type of control. If the surface area that a horned lizard exposes to the sun directly overhead when the lizard sits flat on the ground with its ribs held in a resting position is considered to be 100 percent, then the maximum surface area the lizard can expose by orientation and change in body contour is 173 percent, and the minimum is 28 percent. That is, the lizard can change its radiant heat gain more than sixfold solely by changing its position and body shape.

Color change can further increase a lizard's control of radiative exchange (see the color insert). Objects look dark because they are absorbing energy in the visible part of the solar spectrum, and the radiant energy they absorb warms them. The lightness or darkness of a lizard affects the amount of solar radiation it absorbs, and lizards can darken or lighten by moving dark pigment in their skin. Melanophores are cells that contain the pigment melanin. They are shaped rather like mushrooms, with a broad upper portion connected by a stalk to a lower section. When melanin granules are dispersed into the upper part of the cell, close to the skin surface, the skin appears dark; when the granules are drawn away from the surface into the lower section of the cell, the skin appears light. Lizards heat 10 percent to 75 percent faster when they are dark than they do when they are light.

Combining these mechanisms gives lizards a remarkable independence from air temperature. Lizards occur above the timberline in many mountain ranges, and they are capable of maintaining body temperatures 30°C or more warmer than air temperature during their periods of activity on sunny days. While air temperatures are near freezing, these lizards scamper about with body temperatures as high as those of species that inhabit lowland deserts.

The repertoire of thermoregulatory mechanisms seen in lizards is greater than that of many other ectothermal vertebrates. Turtles, for example, cannot change their body contour or color, and their behavioral thermoregulation is limited to movements between sunlight and shade and in and out of water. Crocodilians are like turtles, although young individuals make minor changes in body contour and color. Most snakes cannot change color, but rattlesnakes and boas lighten and darken as they warm and cool.

Activity Temperature Ranges

The extensive repertoire of thermoregulatory mechanisms employed by ectotherms allows many species of lizards and snakes to keep their body temperature within a range of a few degrees during the active part of their day. Many species of lizards have body temperatures between 33°C and 38°C while they are active (the **activity temperature range**), and snakes often have body temperatures between 28°C and 34°C. This is the region of temperature in which an ectotherm carries out its full repertoire of activities—feeding, courtship, territorial defense, and so on.

These activity temperature ranges have been the focus of much research: Field observations show that thermoregulatory activities may occupy a considerable portion of an animal's time. Less obvious, but just as important, are the constraints that the need for thermoregulation sets on other aspects of the behavior and ecology of ectotherms. For example, some species of lizards and snakes are excluded from certain habitats because it is impossible for them to thermoregulate. In temperate regions, the activity season lasts only during the months when it is warm and sunny enough to permit thermoregulation; at other times of the year, snakes and lizards hibernate. Even during the activity season, time spent on thermoregulation may not be available for other activities, and lizards in temperate regions may have less extensive repertoires of social behavior than do tropical lizards because thermoregulatory behavior in cool climates requires so much time.

Physiological Control of the Rate of Change of Body Temperature

A new dimension was added to studies of ectothermal thermoregulation in the 1960s with the discovery that ectotherms can use physiological mechanisms to adjust their rate of temperature change. The original observations showed that several different kinds of large lizards were able to heat faster than they cooled when exposed to a 20°C difference between body and ambient temperatures. Subsequent studies by other investigators extended these observations to turtles and snakes. From the animal's viewpoint, heating rapidly and cooling slowly prolong the time it can spend in the normal activity range.

The basis of this control of heating and cooling rates lies in changes in peripheral circulation. Heating the skin of a lizard causes a localized expansion of dermal blood vessels (vasodilation) in the warm area. Dilation of the blood vessels, in turn, increases the blood flow through them, and the blood is warmed in the skin and carries the heat into the core of the body. Thus, in the morning, when a cold lizard orients its body perpendicular to the sun's rays and the sunlight warms its back, local vasodilation in that region speeds heat transfer to the rest of the body.

The same mechanism can be used to avoid overheating. The Galápagos marine iguana is a good example. Marine iguanas live on the bare lava flows on the coasts of the Galápagos Islands. In midday, beneath the equatorial sun, the black lava becomes extremely hot—uncomfortably, if not lethally, hot for a lizard. Retreat to the shade of the scanty vegetation or into rock cracks would be one way the iguanas could avoid overheating; however, the males are territorial, and those behaviors would mean abandoning their territories and probably having to fight for them again later in the day. Instead, a male iguana stays where it is, using physiological control of circulation and the cool breeze blowing off the ocean to form a heat shunt that absorbs solar energy on the dorsal surface, carries it throughout the body, and dumps it out the ventral surface.

The process works as follows: In the morning the lizard is chilled from the preceding night and basks to bring its body temperature to the normal activity range. When its temperature reaches this level, the lizard uses postural adjustments to slow the increase in body temperature, finally facing directly into the sun to minimize its heat load. In this posture, the forepart of the body is held off the ground (**Figure 13–16**). The ventral surface is exposed to the cool wind blowing off the ocean, and its body shades a patch of lava under the animal. This lava is soon cooled by the wind. Local vasodilation is produced by warming the blood vessels; it does not matter whether the heat comes from the outside (from the sun) or from inside (from warm blood). Warm blood circulating from the core of the body to the ventral skin warms it and produces vasodilation, increasing blood flow to the ventral surface. The lizard's ventral skin is cooler than the rest of its body because it is shaded and cooled by the wind. The warm

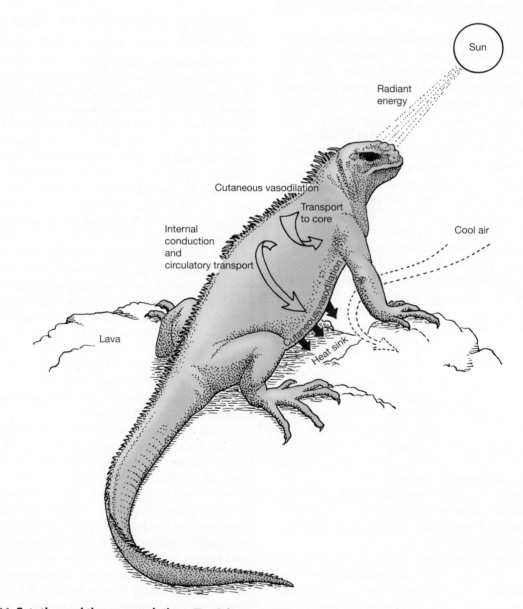

Figure 13–16 Ectothermal thermoregulation. The Galápagos marine iguana uses a combination of behavioral and physiological thermoregulatory mechanisms to shunt heat absorbed by its dorsal surface out its ventral surface.

blood heats the ventral skin, which loses heat by radiation to the cool lava in the shadow created by the lizard's body and by convection to the breeze. In this way, the same cardiovascular mechanism that earlier in the day allowed the lizard to warm rapidly is converted to a regulated heat shunt that rapidly transports solar energy from the dorsal to the ventral surface and keeps the lizard from overheating. In combination with postural adjustments and other behavioral mechanisms, such as the choice of a site where the breeze is strong, these physiological adjustments allow a male iguana to remain on station in its territory all day.

Organismal Performance and Temperature

Minimizing variation in body temperature greatly simplifies the coordination of biochemical and physiological processes. An organism's body tissues are the site of a tremendous variety of biochemical reactions, proceeding simultaneously and depending on one another to provide the proper quantity of the proper substrates at the proper time for reaction sequences to occur. Each reaction has a different sensitivity to temperature, and regulation is greatly facilitated when temperature variation is limited. Thus, coordination of internal processes may be a major benefit of thermoregulation for ectotherms. If the temperature stability that a snake or lizard achieves by thermoregulation is important to its physiology and biochemistry, you might expect that the internal economy of an animal functions best within its activity temperature range, and that is often the case. Ex-

amples of physiological processes that work best at temperatures within the activity range can be found at the molecular, tissue, system, and whole-animal levels of organization.

The wandering garter snake (*Thamnophis elegans vagrans*) provides examples of the effects of body temperature on a variety of physiological and behavioral functions. Wandering garter snakes are diurnal, semiaquatic inhabitants of lakeshores and stream banks in western North America. They hunt for prey on land and in water and feed primarily on fishes and amphibians. Chemosensation, accomplished by flicking the tongue, is an important mode of prey detection for snakes. Scent molecules are transferred from the tips of the forked tongue to the epithelium of the vomeronasal organ in the roof of the mouth. Garter snakes spend the night in shelters, where their body temperatures fall to ambient levels (4°C to 18°C), and emerge in the morning to bask. During activity on sunny days, the snakes maintain body temperatures between 28°C and 32°C.

Temperature affects the speed of crawling and swimming, the frequency of tongue flicks, the rate of digestion, and the rate of oxygen consumption of the snakes (**Figure 13–17**). Crawling, swimming, and tongue flicking are elements of the foraging behavior of garter snakes, and the rates of digestion and oxygen consumption are involved in energy utilization. The ability of garter snakes to crawl and swim was severely limited at the low temperatures they experience during the night when they are inactive. At 5°C snakes often refused to crawl, and at 10°C they were able to crawl only 0.1 meter per second and could swim only

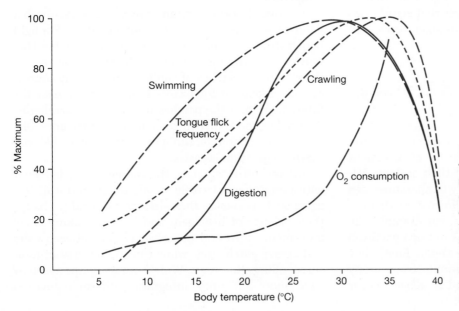

Figure 13–17 Effect of temperature on performance. The ability of a wandering garter snake, *Thamnophis elegans vagrans*, to perform many activities essential to survival depends on its body temperature. The vertical axis shows the percentage of maximum performance achieved at each temperature.

0.25 meter per second. The speed of both types of locomotion increased at higher temperatures. Swimming speed peaked near 0.6 meter per second at 25°C and 30°C, and crawling speed increased to an average of 0.8 meter per second at 35°C. The rate of tongue flicking increased from less than 0.5 flick per second at 10°C to about 1.5 flicks per second at 30°C. The rate of digestion increased slowly from 10°C to 20°C and more than doubled between 20°C and 25°C. It did not increase further at 30°C and dropped slightly at higher temperatures. The rate of oxygen consumption rose as temperature increased from 20°C to 35°C, which was the highest temperature tested because higher body temperatures would have been injurious.

All five measures of performance by garter snakes increased with increasing temperature, but the responses to temperature were not identical. For example, swimming speed did not increase substantially above 20°C, whereas crawling speed continued to increase up to 35°C. The rate of digestion peaked at 25°C to 30°C and then declined, but the rate of oxygen consumption increased steadily to 35°C. More striking than the differences among the functions, however, is the apparent convergence of maximum performance for all the functions on temperatures between 28.5°C and 35°C. This range of temperatures is close to the body temperatures of active snakes in the field on sunny days (28°C to 32°C). Anywhere within that range of body temperatures, snakes would be able to crawl, swim, and tongue-flick at rates that are at least 95 percent of their maximum rates.

The relationship between the body temperatures of active garter snakes and the temperature sensitivity of various behavioral and physiological functions is probably representative of ectotherms in general. That is, in most cases, the body temperatures that ectotherms maintain during activity are the temperatures that maximize organismal performance.

13.6 Temperature and Ecology of Squamates

Squamates, especially lizards, are capable of very precise thermoregulation, and the thermal environment can be a very important feature of their ecology. Microhabitats at which particular body temperatures can be maintained may be one of the dimensions that define the ecological niches of lizards. For example, the five most common species of *Anolis* in Cuba partition the habitat in several ways (**Figure 13–18**). First, they divide the habitat along the continuum from sunny to shady: two species (*A. lucius* and *A. allogus*) occur in deep shade in forests, one (*A. homolechis*) in partial shade in clearings and at the forest edge, and two (*A. allisoni* and *A. sagrei*) in full sunlight. Within habitats in the sunlight-shade continuum, the lizards are separated by the substrates they use as perch sites. In the forest, *A. lucius* perches on large trees up to 4 meters above the ground, whereas *A. allogus* rests on small trees within 2 meters of the ground. *A. homolechis*, which does not share its habitat with another common species of *Anolis*, perches on both large and small trees. In open habitats, *A. allisoni* perches more than 2 meters above the ground on tree trunks and houses, and *A. sagrei* perches below 2 meters on bushes and fence posts. Within this temperature and spatial distribution, the lizards have developed further specializations. Species that live on tree trunks near the ground have long hindlimbs and long tails, whereas species that live on twigs high in the canopy have short hindlimbs and short tails. The body proportions of the different forms appear to be related to locomotion on broad versus narrow surfaces. These relationships among thermal ecology, habitat, and body form have evolved repeatedly on different islands in the Caribbean.

Some species of lizards do not thermoregulate. Lizards that live beneath the tree canopy in tropical forests often have body temperatures very close to air temperature (that is, they are thermally passive), whereas species that live in open habitats thermoregulate more precisely. The relative ease of thermoregulation in different habitats may be an important factor in determining whether a species of lizard thermoregulates or allows its temperature to vary with ambient temperature.

The distribution of sunny areas is one factor that determines the ease of thermoregulation. Sunlight penetrates the canopy of a forest in small patches that move across the forest floor as the sun moves across the sky. These patches of sunlight are the only sources of solar radiation for lizards that live at or near the forest floor, and the patches may be too sparsely distributed or too transient to be used for thermoregulation. In open habitats, sunlight is readily available, and thermoregulation is easier.

The task of integrating thermoregulatory behavior with foraging is relatively simple for sit-and-wait foragers such as *Anolis*. These lizards can readily change their balance of heat gain and loss by making small movements into and out of shade or between calm and breezy perch sites, while continuing to scan their surroundings for prey. Widely foraging species may have more difficulty integrating thermoregulation

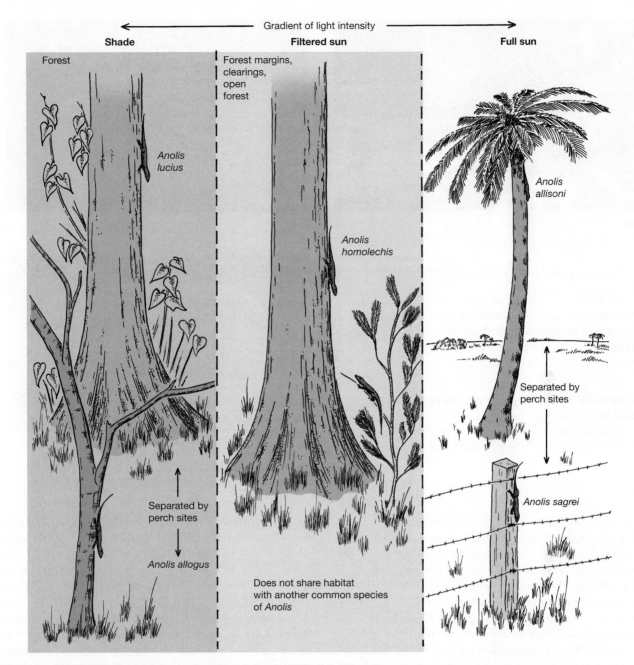

Figure 13–18 Habitat partitioning by Cuban species of *Anolis*. The habitat is divided along a gradient from shaded forest to open pastures and a vertical gradient in perch height.

and predation. They are continuously moving between sunlight and shade and into and out of the wind, and their body temperatures are affected by their foraging activity. These lizards sometimes have to stop foraging to thermoregulate, resuming foraging only when they have warmed or cooled enough to return to their activity temperature range.

Body size is yet another variable that can affect thermoregulation. An example of the interaction of body size, thermoregulation, and foraging behavior is provided by three species of teiid lizards (*Ameiva*) in Costa Rica. *Ameiva* are widely foraging predators that move through the habitat, pushing their snouts beneath fallen leaves and into holes. Three species of

(a)

(b)

(c)

Figure 13–19 Three sympatric species of *Ameiva* from Costa Rica. (a) *Ameiva leptophrys*, which has an average adult mass of 83 grams. (b) *Ameiva festiva*, average adult mass 32 grams. (c) *Ameiva quadrilineata*, average adult mass 10 grams.

Ameiva occur together on the Osa Peninsula of Costa Rica in a habitat that extends from full sunlight (a roadside) to deep shade (forest). The largest of the three species, *A. leptophrys*, has an average body mass of 83 grams; the middle species, *A. festiva*, weighs approximately 32 grams; and the smallest, *A. quadrilineata*, on average weighs 10 grams (**Figure 13–19**). The three species forage in different parts of the habitat: *A. quadrilineata* spends most of its time in the short vegetation at the edge of the road, *A. festiva* is found on the bank beside the road, and *A. leptophrys* forages primarily beneath the forest canopy (**Figure 13–20**). The different foraging sites of the three species may reflect differences in thermoregulation that result from the variation in body size.

The thermoregulatory behavior of the three species is the same: A lizard basks in the sunlight until its body temperature rises to 39°C to 40°C and then moves through the mosaic of sunlight and shade as it searches for food. The body temperature of the lizard drops as it forages, and a lizard ceases foraging and resumes basking when its body temperature has fallen to 35°C. Thus the time that a lizard can forage depends on how long it takes for its body temperature to cool from 39°C or 40°C to 35°C. The rate of cooling for *Ameiva* in the shade is inversely proportional to the body size of the three species: *A. quadrilineata* cools in 4 minutes, *A. festiva* in 6 minutes, and *A. leptophrys* in 11 minutes (**Figure 13–21a** on page 346). That relationship appears to explain part of the microhabitat separation of the three species: The smallest species, *A. quadrilineata*, cools so rapidly that it may not be able to forage effectively in shady microhabitats, whereas *A. leptophrys* cools slowly and can forage in the shade beneath the forest canopy. The species of intermediate body size, *A. festiva*, uses the habitat with an intermediate amount of shade.

The slow rate of cooling of *A. leptophrys* may explain why it is able to forage in the shade. However, field observations indicate that its foraging is actually restricted to shade; it emerges from the forest only to bask. Does some environmental factor prevent *A. leptophrys* from foraging in sunny areas? The answer

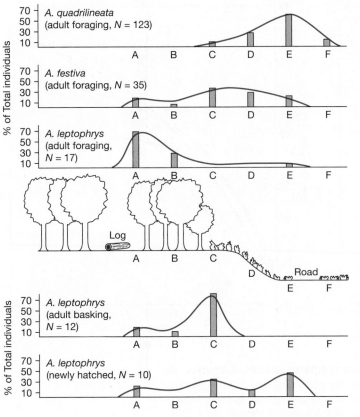

Figure 13–20 Foraging sites of three species of *Ameiva* **in Costa Rica.** The histograms show the number of individuals of the three species seen in each of six locations: (A) a small clearing in the forest; (B) immediately inside the forest edge; (C) outside the edge of the forest; (D) midway between the edge of the forest and open area; (E) low vegetation beside a road; and (F) low vegetation in a large open area without trees.

to that question may lie in the way the body temperatures of the three species increase when they are in open microhabitats. Body size profoundly affects the equilibrium temperature of an organism in the sunlight. A lizard warms by absorbing solar radiation; as it gets warmer, its heat loss by convection, evaporation, and reradiation also increases. When the rate of heat loss equals the rate of heat gain, the body temperature does not increase further. Large lizards reach that equilibrium at higher body temperatures than do small ones. Computer simulations of the heating rates of the three *Ameiva* in sunlight showed that *A. quadrilineata* and *A. festiva* would reach equilibrium at body temperatures of 37°C to 40°C, but *A. leptophrys* would continue to heat until its body temperature reached a lethal 45°C (**Figure 13–21b**). This analysis suggests that *A. leptophrys* would die of heat stress if it spent more than a few minutes in a sunny microhabitat, but that the two smaller species of *Ameiva* would not have that problem.

Thus, as a result of the biophysics of heat exchange, the large body size of *A. leptophrys* apparently allows it to forage in shaded habitats (because it cools slowly) but prevents it from foraging in sunny habitats

(because it would overheat). Field observations of the foraging behavior of hatchling *A. leptophrys* emphasize the importance of heat exchange in the foraging behavior of these lizards. Hatchling *A. leptophrys* forage in open habitats like *A. festiva* and *A. quadrilineata* rather than under the forest canopy like adult *A. leptophrys* (see Figure 13–20). That is, the juveniles of the large species of *Ameiva* behave like adults of the smaller species, probably because of the importance of body size and heat exchange in determining the microhabitats in which lizards can thermoregulate.

The difference in the use of various microhabitats by these three species of lizards looks, at first glance, like an example of habitat partitioning in response to interspecific competition for food. That is, because all three species eat the same sort of prey, they could be expected to concentrate their foraging efforts in different microhabitats to reduce competition. However, this analysis of the thermal requirements of the lizards suggests that interspecific competition for food is, at most, a secondary factor.

If competition for food were important, we would not expect to find hatchling *A. leptophrys* foraging in the same microhabitat as adult *A. quadrilineata*

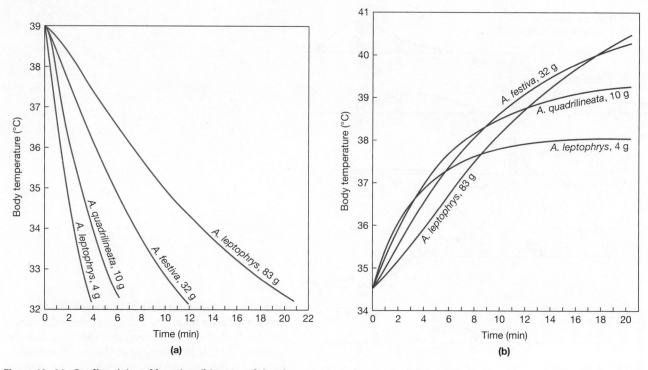

Figure 13–21 Cooling (a) and heating (b) rates of the three sympatric species of *Ameiva*. The largest species, *A. leptophrys*, heats and cools more slowly than the smaller species. If it remained in the sunlight, its body temperature would rise above 40°C. The smaller species heat and cool more rapidly than the large species and reach temperature equilibrium at lower body temperatures.

because the similarity in size of the two lizards would intensify competition for food. The hypothesis that competition for food determines the microhabitat distribution of the animals predicts that the forms most similar in body size should be widely separated in the habitat. In contrast, the hypothesis that energy exchange with the environment is critical in determining where a lizard can forage predicts that species of similar size will live in the same habitat, and that is ap-

proximately the pattern seen. Apparently, the physical environment (radiant energy) is more important than the biological environment (interspecific competition for food) in determining the microhabitat distributions of these lizards. That conclusion reflects the broad-scale ecological significance of the morphological and physiological differences between ectotherms and endotherms, a theme that is developed in the next chapter.

Summary

The extant lepidosaurs include the squamates (lizards and snakes) and their sister group, the Sphenodontidae. The lepidosaurs, with about 9000 species, form the second largest group of extant tetrapods. The tuatara of the New Zealand region is the sole extant sphenodontid. Tuatara are lizardlike animals, about 60 centimeters long, with a dentition and jaw mechanism that give a shearing bite. Sphenodontids were diverse in the Mesozoic and included terrestrial insectivorous and herbivorous species as well as a marine form.

Lizards range in size from tiny geckos and chameleons less than 3 centimeters long to the Komodo monitor lizard, which reaches a length of 3 meters. Most agamid, phrynosomatid, and dactyloid lizards are sit-and-wait predators that maintain territories and detect prey and intruders by vision. These species often employ colors and patterns in visual displays during courtship and territorial defense. Many lacertid, scincid, and teiid lizards are widely foraging predators that detect prey by olfaction and do not maintain territories.

Color change is a temperature-regulating mechanism used by lizards such as the desert iguana (*Dipsosaurus dorsalis*). When they first emerge in the morning, desert iguanas are dark (upper left). By the time it has reached its activity temperature, a lizard has turned light (upper right). This color change reduces heat gained from the sun by almost 25 percent.

Luminescent bacteria in the light organs of fishes emit light as a by-product of their metabolism. (Lower left) A black dragonfish, *Idiacanthus*. A long barbel on the chin bears a luminous lure that is believed to attract prey close enough to be engulfed by the enormous jaws lined with sharp teeth. (Lower right) The flashlight fish, *Photoblepharon*, has a light-emitting organ under each eye. The fish can cover the organ with a pigmented shutter to conceal the light, or open the shutter to reveal it. It uses the light organ in social interactions with other flashlight fish, and in a blink-and-run defense to startle and confuse predators.

Three species of salamanders form a mimicry complex in eastern North America. The red eft (*Notophthalmus viridescens,* top left) and red salamander (*Pseudotriton ruber,* top right) have skin toxins that deter predators. The red-backed salamander (*Plethodon cinereus,* middle left) is not protected by toxins, but predators confuse the erythristic form of that species (middle right) with the toxic species. The experiment described in the text used the mountain dusky salamander (*Desmognathus ochrophaeus,* bottom left) as a palatable control.

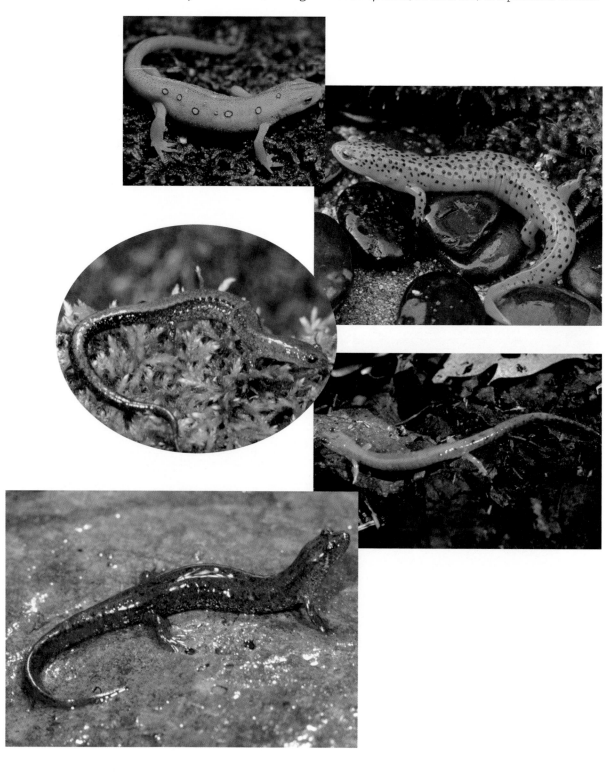

The gular fans of lizards are used in social displays. Color, size, and shape identify the species and sex of an individual. (All the lizards in these photographs are males.) (Upper left) *Anolis grahami*, from Jamaica. (Upper right) Knight anole, *Anolis equestris*, from Cuba. (Lower left) Carolina anole, *Anolis carolinensis*, from Florida. (Lower right) *Anolis chrysolepis* from Brazil.

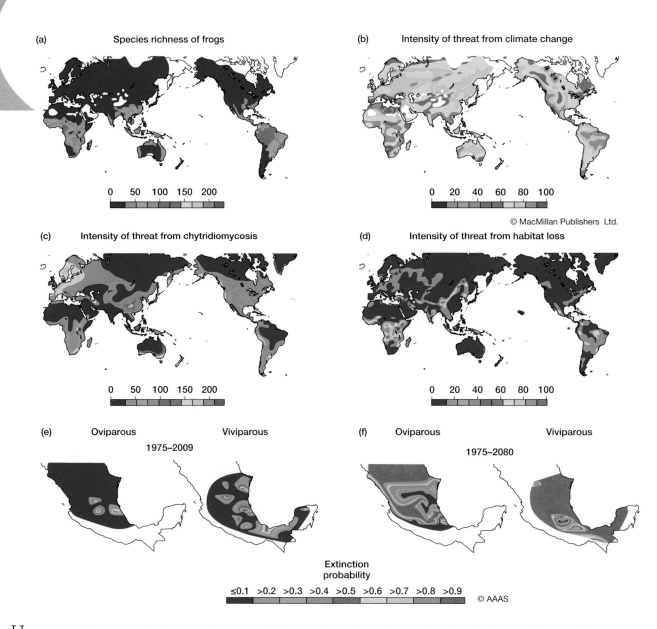

(a) Species richness of frogs

0 50 100 150 200

(b) Intensity of threat from climate change

0 20 40 60 80 100

© MacMillan Publishers Ltd.

(c) Intensity of threat from chytridiomycosis

0 50 100 150 200

(d) Intensity of threat from habitat loss

0 20 40 60 80 100

(e) Oviparous   Viviparous
1975–2009

(f) Oviparous   Viviparous
1975–2080

Extinction probability

≤0.1 >0.2 >0.3 >0.4 >0.5 >0.6 >0.7 >0.8 >0.9

© AAAS

Upper panel: **The geographic distribution of amphibian species and the threats they face.** (a) Species richness of frogs: Frogs are the most diverse group of amphibians, and in tropical regions of South America, Africa, and Southeast Asia more than 100 species of frogs can be found in the same habitats. (b) Global climate change: In most parts of the world climate change is predicted to make the environment unsuitable for at least 50 percent of the species of frogs in the region, and in some areas 100 percent of the species will be affected. (c) Chytridiomycosis: From 50 to 100 percent of frog species are expected to be threatened by chytridiomycosis, with frog communities in South America, western Europe, and southern Australia feeling the greatest impact. (d) Habitat loss: The greatest impact of habitat loss is predicted to occur in South America, Africa, and Asia where logging and agriculture are destroying native forests.

Lower panel: **Extinctions of Mexican *Sceloporus* lizards as a result of climate change.** (e) In 2009 a survey of 200 sites that had been occupied by species of *Sceloporus* in 1975 found that 12 percent of the populations were extinct. Extinctions were grouped in the areas most affected by rising temperatures in northern and central Mexico, and populations at high altitudes showed the greatest levels of extinction. Viviparous species had higher rates of extinction than did oviparous species. (f) By 2080 the temperature increase predicted by global climate models will lead to widespread extinctions of populations of *Sceloporus* in Mexico, and again viviparous species would feel the greatest impact.

Amphisbaenians are specialized burrowing lizards. Their skulls are solid structures that they use for tunneling through soil. Many amphisbaenians have blunt heads, and others have vertically keeled or horizontally spade-shaped snouts. The dentition of amphisbaenians appears to be specialized for nipping small pieces from prey too large to be swallowed whole. The skin of amphisbaenians is loosely attached to the trunk, and amphisbaenians slide backward or forward inside the tube of their skin as they move through tunnels with concertina locomotion.

Snakes are derived lizards. Repackaging the body mass of a vertebrate into a serpentine form has been accompanied by specializations of the mechanisms of locomotion (serpentine, rectilinear, concertina, and sidewinding), prey capture (constriction and the use of venom), swallowing (a highly kinetic skull), and digestion (up- and down-regulation of the digestive system).

Many squamates have complex social behaviors associated with territoriality and courtship, but parental care is only slightly developed. Viviparity has evolved at least 100 times among squamates. Thermoregulation is another important behavior of squamates, and various activities are influenced by body temperature. The ecological niches of some lizards may be defined in part by the microhabitats needed to maintain particular body temperatures.

Discussion Questions

1. Durophagy (eating hard-shelled prey, such as beetles or mollusks) appears as a dietary specialization among lizards, but it is rare among snakes. Why? What specializations of the teeth would you expect to find in durophagous lizards and in snakes?

2. *Anolis* lizards on the island of Jamaica do not show a reduction in signaling behavior during periods of visual "noise" created by the movement of wind-blown vegetation as *Anolis* on Puerto Rico do. The dewlaps of Jamaican *Anolis* are more brightly colored than those of the Puerto Rican lizards, and Jamaican *Anolis* expand and contract their dewlaps more rapidly than the Puerto Rican species do. What difference between the signaling systems of the Jamaican and Puerto Rican *Anolis* might allow the signals of the Jamaican species to be perceived even in visually noisy conditions?

3. Both the American horned lizards (*Phrynosoma*, Phrynosomatidae) and the Australian thorny devil (*Moloch horridus*, Agamidae) eat ants almost exclusively. Suggest an explanation for the convergence of ant-eating specialists in these two families on a short-legged, stocky body form and a spine-covered body.

4. The text mentioned several specializations that allow vipers to eat large prey. The large, heavy-bodied African vipers in the genus *Bitis*, such as the puff adder (*Bitis arietans*), provide particularly clear examples of these specializations. List as many of these specializations as you can, and explain how they work together.

5. Lizards that display sociality are viviparous species; in contrast, oviparous lizards do not appear to form long-term associations with other individuals of their species. What differences between these two groups of lizards might account for the difference in behavior?

Additional Information

Besson, A. A., et al. 2012. Is cool egg incubation temperature a limiting factor for the translocation of tuatara to southern New Zealand? *New Zealand Journal of Ecology* 36:29-37.

Collar, D. C., et al. 2011. Evolution of extreme body size disparity in monitor lizards (*Varanus*). *Evolution* 65:2664–2680.

Corl, A., et al. 2009. Alternative mating strategies and the evolution of sexual size dimorphism in the side-blotched lizard, *Uta stansburiana*: A population-level comparative analysis. *Evolution* 64:79–96.

Costello, E. K., et al. 2010. Postprandial remodeling of the gut microbiota in Burmese pythons. *The ISME Journal* 4:1375–1385.

Davis, A. R. 2012. Kin presence drives philopatry and social aggregation in juvenile desert night lizards (*Xantusia vigilis*). *Behavioral Ecology*, 23:18-24.

Duffield, G. A., and C. M. Bull. 2002. Stable social aggregations in an Australian lizard, *Egernia stokesii*. *Naturwissenschaften* 89:424–427.

Fry, B. J., et al. 2009. A central role for venom in predation by *Varanus komodoensis* (Komodo Dragon) and the extinct giant *Varanus* (*Megalania*) *priscus*. *Proceedings of the National Academy of Sciences* 106:8969–8974.

Glaw, F., et al. 2012. Rivaling the world's smallest reptiles: Discovery of miniaturized and microendemic new species of leaf chameleons (*Brookesia*) from Northern Madagascar. *PLoS ONE* 7(2): e31314. doi:10.1371/journal.pone.0031314/.

Hay, J. M., et al. 2010. Genetic variation in island populations of tuatara (*Sphenodon* spp.) inferred from microsatellite markers. *Conservation Genetics* 11:1063–1081.

Kearney, M., and B. L. Stuart. 2004. Repeated evolution of limblessness and digging heads in worm lizards revealed by DNA from old bones. *Proceedings of the Royal Society of London* B 271:1677–1683.

Lelièvre, H., et al. 2011. Contrasted thermal preferences translate into divergences in habitat use and realized performance in two sympatric snakes. *Journal of Zoology* 284:265–275.

Losos, J. B. 2009. *Lizards in an Evolutionary Tree.* Berkeley, CA. University of California Press.

Mason, R. T., and M. R. Parker. 2010. Social behavior and pheromonal communication in reptiles. *Journal of Comparative Physiology A* 196:729–749.

Ord, T. J., et al. 2011. The evolution of alternative adaptive strategies for effective communication in noisy environments. *The American Naturalist* 177:54–64.

Pryon, R. A., et al. 2011. The phylogeny of advanced snakes (Colubroidea), with discovery of a new subfamily and comparison of support methods for likelihood trees. *Molecular Phylogenetics and Evolution* 58:329–342.

Radder, R. S., et al. 2009. Offspring sex in a lizard depends on egg size. *Current Biology* (2009), doi:10.1016/j.cub.2009.05.027.

Sinervo, B., et al. 2000. Testosterone, endurance, and Darwinian fitness: Natural and sexual selection on the physiological basis of alternative male behaviors in side-blotched lizards. *Hormones and Behavior* 38:222–233.

Townsend, T. M., et al. 2011. Phylogeny of iguanian lizards inferred from 29 nuclear loci, and a comparison of concatenated and species-tree approaches for an ancient, rapid radiation. *Molecular Phylogenetics and Evolution* 61:363–380.

Vidal, N., and S. B. Hedges. 2009. The molecular evolutionary tree of lizards, snakes, and amphisbaenians. *Comptes Rendus Biologies* 332:129–139.

Wang, Y., and S. E. Evans. 2011. A gravid lizard from the Cretaceous of China and the early history of squamate viviparity. *Naturwissenschaften* 98:739–743.

Warner, D. A., and R, Shine. 2011. Interactions among thermal parameters determine offspring sex under temperature-dependent sex determination. *Proceedings of the Royal Society B* 278:256–265.

Wilson, J. A., et al. 2010. Predation upon hatchling dinosaurs by a new snake from the Late Cretaceous of India. *PLoS Biol* 8(3):e1000322. doi:10.1371/journal.pbio.1000322.

Zaher, H., et al. 2009. Molecular phylogeny of advanced snakes (Serpentes, Caenophidia) with an emphasis on South American Xenodontines: A revised classification and descriptions of new taxa. *Papéis Avulsos de Zoologia* 49:115–153.

Websites
General information
The Reptile Database http://www.reptile-database.org/
Herpetological literature
Center for North American Herpetology PDF Library http://www.cnah.org/cnah_pdf.asp
HerpLit Database http://herplit.com/herplit/

CHAPTER 14

Ectothermy: A Low-Cost Approach to Life

Ectothermy is an ancestral character of vertebrates; like many ancestral characters, however, it is just as effective as its derived counterpart, endothermy. Furthermore, the mechanisms of ectothermic thermoregulation are as complex and specialized as those of endothermy. Here we consider the consequences of ectothermy in shaping broader aspects of the lifestyle of fishes, amphibians, and reptiles. The general conclusion from this examination is that success in difficult environments is as likely to reflect the ancestral features of a group as its derived characters.

14.1 Vertebrates and Their Environments

Vertebrates manage to live in the most unlikely places. Amphibians live in deserts where rain falls only a few times a year, and several years may pass with no rainfall at all. Lizards live on mountains at elevations above 4000 meters, where the temperature falls below freezing nearly every night of the year and does not rise much above freezing during the day.

Of course, vertebrates do not seek out only challenging places to live—birds, lizards, mammals, and even amphibians can be found on tropical beaches (sometimes running between the feet of surfers), and fishes cruise the shore. However, even this apparently benign environment is difficult for some animals. Examining the ways that vertebrates live in extreme environments has provided much information about how they function as organisms—that is, how morphology, physiology, ecology, and behavior interact.

In some cases, elegant adaptations allow specialized vertebrates to colonize demanding habitats. More common and more impressive than these specializations, however, is the realization of how minor are the modifications of the ancestral vertebrate body plan that allow animals to endure environmental temperatures ranging from –70°C to +70°C or water conditions ranging from complete immersion in water to complete independence of liquid water. No obvious differences distinguish animals from vastly different habitats—an Arctic fox looks very much like a desert fox, and a lizard from the Andes Mountains looks like one from the Atacama Desert. The adaptability of vertebrates lies in the combination of minor modifications of their ecology, behavior, morphology, and physiology. A view that integrates these elements shows the startling beauty of organismal function of vertebrates.

14.2 Dealing with Dryness—Ectotherms in Deserts

Deserts are produced by various combinations of topography, air movements, and ocean currents and are found from the poles to the equator. But whatever their cause, deserts have in common a scarcity of liquid water. A desert is defined as a region in which the potential loss of water (via evaporation and transpiration of water by plants) exceeds the input of water via precipitation. Dryness is at the root of many features of deserts that make them difficult places for vertebrates to live. The dry air characteristic of most deserts seldom contains enough moisture to form clouds that would block solar radiation during the day or radiative cooling at night. As a result, the daily temperature range in deserts is large compared with that in more humid areas. Scarcity of water is reflected by sparse plant life. With few plants, deserts have low densities of insects for small vertebrates to eat and correspondingly sparse populations of small vertebrates that would be prey for larger vertebrates. Food shortages are chronic and are worsened by seasonal shortages and by unpredictable years of drought when the usual pattern of rainfall does not develop.

Not all deserts are hot; indeed, some are distinctly cold—most of Antarctica and the region of Canada around Hudson Bay and the Arctic Ocean are deserts. The low-latitude deserts north and south of the equator are hot deserts, however, and it is the combination of heat and dryness in low-latitude deserts that creates the most difficult problems.

The scarcity of rain that contributes to the low primary production of deserts also means that sources of liquid water for drinking are usually unavailable to small animals that cannot travel long distances. These animals obtain water from the plants or animals they eat, but plants and insects have sodium and potassium concentrations that differ from those of vertebrates. In particular, potassium is found in higher concentrations in both plants and insects than in vertebrate tissues, and excreting the excess potassium can be difficult if water is too scarce to waste in the production of large quantities of urine.

The low metabolic rates of ectotherms alleviate some of the difficulty caused by scarcity of food and water, but many desert ectotherms must temporarily extend the limits within which they regulate body temperatures or body fluid concentrations, become inactive for large portions of the year, or adopt a combination of these responses.

Terrestrial habitats in deserts are often harsh—hot by day and cold at night. Solar radiation is intense, and air does not conduct heat rapidly. As a result, a sunlit patch of ground can be lethally hot, whereas a shaded area just a few centimeters away can be substantially cooler. Underground retreats offer protection from both heat and cold. The annual temperature extremes at the surface of the ground in the Mohave Desert extend from a low that is below freezing to a maximum above 50°C; but just 1 meter below the surface of the ground, the annual temperature range is only from 10°C to 25°C. Desert animals rely on the temperature differences between sunlight and shade and between the surface and underground burrows to escape both hot and cold.

The Desert Tortoise

The largest ectothermic vertebrates in the deserts of North America are tortoises. The Bolson tortoise (*Gopherus flavomarginatus*) of northern Mexico probably once reached a shell length of a meter, although predation by humans has apparently prevented any tortoise in recent times from living long enough to grow that large. The desert tortoises (*G. agassizii* and *G. morafkai*) of the southwestern United States are smaller than the Bolson tortoise, but are still impressively large turtles (**Figure 14–1**). Adults can reach shell lengths approaching 50 centimeters and may weigh 5 kilograms or more. A study of the annual water, salt, and energy budgets of desert tortoises in Nevada shows the difficulties they face in that desert habitat.

Tortoises in the Mohave Desert construct shallow burrows that they use as daily retreat sites during the summer and deeper burrows for hibernation in winter. The tortoises in one study area emerged from hibernation in spring, and aboveground activity extended throughout the summer until they began hibernation again in November.

Figure 14–2 shows the annual cycle of time spent above ground and in burrows by the tortoises in the study and the annual cycles of energy, water, and salt balance. A positive balance means that the animal shows a net gain, whereas a negative balance represents a net loss. Positive energy and water balances indicate that conditions are good for the tortoises, but a positive salt balance means that ions are accumulating in the body fluids faster than they can be excreted. That situation indicates a relaxation of homeostasis and is probably stressful for the tortoises. The figure shows that the animals were often in negative balance for water or energy, and they accumulated salt during much of the year. Examination of the behavior and dietary habits of the tortoises throughout the year shows what was happening.

(a) **(b)**

Figure 14–1 The desert tortoise, *Gopherus agassizii*. (a) An adult tortoise. (b) A tortoise approaching its burrow.

After they emerged from hibernation in the spring, the tortoises were active for about 3 hours every fourth day; the rest of the time they spent in their burrows. From March through May the tortoises were eating annual plants that had sprouted after the winter rains. They obtained large amounts of water and potassium from this diet, and their water and salt balances were positive. Desert tortoises lack salt glands and their kidneys cannot produce concentrated urine, so they have no way to excrete the salt without losing a substantial amount of water in the process. Instead they retain the salt, and the osmolality of the tortoises' body fluids increased by 20 percent during the spring. This increased concentration shows that they were osmotically stressed as a result of the high concentrations of potassium in their food. Furthermore, the energy content of the plants was not great enough to balance the metabolic energy expenditure of the tortoises, and they were in negative energy balance. During this period, the tortoises were using stored energy by metabolizing their body tissues.

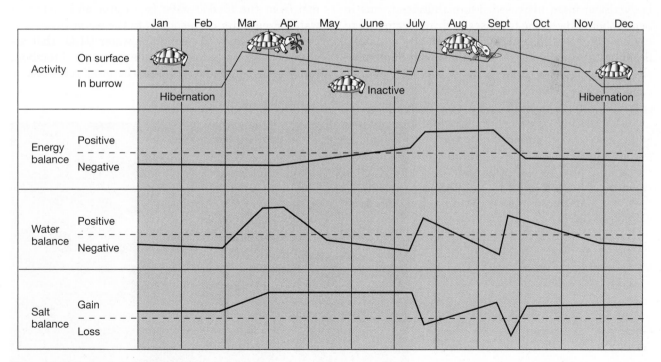

Figure 14–2 Annual cycle of desert tortoises.

As ambient temperatures increased from late May through early July, the tortoises shortened their daily activity periods to about 1 hour every sixth day. The rest of the time the tortoises spent in shallow burrows. The annual plants died, and the tortoises shifted to eating grass and achieved positive energy balances. They stored this extra energy as new body tissue. The dry grass contained little water, however, and the tortoises were in negative water balance. The osmolal concentrations of their body fluids remained at the high levels they had reached earlier in the year.

In mid-July, thunderstorms dropped rain on the study site, and most of the tortoises emerged from their burrows. They drank water from natural basins, and some of the tortoises constructed basins by scratching shallow depressions in the ground that could then trap rainwater. The tortoises drank large quantities of water (nearly 20 percent of their body mass) and voided the contents of their urinary bladders. The osmolal concentrations of their body fluids and urine decreased as they moved into positive water balance and excreted the excess salts they had accumulated when water was scarce. The behavior of the tortoises changed after the rain: they ate every 2 or 3 days and often spent their periods of inactivity above ground instead of in their burrows.

August was dry, and the tortoises lost body water and accumulated salts as they fed on dry grass. They were in positive energy balance, however, and their body tissue mass increased. More thunderstorms in September allowed the tortoises to drink again and to excrete the excess salts they had been accumulating. Seedlings sprouted after the rain, and in late September the tortoises started to eat them.

In October and November the tortoises continued to feed on freshly sprouted green vegetation; however, low temperatures reduced their activity, and they were in slightly negative energy balance. Salts accumulated and the osmolal concentrations of the body fluids increased slightly. In November, the tortoises entered hibernation. Hibernating tortoises had low metabolic rates and lost water and body tissue mass slowly. When they emerged from hibernation the following spring, they weighed only a little less than they had in the fall. Over the entire year, the tortoises increased their body tissues by more than 25 percent and balanced their water and salt budgets, but they did this by tolerating severe imbalances in their energy, water, and salt relations for periods that extended for several months at a time.

The Chuckwalla

The chuckwalla (*Sauromalus ater*) is a herbivorous iguanine lizard that lives in the rocky foothills of desert mountain ranges (**Figure 14–3**). The annual cycle of the chuckwallas, like that of the desert tortoises, is shaped by the availability of water. The lizards face many of the same problems that the tortoises encounter, but their responses are different. The lizards have nasal glands that allow them to excrete salt at high concentrations, and they do not drink rainwater but instead depend on water they obtain from the plants they eat.

Two categories of water are available to an animal from the food it eats: free water and metabolic water. Free water corresponds to the water content of the food—that is, molecules of water (H_2O) that are absorbed across the wall of the intestine. Metabolic water is a by-product of the chemical reactions of

Figure 14–3 A chuckwalla, *Sauromalus ater*.

Table 14–1 Quantity of water produced by metabolism of different substrates

Compound	Grams of Water/Gram of Compound
Carbohydrate	0.556
Fat	1.071
Protein	0.396 when urea is the end product;
	0.499 when uric acid is the end product

metabolism. Protons are combined with oxygen during aerobic metabolism, yielding a molecule of water for every two protons. The amount of metabolic water produced can be substantial; more than a gram of water is released by metabolism of a gram of fat (**Table 14–1**). For animals like the chuckwalla that do not drink liquid water, free water and metabolic water are the only routes of water gain that can replace the water lost by evaporation and excretion.

Chuckwallas were studied at Black Mountain in the Mohave Desert of California. They spent the winter hibernating in rock crevices and emerged from hibernation in April. Individual lizards spent about 8 hours a day on the surface in April and early May (**Figure 14–4**). By the middle of May, air temperatures were rising above 40°C, and the chuckwallas retreated into rock crevices for about 2 hours during the hottest part of the day, emerging again in the afternoon. At this time of year, annual plants that sprouted after the winter rains supplied both water and nourishment, and the chuckwallas gained weight rapidly. The average increase in body mass between April and mid-May was 18 percent (**Figure 14–5**). The water content of the chuckwallas increased faster than the total body mass, indicating that they were storing excess water.

By early June, the annual plants had withered, and the chuckwallas were feeding on perennial plants that contained less water and more ions than the annual plants. Both the body masses and the water contents of the lizards declined. The activity of the lizards

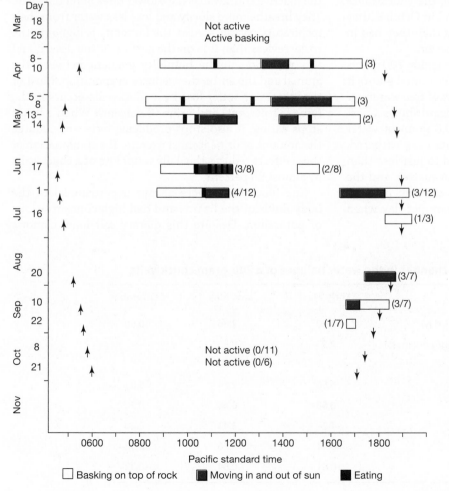

Figure 14–4 Daily behavior patterns in chuckwallas. Arrows indicate sunrise and sunset. Numbers in parentheses for April and May indicate the number of animals whose behavior was recorded. Thereafter the fraction in parentheses indicates the number of lizards active and observed out of the number known to be present.

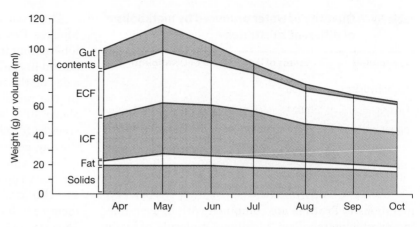

Figure 14–5 Seasonal changes in the body composition of chuckwallas. The total body water is composed of the extracellular fluid (ECF; blood plasma, urine, and water in lymph sacs) and the intracellular fluid (ICF; the water inside cells).

decreased in June and July as the midday temperature increased: Lizards emerged in the morning or in the afternoon but not during the middle of the day. In late June, the chuckwallas reduced their feeding activity; in July, they stopped eating altogether. They spent most of the day in the rock crevices, emerging only in late afternoon to bask for an hour or so every second or third day. From late May through autumn, the chuckwallas lost water and body mass steadily, and in October they weighed an average of 37 percent *less* than they had in April when they emerged from hibernation.

The water budget of a chuckwalla weighing 200 grams is shown in **Table 14–2**. In early May, the annual plants it is eating contain more than 2.5 grams of free water per gram of dry plant material, and the lizard shows a positive water balance, gaining about 0.8 gram of water per day. By late May, when the plants have withered, their free water content has dropped to just less than a gram of water per gram of dry plant matter, and the chuckwalla is losing about 0.8 gram of water per day. The rate of water loss falls to 0.3 gram per day when the lizard stops eating.

Evaporation from the respiratory surfaces and from the skin accounts for about 61 percent of the total water loss of a chuckwalla. As midday temperatures increase and lizards stop eating, they spend most of the day in rock crevices. The body temperatures of inactive chuckwallas are lower than the temperatures of lizards that remain active. Because of their low body temperatures, the inactive chuckwallas have lower rates of metabolism. They breathe more slowly and lose less water from their respiratory passages. Also, the humidity is higher in the rock crevices than it is on the surface of the desert, and this reduction in the humidity gradient between the animal and the air further reduces evaporation. Most of the remaining water loss by a chuckwalla occurs in the feces (31 percent) and urine (8 percent). When a lizard stops eating, it also stops producing feces and reduces the amount of urine it must excrete. The combination of these effects reduces the daily water loss of a chuckwalla by almost 90 percent.

The food plants were always hyperosmolal to the body fluids of the lizards and had high concentrations of potassium. Despite this dietary salt load, osmotic

Table 14–2 Seasonal changes in the water balance of a 200-gram chuckwalla

	Early May	Late May	September
Food intake (g dry mass/day)	2.60	2.86	0.00
Water content of food (g/g dry mass)	2.53	0.96	—
Water gain (g/day)			
Free water	6.56	2.74	0.0
Metabolic water	0.68	0.68	0.20
Total water gain	7.24	3.41	0.20
Water loss (g/day)	6.41	4.26	0.52
Net water flux (g/day)	+0.81	−0.84	−0.32

concentrations of the chuckwallas' body fluids did not show the variation seen in tortoises, because the lizards' nasal salt glands were able to excrete ions at high concentrations. The concentration of potassium ions in the salt-gland secretions was nearly 10 times their concentration in urine. The formation of potassium salts of uric acid was the second major route of potassium excretion by the lizards and was nearly as important in the overall salt balance as nasal secretion. The chuckwallas would not have been able to balance their salt budgets without the two extrarenal routes of ion excretion, but with them they were able to maintain stable osmolal concentrations.

Both the chuckwallas and tortoises illustrate the interaction of behavior and physiology in responding to the characteristics of their desert habitats. The tortoises lack salt-secreting glands and stored the salt they ingested, tolerating increased body fluid concentrations until a rainstorm allowed them to drink water and excrete the excess salt. The chuckwallas were able to stabilize their body fluid concentrations by using their nasal glands to excrete excess salt, but they did not take advantage of rainfall to replenish their water stores. Instead they became inactive, reducing their rates of water loss by almost 90 percent and relying on energy stores and metabolic water production to see them through the period of drought.

Conditions for the chuckwallas were poor at the Black Mountain site during the study. Only 5 centimeters of rain had fallen during the preceding winter, and that low rainfall probably contributed to the early withering of the annual plants that forced the chuckwallas to cease activity early in the summer. Unpredictable rainfall is a characteristic of deserts, however, and the animals living there must be able to adjust to the consequences. Rainfall records from the weather station closest to Black Mountain showed that in 5 of the preceding 10 years, the annual total rainfall was about 5 centimeters. Thus the year of the study was not unusually harsh; conditions are sometimes even worse—only 2 centimeters of rain fell during the winter after the study. However, conditions in the desert are sometimes good. Fifteen centimeters of rain fell in the winter of 1968, and vegetation remained green and lush throughout the following summer and fall. Chuckwallas and tortoises live for decades, and their responses to the boom-or-bust conditions of their harsh environments must be viewed in the context of their long life spans. A temporary relaxation of the limits of homeostasis in bad years is an effective trade-off for survival that allows the animals to exploit the abundant resources of good years.

The intricate interplay of activity, food sources, water balance, and thermoregulation characteristic of tortoises and chuckwallas is threatened by global climate change, which is expected to increase maximum temperatures and decrease precipitation in the desert Southwest. Most desert reptiles are forced to cease midday activity during the summer months, and an increase in the daily maximum temperature would extend the portion of the year when midday activity is not possible, potentially decreasing annual food consumption and reproduction. A reduction in rainfall and the resulting decrease in the growth of desert plants would limit the amount of food available to herbivorous reptiles as well as the availability of prey for insectivorous reptiles.

Desert Amphibians

Permeable skins and high rates of water loss are characteristics that would seem to make amphibians unlikely inhabitants of deserts, but certain species are abundant in desert habitats. Most remarkably, these animals succeed in living in the desert *because* of their permeable skins, not despite them. Anurans are the most common desert amphibians; tiger salamanders are found in the deserts of North America, and several species of plethodontid salamanders occupy seasonally dry habitats in California.

The spadefoot toads are the most thoroughly studied desert anurans (**Figure 14–6**). They inhabit the desert regions of North America—including the edges of the Algodones Sand Dunes in southern California, where the average annual precipitation is only 6 centimeters and in some years no rain falls at all. An analysis of the mechanisms that allow an amphibian to exist in a habitat like that must include consideration of both water loss and gain. A desert anuran must control its water loss behaviorally by its choice of sheltered microhabitats free from solar radiation and wind movement. Different species of anurans utilize different microhabitats—a hollow in the bank of a desert wash, the burrow of a ground squirrel or kangaroo rat, or a burrow the anuran excavates for itself. All these places are cooler and wetter than exposed ground.

Desert anurans spend extended periods underground, emerging on the surface only when conditions are favorable. Spadefoot toads construct burrows about 60 centimeters deep, filling the shaft with dirt and leaving a small chamber at the bottom, which they occupy. In southern Arizona, the spadefoots construct these burrows in September, at the end of the summer rainy season, and remain in them until the rains resume the following July.

Figure 14–6 A desert spadefoot toad, *Spea multiplicata*.

At the end of the rainy season when the spadefoots first bury themselves, the soil is relatively moist. The water tension created by the normal osmolal concentration of a spadefoot's body fluids establishes a gradient favoring movement of water from the soil into the toad. In this situation, a buried spadefoot can absorb water from the soil just as the roots of plants do. With a supply of water available, a spadefoot toad can afford to release urine to dispose of its nitrogenous wastes.

As time passes, the soil moisture content decreases, and the soil moisture potential (the driving force for movement of water) becomes more negative until it equals the water potential of the spadefoot. At this point, there is no longer a gradient allowing movement of water into the toad. When its source of new water is cut off, a spadefoot stops excreting urine and instead retains urea in its body, increasing the osmotic pressure of its body fluids. Osmotic concentrations as high as 600 mmol · [kg H_2O]$^{-1}$ have been recorded in spadefoot toads emerging from burial at the end of the dry season. The low water potential produced by the high osmolal concentration of the spadefoot's body fluids may reduce the water gradient between the animal and the air in its underground chamber so that evaporative water loss is reduced. Sufficiently high internal concentrations should create potentials that would allow spadefoot toads to absorb water from even very dry soil.

The ability to continue to draw water from soil enables a spadefoot toad to remain buried for 9 or 10 months without access to liquid water. In this situation, its permeable skin is not a handicap to the spadefoot—it is an essential feature of the toad's biology. If the spadefoot had an impermeable skin, or if it formed an impermeable cocoon as some other amphibians do, water would not be able to move from the soil into the animal. Instead, the spadefoot would have to depend on the water contained in its body when it was buried. Under those circumstances, spadefoot toads would probably not be able to invade the desert because their initial water content would not see them through a 9-month dry season.

14.3 Coping with Cold— Ectotherms in Subzero Conditions

Temperatures drop below freezing in the habitats of many vertebrates on a seasonal basis, and some animals at high elevations may experience freezing temperatures on a daily basis for a substantial part of the year. Birds and mammals respond to cold by increasing metabolic heat production and insulation, but ectotherms do not have those options. Instead, ectotherms show one of two responses—they avoid freezing by supercooling or synthesizing antifreeze compounds, or they tolerate freezing and thawing by using mechanisms that prevent damage to cells and tissues.

Cold Fishes

The temperature at which water freezes is affected by its osmolal concentration: pure water freezes at 0°C, and increasing the osmolal concentration lowers the freezing point. Body fluid concentrations of marine fishes are 300–400 mmol · [kg H_2O]$^{-1}$, whereas seawater has a concentration near 1000 mmol · [kg H₂O]⁻¹. The osmolal concentrations of the body fluids of marine fishes correspond to freezing points of –0.6°C to –0.8°C, and the freezing point of seawater is –1.86°C. The temperature of Arctic and Antarctic seas falls to –1.8°C in winter, yet the fishes swim in this water without freezing.

A classic study of freezing avoidance of fishes was conducted in Hebron Fjord in Labrador. In summer, the temperature of the surface water at Hebron Fjord is above freezing, but the water at the bottom of the fjord is –1.73°C (**Figure 14–7**). In winter, the surface temperature of the water also falls to –1.73°C, like the bottom temperature. Several species of fishes live in the fjord; some are bottom-dwellers, whereas others live near the surface. These two zones present different problems to the fishes. The temperature near the bottom of the fjord is always below freezing, but ice is not present because ice is lighter than water and remains at the surface. Surface-dwelling fishes live in water temperatures that rise well above freezing in the summer and drop below freezing in winter, and they are also in the presence of ice.

The body fluids of bottom-dwelling fishes in Hebron Fjord have freezing points of –0.8°C year-round. Because the body temperatures of these fishes are –1.73°C, the fishes are supercooled; that is, the water in their bodies is in the liquid state even though it is below its freezing point. When water freezes, the water molecules become oriented in a crystal lattice. The process of crystallization is accelerated by nucleating agents that hold water molecules in the proper spatial orientation for freezing. In the absence of nucleating agents, pure water can remain liquid at –20°C. In the

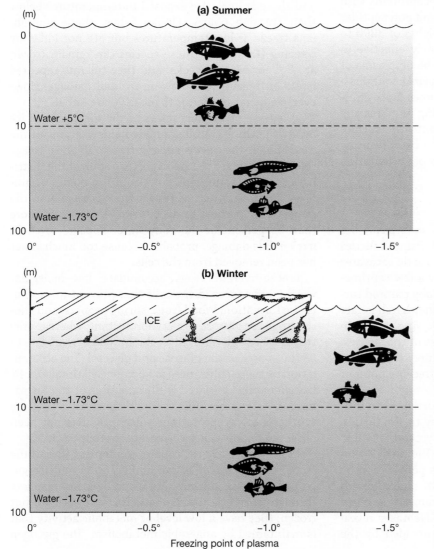

Figure 14–7 Water temperatures and distribution of fishes in Hebron Fjord. The water temperature and occurrence of ice are shown for summer (a) and winter (b). The vertical axis shows the water depth in meters, and the horizontal axis shows the freezing point depression in degrees centigrade. In summer (a) the temperature in the upper 10 meters of water rises to about 5°C, but in winter (b) it falls to –1.73°C. Deeper water remains at –1.73°C year-round. The fishes that live in deep water have freezing points of –0.8°C in summer and winter, whereas the fishes that live in shallow water decrease their freezing points from about –0.8°C in summer to about –1.5°C in winter.

laboratory, the fishes from the bottom of Hebron Fjord can be supercooled to −1.73°C without freezing, but if they are touched with a piece of ice, which serves as a nucleating agent, they freeze immediately. At the bottom of the fjord there is no ice, and the bottom-dwelling fishes exist year-round in a supercooled state.

What about the fishes in the surface waters? They do encounter ice in winter when the water temperature is below the osmotically determined freezing point of their body fluids, and supercooling would not be effective in that situation. Instead, the surface-dwelling fishes synthesize antifreeze substances in winter that lower the freezing point of their body fluids to approximately the freezing point of the seawater in which they swim.

Antifreeze compounds are widely developed among vertebrates (and also among invertebrates and plants). Marine fishes have two categories of organic molecules that protect against freezing: glycoproteins with molecular weights of 2600 to 33,000 and polypeptides and small proteins with molecular weights of 3300 to 13,000. These compounds are extremely effective in preventing freezing. For example, the blood plasma of the Antarctic fish *Trematomus borchgrevinki* contains a glycoprotein that is several hundred times more effective than sodium chloride in lowering the freezing point. The glycoprotein is adsorbed onto the surface of ice crystals and hinders their growth by preventing water molecules from assuming the proper orientation to join the ice-crystal lattice.

Chilly Turtles

Newly hatched painted turtles (*Chrysemys picta*) bet their lives on their ability to supercool. Painted turtles lay their eggs in nests that the female turtle excavates in the soil. A turtle can dig down only a few centimeters, so the nests are not very deep. Baby painted turtles remain in the nest from the time they hatch in late summer or early autumn until the following spring. In northern parts of the geographic range of the species, baby turtles are exposed to temperatures that fall to −10°C or lower, well below the −0.7°C freezing point of their body fluids.

The soil of the nest freezes, so the turtles are in intimate contact with ice crystals. The key to the turtles' survival is avoiding contact with ice crystals that could initiate freezing of their body tissues. The turtles' skin appears to form a barrier that resists penetration by ice crystals; its outer layer consists of alpha keratin with lipid on its inner surface. Recordings of survival of hatchlings in nests in north-central Nebraska reveal the effectiveness of the protection provided by the

skin of painted turtles. All of the turtles survived in nests that cooled to temperatures as low as −8°C, and more than half the turtles survived minimum temperatures as low as −9.7°C. Mortality increased below that temperature, but seven of eight turtles survived in one nest that cooled to −12.7°C.

Frozen Frogs

Terrestrial amphibians that spend the winter in hibernation show at least two categories of responses to low temperatures. One group, which includes salamanders, toads, and aquatic frogs, buries deeply in the soil or hibernates in the mud at the bottom of ponds. These animals apparently are not exposed to temperatures below the freezing point of their body fluids. As far as we know, they have no antifreeze substances and no capacity to tolerate freezing. However, other amphibians apparently hibernate close to the soil surface, and these animals are exposed to temperatures below their freezing points. Unlike fishes, these amphibians freeze at low temperatures but are not killed by freezing (**Figure 14–8**). These species can remain frozen at −3°C for several weeks, and they tolerate repeated bouts of freezing and thawing without damage. However, temperatures lower than −10°C are lethal.

Tolerance of freezing refers to the formation of ice crystals in the extracellular body fluids; freezing of the fluids inside the cells is apparently lethal. Thus, freeze tolerance involves mechanisms that control the distribution of ice, water, and solutes in the bodies of animals. The ice content of frozen frogs is usually in the range of 34 percent to 48 percent. Freezing of more than 65 percent of the body water appears to cause irreversible damage, probably because too much water has been removed from the cells.

Freeze-tolerant frogs accumulate low–molecular weight substances in the cells that prevent intracellular ice formation. Wood frogs, spring peepers, and chorus frogs use glucose as a cryoprotectant, whereas gray tree frogs use glycerol. Glycogen in the liver appears to be the source of the glucose and glycerol. The accumulation of these substances is apparently stimulated by freezing and is initiated within minutes of the formation of ice crystals. This mechanism of triggering the synthesis of cryoprotectant substances has not been observed for any other vertebrates or for insects.

Frozen frogs are, of course, motionless. Breathing stops, the heartbeat is exceedingly slow and irregular or may cease entirely, and blood does not circulate through frozen tissues. Nonetheless, the cells are not frozen; they have a low level of metabolic activity that is maintained by anaerobic metabolism. The glycogen

Figure 14–8 Wood frog (*Lithobates sylvatica*). (a) At normal temperature. (b) Frozen. These photographs were taken in the reverse of the sequence in which they appear here—that is, the image at normal temperature shows the frog after it had thawed out.

content of frozen muscle and kidney cells decreases, and concentrations of lactic acid and alanine (two end products of anaerobic metabolism) increase.

14.4 The Role of Ectothermic Tetrapods in Terrestrial Ecosystems

Life as an animal is costly. In thermodynamic terms, an animal lives by breaking chemical bonds that were formed by a plant (if the animal is an herbivore) or by another animal (if it is a carnivore) and using the energy from those bonds to sustain its own activities. Vertebrates are particularly expensive animals because vertebrates generally are larger and more mobile than invertebrates. Big animals require more energy (i.e., food) than small ones, and active animals use more energy than sedentary ones.

In addition to body size and activity, an animal's method of temperature regulation (ectothermy, endothermy, or a combination of the two mechanisms) is a key factor in determining how much energy it uses and therefore how much food it requires. Because ectotherms rely on external sources of energy to raise their body temperatures to the level needed for activity and endotherms use heat generated internally by their metabolism, ectotherms use substantially less energy than do endotherms. The metabolic rates (i.e., rates at which energy is used) of terrestrial ectotherms are only 10 percent to 14 percent of the metabolic rates of birds and mammals of the same body size. The lower energy requirements of ectotherms mean that they need less food than would an endotherm of the same body size.

Body size is another major difference between ectotherms and endotherms that relates directly to their mode of temperature regulation and affects their roles in terrestrial ecosystems. Ectotherms are smaller than endotherms, partly because the energetic cost of endothermy is very high at small body sizes. As body mass decreases, the mass-specific cost of living (energy per gram) for an endotherm increases rapidly, becoming nearly infinite at very small body sizes (**Figure 14–9**). This is a finite world, and infinite energy requirements are just not feasible. Thus energy requirements, among other factors, apparently set a lower limit to the body size possible for an endotherm.

The mass-specific energy requirements of ectotherms also increase at small body sizes, but because the energy requirements of ectotherms are about one-tenth those of endotherms of the same body size, an ectotherm can be about an order of magnitude smaller than an endotherm. A mouse-size mammal weighs about 20 grams, and few adult birds and mammals have body masses less than 10 grams. The very smallest species of birds and mammals weigh about 3 grams, but many ectotherms are only one-tenth that size (0.3 gram). Amphibians are especially small—20 percent of the species of salamanders and 17 percent of the species of anurans have adult body masses less than 1 gram, and 65 percent of salamanders and 50 percent of anurans are lighter than 5 grams. Squamates are generally larger than amphibians, but 18 percent of the species of lizards and 2 percent of snakes weigh less than 1 gram. Even the largest amphibians and squamates

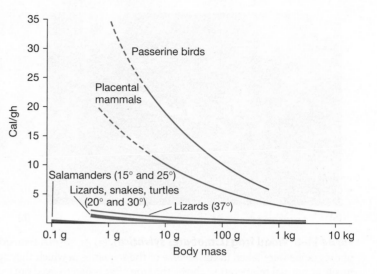

Figure 14–9 Energy cost. Resting metabolic rate (cal · [g · h]$^{-1}$) is shown as a function of body size for terrestrial vertebrates. Metabolic rates for salamanders are shown at 15°C and 25°C as the lower and upper limits of the darkened area, and data for lizards, snakes, and turtles combined are shown at 20°C and 30°C. Data from lizards alone are shown for 37°C. The curve for anurans falls within the lizards-snakes-turtles area, and the relationship for nonpasserine birds is similar to that for placental mammals. The dashed portions of the lines for birds and mammals show hypothetical extensions into body sizes below the minimum for adults of most species of birds and mammals.

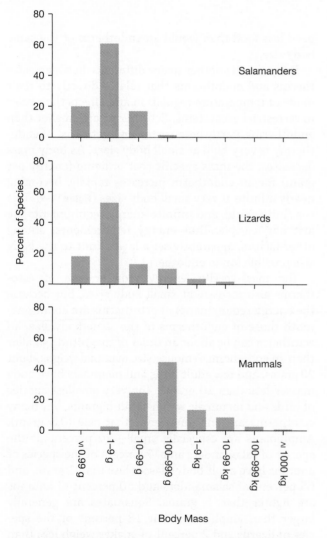

Figure 14–10 Adult body masses of salamanders, lizards, and mammals.

weigh substantially less than 100 kilograms, whereas more than 5 percent of the species of extant mammals weigh 100 kilograms or more (**Figure 14–10**).

Body shape is another aspect of vertebrate body form in which ectothermy allows more flexibility than does endothermy. An animal exchanges heat with the environment through its body surface, and the surface area of the body in relation to the volume of the body (the surface/volume ratio) is one factor that determines how rapidly heat is gained or lost. Small animals have higher surface/volume ratios than do large ones, and that is why endothermy becomes increasingly expensive at progressively smaller body sizes. Metabolic rates of small endotherms must be high enough to balance the high rates of heat loss across their large body surface areas.

Similarly, body shapes that increase the surface/mass ratio have an energy cost that makes them disadvantageous for endotherms. There are no highly elongate endotherms, whereas elongate body forms are widespread among fishes (true eels, moray eels, pipefishes, barracudas, and many more), amphibians (all caecilians and most salamanders, especially the limbless aquatic forms), and reptiles (many lizards and all snakes). Dorsoventral or lateral flattening is another shape that increases the surface/volume ratio. There are no flat endotherms, but flat fishes are common (dorsoventral flattening—skates, rays, flounders; lateral flattening—many coral-reef and freshwater fishes). Some reptiles are also flat (dorsoventral flattening—aquatic turtles, especially soft-shelled turtles, horned lizards; lateral flattening—many arboreal lizards, especially chameleons). Small body sizes and specialized body forms allow ectotherms to fill ecological niches that are not available to endotherms.

Table 14–3 Efficiency of biomass conversion by ectotherms and endotherms. The percentages are net conversion efficiencies calculated as (energy converted/energy assimilated) × 100.

Ectotherm Species	Efficiency (%)	Endotherm Species	Efficiency (%)
Red-backed salamander (*Plethodon cinereus*)	48	Kangaroo rat (*Dipodomys merriami*)	0.8
Mountain salamander (*Desmognathus ochrophaeus*)	76–98	Field mouse (*Peromyscus polionotus*)	1.8
Panamanian anole (*Anolis limifrons*)	23–28	Meadow vole (*Microtus pennsylvanicus*)	3.0
Side-blotched lizard (*Uta stansburiana*)	18–25	Red squirrel (*Tamiasciurus hudsonicus*)	1.3
Hognose snake (*Heterodon platyrhinos*)	81	Least weasel (*Mustela rixosa*)	2.3
Python (*Python curtus*)	33	Savannah sparrow (*Passerculus sandwichensis*)	1.1
Common adder (*Vipera berus*)	49	Marsh wren (*Telmatodytes palustris*)	0.5
Average of 12 species	50	Average of 19 species	1.4

The amount of energy required by ectotherms and endotherms is not the only important difference between them; equally significant is what they do with that energy once they have it. Endotherms expend more than 90 percent of the energy they take in to produce heat to maintain their high body temperatures. Less than 10 percent—often as little as 1 percent—of the energy a bird or mammal assimilates is available for net conversion (that is, increasing the species' biomass by growth of an individual or production of young). Ectotherms do not rely on metabolic heat. The solar energy they use to warm their bodies is free in the sense that it is not drawn from their food. Thus, most of the energy they ingest is converted into the biomass of their species. Values of net conversion efficiencies for amphibians and reptiles range from about 20 percent to more than 90 percent (**Table 14–3**).

Because of the difference in how energy is used, a given amount of chemical energy invested in an ectotherm produces a much larger biomass return than it would in an endotherm. A study of salamanders in the Hubbard Brook Experimental Forest in New Hampshire showed that, although their energy consumption was only 20 percent that of the birds or small mammals in the watershed, their conversion efficiency was so high that the annual increment of salamander biomass was equal to that of birds or small mammals. Similar comparisons can be made between lizards and rodents in deserts.

Small amphibians and squamates occupy key positions in the energy flow through an ecosystem. Because these animals are so small, they can capture tiny insects and arachnids that are too small to be eaten by birds and mammals. Because they are ectotherms, they are efficient at converting the energy in the food they eat into their own body tissues. As a result, the small ectothermic vertebrates in terrestrial ecosystems can be viewed as repackaging energy in a form that avian and mammalian predators can exploit. In other words, when a shrew or a bird searches for a meal in the Hubbard Brook Forest, the most abundant vertebrate prey it will find is salamanders. In this context, frogs, salamanders, lizards, and snakes occupy a position in terrestrial ecosystems that is important both quantitatively (in the sense that they constitute a substantial energy resource) and qualitatively (in that ectotherms exploit food resources that are not available to endotherms).

In a very real sense, small ectotherms can be thought of as living in a different world from that of endotherms. As we saw in the case of the three species of *Ameiva* lizards in Costa Rica (see Chapter 13), interactions with the physical world may be more important in shaping the ecology and behavior of small ectotherms than are biological interactions such as competition. In some cases, these small vertebrates may have their primary predatory and competitive interactions with insects and arachnids rather than with other vertebrates. For example, orb web spiders and *Anolis* lizards on some Caribbean islands are linked by both predation (adult lizards eat spiders, and spiders may eat hatchling lizards) and competition (lizards and spiders eat many of the same kinds of insects). The competitive relationship between these distantly related animals was demonstrated by experiments. When lizards were removed from experimental plots, the abundance of insect prey increased, and the spiders consumed more prey and survived longer than those in control plots where lizards were present.

Ectothermy and endothermy thus represent fundamentally different approaches to the life of a terrestrial vertebrate. An appreciation of ectotherms and endotherms as animals requires understanding the functional consequences of the differences between them. Ectothermy is an ancestral character of vertebrates, but it is a very effective way of life in modern ecosystems.

Summary

All living organisms are mosaics of ancestral and derived characters, interacting to make an organism a functional entity. Ectothermy is an ancestral character that is entirely functional in the modern world. Ectotherms do not use chemical energy from the food they eat to maintain high body temperatures. The results of that ancestral vertebrate characteristic are far-reaching for extant ectothermic vertebrates, and ectotherms and endotherms represent quite different approaches to vertebrate life.

Because of their low energy requirements, ectotherms can colonize habitats in which energy is in short supply. Ectotherms are able to extend some of their limits of homeostasis to tolerate high or low body temperatures and high or low body water contents when doing so allows them to survive in difficult conditions.

When food is available, ectotherms are efficient at converting the energy it contains into their own body tissues for growth or reproduction. Net conversion efficiencies of ectotherms average 50 percent of the energy assimilated compared with an average of 1.4 percent for endotherms.

Ectotherms can be smaller than endotherms because their mass-specific energy requirements are low, and many ectotherms weigh less than a gram, whereas most endotherms weigh more than 10 grams. Because of this difference in body size, many small ectotherms—such as salamanders, frogs, and lizards—eat prey that is too small to be consumed by endotherms. The efficiency of energy conversion by ectotherms and their small body sizes lead to a distinctive role in modern ecosystems, one that is in many respects quite different from that of terrestrial endotherms. Understanding these differences is an important part of understanding the organismal biology of terrestrial ectothermic vertebrates.

Discussion Questions

1. The Lake Eyre dragon (*Ctenophorus maculosus*) is an agamid lizard that lives only on the barren, vegetation-free, salt crust of several dry lake basins in South Australia. The lizards burrow into moist sediment beneath the crust to escape the extreme heat of midday, and they feed on insects and plant debris that blows onto the salt crust from the vegetation surrounding the lake basins. All of the lizard's water is derived from its food; there is no water for it to drink. Describe the ecological and physiological characteristics that allow the Lake Eyre dragon lizard to survive in such an inhospitable environment.

2. Although most fishes are ectotherms, the efficiency of secondary production by fishes is substantially lower than the average of 50 percent measured for terrestrial ectotherms. What difference between aquatic and terrestrial environments might account for this difference?

3. *Callopistes maculatus*, the dwarf tegu, is a teiid lizard that occurs in desert habitats in Chile. A study of its daily activity cycle in summer showed a decrease in activity during midday, although the body temperatures of the lizards that were active in midday were not different from the body temperatures of lizards that were active in the morning or afternoon. That observation is not consistent with the hypothesis that midday temperatures are so high that the lizards must seek shelter to avoid overheating. Apparently dwarf tegus in this habitat are able to maintain their normal activity body temperatures throughout the day. What alternative hypothesis can you suggest to account for the decrease in activity of dwarf tegus in midday?

Additional Information

Baker, P. J., et al. 2010. Winter severity and phenology of spring emergence from the nest in freshwater turtles. *Naturwissenschaften* 97:607–615.

Barrows, C. W. 2011. Sensitivity to climate change for two reptiles at the Mojave-Sonora Desert interface. *Journal of Arid Environments* 75:629–635.

Bartelt, P. E., and W. P. Porter. 2011. Modeling amphibian energetics, habitat suitability, and movements of western toads, *Anaxyrus* (=Bufo) *boreas*, across present and future landscapes. *Ecological Modelling* 221:2675–2686.

Davic, R. H., and H. H. Welsh, Jr. 2005. On the ecological role of salamanders. *Annual Review of Ecology and Systematics* 35:405–434.

Davis, J. R., and D. F. DeNardo. 2010. Seasonal patterns of body condition, hydration state, and activity of gila monsters (*Heloderma suspectum*) at a Sonoran desert site. *Journal of Herpetology* 44:83–93.

Evans, C. W., et al. 2011. How do Antarctic notothenioid fishes cope with internal ice? A novel function for antifreeze glycoproteins. *Antarctic Science* 23:57–64.

Lelièvre, H., et al. 2011. Contrasted thermal preferences translate into divergences in habitat use and realized performance in two sympatric snakes. *Journal of Zoology* 284:265–275.

Peltier, R. 2010. Synthesis and antifreeze activity of fish antifreeze glycoproteins and their analogues. *Chemical Science* 1:538–551.

Rexer-Huber, K. M. J., et al. 2011. Skin ice nucleators and glycerol in the freezing-tolerant frog *Litoria ewingii*. *Journal of Comparative Physiology B* 181:781–792.

Sharp, K. A. 2011. A peek at ice binding by antifreeze proteins. *Proceedings of the National Academy of Sciences* 108:781–782.

Storey, K. B., and J. M. Storey. 2011. Strategies of molecular adaptation to climate change: The challenges for amphibians and reptiles. In K. B. Storey and K. Tannino (eds.), *Temperature Adaptation in a Changing Climate*. Wallingford, UK: CABI Publishers.

Ultsch, G. G. 2006. The ecology of overwintering among turtles: Where turtles overwinter and its consequences. *Biological Reviews* 81:339–367.

Voituron, Y., et al. 2009. Freeze tolerance evolution among anurans: Frequency and timing of appearance. *Cryobiology* 58:241–247.

Websites

Desert ecology

http://digital-desert.com/wildlife/

http://www.ucmp.berkeley.edu/exhibits/biomes/deserts.php

Polar seas

http://www.exploratorium.edu/origins/antarctica/ideas/fish.html

http://www.arctic.uoguelph.ca/cpl/organisms/fish/adaptations/coldwatchem.htm

Ecosystem energy flow

http://www.ecostudies.org/

Freeze tolerance

http://openlearn.open.ac.uk/mod/oucontent/view.php?id=398512&section=1.4.3

Geography and Ecology of the Mesozoic Era

By the early Mesozoic era, Earth's entire land surface had coalesced into a single continent, Pangaea, that stretched from pole to pole. Early Mesozoic (Triassic) faunas and floras showed some regional differentiation due to climate but had no oceanic barriers to dispersal. With the breakup of Pangaea in the later Mesozoic (Jurassic and Cretaceous periods), floras and faunas were geographically isolated and became distinct in different parts of the world.

Many new types of insects appeared in the Mesozoic, including social insects such as bees, ants, and termites. The appearance and rapid radiation of the **angiosperms** (flowering seed plants) during the Cretaceous were important floral changes. A major turnover occurred in terrestrial vertebrates at the end of the Triassic, when the large-animal fauna, including a diverse assemblage of therapsids (nonmammalian synapsids), was replaced by dinosaurs. Jurassic dinosaurs added a new ecological form to the ecosystem—gargantuan herbivorous sauropods. Different kinds of herbivorous dinosaurs, with jaws and teeth specialized for processing tough vegetation, appeared in the Cretaceous.

Mammals, birds, and modern types of amphibians and reptiles appeared in the Mesozoic. The Mesozoic was also the time of a great radiation of marine reptiles—ichthyosaurs, plesiosaurs, mosasaurs, placodonts, and others, none of which survived the end of the Cretaceous. The ecological niches of these Mesozoic marine tetrapods were reinvaded in the Cenozoic era by marine mammals such as whales and seals.

Extinctions were prominent at the end of the Triassic period, which established the dinosaur-dominated

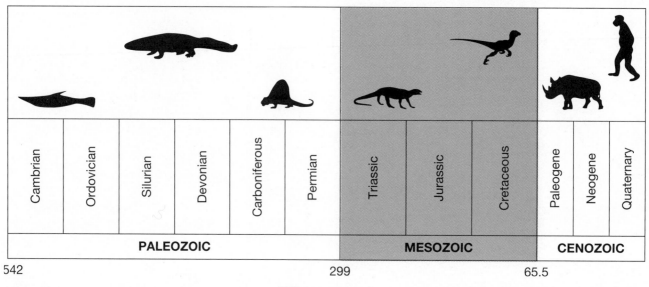

Cambrian	Ordovician	Silurian	Devonian	Carboniferous	Permian	Triassic	Jurassic	Cretaceous	Paleogene	Neogene	Quaternary

PALEOZOIC	MESOZOIC	CENOZOIC

542 299 65.5

Million years ago

faunas of the later Mesozoic, and at the end of the Cretaceous. Faunal changes during the Triassic may be related to low levels of atmospheric oxygen that would favor the type of lungs seen in archosaurs. The end-Cretaceous extinctions are famous as the ones that resulted in the demise of the dinosaurs, but in actuality many types of organisms were affected: invertebrates and plants as well as nondinosaurian vertebrates. Evidence for an impact with an extraterrestrial body (a meteor or asteroid) at the end of the Cretaceous has strengthened the argument for these extinctions being sudden rather than gradual in nature, but problems remain with the resolution of the fossil record and the pattern of the kinds of animals that survived versus the ones that became extinct.

15.1 Mesozoic Continental Geography

By the middle of the Triassic, the entire land area of Earth was concentrated in the supercontinent Pangaea (**Figure 15–1a**). The fragmentation of Pangaea began in the Jurassic with the separation of Laurasia (the northern portion of Pangaea) and Gondwana (the southern portion) by a westward extension of the Tethys Sea. Laurasia rotated away from the other continents, ripping North America from its connection with South America and increasing the size of the newly formed Atlantic Ocean (Figure 15–1b). Epicontinental seas (areas of sea that covered what is now continental land area) were more extensive than they had been in the Triassic.

Separation of the continents and rotation of the northern continents continued, and by the Late Cretaceous, the continents were approaching their current positions, although India was still close to Africa, and Australia, New Zealand, and New Guinea were still well south of their present-day positions (Figure 15–1c).

15.2 Mesozoic Terrestrial Ecosystems

The Mesozoic was marked by a series of large-scale changes in flora and fauna. Terrestrial ecosystems had achieved an essentially modern food web by the end of the Permian, and herbivory had been established in both vertebrates and insects.

The Triassic

The Early Triassic climate was rather unstable and appears to have fluctuated between a warm, wet world with high levels of carbon dioxide and a cool, dry world with low levels of carbon dioxide. The extinctions at the end of the Permian had left a fairly impoverished terrestrial fauna and flora. Triassic ecosystems were slow to recover, and fauna diversity equivalent to that of the Permian did not return until the end of the Triassic. New herbivorous forms of insects appeared in the Triassic, including stick insects. Early Triassic vertebrate faunas were dominated by the herbivorous dicynodont therapsid *Lystrosaurus* as well as carnivorous therapsids (including some early cynodonts, ancestral to mammals) and early archosaurs ("thecodonts"), diapsid reptiles that would eventually give rise to the dinosaurs.

Late in the Triassic, the climate in the places where tetrapods were found shifted from warm and moist to hot and dry; a diversity of therapsids persisted, but the archosaurs increased in diversity and faunal predominance (**Figure 15–2** on page 367). Although dinosaur fossils are unknown until the Late Triassic, evidence from footprints suggests they may have been present as early as the end of the Early Triassic.

Triassic vegetation included familiar modern groups of gymnosperms, such as conifers (pines and other cone-bearing trees, ranging from small bushes to large trees), plus ginkgophytes (relatives of the living ginkgo), cycads, and the now-extinct seed ferns. Other plants included ferns, tree ferns, and horsetails. Although many of these plants exist today, they are not nearly as common and widespread as they were in the Mesozoic. Triassic floras and faunas show regional characters that probably reflect differences in patterns of rainfall and seasonal temperature extremes in areas far from the sea.

Mid-Triassic herbivorous vertebrates (therapsids, archosaurs, and archosaur-related rhynchosaurs) ranged in body mass from 10 to 1000 kilograms (goat size to rhino size), indicative of a woodland habitat rather than closed forest, which supports primarily small-bodied herbivores. All of these herbivores were generalized browsers that would have foraged within a meter of the ground. Higher-level browsers were not present until the evolution of prosauropod dinosaurs in the latest Triassic.

There were virtually no arboreal herbivorous vertebrates. Only when mammals occupied this habitat in the Cenozoic was there a significant radiation of canopy-dwelling herbivorous vertebrates. However, there were plenty of herbivorous insects in the Triassic, and along with them were several arboreal insectivorous diapsid reptiles, including gliding forms.

During the Late Triassic there was a shift in vegetational communities, especially in Gondwana where conifers replaced seed ferns as the dominant plants.

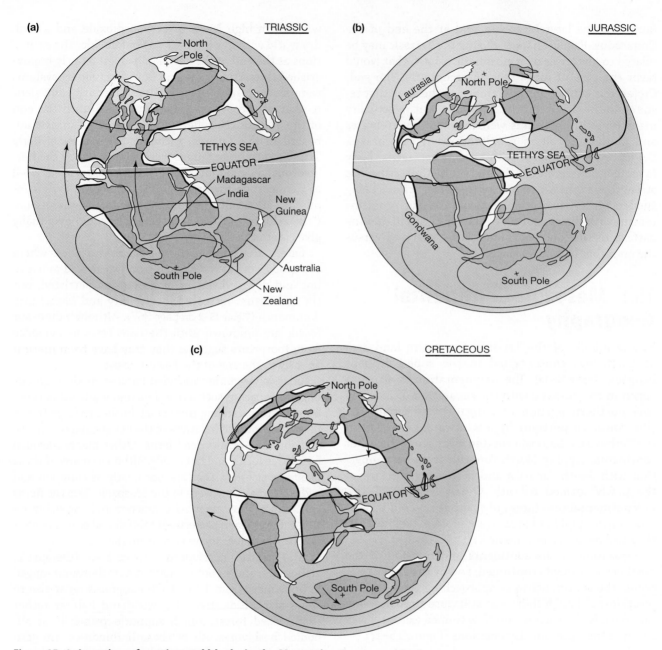

Figure 15–1 Location of continental blocks in the Mesozoic. Continental land areas are shaded, and epicontinental seas are the unshaded areas at the edges of the modern continents. Arrows indicate the direction of continental rotation.

This floral turnover was matched by a major faunal turnover; the original assemblage of large Triassic tetrapods (including therapsids and nondinosaurian archosaurs) became extinct, and new forms took their place. The fossil record shows a complex and protracted faunal turnover, with no evidence of simple competitive replacement.

Both dinosaurs (small carnivorous theropods) and mammals (tiny insectivorous forms) made a first appearance at this time, as did other modern groups such as sphenodontids (the sister group of squamates), turtles, and crocodiles, and now-extinct groups such as pterosaurs (flying reptiles) and various marine reptiles such as ichthyosaurs and plesiosaurs. Toward the end of the Triassic more herbivorous dinosaurs (prosauropods) appeared, but the herbivorous ornithischians were not apparent until the Jurassic.

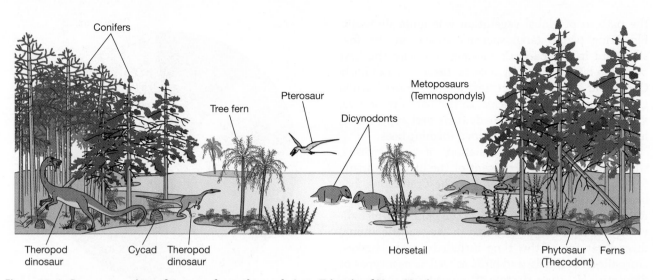

Figure 15–2 Reconstruction of a scene from the early Late Triassic of New Mexico.

We saw in Chapter 7 that low levels of atmospheric oxygen in the Late Permian period may have contributed to heightened levels of extinction before the mass extinction at the end of the Paleozoic era. Oxygen levels continued to drop during much of the Triassic, reaching a low of 15 percent by the middle of the period, and did not approach present-day levels until the Cretaceous.

These low oxygen levels may explain why the recovery of postextinction Triassic faunas was so slow. Low oxygen levels may also explain an evolutionary quandary in the Triassic. The previously successful therapsids were now progressively overshadowed by the rise of a new group, the archosaurs ("thecodonts" at this point in time). The respiratory system of archosaurs, with unidirectional airflow, may have been superior to the tidal flow in the respiratory systems of mammals when oxygen levels were low.

The Jurassic

The conifer- and fern-dominated vegetation of the Late Triassic continued into the Jurassic, although the seed ferns were then reduced in diversity. Several new kinds of insects appeared in the Jurassic, including beetles, thrips, and a variety of bugs (hemipterans).

An abrupt change in terrestrial ecosystems occurred at the start of the Jurassic when the dinosaurs became the predominant element of the large vertebrate fauna, with the addition of larger carnivores (allosaurids) and several kinds of large herbivores, including the high-browsing sauropods. The initial radiation of herbivorous dinosaurs has been causally linked with a contemporaneous radiation of cycads,

but the association does not stand up to statistical scrutiny. Jurassic mammals continued as small insectivores, with the addition of small omnivores, the highly successful multituberculates, that, unlike any other early Mesozoic mammal lineage, persisted into the early Cenozoic. Lizards and the modern groups of amphibians first appeared in the Jurassic, and *Archaeopteryx* (considered by some to be the earliest bird) is known from the Late Jurassic.

Sauropod dinosaurs were gigantic animals, with body masses ranging from 10,000 to 50,000 kilograms or even more, and they may have influenced patterns of vegetational growth and structure as elephants do today. It is not clear how the vegetation of the Jurassic was able to support such a great biomass of herbivores. However, atmospheric carbon dioxide levels were exceptionally high during the Jurassic and may have contributed to high levels of plant productivity.

We have a couple of fantastic windows into the Mesozoic world from localities where soft tissue may be preserved. The Late Jurassic Solnhofen Limestone of Bavaria, Germany, has yielded the feathered fossils of *Archaeopteryx* as well as small dinosaurs, pterosaurs, and many invertebrates. The Jehol Biota from northeastern China preserves an Early Cretaceous ecosystem—plants, invertebrates, fish, and tetrapods. Many feathered dinosaurs, early birds, and early mammals have come from this site.

The Cretaceous

The vegetation of the Early Cretaceous was similar to that of the Late Jurassic. However, by the Late Cretaceous, the pattern of global vegetation was quite different. Flowering plants (angiosperms) appear in the fossil record in the Early Cretaceous and were the first plants to be pollinated by insects. Insects of the Early Cretaceous include butterflies, aphids, short-horned grasshoppers, and gall wasps. In the Late Cretaceous, a variety of social insects made their first appearance, including termites, ants, and hive-forming bees.

The Late Cretaceous radiation of angiosperms, which formed up to 80 percent of the plants in many fossil assemblages by the end of the Cretaceous, has been seen as instrumental in the Cretaceous Terrestrial Revolution (KTR), which marked the point at which diversity on land outstripped that in the sea. The animal diversifications at this time included lizards, snakes, crocodiles, mammals, and birds as well as new types of dinosaurs. However, the role of angiosperms in the vegetational landscape was different from that of today. The large trees were still conifers. The growth forms of angiosperms included small trees and low-growing shrubs and herbs that replaced ferns and cycads as the ground cover. Grasses, which are such a dominant feature of landscapes today, did make an appearance in the latest Cretaceous but were not a predominant feature of the landscape until well into the Cenozoic.

Several new types of herbivorous dinosaurs appeared in the Late Cretaceous. The most diverse of these were the hadrosaurs (duck-billed dinosaurs) and the ceratopsians (horned dinosaurs)—both low-level feeders with complex, shearing cheek teeth, probably for dealing with tough vegetation. However, there is no evidence that they were specialized to feed on the new angiosperm plants, as rare occasions of preserved stomach contents reveal a diet of conifers.

Other later Cretaceous faunal changes include the evolution of mammals with complex molars by the Early Cretaceous and the appearance of mammals belonging to the three modern groups—monotremes, marsupials, and placentals. Although Cretaceous mammals remained small (most were shrew to mouse size, and none was larger than a raccoon), they were now becoming more diverse, with evidence of branching out into a variety of ecological types (diggers, swimmers, gliders, etc.). There was a considerable radiation of birds during the Cretaceous, but few of these forms were closely related to the modern birds that diversified in the Cenozoic. Snakes and modern types of crocodiles are also first known from the Cretaceous.

Thus, although the Jurassic and Cretaceous are usually thought of as the Age of Dinosaurs, there was a considerable diversity of other types of vertebrates present, albeit playing the role of the smaller members of the fauna. The lizards and other reptiles had a size range

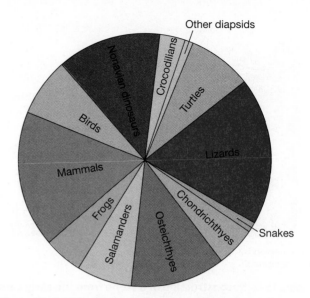

Figure 15–3 The relative abundance of genera at the Late Cretaceous Lance locality in Montana. This locality appears to represent a wooded, swampy habitat with large streams and some ponds.

similar to that of extant species. The very smallest dinosaur would have been about the size of the largest mammal, but, of course, most dinosaurs were considerably larger. Some of the smaller vertebrates probably preyed on dinosaur eggs, and others served as food for juvenile dinosaurs. A fossil of *Repenomamus*, the largest mammal known from that time, has a baby dinosaur (a psittacosaur ceratopsian) in its stomach.

Although dinosaurs are often depicted as inhabiting a strange, extinct world, in fact the dinosaur ecosystem was not so different from our present-day one. The main difference is that the ecological roles for larger vertebrates in the later Mesozoic ecosystems were taken by dinosaurs rather than mammals. All modern tetrapod groups—frogs, salamanders, lizards, snakes, turtles, crocodilians, birds, and mammals—arose in the Late Triassic or Jurassic, and further diversified in the Late Cretaceous. The fauna included far more species of vertebrates that were not dinosaurs than species of dinosaurs, even though dinosaurs may have made up much of the biomass (**Figure 15–3**).

15.3 Mesozoic Climates

Mesozoic climates were equable worldwide, and there were no polar ice caps at any time during the era. Large temnospondyls (aquatic non-amniote tetrapods) are found in Triassic deposits from Australia (where they lasted into the Cretaceous), Antarctica, Greenland, and Spitzbergen, and coal deposits in both the Northern and Southern Hemispheres point to moist climates. In contrast, low and middle latitudes were probably dry—either seasonally or year-round—until the Late Cretaceous and early Cenozoic, when coal deposits of Mesozoic age in middle latitudes indicate the presence of swamps and suggest that the climate had become wetter. These dry lower latitudes had a type of vegetation different from the equatorial vegetation of today, with the absence of wet tropical rain forests.

The Cretaceous plant record also suggests a highly equable world with temperate forests extending into the polar regions, and the Arctic faunas contained a diversity of dinosaurs and large aquatic reptiles (champsosaurs) that would have required a temperate or subtropical climate. A reduction in the diversity of plants at higher latitudes at the end of the Cretaceous signaled an episode of worldwide cooling, however.

15.4 **Mesozoic Extinctions**

The first round of Mesozoic extinctions occurred at the end of the Triassic and affected both terrestrial vertebrates and marine invertebrates. The fauna of the Triassic, which was in many ways a holdover from the late Paleozoic, was replaced by forms that would dominate the Mesozoic, such as the dinosaurs. Eighteen families of tetrapods became extinct at the end of the Triassic. The Triassic extinctions coincided with the initial breakup of Pangaea, but there is now some evidence for an impact with an asteroid or meteor at this time, as well episodes of massive volcanism, and evidence for a rapid increase in the level of atmospheric carbon dioxide with resultant global warming.

The extinctions at the end of the Cretaceous, known as the K-Pg (Cretaceous-Paleogene) boundary, included the demise of the dinosaurs, even though the magnitude of the effect on the global fauna was nowhere near as large as the late Paleozoic (end Permian) extinctions. Many hypotheses have been proposed to explain why dinosaurs became extinct, ranging from the extraterrestrial (their gonads were zapped by radiation from an exploding supernova) to the ridiculous (constipation caused by angiosperms). It must be remembered that dinosaurs were not the only animals that suffered extinction at this time. Thirty-six of the tetrapod families (40 percent) were extinct by the end of the Cretaceous, including not only all nonavian dinosaurs but also all flying reptiles (pterosaurs) and marine reptiles (ichthyosaurs, plesiosaurs, mosasaurs, and others). Birds and mammals suffered lesser, but still significant, extinctions, and insects were among the few groups that survived the Cretaceous relatively intact. There were also extensive extinctions among plants and marine invertebrates. Any explanation for the demise of the dinosaurs must also account for the disappearances of a large variety of other organisms, both on land and in the sea, as well as for the survival of many other lineages.

There has long been debate as to whether the end-Cretaceous extinctions were sudden or gradual in nature. The timing is difficult to gauge from the fossil record because fossils are not usually found in the type of uninterrupted sequences that permit accurate estimation of time intervals. The absence of an animal near the K-Pg boundary might indicate nonpreservation at that point in time rather than extinction, although recently a definitive dinosaur fossil (a ceratopsian horn) has been found very close to the K-Pg boundary in Montana, one of the few places in the world preserving fossils of this age.

In the past few decades, evidence has mounted for the impact of Earth with a large extraterrestrial body, an asteroid or meteor, at the K-Pg boundary. The dust cloud raised by such an impact might have limited the sunlight reaching Earth's surface for several years, resulting in greatly reduced plant photosynthesis and ecosystem collapse. Another potentially catastrophic event also happening on Earth around this time was the eruption of huge volcanoes in India known as the Deccan Traps. Large amounts of sulfur and chlorine gas would have been released from these eruptions, resulting in atmospheric and climatic effects that may have caused extinction of organisms.

While we have excellent evidence that an impact did occur at approximately the same time as the end-Cretaceous extinctions, there is still debate as to what would have been the exact effects of that impact on plant and animal life and whether other prior factors influenced patterns of organismal diversity so that the impact was a final blow rather than a sudden catastrophe. The difference in survival among different types of vertebrates seems to have depended on body size. Almost all vertebrates larger than 10 kilograms became extinct, including all of the nonavian dinosaurs at that time. Many species of smaller vertebrates (including birds and mammals) also became extinct, but of course some survived to reradiate in the Cenozoic. However, we do not know exactly why large crocodiles and turtles survived while other large vertebrates became extinct at this time, nor why modern kinds of amphibians, today so sensitive to environmental perturbations, also survived.

Additional Readings

Archibald, J. D. 2011. *Extinction and Radiation: How the Fall of Dinosaurs Led to the Rise of Mammals*. Baltimore, MD: The Johns Hopkins University Press.

Archibald, J. D., et al. 2010. Cretaceous extinctions: Multiple causes. *Science* 328:973.

Barrett, P. M., et al. 2011. The role of herbivory and omnivory in early dinosaur evolution. *Transactions of the Royal Society of Edinburgh* 101:383–396.

Benton, M. J. 2010. The origins of modern biodiversity on land. *Philosophical Transactions of the Royal Society B* 365:3667–3679.

Berner, R. A., et al. 2007. Oxygen and evolution. *Science* 316:557–558.

Brussatte, S. L., et al. 2011. Footprints pull origin and diversification of dinosaur stem lineage deep into Early Triassic. *Proceedings of the Royal Society B* 278:1107–1113.

Deenen, M. H. L. 2010. A new chronology for the end-Triassic extinctions. *Earth and Planetary Science Letters* 291:112–125.

Fastovsky, D. E., and D. B. Weishampel. 2005. *The Evolution and Extinction of Dinosaurs*, 2nd ed. Cambridge, UK: Cambridge University Press.

Fraser, N.C., and H.-D. Sues. 2011. The beginning of the "Age of Dinosaurs": A brief overview of terrestrial biotic changes during the Triassic. *Transactions of the Royal Society of Edinburgh* 101:189–200.

Hu, Y., et al. 2005. Large Mesozoic mammals fed on young dinosaurs. *Nature* 433:149–152.

Jones, T. D., and N. R. Geist. 2008. Reproductive biology of dinosaurs. In M. K. Brett-Surman et al. (eds.), *The Complete Dinosaur*, 2nd ed. Bloomington, IN: Indiana University Press.

Longrich, N. R., T. Tokaryk, and D. J. Field. 2011. Mass extinction of birds at the Cretaceous-Paleogene (K-Pg) boundary. *Proceedings of the National Academy of Sciences* 108:15253–15257.

Lyson, T. R., et al. 2011. Dinosaur extinction: Closing the '3 m gap.' *Biology Letters*, 7:925–928.

McElwain, J. C., et al. 2009. Fossil plant relative abundances indicate sudden loss of Late Triassic biodiversity in East Greenland. *Science* 324:1554–1556.

Retallack, G. J., et al. 2011. Multiple Early Triassic greenhouse crises impeded recovery from Late Permian mass extinctions. *Palaeogeography, Palaeoclimatology, Palaeoecology* 308:233–251.

Sahney, S., and M. J. Benton. 2008. Recovery from the most profound extinction of all time. *Proceedings of the Royal Society B* 275:759–765.

Sues, H.-D., and N. C. Frazer. 2010. *Triassic Life on Land: The Great Transition*. New York: Columbia University Press.

Tarduno, J. A., et al. 1998. Evidence for extreme warmth from Late Cretaceous Arctic vertebrates. *Science* 282:2241–2243.

Whiteside, J. H., et al. 2010. Compound-specific carbon isotopes from Earth's largest flood basalt eruptions directly linked to the end-Triassic extinctions. *Proceedings of the National Academy of Sciences* 107:6721–6725.

Zhou, Z., and Wang Y. 2010. Vertebrate diversity of the Jehol Biota as compared with other lagerstätten. *Science China Earth Sciences* 53:1894–1907.

Websites

General sites

http://eonsepochsetc.com/index.html
http://www.ucmp.berkeley.edu/mesozoic/mesozoic.html
http://paleobiology.si.edu/geotime/main/index.html

The Jehol and Solnhofen sites

http://www.uua.cn/english/Jehol%20Biota.html
http://www.ucmp.berkeley.edu/mesozoic/jurassic/solnhofen.html

Extinctions

http://www.bbc.co.uk/nature/extinction_events
http://paleobiology.si.edu/geotime/main/htmlversion/cretaceous4.html
http://palaeo.gly.bris.ac.uk/communication/KT.html

CHAPTER 16

Mesozoic Diapsids: Dinosaurs, Crocodilians, Birds, and Others

The Mesozoic saw the first modern vertebrate fauna with all of the extant evolutionary lineages represented and with a full array of vertebrate herbivores and carnivores in the sea, on land, and in the air. Most of these animals were diapsids, and Diapsida is the most diverse lineage of amniotic vertebrates. The huge nonavian dinosaurs of the Mesozoic era are spectacular diapsids, but the lineage also includes most species of extant terrestrial vertebrates. Crocodilians and birds are diapsids, as are lizards, snakes, and turtles. A variety of extinct forms—such as pterosaurs, ichthyosaurs, plesiosaurs, and placodonts—fills the roster of Mesozoic diapsids.

Birds are the most derived dinosaurs. Feathers and flight are the two features commonly associated with birds, but they evolved separately: feathers can be traced back far earlier than birds in the dinosaur lineage, and pterosaurs evolved the ability to fly 70 million years before birds appeared.

16.1 The Mesozoic Fauna

The Mesozoic era extended for more than 180 million years from the close of the Paleozoic 251 million years ago to the beginning of the Cenozoic only 65.5 million years ago. Throughout this vast period evolved a worldwide fauna that diversi-fied and radiated into most of the adaptive zones occupied by all the terrestrial vertebrates living today and some that no longer exist (e.g., the enormous herbivorous and carnivorous tetrapods popularly known as dinosaurs).

Although dinosaurs are the most familiar components of the Mesozoic fauna, they were not present at its start—the earliest dinosaurs appeared in the Late Triassic in the southern part of Pangaea, the landmass that united all of the present continents at that time, and the dinosaurs were not very diverse until the Jurassic. The Permian-Triassic extinction described in Chapter 7 had wiped out 70 percent of the terrestrial species of vertebrates, and during the early Triassic new lineages of herbivorous and carnivorous amniotes radiated into the ecological niches that had been left open. Among these lineages were bipedal diapsid reptiles about the size of a chicken. Paleontologists agree that the two great lineages of dinosaurs, saurischians and ornithischians, arose from these early Triassic diapsids.

The diversification of dinosaurs that began in the Late Triassic appears to have occurred in three stages—their initial appearance and radiation about 231 million years ago were followed by a major radiation in the Middle to Late Jurassic and then several pulses of increased diversity, including a large one in the second half of the Cretaceous. The reason for the dramatic increase in diversity of dinosaurs in the Middle to Late Jurassic is debated: did the extinction of nondinosaur lineages in the first part of the Jurassic open niches that dinosaurs were able to fill, or did dinosaurs prove to be more competitive than nondinosaurian lineages, such as the herbivorous rhynchosaurs, and drive those lineages to extinction?

Although the dinosaurs are the most familiar representatives of the Age of Reptiles, they are only one of many lineages present. By the end of the Triassic, Earth had for the first time an essentially modern fauna that included representatives of all of the vertebrate lineages that exist now—cartilaginous and bony fishes, turtles, crocodilian, mammalian, and (in the form of dinosaurs) avian lineages. Furthermore, these lineages had radiated into all of the kinds of animals that we currently see about us as well as some—the enormous herbivorous and carnivorous dinosaurs—that no longer exist. The play was the same, but the actors were different: in the Mesozoic, diapsid reptiles had the starring roles as the dominant herbivores and apex predators in the sea, on land, and in the air; mammals and birds were bit players.

16.2 Characteristics of Diapsids

The name *diapsid* means "two arches" and refers to the presence of an upper and a lower fenestra in the temporal region of the skull. More distinctive than the openings themselves are the bones that form the arches that border the openings. The upper temporal arch is composed of a three-pronged postorbital bone and a three-pronged squamosal. The jugal and quadratojugal bones form the lower arch.

The lower arch has been lost repeatedly in the radiation of diapsids, and the upper arch is also missing in some forms. Extant lizards and snakes clearly show the importance of those modifications of the skull in permitting increased skull kinesis during feeding, and the same significance may attach to the loss of the arches in some extinct forms. In addition to the two temporal fenestrae, derived diapsids have a fenestra on each side of the head anterior to the eye, and the presence of this opening modifies the relationships among the bones of the palate and the side of the skull.

The diapsids can be split into two groups: Archosauromorpha (Greek *archo* = ruling, *saur* = reptile, and *morph* = form) and Lepidosauromorpha (Greek *lepi* = scale), as shown in **Figures 16–1** and **16–2** on pages 273 and 274. Archosauromorphs include crocodilians and birds, the extinct pterosaurs, dinosaurs, and several Late Permian

and Triassic forms. Lepidosauromorphs include the tuatara and squamates and their extinct relatives as well as three groups of specialized marine tetrapods (placodonts, plesiosaurs, and ichthyosaurs).

Archosauromorphs are the animals most frequently associated with the great radiation of tetrapods in the Mesozoic. Dinosaurs and pterosaurs were distinctive components of many Mesozoic faunas, and other less familiar archosauromorphs also were abundant. The archosaurs (the major lineage of archosauromorphs, including crocodilians, dinosaurs, and birds) are distinguished by the presence of an antorbital (in front of the eye) fenestra. The skull is deep, the orbit of the eye is shaped like an inverted triangle rather than being circular, and the teeth are laterally compressed (**Figure 16–3** on page 374). A trend toward bipedalism was widespread but not universal among archosaurs, and the ventral side of the femur shaft had a distinctive area with a rough surface called the fourth trochanter. The powerful caudofemoral muscle originated on the base of the tail and inserted on the trochanter. When this muscle contracted, it retracted (i.e., pulled back on) the thigh, propelling the animal forward.

Ecological relationships provide a perspective on Mesozoic diapsids. Similar ecological and morphological specializations evolved in the different lineages, and in some cases members of these lineages lived side by side and preyed on, or competed with, one another. **Table 16.1** (on page 376) summarizes the two main phylogenetic groups of Mesozoic faunas from this perspective.

16.3 Marine Lineages

Lepidosauromorph diapsids were more diverse in marine habitats than on land during the Mesozoic, and they include several lineages of secondarily aquatic forms that were derived from terrestrial ancestors. The aquatic lepidosauromorphs were pelagic and coastal predators (plesiosaurs, ichthyosaurs, and mosasaurs) and coastal herbivores (placodonts).

In contrast to lepidosauromorphs, archosauromorphs had a very limited representation in the marine environment. The only lineage of marine archosauromorphs, the metriorhynchid crocodylomorphs, appeared shortly before the middle of the Jurassic and was extinct before the middle of the Cretaceous.

Placodonts

Placodonts are the least specialized of the marine lepidosauromorph lineages. They lived during the Triassic in the Tethys Sea (the body of water that spread from east to west as Laurasia and Gondwana separated during the Mesozoic). The basal forms, such as *Placodus*,

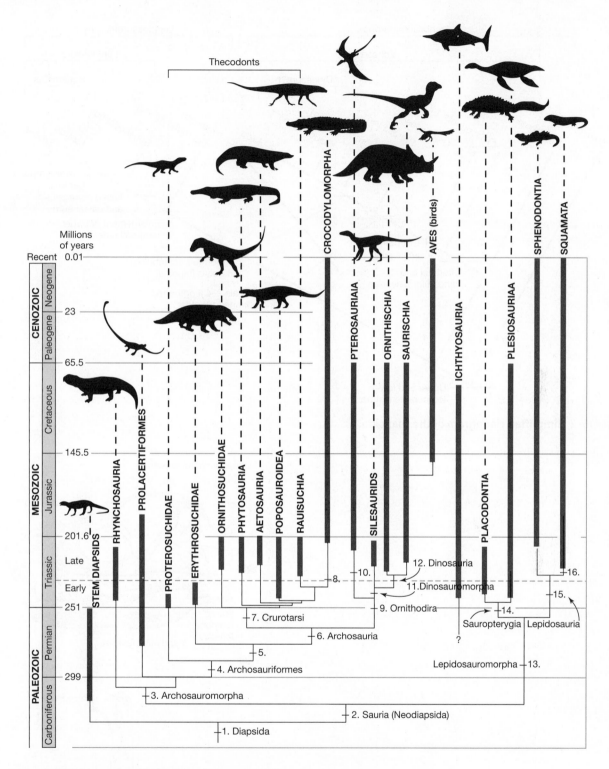

Figure 16–1 Phylogenetic relationships of the Diapsida. This diagram depicts the probable relationships among the major groups of diapsids. (Turtles, which may be diapsids, are not included.) The black lines show interrelationships only; they do not indicate times of divergence or the unrecorded presence of taxa in the fossil record. Only the best-corroborated relationships are shown. The numbers at the branch points indicate derived characters that distinguish the lineages—see the Appendix for a list of these characters.

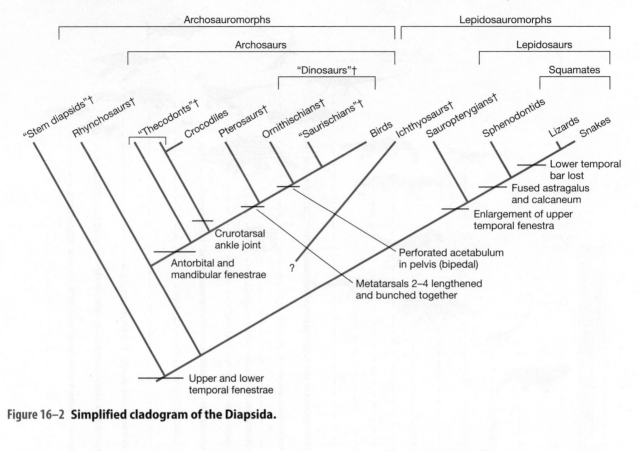

Figure 16–2 Simplified cladogram of the Diapsida.

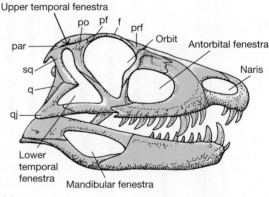

(a)

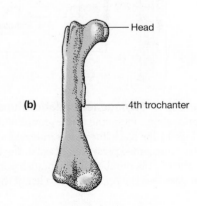

(b)

Figure 16–3 Morphological features of archosaurs.
(a) Skull of *Ornithosuchus* showing the characteristic features of archosaurs: two temporal arches, an orbit shaped like an inverted triangle, and an antorbital fenestra. (b) Femur of *Thescelosaurus* showing the fourth trochanter. (f = frontal, par = parietal, pf = postfrontal, po = postorbital, prf = prefrontal, q = quadrate, qj = quadratojugal, sq = squamosal)

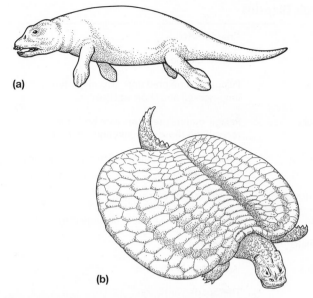

Figure 16–4 Placodonts were coastal herbivores.
(a) *Placodus* was relatively unspecialized, but (b) *Henodus* had dermal armor that was almost as extensive as a turtle's shell.

were stocky and short legged, with extensive development of bone that would have facilitated diving by reducing their buoyancy (**Figure 16–4**). Two lineages of placodonts are recognized. Both of them had bones in the skin (dermal armor). In species like *Henodus* a mosaic of small bones covered the entire dorsal surface of the body and were fused with broadened ribs that produced a carapace almost like that of a turtle. In the other lineage, species like *Placodus*, the armor was only above the vertebral column.

Placodus had forward-projecting anterior teeth, large flat maxillary teeth, and enormous teeth on the palatine bones. This dentition was long considered to indicate that placodonts fed on hard-shelled prey, such as oysters, that they plucked from rocks with their anterior teeth and crushed with the maxillary and palatal teeth. A reinterpretation of that hypothesis has pointed out that the anterior teeth of fossils of *Placodus* are heavily worn, but the posterior teeth show little wear. That is not the pattern one would expect for an animal that used the posterior teeth to crush mollusk shells, and now some scientists propose that *Placodus* was an herbivore, the Mesozoic equivalent of modern dugongs. The anterior teeth of these mammals show heavy wear because they are used to dislodge sea grass from the seafloor, but the posterior teeth are greatly reduced and the sea grass is processed by a horny oral pad.

Plesiosaurs

The plesiosaurs appeared in the Late Triassic and persisted to the very end of the Cretaceous period. The

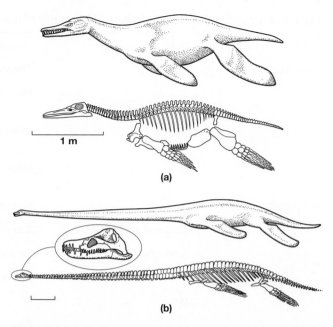

Figure 16–5 Plesiosaurs were pelagic predators. Two radically different forms of plesiosaurs had evolved by the Jurassic: (a) short-necked, fast-swimming forms like *Polycotylus* and (b) long-necked forms like *Elasmosaurus*, which may have pivoted to capture prey with a sweep of the head.

basal forms had slightly elongated necks, and their heads were proportional to their body size. Two ecological specializations are represented among the derived plesiosaurs: pliosauroids had long skulls (more than 3 meters in some forms) and short necks (about 13 cervical vertebrae), whereas plesiosauroids had small skulls and exceedingly long necks with 32 to 76 vertebrae (**Figure 16–5**). Both types had heavy, rigid trunks and appear to have rowed through the water with limbs that acted like oars and may also have served as hydrofoils, increasing the efficiency of swimming. Hyperphalangy (the addition of extra bones lengthening the fingers and toes) increased the size of the paddles, and some plesiosaurs had as many as 17 phalanges per digit. In both types of plesiosaurs, the nostrils were located high on the head just in front of the eyes.

The pliosauroids developed an increasingly streamlined body form during their evolution as the neck became shorter and the paddles larger, whereas plesiosauroids became less streamlined as their necks lengthened and the paddles became smaller in proportion to body size. Pliosauroids were probably speedy swimmers that might have captured swimming cephalopod mollusks and fish by pursuing them the way seals and sea lions hunt their prey. Plesiosauroids may have been ambush hunters, although it is not clear how they would have captured prey because their necks appear to have been quite rigid.

Table 16–1 Ecological and Morphological Diversity of Mesozoic Diapsids

Lineage	Habitat	Group	Characteristics
Lepidosauromorpha	Marine	Placodonts	Coastal marine herbivores
		Plesiosaurs	Pelagic marine predators, two major types: long-necked and short-necked
		Ichthyosaurs	Pelagic marine predators with body forms resembling those of sharks and cetaceans
		Mosasaurs	Pelagic marine predators related to the extant varanid lizards
Archosauromorpha		Metriorhynchids	Coastal marine crocodilians
	Semiaquatic to Terrestrial	Crocodilians	Mostly predators feeding on large prey
	Terrestrial	Ornithischians	Herbivores
		Saurischians	Sauropods, herbivores
	Flying		Theropods, carnivores
		Pterosaurs	The first origin of flight
		Birds	The second origin of flight

A fossil of the Late Cretaceous short-necked plesiosaur *Polycotylus latippinus* contains a single embryo, not yet at full term but already 32 percent of the body length of its mother. That is remarkably large, and its probable size at birth, based on the extent of ossification, is between 35 and 50 percent of its mother's length. Other marine reptiles were viviparous, but these species produced litters consisting of several juveniles, each only 15 to 30 percent of the mother's body length. Production of a single, large offspring appears to be a characteristic that was unique to plesiosaurs among Mesozoic reptiles.

Ichthyosaurs

Ichthyosaurs were dolphinlike diapsids. The Triassic forms were the largest (up to 20 meters long) and retained the elongate body form of their terrestrial ancestors, but the derived ichthyosaurs that lived during the Jurassic were smaller. Regardless of size, ichthyosaurs were the most specialized marine tetrapods of the Mesozoic (**Figure 16–6**). They had a hypocercal tail: the vertebral column bent sharply downward into the ventral lobe of the caudal fin, and the upper lobe was formed of stiff tissue. They also had a dorsal fin that was supported only by stiff tissue, not by a bony skeleton. (We know about the presence of these soft tissues because many ichthyosaur fossils in fine-grained sediments near Holzmaden in southern Germany contain

an outline of the entire body preserved as a carbonaceous film.)

Ichthyosaurs had both forelimbs and hindlimbs (unlike whales and dolphins, which retain only the forelimbs). The limbs of ichthyosaurs were modified into paddles by both hyperphalangy (as in plesiosaurs) and hyperdactyly (the addition of extra fingers and toes).

The streamlining of later forms may have been associated with the development of carangiform locomotion and rapid pursuit of prey. The Jurassic was the high point of ichthyosaur diversity. They were less abundant in the Early Cretaceous, only a single genus remained by the Late Cretaceous, and ichthyosaurs became extinct before the end-Cretaceous extinction event.

Fossil ichthyosaurs with embryos in the body cavity indicate that these animals were viviparous. One fossil appears to be an individual that died in the process of giving birth, with a young ichthyosaur emerging tail first like baby porpoises.

The stomach contents of ichthyosaurs, preserved in some specimens, include primarily cephalopods and fishes, although the remains of a hatchling sea turtle and a bird have been found in fossil ichthyosaurs from the Cretaceous. Most ichthyosaurs had large heads with long, pointed jaws that were armed with sharp teeth in most form. Species of the Triassic genus *Shastasaurus* had short snouts and were toothless. They resembled the extant pygmy and dwarf

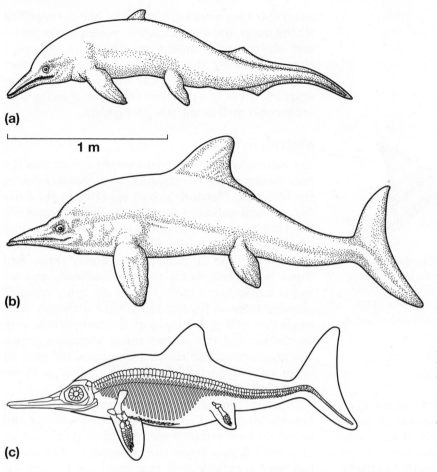

(a)

1 m

(b)

(c)

Figure 16–6 Evolutionary changes in ichthyosaurs. Early ichthyosaurs, such as the Triassic form *Cymbospondylus* (a), had small dorsal and caudal fins and probably swam with undulations of the body. Later ichthyosaurs, represented by *Ophthalmosaurus* from the Jurassic (b), had evolved a sharklike body form and swam with undulations of the caudal fin. The dorsal fin and the upper lobe of the caudal fin included no bone and were supported by only stiff tissue (c).

sperm whales, which use suction feeding to capture cephalopods.

Derived ichthyosaurs had very large eyeballs that were supported by a ring of sclerotic bones. *Ophthalmosaurus*, which had larger eyes than any other vertebrate, is believed to have hunted at great depths—500 meters or more—and detected light emitted by the photophores of its prey.

Mosasaurs

Mosasaurs are a Late Cretaceous radiation of lizards into the shallow epicontinental seas that spread across North America and Europe at that time; they are either varanid lizards or the sister group of varanids. Early mosasaurs, such as *Dallasaurus turneri* from Texas, had body proportions like those of modern varanid lizards and their limbs ended in digits. They probably swam with lateral undulations of the body and tail. The limbs of more derived mosasaurs had elongated digits connected by webbing that converted them into paddles, but their limbs were small and probably played little

role in locomotion, unlike the much larger paddles of plesiosaurs. The most derived mosasaurs, including *Platecarpus tympaniticus*, had evolved a heterocercal tail with the vertebral column bending sharply downward into a bottom lobe and an upper lobe that lacked bony support (**Figure 16–7**). These forms probably used a modified carangiform swimming mode and would have been faster than the earlier forms.

The earliest mosasaurs were about a meter long, but they increased in size during their evolutionary history, and the largest species reached lengths of 17 meters or more. Their skulls were highly kinetic, an ancestral feature, and the teeth of most species were sharp and conical—effective for seizing prey and holding it while the mosasaur swallowed it. Mosasaurs probably ate whatever they were able to catch, and we know something of their diets from fossils that were preserved with their last meals still intact. A single specimen contains bones from a fish, a smaller mosasaur, a flightless seabird, and possibly a shark. A turtle and a plesiosaur have been found in the stomachs of

Figure 16–7 Mosasaurs were sea-going lizards. Early forms were not very different from terrestrial monitor lizards, but derived mosasaurs, such as *Platycarpus* from the Late Cretaceous, were converging on the body form of ichthyosaurs and sharks, with limbs modified to paddles and a hypocercal tail.

other fossil mosasaurs. Fossil shells of ammonites (pelagic cephalopods) with marks that appear to match the shape and arrangement of the teeth of mosasaurs have been described, and a fossil of *Globidens*, a specialized mosasaur with massive blunt teeth, has the remains of clam shells in its abdomen.

Mosasaurs were viviparous, giving birth to litters of four or five young, each about 15 percent of the body length of the mother. A fossil of *Carsosaurus marchesetti*, a mid-Cretaceous mosasaur, contains at least four advanced embryos. Their orientation suggests that they were born tail first, like the reptilian ichthyosaurs and extant marine cetaceans (porpoises and whales) and sirenians (manatees and sea cows), thereby reducing the risk of drowning before they had emerged fully. One of the embryos in the *Carsosaurus* fossil is in the pelvic region, suggesting that the mother died as she was giving birth.

Metriorhynchids

The metriorhynchid crocodylomorphs constituted the only successful marine radiation by archosaurs during the Mesozoic. Metriorhynchids appeared in the Early Jurassic and persisted through the Early Cretaceous. Even the basal members of the group displayed a suite of adaptations to marine life, and the derived forms that lived in the Cretaceous were highly specialized **(Figure 16–8)**. Their heads were streamlined, and air spaces in the bones made their skulls light, probably allowing them to float at the surface with only their heads protruding. The limbs of metriorhynchids were paddlelike, and the posteriormost vertebrae turned sharply downward to create a hypocercal tail with an upper lobe supported by only stiff tissues, as in derived ichthyosaurs and mosasaurs.

Most metriorhynchids were probably fish eaters, but species in the Cretaceous genera *Dakosaurus* and *Geosaurus* were probably apex predators that could attack and kill prey larger than themselves. *Dakosaurus* had bone-crunching jaws that could exert enormous force; the teeth in the upper and lower jaws of *Geosaurus* formed a pair of shearing blades that could slice through tissue. In addition, species in both genera probably employed the crocodilian death-spiral—seizing prey and then rotating rapidly around their long axis, twisting off portions of the prey.

Figure 16–8 Metriorhynchids were marine crocodiles. *Geosaurus* was about 3 meters long and had a short snout, suggesting that it fed on other large marine reptiles rather than specializing on fish. It may have used the crocodilian "death-spiral," twisting pieces from its prey by rotating on its long axis.

16.4 Semiaquatic and Terrestrial Diapsids: Crocodylomorpha

The morphological and ecological diversity of crocodylomorphs during the Mesozoic is far greater than one would guess from an examination of the 23 extant species of crocodilians. All of the extant species are semiaquatic predators, but the history of crocodylomorphs includes many terrestrial species and even a presumed herbivore.

Mesozoic Crocodylomorphs

The evolutionary lineage that includes the modern crocodilians originated in the Triassic from a group of small terrestrial diapsids collectively called sphenosuchids. (Greek *suchus* means "crocodile" and appears frequently in the names of crocodylomorphs.) *Gracilisuchus* and *Terrestrisuchus* are representative sphenosuchids; the former was a bipedal carnivore and about 30 centimeters long, and the latter was quadrupedal and about 50 centimeters long. Terrestrial lineages retained long limbs and moderate body sizes, whereas aquatic lineages evolved shorter limbs and large—in some cases, enormous—body sizes.

All of the extant crocodilians are semiaquatic and short limbed, but long-limbed terrestrial crocodylomorphs persisted into the Cenozoic. *Pristichampsus*, for example, which is known from Eocene deposits in North America and Europe, was a 3-meter-long terrestrial crocodylomorph with hooflike toes that might have pursued and captured mammals.

The Cretaceous was the high point of crocodylomorph diversity, and fossils from Africa and South America (both included in Gondwana during the Cretaceous) have broadened our appreciation of the range of ecological and morphological specializations in the lineage. Paleontological explorations of the Sahara, which was a lush tropical landscape of lakes and meandering rivers in the Cretaceous, have been especially productive.

The diversity of Cretaceous crocodilians included terrestrial species and aquatic forms unlike any vertebrate living today:

- *Araripesuchus rattoides* was a long-legged terrestrial crocodylomorph less than 1 meter long. The species name *rattoides* (Latin *rattus* = rat and *oides* = likeness) means "ratlike" and refers to the protruding teeth in the front of the lower jaw. The cheek teeth resembled the teeth of some herbivorous dinosaurs, and this species may have been herbivorous or omnivorous.

- *Anatosuchus minor* was an aquatic crocodylomorph less than 1 meter long. *Anat*, from a Latin word for duck, refers to the broad, flat snout. *Anatosuchus* had teeth with hook-shaped crowns that could have snared frogs and small fish and a dense array of olfactory structures at the tip of the snout. It had large front feet that it may have used to dig in soft sediments.

- *Pakasuchus kapilimai* was a terrestrial crocodylomorph that lived in the mid-Cretaceous in what is now Tanzania. *Paka* means "cat" in Kiswahili, so *Pakasuchus* translates as "cat crocodile," and with a length of 50 centimeters, *Pakasuchus* was about the size of a small cat. Its mammal-like dentition set *Pakasuchus* apart from other crocodylomorphs; it had caninelike teeth in the front of the jaw, followed by teeth like the premolars and molars of mammals that occlude with a slicing motion that differs from the crushing and piercing bites of other crocodylomorphs.

- *Kaprosuchus saharicus* was a 6-meter-long terrestrial crocodylomorph with two pairs of enormously long caninelike teeth in the upper jaw and another pair in the lower jaw. The teeth projected above and below the snout and resembled the tusks of a boar (Greek *kapro* = boar). *Kaprosuchus* was an apex predator that probably preyed on dinosaurs and other crocodylomorphs.

- Some Cretaceous crocodylomorphs were gigantic. *Sarcosuchus imperator* (Greek *sarco* = flesh), from the Early Cretaceous in Africa, was 11 to 12 meters long and weighed about 8000 kilograms. *Deinosuchus rugosus* (Greek *deino* = terrible), a Late Cretaceous North American form related to alligators, was about the same size. Both of these crocodylomorphs were slightly larger than *Tyrannosaurus rex* and would have been capable of preying on the same herbivorous dinosaurs that *T. rex* ate.

Extant Crocodilians

Crocodilians and birds bracket the nonavian dinosaurs in a cladogram, and we can make inferences about the biology of dinosaurs on the basis of that relationship. For example, we noted in Chapter 1 that crocodilians and birds both have extensive parental care, and the inference that dinosaurs also exhibited parental care is obvious. Inferences about the evolution of endothermy are more complicated because crocodilians are ectotherms and birds are endotherms. Clearly endothermy evolved somewhere between crocodilians and birds, and inferring that some dinosaurs were at least approaching endothermy is not a sure bet, but neither is it wild speculation. Thus, a review of the biology of extant crocodilians sets the stage for nonavian dinosaurs and

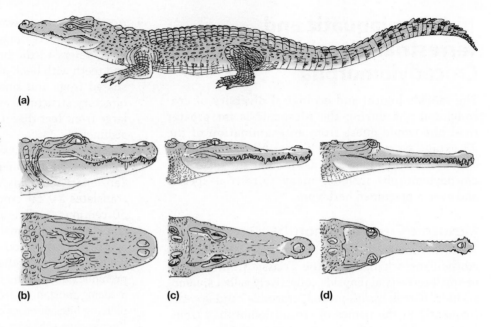

Figure 16–9 Extant crocodilians. Modern crocodilians differ little from one another and from Late Mesozoic forms. The greatest interspecific variation in living crocodilians is seen in the head shape. Alligators and caimans are broad-snouted forms with varied diets. Crocodiles have a range of snout widths. The widest crocodile snouts are almost as broad as those of most alligators and caimans, and these species of crocodilians have varied diets that include turtles, fishes, and terrestrial animals. Other crocodiles have very narrow snouts, and these species are primarily fish eaters. (a) Cuban crocodile; (b) Chinese alligator; (c) American crocodile; (d) gharial.

dispels the mistaken impression of crocodilian behavior that is presented by the sight of well-fed captive alligators and crocodiles resting inert for hours. In fact, crocodilians have extensive and complex predatory and social behaviors.

Diversity and Ecology Only 23 species of crocodiles survive. Most are found in the tropics or subtropics, but three species (the American and Chinese alligators and the American crocodile) have ranges that extend into the temperate zone. Systematists divide living crocodilians (the Crocodylia) into three lineages: Alligatoridae, Crocodylidae, and Gavialidae. The Alligatoridae includes the two species of living alligators and the caimans (**Figure 16–9**). Except for the Chinese alligator, the Alligatoridae is solely a New World group. The American alligator is found in the Gulf Coast states, and several species of caimans range from Mexico to South America and throughout the Caribbean. Alligators and caimans are freshwater forms, whereas the Crocodylidae includes species such as the saltwater crocodile that inhabit estuaries, mangrove swamps, and the lower regions of large rivers. The saltwater crocodile occurs widely in the Indo-Pacific region and penetrates the Indo-Australian archipelago to northern Australia. In the New World, the American crocodile is quite at home in the sea and occurs in coastal regions from the southern tip of Florida through the Caribbean to northern South America.

The saltwater crocodile is probably the largest living species of crocodilian. Until recently, adults may have reached lengths of 7 meters. Crocodilians grow slowly

once they reach maturity, and it takes a long time for them to attain a large size. In the face of intensive hunting during the past two centuries, few crocodilians now attain the sizes they are genetically capable of reaching. Not all crocodilians are giants, however; several diminutive species live in small bodies of water. The dwarf caiman of South America and the dwarf crocodile of Africa are about a meter long as adults and live in swift-flowing streams.

The third family of crocodilians, the Gavialidae, contains only a single species—the gharial, which once lived in large rivers from northern India to Burma. This species has the narrowest snout of any crocodilian; the mandibular symphysis (the fusion between the mandibles at the anterior end of the lower jaw) extends back to the level of the 23rd or 24th tooth in the lower jaw. A very narrow snout of this sort is a specialization for feeding on fish that are caught with a sudden sideward jerk of the head.

Predatory Behavior and Diet Extant crocodilians are semiaquatic, although they have well-developed limbs and some species make extensive overland movements. Several species of crocodiles can gallop, moving the limbs from their normal laterally extended posture to a nearly vertical position beneath the body (**Figure 16–10**). Extant crocodilians hunt in water. The upper and lower jaws are covered with small bulges that are exquisitely sensitive pressure receptors. In complete darkness, American alligators can lunge toward the point of impact of a single drop of water falling on a water surface.

Some crocodilians, such as the Nile crocodile, wait in ambush at the water's edge and attack large mammals

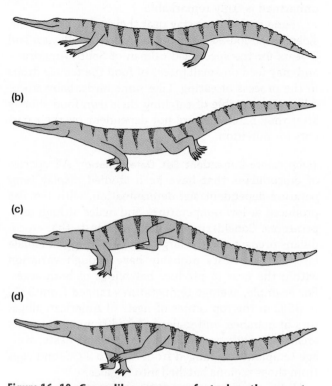

Figure 16–10 Crocodiles can move fast when they want to. Several species of *Crocodylus* can gallop, flexing the vertebral column to extend the length of the stride. These drawings are based on the Australian freshwater crocodile, *Crocodylus johnstoni*, which grows to a length of 3 meters or more and can reach speeds of 17 kilometers per hour. (a) At the start of a stride the crocodile is pushing off with its hind legs, and both forelegs are off the ground and reaching forward. (b) The forelegs are on the ground and the hind legs are off the ground and starting their forward movement. (c) The crocodile's weight is rotating forward over the extended forelegs as the hind legs begin their forward extension. (d) The forelegs are about to leave the ground as the fully extended hind legs are nearly touching the ground.

like antelope and zebra when they come to drink. In tropical areas of Australia, warnings are posted beside rivers and lakes to alert people to the very real danger of attack by crocodiles. After seizing an animal, a crocodilian drags it underwater to drown. When the prey is dead, the crocodilian bites off large pieces and swallows them whole. Sometimes crocodilians wedge a dead animal into a tangle of submerged branches or roots to hold it as the crocodilian pulls chunks of flesh loose. Alternatively, crocodilians can use the inertia of a large prey item to pull off pieces: the crocodilian bites the prey and then rotates rapidly around its own long axis, tearing loose the portion it is holding. Some crocodilians leave

large prey items to decompose for a few days until they can be dismembered easily.

Social Behavior Adult crocodilians, like birds, use sound extensively in their social behavior. Male crocodilians emit a variety of vocalizations during courtship and territorial displays and also slap their heads and tails against the water. Vocal displays are especially important for crocodilians like American alligators that live in dense swamps, because males' territories are often out of sight of other males and females. The roar of a male alligator resounds through the swamp and announces his presence to other alligators up to 200 meters away. Female alligators also roar, but only males produce a subaudible vocalization (i.e., in a frequency below the range of human hearing) that causes drops of water to dance on the water surface and travels for long distances underwater.

Frightened crocodilian hatchlings emit a distress squeak that stimulates adult male and female crocodilians to come to their defense. In addition to summoning the parents, these vocalizations may attract unrelated adults. When staff members at a crocodile farm in Papua New Guinea rescued a hatchling New Guinea crocodile that had strayed from the pond, the hatchling's distress call brought 20 adult crocodiles surging toward it (and toward the staff members!). The dominant male head-slapped the water repeatedly and then charged into the chain-link fence where the staff members were standing, while the females swam about, gave deep guttural calls, and head-slapped the water.

Parental Care The parental care provided by crocodilians appears to be as extensive as the care provided by birds. Adults of both groups protect the nest and eggs, and hatchlings remain with their parents.

Baby crocodilians begin to vocalize before they have fully emerged from their eggs, and these vocalizations are loud enough to be heard some distance away. The calls of the babies stimulate one or both parents to excavate the nest, using their feet and jaws to pull away vegetation or soil (**Figure 16–11**). For example, the female American alligator bites chunks of vegetation out of her nest to release the young when they start to vocalize. Then she picks up the babies in her mouth and carries them—one or two at a time—to water, where she releases them. She repeats this process until all the hatchlings have been carried from the nest to the water. The parents of some species of crocodilians gently break the eggshells with their teeth to help the young escape. The sight of a crocodile, with jaws that could crush the leg of a zebra, delicately cracking the shell of an egg little

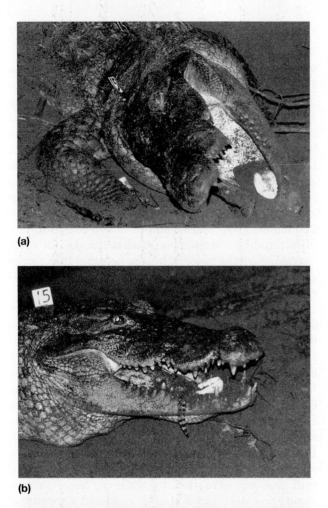

(a)

(b)

(c)

Figure 16–11 Parental care by the mugger crocodile, *Crocodylus palustris*. The numbered tag on the head allows individuals to be recognized. (a) Male parent picking up a hatchling. (b, c) Male parent releasing the hatchling in the water, 9 meters away, where the mother is waiting.

larger than a hen's egg and releasing the hatchling unharmed is truly remarkable.

Young crocodilians stay near their mother for a considerable period—2 years for the American alligator and 3 years for the spectacled caiman of South America—and may feed on small pieces of food the female drops in the process of eating. Like some birds, baby crocodilians are capable of catching their own food shortly after they hatch and are not dependent on their parents for nutrition.

Temperature-Dependent Sex Determination All species of crocodilians that have been studied display temperature-dependent sex determination, with females produced at low temperatures and males at high temperatures. Considering the elaborate nests that crocodilians prepare and the large numbers of eggs a female lays, natural nests probably have enough variation within the nest to produce hatchlings of both sexes. For example, average temperatures ranged from 33°C to 35°C in the top center of nests of American alligators in marshes, and males hatched from eggs in this area. At the bottom and sides of the same nests, average temperatures ranged from 30°C to 32°C, and eggs from those regions hatched into females.

Human activities can distort the ratio of male to female hatchlings. Levees have been constructed in much of the habitat of American alligators, and alligators have readily adopted these artificial sites for nest construction. Nests on levees are hotter than nests in marshes, however, and nearly 100 percent of the hatchlings from nests on levees are males. Global climate change has the potential to disrupt the sex ratios of crocodilians as well as other species of reptiles.

16.5 The First Evolution of Flight: Pterosauria

The archosaurs gave rise to two independent radiations of fliers: pterosaurs and birds. Birds are more familiar, but pterosaurs came first, appearing in the Late Triassic, some 70 million years earlier than birds.

Basal and Derived Pterosaurs

"Ramphorhynchoid" pterosaurs appeared in the Late Triassic and persisted until the Late Jurassic (**Figure 16–12**). They retained a long tail that was stiffened by bony projections extending anteriorly, overlapping half a dozen vertebrae and preventing the tail from bending horizontally or dorsoventrally. A leaf-shaped expansion at the end of the tail may have acted as a rudder. Although

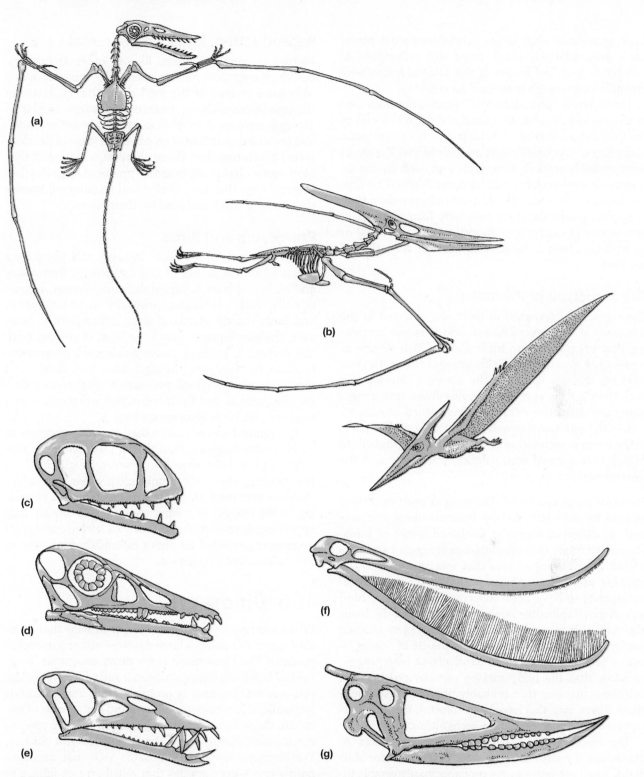

Figure 16–12 Pterosaurs probably filled all the ecological niches for flying vertebrates that birds occupy today. (a) *Rhamphorhynchus*, a long-tailed pterosaur from the Jurassic. (b) *Pteranodon*, a derived (tail-less) pterosaur from the Cretaceous. The skulls of pterosaurs suggest dietary specializations: (c) *Anurognathus* may have been insectivorous; (d) *Eudimorphodon* may have eaten small vertebrates; (e) *Dorygnathus* may have been a fish eater; (f) *Pterodaustro* had a comblike array of teeth that may have been used to sieve plankton; (g) *Dsungaripterus* had a horny beak and broad crushing teeth. Two dietary specializations are plausible: it might have pulled mollusks from rocks or plucked fruit from trees with its horny beak; the broad teeth would have been suitable for either diet.

most pterosaurs were large, rhamphorhyncoid pterosaurs were relatively small, most with wingspans of a meter or less. The fingers of the Triassic forms were probably long enough to be used for climbing.

The derived "pterodactyloid" pterosaurs appeared in the Middle Jurassic and persisted until the end of the Cretaceous. Pterodactyloids lacked tails and teeth. Many forms developed crests on their heads. The crests were probably sexually dimorphic and used during intraspecific interactions, such as courtship and territorial disputes. Pterodactyloids grew to larger sizes than rhamphorhynchoids; the giants were forms like *Quetzalcoatlus* and *Hatzegopteryx* with wingspans of 10 or 11 meters, which is substantially longer than that of any bird.

The Structure of Pterosaurs

The mechanical demands of flight are reflected in the structure of flying vertebrates, and it is not surprising that pterosaurs and birds show a high degree of convergent evolution. In both lineages the teeth were reduced; the tail was lost; the sternum developed a keel; the thoracic vertebrae became fused into a rigid structure called the notarium; postcranial pneumatization of bone was extensive; and the eyes, the parts of the brain associated with vision, and the cerebellum (which is concerned with balance) were large and the olfactory areas were small.

Pterosaur Wings and Flight The wing of pterosaurs was formed by skin stiffened by internal fibers and was entirely different from the feathered wings of birds. The fourth finger of pterosaurs was elongated and supported a membrane of skin that was anchored to the hind leg; another membrane is thought by many paleontologists to have been stretched between the hind legs. A small splintlike bone was attached to the front edge of the finger and probably supported a membrane that ran forward to the neck. The fossils of *Sordes pilosus* ("hairy monster") and *Jeholopterus ningchengensis* show that the body surface was covered by fine, hairlike structures that probably provided insulation. These fibers may also have been colorful, forming patterns used for species and sex recognition.

The aerodynamic characteristics of the wings of pterosaurs suggest that they were best suited for slow flight. Small pterosaurs were probably maneuverable fliers like bats, whereas the large pterosaurs appear to have been specialized for soaring in updrafts like vultures and some hawks and eagles. The maximum lift capability of pterosaur wings was higher than that of birds—an important consideration for taking off and landing because high lift reduces the speed required for these maneuvers.

Reproduction, Eggs, and Parental Care

Pterosaurs were oviparous, like all other archosaurs, but their eggs had flexible shells like those of many squamates instead of the rigid shells that characterize the eggs of crocodilians, nonavian dinosaurs, and birds. The eggs were small in relation to the size of the adults, and pterosaurs hatched at an advanced stage of development. Taken together, those observations suggest that pterosaurs—like many lizards—produced large clutches of small eggs that they buried and abandoned, leaving the young to hatch and fend for themselves.

Pterosaurs and Birds

Birds appeared in the Late Jurassic, and pterosaurs persisted until the end of the Cretaceous. Pterosaurs and birds may have occupied different biomes, at least initially, with pterosaurs primarily in coastal areas and birds mostly at inland sites. Subsequently, however, the two lineages may have been in competition. The derived lineages of pterosaurs show a progressive increase in body size through time, and small- and medium-size species of pterosaurs disappeared during the Jurassic and Cretaceous, just when small- and medium-size birds were appearing.

It is tempting to speculate that the wing structure of birds is more forgiving than that of pterosaurs, because a bird can still fly after losing several flight feathers from a wing, whereas it seems unlikely that a pterosaur could recover from an injury that created a slit extending to the margin of a wing membrane. Probably that hypothesis is too simplistic, however, because birds and pterosaurs coexisted for about 88 million years during the Jurassic and Cretaceous.

16.6 Dinosaurs

When you hear the word *dinosaur*, the image that probably comes to mind is a large animal—either a fearsome predator like *Tyrannosaurus rex* or an enormous, long-necked herbivore like *Apatosaurus*, and whichever image you see, the dinosaur is probably gray. That picture is incomplete: The diversity of sizes and body forms of nonavian dinosaurs extends far beyond those two species; the smallest species were about the size of a chicken. Furthermore, nonavian dinosaurs were not predominantly gray—even species that relied on camouflage to avoid detection by predators had colors and patterns that obscured their outlines, and it is likely that many dinosaurs were as flamboyant as colorful birds are today.

Mobility and social interaction were important components of the biology of nonavian dinosaurs. Their mobility owed much to the structure of their hindlimbs, and their social interactions involved colors, movements, and sounds.

The skeletons of dinosaurs combined the strength required to support animals that weigh thousands or tens of thousands of kilograms with features that minimized the weight of the skeleton. Postcranial pneumatization, the development of open spaces in bones, was prominent in dinosaurs and pterosaurs. These spaces are believed to be the traces of air sacs like those in extant birds that create a one-way flow of air through faveolar lungs.

Hips, Legs, and Ankles

Dinosaurs originated from small (about 1 meter long), agile terrestrial archosaurs that lived in the Late Permian and Early Triassic. These animals were bipedal and already had morphological changes in the hips that allowed the hindlimbs to be held beneath the body in an erect stance (**Figure 16–13**). In addition, the articulation of their ankle joint had been simplified so that it now formed a straight-line hinge rather than the complex articulation of several bones that characterized basal archosaurs. The new ankle structure allowed the hind feet to thrust backward forcefully without twisting. These two changes set the stage for the increase in body size that was so prominent in dinosaurs, but they required additional changes in the pelvis; these changes distinguish the ornithischian and saurischian lineages of dinosaurs.

Among early tetrapods, muscles originating on the pubis and inserting on the femur protracted the leg (moved it forward), and muscles originating on the tail retracted the femur (moved it backward). The ancestral tetrapod pelvis, little changed from *Ichthyostega* through early archosauromorphs, was platelike. The ilium articulated with one or two sacral vertebrae, and the pubis and ischium did not extend far anterior or posterior to the socket for articulation with the femur (acetabulum).

These animals had a sprawling posture, so the femur projected horizontally, the pubofemoral and ischiofemoral muscles extended outward from the pelvis to insert into the femur, and they were long enough to swing the femur through a large arc relative to the ground. But moving the legs under the body made those muscles shorter and less effective in moving the femur because a muscle's maximum contraction is about 30 percent of its resting length. Thus, shorter muscles swung the femur through a smaller arc and reduced the stride to a shuffle.

Dinosaurs solved that problem by moving the origins of the muscles to make them longer, and the ornithischian (bird-hipped) and saurischian (lizard-hipped) dinosaurs did that in different ways: In saurischians, the pubis and ischium both became elongated and the pubis was rotated anteriorly, so that the pubofemoral muscles ran back from the pubis to the femur and were long enough to protract it. Derived ornithischians developed an anterior projection of the pubis that ran parallel to and projected beyond the anterior part of the ilium, providing an anterior origin for protractor muscles.

Although these changes are anatomically different, they are functionally the same—both produced stride lengths that allowed dinosaurs to move over long distances and to support heavy bodies.

Being Noticed

Ask any 6-year-old about dinosaurs and you will be told that they were big. That's not entirely correct—the smallest dinosaurs were about 75 centimeters long and weighed less than 1 kilogram—but most dinosaurs were at least big and many were enormous. A big animal is conspicuous merely because of its size, but dinosaurs had an added array of features that probably made them even more conspicuous in social interactions with conspecifics (members of the same species) and in predator-prey interactions with other species:

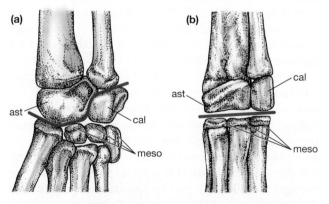

Figure 16–13 The ankle joint of dinosaurs bends like a hinge. (a) In basal archosaurs and crocodilians the axis of bending in the ankle runs at an angle between the astragalus and calcaneum bones. This is called a crurotarsal ankle joint, and it allows the foot to twist sideward as well as to flex forward and back. (b) In both nonavian dinosaurs and birds, bending occurs between the astragalus and calcaneum and their articulations with the mesotarsals, forming a mesotarsal ankle joint that can flex only backward and forward.

- Colors and patterns created by pigments in the skin or by feathers may have identified the species, the sex, and the social status (juvenile or adult) of a dinosaur.
- Long tails and long necks could have been used in social interactions with conspecifics and as deterrents to predators.

- Crests on the head, neck, and back probably identified a dinosaur's species and sex to conspecifics and made it look more formidable during social interactions.
- Complex nasal passages and inflatable tissues are thought to have acted as resonators, producing vocalizations during social interactions and predator-prey encounters.
- Frills and horns were probably used for intraspecific displays and combat and for defense against predators.

Crocodilians and birds form an extant phylogenetic bracket for dinosaurs. The social behaviors of crocodilians on one side of dinosaurs and of birds on the other support the inference that dinosaurs had complex social structures that they maintained with visual and auditory signals. These inferences are supported by fossil discoveries that reveal group behaviors:

- Thanatocoenoses (Greek *thanat* = death and *coeno* = shared) are fossil deposits that contain many individuals of the same species, often both juveniles and adults, suggesting that a group of dinosaurs perished simultaneously as the result of a catastrophe. Determining that all of the individuals actually died at the same time is difficult, however, and several well-known thanatocoenoses are now considered to represent accretions of individuals or small groups over time rather than in a single event. Other examples are convincing, however, such as the remains of at least 10,000 *Maiasaura peeblesorum* (an ornithischian) in a bed of volcanic ash in the Late Cretaceous Two Medicine Formation in Montana.
- Ichnocoenoses (Greek *ichno* = track) are fossilized traces of the activities of animals. Many fossilized trackways of dinosaurs have been described, and some show that groups of adults and juveniles of a single species of dinosaur all moved in the same direction. The difficulty lies in knowing whether they moved at the same time and as a group or independently over a period of time. That determination can be difficult because the paths animals follow are constrained by the habitat—if they are moving along a shoreline, for example, they are likely to move parallel to the water's edge whether they are moving as a group or as individuals. Despite these problems of interpretation, some trackways do appear to record the simultaneous passage of a group of animals traveling in the same direction, with the smaller footprints of juveniles in the middle of the herd.
- Dinosaur nesting areas have been discovered that contain tens or hundreds of dinosaur nests, evenly spaced like the nests in colonies of seabirds. Some of the nests contain eggs, hatchlings, juveniles, and even predators that were feeding on the eggs or hatchlings.

Combining information from observations of extant crocodilians and birds, phylogenetic inferences, and the fossil record reveals that Mesozoic ecosystems were basically similar to modern ecosystems with diapsid reptiles—lepidosauromorphs and archosauromorphs (including early birds)—doing basically the same things that mammals and derived birds do today.

Social Behavior, Reproduction, and Parental Care

Dinosaurs certainly had a repertoire of social behaviors, and some kinds of dinosaurs almost certainly lived and migrated in herds and nested in colonies. We are less sure about the extent of sociality in most dinosaurs, but that does not mean that they did not live or move in groups because inferring behavior from fossils and tracks is difficult.

No fossil of a pregnant dinosaur has ever been found. Considering the number of fossils of pregnant marine reptiles that have been discovered, failure to find a single pregnant dinosaur strongly supports the inference that all dinosaurs were oviparous. We do know that parental care was widespread among dinosaurs, however, and that it included guarding eggs in the nest and caring for the young after they hatched.

The occurrence of parental care among dinosaurs is to be expected because parental care is nearly universal in both of the extant phylogenetic bracketing groups—crocodilians and birds. Furthermore, the number of eggs in a nest suggests that the clutch of eggs was produced by more than one female and that the father was providing parental care. That is the situation we see in ostriches and related species, where a male mates with several females, incubates the eggs, and then cares for the young without assistance from any of the females.

The phylogenetic bracket does not help with a second question about dinosaur reproduction, however—Did dinosaurs have temperature-dependent sex determination? In this case all we can say is possibly, because temperature-dependent sex determination appears to be universal among crocodilians but it is not an ancestral character for birds. It is entirely possible that some lineages of dinosaurs had temperature-dependent sex determination, but there is no way to be sure.

Lineages of Dinosaurs

The Dinosauria includes two evolutionary lineages: the bird-hipped dinosaurs (Ornithischia, Greek *orni* = bird and *ischi* = hip) and the lizard-hipped dinosaurs (Saurischia, Greek *sauro* = lizard). These relationships are shown in **Figures 16–14** and **16–15** on pages 387 and 388.

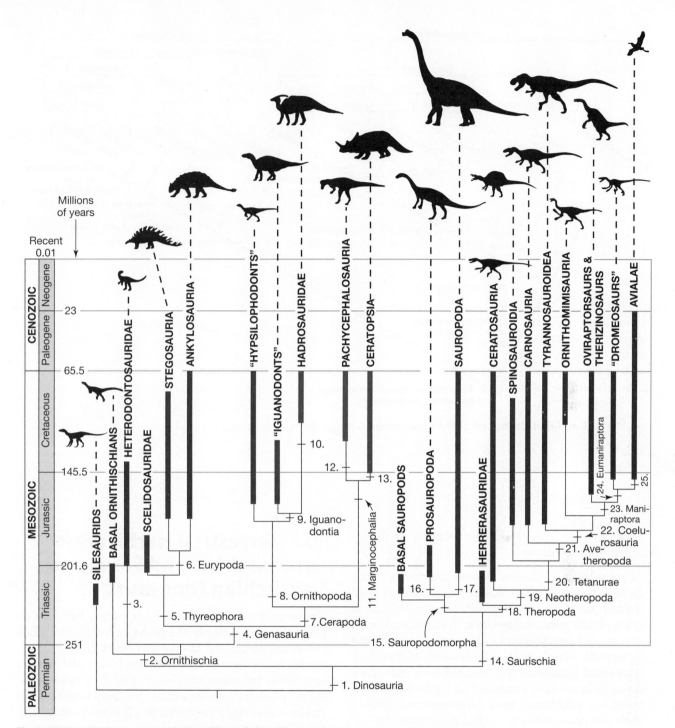

Figure 16–14 Phylogenetic relationships of the Dinosauria. This diagram depicts the probable relationships among the major groups of dinosaurs, including birds. The black lines show interrelationships only; they do not indicate times of divergence or the unrecorded presence of taxa in the fossil record. Only the best-corroborated relationships are shown. The numbers at the branch points indicate derived characters that distinguish the lineages—see the Appendix for a list of these characters. Quotation marks indicate paraphyletic groups.

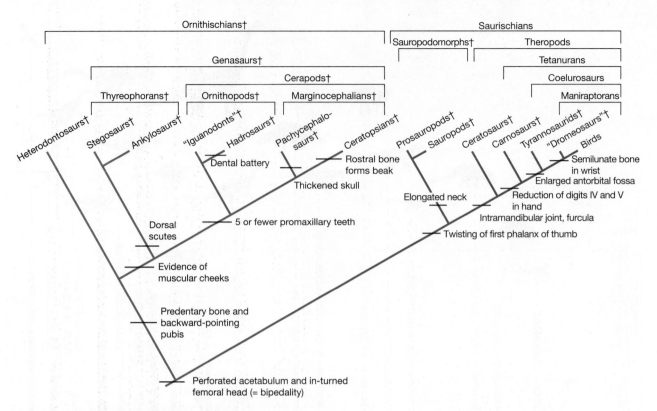

Figure 16–15 **Simplified cladogram of the Dinosauria.** A dagger indicates an extinct lineage.

Each of these lineages has its own set of distinguishing characteristics:

- All ornithischian dinosaurs were herbivores. They include the duck-billed dinosaurs, stegosaurs, and horned dinosaurs, among others. Ornithischians reached their greatest diversity in the Late Cretaceous.
- The saurischian dinosaurs included two lineages: The Sauropoda (Greek *podo* = foot) were long-necked, long-tailed herbivores that reached their maximum diversity in the Late Jurassic. The Theropoda (Greek *therio* = a wild beast) were carnivores and reached their maximum diversity in the Late Cretaceous.
- All pterosaurs were capable of flight. This lineage included insect-eating, fish-eating, mollusk-eating, and flesh-eating forms ranging from species the size of a sparrow to species with wingspans longer than those of any extant birds. Basal pterosaurs had long tails and reached their greatest diversity in the Late Jurassic, whereas derived (tail-less) pterosaurs were dominant in the Cretaceous.

16.7 Terrestrial Herbivores: Ornithischian and Sauropod Saurischian Dinosaurs

Plants do not flee to escape being eaten, but they do place other obstacles in the paths of herbivores: some plants are tough and fibrous and resist attack, for example, and others protect delicate leaves and flowers by growing tall enough to be out of reach of most herbivores. In general, ornithischian dinosaurs evolved teeth and jaws that could cope with tough plant material and they fed on low-growing plants, whereas sauropods evolved long necks that allowed them to reach vegetation that they could process without an elaborate battery of teeth.

Ornithischian Dinosaurs

All ornithischians were herbivorous, and ornithischians displayed far more diversity in body form than did the herbivorous sauropod saurischians. The elaborate crests, frills, and horns that decorated the heads of many species of ornithischians suggest that they had complex social behaviors. Some species probably lived in groups, possibly family groups, and they may have formed herds, at least on a temporary basis; some species nested in colonies.

Three groups of ornithischian dinosaurs are distinguished: Thyreophora, Ornithopoda, and Marginocephalia.

Thyreophora—Stegosaurs and Ankylosaurs The stegosaurs were a group of quadrupedal herbivorous ornithischians that were most abundant in the Late Jurassic, although some forms persisted to the end of the Cretaceous. *Stegosaurus*, a large form from the Jurassic of western North America, is the most familiar of these dinosaurs. It was up to 6 meters long, and its front legs were much shorter than its hind legs (**Figure 16–16a**). A double series of leaf-shaped plates were probably set alternately on the left and right sides of the vertebral column.

The function of the plates of *Stegosaurus* has been a matter of contention for decades. Initially they were assumed to have provided protection from predators, and some reconstructions have shown the plates lying flat against the sides of the body as shields. A defensive function is not very convincing, however. Whether the plates were erect or flat, they left large areas on the sides of the body and the belly unprotected.

A more plausible hypothesis is that the plates were heat exchangers. The plates were extensively vascularized and could have carried a large flow of blood to be warmed or cooled according to the needs of the animal. *Kentrosaurus*, the African counterpart of *Stegosaurus*, had fewer dorsal plates than *Stegosaurus* (**Figure 16–16b**); the plates on *Kentrosaurus* extended only from the neck to the middle of the trunk, and a double row of spikes extended down the tail. It is frustrating not to be able to compare the thermoregulatory behaviors of the two kinds of stegosaurs in a controlled experiment.

The short front legs of stegosaurs kept their heads close to the ground, and their heavy bodies do not give the impression that they were able to stand upright on their hind legs to feed on trees as ornithopods and perhaps sauropods did. Stegosaurs may have browsed on ferns, cycads, and other low-growing plants. The skull was surprisingly small for such a large animal and had a horny beak at the front of the jaws. The teeth show none of the specializations seen in hadrosaurs or ceratopsians. Unlike hadrosaurs and ceratopsians, which appear to have been able to grind or cut plant material into small pieces that could be digested efficiently, stegosaurs may have eaten large quantities of food

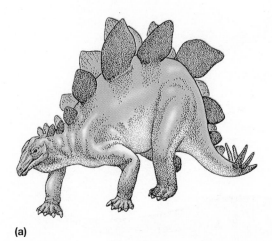

(a)

(b)

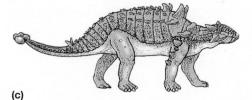

(c)

Figure 16–16 Quadrupedal ornithischians. (a) *Stegosaurus* (Late Jurassic, up to 9 meters) and (b) *Kentrosaurus* (Late Jurassic, up to 5 meters) were stegosaurs. (c) *Euoplocephalus* was an ankylosaur (Late Cretaceous, up to 7 meters).

without much chewing and used gastroliths (Greek *gastro* = stomach and *lith* = stone) in a muscular gizzard to pulverize plant material.

The tails of all the stegosaurs had large spines near their tips. These spines have been assumed to be defensive structures, and a biomechanical analysis of the whipping motion of the tail of *Kentrosaurus* supports that interpretation. Using the tail muscle of an alligator as the basis for calculation, this study concluded that the tip of the tail of an adult *Kentrosaurus* could strike a predator at a velocity of 20 to 40 meters per second. At those speeds the tip of a spine would exert enough force to penetrate skin and muscle to a depth of 30 centimeters and to pierce cortical bone.

The ankylosaurs were a group of heavily armored dinosaurs found in Jurassic and Cretaceous deposits in North America and Eurasia. Ankylosaurs were quadrupedal ornithischians that ranged from 2 to 6 meters long. They had short legs and broad bodies, with osteoderms (bones embedded in the skin) that were fused together on the neck, back, hips, and tail to produce large shieldlike pieces (**Figure 16–16c**). Bony plates also covered the skull and jaws, and even the eyelids of *Euoplocephalus* had bony armor.

Ankylosaurs had short tails, and some species had a lump of bone at the end of the tail that could apparently be swung like a club. The posteriormost caudal vertebrae of these club-tailed forms had elongated neural and hemal arches that touched or overlapped the arches on adjacent vertebrae; they also had ossified tendons running down both sides of the vertebrae. Contraction of the muscles that inserted into these tendons probably pulled the posterior caudal vertebrae together to form a stiff handle for swinging the club head at the end of the tail.

Other species of ankylosaurs had spines projecting from the back and sides of the body. Ankylosaurs must have been difficult animals to attack. Indeed, with a broad, flat body form and an armored back and sides, merely lying flat on the ground might have been an effective defensive tactic for an ankylosaur, and it is possible that the clublike tails were used primarily in male-male combat rather than for defense.

Ornithopoda The first dinosaur fossil to be recognized as such was an ornithopod, *Iguanodon*, found in Cretaceous sediments in England (**Figure 16–17**). Specimens have subsequently been found in Europe and Mongolia,

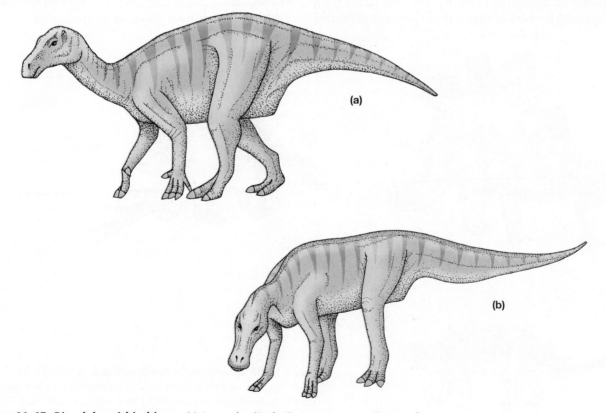

Figure 16–17 Bipedal ornithischians. (a) *Iguanodon* (Early Cretaceous, up to 9 meters).
(b) *Hadrosaurus* (Late Cretaceous, up to 10 meters).

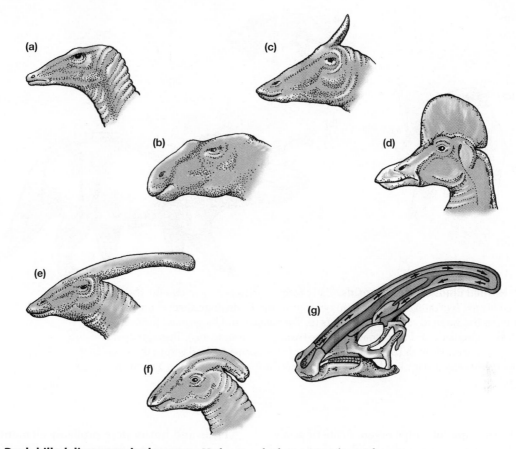

Figure 16–18 Duck-billed dinosaurs, hadrosaurs. Hadrosaurs had a variety of cranial ornaments. *Anatosaurus* (a) and *Kritosaurus* (b) lacked bony crests on their heads but might have had inflatable resonating structures formed by soft tissues. Solid-crested hadrosaurs, such as *Saurolophus*, had a projection formed by a backward extension of the nasal bones that might have supported a frill of skin. In hollow-crested hadrosaurs, such as *Corythosaurus* (d) and *Parasaurolophus* (e), the extensions of the nasal bones were hollow, and (g) the nasal passages ran from the external nares through the crests and down to openings in the roof of the mouth. The crests of male hadrosaurs were larger than those of females ((f) *Parasaurolophus*).

and related forms have been discovered in Africa and Australia. *Iguanodon* grew to a length of 10 meters, although most specimens are smaller.

The derived ornithopods included several specialized forms of hadrosaurs (duck-billed dinosaurs). Hadrosaurs were the last group of ornithopods to evolve, appearing in the middle of the Cretaceous. As their name implies, some duck-billed dinosaurs had a ducklike bill. These were large animals, some reaching lengths of more than 10 meters and weights greater than 10,000 kilograms. The anterior portion of the jaws was toothless, but a remarkable battery of teeth occupied the rear of the jaws. On each side of the upper and lower jaws were four tooth rows, each containing about 40 teeth packed closely side by side to form a massive tooth plate. Several sets of replacement teeth

lay beneath those in use, so a hadrosaur had several thousand teeth in its mouth, of which several hundred were in use simultaneously—perhaps the most advanced vertebrate chewing apparatus ever. Fossilized stomach contents of hadrosaurs consist of pine needles and twigs, seeds, and fruits of terrestrial plants.

The heads of some hadrosaurs were crowned by hollow crests (**Figure 16–18**). These crests were sexually dimorphic—much larger in males than in females. The crest of a male *Parasaurolophus* was a curved structure that extended back over the shoulders. Air flowed from the external nares backward through the crest and then forward to the internal nares, which were located in the palate just anterior to the eyes. These bizarre structures were probably used for visual displays and for vocalizations, with the long nasal passages acting as resonating

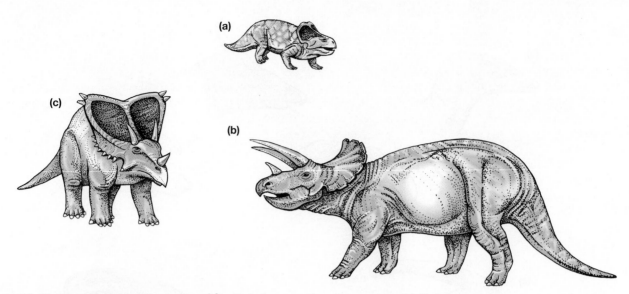

Figure 16–19 Horned dinosaurs, ceratopsids. Early forms, such as *Protoceratops* (a), had short frills and lacked horns. Two categories of derived ceratopsids can be distinguished: short-frilled forms in which the frill extended over the shoulders (b, *Triceratops*), and long-frilled forms in which the frill extended backward to the middle of the trunk (c, *Chasmosaurus*). These groups are not evolutionary lineages; both short-frilled and long-frilled forms evolved within the same lineages. Short-frilled ceratopsians had long horns, whereas the long-frilled species had much shorter horns.

chambers like the pipes of a pipe organ. Acoustic analysis of the ears of hadrosaurs indicates that they would have been able to hear the sound frequencies produced by the resonating columns of air in the nasal passages.

Marginocephalia Ceratopsians, the horned dinosaurs, were the most diverse marginocephalians. Ceratopsians appeared in the Late Jurassic or Early Cretaceous. The distinctive features of ceratopsians were the frill over the neck, which was formed by an enlargement of the parietal and squamosal bones, a parrotlike beak, and teeth that were arranged in batteries in each jaw, somewhat like those of hadrosaurs but with an important difference—they formed a series of knifelike edges rather than a solid surface. The feeding method of ceratopsians probably consisted of shearing vegetation into short lengths rather than crushing it, as hadrosaurs did.

Early ceratopsians, such as *Protoceratops*, had simple frills, unadorned by spikes, and lacked nasal horns, whereas derived ceratopsians had both of these characters (**Figure 16–19**). Two groups of derived ceratopsians can be distinguished: in the short-frilled ceratopsians, such as *Triceratops*, the frill extended backward over the neck, whereas in the long-frilled forms like *Chasmosaurus* the frill extended half the length of the trunk. Both short- and long-frilled ceratopsians had nasal and brow horns developed to varying degrees.

Frills and horns were probably elements of the social behavior of ceratopsians, especially in male-male competition. Observations of modern horned mammals suggest that differences in the size of the frills and the length and arrangement of their horns were associated with different behaviors:

- Species of antelopes with small horns engage in side-on displays with other males. Competing males swing their heads sideward against the flank of their opponent in a comparison of strength. *Protoceratops* and other early ceratopsians may have used displays like this.
- Deer and elk have large antlers, and males engage with each other head on, interlocking their antlers and twisting their necks as each individual attempts to knock the other off his feet. The sturdy horns of *Triceratops* and other short-frilled ceratopsians would have been suitable for trials of strength of this sort.
- Moose have enormous antlers that consist mostly of flat surfaces. Rival male moose face each other head on, twisting and shaking their heads to emphasize the breadth of their antlers. This comparison of antler size is often sufficient to determine dominance, and when moose do engage in trials of strength they are back-and-forth shoving contests. Long-frilled ceratopsians, such as *Chasmosaurus*, may have engaged in these sorts of displays.

Intraspecific combats are generally trials of strength that do not cause physical harm to participants, but the nasal and brow horns of ceratopsians would have been formidable weapons and were probably used for defense against carnivorous dinosaurs. A herd of ceratopsians may have formed a defensive ring when a predator approached, with the adults on the outside, facing the predator, and the juveniles sheltered in the center of the ring.

Nesting and Parental Care by Ornithischians Inferences about parental care by dinosaurs rest on a firmer basis than inferences about their social behavior. Not only do we have observations of parental care by the extant bracketing groups, crocodilians and birds, but nest sites of dinosaurs have been discovered. Many dinosaurs dug hollows in the soil to deposit their eggs, and paleontologists use these four criteria to distinguish fossilized dinosaur nests from other kinds of holes in the ground:

1. The depression cuts through the layers of the surrounding substrate.
2. The depression contains fragments of eggshells or bones.
3. An elevated ridge of sediment surrounds the depression, and the material in the ridge differs from the sediment around and above it in grain size, shape, sorting, or mineral composition.
4. The sediment within the depression differs from the material surrounding it.

Ornithischian dinosaurs laid eggs in an excavation that might have been filled with rotting vegetation to provide both heat and moisture for the eggs. (This method of egg incubation is used by many of the extant crocodilians and by the bush turkeys of Australia.)

Some ornithischians probably provided an extended period of parental care. For example, a nest of the hadrosaur *Maiasaura* in the Late Cretaceous Two Medicine Formation in Montana contains 15 juveniles that were about a meter long—approximately twice the size of hatchlings found in the same area—indicating that the group had remained together after they hatched. The teeth of the baby hadrosaurs showed that they had been feeding; some teeth were worn down to one quarter of their original length. It seems likely that a parent remained with the young. (*Maiasaura* can be translated as "good mother reptile.")

Other fossils suggest that *Maiasaura* and *Hypacrosaurus* (another hadrosaur) grew to one quarter of adult size before they left the nesting grounds, and that another ornithopod found at the same site grew to half its adult size. A nest of the ceratopsian *Protoceratops andrewsi*

from a Late Cretaceous fossil deposit in Mongolia contained 15 juveniles that were larger than a hatchling from the same area, again suggesting that hatchlings remained at the nest and were cared for by a parent.

The discovery of nests with juveniles of several different lineages of ornithischian dinosaurs suggests that parental care was an ancestral characteristic of the lineage. The inference that dinosaurs displayed parental care and prolonged association of parents with their young is entirely plausible in light of the parental care provided by crocodilians and birds.

Sauropods

This group includes the gigantic long-necked, long-tailed dinosaurs that are the centerpieces of paleontology halls at many museums, but those are the derived forms. The earliest sauropodomorph dinosaurs—the prosauropods, a group that was most diverse in the Late Triassic and Early Jurassic—were not as impressive as their descendants. *Plateosaurus*, with a length of 6 meters, was among the largest of the prosauropods.

Prosauropods had long necks and small heads (**Figure 16–20**). *Plateosaurus* had 10 cervical vertebrae, 15 trunk vertebrae, 3 sacral vertebrae (the vertebrae that connect the hips to the vertebral column), and about 46 caudal vertebrae. The long necks of all the prosauropods suggest that they were able to browse on plant material at heights up to several meters above the ground. The ability to reach tall plants might have been a significant advantage during the shift from the low-growing *Dicroidium* flora to the taller bennettitaleans and conifers that occurred in the Late Triassic. The image of sauropods reaching upward to browse is appealing, but there's a problem: Could they have pumped blood to the brain when the head was so far above the heart? That topic is explored in the Discussion Questions at the end of the chapter.

The Structure of Sauropods The derived sauropodomorphs, the sauropods of the Jurassic and Cretaceous, were enormous quadrupedal herbivores. Most fossils consist of fragmentary material, and nearly complete skeletons are known for only about 5 of the 90 genera that have been named. The sauropods were the largest terrestrial vertebrates that have ever existed. The largest of them may have exceeded 30 meters in length and weighed 50,000 kilograms. (For comparison, a large African elephant is about 5 meters long and weighs 4000 to 5000 kilograms.)

The long necks of the derived sauropods resulted both from lengthening the individual cervical vertebrae and from increasing their number, which rose from 10

Figure 16–20 Sauropod dinosaurs. (a) *Plateosaurus* (Late Triassic, up to 12 meters).
(b) *Camarasaurus* (Late Jurassic, up to 18 meters). (c) *Diplodocus* (Late Jurassic, up to 27 meters).

in prosauropods to a minimum of 12 and a maximum of 19 in sauropods. The back became shorter in the derived forms as the number of dorsal vertebrae decreased from 15 in prosauropods to as few as 9 in sauropods. The connection between the vertebral column and the hips was increased to 5 sacral vertebrae, and the long tails contained 80 or more caudal vertebrae.

Sauropods had remarkably small heads in proportion to the size of their bodies, and their dentition consisted of single rows of peglike or spatulate teeth. *Camarasaurus* had teeth along the full length of its jaws, but *Diplodocus* had teeth only in the front of the mouth. Probably sauropods used their teeth to nip and strip twigs and leaves from trees but did little oral processing of the material. The fossilized stomach contents of a sauropod dinosaur, found in Jurassic sediments in Utah, include pieces of twigs and small branches about 2.5 centimeters long and 1 centimeter in diameter.

The question of how sauropods broke down the plant material is still unanswered. Accumulations of

rounded, polished stones associated with a few fossil sauropods have been interpreted as gastroliths that ground food in a gastric mill like the gizzards of birds, but a study of the gastric mill of ostriches has cast doubt on that interpretation. The gastroliths of ostriches differ from the stones found with sauropod fossils in two respects: The surfaces of gastroliths from ostriches are abraded, not polished like the stones found with sauropod skeletons. Furthermore the gastroliths in the gastric mill of ostriches amount to about 1 percent of the weight of the bird, whereas the stones found with sauropod fossils would have been less than 0.1 percent of the dinosaur's weight. At present the most plausible hypothesis is that sauropods had very long passage times and relied on fermentation of their food by symbiotic microorganisms in the gut.

Sauropods were enormously heavy, and the skeletons of the large sauropods clearly reveal selective forces favoring a combination of strength with light weight. The axial skeleton vertebrae and ribs were

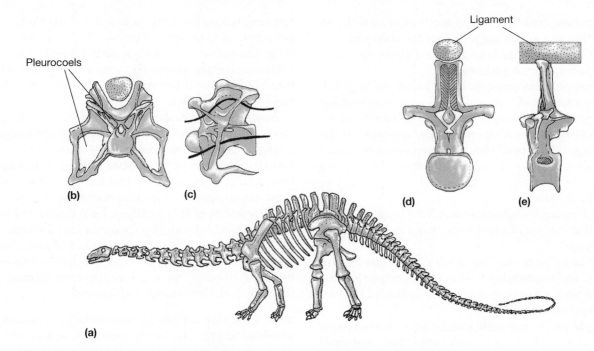

Figure 16–21 Structural features of sauropods. The skeletons of large sauropods combined lightness with strength (a). Vertebrae from the neck region seen from the rear (b) and side (c) show pleurocoels that are thought to have contained air sacs. (The black ribbons in c indicate the position of the pleurocoels.) The neural arches of the dorsal region (d, anterior view; e, lateral view) show a U-shaped depression that accommodated a large ligament that supported the neck and head.

strongly pneumatized (i.e., they contained many air spaces), and the pneumatization moved progressively backward in more derived sauropods, eventually including the hips and tail vertebrae.

The arches of the vertebrae acted like flying buttresses on a large building, while the neural spines held a massive and possibly elastic ligament that helped to support the head and tail (**Figure 16–21**). In cross section, the trunk was deep, like that of an elephant. The tails of sauropods were round in cross section, and in some species they terminated in a long, thin whiplash.

Fossil trackways of sauropods show that the legs were held under the body with the left and right feet only a single foot width apart. The limbs were held straight in an elephant-like pose and moved fore and aft parallel to the midline of the body. This morphology produces the straight-legged locomotion familiar in elephants, and sauropods probably walked with an elephant-like gait, holding their neck erect and their tails in the air.

Understanding the body temperatures of large dinosaurs requires a shift from the concepts of ectothermy and endothermy that apply to extant vertebrates because sauropod dinosaurs were so much larger than any extant species. **Gigantothermy** describes large animals that

maintain body temperatures higher than their surroundings as a result of their low surface/volume ratios. Because their rates of heat loss to the environment are low, a low rate of heat production keeps their body temperatures above the ambient temperature. Some large extant vertebrates are gigantotherms: the leatherback sea turtle can maintain a core body temperature of about 25°C while it is swimming in water with a surface temperature of about 15°C and diving into water with a temperature as low as 0.4°C. A biophysical model that assumes that dinosaurs had low metabolic rates (low rates of internal heat production) predicts stable body temperatures between 25°C for a 12-kilogram juvenile and 41°C for a 13,000-kilogram adult. Thus, gigantothermy allowed large dinosaurs (both ornithischians and saurischians) to maintain high and stable core body temperatures with rates of heat production that are characteristic of extant turtles and crocodilians.

Gigantism as a Way of Life Martin Sander and his colleagues have proposed an interpretation of the biology of the giant sauropods that focuses on the consequences of their morphological characteristics and inferences about their physiology and behavior:

- Gigantism evolved independently in several different lineages, which suggests that the ability to grow to a large body size is an ancestral characteristic of the sauropodomorph lineage.
- In the warm environment of the Mesozoic, large animals maintained high and stable body temperatures by gigantothermy while having low metabolic rates.
- The long neck was one of the key elements in the biology of sauropods because it enlarged the space within which they could feed, probably by allowing them to browse from trees that other dinosaurs could not reach.
- The long neck required a small head because a large head at the end of a long neck would exert too much leverage.
- The large body size required a large quantity of food, and masticating food in the mouth or passing it through a gastric mill would have slowed the rate of food intake.
- Simple teeth were sufficient because all the mouth did was to pluck leaves and twigs that were swallowed without chewing.
- A large gastrointestinal tract allowed for slow rates of passage, giving symbiotic microorganisms time to ferment the food and releasing volatile fatty acids that were absorbed across the wall of the intestine.
- The combination of a low metabolic rate and gigantothermy meant that sauropods had efficient secondary production, which translates into rapid growth of individuals and more energy available for reproduction.

These synergisms among the structural, behavioral, and physiological characters of sauropods produced a type of vertebrate unlike anything that has lived before or since.

Social Behavior of Sauropods Sauropods lacked frills and other sexually dimorphic display structures of the sort seen among ornithischian dinosaurs, but that does not mean that social behavior was entirely absent. After all, extant crocodilians lack sexually dimorphic ornaments, yet they have an extensive repertoire of social behaviors. Still, the evidence for sociality by sauropods is so sparse that we can guess that sauropods had less extensive social interactions than ornithischians did.

An analysis of oxygen isotope ratios in the bones of *Camarasaurus* suggests that these sauropods made seasonal migrations of several hundred kilometers between upland and lowland environments. We cannot tell whether the dinosaurs traveled as individuals moving along the same route at the same time of year or as a cohesive group containing both adults and juveniles. Evidence of possible herd behavior by sauropods may be revealed by a series of tracks found in Early Cretaceous sediments at Davenport Ranch in Texas. These tracks show the passage of 23 sauropods some 120 million years ago; the group appears to have moved in a structured fashion with the young animals in the middle, surrounded by adults.

Another line of evidence comes from sites that preserve multiple individuals of a restricted age range of a single species of dinosaur. These associations suggest that juveniles of the species lived in groups apart from adults, a phenomenon that has been observed in some extant species of grazing mammals. Several age-restricted sites of this sort have been discovered—for example, the Mother's Day Quarry, a Late Jurassic site in Montana, contains numerous fossils of juveniles of a *Diplodocus*-like sauropod, and the Late Cretaceous Big Bend *Alamosaurus* site contains only juveniles of the sauropod *Alamosaurus sanjuanensis*.

Nesting and Parental Care by Sauropods Concentrations of nests and eggs ascribed to sauropods in an Early Jurassic site in South Africa and Cretaceous deposits in southern France and Patagonia suggest that these animals had well-defined nesting grounds to which they returned year after year. Eggs thought to be those of the large sauropod *Hypselosaurus priscus* are found in association with fossilized vegetation similar to that used by alligators to construct their nests. The orientation of the nests suggests that each female dinosaur probably deposited about 50 eggs. The eggs had an average volume of 1.9 liters, about 40 times the volume of a chicken egg. Fifty of these eggs together would weigh about 100 kilograms, or 1 percent of the estimated body mass of the mother. Crocodilians and large turtles have egg outputs that vary from 1 percent to 10 percent of the adult body mass, so an estimate of 1 percent for *Hypselosaurus* seems reasonable.

Some nesting sites of sauropods have revealed astonishing quantities of eggs. During the Late Cretaceous the Auca Mahuevo fossil site in Patagonia was a flood plain drained by shallow stream channels. Thousands of individuals of an unidentified species of large sauropod dinosaur journeyed here to construct nests, creating five layers of egg beds that extend for several kilometers and reach densities of 11 eggs per square meter. The eggs, which were roughly spherical, had diameters between 12 and 15 centimeters, with 20 to 40 eggs in each clutch. The eggs in a clutch are stacked on top of one another in two or three layers. There is no evidence of parental care. Indeed, considering the size of an adult sauropod and the dense spacing of the clutches, any loitering by an adult would have been more likely to crush eggs than to protect them.

We do not yet have direct evidence of parental care by sauropods, but a nest of the prosauropod dinosaur *Massospondylus carinatus* in South Africa may hint at parental care. The embryos in this nest appeared to be close to full term, but the ventral portion of the pelvic girdle was poorly developed, the head was enormous in relation to the body, and teeth were virtually absent. This combination of characters would have made it difficult for the hatchlings to move about or feed themselves, and leads to the inference that adults of this species of sauropodomorph dinosaur might have cared for their young.

16.8 Terrestrial Carnivores: Theropod Dinosaurs

Theropod dinosaurs included three general types of animals: large, probably slow-moving predators that attacked large prey using their jaws as weapons (tyranno-saurs), fast-moving predators that seized small prey with their forelimbs (ornithomimids), and fast-moving predators that used a huge claw on the hind foot to attack prey larger than themselves (dromeosaurs) (**Figure 16–22**).

Tyrannosaurs

Tyrannosaurs are the sister lineage of the coelurosaurs, and like the coelurosaurs early tyrannosaurs were small, lightly built dinosaurs with long arms and legs. Although the derived tyrannosaurs were very large, the distinctive features of the lineage appeared in small dinosaurs, and the evolutionary history of tyrannosaurs proceeded in three stages starting in the Middle Jurassic. In the first stage the skull was strengthened, the anterior upper teeth became specialized for tearing prey, and the jaw-closing muscles grew larger. The second stage, in the Early Cretaceous, produced a miniature tyrannosaur: *Raptorex kriegsteini* was only 3 meters long (compared

Figure 16–22 Theropod dinosaurs. (a) *Coelophysis* (Late Triassic, up to 3 meters). (b) *Ornithomimus* (Late Cretaceous, up to 3.5 meters). (c) *Tyrannosaurus* (Late Cretaceous, up to 12 meters). (d) *Deinonychus* (Early Cretaceous, up to 4 meters).

to 15 meters for *Tyrannosaurus rex*), but it had nearly all the distinctive features of tyrannosaurs, including a large skull, tiny forelimbs, and long hindlimbs. In the third stage, which extended throughout the rest of the Cretaceous, tyrannosaurs increased in size.

The teeth of large tyrannosaurs were as long as 15 centimeters, dagger shaped with serrated edges, and driven by powerful jaw muscles. Marks from the teeth of predatory dinosaurs are sometimes found on fossilized dinosaur bones, and these records of prehistoric predation provide a way to estimate the force of a dinosaur's bite. The pelvis of a horned dinosaur (*Triceratops*) found in Montana bears dozens of bite marks from a *Tyrannosaurus rex*, some as deep as 11.5 millimeters. A fossilized *Tyrannosaurus* tooth was used to make an indentation that deep in the pelvis of a cow. The force required to make the marks on the *Triceratops* pelvis were estimated to range from 6410 to 14,400 newtons. These values exceed the force that can be exerted by several extant predators (dog, wolf, lion, shark). Interestingly, an alligator was the only predator tested that could deliver a bite as powerful as that of the *Tyrannosaurus*, and the jaws and teeth of alligators have many of the same structural characters as the jaws and teeth of *Tyrannosaurus*. A mechanical analysis of the skull of another large theropod, *Allosaurus*, suggests that this species used a slash-and-tear bite that could have killed the prey through loss of blood rather than by crushing.

A coprolite (fossilized dung) the size of a loaf of bread from Saskatchewan, Canada, is believed to have been deposited by a *Tyrannosaurus rex*. It contains bone fragments from a juvenile ornithischian, possibly the head frill of a *Triceratops*. The shattered bone in the coprolite suggests that tyrannosaurs repeatedly bit down on food in the mouth before they swallowed it. This feeding behavior is different from that of extant crocodilians, which swallow large mouthfuls of food without processing it.

Ornithomimids

The ornithomimids were lightly built, cursorial coelurosaurs of the Cretaceous. Despite their name, which means "bird mimic," the ornithomimids are not closely related to birds, but they had evolved into birdlike forms. *Ornithomimus* was like an ostrich in size, shape, and probably ecology as well. It had a small skull on a long neck, and its toothless jaws were covered with a horny bill. The forelimbs were long, and only three digits were developed on the hands. The inner digit was opposable and the wrist was flexible, making the hand an effective organ for capturing small prey. Like ostriches, *Ornithomimus* was probably omnivorous and fed on fruits, insects, small vertebrates, and eggs. Quite possibly it lived in groups, as do ostriches, and its long legs suggest that it inhabited open regions rather than forests.

Apparently not all ornithomimids preyed on small animals. A fossil from the Gobi Desert, *Deinocheirus* ("terrible hand"), had fingers more than 60 centimeters long that appear to have been used for grasping and dismembering large prey. Unfortunately, most of the skeleton had weathered away before the fossil was discovered, so all we have are the hands, arms, pectoral girdle, and a few vertebrae. The proportions of the hands and arms are like those of other ornithomimids, however, and if this theropod had the same body proportions as its relatives, its head might have been 10 meters above the ground—taller than *Tyrannosaurus rex*.

Dromeosaurs

Deinonychus was unearthed by an expedition from Yale University in Early Cretaceous sediments in Montana. It is a small theropod, a little more than 2 meters long. Its distinctive features are the claw on the second toe of the hind foot and the tail. In other theropods, the hind feet were clearly specialized for bipedal locomotion and were very similar to bird feet—the third toe was the largest, the second and fourth were smaller, and the fifth had sometimes disappeared entirely. The first toe was turned backward, as in birds, to provide support behind the axis of the leg. The second toe of dromeosaurs, especially the claw on that toe, was enlarged (**Figure 16–23**). In its normal position, the claw was apparently held off the ground, and it could be bent upward even farther.

It seems likely that dromeosaurs used these claws in hunting, disemboweling prey with a kick. The structure of the tail was equally remarkable. The caudal vertebrae were surrounded by bony rods that were extensions of the prezygapophyses (dorsally) and hemal arches (ventrally) that ran forward about 10 vertebrae from their place of origin. Contraction of muscles at the base of the tail was transmitted through these bony rods, drawing the vertebrae together and making the tail a rigid structure that could be used as a counterbalance or swung like a heavy stick. Possibly the tail was part of the armament of *Deinonychus*, used to knock prey to the ground where it could be kicked, and it may have served as a counterweight for balance as *Deinonychus* made sharp turns. Dromeosaurs probably relied on fleetness of foot to capture active prey. The discovery of five *Deinonychus* skeletons in close association with the skeleton of *Tenontosaurus*, an ornithischian three times their size, suggested that *Deinonychus* hunted in packs or that individuals joined a pursuit in progress. Deinonychosaurs probably used their clawed fore feet to seize prey and then slashed at it with the sicklelike claws on the hind feet. This tactic appears to be illustrated by a remarkable

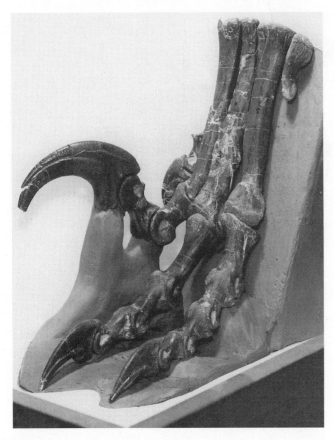

Figure 16–23 The hind foot of _Deinonychus_. The enlarged claw is on the second toe.

discovery in Mongolia of a dromeosaur called _Velociraptor_. It was preserved in combat with a _Protoceratops_, its hands grasping the head of its prey and its enormous claw embedded in the midsection of the _Protoceratops_.

The discovery of _Deinonychus_ stimulated a reexamination of fossils of several other genera of small theropod dinosaurs from the Cretaceous, including _Dromeosaurus_ and _Velociraptor_. All these forms have an enlarged claw on the second toe of the hind foot, and they are now grouped with _Deinonychus_ and birds in the Maniraptora.

Nesting and Parental Care by Theropods

Recognition of parental care by theropods lagged behind discoveries of nests of ornithischian and sauropod dinosaurs because of a mistaken identification. The fossil of a theropod dinosaur that apparently died while attending a nest of eggs was discovered in the Gobi Desert in 1923, but its significance was not recognized until 70 years later. The eggs, which were about 12 centimeters long and 6 centimeters in diameter, were thought to have been deposited by the small ceratopsian _Protoceratops andrewsi_ because adults of that species were by far the most abundant dinosaurs at the

site. The theropod was assumed to have been robbing the nest and was given the name _Oviraptor philoceratops_, which means "egg thief that likes ceratops."

In 1993, paleontologists from the American Museum of Natural History, the Mongolian Academy of Sciences, and the Mongolian Museum of Natural History discovered a fossilized embryo in an egg identical to the supposed _Protoceratops_ eggs. To their surprise, the embryo was an _Oviraptor_ nearly ready to hatch. With the benefit of hindsight it is apparent that the adult dinosaur was resting on its own nest, apparently trying to shelter its eggs from the sandstorm that buried the adult and the nest **(Figure 16–24)**. Nests of a second coelurosaur, _Troodon_, have been found and identified on the basis of embryos, but no eggs or nests of large theropods have yet been positively identified.

Avian Characters in Nonavian Theropods

Extant birds have three distinctive structures: feathers, a furcula, and extensive pneumatization of the postcranial skeleton. All three structures are now known to have been present at least in the earliest theropods, and some extend still further back in the archosaur lineage. Postcranial skeletal pneumatization, for example, was present in sauropod dinosaurs and pterosaurs, and perhaps in the common ancestor of those lineages.

Feathers Four types of featherlike structures (protofeathers and feathers) have been found on fossils of theropods. Protofeathers were especially widespread among coelurosaurs, but two basal tyrannosaurs from China (_Dilong paradoxus_ and _Yutyrannus huali_) had a filamentous covering like that of some coelurosaurs.

The simplest protofeathers were single hollow filaments from 1 to 5 centimeters long. These filaments bore little superficial resemblance to the derived feathers seen in birds, and some paleontologists initially

Figure 16–24 Some coelurosaurs brooded their eggs. An adult oviraptor, _Oviraptor philoceratops_, brooding a nest of eggs. The dinosaur's forearms are spread over the eggs in the same posture used by some extant ground-nesting birds.

Figure 16–25 Sinornithosaurus. This feathered dromeosaur was found in the Liaoning Cretaceous fossil beds of China.

doubted that they were really feathers. Those doubts were relieved by a fossil of *Shuvuuia deserti* in which the filaments retained enough keratin for chemical tests; results showed that the filaments were composed of the form of beta keratin that is unique to feathers.

More complex feathers from nonavian theropods consisted of downlike tufts of filaments and feathers with symmetrical vanes. Their locations varied among species of theropods—along the back, on the forelimbs or hindlimbs or both, and on the tail. The body and tail of *Sinornithosaurus*, a Cretaceous dromeosaur from China, were covered with downy feathers and the forearms bore feathers with symmetrical vanes (**Figure 16–25**).

These feathers were probably used for species and sex identification, in courtship displays, and during male-male displays. The feathers of a specimen of *Anchiornis huxleyi*, a small theropod from the Late Jurassic, are exceptionally well preserved. The shapes and distribution of melanosomes (melanin-containing structures) in the feathers were identified by scanning electron microscopy, and their colors were determined by matching the shapes and density of the melanosomes of *Anchiornis* to the melanosomes that produce black, gray, and reddish-brown colors in the feathers of extant birds. *Anchiornis* was gray, with a crest of reddish-brown feathers on the top of the head and speckles of that

color on the face. The long feathers on the wings and hindlimbs were white with black bands.

In addition to adding colors, patterns, and textures to supplement the colors present in the skin, these early feathers could have provided insulation. Extant birds keep their feet and eggs warm at night by crouching to cover them with their body feathers and by tucking their thinly feathered heads and uninsulated beaks beneath their wings. Feathered theropods could have done these things as well. Furthermore, even protofeathers could have retained the heat produced by a theropod's resting metabolism and muscular activity, starting theropods on the path to endothermal homeothermy, as described in Chapter 11.

The levels of structural complexity we have been considering are not simply a stepwise evolutionary progression in which simple feathers come first and are progressively replaced by more complex feathers. Different types of feathers have different functions, and extant birds have feathers that consist of single strands and tufts of down as well as symmetrical and asymmetrical vaned feathers, each with a different function. Thus, the widespread distribution of feathers among nonavian (i.e., nonflying) dinosaurs emphasizes that the evolution of feathers and the evolution of avian flight were entirely separate phenomena.

Furcula The furcula (wishbone), formed by fusion of the clavicles, is another distinctively avian character that can be traced to basal theropods and appears in nearly all of the theropod lineages. For birds, the furcula is a component of the flight mechanism, but the widespread occurrence of a furcula among theropods shows that the origin of this structure, like the origin of feathers, had nothing to do with flight.

Postcranial Skeletal Pneumatization Early in the evolution of theropods, the cervical and anterior dorsal vertebrae became pneumatized, and expansion of pneumaticity to the posterior dorsal vertebrae, ribs, and long bones appeared independently in as many as a dozen lineages of theropods as well as in sauropods and pterosaurs. Thus, skeletal pneumatization is a third avian character that evolved independently of flight.

Lung Ventilation The ribs of birds have a body projection, the uncinate process, that extends upward and backward from the rib. These structures provide a mechanical advantage for the respiratory muscles by increasing their effectiveness. Oviraptors and dromeosaurs also had uncinate processes, as do basal birds, and a mechanical analysis of the rib cages of these species indicates that the uncinate processes would have been as effective as those in extant birds. These similarities in the skeleton of the rib cage strengthen the inference that nonavian theropods had a flow-through pattern of respiration like that of extant birds.

16.9 The Second Evolution of Flight: Birds

Birds are derived theropod dinosaurs, and few evolutionary transitions are as clearly recorded in the fossil record as the appearance of birds. More than a century ago, Thomas Henry Huxley was an ardent advocate of that relationship, writing that birds are nothing more than "glorified reptiles." Huxley, in fact, was so impressed by their many similarities that he placed birds and reptiles together in his classification as the class Sauropsida. For most of the next century, traditional systematics, with its emphasis on strict hierarchical categories, obscured that evolutionary relationship by placing reptiles and birds at the same taxonomic level (as the class Reptilia and the class Aves). Cladistic systematics emphasizes monophyletic evolutionary lineages, and now birds are again seen as the most derived theropod dinosaurs.

A remarkable biochemical discovery has added a new line of evidence supporting the theropod-bird connection—protein preserved in a fossil of *Tyrannosaurus rex*. Finding organic material in a fossil is an extraordinary occurrence because organic materials are normally replaced by minerals during fossilization. Thus, fossils are rocks that retain the exact shape of the original bone but contain none of the chemicals that were in the bone during the life of the animal. Very rarely, however, areas of soft tissues are left unmineralized. If the organic material has not decayed, the chemical characteristics of tissues from the fossil can be compared with tissues from extant species. In this case, a sample of collagen from a 68-million-year-old *Tyrannosaurus rex* skeleton yielded peptides that were compared to peptides from a variety of extant vertebrates. The closest match was to peptides from a bird, just as would be predicted on the basis of our understanding of the phylogenetic relationships of theropods and birds.

Mosaic Evolution of Avian Anatomical Characters

Major evolutionary changes, such as the transition from nonavian dinosaurs to birds, do not occur all at once. Instead, one character changes, then another, and then another, so the transitional forms present mosaics of ancestral and derived characters. The mosaic nature of evolution is superbly illustrated by the transition from nonavian dinosaurs to birds (**Figure 16–26**).

The dromeosaurs, a derived group of coelurosaurs, had many birdlike characters, including a wrist structure that permitted them to flex the wrists sideways while rotating them. This motion probably allowed coelurosaurs to use their hands to seize prey, and it is recognized in the name Maniraptora (Latin *manus* = hand and *rapt* = seize), which is the lineage that includes dromeosaurs and birds. Birds use this wrist motion to produce a flow of air over the primary feathers of the wings to create lift during flapping flight.

Some still more derived dromeosaurs—such as *Unenlagia*, which was a 2-meter-long terrestrial predator—have a shoulder joint that increased the freedom of movement of the arms. The glenoid fossa (where the humerus articulates with the pectoral girdle) is oriented laterally rather than ventrally in these animals, allowing the arms to be lifted upward and backward and to strike downward and forward to seize prey. This anatomical change, which probably made dromeosaurs more effective as terrestrial predators, is the origin of the up-and-back/down-and-forward arm motion that birds use to flap their wings in flight.

Birdlike Theropods

During the Late Jurassic and throughout the Cretaceous, several lineages of proavians (derived nonavian theropods) lived side by side with the earliest birds.

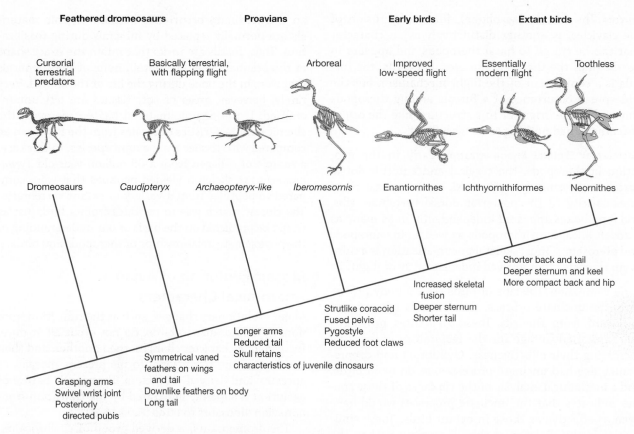

Figure 16–26 Mosaic evolution of the derived characters of birds.

Amber from the Cretaceous of France and Canada contains feathers from both nonavian theropods and birds. The proavians were carnivores, and we have direct evidence of predation—a fossil of *Microraptor*, a proavian from China, contains the remains of a bird.

A substantial reduction in body size characterized these proavians. *Epidexipteryx* was the size of a pigeon and *Scansoriopteryx* was sparrow size, whereas the smallest dromeosaurs, such as *Caudipteryx* and *Sinornithosaurus*, were about a meter long.

This reduction in body size was an important component of the evolution of avian flight. A flying bird is supported by the lift produced by its wings. Wing loading, the weight of the bird divided by the surface area of its wings, is a critical component of flight. Wing loadings of about 2.5 g • cm^{-2} are typical of extant birds that run with flapping wings to take off, and these are plausible values for the earliest birds.

A wing is basically two dimensional, and the surface area of a wing increases as the *square* of its linear dimensions. The weight of a bird, however, is proportional to its volume, which increases as the *cube* of its linear dimensions. As a result, large birds require proportionally larger wings than do small birds to achieve the same

wing loading. The conspicuous reduction in body size that occurred during the transition from dromeosaurs to proavians and birds is probably the result of the importance of wing loading in the evolution of flight.

How and Why Did Birds Get Off the Ground?

The proavians that lived in the Jurassic shared basic features of their body form and ecology: They were terrestrial predators, and their long legs suggest that they ran to pursue prey and to escape from predators. They competed with each other and preyed on each other—it must have been a tough place to make a living, and even an incipient ability to fly might have given one or more of these lineages an advantage.

The earliest flying theropods did not have to be particularly adept—even a brief flight could have foiled an attack by a terrestrial predator. Thus, an enhanced ability to escape predation is a plausible hypothesis for the advantage gained by the first theropods. That is far from the only hypothesis, however, and other scenarios also offer plausible ideas about an advantage that could be gained from feathered wings:

- **A leaping insectivore**—A proavian cursorial predator might have combined an upward leap with wing flapping to capture flying insects.
- **A pouncing proavis**—A proavian ambush predator might have used its wings to glide as it leaped down on prey from an elevated site.
- **A balancing raptor**—A proavian *Deinonychus*-like predator might have used its wings to stabilize itself as it killed prey that it had impaled with the enlarged claw on its second toe.

The Importance of Asymmetrical Flight Feathers

Symmetrical wing feathers are adequate for balancing, gliding, and even leaping into the air in pursuit of insects, but powered flight requires asymmetrical flight feathers to provide thrust. In extant birds these are the primaries (the feathers on the outermost portion of the wing, corresponding to the hand). Symmetrical feathers were widespread among coelurosaurs, but so far *Archaeopteryx* is the only proavian known to have had asymmetrical flight feathers.

Archaeopteryx was certainly not the first theropod to evolve asymmetrical flight feathers, but at this time it is the most basal fossil in which the feathers are preserved well enough to determine that they are asymmetrical (**Figure 16–27**). *Xiaotingia* was another proavian from the Late Jurassic with a skeleton that was very similar to that of *Archaeopteryx*. Unfortunately, the feathers are not preserved clearly in the single specimen of *Xiaotingia* that has so far been described. What *Archaeopteryx* and *Xiaotingia* demonstrate is that a diversity of proavians existed in the Jurassic and that birds (Aves) arose from a lineage of those animals.

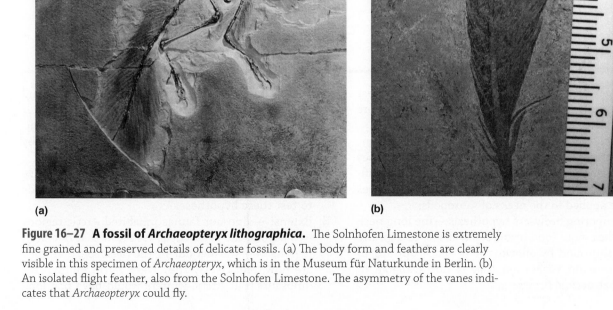

(a)　　　　　(b)

Figure 16–27 A fossil of *Archaeopteryx lithographica*. The Solnhofen Limestone is extremely fine grained and preserved details of delicate fossils. (a) The body form and feathers are clearly visible in this specimen of *Archaeopteryx*, which is in the Museum für Naturkunde in Berlin. (b) An isolated flight feather, also from the Solnhofen Limestone. The asymmetry of the vanes indicates that *Archaeopteryx* could fly.

Summary

The major groups of tetrapods in the Mesozoic were members of the diapsid (two arches) lineage. This group is characterized by the presence of two fenestrae in the temporal region of the skull that are defined by arches of bone.

The lepidosauromorph lineage of diapsids radiated primarily in marine habitats and included both coastal and pelagic forms. Placodonts were herbivores that lived in near-shore habitats in the Tethys Sea during the Triassic. Most were relatively unspecialized—they had paddlelike flippers but were much like their terrestrial ancestors in body form. *Henodus*, however, was a remarkable animal that had converged in some respects on the body form of turtles—it had a broad, flat body covered by bony plates that were probably the functional equivalent of a turtle's shell.

The three other lineages of marine lepidosauromorphs were more specialized. Plesiosaurs and ichthyosaurs were streamlined, and derived ichthyosaurs had converged on the body form of fast-swimming sharks and cetaceans. Mosasaurs were sea-going lizards, closely related to the extant varanid lizards.

The archosauromorph lineage contains crocodilians, pterosaurs and birds as well as the most familiar of the Mesozoic tetrapods, the dinosaurs. Crocodilians were far more diverse in the Mesozoic than they are now and included terrestrial predators as well as aquatic species. The largest crocodilians were as big as the large carnivorous dinosaurs.

Pterosaurs were the first vertebrates to fly, and they were a diverse and successful lineage. Pterosaurs ranged in size from species that were only a little larger than a sparrow to species that had wingspans longer than those of any extant bird.

Two major groups of dinosaurs are distinguished: Ornithischia and Saurischia. The ornithischian dinosaurs were herbivorous, and many had horny beaks on the snout and batteries of specialized teeth in the rear of the jaw. The ornithopods (duck-billed dinosaurs) and pachycephalosaurs (thick-headed dinosaurs) were bipedal, and the stegosaurs (plated dinosaurs), ceratopsians (horned dinosaurs), and ankylosaurians (armored dinosaurs) were quadrupedal.

The saurischians included the sauropod dinosaurs—enormous herbivorous quadrupedal forms like *Apatosaurus* (formerly *Brontosaurus*), *Diplodocus*, and *Brachiosaurus*—and the theropods, which were bipedal carnivores. Large theropods (of which *Tyrannosaurus rex* is the most familiar example) probably preyed on large sauropods. Other theropods were smaller: ornithomimids were probably much like ostriches, and some had horny beaks and lacked teeth. Dromeosaurs were fast-running predators. Birds had evolved by the Jurassic, and the appearance of the derived characters of birds can be traced in nonavian dromeosaurs.

The phylogenetic relationship of crocodilians and birds allows us to draw inferences about some aspects of the biology of dinosaurs. Characters that are shared by crocodilians and birds are probably ancestral for dinosaurs. Social behavior, vocalization, and parental care are the norm for both crocodilians and birds, and increasing evidence suggests that dinosaurs too had elaborate social behavior and vocalizations and that at least some species cared for their young.

Discussion Questions

1. Use the extant phylogenetic bracket method to test the hypothesis that male coelurosaur dinosaurs were the primary providers of care for eggs and young.

2. The necks of giraffes are the closest extant equivalent to the necks of sauropod dinosaurs. Two hypotheses have been proposed to explain the functional significance of the necks of giraffes, and those hypotheses can be applied to the necks of sauropods:
   - Competing browsers hypothesis—The long neck makes more food resources available to a giraffe or sauropod by allowing it to reach upward to browse on vegetation that is not available to short-necked browsers.
   - Necks-for-sex hypothesis—Male giraffes and sauropods use their necks and heads to batter opposing males during intraspecific combat. Thus, long necks increase the fitness of males by increasing their opportunities to mate.

   What predictions can you make about the relative lengths of the necks of male and female sauropods to test these hypotheses?

3. Extensive dinosaur faunas inhabited Arctic regions well beyond the polar circle. We know that northern regions were warmer during the Mesozoic than they are now, and clearly they were warm enough for dinosaurs. But Arctic winters would have had long

periods of dim light (twilight) in the Mesozoic, as they do now. How might dinosaurs have responded to the seasonal changes in day length in the Arctic?

4. Compare the use of inferences about the ecology, behavior, and physiology of extinct animals based on extant phylogenetic brackets with those based on analogies with living forms—for example, the inferences in this chapter about the social behaviors of ceratopsian dinosaurs that were based on analogies with modern herbivorous mammals with horns or antlers. How might you test your conclusions in each case?

5. When a giraffe holds its neck in a vertical position, its brain is about 2 meters above its heart. Contraction of the ventricle must create enough pressure to raise the blood to the level of the head and perfuse the brain (i.e., circulate blood through the brain). The arterial pressure at the aorta of a giraffe is approximately 200 millimeters of mercury (mm Hg), twice that of a human.

The brain of a sauropod dinosaur, such as *Barosaurus*, would have been about 9 meters above the level of its heart if it held its neck vertically. Calculate the aortic pressure required to perfuse the brain using the following information:

- The density of blood is 1.055 g/cm^3 and the density of mercury is 13.554 g/cm^3.
- An aortic pressure of 10 mm Hg is needed to overcome resistance to blood flow in the carotid arteries leading from the heart to the brain.
- A blood pressure of 50 mm Hg at the head is required to perfuse the brain.

What inferences about the biology of sauropods can you draw from your calculation?

6. The concentration of oxygen in the atmosphere is currently 21 percent, but it has been both higher and lower during the Phanerozoic. At the start of the Permian it reached a maximum of 30 percent to 32 percent and then declined throughout the Permian and Triassic, reaching its minimum of 13 percent to 15 percent at the end of the Triassic. Dinosaurs and mammals both appeared in the Triassic, but dinosaurs flourished and radiated into a variety of large-bodied lineages, whereas mammals remained small and did not radiate into multiple lineages. Recall the characteristics of the respiratory systems of sauropsids and synapsids described in Chapter 11 to explain how the low concentration of atmospheric oxygen in the Permian may have favored the development of dinosaurs over mammals.

Additional Information

Asara, J. M., et al. 2007. Protein sequences from mastodon and *Tyrannosaurus rex* revealed by mass spectrometry. *Science* 316:280–285.

Benton, M. J. 2010. The origins of modern biodiversity on land. *Philosophical Transactions of the Royal Society B* 385:3667–3679.

Bhullar, B-A. S. et al. 2012. Birds have paedomorphic dinosaur skulls. *Nature* 487:223–226.

Diedrich, C. J. 2010. Palaeoecology of Placodus gigas (Reptilia) and other placodontids—Middle Triassic macroalgae feeders in the Germanic Basin of central Europe—and evidence for convergent evolution with Sirenia. *Palaeogeography, Palaeoclimatology, Palaeoecology* 285:287–306.

Dodson P. 2003. Allure of El Lagarto—Why do dinosaur paleontologists love alligators, crocodiles, and their kin? *The Anatomical Record Part A* 274A:887–890.

Evans, S. E., and M. E. H. Jones. 2010. The origin, early history and diversification of lepidosauromorph reptiles. In S. Bandyopadhyay (ed.), *New Aspects of Mesozoic Biodiversity*. Heidelberg, Germany: Springer-Verlag, 27–44.

Fastovsky, D. E., et al. 2011. A nest of *Protoceratops andrewsi* (Dinosauria, Ornithischia). *Journal of Paleontology* 85:1035–1041.

Li, Q., et al. 2010. Plumage color patterns of an extinct dinosaur. *Science* 327:1369-1372.

Libby, T., et al. 2012. Tail-assisted pitch control in lizards, robots, and dinosaurs. *Nature* 481:181–184.

Lindgren J, et al. 2011. Landlubbers to leviathans: Evolution of swimming in mosasaurine mosasaurs. *Paleobiology* 37:445–469.

Mallison, H. 1915. Defense capabilities of *Kentrosaurus aethiopicus* Hennig, 1915. *Palaeontologia Electronica* 14(2) 10A:25p; <palaeo-electronica.org/2011_2/255/index.html>.

McKellar, R. C. 2011. A diverse assemblage of Late Cretaceous dinosaur and bird feathers from Canadian amber. *Science* 333:1619–1622. (See also, Dove, C. J. and L. C. Straker L. C. 2012. Comment on "A diverse assemblage of Late Cretaceous dinosaur and bird feathers from Canadian amber." *Science* 335:796, and McKellar R. C. et al. 2012. Response to comment on "A diverse assemblage of Late Cretaceous dinosaur and bird feathers from Canadian amber." *Science* 335:796c.)

Motani, R. 2009. The evolution of marine reptiles. *Evolution: Education and Outreach* 2:224–235.

O'Connor, P. M., et al. 2010. The evolution of mammal-like crocodyliforms in the Cretaceous Period of Gondwana. *Nature* 466:748–751.

Organ, C. L., et al. 2008. Molecular phylogenetics of mastodon and *Tyrannosaurus rex*. *Science* 320:499.

Padian, K., and J. R. Horner. 2011. The evolution of 'bizarre structures' in dinosaurs: Biomechanics, sexual selection, social selection or species recognition? *Journal of Zoology* 283:3–17.

Reisz, R. R., et al. 2012. Oldest known dinosaurian nesting site and reproductive biology of the Early Jurassic sauropodomorph Massospondylus. *Proceedings of the National Academy of Sciences*, in press.

Sander, P. M., et al. 2011. Biology of the sauropod dinosaurs: The evolution of gigantism. *Biological Reviews* 86:117–155.

Schmitz, L., and R. Motani. 2011. Nocturnality in dinosaurs inferred from scleral ring and orbit morphology. *Science* 332:705–708.

Senter, P. 2011. Using creation science to demonstrate evolution: Application of a creationist method for visualizing gaps in the fossil record to a phylogenetic study of coelurosaurian dinosaurs. *Journal of Evolutionary Biology* 23:1732–1743.

Sereno, P. C., and H. C. E. Larsson. 2009. Cretaceous crocodyliforms from the Sahara. *ZooKeys* 28:1–143.

Sookias, R. B., et al. 2012. Rise of dinosaurs reveals major body-size transitions are driven by passive processes of trait evolution. *Proceedings of the Royal Society B*, in press.

Tickle, P. E., et al. 2012. Ventilatory mechanics from maniraptoran theropods to extant birds. *Journal of Evolutionary Biology*, in press.

Varricchio, D. J., et al. 2008. Avian paternal care had dinosaur origin. *Science* 322:1826–1828.

Xu, X. et al. 2012. A gigantic feathered dinosaur from the Lower Cretaceous of Canada. *Nature* 484:92–95.

Young, M. T., et al. 2011. Body size estimation and evolution in metriorhynchid crocodylomorphs: Implications for species diversification and niche partitioning. *Zoological Journal of the Linnean Society* 163:1199–1216.

Websites

The Mesozoic Era
http://eonsepochsetc.com/Mesozoic/mesozoic_home.html
The Paleobiology Database
http://paleodb.org
The Jehol Biota
http://www.uua.cn/english/Jehol%20Biota.html
Crocodilian Biology Database
http://crocodilian.com/cnhc/cbd-evo.htm
DinoBuzz
http://www.ucmp.berkeley.edu/diapsids/dinobuzz.html

Avian Specializations

Flight is a central characteristic of birds. At the structural level, the mechanical requirements of flight shape many aspects of the anatomy of birds. In terms of ecology and behavior, flight provides options for birds that terrestrial animals lack. The ability of birds to make long-distance movements is displayed most dramatically in their migrations. Even small species like hummingbirds travel thousands of kilometers between their summer and winter ranges. Birds use a variety of methods to navigate on these trips, including orienting by the sun and the stars and by using a magnetic sense.

Wing movements are used during swimming by wing-propelled aquatic birds, such as penguins, whereas others (ducks and cormorants are familiar examples) use their feet to swim. Not all birds fly—birds are derived from terrestrial dromeosaurs, and some species of birds (among them ostriches, emus, and the kiwi as well as several lineages of extinct predatory birds) are secondarily flightless. Many more species (both herbivorous birds like grouse and pheasants and predators like roadrunners and bustards) spend most of their time on the ground and fly only short distances to escape from predators.

A second characteristic of birds is diurnality; most species are active only by day. In addition, most birds have excellent vision, and colors and movements play important roles in their lives. Because humans, too, are diurnal and sensitive to color and movement, birds have been popular subjects for behavioral and ecological field studies, and important areas of modern biology draw heavily on studies of birds for data that can be generalized to other vertebrates.

17.1 Early Birds and Extant Birds

Even limited flying abilities would have allowed early birds to move into an adaptive zone that had never been exploited by vertebrates. An animal that could run, launch itself into the air, and then land and run again had new opportunities to attack prey or to escape from predators. The birds that appeared after *Archaeopteryx* quite rapidly developed the range of specializations that we now associate with birds.

Feathered dinosaurs, such as *Caudipteryx*, had long bodies with their center of gravity near the hind legs, shallow trunks (because they lacked the deep sternum of extant birds), long bony tails, front legs that retained claws, and full sets of teeth. They were not very birdlike in appearance; if you saw a *Caudipteryx* while you were out walking, your immediate reaction would not be "There's a bird." However, by the Early Cretaceous a large number of anatomical changes had taken place,

(a)

© 1992 AAAS

(b)

Figure 17–1 Early Cretaceous enantiornithine birds. (a) *Sinornis* is a sparrow-size bird from Early Cretaceous lakebed deposits in China. (b) *Confuciusornis* is a crow-size bird, also from the Early Cretaceous of China.

and you would have no doubt that you were looking at a bird if you saw a *Sinornis* or *Confuciusornis* **(Figure 17–1)**.

The most obvious changes were in the general body form: the center of gravity had shifted forward toward the wings as in modern birds, the bony tail was greatly shortened, and fused vertebrae at the end of the tail formed a pygostyle as they do in modern birds. Less conspicuous changes were the strutlike coracoids (bones that help the shoulder girdle resist the pressures exerted on the chest by the wing muscles), a reduction in the size of the claws on the feet (making them better suited to perching in trees), a larger sternum (giving more area for the origin of flight muscles), and a wrist that could bend back sharply to tuck the wing against the body.

The Mesozoic era saw two independent radiations of birds, and between them they produced a worldwide avian fauna that included most of the ecological types of birds we know today **(Figure 17–2)**. The birds in the earlier

radiation are known as Enantiornithes ("opposite birds") because the articulation between the scapula and coracoid is the reverse of the arrangement of modern birds. Enantiornithines were the dominant birds of the Cretaceous, and they ranged from small to large: *Sinornis* was the size of a sparrow, and *Enantiornis* was the size of a turkey vulture. Most enantiornithines were small to medium sized and probably lived in trees, but some had long legs and appear to have been wading birds, while still others had powerful claws like modern hawks.

The second radiation of birds, the Ornithurae, appeared later in the Cretaceous and radiated into a wide variety of ecological types. Early ornithurines were small, finchlike arboreal species, but by the Late Cretaceous, the lineage had expanded to waders, perchers, and secondarily flightless foot-propelled swimmers and divers such as *Hesperornis* and *Ichthyornis* **(Figure 17–3)**. Modern birds, the Neornithes,

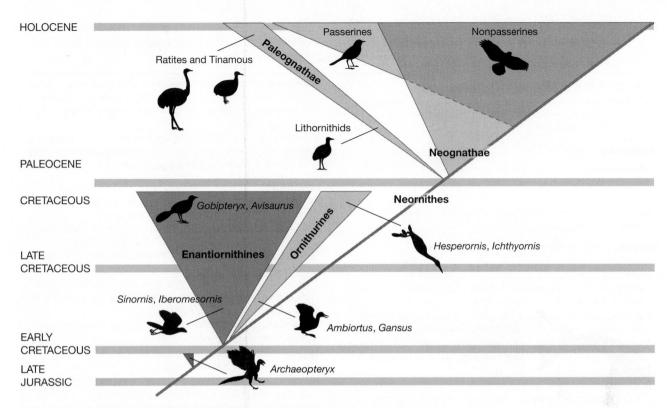

Figure 17–2 Mesozoic and modern birds. The Enantiornithes was the first lineage to radiate and were the dominant group of birds during the Cretaceous period. A second radiation, the basal ornithurines, completed the Mesozoic avian fauna. Neither of these lineages survived the end of the Mesozoic era, and the extant birds (Neornithes) are modern ornithurines. The shaded areas indicate the changes in diversity of the lineages. (Note that the areas representing passerines and nonpasserines overlap.)

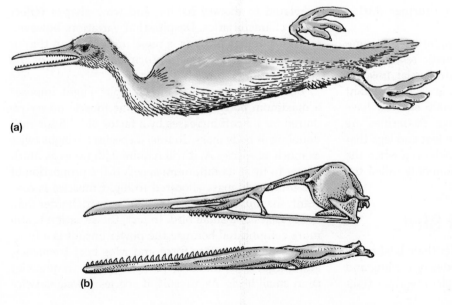

(a)

(b)

Figure 17–3 Hesperornithi-formes. By the Late Cretaceous a lineage of birds had already become specialized as flightless fish-eaters. (a) Reconstruction of *Hesperornis*. Its body form is similar to that of the extant flightless cormorant. (b) Skull of a related species, *Parahesperornis*; note the teeth in the maxilla and dentary bones.

Table 17.1 Major lineages of extant birds (Neornithes). Except as noted, the lineages have representatives on all continents except Antarctica. For additional information, see the Tree of Life Web Project (http://tolweb.org/Neornithes). The table lists the approximate number of species and general characteristics for the major taxa.

Palaeognathae (ratites)

Flightless birds: Five large species (ostrich, emu, and cassowaries) and about 50 smaller species (kiwis, rheas, and tinamous), all in the Southern Hemisphere.

Neognathae

Galloanserae: fowls, ducks, geese and their relatives.

Anseriformes: ducks, geese, and relatives. More than 150 species of semiaquatic birds.

Galliformes: Fowl, quail, megapodes. More than 200 species of ground-dwelling birds.

Neoaves: most extant birds

Apodiformes: hummingbirds and swifts. More than 400 species of arboreal birds with specialized flight. Swifts feed on insects they capture in flight, and hummingbirds are specialized nectar feeders that also capture flying insects.

Cuculiformes: cuckoos. About 150 species of arboreal and ground-dwelling birds that include both herbivorous and predatory forms.

Charadriiformes: stilts, plovers, oystercatchers, and their relatives. Fewer than 100 species of long-billed shorebirds that probe for invertebrates buried in mud or sand.

Gruiformes: cranes, rails, and coots. More than 300 species of mostly herbivorous birds that are associated with freshwater habitats.

Phoenicopteriformes: flamingos. Five species of highly specialized aquatic filter feeders, mostly in tropical regions.

"Water Birds"

Procellariiformes: albatrosses, petrels, and their relatives. More than 100 species of seabirds; many of them are pelagic and fly thousands of kilometers before returning to land.

Sphenisciformes: penguins. Seventeen species of wing-propelled divers that feed on fish. Largely confined to the cold Southern Hemisphere, including Antarctica, but one species lives in the Galápagos Islands.

Ciconiiformes: storks, herons, ibises, pelicans, and their relatives. About 100 species of mostly predatory birds found near water. All except pelicans are long-legged wader

"Land Birds"

Passeriformes: perching birds. Nearly 6000 species of birds with feet that are specialized for holding onto perches.

Suboscines: ovenbirds, antbirds, tyrant flycatchers, and their relatives. Six families with about 1000 species. Suboscines have limited control of the syrinx muscles and do not produce complex vocalizations.

Oscines: songbirds. About 70 families containing about 5000 species. Oscines have great control of their syrinx muscles and produce complex birdsongs.

Psittaciformes: parrots. More than 300 species of arboreal birds, mostly fruit and seed eaters.

Columbiformes: pigeons. More than 300 species of mostly rock dwellers, primarily seed eaters.

Falconiformes: falcons, caracaras, and their relatives. More than 60 species of diurnal birds of prey and scavengers.

Coraciiformes: kingfishers, kookaburras, and relatives. More than 200 species of mostly arboreal, carnivorous, or insectivorous birds.

Piciformes: woodpeckers and relatives. More than 200 species of mostly arboreal insectivorous birds.

Accipitriformes: hawks, eagles, sea eagles, kites, buzzards, New World and Old World vultures, and relatives. More than 100 species of birds of prey and scavengers.

Strigiformes: owls. About 200 species of nocturnal predatory birds that hunt on the wing.

---

probably originated among the ornithurines during the Late Cretaceous.

With about 10,000 extant species, birds are a complex group of animals, and phylogenetic relationships within the Neoaves are especially controversial. **Table 17–1** lists the characteristics of the major lineages of extant birds. The passerines, with about 6000 species, have been a particularly successful lineage. Passerines are characterized by modifications of the feet and legs that allow the toes to hold tightly to a perch even when the bird is asleep; the passerines are commonly called the perching birds.

17.2 **The Structure of Birds**

In many respects, birds are variable: their beaks and feet are specialized for different modes of feeding and locomotion, the morphology of their intestinal tract

is related to dietary habits, and wing shapes reflect flight characteristics. Despite that variation, however, the morphology of birds is more uniform than that of mammals. Much of this uniformity is a result of the specialization of birds for flight.

Consider body size as an example: Flight imposes a maximum body size on birds. The muscle power required for takeoff increases by a factor of 2.25 for each doubling of body mass. That is, if species B weighs twice as much as species A, it will require 2.25 times as much power to fly at its minimum speed. If the proportion of the total body mass allocated to flight muscles is constant, the muscles of a large bird must work harder than the muscles of a small bird. In fact, the situation is still more complicated because the power output is a function of both muscular force and wing-beat frequency, and large birds have lower wing-beat frequencies than small birds. As a result, if species B weighs twice as much as species A, it will develop only 1.59 times as much power from its flight muscles, although it needs 2.25 times as much power to fly. Therefore, large birds require longer takeoff runs than do small birds, and a bird could ultimately reach a body mass at which any further increase in size would move it into a realm in which its leg and flight muscles were not able to provide enough power to take off.

Taking off is particularly difficult for large birds, which have to run and flap their wings to reach the speed needed for liftoff. The largest flying bird known is *Pelagornis chilensis*, a seabird that lived 5 to 10 million years ago in what is now northern Chile. This gigantic predator had a wingspan that is estimated at 5.2 meters, half again the wingspan of the wandering albatross, which has the largest wingspan of any extant species of bird.

Flightless birds are spared the mechanical constraints associated with producing power for flight, but they still do not approach the body sizes of mammals. The largest extant bird is the flightless ostrich, which weighs about 150 kilograms, and the largest bird known, one of the extinct elephant birds, weighed an estimated 450 kilograms. In contrast, the largest terrestrial mammal, the African elephant, may weigh as much as 5000 kilograms.

The structural uniformity of birds is seen even more clearly if their body shapes are compared with those of other sauropsids. There are no quadrupedal birds, for example, nor any with horns or bony armor. Even those species of birds that have become secondarily flightless retain the basic body form of birds.

Feathers

Feathers develop from follicles in the skin, generally arranged in tracts or **pterylae,** which are separated by patches of unfeathered skin, the **apteria** (**Figure 17–4**). Some species—such as ratites, penguins, and mousebirds—lack pterylae, and the feathers are uniformly distributed over the skin.

For all their structural complexity, the chemical composition of feathers is remarkably simple and uniform. More than 90 percent of a feather consists of a specific type of beta keratin, a protein related to the keratin that forms the scales of lepidosaurs. About 1 percent of a feather consists of lipids, about 8 percent is water, and the remaining fraction consists of small amounts of other proteins and pigments, such as melanin. The colors of feathers are produced by structural characters and pigments.

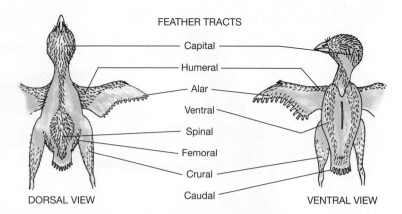

FEATHER TRACTS

Capital — Humeral — Alar — Ventral — Spinal — Femoral — Crural — Caudal

DORSAL VIEW   VENTRAL VIEW

Figure 17–4 Feather tracts of a typical songbird.

Feathers are anchored in the skin by a short, tubular base, the **calamus,** which remains firmly implanted within the follicle until molt occurs (**Figure 17–5**). A long, tapered **rachis** extends from the calamus and bears closely spaced side branches called **barbs. Barbules** branch from the barbs, and proximal and distal barbules branch from opposite sides of the barbs. The ends of the distal barbules bear hooks that insert into grooves in the proximal barbules of the adjacent barb. The hooks and grooves act like Velcro to hold adjacent barbs together, forming a flexible **vane.**

A body-contour feather has several regions that reflect differences in structure. Near the base of the rachis, the barbs and barbules are flexible, and the barbules lack hooks. This portion of a feather has a soft, loose, fluffy texture called plumulaceous or downy. It gives the plumage of a bird its properties of thermal insulation. Farther from the base, the barbs form a tight surface called the vane, which has a pennaceous (sheetlike) texture. This is the part of the feather that is exposed on the exterior surface of the plumage, where it serves as an airfoil, protects the downy undercoat, sheds water, and reflects or absorbs solar radiation. The barbules are the structures that maintain the closed pennaceous character of the feather vanes. They are arranged in such a way that any physical disruption to the vane is easily corrected by preening behavior, in which the bird realigns the barbules by drawing its slightly separated bill over them.

Ornithologists usually distinguish five types of feathers: (1) contour feathers, including typical body feathers and the flight feathers (remiges and rectrices); (2) semiplumes; (3) down feathers of several sorts; (4) bristles; and (5) filoplumes.

The contour feathers include the outmost feathers on a bird's body, wings, and tail. The **remiges** (wing

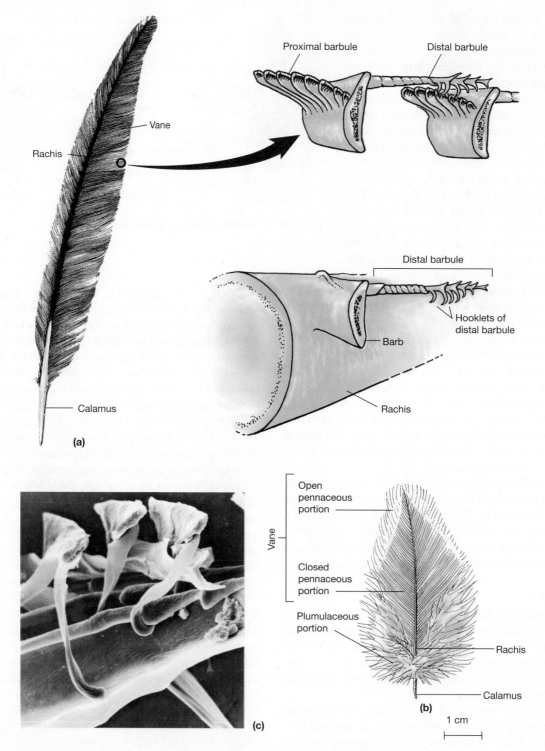

Figure 17–5 Contour feathers. (a) A wing quill, showing its main structural features. (b) A body-contour feather. The inset and electron micrograph (c) show details of the interlocking mechanism of the proximal and distal barbules.

Figure 17–6 A red-tailed hawk. Slotting is visible in the outer primaries.

feathers, singular *remex*) and **rectrices** (tail feathers, singular *rectrix*) are large, stiff, mostly pennaceous contour feathers that are modified for flight. For example, the distal portions of the outer primaries of many species of birds are abruptly tapered or notched so that, when the wings are spread, the tips of these primaries are separated by conspicuous gaps or slots (**Figure 17–6**). This condition reduces the drag on the wing and, in association with the marked asymmetry of the outer and inner vanes, allows the feather tips to twist as the wings are flapped and to act somewhat as individual propeller blades.

Semiplumes are feathers intermediate in structure between contour feathers and down feathers. They combine a large rachis with entirely plumulaceous vanes and can be distinguished from down feathers in that the rachis is longer than the longest barb (**Figure 17–7a**). Semiplumes are mostly hidden beneath the contour feathers. They provide thermal insulation and help to fill out the contour of a bird's body.

Down feathers of various types are entirely plumulaceous feathers in which the rachis is shorter than the longest barb or entirely absent. Down feathers provide insulation for adult birds of all species, and powder down feathers disintegrate into fine particles of keratin that help the contour feathers shed water.

Bristles are specialized feathers with a stiff rachis and without barbs or with barbs on only the proximal portion (**Figure 17–7b**). Bristles occur most commonly around the base of the bill, around the eyes, as eyelashes, and on the head or even on the toes of some birds. Bristles screen out foreign particles from the nostrils and eyes of many birds; they also act as tactile sense organs and possibly as aids in the aerial capture of flying insects, as, for example, do the long bristles at the edges of the jaws in nightjars and flycatchers.

Filoplumes are fine, hairlike feathers with a few short barbs or barbules at the tip (**Figure 17–7c**). In some birds, such as cormorants and bulbuls, the filoplumes grow out over the contour feathers and contribute to

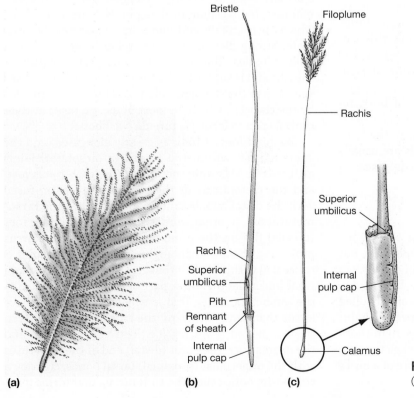

Figure 17–7 Types of feathers.
(a) Semiplume, (b) bristle, (c) filoplume.

the external appearance of the plumage, but usually they are not exposed. Filoplumes are sensory structures that aid in the operation of other feathers. Filoplumes have numerous free nerve endings in their follicle walls, and these nerves connect to pressure and vibration receptors around the follicles and transmit information about the position and movement of the contour feathers. This sensory system probably plays a role in keeping the contour feathers in place and adjusting them properly for flight, insulation, bathing, or display.

Streamlining and Weight Reduction

Birds are the only vertebrates that move fast enough in air for wind resistance and streamlining to be important factors in their lives. Many passerine birds are probably able to fly 50 kilometers per hour or even faster when they must, although their normal cruising speeds are lower. Ducks and geese can fly at 80 or 90 kilometers per hour, and peregrine falcons reach speeds as high as 200 kilometers per hour when they dive on prey. Fast-flying birds have many of the same structural characters as those seen in fast-flying aircraft. Contour feathers make smooth junctions between the wings and the body and often between the head and the body as well, eliminating sources of turbulence that would increase wind resistance. The feet are tucked close to the body during flight, further improving streamlining.

At the opposite extreme, some birds are slow fliers. Many of the long-legged, long-necked wading birds, such as spoonbills and flamingos, fall into this category. Their long legs trail behind them as they fly, and their necks are extended. They are far from streamlined, although they may be strong fliers.

Characteristics of some of the organs of birds reduce body mass. For example, birds lack urinary bladders, and most species have only one ovary (the left). The gonads of both male and female birds are usually small; they hypertrophy during the breeding season and regress when breeding has finished.

Skeleton

Structural modifications can be seen in several aspects of the skeleton of birds. The avian skeleton is not lighter in relation to the total body mass of a bird than is the skeleton of a mammal of similar size, but the distribution of mass is different. Many bones are air-filled (pneumatic, **Figure 17–8**), and the skull is especially light; however, the leg bones of birds are heavier than those of mammals. Thus, the total mass of the skeleton of a bird is similar to that of a mammal, but more of a bird's mass is concentrated in its hindlimbs.

Figure 17–8 Bird bone. The hollow core and reinforcing struts in a long bone from a bird.

Except for the specializations associated with flight, the skeleton of a bird is very much like that of a small coelurosaur (**Figure 17–9**). The pelvic girdle of birds is elongated, and the ischium and ilium have broadened into thin sheets that are firmly united with a **synsacrum,** which is formed by the fusion of 10 to 23 vertebrae. The long tail of ancestral diapsids has been shortened in birds to about five free caudal vertebrae and a **pygostyle** formed by the fusion of the remaining vertebrae. The pygostyle supports the tail feathers (rectrices). The thoracic vertebrae are joined by strong ligaments that are often ossified. The relatively immobile thoracic vertebrae, the synsacrum, and the pygostyle combine with the elongated, rooflike pelvis to produce a nearly rigid vertebral column. Flexion is possible only in the neck, at the joint between the thoracic vertebrae and the synsacrum, and at the base of the tail.

The center of gravity is beneath the wings, and the sternum is greatly enlarged compared with other vertebrates. The sternum of flying birds bears a keel from which the pectoralis and supracoracoideus muscles originate. Strong fliers have well-developed keels and large flight muscles. The scapula extends posteriorly above the ribs and is supported by the coracoid, which is fused ventrally to the sternum. Additional bracing is provided by the clavicles, which, in most birds, are fused at their central ends to form the **furcula** (wishbone).

The hind foot of birds is greatly elongated, and the ankle joint is within the tarsals (a mesotarsal joint) as it was in Mesozoic theropods. The fifth toe is lost, and the metatarsals of the remaining toes are fused with the distal tarsals to form a bone called the **tarsometatarsus.** In other words, birds walk with their toes (tarsals) flat on the ground and the tarsometatarsus projecting upward at an angle. The actual knee joint is concealed within the contour feathers on the body, and what looks like the lower part of a bird's leg is actually the tarsometatarsus. That is why birds appear to have knees that bend backward; the true knee lies between the femur and lower leg (i.e., between the thigh and the drumstick). The outer (distal) end of the tibia fuses with the uppermost (proximal) tarsal bones, creating a composite bone called the **tibiotarsus** that forms most

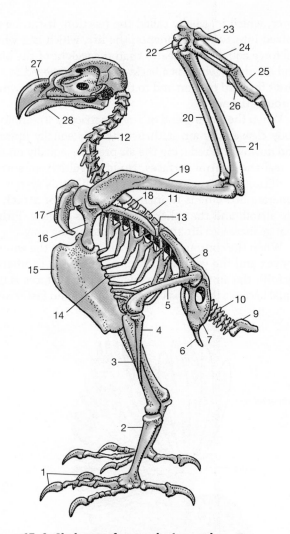

Figure 17–9 Skeleton of an eagle. Legend: 1. Toes.
2. Tarsometatarsus, formed by fusion of the distal tarsals to
the metatarsals. **3.** Tibiotarsus, formed by fusion of the proximal tarsal bones (the calcaneum and astragalus) to the end of
the tibia. **4.** Fibula. **5.** Femur. **6.** Pubis. **7.** Ischium. **8.** Ilium.
9. Pygostyle. **10.** Caudal vertebrae. **11.** Thoracic vertebrae.
12. Cervical vertebrae. **13.** Uncinate process of rib. **14.** Sternal
rib. **15.** Sternum. **16.** Coracoid. **17.** Furcula. **18.** Scapula.
19. Humerus. **20.** Radius. **21.** Ulna. **22.** Carpal bones. **23.** First
digit. **24.** Metacarpal. **25.** Second digit. **26.** Third digit.
27. Upper jaw. **28.** Lower jaw.

of the lower leg (drumstick). The fibula is reduced to a
splint of bone.

Muscles

Power-producing features are equally important components of the ability of birds to fly. The pectoral muscles of a strong flier may account for 20 percent of the total body mass. The power output per unit mass of the pectoralis major of a turtledove during level flight has been estimated to be 10 to 20 times that of most mammalian muscles. Birds have large hearts and high rates of blood flow and complex lungs that use crosscurrent flows of air and blood to maximize gas exchange and to dissipate the heat produced by high levels of muscular activity during flight.

The relative size of the leg and flight muscles of birds is related to their primary mode of locomotion. Flight muscles make up 25 percent to 35 percent of the total body mass of strong fliers such as hummingbirds and swallows. These species have small legs; the leg muscles account for as little as 2 percent of the body mass. Predatory birds such as hawks and owls use their legs to capture prey. In these species, the flight muscles make up about 20 percent of the body mass and the limb muscles 10 percent. Swimming birds—ducks and grebes, for example—have an even division between limb and flight muscles; the combined mass of these muscles may be 30 percent to 60 percent of the total body mass. Birds such as rails, which are primarily terrestrial and run to escape from predators, have limb muscles that are larger than their flight muscles.

Muscle-fiber types and metabolic pathways also distinguish running birds from fliers. The familiar distinction between the light meat and dark meat of a chicken reflects those differences. Fowl, especially domestic breeds, rarely fly, but they are capable of walking and running for long periods. The dark color of the leg muscles reveals the presence of myoglobin in the tissues and indicates a high capacity for aerobic metabolism in the limb muscles of these birds. The white muscles of the breast lack myoglobin and have little capacity for aerobic metabolism. The flights of fowl (including wild species such as pheasants, grouse, and quail) are of brief duration and are used primarily to evade predators. The bird uses an explosive takeoff, fueled by anaerobic metabolic pathways, followed by a long glide back to the ground. Birds that are capable of strong, sustained flight have dark breast muscles with high aerobic metabolic capacities.

17.3 Wings and Flight

Unlike the fixed wings of an airplane, the wings of a bird function both as airfoils (lifting surfaces) and as propellers for forward motion. The primaries, inserted on the hand bones, do most of the propelling when a bird flaps its wings down, and the secondaries along the arm provide lift (**Figure 17–10**). Removing only a few of the primaries from the wings of doves and pigeons greatly reduces their ability to fly, but the secondaries are less critical—a bird can still fly when as much as 55 percent of the total area of the secondaries has been removed.

A bird can alter the area and shape of its wings and their position with respect to the body. These changes in area and shape cause corresponding changes in velocity and lift that allow a bird to maneuver, change direction, land, and take off. The ability of swifts to modify their flight by changes in wing shapes is particularly noteworthy.

The aerodynamic properties of a bird's wing in flight—even in nonflapping flight—are vastly more complex than those of a fixed wing on an airplane or glider. Nevertheless, it is instructive to consider a bird's wing in terms of the basic performance of a fixed airfoil. Although a bird's wing actually moves forward through the air, it is easier to think of the wing as stationary with the air flowing past. The flow of air produces a force, which is usually called the reaction. It can be resolved into two components: the **lift,** which is a vertical force equal to or greater than the weight of the bird, and the **drag,** which is a backward force opposed to the bird's forward motion and to the movement of its wings through the air.

When the leading end of a symmetrically streamlined body cleaves the air, it thrusts the air equally upward and downward, reducing the air pressure equally on the dorsal and ventral surfaces. No lift results from such a condition. There are two ways to modify this system to generate lift. One is to increase the **angle of attack** of the airfoil, and the other is to bend its surface. Either change increases lift at the cost of increasing drag.

When the contour of the dorsal surface of the wing is convex and the ventral surface is concave (a **cambered airfoil**), the air pressure against the two surfaces is unequal because the air has to move farther and faster over

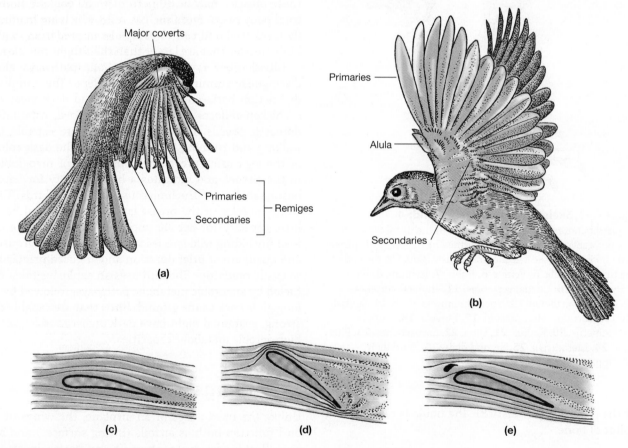

(a)

(b)

(c)   (d)   (e)

Figure 17–10 The wing in flight. (a, b) Drawings from high-speed photographs show the twisting and opening of the primaries during flapping flight. (c-e) Airflow around a cambered airfoil. (c) At a low angle of attack, the air streams smoothly over the upper surface of the wing and creates lift. (d) When the angle of attack becomes steep, air passing over the wing becomes turbulent, decreasing lift enough to produce a stall. (e) A wing slot formed by the alula helps to prevent turbulence by directing a flow of rapidly moving air close to the upper surface of the wing.

the dorsal convex surface than over the ventral concave surface. The result is a lower air pressure above the wing than beneath it. This difference in pressure creates an upward force called lift, and when the lift equals or exceeds the bird's body weight, the bird becomes airborne. The camber of the wing varies in birds with different flight characteristics; it also changes along the length of the wing. Camber is greatest close to the body and decreases toward the wing tip. This change in camber is one reason that the inner part of the wing generates greater lift than the outer part.

If the leading edge of the wing is tilted up so that the angle of attack is increased, the result is increased lift—up to an angle of about 15 degrees, the stalling angle. This lift results more from a decrease in pressure over the dorsal surface than from an increase in pressure below the airfoil. If the smooth flow of air over the wing becomes disrupted, the airflow begins to separate from the wing because of the increased air turbulence over the wing. The wing is then stalled. Stalling can be prevented or delayed by the use of slots or auxiliary airfoils on the leading edge of the main wing. The slots help to restore a smooth flow of air over the wing at high angles of attack and at slow speeds. The bird's **alula** (a small projection on the anterior edge of the wing) has this effect, particularly during landing or takeoff. Also, the primaries act as a series of independent, overlapping airfoils, each tending to smooth out the flow of air over the one behind.

Another characteristic of an airfoil has to do with wing-tip vortexes—eddies of air resulting from the outward flow of air from under the wing and the inward flow from over it. This is induced drag. One way to reduce the effect of these wing-tip eddies and their drag is to lengthen the wing, so that the tip vortex disturbances are widely separated and there is proportionately more wing area where the air can flow smoothly. Another solution is to taper the wing, reducing its area at the wing tip where induced drag is greatest. The ratio of length to width is called the **aspect ratio.** Long, narrow wings have high aspect ratios and high lift-to-drag (L/D) ratios. High-performance sailplanes and albatrosses, for example, have aspect ratios of 18:1 and L/D ratios of about 40:1.

Wing loading is another important consideration. This is the mass of the bird divided by the wing area. The lighter the loading, the less power is needed to sustain flight. Small birds usually have lighter wing loading than do large birds, but wing loading is also related to specializations for powered versus soaring flight. The comparisons in **Table 17–2** illustrate both of these trends. Small species such as hummingbirds, barn swallows, and mourning doves have lighter wing loadings than do large species such as the peregrine, golden eagle, and mute swan; yet the 3-gram hummingbird, a powerful flier, has a heavier wing loading than the more buoyant, sometimes soaring barn swallow, which is more than five times heavier. Similarly, the rapid-stroking peregrine has a heavier wing loading than the larger, often soaring golden eagle.

Flapping Flight

Flapping flight is remarkable for its automatic, unlearned performance. A young bird on its initial flight uses a form of locomotion so complex that it defies precise analysis in physical and aerodynamic terms. The nestlings of some species of birds develop in confined spaces, such as burrows in the ground or cavities in tree trunks, in which it is impossible for them to spread their wings to practice flapping before they leave the nest.

Table 17–2 Wing loading of some representative birds. In general, wing loading increases with body size, but the highest wing loading recorded is found in the thick-billed murre, a sea bird that swims and dives. The murre's wing loading is believed to approach the maximum value possible for a flying bird.

Species	Body Mass (g)	Wing Area (cm²)	Wing loading (g/cm²)
Ruby-throated hummingbird	3	12.4	0.24
Barn swallow	17	118.5	0.14
Mourning dove	130	357	0.36
Thick-billed murre	1033	397	2.6
Peregrine falcon	1222	1342	0.91
Golden eagle	4664	6520	0.72
Mute swan	11,602	6808	1.70

Despite this seeming handicap, many of them are capable of flying considerable distances on their first flights. Diving petrels may fly as far as 10 kilometers the first time out of their burrows. On the other hand, young birds reared in open nests—especially large birds such as albatrosses, storks, vultures, and eagles—frequently flap their wings vigorously in the wind for several days before flying. Such flapping may help to develop muscles, but it is unlikely that these birds are learning to fly. However, a bird's flying abilities do improve with practice for a period of time after it leaves the nest. Landing is especially challenging, and inexperienced young birds may miss the branch they were aiming for and tumble through the tree, grabbing frantically for anything within reach.

There are so many variables involved in flapping flight that it is difficult to understand exactly how it works. A beating wing is both flexible and porous, and it yields to air pressure. Its shape, wing loading, camber, angle relative to the body, and the position of the individual feathers all change remarkably as a wing moves through its cycle of locomotion. This is a formidable list of variables, far more complex than those involved in the analysis of the fixed wing of an airplane, and it is no wonder that the aerodynamics of flapping flight are not yet fully understood. However, the general properties of a flapping wing can be described.

We can begin by considering the flapping cycle of a small bird in flight. A bird cannot continue to fly straight and level unless it can develop a force or thrust to balance the drag operating against forward momentum. The downward stroke of the wings produces this thrust. The wing tips (primary feathers) are the site where thrust is generated, whereas the inner wings (secondaries) generate lift. It is easiest to consider the forces operating on the inner and outer wing separately.

The forces on the wing tips derive from two motions that have to be added together. The tips are moving forward with the bird, but at the same time, they are moving downward relative to the bird. The wing tip would have a very large angle of attack and would stall if it were not flexible. As it is, the forces on the tip cause the individual primaries to twist as the wing is flapped downward and to produce the forces diagrammed in **Figure 17–11**.

The forces acting on the two parts of the wing combine to produce the conditions for equilibrium flight. The positions of the inner and outer wings during the upstroke (dashed lines) and downstroke reveal that vertical motion is applied mostly at the wing tips. Most of the lift is generated by the inner wing and body. Tilting the wing tip during the downstroke produces a resultant force that is directed forward. The movement of the wing tip relative to the air is affected by the forward motion of the bird through the air. As a result of this motion and the tilting of the wing tips during the downstroke, the flow of air across the primaries is different from the flow across the secondaries and the body. When flight speed through the air is constant, the forces acting on the inner wing and the body and on the outer wing combine to produce a set of summed vectors in which thrust exceeds drag and lift at least equals the body mass.

As the wings move downward and forward on the downstroke, the trailing edges of the primaries bend upward under air pressure, and each feather acts as an individual propeller biting into the air and generating thrust. Contraction of the pectoralis major, the large breast muscle, produces the forceful downstroke during level flapping flight. During this downbeat, the thrust is greater than the total drag, and the bird accelerates. In small birds, the return stroke, which is upward and backward, provides little or no thrust. It is mainly a passive recovery stroke, and the bird slows down during this part of the wing-beat cycle.

For large birds with slow wing actions, the upstroke lasts too long to spend in a state of deceleration, and a similar situation exists when any bird takes off—in these situations the bird needs to generate thrust on the upstroke as well as on the downstroke. Thrust on the upstroke is produced by bending the wings slightly at the wrists and elbow and by rotating the humerus upward and backward. This movement causes the upper surfaces of the twisted primaries to push against the air and to produce thrust as their lower surfaces did in the downstroke. In this type of flight, the wing tip describes a rough figure eight through the air. As speed increases, the figure-eight pattern is restricted to the wing tips.

A powered upstroke results mainly from contraction of the supracoracoideus, a deep muscle underlying the pectoralis major and attached directly to the keel of the sternum. It inserts on the dorsal head of the humerus by passing through the foramen triosseum, formed where the coracoid, furcula, and scapula join (**Figure 17–12** on page 420). In most species of birds, the supracoracoideus is a relatively small, pale muscle with low myoglobin content, easily fatigued. In species that rely on a powered upstroke—for fast, steep takeoffs, for hovering, or for fast aerial pursuit—the supracoracoideus is larger. The ratio of weights of the pectoralis major and the supracoracoideus is a good indication of a bird's reliance on a powered upstroke; such ratios vary from 3:1 to 20:1. The total weight of the flight muscles shows the extent to which a bird depends on powered flight. Strong fliers such as pigeons and falcons

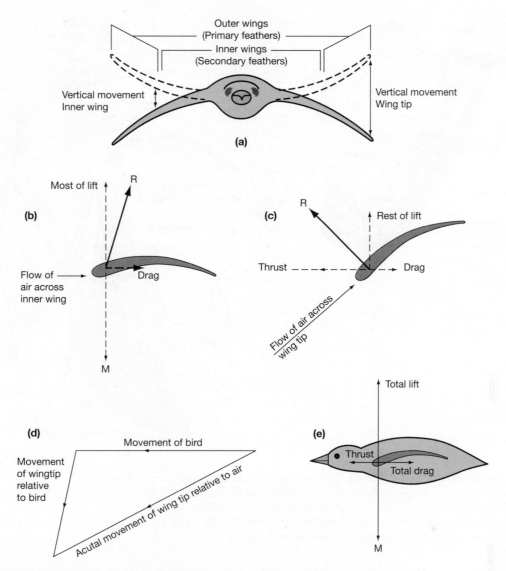

Figure 17–11 Aerodynamic aspects of bird flight. The forces acting on the inner and outer wings and the body of a bird during flapping flight are shown. (a) The positions of the inner wing and wing tips during the upward stroke (dashed outline) and downward stroke (solid outline) reveal that vertical motion is greatest at the wing tips. Thus, the proximal (inner) portion of the wing (the area of the secondaries) acts as if the bird were gliding to generate the forces shown in (b), whereas the distal (outer) portion generates the forces shown in (c). Most of the lift that counteracts the effect of gravity on the body mass (M) is generated by the inner portion of the wing and the body. Canting the outer portion of the wing (the area of the primaries) at a downward angle during the down stroke (c) produces a resultant force (R) that is directed forward to produce thrust. The movement of the wing tip relative to the air is affected by the forward motion of the bird through the air (d). As a consequence of this motion and the canting of the wing tips during the downward stroke, the flow of air across the primaries (c) is different from the flow across the secondaries and the body. When the flight speed through the air is constant, the forces acting on the inner wing and the body (b) and on the outer wing (c) combine to produce a set of summed vectors in which thrust exceeds drag and lift at least equals the gravitational force (e).

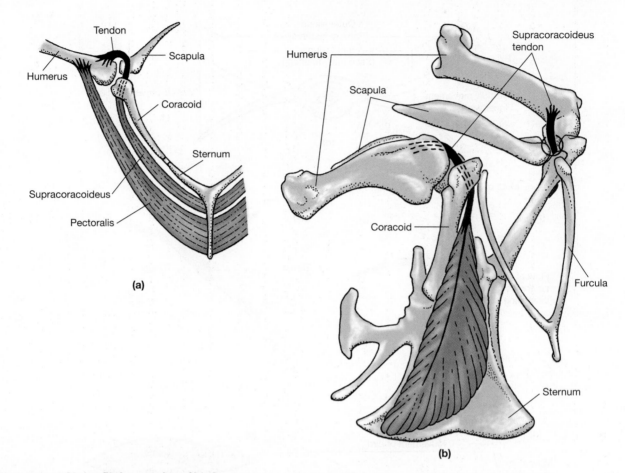

Figure 17-12 Major flight muscles of birds. (a) Cross section through the sternum of a bird showing the relationships of the pectoralis major and supracoracoideus muscles. (b) Frontal and lateral view of the sternum and pectoral girdle of a bird showing the insertion of the supracoracoideus tendon through the foramen triosseum onto the dorsal head of the humerus. The foramen is formed by the articulation of the furcula, coracoid, and scapula.

have breast muscles that make up more than 20 percent of body weight, whereas in some owls, which have very light wing loading, the flight muscles make up only 10 percent of total weight.

Wing Proportions

To a small degree, a bird can change the surface area of the wing by changing the position of its feathers, but most change in wing proportions occurs over evolutionary time. Wings may be large or small in relation to body size, resulting in light wing loading or heavy wing loading. They may be long and pointed, short and rounded, highly cambered or relatively flat, and the width and degree of slotting are additional important characteristics. Depending on whether a bird is primarily a powered flier or a soaring form, the various segments of the wing (hand, forearm, upper arm) are lengthened to different degrees. Hummingbirds have

very fast, powerful wing beats, requiring maximum propulsive force from the primaries. The hand bones of hummingbirds are longer than the forearm and upper arm combined. Most of the flight surface is formed by the primaries, and hummingbirds have only six or seven secondaries. Frigate birds are marine species with long, narrow wings specialized for powered flight as well as for gliding and soaring; they have the lowest wing loading of any extant bird. All three segments of the forelimb are about equal in length. The soaring albatrosses have carried lengthening of the wing to the extreme limit found in birds: the humerus or upper arm is the longest segment, and there may be as many as 32 secondaries in the inner wing (**Figure 17-13**).

Wing Shape and Flight Characteristics

Ornithologists recognize four structural and functional types of wings (**Figure 17-14**). Seabirds, particularly those

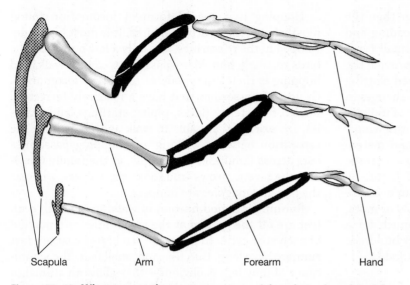

Scapula   Arm   Forearm   Hand

Figure 17–13 Wing proportions. Comparison of the relative lengths of the proximal, middle, and distal elements of the wing bones of a hummingbird (*top*), frigate bird (*middle*), and albatross (*bottom*) drawn to the same size.

such as albatrosses and shearwaters that rely on dynamic soaring, have long, narrow, flat wings lacking slots in the outer primaries. Dynamic soaring is possible only where there is a pronounced vertical wind gradient, with the lower 15 or so meters of air being slowed by friction against the ocean surface. Furthermore, dynamic soaring is feasible only in regions where winds are strong and persistent, such as in the latitudes of the Roaring Forties (between latitudes 40 and 50 degrees south, in which there are strong westerly winds). This is where most albatrosses and shearwaters are found. Starting from the top of the wind gradient, an albatross glides downwind with great increase in ground speed (kinetic energy). Then, as it nears the surface, it turns and gains altitude while gliding into the wind. Because the bird flies into wind of increasing speed as it rises, its loss of airspeed is not as great as its loss of ground speed. Consequently it does not stall until it has mounted back to the top of the wind gradient, where the air velocity becomes stable. At that point, the bird has converted much of its kinetic energy to potential energy, and it turns downwind to repeat the cycle.

Birds that live in forests and woodlands where they must maneuver around obstructions have elliptical wings. These wings have a low aspect ratio, tend to be highly cambered, and usually have a high degree of slotting in the outer primaries. These features are generally associated with rapid flapping, slow flight, and a high degree of maneuverability.

Birds such as swifts, that are aerial foragers, make long migrations, or have a heavy wing loading that is related to some other aspect of their lives, such as diving, have wings with high aspect ratios. These wings have a flat profile (little camber) and often lack slots in the outer primaries. In flight, they show the swept-back attitude of jet-fighter plane wings. All fast-flying birds have converged on this form.

The slotted high-lift wing is a fourth type. It is associated with static soaring typified by vultures, eagles, storks, and some other large birds. This wing has an intermediate aspect ratio between the elliptical wing and the high-aspect-ratio wing, a deep camber, and marked slotting in the primaries. When the bird is in flight, the tips of the primaries turn upward under the influence of air pressure and body weight. Static soarers remain airborne mainly by seeking out and gliding

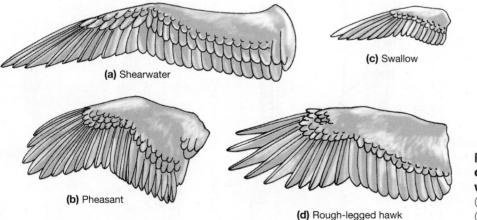

(a) Shearwater

(c) Swallow

(b) Pheasant

(d) Rough-legged hawk

Figure 17–14 Comparison of four basic types of bird wings. (a) Dynamic soaring, (b) elliptical, (c) high aspect ratio, (d) high lift.

in air masses that are rising at a rate faster than the bird's sinking speed. Hence, a light wing loading and maneuverability (slow forward speed and small turning radius) are advantageous. Broad wings provide the light wing loading, and the highly developed slotting enhances maneuverability by responding to changes in wind currents with changes in the positions of individual feathers instead of movements of the entire wing.

In regions where topographic features and meteorological factors provide currents of rising air, static soaring is an energetically cheap mode of flight. By soaring rather than flapping, a large bird can decrease the energy required for flight by a factor of 20 or more, whereas the saving is only one-tenth as much for a small bird. Some condors and vultures cover hundreds of kilometers each day soaring in search of food.

17.4 The Hindlimbs

The ability to perch in trees was an early feature of bird evolution, and by the Cretaceous both the enantiornithine and ornithurine lineages included perching birds. On the ground birds hop, walk, or run, and in water they swim using either the hindlimbs or the wings.

Hopping, Walking, and Running

When they are on the ground, birds hop (make a succession of jumps in which both legs move together), walk (move the legs alternately with at least one foot in contact with the ground at all times), or run (move the legs alternately with both feet off the ground at some times). Modifications of these basic gaits include climbing, wading in shallow water or on insubstantial surfaces such as lily pads, and supporting heavy bodies.

Hopping is a special form of locomotion found mostly in perching, arboreal birds. It is most highly developed in the passerines, and only a few nonpasserine birds regularly hop. Many passerines cannot walk, and hopping is their only mode of terrestrial locomotion. Groups of passerines that have a relatively terrestrial mode of existence (larks, pipits, starlings, and grackles, for example) are able to walk as well as hop. The separation between walking and hopping passerines cuts across families. For example, in the family Corvidae, the ravens, crows, and rooks are walkers, whereas the jays and magpies are hoppers.

Running is a modification of walking in which both feet are off the ground at the same time for a portion of each step cycle. In general, vertebrates that are fast runners have long, thin legs and small feet. The significance of long legs is obvious—they allow an animal to cover more distance with each stride. The reduction in leg and foot mass is a bit more subtle and involves the physics of momentum. When the foot is in contact with the ground, it is motionless; after it pushes off from the ground, the foot and the lower part of the leg must be accelerated to a speed faster than the trunk of the running animal. Reducing the mass of the foot and leg reduces the energy needed for this change in momentum and allows the animal to run faster.

Reducing the number and length of the toes makes the foot smaller. No bird has reduced the length and number of toes in contact with the ground to the extent that the hoofed mammals have, but the large, fast-running ostrich has only two toes on each foot, and many other species have only the three forward-directed toes in contact with the ground (**Figure 17–15**).

Weight-bearing characteristics, such as large, heavy leg bones arranged in vertical columns as supports of great body mass, are well known in large mammals

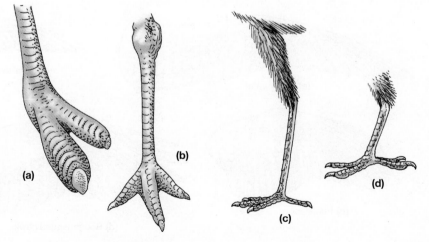

Figure 17–15 Avian feet with specializations for terrestrial locomotion. (a) Ostrich, with only two toes. (b) Rhea, with three toes. (c) Secretary bird, with a typical avian foot. (d) Roadrunner, with zygodactyl foot. (Not drawn to scale.)

(a)

(b)

(c)

(d)

Figure 17–16 Large, flightless, terrestrial birds. (a) The extinct elephant bird of Madagascar. (b) The cursorial ostrich.

such as elephants. No surviving birds show specializations of this kind, although these features are seen in some of the large, flightless terrestrial birds of earlier times. The extinct elephant birds (Aepyornithiformes) of Madagascar and moas (Dinornithiformes) of New Zealand were herbivores that evolved on oceanic islands in the absence of large carnivorous mammals and survived there until contact with humans in the post-Pleistocene period (**Figure 17–16**).

Climbing

Birds climb on tree trunks or other vertical surfaces by using their feet, tails, beaks, and, rarely, their forelimbs. Several distantly related groups of birds have independently acquired specializations for climbing and foraging on vertical tree trunks. Species such as woodpeckers and woodcreepers, which use their tails as supports, begin foraging near the base of a tree trunk and work their way vertically upward, head first, clinging to the bark with strong feet on short legs. The tail is used as a prop to brace the body against the powerful pecking exertions of head and neck, and the pygostyle and free caudal vertebrae in these species are much enlarged and support strong, stiff tail feathers. A similar modification of the tail is found in certain swifts that perch on cave walls and inside chimneys.

Nuthatches and similarly modified birds climb on trunks and rock walls in both head-upward and head-downward directions while foraging. In these species, which do not use their tails for support, the claw on the backward-directed toe (the hallux) is larger than those on the forward-directed toes and is strongly curved.

Perching

The most specialized avian foot for perching on branches is one in which all four toes are free and mobile and of moderate length. Three toes extend forward and are opposable to one toe that extends backward in the same plane; this is an **anisodactyl** foot. It allows a firm grip and is highly developed in the passerine birds. The zygodactylous condition, with two toes extending forward that are opposable to two toes extending backward, is characteristic of birds such as parrots and woodpeckers that climb or perch on vertical surfaces.

The tendons that flex the toes of perching birds (i.e., birds in the Passeriformes) can lock the foot in a tight grip so that the bird does not fall off its perch when it relaxes or goes to sleep. The plantar tendons, which insert on the individual phalanges of the toes, tighten when the legs bend and curl the toes around the perch. Furthermore, the tendons lying underneath the toe bones have hundreds of minute, rigid, hobnail-like projections that mesh with ridges on the inside surface of the surrounding tendon sheath. The projections and ridges lock the tendons in place in the sheaths and help to hold the toes in their grip around the branch. As long as the legs are flexed, the toes are locked around the perch and the bird cannot fall off. Indeed, the bird must extend its legs by standing up before its toes will uncurl.

Swimming on the Surface

Although no birds have become fully aquatic, nearly 400 species of birds are specialized for swimming. Nearly half of these aquatic species also dive and swim underwater. Modifications of the hindlimbs are the most obvious avian specializations for swimming. Other changes include a wide body that increases stability in water; dense plumage that provides buoyancy and insulation; a large preen gland, producing oil that waterproofs the plumage; and structural modifications of the body feathers that retard penetration of water to the skin. The legs are at the rear of a bird's body, where the mass of leg muscles interferes least with streamlining and where the best control of steering can be achieved.

The feet of aquatic birds are either webbed or lobed (**Figure 17–17**). Webbing between the three forward toes (palmate webbing) has been independently acquired at least four times in the course of avian evolution. Totipalmate webbing of all four toes is found in pelicans and their relatives. The hydrodynamic forces acting on the foot of a swimming bird are complex. At the beginning of the power stroke, when the foot is in its forward-most position, the web is nearly parallel to the water surface and moving downward. At this stage, the effect of foot movement is to lift the bird and propel it forward. Later in the stroke, however, when the web passes through the point of being perpendicular to the water surface, hydrodynamic drag is produced on the forward-facing (dorsal) side of the web by the vortex that develops as water flows around the web. At this stage, the drag is the major force propelling the bird forward.

Lobes on the toes have evolved convergently in several phylogenetic lineages of aquatic birds. There are two different types of lobed feet. Grebes are unique in

that the lobes on the outer sides of the toes are rigid and do not fold back as the foot moves forward. Instead of folding the lobes, grebes rotate the third and fourth toes 90 degrees at the beginning of the recovery stroke and move the second toe behind the tarsometatarsus. In this position, the sides of the toes slice through the water like knife blades, minimizing resistance. A simpler mechanism for the recovery stroke occurs in all the other lobe-footed swimmers, where the lobes are flaps that fold back against the toes during forward movement through water and flare open to present a maximum surface on the backward stroke.

Diving and Swimming Underwater

The transition from a surface-swimming bird to a subsurface swimmer has occurred in two fundamentally different ways: by further specialization of a hindlimb already adapted for swimming or by modification of the wing for use as a flipper under water. Highly specialized foot-propelled divers have evolved independently in grebes, cormorants, loons, and the extinct Hesperornithidae. All these families, except the loons, include some flightless forms (**Figure 17–18**). Wing-propelled divers have evolved in the Procellariiformes (the diving petrels), the Sphenisciformes (penguins), and the Charadriiformes (auks and related forms). Only among the waterfowl are there both foot-propelled and wing-propelled diving ducks, but none of these species is as highly modified for diving as specialists like the loons or auks. The water ouzels or dippers (*Cinclus*) are

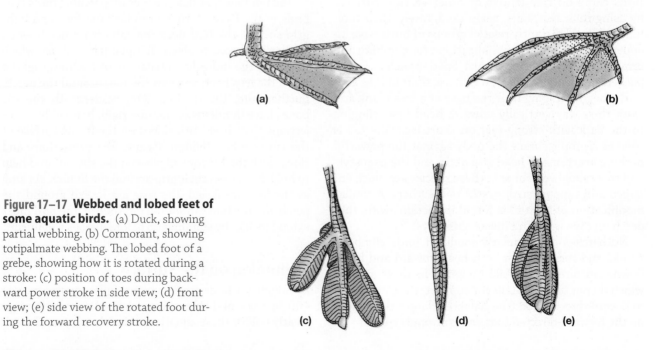

Figure 17–17 **Webbed and lobed feet of some aquatic birds.** (a) Duck, showing partial webbing. (b) Cormorant, showing totipalmate webbing. The lobed foot of a grebe, showing how it is rotated during a stroke: (c) position of toes during backward power stroke in side view; (d) front view; (e) side view of the rotated foot during the forward recovery stroke.

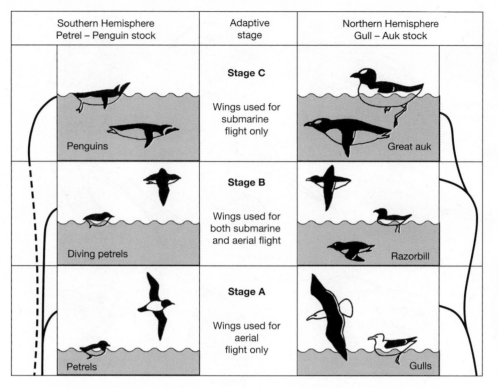

Southern Hemisphere Petrel – Penguin stock	Adaptive stage	Northern Hemisphere Gull – Auk stock
Penguins	**Stage C** Wings used for submarine flight only	Great auk
Diving petrels	**Stage B** Wings used for both submarine and aerial flight	Razorbill
Petrels	**Stage A** Wings used for aerial flight only	Gulls

Figure 17–18 Parallel evolution of swimming and diving birds. Petrels and penguins (Procellariiformes and Sphenisciformes) occupied this niche in the Southern Hemisphere. In the Northern Hemisphere gulls and auks (Charadriiformes) developed parallel specializations.

passerine birds that dive and swim underwater with great facility using their small, round wings, but they lack any other morphological specializations.

17.5 Feeding and Digestion

With the specialization of the forelimbs as wings, which largely prevents any substantial role in capturing prey, birds have concentrated their predatory mechanisms in their beaks and feet. Modifications of the beak, tongue, and intestines are often associated with dietary specializations.

Beaks and Tongues

The presence of a horny beak in place of teeth is not unique to birds—turtles have beaks, and so did the rhynchosaurs, many dinosaurs, pterosaurs, and the dicynodonts. However, the diversity of beaks among birds is remarkable. The range of morphological specializations of beaks defies complete description, but some categories can be recognized (**Figure 17–19**).

Insectivorous birds such as warblers—which find their food on leaf surfaces—usually have short, thin, pointed bills that are adept at seizing insects. Aerial sweepers such as swifts, swallows, and nighthawks—which catch their prey on the wing—have short, weak beaks and a wide gape.

Many carnivorous birds, such as gulls, ravens, crows, and roadrunners, use their heavy pointed beaks to kill their prey. On the other hand, most hawks, owls, and eagles kill prey with their talons and use their beaks to tear off pieces small enough to swallow. Falcons stun prey with the impact of their dive, and then bite the neck of the prey to disarticulate the cervical vertebrae. Fish-eating birds such as cormorants and pelicans have beaks with a sharply hooked tip that is used to seize fish, while mergansers have long, narrow bills with a series of serrations along the sides of the beak, in addition to a hook at the tip. Darters and anhingas have harpoonlike bills that they use to impale fish.

Seeds are usually protected by hard coverings (husks) that must be removed before the nutritious contents can be eaten. Specialized seed-eating birds use one of two methods to husk seeds before swallowing them. One group holds the seed in its beak and slices it by making fore-and-aft movements of the lower jaw. Birds in the second group hold the seed against ridges on the palate and crack the husk by exerting an upward pressure with their robust lower jaw. After the husk has been opened, both kinds of birds use their tongues to remove the contents. Other birds have different specializations for eating seeds: Crossbills extract the seeds of conifers from between the scales of the cones, using the diverging tips of their

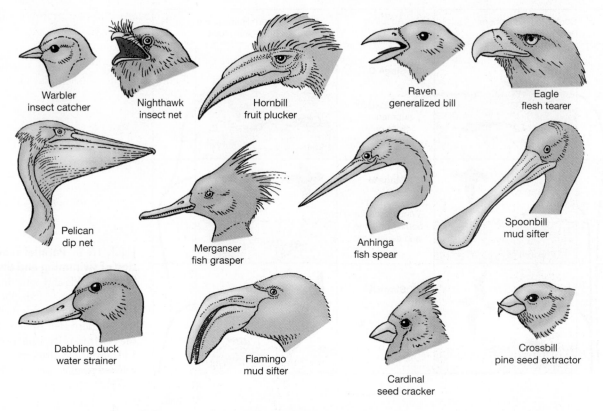

Figure 17–19 Examples of specializations of the beaks of birds.

bills to pry the scales apart and a prehensile tongue to capture the seed inside. Woodpeckers, nuthatches, and chickadees may wedge a nut or acorn into a hole in the bark of a tree and then hammer at it with their sharp bills until it cracks.

The skulls of most birds consist of four bony units that can move in relation to each other. This skull kinesis is important in some aspects of feeding. The upper jaw flexes upward as the mouth is opened, and the lower jaw expands laterally at its articulation with the skull (**Figure 17–20a**). The flexion of the upper and lower jaws increases the bird's gape in both the vertical and horizontal planes and probably assists in swallowing large items. Many birds use their beaks to search for hidden food. Long-billed shorebirds probe in mud and sand to locate worms and crustaceans. These birds

Figure 17–20 Avian skull and jaw kinesis. (a) A yawning herring gull (*Larus argentatus*) shows the kinetic movements of the skull and jaws that occur during swallowing. White arrows show the positions of outward flexion; black arrows show inward flexion. (b) Long-billed wading birds that probe for worms and crustaceans in soft substrates can raise the tips of the upper bill without opening their mouth.

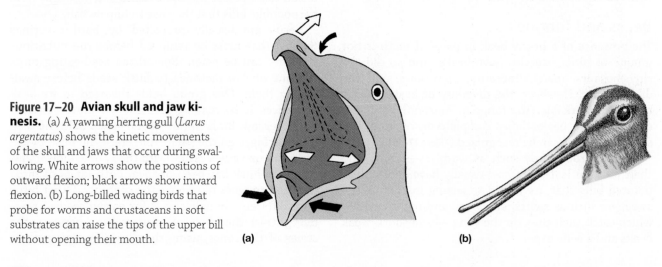

(a)

(b)

display a form of skull kinesis in which the flexible zone in the upper jaw has moved toward the tip of the beak, allowing the tip of the upper jaw to be lifted without opening the mouth (Figure 17–20b). This mechanism enables long-billed waders to grasp prey under the mud.

Tongues are an important part of the food-gathering apparatus of many birds. Woodpeckers drill holes into dead trees and then use their long tongues to investigate passageways made by wood-boring insects. The tongue of the green woodpecker, which extracts ants from their tunnels in the ground, extends four times the length of its beak. The hyoid bones that support the tongue are elongated and housed in a sheath of muscles that passes around the outside of the skull and rests in the nasal cavity (**Figure 17–21**). When the muscles of the sheath contract, the hyoid bones are pushed around the back of the skull and the tongue is projected from the bird's mouth. The tip of a woodpecker tongue has barbs that impale insects and allow them to be pulled from their

tunnels. Nectar-eating birds, such as hummingbirds and sunbirds, also have long tongues and a hyoid apparatus that wraps around the back of the skull. The tip of the tongue of nectar-eating birds is divided into a spray of hair-thin projections, and capillary force causes nectar to adhere to the tongue.

The Digestive System

The digestive systems of birds show some differences from those of other vertebrates. The absence of teeth prevents birds from doing much processing of food in the mouth, and the gastric apparatus takes over some of that role.

Esophagus and Crop Birds often gather more food than they can process in a short time, and the excess is held in the esophagus. Many birds have a **crop,** an enlarged portion of the esophagus that is specialized for temporary storage of food. The crop of some birds is a simple expansion of the esophagus, whereas in others it is a unilobed or bilobed structure (**Figure 17–22**). An additional function of the crop is transportation of food for nestlings. As the foraging adult gathers food, it stores it in the crop. When the adult returns to the nest, it regurgitates the material from its crop and feeds it to the young. In doves and pigeons, the crop of both sexes produces a nutritive fluid (crop milk) that is fed to the young. The milk is produced by fat-laden cells that detach from the squamous epithelium of the crop and are suspended in an aqueous fluid. Crop milk is rich in lipids and proteins but contains no sugar. Its chemical composition is similar to that of mammalian milk, although it differs in containing intact cells.

Stomach The form of the stomachs of birds is related to their dietary habits. Carnivorous and piscivorous (fish-eating) birds need expansible storage areas to accommodate large volumes of soft food, whereas birds that eat insects or seeds require a muscular organ that can contribute to the mechanical breakdown of food. Usually, the gastric apparatus of birds consists of two relatively distinct chambers: an anterior glandular stomach (**proventriculus**) and a posterior muscular stomach (**gizzard**). The proventriculus contains glands that secrete acid and digestive enzymes. The proventriculus is especially large in species that swallow large items such as intact fruit.

The gizzard has several functions, including food storage while the chemical digestion that was begun in the proventriculus continues, but its most important function is the mechanical processing of food. The thick, muscular walls of the gizzard squeeze the contents, and small stones that are held in the gizzards of

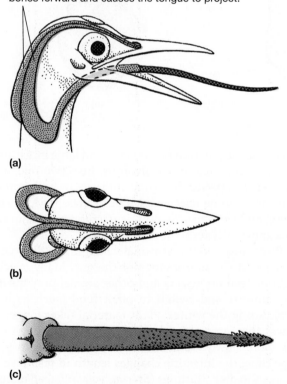

Elongate hyoid bones are enclosed by circular muscles. Contraction of the muscles pushes the bones forward and causes the tongue to project.

(a)

(b)

(c)

Figure 17–21 Tongue projection mechanism of a wood-pecker. The tongue itself is about the length of the bill, and it can be extended well beyond the tip of the beak by muscles that squeeze the posterior ends of the elongated hyoid apparatus, forcing the tongue forward. The detail of the tongue shows the barbs on the tip that impale prey.

Figure 17–22 Anterior digestive tract of birds. (a) The relationship among the parts. The relative sizes of the proventriculus and gizzard vary in relation to diet. Carnivorous and fish-eating birds, like the great cormorant, have a relatively small crop (b) and gizzard (c), whereas seed eaters and omnivores like the peafowl have a large crop (d) and muscular gizzard (e).

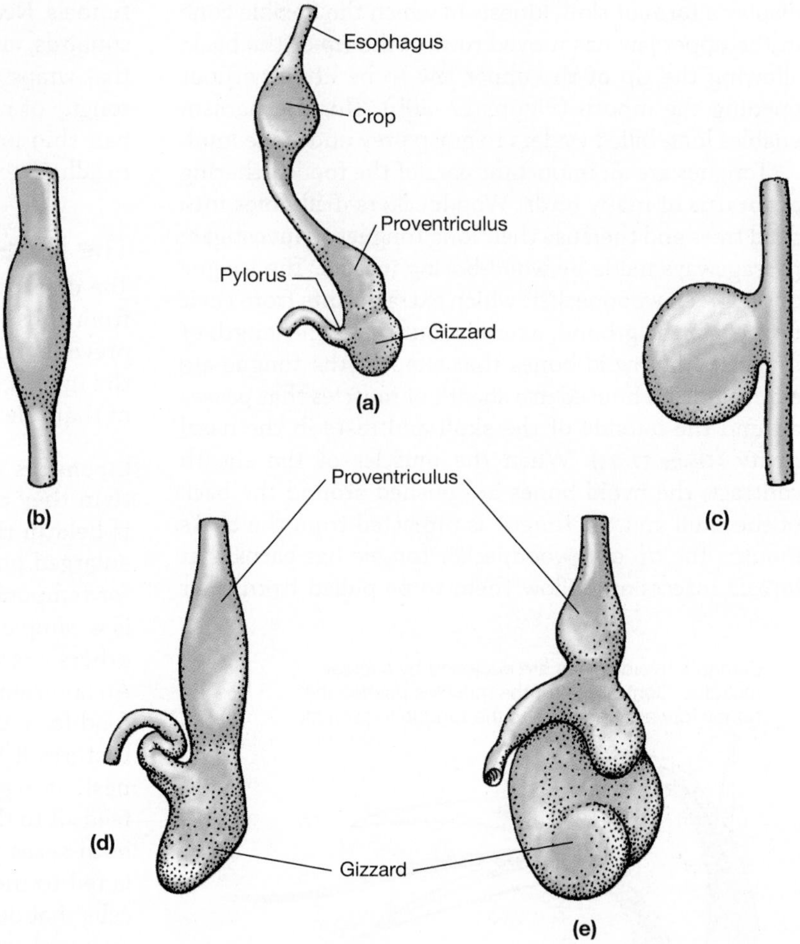

many birds help to grind the food. In this sense, the gizzard is performing the same function as a mammal's teeth. The pressure that can be exerted on food in the gizzard is intense. A turkey's gizzard can grind two dozen walnuts in as little as 4 hours, and it can crack hickory nuts that require as much as 150 kilograms of pressure to break.

Intestine, Ceca, and Cloaca The small intestine is the principal site of chemical digestion, as enzymes from the pancreas and intestine break down the food into small molecules that can be absorbed across the intestinal wall. The mucosa of the small intestine is modified into a series of folds, lamellae, and villi that increase its surface area. The large intestine is relatively short, usually less than 10 percent of the length of the small intestine. Passage of food through the intestines of birds is quite rapid: transit times for carnivorous and fruit-eating species are in the range of a few minutes to a few hours. Passage of food is slower in herbivores and may require a full day. Birds generally have a pair

of ceca at the junction of the small and large intestines. The ceca are small in carnivorous, insectivorous, and seed-eating species, but they are large in herbivorous and omnivorous species such as cranes, fowl, ducks, geese, and ostriches. Symbiotic microorganisms in the ceca apparently ferment plant material.

Birds respond to seasonal changes in diet with changes in the morphology of the gut. Many species of birds feed on insects and other animal prey during the summer and switch to plant food (such as berries) during the winter. Plant material takes longer to digest than animal food, so it must pass through the gut more slowly. To accommodate differences in passage time, the intestine changes length in response to changes in diet. Starlings (*Sturnus vulgaris*) feed mainly on insects from March through June and add progressively more plant material to their diet from late summer through winter. The length of the intestine shows a corresponding cycle, increasing in length by about 20 percent during fall and winter and decreasing by the same amount during spring and early summer. In addition to the anatomical changes in the intestine, the digestive enzymes change to match the reduction in protein and fat and the increase in simple sugars in fruit compared to animal food.

The cloaca temporarily stores waste products while water is being reabsorbed. The precipitation of uric acid in the form of urates (i.e., the potassium and sodium salts of uric acid) frees water from the urine, and this water is returned to the bloodstream. Species of birds that have salt-secreting glands can accomplish further conservation of water by reabsorbing some of the ions that are in solution in the cloaca and excreting them in more concentrated solutions through the salt gland. The mixture of white urates and dark fecal material that is voided by birds is familiar to anyone who has washed an automobile.

17.6 Sensory Systems

A flying bird moves rapidly through three-dimensional space and requires a continuous flow of sensory information about its position and the presence of obstacles in its path. Vision is the sense best suited to provide this sort of information on a rapidly renewed basis, and birds have a well-developed visual system that remains active when the bird is asleep. The importance of vision is reflected in the brain: the optic lobes are large, and the midbrain is an important area for processing visual and auditory information. Olfaction is relatively unimportant for most birds, and the olfactory lobes are small. The cerebellum, which coordinates body movements, is large. The cerebrum is less developed in birds than it is in mammals and is dominated by the corpus striatum.

Vision

The eyes of birds are large—so large that the brain is displaced dorsally and caudally, and in many species the eyeballs meet in the midline of the skull. The eyes of some hawks, eagles, and owls are as large as the eyes of humans. In its basic structure, the eye of a bird is like that of any other vertebrate, but the shape varies from a flattened sphere to something approaching a tube (**Figure 17–23**). An analysis of the optical characteristics of birds' eyes suggests that these differences are primarily the result of fitting large eyes into small skulls. The eyes of a starling are small enough to be contained within the skull, whereas the eyes of an owl bulge out of the skull. An owl would require an enormous, unwieldy head to accommodate a flat eye like that of a starling. The tubular shape of the owl's eye allows it to fit into a reasonably sized skull.

The pecten is a conspicuous structure in the eye of birds. It is formed by blood capillaries surrounded by pigmented tissue and covered by a membrane; it lacks muscles and nerves. The pecten arises from the retina at the point where the nerve fibers from the ganglion cells of the retina join to form the optic nerve. In some species of birds the pecten is small, but in other species the pecten extends so far into the vitreous humor of the eye that it almost touches the lens. The function of the pecten remains uncertain after 200 years of debate. Proposed functions include reduction of glare, a mirror to reflect objects above the bird, production of a stroboscopic effect, and a visual reference point—but none of these seems very likely. The large blood supply flowing to the pecten suggests that it may provide nutrition for the retinal cells and help to remove metabolic waste products that accumulate in the vitreous humor.

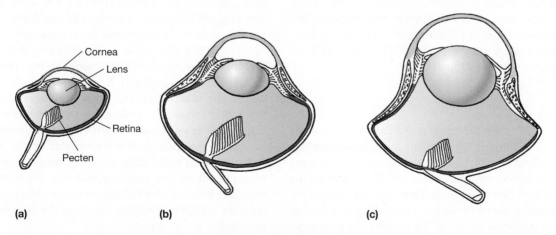

(a) **(b)** **(c)**

Figure 17–23 Variation in the shape of the eye of birds. (a) Flat, typical of most birds. (b) Globular, found in most falcons. (c) Tubular, characteristic of owls and some eagles.

The world must look different to birds than it does to us because most birds have four photosensitive pigments in their retinal cells, whereas humans and other primates have only three. In other words, birds have tetrachromatic vision, whereas we have trichromatic vision. Three of the retinal pigments of birds absorb maximally at wavelengths equivalent to the ones humans perceive—in the red, green, and blue regions of the spectrum. The fourth pigment present in birds ancestrally responds to wavelengths of about 400 nanometers (deep blue). In some species of birds, a mutation in the gene that codes for the deep blue pigment shifted its maximum sensitivity into the ultraviolet, thereby extending the visual sensitivity of birds beyond the range of the human eye. Ultraviolet sensitivity has evolved independently several times among passerines, but it is rare among seabirds.

Sensitivity to wavelengths in the ultraviolet portion of the spectrum is important in many aspects of the behavior of birds. Ultraviolet reflectance distinguishes males from females in some species of birds in which the sexes are indistinguishable to human eyes, indicates the physiological status of prospective mates, and demonstrates the health of nestlings. The colors and patterns of eggs may help parent birds to distinguish their own eggs from the eggs of nest parasites, such as cuckoos, and ultraviolet reflectance may enhance the effectiveness of the warning colors of toxic prey, such as dart-poison frogs.

Oil droplets are found in the cone cells of the avian retina, as they are in other sauropsids. The droplets range in color from red through orange and yellow to green. The oil droplets are filters, absorbing some wavelengths of light and transmitting others. The function of the oil droplets is unclear; it is certainly complex because the various colors of droplets are combined with the four visual pigments. Birds like gulls, terns, gannets, and kingfishers that must see through the surface of water have a preponderance of red droplets. Aerial hawkers of insects (swifts and swallows) have predominantly yellow droplets.

Hearing

In birds, as in other sauropsids, the columella (stapes) and its cartilaginous extension, the extracolumella, transmit vibrations of the tympanum to the oval window of the inner ear. The cochlea of birds appears to be specialized for fine distinctions of the frequency and temporal pattern of sound. The cochlea of a bird is about one-tenth the length of the cochlea of a mammal, but it has about 10 times as many hair cells per unit of length. The space above a bird's basilar membrane (the scala vestibuli) is nearly filled by a folded, glandular tissue. This structure may dampen sound waves, allowing the ear of a bird to respond very rapidly to changes in sounds.

Localization of sounds in space can be difficult for small animals such as birds. Large animals localize the source of sounds by comparing the time of arrival, intensity, or phase of a sound in their left and right ears, but none of these methods is very effective when the distance between the ears is small. The pneumatic construction of the skulls of birds may allow them to use sound that is transmitted through the air-filled passages between the middle ears on the two sides of the head to increase their directional sensitivity. If this is true, then internally transmitted sound would pass from the middle ear on one side to the middle ear on the other side and reach the *inner* surface of the contralateral tympanum. There it would interact with the sound arriving on the external surface of the tympanum via the external auditory meatus. The vibration of each tympanum would be the product of the combination of pressure and phase of the internal and external sources of sound energy, and the magnitude of the cochlear response would be proportional to the difference in pressure across the tympanic membrane.

The sensitivity of the auditory system of birds is approximately the same as that of humans, despite the small size of birds' ears. Most birds have tympanic membranes that are large in relation to the head size. A large tympanic membrane enhances auditory sensitivity, and owls (which have especially sensitive hearing) have the largest tympani relative to their head size among birds. Sound pressures are amplified during transmission from the tympanum to the oval window of the cochlea because the area of the oval window is smaller than the area of the tympanum. The reduction ratio for birds ranges from 11:1 to 40:1. High ratios suggest sensitive hearing, and the highest values are found in owls; songbirds have intermediate ratios (20:1 to 30:1). (The ratio is 36:1 for cats and 21:1 for humans.) The inward movement of the tympanum as sound waves strike it is opposed by air pressure within the middle ear, and birds show a variety of features that reduce the resistance of the middle ear. The middle ear is continuous with the dorsal, rostral, and caudal air cavities in the pneumatic skulls of birds. In addition to potentially allowing sound waves to be transmitted to the contralateral ear, these interconnections increase the volume of the middle ear and reduce its stiffness, thereby allowing the tympanum to respond to faint sounds.

Owls are acoustically the most sensitive of birds. At frequencies up to 10 kilohertz (10,000 cycles per second), the auditory sensitivity of an owl is as great as that of a cat. Owls have large tympanic membranes, large cochleae, and well-developed auditory centers in the brain. Some owls are diurnal, others are crepuscular (active at dawn and dusk), and some are entirely nocturnal. In an experimental test of their capacities for acoustic orientation, barn owls were able to seize mice in total darkness. If the mice towed a piece of paper across the floor behind them, the owls struck the rustling paper instead of the mouse, showing that sound was the cue they were using.

A distinctive feature of many owls is the facial ruff, which is formed by stiff feathers (**Figure 17–24a**). The ruff acts as a parabolic sound reflector, focusing sounds with frequencies higher than 5 kilohertz on the external auditory meatus and amplifying them by 10 decibels. The ruffs of some owls are asymmetrical, and that asymmetry appears to enhance the ability of owls to locate prey because barn owls made large errors in finding targets when their ruffs were removed.

Asymmetry of the aural system of owls goes beyond the feathered ruff. The skull of many species of owls is markedly asymmetrical, and these are the species that have the greatest auditory sensitivity (Figure 17–24b). The asymmetry ends at the external auditory meatus; the middle and inner ears of owls are bilaterally symmetrical. The asymmetry of the external ear openings of owls assists with localization of prey in the horizontal and vertical axes. Minute differences in the time and intensity at which sounds are received by the two ears indicate the direction of the source. The brains of owls integrate time and intensity information with extraordinary sensitivity to produce a map of their environment that integrates auditory and visual information.

Olfaction

The sense of smell is well developed in some birds but poorly developed in most species. The size of the olfactory bulbs is a rough indication of the sensitivity of the olfactory system. Relatively large bulbs are found in ground-nesting and colonial-nesting species, species that are associated with water, and carnivorous and piscivorous species of birds. Some birds use scent to locate prey. The kiwi, for example, has nostrils at the tip of its long bill and finds earthworms underground by smelling them. Turkey vultures follow airborne odors of carrion to the vicinity of a carcass, which they then locate by sight. Shearwaters, fulmars, albatrosses, and petrels were attracted to sponges that had been soaked in fish oil and placed on floating buoys, and they could find the sponges even at night.

Olfaction probably plays a role in the orientation and navigation abilities of some birds. The tube-nosed

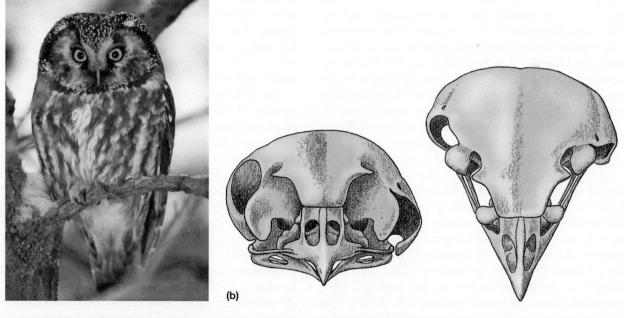

(a)   (b)

Figure 17–24 Anatomical features of owls that increase their auditory sensitivity. (a) A boreal owl (*Aegolius funereus*) showing the asymmetry of its facial ruff. (b) The skull of a boreal owl. The pronounced asymmetry in the position of the ear opening (the external auditory meatus) is characteristic of owls. This asymmetry assists in localization of sound.

seabirds (albatrosses, shearwaters, fulmars, and petrels) nest on islands, and when they return from foraging at sea, they approach the islands from downwind. Homing pigeons use olfaction (as well as other mechanisms) to navigate. Surprisingly, when the right nostrils of homing pigeons were plugged, they had more difficulty finding their way back to their home lofts than did pigeons that had their left nostrils plugged.

17.7 **Social Behavior**

Vision and hearing are the major sensory modes of birds, as they are of humans. One result of this correspondence has been that birds play an important role in behavioral studies. Most birds are active during the day and are thus relatively easy to observe. A tremendous amount of information has been accumulated about the behavior of birds under natural conditions. This background has contributed to the design of experimental studies in the field and in the laboratory.

Colors and Patterns

The colors and patterns of a bird's plumage identify its species and frequently its sex and age as well. Three types of pigments are widespread in birds. Dark colors are produced by melanin—eumelanins produce black, gray, and dark brown, whereas phaeomelanins are responsible for reddish brown and tan shades. Carotenoid pigments are responsible for most red, orange, and yellow colors. Birds obtain these pigments from their diet, and in some cases the intensity of color can be used to gauge the fitness of a prospective mate. Porphyrins are metal-containing compounds chemically similar to the pigments in hemoglobin and liver bile. Ultraviolet light causes porphyrins to emit a red fluorescence. Porphyrins are destroyed by prolonged exposure to sunlight, so they are most conspicuous in new plumage.

Structural colors (purples, blues, and green) result from tiny air-filled structures in the cells on the surface of feather barbs. When daylight is scattered by these structures some wavelengths show constructive interference (i.e., an intensification of that wavelength), whereas other wavelengths show destructive interference (reduction in intensity). The spacing of the air-filled structures determines which colors are intensified. Structural colors can be combined with pigments—green parakeets combine a structural blue with a yellow carotenoid. Blue parakeets are homozygous for an allele that blocks formation of the carotenoid. That allele is a simple recessive, so two blue parakeets can produce only blue offspring.

Some birds use colored areas on the skin, bills, and feet as part of their pattern, and these colors are also produced by the scattering of daylight in a way that causes constructive and destructive interference. Ultraviolet, blue, and green are usually produced by constructive or destructive interference of refracted light by collagen fibers in the dermis and are called structural colors. Yellow, orange, and red colors are produced either by pigments or by a combination of pigments and structural colors.

Vocalization, Sonation, and Visual Displays

Birds use sounds, colors, and movements to recognize other individuals as well as their species, sex, and age. Two categories of sounds play a role in the behavior of birds—vocalization and sonation. Birds produce vocalizations by passing air through the syrinx, a structure that lies at the junction of the trachea and the two bronchi. Sonations are nonvocal sounds that are usually created by feathers knocking against each other or by air moving between or across feathers.

Birdsong The term *birdsong* has a specific meaning that is distinct from a bird call. The song is usually the longest and most complex vocalization produced by a bird. In many species, songs are produced only by mature males and only during the breeding season. Song is controlled by a series of song control regions (SCRs) in the brain. During the period of song learning, which occurs early in life, new connections form between a part of the SCR that is associated with song learning and a region of the brain that controls the vocal muscles. Thus, song learning and song production are closely linked in male birds.

The SCRs are under hormonal control, and in many species of birds, the SCRs of males are larger than those of females and have more and larger neurons and longer dendritic processes. The vocal behavior of female birds varies greatly across taxonomic groups: in some species females produce only simple calls, whereas in other species the females engage with males in complex song duets. The SCRs of females of the latter species are very similar in size to those of males (**Table 17–3**). The function of the SCR in female birds of species in which females do not vocalize has been unclear, but recent experiments suggest that it plays a role in species recognition. When the SCR of female canaries was inactivated, the birds no longer distinguished the vocalizations of male canaries from those of sparrows.

A birdsong consists of a series of notes with intervals of silence between them. Changes in frequency (frequency modulation) are conspicuous components

Table 17–3 Sexual dimorphism in the song control regions (SCR) of the brains of birds. The average ratio of the volumes of the five SCRs in males compared to females (male:female) parallels the difference in the sizes of the song repertoires of males and females.

	Zebra Finch	Canary	Chat	Bay Wren	Buff-Breasted Wren
SCR volume ratio	4.0:1.0	3.1:1.0	2.3:1.0	1.3:1.0	1.3:1.0
Song repertoire	Males only	Males very much greater than females	Males much greater than females	Males the same as females	Males the same as females

of the songs of many birds, and the avian ear may be very good at detecting rapid changes in frequency. Birds often have more than one song type, and some species may have repertoires of several hundred songs.

Birdsongs identify the particular species of bird that is singing, and they often show regional dialects. These dialects are transmitted from generation to generation as young birds learn the songs of their parents and neighbors. In the indigo bunting, one of the best-studied species, song dialects that were characteristic of small areas persisted for as long as 15 years, which is substantially longer than the life of an individual bird. Birdsongs also show individual variation that allows birds to recognize the songs of residents of adjacent territories and to distinguish the songs of these neighbors from those of intruders. Male hooded warblers remember the songs of neighboring males—and recognize them as individuals when they return to their breeding sites in North America after spending the winter in Central America.

The songs of male birds identify their species, sex, and occupancy of a territory. When a recording of a male's song is played back through a speaker that has been placed in the territory of another male, the territorial male responds with vocalizations and aggressive displays, and he may even attack the speaker. These behaviors repel intruders, and even hearing the song of a territorial male keeps intruders at a distance—broadcasting recorded songs in a territory from which the territorial male has been removed delays occupation of the vacant territory by a new male.

Nonvocal Displays In contrast to the precise definition of birdsong, the words *drumming*, *clapping*, and *whistling* have been used to describe the sounds that birds make by striking objects with their beaks or wings and the sounds made as air passes over or through their feathers. These sounds play central roles in the social behavior of many species of birds.

Woodpeckers as well as other species of birds make a drumming sound by pounding with their bills on a resonant object. Drumming could be a territorial signal, informing other woodpeckers that a stand of trees already had a resident, and it could advertise the presence of a male to unmated female woodpeckers. Storks and herons clap their mandibles together, and wing-clapping has been described in many species of birds.

Some species of birds have feathers that are modified to produce sounds. Male club-winged manakins (*Machaeropterus deliciosus*) produce a *tick-ticking* sound during their displays. The rachi of the sixth and seventh secondaries of males are thickened toward their bases and bent, and the rachi of secondaries 1 through 5 are also thickened. During a male's display, these seven feathers vibrate as a unit, producing a vibration with a fundamental frequency of about 1500 hertz, a harmonic with a frequency of about 3000 hertz, and an amplitude higher than that of the fundamental frequency. In contrast to the club-winged manakins, which use muscular movements of their wings to produce sound, male Anna's hummingbirds emit a loud *chirp* by spreading the tail as they dive toward a female. Air flowing across the surface of the vane of the outermost tail feathers causes them to flutter, producing a sound with a fundamental frequency of about 4000 hertz.

Visual Displays Visual displays are frequently associated with songs; for example, a particular body posture that displays colored feathers may accompany singing. Male birds are often more brightly colored than females and have feathers that have become modified as the result of sexual selection. In this process, females mate preferentially with males that have certain physical characteristics. Because of that response by females, those physical characteristics contribute to the reproductive fitness of males, even though they may have no useful function in any other aspect of the animal's ecology or behavior. The colorful areas on the wings of male ducks, the red epaulets on the wings of male red-winged blackbirds, the red crowns on kinglets, and the elaborate tails of male peacocks are familiar examples of specialized areas of plumage that are involved in sexual behavior and display.

17.8 Mating Systems

The mating systems of vertebrates are believed to reflect the distribution of food, breeding sites, and potential mates. The key factor is whether one sex can control resources that the other sex needs. For example, a male could potentially increase his opportunities to mate by defending resources, such as food and nest sites, that females need. If he can exclude other males, he can mate with all the females in the area. However, a male's ability to control access to resources depends on how the resources are distributed in the habitat:

- If resources are more or less evenly distributed, it is unlikely that a male could control a large enough area to monopolize many females. Under those conditions, all males will have access to the resources and to the females.
- If resources are clumped in space, with barren areas between the patches, the females will aggregate in the resource patches, and it will be possible for a male to monopolize several females by defending a patch. Males that are able to defend good patches should have the opportunity to mate with more females than males defending patches of lower quality.

Monogamy, Polygamy, and Extra-Pair Copulations

Perhaps the most important incentive for pair formation for many species of birds is the need for attendance by both parents to raise a brood to fledging (i.e., leaving the nest). Dramatic examples include situations in which continuous nest attendance by one parent is necessary to protect the eggs or chicks from predators while the other parent forages for food. This situation is commonly observed in seabirds, which nest in dense colonies that sometimes include mixtures of two or more species. In the absence of an attending parent, neighbors raid the nest and kill the eggs or chicks. The male and female alternate periods of nest attendance and foraging, and some species engage in elaborate displays when the parents switch duties (**Figure 17–25**).

Social vertebrates exhibit one of two broad categories of mating systems—monogamy or polygamy. **Monogamy** (Greek *mono* = one and *gamy* = marriage) refers to a pair bond between a single male and a single female. The pairing may last for part of a breeding season, an entire season, or a lifetime. **Polygamy** (Greek *poly* = many) refers to a situation in which an individual has more than one mate in a breeding season.

Polygamy can be exhibited by males, females, or both sexes. In **polygyny** (Greek *gyn* = female), a male

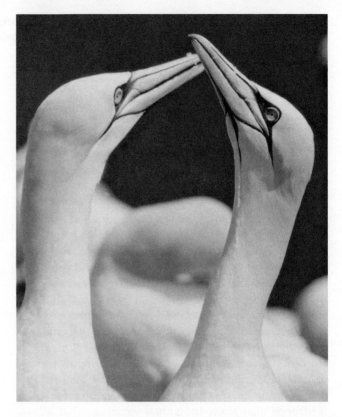

Figure 17–25 Nest exchange display of northern gannets (*Morus bassanus*). The birds engage in an elaborate ritual when one member of the pair returns to the nest after foraging.

mates with more than one female; in **polyandry** (Greek *andr* = male), a female mates with two or more males. **Promiscuity** is a mixture of polygamy and polyandry in which both males and females mate with several different individuals.

Monogamy has been considered the most widespread social system of birds. Both parents in monogamous mating systems usually participate in caring for the young. Monogamy does not necessarily mean fidelity to a mate, however. Genetic studies of monogamous birds have shown that extra-pair copulation (mating with a bird other than the partner) is common. Thus, some of the eggs in a nest may have been fertilized by a male other than the partner of the female who deposited them, and the male partner may have fertilized eggs of other females that are being incubated in their nests. As evidence of extra-pair copulation has accumulated, the term **social monogamy** has been introduced for species in which a male and female share responsibility for a clutch of eggs but do not demonstrate fidelity. **Genetic monogamy** describes a social system in which a male and female share parental responsibilities and do not have extra-pair copulations.

Male Incubation and Polyandry

In polyandrous mating systems, it is the females that control a limited resource, and they use that control to gain access to multiple males. This pattern of breeding among birds seems to be typical of situations in which the cost of each reproductive effort for the female is low (because food is abundant and a clutch contains only a few small eggs) and the probability of successful fledging is small. Spotted sandpipers (*Actitis macularia*) provide an example: Female spotted sandpipers are larger than males, and females compete with other females to establish large nesting territories. A female sandpiper initially mates monogamously with a single male and usually shares parental care. As more males arrive, however, female sandpipers in some populations abandon their first mates and mate with up to three more males.

17.9 Oviparity

Although birds have a great diversity of mating strategies, their mode of reproduction is limited to laying eggs. No other group of vertebrates that contains such a large number of species is exclusively oviparous. Why is this true of birds?

Constraints imposed on birds by their specializations for flight are often invoked to explain their failure to evolve viviparity. Those arguments are not particularly convincing, however, considering that bats have successfully combined flight and viviparity. Furthermore, flightlessness has evolved in at least 15 families of birds, but none of these flightless species has evolved viviparity.

Oviparity is presumed to be the ancestral reproductive mode for archosaurs, and it is retained by both extant groups of archosaurs—crocodilians and birds. However, viviparity evolved in the marine reptiles of the Mesozoic and it has evolved nearly 100 times in the other major lineage of extant sauropsids, the lizards and snakes (lepidosaurs), so the capacity for viviparity is clearly present in sauropsids—why not in birds?

A key element in the evolution of viviparity among lizards and snakes appears to be the retention of eggs in the oviducts of the female for some period before they are deposited. This situation occurs when the benefits of egg retention outweigh its costs. For example, the high incidence of viviparity among snakes and lizards in cold climates may be related to a female ectotherm's ability to speed embryonic development by thermoregulation. A lizard that basks in the sun can raise the temperature of eggs retained in her body and perhaps determine the sex of her offspring, but after depositing the eggs in a nest, she has no control over their temperature and rate of development. Birds are endotherms and can control egg temperature by brooding the eggs. Thus, the reasoning goes, egg retention provides no thermoregulatory advantage for birds so they never took that first step toward viviparity.

Egg Biology

The inorganic part of eggshells contains about 98 percent crystalline calcite, CaCO3, and the embryo obtains about 80 percent of its calcium from the eggshell. An organic matrix of protein and mucopolysaccharides is distributed throughout the shell and may serve as a support structure for the growth of calcite crystals. Eggshell formation begins in the isthmus of the oviduct (**Figure 17–26**). Two shell membranes are secreted to enclose the yolk and albumen, and carbohydrate and water are added to the albumen by a process that involves active transport of sodium across the wall of the oviduct followed by osmotic flow of water. The increased volume of the egg contents at this stage appears to stretch the egg membranes taut. Organic granules are attached to the egg membrane, and these **mammillary bodies** appear to be the sites of the first formation of calcite crystals (**Figure 17–27**). Some crystals grow downward from the mammillary bodies and fuse to the egg membranes, and other crystals grow away from the membrane to form cones. The cones grow vertically and expand horizontally, fusing with crystals from adjacent cones to form the palisade layer.

An eggshell is penetrated by an array of pores that allow oxygen to diffuse into the egg and carbon dioxide and water to diffuse out (**Figure 17–28** on page 437). Pores occur at the junction of three calcite cones, but only 1 percent or less of those junctions form pores; the rest are fused shut. Pores occupy about 0.02 percent of the surface of an eggshell. The morphology of the pores varies in different species of birds: some pores are straight tubes, whereas others are branched, and the pore openings on the eggshell's surface may be blocked to varying degrees with organic or crystalline material—all of these factors affect the rate of evaporation of water from the egg.

Bird eggs must lose water to hatch because the loss of water creates an air cell at the blunt end of the egg. The embryo penetrates the membranes of this air cell with its beak 1 or 2 days before hatching begins, and ventilation of the lungs begins to replace the chorioallantoic membrane in gas exchange. Pipping, the formation of the first cracks on the surface of the eggshell, follows about half a day after penetration

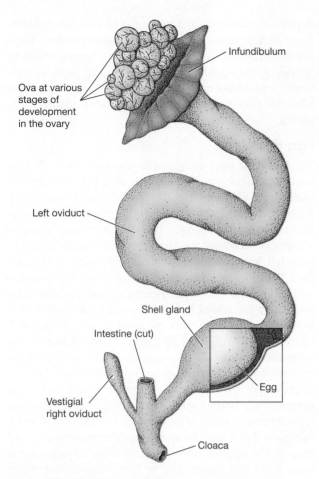

Figure 17–26 Oviduct of a bird. Ova released from the ovary enter the infundibulum of the oviduct. Fertilization occurs at the upper end of the oviduct, then albumen and shell membranes are secreted, and finally the egg is enclosed in calcareous shell. Only the left ovary is functional in most species of birds.

of the air cell, and actual emergence begins half a day later. Shortly before hatching, the chick develops a horny projection on its upper jaw. This structure is called the egg tooth, and it is used in conjunction with a hypertrophied muscle on the back of the neck (the hatching muscle) to thrust vigorously against the shell. The egg tooth and hatching muscle disappear soon after the chick has hatched.

Sex Determination

All birds have heterogametic sex chromosomes, as do mammals, except that in birds the female is the heterogametic sex and the male is homogametic. To avoid confusion with mammals, the sex chromosomes of female birds are designated ZW and male birds are ZZ.

The presence of the W sex chromosome causes the primordial gonad to secrete estrogen, which stimulates the left gonad to develop as an ovary and the left Müllerian duct system to develop into an oviduct and shell gland. (Only the left ovary and oviduct normally develop in birds.) In the absence of estrogen (i.e., when the genotype is ZZ), a male develops. (This is the opposite of the sex determination process in mammals, in which a male-determining gene on the Y chromosome causes an XY individual to develop as a male.)

The evolutionary origin of genetic sex determination (GSD) in birds is unclear because all crocodilians have temperature-dependent sex determination (TSD) and lack heterogametic sex chromosomes. Most birds maintain relatively stable egg temperatures by incubating their eggs during embryonic development, but a lineage of galliform birds, the Megapodiidae, that occurs in the Indoaustralian region is an exception. This family includes about 20 species of ground-dwelling birds that bury their eggs like crocodilians rather than incubating them as birds do. The chicks are fully developed when they hatch and are not dependent on their parents. The Australian brush turkey (*Alectura lathami*) builds a mound of soil and vegetation in which the eggs are deposited (**Figure 17–29**). When eggs of the brush turkey are incubated at 31°C, about 75 percent

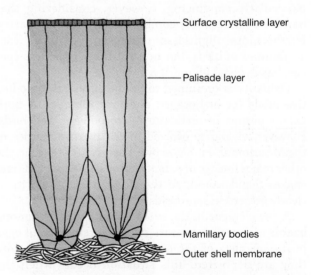

Figure 17–27 Diagram of the crystal structure of an avian eggshell. Crystallization begins at the mammillary bodies. Crystals grow into the outer shell membrane and upward to form the palisade layer. Changes in the chemical composition of the fluid surrounding the growing eggshell are probably responsible for the change in crystal form in the surface crystalline layer.

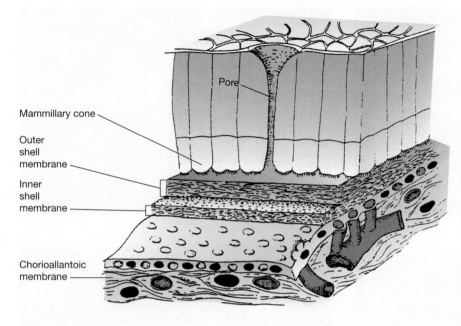

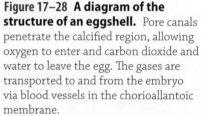

Figure 17–28 A diagram of the structure of an eggshell. Pore canals penetrate the calcified region, allowing oxygen to enter and carbon dioxide and water to leave the egg. The gases are transported to and from the embryo via blood vessels in the chorioallantoic membrane.

of the hatchlings are males, but when the incubation temperature is increased to 34°C, the proportion of males drops to about 28 percent.

Maternal Control of Offspring Sex

To add to the complications of sex determination in birds, female birds can exert some control over the sex of their offspring. Many birds lay one egg per day until all the eggs in a clutch have been produced. Eggs hatch in approximately the sequence in which they were laid, and female birds may be able to adjust the sex of offspring in relation to the laying sequence of

their eggs. A study of house finches (*Carpodacus mexicanus*) found that the eggs laid first by birds in Montana hatched into females, whereas first-laid eggs in an Alabama population hatched into males. Sex-specific growth and survival characteristics appear to be the basis for this difference because babies that hatch first grow faster and reach larger adult sizes than those that hatch later. In Montana large females survive better than smaller ones, whereas in Alabama large males are favored. The sex-biased hatching order increases chick survival by 10 percent to 20 percent.

17.10 Nests

Nests protect the eggs not only from such physical stresses as heat, cold, and rain but also from predators. Bird nests range from shallow holes in the ground to enormous structures that represent the combined efforts of hundreds of individuals over many generations (**Figure 17–30**). The nests of passerines are usually cup-shaped structures because this shape provides support for the eggs and brooding parent. Nests are composed of plant materials that are woven together. Swifts use sticky secretions from their buccal glands to cement the material together to form nests, and grebes, which are marsh-dwelling birds, build floating nests from the buoyant stems of aquatic plants.

Incubation

Some species of birds begin incubation as soon as the first egg is laid, whereas others wait until the clutch

Figure 17–29 An Australian brush turkey on its nest mound.

(a)

(b)

(c)

(d)

Figure 17–30 Diversity of bird nests. Some nests are no more than shallow depressions; other birds build elaborate structures. (a) The piping plover, like many shorebirds, lays its eggs in a depression scraped in the soil. (b) The bald eagle constructs an elaborate nest that is used year after year. (c) Coots build floating nests using the air-filled stems of aquatic plants. (d) Nests of the sociable weaver bird are communal structures that house up to 500 birds and are used for many generations.

is complete. Though starting incubation immediately may protect the eggs, it means that the first eggs in the clutch hatch while the eggs that were deposited later are still developing. A spread in hatching dates forces the parents to divide their time between incubation and gathering food for the hatchlings. Furthermore, the eggs that hatch last produce young that are smaller than their older nestmates, and these young probably have less chance of surviving to fledge. Most passerines, as well as ducks, geese, and fowl, do not begin incubation until the next-to-last or last egg has been laid.

Prolactin, secreted by the pituitary gland, suppresses ovulation and induces brooding behavior, at least in those species of birds that wait until a clutch is

complete to begin incubation. The insulating properties of feathers that are so important in the thermoregulation of birds become a handicap during brooding, when the parent must transfer metabolic heat from its own body to the eggs. Prolactin plus estrogen or androgen stimulates the formation of brood patches in female and male birds, respectively. These brood patches are areas of bare skin on the ventral surface of a bird. The feathers are lost from the brood patch, and blood vessels proliferate in the dermis, which may double in thickness and give the skin a spongy texture. Not all birds develop brood patches, and in some species only the female has a brood patch, although the male may share in incubating the eggs. Ducks and geese create brood patches by plucking the down feathers from their breasts; they use the feathers to line their nests. Some penguins lay a single egg that they hold on top of their feet and cover with a fold of skin from the belly, thus enveloping the egg.

The temperature of eggs during brooding is usually maintained within the range 33°C to 37°C, although some eggs can withstand periods of cooling when the parent is off the nest. Tube-nosed seabirds (Procellariiformes) are known for the long periods that adults spend away from the nest during foraging. Fork-tailed storm petrels (*Oceanodroma furcata*) lay a single egg in a burrow or rock crevice. Both parents participate in incubation, but the adults forage over vast distances, and both parents may be absent from the nest for several days at a time. For storm petrels in Alaska, parents averaged 11 days of absence during an incubation period of 50 days. Eggs were exposed to ambient temperatures of 10°C while the parents were away. Experimental studies showed that storm petrel eggs were not damaged by being cooled to 10°C every four days. The pattern of development of chilled eggs was like that of eggs incubated continuously at 34°C, except that each day of chilling added about a day to the total time required for the eggs to hatch.

Incubation periods are as short as 10 to 12 days for some species of birds and as long as 60 to 80 days for others. In general, large species of birds have longer incubation periods than small species, but ecological factors also contribute to determining the length of the incubation period. A high risk of predation may favor rapid development of the eggs. Among tropical tanagers, species that build open-topped nests near the ground are probably more vulnerable to predators than are species that build similar nests farther off the ground. The incubation periods of species that nest near the ground are short (11 to 13 days) compared to those of species that build nests at greater heights (14 to 20 days). Species of tropical tanagers with roofed-over nests have still longer incubation periods (17 to 24 days).

In those species of birds that delay the start of incubation until all the eggs have been laid, an entire clutch nears hatching simultaneously. Hatching may be synchronized by clicking sounds that accompany breathing within the egg, and both acceleration and retardation of individual eggs may be involved. A low-frequency sound produced early in respiration, before the clicking phase is reached, appears to retard the start of clicking by advanced embryos. That is, the advanced embryos do not begin clicking while other embryos are still producing low-frequency sounds. Subsequently, vocalizations from advanced embryos appear to accelerate hatching by late embryos. Both effects were demonstrated in experiments with bobwhite quail eggs: pairing two eggs, one of which had started incubation 24 hours before the other, accelerated hatching of the late-starting egg by 14 hours and delayed hatching of the early-starting egg by 7 hours.

Parental Care

It is a plausible inference—both from phylogenetic considerations and from fossils of coelurosaurs—that the ancestral form of reproduction in the avian lineage includes depositing eggs in a well-defined nest site, one or both parents attending the nest, young that are able to feed themselves at hatching (**precocial** young), and a period of association between the young and one or both parents after hatching. All of the crocodilians that have been studied conform to this pattern, and evidence is increasing that at least some dinosaurs remained with their nests and young.

Extant birds follow these ancestral patterns, but not all species produce precocial young. Instead, hatchling birds show a spectrum of maturity that extends from precocial young that are feathered and self-sufficient from the moment of hatching to **altricial** forms that are naked and entirely dependent on their parents for food and thermoregulation (**Figure 17–31** and **Table 17–4**). The distinction between precocial and altricial birds includes differences in the amount of yolk originally in the eggs, the relative development of organs and muscles at hatching, and the rates of growth after hatching (**Table 17–5**).

After they hatch, altricial young are guarded and fed by one or both parents. Adults of some species of birds carry food to nestlings in their beaks, but many species swallow food and later regurgitate it to feed the

Table 17.4 Maturity of birds at hatching

Precocial: Eyes open, covered with feathers or down, leave nest after 1 or 2 days
  1. Independent of parents: megapodes
  2. Follow parents but find their own food: ducks, shorebirds
  3. Follow parents and are shown food: quail, chickens
  4. Follow parents and are fed by them: grebes, rails

Semiprecocial: Eyes open, covered with down, able to walk but remain at nest and are fed by parents: gulls, terns

Semialtricial: Covered with down, unable to leave nest, fed by parents
  1. Eyes open: herons, hawks
  2. Eyes closed: owls

Altricial: Eyes closed, little or no down, unable to leave nest, fed by parents: passerines

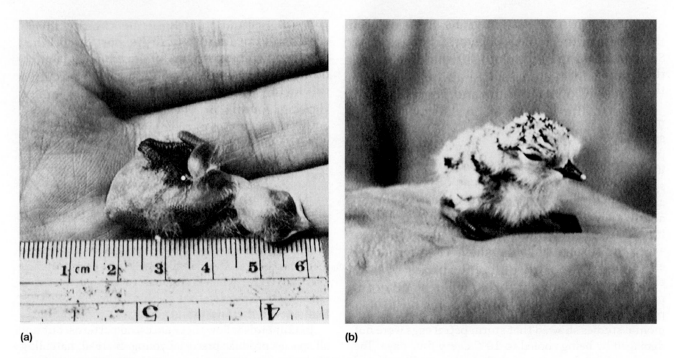

(a) **(b)**

Figure 17–31 Altricial and precocial chicks. (a) Altricial chicks such as that of the tree swallow (*Tachycineta bicolor*) are entirely naked when they hatch and unable even to stand up. (b) Precocial species such as the Kentish plover (*Charadrius alexandrinus*) have chicks that are covered with down when they hatch and can stand erect and even walk. The plover in this photograph has just hatched; it retains the egg tooth on the tip of its bill, whereas the tree swallow chick is 5 days old. The dark color on the leg of the swallow was applied by a researcher to identify individual hatchlings for a study of parental care.

young. Hatchling altricial birds respond to any disturbance that might signal the arrival of a parent at the nest by gaping their mouths widely. The sight of an open mouth appears to stimulate a parent bird to feed it, and the young of many altricial birds have brightly colored mouth linings. Ploceid finches have covered nests, and the mouths of the nestlings of some species are said to have luminous spots that have been likened to beacons showing the parents where to deposit food in the gloom of the nest.

Table 17–5 Comparison of birds with altricial and precocial chicks

Eggs

Amount of yolk in eggs	precocial > altricial
Amount of yolk remaining at hatching	precocial > altricial

Chicks

Size of eyes and brain	precocial > altricial
Development of muscles	precocial > altricial
Size of gut	precocial < altricial
Rate of growth after hatching	precocial < altricial

The duration of parental care is variable. The young of small passerines leave the nest about 2 weeks after hatching and are cared for by their parents for an additional 1 to 3 weeks. Larger species of birds, such as the tawny owl, spend a month in the nest and receive parental care for an additional 3 months after they have fledged, and young wandering albatrosses require a year to become independent of their parents.

17.11 Orientation and Navigation

For as long as people have raised and raced pigeons, we have known that birds released in unfamiliar territory vanish from sight flying in a straight line, usually in the direction of home. The homing pigeon has become a favorite experimental animal for studies of navigation. Experiments have shown that navigation by homing pigeons (and presumably by other vertebrates as well) is complex and is based on a variety of sensory cues. On sunny days, pigeons vanish toward home and return rapidly to their lofts. On overcast days, vanishing bearings are less precise, and birds more often get lost. These observations led to the idea that pigeons use the sun as a compass.

Of course, the sun's position in the sky changes from dawn to dusk. That means a bird must know what time of day it is in order to use the sun to tell direction, and its timekeeping ability requires some sort of internal clock. If that hypothesis is correct, it should be possible to fool a bird by shifting its clock forward or backward. For example, if lights are turned on in the pigeon loft 6 hours before sunrise every morning for about 5 days, the birds will become accustomed to that artificial sunrise. At any time during the day, they will act as if the time is 6 hours later than it actually is. When these birds are released they will be directed by the sun, but their internal clocks will be wrong by 6 hours. That error should cause the birds to fly off on courses that are displaced by 90 degrees from the correct course for home, and that is what clock-shifted pigeons do on sunny days (**Figure 17–32**). Under cloudy skies, however, clock-shifted pigeons head straight for home despite the

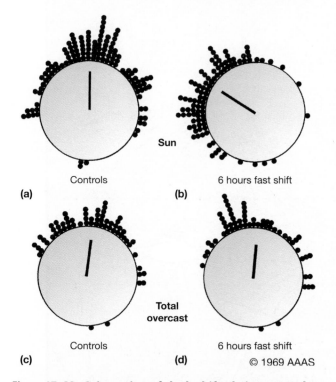

Controls
(a)

Sun

6 hours fast shift
(b)

Controls
(c)

Total overcast

6 hours fast shift
(d)

© 1969 AAAS

Figure 17–32 Orientation of clock-shifted pigeons under sunny and cloudy skies. Each dot in these plots shows the direction in which a pigeon vanished from sight when it was released in the center of the large circle. The home loft is straight up in each diagram. The solid bar extending outward from the center of each circle shows the average direction chosen by the birds.

6-hour error in their internal clocks. Clearly, pigeons have more than one way to navigate: when they can see the sun they use a sun compass, but when the sun is not visible they use another mechanism that is not affected by the clock-shift.

Probably this second mechanism is an ability to sense Earth's magnetic field and use it as a compass. On sunny days, attaching small magnets to a pigeon's head does not affect its ability to navigate, but on cloudy days pigeons with magnets cannot return to their home lofts.

Polarized and ultraviolet lights provide additional cues that pigeons probably use to determine direction, and they can also navigate by recognizing airborne odors and familiar visual landmarks. In addition, pigeons can detect extremely low-frequency sounds (infrasound), well below the frequencies that humans can hear. Those sounds are generated by ocean waves and air masses moving over mountains and can signal a general direction over thousands of kilometers, but their use as cues for navigation remains obscure.

Results of this sort are being obtained with other vertebrates as well and lead to the general conclusion that a great deal of redundancy is built into navigation systems. Apparently, there is a hierarchy of useful cues. For example, a bird relies on the sun and polarized light to navigate on clear days, but it can switch to magnetic direction sensing on heavily overcast days. In both conditions, the bird can use local odors and recognition of landmarks as it approaches home.

Many birds migrate only at night. Under these conditions a magnetic sense of direction might be important. Several species of nocturnally migrating birds use star patterns for navigation. Apparently, each bird fixes on the pattern of particular stars and uses their motion in the night sky to determine a compass direction. As in sun compass navigation, an internal clock is required for this sort of celestial navigation, and artificially changing the time setting of the internal clock produces predictable changes in the direction in which a bird orients.

17.12 Migration

The mobility that characterizes vertebrates is perhaps most clearly demonstrated in their movements over enormous distances. These displacements, which may cover half the globe, require both endurance and the ability to navigate. Other vertebrates migrate, some of them over enormous distances, but migration is best known among birds.

Migratory Movements of Birds

Migration is a widespread phenomenon among birds. About 40 percent of the bird species in the Palearctic are migratory, and an estimated 5 billion birds migrate from the Palearctic every year. Migrations often involve movements over thousands of kilometers, especially in the case of birds nesting in northern latitudes, some marine mammals, sea turtles, and fishes. Short-tailed shearwaters (*Puffinus tenuirostris*), for example, make an annual migration between the North Pacific and their breeding range in southern Australia that requires a round-trip of more than 30,000 kilometers (**Figure 17–33**).

Many birds return each year to the same migratory stopover sites, just as they may return to the same breeding and wintering sites year after year. Migrating birds may be concentrated at high densities at certain points along their traditional migratory routes. For example, species that follow a coastal route may be funneled to small points of land—such as Cape May, New Jersey—from which they must initiate long overwater flights. At these stopovers, migrating birds must find food and water to replenish their stores before venturing over the sea, and they must avoid the predators congregating at these sites. Development of coastal areas for human use has destroyed many important resting and refueling stations for migratory birds. The destruction of coastal wetlands has caused serious problems for migratory birds on a worldwide basis.

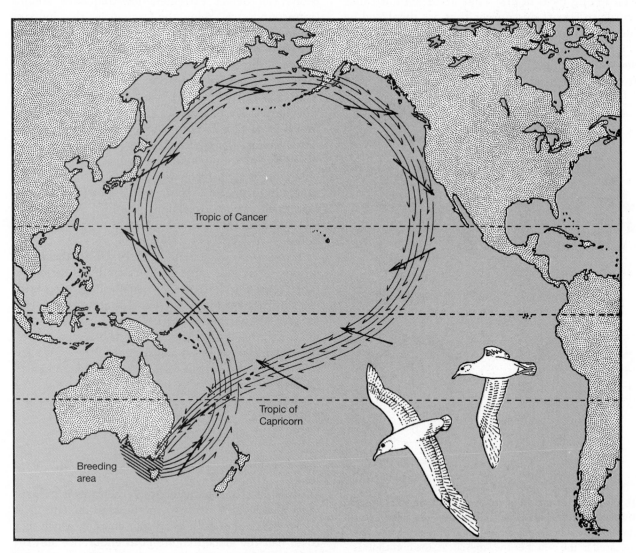

Figure 17–33 The migratory path of the short-tailed shearwater. As this species migrates from its Australian breeding area to its northern range, it takes advantage of prevailing winds in the Pacific region to reduce the energetic cost of migration.

Loss of migratory stopover sites may remove a critical resource from a population at a particularly stressful stage in its life cycle.

Costs and Benefits of Migration

The high energy costs of migration must be offset by energy gained as a result of moving to a different habitat. The normal food sources for some species of birds are unavailable in the winter, and the benefits of migration for those species are starkly clear. Other species may save energy mainly by avoiding the temperature stress of northern winters. In other cases, the main advantage of migration may come from breeding in high latitudes in the summer where resources are abundant and long days provide more time to forage than the birds would have if they remained closer to the equator.

Physiological Preparation for Migration

Migration is the result of a complex sequence of events that integrate the physiology and behavior of birds. Fat is the principal energy store for migratory birds, and birds undergo a period of heavy feeding and premigratory fattening (**Zugdisposition,** migratory preparation) in which fat deposits in the body cavity and subcutaneous tissue increase 10-fold, ultimately reaching 20 percent to 50 percent of the nonfat body mass. Fat is metabolized rapidly when migration begins, and many birds migrate at night and eat during the day. Even diurnal migrants divide the day into periods of migratory flight (usually early in the day) and periods of feeding. In addition, pauses of several days to replenish fat stores are a normal part of migration. *Zugdisposition* is followed by **Zugstimmung** (migratory mood), in which the bird undertakes and maintains migratory flight. In caged birds, which are prevented from migrating, this condition results in the well-known phenomenon of **Zugunruhe** (migratory restlessness).

Timing of Migration

Preparation for migration must be integrated with environmental conditions, and this coordination appears to be accomplished by the interaction of internal rhythms with an external stimulus. Day length is the most important cue for *Zugdisposition* and *Zugstimmung* for birds in northern temperate regions. Northward migration in spring is induced by increasing day length (**Figure 17–34**). The direction in which migratory birds orient during *Zugunruhe* depends on their physiological condition. In one experiment, the photoperiod was manipulated to bring one group of indigo buntings into their autumn migratory condition at the same time that a second group of birds was in its spring migratory condition. When the birds were tested under an artificial planetarium sky, the birds in the spring migratory condition oriented primarily in a northeasterly direction, whereas the birds in the autumn migratory condition oriented in a southerly direction. In Figure 17–34 dark circles show the mean nightly headings, pooled for several observations, for each of six birds in the spring migratory condition and five birds in the autumn condition.

Underlying the responses of birds to changes in day length is an endogenous (internal) rhythm. This circannual (about a year) cycle can be demonstrated by keeping birds under constant conditions. Fat deposition and

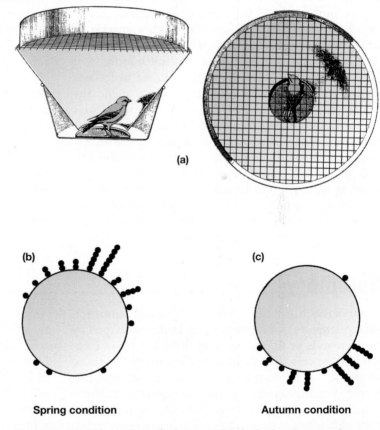

(a)

(b) Spring condition

(c) Autumn condition

Figure 17–34 Orientation of migratory birds during Zugunruhe.
(a) The birds were tested in circular cages that allow a view of the sky. The bird stands on an ink pad, and each time it hops onto the sloping wall it leaves a mark on the blotting paper that lines the cage. (b) In spring the birds oriented toward the north, and (c) in autumn toward the south.

migratory restlessness coincide in most species and alternate with gonadal development and molt, as they do in wild birds. When the rhythms are free running (that is, when they are not cued by external stimuli), they vary between 7 and 15 months. In other words, the birds' internal clocks continue to run, but in the absence of the cue normally provided by changing day length, the internal rhythms drift away from precise correspondence with the seasons.

Synchronizing Migration

Mate fidelity appears to be important to the lifetime reproductive success of long-lived migratory birds, and a change of partners is often followed by a reduction in reproductive success. Consequently, migratory birds that are pair-bonded to their mates must synchronize their internal rhythms so that both members of the pair migrate at the same time.

If that degree of coordination seems complicated for species of birds that migrate together and live as a pair at both ends of their migratory journey, think of the problem that confronts the Icelandic black-tailed godwit (*Limosa limosa islandica*), a species in which the males and females migrate separately and spend the winter in locations that are hundreds of kilometers apart! Icelandic godwits breed in Iceland during the summer and migrate south to spend the winter in England, Ireland, France, or Spain. Males and females depart from Iceland separately and winter in different locations: the average distance between the members of 14 pairs during the winter was 955 kilometers, and the most distant pair was separated by almost 2000 kilometers. Returning godwits start to appear in Iceland in mid-April, and all of the birds have returned by mid-May. Although the return dates for the population as a whole extend over a period of 30 days, the average interval between the arrivals of the members of a pair was only 3 days. Synchrony in arrival dates appears to contribute to the maintenance of pair bonds; the only changes of partners recorded during one study occurred in two of the three pairs in which the individuals arrived more than 8 days apart.

Climate Change and Migration

Global warming is a reality, and its effects on plants and animals have already been detected. Growing seasons and flowering times of plants in high latitudes now begin earlier in the spring and extend later into the fall than in previous years. Migratory birds are arriving earlier at their summer breeding ranges, and birds are laying eggs earlier. Complications are occurring for some species because the breeding cycles of birds are matched to the breeding cycles of their prey species, so that the birds' eggs hatch at the time of maximum availability of food for nestlings. Changing temperatures affect both the birds and their prey and can throw their cycles out of synchrony.

A mismatch of this sort appears to be responsible for a decrease in the population of pied flycatchers (*Ficedula hypoleuca*) during the past two decades. The decline in flycatchers corresponds to the degree of mismatch between the time caterpillar densities reach their peak and the time the flycatchers' eggs hatch. In areas where the mismatch is small, the populations of flycatchers have declined only 10 percent, whereas in the areas of greatest mismatch the decline has reached 90 percent. The farther a species travels on its migratory journey, the more likely it is to suffer from the effects of global climate change.

Summary

Flight is the distinctive mode of avian locomotion, and many elements of the anatomy of birds are shaped by the demands of flight. Flapping flight is a more complicated process than flight with fixed wings like those of aircraft, but it can be understood in aerodynamic terms. Feathers make up the aerodynamic surfaces responsible for lift and propulsion during flight, and they also provide streamlining. There are many variations and specializations within the four basic types of wings, but, in general, high speed and elliptical wings are used in flapping flight and high-aspect-ratio and high-lift wings are used for soaring and gliding.

The hollow bones of birds, an ancestral character of the coelurosaur lineage, combine lightness and strength. Some parts of the skeletons of birds are light in relation to the sizes of their bodies, and some of this lightness has been achieved by modification of the skull. Several skull bones that are separate in other diapsids are fused in birds, and teeth are absent. The function of teeth in processing food has been taken over by the muscular gizzard, which is part of the stomach of birds. The gizzard is well developed in birds that eat hard items such as seeds; it may contain stones that probably assist in grinding food.

The basic structure of the bird foot is very similar to the feet of coelurosaurs. Modifications of the feet are related to the lifestyle of a species—many flightless birds have reduced the number of toes, and aquatic birds have developed lobes on the toes or webs between the toes to increase the surface area acting on the water.

All birds are oviparous, perhaps because egg retention is the first step in the evolution of viviparity, and the specializations of the avian way of life do not make egg retention advantageous for birds. Female birds can exert a degree of control over the number of eggs they lay and the sex of the hatchlings. Most birds have extended periods of parental care, and during this period young birds learn species-specific behaviors such as song.

Many of the complex social behaviors of birds are associated with reproduction, and birds have contributed greatly to our understanding of the relationship between ecological factors and the mating systems of vertebrates. Social monogamy, with both parents caring for the young, is the most common mating system for birds, but extra-pair copulations are a common feature of monogamous mating systems and provide opportunities for both males and females to increase their fitness.

Migration is the most dramatic manifestation of the mobility of birds, and some species travel tens of thousands of kilometers in a year. Migrating birds use a variety of cues for navigation, including the position of the sun, stars, polarized light, and Earth's magnetic field.

Discussion Questions

1. Imagine that you are walking in a penguin colony and you find a bird's humerus. The skin and feathers disappeared long ago, leaving only the bleached bone. You know that the bone must have come from either a penguin or a gull because those are the only two birds of appropriate size that live in the area. How can you determine which kind of bird the bone came from?

2. Megapode birds, such as the brush turkey, bury their eggs in soil as crocodilians do rather than brooding them in a nest like other birds. How could you determine whether this mode of incubation is a retained ancestral trait or a derived condition that happens to have reverted to the crocodilian mode?

3. Many birds eat plant material, but the hoatzin (*Opisthocomus hoazin*), a South American bird, is the only avian species known to employ foregut fermentation. Hoatzins are herbivorous; green leaves make up more than 80 percent of their diet. More than a century ago, naturalists observed that hoatzins smell like fresh cow manure, and a study of the crop and lower esophagus of hoatzins revealed that bacteria and protozoa like those found in the rumen of a cow break down plant cell walls, releasing volatile fatty acids that are absorbed in the gut. Considering the abundance of plant material that is available to birds and the widespread use of fermentative digestion by mammals, why do you think the hoatzin is the only foregut fermenter among birds?

4. Crows are notorious nest robbers, stealing eggs from the nests of passerine birds, and they will even eat other birds if they can catch them. Adult passerines attack crows that approach their nests, darting at them and pursuing them when they fly away. What anatomical difference in the wings of passerines and crows might allow passerines to engage in such seemingly risky behavior?

5. The need to have two parents caring for hatchlings appears to be the reason that many species of birds are monogamous. But if the value of monogamy lies in rearing a brood successfully, why is extra-pair copulation so common among monogamous species of birds?

6. Some species of birds have a distinctive mode of courtship called "lekking." Males of lekking species assemble in groups (leks) and display. Females come to the leks and observe the displays by the males. Eventually a female selects a male and mates with him. What is the fundamental difference between this situation and the formation of territories by males that call from their territories to attract females? (Hint: What does a female base her choice on in each situation?)

Additional Information

Bennett, P. M., and I. P. F. Owens. 2002. *Evolutionary Ecology of Birds: Life Histories, Mating Systems, and Extinction.* Oxford, UK: Oxford University Press.

Bostwick, K. S., et al. 2010. Resonating feathers produce courtship song. *Proceedings of the Royal Society B* 277:835–841.

Both, C., et al. 2010. Avian population consequences of climate change are most severe for long-distance migrants in seasonal habitats. *Proceedings of the Royal Society B* 277:1259–1266.

Chiappe, L. M., and G. J. Dyke. 2002. The Mesozoic radiation of birds. *Annual Review of Ecology and Systematics* 33:91–124.

Cox, G. W. 2010. *Bird Migration and Global Change*. Washington, DC: Island Press.

Doutrelant, C., et al. 2012. Female plumage coloration is sensitive to the cost of reproduction. An experiment in blue tits. *Journal of Animal Ecology* 81:87–96.

Dyke, G., and G. Kaiser (eds.). 2011. *Living Dinosaurs: The Evolutionary History of Modern Birds*. Hoboken, NJ: Wiley-Blackwell.

Gagliardo, A., et al. 2011. Olfactory lateralization in homing pigeons: A GPS study on birds released with unilateral olfactory inputs. *The Journal of Experimental Biology* 214:593–598.

Göth, A. 2007. Incubation temperatures and sex ratios in Australian brush-turkey (*Alectura lathami*) mounds. *Austral Ecology* 32:378–385.

Hunt, D. W., et al. 2009. Evolution and spectral tuning of visual pigments in birds and mammals. *Philosophical Transactions of the Royal Society B* 364:2941–2955.

Jones, T., and W. Cresswell. 2010. The phenology mismatch hypothesis: Are declines of migrant birds linked to uneven global climate change? *Journal of Animal Ecology* 79:98–108.

Kenward. B., et al. 2011. On the evolutionary and ontogenetic origins of tool-oriented behaviour in New Caledonia crows (*Corvus moneduloides*). *Biological Journal of the Linnean Society* 102:870–877.

Louchart, A., and L. Viriot. 2011. From snout to beak—The loss of teeth in birds. *Trends in Ecology and Evolution* 26:663–673.

Lovejoy, T. E., and J. Elphick. 2007. *Atlas of Bird Migrations: Tracing the Great Journeys of the World's Birds*. Buffalo, NY: Firefly Books.

Knott, B., et al. 2010. Avian retinal oil droplets—Dietary manipulation of colour vision? *Proceedings of the Royal Society B* 277:953–962.

Maan, M. E., and M. E. Cummings. 2012. Poison frog colors are honest signals of toxicity, particularly for bird predators. *The American Naturalist*, in press.

Martin, G. R. 2012. Through birds' eyes: Insights into avian sensory ecology. *Journal of Ornithology*, in press.

Odeen, A. 2011. Evolution of ultraviolet vision in the largest avian radiation—The passerines. *BMC Evolutionary Biology* 11:313.

Rijke, A. M., and W. A. Jesser. 2011 The water penetration and repellency of feathers revisted. *The Condor* 113: 245–254.

Unwin, M. 2011. *The Atlas of Birds: Diversity, Behavior, and Conservation*. Princeton, NJ: Princeton University Press.

Vegvari, Z., et al. 2010. Life history predicts advancement of avian spring migration in response to climate change. *Global Change Biology* 16:1–11.

Wang, X. 2011. Avian wing proportions and flight styles—First step towards predicting the flight modes of Mesozoic birds. *PLoS ONE* 6(12): e28672. doi:10.1371/journal.pone.0028672.

Websites

Bird families of the world
http://creagrus.home.montereybay.com/list.html
Birdnet: The biology and diversity of birds
http://www.nmnh.si.edu/BIRDNET/birds/aboutbirds.html
The Cornell Lab of Ornithology
http://www.birds.cornell.edu/Page.aspx?pid=1478
The Tree of Life
http://tolweb.org/Neornithes/15834

PART IV

Synapsida: The Mammals

Because we *are* mammals, humans tend to think of mammals as the dominant kinds of vertebrates. That perspective does not withstand examination— there are barely more than half as many extant species of mammals as of birds, for example, and ray-finned fishes include as many extant species as all of the tetrapods combined—but it is pervasive. In some contexts, mammals are exceptionally successful derived vertebrates. The size range of mammals—from shrews to whales—is impressive, as is the development of flight by bats and echolocation by bats and cetaceans (whales and dolphins). Mammals probably have more complex social systems than any other kinds of vertebrates, and many features of their biology are related to interactions with other individuals of their species, ranging from the often prolonged dependence of young on their mother to lifelong alliances between individuals that affect their social status and reproductive success in a group.

The anatomical and physiological characteristics of mammals—respiration with a diaphragm, hair that provides insulation, high metabolic rates, teeth with complex surfaces that process food efficiently—make them successful in a wide variety of habitats. The progressive appearance of these characteristics can be traced clearly through the stem groups in the mammalian lineage.

Humans differ from other vertebrates in the extent to which they have come to dominate all of the habitats of Earth and in their effects—direct and indirect—on other vertebrates. In this portion of the book we will explore the evolution of mammals, the adaptive zones opened to them by their distinctive characteristics, the origin and radiation of humans, and the impact of humans on other species.

CHAPTER 18

The Synapsida and the Evolution of Mammals

We must backtrack to the end of the Paleozoic era to find the origin of the second lineage of terrestrial vertebrates, the synapsids. The synapsids actually had their first radiation (pelycosaurs) and a significant portion of their second radiation (therapsids) in the Paleozoic, before the radiations of the diapsids we have already discussed. The third radiation of the synapsid lineage (mammals) did not begin until the end of the Triassic and did not reach its peak until the Cenozoic era. Nonetheless, throughout the late Paleozoic and early Mesozoic, the synapsid lineage was becoming increasingly mammal-like.

The three groups of extant mammals—monotremes, marsupials, and placentals—had evolved by the late Mesozoic era, and they were accompanied by several groups of mammals that are now extinct.

18.1 The Origin of Synapsids

Synapsids include mammals and their extinct predecessors, commonly called "mammal-like reptiles" (**Figures 18–1** and **18–2**). In Chapter 11 we considered the ways in which synapsids differed in various aspects of their biology from the other major group of

amniotes, the sauropsids. Synapsids are distinguished from other amniotes by the presence of a lower temporal (synapsid) fenestra plus a few other skull features. Changes in the structure of the skull and skeleton of nonmammalian synapsids, and their probable relation to metabolic status and the evolution of the mammalian condition, will be described later.

The term *synapsid* is often used incorrectly to refer to only the extinct nonmammalian forms, but it in fact includes all amniotes descended from a common ancestor with the synapsid type of temporal fenestration. *Mammal-like reptile* is an appealing term for the ancestors of mammals, yet it is a misleading one and is now seldom used in technical publications. As we learned in Chapters 9 and 11, mammals are not the descendants of any animals closely related to modern reptiles, and to think of early synapsids as some sort of large, peculiar lizardlike beasts would indeed be inaccurate. Moreover, since mammals originated within the group of animals called "mammal-like reptiles," it is a paraphyletic assemblage rather than a true evolutionary group. (This is the same problem we faced in Chapter 16 with "dinosaurs," which is also a paraphyletic assemblage

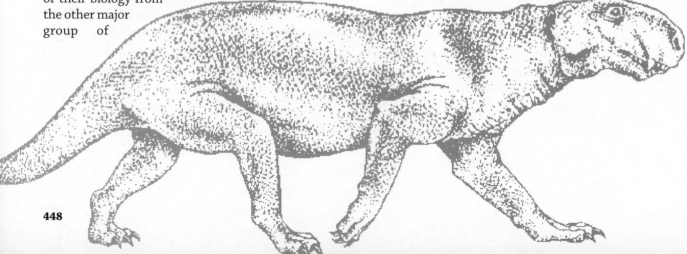

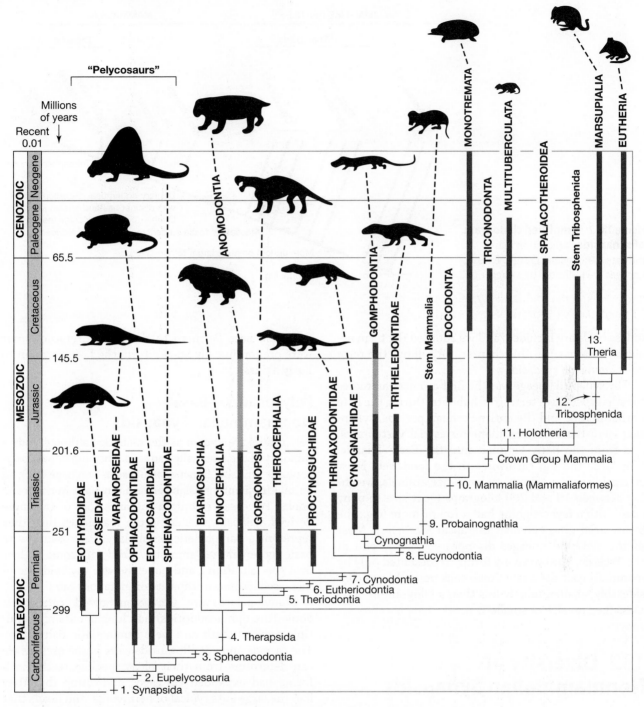

Figure 18–1 Phylogenetic relationships of the Synapsida. This diagram depicts the probable relationships among the major groups of synapsids. The black lines show relationships only; they do not indicate times of divergence or the unrecorded presence of taxa in the fossil record. Lightly shaded bars indicate ranges of time when the taxon is believed to be present, because it is known from earlier and later times, but is not recorded in the fossil record during this interval. Only the best-corroborated relationships are shown. Note that some researchers place the tritylodontids (here included with the Gomphodontia) as the sister taxon to Mammalia. The numbers at the branch points indicate derived characters that distinguish the lineages—see the Appendix for a list of these characters.

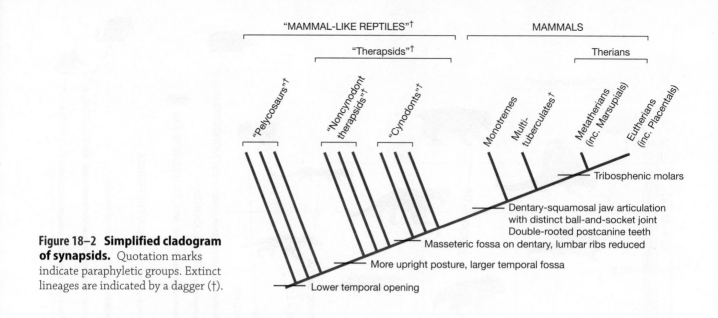

Figure 18–2 Simplified cladogram of synapsids. Quotation marks indicate paraphyletic groups. Extinct lineages are indicated by a dagger (†).

Labels in figure:
"MAMMAL-LIKE REPTILES"†
"Therapsids"†
MAMMALS
Therians
"Pelycosaurs"†
"Noncynodont therapsids"†
"Cynodonts"†
Monotremes
Multituberculates†
Metatherians (inc. Marsupials)
Eutherians (inc. Placentals)
Tribosphenic molars
Dentary-squamosal jaw articulation with distinct ball-and-socket joint
Double-rooted postcanine teeth
Masseteric fossa on dentary, lumbar ribs reduced
More upright posture, larger temporal fossa
Lower temporal opening

unless birds are included, and we adopted the term *noavian dinosaur*.) In this chapter, we will use the term *nonmammalian synapsid*.

The synapsid lineage was the first group of amniotes to radiate widely in terrestrial habitats. During the Late Carboniferous and the entire Permian periods, synapsids were the most abundant terrestrial vertebrates, and from the Early Permian into the Early Triassic, they were the top carnivores in the food web. Most synapsids were medium- to large-size animals, weighing between 10 and 200 kilograms (goat size to pony size), with a few weighing half a ton or more (e.g., the bison-sized dinocephalian therapsid *Moschops*). Most of the synapsid lineages disappeared by the end of the Triassic. The surviving forms (represented only by mammals past the Early Cretaceous period) were considerably smaller, mostly less than a kilogram in body mass (i.e., rat size or smaller).

18.2 **Diversity of Nonmammalian Synapsids**

The two major groups of nonmammalian synapsids were the pelycosaurs and the therapsids; **Table 18–1** lists the major subgroups within each. Pelycosaurs, the more basal of the two groups, were found mainly in the paleoequatorial latitudes of the Northern Hemisphere (Laurasia) and were predominantly known from the Early Permian. The more derived therapsids were found mainly in higher latitudes in the Southern Hemisphere (Gondwana). Therapsids ranged in age

from the Late Permian to the Early Cretaceous but were predominantly known from the Late Permian to Early Triassic.

Pelycosaurs—Basal Nonmammalian Synapsids

Pelycosaurs are known as the sailbacks, although only a minority of them actually had sails. The best-known pelycosaur is probably *Dimetrodon* (**Figure 18–3** on page 452), an animal frequently mislabeled as a dinosaur in children's books. Pelycosaurs contain the ancestors of the more derived synapsids, including the mammals; thus they represent a paraphyletic assemblage. Pelycosaurs were basically generalized amniotes, albeit with some of their own specializations, and none showed any evidence of increased locomotor capacity or metabolic rate.

Many pelycosaurs were generalized carnivores. Some (the ophiacodontids) had long snouts and multiple pointed teeth and were semiaquatic fish eaters. The caseids and edaphosaurids were herbivores, as we can determine from their blunt, peglike teeth. Both forms had expanded rib cages, indicating that they had the large guts typical of herbivores, and heads that look surprisingly small in comparison to their barrel-shaped bodies.

The most derived pelycosaurs were the sphenacodonts, such as *Dimetrodon*—mainly large, carnivorous forms with large, sharp teeth. An enlarged caninelike tooth in the maxillary bone and an arched palate (a forerunner of the later separation of mouth and nasal passages) are key features linking sphenacodonts to more derived synapsids.

Table 18.1 Major groups of nonmammalian synapsids*

Pelycosaurs

Eothyrididae: Basal pelycosaurs, small (cat size) and probably insectivorous. Early Permian of North America (e.g., *Eothyris*).

Caseidae: Large (pig size) herbivorous forms. Middle Permian of North America and Europe (e.g., *Casea*, *Cotylorhynchus* [see Figure 18–3b]).

Varanopseidae: Generalized, medium-size forms. Early Permian of North America (e.g., *Varanops*) and Late Permian of South Africa and South America.

Ophiacodontidae: The earliest known pelycosaurs (but not the most basal). Medium size with long, slender heads, reflecting semi-aquatic fish-eating habits. Late Carboniferous to Early Permian of North America and Europe (e.g., *Ophiacodon* [see Figure 18–4a]).

Edaphosauridae: Large herbivores, some with a sail. Early Permian of North America (e.g., *Edaphosaurus*).

Sphenacodontidae: Large carnivores, some with a sail. Early Permian of North America and Europe (e.g., *Haptodus*, *Dimetrodon* [see Figures 18–3a and 18–4b]).

Noncynodont Therapsids

Biarmosuchia: The most basal therapsids, medium-size (dog size) carnivores. Late Permian of Eastern Europe (e.g., *Biarmosuchus*).

Dinocephalia: Medium to large (cow size) carnivores and herbivores. Some large herbivores had thickened skulls, possibly for head-butting in intraspecific combat. Late Permian of Eastern Europe and South Africa (e.g., *Titanophoneus*, *Moschops* [see Figure 18–3c]).

Anomodontia: Small (rabbit size) to large herbivores, the most diverse of the therapsids. Includes the dicynodonts, which retained only the upper canines and substituted the rest of the dentition with a turtlelike horny beak. Includes burrowing and semiaquatic forms. Late Permian to Late Triassic worldwide (including Antarctica) and Early Cretaceous of Australia (e.g., *Dicynodon* [see Figure 18–4c], *Lystrosaurus* [see Figure 18–5]).

Gorgonopsia: Medium to large carnivores. Late Permian of Eastern Europe and South Africa (e.g., *Scymnognathus* [see Figure 18–4d], *Lycaenops* [see Figure 18–3d]).

Therocephalia: Small to medium-size carnivores and insectivores. Similarities to cynodonts include a secondary palate, complex postcanine teeth, and evidence of nasal turbinate bones. Late Permian to mid-Triassic of Eastern Europe and South and East Africa (e.g., *Pristerognathus*).

Cynodont Therapsids

Procynosuchidae: The most basal cynodonts, medium-size (rabbit size) carnivores and insectivores. Late Permian of Eastern Europe and South and East Africa (e.g., *Procynosuchus*).

Thrinaxodontidae: Medium-size (rabbit size) carnivores and insectivores. Early Triassic of Eastern Europe, South Africa, South America, and Antarctica (e.g., *Thrinaxodon* [see Figure 18–7b]).

Cynognathidae: Medium to large (dog size) carnivores. Early Triassic of South Africa and South America (e.g., *Cynognathus* [see Figure 18–4e]).

Gomphodontia: Small (mouse size) to medium-size (dog size) herbivores. Includes the large diademodontids (e.g., *Diademodon*, Early Triassic of Africa and East Asia) and transversodontids (e.g., *Traversodon*, Middle Triassic of South and East Africa, South America, East Asia) and the small-size tritylodontids (e.g., *Tritylodon*, *Oligokyphus* [see Figure 18–4f]). Late Triassic to Middle Jurassic of North America, Europe, East Asia and South Africa, plus Early Cretaceous of Russia. Tritylodonts were rodentlike forms that paralleled the mammalian condition of the postcranial skeleton. Some researchers consider them as separate from the other gomphodonts and as the sister group to mammals.

Chiniquodontidae and Probainognathidae: Small carnivores and insectivores, closely related to Tritheledontidae (not included in Figure 18–1). Middle to Late Triassic of North and South America (e.g., *Probelesodon* [see Figures 18–3e and 18–7c]).

Tritheledontidae: Small carnivores and insectivores. Approached mammalian condition in form of jaw joint and postcranial skeleton. Late Triassic of South America to Early Jurassic of South Africa (e.g., *Diarthrognathus*).

*Note that large, medium, and small refer to size within a particular group. A medium-size noncynodont therapsid is much larger than a medium-size cynodont.

Elongation of the neural spines of the trunk into the well-known pelycosaur sail was a remarkable feature of some edaphosaurids and sphenacodontids. These sails must have been evolved independently in the two groups because they are absent from early members of both groups and differ in detailed structure when they are present. Marks of blood vessels on the spines, and spines that had been broken and then healed, suggest that a web of skin covered the spines. There is no evidence of sexual dimorphism in these sails, so it is unlikely that they were devices evolved primarily for display. Such a large increase in the surface area of an animal would affect its heat exchange with the environment, and it seems likely that the sails were temperature-regulating devices, picking up warmth from the sun's rays, which could then be transferred to the body by blood vessels.

Therapsids—More Derived Nonmammalian Synapsids

A flourishing fauna of more derived synapsids—grouped under the general name of therapsids—extended from the Middle Permian to the Early Cretaceous. (Note that the term *therapsid* properly

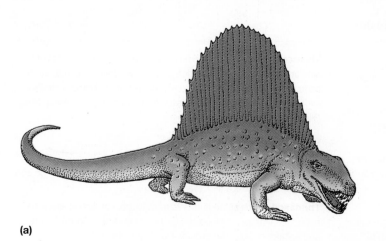

(a)

Figure 18–3 Diversity of nonmammalian synapsids. (a) The sphenacosaurid pelycosaur *Dimetrodon* (about the size of a Saint Bernard dog). (b) The caseid pelycosaur *Cotylorhynchus* (similar size to *Dimetrodon* or a little smaller). (c) The dinocephalian therapsid *Moschops* (about the size of a cow). (d) The gorgonopsid therapsid *Lycaenops* (about the size of a Labrador retriever). (e) The cynodont *Probelesodon* (about the size of a Jack Russell terrier). (The species are shown approximately to scale.)

(b)

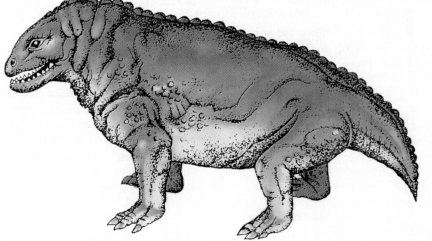

(c)

(d)

(e)

includes not only these beasts but also mammals.) Therapsids all had modifications suggesting an increase in metabolic rate over the more generalized pelycosaurs; thus they are often portrayed as having hair. They were all fairly heavy-bodied, large-headed, stumpy-legged forms. The image of this body form, combined with incipient hairiness, prompted one cartoonist to declare them "too ugly to survive."

Derived Features of Therapsids The earliest therapsids had various new features suggesting that they had a higher metabolic rate than the pelycosaurs. A larger temporal fenestra provided space for the origin of larger external adductor muscles on the skull roof, the upper canines were longer, and the dentition now showed a distinct differentiation into incisors, canines, and postcanine teeth (**Figure 18–4**). The choanae (internal nostrils) were enlarged, and a trough in the roof of the mouth (possibly covered by a soft-tissue secondary palate in life) indicates the evolution of a dedicated airway passage separate from the rest of the oral cavity. The entire skull was more rigid, the head was capable of

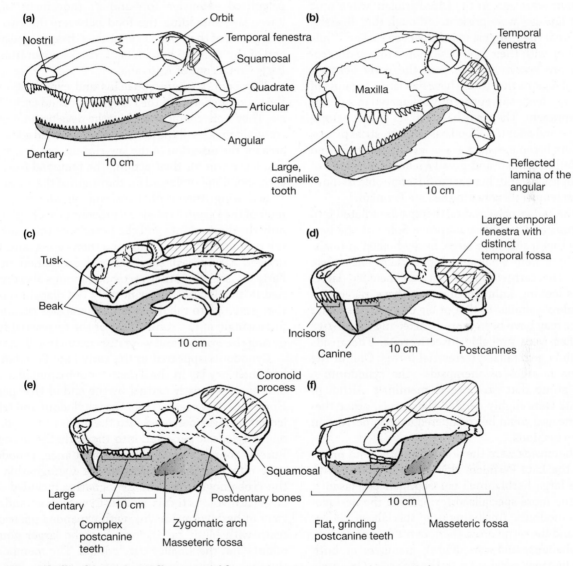

Figure 18–4 Skulls of nonmammalian synapsids. (a) The ophiacodontid pelycosaur *Ophiacodon*. (b) The sphenacodontid pelycosaur *Dimetrodon*. (c) The dicynodont therapsid *Dicynodon*. (d) The gorgonopsid therapsid *Scymnognathus*. (e) The cynodont *Cynognathus*. (f) The tritylodontid cynodont *Oligokyphus*. Light-colored stippling = the dentary bone; hatching = the temporal fossa. (The opening in the skull is the temporal fenestra, and the temporal fossa is the depression on the skull exposed by the temporal fenestra where the jaw muscles originate.)

increased dorsoventral flexion on the neck (i.e., a nodding action), and the neck as a whole was more flexible, allowing the head to be turned left and right. The pectoral and pelvic girdles were less massive than those of pelycosaurs, the limbs were more slender, and the shoulder joint appears to have allowed more freedom in movement of the forelimb, which would have made a longer stride possible. Therapsid bones were more vascularized than those of pelycosaurs, possibly indicative of more mammal-like growth rates.

Therapsid Diversity Therapsids first appeared around 267 million years ago, in the Late Permian, when nine different lineages were present, although this diversity suggests that their origins may have been earlier. While pelycosaurs, with their food base rooted in aquatic ecosystems, are known mainly from the Early Permian equatorial regions that would have had large amounts of rainfall, the therapsids apparently originated in a different environment. They may have evolved in drier tropical regions and adopted a food base of terrestrial plants. Pelycosaurs became extinct as sea levels rose at the end of the Early Permian and greatly reduced the area of moist tropical habitat, but therapsids were able to move into more temperate zones in the Late Permian.

Like the earlier pelycosaurs, therapsids radiated into herbivorous and carnivorous forms. Some of the herbivorous therapsids were large, heavy-bodied animals, while others were small and superficially rather like rodents. The carnivorous therapsids included large, ferocious-looking animals that may have played an ecological role similar to that of big cats today, smaller ones that may have been more foxlike, and rabbit-size forms that were probably insectivorous. Therapsids were mainly generalized terrestrial forms. One of the numerous lineages of therapsids—the cynodonts—was the group that gave rise to mammals. Although cynodonts were highly derived therapsids, other therapsid lineages could be considered equally derived in their own fashions.

The therapsids were the dominant large land mammals in the Late Permian, sharing center stage with only the large herbivorous parareptilian pareiasaurs. Among the more specialized forms were the herbivorous anomodonts (including the specialized dicynodonts) and the carnivorous theriodonts (gorgonopsids, therocephalians, and cynodonts); members of both of these lineages survived into the Triassic. One early anomodont (*Tiarajudens* from the Middle Permian of Brazil) had huge, elongated saberlike teeth, possibly used for display as in the males of some living herbivorous mammals, like mouse deer and musk deer. Another anomodont, *Suminia*, had long fingers and toes and was probably arboreal (the first known arboreal tetrapod).

Dicynodonts were the most abundant large terrestrial animals in the Late Permian and diversified into several different ecological types, including long-bodied burrowers and probable semiaquatic forms such as *Lystrosaurus* (**Figure 18–5**). The specializations of dicynodont skulls for herbivory include the loss of the marginal teeth (with the retention of tusklike canines), and there is evidence that the toothless jaw margins were covered by a turtlelike horny beak for shearing vegetation. The structure of the dicynodont jaw articulation permitted extensive fore-and-aft movement of the lower jaw, shredding the food between the two cutting plates of the beak. Dicynodonts disappeared at the end of the Triassic, but a relict form is known from the Early Cretaceous of Australia.

Theriodonts (therocephalians and cynodonts) were the major predators of the Late Permian and Early Triassic. They were characterized by the development of the coronoid process of the dentary that provided additional area for the insertion of the jaw-closing muscles, as well as a lever arm for their action. The temporal fossa was correspondingly enlarged for the origin of these muscles.

Following (therocephalians and cynodonts) the tumult of the Permo-Triassic extinctions (see Chapter 15), only the dicynodonts and the cynodonts (plus some of the cynodonts' sister group, the therocephalians, both derived carnivores) survived and diversified in the Early Triassic. But other vertebrate groups also diversified in the Triassic, most notably the diapsid reptiles that gave rise to the dinosaurs. Therapsids became an increasingly minor component of the terrestrial fauna during the Triassic and were near extinction at its end.

Cynodonts appeared in the early Late Permian and had their heyday in the Triassic; nonmammalian cynodonts were mostly extinct by the end of that period. However, a few species from the tritylodont and tritheledont lineages persisted into the Jurassic period, and one tritylodont persisted into the Early Cretaceous of Russia. Soon after their first appearance, cynodonts split into two major lineages: the Cynognathia and the Probainognathia. The Cynognathia included some large carnivorous forms, such as *Cynognathus*, and a variety of herbivorous forms, with expanded postcanine teeth with blunt cusps, including the larger gomphodonts and the smaller tritylodonts. The members of the Probainognathia were in general smaller, less specialized carnivorous and insectivorous forms, such as the tritheledontids; however, this was the lineage that gave rise to mammals.

The evolution of cynodonts was in general characterized by a reduction of body size. In fact, several

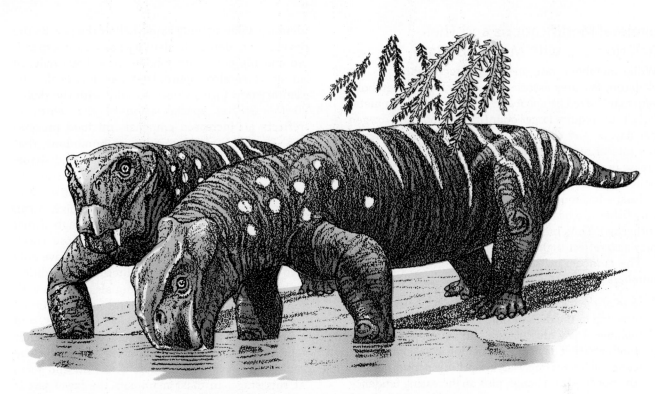

Figure 18–5 A reconstruction of the Early Triassic dicynodont therapsid *Lystrosaurus*.

Middle Triassic lineages of cynodonts were independently evolving smaller body size and, along with this, attaining various mammal-like features related to this miniaturization. Some early cynodonts were the size of large dogs, but by the Middle Triassic, the carnivorous cynodonts were only about the size of a rabbit. The earliest mammals of the latest Triassic were smaller than 100 millimeters, about the size of a shrew, while the contemporaneous tritylodontid cynodonts were a only little larger, about the size of a mouse.

Cynodonts had a variety of derived features, making them more mammal-like than other therapsids: All cynodonts had multicusped cheek teeth (i.e., with small accessory cusps anterior and posterior to the main cusp), with evidence of some occlusion (contact between teeth during biting), suggesting a more precise and more powerful bite. An enlarged infraorbital foramen—the hole under the eye through which the sensory nerves from the snout pass back to the brain—suggests a highly innervated face, perhaps indicative of a mobile, sensitive muzzle with lips and whiskers. Cynodonts and therocephalians also had evidence of maxillary turbinates, scroll-like bones in the nasal passages that warm and humidify the incoming air and help to prevent respiratory water and heat loss.

18.3 Evolutionary Trends in Synapsids

The synapsid lineage crossed a physiological boundary as the animals moved from ectothermy to endothermy, and this change was accompanied by changes in ecology and behavior. One major change took place between pelycosaurs and therapsids, and a second with the origin of cynodonts. Physiology, ecology, and behavior do not fossilize directly, but some of the changes that were occurring in metabolism, ventilation, and locomotion can be traced indirectly through changes in the skull and the postcranial skeleton. However, in tracing the evolution of a feature like endothermy, we must remember that evolution is not goal-oriented; the changes toward a more mammal-like condition proceeded by a series of small steps. Presumably, each small change was advantageous for the animal that had it at that time, and only in retrospect can we detect a pattern extending back from the modern condition to the ancestral state. The nonmammalian synapsids were well-adapted animals in their own right, not merely evolutionary stages on the road to full mammalhood.

Skeletal Modifications and Their Relationship with Metabolic Rate

While metabolic rate does not directly show on the skeleton, features associated with a higher metabolic rate may indeed be apparent. Animals with high metabolic rates require larger quantities of food and oxygen per day, so any changes suggesting improvements in the rate of feeding or respiration may indicate an increase in metabolic rate. Only animals with high metabolic rates are capable of sustained activity (on land at least—locomotion is cheaper in the water, which is why fishes can be continuously active despite being ectotherms). Thus indicators of greater levels of activity may also reflect higher metabolic rates.

The numbers of the features listed here match the numbers in **Figure 18–6**, which illustrates such changes.

1. *Size of the temporal fenestra*—A larger fenestra (the opening in the dermal skull roof) indicates a greater volume of jaw musculature and hence implies more food eaten per day. A small opening in pelycosaurs becomes an increasingly larger opening in more derived therapsids. This change, plus an increasing tendency to enclose the braincase with dermal bone, results in a distinct temporal fossa (an area on the skull roof itself) for the origin of a larger volume of jaw musculature (see also Figure 18–4). The external adductor (jaw-closing) muscles pass from their origin on the temporal fossa, through the temporal fenestra, to insert on the lower jaw. In mammals, and also in a few derived cynodonts, the fenestra is enlarged further with the loss of the postorbital bar so that the orbit is now confluent with the temporal fenestra.

2. *Condition of the lower temporal bar*—A bar of bone bowed out from the skull in the region of the orbit indicates the presence of a **masseter muscle** originating from this bony bar and inserting on the lower jaw, again implying more effective food processing (**Figure 18–7** on page 458). The temporal bar was originally very close to the upper border of the lower jaw, leaving no room for muscle insertion on the outside of the lower jaw. In cynodonts and mammals, the bar is bowed outward, forming the **zygomatic arch** and indicating the presence of a masseter muscle. A corresponding masseteric fossa (depression) on the dentary in these animals also indicates the presence of this muscle.

3. *Lower jaw and jaw joint*—Changes reflect the increased compromise between food processing and hearing in synapsids, as we will explain later. The lower jaw in pelycosaurs resembles the general amniote condition. The tooth-bearing portion, the dentary, takes up only about half of the jaw. By the level of cynodonts, the dentary has greatly expanded and the postdentary bones have been reduced in size (see also Figure 18–4). In mammals, the dentary now forms a new jaw joint with the skull.

4. *Teeth*—Greater specialization of the dentition reflects an increased emphasis on food processing. The teeth of pelycosaurs are **homodont;** that is, they are virtually all the same size and shape, with no evidence of regionalization of function. In more derived synapsids, the teeth become increasingly **heterodont,** or differentiated in size, form, and function (see Figure 18–4), although actual mastication with precise occlusion of the cheek teeth is a mammalian feature. Mammals have only two sets of teeth during their lifetime, a condition called **diphyodonty.** Most mammals are born with a set of small teeth (the "milk teeth" or "deciduous" teeth) that consists of the incisors, canines, and premolars. The milk teeth are replaced by the adult dentition as the jaws grow to their adult size. The molars are part of the adult dentition and form at this stage. In most mammals the lower jaw is a bit narrower than the upper jaw, so the teeth of the lower jaw lie slightly lingual (i.e., closer to the tongue) than the teeth of the upper jaw. This tooth arrangement allows mammals to chew with a rotary jaw motion. Mammals also have double-rooted molars and prismatic enamel capping their teeth.

5. *Development of a bony secondary palate*—A secondary palate separates the nasal passages from the mouth and allows an animal to breathe and eat at the same time. It also reinforces the skull against stresses from increased amounts of food processing. No bony secondary palate is apparent in pelycosaurs. An incipient, incomplete one is present in some noncynodont therapsids, although as previously mentioned this may have been completed by soft tissue, and a complete one is present in derived cynodonts and in mammals. A ridge inside of the nasal passage, suggesting the presence of nasal turbinates for reclaiming respiratory water (see Chapter 11), is found in cynodonts and therocephalians. Mammals have merged the originally double nasal opening into a single median one, probably reflecting an increase in size of the nasal passages and a higher rate of ventilation.

6. *Presence of a parietal foramen*—A hole in the skull for the pineal eye reflects temperature regulation by behavioral means. It is present in pelycosaurs and in most therapsids but lost in derived cynodonts and mammals.

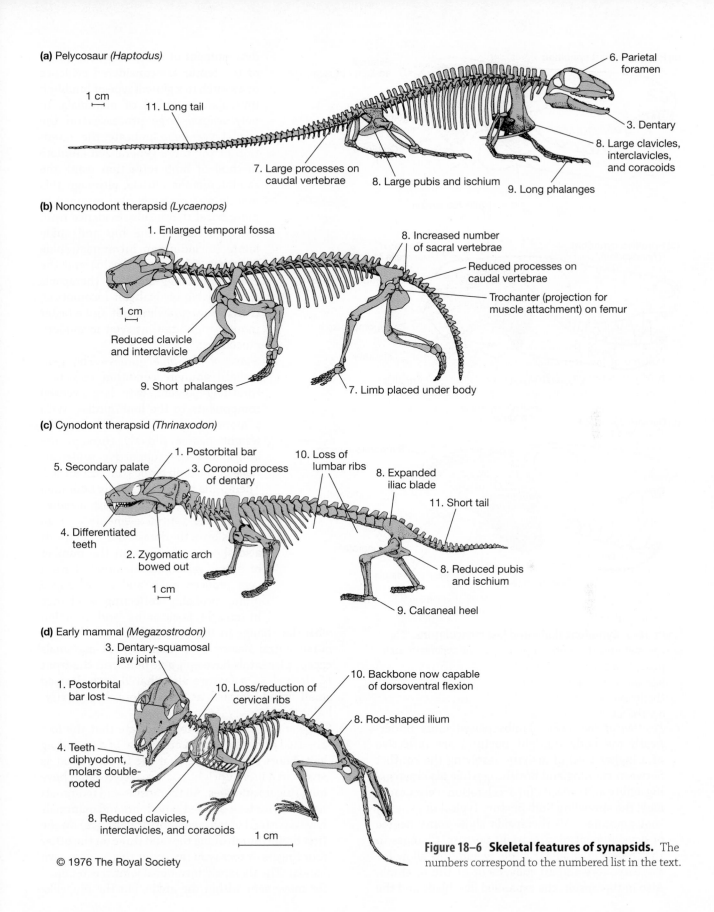

(a) Pelycosaur *(Haptodus)*

1 cm

11. Long tail

6. Parietal foramen

3. Dentary

8. Large clavicles, interclavicles, and coracoids

7. Large processes on caudal vertebrae

8. Large pubis and ischium

9. Long phalanges

(b) Noncynodont therapsid *(Lycaenops)*

1. Enlarged temporal fossa

8. Increased number of sacral vertebrae

Reduced processes on caudal vertebrae

Trochanter (projection for muscle attachment) on femur

1 cm

Reduced clavicle and interclavicle

9. Short phalanges

7. Limb placed under body

(c) Cynodont therapsid *(Thrinaxodon)*

1. Postorbital bar

5. Secondary palate

3. Coronoid process of dentary

10. Loss of lumbar ribs

8. Expanded iliac blade

11. Short tail

4. Differentiated teeth

2. Zygomatic arch bowed out

1 cm

8. Reduced pubis and ischium

9. Calcaneal heel

(d) Early mammal *(Megazostrodon)*

3. Dentary-squamosal jaw joint

10. Backbone now capable of dorsoventral flexion

1. Postorbital bar lost

10. Loss/reduction of cervical ribs

8. Rod-shaped ilium

4. Teeth diphyodont, molars double-rooted

8. Reduced clavicles, interclavicles, and coracoids

1 cm

© 1976 The Royal Society

Figure 18–6 Skeletal features of synapsids. The numbers correspond to the numbered list in the text.

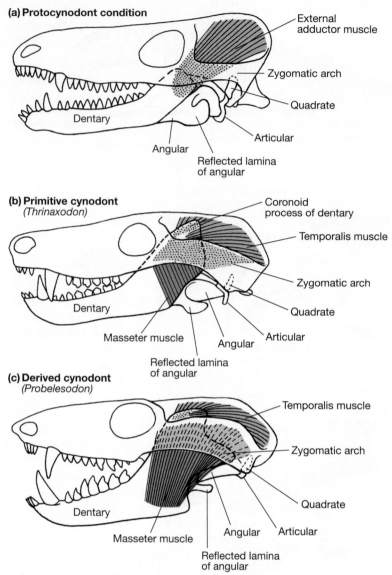

(a) Protocynodont condition

External adductor muscle

Zygomatic arch

Quadrate

Dentary

Articular

Angular

Reflected lamina of angular

(b) Primitive cynodont (*Thrinaxodon*)

Coronoid process of dentary

Temporalis muscle

Zygomatic arch

Quadrate

Dentary

Articular

Masseter muscle

Angular

Reflected lamina of angular

(c) Derived cynodont (*Probelesodon*)

Temporalis muscle

Zygomatic arch

Quadrate

Dentary

Articular

Angular

Masseter muscle

Reflected lamina of angular

Figure 18–7 Cynodont skulls and jaw musculature. The squamosal forms the posterior portion of the zygomatic arch.

7. *Position of the limbs*—Limbs placed more underneath the body (upright posture) are reflective of a higher level of activity, resolving the conflict between running and breathing while also increasing agility and capacity for acceleration. Pelycosaurs have the sprawling limb posture typical of generalized amniotes. All therapsids show some degree of development of an upright posture. Changes in the therapsid shoulder girdle would now allow for a greater fore-and-aft movement of the forelimb. Also in therapsids, the expanded iliac blade and the

development of the greater trochanter of the femur are considered evidence of a switch to a gluteal type of hindlimb musculature typical of mammals. In pelycosaurs, large processes on the caudal vertebrae indicate the retention of the more generalized amniote method of limb retraction using the caudofemoralis muscle, although this muscle was probably also retained in more basal therapsids. Evidence from the structure of the hip and ankle joints (including an intratarsal ankle joint between the astragalus and the calcaneum) suggests that therapsids were capable of dual-gait locomotion, with a slow sprawling gait and a faster more upright gait, as seen in modern crocodiles.

8. *Shape of the limb girdles*—The generalized amniote condition reflects a sprawling posture, with large ventral components to the limb girdles. With a more upright posture, more of the weight passes directly through the limbs, and the supportive undercarriage of the limb girdles can be reduced. Therapsids have more lightly built limb girdles than pelycosaurs, with a reduction of the ventral elements and an expansion of the dorsal ones. Therapsids also show an increase in the number of sacral vertebrae. Mammals have a very reduced pubis and a rod-shaped ilium, probably reflecting a change in muscle positioning and function with the change in the vertebral column to allow dorsoventral flexion (see item 10). All mammals except placentals have epipubic bones on the front of the pelvis, a feature also shared with derived small cynodonts such as tritheledontids and tritylodontids.

9. *Shape of the feet*—Long toes indicate that the feet are used as holdfasts and are typical of a sprawling gait. Short toes indicate that the feet are used as levers in a more upright posture. Pelycosaurs have long fingers and toes. All therapsids and mammals have shorter feet; derived cynodonts and mammals have a pattern of two segments (phalanges) on the first digit (thumb or big toe) and three on the other four fingers or toes (you can count this on your own hands). The therapsid intratarsal joint was retained for movement within the ankle, but the hingelike

joint between the tibia and astragalus, more specialized for fore-and-aft motion of the foot, was emphasized in derived cynodonts. Along with this, a distinct calcaneal heel is seen in cynodonts and mammals, providing a lever arm for a greater degree of push-off from the gastrocnemius (calf) muscle. Mammals also have an opposable big toe.

10. *Form of the vertebral column*—The reduction of the lumbar ribs may indicate the presence of a muscular diaphragm, evidence of a higher rate of respiration. Extant mammals use the diaphragm in addition to the ribs to inhale air into the lungs, and the diaphragm is especially important in obtaining additional oxygen during activity.

Some of the early mammals retain lumbar ribs, but with their reduction and eventual loss mammals have evolved distinctive differences between the thoracic and lumbar vertebrae, indicative of the mammalian mode of dorsoventral flexion, although the distinctive bounding gait of modern mammals is probably limited to therians (marsupials and placentals). Most therapsids have restricted the number of neck vertebrae to seven, a characteristic feature of modern mammals. Mammals have also reduced or lost the ribs on the cervical vertebrae, which may reflect the conversion of the sphincter colli muscle into the muscles of facial expression.

11. *Tail*—A long, heavy tail is the generalized amniote condition and is retained in pelycosaurs. A shorter tail, as in most therapsids and mammals, indicates a more upright posture in which limb propulsion is more important than axial flexion.

Evolution of Jaws and Ears

In the original synapsid condition a tooth-bearing dentary bone formed the anterior half of the jaw, with a variety of bones (known collectively as postdentary bones) forming the posterior half. (This is the generalized condition that is present in other tetrapods and bony fishes.) The jaw articulated with the skull via the articular bone in the lower jaw and the quadrate bone in the skull. Within cynodonts we can see a progressive enlargement in the size of the dentary and a decrease in the size of the postdentary bones. This trend was probably related to the increase in the volume of jaw adductor musculature, which inserted on the dentary (see Figure 18–7). In the most derived cynodonts, a condylar process of the dentary grew backward and eventually contacted the squamosal bone of the skull. In mammals and some very derived cynodonts, this contact between the dentary and the squamosal

formed a new jaw joint, the dentary-squamosal jaw joint. In these derived cynodonts, and in the earliest mammals, this new jaw joint coexisted with the old one, but in later mammals the dentary-squamosal jaw joint was the sole one. The bones forming the old jaw joint were now part of the middle ear: the articular became the malleus, and the quadrate became the incus, and they joined with the original stapes (the former hyomandibula) to form the distinctive three-boned mammalian middle ear.

The bare facts of this transition have been known for a century or so, but the evolutionary interpretation of these facts has changed over time. Almost two centuries ago, embryological studies demonstrated that the malleus and incus of the middle ear of mammals are homologous with the articular and quadrate bones that formed the ancestral jaw joint of other gnathostomes (**Figure 18–8**). Originally it was assumed that a lizardlike middle ear, with the stapes alone forming the auditory ossicle (the bone that transmits vibrations from the eardrum to the inner ear), was the basal condition for all tetrapods. We now have good evidence that an enclosed middle ear evolved separately in modern amphibians and amniotes and probably at least three times convergently within amniotes. However, before anatomists had this information, they assumed that the mammal ancestors had a lizardlike middle ear, and that when the new jaw joint was formed, an originally single-boned middle ear was transformed into a three-boned one using some leftover jawbones. (This evolutionary scenario also carried the tacit assumption that the mammalian condition of both ear bones and jaw articulation was inherently superior to that of other tetrapods.)

On reconsideration, there are some problems with this traditional story. Even if it were true that the mammalian type of jaw joint *was* somehow superior, how could researchers explain the millions of years of mammal ancestors' evolution during which the dentary was enlarging prior to contacting the skull? (Recall that evolution has no foresight.) Why is the original jaw articulation an inherently weak one? It appears to work well enough in other vertebrates; no one has ever accused *Tyrannosaurus rex* of having a weak jaw! And why disrupt a perfectly functional middle ear to insert some extra bones that happened to be available? Even if a three-boned middle ear were ultimately superior, there would be a period of adjustment to a new condition that would probably be less effective than the earlier condition.

In the mid-twentieth century, as more fossils of nonmammalian synapsids were discovered, the evolution of the middle ear could be traced from its beginning in

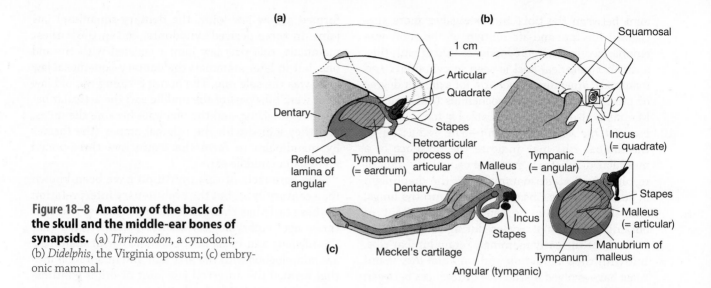

Figure 18–8 Anatomy of the back of the skull and the middle-ear bones of synapsids. (a) *Thrinaxodon*, a cynodont; (b) *Didelphis*, the Virginia opossum; (c) embryonic mammal.

basal synapsids through therapsids to early mammals. The first indication of a mammalian type of middle ear is seen in the sphenacodontid pelycosaurs, although they retained the basal amniote condition of a large stapes bracing the dermal skull roof to the braincase. Derived sphenacodontids had a structure called the "reflected lamina" on the angular bone of the lower jaw, which was a little flange on the bone that jutted out and backward (see Figures 18–4b and 18–7). This structure probably originated as the insertion for an enlarged and more elaborated pterygoideus muscle (one of the jaw-closing muscles). However, the reflected lamina became smaller but more clearly defined in derived therapsids and eventually became the structure holding the eardrum (a homology that we can trace in mammalian development today; see Figure 18–8).

In therapsids the bracing role of the stapes was lost, and the stapes was reduced in size. In all gnathostomes the stapes articulates with the otic capsule close to the housing of the inner ear. If this bone can vibrate it will transmit sound waves: thus, making this bone lighter so that it vibrates more readily renders hearing more acute. The quadrate-stapes contact in therapsids resembles the incus-stapes articulation of mammals. This anatomy suggests that the chain of bones that make up the middle ear of mammals had been used for hearing all along in synapsids, although with the relatively large size of these bones in nonmammalian synapsids, hearing would have been restricted to low-frequency sound.

Using the same set of bones for two different functions—as a jaw joint and as a hearing device—might seem like a rather clumsy, makeshift arrangement. This trade-off between hearing and jaw function would probably be adequate for the early synapsids, which, as ectotherms with low food intakes, would not have been using their jaw joints in complex chewing movements. A problem would arise in more derived synapsids because their higher metabolic rates would demand greater food intake and thus heavier jaw musculature, more use of the jaw in feeding, and greater stress on the jaw joint that would be incompatible with the role of these bones in hearing. Thus, the evolutionary history of the jaw in synapsids has been interpreted as representing a conflict between feeding and hearing. It is also likely that, as long as the articular and quadrate bones formed the primary jaw joint, no eardrum would have been present. An eardrum functions when there is an enclosed, air-filled, middle-ear cavity, which would be difficult to sustain around a working jaw joint. The reflected lamina may have provided support for connective tissue that acted to help transmit low-frequency sounds to the ear ossicles and thus to the inner ear.

In cynodonts the dentary bone was enlarged, and all of the jaw muscle insertions were transferred to this bone. This change would have isolated the auditory postdentary bones somewhat from the feeding apparatus, allowing them to become smaller. Note that the reason earlier researchers had interpreted the jaw of synapsids as being weak is that the postdentary bones remained loose and wobbly, as they did not fully fuse with one another and with the dentary as in many other tetrapods. However, there was a very good reason for these bones *not* to fuse: if they had done so, they would have compromised their role as vibrating auditory ossicles.

A fossa on the outside of the dentary shows that the mammalian type of masseter muscle was first apparent in cynodonts. In mammals, the masseter can move the lower jaw laterally and can be thought of as holding the lower jaw in a supportive sling. The original role of the masseter muscle may have been to help in relieving stresses at the jaw joint by this slinglike action, and thereby helping to resolve the conflict between jaw use and hearing.

The dentary increased in size in progressively more derived cynodonts. Finally, the condylar process of the dentary was large enough to contact the squamosal in the skull and to form a new jaw joint with this bone, freeing the original jaw joint from its old function and allowing it to be devoted entirely to hearing. The original reason for the backward extension of the condylar process may have been to prevent dislocation between the dentary and the postdentary bones when the animal bit hard with its back teeth. (Biomechanical analyses show that this biting action would tend to rotate the postdentary bones upward, a movement blocked by the condylar process.)

The decreasing size of later cynodonts would have accelerated this process. Smaller animals have relatively larger brains than larger animals, so a decrease in size would result in a relatively larger braincase. In turn, that change in proportions would move the posterior end of the dentary closer to the squamosal. Smaller animals also have smaller, lighter ear ossicles that would respond to higher frequency sounds.

The transition from the nonmammalian to the mammalian condition can be visualized by comparing the posterior half of the skull of *Thrinaxodon*, a fairly generalized cynodont, with that of *Didelphis*, the Virginia opossum (see Figure 18–8). The mammalian jaw joint is a new structure, formed by the dentary of the lower jaw and the squamosal bone of the skull. The tympanum (eardrum) and middle-ear ossicles lie behind the jaw joint and are much reduced in size, but the bones retain the same relation to each other as they had in cynodonts. The lower jaw of a fetal mammal viewed from the medial side shows that the angular (tympanic), articular (malleus), and quadrate (incus) develop in the same positions they had in the cynodont skull. The homologue of the angular, the tympanic bone, supports the tympanum of mammals. The mammalian manubrium of the malleus is a new feature that also provides support for the tympanum, and the ancestral jaw joint persists as the articulation between the malleus and the incus.

A classic example of an evolutionary intermediate is provided by the tritheledontid cynodont *Diarthrognathus*. This animal gets its name from its double jaw joint (Greek *di* = two, *arthro* = joint, and *gnath* = jaw).

It has both the ancestral articular-quadrate joint and, next to it, the mammalian dentary-squamosal articulation. The postdentary bones are still retained in a groove in the lower jaw. Note that this anatomy is still seen in the earliest mammals. This condition can be seen in fossils because even when these small, loose bones fall out and are not preserved, the distinctive trough in which they lay remains on the internal side of the lower jaw. In fact, the jaw and ear region of an adult Jurassic mammal resembles the condition seen in fetal mammals today, as shown in Figure 18–8c.

The type of middle ear seen in all mammals today (called the definitive mammalian middle ear, DMME) involves the loss of the embryonic Meckel's cartilage (the original gnathostome lower jaw bone) in the adult, and the separation of the middle-ear bones from the mandible. The reason for this shift may be the increase in the size of the mammalian forebrain, especially the olfactory region. The expansion of the skull during embryonic development to accommodate the expansion of the brain would dislocate the middle bones from their original position and thus provide the initial condition for their subsequent enclosure in a discrete middle-ear cavity distinct from the lower jaw.

The evolution of the definitive mammalian middle ear was originally assumed to be a one-time occurrence, although subtle differences in the middle-ear anatomy between monotremes and therian mammals (marsupials and placentals) initially led to some speculations that this evolution had occurred independently in the two groups. We now have fossils of several different species of Mesozoic mammals with this telltale trough in the dentary, and some with even the postdentary bones preserved or with an ossified Meckel's cartilage that functioned as a transitional way of supporting the middle ear bones. From what we know about the interrelationships of these mammals, it seems clear that the DMME evolved convergently several times (including independently in monotremes and therians) and that there may even have been reversals from a DMME to a condition in which the ear bones are still attached to the mandible.

How could such a complex evolutionary pattern have come about? The evolution of the DMME from the condition in the earliest mammals is primarily a developmental issue. All mammals start their life with middle-ear bones attached to the mandible, and the genetic basis for the timing of the resorption of Meckel's cartilage that frees the ear bones is becoming known. An apparent reversal to a less derived condition of the middle ear would require merely a shift in developmental timing. That developmental plasticity seems to have ceased with the therians, following the evolution of the coiled cochlea in the inner ear. (The coiling of the cochlea permits a longer structure, one that is capable of better pitch discrimination, to fit within the inner ear.) This stage heralded a central improvement in hearing, rendering a stable form of the DDME a necessity.

18.4 The First Mammals

The earliest mammals are identified by some derived features of the skull that reflect enlargement of the brain and inner-ear regions, a dentary-squamosal jaw joint that now has a distinct ball-and-socket articulation, and postcanine teeth with divided roots. These mammals were tiny, a couple of orders of magnitude smaller than cynodonts (a body mass of less than 100 grams—mouse size). In contrast, the smallest cynodonts would have weighed about 500 to 1000 grams, rat size.

The oldest well-known mammals from the latest Triassic and earliest Jurassic include *Morganucodon* (also known as *Eozostrodon*) from Wales (**Figure 18–9**), *Megazostrodon* from South Africa (see Figure 18–6d), and *Sinoconodon* from China. A somewhat older possible mammal is *Adelobasileus* (about 225 million years old), known only from an isolated braincase from the Late Triassic of Texas. Some workers prefer to limit the group "Mammalia" to those animals bracketed by the interrelationships of surviving mammals (i.e., to the Crown Group Mammalia in Figure 18–1); the name of the crown-group plus stem mammals is then *Mammaliaformes*.

Extant mammals are characterized by two salient features: hair and mammary glands. Neither of these features is directly preserved in the fossil record, although we may be able to infer the point in synapsid evolution at which these features were acquired. Another feature typical of most living mammals is viviparity—giving birth to young rather than laying eggs. However, we know that this was not a feature of the earliest mammals because some living mammals, the monotremes, still lay eggs. Mammals are also well known for having brains larger than those of other amniotes (except for birds); brains themselves do not fossilize, but clues can be obtained from the skull anatomy.

Features of the Earliest Mammals

How much like modern mammals would these early forms have been? We know about their bony anatomy, but what can we know about their soft anatomy, physiology, and behavior? Any feature shared by all extant mammals (e.g., lactation) would have been present in their common ancestor, but what can we infer about the biology of the stem mammals that predate the split between the **monotremes** (the platypus and echidna of Australia and New Guinea) and the more derived **therians** (marsupials and placentals)?

We can deduce some features of the soft-tissue anatomy of the earliest mammals from the condition in monotremes. For example, monotremes lay eggs, and because egg-laying is the generalized amniote condition, we can be fairly confident that this was true of the earliest mammals. However, monotremes are not good examples for making deductions about the behavior and ecology of the earliest mammals. Monotremes are relatively large mammals, and they are specialized in their habits.

Metabolic Rate We have already argued that at least some cynodonts had higher metabolic rates than the other therapsids, and this would have been the minimal condition for the earliest mammals. Monotremes, although endothermal, have a metabolic rate lower than that of therians and are not good at evaporative cooling; early mammals were probably similar.

Feeding and Mastication Most of our information about mammals comes from their teeth because hard, enamel-containing teeth are more likely to fossilize than are bones. Fortunately, mammalian teeth turn out to be very informative about their owners' lifestyle.

Most vertebrates have multiply-replacing sets of teeth and are hence termed **polyphyodont**. In contrast, the general mammalian condition is diphyodonty; that is, they have only two sets of teeth (like our milk teeth and our permanent teeth), and the molars are not replaced at all but instead erupt fairly late in life. This seems to have been the condition in almost all of the earliest known mammals. Mammals also have molars with precise occlusion that is made possible by an interlocking arrangement of the upper and lower teeth, producing characteristic wear patterns on the teeth.

Figure 18–9 Reconstruction of the Early Jurassic mammal *Morganucodon.*

Precise occlusion enables the cusps on the teeth to cut up food very thoroughly, creating a large surface area for digestive enzymes to act on and thereby promoting rapid digestion. Only mammals **masticate** (thoroughly chew) their food in this fashion. The cheek teeth of mammals also are set in an alternating fashion, and the lower teeth are closer together than the uppers. In addition, the jaws are moved in a rotary fashion—in contrast with the simple up-and-down movement of the jaws in most other tetrapods—which means that mammals can chew on only one side of the jaw at any one time (**Figure 18–10**).

Fossils of early mammals have tooth wear indicative of precise occlusion, interlocking upper and lower teeth, and narrow-set lower jaws that would require sideways movement of the lower jaw to occlude the

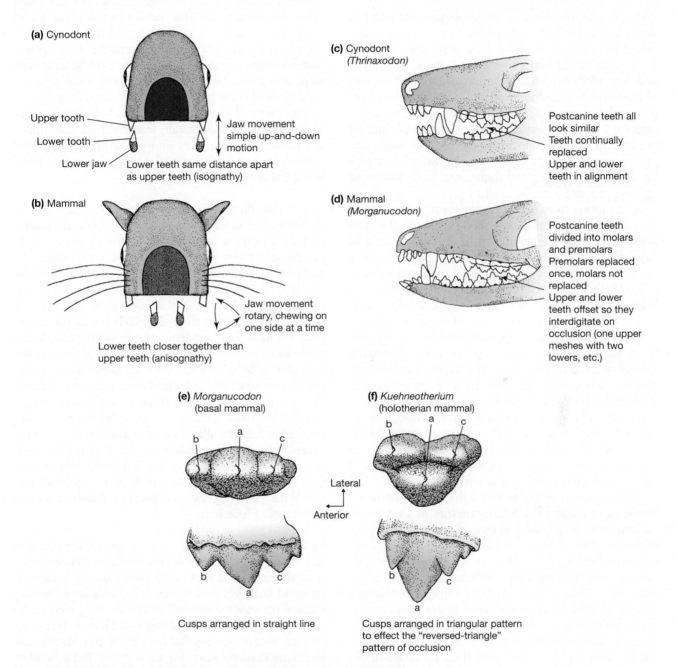

(a) Cynodont

Upper tooth

Lower tooth

Lower jaw

Jaw movement simple up-and-down motion

Lower teeth same distance apart as upper teeth (isognathy)

(b) Mammal

Jaw movement rotary, chewing on one side at a time

Lower teeth closer together than upper teeth (anisognathy)

(c) Cynodont (*Thrinaxodon*)

Postcanine teeth all look similar
Teeth continually replaced
Upper and lower teeth in alignment

(d) Mammal (*Morganucodon*)

Postcanine teeth divided into molars and premolars
Premolars replaced once, molars not replaced
Upper and lower teeth offset so they interdigitate on occlusion (one upper meshes with two lowers, etc.)

(e) *Morganucodon* (basal mammal)

b a c

Lateral

Anterior

b c

a

Cusps arranged in straight line

(f) *Kuehneotherium* (holotherian mammal)

b a c

b c

a

Cusps arranged in triangular pattern to effect the "reversed-triangle" pattern of occlusion

Figure 18–10 Occlusion and molar form in cynodonts and early mammals. (a and b) Cross-sectional view through the muzzle. (c and d) Side view of the jaws. (e and f) Schematic upper molars in occlusal ("tooth's eye") and lateral views.

teeth. These features were not present in any cynodont. Thus we can infer that the earliest mammals used the basic mammalian pattern of jaw movement. Mammalian teeth must be durable because mammals have only two sets of teeth per lifetime, and teeth with prismatic, wear-resistant enamel are a distinctive mammalian feature (also shared with the derived tritheledontid cynodonts), as are double-rooted molars.

Brains, Senses, and Behavior Brains do not fossilize, but the original size and shape of the brain may be determined from the inside of the skull, the endocranial cavity. In reptiles the brain does not fill the entire cavity, and thus, while we can get an approximation of the size of the brain, we cannot determine details of the structure of the brain itself. If the brain fills the endocranial cavity it leaves impressions on the cavity wall, and an impression of the endocranial space (and endocast) provides details of its structure. (Nowadays complex computerized imaging techniques can do this without any risk of damage to the skull.)

An endocast from a derived cynodont shows a brain of a similar size to that of a modern reptile (i.e., the generalized amniote condition), but with a couple of key differences. There were better-developed olfactory lobes (the portion of the brain concerned with smell) and a better-developed cerebellum (the portion of the brain concerned with neuromuscular control). Olfaction has always been an important mammalian sensory mode—it is rather ironic that we consider ourselves to be the most derived mammals, yet we have secondarily greatly reduced this function. The larger cynodont cerebellum was consistent with the more mammal-like postcranial skeleton of cynodonts, indicative of more sophisticated control of locomotion. However, the cerebral hemispheres were still small: even derived cynodonts would not have had behavior as complex and flexible as most modern mammals, and they would also have lacked good vision and hearing.

Computerized X-ray tomography has recently allowed us to discern the brain structure of a couple of early mammals, showing stages in mammalian brain evolution. One of the most basal mammals (*Morganucodon*) had enlarged the olfactory lobes and the cerebellum beyond the cynodont condition and had a larger neocortex. Enlargement of these regions continued, and *Hadrocodium,* a mammal closer to the crown group, had a brain almost the size of some living mammals. Then, in crown-group mammals we see further evidence for olfactory sophistication, with the ossification of the ethmoid turbinal bones in the nose that provided a 10-fold increase in the surface area of the olfactory epithelium. Thus, the earliest mammals showed a distinctive shift in brain anatomy from the cynodont condition. Part of this change may be due to their great decrease in size, because smaller animals have relatively larger brains and the process of miniaturization would have provided an opportunity for brain reorganization.

The enlargement of the neocortex in early mammals may have heralded some improvement of vision, but mammals in general do not rely on vision as much as other amniotes do, and most lineages lack good color discrimination. The cochlea in the inner ear of early mammals is not elongated, which indicates that their hearing was not acute. Monotremes lack a pinna (external ear), and it is likely that the earliest mammals were also pinna-less, even though this makes reconstructions of them look rather nonmammalian.

The teeth of the earliest mammals indicate an insectivorous diet. Like small, insectivorous marsupials and placentals today, they were probably nocturnal and solitary in their behavior, with the mother-infant bond being the only strong social bond.

Hair If cynodonts were indeed endothermal, it is possible they had at least some hair for insulation. However, hair appears to be primarily a sensory structure, as can be seen today in the whiskers of many mammals, and only secondarily co-opted for insulation. Mammalian hairs develop in association with migrating neural crest cells (see Chapter 2) that induce the formation of mechanoreceptors in the hair follicle for tactile sensation. We have fossil evidence of a dense fur coat in the fairly basal mammal *Castorocauda*. The increase in the size of the neocortex recorded in *Morganucodon* may reflect increased sensory input from a coating of hairs. The possible presence of a **Harderian gland** in *Morganucodon* may also indicate a coat of fur because in extant mammals this gland secretes a substance used for preening and waterproofing the fur. Perhaps the need to evolve fur as insulation was not really pressing before the earliest mammals, which were very small and would have had high ratios of body surface to mass and high rates of heat loss.

Lactation Milk is the source of nutrition for the young in all mammals, but how could this condition have evolved? A small amount of proto-milk would not be sufficient, so could lactation and mammary glands have initially evolved for another reason? Obviously, the precursor to any type of feeding of the young would have to be parental care. Few extant reptiles (except for crocodilians and birds) are known to care for their young, but a fossil of an adult basal pelycosaur from the late Middle Permian of South Africa is accompanied by several juveniles, suggesting that parental care may be a general synapsid trait.

Monotremes today lay parchment-shelled eggs, like the eggs of most lizards and very unlike the calcareous-shelled eggs of birds. Such thin-shelled eggs are likely the basal amniote condition and so were probably the condition in the nonmammalian synapsids. The original function of proto-mammary glands might have been to keep the eggs moist. Mammary glands appear to be rather closely linked to one of the secretory glands associated with hair (apocrine glands). Thus, they probably originally had the ability to secrete small quantities of organic materials, as do the apocrine glands of mammals today.

In addition to nutrition, milk has an important protective function in all mammals, passing on innate immunity to the newly born (or newly hatched) young. Milk contains proteins that are related to the lysozyme enzymes that attack bacteria—human milk has antimicrobial properties—and protection of eggs from microorganisms could have been another function of the proto-mammary glands.

Once proto-mammary secretion had evolved, whether for moisture production or immune defense or both, a more copious, more nutritive secretion that was accidentally ingested by the young could have been beneficial.

Although monotremes produce milk, their mammary glands lack nipples, and nipples were probably lacking in the earliest mammals as well. However, monotreme mammary glands are associated with hairs, which reinforces the hypothesis that mammary glands derived from hair-associated skin glands. (This hypothesis is very different from the popular notion that mammary glands came from "sweat glands"—human sweat glands are not associated with hairs.)

When in synapsid evolution would these secretions have become essential for feeding the young? Mammalian teeth occlude precisely, unlike the teeth of reptiles, but precise occlusion cannot develop until the jaws have reached their adult size. Thus, mammals are diphyodont—they have a set of milk teeth while their jaws are growing and a second set of adult teeth that do not appear until the jaws have reached their full size. However, an animal can be diphyodont only if it is fed milk during its early life. With a liquid diet, the jaw could grow while it had no need of teeth, and permanent teeth could erupt in a near adult-size jaw. Thus, if an animal has precise occlusion and diphyodonty, it must first have evolved lactation. However, we don't know if lactation originated with mammals, or if it was present in some or all nonmammalian synapsids.

What is the evolutionary advantage of lactation for mammals? Lactation allows the production of offspring to be separated from seasonal food supply. Unlike birds, which must lay eggs only when there is the appropriate food for the fledglings, mammals can store food as fat and convert it into milk at a later date. Provision of food in this manner by the mother alone also means that she does not have to be dependent on paternal care to rear her young. Finally, lactation makes viviparity less strenuous on the mother because the young can be born at a relatively undeveloped stage and cared for outside of the uterus.

Mammals Suckle The ability to suckle is a unique mammalian feature. Mammals can form fleshy seals against the bony hard palate with the tongue and with the epiglottis, effectively isolating the functions of breathing and swallowing (**Figure 18–11**). Mammals use these seals to suckle on the nipple while breathing through the

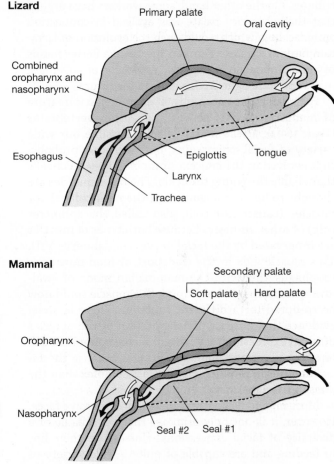

Figure 18–11 Longitudinal-section views of the oral and pharyngeal regions. Seal #1 prevents substances from entering the pharynx by appressing the back of the tongue against the soft palate. Seal #2 (lacking in postinfant humans) appresses the epiglottis against the back of the soft palate: this allows air to enter the trachea but blocks the entrance of material from the oropharynx. However, liquids can pass from the oropharynx, around the trachea, and into the esophagus while this seal is in place.

nose. Adult humans have lost the posterior seal that allows this action (because the larynx shifts ventrally in early childhood). This makes us more liable to choke on our food, but also enables us to breathe voluntarily through the mouth as well through the nose. We retain the anterior seal into adulthood; this is what stops us from swallowing water when we gargle.

This mammalian pharyngeal anatomy is also important for our mode of swallowing a discrete bolus of food that has been chewed into fine particles (deglutination) rather than swallowing large items intact like a snake swallowing a mouse. Changes in the bony anatomy of the palate and surrounding areas indicate that these functions came into use only with the most derived cynodonts, suggesting that this type of swallowing and the capacity for suckling are fundamentally mammalian attributes. On the other hand, some workers have argued that the secondary palate was evolved in connection with suckling, which would push the evolution of lactation back to the Late Permian therapsids. Better fossils of nonmammalian synapsids, especially of juveniles, may shed more light on this issue in the future.

Facial muscles are another characteristic feature of mammals that is absent from other vertebrates (**Figure 18–12**). These muscles make possible our wide variety of facial expressions, but they were probably first evolved in the context of mobile lips and cheeks that enable the young to suckle. The facial muscles are thought to be homologous with the neck constrictor muscles (constrictor colli, also called the sphincter colli) of other amniotes because both types of muscles are innervated by the facial nerve (cranial nerve VII). This muscle aids in the transport of food down the esophagus and, with the mammalian mode of swallowing a discrete bolus of food, this muscle could now be co-opted for a different function. There is some evidence that at least the elaboration of the muscles of facial expression may have occurred somewhat differently in various mammal lineages because the facial muscles in monotremes are less complex than the ones in therians.

Most mammals do not have highly expressive faces. However, it is not only primates that are capable of a diversity of facial expressions. Horses use their lips in feeding and are capable of quite a wide variety of expressions, whereas cows use their tongues and are poker-faced. This is why Mr. Ed (the talking horse on television) seems plausible to us, whereas a cow could never play that role. All mammals with well-developed facial muscles display similar expressions for similar emotions; for example, the snarl of an angry human is like the snarl of an angry dog or even an angry horse.

Lizard (reptile)—No muscles of facial expression

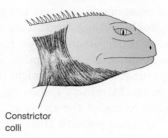

Constrictor colli

Rodent (mammal)—Moderate development of muscles of facial expression

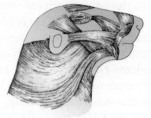

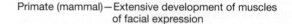

Primate (mammal)—Extensive development of muscles of facial expression

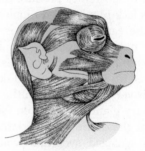

Figure 18–12 Muscles of facial expression.

18.5 The Radiation of Mesozoic Mammals

The Cenozoic is often called the Age of Mammals, but this description overlooks a tremendous amount of mammalian evolution that occurred during the Mesozoic. While it is true that mammals were small (none bigger than a raccoon, and most the size of a mouse or shrew) until the end of the Cretaceous, the radiation of Mesozoic mammals nonetheless represents two thirds of mammalian history. Mesozoic mammals were diverse taxonomically: more than 310 Mesozoic genera are now known, the majority discovered in the past few decades. Some of these new finds (many of them from the same spectacular Early Cretaceous Jehol Biota in China that produced many feathered dinosaurs and early birds) showed a range of body forms that was previously unknown, mammals with skeletons indicative of swimming, digging, and gliding habits (**Figure 18–13**). Another recent discovery is of a relatively large predatory form, *Repenomamus*, famous for being preserved with a juvenile dinosaur in its stomach. The

Figure 18–13 Newly discovered Mesozoic mammals. (a) *Repenomamus*, a large (for a Mesozoic mammal!) predator from the Early Cretaceous of China (about the size of a corgi dog). (b) *Volaticotherium*, a gliding form from the Late Jurassic or earliest Cretaceous of China (about the size of a flying squirrel). (c) *Castorocauda*, a swimming form from the Middle Jurassic of China (about the size of a rat). (d) *Fruitafossor*, a digging form that likely ate social insects from the Late Jurassic of North America (about the size of a chipmunk). Note that *Repenomamus* and *Volaticotherium* both lacked a therian-like coiled cochlea, so they probably lacked external ears.

teeth of Mesozoic mammals reveal mainly insectivorous or carnivorous diets with some omnivory, but these mammals were too small to be herbivores with a highly fibrous diet.

Although it is unlikely there was direct competition between mammals and dinosaurs, mammals did not diversify into large forms with more varied diets until the dinosaurs' extinction. The presence of dinosaurs must in some way have been preventing the evolution and radiation of larger-sized mammals. However, it is worth noting that even today the majority of mammals are no bigger than the ones known in the Mesozoic.

There were two major periods of mammalian diversification during the Mesozoic. The first, spanning the Jurassic to Early Cretaceous, produced an early radiation of forms that in the main did not survive past the Mesozoic. Many of these animals retained features of the nonmammalian synapsids, such as the lack of a fully enclosed middle ear and short ribs on the lumbar vertebrate. This radiation included one clade in Gondwana (the australosphenidans, Latin *austral* = southern and *sphen* = a wedge) that contained the origin of monotremes, and another in Laurasia (the boreosphenidans, Greek *boreo* = northern) that contained the origin of therians. At this time the northern and southern continental blocks were separated, so these two radiations of mammals occurred somewhat independently,

both groups "experimenting" with more complex types of cheek teeth.

The second radiation, which got under way in the later Early Cretaceous, was composed of more derived northern mammals, including the therians (relatives of today's marsupials and placentals) and the rodentlike multituberculates (which survived into the Cenozoic but are now extinct). However, note that both the earliest multituberculates and the earliest eutherians (the group including lineages of extant placentals and their Mesozoic relatives) first appeared at the end of the Middle Jurassic, 160 million years ago. The discovery of *Juramaia sinensis*, a very early eutherian from China, shows that the split between marsupials and placentals must have occurred by this time. However, fossils of marsupials and monotremes are not known until the Early Cretaceous. The radiation of these mammals will be discussed more extensively in Chapter 20.

Changes in other aspects of the terrestrial ecosystem are also apparent at this time. The Early Cretaceous marks the time of the initial radiation of the angiosperms, the flowering plants, and the radiation of the more modern mammals seems to be tightly tied to the radiation of angiosperm vegetation. Among other tetrapods, the Cretaceous dinosaur faunas were distinctively different from earlier ones, and snakes made their first appearance, possibly in association with a diversity of small mammals on which they fed.

Summary

The synapsid lineage is characterized by a single lower temporal fenestra on each side of the skull. The first synapsids were the pelycosaurs of the Late Carboniferous and Early Permian, including the familiar sailbacks. Many pelycosaurs were large animals for their time, up to the size of a large pig, and included the dominant carnivores and herbivores of their day.

The earliest therapsids were derived in having a more upright posture with the feet and limbs placed more under the body and cranial evidence of a greater volume of jaw musculature. The most derived therapsids, the Triassic cynodonts, had many features suggesting that they were endothermal, including turbinate bones in the nose, a secondary palate, and a reduction of the lumbar rib cage, indicating the presence of a diaphragm. Many features of the cynodont skull that changed over its history can be understood in the context of an evolutionary conflict between

chewing and hearing, because two of the bones of the mammalian middle ear formed the original synapsid jaw joint.

Most nonmammalian synapsids were extinct by the end of the Triassic. The first true mammals are known from the latest Triassic or earliest Jurassic. Most of these early mammals had teeth that precisely interlocked to chew food and were replaced only once. Evolution of mammals from cynodonts was accompanied by a substantial reduction in body size—early mammals were the size of shrews. We can infer the probable biology of these early mammals from their skeletal remains and from the biology of the living monotremes; like monotremes they must have laid eggs, and from patterns of tooth replacement we can infer that lactation must have evolved by this time.

Mesozoic mammals were mainly small insectivorous or omnivorous forms, although we know of greater diversity in ecological types now than we did at the end of the last century. The modern groups of mammals—monotremes and therians—can trace their origin back to the Late Jurassic or Early Cretaceous, a time of evolutionary turnover not only in mammals but also in other tetrapods as well as in plants. Mammals did not diversify into larger, more specialized forms until the Cenozoic, after the extinction of the dinosaurs.

Discussion Questions

1. Why does a more upright posture (i.e., limbs more under the body) in therapsids indicate a higher metabolic rate?
2. Why does the inferred appearance of a diaphragm in cynodonts make sense in terms of other attributes of these synapsids?
3. Why might a larger brain in early mammals be related (at least in part) to their "miniaturization" from the cynodont condition?
4. Why can we infer that lactation must have evolved at least by the time of the first mammal?
5. We now have a series of transitional fossils documenting the evolution of the mammalian ear. How could people be so sure, before these fossils were known, that the "new" ear ossicles in mammals (malleus and incus) were derived from the bones that formed the original gnathostome jaw joint?
6. Why do some people think that early mammals (including most Mesozoic forms) lacked a pinna (external ear)?

Additional Information

Angielczyk, K. D. 2009. *Dimetrodon* is not a dinosaur: Using tree thinking to understand the ancient relatives of mammals and their evolution. *Evolution, Education and Outreach* 2:257–271.

Blackburn, D. G. 1991. Evolutionary origins of the mammary gland. *Mammal Review* 21:81–96.

Blackburn, D. G., et al. 1989. The origins of lactation and the evolution of milk: A review with new hypotheses. *Mammal Review* 19:1–26.

Botha-Brink, J., and S. P. Modesto. 2007. A mixed-age classed 'pelycosaur' aggregation from South Africa: Earliest evidence of parental care in amniotes? *Proceedings of the Royal Society B* 274:2829–2834.

Chinsamy-Turan, A. (ed.). 2011. *Forerunners of Mammals.* Bloomington, IN: Indiana University Press.

Cisneros, J. C. 2011. Dental occlusion in a 260-million-year-old therapsid with saber canines from the Permian of Brazil. *Science* 331:1603–1605.

Fröbisch, J., and R. R. Reisz. 2009. The Late Permian herbivore *Suminia* and the early evolution of herbivory in terrestrial vertebrate ecosystems. *Proceedings of the Royal Society B* 276:3611–3618.

Gonick, L. 1990. *The Cartoon History of the Universe*, vols. 1–7. New York: Doubleday.

Hillenius, W. J. 2000. The septomaxilla of nonmammalian synapsids: Soft tissue correlates and a new functional interpretation. *Journal of Morphology* 245:207–229.

Hu, Y., et al. 2005. Large Mesozoic mammals fed on young dinosaurs. *Nature* 433:149–152.

Ji, Q., et al. 2002. The earliest-known eutherian mammal. *Nature* 416:816–822.

Ji, Q., et al. 2009. Evolutionary development of the middle ear in mammals. *Science* 326:278–281.

Kemp, T. S. 2005. *The Origin and Evolution of Mammals.* Oxford, UK: Oxford University Press.

Kemp, T. S. 2006a. The origin and early radiation of the therapsid mammallike reptiles: A palaeobiological hypothesis. *Journal of Evolutionary Biology* 19:1231–1247.

Kemp, T. S. 2006b. The origin of mammalian endothermy: A paradigm for the evolution of complex biological structure. *Zoological Journal of the Linnean Society* 147:473–488.

Kemp, T. S. 2007. Acoustic transformer function of the postdentary bones and quadrate of a nonmammalian cynodont. *Journal of Vertebrate Paleontology* 27:431–441.

Kemp, T. S. 2009. The endocranial cavity of a nonmammalian cynodont, *Chiniquodon theotenicus*, and its implications for the origin of the mammalian brain. *Journal of Vertebrate Paleontology* 29:1188–1198.

Kielan-Jaworoswka, Z., et al. 2004. *Mammals from the Age of Dinosaurs.* New York: Columbia University Press.

Luo, Z.-X. 2007. Transformation and diversification in early mammal evolution. *Nature* 450:1011–1019.

Luo, Z.-X. 2011. Developmental patterns in Mesozoic evolution of mammal ears. *Annual Reviews of Ecology, Evolution, and Systematics* 42:355–380.

Luo, Z.-X., et al. 2007. A new eutriconodont mammal and evolutionary development in early mammals. *Nature* 446:288–293.

Lou, Z.-X., et al. 2011. A Jurassic eutherian mammal and divergence of marsupials and placentals. *Nature* 476:442–445.

Meng, J., et al. 2011. Transitional mammalian middle ear from a new Cretaceous Jehol eutriconodont. *Nature* 472:181–185.

Oftedal, O. T. 2011. The evolution of milk secretion and its ancient origins. *Animal* 6:355–368.

Rich, T. H., et al. 2005. Independent origins of middle ear bones in monotremes and therians. *Science* 307:910–914.

Rowe, T. B., et al. 2011. Fossil evidence on origin of the mammalian brain. *Science* 332:955–957.

Ruben, J. A., and T. D. Jones. 2000. Selective factors associated with the origin of fur and feathers. *American Zoologist* 40:585–596.

Vorbach, C., et al. 2006. Evolution of the mammary gland from the innate immune system? *BioEssays* 28:606–616.

Weil, A. 2011. A jaw-dropping ear. *Nature* 472:174–175.

Websites
Synapsids
http://tolweb.org/Synapsids and http://tolweb.org/Therapsids
http://www.palaeos.org/Synapsida
Mammal ear and brain evolution
http://beta.revealedsingularity.net/article.php?art=mammal_ear
Mammal origins and Mesozoic mammals
http://dinosaurs.about.com/od/otherprehistoriclife/a/earlymammals.htm

CHAPTER 19

Geography and Ecology of the Cenozoic Era

The role of Earth's history in shaping the evolution of vertebrates is difficult to overestimate. The positions of continents have affected climates and the ability of vertebrates to migrate from region to region. The geographic continuity of Pangaea in the late Paleozoic and early Mesozoic eras allowed tetrapods to migrate freely across continents, and the faunas were fairly similar in composition across the globe. By the late Mesozoic, however, Pangaea no longer existed as a single entity. Epicontinental seas extended across the centers of North America and Eurasia, and the southern continents were separating from the northern continents and from one another. This isolation of different continental blocks resulted in the isolation of their tetrapod faunas and limited the possibility for migration. As a result, tetrapod faunas became progressively more different on different continents.

Distinct regional differences became apparent in the dinosaur faunas of the late Mesozoic, and they have been a prominent feature of Cenozoic mammalian faunas, although the uniqueness of the Cenozoic mammalian faunas on different continents results in large part from early migration events, with subsequent isolation.

Continental drift in the late Mesozoic and early Cenozoic moved the northern continents that had formed the old Laurasia (North America and Eurasia) from their early Mesozoic near-equatorial position into higher latitudes. The greater latitudinal distribution of continents, plus changes in the patterns of ocean currents that were the result of continental movements, caused cooling trends in high northern and southern latitudes during the late Cenozoic era. An ice sheet formed in the Arctic in the late Eocene (about 46 million years ago) but it lasted only about 2 million

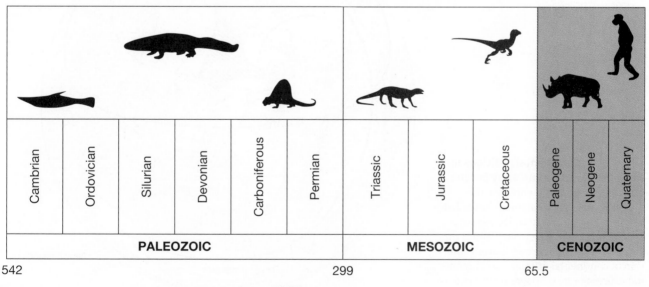

Cambrian	Ordovician	Silurian	Devonian	Carboniferous	Permian	Triassic	Jurassic	Cretaceous	Paleogene	Neogene	Quaternary
PALEOZOIC						**MESOZOIC**			**CENOZOIC**		

542   299   65.5

Million years ago

years. The current Arctic ice cap developed some 5 million years ago, this cooling led to a series of ice ages that began in the Pleistocene and continued to the present. We are currently living in a relatively ice-free interglacial period. Both the fragmentation of the landmasses and the changes in climate during the Cenozoic have been important factors in the evolution of mammals.

19.1 Cenozoic Continental Geography

The breakup of Pangaea in the Jurassic period began with the movement of North America that opened the ancestral Atlantic Ocean. Rifts formed in Gondwana, and India then moved northward on its separate oceanic plate (**Figure 19–1**), eventually to collide with Eurasia. The collision of the Indian and Eurasian plates in the mid-Cenozoic produced the Himalayas, the highest mountain range in today's world.

South America, Antarctica, and Australia separated from Africa during the Cretaceous but maintained connections into the early Cenozoic. Africa was separated from Europe and Asia by the Tethys Sea until the Miocene, and North and South America did not have a solid land connection until the Pleistocene. In the middle to late Eocene, Australia separated from Antarctica and, like India, drifted northward. Intermittent land connections between South America and Antarctica were retained until the middle Cenozoic, and Eocene Antarctic mammals are basically a subset of the South American ones. New Zealand separated from Australia sometime in the middle Mesozoic and has a diversity of endemic tetrapods—most notably the generalized diapsid reptile *Sphenodon* (tuatara), but the only native mammal is a bat. Other nonflying mammals that are found in New Zealand today, such as possums, deer, and hedgehogs, were brought in by humans during the last few centuries.

In the Northern Hemisphere there were two early Cenozoic routes connecting the northern continents at high, but relatively ice-free, latitudes. The trans-Bering Bridge (situated where the Bering Strait is today between Alaska and Siberia) connected western North American and Asia intermittently throughout the Cenozoic. A connection between eastern North America and Europe via Greenland and Scandinavia was apparent during the early Eocene. Europe and Asia have been a single continental mass since the end of the Eocene, but prior to that were separated by a sea called the Turgai Straits. Other tectonic movements also influenced Cenozoic climates. For example, the uplift of mountain ranges such as the Rockies, the Andes, and

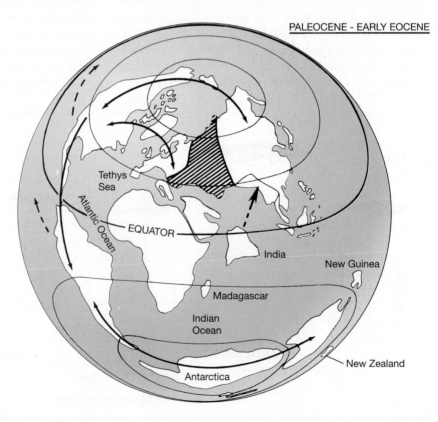

PALEOCENE - EARLY EOCENE

Figure 19–1 Continental positions in the late Paleocene and early Eocene epochs. An epicontinental sea, the Turgai Straits (cross-hatching), extended between Europe and Asia. The sea separating Europe and Asia from Africa is the Tethys Sea. Dashed arrows show the direction of continental drift, and solid arrows indicate major land bridges mentioned in the text.

the Himalayas in the middle Cenozoic led to alterations of global and local rainfall patterns, resulting in the spread of grasslands—a new habitat type—in the higher latitudes.

19.2 Cenozoic Terrestrial Ecosystems

The evolution of animals and plants during the Cenozoic is closely connected with climatic changes, which in turn are related to changes in the positions of continents. The key to understanding changes in climatic conditions at this time lies in knowing that the major landmasses were moving *away* from the equatorial region and toward the poles. The passage of landmasses over the polar region—Antarctica in the early Cenozoic and Greenland in the later Cenozoic—allowed the formation of polar ice caps and culminated in the periods of glaciation (ice ages) of the Pleistocene. The story of Cenozoic terrestrial ecosystems is also the story of the temperate regions of the higher latitudes, which became cooler and drier with accompanying changes in the types and productivity of the vegetation, such as the replacement of tropical-like forests with woodland and grassland starting around 45 million years ago. (The term *tropical-like* indicates a forest that had a multilayered structure like modern tropical forests but with a somewhat different assortment of species.)

The radiation of modern types of mammals is a prominent feature of the Cenozoic, and the Cenozoic is commonly known as the Age of Mammals. This name is misleading in two respects: the time elapsed since the start of the Cenozoic represents only about a third of the time spanned by the total history of mammals and the most diverse vertebrates of the Cenozoic, in terms of numbers of species, are the teleost fishes. The radiation of larger mammals is almost certainly related to the extinction of the dinosaurs, which left the world free of large tetrapods and provided a window of opportunity for other groups. The diversification of angiosperm plants in the Late Cretaceous was also critical for the late Mesozoic radiation of many types of mammals (such as the herbivorous multituberculates), setting the stage for their later Cenozoic diversification.

In the warm world of the early Cenozoic, tropical-like forests were found in high latitudes—even extending into the Arctic Circle, indicating ice-free conditions year round with summer temperatures reaching 20 °C. These forests were composed of broad-leaved plants, a little like the swamp-cypress forests and broadleaved floodplain forests seen today in the southeastern United States. These forests evidently could withstand 3 months of continuous light and 3 months of continuous darkness. Vertebrates in these polar forests included reptiles such as turtles and crocodiles as well as mammals resembling—though not closely related to—the tree-dwelling primates, small tapirs and small hippos of present-day tropical areas, while mammals common at lower latitudes were absent. **Figure 19–2** is a reconstruction of the subtropical conditions seen in Europe in the early Eocene.

Many of the early Cenozoic lineages of mammals are now extinct. These groups are often called *archaic mammals,* a rather pejorative term coming from our own perspective in the comfort of the Recent epoch. Archaic mammals were mainly small- to medium-size generalized forms, many of them arboreal. Larger specialized predators and herbivores with teeth suggesting a high-fiber herbivorous diet did not appear until the late Paleocene. Members of most present-day orders did not make their first appearance until the Eocene, but there are some notable exceptions, such as the order Carnivora (dogs, cats, and others) whose first members (miacoids) were present in the early Paleocene.

It appears that browsing by herbivorous dinosaurs kept forests at bay during the Cretaceous, much as large herbivorous mammals such as elephants maintain savanna habitats today. High-latitude tropical-like forests were established a mere million years or so into the Cenozoic, but it was not until the Eocene that more open canopied forests capable of supporting denser undergrowth vegetation appeared, and this is when the larger mammals began to radiate into the diversity of terrestrial niches they occupy today.

Modern types of birds also diversified at the start of the Cenozoic, including the passerines (songbirds), first known from the early Eocene, and large terrestrial birds including carnivorous forms and herbivores like the present-day ostrich. Many of the present-day groups of non-avian reptiles and amphibians had their origins in the latest Cretaceous or early Cenozoic. There was a modest radiation of terrestrial crocodiles in the Southern Hemisphere during the Paleocene and Eocene, but only aquatic crocodilians survived into the later Cenozoic. Among the insects, modern butterflies and moths first appeared in the middle Eocene.

The later Cenozoic world was generally cooler and drier than that of the early Cenozoic. Changes in terrestrial ecosystems reflected these climatic changes. Temperate forests and woodlands replaced the tropical-like vegetation of the higher latitudes, and tropical forests were confined to equatorial regions. Extensive grasslands first appeared in the Miocene epoch in the northern latitudes, forming swaths of savanna (grassland with scattered trees) across North America and

Figure 19–2 A reconstruction of a scene from the early Eocene of Europe. The trees in this tropical-like rain forest include sequoia, pine, birch, palmetto, swamp cypress, and tree ferns. In the foreground are cycads and magnolia. The birds are ibises, with a *Gastornis* (an extinct flightless predatory bird) in the background. The hippolike mammals are *Coryphodon,* belonging to the extinct ungulate-like order Pantodonta. An early primate (*Cantius,* an omomyid) climbs up a liana vine. Crouching among the ferns is the oxyaenid *Palaeonictis,* a catlike predator belonging to the extinct order Creodonta. Its potential prey are the early artiodactyls, *Diacodexis,* in the foreground.

central Asia. In the late Pliocene, 2 to 3 million years ago, savannas appeared in more tropical areas such as East Africa, and the more temperate grasslands turned into treeless prairie or steppe. This vegetational change coincided with the emergence of our own genus *Homo,* a hominid well adapted to life on the tropical savannas. New types of vegetation also appeared in the Plio-Pleistocene: tundra (treeless shrubland) and taiga (boreal evergreen forests) in the Arctic regions and deserts in the tropical and temperate regions.

The radiation of mammals in the late Cenozoic reflected these vegetational changes. Large grazing mammals such as horses, antelope, rhinoceroses, and elephants evolved along with the emerging grasslands—and with them the carnivores that preyed on them, such as large cats and dogs. Some small mammals also diversified, most notably modern types of rodents, such as rats and mice. This diversification of small mammals may explain the concurrent late Cenozoic diversification of modern types of snakes. Late Cenozoic lizards included some very large varanoids, including not only the largest lizard known today, the Komodo dragon, but also a crocodile-size predator, *Megalania,* in the Pleistocene of Australia. Many modern types of birds first appeared in the late Cenozoic, including birds of prey such as eagles, hawks, and vultures. The more open habitats of the late Cenozoic also favored the diversification of social insects that live in grasslands, such as ants and termites.

19.3 Cenozoic Climates

Today's world is a very cool and dry place in comparison with most of the Cenozoic (and, indeed, with most of Earth's history), and it is also quite varied in terms of habitats and climatic zones. At the start of the Cenozoic, the world was covered with tropical-like forests and with the types of animals that live in such habitats. We still have forests of this type, now confined to the equatorial regions, but we have added other types of environments at other latitudes (temperate woodland and grasslands) and also tropical grasslands plus very new types of habitats, such as deserts and arctic tundra. Thus the numbers of different types of animals have

increased enormously with this great diversity of habitat types. In addition, the division of the world into separate continents has meant that many faunas and floras have evolved in geographic isolation, thus boosting the total global diversity.

Paleogene (Paleocene–Oligocene) Climates

The world of the early Cenozoic still reflected the "hothouse world" of the Mesozoic: the tropics appear to have been hotter than those of the present day, as evidenced by the gigantic Paleocene South American snake, *Titanoboa.* There was an increase in the levels of atmospheric carbon dioxide at the start of the Eocene, which would have resulted in warming via a greenhouse effect, plus a rise in oxygen levels to present-day amounts that may have favored the larger placental mammals.

There was a transitory spike in temperature at the Paleocene-Eocene boundary (**Figure 19–3**), lasting around 100,000 years, known as the Paleocene-Eocene thermal maximum (PETM). The probable cause was the release of methane, a greenhouse gas, into the atmosphere from shallowly buried sediments on the ocean continental shelf. This warming had profound effects on mammalian evolution, and many of the modern orders of mammals appeared at this time. A more prolonged period of warm temperatures followed, peaking at around the early-middle Eocene boundary 50 million years ago (see Figure 19–3). From this high point, the higher latitude regions started to cool, with a rather precipitous drop in mean annual temperature (and in atmospheric CO_2 levels) in the latest Eocene and earliest Oligocene epochs, plunging the Earth into the start of the colder "icehouse" world of the later Cenozoic.

A primary cause of this climatic cooling was probably a reverse greenhouse effect related to sharply falling levels of atmospheric carbon dioxide, which plummeted at this time from around four times today's levels to a level resembling that of the present day. Additionally, cold polar-bottom water massed over the poles when Australia broke away from Antarctica and Greenland broke away from Norway. Ocean circulation carried the cold water toward the equator, cooling the temperate latitudes. The Antarctic ice cap probably formed by the

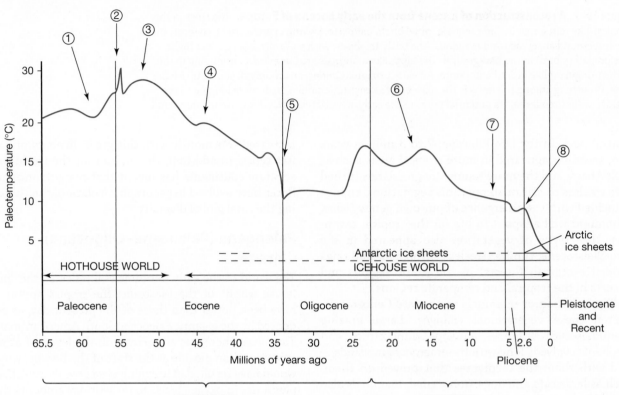

Figure 19–3 Mean annual paleotemperatures in the northern higher latitudes during the Cenozoic. The lines indicating the presence of ice sheets are dashed when representing minimal (partial or ephemeral) ice coverage and solid when representing maximal (full-scale and permanent) ice coverage. **1.** Tropical forests within the confines of the Arctic and Antarctic Circles. Archaic mammals predominate. **2.** Paleocene-Eocene Thermal Maximum. Transitory (100,000 years) time of high temperatures; many modern types of mammals appear. **3.** Early Eocene Climatic Optimum. Temperatures reach a Cenozoic maximum. Modern types of mammals diversify, and archaic mammals decline. **4.** Shortly following Middle Eocene Climatic Optimum. Cooling brings winter frosts to higher latitudes. Tropical forests replaced by woodlands in higher latitudes, and polar forests disappear; more tropical types of vertebrates (e.g., primates, many kinds of reptiles) now disappear from higher latitudes, and remaining archaic mammals become extinct. **5.** Eocene-Oligocene temperature plunge results in extinctions and an overall cooler world. **6.** Middle Miocene Climatic Optimum. Warming and initial drying bring wooded grasslands to higher latitudes. **7.** Late Miocene cooling and further drying result in spread of grasslands and loss of woodlands. In higher latitudes, grazing mammals predominate, browsers decline in numbers, all crocodiles and large turtles now disappear. **8.** Significant further cooling at 2.5 million years ago plunges the world into an ice age.

end of the Eocene, although a lasting Arctic ice cap did not form until some 30 million years later, in the latest Miocene. Following a rather cool early Oligocene, temperatures rose again in the higher latitudes in the late Oligocene.

Neogene (Miocene–Recent) Climates

Temperatures fell a little in the earliest Miocene, but then rose again and reached a late Cenozoic maximum in the middle Miocene, around 16 million years ago. This warming event may be due to the opening of Drake's Passage between Antarctica and South America, which isolated the cold polar water around Antarctica. However, expansion of the Antarctic ice cap in the late Miocene once again brought cooling to the higher latitudes, a trend that has persisted to the present day with occasional remissions. The Miocene world was in general drier than in the Paleogene, and the combination of warmth and dryness promoted the spread of grasslands.

The closing of the Isthmus of Panama that joined North and South America in the Pliocene epoch also profoundly affected global climate. Warm waters could no longer circle the equator, and it is during this time that the Gulf Stream was formed, carrying warm water from the subtropical American regions to Western Europe. Elsewhere in the world, disruption of oceanic circulation resulted in cooling and led to the formation of the Arctic ice cap. The polar ice caps have also been linked with declining levels of atmospheric carbon dioxide, apparent with the formation of the Antarctic ice cap in the late Eocene, its expansion in the late middle Miocene, and the formation of the Arctic ice cap in the late Pliocene.

The extensive episodic continental glaciers that characterize the Pleistocene epoch were events that had been absent from the world since the Paleozoic. These ice ages had an important influence not only on Cenozoic mammal evolution in general—almost all of today's species have their origin in the Pleistocene—but also on our own evolution and even on our present civilizations. We still live in a world with abundant polar ice, but right now we inhabit a warmer, interglacial period. An enormous volume of water is still locked up in glaciers and polar ice caps. The melting of the glaciers of the last glacial episode at the end of the Pleistocene, around 10 thousand years ago, caused the sea level to rise by about 140 meters (almost half the height of the Empire State Building) to its present relatively stable condition. If the present-day glaciers were to melt, sea level would rise at least another 50 meters, submerging most of the world's coastal cities and some countries. No wonder that present-day global warming is the focus of so much concern!

Today glaciers cover 10 percent of Earth's land surface, mostly in polar regions but also on high mountains. At times in the Pleistocene, an ice mass that was probably between 3 and 4 kilometers thick covered as much as 30 percent of the land and extended southward in North America to 38°N latitude (southern Illinois). A similar ice sheet covered northern Europe. However, much of Alaska, Siberia, and Beringia (Pleistocene land that is now underwater between Alaska and northeast Asia) were free of ice and housed a biome known as the Mammoth Steppe or the steppe-tundra, which has since disappeared. This steppe-tundra habitat was obviously much more productive than present-day high-latitude habitats: it contained highly diverse fauna of large mammals, combining present-day Arctic mammals such as reindeer and musk ox with now-tropical mammals such as lions and rhinoceroses.

These continental glaciers advanced and retreated several times during the Pleistocene. (The Southern Hemisphere was less affected because at that time the southern continental landmasses were farther from the poles than the northern ones, as they are today.) There were four major episodes of glaciation, interspersed with many (perhaps 20 or more) minor ones. Although the glaciers would have advanced slowly enough for animals to migrate toward the equator, problems would occur if routes were blocked by mountain ranges or seaways. Eurasian animals had the advantage of broad connections between temperate and tropical zones, both in Asia and in Africa. In contrast, North American animals would have to traverse the narrow Isthmus of Panama to reach the more tropical areas of South America. This geographic bottleneck may have limited the migration of certain types of mammals.

Drying of the ice-free portions of Earth due to the volume of water tied up in glaciers was at least as important for terrestrial ecosystems as the glaciers themselves. Many of the equatorial areas that today are covered by lowland rain forests were much dryer then, even arid. Today's relatively mild interglacial period is apparently colder and drier than other interglacial periods in the Pleistocene. For example, during other interglacial periods hippopotamuses were found in what is now the Sahara Desert and in England.

The episodes of glaciation were caused by Milankovitch cycles (named after the astronomer who proposed them in the 1930s). There are three different cycles of variation in the eccentricity of the Earth's orbit around the sun or in the tilt of Earth's axis, lasting from 23,000 years to 100,000 years, that affect the amount of solar radiation reaching Earth. Normally these cycles are out of phase with each other, like discordant keys played on a piano, but every so often they line up together like notes making a chord, with a large resultant effect on Earth's climate. It appears that glacial episodes get their start not from the world as a whole getting colder year-round, but from cool summers that prevent the melting of winter ice. It is important to realize that these Milankovitch cycles must have been in existence throughout Earth's history. However, it was only after the late Neogene formation of the Arctic ice cap that enough polar ice existed to plunge the Northern Hemisphere into a series of ice ages.

19.4 Cenozoic Extinctions

The best-known extinction of the Cenozoic occurred at the end of the Pleistocene, although this was by no means the extinction of greatest overall magnitude. The Pleistocene extinction (also known as the Quaternary extinction) appears dramatic because of the extinction of the megafauna (mammals larger than 20 kilograms in body mass). This included many very large mammals that are not only extinct but also not represented by any similar types today, such as glyptodonts (giant armored mammals related to armadillos) and ground sloths in North and South America, and diprotodontids (the largest of all marsupials, some as big as a hippopotamus) in Australia. The extinction also included many larger and exotic forms of more familiar mammals, such as the saber-toothed cats and the mammoths of the Northern Hemisphere and Africa; the Irish elk, cave bears, and woolly rhinoceroses of Eurasia; and giant kangaroos and wombats in Australia. Large terrestrial birds also suffered in these extinctions—including herbivores such as the moas of New Zealand and the elephant bird of Madagascar and carnivores such as the New World phorusrhacids. Although smaller mammals did not suffer as many extinctions, there was nonetheless a decline in the diversity of their communities.

There is much debate about the cause of these extinctions. The main extinctions occurred at the end of the last glacial period, which ended about 12,000 years ago. (Surprisingly enough, animals appear to be more vulnerable to extinction when the climate changes from glacial to interglacial rather than the other way around, probably because the effects of warming occur faster than the effects of cooling.) Thus climatic change is an obvious explanation for the extinctions.

However, many scientists have noted that it is only the last glacial period—rather than any of the previous ones—that brought extinctions of such magnitude. This observation suggests that part, if not all, of the blame for megafaunal extinctions should be placed on the spread of modern humans and modern hunting techniques, which occurred at that time. This is the overkill hypothesis, and the survival of mammoths until only a few thousand years ago on human-free Wrangel Island, off the coast of Siberia, appears to support this view. However, although the fossil record shows a rather sudden disappearance of large mammals in North America between 13,500 and 11,500 years ago, DNA from fossil sediments shows that mammoth and horses survived until 10,500 years ago in Alaska—well past the date when humans arrived. The extinctions were prolonged in Europe, and megafaunal animals persisted along with immigrant humans for a thousand years or more.

Researchers who favor climate change as the cause of the extinctions point out that the historic examples of extinction following the appearance of humans (whether by hunting or habitat destruction) apply to island situations rather than to continental ones.

The overkill and climate change hypotheses are not mutually exclusive, of course, and the two forces may have acted synergistically, with hunting pressure providing the last straw that drove already unstable populations to extinction, or a climate change proving catastrophic to a fine balance between human hunting and survival.

About 30 percent of mammalian genera became extinct at the end of the Pleistocene. That is approximately the magnitude of the other two major Cenozoic extinctions (late Eocene and late Miocene). However, the preceding two extinctions differ in several critical ways from that of the Pleistocene. Mammals of all body sizes (not just large ones) were affected in both the Eocene and Miocene. Other organisms, both terrestrial and marine, also experienced profound extinctions. That late Pleistocene extinction affected primarily large terrestrial mammals and birds, which are the species most likely to be viewed as prey or competitors by human hunters, is evidence in favor of the overkill hypothesis.

The late Eocene extinctions were associated with the dramatic fall in higher-latitude temperatures and the resulting environmental changes. The archaic mammals disappeared at this time, as did more modern types of mammals adapted to tropical-like forests, such as the early primates. The early Cenozoic diversity of amphibians and reptiles in higher latitudes was also greatly reduced during the late Eocene.

The late Miocene extinctions were associated with global drying as well as falling higher-latitude temperatures. The major extinctions were of browsing mammals (including a variety of large browsing horses), which suffered habitat loss as the savanna woodlands turned into open grasslands and prairie. North America was especially hard hit by the climatic events of the late Miocene because of its relatively high latitudinal position and the fact that animals could not migrate to more tropical areas in South America before the formation of the Isthmus of Panama in the Pleistocene.

Additional Information

Archibald, J. D. 2011. *Extinction and Radiation: How the Fall of Dinosaurs Led to the Rise of Mammals*. Baltimore, MD: Johns Hopkins University Press.

Barnosky, A. D., and B. P. Kraatz. 2007. The role of climatic change in the evolution of mammals. *BioScience* 57: 523–532.

Barnosky, A. D., and E. L. Lindsey. 2010. Timing of Quaternary megafaunal extinctions in South America in relation to human arrival and climate change. *Quaternary International* 217:10–29.

Blois, J. L., and E. A. Hadley. 2009. Mammalian response to Cenozoic climatic change. *Annual Review of Earth and Planetary Sciences* 37:181–208.

Blois, J. L., et al., 2010. Small mammal diversity loss in response to late-Pleistocene climatic change. *Nature* 465:771–774.

Burney, D. A., and T. F. Flannery. 2005. Fifty millennia of catastrophic extinctions after human contact. *Trends in Ecology and Evolutionary Biology* 20:395–401.

Eberle, J. J. and D. R. Greenwood. 2012. Life at the top of the greenhouse Eocene world—A review of the Eocene flora and vertebrate fauna from Canada's High Arctic. *Geological Society of America Bulletin* 124:3–23.

Eldrett, J. S., et al. 2009. Increased seasonality through the Eocene to Oligocene transition in northern high latitudes. *Nature* 459:969–974.

Faith, J. T., and T. A. Surovell. 2009. Synchronous extinction of North America's Pleistocene mammals. *Proceedings of the National Academy of Sciences* 106:20641–20645.

Falkowski, P. G., et al. 2005. The rise of oxygen over the past 205 million years and the evolution of large placental mammals. *Science* 309:2202–2204.

Godfrey, L. R., and M. T. Irwin. 2007. The evolution of extinction risk: Past and present anthropogenic impacts on primate communities of Madagacar. *Folia Primatologia* 78:405–419.

Guthrie, R., 2004. Radiocarbon dating of mid-Holocene mammoths stranded on an Alaskan Bering Sea island. *Nature* 441:746–749.

Haile, J., et al. 2009. Ancient DNA reveals late survival of mammoth and horse in interior Alaska. *Proceedings of the National Academy of Sciences* 196:22353–22357.

Head, J. J., et al. 2009. Giant boid snake from the Palaeocene neotropics reveals hotter past equatorial temperatures. *Nature* 457:715–718.

Janis, C. M. 1993. Tertiary mammal evolution in the context of changing climates, vegetation, and tectonic events. *Annual Review of Ecology and Systematics* 24:467–500.

Koch, P. L., and A. D. Barnosky. 2006. Late Quaternary extinctions: State of the debate. *Annual Review of Ecology, Evolution and Systematics* 37:215–250.

Lister, A. M. 2004. The impact of Quaternary Ice Ages on mammalian evolution. *Philosophical Transactions of the Royal Society* 359:221–241.

McGlone, M. 2012. The hunters did it. *Science* 335:1452–1453.

Pearson, P. N., et al. 2009. Atmospheric carbon dioxide through the Eocene-Oligocene climate transition. *Nature* 461:1110–1114.

Prideaux, G. J., et al. 2010. Timing and dynamics of Late Pleistocene mammal extinctions in southwestern Australia. *Proceedings of the National Academy of Scicnces* 107:22157–22162.

Prothero, D. R. 2006. *After the Dinosaurs: The Age of Mammals*. Bloomington, IN: Indiana University Press.

Stickley, C. E., et al. 2009. Evidence for middle Eocene Arctic sea ice from diatoms and ice-rafted debris. *Nature* 460:376–380.

Tripati, A. K., et al. 2009. Coupling of CO_2 and ice sheet stability over major climatic transitions of the last 20 million years. *Science* 326:1394–1397.

Wilson, G. P. 2012. Adaptive radiation of multituberculates before the radiation of dinosaurs. *Nature* 483:457–483.

Zachos, J., et al. 2008. An early Cenozoic perspective on greenhouse warming and carbon-cycle dynamics. *Nature* 451:279–283.

Websites

Cenozoic era

http://www.bobainsworth.com/fossil/cenozoic.htm
http://www.fossilmuseum.net/Paleobiology/Cenozoic_Paleobiology.htm

Megalania and other Pleistocene giants

http://www.pbs.org/wgbh/nova/bonediggers/vani-01.html
http://australianmuseum.net.au/Megalania-prisca

Pleistocene extinctions

http://cpluhna.nau.edu/Biota/megafauna_extinctions.htm
http://scienceblogs.com/laelaps/2008/01/eurasian_pleistocene_extinctio.php

CHAPTER 20

Mammalian Diversity and Characteristics

Cenozoic mammals are adapted to a wide variety of lifestyles and display great diversity in body size, form, and ecology. In addition to the familiar features of lactation and hair, mammals have other derived characters that set them apart from other vertebrates. An understanding of mammalian diversity and specializations requires an understanding of how placentals differ from marsupials as well as how therians (marsupials and placentals) differ from monotremes. Much of the diversity seen among Cenozoic therian mammals reflects convergent evolution due to the isolation of different groups of mammals on different continental landmasses. The changing climates of the higher latitudes during the Cenozoic era also resulted in a wider diversity of mammals adapted to new habitats, such as grasslands, and changes in sea level and continental positions resulted in various mammalian intercontinental migrations.

20.1 Major Lineages of Mammals

Although mammals are perceived as the dominant terrestrial animals of the Cenozoic, their extant species diversity (about 5488 species) is only a little more than half that of birds (about 10,000 species) and is considerably less than that of lepidosaur reptiles (about 9000 species). Mammalian species diversity is in fact about the same as that of amphibians (about 6800 species), making mammals and amphibians the smallest lineages of tetrapods (at least in terms of numbers of species).

Mammals do, of course, include the largest living terrestrial and aquatic vertebrates (the blue whale, which weighs about 110,000 kilograms, is twice the estimated weight of the largest dinosaur). Mammals have great morphological diversity; no other vertebrate group has forms as different from each other as a whale is from a bat. Even among strictly terrestrial mammals, there is a tremendous morphological difference between, for example, a mole and a giraffe. However, it is prudent to consider that mammals do not rise above other vertebrates when some measures of evolutionary success, such as species diversity, are considered.

Traditionally, the class Mammalia has been divided into three subclasses: **Allotheria** (multituberculates, now extinct), **Prototheria** (monotremes), and **Theria.** This classification does not take into account the large diversity of Mesozoic mammals (see Chapter 18), but these three groups do reflect basic divisions in body plans among those mammals that survived into the Cenozoic. Therians are subdivided into two infraclasses: Metatheria, including marsupials and their extinct Mesozoic stem-group relatives, and Eutheria, including placentals and their extinct stem-group relatives. The term *placental* mammal is a bit of a misnomer because marsupials also have a placenta (see Section 21.1 in Chapter 21), but this is a legacy of earlier terminology. "Marsupials" and "placentals" are the commonly-used terms that refer to the radiation of extant forms.

Multituberculates and monotremes were originally seen as being very distantly related to therians. Living monotremes retain some ancestral anatomical features in addition to their egg-laying habits and were once considered to be an independent radiation from the cynodonts. The teeth of multituberculates are so different from those of other early mammals that they, too, were thought to have a very early offshoot from the main lineage. However, we now know of a diversity of Mesozoic mammals basal to these groups. Most workers consider multituberculates to be more closely related to monotremes than to therians based on the more derived nature of their shoulder girdle, although their brains (as determined from skull endocasts) were not much changed from the basal mammalian condition.

Living monotremes are toothless as adults, so the characters of dental anatomy on which much of mammalian phylogeny is based cannot easily be used to evaluate their relationships. However, teeth are found in juvenile platypuses, and teeth are known from fossils. A single jaw of a fossil (*Steropodon*) from the Early Cretaceous of Australia shows fully formed teeth that are clearly triangular, unlike the less complex teeth of many Mesozoic mammals.

Multituberculates—Rodentlike Mammals of the Mesozoic Era

Multituberculates were a very long-lived group, known from the Late Jurassic period to the early Cenozoic (late Eocene epoch). Multituberculates were probably small terrestrial and semiarboreal omnivores like extant rodents. Their very narrow pelvis suggests that they did not lay eggs but may have given birth to extremely altricial young.

Some multituberculates were rather squirrel-like: the structure of their ankle bones shows that they could rotate their foot backward to descend trees head first, like a squirrel, and their caudal vertebrae indicate a prehensile tail (**Figure 20–1**). These squirrel-like multituberculates also had an enlarged lower posterior premolar that formed a shearing blade, perhaps used to open hard seeds. Other multituberculates were more terrestrial and rather like a ground hog or a wombat.

Figure 20–1 **The Early Cenozoic multituberculate** *Ptilodus*.

The extinction of multituberculates in the Eocene might have been due to competition with rodents, which first appeared in the late Paleocene epoch.

Multituberculates get their name from their molars, which are broad, multicusped (multituberculed) teeth specialized for grinding rather than shearing. Wear on the teeth indicates that multituberculates moved their lower jaw backward while bringing their teeth into occlusion. The teeth and jaw movements of multituberculates were also similar to those of rodents, except rodents move their lower jaws *forward* into occlusion.

20.2 Mammalian Ordinal Diversity

One current hypothesis of interrelationships of living therian mammal orders is shown in **Figure 20–2**. Controversies about the interrelationships among the various placental mammal orders are discussed later.

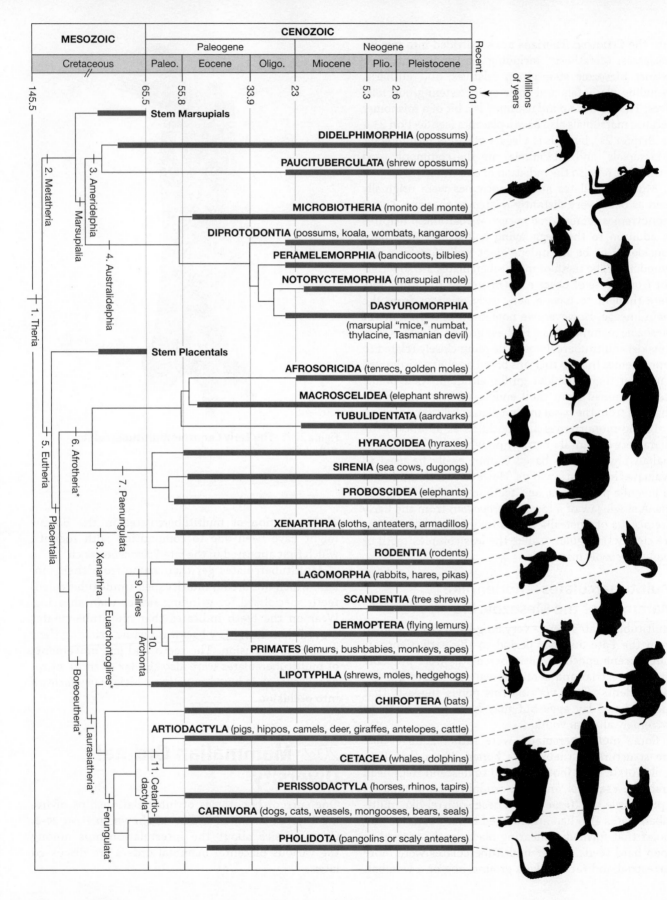

Figure 20–2 Phylogenetic relationships of extant mammalian orders, excluding Monotremata. This diagram depicts the probable relationships among the major groups of living therian mammals. The black lines show interrelationships only; they do not indicate times of divergence or the unrecorded presence of taxa in the fossil record. Only the best-corroborated relationships are shown. Numbers at the branch points indicate derived characters that distinguish the lineages—see the Appendix for a list of these characters. This cladogram is based on a combination of morphological and molecular characters. Higher taxa for which the primary (or only) evidence is molecular are indicated by an asterisk (*). Afrotheria is placed here as the basal placental clade; note, however, that other researchers promote Afrotheria and Xenarthra as sister groups or one or both groups in a more nested position within the other placentals. Also note that some recent molecular analyses place the bats (Chiroptera) as the sister group to the ungulates (Cetartiodactyla + Perissodactyla).

Monotremes

Monotremes are grouped in the infraorder Ornithodelphia (Greek *ornitho* = bird and *delphy* = womb, referring to the single functional oviduct in the platypus and many birds; contrast with the *didelphid* opossums) and in the order Monotremata (Greek *mono* = one and *trema* = hole, referring to the cloaca—the single opening of the excretory and reproductive tracts). (An old term for therians was *Ditremata,* referring to the two separate openings.) There are two families: the Ornithorhynchidae (Greek *rhynchus* = beak) contains the platypus, a semiaquatic animal that feeds on aquatic invertebrates in the streams of eastern Australia and Tasmania, and the Tachyglossidae (Greek *tachy* = fast and *glossa* = tongue) contains two types of echidnas, the short-nosed echidna of Australia (which eats mainly ants and termites) and three species of long-nosed echidnas in New Guinea that include earthworms in their diet (**Figure 20–3**).

The small teeth of Cretaceous Australian monotremes suggest a previously unsuspected Mesozoic diversity of these animals, but all the animals were very small. The few known Cenozoic monotreme fossils represent only echidnas or platypuses, and monotreme diversity overall has probably always been fairly limited. That is to say, the platypus and echidna are not the relicts of a once much larger radiation but rather have been its mainstay. A handful of teeth from the Paleocene of Patagonia is the only evidence of monotremes outside of Australasia. Molecular data show a relatively recent (19–48 million years) split of the echidna from the platypus. Since early Cenozoic platypuses seem to be much like the modern form, it has been suggested that echidnas may have originated from a semiaquatic lineage that became more terrestrial.

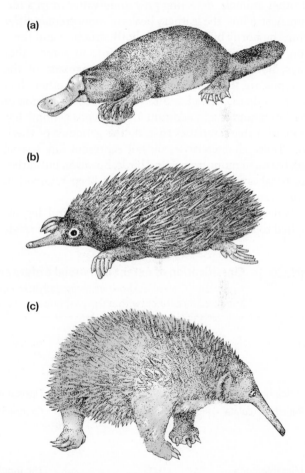

(a)

(b)

(c)

Figure 20–3 Diversity of living monotremes. (a) The platypus, *Ornithorhynchus anatinus.* (b) The short-nosed echidna (Australia), *Tachyglossus aculeatus.* (c) The long-nosed echidna (New Guinea), *Zaglossus bruijni.* (The species are drawn approximately to scale; the platypus is the size of a large house cat.)

Figure 20–4 Diversity of living marsupials. (a) The common North American opossum, *Didelphis virginiana* (Didelphidae: Didelphimorphia). (b) The shrew opossum, *Lestoros inca* (Caenolestidae: Paucituberculata). (c) The monito del monte, *Dromiciops australis* (Microbiotheriidae: Microbiotheria). (d) The Tasmanian devil, *Sarcophilus harrisii* (Dasyuridae: Dasyuromorphia). (e) The marsupial mole, *Notoryctes typhlops* (Notoryctidae: Notoryctemorphia). (f) The bilby, or rabbit-eared bandicoot, *Macrotis lagotis* (Thylacomyidae: Peramelemorphia). (g) The honey possum, *Tarsipes rostratus* (Tarsipedidae, Phalangeroidea, Diprotodontia). (h) The koala, *Phascolarctos cinereus* (Phascolarctidae, Phascolartoidea, Diprotodontia). (i) The long-nosed potoroo (rat kangaroo), *Potorous tridactylus* (Macropodidae, Macropodoidea, Diprotodontia). (The species are drawn approximately to scale; the North American opossum is the size of a large house cat.)

The sprawling stance of monotremes, reminiscent of reptiles, may be a specialization for swimming or digging rather than a basal mammalian condition. Monotremes also have their own unique specializations. Both platypuses and echidnas lack teeth as adults and have a leathery bill or beak. This beak contains receptors that sense electromagnetic signals from the muscles of other animals, detecting prey underwater or in a termite nest. Thus, the leathery beaks of monotremes differ from the horny bills of birds in both structure and function. Monotremes are similar to birds in some other ways, however. For example, the spermatozoa of the platypus are threadlike, like those of birds, rather than having a distinct head and tail like the spermatozoa of therian mammals. In addition, the platypus genome has more bird characteristics than do the genomes of therians. These characteristics do not represent any special link between monotremes and birds, however, but rather are basal amniote features that have been retained in monotremes and birds but lost in therians.

Male platypuses have a spur on the hind leg attached to a venom gland, which is used to poison rivals or predators. A similar spur is seen in some Mesozoic mammals, so this may not be a unique monotreme feature. The possession of venom is a rare feature in mammals. Other extant venomous mammals are found in the order Lipotyphla and include a few species of shrews and the cat-size *Solenodon* that occurs on some Caribbean islands. These mammals have venom in their saliva, which they inject via a bite with their canines. *Solenodon* (but not shrews) has a groove on the lower canine to assist with the venom delivery, and a similarly grooved upper canine is seen in *Bisonalveus*, a Paleocene insectivorous mammal that is not closely related to any living mammal.

Marsupials

Extant marsupials can be divided into at least four lineages that are equivalent in morphological and genetic diversity to placental orders. The current scheme of marsupial interrelationships is shown in Figure 20–2 and **Table 20–1**, and **Figure 20–4** illustrates the diversity of living marsupials.

Table 20–1 Classification of extant marsupial orders and approximate numbers of families, genera, and species.
Note that different authorities recognize different numbers of families within an order. This table lists all the extant and recently extinct (within the past hundred years) taxa.

Order	Number of Families/ Genera/Species	Major Examples and Higher-Level Classification
		Ameridelphia
Didelphimorphia	1/17/87	Opossums; 20 g to 6 kg; Neotropical region (plus one North American species)
Paucituberculata	1/3/6	Caenolestids or rat opossums; 15 to 40 g; Neotropical region
		Australidelphia
Microbiotheria	1/1/1	The monito del monte; about 25 g; Neotropical region
Dasyuromorphia	3/20/69	Marsupial mice, native cats, Tasmanian devil, Tasmanian tiger (thylacine), and marsupial anteater (numbat; 5 g to 20 kg; Australian region
Notoryctemorphia	1/1/2	Marsupial mole; 50 g; Australian region
Peramelemorphia	3/8/23	Bandicoots and bilbies; 100 g to 5 kg; Australian region
Diprotodontia	11/40/143	Possums, flying phalangers, cuscuses, honey possum (noolbenger), koala, wombats, potoroos, wallabies, and kangaroos; 12 g to 90 kg; Australian region; possums and wallabies introduced into New Zealand by humans

There is a fundamental split in the living marsupials into the Ameridelphia of the New World and the Australidelphia of (mainly) Australia. This division was originally proposed based on the ankle morphology and has now been supported by molecular and genetic data. A third group, now extinct and possibly more basal than either of these living groups, was the Deltatheroida of the Late Cretaceous of Asia. Note that the original metatherian/eutherian divergence was most likely in Asia.

The ameridelphian order Didelphimorphia includes didelphoids (opossums plus related extinct forms, such as the Northern Hemisphere marsupials of the early Cenozoic) and some extinct Cretaceous forms. Present-day opossums are quite a diverse group of small- to medium-size marsupials, mainly arboreal or semiarboreal omnivores, including animals such as the herbivorous woolly opossum and the otterlike yapok. A much greater diversity of didelphoids existed in the early to middle Cenozoic, including jerboa-like hopping forms and molelike burrowing forms. Other ameridelphians are the Paucituberculata (caenolestids or rat opossums), which are small, terrestrial, shrewlike forms. The borhyaenoids (Eocene-Pliocene), which are now placed in the order Sparassodonta, which is outside of the phylogenetic grouping of extant marsupials, formed a large radiation of carnivorous forms, including ferretlike, doglike, and bearlike forms, and even a Neogene saber-tooth parallel, *Thylacosmilus*.

The one remaining type of South American marsupial is *Dromiciops,* the *monito del monte,* a tiny mouse-like animal living in the montane forests of Chile and Argentina, placed in its own distinct order Microbiotheria. However, this animal is more closely related to the Australian marsupials than to the other South American ones, and it may be a distant relict of the stock of originally South American marsupials that migrated across Antarctica to Australia in the early Cenozoic.

The Australian Australidelphia fall into three major orders. The Dasyuromorphia is mainly composed of the carnivorous dasyurids: the marsupial cats and marsupial mice (which would be better called marsupial shrews, since they are carnivorous and insectivorous rather than omnivorous). Some larger dasyurids include rather doglike forms known today only from Tasmania, an island off the south coast of Australia: the Tasmanian devil and the Tasmanian tiger or thylacine (also known as the "marsupial wolf" and sometimes placed in its own family, Thylacinidae), which has been extinct since 1936. The numbat (the marsupial anteater) is related to the dasyurids (included with Dasyuromorphia in Figure 20–2), and the tiny marsupial mole *Notoryctes* (in its own order Notoryctemorphia) may be a more distant relation.

The Peramelemorphia includes the peramelids, the bandicoots and bilbies. The long-eared bilbies look a little like rabbits, and Australians celebrate an "Easter Bilby" rather than an "Easter Bunny," but they are insectivorous rather than herbivorous. Peramelids share with the final group of marsupials (the diprotodontians) a condition of the hind feet known as syndactyly, in which the second and third toes are reduced in size and enclosed within the same skin membrane so that they appear to be a single, double-clawed toe. The syndactylous toes are used for grooming.

The largest group of marsupials is the Diprotodontia. Diprotodontians get their name from their modified lower incisors, which project forward rather like the incisors of rodents (Greek *di* = two, *proto* = first, and *dont* = tooth). This lineage includes herbivorous and omnivorous forms today, although marsupial "lions" such as the Pleistocene *Thylacoleo* appear to represent an independent evolution of carnivory from an herbivorous ancestry. *Thylacoleo* shows some interesting specializations: coming from an herbivorous ancestry, it had lost its canines but modified the incisors into caninelike teeth.

The three major radiations within the diprotodontians are the phalangeroids, phascolarctoids, and macropodoids. Phalangeroids represent an arboreal radiation of somewhat primatelike animals, also including gliders. They comprise six families, including possums, phalangers, ringtails, cuscuses, and the diminutive honey possum or noolbenger, which is one of the few nectar-eating mammals apart from the bats. Phascolarctoids or vombatiformes include the arboreal koala and the terrestrial, burrowing wombats. Extinct phascolarctoids include the sheep- to rhinoceros-size diprotodontids, which looked like giant wombats and grazed on the Plio-Pleistocene Australian savannas. Macropodoids, with a few exceptions, are the hopping marsupials: they include the small, omnivorous rat kangaroos (or potoroos) and the larger, herbivorous true kangaroos (including wallabies and tree kangaroos, the latter having abandoned hopping for climbing). The largest kangaroos today have a body mass of about 90 kilograms, but larger ones (up to three times that size) existed in the Pleistocene, including a radiation of one-toed, short-faced browsing sthenurine kangaroos, now all extinct. Sthenurines, with their short faces and long forearms, could be the source of legends of giant rabbits in Australia's interior. Perhaps, as desert forms, they also had large rabbitlike ears for cooling.

Placentals

Extant placental mammals can be grouped into a number of distinct lineages, but their interrelationships have been a source of controversy, which suggests that the diversification of these groups from an ancestral stock occurred very rapidly without leaving many morphological clues. The current scheme of the interrelationships among placental mammals is shown in Figure 20–2 and **Table 20–2. Figure 20–5** on page 490 illustrates the diversity of placental mammals.

Over the past decade, molecular studies have resulted in a very different view of placental interrelationships from phylogenies based on morphological data alone. Probably the most radical difference is the creation of an endemic grouping of African mammals, the Afrotheria. A grouping of Paenungulata—the rodentlike hyraxes (the conies of the Bible), the sirenians (manatees and dugongs), and the proboscideans (elephants and extinct relatives)—had long been supported by morphology, but molecular data show that aardvarks, elephant shrews, tenrecs (Madagascan insectivorous forms), and the golden mole are also closely related to these larger African mammals. This endemic grouping, which usually appears to fall at or near the base of the placental phylogeny, is indicative of a long period of independent isolated evolution of mammals on the African continent. This fits with the patterns of Cenozoic continental movements as Africa was isolated from Eurasia during the first half of the Cenozoic.

The grouping of sirenians, proboscideans, and the now extinct desmostylians is termed the Tethytheria. Both sirenians and desmostylians are aquatic forms, and some early proboscideans have anatomical features that suggest a semiaquatic existence; this is also borne out by isotopic signatures in their teeth. Later proboscideans may represent a secondary return to a more terrestrial existence from an originally semiaquatic stock. However, popular speculation that the elephant's trunk was originally a snorkel is mistaken; the skulls of early fossil proboscideans show that these animals lacked trunks.

The next major grouping of placental mammals, long considered basal on morphological grounds, comprises the order Xenarthra: sloths, anteaters, and armadillos. These animals are sometimes known as *edentates,* meaning "without teeth." Only anteaters are completely toothless, but all of these animals have simplified their dental pattern. Xenarthrans are South American endemics and have been found on the North American continent only since the late Miocene epoch. The pangolins or scaly anteaters (order Pholidota) were originally thought to be related to the xenarthrans based on morphological features, but these features represent convergences due to a similar lifestyle, and molecular data place them as the sister taxon to the Carnivora.

All other placental orders are placed in the supergrouping Boreoeutheria (meaning "northern mammals"). This grouping can be subdivided into the Euarchontoglires (rodents, primates, and relatives) and the Laurasiatheria (insectivores, bats, carnivores, and ungulates, including whales). Bats (order Chiroptera) were long considered to belong with primates in the grouping Archonta, an assemblage that also includes tree shrews (once thought to be basal primates) and dermopterans (flying lemurs—actually gliders and not lemurs at all). However molecular evidence has reassigned bats as closer to the insectivores, and the remaining archontans are now termed the Euarchonta. The Glires includes the rodents and rabbits. Elephant shrews were originally placed in this grouping, but molecular data have assigned them to the Afrotheria.

Insectivores (originally considered to belong to a single order, "Insectivora") were often considered to be the basal stock from which other placentals were derived. However, although modern insectivores such as shrews may superficially resemble ancestral placental mammals, they are not closely related to them. With the loss of tenrecs and golden moles to the Afrotheria, the remaining insectivores are now placed in the order Lipotyphla. The Lipotyphla is sometimes divided into two orders: the Erinaceomorpha (hedgehogs and relatives) and the Soricomorpha (shrews and relatives).

The largest grouping within the Laurasiatheria is the Ferungulata, mainly comprising carnivores and ungulates. The Carnivora (informally termed *carnivorans* to distinguish them from other, unrelated carnivorous mammals) are distinguished by having a pair of specialized shearing teeth, or **carnassials,** formed from the upper last (fourth) premolar and the lower first molar. Carnivorans include not only specialized carnivores, such as cats and dogs, but also secondarily herbivorous forms, such as the panda bear, and one of the three living groups of secondarily aquatic mammals, the pinnipeds (seals, sea lions, and walruses), which are related to bears.

The extant ungulates (hoofed mammals), the **perissodactyls** (odd-toed ungulates) and **artiodactyls** (even-toed ungulates), were previously placed with some extinct groups in the higher taxon Ungulata, but molecular data imply that these two orders, although closely related, may not be each other's closest relatives. Some archaic ungulates in the early Cenozoic were also omnivorous or even carnivorous, including the mesonychids, once thought to be ancestral to whales. In South America, there were several hoofed orders that are now entirely extinct. These extinct orders are probably related to the living ungulates, but of course they cannot be slotted into a molecular phylogeny.

Table 20–2 **Classification of extant placental orders and approximate numbers of families, genera, and species.**
Note that different authorities recognize different numbers of families within an order. The geographic regions (shown in Figure 20–17) represent the distribution of the entire orders (and for the present day only); families within an order often have smaller distributions.

Order	Number of Families/ Genera/Species	Major Examples and Higher-Level Classification
	Afrotheria	
Afrosoricida	2/19/51	Tenrecs, otter shrews, and golden moles; 5 g to 2 kg; Afrotropical region
Macroscelidea	1/4/16	Elephant shrews; 25 to 500 g; Afrotropical region with one species in Morocco and northern Africa
Tubulidentata	1/1/1	Aardvark; 64 kg; Afrotropical region
Hyracoidea	1/3/4	Hyraxes (conies or dassies); 4 kg; Afrotropical region and Asia Minor
Sirenia	2/2/4	Dugongs and manatees; 140 to more than 1000 kg; coastal waters and estuaries of all tropical and subtropical oceans except the eastern Pacific (in the Atlantic drainage they enter rivers)
Proboscidea	1/2/3	Elephants and fossil relatives; 4500 to 7000 kg; Afrotropical and Oriental regions
	Xenarthra	
Cingulata	1/9/21	Armadillos; 120 g to 60 kg; Neotropical region (plus some in southern United States)
Pilosa	4/5/10	Sloths and anteaters: 350 g to 40 kg; Neotropical region
	Boreoeutheria	
	Euarchontoglires	
	Glires	
Lagomorpha	2/12/91	Rabbits, hares, and pikas; 180 g to 7 kg; worldwide except Antarctica, introduced in Australia by humans
Rodentia	33/481/2277	Rats, mice, squirrels, guinea pigs, and capybara; 7 g to more than 50 kg; worldwide except Antarctica

(Continued)

A surprise in the reexamination of mammalian interrelationships over the past couple of decades has been the realization that whales and dolphins (order Cetacea) are technically ungulates (even though they lack hooves), related to the artiodactyls. This relationship had occasionally been proposed on anatomical groups, but it was not universally accepted until it was confirmed by molecular studies. Discoveries of the limbs of early whales show that they had ankles with the characteristic artiodactyl "double pulley" astragalus. Molecular phylogenetic studies now place cetaceans within the artiodactyls, with hippos as their closest relatives. This placement renders the traditional notion of Artiodactyla paraphyletic, and whales and artiodactyls are now grouped as the single order *Cetartiodactyla*.

Chapter 21 will discuss the evolutionary history of whales in more detail.

20.3 Features Shared by All Mammals

As diverse as mammals are, they are united by a suite of characteristics that define them and the ecological and behavioral zones that they can occupy. Lactation has already been identified as a critical characteristic of mammals because it is an essential precursor to the precise occlusion of teeth that characterizes mammals (see Chapter 18). Skeletomuscular characteristics of mammals underlie functions as diverse as facial expressions used in social behavior and the

Table 20–2 Continued

Order	Number of Families/ Genera/Species	Major Examples and Higher-Level Classification
		Boreoeutheria
		Euarchonta
Scandentia	2/5/20	Tree shrews; 400 g; Oriental region
Primates	15/69/376	Lemurs, monkeys, apes, and humans; 85 g to more than 275 kg; primarily Oriental, Afrotropical and Neotropical regions, humans are now worldwide
Dermoptera	1/2/2	Flying lemur; 1 to 2 kg; Oriental region
		Laurasiatheria
Lipotyphla*	3/54/542	Hedgehogs, moles, and shrews; 2 g to 1 kg; worldwide except Australia and Antarctica (although only a single species of shrew is known from South America, a Pleistocene immigrant)
Chiroptera	19/202/1120	Bats; 4 g to 1.4 kg; worldwide (including New Zealand) except Antarctica
		Ferungulata
Carnivora	15/132/286	Dogs, bears, raccoons, weasels, hyenas, cats, sea lions, walruses, and seals (these last three are often assigned to the suborder Pinnipedia); 70 g to 760 kg, some marine forms over 100 kg; worldwide
Pholidota	1/1/8	Pangolins (scaly anteaters); 2 to 33 kg; Afrotropical and Oriental regions
Perissodactyla	3/17	Odd-toed ungulates: horses, tapirs, and rhinoceroses; 150 to 3600 kg; worldwide except Antarctica (domestic horses introduced by humans into North America and Australia)
Artiodactyla[†]	10/90/241	Even-toed ungulates: swine, hippopotamuses, camelids, deer, giraffes, antelopes, sheep, and cattle; 2 to 2500 kg; worldwide except Antarctica (introduced into Australia and New Zealand by humans)
Cetacea[†]	11/40/82	Porpoises, dolphins, sperm whales, and baleen whales; 20 to 120,000 kg; worldwide in oceans and in some rivers and lakes in Asia, South America, North America, and Eurasia

Originally Insectivora, now sometimes split into two separate orders, Erinaceomorpha and Soricomorpha.
[†]*These two orders are now combined into the Cetartiodactyla.*

types of locomotion they employ. Characteristics of the skin and pelage control interactions with the physical environment, and modifications of the integument form structures that extend from head (horns) to toe (hooves). The cardiovascular, respiratory, and urogenital systems are closely coordinated, and the sensory systems still retain traces of the origin of mammals as nocturnal animals.

Lactation

The females of all mammalian species lactate, feeding their young by producing milk. Mammary glands are entirely absent from the males of marsupials, but they are present and potentially functional in male monotremes and placentals. There are two species of bats in which the males lactate, and there have been examples of human males producing milk under certain circumstances. It has long been a mystery why male mammals retain mammary glands but do not lactate; indeed, breast cancer affects human males as well as females.

Although all mammals lactate, only therians have nipples so that the young can suck directly from the breast rather than from the mother's fur. Mammary hairs, probably a basal mammalian feature, are present in monotremes and marsupials, and the mammae de-

Figure 20–5 Diversity of living placentals. (a) Two-toed sloth, *Choloepus didactylus* (Megalonychidae, Xenarthra); cat size. (b) Common tenrec, *Tenrec ecaudatus* (Tenrecidae, Tenrecoidea, Afrotheria); rat size. (c) Golden-rumped elephant shrew, *Rhynchocyon chrysopygus* (Macroscelididae, Macroscelidae, Afrotheria); rat size. (d) Naked mole rat, *Heterocephalus glaber* (Bathyergidae, Rodentia, Euarchontoglires); mouse size. (e) Flying lemur, *Cynocephalus volans* (Cynocephalidae, Dermoptera, Euarchontoglires); cat size. (f) Spotted hyena, *Crocuta crocuta* (Hyaenidae, Carnivora, Laurasiatheria); wolf size. (g) Asiatic tapir, *Tapirus indicus* (Tapiridae, Perissodactyla, Laurasiatheria); pony size. (h) African water chevrotain, *Hyemoschus aquaticus* (Tragulidae, Artiodactyl, Laurasiatheria); rabbit size. (i) Rock hyrax, *Procavia capensis* (Procaviidae, Hyracoidea, Afrotheria); rabbit size. (The species are not drawn to scale.)

velop from areola patches confined to the abdominal region. Placentals lack mammary hairs, and the mammae develop from mammary lines that form along the entire length of the ventral body surface.

Skeletomuscular System

In Chapter 18 we discussed the evolution of the mammalian skeletal system within the synapsid lineage. Here we will focus on those features that are uniquely mammalian in comparison with other extant amniotes.

Cranial Features In the mammalian skull, the dermal bones that originally formed the skull roof have grown down around the brain and completely enclose the braincase (**Figure 20–6**). The bones that form the lower border of the synapsid temporal opening are bowed out into a zygomatic arch, which we commonly refer to as the cheekbone. This is the bony bar that you can feel just below your eyes.

The dentition of mammals is divided into several types of teeth (a condition called heterodonty): incisors, canines, premolars, and molars. Most placental mammals have two sets of dentitions in their lifetime (marsupials replace only one tooth, the last premolar). The first set—or the milk teeth—consists of incisors, canines, and premolars only, although the form of these premolars may be like that of the adult molars. The permanent, adult dentition consists of the second set of the original teeth with the addition of the later-erupting molars. (Our last molars are known as wisdom teeth, so called because they erupt at the age— late teens—at which we supposedly have attained wisdom.) Mammals are the only animals that masticate (chew the food) and swallow a discrete bolus of food comprised of small food particles mixed with saliva (the entire act being termed *deglutination*).

Therian mammals have unique types of molars called **tribosphenic molars.** The basal type of mammalian molar, exemplified by the basal mammal

Morganucodon, has the three main molar cusps in a more or less straight line. In the more derived type, exemplified by the contemporaneous holotherian *Kuehneotherium,* the principal, middle cusp is shifted so that the teeth assume a triangular form in occlusal (food's-eye) view (**Figure 20–7** on page 493). The apex of the triangle formed by the upper tooth points inward, while that of the lower tooth points outward, forming an intermeshing relationship called reversed triangle occlusion. The longer sides of these triangular teeth result in a greater amount of available area for shearing action.

In the more derived type of tribosphenic molar, a new cusp, the protocone, is added in the uppers, which occludes against a basined addition to the lowers called the talonid. The tribosphenic molar adds the function of crushing and punching to the original tooth, which acted mainly to cut and shear. The presence of this tooth presumably reflects a greater diversity of dietary items taken.

Postcranial Features Unlike the sprawling posture of extant reptiles, mammals have a more upright posture with the limbs positioned underneath the body (**Figure 20–8** on page 494). However, the fully upright posture of familiar mammals such as cats, dogs, and horses is a derived one; the semisprawling stance of a mammal such as the opossum probably represents the generalized mammalian condition.

Mammals have an ankle joint that differs from that of other amniotes. The main site of movement is not within the bones of the ankle joint (the mesotarsal joint); it is between the tibia and one of the proximal ankle bones, the astragalus. This new ankle joint allows only fore-and-aft movement of the foot on the leg, which is good for rapid locomotion. The original intratarsal joint is retained in some mammals to allow for movement within the foot, such as inversion and eversion.

Along with this new joint, mammals have a projection of the other proximal ankle bone, the calcaneum,

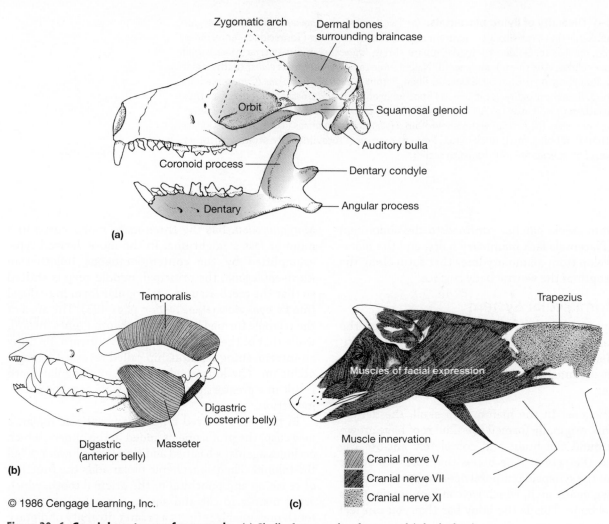

Figure 20–6 Cranial anatomy of mammals. (a) Skull of a generalized mammal (a hedgehog): the condyle on the dentary fits into the glenoid on the squamosal to form the jaw joint. (b) The muscles of mastication. (c) Superficial view of the muscles of facial expression plus shoulder muscles innervated by cranial nerves.

to form the calcaneal heel. The heel is the point of insertion of the gastrocnemius (calf) muscle. In the pelvic girdle, the ilium is rod shaped and directed forward, and the pubis and ischium are short in contrast to the more platelike pubis and ischium of reptiles. The femur has a distinct trochanter on the proximal lateral side (the greater trochanter) for the attachment of the gluteal muscles, which are now the major retractors of the hindlimbs. It is the gluteals that give mammals their characteristic rounded rear ends.

With very few exceptions, all mammals have seven cervical (neck) vertebrae. Manatees and one type of tree sloth (the two-toed form, *Choloepus*) have six cervical vertebrae. The three-toed tree sloth (*Bradypus*)

appears to have 8 to 10 cervical vertebrae, but developmental studies show that the "extra" ones (numbers 8–10) are in fact modified thoracic vertebrae.

Mammals also have a uniquely specialized atlas-axis complex of the first two cervical vertebrae; a mammal can rotate its head on its neck in two places: not only in the more general up-and-down fashion (at the joint between the skull and the atlas) but also in the more derived side-to-side fashion (at the joint between the atlas and the axis). Mammals have restricted the ribs to the anterior (thoracic) trunk vertebrae (although small lumbar ribs are still present in some Mesozoic mammals).

The lumbar vertebrae now have zygapophyseal connections that allow dorsoventral flexion; they also have

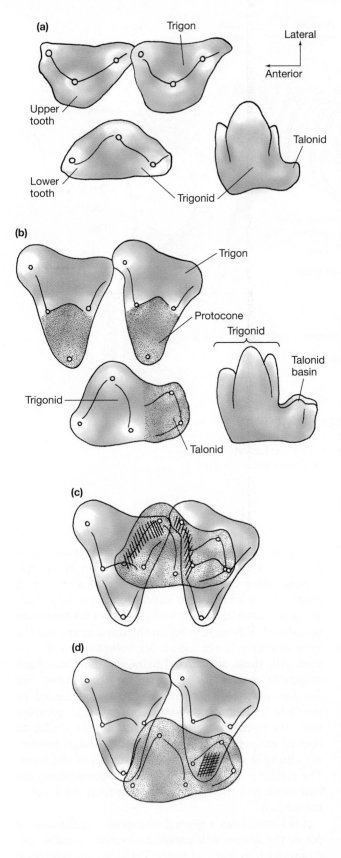

(a)

Trigon

Lateral

Anterior

Upper tooth

Talonid

Lower tooth

Trigonid

(b)

Trigon

Protocone

Trigonid

Talonid basin

Trigonid

Talonid

(c)

(d)

Figure 20–7 Evolution of mammalian molars. (a) Schematic occlusal view of reversed triangle molars of a nontherian holotherian mammal (e.g., *Kuehneotherium*). The lower molar is also illustrated in side view. (b) Similar view of the tribosphenic molars of a therian. The new portions (the protocone in the uppers and the basined talonid in the lowers) are shaded. Parts (c) and (d) show the action of the lower molars in occlusion with the uppers: the lower molar is shaded, and areas of tooth contact are cross-hatched. (c) Initial contact between the teeth at the start of occlusion. The trigonid cusps produce a shearing action alongside the cusps at the back of the trigon of the anterior upper tooth and at the front of the trigon of the posterior upper tooth. (d) Mortar-and-pestle action of the molars at the end of the occlusal power stroke. The protocone (the pestle of the combination) fits into the basin formed between the cusps in the trigonid (the mortar).

large transverse processes for the attachment of the longissimus dorsi (one of the epaxial muscles) that produces this movement during locomotion. The capacity to twist the spine in both lateral and dorsoventral directions in mammals may relate to their ability to lie down on their sides, something that other vertebrates cannot do easily. This ability may have been important in the evolution of suckling because the nipples are on the ventral surface of the trunk.

Epiphyses on the long bones reflect the mammalian feature of determinate growth. The **epiphyses** are the ends of the bones that are separated from the shaft (the diaphysis) by a zone of growth cartilage in immature mammals. At maturity, the ossification centers in the epiphyses and the shaft of the bone fuse, and the epiphyses no longer appear as distinct structures.

The Integument

The variety of mammalian integuments is enormous and reflects the importance of the skin in maintaining the internal body temperature in an endothermic animal, especially one living in a harsh climate. Some small rodents have a delicate epidermis only a few cells thick. Human epidermis varies from a few dozen cells thick over much of the body to more than a hundred cells thick on the palms and soles. Elephants, rhinos, hippos, and tapirs were once classified together as pachyderms (Greek *pachy* = thick and *derm* = skin) because their epidermis is several hundreds of cells thick. The texture of the external surface of the epidermis varies from smooth (in fur-covered skins and the hairless skin of whales and dolphins) to rough and crinkled (many hairless terrestrial mammals). The tails of opossums and many rodents are covered by epidermal

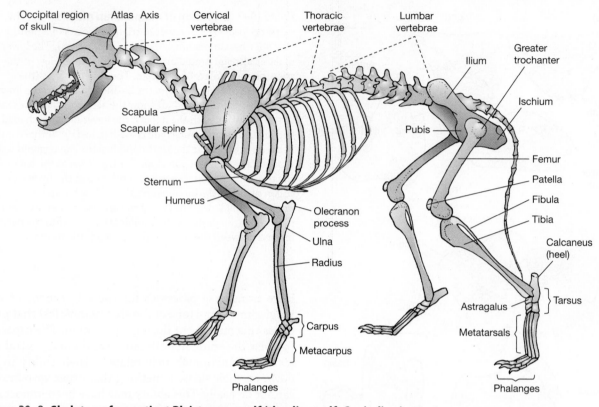

Figure 20–8 Skeleton of an extinct Pleistocene wolf (the dire wolf, *Canis dirus*). The long neural spines of the thoracic vertebrae serve as the attachment area of the nuchal ligament, which runs from the back of the head and helps to hold the head up. This animal is standing on the phalanges in a digitigrade form of foot posture, in contrast with the more generalized plantigrade foot posture of the opossum in Figure 20–14b.

scales similar to those of lizards but lacking the hard beta keratin found in birds and reptiles.

Figure 20–9 illustrates the typical structure of mammalian skin. Note that, although mammalian skin is like that of other vertebrates in basic form, with epidermal, dermal, and hypodermal layers, there are also unique components. Mammalian skin has hair and lubricant- and oil-producing glands that secrete volatile substances, water, and ions. Typical mammalian structures derived from the keratinous layer of the epidermis include nails, claws, hoofs, and horns. Sensory nerve endings include free nerve endings (probably pain receptors), beaded nerve nets around blood vessels, Meissner's corpuscles (touch receptors), Pacinian corpuscles (pressure receptors), nerve terminals around hair follicles, and heat and cold receptors. Vascular plexuses (intertwined blood vessels) of the skin are the basis for countercurrent blood flow that allows for heat to be gained or lost via the skin.

Hair Hair is composed of keratin, and it grows from a deep invagination of the germinal layer of the epidermis called the hair follicle. Hair has a variety of functions in addition to insulation, including camouflage, communication, and sensation via **vibrissae** (whiskers). Vibrissae grow on the muzzle, around the eyes, or on the lower legs; touch receptors are associated with these specialized hairs, and hair may have originally evolved for these sensory properties. Fur consists of closely placed hairs, often produced by multiple hair shafts arising from a single complex root. Its insulating effect depends on its ability to trap air within the fur coat (or pelage), and its insulating ability is proportional to the length of the hairs. The erector pili muscles that attach midway along the hair shaft pull the hairs erect to deepen the layer of trapped air.

Cold stimulates a general contraction of the erector pili via the sympathetic nerves, as do other stressful conditions such as fear and anger. This autonomic reaction

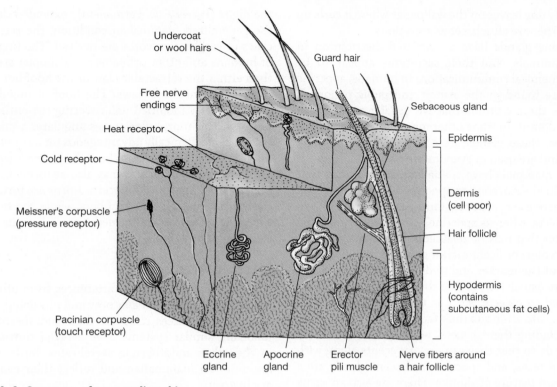

Figure 20–9 **Structure of mammalian skin.**

still occurs in hairless mammals, such as ourselves, where it produces dimples (goose bumps) on the skin's surface. Hair erections can serve for communication as well as for thermoregulation; mammals use them to send a warning of fear or anger as seen in the display of a puffed-up cat or the raised hackles of a dog.

The color of hair depends on the quality and quantity of melanin injected into the forming hair by melanocytes at the base of the hair follicle. The color patterns of mammals are built up by the colors of individual hairs. Because exposed hair is nonliving, it wears and bleaches. Replacement occurs by growth of an individual hair or by **molting,** in which old hairs fall out and are replaced by new hairs. Most mammals have pelage that grows and rests in seasonal phases; molting usually occurs only once or twice a year.

Glandular Structures Secretory structures of the skin develop from the epidermis. Modern amphibians (but not fish) have a glandular skin, and the condition in mammals probably represents the basal tetrapod condition, reduced or restricted in sauropsids. There are three major types of skin glands in mammals: **eccrine, sebaceous,** and **apocrine glands.** Except for the eccrine glands, skin glands are associated with hair follicles, and

the secretion in all of them is under neural and hormonal control. A full component of these skin glands is found in monotremes as well as in therians, and thus they may be assumed to be a basic feature of all mammals.

Eccrine glands produce a secretion that is mainly watery, with little organic content. In most mammals, eccrine glands are restricted to the soles of the feet, prehensile tails, and other areas that contact environmental surfaces, where they improve adhesion or enhance tactile perception. Eccrine glands are found over the body surface only in primates and especially in humans, where they secrete copious amounts of fluid for evaporative cooling. Profuse thermoregulatory sweating has evolved convergently in different lineages of mammals: eccrine glands function as sweat glands in humans, whereas ungulates sweat via apocrine glands. The sweat glands of humans are not a general mammalian feature. Other mammals can induce evaporative cooling by other means; for example, dogs pant and kangaroos lick their forearms.

Sebaceous glands are found over the entire body surface. They produce an oily secretion, sebum, which lubricates and waterproofs the hair and skin. Sheep lanolin, our own greasy hair, and the grease spots that the family dog leaves on the wallpaper where it curls up in the corner are all sebaceous secretions.

Apocrine glands have a restricted distribution in most mammals, and their secretions appear to be used in chemical communication. In humans, apocrine glands are found in the armpit and pubic regions—these are the secretions that we usually try to mask with deodorant. In some other mammals, such as large ungulates, these glands are scattered over the body surface and are used in evaporative cooling.

Many mammals have specialized scent glands that are modified sebaceous or apocrine glands. Sebaceous glands secrete a viscous substance, usually employed to mark objects, whereas apocrine glands produce volatile substances that may be released into the air as well as placed on objects. Scent marking is used to indicate the identity of the marker and to define territories. Scent glands are usually on areas of the body that can be easily applied to objects, such as the face, chin, or feet. Domestic cats often rub their face and chin to mark objects, including their owners. Many carnivorans have anal glands so that they can deposit scents along with urine and feces, and apocrine anal scent glands are a well-known feature of skunks. There are also apocrine glands in the ear that produce earwax.

Mammary glands have a more complex, branching structure than do other skin glands, but they appear to have been derived from some sort of basal apocrine gland because the detailed mode of secretion (exocytosis) is similar to these glands, as are some aspects of their development. Note that the common notion that mammary glands are "modified sweat glands" is a serious misnomer.

Claws, Nails, Hooves, and Horns Some integumentary appendages are involved in locomotion, offense, defense, and display. Claws, nails, and hooves are accumulations of keratin that protect the terminal phalanx of the digits (**Figure 20–10**). Permanently extended claws are the generalized mammalian condition; the retractable claws of catlike carnivores are derived. The fingernails of humans and other primates are a simpler structure than either the retractable claw or the hoof but are derived from ancestral claws. The hoof of ungulates is an extensively modified nail covering the entire third phalanx. This morphology gives ungulates a small foot (which is mechanically advantageous for a running animal) that is solid enough to bear the animal's weight.

The horns of ungulates may also be formed, at least in part, from keratin. Rhinoceros horns are formed entirely from matted keratin fibers, while the horns of cows and antelope are formed from a keratin sheath covering a bony core.

Internal Anatomy

Mammals have numerous differences from other amniotes in their internal anatomy and physiology. Some of these differences relate to their endothermic metabolism; similar systems have evolved convergently in the other endothermic vertebrates, birds. Others are uniquely mammalian and reflect their evolutionary history.

Adipose Tissue Mammalian adipose tissue (white fat) is distributed as subcutaneous fat, as fat associated with various internal organs (e.g., heart, intestines, kidneys), as deposits in skeletal muscles, and as cushioning for joints. Adipose tissue is not simply inert material used only as an energy store when fasting or as an insulating layer of blubber in marine mammals, as has often been assumed. More recent studies have revealed that adipocytes (fat cells) secrete a wide variety of messenger molecules that coordinate important metabolic processes. Some small fat storage sites have specific properties that equip them to interact locally with the immune system and possibly other organs.

Figure 20–10 Skin appendages associated with terminal phalanges. The unguis is the fully keratinized portion of the appendage (claw, nail, or hoof), and the subunguis is the less keratinized portion underlying it. (a) Retractable claws. *Left:* Hair and thick epidermal pads associated with the base of the claws (generalized mammalian condition). *Center:* Longitudinal section of a claw showing its close relationships with the tendons, dermis, and bone of the third (terminal) phalanx. *Right:* Claw retraction mechanism characteristic of cats. (b) The hoof of a horse. *Left:* Normal appearance of the hoof of a shod horse. (Horseshoes are devices used to minimize wear of the hoof on unnaturally hard and abrasive surfaces.) *Right:* Longitudinal section of lower front foot showing relationship of phalanges to hoof. (c) The human nail. *Left:* The external appearance of the nail. *Right:* A longitudinal section of the end of a finger showing the association of the nail with the epidermis and terminal phalanx of the digit. (d) Horns of ungulates. *Left:* The horn of a cow (animal facing toward the left), a bony core covered with a keratin sheath. *Right:* The horn of a rhinoceros (animal facing toward the right), made entirely from keratin fibers.

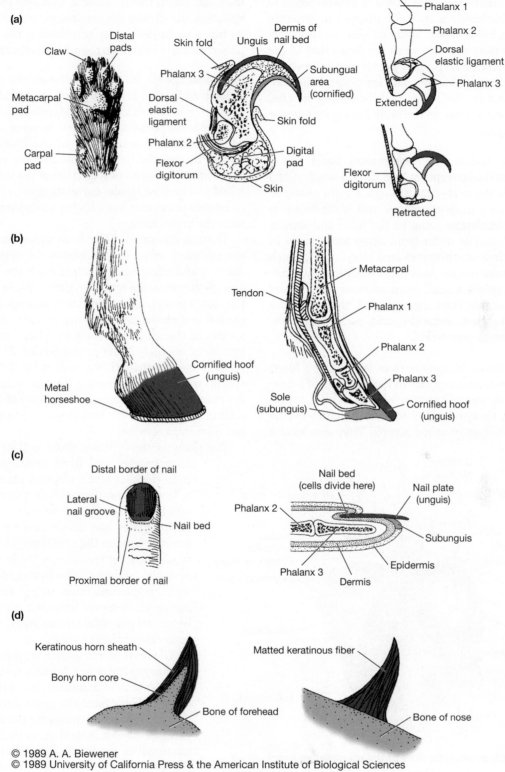

(a)

Claw · Distal pads
Metacarpal pad
Carpal pad

Skin fold · Unguis · Dermis of nail bed
Phalanx 3
Dorsal elastic ligament
Phalanx 2
Flexor digitorum
Subungual area (cornified)
Skin fold
Digital pad
Skin

Phalanx 1
Phalanx 2
Dorsal elastic ligament
Phalanx 3
Extended

Flexor digitorum
Retracted

(b)

Metal horseshoe
Cornified hoof (unguis)

Tendon
Metacarpal
Phalanx 1
Phalanx 2
Phalanx 3
Sole (subunguis)
Cornified hoof (unguis)

(c)

Distal border of nail
Lateral nail groove
Nail bed
Proximal border of nail

Nail bed (cells divide here)
Nail plate (unguis)
Phalanx 2
Subunguis
Phalanx 3
Dermis
Epidermis

(d)

Keratinous horn sheath
Bony horn core
Bone of forehead

Matted keratinous fiber
Bone of nose

© 1989 A. A. Biewener
© 1989 University of California Press & the American Institute of Biological Sciences

Mammals also have a unique type of adipose tissue—brown fat. This tissue is specially adapted to generate heat and can break down lipids or glucose to generate energy as heat at a rate up to ten times that of the muscles. Brown fat is especially prominent in newborn mammals, where it is important for general thermoregulation, and in the adults of hibernating species, where it is used to rewarm the body during emergence from hibernation.

Cardiovascular System The mammalian heart has a complete ventricular septum and only a single (left) systemic arch (**Figure 20–11**), although a right systemic arch is apparent in development and contributes to part of the circulation going to the head and arm in the adult. Mammals differ from other vertebrates in the form of their erythrocytes (red blood cells), which lack nuclei in the mature condition. Additionally, while monotremes retain a small sinus venosus as a distinct chamber, therians have incorporated this structure into the right atrium as the sinoatrial node, which now acts as the heart's pacemaker.

Respiratory System Mammals have large, lobed lungs with a spongy appearance due to the presence of a finely branching system of bronchioles in each lung, terminating in tiny thin-walled, blind-ending chambers (the sites of gas exchange) called alveoli. They also have a

muscular sheet, the diaphragm, that aids the ribs in inspiration and divides the original pleuroperitoneal cavity into a peritoneal cavity surrounding the viscera and paired pleural cavities surrounding the lungs.

Urogenital System All mammals retain the original tetrapod bladder that is lost in many sauropsids, and they excrete relatively dilute urine. Mammals have entirely lost the renal portal system seen in other vertebrates, which supplies venous blood to the kidney in addition to the arterial blood supplied by the renal artery. Mammals also have a new portion of the kidney tubule called the loop of Henle, correlating with their ability to excrete urine that has a higher concentration of salt than the body fluids.

Therian mammals differ from other vertebrates in various ways. In most vertebrates, the urinary, reproductive, and alimentary systems reach the outside via a single common opening, the cloaca, while in therians the cloaca is replaced by separate openings for the urogenital and alimentary systems. In most (but not all) species of therians, the testes are placed in a scrotum outside of the body in males (see Section 21.1), and the penis is used for urination as well as for the passage of sperm, with the ducts leaving the testes and the bladder combined into a single urethra (**Figure 20–12**). The clitoris in the female is the homologue of the penis but is not used to pass urine.

The glans is the bulbous distal end of the penis. Monotremes and most marsupials have a bifid (forked) glans, whereas placentals have a single glans. Some male placentals have a bone in their penis, the os penis or **baculum**; females may also have a corresponding structure, called the os clitoris. This structure is seen among certain species in nonhuman primates, rodents, insectivores, carnivorans, and chiropterans (more astute students may notice a mnemonic here) but is presumed to be a basal placental feature and thus homologous among these groups.

The urethra and vagina are joined in a single **urogenital sinus** leading to the outside in most female mammals, but primates and some rodents have the more derived condition of separate openings for the urinary and genital systems. Note that, in the more usual mammalian condition, the clitoris is within the urogenital sinus, where it receives direct stimulation from the penis during copulation, a rather more practical arrangement than in humans. Primates also are unusual in having a pendulous penis that cannot be

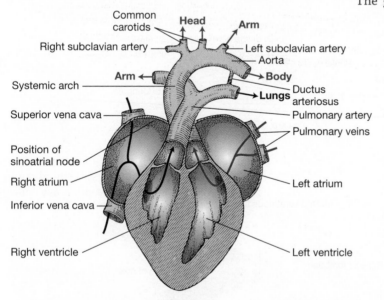

Figure 20–11 Diagrammatic view of the mammalian heart. Black arrows indicate the passage of deoxygenated blood, colored arrows the passage of oxygenated blood. The ductus arteriosus (ligamentum arteriosum in the adult condition) is the remains of the dorsal part of the pulmonary arch; it is functional in the amniote fetus, where it is used as a bypass shunt for the lungs, but closes after birth.

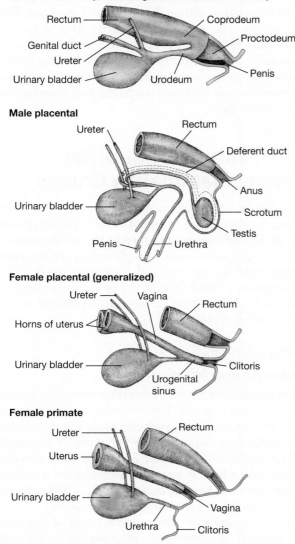

Male monotreme (similar to general amniote condition)

Rectum

Coprodeum

Proctodeum

Genital duct

Ureter

Penis

Urinary bladder

Urodeum

Male placental

Ureter

Rectum

Deferent duct

Anus

Scrotum

Urinary bladder

Testis

Penis

Urethra

Female placental (generalized)

Ureter

Vagina

Rectum

Horns of uterus

Clitoris

Urinary bladder

Urogenital sinus

Female primate

Ureter

Rectum

Uterus

Urinary bladder

Vagina

Urethra

Clitoris

Figure 20–12 Anatomy of the urogenital ducts in mammals. The head is to the left.

retracted into the body. Most male mammals extend the penis from an external sheath (normally the only visible portion) only during urination and copulation, again a seemingly more practical arrangement.

Sex Determination and Sex Chromosomes

Mammals always have distinct sexes, and sex determination is by distinctive sex chromosomes, the X and Y chromosomes: females have two X chromosomes, and males have an XY combination. However, the platypus is peculiar in having multiple sex chromosomes: males

have five X and five Y chromosomes, and the mode of sex determination is unclear. In therians it appears that a gene located on the Y chromosome in mammals initiates male gonadal development, and female gonadal development results from its absence. Once a gonad has had its primary sex declared as female or male, one of the sex hormones—estrogen or testosterone—is produced. These hormones affect development of the secondary sex characteristics. In humans, the genitalia, breasts, hair patterns, and differential growth patterns are secondary sex characteristics. Horns, antlers, and dimorphic color patterns are familiar differences that we associate with gender in other mammals.

Sensory Systems

The sensory systems of mammals differ from those of other tetrapods in various ways. Mammals have exceptionally large brains among vertebrates, and their brains evolved along a pathway somewhat different from that of other amniotes. Mammals are more reliant on hearing and olfaction and less reliant on vision than are most other tetrapods.

The Brain The enlarged portion of the cerebral hemispheres of mammals, the neocortex or neopallium, forms differently from the enlarged forebrain of derived sauropsids (see Section 11.6). Other unique features of the mammalian brain include an infolded cerebellum and a large representation of the area for cranial nerve VII, which is associated with the facial musculature. In addition to the anterior commissure, a structure that links the hemispheres in all amniotes, placentals have a new nerve tract linking the two cerebral hemispheres—the corpus callosum.

Olfaction The keen sense of smell of most mammals is probably related to their primarily nocturnal behavior. The olfactory receptors are located in specialized epithelium on the scroll-like nasoturbinal and ethmoturbinal bones in the nose. The olfactory bulb is a prominent portion of the brain in many mammals, but primates have a relatively small olfactory portion of the brain and a poor sense of smell, probably related to their diurnal and arboreal habits. The sense of smell is also reduced or completely absent in the toothed whales. Interestingly, both humans and whales retain a large number of genes for the sense of olfaction, but many of these genes (all of them in the toothed whales) are nonfunctional pseudogenes, providing confirmation that these mammals had ancestors with a better sense of smell than their extant descendants.

Vision Mammals evolved as nocturnal animals, and visual sensitivity (forming images in dim light) was more important than visual acuity (forming sharp images). Mammals have retinas composed of primarily rod cells (a derived condition, most vertebrate retinas are cone-dominated), which have high sensitivity to light (thus providing good night vision) but are relatively poor at acute vision. Acute vision is possible in only one small region of the retina, the all-cone fovea.

In addition to providing high visual acuity, cones are the basis for color vision. However, most mammals can perceive some color and have monochromatic (one visual color pigment) or dichromatic (two pigments) vision. Genetic data show that at least one gene coding for a visual pigment in other vertebrates has been lost in mammals.

Anthropoid primates are unique among mammals in having good color vision and a brain that is specialized for the visual sensory mode. Loss of the olfactory genes by anthropoids coincided with the duplication of one of the visual pigment genes, resulting in trichromatic color vision.

Hearing The evolution of the mammalian three-boned middle ear was discussed in Chapter 18. Several other features contribute to increased auditory acuity in therian mammals. These include a long cochlea, coiled so that it can fit inside the otic capsule, capable of precise pitch discrimination (**Figure 20–13**). In addition, the external ear (**pinna**) helps to determine sound direction. The pinna, in combination with the narrowing of the external auditory meatus (earhole) of mammals, concentrates sound from the relatively large area encompassed by the external opening of the pinna to the small, thin, tympanic membrane. The pinna is unique to therian mammals, although it has a feathery analogue

in certain owls. Most mammals can move their pinnae to detect sound, although anthropoid primates lack this ability. The auditory sensitivity of a terrestrial mammal is reduced if the pinnae are removed. Aquatic mammals use entirely different systems to hear underwater and have reduced or lost the pinnae. Cetaceans, for example, use the lower jaw to channel sound waves to the inner ear.

20.4 **Features That Differ Among Mammalian Groups**

Each lineage of extant mammals has distinctive characteristics that define elements of the lifestyles of each group. The differences in reproductive modes among monotremes, marsupials, and placentals are the most conspicuous differences, and they have far-reaching consequences. Differences in the structure of the skeleton and the dentition are also deeply rooted in the biology of each group.

Differences Between Therians and Nontherians

Therians are distinguished from monotremes by a number of derived features. The most obvious one is giving birth to young rather than laying eggs. The differences in reproduction between different types of mammals will be considered in Chapter 21. Therians are also distinguished by having mammae with nipples, a cochlea in the inner ear with at least two and a half coils, an external ear (pinna), and tribosphenic molars.

Therians also have distinctive features of the skull and skeleton. Therians have completely lost the sclerotic bony rings around the eyes. Monotremes retain sclerotic cartilages, although they do not ossify to form a bony ring, as seen in many other amniotes, including non-mammalian synapsids. While all mammals have a specialized ankle joint between the tibia and the astragalus, in therians the astragalus bone in the ankle sits directly on top of the calcaneum, rather than being placed side by side with it as in the generalized amniote condition (**Figure 20–14**), although this was apparently evolved convergently in placentals and marsupials. Monotremes have only partial superposition of the astragalus on the calcaneum and in addition retain the contact between the fibula and calcaneum as part of the functional joint. This more derived ankle joint in therians probably made possible new forms of locomotion, such as bounding and hopping.

The therian shoulder girdle is also extremely derived. While monotremes have the derived mammalian

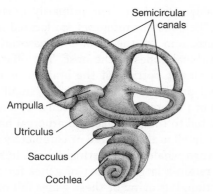

Semicircular canals

Ampulla

Utriculus

Sacculus

Cochlea

© 1998 The McGraw-Hill Companies, Inc.

Figure 20–13 Design of the vestibular apparatus in therian mammals. The long, coiled cochlea is equivalent to the lagena of other vertebrates.

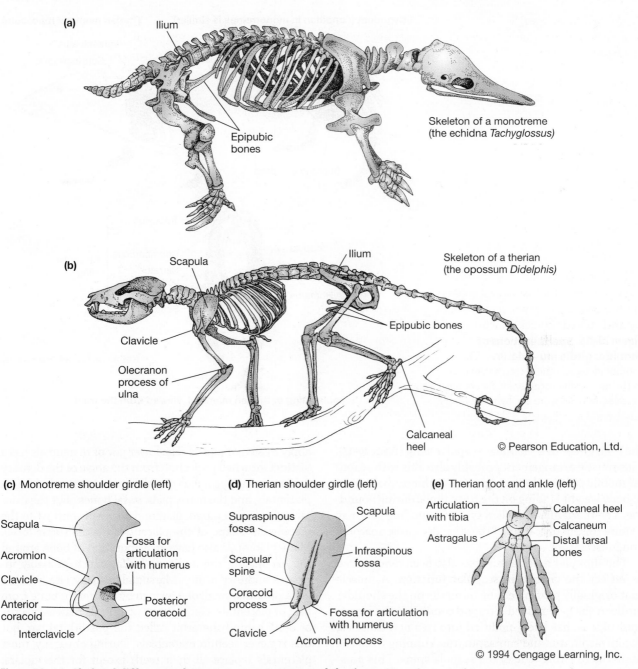

(a)

Ilium

Epipubic
bones

Skeleton of a monotreme
(the echidna *Tachyglossus*)

(b)

Scapula

Ilium

Skeleton of a therian
(the opossum *Didelphis*)

Clavicle

Epipubic bones

Olecranon
process of
ulna

Calcaneal
heel

© Pearson Education, Ltd.

(c) Monotreme shoulder girdle (left)

Scapula

Acromion

Clavicle

Anterior
coracoid

Interclavicle

Fossa for
articulation
with humerus

Posterior
coracoid

(d) Therian shoulder girdle (left)

Supraspinous
fossa

Scapula

Scapular
spine

Infraspinous
fossa

Coracoid
process

Clavicle

Fossa for articulation
with humerus

Acromion process

(e) Therian foot and ankle (left)

Articulation
with tibia

Calcaneal heel

Astragalus

Calcaneum

Distal tarsal
bones

© 1994 Cengage Learning, Inc.

Figure 20–14 Skeletal differences between monotremes and therians. Note that epipubic
bones, present in monotremes and marsupials, have been lost in placentals.

form of pelvis, their shoulder girdle is more reminiscent of the typically reptilian condition. Therians have lost the ventral elements of the shoulder girdle, the coracoid and interclavicle bones, although these bones are retained in marsupial newborns (see Chapter 21). Therians also have expanded the scapula with the addition of a new area of bone (the supraspinous fossa). This sits in front of the scapular spine, which is in part homologous with the original anterior border of the

scapula, although some portion of it is also a new addition. The clavicle (collarbone) is retained in most therians but is lost in many running-adapted placentals (e.g., dogs and horses).

This reduction of the ventral elements of the shoulder girdle allows the scapula to move as an independent limb segment around its dorsal border, adding to the length of the stride during locomotion. Additionally, certain hypaxial muscles have become modified to hold

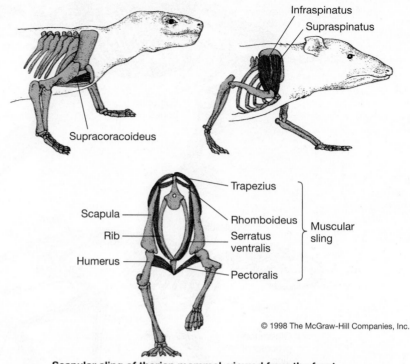

Cynodont (condition in monotremes is similar)

Therian mammal (opossum)

Infraspinatus

Supraspinatus

Supracoracoideus

Trapezius

Scapula

Rhomboideus

Rib

Serratus ventralis

Muscular sling

Humerus

Pectoralis

© 1998 The McGraw-Hill Companies, Inc.

Scapular sling of therian mammal, viewed from the front

Figure 20–15 Specializations of shoulder girdle musculature. The rhomboideus and the serratus ventralis in the mammalian scapular sling are new muscles, derived from the hypaxial layers.

the limb girdle in a muscular scapular sling (**Figure 20–15**). This muscle arrangement probably also aids with scapular mobility during locomotion and cushions the impact of body weight landing on the front limbs during bounding. The therian mammalian type of bounding gait may require this modification of the shoulder girdle anatomy and musculature.

The shoulder musculature has also been reorganized to reflect the change in scapular function. A muscle that originally protracted the humerus on the shoulder girdle in the generalized tetrapod condition, the supracoracoideus, has been modified into two new muscles, the infraspinatus and supraspinatus, running from the humerus to either side of the scapula spine. This muscular arrangement likely helps to stabilize the limb on the shoulder girdle during bounding.

Differences Between Marsupials and Placentals

Although there are numerous differences between marsupials and placentals in their reproductive biology, which will be discussed in Chapter 21, there are few major differences in their reproductive anatomy.

Features in the skull and dentition distinguish marsupials from placentals, although not all of these features

apply to all marsupials. The lower jaw of marsupials has a distinct inturned projection from the angle of the dentary (where the pterygoideus muscle inserts) that is lacking in placentals, and the marsupials' nasal bones abut the frontal bones with a flared, diamond shape in contrast to the rectangular shape of the placentals' nasals (**Figure 20–16**). Many placentals also have an elaboration of bone around the ear region, the **auditory bulla,** which probably increases auditory acuity. Marsupials usually either lack a bulla or have a small one formed by a different bone from that of placentals. Herbivorous placentals may have a bar of bone behind the orbit called the postorbital bar, but this is never seen in marsupials. During ontogeny, most placentals replace all their teeth except for the molars, whereas marsupials replace only the last premolar.

Marsupials also differ from placentals in their dental formula, or numbers of different types of teeth. The basal dental formula of placentals (i.e., the maximum number of teeth usually seen, as counted on one side of the skull only) is three upper and three lower incisors (a total of 12 incisors, counting both sides), one upper and one lower canine (a total of 4), four upper and four lower premolars (a total of 16), and three upper and three lower molars (a total of 12). Many placentals have fewer teeth than this, but only a few mammals with highly specialized diets, such as armadillos and

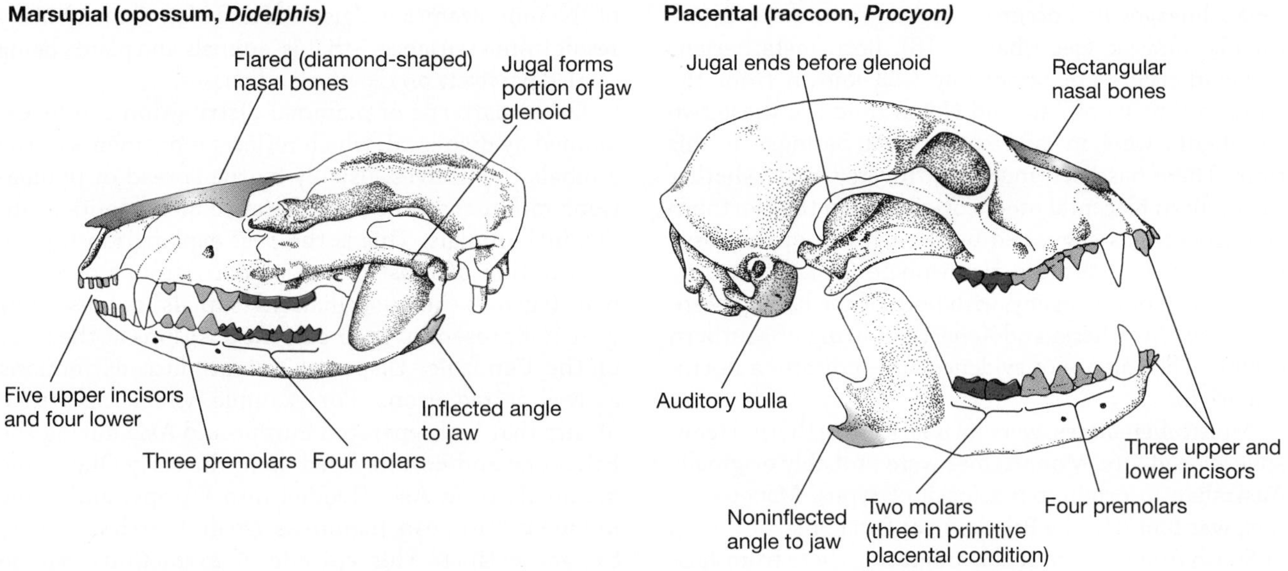

Marsupial (opossum, *Didelphis*)

Flared (diamond-shaped) nasal bones

Jugal forms portion of jaw glenoid

Five upper incisors and four lower

Three premolars   Four molars

Inflected angle to jaw

Placental (raccoon, *Procyon*)

Jugal ends before glenoid

Rectangular nasal bones

Auditory bulla

Noninflected angle to jaw

Two molars (three in primitive placental condition)

Four premolars

Three upper and lower incisors

Figure 20–16 Skull differences between marsupials and placentals.

porpoises, have more teeth than the standard placental formula. Humans have lost a pair of incisors and two pairs of premolars from each side, so we have only 32 teeth instead of the generalized placental component of 44. Marsupials have more incisors (five uppers and four lowers) and molars (four uppers and lowers) than do placentals and have fewer premolars (three uppers and lowers) (see Figure 20–16).

The postcranial skeleton of marsupials can be distinguished from the skeleton of placentals primarily by the **epipubic bones** that project forward from the pubis. Epipubic bones were once considered a unique feature of marsupials, but they are now recognized as a basal mammalian feature that is retained in a few Cretaceous eutherians. Epipubic bones may have had the original function of supporting some sort of pouch used to house incubating eggs (as in echidnas) or immature young. While both marsupial sexes have epipubic bones, the bones tend to be more robust in females. Longer retention of the young in the uterus would make a pouch redundant in eutherians.

Epipubic bones also have a role in locomotion: in marsupials they act in concert with the thigh and abdominal muscles to stiffen the torso and to resist bending of the trunk during locomotion, and they also may limit asymmetrical movements of the hindlimbs in gaits such as the gallop. This may be the reason why the only quadrupedal cursorial marsupial, the recently extinct thylacine, has lost these bones, as did the long extinct borhyaenids that took the role of marsupial doglike predators in South America. The loss of epi-

pubic bones in most placentals may be associated with the locomotor mechanics outlined above or may relate to the fact that these bones are rigid components of the abdominal wall that would interfere with the expansion of the abdomen during pregnancy. (These two hypotheses are not mutually exclusive.)

20.5 Cenozoic Mammal Evolution

During the early Cenozoic, the continents had not moved as far from their equatorial positions as they are today, but the major landmasses were more isolated than they are at present. Australia had broken free from the other southern continents by the mid-Eocene, but North and South America did not come into contact until the Pleistocene. India collided with Asia in the Miocene and formed the Himalayan mountain ranges. Africa made contact with Eurasia in the late Oligocene or early Miocene epoch, closing off the original east-west expanse of the Tethys Sea to form the now enclosed Mediterranean basin.

Time and Place of Modern Mammal Origins

The more derived mammals with tribosphenic molars (including therians) were originally a Northern Hemisphere radiation, and the fossil record shows that the evolutionary split between marsupial and placental lineages had occurred in Asia by the end of the Middle Jurassic (see Chapter 18). Both metatherians and eutherians are reasonably well known from the Late Cretaceous of Asia and North America; these two continents were in land contact via Beringia at this time. There has been much controversy as to whether the modern placental orders originated in the Northern Hemisphere (as suggested by the pattern of the fossil record) or the Southern Hemisphere (as suggested by the pattern of phylogeny, with both of the basal placental clades, Afrotheria and Xenarthra, being of southern origin). The balance of evidence now indicates a northern origin.

Multituberculates were primarily Northern Hemisphere residents. Monotremes were probably originally Australian in origin, but a fossil platypus, *Monotrematum,* was found in the Paleocene of Patagonia, at the tip of South America—probably migrating there from Australia across Antarctica, which was warm at that time.

The radiation of extant orders of placentals has long been assumed to correlate with the disappearance of the dinosaurs, and there is little fossil evidence of members of extant orders until the start of the Cenozoic, 65 million years ago. Molecular analyses, using the concept of the molecular clock, place the divergence of placentals considerably earlier than the fossil record does: some early studies came up with divergence times for the modern placentals as long as 140 million years ago. That timing would place placental origins in the Early Cretaceous, which implies an unreasonably long absence from the fossil record. More recent studies, with greater sampling of mammal diversity, have yielded a divergence date of about 80 million years, during the latest Cretaceous. Paleontologists are more comfortable with this date: obviously placental mammals must have started diverging some time before their first fossil record appearance at 65 million years ago, and a fossil record gap of 15 million years is more realistic than one of 70 million years or more.

Biogeography of Cenozoic Mammals

At the start of the Cenozoic, all mammals were small and fairly unspecialized. The marsupials appear to have been omnivorous and arboreal, like modern opossums, while the placentals were mostly terrestrial, shrewlike insectivorous forms or archaic ungulates.

The radiation of mammals occurred during the fractionation of the continental masses, when different stocks became isolated on different continents. This separation of ancestral stocks from one another by Earth's physical processes, rather than by their own movements, is called **vicariance.** However, only a few instances of the distribution of present-day mammals (e.g., the isolation of the monotremes in Australia and New Guinea) likely result from vicariance—that is, animals and plants being carried passively on moving landmasses.

Other patterns of mammal distribution can be explained by **dispersal,** which reflects movements of the animals themselves, usually by the spread of populations rather than the long-distance movements of individual animals. This is the best explanation for the distribution of most of the modern groups of mammals (such as the Australian marsupials) because they were not present in their current positions at the start of the Cenozoic. Dispersal can produce extinctions as well as radiations. For example, when the Turgai Straits that had separated Europe and Asia during the Paleocene and Eocene dried up in the early Oligocene, mammals from Asia flooded into Europe, and some uniquely European mammals (mainly archaic types) became extinct. This episode of extinctions was so dramatic that it is known as *La Grande Coupure* ("the Great Separation").

Although today marsupials are considered the quintessential Australian mammals, they did not reach that continent until the early Cenozoic. Marsupials reached North America (from Asia) in the Late Cretaceous, where they enjoyed a modest radiation of small forms, and the initial Cenozoic diversification of marsupials occurred primarily in South America, where the more basal types of marsupials persist today. Both anatomical and molecular data show that the Australian marsupials derive from a common stock from within the radiation of the South American ones, and their diversification is discussed later in the chapter.

During the Cenozoic, marsupials dispersed not only across the southern continents but also across the northern ones. Generalized, rather opossum-like marsupials are known from the Paleogene of North America, where they are fairly numerous as fossils, and the Paleogene of Europe, Asia, and northern Africa, where they are much more rare. Some fragmentary remains of marsupials are known from as late as the mid-Miocene in North America and Europe, but their mid-Cenozoic extinction in the Northern Hemisphere may be just part of the extinction of many archaic mammals taking place at that time. There is no need to invoke a complex scenario of competition with placentals. The present-day North American native marsupial, the common opossum *Didelphis virginiana,* is a Pleistocene immigrant from South America. Opossums are still moving north and were first recorded in Canada in the 1950s.

Today the mammals of the Northern Hemisphere (Holarctic) and of Africa, Madagascar, South America, and Australia are strikingly different from one another. The fauna of Africa was even more distinct from the fauna of Holarctica before Africa collided with Eurasia in the Oligocene-Miocene times. For example, hyraxes, which now survive only as small rodentlike forms, assumed ecological roles taken today by antelope and pigs. Similarly, the South American fauna was more distinct from that of Holarctica before South America was connected with North America in the Pleistocene, with the role of large carnivores taken by the now-extinct borhyaenoid marsupials (**Figure 20–17**) and the role of large herbivores by the now-extinct native ungulates of uncertain phylogenetic affinities, such as litopterns and notoungulates.

Even the North American fauna was much more distinct from that of the rest of Holarctica for most of the Cenozoic than it is today. Much of the present North American fauna (e.g., deer, bison, and rodents such as voles) emigrated from Eurasia in the Pliocene and Pleistocene via the Bering land bridge. Today only Madagascar and Australia retain distinctly different faunas from the rest of the world.

Convergent Evolution of Mammalian Ecomorphological Types

The term *ecomorphology* describes the way in which an animal's form (its morphology) is related to its activities in its environment (its ecology). The separate evolution of different basic mammalian stocks on different continents illustrates the convergent evolution of ecomorphs in Cenozoic mammals.

Specialized running herbivores include the extinct litopterns of South America, the Holarctic artiodactyl antelope of Eurasian origin, and the perissodactyl horses (most of the evolution of horses took place in North America). These ungulates are strikingly similar in many respects. Even the Australian marsupials such as kangaroos demonstrate convergences to placental herbivores in jaws, teeth, and feeding behavior. Convergent evolution also took place among large-bodied, slower-moving herbivores, such as the Holarctic rhinoceroses (perissodactyls), the elephants (which originated in Africa), the extinct notoungulates of South America, and the extinct marsupial diprotodontids of Australia.

Carnivorous mammals show similar convergences. The wolves of Holarctica, the recently extinct thylacine of Australia, and the extinct marsupial borhyaenids of South America have similar body shapes and tooth forms, although only the wolves have the long legs that distinguish fast-running predators. Similar types of intercontinental convergences, again involving both marsupials and placentals, occurred with a carnivorous ecomorph that does not exist today—the saber-toothed catlike predator.

Mammals specialized for feeding on ants and termites (myrmecophagy) include the giant anteater in South America, the aardvark in Africa, the pangolin in tropical Asia and Africa, and the spiny anteater (a monotreme) in Australia. Ants and termites are social insects that employ group defense. They often build impressive earthen nests containing thousands of individuals, some of which are strong-jawed soldiers. The convergent specializations of these unrelated or distantly related mammals include a reduction in the number and size of teeth, changes in jaw and skull shape and strength, and forelimbs modified for digging.

Other examples of convergent evolution of ecomorphs abound, such as burrowing moles or mole-like animals and gliders. Gliding ecomorphs include the flying squirrels of the Northern Hemisphere and the marsupial flying phalangers in Australia, as well as the flying lemurs (not true primates) of Southeast Asia and a completely different type of flying rodent, the "scaly-tailed squirrel" of Africa, which is not closely related to northern squirrels. One of the most striking examples of convergence is between the marsupial mole *Notoryctes* from desert regions in Western Australia and the golden moles (Chrysochloridae) from Africa; both of these phylogenetically distant mammals have adaptations that allow them to burrow rapidly through soft sand.

Cenozoic Mammals of the Southern Continents

The mammals of Australia, the island of Madagascar, and South America differ from those of the Northern Hemisphere, providing clear examples of the effects of biogeographical isolation.

Cenozoic Isolation of Australia The terrestrial mammalian fauna of Australia has always been composed almost entirely of marsupials and monotremes. The earliest Australian marsupial fossils are of early Eocene age, and there are abundant remains from the late Oligocene onward. South America, Antarctica, and Australia were still close together in the early Cenozoic. Australia was populated by marsupials that probably moved from South America across Antarctica, which was warm and ice-free until about 45 million years ago. Marsupial fossils are known from the Eocene of western Antarctica, and, in fact, the major barrier to dispersal to Australia was probably crossing the midcontinental mountain ridge between western and eastern Antarctica.

Once marsupials reached Australia, they enjoyed the advantages of long-term isolation, evolving to fill a variety of niches with food habits ranging from complete

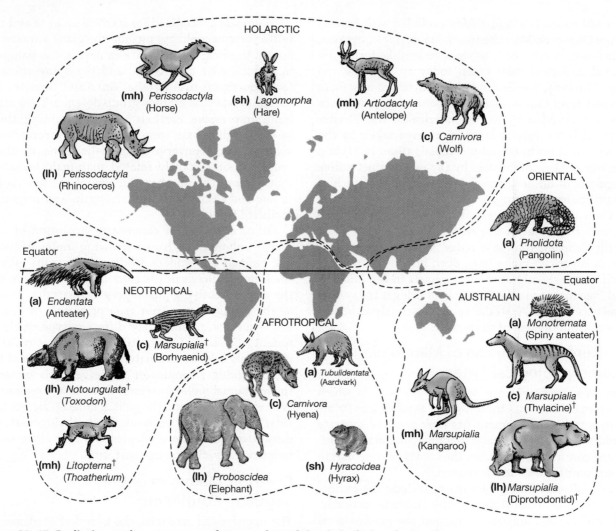

Figure 20–17 Radiation and convergence of mammals evolving in isolation during the Cenozoic era. Mammals are grouped with landmasses on which they probably originated. The major biogeographical regions of the world today are shown. The current convention, reflecting the historical patterns of colonization by Europeans, is to call American continents the New World and other continents (including Australia) the Old World. The northern continents, North America and northern Eurasia, can be grouped together as Holarctica; they can be further subdivided into the Nearctic (North America, including Greenland) and the Palaearctic (Europe and northern Asia, including Asia Minor). India and Southeast Asia fall within the tropics and together are called the Oriental region. Africa (including Madagascar) forms the Afrotropical or African region. The Afrotropical and Oriental regions are sometimes grouped together as the Old World Tropics, or Paleotropics. South and Central America form the Neotropical region, or New World Tropics. Australia and associated islands (including New Guinea, Tasmania, and New Zealand) make up the Australian region. Key to ecomorphological types of mammals: (a) = ant or termite eater, (c) = carnivore (doglike form), (lh) = large herbivore, (mh) = medium-sized herbivore, (sh) = small herbivore. A dagger (†) indicates an extinct taxon.

herbivory to carnivory. **Figure 20–18** shows some Australian marsupials and notes their convergences with northern placentals. Possums and phalangers are a diverse radiation of arboreal mammals that are convergent not only with squirrels but also with primates.

A particularly striking example of convergence is the aye-aye of Madagascar (a lemur, a type of primate) and the striped possum of the York Peninsula of northeastern Australia. Both pry wood-boring insects out from under tree bark with an elongated

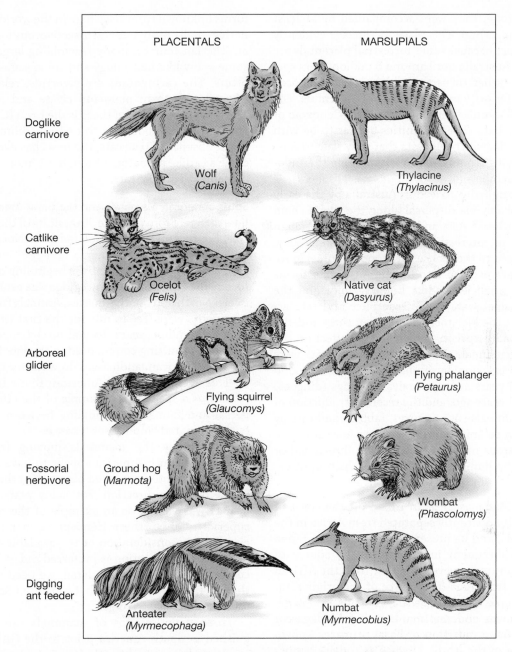

PLACENTALS | MARSUPIALS

Doglike carnivore

Wolf
(*Canis*)

Thylacine
(*Thylacinus*)

Catlike carnivore

Ocelot
(*Felis*)

Native cat
(*Dasyurus*)

Arboreal glider

Flying squirrel
(*Glaucomys*)

Flying phalanger
(*Petaurus*)

Fossorial herbivore

Ground hog
(*Marmota*)

Wombat
(*Phascolomys*)

Digging ant feeder

Anteater
(*Myrmecophaga*)

Numbat
(*Myrmecobius*)

Figure 20–18 Convergences in body form and habits between placental and marsupial mammals. Note that the animals here are not drawn to scale. For example, the placental ant-eater is the size of a large dog, while the numbat is about the size of a squirrel.

finger used as a probe, and both have chisel-like incisors. Certain insectivores (apatemyids) from the early Cenozoic of Holarctica had similar specializations but became extinct in the early Miocene when woodpeckers evolved—birds that also feed on wood-boring insects. Madagascar and Australia are the only places in the world today that lack woodpeckers, which sug-

gests that the woodpeckers elsewhere outcompeted the wood-boring mammals.

The early Eocene fauna of Australian mammals revealed a remarkable discovery: in addition to several marsupial species, it preserved the molar that apparently belonged to a placental mammal. This fossil suggests that marsupials did not make the journey across

Antarctica alone; they were accompanied by at least one type of placental. This placental lineage evidently became extinct because other terrestrial placentals are unknown in Australia until around 5 million years ago. This animal rather turns the tables on the supposed marsupial inferiority in the presence of placentals. The only other placentals known in the early Cenozoic of Australia are bats. Their affinities appear to be with the bats of Asia, which suggests that they flew to Australia by moving from island to island along the Indo-Australian Archipelago.

Rodents originally arrived in Australia by the early Pliocene probably via dispersal along the island chain between Southeast Asia and New Guinea. Australian rodents are an interesting endemic radiation today; they are related to the mouse-rat group of Eurasian rodents but have evolved into unique Australian forms such as the small jerboa-like hopping mice and the large (otter size) water rats. True mice and rats arrived later, in the Pleistocene. However, these rodents apparently had surprisingly little overall effect on the Australian marsupials. Far more serious threats were the original invasion by humans from Asia around 50 thousand years ago, the arrival of dogs (dingoes) around 4000 years ago, and the recent introduction of domestic mammals such as foxes, rabbits, and cats by the European colonists during the past few centuries. Today, numerous marsupial species are threatened or endangered by humans and their introduction of exotic species.

Madagascar Madagascar is an island off the coast of East Africa. It apparently separated from Africa in the mid-Mesozoic, and its present-day mammalian fauna represents subsequent immigration. The source of these immigrants, however, whether from the African mainland or from Asia (the source of the original indigenous people of Madagascar), remains in debate. The best-known endemic mammals of Madagascar are the lemurs, a radiation of basal primates known nowhere else in the world. Diverse as today's lemurs are, their diversity was much greater in the recent past. Much larger (up to gorilla size) lemurs—both terrestrial and arboreal—paralleled the radiation of the great apes among the anthropoid primates. The extinction of these giant lemurs appears to be related to the arrival of humans a couple of thousand years ago (although reports suggest that some species of giant lemurs may have survived until only a few hundred years ago) and is probably due to habitat destruction rather than to hunting.

The other native Madagascan mammals belong to the Carnivora and the Afrotheria and are composed of forms unknown from elsewhere in the world today. The native small omnivores and insectivorous forms are the afrotherian tenrecs, many resembling large shrews—some spiny like hedgehogs and some semiaquatic like otters. The carnivorans are viverrids, related to the African and Asian mainland civets and mongooses. The most remarkable of these is the fossa (*Cryptoprocta ferox*), which is about the size of a medium-sized dog. In the absence of true cats, the fossa has evolved to become a catlike predator, although it is more arboreal in its habits than true cats.

South American Mammals and the Great American Biotic Interchange From the Late Jurassic until the late Cenozoic, South America was isolated from North America by a seaway between Panama and the northwestern corner of South America. At the beginning of the Pleistocene the Panamanian land bridge was established between North and South America. Animals from the two continents were free to mix for the first time in more than 100 million years. Faunal interchange by island dispersal or rafting commenced in the late Miocene as the two American landmasses drew nearer to each other. This event, the Great American Biotic Interchange (GABI), is a spectacular example of the effects of dispersal and faunal intermingling between two previously separated landmasses (**Figure 20–19**).

Superficially the mammals moving from North America to South America appear to have been more numerous and to have fared better than those moving in the opposite direction. For many years, the interchange was viewed as an example of the competitive superiority of Northern Hemisphere mammals. However, with reconsideration of the available evidence, a different interpretation is preferred today. Before discussing the GABI in more detail, we need to consider the diversity of mammals that inhabited South America before the interchange.

Three major groups of mammals can be distinguished in South America prior to the GABI. In fact, much of the original South American mammal fauna was extinct by the time of the GABI, and competition with northern immigrants was not responsible for their demise.

- *Early inhabitants, known from the Paleocene, evolving in situ or originating from North America: marsupials, xenarthrans, and archaic ungulates.* Xenarthrans were known only from South America prior to the late Miocene. The past diversity of xenarthrans was considerably greater than that of today, including the armadillo-related glyptodonts (cow-size beasts encased in a turtlelike carapace of dermal bone) and

Figure 20–19 Mammalian taxa involved in the Great American Biotic Interchange. All of the species shown in South America migrated from North America, and the ones shown in North America originated in South America and moved northward. A dagger (†) indicates taxa that are now totally extinct; an asterisk (*) indicates taxa that are now extinct in that area.

ground sloths (ranging from the size of a large dog to the size of a rhinoceros).

The endemic South American ungulates, now all extinct, radiated into several orders. The ones surviving at the time of the GABI were the litopterns and the notoungulates. Litopterns were lightly built, cursorial animals that were small and ponylike or larger and camel-like. Notoungulates were stockier and diverged into small rodentlike forms or large rhinolike forms. In the absence of rodents and carnivorans, marsupials radiated into these niches during the isolation of South America in the early Cenozoic.

• *Late Eocene or early Oligocene colonizers, probably arriving by rafting from Africa.* These colonizers include the caviomorph rodents (e.g., guinea pigs, agoutis, and capybaras) and the South American monkeys (discussed further in Chapter 24). The caviomorphs diversified during the Miocene and Pliocene to include the largest rodents that have ever lived. One of these, *Phoberomys* from the late Miocene of Venezuela, was the size of a small rhinoceros, and even today the pig-size capybara is extremely large for a rodent.

- *Late Miocene arrivals, coming from North America via an island arc linking the two continents.* These include raccoons (moving from north to south) and some small ground sloths (moving from south to north).

With the establishment of the Panamanian land bridge in the Pleistocene, some animals from North America moved southward and some South American forms moved northward. The interchange actually occurred in four major pulses that may reflect climatic events that rendered the habitats of Panama suitable for migration. The major pulse was at the start of the Pleistocene, 2.6 to 2.4 million years ago. Subsequent pulses of migrations (in both directions) occurred at 1.8 million years ago, 700,000 years ago, and 12,000 years ago. It is often assumed that the now-extinct native South American mammals such as the borhyaenids (marsupial carnivores) and the endemic ungulates were outcompeted by the northern immigrants. However, the fossil record shows that these groups had declined in numbers prior to the Pleistocene, probably as the result of late Neogene climatic changes.

On each continent the newcomers and the native fauna appeared to coexist, and for the most part the immigrants enriched the existing fauna rather than displaced it. However, a disparity appeared during the Pleistocene extinctions. Although these extinctions affected the largest mammals on both continents, the southern immigrants in North America were affected more profoundly than were the North American forms in South America.

Today about half the generic diversity of South American mammals consists of forms with a North American origin. Some notable radiations of mammals of northern origin include deer, canids (dogs), and felids (cats). However, some notable northern immigrants to South America are now extinct there—for example, horses.

The southern species that persist in North America are mostly confined to Central America (e.g., capybaras) and the southern United States (e.g., armadillos). Opossums and porcupines are exceptions—they have extended their geographic ranges to northern North America and have remained successful there.

A key to the apparent greater success of the immigrants to South America lies in understanding the importance of biogeography. The equator and much of tropical America lie within South America. During climatic stress, such as the glacial periods, South America retains more equable habitats than North America, so fewer extinctions would be expected. Vegetational changes associated with climatic changes obviously also played a role in the survival or extinction of species and in the hospitality of the Panamanian environment for migrating animals.

By the late Pleistocene, the Central American corridor was evidently closed to migration of savanna-adapted mammals, possibly because of the development of tropical forests. Mammals that arrived in North America only in the late Pleistocene, such as bison and mammoths, never reached South America. The glyptodonts and ground sloths that had migrated northward would also have been unable to return to South America. When the North American savannas disappeared at the end of the Pleistocene, these southern-originating edentate mammals were unable to adapt to the cooler prairies that housed the bison and so became extinct.

A final point of consideration in the GABI is that counts of who moved where—and when—usually consider only species known from the fossil record. Because tropical habitats rarely preserve fossils, we have little information about the fossil history of Central America. Yet the large diversity of opossum-like marsupials, edentates, monkeys, and caviomorph rodents in Central America today can only have come from South America. Because of our present-day political boundaries, we often forget that Central America is geologically part of North America, not South America, even though we group its flora and fauna together with those of South America as the Neotropics. A proper tally of the immigrants from South America to Central America is necessary before we can write the final chapter on the Great American Biotic Interchange.

Summary

The major groups of living mammals are monotremes and therians (marsupials and placentals); another important group, the multituberculates, did not survive the early Cenozoic. Placentals are by far the most diverse group of living mammals, and they are found worldwide, while monotremes and marsupials are confined to the Southern Hemisphere (with the exception of the North American opossum). New molecular techniques have altered previous notions of the phylogenetic relationships of placental mammals: most notable is the recognition of an endemic group of African mammals, the Afrotheria.

All mammals share uniquely derived features. Lactation is the most obvious mammalian feature, and all mammals have hair and a variety of skin glands used for hair lubrication and waterproofing, olfactory communication, and thermoregulation. Mammary glands evolved from these types of skin glands. Therians are not only more derived than monotremes in their mode of reproduction (viviparity as opposed to egg laying) but also more derived in many features of the skull, dentition, skeleton, and soft anatomy. Only a few dental and skeletal differences distinguish marsupials and placentals, most notably the loss of the epipubic bones in placentals.

The diversity of Cenozoic mammals can be best understood in the context of changing patterns of biogeography. The distribution of some mammalian groups reflects vicariance, as in the monotremes of Australia. Other patterns of distribution reflect dispersal, such as the migration of marsupials to Australia from South America. Isolation of the various continents has resulted in unique mammalian faunas in Australia, Madagascar, and South America. The South American fauna was even more different from the rest of the world until the formation of the Isthmus of Panama about 2 million years ago.

Discussion Questions

1. Why do we now think that the Cenozoic radiation of monotremes had semiaquatic origins?
2. Magazines and television shows like to say that the platypus is a mix of different types of animals— part mammal, part reptile, part bird. How is that incorrect (at least in terms of morphology)?
3. How does new information about the little mountain opossum *Dromiciops* help us understand past patterns of marsupial diversification?
4. The epipubic bones of nonplacental mammals have long been thought to have had a reproductive function (e.g., support of the pouch in marsupials). What evidence do we have that some of their function may be related to locomotion and support of the abdominal wall?
5. The first eutherian mammal and the first metatherian mammal are both found in China. Is this an enormous coincidence?
6. Various types of Australian marsupials have often been compared with similar types of placentals (e.g., gliding squirrels and gliding possums) as examples of convergent evolution. However, there is no equivalent among marsupials to a placental whale or even to a seal. What is a possible reason for this?

Additional Information

Bininda-Emonds, O. R. P., et al. 2007. The delayed rise of present-day mammals. *Nature* 446:507–511.

Fischer, M. S., et al. 2002. Basic limb kinematics of small therian mammals. *Journal of Experimental Biology* 205:1315–1338.

Fox, R. C., and C. S. Scott. 2005. First evidence of a venom delivery system in extinct mammals. *Nature* 435: 1091–1093.

Gilad, Y., et al. 2004. Loss of olfactory receptor genes coincides with the acquisition of full trichromatic vision in primates. *PLoS Biology* 2:e0120.

Godthelp, H., et al. 1992. Earliest known Australian Tertiary mammal fauna. *Nature* 356:514–516.

Hautier, L., et al. 2010. Skeletal development in sloths and the evolution of mammalian vertebral patterning. *Proceedings of the National Academy of Sciences* 107:18903–18908.

Holt, C., and J. A. Carver. 2012. Darwinian transformation of a 'scarcely nutritious fluid' into milk. *Journal of Evolutionary Biology* 25:1253–1263.

Hurum, J. H., et al. 2006. Were mammals originally venomous? *Acta Palaeontologica Polonica* 51:1–11.

Jacobs, G. H. 2012. The evolution of vertebrate color vision. In C. Lopez-Larrea (ed.), *Sensing in Nature.* Austin, TX: Landes Bioscience, pp. 156–172.

Kishida, T., et al. 2007. The olfactory receptor gene repertoires in secondary-adapted marine vertebrates: Evidence for reduction of the functional proportions in cetaceans. *Biology Letters* 3:428–430.

Kitazoe, V., et al. 2007. Robust time estimation reconciles views of the antiquity of placental mammals. *PLoS ONE* 4:e834.

Kunz, T. H., and D. J. Hosken. 2009. Male lactation: Why, why not and is it care? *Trends in Ecology and Evolutionary Biology* 24:80–85.

Lee, M. S. Y., and A. B. Camens. 2009. Strong morphological support for the molecular evolutionary tree of placental mammals. *Journal of Evolutionary Biology* 22:2243–2257.

Liu, A. G. S. C., et al. 2008. Stable isotope evidence for an amphibious phase in early proboscidean evolution. *Proceedings of the National Academy of Sciences* 105: 5786–5791.

Long, J., et al. 2002. *Prehistoric Mammals of Australia and New Guinea*. Baltimore, MD: Johns Hopkins University Press.

Meredith, R. W., et al. 2011. Impacts of the Cretaceous Terrestrial Revolution and KPg extinction on mammal diversification. *Science* 334:521–524.

Murphy, W. J., et al. 2007. Using genomic data to unravel the root of placental mammal phylogeny. *Genome Research* 17:413–421.

Nilsson, M. A., et al. 2010. Tracking marsupial evolution using archaic genomic retroposon insertion. *PLoS Biology* 8:e1000436.

Oftedal, O. T. 2011. The evolution of milk secretion and its ancient origins. *Animal* 6:355–368.

Phillips, M. J., et al. 2009. Molecules, morphology, and ecology indicate a recent, amphibious ancestry for echidnas. *Proceedings of the National Academy of Sciences* 106:17089–17094.

Pond, C. M. 1998. *The Fats of Life*. Cambridge, UK: Cambridge University Press.

Reilly, S. M., and T. D. White. 2003. Hypaxial motor patterns and the function of epipubic bones in primitive mammals. *Science* 299:400–402.

Rose, K. D. 2006. *The Beginning of the Age of Mammals*. Baltimore, MD: Johns Hopkins University Press.

Rose, K. D., and J. D. Archibald (eds.). 2005. *The Rise of Placental Mammals*. Baltimore, MD: Johns Hopkins University Press.

Sánchez-Villagra, M. R., et al. 2003. The anatomy of the world's largest extinct rodent. *Science* 301:1708–1710.

Spaulding, M., et al. 2009. Relationships of Cetacea (Artiodactyla) among mammals: Increased taxon sampling alters interpretations of key fossils and character evolution. *PLoS ONE* 4:e7062.

Vaughan, T. A., et al. 2011. *Mammalogy*, 5th ed. Sudbury, MA: Jones and Bartlett Publishers.

Warren, W. C., et al. 2008. Genome analysis of the platypus reveals unique signatures of evolution. *Nature* 453: 175–184.

Webb, S. D. 2006. The Great American Biotic Interchange: Patterns and processes. *Annals of the Missouri Botanic Garden* 93:245–257.

Wible, J. R., et al. 2007. Cretaceous eutherians and Laurasian origin for placental mammals near the K/T boundary. *Nature* 447:1003–1006.

Woodburne, M. P. 2010. The Great American Biotic Interchange: Dispersals, tectonics, climate, sea level, and holding pens. *Journal of Mammalian Evolution* 17: 245–264.

Websites

General

http://animals.nationalgeographic.com/animals/mammals/

Monotremes

http://www.ucmp.berkeley.edu/mammal/monotreme.html

Marsupials

http://palaeo.gly.bris.ac.uk/palaeofiles/marsupials/Index.htm

http://www.australianwildlife.com.au/riversleigh.htm

Ungulates

http://www.ultimateungulate.com/

Mammalian Specializations

Mammals are a highly diverse group of organisms, adapted to a wide variety of lifestyles and displaying broad ecological and morphological diversity. The three major types of living mammals—monotremes, marsupials, and placentals—can be distinguished by profound differences in their reproduction. Differences in morphological specializations can also be seen in the skull and teeth for feeding and in the postcranial skeleton for locomotion.

21.1 Mammalian Reproduction

The mode of reproduction is the major, and most obvious, difference among the three major groups of extant mammals. Although all mammals lactate and care for their young, monotremes are unique among living mammals in laying eggs, which is the ancestral amniote condition. With the evolution of viviparity in therians, the uterine glands that add the shell and other egg components were lost, making a return to oviparity difficult or impossible. Among the therians, marsupials and placentals have profound differences in their lengths of gestation; marsupials are familiar to many people as pouched mammals, the pouch housing the young that would still be inside the uterus of a comparable placental mammal.

All mammals grow from an initial embryonic ball of cells (the blastocyst) that forms both the embryo and the trophoblast. The trophoblast is a dif-

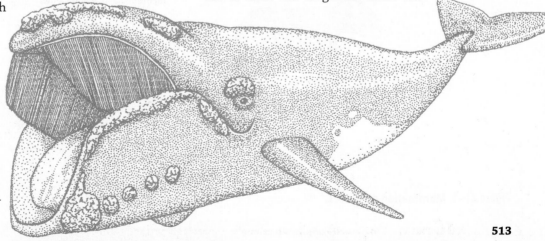

ferentiation of extraembryonic tissue specialized for obtaining nutrition in the uterus, for producing hormones to signal the state of pregnancy to the mother, and (in therians) for helping the embryo implant into the uterine wall.

Additionally, all mammals have a glandular uterine epithelium (**endometrium**), which secretes materials that nourish embryos in the uterus, and a **corpus luteum** (Latin *corpus* = body and *lut* = yellowish; plural *corpora lutea*), a hormone-secreting structure formed in the ovary from the follicular cells that remain after the egg is shed. Hormones secreted by the corpus luteum are essential for the establishment and at least the initial maintenance of pregnancy, although only in placentals is there feedback from the placenta that maintains the life of the corpus luteum. This system suppresses ovulation and allows the extended gestation of placentals.

Reproductive Mode of Monotremes

The reproductive tract of monotremes retains the generalized amniote condition. The two oviducts remain separate and do not fuse in development except at the base, where they join with the urethra from the bladder to form the urogenital sinus. The oviducts swell to form two uteri, in which the fertilized egg is retained (only the left oviduct is functional in the platypus). In all mammals, the eggs are fertilized in the anterior portion of the oviduct, the **fallopian tube,** before they enter the uterus.

The ovaries of monotremes are bigger than those of therians, and monotremes provide the embryo with yolk. However, the eggs of monotremes are much smaller at ovulation than are those of reptiles or birds of similar body size. The amount of yolk is not sufficient to sustain the embryo until hatching, and the eggs are retained in the uterus, where they are nourished by maternal secretions and increase in size before the shell is secreted. The eggshell is leathery, like that of lizards, rather than rigid like the calcareous eggshells of birds.

Monotremes lay one or two eggs, and the young hatch soon after the egg is laid (after only 12 days for the platypus). The young hatch at an extremely altricial stage (**Figure 21–1**), and brooding by the mother continues for another 16 weeks. The platypus usually lays its eggs in a burrow, whereas echidnas keep their eggs in a ventral pouch. All monotremes have a low reproductive rate—they breed no more than once a year.

Reproductive Mode of Therians

All therians have a placenta, which is formed from the extraembryonic membranes of the fetus. Marsupials and placentals are often thought to differ in their type of placentation, and it is sometimes stated that marsupials lack a placenta. However, there is more similarity in placentation between these two types of mammals than is commonly believed.

All marsupials and some placental mammals have an initial **choriovitelline placenta,** developed from the yolk sac (although this structure is vestigial in humans). Placentals also have a later-developing **chorioallantoic placenta,** developed from the combination of the chorionic and allantoic amniote membranes. Some placentals retain a choriovitelline placenta even after the chorioallantoic placenta has developed. Six distinct layers of tissues separate fetal and maternal blood in most placentals, but the placenta of some placental mammals (including anthropoid primates and some rodents) penetrates so deeply into the uterine lining that only a layer or two of tissues separate the fetal and maternal blood systems. There is much variety in the form of the mammalian placenta as well as in the types of placentation.

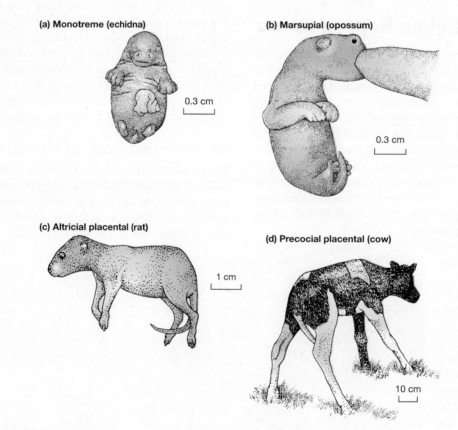

(a) Monotreme (echidna)   0.3 cm

(b) Marsupial (opossum)   0.3 cm

(c) Altricial placental (rat)   1 cm

(d) Precocial placental (cow)   10 cm

Figure 21–1 Mammalian neonates.

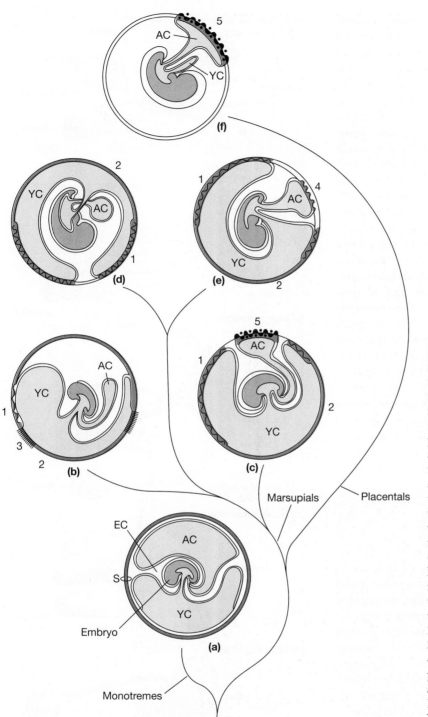

Figure 21–2 Types of placentation in marsupials and placentals. (a) Egg-laying monotreme. (b) Dasyurid—the allantois reaches the chorion and then retreats from it without forming a placental structure. (c) Bandicoot—a complex chorioallantoic placenta is formed at the close of gestation, and the choriovitelline placenta remains functional until the young are born. (d) Possum and kangaroo—the allantois may grow to a large size but remains enshrouded in the folds of the yolk-sac wall. (e) Koala and wombat—the allantois reaches the chorion, forming an apposed chorioallantoic placenta. (f) Placental—choriovitelline placenta is short lived, and a complex chorioallantoic placenta is the functional one for most of the gestation. Dark shading indicates areas of placentation. (1 = vascular choriovitelline placenta, 2 = nonvascular choriovitelline placenta, 3 = syncytialized choriovitelline placenta, 4 = apposed (nonvascular) chorioallantoic placenta, 5 = syncytialized (vascular) chorioallantoic placenta; AC = allantoic cavity, EC = extraembryonic coelom, S = shell, YC = yolk-sac cavity)

Most marsupials have only a choriovitelline placenta, but some (bandicoots, koalas, and wombats) show a transitory chorioallantoic placenta near the end of their short gestation (**Figure 21–2**). The lack of a chorioallantoic placenta in most marsupials represents suppression of the chorioallantoic membrane during development; the rate of outgrowth of this membrane in development is slower than that seen in all other amniotes, not just other mammals.

The ability to maintain the embryo in a state of arrested development prior to implantation (**embryonic diapause**) is an important reproductive feature of some therians. This capacity enables the mother to space successive litters and to separate the time of mating and fertilization from the start of gestation. Thus, diapause allows mating and the birth of young to occur at optimal times of the year. Embryonic diapause is particularly well developed in kangaroos and has often been perceived as a derived marsupial feature. However, embryonic diapause also occurs in a wide variety of placentals.

Monotremes are like other vertebrates in having testes that are retained within the abdomen. Some therians, both marsupials and placentals, also retain the testes in the abdomen, either in the original position high within the body or partially descended and housed subcutaneously at the base of the abdomen. However, the testes of most therian mammals descend into a scrotum during development.

The value of a scrotum is obscure. The traditional idea is that a scrotum provides a cooler environment for sperm production, but there is no simple correlation between core body temperature and testicular position among mammals. The scrotum probably evolved convergently in marsupials and placentals because the control of scrotal development is different in the two groups. In marsupials testicular descent is under direct genetic control, whereas in placentals it is hormonally determined. Additionally, the scrotum is in front of the penis in marsupials and behind the penis in most placentals, although there are some exceptions (e.g., rabbits have a prepenile scrotum).

Reproduction of Placental Mammals

In all therians, the ureters draining the kidney enter the base of the bladder rather than the cloaca or urogenital sinus, as in most other vertebrates (**Figure 21–3**). In placentals, the ureters pass laterally around the developing reproductive ducts to enter the bladder. This arrangement allows the oviducts of females to fuse in the midline anterior to the urogenital sinus for much of their length. In males, this anatomical arrangement results in the **vasa deferentia** (Latin *vas* = vessel; singular *vas deferens*), the male reproductive tracts, looping around the ureters in their passage from the scrotum to the urogenital sinus. All placentals have a single, midline vagina, but only a few have a single median uterus as seen in humans. Most placentals have a uterus that is bipartite (divided lengthwise into left and right sides) for some or all of its length, and a bipartite uterus sometimes occurs as a developmental abnormality in humans.

In most placentals, the urogenital sinus and the alimentary canal have separate openings with a distinct external space (the perineum) between them. The perineum is not so apparent in marsupials, and a well-defined separation of female urogenital openings

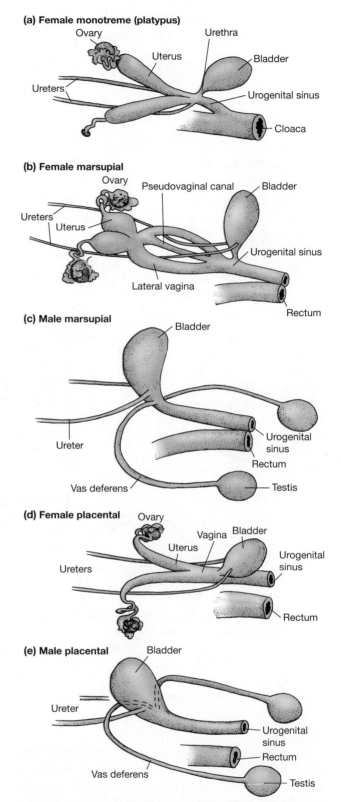

Figure 21–3 Mammalian reproductive tracts. The structures are shown as if the animal were lying on its back with its head to the left.

into distinct external urethral and vaginal openings is seen only in primates and some rodents.

Some placentals (e.g., rodents and insectivorans) are born in a highly altricial state in which they are only slightly more developed than some marsupial young; others are born in more developed stages, extending to a highly precocial state (most ungulates) in which the young can run within a few hours of birth. All placentals, however precocial, still require a period of lactation for the transfer of essential antibodies from the mother as well as for nutrition. The period of lactation in most placentals is relatively short in comparison with other mammals—usually shorter than the period of gestation. In all mammals, larger-bodied species tend to have fewer young per litter than do smaller ones—but the total number of young that the mother produces in her lifetime may not be so different because larger mammals have a longer life span than do smaller ones.

Reproduction of Marsupials

In marsupials, the ureters pass medial to the developing reproductive ducts to enter the bladder. This arrangement prevents the oviducts of the females from fusing in the midline, at least near the base, and means that the vasa deferentia of the males do not have to loop around the ureters. The female reproductive tract consists of two lateral vaginae that unite anteriorly, from which point the two separate uteri diverge (see Figure 21–3b). The lateral vaginae are for the passage of sperm only. Male marsupials have a bifed (forked) penis to complement this anatomy. Birth of the young is through a midline structure, the median vagina or **pseudovaginal canal,** which develops at the first parturition.

The length of gestation in marsupials is relatively independent of body size, and the young are ejected at the end of the estrous cycle. However, larger marsupials take longer to rear their young than smaller ones, having a longer period of lactation. In contrast to placentals, marsupials retain direct evidence of their oviparous ancestry; a transient shell membrane appears, and the eggs still contain a small amount of yolk.

The development and growth of marsupial young is very different from that of both monotremes and placentals and represents the derived condition. Marsupial neonates have well-developed forelimbs in comparison with other altricial neonates, and their lungs are relatively large at birth, so that the newborn marsupial can attach itself to a nipple and begin suckling. Development of the jaws, secondary palate, facial muscles, and tongue is advanced, while that of the central nervous system is retarded.

Most, but not all, marsupials enclose the nipples within a pouch—some dasyurids (marsupial mice) and some didelphids (opossums) lack a pouch. When young marsupials are born, they evidently have an instinct to climb upward until they attach themselves to a nipple to complete their development. Marsupial neonates have a novel condition of their shoulder girdle, called a shoulder arch, that allows them to make this trip (**Figure 21–4**). In making this journey, the young marsupials do not lever themselves up with their limbs like a gymnast on the parallel bars. Rather, they wriggle their bodies, using the front claws as holdfasts, and this shoulder arch aids in the functional requirements for the crawl to the nipple. The marsupial shoulder arch includes the coracoid and interclavicle bones,

Newborn

Adult

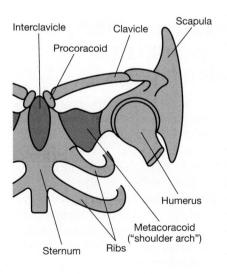
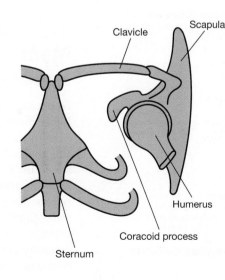

Figure 21–4 Shoulder girdles of newborn and adult marsupials. Viewed from the ventral side with only the left limb shown.

which are portions of the tetrapod shoulder girdle that are present in most tetrapods, including monotremes, but absent from all therians except for their transient retention in newborn marsupials. Here they provide a strong brace for the front limbs during the crawl to the nipple, but this specialized neonatal anatomy limits the subsequent development of the shoulder girdle in the adults, and marsupials have less variation than placentals in the form of their shoulder girdles.

In kangaroos, as in most other species, the mother adopts a distinct sitting birth posture (**Figure 21–5**); she licks a path from the vagina to the pouch but does not otherwise aid the young in its journey. (A kangaroo's pouch opens anteriorly, but the pouches of many other marsupials open posteriorly, shortening the distance the young must travel.) Lactation also continues for some time after the young have become sufficiently mature to detach from the nipple. This is when we typically see the young-at-foot, hopping in and out of the pouch.

Although the composition of the milk varies little during pregnancy in placentals, the milk changes markedly in marsupials and monotremes. The first milk is dilute and rich in protein, while the later milk is more concentrated and richer in fats. Concurrent asynchronous lactation has been observed in some kangaroos; that is, an immature pouch young is attached to one nipple while a more mature, independent pouch young drinks from another nipple, and the mother produces different kinds of milk at the two nipples. The composition of milk is probably determined by how long the young spend suckling on the nipple per day.

The Earliest Therian Condition

The reproductive mode of the earliest therians is not clear, although the birth (or hatching) of highly altricial young is probably the basal condition. The specializations of the reproductive anatomy of marsupials and placentals cannot easily be derived from each other, and it is possible that viviparity evolved independently in the two groups. The early therians of the Cretaceous period were small animals, most weighing less than a kilogram. They are likely to have had the same life-history features as extant small living therians—short life spans, several litters produced in rapid succession or a single large litter, and a short gestation period. Some fetal placentals show evidence of mouth seals—tissue that develops around the lateral margins of the mouth of neonatal marsupials to aid in attachment to the nipples. This feature suggests that attachment to a nipple is a basal therian feature, whether or not the earliest therians had a marsupial-like pouch.

Birth posture of grey kangaroo

Red kangaroo with 3 different young at 3 different developmental stages

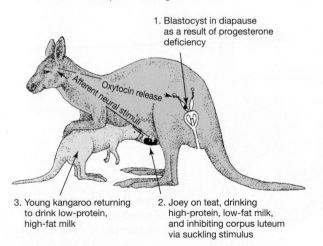

1. Blastocyst in diapause as a result of progesterone deficiency

Oxytocin release

Afferent neural stimuli

3. Young kangaroo returning to drink low-protein, high-fat milk

2. Joey on teat, drinking high-protein, low-fat milk, and inhibiting corpus luteum via suckling stimulus

Figure 21–5 Birth posture and embryonic diapause in kangaroos.

21.2 Some Extreme Placental Mammal Reproductive Specializations

Numerous reproductive specializations exist among living placentals. Some of the most extreme are those seen in the naked mole rat (*Heterocephalus glaber*) and the spotted hyena (*Crocuta crocuta*). Naked mole rats are found in arid areas in sub-Saharan Africa and live as underground burrowers feeding on plant roots and tubers. They are **eusocial,** living in a type of social system otherwise seen only in social insects such as ants, termites, and bees. Like these insects, animals within a colony are closely related, but they lack the genetic characteristics of insect colonies.

Mole rats live in colonies of up to 40 individuals with only one breeding female, the queen, who produces one to four litters per year with up to two dozen young in each litter. Other colony members are divided into different social castes, including workers who help maintain the colony and carers for the young. All the males in a colony produce sperm, but only the nonworkers are large enough to copulate successfully with the queen. If the queen dies, one of the faster-growing female infrequent workers may become the new queen.

Spotted hyenas, found in the African savannas, are the only hyenas that regularly hunt in packs, and they also have females with masculinized genitalia. The clitoris is so enlarged that it resembles a male penis (complete with the capacity for erection), and the labia are fused to form a structure resembling a scrotum. For many years it was mistakenly thought that these hyenas were hermaphrodites. Although hyena females may engage in display behavior with these structures, we now think that the masculinization of the females' genitalia is primarily a side effect during development due to high levels of male hormones. The social systems of spotted hyenas are female-dominated with high frequencies of aggressive interactions. High levels of testosterone are correlated with high levels of aggressive behavior in female spotted hyenas, and aggressive individuals have high social status in their packs.

This masculinization has clear disadvantages, however. All the functions of the original urogenital sinus must now be transmitted through this penislike structure; thus female hyenas urinate, copulate, and give birth through the enlarged clitoris. It is perhaps not surprising that there is high mortality among females giving birth for the first time.

21.3 Are Placental Mammals Reproductively Superior to Marsupials?

The marsupial mode of reproduction was once thought to be inferior to that of placentals. This opinion was based primarily on the assumption that marsupials had been unable to compete with placentals and were able to radiate only in Australia, where they apparently evolved in isolation from placental mammals. The low diversity of marsupial species (about 6 percent that of placentals) was also cited as evidence of their evolutionary inferiority. That analysis is faulty, however, and the low diversity of marsupials is probably no more than an accident of history. The continents on which marsupials radiated had a smaller total land area

during the Late Cenozoic than the continents that placentals inhabited. Marsupials would thus be expected to have less species diversity than placentals.

Placental chauvinists have argued that marsupial reproductive physiology and anatomy are inferior to the reproductive system of placentals, but those arguments also are faulty. For example, it has been suggested that marsupials are unable to give birth to large offspring because they cannot maintain embryos in the uterus past one estrous cycle. The physiological basis for this deficiency was said to be a lack of hormonal feedback between the developing young and the maternal brain. This idea sounded good at the time, but we now know that hormonal feedback of this nature does exist in marsupials.

The lack of midline fusion of the oviducts in marsupials is an anatomical character that has been held responsible for their supposed reproductive inferiority, because the restricted space in the marsupial reproductive tract would make it impossible to carry large young to full term. Once again the reasoning is faulty: many placentals have uteri that are almost completely separate, with midline fusion occurring only at the base. This type of duplex uterus is seen in cows, for example, which give birth to large young.

It seems more likely that marsupials and placentals evolved different but equivalent reproductive strategies. The most significant difference may be the timing of the largest energy investment in reproduction. Placentals pay upfront with high investments of energy during intrauterine development, whereas energy investment by marsupials is greatest during the extended period of postgestation lactation.

The difference in the timing of energy investment may render the marsupial reproductive mode superior to that of placental mammals under some conditions. Although marsupial and placental mothers have invested similar total amounts of energy in their young by the time they become independent of maternal care, marsupials supply energy at slower rates. Thus, the death of a nursing young during a drought, for example, is less costly for a marsupial than it is for a placental because the marsupial has invested less energy in its young. In these circumstances a marsupial mother might have retained enough stored energy to conceive again immediately, whereas a placental would have to wait until the following year.

The death of an offspring might even be part of a strategy that increases the lifetime reproductive success of a female marsupial. Some female kangaroos eject pouch young when predators are pursuing them. This behavior could be adaptive if sacrificing a baby allows the mother to escape from a predator. Kangaroos are long-lived animals with high reproductive potential, and the loss of one baby has little impact on the lifetime reproductive success of a female kangaroo.

That hypothesis leads to the prediction that mothers would more readily sacrifice small babies than larger ones because a small baby represents less maternal investment than a larger baby does. Observations of wallabies do not support that prediction, however, because these kangaroos are more likely to eject large pouch young than small ones. Furthermore, the fossil record reveals no evidence of any pursuit predators in Australia before humans and their dogs arrived, some 40,000 years ago. Perhaps ejection of pouch young is accidental and larger babies are simply more likely than small ones to be bounced out of the pouch as the mother bounds away from a predator.

These scenarios are based on current conditions, however, and an understanding of the evolutionary reasons for the development of marsupial and placental reproductive strategies must consider the animals and conditions that existed at the time of their divergence in Asia, back in the Early Cretaceous period. At this time both marsupials and placentals were small and lived in a world of equable climates, so killing droughts were probably not frequent occurrences. The arid interior of Australia did not develop until the late Cenozoic, and in any case today many marsupials live outside of Australia. Furthermore, these small species probably produced only a single litter per lifetime, so the ability to dump the pouch young is unlikely to have been adaptive.

Both marsupial and placental reproductive strategies probably represented equally good ways of achieving the same goal for the small therians of the Early Cretaceous. However, the marsupial reproductive system may have limited the mode of life of larger animals that evolved during the Cenozoic. For example, there is no marsupial equivalent of a whale. A fully aquatic marsupial could not carry altricial young in a pouch because they would be unable to breathe air; nor could a marsupial give birth underwater because the tiny neonates would be swept away by currents. There is only one semiaquatic marsupial, the South American yapok or water opossum, which seals up its pouch during its short underwater forays.

There are no marsupial equivalents to bats, but this might just be a matter of evolutionary chance. Gliding has evolved several times convergently among both marsupials and placentals, but flight has evolved only once in synapsids. It is also probably true that a marsupial could not afford to reduce its front feet to nonclasping limbs with few fingers, like the front limbs of hoofed ungulates, because the newborn young would be unable to climb to the pouch. (This line of reasoning probably also explains why the marsupial equivalents

of horses and antelope, the large kangaroos, have specialized only their back legs for locomotion.)

Although the marsupial mode of reproduction may not be adaptively inferior overall, it is true that placentals have a faster reproductive rate than marsupials, especially at smaller body sizes. Certainly one reason that feral placental mammals are diversifying at the expense of the native Australian marsupials is because they can reproduce faster and more often.

21.4 Specializations for Feeding

Mammals need large quantities of food, and starting to process food in the mouth rather than waiting until it reaches the stomach speeds digestion. All mammals masticate their food, and all mammals have a distinct swallowing reflex whereby they ingest a discrete bolus of finely chewed food (deglutination). The mammalian tongue, important in both oral food processing and swallowing, has a unique system of intrinsic muscles, and the muscular cheeks of mammals—which keep food in the mouth as it is chewed—are derived from their unique facial muscles.

Dentition

Mammalian teeth are shown in **Figure 21–6**. The incisors are used to seize food. The incisors of mammals that gnaw, such as rodents and rabbits, may be enormously enlarged and grow continuously throughout life. Canines are used to stab prey and are often lost in herbivores, which have no need to subdue their food items, but the canines are also used in social signals and may be retained in modified form. The tusks of pigs and walruses are modified canines, but the tusks of elephants are modified incisors. Upper canines are generally larger in male primates than in females, even slightly so in humans. Hornless ruminants, such as the mouse deer, may retain large upper canines in the males for fighting and display.

Premolars pierce and slice food, and molars break food into fine particles. Premolars and molars are usually different in form. Premolars generally have a single cusp, whereas molars have three or more cusps. Many herbivores use the entire postcanine tooth row for mastication, and in these mammals the premolars resemble the molars (i.e., are molarized).

Omnivorous and fruit-eating mammals have reduced the originally pointed cusps of their molars to rounded, flattened structures suitable for crushing and pulping. They have also added a fourth cusp to the upper molars and increased the size of the

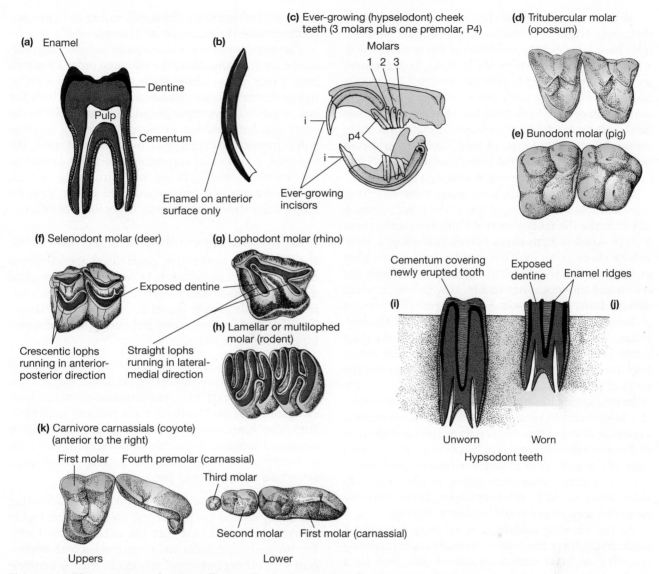

(a) Enamel

Dentine

Pulp

Cementum

(b)

Enamel on anterior surface only

(c) Ever-growing (hypselodont) cheek teeth (3 molars plus one premolar, P4)

Molars
1 2 3

i

p4

i

Ever-growing incisors

(d) Tritubercular molar (opossum)

(e) Bunodont molar (pig)

(f) Selenodont molar (deer)

Exposed dentine

Crescentic lophs running in anterior-posterior direction

(g) Lophodont molar (rhino)

Straight lophs running in lateral-medial direction

(h) Lamellar or multilophed molar (rodent)

Cementum covering newly erupted tooth

Exposed dentine

Enamel ridges

(i)

(j)

Unworn

Worn

Hypsodont teeth

(k) Carnivore carnassials (coyote) (anterior to the right)

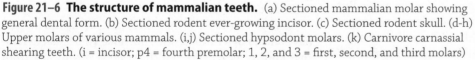

First molar   Fourth premolar (carnassial)

Third molar

Second molar   First molar (carnassial)

Uppers   Lower

Figure 21–6 The structure of mammalian teeth. (a) Sectioned mammalian molar showing general dental form. (b) Sectioned rodent ever-growing incisor. (c) Sectioned rodent skull. (d-h) Upper molars of various mammals. (i,j) Sectioned hypsodont molars. (k) Carnivore carnassial shearing teeth. (i = incisor; p4 = fourth premolar; 1, 2, and 3 = first, second, and third molars)

talonid basin in the lower molars so that these teeth now appear square (quadritubercular) rather than triangular. These teeth are called **bunodont** (Greek *buno* = a hill or mound and *dont* = tooth) in reference to the rounded cusps. We have bunodont molars, as do most primates and other omnivores such as pigs and raccoons.

In herbivores, the simple cusps of the bunodont tooth run together into ridges, or **lophs.** Lophed teeth work best when the enamel has been worn off the top of the ridges to expose the underlying dentine. Each ridge then consists of a pair of sharp enamel blades lying on either side of the interven-ing dentine. When these teeth occlude and the jaws move sideways, the food is grated between multiple sets of flat, shearing blades. Such lophed teeth have been evolved convergently many times among mammals, and there are several different ways of running the cusps together to form lophs. **Selenodont** molars (teeth with crescentic ridges) are seen in ruminants, camelids, and koalas. **Lophodont** molars (teeth with ridges that run in a predominantly internal-external direction) are seen in perissodactyls like horses and rhinos, and in kangaroos. Elephants, some rodents, and wombats have a more complex type of molar called multilophed or lamellar.

All herbivorous mammals face a similar problem with their teeth—that of dental durability, because all mammals have inherited the condition of diphyodonty (only a single set of replacement teeth) from their common ancestor. Diphyodonty is probably essential for precise occlusion of the adult teeth, but it presents a problem in that one set of adult teeth must last a lifetime. Herbivores face a particular problem because vegetation is more abrasive than other forms of food. Grazing herbivores feed close to the ground and have an additional problem because they ingest dust and grit with their food.

Herbivorous mammals have made their dentition more durable in a variety of ways. The most common way to make the molars more durable is to make them high crowned or **hypsodont** (Greek *hyps* = high). Hypsodont cheek teeth look like regular teeth when seen in the jaw, but the division at the base between the crown and the root is not visible, as with low-crowned (**brachydont;** Greek *brachy* = short) teeth. The crowns of hypsodont teeth extend into the depth of the jawbones. Species with hypsodont teeeth have very deep lower jaws and deep cheek regions, and the orbits may have moved posteriorly, thus accommodating the roots of the upper teeth in the maxilla.

In ungulates with hypsodont teeth, a layer of cementum is extended to cover the entire tooth. Cementum is a bonelike material that covers only the root and the base of the crown in most mammals, but in ungulates the high lophs of the teeth must be laid down during tooth development. Without cementum acting as filler, the individual lophs would be tall, freestanding blades once the tooth had erupted and would be likely to fracture.

As the chewing surface is worn away, hypsodont teeth erupt from the base to provide a continuously renewing occlusal surface, much as the lead in a mechanical pencil is pushed up as it is worn away. But when the tooth is worn out, the animals can no longer eat. (Most mammals in the wild will have died of natural causes long before their teeth wear out, but domestic horses surviving into their late twenties and thirties often must be fed soft food because they have virtually no molars left.) However, some mammals have molars in which the roots do not close and the tooth is functionally ever-growing, or **hypselodont.** For a variety of reasons, this appears to be an option primarily for small mammals, such as rabbits and some rodents.

Elephants do not have ever-growing cheek teeth but instead employ the novel feature of molar progression. Each molar is not only hypsodont but also greatly enlarged, the size of the entire original tooth row. Each molar is erupted and worn in turn: when it becomes worn down, its remaining stub falls out of the front of the jaw, and the molar behind erupts from the back to replace it. Elephants have a total of six molars (three milk molars and three permanent molars) in each upper and lower jaw half.

Carnivorous mammals have specialized shearing postcanine teeth. Mammals in the placental order Carnivora have a pair of specialized teeth modified into a set of tightly shearing blades, the carnassials, formed from the last premolar in the upper jaw and the first molar in the lower jaw. Some extinct carnivores belonging to other orders formed their carnassials from different teeth. The thylacine, a marsupial carnivore (sometimes known as the "marsupial wolf"), lacked true carnassials; instead, each molar was somewhat specialized into a bladelike shape, but none was significantly larger than the others.

Craniodental Specializations of Mammals

The hedgehog shows the generalized insectivorous condition that can be taken to represent the original mammalian mode (**Figure 21–7**). The molars of generalized mammals usually retain the unmodified triangular shape and pointed individual cusps that are useful for puncturing insect cuticle.

Anteaters Mammals that specialize in eating ants and termites are called **myrmecophagous** (Greek *myrme* = ant and *phago* = to eat). Myrmecophagous mammals have elongated jaws and teeth that are reduced or absent. They also have enlarged salivary glands and highly elongated tongues. Reduction of the teeth is also seen in mammals that specialize in getting nectar (e.g., some bats and the honey possum).

Aquatic Feeders Aquatic fish-eating and squid-eating mammals such as porpoises and dolphins have highly elongate jaws that have lost the anteriormost teeth. The form of their skulls and teeth has become convergent with other piscivorous tetrapods, such as ichthyosaurs and crocodiles.

The baleen whales (mysticetes) lack teeth, and instead have sheets of a fibrous, stiff, hornlike epidermal derivative known as **baleen,** which extends downward from the upper jaw. These whales use the baleen to strain planktonic organisms from the water. (Some Oligocene baleen whales apparently had both teeth and baleen.)

Differences Between Carnivorous and Herbivorous Mammals The basic mode of mammalian mastication is best understood by considering the difference between carnivorous and herbivorous mammals (**Figure 21–8** on page 524). All mammals use a combination of masseter, temporalis, and pterygoideus muscles to close the jaws, and the digastric muscle (in therians) to open the jaw. The relative sizes of the muscles and the shapes of the skulls reflect the different demands of masticating flesh and vegetation. The temporalis has its greatest mechanical

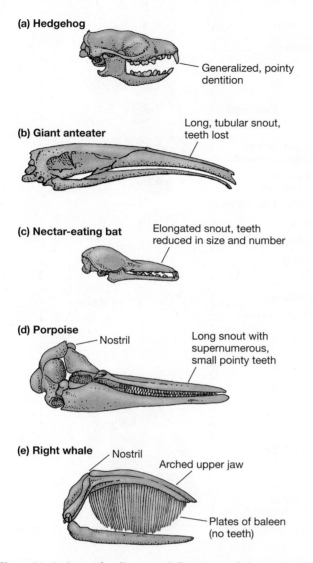

(a) Hedgehog

Generalized, pointy dentition

(b) Giant anteater

Long, tubular snout, teeth lost

(c) Nectar-eating bat

Elongated snout, teeth reduced in size and number

(d) Porpoise

Nostril

Long snout with supernumerous, small pointy teeth

(e) Right whale

Nostril

Arched upper jaw

Plates of baleen (no teeth)

Figure 21–7 Some feeding specializations of the teeth and skulls of mammals.

advantage at initiating jaw closure at moderate to large gapes, when the incisors and canines are likely to be used. Thus a large temporalis is typical of carnivores, which use a forceful bite with their canines to kill and subdue prey. In the generalized mammalian condition the jaw joint is on the same level as the tooth row so the teeth come into contact sequentially as the jaw closes, like the blades of a pair of scissors—a morphology well suited for teeth that primarily cut and shear. The postglenoid process prevents the strong temporalis muscle from dislocating the lower jaw. Strong muscles run from the high occipital region of the head to the cervical vertebrae. These muscles are probably important for resisting struggling prey.

The skulls of herbivorous mammals are modified to grind tough, resistant food. The protein content

of leaves and stems is usually low, and the protein is enclosed by a tough cell wall formed by **cellulose** (a complex carbohydrate); thus herbivores must process large amounts of food per day. The masseter exerts crushing force at the back of the tooth row and also moves the jaw from side to side, aiding the complex molars in breaking down the vegetation.

Herbivorous mammals have a large masseter and small temporalis in comparison to the generalized mammalian condition. This morphology is reflected in the increased size of the angle of the lower jaw (where the masseter inserts) and the reduced size of the coronoid process and temporal fossa (where the temporalis inserts).

The jaw joint has been shifted in herbivores so that it is high on the skull, offset from the tooth row. This offset brings all the teeth in the upper and lower jaws together simultaneously, with a grinding action that shreds plant material between the lophs of the upper and lower teeth, much as the offset handle of a cooking spatula allows you to apply the entire blade of the spatula to the bottom of the frying pan while keeping your hand above the pan's rim.

Herbivores also usually have elongated snouts, resulting in a gap between the cheek teeth and the incisors called the **diastema.** The function of the diastema is uncertain. It may allow extra space for the tongue to manipulate food, or it may be only a reflection of the elongation of the jaw for other reasons. A long jaw allows an animal to reach into narrow spaces to nip off a leaf or fruit with its incisors.

Many herbivorous placental mammals ossify the cartilaginous partition at the back of the orbit to form a bony postorbital bar. This bar is probably important in absorbing stress from the jaws during the constant chewing of herbivores, thus protecting the braincase. Herbivores usually have a fairly low occipital region because they do not need to have a carnivore-like attachment for muscles that help brace the head on the neck. An exception can be found in pigs, which root with their snouts and have strong muscles linking the back of the head to the neck.

Many rodents have a highly specialized type of food processing. Their upper and lower tooth rows are the same distance apart—unlike the condition in most mammals, in which the lower tooth rows are closer together than the uppers. This derived condition in rodents is combined with a rounded jaw condyle, which allows forward and backward jaw movement, and the insertion of a portion of the masseter muscle far forward on the skull so that the lower jaw can be pulled forward into occlusion. This jaw apparatus allows rodents to chew on both sides of the jaw at once, presumably a highly efficient mode of food processing. Note that this mode

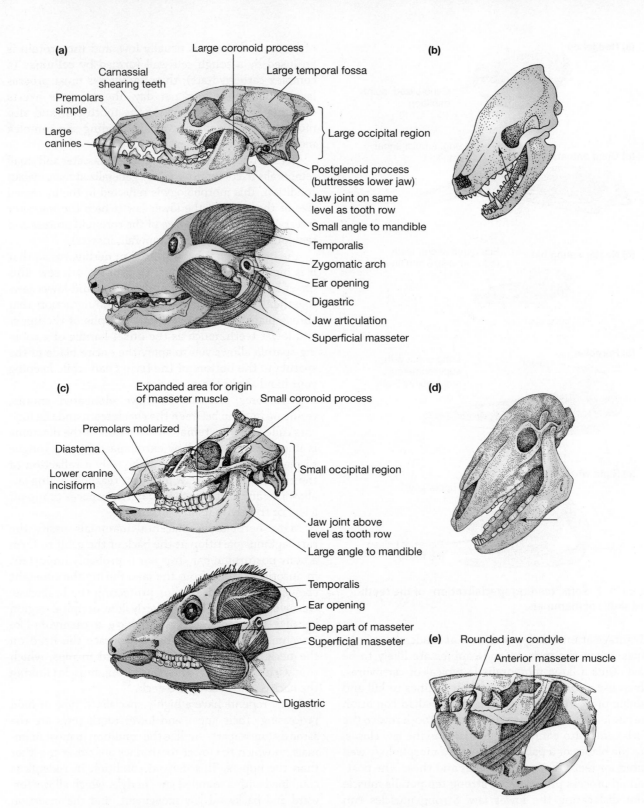

Figure 21–8 Craniodental differences between carnivorous and herbivorous mammals.
(a) Carnivore skull and musculature (a dog). (b) Action of carnivore jaws. (c) Herbivore skull and musculature (a deer). (d) Action of herbivore jaws. (e) Rodent skull and musculature (beaver).

of chewing can be achieved only with flattened, lamellar teeth because the high ridges of other types of teeth would prevent this jaw motion.

Herbivores, Microbes, and the Ecology of Digestion

The specialized teeth of herbivores can rupture the cell walls of vegetation and expose the cell contents, but only special enzymes (**cellulases**) can digest the cellulose that forms the cell wall and constitutes a large percentage of the plant material. However, no multicellular animal has the ability to synthesize cellulase. Thus the efficient use of plants as food requires cellulase enzymes produced by microorganisms that live as symbionts in the guts of herbivorous animals. While all mammals have gut symbiotic microorganisms of some kind, herbivorous mammals have independently evolved specialized chambers within the digestive tract to house symbiotic microorganisms that convert the cellulose and lignin of plant cell walls into digestible nutrients (volatile fatty acids) that can be absorbed through the gut wall (**Figure 21–9**).

Distinctly different solutions to the problems posed by plants as food can be seen in different mammals.

Horses and other perissodactyls are examples of **hindgut (monogastric) fermenters.** These animals have a simple stomach and have enlarged both the large intestine and the cecum as fermentation chambers. Other hindgut fermenters include elephants, hyraxes, New World howler monkeys, wombats, koalas, rabbits, and many rodents. Some degree of hindgut fermentation is probably a generalized vertebrate character: it occurs among birds, lizards, turtles, and fishes as well as omnivorous and carnivorous mammals such as humans and dogs.

Cows and other ruminant artiodactyls are examples of **foregut (ruminant) fermenters,** in which the nonabsorptive forestomach is divided into three chambers that store and ferment food, followed by a fourth chamber (the true stomach) in which digestion occurs. Ruminants are so called because they ruminate, or chew, the cud. Camels resemble other ruminants in many respects but have only three-chambered stomachs (the omasum is lacking). A simpler type of foregut fermentation, without extensive stomach division or cud chewing, is found in many other mammals, including Old World colobine monkeys, kangaroos, hippos, and some rodents. Interestingly, the gut symbionts of hindgut fermenters and foregut fermenters are somewhat distinctive in each group,

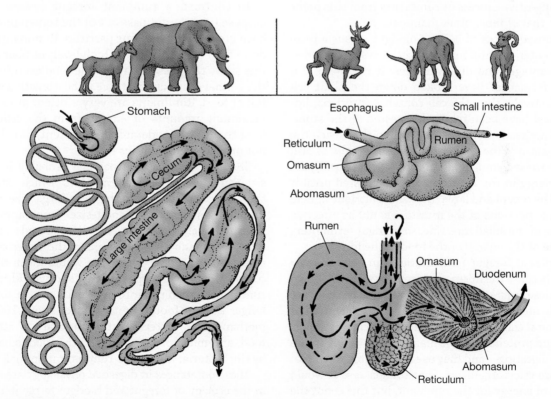

Figure 21–9 Hindgut and ruminant foregut digestive systems. *Left:* The hindgut fermenting system. Fermentation occurs in the enlarged cecum and colon (large intestine). *Right:* The ruminant system, showing the four-chambered stomach.

regardless of phylogenetic associations (e.g., the sheep's gut flora is more like that of a kangaroo than of a horse).

Hindgut fermenters chew their food thoroughly as they eat, fracturing the plant cell walls with their teeth so that the cell contents are released. These cell contents are processed and absorbed in the stomach and small intestine. The cellulose of the plant cell wall is not digested until it reaches the cecum and large intestine, where it is attacked by the symbiotic microorganisms.

Some small hindgut fermenters, such as rabbits and rodents, ferment the food largely in the cecum and do not absorb much of the initial products of fermentation. Instead they rely on **coprophagy,** re-eating the first set of feces that are produced and recycling the nutrients.

Ruminant foregut fermenters do not need to chew their food as thoroughly on initial mastication, partly because the plant cell walls will be chemically disrupted in the stomach and partly because boluses of food (the cud) will be returned to the mouth and chewed again. Ruminants have less extensive modifications of the skull and teeth than do hindgut fermenters. Food cannot pass into the true stomach until it has been reduced by remastication to small particles. Most or all of the cellulose has been broken down and absorbed before the food reaches the true stomach, and the digestive process of ruminants from this point on is like that of most other mammals.

The consequences of being a hindgut versus a foregut fermenter are profound, and each system has distinct advantages and disadvantages. If the teeth of a hindgut fermenter have broken down the plant cell walls effectively, then the cell contents (proteins, lipids, sugars) have been released, digested in the stomach, and are ready to be absorbed when they reach the small intestine. The cellulose in the cell walls is still intact, however, and it represents a substantial part of the energy in the food that is not digested until it reaches the cecum and large intestine. Nutrient uptake from these portions of the intestine is not as effective as uptake in the small intestine, so hindgut fermenters lose some of the energy of the food in the feces.

In contrast, foregut fermentation can be extremely efficient because the microorganisms have broken down the cellulose *before* it reaches the small intestine. The symbiotic microorganisms in the rumen ferment *all* of the chemical compounds in the food—including lipids, sugars, and proteins that the mammalian system has no problem digesting—and they use these materials to produce more microorganisms. That sounds as if it would be a loss of energy for the ruminant, but this is not the case because surplus microorganisms subsequently pass into the true stomach where they are digested, so the ruminant ultimately gets those nutrients.

In fact, this protein fermentation actually ends up being beneficial to the ruminant in a process called nitrogen cycling. Microorganisms ferment the protein into ammonia, which is then taken via the circulation to the liver, where it is converted into urea. This urea is transported by the circulatory system to the rumen, where the microorganisms use it for their own growth. Thus all the protein that the ruminant digests is microbial protein. An advantage of this system is that the microorganisms make all of the essential amino acids needed in the diet. Thus, a ruminant can be more limited in its selection of plant species than a hindgut fermenter, which must find all of its essential amino acids by eating a variety of plant sources. Waste urea can also be fed into this recycling, meaning that less urea must be excreted in the urine, thereby conserving water.

A hindgut fermentation system processes material relatively rapidly, whereas ruminants process food more slowly. Food moves through the gut of a horse in 30 to 45 hours, compared to 70 to 100 hours for a cow. While hindgut fermenters are not so efficient at extracting energy from the cellulose, they can process a large volume of food fairly rapidly. Hindgut fermenters can thus bulk process less nutritious, fibrous food, obtaining their energy primarily from the cell contents.

In contrast, a ruminant foregut system is slow because food cannot pass out of the rumen until it has been ground into very fine particles. Ruminants do not do well on diets containing high levels of fiber because this slows the passage rate of the food even further—the animal can literally starve to death with a stomach full of food. Ruminants are very efficient at extracting maximum amounts of energy from the cellulose in food that has a moderate fiber content, but they cannot process highly fibrous food.

The slow rate of food passage probably limits the size range of ruminant fermenters, which attain neither the very small nor the very large size of some hindgut fermenters. Rabbit-size ruminants would have relatively greater metabolic demands (a simple effect of metabolic scaling), and the necessary long retention of food in the rumen would mean that they couldn't eat enough per day to survive. Rhino-sized ruminants would run into problems with an absolutely longer time of food retention: after about 100 hours, methanogenic bacteria in the gut start to attack the food, and much of the energy in the food is harvested by the bacteria, not by the ruminant mammal.

These differences in digestive physiology are reflected in the ecology of foregut and hindgut fermenters. Hindgut fermenters can survive on very low-quality food such as fibrous grasses, as long as it is available in large quantities, but they cannot survive so well in areas where the absolute quantity of food is the limiting factor. Ruminants are the main herbivores in places like the Arctic and deserts, where the food is of moderately good quality but severely limited in quantity; such food best supports an animal that can make the most efficient use of its dietary intake. Ruminants also have an advantage in desert conditions because of their water-conserving nitrogen cycling, and they are thus better able to go without water for a few days than are hindgut fermenters. However, hindgut fermenters do better in situations where food is abundant but of poor quality, as in some grassland habitats (where horses do better than cows, for example).

21.5 Locomotor Specializations

A tree shrew illustrates the bounding and scrambling that is the basic mode of mammalian locomotion (Figure 21–10). The limbs and back are flexed during locomotion, and the basic therian gait appears to be highly dependent on this flexed spine and limbs, which bend at the elbow-ankle, shoulder-knee, and top of scapula-hip joints in a three-part zigzag fashion. The independent movement of the scapula (a new feature in therians) is critical to this arrangement: the major pivot points of the limbs during locomotion are formed by the dorsal border of the scapula (rotating around its own axis) in the forelimb and by the hip joint (the articulation of the femur with the pelvic girdle) in the hindlimb. This morphology is probably an adaptation to locomotion at small body size on an irregular ground surface. In contrast, larger mammals, such as the horse, move with a stiffer back and straighter legs, and they gallop rather than bound.

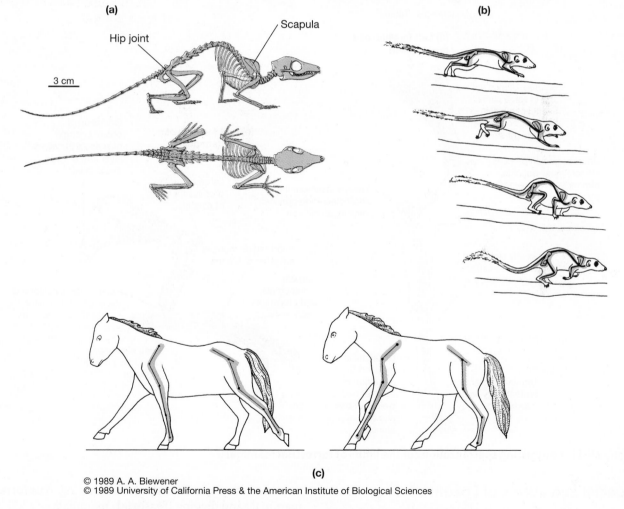

Figure 21–10 Gait and locomotion in mammals. (a) Skeleton of tree shrew (*Tupaia glis*) in typical posture with flexed limbs and quadrupedal stance. (b) Sequential phases of the bounding run in a tree shrew. Note the relatively flexed limbs and mobile back. (c) Sequential phases of the gallop in a horse. Note the relatively straight limb angles and immobile back.

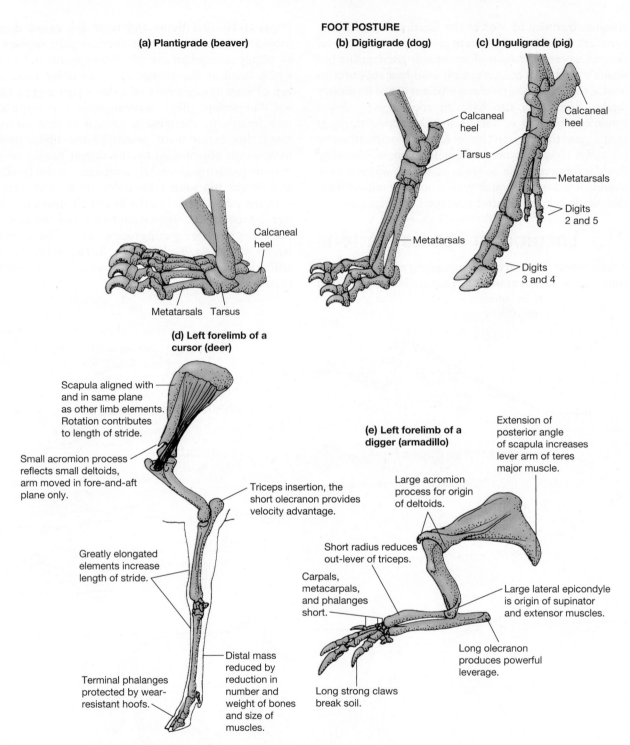

FOOT POSTURE

(a) Plantigrade (beaver)

Calcaneal heel

Metatarsals   Tarsus

(b) Digitigrade (dog)

Calcaneal heel

Tarsus

Metatarsals

(c) Unguligrade (pig)

Calcaneal heel

Metatarsals

Digits 2 and 5

Digits 3 and 4

(d) Left forelimb of a cursor (deer)

Scapula aligned with and in same plane as other limb elements. Rotation contributes to length of stride.

Small acromion process reflects small deltoids, arm moved in fore-and-aft plane only.

Triceps insertion, the short olecranon provides velocity advantage.

Greatly elongated elements increase length of stride.

Terminal phalanges protected by wear-resistant hoofs.

Distal mass reduced by reduction in number and weight of bones and size of muscles.

(e) Left forelimb of a digger (armadillo)

Extension of posterior angle of scapula increases lever arm of teres major muscle.

Large acromion process for origin of deltoids.

Short radius reduces out-lever of triceps.

Carpals, metacarpals, and phalanges short.

Large lateral epicondyle is origin of supinator and extensor muscles.

Long olecranon produces powerful leverage.

Long strong claws break soil.

Figure 21–11 Contrasting specializations of the limbs of terrestrial mammals.

Specialized Forms of Locomotion

Larger animals experience the world differently from smaller ones because of physical size and scaling effects, and larger mammals are usually modified for more specialized forms of locomotion. **Figure 21–11** contrasts the specializations of running (**cursorial**) mammals and digging (**fossorial**) mammals.

Cursorial Limb Morphology The number of strides it takes to travel a given distance determines the cost of locomotion, and a long-legged mammal can cover a given distance in fewer strides than a shorter-legged one. Long legs also provide a long out-lever arm for the major locomotor muscles, such as the triceps in the forelimb and the gastrocnemius in the hindlimb. This arrangement favors speed of motion rather than power.

Elongation is limited mostly to the lower portions of the limb: the radius and ulna in the forelimb, the tibia and fibula in the hindlimb, and the metapodials (a collective term used to describe the metacarpal and metatarsal bones). The humerus and femur are not elongated, nor are the phalanges.

Muscles are limited to the proximal portion of the limb, reducing the mass in the lower limb. There is almost no muscle in a horse's leg below the wrist (the horse's so-called knee joint) or the ankle (the hock). That anatomy makes sense in mechanical terms because the foot is motionless at the start of each stride (as it pushes against the ground); when it leaves the ground, it must be accelerated from zero velocity to a speed greater than the body speed as it moves forward ready for the next contact with the ground. The lighter the foot, the less effort is needed.

The force of muscular contraction by the muscles in the upper limb is transmitted to the lower limb via long elastic tendons. These tendons are stretched with each stride, storing and then releasing elastic energy and contributing to locomotor efficiency (**Figure 21–12**).

The leg tendons of a hopping kangaroo are an obvious example of elastic storage, as the animal bounces on landing as if using a pogo stick. However, all cursorial animals rely on energy storage in tendons for gaits faster than a walk. Even humans rely on elastic energy storage in tendons, especially in the Achilles tendon that attaches the gastrocnemius (calf) muscle to the calcaneal heel. People who have damaged Achilles tendons (a common sports injury) find running difficult or impossible. Lengthening these tendons to increase the amount of stretch and recoil may be part of the evolutionary reason for limb elongation and changes in foot posture.

Other cursorial modifications restrict the motion of the limb to a fore-and-aft plane so that most of the thrust on the ground contributes to forward movement. The clavicle is reduced or lost, the wrist and ankle bones allow motion only in a fore-and-aft plane, and the forelimb cannot be supinated. (Note how easily you can turn your hand at the wrist so that your palm faces upward; a dog can't turn its forepaw that much, and a horse has almost no ability to rotate this joint.)

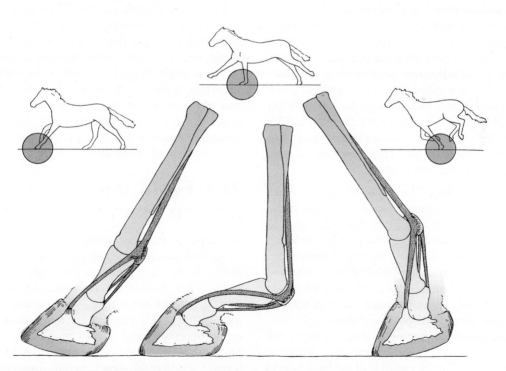

Figure 21–12 Springing action of the elastic tendons in the foot of a horse. The tendon is stretched as the horse's body moves forward over the leg and shortens as the foot leaves the ground, providing additional propulsive force.

The number of digits may also be reduced, perhaps to reduce the weight of the foot so that it can be accelerated and stopped more easily. Carnivores often reduce digit 1 but otherwise compress the digits together rather than reducing their number. Artiodactyls reduce or lose digits 1, 2, and 5, becoming effectively four-toed like a pig or two-toed like a deer. Perissodactyls lose digits 1 and 5 and reduce digits 2 and 4, becoming three-toed like a rhinoceros or single-toed like a horse.

Such cursorial specializations occurred convergently among many different mammalian lineages. Early Cenozoic ungulates and carnivorans were not highly cursorial, and it has long been assumed that these specializations, especially the evolution of longer legs, must have arisen in the context of predator-prey relationships. Longer legs would have given a carnivore a little more speed to pursue the herbivore, resulting in selection for ungulates with longer legs to make a faster escape.

This idea of a coevolutionary arms race between predator and prey is appealing, but the fossil record does not support it. If coevolution were the driving force, ungulates and carnivores would have evolved their longer limbs at the same time, in lockstep fashion, but this is not the case. Cursorial ungulates are known from 30 million years ago, whereas cursorial carnivores became apparent only within the past 5 million years. Why would ungulates evolve cursorial specializations if not to flee predators? Probably because all of the limb modifications that make a mammal a faster runner also make it more efficient at slower gaits, such as a trot. These cursorial adaptations appeared in the mid-Cenozoic when open habitats started to become prominent (see Chapter 19), and the ungulates would have had to forage farther each day for food (as do ungulates in more open habitats today). However, modifications for endurance at slow gaits also provide speed at fast gaits, and speed became valuable when cursorial predators later evolved.

Fossorial Limb Morphology Mammals do several types of digging. The most common is scratch digging at the surface—that is, the type of digging that a dog uses to bury a bone. Animals that are truly fossorial (burrowing under the ground surface) have a variety of anatomical specializations. No mammal is limbless and elongated like a burrowing lizard or caecilian, although mammals that follow their prey down burrows (such as weasels and ferrets) have elongated bodies and short legs.

A limb specialized for digging is almost exactly the opposite of a limb specialized for running. Running limbs maximize speed at the expense of power, whereas digging limbs maximize power at the expense of speed. Fossorial mammals achieve this mechanical advantage in the forelimb with a long olecranon process (for the triceps muscle to retract the hand) and a relatively short forearm.

Some subterranean diggers are called rapid scratch diggers. These animals, such as the African golden mole and the Australian marsupial mole, dig with both fore and hind feet and move through sandy soil by pushing the grains aside and back without constructing an open burrow. True moles, which live in more compact soil, rotate their forelimb with a laterally facing palm rather than simply retracting it. This form of digging is termed *rotation thrust digging,* and the anatomical adaptations are somewhat different from those of scratch diggers because they employ the teres major muscle to rotate their arms rather than the triceps to retract it. Moles burrow just below the surface of the ground, seeking worms and insect larvae in the roots of plants, and they push the soil upward as they tunnel. Finally, many rodents (gophers are an example) use their ever-growing incisors rather than their limbs to dislodge soil; this process is termed *chisel-tooth digging.* Gophers have a pronounced diastema and can pull their lips together behind the incisors to prevent soil from entering the mouth. Gophers push the soil they excavate out onto the surface of the ground, forming the mounds that are a familiar feature of the landscape wherever gophers live.

Digging mammals retain all five digits, tipped with stout claws for breaking the substrate. They also have large projections on their limb bones for attachment of strong muscles, such as the enlarged acromion process on the scapula. Scratch diggers at the surface, such as anteaters, have a very stout pelvis, with many vertebrae involved in the sacrum, for bracing the hindlimb while digging with the forelimb. However, underground burrowers such as moles do not have this type of strengthened pelvis and sacrum.

21.6 Evolution of Aquatic Mammals

Semiaquatic mammals are not very different from terrestrial mammals except for somewhat more paddlelike limbs and a denser fur coat, and lineages of semiaquatic mammals have evolved numerous times. Among extant mammals, we can see examples in monotremes (platypus) and marsupials (yapok, or water opossum) and within placentals in the orders Lipotyphla (water shrew, desman), Tenrecoidea (otter tenrec), Rodentia (beaver, coypu, muskrat, Australian water rat), Carnivora (otters, mink, polar bear), and Artiodactyla (hippopotamus). The fossil record adds hippopotamus-like rhinoceroses, several independent evolutions of otterlike animals (including the recently extinct sea mink), and even a semiaquatic sloth.

Specialization for fully aquatic life is a different matter, however, and fully aquatic mammals—primarily marine forms that never or rarely come out onto land—have evolved only three times: in the orders Cetacea (whales, porpoises, and dolphins), Sirenia (dugongs and manatees), and Carnivora (seals, sea lions, and walruses) (**Figure 21–13**). Cetaceans and sirenians cannot come onto land, but pinnipeds are more amphibious, emerging onto land to court, mate, and give birth. Cetaceans and pinnipeds are carnivorous, but sirenians are herbivores. All use blubber (a thick layer of subcutaneous fat) for insulation as well as (or instead of) hair, and show various adaptations of their physiology to allow them to stay underwater and in some cases (whales and seals) to dive to great depths.

Morphological Adaptations for Life in Water

Most semiaquatic mammals use the limbs to swim (paraxial swimming), as we do ourselves. This type of swimming is fairly inefficient in terms of the drag forces created in the water. Most fully aquatic mammals use axial swimming, that is, undulations of the body and tail (or the tail-like hind limbs of pinnipeds) via dorsoventral flexion, rather than laterally like fishes and aquatic sauropsids. This swimming motion is a modification of the flexion of the vertebral column that is used by terrestrial mammals. Fully aquatic mammals have short paddlelike limbs, with a short proximal portion and elongated phalanges, and these limbs are used like fish fins for braking and steering, not for propulsion. **Figure 21–14** on page 533 illustrates the aquatic modifications of a generalized whale.

Early cetaceans and sirenians had hind legs, but these were lost as later forms became specialized for axial swimming. Pinnipeds, which still come out onto land at times, retain hind legs and also retain zygapophyses in their trunk vertebrae. What appears to be a tail in pinnipeds is actually a modified pair of hindlimbs that have been turned backward. True seals are unable to change the position of the hindlimbs and are clumsy on land, but sea lions and walruses can turn the hind legs forward and move on land with vertical flexions of the vertebral column in an effective, although ungainly, humping fashion. Sea lions additionally have a derived type of paraxial swimming in which the forelimbs are used in synchrony for underwater flying, creating lift in the water in concert with dorsoventral movements of the back and hind legs.

The Evolution of Whales

An almost perfect fossil record sequence of early cetacean evolution has been assembled, starting in the Eocene of Pakistan along the shores of the ancient Tethys Sea. This sequence shows a progression among early whales from terrestrial forms with a full set of legs to aquatic forms with reduced and modified limbs (**Figure 21–15** on page 534). Whales are now known to be derived from an ancestral type of artiodactyl, and are most closely related to the hippopotamuses among the modern forms.

Indohyus from the middle Eocene of India may be closely related to the ancestral lineage of whales. *Indohyus* was small (raccoon size) and probably rather like a present-day mouse deer in appearance, but with shorter legs and a longer tail. Its ear region shows the characteristic cetacean derived condition (a thickened involucrum of the auditory bulla), and its heavy (osteosclerotic) bones indicate a semiaquatic lifestyle.

The Eocene fossil whales all belong to the suborder Archaeoceti, which was largely extinct by the end of the epoch. The modern suborders of whales, Odontoceti (toothed whales) and Mysticeti (baleen whales), arose from a common ancestor among the archaeocetes and first appeared in the latest Eocene or early Oligocene.

The earliest archaeocetes are in the family Pakicetidae from the late early Eocene. These animals were coyote to wolf size, and they have many features indicating that they were amphibious. Their teeth were like those of modern fish eaters, and they had orbits situated on the top of their head, like a hippopotamus, allowing them to observe their surroundings as they floated at the surface. (Later forms [protocetids] had more laterally placed eyes, indicating that they were no longer peering out above the water's surface.) Pakicetid postcranial remains show heavy bones for ballast and robust tails like those of otters, which swim with flexion of the vertebral column. The oxygen isotope ratios in the teeth of pakicetids show that they drank freshwater, not salt water, indicating that their amphibious life was conducted in rivers or at least near-shore environments. The middle ears of pakicetids were slightly, but significantly, modified for underwater hearing, including the diagnostic whale features of the ear region in the skull, which is how we can trace the whale pedigree to these terrestrial forms.

The Ambulocetidae were somewhat later forms from the early middle Eocene. They were about the size of a sea lion and had hindlimbs with enormous feet, possibly a specialization in this group for a mode of locomotion involving dorsoventral flexion and paddling with the

Figure 21–13 Diversity of marine mammals.
(a) Toothed whale: bottlenose dolphin, *Tursiops truncatus* (Cetacea, Odontoceti, Delphinidae). (b) Baleen whale: North Atlantic right whale, *Eubalaena glacialis* (Cetacea, Mysticeti, Balaenidae). (c) Sea cow: dugong, *Dugong dugon* (Sirenia, Dugongidae). (d) Sea lion: Cape fur seal, *Arctocephalus pusillus* (Carnivora, Otariidae). (e) True seal: harbor seal, *Phoca vitulina* (Carnivora, Phocidae). The drawings are approximately to scale except for the baleen whale, which is about one quarter of its normal size in comparison to the other animals.

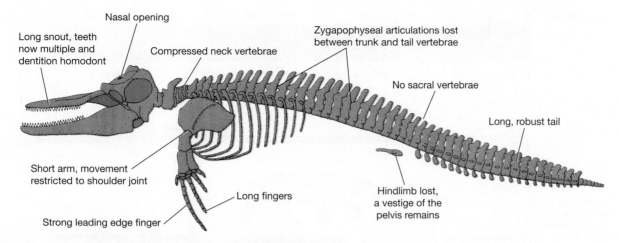

Figure 21–14 Specializations for aquatic locomotion in whales.

Labels (clockwise from upper left): Nasal opening; Long snout, teeth now multiple and dentition homodont; Compressed neck vertebrae; Zygapophyseal articulations lost between trunk and tail vertebrae; No sacral vertebrae; Long, robust tail; Hindlimb lost, a vestige of the pelvis remains; Long fingers; Strong leading edge finger; Short arm, movement restricted to shoulder joint

hind feet pointed backward like those of a seal. However, although their skeletons were more modified for aquatic life than those of the pakicetids, their fingers were not embedded in a webbed flipper, and both fingers and toes had little hooves at the tips, betraying their ungulate origins. Their fossil remains are found in coastal environments, but their tooth isotopes show that they still drank freshwater. The robust skull and teeth of ambulocetids suggest that they were specialized crocodile-like ambush predators feeding on large prey.

The Remingtonocetidae was a more derived group. They were the size of seals and had relatively robust hindlimbs that were probably still capable of bearing their weight on land. Unlike the ambulocetids, they had delicate skulls with long, narrow snouts and very small eyes, indicating that they fed underwater on fish. Both remingtonocetids and the later protocetids had an ear anatomy that improved underwater hearing, including an enlarged canal in the lower jaw that houses a sound-conducting fat pad in modern whales. They also had a reduction in the size of their semicircular canals, which is associated with underwater navigation in modern whales. In addition, the tooth isotopes of remingtonocetids (and all later whales) show that they did not drink freshwater; they must have obtained all of their water from their food, as modern whales do.

Animals that appear to be still more whalelike in appearance and behavior appeared in the late middle Eocene. These were the Protocetidae, which had more reduced hindlimbs than earlier whales. Most retained a connection of the hindlimb to the sacrum but would probably have been clumsy on land. However, evidence that at least some protocetids came on land to give birth, as do pinnipeds today, can be found in the fossil *Maiacetus,* which was preserved with a late-stage embryo in its body. This embryo was positioned so as to be born head first, like terrestrial mammals, rather than in the tail-first mode of modern whales.

The earlier protocetids probably still relied on their hindlimbs for paddling, but the later ones may have had an axial swimming locomotion, with oscillations of the lumbar spine. Protocetids were the first whales to be found in offshore marine habitats and beyond the Indo-Pakistan region—as far afield as the coasts of North America and Africa. They may have had a lifestyle like modern seals: fully aquatic, but not obligatorily so, and still able to come onto land at least for reproductive purposes.

Finally, in the later Eocene, the Basilosauridae appeared. These whales had lost the hindlimb-sacral connection and had greatly reduced hindlimbs, although all of the skeletal elements were still present and the reduced hindlimbs might have been used as copulatory guides. (In all mammals certain muscles run from the pelvis to the genitals, which explains why a remnant of the pelvis is retained in whales and sea cows.)

The necks of basilosaurids were short, their forelimbs were flipperlike with an immobile elbow, and the morphology of their tail vertebrae suggests that they had a tail fluke like modern whales. The ears of basilosaurids were essentially like those of modern whales, with the entire bony ear region isolated from the rest of the skull by air-filled sinuses, although they still retained an external auditory meatus (ear hole). All of these features suggest that they were obligatorily aquatic. The basilosaurines, known from the Northern Hemisphere, had long bodies (up to 16 meters), with greatly elongated trunk vertebrae and small heads. The dorudontines, known from both the Northern and Southern Hemispheres, were more dolphinlike in appearance and contained the ancestry of the modern whales.

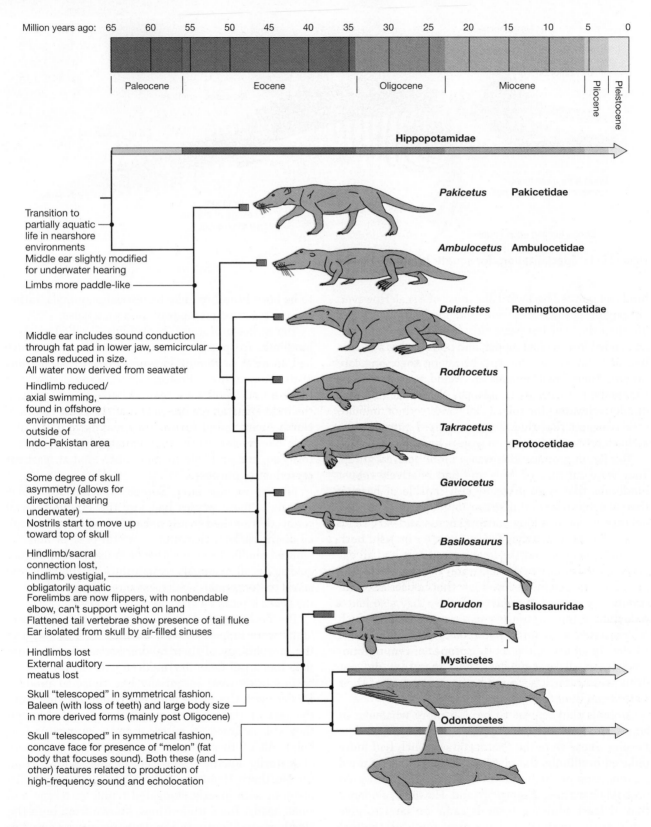

Figure 21–15 A phylogeny of whales showing the sequence of appearance of aquatic specializations.

The initial radiation of modern whales more or less coincides with the extinction of the archaeocetes at the Eocene-Oligocene boundary, a time when higher-latitude temperatures fell dramatically, although some archaeocetes may have survived in the New Zealand region until the late Oligocene. Toothed whales and baleen whales are separate radiations, even though the earliest baleen whales retained teeth. Modern whales differ from archaeocetes in having a "telescoped" skull, in which the nostrils have moved to the top of the head, where they form the blowhole, and the upper and lower jaws extend out into a "beak," although the mode of skull telescoping is different in the two suborders.

The modern whale radiation is probably related to changes in oceanic circulation that increased the productivity of the oceans, resulting in the novel feeding strategies of echolocation-assisted predation (toothed whales) and filter feeding (baleen whales). Another pulse of cetacean radiation occurred in the late Miocene, again concurrent with a lowering of higher-latitude temperatures and changes in oceanic circulation. At this time, many existing families became extinct, and some modern forms (such as dolphins and porpoises) made their first appearance.

Thus the fossil record shows that whales have long been an extremely diverse and successful group of mammals. Sadly, recent molecular analyses suggest that North Atlantic baleen whales were up to 20 times more abundant prior to the start of human commercial exploitation than they are today.

Summary

The major groups of living mammals can be distinguished by their mode of reproduction. Therians (marsupials and placentals) are more derived than monotremes and give birth to live young (viviparity) as opposed to laying eggs. Marsupials give birth to very immature young that complete their development attached to nipples that are usually, but not always, enclosed in a pouch. Placentals give birth to more mature young and have a shorter period of lactation than marsupials. The marsupial mode of reproduction has often been considered inferior to the placental one, but the differences may reflect only their separate evolutionary histories, with neither method being inherently superior to the other.

Cenozoic mammals have diversified into a variety of feeding types, reflected in different anatomies of their skulls and dentitions. The skulls and teeth of herbivores are in general more specialized than those of omnivores and carnivores because fibrous plant material is difficult to chew and abrades the teeth. Herbivores have teeth that form low-relief cutting blades, and these teeth may be high crowned (hypsodont) to ensure that the dentition resists a lifetime of abrasion. Carnivorous mammals usually have a pair of specialized bladelike cutting teeth, the carnassials.

Vegetation is also more difficult to digest than other diets, and many herbivores have evolved a symbiotic association with microorganisms in their guts, which ferment the plant fiber and aid in its chemical breakdown.

The basal mammalian mode of locomotion is probably some sort of bounding. With the radiation of larger mammals in the Cenozoic, more specialized types of locomotion evolved. Cursorial (running-adapted) mammals have elongated their legs and changed their foot posture, restricting the range of limb motion and sometimes reducing the number of digits. Fossorial limbs, modified for digging, are short and stout with heavy muscles and large claws. Both types of adaptations have evolved convergently in different mammalian groups.

Many mammals have returned to the water to become semiaquatic, but only three modern groups of mammals (pinnipeds, sirenians, and cetaceans) are fully aquatic, living in marine environments. Marine mammals share a number of morphological features relating to the demands of underwater locomotion as well as physiological adaptations for diving and other behaviors. We have an excellent fossil record of whale evolution showing how they evolved from terrestrial species, through amphibious, near-shore forms, to obligatorily aquatic fully marine forms.

Discussion Questions

1. The eggs of monotremes are much smaller than those of lizards of comparable body size. What is a possible reason for this?

2. Some people have speculated that marsupials and placentals may have evolved viviparity independently. What evidence could be used to support this?

3. Why do some herbivorous mammals have hypsodont (high-crowned) molars? Would herbivorous dinosaurs have encountered a similar problem?
4. Why are horses more dependent on drinking water than ruminants or camels?
5. How and why is the mobile scapula of therians essential for their form of locomotion?
6. Why do seals and sea lions retain zygapophyses in their backbones, while whales and sea cows do not? How could we use this anatomy to determine the behavior of extinct whales (and what additional evidence do we have in one fossil whale to back up behavioral predictions)?

Additional Information

Clauss, M., et al. 2003. The maximum attainable body size of herbivorous mammals: Morphophysiological constraints on foregut, and adaptations of hindgut fermenters. *Oecologica* 136:14–27.

Damuth, J., and C. M. Janis. 2011. On the relationship between hypsodonty and feeding ecology in ungulate animals, and its utility in palaeoecology. *Biological Reviews* 86:733–758.

Domning, D. P. 2001. The earliest known fully quadrupedal sirenian. *Nature* 413:625–627.

Fischer, M. S., et al. 2002. Basic limb kinematics of small therian mammals. *Journal of Experimental Biology* 205:1315–1338.

Gingerich, P. D., et al. 2009. New protocetid whale from the Eocene of Pakistan: Birth on land, precocial development, and sexual dimorphism. *PLoS ONE* 4:e4366.

Janis, C. M. 1976. The evolutionary strategy of the Equidae and the origins of rumen and cecal digestion. *Evolution* 30:757–774.

Janis, C. M., and P. B. Wilhelm. 1993. Were there mammalian pursuit predators in the Tertiary? Dances with wolf avatars. *Journal of Mammalian Evolution* 1:103–126.

Ley, R. E., et al. 2008. Evolution of mammals and their gut microbes. *Science* 320:1647–1651.

Muller, M. N., and R. Wrangham. 2002. Sexual mimicry in hyenas. *Quarterly Review of Biology* 77:3–16.

Nummela, S., et al. 2004. Eocene evolution of whale hearing. *Nature* 430:776–778.

Reidenberg, J. S. 2007. Anatomical adaptations of aquatic mammals. *The Anatomical Record* 290:507–513.

Renfree, M. B., and G. Shaw. 2000. Diapause. *Annual Review of Physiology* 62:353–375.

Roman, J., and S. R. Palumbi. 2003. Whales before whaling in the North Atlantic. *Science* 301:508–510.

Sears, K. E. 2004. Constraints on the morphological evolution of marsupial shoulder girdles. *Evolution* 58:2352–2370.

Selwood, L., and R. Johnson. 2006. Trophoblast and hypoblast in the monotreme, marsupial, and eutherian mammal: Evolution and origins. *BioEssays* 28:128–145.

Smith, K. K. 2001. The evolution of mammalian development. *Bulletin of the Museum of Comparative Zoology* 156:119–135.

Spoor, F., et al. 2002. Vestibular evidence for the evolution of aquatic behaviour in early cetaceans. *Nature* 417:163–166.

Steadman, M. E., et al. 2009. Radiation of extant cetaceans driven by restructuring of the oceans. *Systematic Biology* 58:573–585.

Thewissen, J. G. M., et al. 2009. From land to water: The origin of whales, dolphins, and porpoises. *Evolution, Education and Outreach* 2:272–288.

Tyndale-Biscoe, C. H., and M. B. Renfree. 1987. *Reproductive Physiology of Marsupials*. Cambridge, UK: Cambridge University Press.

Uhen, M. D. 2010. The origin(s) of whales. *Annual Reviews of Earth and Planetary Sciences* 38:189–219.

Werdelin, L., and A. Nilsonne. 1999. The evolution of the scrotum and testicular descent in mammals: A phylogenetic view. *Journal of Theoretical Biology* 196:61–72.

Websites
Anatomy
http://animaldiversity.ummz.umich.edu/site/topics/mammal_anatomy/
Marine mammals
http://www.wired.com/wiredscience/2011/10/evolutionary-treasures-locked-in-the-teeth-of-early-whales/
http://collections.tepapa.govt.nz/exhibitions/whales/Segment.aspx?irn=161
http://www.gma.org/marinemammals/index.html
Afrotheria
http://afrotheria.net/ASG.html
Australian mammals
http://australianmuseum.net.au/Mammals
Digestive physiology of herbivores
http://www.vivo.colostate.edu/hbooks/pathphys/digestion/herbivores/index.html

CHAPTER 22

Endothermy: A High-Energy Approach to Life

Endothermy is a derived character of mammals (synapsids) and birds (sauropsids). The two lineages evolved endothermy independently, but the costs and benefits are the same for both. Endothermy is a way to become relatively independent of many of the challenges in the physical environment, especially cold. Large birds and mammals can live in the coldest habitats on Earth, assuming they can find enough food. That qualification expresses the major problem of endothermy: it is energetically expensive, and small endotherms sometimes must allow their body temperatures to fall because they do not have enough energy to remain homeothermic.

At the opposite end of the temperature range—in hot environments—endotherms face the challenge of keeping cool. Evaporation of water by sweating or panting is an effective cooling mechanism, but as a strategy it works only when water is readily available. Hot deserts combine temperature extremes with scarcity of water. Birds and large mammals may be able to travel to waterholes, and some species have remarkable abilities to obtain, transport, and conserve water.

22.1 Endothermic Thermoregulation

Birds and mammals are endotherms. They regulate their body temperatures by balancing heat production with heat loss. An endotherm can increase or decrease its heat production by varying its metabolic rate and modify its rate of heat loss by adjusting its insulation.

Endotherms produce metabolic heat in several ways.

Besides the obligatory heat production derived from the basal or resting metabolic rate, there is the heat increment of feeding, often called the **specific dynamic action** or thermic effect of the food. This added heat production after eating apparently results from the energy used to assimilate molecules and synthesize protein, and it varies depending on the type of foodstuff being processed. It is highest for a meat diet and lowest for a carbohydrate diet.

Contraction of skeletal muscle produces large amounts of heat. This is especially true during locomotion, which can result in a heat production exceeding the basal metabolic rate by 10-fold to 15-fold. This muscular heat can compensate for heat loss in a cold environment, but it can be a problem for animals in warm places. Cheetahs, for example, show a rapid increase in body temperature when they chase prey, and it is usually overheating rather than exhaustion that causes a cheetah to break off a pursuit. **Shivering**, the generation of heat by muscle-fiber contractions in an asynchronous pattern that does not result in gross movement of the whole muscle, is an important mechanism of heat production.

Endotherms usually live under conditions in which air temperatures are lower than the regulated body temperatures of the animals themselves, and in this situation heat is lost to the environment. Hair and feathers reduce the rate of heat loss by trapping air, and a mammal or bird can change its insulation by raising and lowering the hair or feathers to change the thickness of the layer of trapped air. We humans have goose bumps on our arms and legs when we are cold because our few remaining hairs rise to a vertical position in an ancestral mammalian attempt to increase our insulation.

Costs and Benefits of Endothermic Thermoregulation

Endothermy has both benefits and costs compared to ectothermy. On the positive side, endothermy allows birds and mammals to maintain high body temperatures when solar radiation is not available or is insufficient to warm them—at night, for example, or in the winter. The thermoregulatory capacities of birds and mammals allow them to live in climates that are too cold for ectotherms. On the negative side, endothermy is energetically expensive. The metabolic rates of birds and mammals are nearly an order of magnitude greater than those of amphibians and reptiles. The energy to sustain those high metabolic rates comes from food, and endotherms need more food than do ectotherms.

Mechanisms of Endothermic Thermoregulation

Body temperature and metabolic rate must be considered simultaneously to understand how endotherms maintain their body temperatures at a stable level in the face of environmental temperatures that may range from -70°C to +56°C. Most birds and mammals conform to the generalized diagram in **Figure 22–1**.

Zone of Tolerance Each species of endotherm has a range of environmental temperatures over which it can keep its body temperature stable by adjusting heat loss and heat production. This temperature range is called the **zone of tolerance**. Above this range, the animal's ability to dissipate heat is inadequate; both the body temperature and metabolic rate increase as the environmental temperature increases until the animal dies. Below the zone of tolerance even the maximum metabolic heat production does not match heat loss. As the body temperature falls, the metabolic rate decreases, leading to an increasingly rapid decrease in body temperature and death.

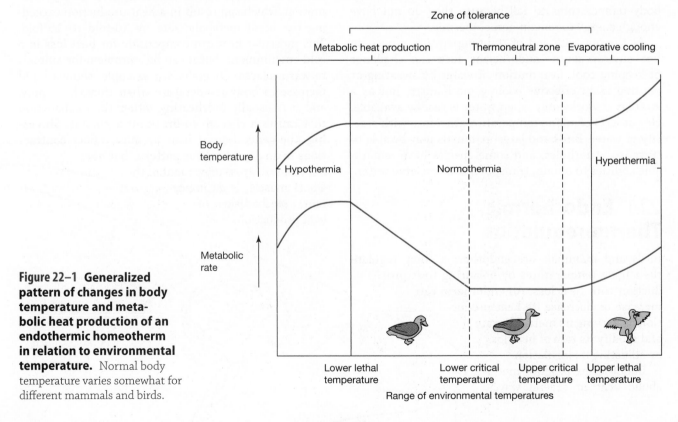

Figure 22–1 Generalized pattern of changes in body temperature and metabolic heat production of an endothermic homeotherm in relation to environmental temperature. Normal body temperature varies somewhat for different mammals and birds.

The zones of tolerance of large animals usually extend downward to lower temperatures than those of smaller animals because heat is lost from the body surface, and large animals have lower surface/volume ratios than those of small animals. Thus, large animals retain heat better at lower environmental temperatures. Similarly, well-insulated species have lower zones of tolerance than those of poorly insulated ones.

Thermoneutral Zone The **thermoneutral zone** (TNZ) is the range of environmental temperatures within which the metabolic rate of an endotherm is at its standard level and the animal controls body temperature by changing its rate of heat loss without increasing heat production. The thermoneutral zone is also called the zone of physical thermoregulation because an animal adjusts its heat loss using nonmetabolic processes such as fluffing or sleeking its hair or feathers, postural changes such as huddling or stretching out, and changes in blood flow (vasoconstriction or vasodilation) to exposed parts of the body (feet, legs, face).

Lower Critical Temperature The **lower critical temperature** (LCT) is the point below which an animal must increase its metabolic heat production to maintain a stable body temperature.

Zone of Chemical Thermogenesis The **zone of chemical thermogenesis** ("heat production") lies below the lower critical temperature. In this zone, the metabolic rate increases to increase heat production as the ambient temperature falls. The quality of an animal's insulation determines how much additional metabolic heat production is required to offset a change in environmental temperature. Well-insulated animals have relatively shallow slopes for the graph of increasing metabolism below the lower critical temperature, and poorly insulated animals have steeper slopes.

Lower Lethal Temperature Ultimately an animal reaches its **lower lethal temperature**. At this point, metabolic heat production is at its maximum rate and is still insufficient to balance the heat lost to the environment. The body temperature falls, and because the rates of the chemical reactions that produce metabolic heat are sensitive to temperature, heat production falls as well. A positive feedback loop is initiated in which falling body temperature reduces heat production, causing a further reduction in body temperature. Death from **hypothermia** (a body temperature lower than normal) follows.

Upper Critical Temperature Endotherms are remarkably good at maintaining stable body temperatures in cool environments, but they have difficulty at high ambient temperatures. The **upper critical temperature** (UCT) is the point at which heat loss via convection and conduc-

tion to the animal's surroundings has been maximized by using all of the physical processes available to an animal—exposing the poorly insulated areas of the body and maximizing cutaneous blood flow to increase nonevaporative heat loss. When these mechanisms are insufficient to balance heat gain, the only option vertebrates have is to use evaporation of water to cool the body.

Zone of Evaporative Cooling The temperature range from the upper critical temperature to the upper lethal temperature is the **zone of evaporative cooling**. Some mammals sweat, a process in which water is released from sweat glands on the surface of the body and evaporation of the sweat cools the body surface. Other animals pant, breathing rapidly and shallowly so that evaporation of water from the respiratory system provides a cooling effect. Many birds use a rapid fluttering movement of the gular region to evaporate water for thermoregulation. Panting and gular flutter require muscular activity, and some of the evaporative cooling they achieve is used to offset the increased metabolic heat production they require.

Upper Lethal Temperature At the **upper lethal temperature,** evaporative cooling cannot balance the heat flow from a hot environment. As an animal's body temperature rises, the metabolic rate increases, and metabolic heat production raises the body temperature still more, further increasing the metabolic rate. This process can lead to an explosive rise in body temperature and death from **hyperthermia** (a body temperature higher than normal).

The difficulty that endotherms experience in regulating body temperature in high environmental temperatures may be one of the reasons that the body temperatures of most endotherms are in the range of 35°C to 40°C. Most habitats seldom have air temperatures that exceed 35°C. Even the tropics have average yearly temperatures below 30°C. Thus the high body temperatures maintained by endotherms ensure that, in most situations, the heat gradient is from animal to environment. (Still higher body temperatures—about 50°C, for example—could ensure that mammals were always warmer than their environment. There are upper limits to the body temperatures that are feasible, however. The fluidity of the plasma membranes of cells increases with increasing temperature, disrupting physiological functions such as ion transport and nerve conduction, and many proteins denature at temperatures above 50°C.)

Effectiveness of Endothermic Thermoregulation

The low critical temperatures of Arctic birds and mammals dramatically illustrate how effective insulation

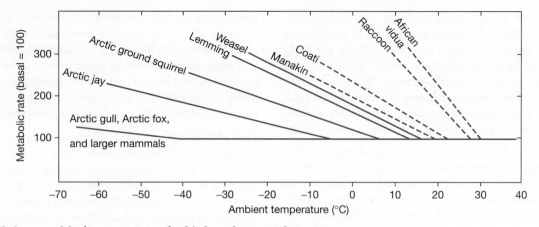

Figure 22–2 Lower critical temperatures for birds and mammals. Solid lines are for Arctic birds and mammals; dashed lines are for tropical birds and mammals. The basal metabolic rate for each species is considered to be 100 units to facilitate comparisons among species. The lower end of the zone of thermoneutrality is at the intersection of the basal metabolic rate (horizontal line) and the increased metabolic rate in the zone of chemical thermogenesis. The steepness of the slope shows how rapidly metabolic heat production increases as ambient temperature falls. The slopes for tropical animals are steeper than those for Arctic animals. Within a climate type, large animals have lower LCTs and shallower slopes than small animals because large animals have smaller surface/volume ratios.

can be (**Figure 22–2**). Arctic birds and mammals maintain resting metabolic rates at environmental temperatures well below freezing, and they show only small increases in metabolism (i.e., flatter slopes) below the LCT. The Arctic fox, for example, has an LCT of –40°C, and at –70°C (approximately the lowest air temperature ever recorded) a fox has elevated its metabolic rate only 50 percent above its standard level. Under those conditions the fox is maintaining a body temperature approximately 110°C higher than air temperature. Arctic birds are equally impressive, and the Arctic glaucous gull, like the Arctic fox, has a LCT near –40°C and can withstand –70°C with only a modest increase in metabolism.

The effect of the long hair and dense pelage of Arctic mammals and the layer of down beneath the outer feathers of Arctic birds can be understood by comparing the LCTs of Arctic animals with those of tropical species: tropical mammals have lower LCTs, between 20°C and 30°C. As air temperatures fall below their LCTs, these animals are no longer in their thermoneutral zones and must increase their metabolic rates to maintain normal body temperatures. For example, a tropical raccoon has increased its metabolic rate approximately 50 percent above its standard level at an environmental temperature of 25°C.

22.2 Endotherms in the Cold

Most endotherms (especially small ones) expend most of the energy they consume just keeping themselves

warm—even in the moderate conditions of tropical and subtropical climates. Nonetheless, endotherms have proved themselves very adaptable in extending their thermoregulatory responses to allow them to inhabit even Arctic and Antarctic regions.

An animal can maintain a stable body temperature in extreme cold by increasing heat production or by decreasing heat loss. On closer examination, the option of increasing heat production does not seem particularly attractive. Any significant increase in heat production would require an increase in food intake. This scheme poses obvious ecological difficulties in terrestrial Arctic and Antarctic habitats where primary production is extremely low, especially during the coldest parts of the year. For most polar animals the quantities of food necessary would probably not be available.

Because they lack the option of increasing heat production significantly, conserving heat within the body is the primary thermoregulatory mechanism of polar endotherms. Insulative values of pelts from Arctic mammals are two to four times those of tropical mammals. In Arctic species, insulative value is closely related to fur length (**Figure 22–3**). Small species such as the least weasel and the lemming have fur only 1 to 1.5 centimeters long. Presumably, the thickness of their fur is limited because longer hair would interfere with the movement of their legs. Large mammals (caribou, polar and grizzly bears, Dall sheep, and Arctic fox) have hair 3 to 7 centimeters long. There is no obvious reason why their hair

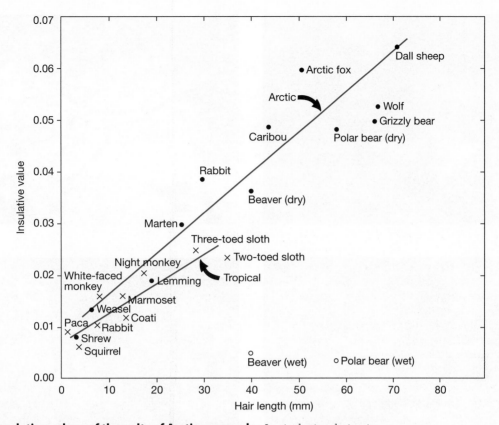

Figure 22–3 Insulative values of the pelts of Arctic mammals. In air the insulation is proportional to the length of the hair. The pelts of tropical mammals (×) have approximately the same insulative value as those of Arctic mammals (•) at short hair lengths, but long-haired tropical mammals like sloths have less insulation than Arctic mammals with hair of the same length. Immersion in water greatly reduces the insulative value of hair, even for such semiaquatic mammals as the beaver and polar bear (o).

could not be longer; apparently, they do not need more insulation. The insulative values of pelts of short-haired tropical mammals are similar to those measured for the same hair lengths in Arctic species. Long-haired tropical mammals, like sloths, have fewer hairs per square centimeter of skin and thus less insulation than Arctic mammals with hair of similar length.

22.3 Avoiding Cold and Sharing Heat

Small mammals can buffer the effects of winter conditions by changing their own behavior and by sharing a nest with other individuals. Snow provides insulation; the ground surface beneath a layer of snow can be more than 20°C warmer than air temperature. Small mammals can forage in this protected subnivean (beneath the snow) zone. When they are inactive they can retreat to nests that provide insulation, and by huddling in groups small mammals can reduce their own heat loss and capture some of the heat that other individuals lose.

Modifying Activity Patterns and Seeking Favorable Microclimates

Shrews are among the smallest species of mammals (**Figure 22–4**). Their food requirements are legendary, and they have such high surface/volume ratios that their thermoneutral zones are very narrow. Nonetheless, two species of shrews, the short-tailed shrew (*Blarina brevicauda,* adult body mass about 15 grams) and the masked shrew (*Sorex cinereus,* adult body mass about 4 grams), are active throughout the winter when the air temperature is as low as –29°C. The shrews are not exposed to those frigid temperatures, however, because they confine their activity to sheltered microhabitats, such as spaces within the layer of leaves on the forest floor where the temperature is –4°C and tunnels that have a comparatively balmy temperature of 1°C. The shrews also reduce their exposure to cold conditions by spending less time

(a)

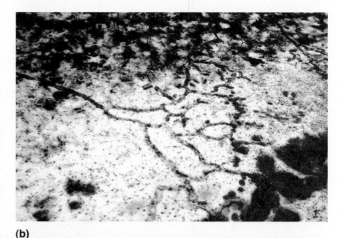
(b)

(c)

(d)

Figure 22–4 Small mammals use behavioral adjustments to avoid cold. (a) The short-tailed shrew is too small to grow a deep pelage. (b) Shrews minimize their exposure to cold in the winter by foraging on the ground surface beneath the snow and in tunnels through the leaf litter. These tunnels are revealed when the snow melts. (c) White-footed deer mice huddle in groups in nests. (d) A deer mouse nest is made of dried leaves loosely woven together.

outside their nests; short-tailed shrews forage for less than 3 hours per day in winter compared to 8 hours per day in summer.

Construction of Nests and Social Thermoregulation

Many small mammals construct nests underground or in hollows in trees and form groups that occupy them together during the winter. White-footed deer mice (*Peromyscus leucopus*), for example, construct elaborate nests during the winter, lining them with dried grasses and herbaceous vegetation. Like shrews, mice spend more time in their nests during the winter than in the summer. Several individuals huddle together in a nest,

and the group has a smaller surface/volume ratio than an individual mouse has, thus reducing heat loss. The mice in the center of the pile are warmer than those on the outside, of course, and there is a continuous shifting of positions as mice from the outside of the pile grow cold and push their way into the center.

Social thermoregulation has a substantial effect on the temperatures that taiga voles (*Microtus xanthognathus*) are exposed to during the winter. Groups of 5 to 10 individuals nesting together can maintain the nest temperature between 4°C and 10°C when the soil around the nest is between –3°C and –6°C and air temperatures on the surface are between –10°C and –20°C. Furthermore, the voles may forage on different schedules so that the nest is never empty. As a result, the voles return to a warm nest after foraging.

22.4 Facultative Heterothermy

Winters pose a challenge to endotherms because low environmental temperatures increase the need for heat production—and for the food to sustain higher rates of heat production—at the same time that food becomes scarce. Birds and mammals that weigh less than 100 grams are particularly vulnerable to low temperatures for three reasons:

- They have large surface/volume ratios and high rates of heat loss.
- They have large food requirements because their mass-specific metabolic rates (i.e., energy consumed per gram of body mass) are higher than those of large animals.
- Their layers of insulation are thinner than those of large animals. (An animal the size of a reindeer can have pelage 3 centimeters deep, but if a mouse had hair that long, its legs would not reach the ground!)

The rate at which an animal loses heat to the environment can be reduced by decreasing the heat gradient between the animal and the environment. The sheltered microclimates described in the preceding section adjust the environmental side of that relationship, and intentionally lowering body temperature (facultative hypothermia) is another way to reduce the temperature gradient between an animal and its environment. Patterns of reduced body temperature form a continuum from species that have slightly lower body temperatures in winter than in summer while continuing their normal activities (seasonal hypothermia), through those that allow their body temperature to drop a few degrees and become inactive for hours, to those that have profound reductions of body temperature that extend for weeks or months (hibernation).

Shallow (Seasonal) Hypothermia

Some animals lower their body temperature as much as 5°C in winter while maintaining normal activity. North American gray squirrels (*Sciurus carolinensis*), Abert's squirrels (*S. aberti*), and red squirrels (*Tamiasciurus hudsonicus*) are active through the winter with body temperatures that are 1 to 6°C lower than in summer.

Barnacle geese use hypothermia in preparation for migration. Temperature sensors implanted in the abdominal cavities of barnacle geese in Norway showed a month-long period of hypothermia that began about a week before migration started and extended through the period of recovery after the birds arrived at the overwintering site. Body temperature declined from a daily average of about 40°C to about 36°C, and then rose to the normal levels. The energy saved by this period of hypothermia is believed to reduce the rate at which fat is used during migration and to speed the replacement of fat stores following migration.

Rest-Phase Hypothermia

Rest-phase hypothermia is a drop in body temperature that coincides with an animal's normal period of inactivity—at night for most birds and during the day for most small mammals. Metabolic rate and thus heat production decrease, and the body temperature falls 5°C to 10°C lower than normal (**Figure 22–5**). Then the normal

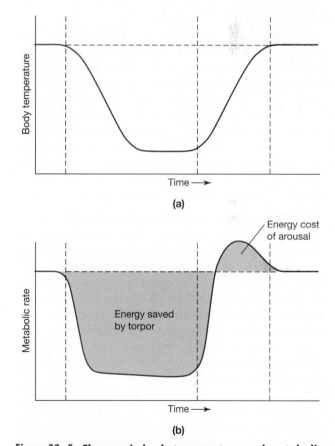

Figure 22–5 Changes in body temperature and metabolic rate during daily torpor. A decrease in metabolic rate (b) precedes a fall in body temperature (a) to a new set point. An increase in metabolism produces the heat needed to return to normal body temperatures; the metabolic rate during arousal briefly overshoots the resting rate.

Figure 22–6 The black-capped chickadee, *Poecile atricapillus*. This is one of the smallest species of birds that overwinters in Canada and the northern United States.

thermoregulatory mechanisms—thermogenesis and insulation—balance heat gain and heat loss at this new level, saving energy because the temperature gradient between the animal and its environment is smaller. An increase in metabolic rate and heat production initiates warming, and the metabolic rate rises briefly above the resting rate. This period of elevated metabolism is the energy cost of arousal, and it is small compared to the energy that was saved during the time the animal was hyperthermic. Calculations indicate that even a short period of rest saves energy for a small bird or mammal.

Chickadees on Winter Nights Rest-phase hypothermia is widespread among small birds and mammals and has been reported in more than two dozen species of passerine birds—chickadees are an example (**Figure 22–6**). These are small (10 to 12 grams) passerine birds that are winter residents in northern latitudes, where they regularly experience environmental temperatures that do not rise above freezing for days or weeks.

In winter, chickadees around Ithaca, New York, maintain body temperatures between 40°C and 42°C during the day and cool to 29°C or 30°C at night. This reduction in body temperature permits a 30 percent reduction in energy consumption. The chickadees rely primarily on fat stores they accumulate as they feed during the day to supply the energy needed to carry them through the following night. Thus the energy available to them and the energy they use at night can be estimated by measuring the fat content of birds as they go to roost in the evening and as they begin activity in the morning. Chickadees have an average of 0.80

gram of fat when they cease activity in the evening, and by morning the fat store has decreased to 0.24 gram. The fat metabolized during the night (0.56 gram per bird) corresponds to the metabolic rate expected for a bird with a body temperature of 30°C.

Nighttime hypothermia is essential for the chickadees because they would require 0.92 gram of fat per bird to maintain a body temperature of 40°C through the night—more fat than the birds have when they go to roost in the evening. If chickadees did not become torpid, they would starve before morning. Even with torpor, they use 70 percent of their fat reserve in one night. They do not have an energy supply to carry them far past sunrise, and chickadees are among the first birds to begin foraging in the morning. They also forage in weather so foul that other birds, which are not in such a precarious energy balance, remain on their roosts. The chickadees must reestablish their fat stores each day if they are to survive the next night.

Migratory Stopovers Another function of rest-phase hypothermia may be to speed the rebuilding of energy stores during migratory stopovers. Garden warblers (*Sylvia borin*) and icterine warblers (*Hippolais icterina*), like many other species of passerines, migrate at night. Both species normally have body temperatures near 39.5°C, and they allow their temperatures to fall to 33.5°C at night while they remain at a stopover site, feeding and replenishing their fat stores. Hypothermic migrating blackcaps (*Sylvia atricapilla*) also allow their body temperatures to fall from a daytime average of 42.5°C to 35.5°C at night, and that reduction in body temperature corresponds to an energy savings of more than 30 percent.

Brooding Hummingbirds Hummingbirds in the Rocky Mountains illustrate the flexibility of rest-phase hypothermia. Hummingbirds depend on the energy they gather from nectar during the day to sustain them through the following night. These very small birds (3 to 10 grams) have extremely high energy expenditures and yet are found during the summer in northern latitudes and at high altitudes. Studies of nesting broadtailed hummingbirds at an altitude of 2900 meters near Gothic, Colorado, showed that the birds adjust their use of rest-phase hypothermia on a day-by-day basis. The air temperature drops to nearly freezing at night, and hummingbirds that are not brooding eggs become hypothermic at night.

Hummingbirds that are brooding eggs behave differently, however. Brooding hummingbirds normally do not become hypothermic. The lowered egg temperature that results from the parent bird's becoming torpid does not damage the eggs, but it slows development and delays hatching. Presumably, there are advantages to hatching the eggs as quickly as possible. As a result, brooding hummingbirds expend energy to keep themselves and their eggs warm through the night, provided they have the energy stores necessary to maintain the high metabolic rates needed.

On some days, bad weather interferes with foraging by the parent birds, and they go into a night with insufficient energy supplies to maintain normal body temperatures. In this situation, the brooding hummingbirds do become torpid for part of the night. One bird that had experienced a 12 percent reduction in foraging time during the day became torpid for 2 hours, and a second bird that had lost 21 percent of its foraging time was torpid for 3.5 hours. Torpor can thus be a flexible response that integrates the energy stores of a bird with environmental conditions and such biological requirements as brooding eggs.

Deep Hypothermia: Hibernation

The deep hypothermia that occurs during **hibernation** is a comatose condition, much more profound than the deepest sleep. Voluntary motor responses are reduced to sluggish postural changes, but some sensory perception of powerful auditory and tactile stimuli and environmental temperature changes is retained. Perhaps most dramatically, a hibernating animal can arouse spontaneously from this state using heat production by brown fat, a tissue metabolically specialized for heat production. Some endotherms can rewarm under their own power from the lowest levels of hibernation; others must warm passively with an increase in environmental temperature until some threshold is reached at which arousal starts.

Body Size and Hibernation The largest mammals that hibernate are marmots, which weigh about 5 kilograms, and deep hypothermia and body size are closely related. Deep hypothermia is not as advantageous for a large animal as for a small one. In the first place, the energy cost of maintaining a high body temperature is lower for a large animal than for a small one, and as a consequence, a small animal has more to gain from hyperthermia. Second, a large animal cools off more slowly than a small animal, so its metabolic rate does not decrease as rapidly. Furthermore, large animals have more body tissue to rewarm on arousal, and their costs of arousal are correspondingly greater than those of small animals. Bears in winter dormancy, for example, lower their body temperatures only about 5°C from normal levels, and their metabolic rate decreases about 50 percent. That small reduction in body temperature, combined with the large fat stores bears accumulate

before retreating to their winter dens, is sufficient to carry them through the winter, but the hypothermia is not deep enough to be considered hibernation.

Physiological Changes During Hibernation Profound changes occur in a variety of physiological functions of animals in deep hypothermia. Body temperature drops to within 1°C or less of the surrounding temperature, and in some cases (bats, for example) extended survival is possible at body temperatures just above the freezing point of the tissues. In Arctic ground squirrels (*Urocitellus parryii*) the temperature of parts of their bodies actually falls as low as −2.9°C. Oxidative metabolism and energy use are reduced to as little as one-twentieth of their rates at normal body temperatures. Although body temperatures fall very low during hibernation, temperature regulation does not entirely cease. Instead the hypothalamic thermostat is reset. If the body temperature of a hibernating animal falls below the new set point, thermogenic processes bring the hibernating animal back to the regulated level.

Respiration is slow during hibernation, and an animal's overall breathing rate can be less than one breath per minute. Heart rate is drastically reduced, and blood flow to peripheral tissues is virtually shut down, as is blood flow posterior to the diaphragm. Most of the blood is retained in the body's core.

Arousal from Hibernation Hibernation is an effective method of conserving energy during long winters, but hibernating animals do not remain at low body temperatures for the whole winter. Periodic arousals are normal, and these arousals consume a large portion of the total amount of energy used by hibernating mammals. An example of the magnitude of the energy cost of arousal is provided by Richardson's ground squirrels (*Urocitellus richardsonii*) in Alberta, Canada (**Figure 22–7**).

The activity season for ground squirrels in Alberta is short: they emerge from hibernation in mid-March, and adult squirrels reenter hibernation 4 months later, in mid-July. Juvenile squirrels begin hibernation in September. When the squirrels are active, they have body temperatures of 37°C to 38°C, and their temperatures fall as low as 3°C to 4°C when they are torpid.

Figure 22–8 shows the body temperature of a juvenile male ground squirrel from September through March; periods of torpor alternate with arousals throughout the winter. Hibernation begins in mid-September with short bouts of torpor followed by rewarming. At that time, the temperature in the burrow is about 13°C. As the winter progresses and the temperature in the burrow falls, the intervals between arousals lengthen and the body temperature of the torpid animal

Figure 22–7 Richardson's ground squirrel, *Urocitellus richardsonii*.

bined metabolic expenditures for those three phases of the torpor cycle account for an average of 83 percent of the total energy used by the squirrel.

Surprisingly, we have no clear understanding of why a hibernating ground squirrel undergoes these arousals that increase its total winter energy expenditure nearly fivefold. Ground squirrels do not store food in their burrows, so they are not using the periods of arousal to eat. They do urinate during arousal, so eliminating accumulated nitrogenous wastes may be the reason for arousal, and spending some time at a high body temperature may be necessary to carry out other physiological or biochemical activities, such as resynthesizing glycogen, redistributing ions, or synthesizing serotonin. Arousal may also allow a hibernating animal to determine when environmental conditions are suitable for emergence. Whatever their function, the arousals must be important because the squirrel pays a high energy price for them during a period of extreme energy conservation.

Communal Hibernation Many species of ground squirrels and marmots hibernate in groups. For example, alpine marmots (*Marmota marmota*) hibernate in groups consisting of a male, female, and their young from the past several years—some groups contain as many as 20 individuals. The animals huddle together, producing a heap of animals that has a smaller surface/volume ratio than a single individual. The body temperatures of the marmots in the heaps remain 3°C or 4°C above freezing, although the air temperature in the burrow is below freezing. Communal hibernation increases survival, especially of the youngest individuals, the ones that were born the previous spring.

Bouts of arousal by communally hibernating marmots are synchronized, which saves energy because heat is shared by all of the individuals. The older individuals begin arousal first so that heat transferred from them assists the younger individuals in warming up. The effectiveness of this synchrony in arousal is demonstrated by comparing weight loss during hibernation: marmots in highly synchronized groups lose 20 to 25 percent of their initial body mass, whereas

decreases. By late December, the burrow temperature has dropped to 0°C, and the periods between arousals are 14 to 19 days. In late February, the periods of torpor become shorter, and in early March the squirrel emerges from hibernation.

A torpor cycle consists of entry into torpor, a period of torpor, and an arousal (**Figure 22–9**). In this example of the male ground squirrel, entry into torpor begins shortly after noon on February 16; 24 hours later, the body temperature has stabilized at 3°C. This period of torpor lasts until late afternoon on March 7, when the squirrel starts to arouse. In 3 hours the squirrel warms from 3°C to 37°C. It maintains that body temperature for 14 hours and then enters into torpor again.

These periods of arousal account for most of the energy used during hibernation (**Table 22–1**). The energy costs associated with arousal include the cost of warming from the hibernation temperature to 37°C, the cost of sustaining a body temperature of 37°C for several hours, and the metabolism above torpid levels as the body temperature slowly decreases during reentry into torpor. For the entire hibernation season, the com-

Figure 22–8 Record of body temperature during a complete torpor season for a Richardson's ground squirrel.

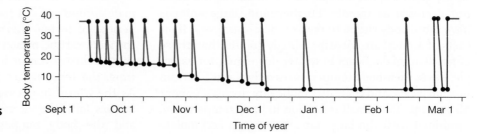

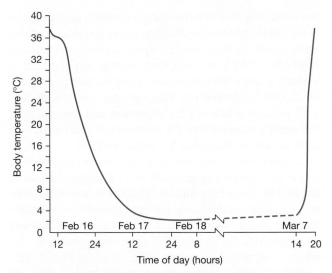

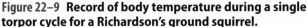

Figure 22–9 Record of body temperature during a single torpor cycle for a Richardson's ground squirrel.

those in poorly synchronized groups lose 40 to 45 percent—almost twice as much.

22.5 Migration to Avoid Cold

Migrations are usually a response to seasonal changes in climatic factors such as temperature or rainfall. In turn, these conditions influence food supply and the occurrence of suitable breeding conditions.

Long-distance migration is more feasible for birds and marine animals than for terrestrial species, partly because geographic barriers are less of a problem and partly because the energy cost of transport is less for swimming fish and flying birds than for walking mam-

Table 22–1 Use of energy during different phases of the hibernation cycle by Richardson's ground squirrel

| | Percentage of Total Energy per Month | | | |
Month	Torpor	Warming	Intertorpor homeothermy	Reentry
July	8.5	17.2	56.5	17.8
September	19.2	15.2	49.9	15.7
November	20.8	23.1	43.1	13.0
January	24.8	24.1	40.0	11.1
March	3.3	14.0	76.4	6.3
Average for season	**16.6**	**19.0**	**51.6**	**12.8**

mals. We can consider the costs and benefits of migrating by considering two kinds of animals that represent extremes of body size. The baleen whales are the largest animals that have ever lived, and hummingbirds are among the smallest vertebrate endotherms, yet both whales and hummingbirds migrate.

Whales

The annual cycle of events in the lives of the great baleen whales is particularly instructive in showing how migration relates to the use of energy and how it correlates with reproduction in the largest of all animals. Most baleen whales spend summers in polar or subpolar waters of either the Northern or the Southern Hemisphere, where they feed on krill or other crustaceans that are abundant in those cold, productive waters. For 3 or 4 months each year, a whale consumes a vast quantity of food that is converted into stored energy in the form of blubber and other kinds of fat. Pregnant female whales are using energy to support the development of their unborn young, which may grow to one-third the length of their mothers before birth.

Near the end of summer, whales begin migrating toward tropical or subtropical waters where the females bear their young. The young grow rapidly on the rich milk provided by their mothers, and by spring the calves are mature enough to travel with their mothers back to Arctic or Antarctic waters. The calves are weaned about the time they arrive in their summer quarters.

From a bioenergetic and trophic point of view, the remarkable feature of this annual migration is that virtually all of the energy required to fuel it comes from ravenous feeding and fattening during the 3 or 4 months spent in polar seas. Little or no feeding occurs during migration or during the winter period of calving and nursing. Energy for all these activities comes from the abundant stores of blubber and fat.

The gray whale (*Eschrichtius robustus*) of the Pacific Ocean has one of the longest and best-known migrations (**Figure 22–10**). The summer feeding waters are in the Bering Sea and the Chukchi Sea north of the Bering Strait in the Arctic Ocean. Gray whales feed on amphipods (small crustaceans) they gather by filtering mouthfuls of mud from the sea floor. Most gray whales follow the Pacific Coast of North America, moving south to Baja California and adjacent parts of western Mexico. They arrive in December or January, the females bear their young in warm shallow lagoons (**Figure 22–11** on page 549), and then they mate with the males, which also make the trip. They depart in March to return to the North Pacific, completing a round trip of at least 9000 kilometers.

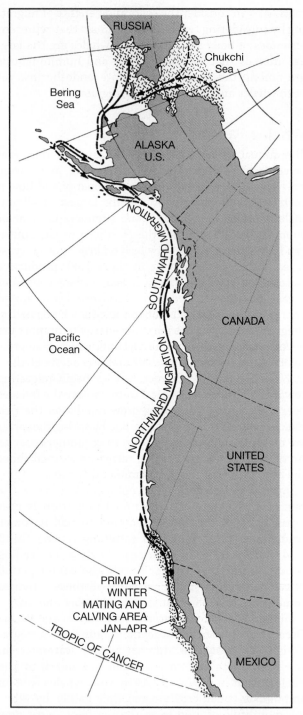

Figure 22–10 Migratory route of the gray whale between the Arctic Circle and Baja California.

the metabolic rate of a free-ranging whale, including the locomotion involved in feeding and migrating, is about three times the basal rate (a typical level for mammals), then the whale's average daily energy expenditure is about 3 million kilojoules. Assuming an energy content of 20,000 kilojoules per kilogram for amphipods and a 50 percent efficiency for digestion and assimilation, the energy requirement for existence is equivalent to a daily intake of about 300 kilograms of food.

In addition to satisfying its daily energy needs, a whale on the feeding grounds must accumulate a store of blubber. Body fat contains 38,500 kilojoules per kilogram, so the whale's daily energy expenditure is equivalent to metabolizing almost 80 kilograms of blubber or fat per day. To live for 245 days without eating, the whale must metabolize about 19,000 kilograms of fat.

Accumulating that amount of fat in 120 days of active feeding in Arctic waters at a conversion efficiency of 50 percent requires the consumption of more than 75,500 kilograms of amphipods, or more than 600 kilograms per day. Thus a gray whale must consume approximately 300 + 600 = 900 kilograms of amphipods per day to meet its daily metabolic needs and store energy for its migration.

This is a minimum estimate for females because the calculations do not include the energetic costs of the developing fetus or the cost of milk production. Nor do they include the cost of transporting 20,000 kilograms of fat through the water. However, a large whale can do all this work and more without exhausting its insulating blanket of blubber because nearly half the total body mass of a large whale consists of blubber and other fats.

Why does a gray whale confine feeding to only 4 months and expend all this energy to migrate during the other 8 months? An adult gray whale is too large and too well insulated to be stressed by the cold Arctic and sub-Arctic waters, which do not vary much from 0°C between summer and winter. It seems strange for an adult whale to abandon an abundant source of food and swim off on a forced starvation trek into warm waters that may cause stressful overheating.

The advantage of migration probably accrues to the newborn calf, which, although it is large, lacks an insulative layer of blubber. If the calf were born in cold northern waters, it would have to use a large fraction of its energy intake (milk produced from its mother's stored fat) to generate metabolic heat to regulate its body temperature. That energy could otherwise be used for rapid growth. Apparently it is more effective, and perhaps energetically more efficient, for the mother whale to migrate thousands of kilometers into warm waters to give birth and nurse in an environment where the young whale can devote most of its energy intake to rapid growth.

The amount of energy expended by a whale in this annual cycle is phenomenal. The basal metabolic rate of a gray whale with a fat-free body mass of 50,000 kilograms is approximately 1 million kilojoules per day. If

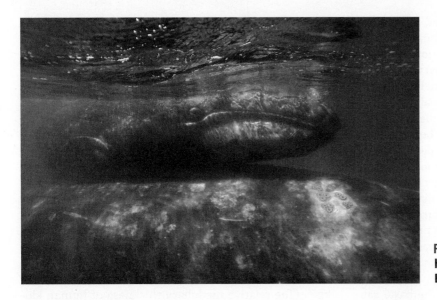

Figure 22–11 A gray whale calf resting on his mother's back in San Ignacio Lagoon, Baja California, Mexico.

Hummingbirds

At the opposite end of the size range of endotherms, hummingbirds are the smallest birds that migrate. Ornithologists have long been intrigued by the ability of the ruby-throated hummingbird (*Archilochus colubris*), which weighs only 3.5 to 4.5 grams, to make a nonstop flight of 800 kilometers during migration across the Gulf of Mexico from Florida to the Yucatán Peninsula.

Like most migratory birds, ruby-throated hummingbirds store subcutaneous and body fat by feeding heavily prior to migration. A hummingbird with a lean mass of 2.5 grams can accumulate 2 grams of fat. Measurements of a hummingbird hovering in the air in a respirometer chamber in the lab indicate an energy consumption of 3 kilojoules per hour. Hovering is energetically more expensive than forward flight, so these values represent the maximum energy used in migratory flight. Even so, 2 grams of fat produce enough energy to last for 24 to 26 hours of sustained flight.

Hummingbirds fly at speeds of about 40 kilometers per hour, so crossing the Gulf of Mexico requires about 20 hours. Thus, by starting with a full store of fat, they have enough energy for the crossing with a reserve for unexpected contingencies such as a headwind that slows their progress. In fact, most migratory birds wait for weather conditions that will generate tailwinds before they begin their migratory flights, thereby further reducing the energy cost of migration.

22.6 Endotherms in the Heat

Hot environments place more severe physiological demands on endotherms than do polar conditions.

Endotherms encounter two problems in regulating their body temperature in hot deserts. The first results from a reversal of the normal relationship of an animal to the environment. In most environments, an endotherm's body temperature is warmer than the air temperature. In this situation, heat flows from the animal to its environment, and thermoregulatory mechanisms achieve a stable body temperature by balancing heat production and heat loss. Very cold environments merely increase the gradient between an animal's body temperature and the environment. The example of Arctic foxes with lower critical temperatures of −40°C illustrates the success that endotherms have had in providing sufficient insulation to cope with enormous gradients between high core body temperatures and low environmental temperatures.

In a hot desert, the temperature gradient is not increased—it is reversed. Desert air temperatures can climb to 40°C or 50°C during summer, and the ground temperature may exceed 60°C or 70°C. Instead of losing heat to the environment, an animal is continuously absorbing heat, and that heat plus metabolic heat must somehow be dissipated to maintain the animal's body temperature in its normal range. Maintaining a body temperature 10°C below the ambient temperature can be a greater challenge for an endotherm than maintaining it 100°C above the ambient temperature.

Temperature Stress and Scarcity of Water Interact in Deserts

Evaporating a kilogram of water dissipates approximately 2400 kilojoules. (The exact value varies slightly with temperature.) Thus, evaporative cooling is an effective cooling mechanism as long as an animal has an unlimited supply of water. In a hot desert, however, where thermal stress is greatest, water is a scarce commodity and its use must be carefully rationed. Calculations show, for example, that if a kangaroo rat (*Dipodomys*) were to venture out into the desert sun, it would have to evaporate 13 percent of its body water per hour to maintain a normal body temperature. Most mammals die when they have lost 10 percent to 20 percent of their body water, so evaporative cooling is of limited utility in deserts except as a short-term response to a critical situation.

Disposing of waste products in the desert is a separate problem. Birds and mammals eat a lot to maintain their high metabolic rates, and they produce correspondingly large volumes of waste. Urine is more of a problem than feces for mammals because nitrogenous wastes are dissolved in water when they are excreted from the kidney.

Birds have less difficulty conserving urinary and fecal water than mammals do because the basal sauropsid excretory system reabsorbs water in the bladder or cloaca and precipitates both nitrogenous wastes and excess ions as sodium and potassium urates.

Mammals, too, maintain water balance in deserts thanks to the ability of the mammalian kidney to produce urine that has an osmotic concentration several times higher than that of the blood. An ultrafiltrate of the blood is processed in the nephron to recover needed compounds and secrete waste products into the forming urine. Most mammals have two types of nephrons: those with a cortical glomerulus and abbreviated loops of Henle that do not penetrate far into the medulla, and those with juxtamedullary glomeruli with loops that penetrate as far as the papilla of the renal pyramid. The long loops of Henle experience large osmotic gradients along their lengths.

Form and function are intimately related in mammalian kidneys. A thick medulla forms a long renal pyramid with great concentrating power, and the maximum urine osmolalities of mammals are proportional to the relative medullary thickness of their kidneys (the ratio of the depth of the medulla to the thickness of the cortex). Some desert rodents with exceptionally long renal pyramids have relative medullary thicknesses of 15 to 20, and the maximum urine concentrations of these animals exceed 7000 mmol · kg^{-1}. (The relative medullary thickness of human kidneys is an unimpressive 3.0, and the maximum urine concentration is only 1430 mmol · kg^{-1}.) **Figure 22–12** shows the strong correlation between relative medullary thickness and maximum urine concentration, but substantial variation is apparent, indicating that other anatomical or physiological factors are involved. For mammals as a group, relative medullary thickness accounts for 59 percent of interspecific variation in maximum urine concentration.

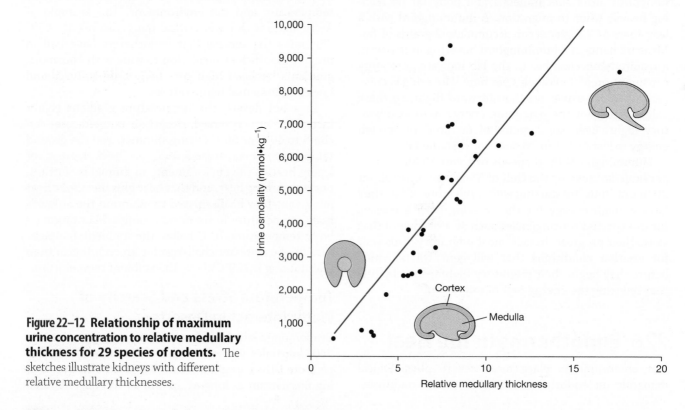

Figure 22–12 Relationship of maximum urine concentration to relative medullary thickness for 29 species of rodents. The sketches illustrate kidneys with different relative medullary thicknesses.

Strategies for Desert Survival

Deserts are challenging environments, especially for endotherms, but they contain a mosaic of micro-environments that animals can use to find the conditions they need. Broadly speaking, endotherms have three general methods for coping with desert conditions:

- **Avoidance**—Some endotherms manage to avoid desert conditions by behavioral means. They live in deserts but are rarely exposed to the full rigors of desert life.
- **Relaxation of homeostasis**—Some endotherms have relaxed the limits of homeostasis. They survive in deserts by tolerating hyperthermia (high body temperature) or dehydration (low body-water content).
- **Specializations**—Anatomical specializations allow some endotherms to evade the problems of life in the desert.

These categories are not mutually exclusive; many desert endotherms combine elements of all three responses.

Avoidance—Nocturnal Rodents Rodents are the pre-eminent small mammals of arid regions. It is a commonplace observation that population densities of rodents are higher in deserts than in moist habitats. Several ancestral features of rodent biology allow them to extend their geographic ranges into hot, arid regions. Among the most important of these characters are the normally nocturnal habits of many rodents and their practice of living in burrows. A burrow provides escape from the heat of a desert, giving an animal access to a sheltered microenvironment while soil temperatures on the surface climb above lethal levels. In addition, rodents in general have kidneys that produce concentrated urine—even a laboratory white rat produces urine that is twice as concentrated as human urine.

Kangaroo rats are among the most specialized desert rodents in North America (**Figure 22–13**). Merriam's kangaroo rat (*Dipodomys merriami*) occurs in desert habitats from central Mexico to northern Nevada. A population of this species lives in extreme conditions in the Sonoran Desert of southwestern Arizona. During the summer, daytime temperatures at the ground surface approach 70°C, and even a few minutes of exposure would be deadly. Kangaroo rats spend the day in burrows 1 to 1.5 meters underground, where air temperatures do not exceed 35°C even during the hottest parts of the year. In the evening, when the kangaroo rats emerge to forage, external air temperatures have fallen to about 35°C.

Avoidance—Birds Because birds fly, they are much more mobile than mammals. A kangaroo rat or ground squirrel is confined to a home range less than 100 meters in diameter, but it is quite possible for a desert bird with the same body size as those rodents to fly many kilometers to reach water. For example, mourning doves (*Zenaida macroura*) in the deserts of North America congregate at dawn at waterholes, with some individuals flying 60 kilometers or more to reach them.

The normally high and variable body temperatures of birds give them another advantage that is not shared by mammals. With body temperatures normally about 40°C, birds face the problem of a reversed temperature gradient between their bodies and the environment for a shorter portion of each day than does a mammal. Furthermore, birds' body temperatures are normally variable, and birds tolerate moderate hyperthermia without apparent distress. These are all ancestral characters that are present in virtually all birds. Neither the body temperatures nor the lethal temperatures of desert birds are higher than those of related species from non-desert regions.

The mobility provided by flight does not extend to fledgling birds, and the most conspicuous adaptations of birds to desert conditions are those that ensure a supply of water for the young. Altricial fledglings, those that need to be fed by their parents after hatching, receive the water they need from their food. One pattern of adaptation in desert birds ensures that reproduction will occur at a time when succulent food

Figure 22–13 Merriam's kangaroo rat, *Dipodomys merriami*. Kangaroo rats are nocturnal, and because deserts cool off rapidly at night, low environmental temperatures can be more of a problem for them than heat.

is available for fledglings. In the arid central region of Australia, bird reproduction is precisely keyed to rainfall. The sight of rain is apparently sufficient to stimulate courtship, and mating and nest building commence within a few hours of the start of rain. This rapid response ensures that the baby birds will hatch in the flush of new vegetation and insect abundance stimulated by the rain.

A different approach, very like that of mammals, has been evolved by columbiform birds (pigeons and doves), which are widespread in arid regions. Fledglings are fed on pigeon's milk, a liquid substance produced by the crop under the stimulus of prolactin, which is a hormone secreted by the anterior pituitary gland. The chemical composition of pigeon's milk is very similar to that of mammalian milk; it is primarily water plus protein and fat, and it simultaneously satisfies both the nutritional requirements and the water needs of the fledgling. This approach places the water stress on the adult, which must find enough water to produce milk as well as meet its own water requirements.

Relaxation of Homeostasis—Diurnal Rodents Not all rodents that live in deserts are nocturnal. Ground squirrels forage during the day, running across the desert surface. The almost frenetic activity of desert ground squirrels on intensely hot days is a result of the thermoregulatory problems that small animals experience under these conditions. Studies of the antelope ground squirrel (*Ammospermophilus leucurus*) at Deep Canyon—near Palm Springs, California—illustrate the short-term relaxation of body temperature homeostasis (**Figure 22–14**).

The heat on summer days at Deep Canyon is intense, and the squirrels are exposed to high heat loads for most of the day. They have a bimodal pattern of activity that peaks in the midmorning and again in the late afternoon. Relatively few squirrels are active in the middle of the day. The body temperatures of antelope ground squirrels are labile, and the body temperatures of individual squirrels vary as much as 7.5°C (from 36.1°C to 43.6°C) during a day. The squirrels use this variability of body temperature to store heat during their periods of activity.

High temperatures limit the squirrels' bouts of activity to no more than 9 to 13 minutes. They sprint furiously from one patch of shade to the next, pausing only to seize food or to look for predators. The squirrels minimize exposure to the highest temperatures by running across open areas, and they seek shade or their burrows to cool off. On a hot summer day, a squirrel can maintain a body temperature lower than 43°C (the maximum temperature it can tolerate) only by retreating every few minutes to a burrow deeper than 60 centimeters, where the soil temperature is 30°C to 32°C.

The body temperature of an antelope ground squirrel shows a pattern of rapid oscillations, rising while the squirrel is in the sun and falling when it retreats to its burrow (**Figure 22–15**). Ground squirrels do not sweat or pant; instead, they combine transient heat storage with passive cooling in a burrow to permit diurnal activity.

Relaxation of Homeostasis—Dromedary Camels Large animals, including humans, have specific advantages and disadvantages in desert life that are directly related to body size. A large animal has nowhere to hide from desert

Figure 22–14 The antelope ground squirrel, *Ammospermophilus leucurus*.

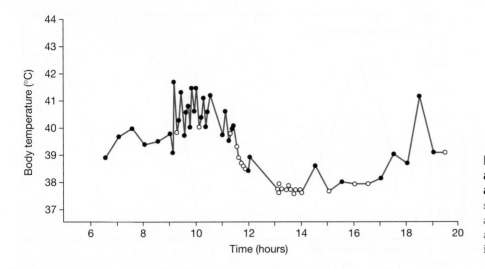

Figure 22–15 Short-term cycles of activity and body temperature of an antelope ground squirrel. The squirrel warms up during periods of activity on the surface (dark circles) and cools down when it retreats into its burrow (open circles).

conditions. It is too big to burrow underground, and few deserts have vegetation large enough to provide useful shade to an animal much larger than a jackrabbit. On the other hand, large body size offers some options not available to smaller animals. Large animals are mobile and can travel long distances to find food or water, whereas small animals may be limited to home ranges only a few meters or tens of meters in diameter. Large animals have low surface/volume ratios and can be well insulated. Consequently, they absorb heat from the environment slowly. A large body mass gives an animal a large thermal inertia; that is, it can absorb a large amount of heat energy before its body temperature rises to dangerous levels.

The two extant species of camels, the Asian (two-humped) bactrian camel (*Camelus ferus*) and the Arabian (one-humped) dromedary camel (*Camelus dromedarius*) are the classic large desert animals (**Figure 22–16**). Dromedaries make journeys in excess of 500 kilometers, lasting 2 or 3 weeks, during which they do not have an opportunity to drink. Their longest trips take place in winter and spring, when air temperatures are relatively low and scattered rainstorms have produced fresh vegetation that provides them with a little food and water.

Camels are large animals—adult body masses of dromedary camels are 400 to 450 kilograms for females and up to 500 kilograms for males. The camel's adjustments to desert life are revealed by comparing the daily cycle of body temperature in a camel that receives water daily with the cycle of one that has been deprived of water (**Figure 22–17**). The watered camel shows a small daily cycle of body temperature with a minimum of 36°C in the early morning and a maximum of 39°C in midafternoon. When a camel is deprived of water, the daily temperature variation triples. Body temperature is allowed

to fall to 34.5°C at night and climbs to 40.5°C during the day.

The significance of this increased daily fluctuation in body temperature can be assessed in terms of the water that the camel would expend in evaporative cooling to prevent the 6°C rise. With a specific heat of 4.2 kJ/(kg • °C), a 6°C increase in body temperature for a 500-kilogram camel represents storage of 12,600 kilojoules of heat. Evaporation of a kilogram of water dissipates approximately 2400 kilojoules. Thus, a camel would have to evaporate slightly more than 5 liters of water to maintain a stable body temperature at the nighttime level, and it can conserve that water by tolerating hyperthermia during the day.

Figure 22–16 Bactrian camels. In the heat of the day, most of these camels have faced into the sun to reduce the amount of direct solar radiation they receive. Now they are pressed against one another to reduce the heat they gain by convection and reradiation.

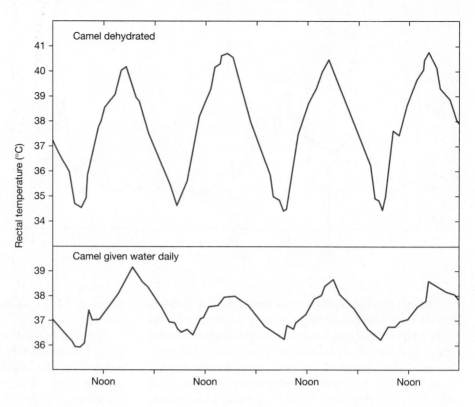

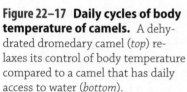

Figure 22–17 Daily cycles of body temperature of camels. A dehydrated dromedary camel (*top*) relaxes its control of body temperature compared to a camel that has daily access to water (*bottom*).

In addition to saving water that is not used for evaporative cooling, the camel receives an indirect benefit from hyperthermia via a reduction of energy flow from the air to the camel's body. As long as the camel's body temperature is lower than the air temperature, a gradient exists that causes the camel to absorb heat from the air. At a body temperature of 40.5°C, the camel's temperature is equal to that of the air for much of the day, and no net heat exchange takes place. Thus the camel saves an additional quantity of water by eliminating the temperature gradient between its body and the air. The combined effect of these measures on water loss is illustrated by data from a young dromedary camel (**Table 22–2**). When deprived of water, the camel reduced its evaporative water loss by 64 percent and reduced its total daily water loss by half.

Behavioral mechanisms and the distribution of hair on the body aid dehydrated camels in reducing their heat load. In summer, camels have hair 5 or 6 centimeters long on the back and up to 11 centimeters long over the hump. On the ventral surface and legs the hair is only 1.5 to 2 centimeters long. Early in the morning, camels lie down on surfaces that have cooled overnight by radiation of heat to the night sky. They tuck their legs beneath the body, and this places the ventral surface, with its short covering of hair, in contact with the cool ground. In this position a camel exposes only its well-protected back and sides to the sun and places its lightly furred legs and ventral surface in contact with cool sand, which may be able to conduct away some body heat. Camels may assemble in small groups and lie pressed closely together throughout the day. Spending a day in the desert sun squashed between two sweaty camels may not be your idea of fun, but in this posture a camel reduces its heat gain because it keeps its sides in contact with other camels (both at about 40°C) instead of allowing solar radiation to raise its fur surface temperature to 70°C or higher.

Despite camels' ability to reduce water loss and to tolerate dehydration, the time eventually comes when even camels must drink. These large, mobile animals can roam across the desert seeking patches of vegetation

Table 22–2 Daily water loss of a 250-kilogram camel

Condition	Water Loss (L/day) by Different Routes			
	Feces	Urine	Evaporation	Total
Drinking daily (8 days)	1.0	0.9	10.4	12.3
Not drinking (17 days)	0.8	1.4	3.7	5.9

produced by local showers and move from one oasis to another; when they drink, however, they face a problem they share with other grazing animals: waterholes can be dangerous places. Predators frequently center their activities around waterholes, where they are assured of water as well as a continuous supply of prey animals. Reducing the time spent drinking is one method of reducing the risk of predation, and camels can drink remarkable quantities of water in very short periods of time. A dehydrated camel can drink as much as 30 percent of its body mass in 10 minutes. (A very thirsty human can drink about 3 percent of body mass in the same time, and consuming 10 percent of body mass in 10 minutes would be lethal!)

The water a camel drinks is rapidly absorbed into its blood. The renal blood flow and glomerular filtration rate increase, and urine flow returns to normal within a half-hour of drinking. The urine changes from dark brown and syrupy to colorless and watery. Aldosterone, a hormone produced by the adrenal cortex, helps to counteract the dilution of the blood by the water the camel has drunk by stimulating sodium reabsorption in the kidney. Nonetheless, dilution of the blood causes the red blood cells to swell as they absorb water by osmosis. Camel erythrocytes are resistant to this osmotic stress, in part because their hemoglobin binds water and reduces its osmotic activity, but other desert ruminants have erythrocytes that would burst under these conditions. Bedouin goats, for example, have fragile erythrocytes, and the water a goat drinks is absorbed slowly from the rumen. Goats require 2 days to return to normal kidney function after dehydration.

Relaxation of Homeostasis—Large African Antelope The 100-kilogram oryx (*Oryx beisa*) and the 200-kilogram eland (*Taurotragus oryx*) use heat storage like the dromedary, but they allow their body temperatures to rise considerably higher than the 40.5°C level recorded for the camel. Rectal temperatures of 45°C have been recorded for the oryx. Body temperatures higher than 43°C rapidly produce brain damage in most mammals, but Grant's gazelles can maintain rectal temperatures of 46.5°C for as long as 6 hours with no apparent ill effects. These antelope keep their brain temperature lower than body temperature by using a countercurrent heat exchange to cool blood before it reaches the brain. The blood supply to the brain passes via the external carotid arteries, and at the base of the brain in these antelope, the arteries break into a rete mirabile that lies in a venous sinus **(Figure 22–18)**. The blood in the sinus is venous blood, returning from the walls of the nasal passages where it has been cooled by the evaporation of water. This

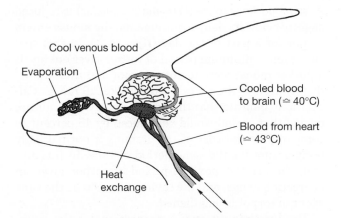

Figure 22–18 The countercurrent heat-exchange mechanism that cools blood going to a gazelle's brain. Blood leaves the heart at about 43°C (light color) and is cooled to about 40°C by cool venous blood (dark color) returning from the nasal passages where it was cooled by evaporation of water.

chilled venous blood cools the warmer arterial blood before it reaches the brain.

A mechanism of this sort is widespread among mammals, but the anatomical details vary. Horses, for example, do not have a venous sinus and carotid rete mirabile but cool the blood in the internal carotid arteries by passing it through the guttural pouches. The guttural pouches are outgrowths from the auditory tubes that envelope the internal carotid arteries and are filled with air that is cooler than the blood.

Large animals illustrate one approach to desert life. Too large to escape the rigors of the environment, they survive by tolerating a temporary relaxation of homeostasis. Their success under the harsh conditions in which they live is the result of complex interactions among diverse aspects of their ecology, behavior, morphology, and physiology. The strategy the antelope ground squirrel uses is basically the same as that employed by a camel—saving water by allowing the body temperature to rise until the heat can be dissipated passively. The difference between the two animals is a consequence of their difference in body size: a camel weighs 500 kilograms and can store heat for an entire day and cool off at night, whereas an antelope ground squirrel weighs about 100 grams and heats and cools many times during the course of a day.

Relaxation of Homeostasis—Desert Rodents The significance of daily torpor as an energy conservation mechanism in small birds was described earlier. Many desert rodents also have the ability to become torpid. In most cases, limiting the food available to an animal can induce the torpor. When the food ration of the

Arizona pocket mouse (*Perognathus amplus*) is reduced slightly below its daily requirements, the mouse enters torpor for a part of the day (**Figure 22–19**). In this species, even a minimum period of torpor saves energy. If a pocket mouse were to enter torpor and then immediately arouse, the process would take 2.9 hours. Calculations indicate that the overall energy expenditure during that period would be reduced by 45 percent as compared to the cost of maintaining a normal body temperature for the same period of time. In this animal, the briefest possible period of torpor gives an energetic savings, and the saving increases as the time spent in torpor is lengthened.

The duration of torpor is proportional to the severity of food deprivation for the pocket mouse. As its food ration is reduced, the mouse spends more time each day in torpor and conserves more energy. Adjusting the time spent in torpor to match the availability of food may be a general phenomenon among seed-eating desert rodents. These animals appear to assess the rate at which they accumulate food supplies during foraging rather than their actual energy balance. Species that accumulate caches of food enter torpor even with large quantities of stored food on hand if they are unable to add to their stores by continuing to forage. When seeds are deeply buried in the sand, and thus hard to find, pocket mice spend more time in torpor than they do when the same quantity of seed is close to the surface. This behavior is probably a response to the chronic food shortage that desert rodents may face because of the low primary productivity of desert communities and the effects of unpre-dictable variations from normal rainfall patterns, which may almost completely eliminate seed production by desert plants in dry years.

Specializations—Male Sandgrouses Seed-eating desert birds with precocial young, like the sandgrouse (more than a dozen species of *Pterocles*) inhabiting the deserts of Africa and the Near East, face particular problems in providing water for their young. Sandgrouse chicks begin to find seeds for themselves within hours of hatching. However, they are unable to fly to water-holes as their parents do, and seeds do not provide the water they need. Instead, adult male sandgrouse transport water to their broods. The belly feathers of males have a unique structure in which the proximal portions of the barbules are coiled into helices. When the feather is wetted, the barbules uncoil and trap water. The feathers of male sandgrouse hold 15 to 20 times their weight of water, and the feathers of females hold 11 to 13 times their weight.

Male sandgrouses in the Kalahari Desert of southern Africa fly to waterholes just after dawn and soak their belly feathers, absorbing 25 to 40 milliliters of water. Some of this water evaporates on the flight back to their nests, but calculations indicate that a male sandgrouse can fly 30 kilometers and arrive with 10 to 28 milliliters of water still adhering to its feathers. As the male sand-grouse lands, the chicks rush to him and, seizing the wet belly feathers in their beaks, strip the water from them with downward jerks of their heads (**Figure 22–20**). In a few minutes, the young birds have satisfied their thirst, and the male rubs himself dry on the sand.

Figure 22–19 An Arizona pocket mouse, *Perognathus amplus*, in torpor.

Figure 22–20 A male chestnut-bellied sandgrouse, *Pterocles exustus*, providing water to a chick. The chick is stripping water from the modified breast feathers of the male.

Summary

Endothermy is an energetically expensive way of life. It allows birds and mammals considerable freedom from the physical environment, especially low temperatures, but it requires a large base of food resources to sustain high rates of metabolism. Endothermy is remarkably effective in cold environments; some species of birds and mammals can live in the coldest temperatures on Earth. Small mammals can minimize exposure to severe conditions by remaining in sheltered microhabitats and adjusting their behavior patterns. For large mammals and birds, the insulation provided by hair or feathers is so good that little increase in metabolic heat production is needed to maintain a gradient of 50°C or more between the animal's internal temperature and the temperature of the air surrounding it.

Endothermy is energetically expensive, however, and maintaining a high body temperature requires a steady supply of food. When food is scarce, many birds and mammals become hypothermic—that is, they allow their body temperatures to fall for periods ranging from hours to months. Lower body temperatures reduce heat loss to the environment and the amount of food an animal must find.

Hot environments are more difficult for endotherms than are cold environments because endothermal thermoregulation balances internal heat production with heat loss to the environment. When the environment is hotter than the animal, the movement of heat is reversed. Evaporative cooling is effective as a short-term response to overheating, but it depletes the body's store of water and creates new problems. Small animals—nocturnal rodents, for example—can often avoid much of the daily heat load in hot environments by spending the day underground in burrows and emerging only at night.

Larger animals have nowhere to hide and must meet the heat load head on. Camels and other large mammals of desert regions relax their limits of homeostasis when confronted by the twin problems of high temperatures and water shortage: they allow their body temperatures to rise during the day and fall at night. This physiological tolerance is combined with behavioral and morphological characters that reduce the amount of heat that actually reaches their bodies from the environment.

Environments that are both hot and dry—deserts— pose a dual challenge. Animals must have a way of cooling themselves, but water is in short supply. Minimizing the water used to excrete metabolic wastes is an important consideration for desert animals, and birds and mammals conserve water in different ways. Birds have the ancestral sauropsid character of excreting nitrogenous wastes as salts of uric acid; this process releases water as the urate salts precipitate from solution in the urine. Mammals have a unique structure in the kidney, the loop of Henle, that enables them to produce urine with high concentrations of urea and salts.

Mobility is an important part of the response of large endotherms to both hot and cold environments. Seasonal movements away from unfavorable conditions (migration) or regular movements between scattered oases that provide water and shade are options available to medium-sized and large mammals, and the mobility of birds makes long-distance movements feasible even for small species. Some small mammals in the desert conserve water and energy by retreating underground and becoming hypothermic for periods ranging from hours to days.

The effectiveness of the same response—hypothermia—at both extremes of the environmental temperature range emphasizes the versatility of the basal mechanisms of endotherms that permit them to inhabit the full range of environmental conditions on Earth.

Discussion Questions

1. Why does an endotherm's metabolic rate increase above the upper lethal temperature? And why does the metabolic rate fall below the lower lethal?
2. What Arctic animals use countercurrent circulation? Where in those animals is it most conspicuous?
3. Compare the advantages and disadvantages of large body size for mammals that live in cold environments. Go beyond the information presented in the chapter.
4. Most mammals have a layer of cutaneous fat beneath the skin, but dromedary camels concentrate fat storage in their hump. What is the significance of that difference for the thermoregulation of camels?

5. How can you estimate the concentration of the urine that a species of rodent produces if the only information you are given is a photograph of an intact kidney from that species? Explain.

6. Are birds more like humans or kangaroo rats in the challenges they face in deserts and the way they cope with these challenges?

Additional Information

Beuchat, C. A. 1990. Body size, medullary thickness, and urine concentrating ability in mammals. *American Journal of Physiology* 258:R298–R308.

Florant, G. L, and J. E. Healy. 2012. The regulation of food intake in mammalian hibernators—A review. In press.

Gilbert, C., et al. 2010. One for all and all for one: The energetic benefits of huddling in endotherms. *Biological Reviews* 85:545–569.

Gilbert, C., et al. 2012. Private heat for public warmth: How huddling shapes individual thermogenic responses of rabbit pups. PloS ONE 7(3): e33553. doi:10.1371/journal.pone.0033553

Glanville, E. J., and F. Seebacher. 2010. Advantage to lower body temperatures for a small mammal (*Rattus fuscipes*) experiencing chronic cold. *Journal of Mammalogy* 91:1197–1204.

Jefimow, M., et al. 2011. Social thermoregulation and torpor in the Siberian hamster. *The Journal of Experimental Biology* 214:1100–1108.

Lovegrove, B. G. 2005. Seasonal thermoregulatory responses in mammals. *Journal of Comparative Physiology B* 175:231–247.

McKechnie, A. E., and G. G. Lovegroce. 2003. Avian facultative hypothermic responses: A review. *The Condor* 104:705–724.

Schubert, K. A., et al. 2010. Daily torpor in mice: High foraging costs trigger energy-saving hypothermia. *Biology Letters* 6:132–135.

Schwimmer, H., and A. Haim. 2009. Physiological adaptations of small mammals to desert ecosystems. *Integrative Zoology* 4:357–366.

Smit, B., and A. E. McKechnie. 2010. Avian seasonal metabolic variation in a subtropical desert: Basal metabolic rates are lower in winter than in summer. *Functional Ecology* 24:330–339.

Storey, K. B. 2010. Out cold: Biochemical regulation of mammalian hibernation—A mini-review. *Gerontology* 56:220–230.

Wojciechowski, M. S., and B. Pinshow. 2009. Heterothermy in small, migrating passerine birds during stopover: Use of hypothermia at rest accelerates fuel accumulation. *The Journal of Experimental Biology* 212:3068–3075.

Websites

Desert biology

http://openlearn.open.ac.uk/mod/oucontent/view.php?id=398612&direct=1

Hibernation and torpor

http://openlearn.open.ac.uk/mod/oucontent/view.php?id=398616&direct=1

Daily torpor by a hummingbird

http://scienceblogs.com/grrlscientist/2006/04/hummingbirds_and_torpor.php

Metabolism and thermoregulation

http://openlearn.open.ac.uk/mod/oucontent/view.php?id=398668&section=5.2

Polar biology

http://openlearn.open.ac.uk/mod/oucontent/view.php?id=398620

CHAPTER 23

Body Size, Ecology, and Sociality of Mammals

The origin of some of the derived features of mammalian behavior might lie in the nocturnal habits that are postulated for Mesozoic mammals. If these animals had to rely on scent or hearing instead of vision to interpret their surroundings, ancestral mammals may have benefited from an increased ability to associate information received via the ears and nose and to compare the intensity of stimuli over intervals of time—for example, Are the footsteps of a predator getting louder? Is the scent of the prey getting stronger? In turn, greater associative capacity might contribute to more complex social behavior, and the contact between mother and young during infancy could provide an opportunity to modify behavior by learning.

Social behaviors and interactions between individuals play a large role in the biology of extant mammals. These behaviors are modified by the environment, and relationships among energy requirements, resource distribution, and social systems can often be demonstrated. In this chapter we consider some examples of those interactions that illustrate the complexity of the evolution of mammalian social behavior. In addition, we consider the social behavior of several species of primates. The social behavior of many primates is elaborate but not necessarily more complex than the behavior of some other kinds of mammals, including cetaceans, elephants, and canids. However, primates have been the subjects of more field studies than have other mammals, and we know a great deal about their social behavior and its consequences for the fitness of individuals, and even a little about the way some species of primates view their own social systems.

23.1 Social Behavior

Sociality means the state of living in structured groups, and some form of group living is found among nearly all kinds of vertebrates. However, the greatest development of sociality is found among mammals. Much of the biology of mammals can best be understood in the context of what sorts of groups form, the advantages of group living for the individuals involved, and the behaviors that stabilize groups. Mammals may be particularly social animals due to the interaction of several mammalian characteristics, no single one of which is directly related to sociality but which, in combination, create conditions in which sociality is likely to evolve. Thus, the relatively large brains of mammals (which presumably facilitate complex behavior and learning), the prolonged association of parents and young, and high metabolic rates and endothermy (with the resulting high resource requirements) may be viewed as conditions that are conducive to the development of interdependent social units.

Of course, not all mammals are social; in fact, there are more species of solitary mammals than social ones. Both solitary and social species are known among marsupials and placentals. Monotremes appear to be solitary (as are most placentals of that body size), but the three living monotreme species are too small a sample to form a basis for speculations about the phylogenetic origins of sociality among mammals. Of course, the social behavior of mammals does not operate in a vacuum; it is only one part of the biology of a species. Social behavior interacts with other kinds of behavior (such as food gathering, predator avoidance, and reproduction), with the morphological and physiological characteristics of a species, and with the distribution of resources in the habitat. Our emphasis in this chapter is on those interactions, and we illustrate the interrelationships of behavior and ecology with examples drawn from both predators and their prey.

23.2 **Population Structure and the Distribution of Resources**

From an ecological perspective, the distribution of resources needed by a species is usually a major factor in determining its social structure. If resources are too limited to allow more than one individual of a species to inhabit an area, there is little chance of that species developing social groupings. Thus the distribution of resources in the habitat and the amount of space needed by an individual to meet its resource requirements are important factors influencing the sociality of mammals.

Most animals have a **home range,** an area in which they spend most of their time and find the food and shelter they need. Home ranges are not defended against the incursions of other individuals—an area that is defended is called a **territory.** The value of staying within a home range probably lies in the familiarity of an individual animal with the locations of food and shelter. Many species of vertebrates employ a type of foraging known as traplining, in which they move over a regular route and visit specific places where food may be available. For example, a mountain lion may carefully approach a burrow where a marmot lives, beginning its stalk long before it can actually see whether the marmot is outside its burrow; a hummingbird may return to patches of flowers at intervals that match the rate at which nectar is renewed. This kind of behavior demonstrates a familiarity with the home range and with the resources likely to be available in particular places.

The **resource dispersion hypothesis** predicts that the size of the home range of an individual animal depends primarily on two factors: the resource needs of the individual and the distribution of resources in the environment. That is, individuals of species that require large quantities of a resource such as food should have larger home ranges than individuals of species that require less food. Similarly, the home ranges of individuals should be smaller in a rich environment than in one where resources are scarce. The resource dispersion hypothesis is a very general statement of an ecological principle. It applies equally well to any kind of animal and to any kind of resource. The resources usually considered are food, shelter, and access to mates. In Section 17.7 we considered the role of the monopolization of resources by individuals in relation to the mating systems of birds, and here we discuss the role of resource dispersion in relation to the home range size and sociality of mammals.

Body Size and Resource Needs

Studies of mammals have concentrated on food as the resource of paramount importance in determining the sizes of home ranges. The energy consumption of vertebrates increases in proportion to body mass raised to a power that is usually between 0.75 and 1.0. If we assume that energy requirements determine home range size, we can predict that home range size will also increase in proportion to the 0.75 or 1.0 power of body mass. That prediction appears to be correct in general but perhaps wrong in detail (**Figure 23–1**). That is, the sizes of the home ranges of mammals do increase with increasing body size, but the rates of increase (the slopes in Figure 23–1) are somewhat greater than expected. Home ranges appear to be proportional to body mass raised to powers between 1.0 and 2.0. This relationship between energy requirements and home range sizes suggests that energy needs are important in determining the size of the home range but that additional factors are involved. One possibility is that the efficiency with which an animal can find and use resources decreases as the size of a home range or the fragmentation of resources increases. If that hypothesis is correct, then the sizes of home ranges would be expected to increase with the body sizes of animals more rapidly than energy requirements increase with body size.

The failure of the resource dispersion hypothesis to predict the exact relationship between body size and the size of home ranges indicates that we have more to learn about how animals use the resources of their home ranges. So far we have been assuming that resources are distributed evenly throughout the home

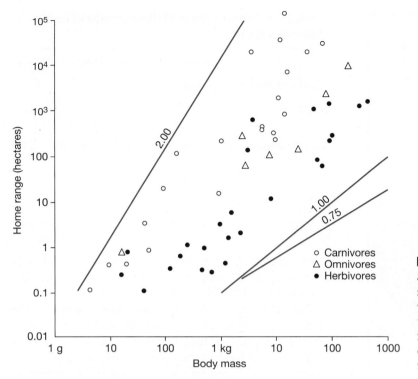

Figure 23–1 Home range size of mammals as a function of body mass on a log-log scale. All groups have slopes greater than 1.00, which indicates a disproportionate (i.e., allometric) increase in home range size as body size increases. Metabolic rates increase with a slope of about 0.75; thus, home range size increases more rapidly than food requirements.

range, but that assumption overlooks the structural complexity of most habitats. What insights can we gain from a more realistic consideration of how mammals gather food?

The Availability of Resources

Three factors seem likely to be important in determining the availability of food to mammals: what they eat, whether their food is evenly dispersed throughout the habitat or is found in patches, and how they gather their food. We will consider examples of each of these factors.

Dietary Habits Figure 23–1 shows that home range size increases with body size and that dietary habits also affect home range size. Herbivores have smaller home ranges than do omnivores of the same body size, and carnivores have larger home ranges than do herbivores or omnivores. For example, the home range of a red deer (an herbivore) that weighs 100 kilograms is approximately 100 hectares. A bear (an omnivore) of the same body size has a home range larger than 1000 hectares, and a tiger (a carnivore) has a home range of more than 10,000 hectares.

The relationship between home range size and the dietary habits of mammals probably reflects the abundance of different kinds of food. The grasses and leaves eaten by some herbivores are nearly ubiquitous, and a small home range provides all the food an individual

requires. The plant materials (seeds and fruit) eaten by omnivores are less abundant than leaves and grasses, and their distribution is more fragmented in space and time because different species of plants produce seeds and fruit at different seasons. Thus, a large home range is probably necessary to provide the food resources that an individual omnivore needs. The vertebrates that are eaten by carnivores are still less abundant and more fragmented, and a correspondingly larger home range is apparently needed to ensure an adequate food supply. Seasonal changes in food supply add another level of variation. When food is scarce, animals must expand their home ranges.

Distribution of Resources We have assumed that one part of a home range is equivalent to another part in terms of the availability of food. This assumption may be valid for some grazing and browsing herbivorous mammals, but it is clearly not true for mammals that seek out fruiting trees (which represent patches of food) or for any carnivorous mammal that preys on animals living in groups. The sizes of the home ranges of animals that use food occurring in patches should reflect the distribution of patches: home ranges should be small if patches are abundant and large if patches are widely dispersed.

That relationship is well illustrated by the home ranges of Arctic foxes (*Alopex lagopus*) in Iceland. The

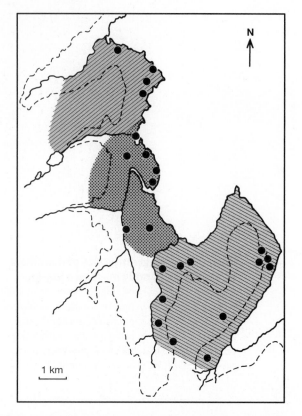

Figure 23–2 Map of the territorial boundaries of three groups of Arctic foxes in Iceland. The 200-meter contour line is shown. Black dots mark the sites of dens used by the foxes.

Table 23–1 Home ranges of three groups of Arctic foxes in Iceland

	Group 1	Group 2	Group 3	Average
Total area (km²)	10.3	8.6	18.5	12.5
Length of coastline (km)	5.6	5.4	10.5	7.2
Length of productive coastline (km)	5.6	5.4	6.0	5.7
Driftwood productivity (logs/ year)	1800	1800	2100	1900

foxes live in social groups consisting of one male and two females plus the cubs from the current year. The home ranges of the individuals in a group overlap widely, and there is very little overlap between the home ranges of members of different groups. The home ranges of the foxes are located along the coast and do not extend far into the uplands (**Figure 23–2**). Between 60 and 80 percent of the diet is composed of items the foxes find on the shore—carcasses of seabirds, seals, and fishes, and invertebrates from clumps of seaweed washed up on the beach. Little food is available for foxes in the uplands. The foxes concentrate their foraging on the beach during the 3 hours before low tide, which is the best time for beachcombing. The foxes approach the shore carefully, stalking along gullies. They creep out on the beach carefully, apparently looking for birds that are resting or feeding. If birds are present, the foxes stalk them. If no birds are present, the foxes search the beach for carrion.

Researchers studied three groups of foxes using radiotelemetry to follow the movements of individuals. The areas of the home ranges vary more than twofold, from 8.6 to 18.5 square kilometers (**Table 23–1**). The sizes

of the home ranges are slightly more similar when only the coastline is considered: each territory includes between 5.4 and 10.5 kilometers of coastline.

The coastline is, of course, the main source of the foxes' food, but not all areas of the coastline accumulate floating objects. The distribution of food on the beaches is patchy and depends on the directions of currents. As a result, some parts of the shore are more productive than others. The length of productive coastline occupied by each group of foxes is quite similar—from 5.4 to 6.0 kilometers. Farmers in Iceland use driftwood to make fence posts, and the amount of driftwood that is harvested by the farmers from the coasts in the home ranges of the three groups of foxes varies only from 1800 to 2100 logs per year. Because both driftwood and carrion are moved by currents and deposited on the beaches, the harvest of driftwood by farmers probably reflects the harvest of carrion by the foxes. Thus, the home range sizes of the three groups appear to match the distribution of their most important food resource, and the productive areas of the home ranges of the three groups are very similar despite the more than twofold difference in total areas of their home ranges.

The social structure and ecology of Arctic foxes are being disrupted by the northward movement of red foxes (*Vulpes vulpes*) as the climate of polar regions is becoming warmer. Red foxes are larger than Arctic foxes, and they appear to be more versatile predators as well. Human activities accompanying climate change are also playing a role in the northward movement of red foxes. Construction of roads and houses has increased the availability of road-killed animals and garbage, and red foxes are exploiting both of those sources of food.

Group Size and Hunting Success It is readily apparent that the average home range size of a species of mammal can influence the social system of that species. Individuals of a species probably encounter each other frequently when home ranges are small, whereas individuals of species that roam over thousands of hectares may rarely meet. Thus the distribution of resources in relation to the resource needs of a species is one factor that can limit the degree to which social groupings occur. However, sociality may influence resource distribution if groups of animals are able to exploit resources that are not available to single individuals.

The influence of sociality on resource distribution may be seen among predatory animals that can hunt individually or in groups. Some species of prey are too large for an individual predator to attack but are vulnerable to attack by a group of predators. For example, spotted hyenas (*Crocuta crocuta*) weigh about 50 kilograms. When hyenas hunt individually, they feed on Thomson's gazelles (*Gazella thomsoni*, 20 kilograms) and juvenile

wildebeest (*Connochaetes taurinus*, about 30 kilograms) (**Figure 23–3**). However, when hyenas hunt in packs, they feed on adult wildebeest (about 200 kilograms) and zebras (*Equus burchelli*, about 220 kilograms). Some species of prey have defenses that are effective against individual predators but less effective with groups of predators. For example, the success rate for solitary lions (*Panthera leo*) hunting zebras and wildebeest is only 15 percent, whereas lions hunting in groups of six to eight individuals are successful in up to 43 percent of their attacks. Groups of lions make multiple kills of wildebeest more than 30 percent of the time, but individual lions kill only a single wildebeest.

The relationship of the sociality and body size of a predator to the size of its prey is shown in **Figure 23–4**: social predators (defined as those that hunt in groups of 8 to 10 individuals) attack larger prey than do weakly social predators (average group sizes of 1.6 to 3.1 indi-

(a)

(b)

(c)

(d)

Figure 23–3 Spotted hyenas. (a) Spotted hyenas may hunt individually or in packs. They prey on small animals like the Thomson's gazelle (b) when they hunt individually. When they hunt in packs, they attack larger prey such as wildebeest (c) and zebras (d).

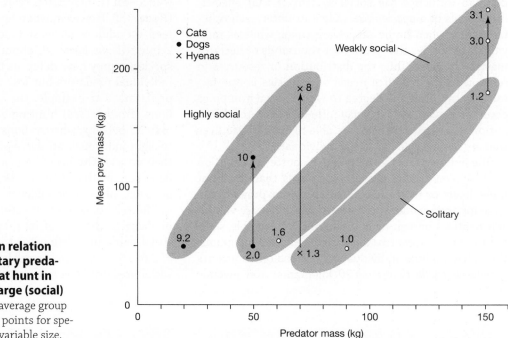

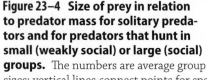

Figure 23–4 Size of prey in relation to predator mass for solitary predators and for predators that hunt in small (weakly social) or large (social) groups. The numbers are average group sizes; vertical lines connect points for species that hunt in groups of variable size.

viduals), and these weakly social predators attack larger prey than do solitary predators (average group sizes of 1.0 to 1.3 individuals). One consequence of sociality for predatory mammals thus appears to be an increase in the potential food resources of an environment: individual predators may be able to extend the range of prey species they can attack by hunting in groups.

Of course, the major disadvantage to hunting in a group is that there are more mouths to feed when a kill is made. The food requirement of a group of predators is the sum of the individual requirements of the group members, and for group hunting to be advantageous, the per capita amount of food obtained by hunting in a group must exceed the amount caught by a solitary hunter.

What factors could contribute to the evolution of cooperative hunting? That is, do predators form groups *because* that allows them to hunt large prey, or *must* they hunt large prey because they live in groups for some other reason? A study of lions suggests that the second hypothesis is correct for this species. Female lions are the only female felids to live in social groups. Female lions defend a group territory and protect their cubs from other groups of female lions. The high population densities that are characteristic of lions may have favored group defense of a territory. The presence of large prey makes it possible for lions to hunt in groups, but group hunting does not increase

the amount of food available per lion—a female lion hunting by herself can catch as much prey as her share of a group capture. So, groups form because of the advantages they provide in the social structure of the population of lions on the Serengeti, and group hunting is a by-product of that social structure.

A similar interpretation has been suggested for the formation of groups of male cheetahs. Male cheetahs may live alone or form permanent coalitions of two or three individuals that live and hunt together. These coalitions are often composed of littermates, and a coalition is more successful in occupying a territory than is a single male. Competition for territories is intense, and territorial disputes are an important cause of mortality for male cheetahs. Cheetahs hunting singly concentrate on small prey such as Thomson's gazelles, whereas coalitions attack larger prey such as wildebeest. Overall foraging success increases with group size for male cheetahs, but the benefit of having the assistance of a coalition in holding a territory and controlling access to females is probably more important than the effect of the coalition on food intake.

23.3 **Advantages of Sociality**

Sociality is not limited to species of mammals that hunt in groups, and the potential advantages of sociality are not confined to predatory behavior.

Mammals may derive benefits from sociality in terms of reproduction and care of young, avoiding predation, and facilitating feeding.

Defenses Against Predators

One probable advantage of sociality is the reduced risk of predation for an individual that is part of a group compared to the risk for a solitary individual. The proposed benefits of sociality in avoiding predation take many forms, but most of them can be grouped into three categories.

More Eyes, More Time A group of animals may be more likely to detect the approach of a predator than is an individual, simply because a group has more eyes, ears, and noses to keep watch. As a result of the extra watchers, an individual in a group may be able to devote a larger proportion of its time to feeding and less time to watching for predators than a solitary individual can. Mammals in groups generally live in open habitats, whereas solitary species are usually found in forests; that relationship may partially reflect the antipredator aspects of group living.

Dilution, Confusion, and the Selfish Herd Some of the benefits of sociality in predator avoidance result from a reduced risk of predation for an individual when it is part of a group. The presence of a large number of potential prey animals may exceed the predatory capacity of a limited number of predators if the prey are present for only a limited time. For example, the huge herds of nomadic wildebeest that follow the shifting rains across the African savanna contain far more individuals than a pack of wild dogs can eat during the time it takes the herd to cross the pack's home range—and, when the dogs do attack a herd, the confusion that arises as they try to single out and pursue one animal among a crowd of animals that all look the same may increase a wildebeest's chance of escape. The benefit could be especially strong for individuals that manage to remain in the center of the herd; predators pursue and capture individuals on the periphery of the herd before they reach the ones in the middle. (This is the selfish herd hypothesis, which proposes that the protection an individual receives from being in a group depends on the behavior of the other individuals in the group—not everyone can be in the center.)

Group Defenses Large social mammals can form a defensive wall when a predator approaches. Typically, the adults confront the predator, with young animals and sometimes females sheltered behind them. Musk oxen (*Ovibos moschatus*) are one of the most familiar examples of this behavior, sometimes forming a complete circle, with the adults facing outward to confront a pack of wolves and the young oxen sheltered on the inside of the circle.

Sociality and Reproduction

The extensive period of dependence of many young mammals on their parents provides a setting in which many benefits of sociality can be manifested. Maternal care of the offspring is universal among mammals, and males of many species also play a role in parental care. Group living provides opportunities for complex interactions among adults and juveniles that involve various sorts of **alloparental behavior** (care provided by an individual that is not a parent of the young receiving the care). Collaborative rearing of the young of several mothers is characteristic of lions and of many canids. Frequently, nonbreeding individuals join the mothers in protecting and caring for the young. Among dwarf mongooses (*Helogale parvula*), this kind of behavior extends to the care of sick adults; similar behaviors are reported for mammals as diverse as elephants and cetaceans. Many social groups of mammals consist of related individuals, and these nonbreeding helpers may increase their own fitness by assisting in rearing the offspring of their kin.

23.4 Body Size, Diet, and the Structure of Social Systems

The complex relationships among body size, sociality, and other aspects of the ecology and behavior of herbivorous mammals are illustrated by the variation in social systems of African antelope (family Bovidae). The smallest species of antelope have adult weights of 3 to 4 kilograms (the dik-diks, *Madoqua,* and some duikers, *Cephalophus*); one of the largest (the Cape buffalo, *Syncerus caffer*) weighs 400 kilograms (**Figure 23–5**). The smallest species are forest animals that browse on the most nutritious parts of shrubs, live individually or in pairs, defend a territory, and hide from predators (**Table 23–2** on page 567). The largest species (including the 500-kilogram eland, *Taurotragus oryx,* and the Cape buffalo) are grassland animals that feed unselectively, live in large herds, are migratory, and use group defense to deter predators. Species with intermediate body sizes are also intermediate in these ecological and behavioral characteristics. It seems likely that the correlated variation in body size, ecology, and behavior among these antelope reveals functional relationships among these aspects of their biology. How might such diverse features of mammalian biology interact?

(a)

(b)

(c)

(d)

Figure 23–5 Diet and body size are correlated in African bovids. The smallest species, such as the dik-dik (a), eat leaves and fruit and are highly selective in their choice of the parts of plants they consume (type I diet). Medium-sized antelope, such as the impala (b), are less selective in their browsing (type II diet). Larger species, such as the hartebeest (c), are grazers that select new growth (type III diet). The largest species, such as the Cape buffalo (d), browse and graze unselectively (type IV diet).

An antelope's feeding habits appear to provide a key that can be used to understand other aspects of its ecology and behavior. The diets of different species are closely correlated with their body size and habitats. In turn, those relationships are important in setting group size. The size of a group determines the distribution of females in time and space, and this is a major factor in establishing the mating system used by males of a species. Group size also plays an important role in determining the appropriate antipredator tactics for a species. Mating systems and antipredator

mechanisms are central factors in the social organization of a species.

Body Size and Food Habits

Antelope are ruminants, relying on symbiotic microorganisms in the rumen to convert cellulose from plants into compounds that can be absorbed by the vertebrate digestive system. The effectiveness of ruminant digestion is proportional to body size. Body size is the primary factor determining the anatomical characteristics of the gut of ungulates. The volume of the rumen is pro-

Table 23–2 Elements of the ecology and social systems of African ungulates

Diet Type	Examples	Body Mass (kg)	Food Habits	Group Size	Mating System	Predator Avoidance
I	Dik-dik, some duikers	3–20	Highly selective browsers	1–2	Stable pair, territorial	Hide
II	Thomson's gazelle, impala	20–100	Moderately selective browsers and grazers	2–100	Male territorial in breeding season, temporary harems of females	Flee
III	Wildebeest, hartebeest	100–200	Grazers, selective for growth stage	Large herds	Nomadic, temporary harems	Flee, hide in herd, threaten predator
IV	Eland, buffalo	300–900	Unselective browsers and grazers	Large herds	Male hierarchy	Group defense

portional to body mass in species with different body sizes, whereas metabolic rates are proportional to the 0.75 power of body mass. The ecological consequence of this difference in allometric slopes is illustrated in **Figure 23–6**: a large ruminant has proportionately more capacity to process food than does a small ruminant. For animals of very small body size, the metabolic requirements become high in relation to the volume of rumen that is available to ferment plant material.

Because of this relationship, small ruminants must be more selective feeders than large ruminants. That is, a large ruminant has so much volume in its rumen that it can afford to eat large quantities of food of low nutritional value. It does not extract much energy from a unit volume of this food, but it is able to obtain its daily energy requirements by processing a large volume of food. Small ruminants, in contrast, must eat higher-quality food and rely on obtaining more energy per unit volume from the smaller volume of food that they can fit into their rumen in a day. In fact, 40 kilograms is the approximate lower limit of body size at which an unselective ruminant can balance its energy budget; species larger than 40 kilograms can be unselective grazers, whereas smaller species must eat only the most nutritious parts of plants.

The species of antelope in this example can be divided into four feeding categories:

- **Type I species** are selective browsers. They feed preferentially on certain species of plants, and they choose the parts of those plants that provide the highest-quality diet—new leaves (which have a higher nitrogen and lower fiber content than mature leaves) and fruit. Dik-diks and duikers fall into

this category; they have adult body masses between 3 and 20 kilograms. These animals show little sexual dimorphism in body size and appearance. The males have small horns, while the females are hornless or may also have small horns.

- **Type II species** are moderately selective grazers and browsers. They eat more parts of a plant than do the type I species, and they may have seasonal changes in diet as they exploit the availability of fresh shoots or fruits on particular species of plants. Thomson's gazelles (*Gazella thomsoni*) and impalas (*Aepyceros melampus*) weigh 20 to 100 kilograms and have

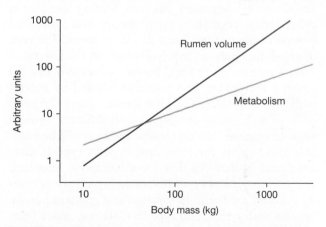

Figure 23–6 Rumen volume and energy requirements in relation to body mass. Rumen volume increases in proportion to body size (a slope of 1), whereas energy requirements are proportional to metabolism (a slope of 0.75). Thus, large species are more effective ruminants than are small species. Both axes are drawn with logarithmic scales, and the scale of the vertical axis is in arbitrary units.

type II diets. These animals show substantial sexual dimorphism in body size, with the males being about a third again larger than the females in body mass. They are highly dimorphic in appearance: Males have large, elaborate horns and may have a different coat color from females. Females are either hornless or have horns that are much smaller than those of males.

- **Type III species** are primarily grazers that are not selective for species of grass but selective for the parts of the plant; that is, they eat the leaves and avoid the stems. Hence, they are selecting for a growth stage: They avoid grass that is too short because that limits their food intake. They also avoid grass that is too long because it has too many stems that are low-quality food. Wildebeest (*Connochaetes taurinus*) and hartebeest (*Alcelaphus buselaphus*), which weigh about 200 kilograms, are type III feeders. Type III species show little sexual dimorphism in size or appearance, and the horns of females are nearly as large as those of males.

- **Type IV species** are very large and are unselective grazers and browsers. They eat all species of plants and all parts of the plant. Elands (*Taurotragus oryx*, 500 kilograms) and buffalo (*Syncerus caffer*, 400 kilograms) are type IV species. Males are substantially larger than females, but there is little dimorphism in horns.

Food Habits and Habitat

The food habits of different antelope species are important in determining what sorts of habitats provide the resources they need. Selective feeding operates at three levels: vegetation type, species and individual groups of plants, and parts of plants eaten. The type of vegetation present largely depends on the habitat—forests contain shrubs and bushes, whereas plains are covered with grass. The resources needed by species with type I diets are found in forests where the presence of a diversity of species with different growth seasons ensures that new leaves and fruit will be available throughout the year. Species with type II diets are found in habitats that are a mosaic of woodland and grassland, and type III species (which are primarily grazers) are found in savanna and grassland areas. Species with type II and type III diets may move from place to place in response to patterns of rainfall. For example, wildebeest require grass that has put out fresh new growth but that has not had time to mature. To find grass at this growth stage, wildebeest have extensive nomadic movements that follow the seasonal pattern of rain on the African plains.

Type IV feeders eat almost any kind of plant material, and they can find something edible in almost any habitat. They occupy a range of habitats, including grassland and brush, and do not have nomadic movements.

Habitat and Group Size

The habitats in which antelope feed and the types of food they eat put certain constraints on the sorts of social groupings that are possible. For example, species with type I diets live in forests and feed on scattered, distinct items. They eat an entire new leaf or fruit at a bite, and they must move between bites. A type I feeder completely removes the items it eats, so it changes the distribution of resources in its habitat. The upper part of **Figure 23–7** shows the effect of selective browsing on the new leaves on a bush—after the first individual has fed, the bush no longer has new leaves. A second individual cannot feed close behind the first because the food resources of an individual bush are entirely consumed by the first individual to feed there. As a result, the feeding behavior of species with type I diets makes it impossible for a group of animals to feed together. If one individual attempts to follow behind another to feed, the second animal must search to find food items overlooked by the individual ahead; consequently, it falls behind. Alternatively, the second individual can move aside from the path of the first animal to find an area that has not already been searched. In either case, small animals in dense vegeta-

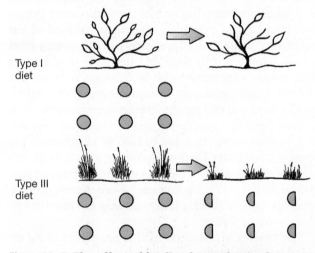

Figure 23–7 The effect of feeding by a selective browser with a type I diet and a grazer with a type III diet. *Top:* The browser removes entire food items (new leaves or fruit), thereby changing the distribution as well as the abundance of food in the habitat. *Bottom:* The grazer removes part of a grass clump, changing the abundance of food in the habitat but not its distribution.

tion rapidly lose track of each other, and no cohesive group structure is maintained.

Instead of feeding as a group, type I species are solitary or occur in pairs, and the individuals of a pair are only loosely associated as they feed. A type I diet places a premium on familiarity with a home range because a tree or bush is a patch of food that must be visited repeatedly to harvest fruit or new leaves as they appear.

Species of antelope with type II and type III diets are less selective than species with type I diets, and their feeding has less impact on the distribution of resources. These species do not remove all of the food resources in an area, and other individuals can feed nearby. Type III feeders, in particular, graze as they walk—taking a bite of grass, moving on a few steps, and taking another bite. This mode of feeding changes the abundance of food, but not its distribution in space (see Figure 23–7); herds of wildebeest graze together, all moving in the same direction at the same speed and maintaining a cohesive group. Rainfall is a major determinant of the distribution of suitable food in the habitat of these species. The rainstorms that stimulate the growth of grass are erratic, and the patch sizes in which food resources occur are enormous—hundreds of square kilometers of new grass where rain fell are separated by hundreds of kilometers of old, dry grass that did not receive rain. Instead of having home ranges or territories, species with type II and type III diets move nomadically with the rains. Group sizes change as the distribution of resources changes, from half a dozen to 60 individuals for species with type II diets and from herds of 300 or 400 to superherds of many thousands of individuals during the nomadic movements of wildebeest.

Species of antelope with type IV diets are so unselective in their choice of food that they can readily maintain large groups. Herds of buffalo may number in the hundreds. Because these species can eat almost any kind of vegetation, the distribution of resources does not change seasonally, and the size of the herds is stable.

Group Size and Mating Systems

The mating systems of African antelope are closely related to the size of their social groups and the distribution of food because those are the major factors that determine the distribution of females and the potential for males to obtain opportunities to mate by controlling resources that females need. The females of species with type I diets are dispersed because the distribution of resources in the habitat does not permit groups of individuals to form. A male of a type I species can defend food resources, but individuals must disperse through the territory to feed, and it is not feasible for a male to maintain a territory that attracts a group of females. Males of type I species pair with one female; the male defends its territory year-round, the pair bond with an individual female appears to be stable, and offspring are driven out of the territory as they mature.

A group of individuals of type II species contains several males and females. The evenly distributed nature of food of these species makes it difficult for a male to monopolize resources. Only some of the males are territorial, and even this territoriality is manifested for only part of the year. A territorial male tries to exclude other males from its territory and to gather groups of females. Exclusive mating rights are achieved by holding a patch of ground and containing females within it, driving them back if they try to leave. These species have no long-term association between a male and a particular female.

Type III species are nomadic, and males establish territories only when the herd is stationary. During these periods male wildebeest have access to groups of females within their territories, but the association between a male and females is broken when the herd moves on. However, mothers and their daughters maintain associations for 2 or 3 years. Unmated male wildebeest form bachelor herds with hierarchies, and individuals at the top of a bachelor hierarchy try to displace territorial males. If a territorial male is displaced, it joins a bachelor herd at the bottom of the hierarchy and must work its way up to the top before it can challenge another territorial male.

The social structure of buffalo (a type IV species) differs in two respects from that of wildebeest (type III). First, each herd includes many mature males that form a dominance hierarchy. The ability to attain dominance over other males is largely related to body size, and the males of type IV antelope grow throughout life. A mature male may thus be twice as heavy as a female. The individuals near the top of the hierarchy court receptive females, but no territoriality or harem formation is seen. Second, the female membership of the herd is fixed, and this situation results in a degree of genetic relationship among all the members of a buffalo herd. That genetic relationship among individuals creates situations in which the fitness an individual gains from assisting its relatives may be a factor in the social behavior of the Cape buffalo.

Mating Systems and Predator Avoidance

Prey species have various ways to avoid predators, but only some will work in a given situation. In general, a prey species can (1) avoid detection by a predator, (2) flee after it has been detected but before the predator attacks, (3) flee after the predator attacks, or (4) threaten or attack the predator. Body size, habitat, group size, and the mating system all contribute to determining the risk of predation faced by a species as well as which predator avoidance methods are most effective.

Predators usually attack prey that are the same size as the predator or smaller. Thus, small species of prey animals potentially have more species of predators than do large species. Species of antelope with type I diets are small; consequently, they are at risk from many species of predators. Furthermore, small antelope may not be able to run fast enough to escape a predator after it has attacked. On the other hand, these small antelope live in dense habitats where they are hard to see. They are cryptically colored and secretive, and they rely on being inconspicuous to avoid detection by predators. If they are pursued, they may be able to use their familiarity with the geography of their home range to avoid capture.

Groups of animals are more conspicuous to predators than are individuals, but groups also have more eyes to watch for the approach of a predator. Species of antelope with type II diets live in small groups in open habitats, where they can detect predators at a distance. These antelope avoid predators by fleeing either before or after the predator attacks. The antelope may attack small predators but usually only when a member of the group has been captured. This sort of defense is normally limited to a mother protecting her young; the rest of the group does not participate.

Species of antelope in the type III diet category are large enough to have relatively few predators, and in a group they may be formidable enough to scare off a predator. Wildebeest sometimes form a solid line and walk toward a predator, a behavior that is effective in deterring even lions from attacking. Many predators of wildebeest focus their attacks on calves, and defense of a calf is usually undertaken only by the mother. Much of the antipredator behavior of wildebeest depends on the similarity of appearance of individuals in the herd to each other. Field observations have shown that the individuals in a group of animals that are distinctive in their markings or behavior are most likely to be singled out and captured by predators.

One of the unavoidable events that distinguishes a female wildebeest is giving birth to a calf, and the reproductive biology of the species has specialized features that appear to minimize the risks associated with giving birth. For wildebeest, the breeding season and births are highly synchronized. Mating occurs in a short interval; consequently, 80 percent of the births occur within a period of 2 or 3 weeks. Furthermore, nearly all of the births that will take place on a day occur in the morning in large aggregations of females, all giving birth at once. A female wildebeest that is slightly out of synchrony with other members of her group can interrupt delivery at any stage up to the emergence of the calf's head in order to join the mass parturition. Presumably, this remarkable synchronization and control of parturition reflect the advantage of presenting predators with a homogeneous group of cows and calves rather than a group with only a few calves that could readily be singled out for attack.

Type IV species, such as buffalo, are formidable prey even for a pride of lions. They escape much potential predation simply because of their size. When buffalo are attacked, they engage in group defense; if a calf is captured, its distress cries bring many members of the group to its defense. This altruistic behavior probably represents kin selection because the stability of the female membership of buffalo herds results in genetic relationships among the individuals.

23.5 Horns and Antlers

Horns and antlers are conspicuous features of the antelope we have been discussing, and they are characteristic of many large ungulates. Horns and antlers appear to play a primary role in social recognition, sexual display, and jousting between males, although they may also be used for defense. **Figure 23–8** illustrates

Figure 23–8 Horns and antlers. (a) Horns of a bighorn sheep. The horn has a bony core covered by epidermis that produces keratin, which forms the visible part of the horn. Horns grow throughout the entire lifetime of an individual, and the horns of old males form more than a complete circle. Females have much smaller horns with little or no curvature. (b) Rhinoceros horns. One or two horns grow in the midline of the skull; the anterior horn grows above the nasal bones, and if a second horn is present, it is above the frontal bones. When horns are present they are found in both males and females, but not all species of rhinos have horns. (c) Giraffe ossicones. Males and females have horns of approximately the same size. An ossicone has a bony core that is covered by skin and hair. The hair at the tip of the ossicone wears away in old individuals, but the skin covers the bone throughout the giraffe's life. (d) Antlers of a caribou. The antlers of male caribou are much larger than those of females, and females of most species of deer do not have any antlers. Antlers are solid bone and are grown and shed annually. At the end of the breeding season, bone is reabsorbed from the junction region between the base of the antler and the pedicle of the skull, and the antler then falls off.

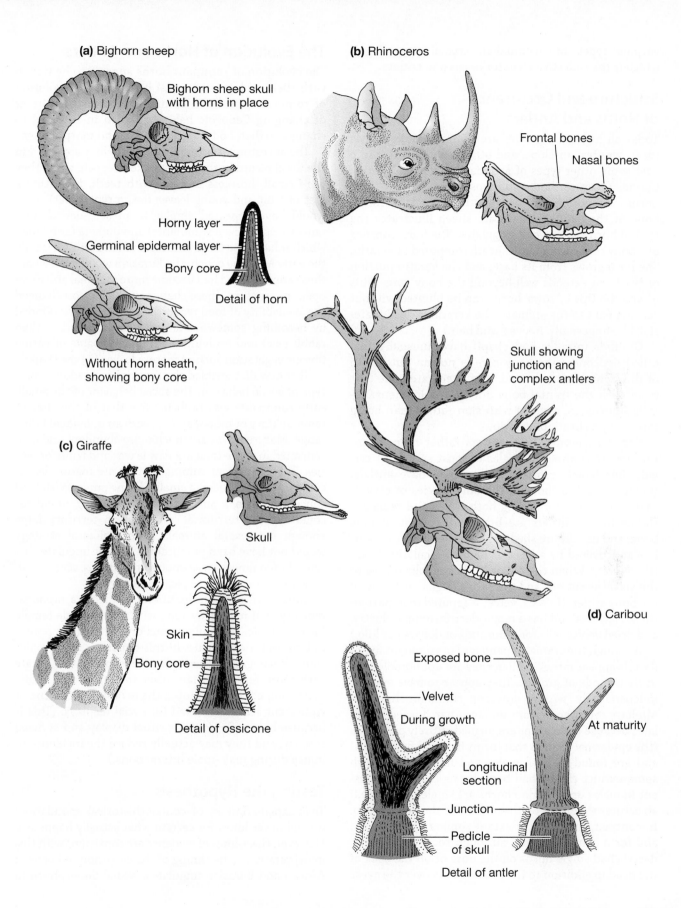

(a) Bighorn sheep

Bighorn sheep skull with horns in place

Horny layer
Germinal epidermal layer
Bony core

Detail of horn

Without horn sheath, showing bony core

(b) Rhinoceros

Frontal bones
Nasal bones

Skull showing junction and complex antlers

(c) Giraffe

Skull

Skin
Bony core

Detail of ossicone

(d) Caribou

Exposed bone

Velvet
During growth

At maturity

Longitudinal section

Junction

Pedicle of skull

Detail of antler

various types of mammalian cranial appendages, which is the collective term for horns and antlers.

Structure and Occurrence of Horns and Antlers

Today all horned ungulates are ruminant artiodactyls (deer, giraffes, antelope, and other bovids), but in the past some other types of artiodactyls also had horns. Cervids (deer, caribou, and moose) have antlers. Both horns and antlers are outgrowths from the frontal bone of the skull. A mountain sheep illustrates the typical horn structure of ungulates. The horn consists of a bony core covered by a sheath composed of keratin. The horn grows from its base, and the keratin portion of the horn extends well beyond the bony core. (This is why the tips of cows' horns can be blunted without causing pain to the animal—the keratin is dead material, like fingernails, hooves, and hair.)

Giraffids (giraffes and okapi) have unusual horns called ossicones. The bony core is not an outgrowth of the frontal bone; instead, it is a separate bone that fuses with the frontal bone during development. Giraffes' horns are covered with skin rather than being formed by a sheath of keratin.

Caribou (cervids) have antlers rather than horns. Antlers are confined to males of most species of cervids, but female reindeer and caribou have antlers. (Reindeer and caribou are the only species of cervids that form herds containing both males and females.) Unlike horns, antlers are branched, consist of only bone, and usually are shed annually. As they grow, antlers are covered by a layer of highly vascularized skin (the velvet). When the antler is mature, blood flow to the velvet is cut off, and the skin sloughs off to reveal the bony antler. (In Asia, velvet is reputed to impart virility. New Zealand has a major deer-farming industry, and dried velvet sells for thousands of dollars per kilo.)

Although the cranial appendages of ruminant artiodactyls appear rather similar, they are not homologous in their mode of growth. They appear to have evolved independently within different ruminant lineages. Modern rhinoceroses are unlike other horned ungulates because their horns are formed entirely of keratin (the epidermal protein that forms hair and fingernails) and are found in both males and females (although some extinct rhinos had bony horns that were present in males only). Rhino horns are single (not paired) structures that form on the midline of the nose region. In contrast, the horns of extant ruminants are paired and form above the eyes, although some fossil artiodactyls had single horns on the nose or on the back of the head in addition to the paired horns over the eyes.

The Evolution of Horns and Antlers

The evolution of ruminant horns appears to be tied in with their ecology and social behavior. The evolution of ruminant horns can be understood in the context of changing Cenozoic habitats, which in turn led to changes in diet, body size, behavior, and morphology.

The ancestors of horned ruminants first appeared in the fossil record in the late Oligocene epoch, when they were small, hornless animals with teeth, indicating a diet of fruit and young leaves (i.e., a type I diet). They would have appeared similar to present-day chevrotains (unspecialized ruminants) and duikers (antelope), which inhabit the tropical forests of Asia and Africa. By the early Miocene epoch, the Eurasian woodlands where these animals lived had become more seasonal and more open in structure. These changes in vegetation changed the availability of food resources. Ruminants responded by becoming somewhat larger (goat size rather than rabbit size) and evolving teeth more capable of eating fibrous vegetation such as mature leaves (a type II diet).

This new diet enabled the ruminants to adopt a new type of social behavior. The social behavior of the small, early ruminants was probably like that of the chevrotains: solitary or monogamous, with an individual home range. Mature leaves are much more abundant and concentrated in space than are new leaves and fruits or berries. The new, larger ruminants that ate mature leaves could find their food in a smaller home range. With food concentrated in a smaller area, leaf-eating ungulates could become territorial, defending a territory large enough for several animals. This ecological strategy would not have been practical for smaller ungulates because home ranges large enough to support several animals would be too big to patrol effectively.

Thus territorial ruminants moved from a monogamous type of mating system, with only a single female for each male, to a polygynous one with many potential mates for each male. In this situation, some males could have greater reproductive success (i.e., mate with more females) than other males. Intense male-male competition promoted the evolution of horns or equivalent structures used for social displays. (This is because horns are used for ritual display and stylized combat, and they may actually reduce the incidence of injury during male-male interactions.)

Testing the Hypothesis

That explanation is, of course, historical speculation; we will never know for certain what actually happened. However, three lines of evidence are consistent with this interpretation—the timing of the evolution of horns in African and Eurasian ungulates, sexual dimorphism in the occurrence of horns and antlers, and the absence of horns among North American ungulates.

When Horns Appeared Horns or their equivalents appeared in different families of ruminants in Africa and Eurasia at about the same time in the early Miocene. The evolution of horns was correlated with a change in habitat (seen from the plant fossil record) and an increase in body size (seen in the fossil record of the animals). Among present-day ruminants, the smaller, solitary forms are hornless. Larger forms, in which the females live in groups and the males are territorial, have horns. African and Eurasian ungulates crossed this size threshold as the climate and habitat changed in the Miocene. Thus the evolution of horns correlates with a likely change in behavior, from solitary and monogamous to group forming and polygamous. What we see today as an ecomorphological shift along a habitat gradient (forest to woodland) may be what happened in evolution over a temporal shift of changing habitats in the Miocene.

Sexual Dimorphism Horns appear to have evolved initially only in males. Fossils show that early members of all horned ruminant lineages included both horned and hornless individuals. Presumably, those with horns were males and those without horns were females. This sexual dimorphism suggests that horns were initially used in male-male interactions. If horns had originally been used for some activity that both sexes engage in, such as defense against predators, both sexes should have evolved them simultaneously.

Large, grazing ruminants (type III and IV species such as wildebeest and buffalo) now have home ranges too large to be defended as territories. These animals are no longer sexually dimorphic, although, instead of the males losing their horns, the females have evolved horns as well. The females use these horns in competition with the males for feeding resources, now that they live with males year-round.

North American Ungulates The absence of horns among North American ungulates such as camels and horses may be partially explained by the different pattern of vegetational change on that continent. Grasslands rapidly replaced forests in North America, without a persistent stage of open forests. In grassland habitats, camels and horses were unlikely to have passed through an evolutionary stage in which territorial defense would have been a worthwhile ecological strategy. Perhaps this is why they never evolved the type of sexual dimorphism seen in antelope. Additionally, hindgut fermenters like horses are less efficient feeders than ruminants and so may always have required a home range area that is too large to defend as a territory.

Horses and camelids (camels and llamas) have a different type of social system, called harems. Both these types of ungulates form permanent associations of females and their young, usually accompanied by only a single male. Males that are not part of a harem association form bachelor herds. In this social system, the male defends a group of females from other males rather than defending a piece of real estate. The term *harem* conjures up visions of a male controlling and dominating a group of females, and that was the original interpretation applied by behavioral ecologists. However, more careful observation (and more female ecologists who have brought new perspectives to the field) has revealed that it is the bond between the females that is the basis of the harem. The females then allow a male to join their social grouping because he keeps other males from constantly pestering them and interfering with the time they can spend feeding.

23.6 Primate Societies

The phylogenetic relationship of humans to other primates has led some biologists to assume that primates are the animals that should have the most elaborate social systems and that studying the social systems of primates will provide information about the evolution of human behavior. Both assumptions are controversial. A growing base of information indicates not only that complex social systems exist among many kinds of vertebrates other than primates but also that interpreting primate behavior in the context of human evolution is fraught with difficulty and must be cautiously approached. Nonetheless, some primates do have elaborate and complex social systems, and more long-term research has focused on the social systems of primates than on any other vertebrates (Table 23–3). A review of primate behavior emphasizes its variety and complexity and sets the stage for considering the evolution of humans in the next chapter.

The approximately 200 species of primates are ecologically diverse. They live in habitats ranging from lowland tropical rain forests, to semideserts, to northern areas that have cold, snowy winters. Some species are entirely arboreal, whereas others spend most of their time on the ground. Many are generalist omnivores that eat fruit, flowers, seeds, leaves, bulbs, insects, bird eggs, and small vertebrates. However, many of the colobus monkeys (*Colobus*) and howler monkeys (*Alouatta*) are specialized folivores (leaf eaters) with digestive tracts in which bacteria and protozoans ferment cellulose, and some of the small monkeys are insectivores.

Table 23–3 Social organization of extant primates

Taxon	Social Organization
Prosimii	
Lemuriformes	
Lemuroidea (aye-ayes, lemurs, indris, and sifakas)	Largely solitary or monogamous pairs
Lorisoidea (bushbabies, lorises, pottos, and angwantibos)	Largely solitary
Tarsiiformes (tarsiers)	Solitary or monogamous pairs
Anthropoidea	
Platyrrhini (New World monkeys)	Monogamous pairs to large groups
Cebidae (marmosets, tamarins, capuchins, squirrel monkeys, and owl monkeys)	
Atelidae (howler monkeys, spider monkeys, sakis, ukaris, and titis)	
Catarrhini (Old World monkeys and apes)	Small to large groups
Cercopithecidae (vervet monkeys, guenons, mangabeys, macaques, baboons)	
Colobinae (colobus monkeys and langurs)	
Hominidae (apes and humans)	
Gibbons	Monogamous pairs
Orangutans	Solitary
Gorillas	Small groups with a variable number of resident males
Chimpanzees and bonobos	Closed social network containing several breeding males and females
Humans	Closed social network containing several breeding males and females

Social Systems of Primates

The social systems of primates can be classified by the movement of males and females between groups (**Table 23–4**). Four social systems can be defined on this basis:

- **Female transfer systems**—In species with this type of social organization, most females move away from the group in which they were born to join another social group. Because of this migration of females among groups, the females in a group are not closely related to each other. In contrast, males often remain with their natal groups, and associations of male kin may be important elements of the social behaviors of these species of primates. Male chimpanzees, for example, cooperate in defending their territories from invasion by neighboring males. Most species of primates with female transfer systems live in relatively small social groups.
- **Male transfer systems**—Most females of these species spend their entire lives in the group in which they were born. Social relations among the females in a group are complex and based on kinship. Males of these species emigrate from their natal group as adolescents, and they may continue to move among groups as adults. In some of these species, a single male lives with a group of females until he

is displaced by a new male. In other species, several males may be part of the group and maintain an unstable dominance hierarchy among themselves. Cooperation by several adult males may allow them to resist challenges from younger, stronger males that they would not be able to subdue if they acted as individuals. Group size is usually larger for male transfer species than for female transfer species.
- **Monogamous species**—A single male and female form a pair, sometimes accompanied by juvenile offspring. These species of primates show little sexual dimorphism, the sexes share parental care and territorial defense, and the offspring are expelled from the parents' territory during adolescence.
- **Solitary species**—These species live singly or as females with their infants and juvenile offspring. Males of some species of prosimians maintain territories that include the home ranges of several females and exclude other males from their territories, whereas male orangutans do not defend their territories. Instead, they repulse other males when a female within the male's home range comes into estrus.

Ecology and Primate Social Systems

Three ecological factors appear to be particularly important in shaping the social systems of primates, as they are for other vertebrates:

Table 23–4 Characteristics of the social systems of primates

System	Group Size	Number of Males in Group	Male Behavior	Example
Female transfer	Small	One or many	Territorial, harems, sometimes male kinship groups	Chimpanzees, gorillas, howler monkeys, hamadryas baboons, colobus monkeys, some langurs
Male transfer	Large	One or several	Male hierarchy, whole group (males and females) may exclude conspecifics from food sources	Most cercopithecines: yellow baboons, mangabeys, macaques, guenon monkeys
Monogamous	Male and female, plus juvenile offspring	One	Both sexes participate in territorial defense and parental care	Gibbons, marmosets, some tamarins, indris, titis
Solitary	Individual, or female plus juvenile male offspring	—	Range of male overlaps ranges of several females	Bushbabies, tarsiers, lorises, orangutans

- **Distribution of food resources**—The defensibility of food resources appears to determine whether individuals will benefit from not attempting to defend resources, defending individual territories, or forming long-term relationships with other individuals and jointly defending resources.
- **Group size**—The distribution of food in time and in space may determine how large a group can be and whether the group can remain stable or must break up into smaller groups when food is scarce.
- **Predation**—The risk of predation may determine whether individuals can travel alone or require the protection of a group, whether the benefit of the additional protection provided by a large versus a small group outweighs the added competition among individuals in a large group, and whether the presence of males is needed to protect the young.

Behavioral Interactions

Life within a group of primates is a balance between competition and cooperation (**Figure 23–9**). Competition is manifested by aggression. Some aggression—for example, the defense of food, sleeping sites, or mates—is closely linked to resources. Other types of aggression involve establishing and maintaining dominance hierarchies, which can be an indirect form of resource competition if high-ranking individuals have preferential access to resources.

Cooperation, too, is diverse. Grooming behavior in which one individual picks through the hair of another, removing ectoparasites and cleaning wounds, is the most common form of cooperation. Other types of cooperation include sharing food or feeding sites, collective defense against predators, collective defense of a territory or a resource within a home range, and formation of alliances between individuals. Two-way, three-way, and even more complex alliances that function during competition within a group are common among primates.

Kinship and the concept of inclusive fitness play important roles in the interpretation of primate social behavior. Because fitness is nearly impossible to demonstrate in wild populations, behaviorists normally search for effects that are likely to be correlated with fitness, such as access to females (for males), interbirth interval (for females), or the probability that offspring will survive to reproductive age. Behaviors that increase these measures are assumed to increase fitness. The behaviors may directly benefit the individual displaying the behavior (personal fitness), or they may be costly to the personal fitness of the individual but sufficiently beneficial to its close relatives to offset the cost to the individual (inclusive fitness).

Social Relationships Among Primates

Four general types of relationships among individuals have been described in the social behavior of primates: adult-juvenile, female kinship, male-male, and male-female.

Adult-Juvenile Associations Primates are born in a relatively helpless state compared to many mammals, and

(a) **(b)**

Figure 23–9 Social behaviors of savanna baboons (*Papio cynocephalus*). (a) A male friend grooming a female baboon in estrus. (b) Aggression among male baboons.

they depend on adults for unusually long periods. The relationship of a mother to her infant is variable within a species—some mothers are protective, whereas others are permissive. Permissive mothers often wean their offspring earlier than protective mothers and may have shorter intervals between the births of successive offspring, although this relationship has not been observed in all species. The offspring of permissive mothers may suffer higher rates of mortality than the offspring of protective mothers, and the incompetence of some inexperienced mothers appears to lead to high mortality among firstborn offspring.

Older siblings often participate in grooming and carrying an infant, but they may also assault, pinch, and bite the infant while it is being fed or groomed by the mother. Allomaternal behavior provided by an adult female who is not the mother includes cuddling, grooming, carrying, and protecting an infant. Several factors seem to influence allomaternal behavior: young infants are preferred to older ones, infants of high-ranking mothers receive more attention and less abuse than infants of low-ranking mothers, and siblings may participate more than unrelated females in allomaternal behavior. Males of the monogamous New World primates participate extensively in caring for infants, carrying them for much of the day and sharing food with them, whereas the relationships of males of Old World primates with infants are more often characterized by proximity and friendly contact than by care.

Female Kinship Bonds The females of some species of semiterrestrial Old World primates live in groups that include several males and females. This social organization is typical of savanna baboons (*Papio*

cynocephalus), several species of macaques (*Macaca*), and vervet monkeys (*Cercopithecus aethiops*). Females of these species remain for their entire lives in the troops in which they were born. Female kinship bonds and kin selection play important roles in the behaviors of females in male transfer systems because the females in a group are related to one another.

The females form a dominance hierarchy and compete for positions in the hierarchy. Related females within a group are referred to as **matrilineages**. Females consistently support their female relatives during encounters with members of other matrilineages. The supportive relationship among females within a matrilineage is an important element of a group's social structure. For example, when their female kin are nearby, young animals can dominate older and larger opponents from subordinate matrilineages. Furthermore, high-ranking females retain their position in the hierarchy even when age or injury reduces their fighting ability. An adolescent female savanna baboon normally attains a rank in the group just below that of her mother, and this inheritance of status provides stability in the dominance relationships among the females of a group. However, the social rank of the matrilineage is not fixed: low-ranking female savanna baboons, with their female kin, may challenge higher-ranking individuals; if they are successful, their entire matrilineage may rise in rank within the group.

Female kinship bonds are clearly important elements of the social structure of male transfer systems, but the exact contribution of the long-term relationships among females to the fitness of individual females is not clear. In some species high-ranking females reproduce earlier than low-ranking individuals and have shorter interbirth intervals and higher infant survival, but those correlations are not present in all the species that have been studied. Furthermore, female kinship bonds are manifested weakly, if at all, in female transfer systems, which include most species of apes and many species of monkeys.

Male-Male Alliances Male primates in male transfer species often form dominance hierarchies, but male rank depends mainly on individual attributes and is therefore less stable than female dominance systems based on matrilineage. Young adult males, which are usually recent immigrants from another group, have the greatest fighting ability and usually achieve the highest rank. Some older males achieve stable alliances with each other that enable them to overpower younger and stronger rivals in competition for opportunities to court receptive females. These males probably attain greater mating success by engaging in these reciprocal alliances than they would achieve on the basis of their individual ranks in the hierarchy.

Cooperative relationships among males are most common in female transfer systems because the males of these species remain in their natal group. As a result, kin relationships exist among the males in a group. In red colobus monkeys (*Colobus badius*), for example, only males born in the group appear to be accepted by the adult male subgroup, and the membership of this subgroup can remain stable for years.

Adult males spend much of their time in close proximity to each other and cooperate in aggression against males of a neighboring group. Within a group, male chimpanzees spend more time together than do females, and they engage in a variety of cooperative behaviors, including greeting, grooming, and sharing meat. However, this apparent cooperation is simply a way of cementing relationships that are based on intense and sometimes violent competition over females.

Male-Female Friendships Interactions between individual male and female baboons are not randomly distributed among members of the group. Instead, each female has one or two particular males called friends. Friends spend much time near each other and groom each other often. These friendships last for months or years, including periods when the female is not sexually receptive because she is pregnant or nursing a baby. Male friends are solicitous of the welfare of their female friends and of their infants. Similar male-female friendships have been described in mountain gorillas (*Gorilla gorilla beringei*), gelada baboons (*Theropithecus gelada*), hamadryas

baboons (*Papio hamadryas*), rhesus macaques (*Macaca mulatta*), and Japanese macaques (*Macaca fuscata*).

Friendship with a male protects a female and her offspring from predators and from other members of the group. The advantage of friendship for a male is less apparent. If the female's offspring had been sired by the male, protecting the offspring would contribute to the male's fitness. However, only half the friendships between male and infant savanna baboons involved relationships in which the male was the likely sire of the infant. The other friendships involved males that had never been seen mating with the mother of the infant. Perhaps males who participate in a friendship with a female have a chance of mating with that female many months later, when she is again receptive.

Sociality and Survival

The proportion of infants that survive the first year of life is higher for females that engage in extensive social behavior than for those that are not as socially integrated in the group (**Figure 23–10**). The influence of sociality on infant survival can be distinguished statistically from the effect of the dominance status of a female. Although it is true, as would be expected, that high-ranking females have high sociality scores, the correlation between sociality and infant survival remains when the effect of dominance rank is removed in the statistical analysis.

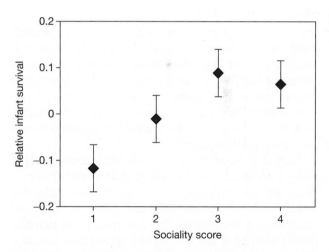

Figure 23–10 Sociality and infant survival. Relative infant survival is an index of how well the infants of an individual female baboon survive their first year of life compared to the average survival of all infants in the troop during that year. (Average survival varies substantially from year to year.) High sociality scores are females that are more socially integrated than the median female, and low scores are females that are less integrated than the median.

Why do social interactions among female baboons contribute to their infants' survival? Studies of humans may provide insights because they indicate that social support moderates the effects of stress. Women who have extensive social networks give birth to heavier babies and have lower incidences of disease, accidents, and mental disorders. In contrast, feelings of loneliness are correlated with higher rates of illness and death.

One of the mechanisms by which sociality may enhance fitness is physiological. Supportive social interactions among humans stimulate the release of neurohormones called endorphins, which produce a sense of relaxation, and similar relationships may exist among nonhuman primates. Laboratory studies of other nonhuman vertebrates, including primates, have shown that the presence of familiar individuals of the same species lowers heart rate and cortisol levels (indicating reduced stress), delays reproductive senescence, and lengthens the life span. Thus, social interactions may contribute to a benign environment for infant baboons.

Behavioral mechanisms may also be responsible for some of the effects of sociality on the survival of infant baboons: social interactions among adult baboons increase tolerance from high-ranking individuals and may provide protection from harassment by females from other matrilineages as well as access to valuable resources such as desirable feeding and drinking sites.

How Do Primates Perceive Their Social Structure?

The preceding summary of primate social structures represents the results of tens of thousands of hours of observations of individual animals over many years. Statistical analyses of interactions between individuals—grooming sessions, aggression, defense—reveal correlations associated with factors such as age, personality, kinship, and rank. Do the animals themselves recognize those relationships?

That is a fascinating but difficult question, especially with studies of free-ranging animals. Observations are accumulating that suggest that primates probably do recognize different kinds of relationships among individuals. For example, when juvenile rhesus macaques are threatened by another monkey, they scream to solicit assistance from other individuals who are out of sight. The kind of scream they give varies depending on the intensity of the interaction (threat or actual attack) and the dominance rank and kinship of their opponent. Furthermore, a mother baboon interprets the screams of her juvenile and responds more or less vigorously depending on the

nature of the threat her offspring faces. When tape-recorded screams were played back to the mothers, the mothers responded most strongly to screams that were given during an attack by a higher-ranking opponent, less strongly to screams that were given in response to interactions with lower-ranking opponents, and least strongly to screams that were given in interactions with relatives. Baboons were more interested in playbacks that artificially reversed the dominance status of two individuals—that is, when a higher-ranking animal sounded as if it was giving a subordinate response to a lower-ranking individual.

In experiments with free-ranging vervet monkeys, the screams of a juvenile were played back to three females, one of them the mother of the juvenile. The mother responded more strongly to the screams than did the other two monkeys, as might be expected if females can recognize the voices of their own offspring. However, the other two monkeys responded to the screams by looking toward the mother, suggesting that they were able not only to associate the screams with a particular juvenile but also to associate that juvenile with its mother.

Observations of redirected aggression also suggest that some primates classify other members of a group by matrilineage and friendships. When a baboon or macaque has been attacked and routed by a higher-ranking opponent, the victim frequently attacks a bystander who took no part in the original interaction. This behavior is known as redirected aggression—and the targets of redirected aggression are relatives or friends of the original opponent more frequently than would be expected by chance. Vervet monkeys show still more complex forms of redirected aggression: They are more likely to behave aggressively toward an individual when they have recently fought with one of that individual's close kin. Furthermore, an adult vervet is more likely to threaten a particular animal if that animal's kin and one of its own kin fought earlier that same day. This sort of feud is seen only among adult vervets, suggesting that it takes time for young animals to learn the complexities of the social relationships of a group.

These sorts of observations suggest that adult primates have a complex and detailed recognition of the genetic and social relationships of other individuals in their group. Furthermore, they may be able to recognize more abstract categories—such as *relative* versus *nonrelative, close relative* versus *distant relative,* or *strong friendship bond* versus *weak friendship bond*—that share similar characteristics independent of the particular individuals involved.

Summary

Sociality, the state of living in structured groups, is a prominent characteristic of the behavior of many species of mammals. However, social behavior is only one aspect of the biology of a species, and social behaviors coexist with other aspects of behavior and ecology, including finding food and escaping from predators. The size and geography of an animal's home range are related to the distribution and abundance of resources, the body size of the animal, and its feeding habits. Large species have larger home ranges than do small species, and at any body size carnivores have the largest home ranges and herbivores the smallest.

Social systems are related to the distribution of food resources and to the opportunities for an individual (usually a male) to increase access to mates by controlling access to resources. Dietary habits, the structural habitat in which a species lives, and its means of avoiding predators are closely linked to body size and mating systems. These aspects of biology form a web of interactions, each influencing the others in complex ways.

The social systems of primates have been the subjects of field studies, and more information about social behavior under field conditions is available for primates than for other mammals. The social systems of primates are complex but not unique among mammals. Some primates are solitary or monogamous; others live in groups and display behaviors that suggest not just recognition of other individuals but also recognition of the genetic and social relationships among other individuals. Studies of other kinds of mammals will probably reveal similar phenomena. Understanding the behavior of mammals requires a broad understanding of their ecology and evolutionary histories.

Discussion Questions

1. The chapter cited a few advantages and disadvantages of sociality but did not provide a complete list. Begin with the examples in this chapter and compile a more complete list of the pros and cons of group living.
2. How could you determine whether wolves form packs in order to prey on large animals or wolves have to prey on large animals because they live in groups?
3. Why don't zebras fit into the relationship of body size, diet, and social system seen among antelope?
4. Many hadrosaur dinosaurs had elaborate structures on their heads that may have been social signals like the horns and antlers of mammals. What predictions can you make about the social behavior of hadrosaurs, and how could you use the fossil record to test those predictions?
5. Why is alloparental care common among male-transfer species of primates but not among female-transfer species?

Additional Information

Byrd, D. E., et al. 2011. Evolution of ruminant headgear: A review. *Proceedings of the Royal Society B* 278:2857–2865.

Caro, T. M. 1994. *Cheetahs of the Serengeti Plains*. Chicago: University of Chicago Press.

Cheney, D., and R. Seyforth. 1990. *How Monkeys See the World*. Chicago: University of Chicago Press.

Clutton-Brock, T. 2009. Structure and function in mammalian societies. *Philosophical Transactions of the Royal Society B* 364:3229–3942.

Davis, E. B., et al. 2011. Evolution of ruminant headgear: A review. *Proceedings of the Royal Society B* 278:2857–2865.

de Jong, H. H., et al. 2011. Resource partitioning among African savanna herbivores in North Cameroon: The importance of diet composition, food quality and body mass. *Journal of Tropical Ecology* 27:503–513.

Gilbert, C., et al. 2012. Private heat for public warmth: How huddling shapes individual thermogenic responses of rabbit pups. *PLoS ONE* 7(3): e33553. doi:10.1371/journal.pone.0033553.

Hayward, M. W., et al. 2011. Do lions *Panthera leo* actively select prey or do prey preferences simply reflect chance responses via evolutionary adaptations to optimal foraging? *PLoS ONE* 6(9):e23607. doi:10.1371/journal.pone.0023607.

Henden, J-A., et al. 2009. Strength of asymmetric competition between predators in food webs ruled by fluctuating prey: The case of foxes in tundra. *Oikos* 119:27–34.

Höner, O. P., et al. 2012. The impact of a pathogenic bacterium on a social carnivore population. *Journal of Animal Ecology* 81:36–46.

Hopcraft, J. C., et al. 2012. Body size and the division of niche space: Food and predation differentially shape the distribution of Serengeti grazers. *Journal of Animal Ecology* 81:201–213.

Huchard, E., et al. 2010. More than friends? Behavioural and genetic aspects of heterosexual associations in wild chacma baboons. *Behavioral Ecology and Sociobiology* 64:769–781.

Jarman, P. J. 1974. The social organization of antelope in relation to their ecology. *Behaviour* 58:215–267.

Mosser, A., and C. Packer. 2009. Group territoriality and the benefits of sociality in the African lion, *Panthera leo. Animal Behaviour* 78:359–370.

Rodnikova, A., et al. 2011. Red fox takeover of arctic fox breeding den: An observation from Yamal Peninsula, Russia. *Polar Biology* 34:1609–1614.

Schultz, S., et al. 2011. Stepwise evolution of stable sociality in primates. *Nature* 479:219–224.

Silk, J. B. 2007. The adaptive value of sociality in mammalian groups. *Philosophical Transactions of the Royal Society B* 362:539–559.

Silk, J. B., et al. 2010. Strong and consistent social bonds enhance the longevity of female baboons. *Current Biology* 20:1359–1361.

VanderWaal, K. L., et al. 2009. Optimal group size, dispersal decisions and postdispersal relationships in female African lions. *Animal Behaviour* 77:949–954.

Yeakel, J. D., et al. 2009. Cooperation and individuality among man-eating lions. *Proceedings of the National Academy of Science* 106:19040–19043.

Websites

Baboons
http://www.psych.upenn.edu/~seyfarth/Baboon%20research/

Chimpanzees
http://www-bcf.usc.edu/~stanford/chimphunt.html

Hyenas
http://www.hyaenidae.org/the-hyaenidae/spotted-hyena-crocuta-crocuta/crocuta-social-behavior.html

Lions
http://www.cbs.umn.edu/lionresearch/about/socialbehavior.shtml

Vervet monkeys
http://pin.primate.wisc.edu/factsheets/entry/vervet/behav

Wolves
http://www.pbs.org/wgbh/nova/wolves/

CHAPTER 24

Primate Evolution and the Emergence of Humans

Primates have been a moderately successful group for most of the Cenozoic era, although since the end of the Eocene epoch they have been primarily confined to tropical latitudes (with the obvious exception of humans). Primates include not only the anthropoids, the group of apes and monkeys to which humans belong, but also the prosimians, animals such as bush babies and lemurs. Genomic techniques show that chimpanzees are the closest extant relatives of humans, and both genetics and the fossil record indicate that the separation of humans from the African great apes occurred less than 10 million years ago. Fossils of the genus *Australopithecus,* the sister taxon to our own genus (*Homo*), clearly show that bipedal walking arose before the acquisition of a large brain. A great diversity of new fossils shows that the picture of early human evolution was much more complex and diverse than previously believed. *Homo* and *Australopithecus* lived together in Africa for more than a million years, and the extinction of the australopithecines was probably related to climatic changes rather than to competition with our ancestors. Our current situation, in which we—*Homo sapiens*—are the only species of hominin on Earth, is new; as recently as 30,000 years ago, we shared the planet with *Homo erectus*, *Homo floresiensis*, and *Homo neanderthalensis*.

24.1 Primate Origins and Diversification

Humans share many biological traits with the animals that are variously called apes, monkeys, and prosimians: we are all members of the order Primates (Latin *prima* = first). The first primates were arboreal forms living in early Cenozoic forests. Humans are late-appearing primates, and complex social systems such as those discussed in Chapter 23 are an ancestral feature of the primate lineage.

Characteristics of Primates

Features that are typical of primates are listed in **Table 24–1**. Note that many of these characters are not unique to primates; for example, many mammals retain the clavicle, pigs have bunodont molars similar to those of primates, and many ungulates and kangaroos have only a single young per pregnancy.

Most of these traits have been attributed to an arboreal life. All the basic modifications of the limbs can be seen as contributing to arboreal locomotion, as can the stereoscopic depth perception that results from binocular vision, and the enlarged brain that coordinates visual perception and locomotory response. The bunodont teeth imply an originally frugivorous diet. Primates today

Table 24.1 Characteristics of primates

Retention of the clavicle (which is reduced or lost in many mammalian lineages) as a prominent element of the pectoral girdle

A shoulder joint allowing a high degree of limb movement in all directions and an elbow joint permitting rotation of the forearm

General retention of five functional digits on the fore- and hindlimbs; enhanced mobility of the digits, especially the thumb and big toes, which are usually opposable to the other digits

Claws modified into flattened and compressed nails

Sensitive tactile pads developed on the distal ends of the digits

A trend toward a reduced snout and olfactory apparatus, with most of the skull posterior to the orbits

Reduction in the number of teeth compared to the generalized mammalian condition but with the retention of simple bunodont molar cusp patterns

A complex visual apparatus with high acuity, and a trend toward development of forward-directed binocular eyes and tricolor perception

Large brain relative to body size, in which the cerebral cortex is particularly enlarged

A trend toward derived fetal nourishment mechanisms

Only two mammary glands (some exceptions)

Typically only one young per pregnancy associated with prolonged infancy and pre-adulthood

A trend toward holding the trunk of the body upright, leading to facultative bipedalism

---

are mainly herbivorous, with smaller forms including insects in the diet and larger forms including leaves. Most primates are arboreal, but some have become secondarily terrestrial (baboons, for example), and humans are the most terrestrial of all. Even so, many of the traits that are most distinctively human are derived from earlier arboreal specializations.

Arboreality cannot be the entire basis for these primate features because they are not shared with many other arboreal mammals. For example, squirrels lack big brains and an opposable thumb, and they retain claws. Squirrels are rather generalized, all-purpose climbers, however, and skeletons of carpolestids, animals belonging to the primate-related plesiadapiforms of the early Cenozoic, reveal specialized climbers with long fingers and an opposable thumb that were probably specialized for feeding at the very ends of branches (**Figure 24–1**). True primates may have been derived from a similarly specialized animal.

Evolutionary Trends and Diversity in Primates

Figure 24–2 is a simplified representation of the interrelationships among primates, and Table 23–3 presented a traditional classification of extant primates.

Plesiadapiforms The first primatelike mammals (plesiadapiforms; Greek *plesi* = near, Latin *adapi* = rabbit and *form* = shape) appeared in the earliest Cenozoic in the Northern Hemisphere and persisted until the end of the Eocene epoch. Plesiadapiforms were rather squirrel-like and ranged from chipmunk size to marmot size; different lineages had teeth indicating that they had different diets—omnivorous, gum eating, and insectivorous (**Figure 24–3** on page 584). The decline in plesiadapiform diversity coincided with the evolution and radiation of rodents in the late Paleocene, perhaps reflecting competition.

Plesiadapiforms shared some derived features of the teeth and skeleton with true primates, but they differed in having smaller brains and longer snouts, in lacking a postorbital bar and an opposable hallux (big toe), and in the presence of rodentlike incisors in some forms. Plesiadapiforms also retained claws. In contrast,

Figure 24–1 Reconstruction of the plesiadapiform *Carpolestes simpsoni.*

© 2002 AAAS

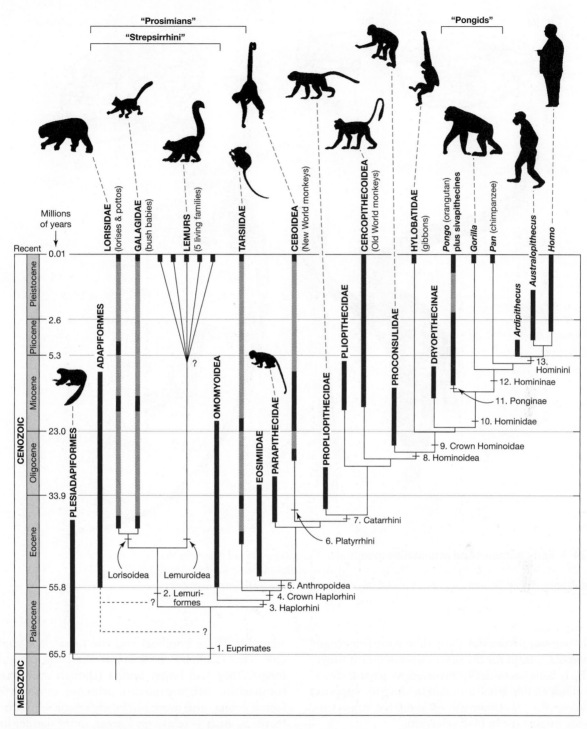

Figure 24–2 Phylogenetic relationships of the primates. This diagram depicts the probable relationships among the major groups of primates. The black lines show interrelationships only; they do not indicate times of divergence or the unrecorded presence of taxa in the fossil record. Lightly shaded bars indicate ranges of time when the taxon is believed to be present, because it is known from earlier times, but is not recorded in the fossil record during this interval. The numbers at the branch points indicate derived characters that distinguish the lineages—see the Appendix for a list of these characters. Question marks indicate uncertainties about relationships; quotation marks indicate paraphyletic groupings.

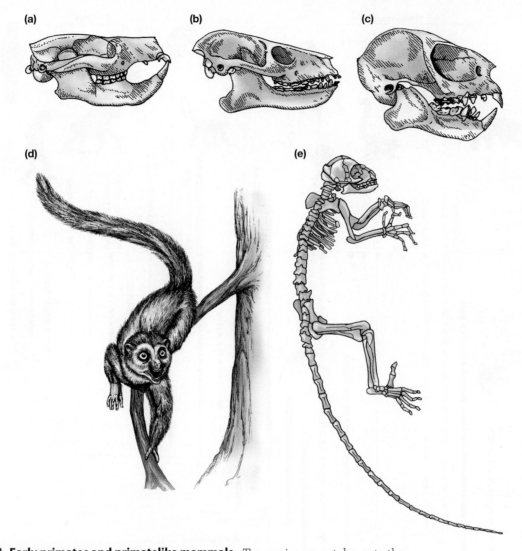

Figure 24–3 Early primates and primatelike mammals. The species are not drawn to the same scale. (a) Skull of the plesiadapiform *Plesiadapis*. (b) Skull of the adapiform *Notharctus*. (c) Skull of the omomyoid *Tetonius*. (d) Skeleton and life restoration of the adapiform *Darwinius*.

all true primates have flat nails (like your fingernails and toenails) except for the marmosets of South America, which have secondarily reverted to having claws. While claws enable small animals to cling for support, nails allow the development of sensitive fingertips, which are important in food selection.

Prosimians The first true primates, or Euprimates (Greek *eu* = good), are known from the latest Paleocene or earliest Eocene of North America, Eurasia, and northern Africa. These early primates belong to a group that has traditionally been called prosimians (Greek *pro* = before; Latin *simi* = ape). Most Eocene prosimians, including the lemurlike adapiforms (or

adapoids, three families) and the tarsierlike omomyoids (two families), were larger than the plesiadapiforms. They had larger brains (though still small in comparison with any modern primate), more forward-facing orbits, and more obviously specialized arboreal features, such as relatively longer, more slender limbs.

The extant prosimians include the bush babies of Africa; the lemurs of Madagascar; and the lorises, pottos, and tarsiers of Southeast Asia (**Figure 24–4**). Both molecular characters and a sparse fossil record support a deep split between the lorisoids and lemuriformes that occurred at least 40 million years ago. Prosimians are in general small, nocturnal, long snouted, and relatively small brained compared with the more derived

Figure 24–4 Diversity of living prosimians. Lemurs: (a) ring-tailed lemur, *Lemur catta*; (b) indri, *Indri indri*; (c) aye-aye, *Daubentonia madagascariensis*. Loris: (d) Demidoff's bush baby, *Galagoides demidovii*. Potto: (e) *Perodicticus potto*. Tarsier: (f) *Tarsius spectrum*.

anthropoids (Greek *anthrops* = man) or apes and monkeys. Prosimians are mostly generalized omnivores, with a few specialized herbivores.

The group called prosimians is paraphyletic because several derived features (for example, a short snout with a dry nose rather than a wet, doglike one) indicate that tarsiers are more closely related to the anthropoids than are the other prosimians. Molecular data support the association of tarsiers with anthropoids, and morphological characters cluster the extinct omomyoids with this lineage. An alternative division of the primates is into the Strepsirhini (lemuroids and lorisoids; Greek *strepsi* = twisted and *rhin* = nose) and the Haplorhini (tarsiers and anthropoids; Greek *haplo* = simple). Living strepsirrhines form a monophyletic grouping, the Lemuriformes, but there is debate about whether the adapiforms fit within this clade or rather form the sister group to all other primates.

In 2009 a fossil adapiform, *Darwinius masillae*, attracted media attention. This animal, nicknamed "Ida," is known from a beautifully preserved juvenile specimen from the early middle Eocene (47 million years ago) locality of Messel in Germany. Messel is a World Heritage Site that preserves in wonderful detail a fauna representing a tropical-like rain forest. This specimen is the most complete fossil primate skeleton known and also preserves nonbony elements such as the gut contents. The researchers who described *Darwinius* considered that new aspects of its anatomy warranted reevaluation of primate phylogeny—placing adapiforms as basal haplorhines (rather than as strepsirrhines or basal primates). This phylogenetic rearrangement places *Darwinius* close to the origin of anthropoids, and it was widely hailed as a new "missing link." Other researchers have disputed this claim, relegating *Darwinius* to a side branch in terms of human origins. Opinions remain divided, but whatever the systematic position of *Darwinius*, it is not especially close to human origins, despite the unfortunate claims published in popular media.

The diversification of the early Cenozoic primates throughout the Northern Hemisphere reflects the tropical-like climates of the higher latitudes at that time. With the late Eocene climatic cooling in the

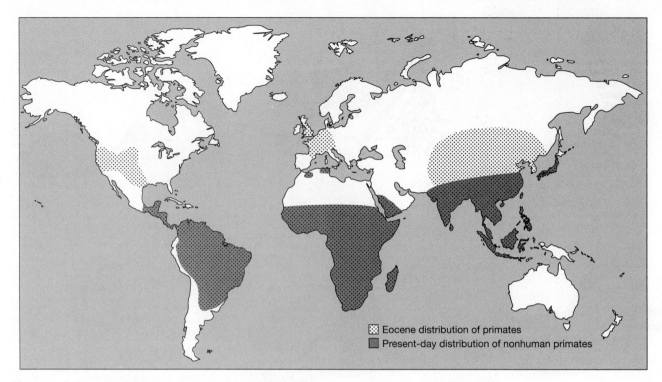

Eocene distribution of primates

Present-day distribution of nonhuman primates

Figure 24–5 Distribution map of primates in Recent and Eocene times. This map shows the locations of fossil sites that contain Eocene primates. As these primates evidently dispersed across the Northern Hemisphere, they must actually have ranged into much higher latitudes at this time.

temperate latitudes, these early primates declined and were virtually extinct by the end of the Eocene. Even today, almost all nonhuman primates are restricted to the tropics (**Figure 24–5**) as they have been for most of the later Cenozoic, except for some excursions of apes into northern portions of Eurasia during the warming period in the late Miocene.

Present-day prosimians are a moderately diverse Old World tropical radiation, first known from the late middle Eocene of Africa (about 40 million years ago). The lemurs of the island of Madagascar have undergone an evolutionary diversification into five different families. These include some large (raccoon size) diurnal specialized herbivores, such as the rather koala-like indri (*Indri indri*) and the peculiar aye-aye (*Daubentonia madagascariensis*), which uses its specialized middle finger, which is bony and devoid of soft tissues, to probe grubs out of tree bark.

Perhaps as recently as 2000 years ago lemurs were far more diverse and included giant arboreal species resembling bear-size koalas and sloths. It seems that the lemurs, in isolation from the rest of the world, evolved their own version of primate diversity, including parallels to the anthropoid apes. Unfortunately, much of this diversity is now gone, probably because of the immigration of humans to the island from Asia a few thousand years

ago. Accounts from early European explorers suggest that some of the giant lemurs may have been alive as recently as 500 years ago. The decline of lemurs is continuing, and many extant species are threatened with extinction as a result of continued destruction of forests on Madagascar.

Anthropoids Modern anthropoids are mostly larger than prosimians, with larger brains housing relatively small olfactory lobes, and they are frugivorous or folivorous rather than omnivorous or insectivorous. They also are usually diurnal, with complex social systems. Anthropoids either walk quadrupedally along the tops of branches or hang suspended by all four limbs below branches, whereas prosimians are usually clinging and leaping forms. Large anthropoids typically employ a form of suspensory climbing in which the animals move relatively slowly by clinging underneath branches. **Brachiation**, a specialized version of arboreal locomotion in which the animals swing rapidly from the underside of one branch to the underside of the next using their hands to grasp the branches, is seen today in gibbons and spider monkeys, where it has evolved independently.

Modern anthropoids can be distinguished from prosimians by a variety of skull features that reflect their large brain sizes and their fibrous diets that require extensive chewing (**Figure 24–6**). A distinguishing anthropoid feature, partially developed in tarsiers, is the bony postorbital septum. The functional significance of this character is related to the evolution of large, forwardly oriented orbits and the need to prevent mechanical disturbance of the eye when the temporalis muscle contracts. Anthropoids also lack the grooming claw on the second toe that is seen in modern prosimians.

The origin of anthropoids appears to be related to a shift from foraging at night to being diurnal, and this change in behavior was accompanied by a drastic reduction in body size. The earliest-known anthropoids are all tiny (weighing no more than a couple hundred grams) and must have been insectivorous because only mammals larger than about 500 grams can include fibrous, low-quality food like leaves in their diet.

These early anthropoids are known from the Eocene of both Asia and Africa. However, the anthropoids in Asia today are clearly derived from a relatively recent (middle Miocene) African ancestry (note that Africa was an island continent until about 25 million years ago). There is considerable debate as to whether anthropoids originated in Asia or Africa. The primate radiation as a whole most likely had Asian roots, however, because tree shrews and flying lemurs (mammals related to primates) are both Asian natives.

The earliest Asian anthropoids are the Eosimiidae, known from the early Eocene to the early Oligocene of China, India, and Pakistan. *Eosimias* (Greek *eos* = dawn), from the middle Eocene of China, was a tiny species that was only about 5 centimeters long and weighed about 10 grams.

The earliest known eosimiid, *Anthrasimias* (Greek *anthra* = coal, because the specimen was found in a coal mine), is from the early Eocene of India. Some larger anthropoid primates (amphipithecines) are known from the late middle Eocene to the early Oligocene of Myanmar, Thailand, and Pakistan. Eosimiids have been established as stem anthropoids, predating the catarrhine/platyrrhine split.

Another diverse radiation of stem anthropoids is known from a similar time range in the Fayum Formation of Egypt. These forms (e.g., the parapithecids) were all small, generalized squirrel-like forms that retained relatively small brains.

Platyrrhine and Catarrhine Anthropoids Modern anthropoids can be divided into the Platyrrhini (Greek *platy* = broad)—the New World (broad-nosed) monkeys—and the Catarrhini (Greek *cata* = downward)—the Old World (narrow-nosed) monkeys and apes (**Figure 24–7**). These two groups differ in important characters:

- **Color vision**—All catarrhines have trichromatic color vision, which is also seen in a few platyrrhines, such as the howler monkeys. Trichromatic color vision is produced by a duplication of the opsin gene coding for color reception in the red-green part of the spectrum. (An independent gene duplication event leading to trichromatic vision occurred among some marsupials, such as the honey possum.)

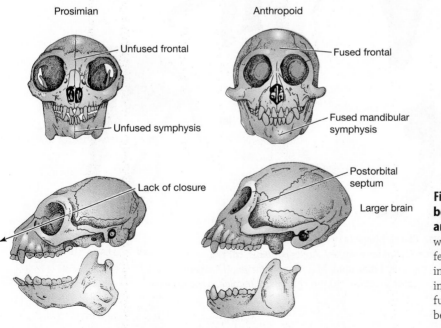

Prosimian — Unfused frontal — Unfused symphysis — Lack of closure

Anthropoid — Fused frontal — Fused mandibular symphysis — Postorbital septum — Larger brain

Figure 24–6 Cranial differences between prosimians and anthropoids. Anthropoids have a bony wall behind the orbit (as do tarsiers, a feature linking tarsiers and anthropoids in the Haplorhini), and the bones joining the two halves of the lower jaws are fused, as are the paired frontal bones between the eyes.

Platyrrhines—New World

Catarrhines—Old World

Figure 24–7 Diversity of living monkeys. Platyrrhines: (a) pygmy marmoset, *Callithrix pygmaea*; (b) squirrel monkey, *Saimiri sciureus*; (c) spider monkey, *Ateles paniscus*; (d) red howler monkey, *Alouatta seniculus*. Catarrhines: (e) pig-tailed macaque, *Macaca nemestrina*; (f) savanna baboon, *Papio anubis*; (g) Hanuman langur, *Presbytis vetulus*; (h) red colobus, *Piliocolobus badius*.

- **Olfaction**—While all anthropoids have small olfactory bulbs, platyrrhine primates seem to have retained a relatively good sense of smell, and most of the genes associated with olfaction are still functional. In contrast, only 50 percent of the olfactory genes of catarrhines code for functional receptor proteins, and catarrhines have also lost a functional vomeronasal organ.
- **Skulls, brains, and teeth**—Platyrrhines and catarrhines also differ in some details of the skull, especially in the ear region. Additionally, while all modern anthropoids have relatively large brains, the fossil record shows that enlargement of the brain from a prosimian-like condition occurred independently in the two groups. Furthermore, platyrrhines retain the generalized primate condition of having three premolars on each side of the jaw, whereas catarrhines have only two premolars.

Platyrrhines Platyrrhines first appeared in South America in the Oligocene epoch and are an exclusively New World radiation. They are presumed to have rafted from Africa across the Atlantic Ocean. (Rodents of probable African origin also reached South America a little earlier than platyrrhines, probably also by rafting.) Platyrrhines can be divided into the cebids and the atelids.

Cebids include the cebines (e.g., the familiar capuchin or organ grinder monkey and the squirrel monkey), the callitrichines (marmosets and tamarins), and the aotines (the owl monkey). Marmosets and tamarins are small and squirrel-like and have secondarily clawlike nails on all digits except for the big toe. They have simplified molars, and most species eat gum exuded from trees. They are also unusual among anthropoid primates, and more like the prosimians, in producing twins. The owl monkeys are the only nocturnal anthropoids.

Atelids include the atelines (wooly, howler, and spider monkeys), the callicebines (titi monkeys), and the pithecines (uakaris and saki monkeys). The atelines are distinguished by their prehensile tail and a specialized suspensory mode of arboreal locomotion aided by the tail.

The cebid radiation paralleled that of Old World monkeys to a certain extent, although there is no terrestrial radiation of cebids equivalent to that of baboons and macaques, nor was there ever a cebid radiation equivalent to the anthropoid great apes. Perhaps the extensive radiation of ground sloths in South America inhibited a terrestrial radiation among the primates. However, a striking parallel does exist between the spider monkey and the gibbon—both are specialized brachiators with exceptionally long arms that they use to swing through the branches, and they have evolved a remarkable convergence in a wrist joint modification that allows exceptional hand rotation. Spider monkeys can be distinguished from gibbons chiefly by their use of a prehensile tail as a fifth limb during locomotion. (Gibbons, like all apes, lack a tail entirely.)

Catarrhines The catarrhines include the Old World monkeys, apes, and humans. Stem catarrhines (i.e., forms that predate the divergence of Old World monkeys and apes) include the propliopithecids, small forms from the late Eocene and the early Oligocene of Africa, and the pliopithecids, larger Miocene forms that ranged from Africa into Eurasia. The split between apes and monkeys has been estimated from molecular studies to have occurred about 28 million years ago. This date matches a catarrhine fossil from the late Oligocene of Arabia, *Saadanius,* which appears to be close to this split.

Catarrhines have nostrils that are close together and open forward and downward, and they have a smaller bony nasal opening from the skull than do platyrrhines. The great apes (orangutans, gorillas, and chimpanzees) and humans are the largest extant primates, rivaled only by some of the extinct lemurs. Catarrhines' tails are often short or absent, and they never evolved prehensile tails. The group consists of two clades: the Old World monkeys (Cercopithecoidea; Greek *cerco* = tail and *pithecus* = ape) and the gibbons, great apes, and humans (Hominoidea; Latin *homini* = man).

Present-day Old World monkeys include two groups: colobines (Greek *colobo* = shortened) and cercopithecines. Colobines are found in both Africa and Asia, including colobus monkeys, langur monkeys, proboscis monkeys, and the golden monkey. They are more folivorous than are the cercopithecines, and they have more lophed, higher-cusped molars and a complex forestomach in which plant fiber is fermented. Colobines are primarily arboreal, with long tails and hind legs that are longer than their fore legs.

Cercopithecines are primarily an African radiation and include macaques, mangabeys, baboons, guenons, the patas monkey, and the so-called Barbary ape of Gibraltar. Cercopithecines are more omnivorous and frugivorous than are colobines, and this dietary difference is reflected in their broad incisors and their flat, bunodont molars. Cercopithecines are also more terrestrial than colobines, with short tails and fore- and hindlimbs of equal length. Cercopithecines have cheek pouches in which they carry food, and hands with a longer thumb and shorter fingers than the hands of colobines.

The first Old World monkeys are known from the middle Miocene, a slightly later date than the first true apes, which are the generalized proconsulids from the early Miocene of Africa. The radiation of monkeys in the late Miocene and Pliocene coincided with the reduction in diversity of the earlier radiation of generalized apes and apelike forms. Because we ourselves are apes, we often think of the monkeys as being the earlier, less derived members of the anthropoid radiation. However, among the Old World anthropoids, the converse is actually true: Apes were originally the more generalized forms, although the extant forms are specialized. The extant radiation of cercopithecoid monkeys is more derived in many respects than that of apes and more successful in terms of species diversity.

24.2 **Origin and Evolution of the Hominoidea**

Apes and humans are placed in the Hominoidea. Hominoids are distinguished morphologically from other recent anthropoids by a pronounced widening and dorsoventral flattening of the trunk relative to body length, so that the shoulders, thorax, and hips have become proportionally broader than in monkeys.

In all hominoids the clavicles are elongated, the iliac blades of the pelvis are wide, and the sternum is broad. The shoulder blades of hominoids lie over a broad, flattened back in contrast to their lateral position next to a narrow chest in monkeys (**Figure 24–8**) and most other quadrupeds. The lumbar region of the vertebral column is short (**Figure 24–9**). The caudal vertebrae have become reduced to vestiges (called the coccyx) in all recent hominoids, and normally no free tail appears postnatally. Balance in a bipedal pose is assisted by the flat thorax, which places the center of gravity near the vertebral column. These and other anatomical specializations of the trunk are common to all hominoids and help to maintain the erect postures that these primates assume during sitting, vertical climbing, and bipedal walking.

The skulls of hominoids differ from those of other catarrhines in their extensive formation of sinuses—hollow, air-filled spaces lined with mucous membranes that develop between the outer and inner surfaces of skull bones. Chimpanzees, gorillas, and humans share the derived character of true frontal sinuses.

Monkey

Ape

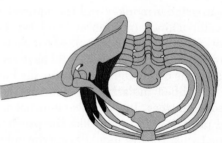

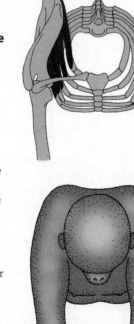

Figure 24–8 Differences in the form of the shoulder girdle in monkeys and apes. The upper portion of the picture shows the skeleton with the skull and neck vertebrae removed and also the position of the serratus muscle that links the scapula to the ribs. Note the broader chest and the dorsal position of the scapula in the ape; the curvature of the ribs is also greater, with the vertebral column lying more in the middle of the rib cage, closer to the center of gravity. These features all make it easier for a hominoid to balance in an upright position—this is in contrast to monkeys, which must bend their knees and lean forward to avoid tipping over backward if they are balancing on their hind legs.

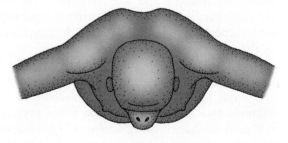

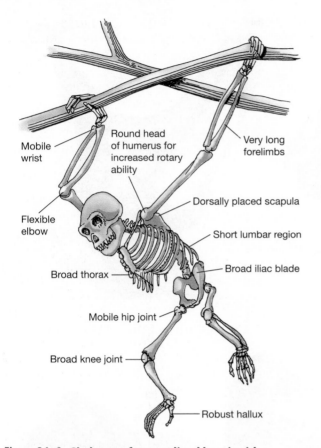

Mobile wrist

Round head of humerus for increased rotary ability

Very long forelimbs

Flexible elbow

Dorsally placed scapula

Short lumbar region

Broad iliac blade

Broad thorax

Mobile hip joint

Broad knee joint

Robust hallux

Figure 24–9 Skeleton of generalized hominoid, showing morphological specializations for suspensory locomotion.

Diversity and Evolution of Nonhuman Hominoids

Primates that we can call apes in the broad sense of the word have been around since the late Eocene. However, primates that can be included in a monophyletic Hominoidea date only from the early Miocene, when the anthropoid lineage diverged into the hominoids and the cercopithecoids (Old World monkeys). Modern apes are a highly specialized radiation of large tropical animals. The apes that radiated in the Miocene were more generalized animals, and they radiated into temperate parts of the Old World as well as the tropics. Apes and monkeys can be distinguished by their teeth, which is fortuitous because teeth are frequently the only remains of fossil species. Monkeys have lower molars with four cusps, whereas those of hominoids have five cusps. Additionally, the teeth of apes are usually flatter in relief than those of monkeys, and the grooves between the posterior cusps form a distinct Y pattern.

Diversity of Present-Day Apes

Extant apes include the Asian gibbons (including siamangs) and the orangutan as well as the African chimpanzees and the gorilla (**Figure 24–10**). Evidence of a type of culture, defined as social learning (for example, of tool use) with distinct differences among different biogeographic regions, has been observed in both chimps and orangutans, and chimpanzees have been shown to be capable of foresight and planning for future events. All of the great apes are critically endangered. Recent estimates show that the numbers of chimpanzees and gorillas in western equatorial Africa, a region considered to be the last refuge of their tropical habitat, have declined by more than half in the past few decades, primarily as a result of commercial hunting, mechanized forest logging, and the spread of the Ebola virus.

Nine species of gibbons (*Hylobates*) occur in Southeast Asia, both on the mainland (from India to China) and in the islands (Borneo, Sumatra, Java, and other nearby islands). They are the smallest apes, and they differ from other apes (and, indeed, the great majority of mammals) in their monogamous social system. Gibbons move through the trees most frequently by brachiation. They become entirely bipedal when moving on the ground, holding their arms outstretched for balance like a tightrope walker using a pole.

There are two subspecies of the orangutan (*Pongo pygmaeus*); one lives on Borneo and the other on Sumatra, although their ranges were larger in prehistoric times. Orangutans are about the same size as humans but are extremely sexually dimorphic, with males weighing twice as much as females. They are generally solitary, with groups consisting of females and their offspring. Orangutans are arboreal but rarely swing by their arms, preferring slow climbing among the branches of trees. They usually hang with all four limbs below the branch but sometimes walk on their hindlimbs on top of the branch, supporting themselves by grasping tree limbs above them.

Gorillas and chimpanzees live in the tropical forests of central Africa. Both are more terrestrial than gibbons and orangutans. On the ground they move quadrupedally by knuckle walking, a derived mode of locomotion in which they support themselves on the dorsal surface of digits 3 and 4 with the hand making a fist, rather than placing their weight flat on the palm of the hand, as we do when we walk on all fours.

Gorillas are the largest and most terrestrial of the apes. Unlike orangutans, they are highly social and live in groups, but like orangutans, gorillas are highly sexually dimorphic in body size—the males may weigh up

Figure 24–10 Diversity of extant apes.
(a) Siamang (a type of gibbon), *Hylobates syndactylus*. (b) Orangutan, *Pongo pygmaeus*. (c) Gorilla, *Gorilla gorilla*. (d) Common chimpanzee, *Pan troglodytes*.

to 200 kilograms, twice the mass of the females—and they are the most folivorous of the apes. Three geographically isolated subspecies of gorilla (*Gorilla gorilla*) have been described: the western lowland gorilla, the eastern lowland gorilla, and the mountain gorilla.

There are two (possibly three) species of chimpanzees. The larger and more widely distributed common chimpanzee (*Pan troglodytes*) lives primarily in Central and East Africa. The western subspecies (*Pan troglodytes verus*), from Nigeria and the Cameroon, may be a separate species. There is also a smaller species of chimpanzee, the bonobo (*Pan paniscus*), in Central Africa south of the Congo River. Bonobos live in more forested habitats than the other chimpanzees. Chimpanzees are more omnivorous than are the more strictly herbivorous gorillas; they are also more arboreal, exhibiting a greater degree

of suspensory locomotion. They are only moderately sexually dimorphic and, like gorillas, live in groups.

Relationships Within the Hominoidea

Hominoid classification has changed considerably over the past decade or so. Traditionally the Hominoidea was considered to include three families: Hylobatidae (gibbons), Pongidae (other apes, or "great apes"), and Hominidae (humans). The Hylobatidae, as originally conceived, remains a valid family. However, the Pongidae, a grouping of the great apes with the exclusion of humans, is paraphyletic: molecular data show that chimpanzees are more closely related to humans than are the gorilla and the orangutan, and in turn the gorilla is more closely related to the humans and chimps than is the orangutan.

Despite the fact that humans look rather different from the great apes, and we are certainly different in terms of culture and language, molecular studies show that we are very close to them genetically and that we diverged from them very recently in geological terms. Many researchers now consider that the great apes and humans belong in a single family, and the term "Hominidae" now includes these apes as well as ourselves. Humans and their immediate ancestors (i.e., extinct taxa more closely related to *Homo sapiens* than to chimpanzees) are now considered to be in the tribe Hominini within the subfamily Homininae (which contains humans, chimpanzees, and gorillas). (The orangutan and its fossil relatives are in the subfamily Ponginae.) Thus, what was called a hominid in fossil studies only a few years ago (and in earlier editions of this book), meaning a fossil human, is now called a **hominin.**

Both molecular evidence and the fossil record suggest that the Hylobatidae and the Hominidae split about 17 million years ago, and the Ponginae and the Homininae separated about 13 million years ago. Molecular evidence indicates that gorillas separated from the common ancestor of chimpanzees and humans between 6 and 8 million years ago, and that humans separated from their common ancestor with chimpanzees between 5 and 6 million years ago. This latter date fits well with the earliest definitive hominin fossil, *Ardipithecus,* known from 5.8 million years ago. An intriguing twist to this tale is that genetic data suggest that chimps and humans may have been interbreeding, and thus hybridizing their genomes, before the split between the two lineages became final. There is also evidence that chimpanzees have been evolving genotypic changes faster than humans have, especially in the Y chromosome. The gorillas separated into eastern and western populations about 3 million years ago. The three chimpanzee species are thought to have separated from each other more recently—between about 2.5 and 1.6 million years ago.

Diversity of Fossil Hominoids

Molecular data indicate that the split between apes and Old World monkeys occurred at the start of the Miocene, about 23 million years ago. The first true hominoids are from the early Miocene of East Africa at about 20 million years ago. These include proconsulids and various other taxa such as *Morotopithecus.* These early hominoids were primarily arboreal forms living in forested habitats. They had bunodont molars suggesting a frugivorous diet, and they were more derived than the propliopithecid stem anthropoids of the Fayum Formation (such as *Aegyptopithecus*); for example, like all apes, they lacked a tail.

Proconsulids ranged from the size of a small monkey to the size of a modern human. Although they remained generalized arboreal quadrupeds, not yet showing the specialized suspensory locomotion of many later apes, their hands and feet were more capable of gripping than those of monkeys, and the elbow joint was more stable. *Morotopithecus* was the size of a small human and is the first hominoid to have derived features of skeletal anatomy that show a capacity for suspensory locomotion. These features include a highly mobile shoulder joint, a short stiff back resisting lumbar flexion, and a moderately mobile hip joint. New material indicates that this animal may be more derived than previously thought, perhaps even more closely related to the Hominidae than are the gibbons.

By the middle Miocene, more derived hominoids had diversified into a variety of ecological types. Some of these apes, such as *Afropithecus, Kenyapithecus,* and *Equatorius,* remained in Africa, although a specimen of *Kenyapithecus* is known from Turkey. These genera are of controversial affinities, probably basal hominoids (or even hominids) of some sort, and are sometimes placed within the family Afropithecidae (which may well be paraphyletic, and so is not shown on Figure 24–2).

Other apes spread broadly into Eurasia, following the general middle Miocene warming trend and the connection of Africa to the Eurasian mainland (see Chapter 19). Both cercopithecine and colobine monkeys are also known from Eurasia in the late Miocene and Pliocene. The later Cenozoic Eurasian apes include the dryopithecids (Greek *dryo* = tree) and sivapithecids (after the Hindu god Shiva), which shared anatomical evidence of an upright posture. Both of these groups are more derived than the gibbons, and so are included in the Hominidae. Gibbons themselves were unknown in the fossil record until recently, when a possible early member of the lineage, *Yuanmoupithecus,* was described from a site in China that is dated to about 10 million years ago.

The dryopithecines diversified in Europe (fossils are known from France and Spain); they are now generally considered to be the sister group to the Hominidae. The sivapithecines were mainly an Asian radiation and were the stem stock for the modern orangutan (i.e., all members of the subfamily Ponginae). These Eurasian apes primarily occupied forested habitats and were somewhat modified for suspensory locomotion, although not to the extent seen in their modern relatives. Sivapithecines flourished from about 16 million years ago (the Middle Miocene Climatic Optimum) to 9 million years ago, at the start of the late Miocene, when the climate became cooler and drier in regions across the middle latitudes of the Northern Hemisphere. Both the dryopithecines and the sivapithecines included several genera, and the most interesting may be the sivapithecine *Gigantopithecus* that lived from the late Miocene to the Pleistocene. The Pleistocene species of *Gigantopithecus blacki* is the largest primate that has ever lived. With an estimated body mass of 300 kilograms, it would have been twice the size of an average gorilla. Some people have speculated that a surviving lineage of *Gigantopithecus* is behind the legends of the Yeti in Tibet and Bigfoot or Sasquatch in northwestern North America. Bigfoot, if it exists, could have reached North America by migrating across Beringia during the Pleistocene epoch, as did so many other mammals.

The various Eurasian dryopithecine genera include *Dryopithecus* itself, first known from 12 million years ago, and later forms such as *Hispanopithecus, Ouranopithecus,* and *Pierolapithecus,* which show features of both the skull and the postcranial skeleton that resemble the African apes (*Ouranopithecus* [Greek *ourano* = heaven] may actually be more closely related to the Homininae). This abundance of Eurasian apes and the absence of hominoids from Africa in the period between 13.5 and 10 million years ago have been used to argue for a Eurasian origin of the Homininae. However, more recent finds of late Miocene African apes that appear to be stem Homininae are challenging this viewpoint.

Fossils of gorillas and chimpanzees have been found only very recently. Possible basal gorillas include *Chororapithecus* and *Nakalipithecus,* both known from Kenya from about 10 million years ago. However, debate exists about the affinities of these animals, especially because they predate the molecular estimates of the split between gorillas and chimpanzees (plus humans), and they may be stem Homininae. The only fossil evidence of chimpanzees is some teeth from the middle Pleistocene of Kenya.

Although humans have evolved many of their own specializations, we have retained the ancestral characteristics of our clade in numerous dental and associated cranial features, while the lineages of great apes evolved their own unique derived characters. It is important to note, however, that all of the living hominoids are derived in comparison with the Mio-Pliocene ape radiation. The fact that gibbons and orangutans are in more basal positions on the cladogram than are humans or African apes (see Figure 24–2) does not imply that earlier apes looked like these modern forms.

24.3 Origin and Evolution of Humans

Anatomical differences between humans and apes appear in the skull and jaws, the trunk and pelvis, and, to a lesser extent, the limbs. Apes have long

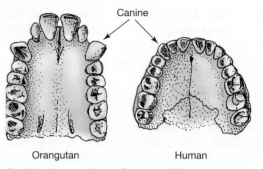

Figure 24–11 Comparison of ape and human upper jaws. The premolars and molars form parallel rows in apes, whereas they diverge posteriorly in humans.

jaws that are rectangular or U shaped, with the molar rows parallel to each other; the canines are large and pointed; and there is a gap between the canines and the incisors. Hominins have short jaws in association with the shortening of the entire muzzle. The human jaw is V shaped or bow shaped, with the teeth running in a curve that is widest at the back of the jaw; the canines are small and blunt; and the entire dentition is relatively uniform in size and shape without any gaps between the teeth (**Figure 24–11**). In addition, the hominin palate is prominently arched, whereas the ape palate is flatter between their parallel rows of cheek teeth.

Several evolutionary trends can be identified within the hominins. The points of articulation of the skull with the vertebral column (the occipital condyles) and the foramen magnum (the hole for the passage of the spinal cord through the skull) shifted from the ancestral position at the rear of the braincase to a position under the braincase. This change balances the skull on top of the vertebral column and signals the appearance of an upright, vertical posture. The braincase itself became greatly enlarged in association with an increase in forebrain size. By the end of the middle Pleistocene, a prominent vertical forehead developed in contrast to the sloping forehead of the apes. The brow ridges and crests for muscular attachments on the skull became reduced in size, in association with the reduction in size of the muscles that once attached to them. The human nose became a more prominent feature of the face, with a distinct bridge and tip.

Early Hominins

The earliest well-known hominins are the australopithecines (Latin *austral* = southern), known primarily from East and South Africa in the Pliocene and early

Pleistocene. Many new hominins have been described in the past decade, which has resulted in a confusing plethora of names and arguments about who is related to whom and when bipedality first evolved.

Extremely Early Possible Hominins We discussed possible late Miocene hominids in an earlier section. As yet, no Miocene taxon is presumed to be a true hominin (that is, a bipedal animal on the direct human lineage), but a couple of candidates for that designation have emerged within the past decade: *Orrorin tugenensis* from Kenya, dated at 6 million years, and *Sahelanthropus tchadensis* from the central African country of Chad, dated at 6.5 million years. Both of these possible early hominins are found in areas that were woodland or forest.

Spectacular as these finds are, there is still debate as to their hominin status. The preserved fragments of *Orrorin* combine an australopithecine-like femur that has features indicating bipedality with large, apelike canines and finger bones indicating climbing abilities. *Sahelanthropus* is known only from a skull, although the ventral position of the foramen magnum suggests that the head was balanced on top of the vertebral column, as in bipedal hominins. Unlike *Orrorin*, this animal has small, humanlike canines and hominin-like molars, but it also has a massive brow ridge and various cranial features that are more apelike than humanlike.

The Generalized Australopithecine Condition In terms of their biology, australopithecines are perhaps best thought of as being like bipedal apes with a modified dentition that includes the derived hominin features of thick enamel and small canines. Males were usually larger than females, and individuals grew and matured rapidly, unlike the prolonged childhood that characterizes our own species. Microwear analysis of their teeth suggests that early australopithecines, at least, were primarily fruit eaters, perhaps including some meat in their diet as do present-day chimpanzees.

All australopithecines appear to have been capable of bipedal walking. Humanlike footprints were found by Mary Leakey and her associates at Laetoli in Tanzania, in volcanic ash beds radiometrically dated between 3.6 and 3.8 million years ago. The tracks were probably made by *Australopithecus afarensis,* which is known from fossils at the same site. Analysis of these footprints indicates that they do not differ substantially from modern human trails made on a similar substrate, demonstrating the antiquity of bipedalism in hominin ancestry, far earlier than the appearance of an enlarged brain.

The most completely known early australopithecine is *Australopithecus afarensis*. This species is best known from a substantial part of a single young adult female skeleton, popularly known by the nickname Lucy. Lucy was found in the Afar region of Ethiopia, not far from the Red Sea, in a deposit dated at 3.2 million years. Lucy is the most complete pre-*Homo* hominin fossil ever found, consisting of more than 60 pieces of bone from the skull, lower jaw, arms, legs, pelvis, ribs, and vertebrae. Her overall body size was small. Young but fully grown when she died, Lucy was only about 1 meter tall and weighed perhaps 30 kilograms. Other finds indicate that males of her species were larger, averaging 1.5 meters tall and weighing about 45 kilograms. Lucy's teeth and lower jaw are also clearly humanlike. Her diet may have been rich in hard objects, probably fruits that would have been unevenly distributed in space and time. *Australopithecus afarensis* had a brain size of 380 to 450 cubic centimeters, quite close to that of modern chimpanzees and gorillas.

Despite modifications for bipedality, australopithecines retained some apelike features both of limb anatomy and of the semicircular canals in the inner ear (structures responsible for orientation and balance), suggesting the retention of apelike orientation in an arboreal environment. A degree of arboreality is also reflected in the hands and feet, with the bones of the fingers and toes significantly longer and more curved than those of modern humans (**Figure 24–12**), although a recently discovered foot bone (the fourth metatarsal) also shows that *Australopithecus afarensis* had a humanlike arch to the foot. The hands of australopithecines were more humanlike than those of fully arboreal apes such as gibbons and orangutans, and they lacked the robust fingers of the knuckle-walking chimpanzees and gorillas. It appears that australopithecines were able to stand and walk bipedally but still spent much of their time in the trees and probably were not capable of sustained running. Biomechanical studies of australopithecines show that they walked as we do. (The caveman image of walking with a stoop and bent hips and knees has its basis in Hollywood, not in science—it is someone's idea of an intermediate gait between chimpanzees and humans and it probably never existed.)

Other Early Australopithecines Recently, hominins earlier and more basal than Lucy have been discovered. *Ardipithecus ramidus* was originally described in 1994 on the basis of a few jaws and teeth. *Ardipithecus* (*ardi* = ground, floor) is known from Ethiopian sediments about 4.4 million years old, with associated

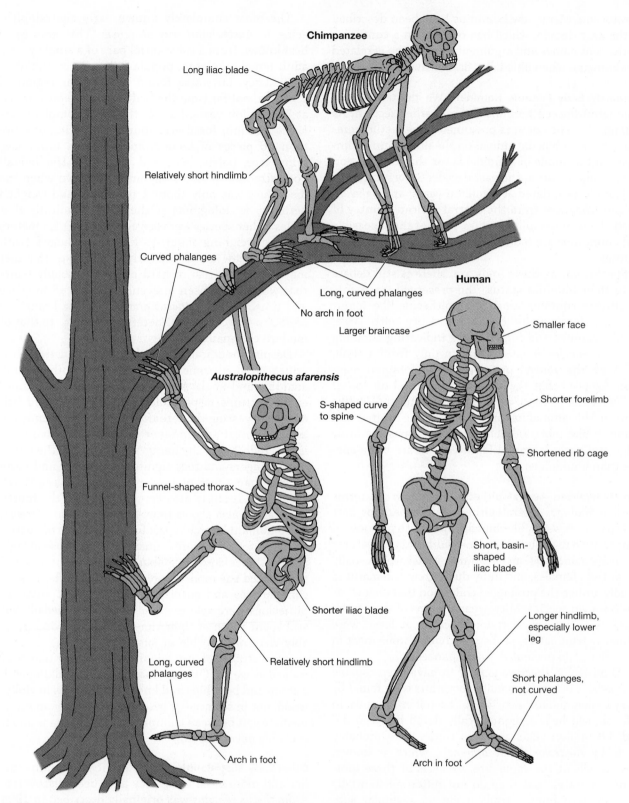

Chimpanzee

Long iliac blade

Relatively short hindlimb

Curved phalanges

Long, curved phalanges

No arch in foot

Larger braincase

Human

Smaller face

Australopithecus afarensis

S-shaped curve to spine

Shorter forelimb

Shortened rib cage

Funnel-shaped thorax

Short, basin-shaped iliac blade

Shorter iliac blade

Longer hindlimb, especially lower leg

Long, curved phalanges

Relatively short hindlimb

Short phalanges, not curved

Arch in foot

Arch in foot

Figure 24–12 Comparison of skeletons of a chimpanzee, an australopithecine, and a human. Note the shorter iliac blade, the less funnel-shaped rib cage, the longer legs, and the shorter fingers and toes in *Homo*.

animals indicating a wooded habitat; a new specimen, *Ardipithecus ramidus kadabba,* extends the range of this genus back to 5.5 million years ago. From the original description, one could conclude that *Ardipithecus* was an upright biped, judging from the position of the foramen magnum, and that it might have had a more humanlike social system, judging from the small canines with less sexual dimorphism, than modern African apes. A lesser degree of sexual dimorphism is associated with a pair-bonding type of sociality, as opposed to one where the males compete for access to females, as in the African apes today.

A full description of *Ardipithecus,* including the postcranial skeleton, did not appear until 2009, when 11 papers appeared simultaneously in a special issue of the journal *Science* that described several important characters of *Ardipithecus* (**Figure 24–13**):

- The structure of the pelvis of *Ardipithecus* indicates that it had at least a degree of bipedal locomotion.
- The foot had long toes and a highly divergent big toe, both features indicating that *Ardipithecus* was more arboreal than any species of *Australopithecus.*
- The arms and hands of *Ardipithecus* were longer than those of other hominins but did not closely resemble those of the living African apes: there is no evidence for knuckle walking, and the wrist and finger joints were much more flexible, suggesting that the hands were used for support while walking bipedally along tree branches.
- *Ardipithecus* had a brain that was smaller in relation to its body size than the brain of any species of *Australopithecus.*
- *Ardipithecus* had a more sloping face than any species of *Australopithecus,* but less sloping than in the face of a chimpanzee.

In general, the anatomy of *Ardipithecus* shows that the common ancestor of chimpanzees and humans was not simply like a chimpanzee, and that chimps have undergone substantial evolution since their split from hominins.

Several new species of *Australopithecus* have been found in the past decade, which fill in more gaps in the story of human evolution. **Figure 24–14** shows one current hypothesis of how the australopithecine species were related to each other and to our own genus *Homo.* The earliest known member of the genus *Australopithecus* is now *A. anamensis,* described in 1995, from sites in Kenya and Ethiopia ranging from 4.2 to 3.9 million years old. This hominin appears to be intermediate in anatomy between *Ardipithecus* and *Australopithecus afarensis,* with an estimated body

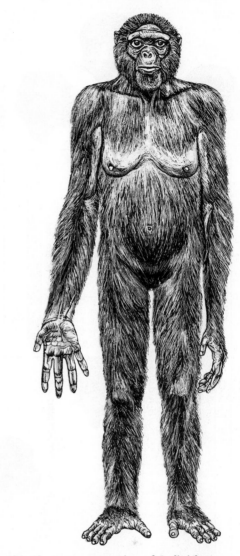

Figure 24–13 A reconstruction of *Ardipithecus ramidus.* Features indicating that *Ardipithecus* was bipedal include the position of the foramen magnum on the base of the skull and the structure of the pelvis and hindlimbs. Arboreal characteristics visible in this reconstruction include long arms and hands with long fingers, and a foot with long toes and a big toe that diverges widely from the remaining toes.

mass of about 50 kilograms. The fossils are associated with those of woodland types of mammals, reinforcing the notion that early hominin evolution took place in the woodlands rather than on the savanna, as was once assumed.

Australopithecus bahrelghazali, known from a single jaw, is slightly younger than *A. anamensis* and contemporaneous with *A. afarensis.* It was described in 1995, from a site in Chad, in central Africa, demonstrating

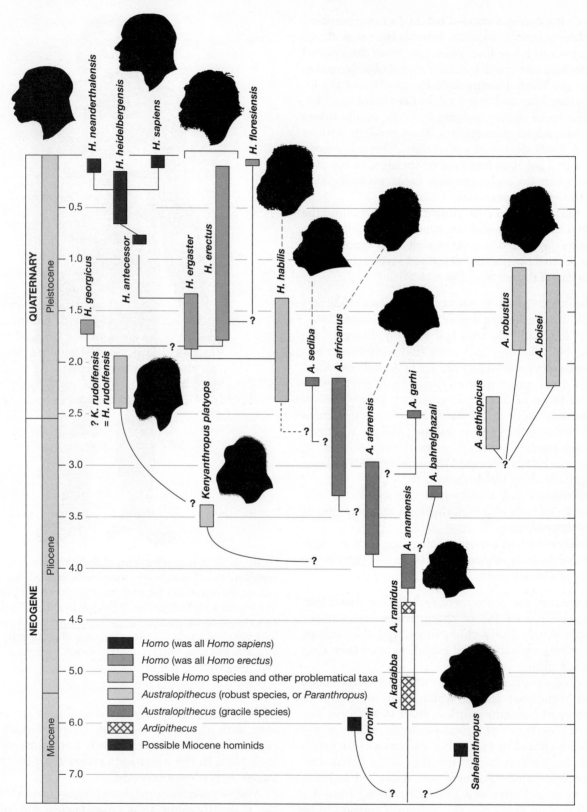

Figure 24–14 A hypothesis of the phylogenetic relationships within the Hominini. Question marks and dashed lines indicate uncertainties about relationships.

that early hominins were more widespread in Africa than had previously been supposed. A still younger species from Ethiopia, *Australopithecus garhi,* known from a single fragmentary skull, was described in 1999. Dated at 2.5 million years old, it is only slightly older than the earliest known specimen of the genus *Homo,* and its association with some fossilized butchered animal bones has led to the speculation that it may have been the first species in our lineage to eat meat and use tools.

Australopithecus africanus, which lived between 2.1 and 3.3 million years ago, had very robust arm bones, suggesting that it may have spent even more time in trees than *A. afarensis.* This greater degree of arboreality implies that it may not be in the direct evolutionary line to *Homo. Australopithecus africanus* apparently also differed from *A. afarensis* by including more meat in its diet. A final species, *Australopithecus sediba,* was described in 2010 from about 1.8 million years ago in South Africa. This species is known from four partial skeletons; in both its cranial and postcranial features it appears to be the closest of the australopithecines to the genus *Homo.*

Another early hominin, known from 3.5 million years ago in Kenya, represents a completely new genus, *Kenyanthropus platyops.* This hominin is markedly different from its contemporary, *Australopithecus afarensis,* combining the unique features of a derived face with a less derived cranium. Although its relationships remain uncertain, there is speculation that it may represent an entirely separate lineage of early hominins, perhaps ancestral to the hominin *Kenyanthropus* [*Homo*] *rudolfensis.*

The Robust Australopithecines These australopithecines, appearing later than most of the other forms, were distributed in East and South Africa from 2.5 to about 1.2 million years ago. Called robust australopithecines, in contrast with the earlier, smaller forms dubbed the gracile australopithecines, they are usually placed in their own genus *Paranthropus*—*Paranthropus robustus* of South Africa and *P. aethiopicus* and *P. boisei* of East Africa. However, it is not entirely clear how these robust australopithecines were interrelated or even whether they represent a single radiation from within the gracile australopithecines. The robust type of australopithecine may have arisen independently on more than one occasion from different gracile australopithecine species.

The robust australopithecines were relatively large, powerfully built forms with pronounced sagittal crests on the skulls and the body proportions of a football player (although they were no more than about 1.5 meters tall). They were terrestrial, savanna-dwelling vegetarians,

and their huge molars exhibited heavy wear, suggesting a coarse and fibrous diet. These hominins lived sympatrically with early *Homo* and were a highly successful radiation during the Pleistocene. It is likely that their extinction in the mid-Pleistocene was related to climatic changes in Africa rather than to any competition with early humans.

Ecological and Biogeographic Aspects of Early Hominin Evolution

For many years the evolution of the hominin lineage was assumed to be related to the appearance of the African savannas, which are grasslands with widely spaced bushes and trees. The spread of African savanna environments was probably related to the formation of the Isthmus of Panama 2.5 million years ago, which blocked the flow of water between North and South America, leading to profound global climatic changes. Paleontological evidence from both the flora and fauna of eastern Africa suggests that the environment changed to a savanna habitat at this time. Grazing antelopes increased in abundance, and new types of carnivores appeared.

It used to be thought that the development of savanna habitats coincided with the split of the human lineage from that of the other apes. The traditional view has long been that the development of the Rift Valley, which extends from north to south in eastern Africa, isolated the human lineage from that of the other apes. Subsequently humans became adapted for these new, open grassland habitats by adopting a bipedal gait, while the apes were relegated to the tropical forests to the west of the Rift Valley and remained primarily arboreal.

We now know that the emergence of broad expanses of savanna occurred 2 to 3 million years ago, whereas hominin bipedality certainly extends back to at least 4.5 million years and possibly 7 million years ago. Thus, it seems probable that the origins of humans and bipedality took place in forested environments, although some isolation of humans from other hominoids undoubtedly did occur.

The gracile australopithecines were primarily a Pliocene radiation, preceding the expansion of savanna habitats. At the start of the Pleistocene, about 2.5 million years ago and coincident with major climatic changes, the hominin lineage split into two—one lineage leading to our direct ancestors, the genus *Homo,* and the other leading to the robust australopithecines. Thus in an earlier part of the Pleistocene there were two lineages of hominins: early true humans and robust australopithecines. Further climatic cooling and drying resulted in the reduced abundance of robust australopithecines about 1.8 million years ago, and their extinction later in the epoch. However, had the Pleistocene climatic changes been different—such as reverting to a wetter and warmer regime—it might have been our ancestors who became extinct and the robust australopithecines that survived.

24.4 Derived Hominins (the Genus *Homo*)

The earliest species of *Homo*, *H. habilis* (Latin *habilis* = able or "handyman"), existed in East Africa from 2.33 to 1.44 million years ago. This taxon is rather poorly known and has been the subject of intense debate. *Homo habilis* is best distinguished from australopithecines by its larger cranial capacity (between 500 and 750 cubic centimeters in contrast to 380 to 450 cubic centimeters for *Australopithecus afarensis*). However, this brain size is still small in comparison with later species of *Homo*. *Homo habilis* also differed from *Australopithecus* in having a smaller face and a smaller jaw and dentition, with smaller cheek teeth and larger front teeth. Like *Australopithecus,* it had a relatively small body and retained some specializations for climbing. *Homo habilis* has been found in association with stone artifacts and fossil bones with cut marks, suggesting the use of tools and hunting (or at least scavenging) behavior.

Some paleoanthropologists think that *H. habilis* does not have enough derived characters to be included in the genus *Homo* and instead place it within the australopithecines. *Homo habilis* co-occurred with the more derived *H. erectus* in East Africa for nearly half a million years.

Many specialists have split the original *H. habilis* into two species: *H. rudolfensis,* known from a single skull from about 1.9 million years ago, and *H. habilis* (redefined, known from 2.4 to 1.4 million years ago). *Homo rudolfensis* was somewhat larger-brained than *H. habilis,* and some researchers have concluded that it was not a member of our genus at all, but a descendant of *Kenyanthropus platyops.* This view is reflected in Figure 24–14, where the taxon is called *Kenyanthropus rudolfensis.*

About 1.9 million years ago a new hominin appeared in the fossil record—*Homo erectus* (Latin *erect* = upright)—originally described in the late nineteenth century as *Pithecanthropus erectus* and known at that time as Java Man or Peking Man (**Figure 24–15**). Like modern humans, this hominin had a large body, lacked adaptations for climbing, and had relatively small teeth and jaws: there is no doubt that this taxon belongs in the genus *Homo.*

Homo erectus and *Homo ergaster*

Homo erectus originated in East Africa, where it co-existed for at least several hundred thousand years with two of the robust australopithecines and overlapped in time with *H. habilis. Homo erectus* was the first intercontinentally distributed hominin. It appears to have spread to Asia at least 1.7 million years ago and subsequently perhaps into Europe. This appearance of *Homo* in Asia was once hailed as a landmark in human evolution, with the notion of "out of Africa" being some kind of early human achievement similar to the first man on the moon. In fact, all kinds of other mammals were moving between Africa and Asia at this time, and the discovery of a greater variety of early hominins in Asia merely indicates that hominins, too, were migrating into Asia in the early Pleistocene.

Currently the older African form of *H. erectus* is usually called *Homo ergaster* (**Figure 24–16** on page 602), and the name *erectus* is reserved for the Asian hominin. *Homo ergaster* is thought to be more closely related to later hominins. The differences between *H. erectus* and *H. ergaster* are subtle, however, and the following description applies to both species.

Four characteristics of *Homo erectus* and *H. ergaster* represent a significant change in the evolutionary history of humans:

- They were substantially larger than earlier hominins (up to 1.85 meters tall and weighing at least 65 kilograms—the same size as modern humans), with a major increase in female size that reduced sexual dimorphism so that the males were only about 20 percent to 30 percent larger than the females, as in our own species. The reduction in sexual dimorphism in these and later species of *Homo* implies a change from a polygynous mating system (in which males competed with each other for access to females) to monogamous pair bonding in which female choice among potential mates played a larger role.
- *Homo erectus* and *H. ergaster* had body proportions like those of humans. There is debate about the nature of the postcranial skeleton of *H. habilis* and *K. rudolfensis* because of the fragmentary nature of the fossil material, but there is no doubt that *H. erectus* and *H. ergaster* had the short arms, long lower legs, narrow pelvis, and barrel-shaped chest that are characteristic of modern humans.
- *Homo erectus* and *H. ergaster* also had a larger brain than earlier *Homo* species, with cranial capacities ranging from 775 to 1100 cubic centimeters. Because the brain is a metabolically demanding organ, their larger brains imply that *H. erectus* and

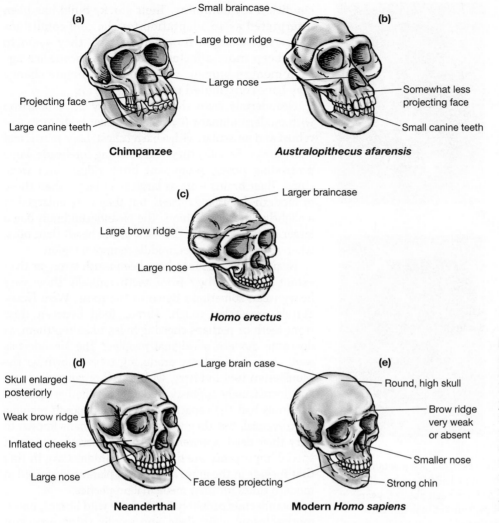

Figure 24–15 Progressive changes in the skull of hominins. Basal hominoids had small brains and large jaws, which gave them sloping faces. As the volume of the brain increased and the size of the jaws decreased, the lower portion of the skull no longer projected beyond the braincase and the face became flatter.

H. ergaster had greater nutritional needs than earlier hominins.

- *Homo erectus* and *H. ergaster* were the first hominins to have a humanlike nose—broad and flat, but with downward-facing nostrils. Their other facial features were less like modern humans: the jaw projected beyond the plane of the upper face (prognathous), the teeth were relatively large, there was almost no chin, the forehead was flat and sloping, and the bony eyebrow ridges were prominent.

- *Homo erectus* and *H. ergaster* were also the first hominins to have delayed tooth eruption and relatively small teeth for their body size. The smaller teeth suggest that these species may have cooked their food because cooked food is easier than raw food to chew. The delayed tooth eruption suggests a humanlike extended childhood, which would also imply a humanlike extended life span and humanlike passage of learned information from one generation to the next.

Precursors of *Homo sapiens*

The species *Homo sapiens* (Latin *sapien* = wise), as originally defined, included not only the modern types of humans but also the Neandertals and some earlier forms. More recent evidence has complicated this picture, with the result that the forms originally included in *Homo sapiens* have been split into several different species; the term *H. sapiens* is now reserved for modern humans.

Homo heidelbergensis, known from both Africa and Europe between 700,000 and 200,000 years ago, and *H. antecessor,* known from about 800,000 years ago in Spain, are precursors of modern humans. These hominins had slightly larger brains, thicker and more robust skulls, larger teeth, and a less prognathous face than *Homo erectus.*

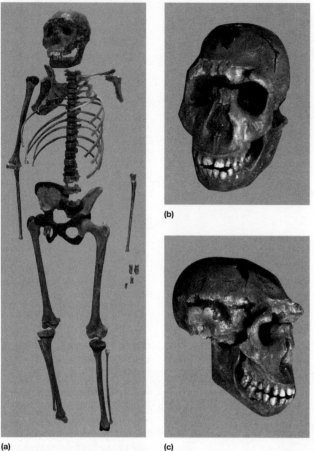

(a)

(b)

(c)

Figure 24–16 *Homo ergaster.* This skeleton of a juvenile male, known as Turkana boy, was found in 1984 near Lake Turkana in Kenya. He is thought to have been about 8 years old when he died. He was 154 centimeters tall at death, and his estimated adult height would have been about 163 centimeters.

The Neandertals

The first recognized fossils of *Homo neanderthalensis* were found in the Neander Valley in western Germany in 1856, and fossils with Neandertal features first appear about 200,000 years ago, roughly contemporaneous with the earliest members of the modern form of *Homo sapiens.* Recent analysis of ancient DNA from the bone of a Neandertal shows considerable genetic difference from modern humans, suggesting that Neandertals were not directly ancestral to modern humans and that the two lineages diverged from *Homo heidelbergensis* about 500,000 years ago. Neandertals are often popularly displayed as primitive cavemen, with the unspoken assumption that this is what nonhuman hominins looked like. In fact, the Neandertal features represent a derived

condition for hominins. Their stocky build has been interpreted as an adaptation for the cold conditions of Ice Age Europe; however, in fact, they seem to have been more affected by the Pleistocene ice age than modern humans were, and this climate change may have contributed to their extinction.

Neandertals were short and stocky in comparison with modern humans (**Figure 24–17**). Their body form was robust and muscular, with a barrel chest, large joints, and short limbs. Facially, they had receding foreheads, large protruding noses, prominent brow ridges, and weak chins. Their brains were as large as or larger than those of modern-day *Homo sapiens,* but they were enlarged in a slightly different fashion. The Neandertal brain had a larger occipital area (at the back of the head) than ours, whereas we have a larger, middle temporal region.

Neandertals appear to have been much stronger than extant humans. Their front teeth typically show very heavy wear, sometimes down to the roots. Were Neandertals processing tough, fibrous food between their front teeth or perhaps chewing hides to soften them, as do some modern aboriginal peoples? The Neandertals were stone toolmakers, producing tools known as the Mousterian tool industry, with a well-organized society and increasingly sophisticated tools. Whether the Neandertals had the capacity for complex speech remains controversial, but they were the first humans known to bury their dead, apparently with considerable ritual. Of special importance are burials at Shanidar Cave in Iraq that include in the grave a variety of plants recognized in modern times for their medicinal properties.

Neandertals probably hunted the wild horses, mammoths, bison, giant deer, and woolly rhinoceroses of the Eurasian high-latitude plains. Mousterian-style hunting tools were of the punching, stabbing, and hacking type—throwing spears and bows and arrows are unknown. Many skeletal remains of Neandertals show evidence of serious injury during life, and their patterns of injury resembled that of present-day rodeo bull riders. That similarity suggests that, like bull riders, Neandertals were in close contact with large, frightened animals. Despite the high incidence of injuries, 20 percent of Neandertals were more than 50 years old at the time of death. It was not until after the Middle Ages that human populations again achieved this longevity.

Modern *Homo sapiens* reached Europe and Asia between 40,000 and 50,000 years ago, and European Neandertals and the populations of *Homo erectus* remaining in Asia vanished between 40,000 and 30,000 years ago, with a relict population of Neandertals remaining in Gibraltar (the southernmost point in Europe) until 28,000 years ago. There is much debate about the role

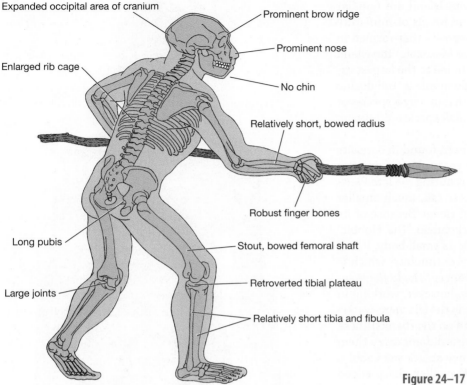

Expanded occipital area of cranium

Enlarged rib cage

Long pubis

Large joints

Prominent brow ridge

Prominent nose

No chin

Relatively short, bowed radius

Robust finger bones

Stout, bowed femoral shaft

Retroverted tibial plateau

Relatively short tibia and fibula

Figure 24–17 Reconstruction of a Neandertal.

of *H. sapiens* in the disappearance of the other species of humans. Did our species gradually outcompete the others in a noncombative fashion, or was there some type of direct conflict? The climate was changing rapidly between about 45,000 to 30,000 years ago when the two species intermingled in Europe, and the population densities of humans increased 10-fold during that time. These two observations suggest that modern humans were better adapted to the new conditions and that direct or indirect competition between the species contributed to the decline of Neandertals.

Additional Derived Hominins

The increasing number and the geographic range of sites with fossils of derived hominins indicate that the radiation of humans in the Pleistocene and Holocene was far more dynamic than we realized a decade ago. Evidence is accumulating that groups of *Homo* moved out of Africa repeatedly between 2 million and 60,000 years ago and migrated northward into Europe and eastward into Asia. Some of these migrations were probably aided by periods of low sea levels that connected islands, but open water was not an insurmountable barrier. *Homo erectus* moved eastward along the Indonesian archipelago across a series of islands that were separated by deep water at the time, and a toe

bone found in Callao Cave on the island of Luzon indicates that *Homo sapiens* reached the Philippines at least 67 thousand years ago.

The Dmanisi Hominins A hominin, found at Dmanisi, Georgia, is dated at about 1.8 million years ago at the beginning of the Pleistocene. (Georgia, formerly part of the Soviet Union, is a country near Turkey.) This is the earliest date for a hominin in Europe, and it is close to the date of the first *Homo erectus* found outside of Africa. It is not clear whether this hominin (called *Homo georgicus* by some people) is an early offshoot from *H. erectus* or from an even earlier hominin species, such as *Homo habilis*. Although it resembles *H. erectus* in a number of ways, its smaller stature (about 1.5 meters) and smaller brain (600 to 775 cubic centimeters) lead researchers to suspect that it might have evolved from a less derived hominin. Aspects of the elbow and shoulder joints support the latter interpretation, yet the lower limbs are long—suitable for running and distance travel as seen in early *Homo*.

Homo floresiensis The "island rule" in evolutionary biology refers to a strange phenomenon that occurs when a lineage of animals is isolated on an island: small species evolve to larger body sizes and large species become miniaturized. The giant land tortoises found on the

Galápagos Islands and on Aldabra Island are familiar examples of island gigantism, and fossils of miniature dinosaurs have been found in deposits that formed in areas that were islands during the Mesozoic. The island of Flores in eastern Indonesia is home to the largest extant species of lizard, the Komodo monitor, and during the late Pleistocene Flores was inhabited by a species of miniature elephant and a very small species of human, *Homo floresiensis* (**Figure 24–18**).

Fossils of *H. floresiensis* have been found in deposits ranging from 12,000 to 74,000 years old, making it a contemporary of modern humans during the late Pleistocene. It was only about 1 meter tall, much smaller than any other known species of *Homo*. Because of its small size this creature was nicknamed "the Hobbit" by the popular press. Along with its small body, it had a small brain—about 400 cubic centimeters, which is no bigger than that of a chimpanzee. *Homo floresiensis* made a variety of stone tools, however, working in a logical sequence. The starting materials were water-worn stone cobbles that are found on riverbanks. These cobbles are large, and *H. floresiensis* did not carry them back to their shelters. Instead they struck the cobbles with hammer stones to break off smaller, sharp-edged flakes and then took the flakes back to their shelters, where they used a variety of knapping techniques to produce tools called blades, scrapers, and penetrators.

Homo floresiensis is considered a distinct species of human with a relatively derived, though miniature, brain. The origin of *H. floresiensis* is still a subject of debate; the most widely accepted hypothesis is that it is derived from *H. erectus,* but some researchers believe that it might be derived from a more basal hominin than *H. erectus.*

The Denisovans Fragmentary remains of hominins who lived between 50,000 and 30,000 years ago have been found in Denisova Cave in the Altai Mountains of southern Siberia. Although the materials consist of isolated teeth and finger and toe bones, they contain DNA that indicates that the Denisovan lineage split from the Neandertal lineage about 200,000 years ago. The Denisovans were genetically distinct from Neandertals, but they were contemporaneous with both Neandertals and modern humans. Indeed, a preliminary analysis of the Denisova Cave indicated that it was occupied by Denisovans 50,000 years ago, by Neandertals 45,000 years ago, and by modern humans soon after that.

Origin of Modern Humans

A single African origin of *Homo sapiens* is now supported both by the fossil record and by genetic studies of modern humans, especially the evolution of mitochondrial

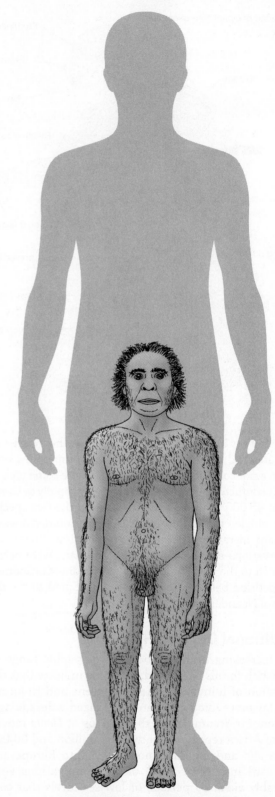

Figure 24–18 Reconstruction of *Homo floresiensis*. The outline of a modern human shows the relative sizes of *H. sapiens* and *H. floresiensis.*

DNA and the Y chromosome. Few people still adhere to the older multiregional model of *H. sapiens* evolving independently in different areas, each from an already distinctive local population of *H. erectus*. Mitochondrial DNA is inherited only from the mother because it resides in the cytoplasm of the egg, not in the nucleus, and the genome is small—only about 16,000 base pairs. Analysis of mitochondrial DNA allows one to trace the maternal lineage of an individual. A study of mitochondrial DNA from people all over the world showed that all living humans can trace their mitochondria to a woman who lived in Africa about 170,000 years ago. This hypothetical common ancestor has been called the African Eve, a phrase that obscures the biological meaning of the discovery. It does not mean that there was only one woman on Earth 170,000 years ago; instead, it means that only one woman has had an unbroken series of daughters in every generation since then.

A similar approach can be used with the Y chromosome, which is passed only from father to son. The Y chromosome has about 60 million base pairs, so it is much more difficult to study than mitochondrial DNA. An analysis of 2600 base pairs from the Y chromosome indicates that all human males are descended from a single individual, who is estimated to have lived in Africa 59,000 years ago. Naturally, this individual has been called the African Adam.

The difference between the estimates—170,000 years versus 59,000 years—results from uncertainty about the rates of mutation in mitochondrial DNA and the Y chromosome. What is most significant is that both studies indicate that the common ancestor of modern humans lived in Africa, and this conclusion is reinforced by other genetic information. For example, there is more variation in the human genome in Africa than in the rest of the world combined, which is exactly what would be expected if humans originated in Africa. Furthermore, humans have only about one-tenth the genetic variation of chimpanzees, and that observation might indicate that human populations were once very small, passing through a genetic bottleneck.

Although they have met with some criticism, these molecular studies generally agree with the fossil record, which shows that modern *Homo sapiens* originated in Africa about 200,000 years ago. Earlier forms from as far back as 500,000 years ago, known as "archaic" *Homo sapiens,* may be *Homo heidelbergensis.* Three skulls of hominins found in Ethiopia have been dated at about 160,000 years; they appear to be at the very base of the modern human lineage. By 125,000 years ago, anatomically modern humans were widespread across Africa. They crossed over into the Levant region of Asia (the area termed the Middle East today) about 120,000 years ago, and they first appeared in more northern Eurasia (and also in Australia) between 40,000 and 50,000 years ago. The data that were available a decade ago supported the "replacement hypothesis"—that is, an interpretation of human evolution by which modern humans completely replaced the human populations they encountered, and the older populations vanished without a trace.

The replacement hypothesis has been modified by genetic studies that reveal a low level of interbreeding between modern humans and the populations of humans that they met as they spread from Africa. The complete Neandertal genome was determined from DNA extracted from fossils found in Croatia. Comparison of the Neandertal genome with the genome of living humans reveals that between 1 percent and 4 percent of the nuclear DNA of extant Eurasians came from Neandertals, not from our African ancestors. The most plausible interpretation of that observation is that modern humans interbred with Neandertals in the Middle East, where the two species coexisted between 80,000 and 50,000 years ago. Sometime later a subgroup of modern *Homo sapiens* moved eastward into Asia, where they interbred with Denisovans in Asia. The descendants of this group of modern humans are found in Melanesia (the islands in the South Pacific east of Australia), and about 8 percent of their genome is derived from archaic species of *Homo.*

A 4 percent to 8 percent representation of archaic species of *Homo* in the genome of non-African modern humans does not discredit the replacement hypothesis—after all, more than 90 percent of the genome of human populations outside of Africa does represent a recent African origin—but it calls for modification of the hypothesis. Paleoanthropologists now describe the spread of modern humans as "replacement with hybridization" or "leaky replacement" to recognize the genetic contribution of archaic human populations to the genome of modern humans.

24.5 Evolution of Human Characteristics

Humans are classically distinguished from other primates by three derived features: a bipedal stance and mode of locomotion, an extremely enlarged brain, and the capacity for speech and language. Here we examine possible steps in the evolution of each of these key features and also consider the loss of body hair and the evolution of tool use.

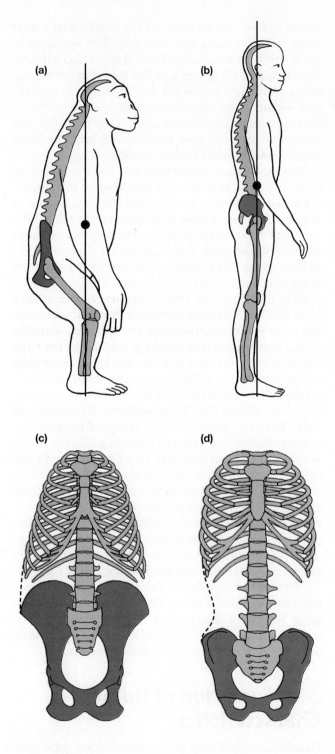

Figure 24–19 Structural and postural differences between chimpanzees and humans. The trunk regions and vertebral columns of chimpanzees (a and c) and humans (b and d) show structural characters associated with quadrupedal and bipedal locomotion. (a) When a chimpanzee stands bipedally, its center of gravity (shown by a dot) is anterior to its hip joints and it must bend its legs at the hip and knee to place its center of gravity over its feet. (b) In contrast, the S-shaped curve of the human vertebral column places the center of gravity directly over the hips and feet when the legs are held straight. A chimpanzee must use muscular contractions to support its weight when it stands erect, whereas most of the weight of a human is transmitted directly through the limb bones to the ground and the leg muscles are mostly inactive. Humans (d) also differ from chimpanzees (c) in having a longer trunk region with a more barrel-shaped (versus funnel-shaped) rib cage, resulting in a distinct waist.

and hindlimbs, thereby freeing the forelimbs from obligatory functions of support, balance, or locomotion (**Figure 24–19**). The most radical changes in the hominin postcranial skeleton are associated with the assumption of a fully erect, bipedal stance in the genus *Homo*. Anatomical modifications include the S-shaped curvature of the vertebral column, the modification of the pelvis and position of the acetabulum (hip socket) in connection with upright bipedal locomotion, and the lengthening of the leg bones and their positioning as vertical columns directly under the head and trunk. Humans also differ from apes in having a longer trunk region with a more barrel-shaped (versus funnel-shaped) rib cage, resulting in a distinct waist. The humanlike waist may be a specific adaptation for bipedal walking, allowing rotation of the pelvis in striding without also involving the upper body.

The secondary curve of the spine in humans is a consequence of bipedal locomotion and forms only when an infant learns to walk. We have by no means perfected our spines for the stresses of bipedal locomotion, which are quite different from those encountered by quadrupeds. One consequence of these stresses is the high incidence of lower-back problems in modern humans.

Humans stand in a knock-kneed position, which allows us to walk with our feet placed on the midline and reduces rolling of the hips from side to side. This limb position leaves some telltale signatures at the articulation of the femur with the hip and at the knee joint. This type of bony evidence can aid researchers in deducing whether fossil species were fully bipedal. An

Bipedality

Although all modern hominoids can stand erect and walk to some degree on their hind legs, only humans display an erect bipedal mode of striding locomotion involving a specialized structure of the pelvis

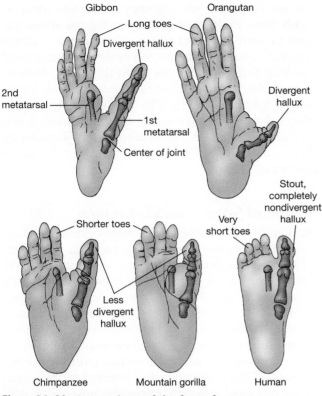

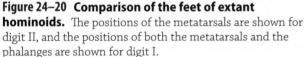

Figure 24–20 Comparison of the feet of extant hominoids. The positions of the metatarsals are shown for digit II, and the positions of both the metatarsals and the phalanges are shown for digit I.

unfortunate consequence of this limb position is that humans, especially athletes, are rather prone to knee dislocations and torn knee ligaments. Because women have wider hips than men, their femurs are inclined toward the knees at a more acute angle than men's femurs, and female athletes are especially prone to knee injuries.

The feet of humans show drastic modifications for bipedal, striding locomotion. The feet have become flattened except for a tarsometatarsal arch, with corresponding changes in the shapes and positions of the tarsals and with close, parallel alignment of all five metatarsals and digits. In addition, the big toe is no longer opposable, as in apes and monkeys (**Figure 24–20**), although it may still have had some capacity to diverge from the rest of the toes in early hominins, especially in *Ardipithecus*.

There are almost as many hypotheses about the reasons for human bipedality as there are anthropologists. Among the suggested reasons are improved predator avoidance (being able to look over

tall grass), freeing the hands for carrying objects (for either hunting or collecting food), thermoregulation (an upright ape presents a smaller surface area to the sun's rays), and energy efficiency of locomotion. An obvious problem with all such hypotheses is that they are difficult to test and not mutually exclusive. Although humans are extremely efficient at bipedal locomotion, especially at walking, it seems unlikely that bipedality evolved specifically for efficient, striding locomotion. Other apes are not nearly as efficient as humans at bipedal walking, but this is the evolutionary condition from which human bipedality must have commenced. That is, a protohominin or early hominin must have walked bipedally in an inefficient way first, before selection could act to increase efficiency, and in fact human locomotion is not particularly efficient when compared to that of quadrupedal mammals.

A tendency to walk bipedally appears to be a basal hominoid feature, but its antiquity is obscured by the knuckle walking of gorillas and chimpanzees. Knuckle walking is a derived trait within primates, but the situation for the great apes is not clear. Knuckle walking might have evolved independently in chimpanzees and gorillas, or it might have been present in the common ancestor of great apes and hominins and been lost from the hominin lineage.

More basal apes—gibbons and orangutans—tend to walk in a clumsy bipedal stance when they are on the ground. Their upright trunk, adapted for arboreal locomotion, predisposes them to do this. Wild orangutans walk bipedally in the trees, supporting themselves by grasping overhead branches above with their hands. This behavior allows them to move out onto narrower supports than they could reach quadrupedally or even by hanging underneath the branch. Although orangutans are not on the direct line to humans, this behavior shows that bipedal walking could have evolved in trees instead of on the ground.

Origin of Large Brains

The human brain increased in size threefold over a period of about 2.5 million years. Human brains are not simply larger versions of ape brains but have a number of key differences, such as a relatively much larger prefrontal cortex and relatively smaller olfactory bulbs. We still do not know what selective pressures led to humans evolving such large brains. Speculations include increasing ability for social interactions, conceptual complexity, tool use, dealing with rapidly changing ecological conditions, language, or a mixture of these elements.

Brain tissue is metabolically expensive to grow and to maintain; on a gram-for-gram basis, brain tissue has a resting metabolic rate 16 times that of muscle. Most of the growth of the brain occurs during embryonic development, but human brains continue to grow after birth and require continued energy input from the mother. Thus, selective pressures for larger brains can be satisfied only in an environment that provides sufficient energy, especially to pregnant and lactating females. The evolution of larger brains may have required increased foraging efficiency (partially achieved through larger female size and mobility) and high-quality foods in substantial quantities (partially achieved through the use of tools and fire). The increase in brain size commencing in *Homo erectus* has been ascribed to the development of cooking, because cooked food requires less energy to digest and thus frees up energy for the brain. However, there is no evidence for cooking with fire until about 500,000 years ago, more than a million years later than the first *H. erectus*.

Larger brains also would have required a change in life-history pattern that probably exaggerated the ancestral primate character of slow rates of pre- and postnatal development, thus lowering daily energy demands and also a female's lifetime reproductive output. Our prolonged period of childhood may allow children to stay with the family long enough to acquire the necessary knowledge for survival from their parents. (Imagine the chaos if human children became sexually mature at about 3 years old, as horses do.) There is also speculation that human menopause, a post-reproductive time span that is not seen in other female mammals, may be a more effective way for a woman to increase the representation of her alleles in the population than having more children of her own near the end of her life span.

The origin of bipedality may be linked to the origin of large brain size and a change in human reproductive biology and life history: only after bipedality had evolved would a female hominin be able to carry the highly altricial type of human newborn. Whereas apes give birth to young that can cling to their mother's fur a few weeks after birth, our large brain necessitates that we are born relatively helpless, with much more brain growth after birth than in other mammals. (If we were born with brains almost fully formed, birth would be even more of a problem for human mothers than it is now.)

An alternative hypothesis proposes that the change in the structure of the foot that is associated with bipedal walking was the key element that made infant hominins more helpless than infant chimpanzees. The big toes of hominins, unlike those of chimpanzees, are not divergent—that is, they do not project at an angle from the foot. A projecting big toe allows the foot to grasp branches when a chimpanzee is climbing, but it does not permit the heel-to-toe rotation of the foot that hominins use when they walk. Without a projecting big toe, infant hominins could no longer use their hind feet to cling to their mother's hair, and the mother would have had to carry the infant. If this hypothesis is correct, the helplessness of hominin infants dates back well before the increase in the size of the brain.

Origin of Speech and Language

Although other animals can produce sounds and many mammals communicate by using a specific vocabulary of sounds, as anyone who has kept domestic pets well knows, the use of a symbolic language is a uniquely human attribute. Although gorillas and chimpanzees have been taught to use some human words and form simple sentences, this is a long way from the complexity of human language. Where in human evolution did language evolve, and how can we find out from the fossil record? The first evidence of human writing is only a few thousand years old; obviously, language evolved before this, but how long before?

Controlled speech might not not have been possible until a later stage than *Homo erectus*. In *H. erectus,* the spinal cord in the region of the thorax is much smaller than it is in modern humans. This observation suggests that *H. erectus* lacked the capacity for the complex neural control of the intercostal muscles that allows modern humans to control breathing in such a way that we can talk coherently. Furthermore, the hypoglossal canal (the exit from the skull for cranial nerve XII, which innervates the tongue muscles) is smaller in other hominins (and is also smaller in chimpanzees and gorillas) than in the modern humans and Neandertals. Additionally, a specific gene involved in the production of language in humans, *FOXP2*, has the same two differences from the chimpanzee condition in both modern humans and Neandertals.

Even if more derived *Homo* species had evolved the capacity for language, they would not have been able to produce the range of vowel sounds that we can produce until a change in the anatomy of the pharynx and vocal tract had taken place. The original position of the mammalian larynx is high in the neck, right behind the base of the tongue. However, in modern humans, the larynx shifts ventrally at the age of 1 to 2 years, resulting in the creation of a much larger resonating chamber for vocalization. This change in the vocal tract anatomy is associated with a change in the shape of the base of the skull, so we can infer from fossil skulls when the shift in larynx position occurred. Although there are some differences of opinion, a fully modern condition of the vocal tract was probably not a feature of the genus *Homo* until *H. sapiens* about 50,000 years ago. Speech also requires a change in neural capacities in order to process the rapid frequency of transmitted sounds, which are decoded at a rate much faster than other auditory signals.

Some evolutionary problems with the origin of the vocal tract trait still exist. The descent of the larynx means that the original mammalian seal between the palate and the epiglottis is lost, making humans especially vulnerable to choking on their food. It seems likely that a risk of choking would have been a powerful selective force acting in opposition to repositioning the larynx. In addition, the death of infants from SIDS (sudden infant death syndrome) is associated with the developmental period when the larynx is in flux. It is difficult to imagine that the ability to produce a greater range of vowel sounds could counteract these antagonistic selective forces. Perhaps there was another, more immediately powerful reason for repositioning the larynx.

One advantage that repositioning the larynx affords us is the ability to voluntarily breathe through our mouths, an obviously important trait in the production of speech. However, perhaps a more important function of mouth breathing is apparent to anyone who has a bad cold. Many of us would suffocate every winter if we were unable to breathe through our mouths. It is tempting to speculate that the human capacity for well-enunciated speech owes its existence to a prior encounter of our species with the common cold virus.

Loss of Body Hair and Development of Skin Pigmentation

Humans are unique among primates in their apparent loss, or at least reduction, of body hair and their development of heavily pigmented skin. (Humans have the same number of hair follicles as other apes, but the hairs themselves are minuscule.) These features are related, as fur protects animals from deleterious effects of the sun's rays. Chimpanzees have relatively unpigmented skin, but with the loss of body hair there would be a need to gain skin pigments. It is not clear why humans lost the majority of their body hair. Speculations include increased use of eccrine sweat glands in evaporative cooling (possibly important in hunting) and increased problems with skin parasites such

as lice, ticks, and fleas (perhaps in association with a more sedentary way of life with groups of people living together in confined quarters). However, if cooling during hunting was the driving force, then we would expect men to be less hairy than women, rather than the reverse.

Although we will probably never know precisely why humans lost their covering of hair, genetics enables us to figure out when this might have happened. The genes involved in human skin pigmentation appear to date from at least 1.2 million years ago. This implies that hairlessness first came about with the lineage leading to *Homo sapiens*, perhaps also including the *H. erectus* lineage that branched off about 1.7 million years ago. This corresponds with the time when hominins adopted a home base, which might have made them especially prone to parasitic infections and suggests that we shared hairlessness with other species, such as the Neandertals. This hypothesis would also explain the tendency for females to be less hairy than males, as they would most likely spend more of their time at the home base.

A study of the genetics of skin parasites provides information about when humans started to wear clothes. The human body louse is different from head lice and pubic lice in that it clings to human clothing rather than to hair. This parasite presumably evolved from the head louse after humans started to wear clothes, and this separation occurred between 40,000 and 70,000 years ago, broadly coincident with the emergence of modern *Homo sapiens* in Europe.

Humans living close to the equator have dark skin, whereas people who live farther from the equator have lighter skin. A balance between protection against the damaging effects of ultraviolet light and the need for vitamin D synthesis appears to offer the best general explanation of this phenomenon. Folic acid is essential for normal embryonic development, and ultraviolet light breaks down folic acid in the blood. Melanin in the skin blocks penetration by ultraviolet light, thereby protecting folic acid. Too much melanin in the skin creates a different problem, however, because vitamin D is converted from an inactive precursor into its active form by ultraviolet light in blood capillaries in the skin. Thus, human skin color probably represents a compromise—enough melanin to protect folic acid while still permitting enough ultraviolet penetration for vitamin D synthesis.

Origins of Human Technology and Culture

Tool use has been recorded in three lineages of vertebrates (fishes, birds, and mammals), but only passerine birds and primates use tools for a variety of tasks.

New Caledonian crows tear stiff leaves to make probes they use to extract wood-boring insect larvae from cavities, matching the size of the probe they manufacture to the size of the opening. In laboratory tests, two crows, Betty and Abel, selected a wire bent into a hook in preference to a straight wire to pull a piece of food from a vertical pipe. When Abel stole the hook, Betty bent the straight wire into a hook and used it to retrieve the food item.

Passerines use tools primarily to capture, prepare, or transport food, and only 3 percent of passerine species use tools in multiple ways. In contrast, primates use tools for the same activities as birds and in other ways as well. Both baboons and chimpanzees use sticks and stones as weapons very much in the way that ancestral humans must have done, for example, and chimpanzees use these same types of materials as tools in obtaining food.

The earliest recognized simple stone tools are found throughout East Africa and date to 2.5 to 2.7 million years ago; they may have been made by either *Homo habilis* or *Australopithecus robustus*. Both of these hominins, but not earlier australopithecines, had a thumb joint that would have allowed a precision grip. The so-called Oldowan tools are generally circular and worked on only one side. They remained relatively unchanged for a million years, and similar tools are made by extant hunter gatherer societies such as the Australian Aborigines. Acheulean tools, which differ from Oldowan tools in having a distinct long axis and being chipped on both sides, appeared in East Africa about 1.4 million years ago. These tools, which include cleavers and so-called hand axes, were apparently made by *Homo erectus* and changed very little in style for the next 1.2 million years. The lack of any dramatic advances in tool manufacture over this immense time period is surprising, especially in light of the spread of these tools to southwest Asia and Western Europe. A remarkable conclusion from paleoanthropological findings is that the use of tools precedes the origin of the big-brained *Homo sapiens* by at least 1.5 million years. The use of tools by earlier hominins may have been a major factor in the evolution of the modern *Homo* type of cerebral cortex; in fact, the elaborate brain of *Homo sapiens* may be the consequence of culture as much as its cause.

By at least 750,000 years ago, *Homo erectus* was making advanced types of stone tools and had also apparently learned to control and use fire. With fire, humans could cook their food, which increases its digestibility and decreases the chance of bacterial infection, and preserve meat for longer periods of time than it would remain usable in a raw state. Humans could also keep themselves warm in cold weather, ward off predators, and light up the dark to see, work, and socialize.

It is often assumed that much of the evolution of human tool use and culture developed in the context of humans hunting other animals. The term "Man the Hunter" was coined in the 1960s. Anthropologists have pointed out that much of our perception of human evolution as being an upward and onward quest has more to do with Western cultural myths of the "Hero's tale" than with anthropological data. The archeological evidence for hunting by early hominins, such as stone tools and bones with cut marks, suggests that early humans were scavengers rather than hunters. Furthermore, the corollary phrase—"Woman the Gatherer"—has been discredited. Studies of the few remaining hunter gatherer societies have shown that tasks are shared by both sexes: men do substantial amounts of gathering, and men and women both contribute to the heavy tasks of butchering large prey, cutting it into manageable pieces, and carrying it back to the home base.

24.6 How Many Species of Humans Were Contemporaneous?

We now live in an unusual time for hominins because only in the past 12,000 years has there been just one species of hominin. Throughout hominin history (possibly with the exception of the first 500,000 years) several species of hominins have coexisted. As recently as 30,000 years ago our species, *Homo sapiens*, shared the planet with at least three other species—*H. neanderthalensis*, the Denisovans, and *H. floresiensis*—and *H. erectus* might have been present as well.

Homo erectus was long believed to have disappeared between 200,000 and 300,000 years ago, but recent discoveries have shown that it survived at least until 150,000 years ago and possibly until 30,000 or 40,000 years ago. Remains of *H. erectus* from Java have been dated to between 27,000 and 53,000 years old by two methods and to between 40,000 and 60,000 years old by a third method. The sediments in which these fossils are found are complex and hard to decipher and these young dates are controversial, but if they are correct late-surviving *H. erectus* were contemporaries of Neandertals, *H. floresiensis*, Denisovans, and modern *H. sapiens*. It is hard to imagine how it would feel to live with other species of humans to whom we were as closely related as dogs are to wolves and coyotes.

Summary

Evidence for the origin of *Homo sapiens* comes from the Cenozoic fossil record of primates and from comparative studies of living monkeys, apes, and humans. The first primatelike mammals are known from the early Paleocene, but the earliest true primates are not known until the early Eocene. These early primates were similar to the extant lemurs and were initially present in North America and the Old World. All were arboreal, and some had larger brains in relation to their body size than had other mammals of that time.

After the climatic deterioration of the late Eocene, primates were confined to the tropical regions until the middle Miocene. The anthropoids evolved in Asia by the middle Eocene and soon spread to Africa. By the Oligocene, two distinct groups of extant anthropoids had evolved: the platyrrhine monkeys of the New World tropics and the catarrhine monkeys and apes of the Old World. The apes and humanlike species, including *Homo sapiens*, are grouped in the Hominoidea. Many morphological features distinguish the hominoids from other catarrhines, and enlargement of the brain has been a major evolutionary force molding the shape of the hominoid skull, especially in the later part of human evolution.

The first known hominoids occur in the early Miocene, about 25 million years ago. By the late Miocene, hominoids had diversified and spread throughout Africa, Europe, and Asia. The genetic closeness of humans and great apes (orangutans, gorillas, and chimpanzees) has led to a regrouping of the traditional family Hominidae to include these apes as well as humans.

Humans and their fossil relatives have now been assigned to a tribe within the Hominidae, the Hominini. A variety of hominin fossils occurs in late Pliocene and early Pleistocene deposits of Africa. The earliest well-known hominins were the australopithecines, known from 4.4 to 1.2 million years ago. Australopithecines were bipedal but retained many arboreal characters and had a relatively small apelike cranial volume.

The earliest member of the genus *Homo* dates from about 2.5 million years ago, concurrent with the earliest stone tools found in East Africa. *Homo* appears to be a primarily terrestrial genus. A global climatic change at around this time may have prompted the evolution both of *Homo* and of the robust australopithecines.

Homo erectus lived in Africa and Eurasia from about 1.8 million years ago to at least as recently as 150,000 years ago and possibly up to 30,000 years ago. This hominin had a brain capacity approaching the lower range of *Homo sapiens*, made stone tools, and used fire. *Homo sapiens*, the only surviving species of the tribe Hominini, came into existence about 200,000 years ago. By about 40,000 years ago, some populations of *Homo* had a well-organized society with a rapidly developing culture, especially obvious in their use of stone tools and the development of art. For much of human history, several species of the genus *Homo* have coexisted: at least three archaic species of *Homo* were in existence 50,000 years ago, and two of them interbred with modern humans. The present-day situation, with *Homo sapiens* as the sole existing hominin, is a highly unusual one.

Discussion Questions

1. There are no (nonhuman) primates today in North America. What elements of the Earth's climate in the Eocene allowed primates to live in North America, and what climatic changes accompanied their disappearance from this continent?

2. During the Miocene, when temperatures became warm at higher latitudes, monkeys and apes radiated from Africa into Eurasia. What prevented primates from reappearing in North America during this time?

3. What is the functional significance of the S-shaped bend in the vertebral column of *Homo sapiens*?

4. The South American monkeys (platyrrhines) first appear in the fossil record of that continent in the Oligocene, and we deduce that they must have arrived there at least by the late Eocene. This reasoning involves some rather complex arguments about crossing the Atlantic Ocean. Why don't we think that the platyrrhines might have reached South America a lot earlier when the Atlantic Ocean was narrower and traversing it would have been easier?

5. Humans originated in Africa, and their closest relatives, the chimps and gorillas, are known only from Africa. So why is there speculation that this lineage (the Homininae) had its origins in Eurasia?

6. Suppose that Bigfoot (or Sasquatch, the legendary primate that is imagined to live in the Pacific Northwest) or the Yeti (the equally imaginary primate that is supposed to live in the Himalayas) were found to be real and derived from *Homo erectus*. What moral responsibility would we (*Homo sapiens*) have to a species that was as closely related to us as a wolf is to a dog?

Additional Information

Multiple authors. 2009. *Ardipithecus ramidus*. *Science* 326 (5949):36–43 and 60–74.

Andrews, P., and E. Delson. 2000. Hominoidea. In E. Delson et al. (eds.), *Encyclopedia of Human Evolution and Prehistory*, 2nd ed. New York: Garland Publishing.

Bajpai, S., et al. 2008. The oldest Asian record of Anthropoidea. *Proceedings of the National Academy of Sciences* 105:11093–11098.

Bakewell, M. A., et al. 2007. More genes underwent positive selection in chimpanzee evolution than in human evolution. *Proceedings of the National Academy of Sciences* 104:7489–7494.

Begun, D. R. 2010. Miocene hominids and the origins of the African apes and humans. *Annual Review of Anthropology* 39:67–84.

Bentley-Condit, V. K. and E. O. Smith. 2010. Animal tool use: current definitions and an updated comprehensive catalog. *Behaviour* 147:185–221.

Berger, L. R., et al. 2010. *Australopithecus sediba*: A new species of *Homo*-like australopith from South Africa. *Science* 328:195–204.

Bokma, F., et al. In press. Unexpectedly many extinct hominins. *Evolution* 66.

Brown, P. 2012. LB1 and LB6 *Homo floresiensis* are not modern human (*Homo sapiens*) cretins. *Journal of Human Evolution* 62:201–204.

Brown, P., et al. 2004. A new small-bodied hominin from the Late Pleistocene of Flores, Indonesia. *Nature* 431:1055–1061.

Brunet, M., et al. 2005. New material of the earliest hominid from the Upper Miocene of Chad, Central Africa. *Nature* 434:752–755.

Cerling, T. E., et al. 2011. Diet of *Paranthropus boisei* in the early Pleistocene of East Africa. *Proceedings of the National Academy of Sciences* 108:9337–9341.

Chaimanee, Y., et al. 2012. Late Middle Eocene primate from Myanmar and the initial anthropoid colonization of Africa. *Proceedings of the National Academy of Sciences* In press.

Crompton, R. H., et al. 2010. Arboreality, terrestriality and bipedalism. *Proceedings of the Royal Society B* 365:3301–3314.

Enard, W., et al. 2002. Molecular evolution of *FOXP2*, a gene involved in speech and language. *Nature* 418:869–872.

Finlayson, C. 2005. Biogeography and evolution of *Homo*. *Trends in Ecology and Evolutionary Biology* 20:457–463.

Finlayson, C., et al. 2006. Late survival of Neandertals at the southernmost extreme of Europe. *Nature* 443:850–853.

Finlayson, C., and J. S. Carrión. 2007. Rapid ecological turnover and its impact on Neandertal and other human populations. *Trends in Ecology and Evolutionary Biology* 22:213–222.

Franzen, J. L., et al. 2009. Complete primate skeleton from the middle Eocene of Messel in Germany. *PLoS ONE* 4:e5723.

Gibbons, A. 2009. Celebrity fossil primate: Missing link or weak link? *Science* 324:1124–1125.

Gibbons, A. 2010. Close encounters of the prehistoric kind. *Science* 328:680–684.

Gibbons, A. 2011. A new view of the birth of *Homo sapiens*. *Science* 331:392–394.

Gibbons, A. 2011. Who were the Denisovans? *Science* 333:1084–1087.

Graves, R. R., et al. 2010, Just how strapping was KNM-WT 15000? *Journal of Human Evolution* 59:542–554.

Green, R. E. 2010. A draft sequence of the Neandertal genome. *Science* 328:710–722.

Harrison, T. 2010. Apes among the tangled branches of human origins. *Science* 327:532–534.

Jablonski, N. G., and G. Chaplin. 2003. Skin deep. *Scientific American* 13(2):72–79.

Kay, R. F., et al. 1998. The hypoglossal canal and the origin of human vocal behavior. *Proceedings of the National Academy of Science, USA* 95:5417–5419.

Kenward, K. et al. 2011. On the evolutionary and onotgenetic origins of tool-oriented behaviour in New Caledonian crows (*Corvus moneduloides*). *Biological Journal of the Linnean Society* 102:870–877.

Kittler, R., et al. 2003. Molecular evolution of *Pediculus humanus* and the origin of clothing. *Current Biology* 13:1414–1417.

Krause, J., et al. 2007. The derived *FOXP2* variant of modern humans was shared with Neandertals. *Current Biology* 17:1908–1912

Lieberman, P. 2007. The evolution of human speech: Its anatomical and neural bases. *Current Anthropology* 48:39–66.

Lordkipanidze, D., et al. 2007. Postcranial evidence from early *Homo* from Dmanisi, Georgia. *Nature* 449:305–310.

McLean, C. Y., et al. 2011. Human-specific loss of regulatory DNA and the evolution of human-specific traits. *Nature* 471:216–219.

McNulty, K. P. 2010. Apes and tricksters: The evolution and diversification of human's closest relatives. *Evolution and Education Outreach* 3:322–332.

Mellars, P., et al. 2011. Tenfold population increase in Western Europe at the Neandertal-to-modern human transition. *Science* 333:623–627.

Morwood, M. J. 2004. Further evidence for small-bodied hominins from the Late Pleistocene of Flores, Indonesia. *Nature* 437:1012–1071.

Patterson, N., et al. 2006. Genetic evidence for complex speciation of humans and chimpanzees. *Nature* 441:1103–1108.

Pennisi, E. 2012. The burdens of being a biped. *Science* 336:974.

Prüfer, K., et al. 2012. The bonobo genome compared with the chimpanzee and human genomes. *Nature* 486:527–531.

Rantala, M. J. 2007. Evolution of nakedness in *Homo sapiens*. *Journal of Zoology* 273:1–7.

Reich, D., et al. 2010. Genetic history of an archaic hominin group from Denisova Cave in Siberia. *Nature* 468:1053–1060.

Richmond, B. G., and W. L. Jungers. 2008. *Orrorin tugenensis* femoral morphology and the evolution of hominin bipedalism. *Science* 319:1662–1665.

Rogers, A. R., et al. 2004. Genetic variation at the *MC1R* locus and the time since loss of body hair. *Current Anthropology* 45:105–108.

Seiffert, E. R., et al. 2009. Convergent evolution of anthropoid-like adaptations in Eocene adapiform primates. *Nature* 461:1118–1122.

Shoenemann, P. T. 2006. Evolution of the size and functional areas of the human brain. *Annual Review of Anthropology* 35:379–406.

Strait, D. S. 2010. Evolutionary history of the australopithecines. *Evolution and Education Outreach* 3:341–352.

Tattersall, I., and J. H. Schwartz. 2009. Evolution of the genus *Homo*. *Annual Review of Earth and Planetary Sciences* 37:67–92.

Taylor, A. H. et al. 2011. New Caledonian crows learn the functional properties of novel tool types. *PLoS ONE* 6(12): e26887. doi:10.1371/journal.pone.0026887

Thorpe, S. K. S., et al. 2007. Origin of human bipedalism as an adaptation for locomotion on flexible branches. *Science* 316:1328–1331.

Tocheri, M. W., et al. 2007. The primitive wrist of *Homo floresiensis* and its implications for hominin evolution. *Science* 317:1743–1745.

Ward, C. V., et al. 2011. Complete fourth metatarsal and arches in the foot of *Australopithecus afarensis*. *Science* 331:750–753.

White, T. D., et al. 2006. Asa Issie, Aramis, and the origin of *Australopithecus*. *Nature* 440:883–889.

White, T. D., et al. 2009. *Ardipithecus ramidus* and the paleobiology of early hominids. *Science* 326:75–86. (See also 10 other papers in the same issue.)

Williams, B. A., et al. 2010. New perspectives on anthropoid origins. *Proceedings of the National Academy of Sciences* 107:4797–4804.

Wood, B., and T. Harrison. 2011. The evolutionary context of the first hominins. *Nature* 470:347–352.

Yokoyama, Y., et al. 2008. Gamma-ray spectrometric dating of late *Homo erectus* skulls from Ngandong and Sambungmacan, Central Java, Indonesia. *Journal of Human Evolution* 55:274–277.

Websites

Diversity of living primates

Palomar College http://anthro.palomar.edu/primate/

Human evolution

American Museum of Natural History http://www.amnh.org/exhibitions/permanent/humanorigins/

Australian Museum of Natural History http://australianmuseum.net.au/Human-Evolution/

Institute of Human Origins http://iho.asu.edu/

John Hawks Weblog http://johnhawks.net/weblog

National Museum of Natural History http://humanorigins.si.edu/sitemap

Natural History Museum http://www.nhm.ac.uk/nature-online/life/human-origins/

CHAPTER 25

The Impact of Humans on Other Species of Vertebrates

A review of the biology of vertebrates must include consideration of the effect of the current dominance of one species, *Homo sapiens,* on other members of the vertebrate clade. Never before in Earth's history has a single species so profoundly affected the abundance, and even the prospects for survival, of other species.

Some of the influence of humans derives from the size of our population—almost 7 billion now and projected to increase to more than 9 billion in 2050—and from our worldwide geographic distribution. However, other species of vertebrates match or surpass humans in numbers, and some at least come close to matching us in geographic distribution. Technology and the consequences of advances in technology are what set humans apart from other vertebrates. Human technology began with stone tools some 2.5 million years ago, and increased consumption of resources has accompanied increased technology. Even existing human hunter gatherer societies use more resources than do other species of animals. Extraction and consumption of energy and other resources by human societies have worldwide impacts ranging from oil spills and contamination of soil and water with heavy metals to the loss of wild habitats to agriculture and urbanization. Per capita energy use has tripled since 1850, and most of the increase has occurred in the past few decades. The most highly industrialized societies have the highest per capita rates of consumption of energy and other resources. The United States, for example, has about 4.5 percent of the world's population and accounts for about 25 percent of the world's consumption of resources. The developed nations as a group have 20 percent of the world's population and are responsible for 60 percent of resource consumption. Societies with high rates of consumption have correspondingly high rates of production of greenhouse gases and waste products.

Less technologically developed societies are not necessarily ecologically benign. They present special problems as growing human populations increasingly impinge on areas that have so far been relatively undisturbed. Wealthy nations at least have the resources to control pollution and create national parks, monuments, and wildlife reserves. Poorer societies struggle daily to meet their basic survival needs and understandably regard conservation as a luxury beyond their reach. Perhaps the most insidious process affecting the relationship between humans and other vertebrates is the spread of Western cultural values that emphasize material possessions, thereby increasing demand for consumer products and expanding the geographic influence of high-impact societies.

This book has emphasized the evolutionary history of vertebrates and their characteristics as organisms, and both kinds of information are essential parts of efforts to conserve natural habitats and to protect endangered species. We tend to assume that the natural state of an environment is the way it was at the time of the first written record, but fossil evidence shows that humans had enormous impacts on vertebrate faunas long before writing was invented some 5000 years ago. Extinctions of vertebrates due to human activities began about 50,000 years ago, and they have accelerated ever since.

The scientific study of vertebrates dates back only a few hundred years, but it, too, has accelerated rapidly. Information about the biology of vertebrates can be used to identify causes of population declines and extinctions and possibly to prevent some species that are currently endangered from becoming extinct.

25.1 Humans and the Pleistocene Extinctions

Starting from the appearance of the earliest vertebrates in the Late Cambrian or Ordovician period, the diversity of vertebrates increased slowly throughout the Paleozoic and early Mesozoic eras, and then more rapidly during the past hundred million years. This overall increase has been interrupted by eight periods of extinction for aquatic vertebrates and six for terrestrial forms. Extinction is as normal a part of evolution as species formation, and the duration of most species in the fossil record appears to be from 1 million to 10 million years. Periods of major extinction (a reduction in diversity of 10 percent or more) are associated with changes in climate and the consequent changes in vegetation. But that pattern of reasonably long-lived species and extinctions correlated with shifts in climate and vegetation changed at about the time that humans became a dominant factor in many parts of the world.

For example, the number of genera of Cenozoic mammals reached a peak in the mid-Miocene and a second peak in the early Pleistocene epoch. Only 60 percent of the known Pleistocene genera are living now, and the extinct forms include most species of large terrestrial mammals—those weighing more than 20 kilograms. These large mammals, plus some enormous species of birds and reptiles, are collectively called the Pleistocene megafauna. The term is most commonly applied to North American species, such as ground sloths, mammoths, mastodons, and the giant Pleistocene beaver that was the size of a bear, but other continents also had megafaunas that became extinct during the Pleistocene.

The first humans to reach Australia were met by a megafauna that included representatives of four groups: marsupials, flightless birds, tortoises, and echidnas. The largest Australian land animals in the Pleistocene were several species of herbivorous mammals in the genus *Diprotodon*, the largest of which probably weighed about 2000 kilograms. The largest kangaroos weighed 200 kilograms, and one of them may have been carnivorous. The largest echidnas reached weights of 20 to 30 kilograms and were waist high to an adult human. The flightless bird *Genyornis newtoni* was twice the height of a human, and the horned turtles in the family Meiolaniidae were nearly as large as Volkswagen Beetle cars. Perhaps the most dramatic species in the Australian megafauna was a monitor lizard at least 6 meters long (as large as a medium-sized *Allosaurus*). The turtles, the monitor lizard, *Genyornis,* and all marsupial species weighing more than 100 kilograms became extinct in the late Pleistocene.

The role that humans played in the extinctions of large animals in the Pleistocene was discussed in Section 19.4. Specialists agree that the arrival of a new, technologically advanced species of predator would inevitably have an impact on other species, but the relative importance of the direct effect of hunting versus indirect effects resulting from changes in the habitat and the introduction of new pathogens remains a subject of debate.

The most striking evidence of the impact of humans on other vertebrates is the apparent correspondence in the times of the arrival of humans on continents and islands and the extinction of the megafauna (**Figure 25–1**). At least 20 genera of giant mammals (marsupials and monotremes), birds, and non-avian reptiles were extinct by 40,000 years ago. The reduced herbivore pressure that resulted from the loss of these species triggered the transition from a mixed tropical forest vegetation to the dry **sclerophyll** vegetation that characterizes much of Australia today. In North and South America, at least eight species of large mammals survived until about 10,000 years ago. Humans colonized islands later than continents, and extinctions occurred between 10,000 and 4000 years ago on islands in the Mediterranean Sea, 4000 years ago on islands in the Arctic Ocean north of Russia, 2000 years ago on Madagascar, and only a few hundred years ago on islands in the Pacific Ocean. In each of these cases, the dates of extinctions closely follow the dates when humans are believed to have arrived.

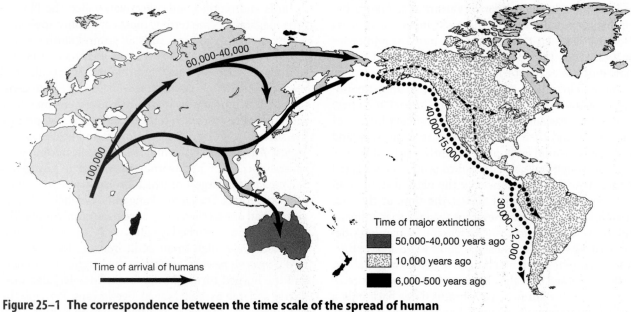

60,000–40,000

100,000

40,000–15,000

30,000–12,000

Time of major extinctions

■ 50,000–40,000 years ago

▒ 10,000 years ago

■ 6,000–500 years ago

Time of arrival of humans

Figure 25–1 The correspondence between the time scale of the spread of human populations and the extinction of native species of vertebrates. Modern humans left Africa about 100,000 years ago and spread eastward across Eurasia and into the Americas. The dates of extinctions of Pleistocene megafauna correspond with estimates of when humans arrived in Australia, the Americas, and major islands such as New Zealand, Madagascar, and the islands north of Siberia.

The Overkill Hypothesis

Overhunting has been proposed as the primary reason for megafaunal extinctions, and some evidence for overhunting is dramatic. There were 11 species of moa (giant flightless birds) on New Zealand when the Maori arrived in the late thirteenth century, and all appear to have become extinct within 100 years. The role of Maori hunters in the extinction of moa is amply documented. On the North Island of New Zealand a butchering site was discovered on the sand dunes at Koupokonui. The remains of hundreds of individuals of three species of moa were found in and around ovens. Uncooked moa heads and necks had been left in piles to rot, while the legs were roasted. At Wairau Bar on the South Island the ground is littered with the bones of moa, an estimated 9000 individuals plus 2400 eggs, and at Waitaki Mouth are the remains of an estimated 30,000 to 90,000 moa.

The best-documented examples of the impact of overhunting come from islands and are relatively recent. It is less certain that hunting alone was responsible for continent-wide megafaunal extinctions. Although hunting was probably part of the equation, diseases and habitat changes also played a role.

Disease

Although there is no paleontological evidence of disease in the Pleistocene megafauna, modern examples of transmission of disease from domestic animals to wild species abound, and emerging infectious diseases now threaten both wild animals and humans. Within the past two decades, lions and wild dogs in Africa have been infected by canine distemper transmitted from domestic dogs. Wild dogs (*Lycaon pictus*) are now nearly extinct in the Serengeti Plains. Less than a century ago, packs contained 100 or more animals, but today packs consist of only 10 or so adults. The population of wild dogs on the Serengeti is less than 60 animals, and the entire surviving population of the species is no more than 5000 individuals.

Other examples of the transmission of diseases from humans and our pets to wild animals can be cited: Rabies has killed a substantial proportion of the approximately 500 remaining Ethiopian wolves, the parasite *Toxoplasma gondii* from domestic cat feces has killed large numbers of sea otters, and Ebola virus has been responsible for the deaths of thousands of gorillas and chimpanzees. Wild mountain gorillas in Uganda have contracted mange from parasitic mites in clothing discarded by tourists, and if measles or tuberculosis gets into the gorilla population, the results will be devastating. A new twist has been added by the discovery that human influenza B virus caused a respiratory infection in harbor seals on the Dutch coast, and the seals are now a reservoir for the virus that could pose a threat to human health.

A newly introduced disease does not kill every individual of a previously healthy population. Some individuals survive, perhaps because they are resistant or maybe just because they are lucky enough to avoid infection. Nonetheless, a disease that drastically reduces the number of individuals of a species may start a process that leads to extinction, and this may be happening to some populations of African wild dogs. The hunting method of wild dogs—prolonged pursuit of antelope until an individual antelope is captured—is energetically expensive. A pack of wild dogs hunts cooperatively, with one individual taking up the chase as another tires. The drastic reduction that has occurred in pack size means that each dog must work harder. To make things worse, spotted hyenas (*Crocuta crocuta*) steal the kills made by wild dogs, and a small pack of dogs probably has more difficulty defending its kills from hyenas than does a larger pack. Wild dogs normally hunt for about 3.5 hours per day, and calculations of the energy cost of hunting and the energy gained from prey show that the dogs just meet their daily energy needs on this schedule. If hyenas steal some of the kills, the wild dogs must increase the time they spend hunting. A 10 percent loss of prey would force the wild dogs to double their hunting time, and a 25 percent loss would force them to spend 12 hours a day hunting. Wild dogs are already working at nearly their physiological limits when they hunt for 3.5 hours per day, and they probably cannot survive if they lose much food to hyenas. Thus, a drastic reduction in pack size sets the stage for a competitive interaction with hyenas, and this interaction could be the factor that drives a pack of wild dogs to extinction.

Fire

With the arrival of modern humans came the use of fire to manipulate the habitat. *Genyornis newtoni* was a flightless bird that inhabited inland plains and some coastal areas of Australia when the first humans arrived about 50,000 years ago. Although *Genyornis* was a ponderous bird and was probably less fleet footed than emus (*Dromaius novaehollandiae*), there is only one site known with evidence that humans hunted *Genyornis*. The reason that *Genyornis* became extinct and emus survived may lie in their feeding habits. Emus eat a wide variety of items including grasses, whereas the chemical composition of *Genyornis* eggshells suggests that they were browsers, eating leaves from shrubs. Wildfires were a part of the Australian landscape long before humans arrived, but humans changed the fire regime. Natural fires occur during the dry season and do not recur until the vegetation in a burned area has regenerated. The early human inhabitants of Australia may have set fires at other times

of the year and at shorter intervals than the natural fire cycle. A regime of more frequent burning would have converted the shrub lands that *Genyornis* depended on to the grasslands and spinifex that characterize the inland Australian plains today. Thus, habitat change produced by the new fire regime created by humans may have been responsible for the extinction of *Genyornis*. The large herbivorous mammals that became extinct in Australia were also browsers, and the same habitat changes may have been responsible for their disappearance.

25.2 Humans and Recent Extinctions

As we move closer to the present, the role of humans as the major cause of extinction becomes unambiguous. Excavations of fossils preserved in tubelike lava caves formed by volcanoes show that the Hawaiian Islands probably had more than 100 species of native birds when Polynesian colonists arrived about 1700 years ago. By the time European colonists reached Hawaii in the late eighteenth century, that number had been reduced by half.

The Age of Exploration, which began in the fifteenth century, brought sophisticated weapons and commercial trade to areas that had known only stone tools and hunter gatherer economies. Ships sailed from Europe to all corners of the world, stopping en route to renew their supplies of food and water from oceanic islands. Not surprisingly, extinctions on islands began about two centuries before extinctions on continents (**Figure 25–2**). One notable example is the dodo (*Raphus cucullatus*), a flightless bird related to pigeons that lived on the island of Mauritius in the Indian Ocean. The dodo was last seen alive in 1662 and was almost certainly extinct by 1690. In the Hawaiian Islands about one-third of the native species of birds that were still surviving when Captain Cook arrived in the late eighteenth century are now extinct.

Animal species continue to become extinct today, and a worldwide survey of extinctions since the start of European colonization reveals two trends: island extinctions began almost two centuries earlier than continental extinctions, and both island and continental extinctions have increased rapidly from the early or mid-nineteenth century through the twentieth century. More than 800 species have become extinct in the past 500 years, a rate of extinction that is thought to be 1000 to 10,000 times higher than it would have been without the effect of humans. The increasing application of biotechnology and genetic engineering is creating a new category of risks that go beyond the historical concerns with pollution and disease to include direct interference with natural selection and evolution.

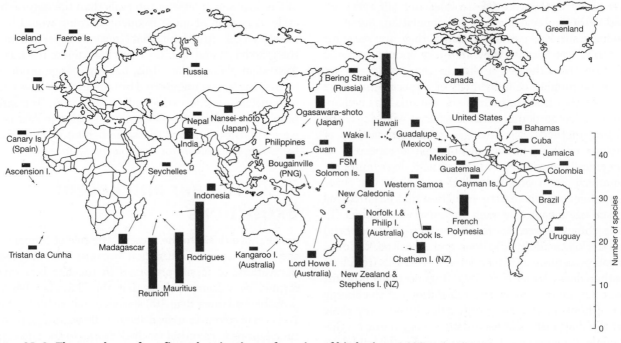

Figure 25–2 The numbers of confirmed extinctions of species of birds since 1600. Islands
have suffered more extinctions than continental areas.

The IUCN (International Union for Conservation of Nature) has summarized the best information available about the conservation status of animals. The 2011 edition of the *IUCN Red List of Threatened Species* lists a total of 20,426 species of plants and animals currently at risk of extinction. They place species in categories of risk by using criteria that focus on the absolute size of wild populations and changes in the populations during the past 10 years.

TWO CATEGORIES OF *EXTINCT* SPECIES ARE DEFINED:

- **Extinct**—A species is Extinct when no living individuals exist. The golden toad (*Bufo periglenes*) from the montane cloud forest of Costa Rica was first described in 1967 and had vanished by 1989. Climate change and infection by chytrid fungi have been proposed as the causes of its extinction.
- **Extinct in the Wild**—A species is Extinct in the Wild when it is known to survive only in cultivation, in captivity, or as a naturalized population (or populations) well outside the past range. The black-footed ferret (*Mustela nigripes*) was a resident of prairie dog towns that once covered thousands of hectares in North America. Habitat loss as land was converted to agricultural use, prairie dog extermination programs, and bubonic plague have reduced the extent of prairie dog

colonies to less than 2 percent of their original area, and black-footed ferrets have suffered from those changes. The last wild ferrets were taken into captivity in the mid-1980s. A captive breeding program was initiated, and more than 200 captive-bred ferrets have been released since 1991. The reintroduced populations of ferrets are growing, and the prospects for successful reestablishment appear to be good.

In contrast, reintroduction of California condors has been less successful. The geographic range of California condors once extended over much of North America, but by 1982 the population had declined to 22 individuals. The last wild condor was brought into captivity in 1987. By 2009 a captive breeding program had raised the number of condors to 350, and 180 of them had been released at sites in California, Utah, Arizona, and Baja California (Mexico). The released birds are not thriving, however. Adult condors are poisoned by lead from bullets and shotgun pellets in carcasses of animals shot by hunters, and nestlings are killed by ingesting anthropogenic microtrash, such as metal nuts and bolts and pieces of wire, glass, and plastic. A review of the condor program published in 2010 concluded that establishing self-sustaining populations of wild California condors is possible, but will require solving a host of interlocking problems, of which lead poisoning is the most urgent.

THREE CATEGORIES OF *THREATENED* SPECIES ARE DEFINED:

- **Critically Endangered**—A species is Critically Endangered when it is facing an extremely high risk of extinction in the wild in the immediate future. The common sturgeon (*Acipenser sturio*) is a large fish (up to 3 meters and 300 kilograms). Its historic range included the entire coastline of Europe from the North Cape to the Baltic, Mediterranean, and Black Seas. The species is now extinct in some of its former spawning rivers, including the Elbe, Rhine, and Vistula. Breeding populations are restricted to a few European rivers—the Gironde in France, the Guadalquivir in Spain, and the lower Danube in Romania and Bulgaria. Overharvesting is the primary cause of its decline. The flesh is prized and the roe used to make caviar, so gravid females are especially sought and are killed before they can reproduce.
- **Endangered**—A species is Endangered when it is facing a very high risk of extinction in the wild in the near future. One example of an endangered species is the giant panda (*Ailuropoda melanoleuca*), which was once distributed throughout Myanmar, northern Vietnam, and a large part of eastern and southern China. Wild pandas now occur only in fragmented populations in mountain ranges in western China; the total population is thought to be about 1200 animals. Attempts to breed giant pandas in captivity outside of China have been generally unsuccessful.
- **Vulnerable**—A species is Vulnerable when it is facing a high risk of extinction in the wild in the medium-term future. The great white shark (*Carcharodon carcharias*) has a worldwide distribution in warm and temperate seas. As a top predator, it has always

had a low population density, and overfishing is considered a potential threat.

THREE CATEGORIES OF *LOWER RISK* ARE RECOGNIZED:

- **Conservation Dependent**—These are species that are being sustained by ongoing conservation programs; without those programs the species would likely qualify for one of the threatened categories within 5 years.
- **Near Threatened**—These are species that are close to Vulnerable status for which no conservation measures are in place.
- **Least Concern**—These species do not qualify for Conservation Dependent or Near Threatened classification.

We simply lack enough information to reach a conclusion about the status of many species that have been evaluated, and these are placed in a category of their own:

- **Data Deficient**—A species is listed as Data Deficient when information about the species' distribution and abundance is insufficient to assess its risk of extinction. Listing a species in this category emphasizes the need for more information. Only species that have been evaluated can be included in this category, and the selection of a species for evaluation indicates that it may be at risk.

A summary of the status of vertebrates in the *2011 IUCN Red List of Threatened Species* shows that from 14 percent to more than 40 percent of the species of vertebrates evaluated are Extinct, Extinct in the Wild, Critically Endangered, Endangered, or Vulnerable (**Table 25–1**). The proportions of mammals, birds, reptiles,

Table 25–1 Status of vertebrates included in the 2011 edition of the IUCN Red List

Group	Number of Species Evaluated (species for which data are deficient are omitted)	Extinct or Extinct in the Wild	Critically Endangered or Endangered	Vulnerable	Percent of Species That Are Extinct, Extinct in the Wild, Critically Endangered, Endangered, or Vulnerable
Mammalia	4665	79	641	497	26
Aves	9990	136	571	682	14
Reptilia	2755	22	421	351	29
Amphibia	4723	39	1262	655	41
Actinopterygii	6791	67	819	1014	28
Sarcopterygii	5	0	1	1	40
Chondrichthyes	812	0	66	115	22

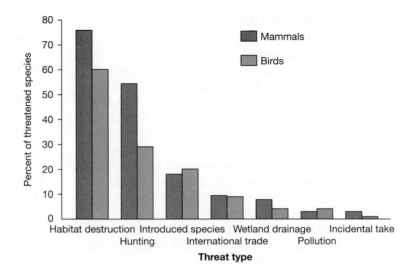

Figure 25–3 **The major threats affecting birds (on a worldwide basis) and mammals (Australasia and the Americas).** Habitat destruction is the single most important threat for both kinds of animals, affecting 60 percent of birds and 76 percent of mammals. Hunting (for food and sport) is a greater threat to mammals than to birds. Introduced species may be predators or competitors, and international trade refers to commercial exploitation for fur, feathers, and the pet trade. Incidental take is the term used to designate accidental mortality, such as dolphins that are drowned by boats fishing for tuna.

and fishes in these categories have changed little since the 2003 issue of the Red List, but the number of amphibians increased more than 10-fold, from 164 species in 2003 to 1956 species in 2011. This dramatic increase is largely the result of the spread of chytrid infections described in Chapter 10.

It is not easy to calculate the rate at which extinctions are occurring, and different assumptions can produce different values. What is clear is that destruction of habitat is the major threat, affecting 60 percent of threatened species of birds and nearly 80 percent of threatened species of mammals (**Figure 25–3**).

25.3 Global Climate Change and Vertebrates

The first worldwide wave of extinctions caused by humans began with the spread of modern humans from Africa about 100,000 years ago and accelerated about 50,000 years ago as humans reached Australia and the Americas. A second wave of extinctions that can be traced to humans began about 500 years ago with the spread of Europeans around the globe during the Age of Exploration. A third wave of anthropogenic extinctions promises to dwarf the two waves that preceded it. *Anthropogenic* means "originating with humans," and that term covers the enormous variety of insults that humans have inflicted on the environment, which extend from habitat destruction to the release of toxic chemicals and bioactive compounds that interfere with embryonic development and alter sex determination. And now another issue has emerged—global climate change—and it will be the dominant environmental problem of this century.

As recently as 10 years ago, the questions about global climate change that were circulating in the scientific community were: Is it real? and Are humans responsible for it? Now scientists know beyond any reasonable doubt that the answer to both of those questions is yes, and discussion focuses on predicting how fast the climate will change and what biological, social, and economic changes will accompany changes in the climate.

Working Groups I and II of the Intergovernmental Panel on Climate Change (IPCC), established by the World Meteorological Organization and the United Nations Environment Programme, have produced a comprehensive analysis of what we can expect: In general, maximum and minimum temperatures will increase, precipitation will decrease, and extreme weather events will become more frequent. All of the continents will experience drought, and its effects will be especially severe in regions that are already on the edge of aridity, including Australia, sub-Saharan Africa, and the American Midwest. The rate of change in the Arctic has been astonishingly rapid, with the ice cover of the Arctic Ocean diminishing by measurable amounts every summer. The Bering Sea is already shifting from an Arctic to a sub-Arctic ecosystem.

Climate has changed many times in Earth's history, and the Pleistocene in particular was marked by periods of glacial cooling and interglacial warming. It is the speed of change in the current episode that is alarming—the mean global temperature is increasing at a rate that exceeds the ability of most plants and animals to adjust.

Predicting the effects of changes in the mean annual air temperature and precipitation on individual species or communities is extraordinarily complicated. The responses of individual species are merely the start of a cascade of interacting events that have positive and negative repercussions on other elements of the community. In an experimental study of a prairie grassland habitat, changes in rainfall that *increased* biodiversity and productivity in the initial 2 years of the project led to a simplification of the food web, so that the final result was *reduced* biodiversity and productivity.

Despite the difficulties of making predictions about individual species, the IPCC paints a bleak picture in its forecasts of the magnitude of the changes in habitats. An analysis by an international team predicted that a minimum of 18 percent of the species in the regions sampled—and perhaps as many as 35 percent—will be in irreversible decline toward extinction by 2050 as a result of global warming.

What Biomes Are Most Affected?

Climate change affects the entire Earth. The global average temperature reached record highs in 2005 and again in 2010—0.62°C above the average for the twentieth century—but some biomes are warming more rapidly than others. The far north shows the most rapid rate of warming. In 2011 the average temperature within the Arctic Circle was 1.5°C higher than the 30-year average (1981–1910), and the temperature over the Arctic Ocean was 3°C higher than that average.

Coral reefs have also been severely affected by climate change. Global sea temperature in the twenty-first century has been about 0.5°C higher than the average temperature for the twentieth century, and temperatures in the Caribbean have risen by about 0.8°C. These temperature changes, combined with increasing acidity, are stressing coral reefs around the world.

Polar Regions The sea ice in the Arctic Ocean has been shrinking rapidly in the twenty-first century, covering the smallest area that has ever been recorded. The rate at which summer ice has been shrinking has exceeded projections, and a nearly ice-free Arctic summer could occur in the next several decades.

Many Arctic marine mammals depend on the sea ice, particularly the sea ice that forms over the continental shelf (that is, in relatively shallow water close to shore), and this is the ice that is disappearing fastest. Ringed seals (*Pusa hispida*) and bearded seals (*Erignathus barbatus*) dive from the ice in pursuit of fish, and invertebrates and walruses (*Odobenus rosmarus*) descend to the seafloor to feed on clams. All of these pinnipeds go out onto the surface of the ice to rest, molt, and give birth. Sea ice is the primary habitat of polar bears; that is where they capture the seals that make up nearly all of their diet.

The timing of the sea ice breakup in the late summer and the formation of new sea ice in autumn are intimately linked to the reproductive cycle of polar bears. Mating occurs in April and May, when the bears are on the sea ice. Implantation of the eggs in the wall of the uterus is delayed, and the fertilized eggs remain in embryonic diapause for about 4 months. During this time the female hunts seals on the ice, accumulating fat—in a good season she can more than double her body weight during this period. In August or September, as the sea ice is breaking up, the female constructs a maternity den in the snow. Gestation begins when she enters the den and lasts about 2 months. Polar bear cubs weigh only 600 to 800 grams at birth, which is less than 0.5 percent of the weight of their mother. Between February and May, when the cubs have grown large enough to follow their mother (about 10 kilograms), she leaves the den and moves out onto the ice to hunt for seals.

As the Arctic has warmed, sea ice has been melting earlier in the summer, which reduces the time that female polar bears have to build the fat store they need to carry them through an 8-month period of fasting in the maternity den. A 5-year study of female polar bears in the Beaufort Sea found that their annual survival rate was 96 to 99 percent from 2001 to 2003 when the ice-free period was 101 days, but fell to 73 to 79 percent when the ice-free period had increased to an average of 135 days. A projection based on these data indicates that the survival rate of female polar bears would drop to 50 percent if the ice-free period lengthened to 150 days.

Male polar bears do not face the energy demands of pregnancy and nursing, but males are also affected by the progressively earlier melting of sea ice in summer and the later reforming of sea ice in the winter. One projection based on observations of current conditions is that 3 percent to 6 percent of adult males in Western Hudson Bay would die of starvation if the summer ice-free period lasted for 120 days, and mortality would increase to between 28 and 48 percent if the ice-free period extended to 180 days.

Warming affects polar bears in yet another way—mating success. Male and female polar bears are solitary, and mating occurs when a male and female encounter each other by chance during the period in April and May when the female is fertile. As the area of continuous ice grows smaller, the bears' habitat is fragmented and their mate-searching efficiency decreases. If searching efficiency decreases four times as fast as the area of ice, female mating success will fall to 72 percent, further accelerating the decline in the polar bear population.

1996 - Healthy   1997 - Bleached

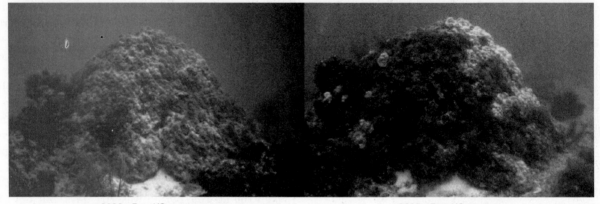

2000 - Dead/Overgrown   2005 - Dead/Overgrown

Figure 25–4 Coral bleaching.

Coral Reefs Half a world away from the Arctic, tropical seas are feeling the impact of global climate change and coral reefs are suffering: the amount of coral in the Caribbean has decreased by 80 percent since the 1960s, and the Great Barrier Reef of Australia has lost 50 percent of its coral. Coral bleaching is the most conspicuous effect of warmer water temperatures, and it occurs when coral polyps (sea anemone-like invertebrates that build coral reefs) expel their endosymbiotic zooxanthellae (photosynthetic dinoflagellates). When the pigmented zooxanthellae are absent, the white color of the calcium carbonate structure of the reef is exposed (**Figure 25–4**).

Bleaching is a stress response by coral polyps. Short periods of bleaching do not kill coral polyps, but prolonged exposure to elevated water temperature can be lethal. A water temperature 1°C higher than the normal maximum temperature is sufficient to initiate bleaching. Temperature-induced coral bleaching that affects entire ocean basins is a new phenomenon: The first event of this sort occurred in the Caribbean in 1979. A worldwide bleaching event started in the

eastern Pacific in 1997 and had spread to most of the world's coral reefs by 1998. Overall mortality of corals from this event is estimated to be 16 percent, and the western Indian Ocean lost 50 percent of its coral.

Coral is a foundation organism—coral reefs provide the three-dimensional structure that is the basis of the entire coral reef ecosystem. Bleaching events can initiate a coral death spiral when seaweed finds a foothold on the dead coral. As seaweed spreads, it interferes with the reproduction of coral polyps both physically (by occupying places the polyps would settle) and chemically (by releasing toxins that kill coral polyps). Herbivorous fishes that normally keep seaweed in check cannot cope with the massive invasion of seaweed, so more seaweed invades, further interfering with the regeneration of coral. This process is exacerbated by waste from shoreline communities containing pathogens that infect the coral polyps.

In addition to the damage caused by warming of the oceans and diseases, corals face the threat of ocean acidification. Seawater is the largest sink for atmospheric carbon dioxide, and when CO_2 dissolves in water it dissociates to form a proton (H⁺) and a bicarbonate ion (HCO_3^-). The proton combines with carbonate (CO_3^{2-}) to form another bicarbonate ion, which reduces the concentration of carbonate in the water. A carbonate concentration of 200 μmol • kg^{-1} of seawater is the lowest level of carbonate at which corals, mollusks, and other carbonate-depositing organisms can accrete new material, and that concentration of carbonate will occur when the atmospheric CO_2 level reaches 450 parts per million. That day may not be far distant; in 53 years levels of atmospheric CO_2 rose from about 315 parts per million in 1958 to about 392 parts in 2011. If the rate of increase in atmospheric CO_2 remains constant, we will reach 450 parts per million in about 40 years, and the rate of increase is rising.

Coral reefs are one of the most diverse habitats on Earth, with thousands of species of invertebrate and vertebrate animals, and their decline will have profound impacts on the entire global environment.

What Vertebrates Are Most Affected?

A second perspective on global climate change comes from asking "What group of vertebrates will be affected most severely?" In earlier chapters we described some of the effects of climate change, and there are many candidates for the most affected group. Marine fishes and the birds and mammals that feed on them will feel the effects of ocean warming and acidification. All of the species of vertebrates (including humans) that depend on coastal habitats will be displaced as sea levels rise. Long-distance migrants face double jeopardy because they are vulnerable to changes that occur at both ends of their migratory journey. A strong case can be made for any of these organisms, but yet another group—lizards—may be at even greater risk.

Lizards are ectotherms, and that characteristic lies at the heart of the effects of global climate change for this group. Both ectothermic and endothermic thermoregulatory mechanisms allow vertebrates to control their body temperatures, but there is an important difference: by using metabolic heat to raise the body temperature, endotherms achieve substantial independence from the temperature of their immediate surroundings. If a mountain lion needs to crouch in the shadow of a rock while it waits to ambush a deer, it can do that. Life is not so simple for a lizard because it must select an ambush site that not only allows it to see and attack prey but also has an exposure to sun and wind so that the lizard can thermoregulate. As the global climate changes, lizards in some habitats are having difficulty integrating thermoregulation with their other activities.

A study of populations of 48 species of spiny swifts (*Sceloporus*) in Mexico revealed the vulnerability of lizards to climate change. Spiny swifts bask in the sunlight to raise their body temperatures to activity levels and then move to shaded crevices to avoid overheating. As temperatures rise, the lizards spend more time in crevices, thereby reducing the time they can spend feeding. This limitation on energy intake would be exacerbated if higher nighttime temperatures increased their metabolic rates when the lizards were sleeping, consuming energy that could otherwise be devoted to reproduction.

Increasing temperatures might also affect embryonic development directly. Many high-altitude species of *Sceloporus* are viviparous, and female *Sceloporus* maintain lower body temperatures when they are pregnant than they do at other times of the year. Higher temperatures could force pregnant females to choose between maintaining body temperatures that are higher than the optimum for their embryos and consuming less food and potentially depriving the embryos of nutrients. Egg-laying species of *Sceloporus* might have difficulty finding nest sites with suitable temperature or water conditions for egg development. And temperature-dependent sex determination is another potential complication; TSD is not known to occur in *Sceloporus*, but it is hard to detect without detailed study and we cannot say that it is absent from the genus.

Data from Mexican weather stations show that the maximum daily air temperature has increased most during the period from January to May, which corresponds to the breeding season for many species of *Sceloporus* in Mexico. The increase in temperature has been greatest in northern and central Mexico and at high altitudes. The biophysical model used this information about the rate of change in maximum air temperature to make the predictions about the risk of extinction of populations of lizards that are shown on the fourth page of the color insert.

The researchers tested their model's predictions by returning to 200 sites that had populations of *Sceloporus* in 1975. Twelve percent of the populations had become extinct by 2009, and the rate of change in maximum temperature was positively correlated with extinctions of local populations. For example, two populations of *Sceloporus serrifer*, the species shown on the cover, were extinct and two others were still present. The restriction on the number of hours a lizard could be active during the reproductive season at the two extinct sites was dramatic: by April 2009 a lizard at those sites would have had to spend most of the daylight hours in a retreat site to escape high temperatures.

The risk experienced by mountain-dwelling species is counterintuitive because one would think that these species could move upward to escape higher temperatures. In practice that response does not work for two related reasons: First, mountains end and once a population of lizards has been forced to the top of a mountain it has nowhere else to go. Furthermore, mountains grow narrower as one moves upward so there is less area available, and this problem is exacerbated by interspecific competition if species from lower on the mountain are also moving upward.

Extending the model to 2050 and 2080 expands the area of Mexico in which extinctions are expected to occur and raises the probability of extinction for many populations above 50 percent. Applying the model on a global level and extending it to include 34 families of lizards, researchers predict that extinctions of local populations will reach 39 percent by 2080 and that 20 percent of the species of lizards in those families may be extinct.

25.4 Organismal Biology and Conservation

The descriptions of the plights of polar bears, corals, and lizards in the preceding section show that our recognition of the effects of even a global-level phenomenon like climate change depends on understanding the biology of the organisms affected. Ultimately we must know about the anatomical and physiological characteristics of a species and how they are reflected in the ecology and behavior of the species to understand how individuals of that species work as organisms and how environmental changes will affect them.

Developing management plans for endangered species requires enormous amounts of information about the basic biology of the species concerned. Without a thorough understanding of all aspects of a species' biology, well-intentioned management efforts can be ineffective or even have negative effects on the species. In the following sections we describe several examples of situations in which an understanding of the ecology, behavior, or physiology of a species is central to effective management.

What Is Critical in a Critical Habitat?

Federal law and some state laws require assessment of the habitat requirements of species that are considered at risk, and wildlife biologists may be charged with the responsibility of determining the critical habitat for a species. In biological terms, the elements of its habitat that are critical for the success of a species are likely to be complex, and the differences between legislative and biological perspectives can create tension and conflict.

Subtle Elements of Critical Habitat Complex interactions between the elements of a habitat and the needs of a species may determine the survival or extinction of populations of threatened species, but they can be detected only by careful study and application of basic biological information. The desert tortoise (*Gopherus agassizii*) of North America provides an example of this sort of interaction (**Figure 25–5**).

Desert tortoises are herbivores and they lack salt-excreting glands. Plants contain higher concentrations of potassium than do animals, so herbivorous animals must excrete some of the excess potassium they get in their food. Desert tortoises excrete potassium along with nitrogenous wastes in the form of salts of uric acid. Because potassium excretion is chemically linked to nitrogen excretion, an excess of potassium in a tortoise's diet robs the tortoise of nitrogen. That is, a tortoise on a high-potassium diet uses so much nitrogen getting rid of the excess potassium that it does not have enough nitrogen for protein synthesis.

Feeding-choice trials with captive tortoises showed that tortoises selected the diet with the lowest potassium. When tortoises were offered only high-potassium diets, they reduced the amount of food they ate. Wild tortoises may be able to survive only if plants with favorable ratios of nitrogen to potassium are available.

Plant species vary in potassium content, and two changes that have occurred in the tortoises' habitat in the past 200 years may have reduced the availability of low-potassium plants:

- **Commercial grazing**—Prior to European settlement, tortoises, rabbits, hares, deer, and antelope

Figure 25–5 A desert tortoise in the Mojave Desert. This tortoise is feeding on yellow fiddlehead (*Amsinckia menziesii*), a native species of plant that is an important component of the tortoises' diet in March and April.

shared the open range, and their population densities were low because there was not much food in the habitat. This situation changed in the 1800s, when ranchers introduced herds of cattle and sheep at densities far higher than the arid grasslands and deserts could support. The damage that livestock grazing causes in the arid Southwest is obvious to any ecologist, although lobbyists for the ranching industry vehemently deny it at legislative and regulatory hearings. Probably cattle, sheep, and tortoises all prefer plants that have low potassium and high nitrogen contents; if this is the case, then high densities of grazing mammals may reduce the availability of high-quality plants for tortoises.

- **Alien plants**—In many parts of North America, including the southwestern deserts where desert tortoises occur, Eurasian plants that accompanied the spread of European settlers have replaced some native species of plants. These Eurasian plants appear to have higher potassium and lower nitrogen levels than the native plants they have replaced. If this is true, a desert habitat with even a dense growth of Eurasian plants would not provide a diet on which tortoises could survive and grow. Furthermore, the alien plants are shading out the native annuals and increasing the frequency of fires, changing the vegetation of the desert.

Both of these concerns are in the hypothesis stage now—they are potential problems, but we lack data to decide whether they are real problems. Measuring the impact of commercial grazing and alien plants on tortoises and other native desert herbivores will require the combined efforts of ecologists and botanists working in the field as well as nutritionists and physiologists working in the laboratory.

Critical Habitat for Migratory Songbirds

The problems of identifying and then protecting critical habitat are particularly severe for migratory species, for which even the most superficial definition of critical habitat must include both ends of their migratory paths and the areas in which they stop to rest and feed as they move back and forth. To complicate conservation efforts even more, migratory animals usually move through several national jurisdictions and may pass over or through international waters where no national jurisdiction exists. The legislative and diplomatic effort required to protect these species is enormous and must be based on sound biological information.

Neotropical Migrants Birds are the best-known migratory animals, and migration is a central feature of the

biology of many species. Loss of habitat in their summer and winter ranges and at stopping places along the migratory routes appears to be contributing to declining populations of shore birds and songbirds that migrate between summer ranges in North America and winter ranges in the tropics.

About 350 species of songbirds occur in North America, and about 250 of them spend their winters in the New World tropics, which extend from southern Mexico through Central and northern South America and into the West Indies. The remaining 100 species of North American songbirds either are year-round residents in northern habitats or migrate only short distances south of their summer ranges. Neotropical migrants make up the majority of species of songbirds in most habitats in North America. In most wooded areas, half of the breeding species are Neotropical migrants, and in some northern regions of North America, more than 90 percent of the songbirds are migrants.

Bird watchers have a nationwide system of local bird censuses, and starting in the 1970s and 1980s, these counts revealed dramatic decreases in the numbers of some Neotropical migrants such as wood thrushes (*Hylocichla mustelina*) and cerulean warblers (*Setophaga cerulea*) in certain areas—as much as 1 percent annually over the past 30 years. In contrast, populations of year-round residents and short-distance migrants, such as chickadees (*Poecile atricapillus*) and northern cardinals (*Cardinalis cardinalis*), were stable or even increasing. The pattern of decline is not consistent across species or regions of North America, however. Populations of some species of Neotropical songbirds are declining, but others are increasing. Populations of songbirds in the forested areas of the eastern United States are declining, but those in the West are not.

Summer Range This complex picture results from the variety of factors that affect migratory birds in their summer and winter ranges and in stopover points on their routes of migration. Changes in land use are probably the primary reason for declining bird populations. Neotropical migrants breed in their summer ranges, and fragmentation of forests into smaller and smaller patches has reduced the total amount of breeding habitat available and changed the nature of the habitat that remains. Nests near the edges of woodlands generally suffer higher rates of failure than do those nearer the center, and as forest patches grow smaller, a greater proportion of the habitat is near an edge.

Part of the increased rate of nest failure near the edges of woodlands may result from nest parasitism by brown-headed cowbirds (*Molothrus ater*), which

(a)

(b)

(c)

Figure 25–6 Nest parasitism by brown-headed cow-birds. (a) A female cowbird about to deposit an egg in the nest of a blue grosbeak. (b) Two cowbird eggs in a robin's nest. The parent birds do not remove the cowbird eggs, even though they are speckled brown and white and look nothing like the plain blue robin egg. (c) An adult yellow warbler feeding a cowbird chick that is nearly as large as the foster parent.

are obligate nest parasites (**Figure 25–6**). Cowbirds do not build their own nests but lay their eggs in the nests of other species; some 200 species of birds have been reported to be parasitized by cowbirds. A female cowbird lays 20 to 40 eggs, one or two in each nest, and a few pairs of cowbirds can parasitize all the nests in a small woodland. A female cowbird often removes an egg of the host species from a nest when she lays her own, and cowbird eggs develop and hatch more rapidly than the eggs of the host species, giving the cowbird nestling an advantage. Larger and pushier than the nestlings of the host species, cowbird nestlings take so much of the food their unwitting foster parents bring to the nest that the host's nestlings may starve.

Cowbird parasitism is insidious. When a nest with eggs or fledglings is lost to a predator, the parent birds usually build another nest and start a second clutch. They may succeed in reproducing that season despite the loss of their first clutch. In contrast, the parent birds that serve as hosts for the cowbirds do not dis-

tinguish the cowbird eggs or nestlings from their own. Thus, when a pair of birds has raised a cowbird to fledging, they behave as if they had successfully fledged their own young and do not nest again that year.

Cowbirds are insectivorous and feed primarily in open fields. Because their movements are not limited by the need to care for their young, they can travel long distances between the open fields where they feed and the woodland edges where they find the nests of other birds. Populations of cowbirds have increased dramatically as open fields have replaced woodland habitats in North America, and the small remaining woodlands offer songbirds too little interior space to conceal their nests from cowbirds. Thus habitat fragmentation in the summer breeding range of Neotropical migrants appears to be responsible directly and indirectly for some of the decrease in their populations.

Winter Range Habitat change in the winter range is probably also responsible for some population declines.

The huge landmass of North America funnels down to a much smaller area in Central America, where many migrant birds overwinter. Competition for food and space in the small landmass of Central America may be one of the reasons that migrants do not breed in their winter ranges. Because the land area in the winter range is small, habitat changes caused by agricultural practices on even a relatively small scale could affect large numbers of birds. Coffee originated in Ethiopia and was brought to Latin America by Spanish colonists. Traditionally, coffee trees have been grown beneath a canopy of taller trees. This method produces shade-grown coffee. Coffee plantations of this sort are similar in structure to natural forests, although they are less complex and have fewer species of trees, and these traditional coffee plantations provide important habitats for birds. A survey of traditional plantations in Chiapas, Mexico, by the Smithsonian Migratory Bird Center found that more than 150 species of birds live in them, including Neotropical migrants.

In the past 20 years, hybrid coffee trees that grow in full sun have replaced traditional plantations. Sun-grown coffee produces substantially higher yields than shade-grown, and its use has spread rapidly. Currently about 20 percent of cropland planted in coffee in Mexico, 40 percent in Costa Rica, and 70 percent in Colombia has no shade canopy. Sun-grown coffee requires intensive cultivation, with heavy use of chemical fertilizers, insecticides, herbicides, and fungicides. These methods increase soil erosion, acidification, and the amount of toxic runoff. Furthermore, because sun-grown coffee plantations lack the complex structure of shade-grown coffee plantations, they do not provide habitats for forest birds. The diversity of birds plummets when a coffee plantation is converted from traditional shade-grown into sun-grown coffee. Studies in Colombia and Mexico found 94 percent to 97 percent fewer bird species in sun-grown coffee plantations than in shade-grown plantations.

Migratory birds depend on forested habitats in their winter ranges, but the land that can be set aside in parks and reserves in Central and northern South America is insufficient to maintain healthy bird populations. If migratory birds are to survive, human-dominated landscapes in the Neotropics must provide suitable habitats for them, such as traditional shade-grown coffee plantations. The problem is that sun-grown coffee produces a faster and higher return on a financial investment; shade-grown coffee is environmentally friendly but not economically attractive. To promote shade-grown coffee, the Smithsonian Migratory Bird Center has developed the Bird-Friendly® program, which certifies coffee growers that adhere to environmentally sound practices (Figure 25–7).

Critical Habitat for Large Mammals

Large animals need a lot of room—they have home ranges that cover hundreds or thousands of hectares, especially in the case of large predatory mammals. Furthermore, animals need different habitats for different activities, and the critical habitat must include the resources needed for hunting, mating, and raising young.

For species that have a well-developed social system, the ages and sexes of the individuals in a population may affect the success of a conservation plan. Populations of elephants in African game parks have increased substantially since the parks were established. This population growth testifies to the effectiveness of the parks in protecting the elephants, but it creates serious problems with overcrowding. As their populations increase, foraging elephants tear down mature trees to eat the upper branches and often move outside of the parks, where they raid gardens and orchards. To reduce these problems South Africa has developed a process of selective removal of elephants. Some individuals are trapped and moved to other parks that are below their capacity. It turns out that creating a mixture of juvenile and adult elephants is the key to making this plan work.

Adolescent Elephants Between 1981 and 1993, young male and female African elephants (*Loxodonta africana*) that had been orphaned when their mothers were killed were relocated from Kruger National Park to Pilanesburg. There were no elephants at Pilanesberg, and the orphans matured in the absence of older elephants. When the first elephant calf was born in

Figure 25–7 **The Bird-Friendly® label for shade-grown coffee.**

1989, the relocation program looked like a success story, except for one problem—the young male elephants were attacking and killing white rhinoceroses (*Ceratotherium simum*). By 1997, the elephants had killed more than 40 rhinoceroses.

The African elephant is an endangered species, and the white rhinoceros is listed as vulnerable. Both species face sufficient risk of extinction without the added complication of having one species attacking and killing the other. The aggressive behavior of the young male elephants at Pilanesburg is not typical of African elephants, but it was not unique to this population. It occurred in some other populations, especially in Hluhluwe-iMfolozi Park, which is in northern KwaZulu-Natal province. This was the second population to be established with orphans from Kruger National Park, and the young male elephants at Hluhluwe-Umfolozi began to kill rhinoceroses about 2 years after the behavior appeared in the Pilanesberg population.

The attacks on rhinoceroses occurred when the male elephants were in musth. Musth is an annual period during which the level of testosterone circulating in the bloodstream of male elephants rises dramatically. Elephants in musth adopt a distinctive posture that allows them to be recognized from a distance. Their temporal glands swell and secrete an oily material, and sexual and aggressive behavior increases (**Figure 25–8**). In natural populations, males first enter musth when they are 25 to 30 years old. The duration of musth lengthens as the elephants get older—from a few days to a few weeks for animals between 25 and 30 to 2 to 4 months for animals older than 40. Thus, the young males at Pilanesberg had developed musth earlier than would be expected, and their periods of musth lasted far longer than normal for such young males.

The populations of elephants that are killing rhinoceroses have one feature in common—all were established by moving only young animals to locations where they matured in the absence of adults. That abnormal situation turned out to be the key to the early and prolonged musth and aggressive behavior of the young males. In natural populations of elephants that include older males, young males are unlikely to be in musth, and when they do enter musth, the periods are short. Social interactions are responsible for this situation—young males in musth engage in aggressive interactions with the older bulls in musth, and the young males are defeated and driven away by the bulls. Within minutes to hours after being defeated in an aggressive encounter, a young male loses the signs of musth. Repeated interactions with older bulls drive young males out of musth, and the presence of older males in a population may delay its onset in young males.

In 1998, six adult bull elephants were moved from Kruger National Park to Pilanesberg, and their presence had a dramatic effect on the young males. The duration of musth in young males dropped sharply, in most cases falling from weeks to just a few days. Gratifyingly, the young males also stopped killing rhinoceroses.

25.5 Captive Breeding

In some situations, the threat to a species is so acute that the only alternative to watching the species

(a)

(b)

Figure 25–8 African elephants in musth. (a) Secretion from the temporal gland of a male in musth (*arrow*). (b) A large bull sparring with a smaller male.

become extinct is to gather as many individuals as possible and use them to establish a captive breeding program. The Amphibian Ark program is an example of this approach: in the face of the rapid southward advance of chytrid fungi, a team of scientists collected founder populations of 35 species of amphibians from El Valle de Antón, Panama, and transported them to Zoo Atlanta and the Atlanta Botanical Garden to establish breeding programs.

Goals of Captive Breeding

More than 500 species and subspecies of animals are being bred in captivity, and all of the major types of vertebrates are included. About 60 of these breeding programs are guided by Species Survival Plans (SSPs) developed by the Association of Zoos and Aquariums (AZA). A successful captive breeding program is far more than a group of animals that are reproducing in a zoo. The SSP defines a breeding plan that minimizes inbreeding and equalizes the genetic representation of each of the founding members of the captive population.

Genetics of Small Populations If you draw individuals randomly from a population of animals, you will get more individuals with common alleles than with rare ones, and alleles that are very rare may be absent from your sample. This is the familiar genetic bottleneck effect, and it is an inevitable element of captive breeding for most species because only a few wild-caught individuals are available to be the founders of a captive breeding program.

The first requirement of an SSP is that the animals in the captive population must represent a natural genetic component of the population because preserving as much of the genetic diversity of a species as possible is essential for long-term success. Most species show geographic variation in size, color, and behavior that reflects underlying genetic differences. These local varieties are often called subspecies, and the differences among subspecies are the result of local selective pressures. Captive breeding programs can retain this genetic diversity only by breeding pure subspecies lineages.

In addition to geographic variation in genetic characters, there is genetic variation among individuals. Many genetic loci have multiple alleles, and this heterozygosity is the raw material on which natural selection acts. Studbooks record the pedigrees and genetic characteristics of all the individuals of a species in captivity. The AZA currently maintains studbooks for almost 450 species and subspecies; when SSPs are developed, this information is used to decide which individuals should be bred to each other.

Several criteria are used to make these pairings: the relatedness of the individuals that are to be mated should be minimized, the bloodlines of different founders should be equally represented, and the rarest bloodlines should have highest priority for breeding.

The relatedness of an individual to the other individuals in the population is expressed as its kinship value. A low kinship value means that an individual's genes are not well represented in the population, and these individuals are good candidates for mating. It often turns out that the best mating involves animals that are in zoos thousands of kilometers apart; when this is the case, one partner must be shipped to the zoo where the other partner lives.

Inbreeding Depression When the offspring of a small founder population are bred for generation after generation, the average fitness of the population is reduced by two mechanisms. First, every population carries some recessive alleles that have deleterious effects. In a large, genetically diverse population, these genetic loci are usually heterozygous—that is, the deleterious recessive allele is usually paired with a normal dominant allele, so the deleterious character is not expressed. As the captive population becomes more inbred, however, more and more of these deleterious recessive alleles appear in the homozygous condition and are expressed.

Second, individuals that are heterozygous at most of their genetic loci are generally more robust than those that are mostly homozygous. The progressive reduction in heterozygosity that occurs as a captive population becomes more inbred lowers the average fitness of the animals in the population. The consequence of these two processes is known as inbreeding depression, and it often appears as reduced rates of reproduction and increased infant mortality in breeding colonies.

Domestication Inadvertent domestication is another genetic pitfall of captive breeding programs. The individuals that adjust well to captivity are more likely to reproduce than the ones that don't adjust, and some of the tendency to adjust is genetically determined. As a result, any captive breeding program selects for genetic characteristics that increase reproductive success in a captive situation, but these characteristics may be quite different from the ones that are beneficial to wild animals. Among antelope, for example, some individuals quickly learn that when a keeper approaches their pen it means they are about to be fed, and they gather eagerly at the feeding area. Other individuals, however, run away when they see the keeper approaching, possibly injuring themselves when they crash into the fence around their enclosure. Extreme wariness is a detrimental character in captivity. However, in the wild, the wary individuals may be less vulnerable to predators than their bolder companions, and loss of wariness in a captive-bred population can spell trouble when captive-bred animals are released into the wild.

Animal Behavior and Captive Breeding Programs

Animals have complex suites of behaviors that enable them to find appropriate food, evade predators, function in a group, and identify suitable mates. Some of these behaviors are innate—that is, an animal is born with the ability to produce the correct response to certain stimuli—but many behaviors are learned by trial and error. Animals raised in captivity do not experience the same stimuli they would in their natural environment and may not be able to function effectively if they are released. As captive breeding programs are developed to restore species to the wild, increasing effort is being devoted to ensuring that the animals to be released are competent to survive in the wild.

Imprinting, Learning, and the Captive Husbandry of Birds The process known as imprinting is an important feature of the behavior of birds and mammals. Imprinting is a special kind of learning that occurs only during a restricted time in ontogeny called the critical period. Once imprinting is established, it is permanent and cannot be reversed.

Colors, patterns, sounds, and movements are the major stimuli for imprinting among birds, whereas scent is the most important stimulus for mammals. Two types of imprinting can be distinguished:

- **Filial imprinting** is the process of learning to recognize the individual characteristics of the parents, and it is responsible for keeping the young with the mother after they move away from the nest or den.
- **Sexual imprinting** refers to learning the characteristics of other members of the species. Sexual imprinting during infancy allows an individual to recognize a mate of its own species when it matures. Birds and mammals normally imprint on their parents and siblings because those are the only objects in the den or nest that are emitting visual, auditory,

and olfactory stimuli. In the absence of their parent, however, infants may imprint on inanimate objects or on members of another species, including humans.

The confusion of species identification by birds that have imprinted on a foster parent or a keeper can be disastrous for programs in which endangered species are reared in captivity and then released. Young birds must recognize appropriate mates if they are to establish a breeding population, and captive-rearing programs go to great lengths to ensure that the young birds are properly imprinted.

For example, some hatchling California condors (*Gymnogyps californianus*) in the breeding facility at the San Diego Wild Animal Park are reared in enclosed incubators and fed by a technician who inserts her hand into a rubber glove modeled to look like the head of an adult condor (**Figure 25–9**). Puppet-rearing and parent-rearing have advantages and disadvantages: Puppet-rearing increases the productivity of a pair of condors because eggs can be removed from the nest to an incubator, stimulating the female to produce additional eggs, but parent-rearing probably helps young condors to develop natural behaviors.

Still more training may be necessary to produce captive-reared young that can survive after they have been released. A husbandry program for the northern bald ibis (*Geronticus eremita*) is an example of how complicated this process can be. Within historic times, the geographic range of the bald ibis extended from the Middle East and North Africa north to Switzerland and Germany, but the wild population has dwindled to fewer than 220 birds in a reserve in Morocco. Bald ibises flourish in captivity, and there are more than 700 captive individuals. This species seems ideal for reintroduction—it is prolific, and its disappearance from the wild seems to have been caused by human predation rather than by pollution or loss of habitat. Yet two attempts to establish populations by releasing captive-reared birds have failed.

The reason for the failures seems to have been the absence of normal social behaviors in the captive-reared birds. Bald ibises are social birds with extended parental care, and it seems that juveniles learn appropriate behaviors from adults. For some reason, this learning did not occur in captivity. An attempt is now under way to instruct young bald ibises in these social skills: human foster parents are hand-rearing the birds, teaching them to find their way to fields where they can forage, to recognize predators and other dangers such as automobiles,

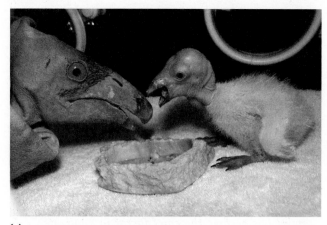

(a)

(b)

Figure 25–9 Captive husbandry of endangered species of birds. (a) A hatchling California condor in its incubator with the model condor head used to feed it. (b) A northern bald ibis. The trainer, wearing a bicycle helmet that has been modified to look like an adult bald ibis, is teaching a young captive-bred ibis how to search for food.

and to engage in mutual preening, which is an important social behavior.

Tameness can be a problem with captive-bred animals, even when extraordinary care is used to avoid imprinting on humans. Asiatic houbara bustards (*Chlamydotis undulata macqueenii*) being reared in Saudi Arabia succeeded in finding food after they were released, but many of them fell victim to predators, apparently because their sheltered rearing conditions had not taught them to be afraid. That problem may have been solved with the assistance of a young fox named Sophie who is let into a cage with the birds and allowed to chase them. Sophie thinks it's play, but three training sessions apparently convinced the birds to be more wary because survival after release was greater in the trained group than in a control group that had not been exposed to Sophie.

25.6 The Paradoxes of Conservation

Although conservation must address biological issues, conservation is not a purely biological issue. Human societies that burn fossil fuels and manufacture products will inevitably produce pollutants that travel far from their sources. It is relatively easy to believe that it is wrong for a multinational corporation to build a salt-extraction plant that will destroy a protected habitat or for a paper mill to release pollutants to save itself the cost of removing them from its waste discharge. Is it equally clear that a subsistence-level farming family in a developing country should not disrupt the habitat of an endangered species to grow the food it needs to survive?

Rich and poor nations respond differently to the often conflicting demands of earning a living versus conserving natural resources. A wealthy nation, especially a large country like the United States, can afford to set aside land to protect habitats and organisms, but that option may not be possible for a poor nation. Effective conservation efforts cannot focus on only biological questions. In the real world, conservation requires an intricate balance among biological, political, economic, and cultural values, and no one response is right for all species and all habitats. Some species need protected areas, but many of the most important areas of relatively undisturbed habitat are in countries where poverty and pressing social problems make complete protection an unrealistic goal. Conservation programs must address social and economic issues as well as biological ones if they are to succeed. The people living near parks and management areas must believe that protecting those habitats and the species they contain will contribute more to their own standard of living than they could gain by clearing the land and killing the animals.

Efforts of this sort are under way. The Natural Resources Defense Council is working with the villages around San Ignacio Lagoon to develop sustainable fisheries that are environmentally friendly. These fisheries also will provide an economic basis to make it easier to resist future proposals for industrial development that would damage the lagoon. Project Piaba in Brazil began as a study of fish diversity in the middle Rio Negro basin and has grown into a community-based interdisciplinary project that fosters ecologically sound economic development. The Rio Negro is the primary source of ornamental aquarium fish; more than 20 million fish are exported annually, with a retail value of more than $100 million.

If it were not for the fishery, the area would be developed for mining, forestry, and agriculture. Since sustaining the fishery requires sustaining the ecosystem, scientists working with Project Piaba are studying the structure and function of the aquatic systems. They are using that information to develop fishery management procedures to give the local people an incentive to preserve the integrity of the environment that supports the fishery.

The magnitude and complexity of the problems facing conservation biologists are daunting, and the scale on which remedial efforts must be attempted is nearly beyond comprehension. Alison Jolly has calculated that the total world population of all wild primates combined is less than that of any of the Earth's major cities. The entire extant populations of many species of primates are no larger than the population of a small town. Conditions are equally critical for many other vertebrates. More than three decades ago, Alison Jolly expressed the problems we face as biologists today:

> *This realization has been painful. It began for me in Madagascar, where the tragedy of forest felling, erosion, and desertification is a tragedy without villains. Malagasy peasant farmers are only trying to change wild environments to feed their own families, as mankind has done everywhere since the Neolithic Revolution. The realization grew in Mauritius, where I watched the world's last five echo parakeets land on one tree and knew they will soon be no more. It has come through an equally painful intellectual change. I became a biologist through wonder at the diversity of nature. I became a field biologist because I preferred watching nature go its own way to messing it about with experiments. At last I understood that biology, as the study of nature apart from man, is a historical exercise. From the Neolithic Revolution to its logical sequels of twentieth-century population growth, biochemical engineering of life forms, and nuclear mutual assured destruction, the human mind has become the chief factor in biology. . . . the urgent need in [vertebrate] studies is conservation. It is sheer self-indulgence to write books to increase understanding if there will soon be nothing left to understand.*

Reviews of the biodiversity crisis agree that an essential first step in coming to grips with the problem is a clear understanding of what species exist, where they are, and what the critical elements in their survival are. This is an enterprise that must enlist biologists from specialties as diverse as systematics, ecology, behavior, physiology, genetics, nutrition, and animal husbandry. There are so many species about which we know almost nothing, and there is so little time for us to learn.

Summary

The diversity of vertebrates has increased steadily (albeit with several episodes of extinction) for the past 500 million years, peaking in the mid-Miocene, about 15 million years ago. Much of the decline in diversity of vertebrates (and other forms of life) since then can be traced to the direct and indirect effects of humans on the other species with which we share the planet. Major threats to the continued survival of species of vertebrates include habitat destruction, pollution, and overhunting. At the base of all these phenomena is the enormous increase in the use of resources by humans. The world's human population is currently almost 7 billion people, double what it was only 50 years ago, and is predicted to grow to 9 billion by 2050. The net increase in the human population (i.e., births minus deaths) is more than 200,000 each day, and 75 million additional people demand resources each year. Consumption of resources and pollution of the environment increase at rates far greater than the rate of population growth. This differential is increasing as global communication (especially television) exposes the overwhelming majority of humans to a Western lifestyle, raising expectations and aspirations worldwide. Typically the use of resources in a modern technological society increases four to five times faster than population growth, and the release of pollutants rises in proportion to resource use.

Conservation efforts are driven by political and economic issues, and strategies that are appropriate for developed countries may be impractical in developing nations. Programs that integrate the needs of humans and wildlife have the best chances for long-term success. Information about the basic biology of organisms plays an essential role in conservation by defining the critical elements of a habitat that must be preserved to ensure that a species can survive, by identifying sources of problems, and by guiding management of wild populations and reintroduction programs.

Discussion Questions

1. Diagram the sequential changes in the coral reef death spiral, and explain the positive feedback process that promotes the accelerating growth of seaweed following the death of coral polyps in a bleaching episode.

2. Select a species of animal that interests you and outline the critical elements of a captive breeding and release program for that species. What measures will you take to ensure that your captive-bred individuals are physiologically and behaviorally competent to survive in the wild?

3. What are the pros and cons of releasing captive-bred individuals of a species into an existing population of that species or into a vacant habitat where the species once lived? (Assume that the existing population is below the carrying capacity of the habitat and that the conditions that were responsible for the extinction of the species in the vacant habitat no longer exist.)

4. What effect might global climate change have on the tuatara (*Sphenodon punctatus*), a cold-climate lepidosaur with temperature-dependent sex determination?

5. Do you agree with the argument that global climate change will have little impact on biodiversity because montane species can move to higher elevations as temperature increases and continental species can move northward (or southward for species in the Southern Hemisphere)?

6. The table lists characteristics of species of vertebrates that increase or decrease their potential for survival. Select a species of vertebrate that you think is likely to be at risk and another that is likely to be safe, and use the table to determine how well the characteristics of your species fit the categories. (Selecting species that are taxonomically similar—two mammals or two lizards, for example—simplifies the comparison.)

At Risk	Safe
Large body size	Small body size
Carnivore	Herbivore, scavenger, insectivore
Narrow habitat tolerance	Wide habitat tolerance
Valued for a product—meat, fur, oil, etc.	Not a source of a commercially valuable product
Restricted geographic distribution	Wide geographic distribution
Intolerant of the presence of humans	Tolerant of the presence of humans
Apex predator that bioaccumulates toxins from its prey	Lower tropic level predator

Additional Information

Barnosky, A. D., et al. 2011. Has the Earth's sixth mass extinction already arrived? *Nature* 471:51–57.

Burney, D. A. and T. F. Flannery. 2005. Fifty millennia of catastrophic extinctions after human contact. *TRENDS in Ecology and Evolution* 120:395-401.

Butchart, S. H. M., et al. 2010. Global biodiversity: Indicators of recent declines. *Science* 328:1164–1168.

Daszak, P., et al. 2000. Emerging infectious diseases of wildlife—Threats to biodiversity and human health. *Science* 287:443–449.

Derocher, A. E., et al. 2011. Sea ice and polar bear den ecology at Hopen Island, Svalbard. *Marine Ecology Progress Series* 441:273–279.

Derocher, A. E. 2012. *Polar Bears: A Complete Guide to their Biology and Behavior*. Baltimore, MD: Johns Hopkins University Press.

Graham, N. A. J., et al. 2011. Extinction vulnerability of coral reef fishes. *Ecology Letters* 14:341–348.

Grebmeier, J. M., et al. 2006. A major ecosystem shift in the northern Bering Sea, *Science* 311:1461–1464.

Gruber, N., et al. 2012. Rapid progression of ocean acidification in the California Current System. *Science* 337:220-223.

Hare, W. L., et al. 2011. Climate hotspots: Key vulnerable regions, climate change and limits to warming. *Regional Environmental Change* 11(Suppl. 1):S1–S13.

Hoegh-Guldberg, O. 2011. Coral reef ecosystems and anthropogenic climate change. *Regional Environmental Change* 11(Suppl. 1):S217–S227.

Hoffmann, M., et al. 2010. The impact of conservation on the status of the world's vertebrates. *Science* 330:1503–1509.

Holdaway, R. N., and C. Jacomb. 2000. Rapid extinction of the moas (Aves: Dinornithiformes): Model, test, and implications. *Science* 287:2250–2254.

Huey, R. B., et al. 2009. Why tropical forest lizards are vulnerable to climate warming. *Proceedings of the Royal Society B* 276:1939–1948.

Johannessen, O. M., and M. W. Miles. 2011. Critical vulnerabilities of marine and sea ice–based ecosystems in the high Arctic. *Regional Environmental Change* 11(Suppl. 1):239–248.

Jolly, A. 1980. *A World Like Our Own: Man and Nature in Madagascar*. New Haven, CT: Yale University Press.

Miller, G. H., et al. 1999. Pleistocene extinction of *Genyornis newtoni:* Human impact on Australian fauna. *Science* 283:205–208.

Oftedal, O. T., and M. E. Allen. 1996. Nutrition as a major facet of reptile conservation. *Zoo Biology* 15:491–497.

Regher, E. V., et al. 2010. Survival and breeding of polar bears in the southern Beaufort Sea in relation to sea ice. *Journal of Animal Ecology* 79:117–127.

Rule, S., et al., 2012. The aftermath of megafaunal extinction: Ecosystem transformation in Pleistocene Australia. *Science* 335:1483–1486.

Service, R. F. 2012. Rising acidity brings an ocean of trouble. *Science* 337:146–148.

Sinervo, B., et al. 2010. Erosion of lizard diversity by climate change and altered thermal niches. *Science* 328:894–899.

Slotow, R., et al. 2000. Older bull elephants control young males. *Nature* 408:425–426.

Suttle, K. B., et al. 2007. Species interactions reverse grassland response to climate. *Science* 315:640–642.

Walsh, P. D. 2003. Catastrophic ape declines in western equatorial Africa. *Nature* 422:611–614.

Walters, J. R., et al. 2010. Status of the California condor (*Gymnogyps californianus*) and efforts to achieve its recovery. *The Auk* 127:969-1001.

Walther, G.-R. 2007. Tackling ecological complexity in climate impact research. *Science* 315:606–607.

Wang, M., and J. E. Overland. 2009. A sea ice free summer Arctic within 30 years? *Geophysical Research Letters* 36, L07502, doi:10.1029/2009GL037820.

Warren, R., et al. 2010. Increasing impacts of climate change upon ecosystems with increasing global mean temperature rise. *Climatic Change* 106:141–177.

Websites
Biodiversity http://www.biodiversityscience.com/
Climate change
Arctic Sea Ice, National Snow and Ice Data Center http://nsidc.org/arcticseaicenews/
ClimateWatch http://www.climatewatch.noaa.gov/
Intergovernmental Panel on Climate Change http://www.ipcc-wg2.gov/
Coral reefs
http://www.reefbase.org/main.aspx
IUCN Red List of Threatened Species
http://www.iucnredlist.org/
Migratory birds
http://nationalzoo.si.edu/ConservationAndScience/MigratoryBirds/Coffee/
Overkill hypothesis
http://www.amnh.org/science/biodiversity/extinction/Day1/overkill/Bit1.html
http://austhrutime.com/megafauna_dreamtime.htm

Appendix

FIGURE 1–4 Phylogenetic relationships of extant vertebrates. This diagram depicts the probable relationships among the major groups of extant vertebrates. Note that the cladistic groupings are nested progressively; that is, all placental mammals are therians, all therians are synapsids, all synapsids are amniotes, all amniotes are tetrapods, and so on. The lines show interrelationships only; they do not indicate times of divergence or the unrecorded presence of taxa in the fossil record. The numbers at the branch points indicate derived characters that distinguish the lineages.

Legend: Selected derived characters of the groups identified by numbers. More details are provided in subsequent chapters. **1. Vertebrata**—vertebrae or vertebral elements (arcualia). **2. Gnathostomata**—all living vertebrates except hagfishes and lampreys: jaws formed from mandibular arch, teeth containing dentine. **3. Osteichthyes**—all living gnathostomes except the Chondrichthyes: lung or swim bladder derived from the gut, unique pattern of dermal bones of the head and shoulder region. **4. Sarcopterygii**—all living osteichthyans except ray-finned fishes: unique supporting skeleton in fins. **5. Rhipidistia**—all living sarcopterygians except coelacanths: teeth with a distinctive pattern of folded enamel, a distinctive pattern of dermal skull bones. **6. Tetrapoda**—all living rhipidistians except lungfishes: limbs with carpals, tarsals, and digits. **7. Lissamphibia**—all living amphibians: structure of the skin and elements of the inner ear. **8. Batrachia**—salamanders and frogs: characteristics of the ears and loss of dermal scales. **9. Amniota**—all living tetrapods except amphibians: a distinctive arrangement of extraembryonic membranes (the amnion, chorion, and allantois). **10. Sauropsida**—all living amniotes except synapsids (mammals and their extinct relatives): tabular and supratemporal bones small or absent, beta keratin present. **11. Diapsida**—all living sauropsids (probably including turtles): skull with both dorsal and ventral temporal openings (fenestrae). **12. Archosauria**—all living diapsids except turtles and lepidosaurs: a fenestra anterior to the orbit of the eye. **13. Synapsida (including Mammalia)**—all living mammals and their extinct relatives: only a lower temporal fenestra. **14. Theria**—all living mammals except monotremes: tribosphenic molar, a mobile scapula with loss of the coracoid bone.

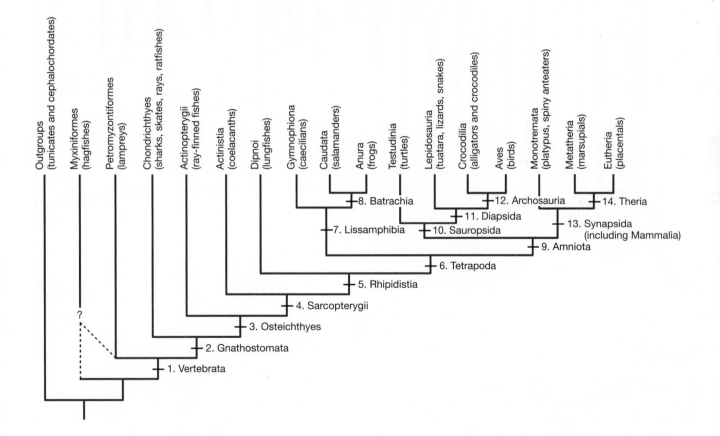

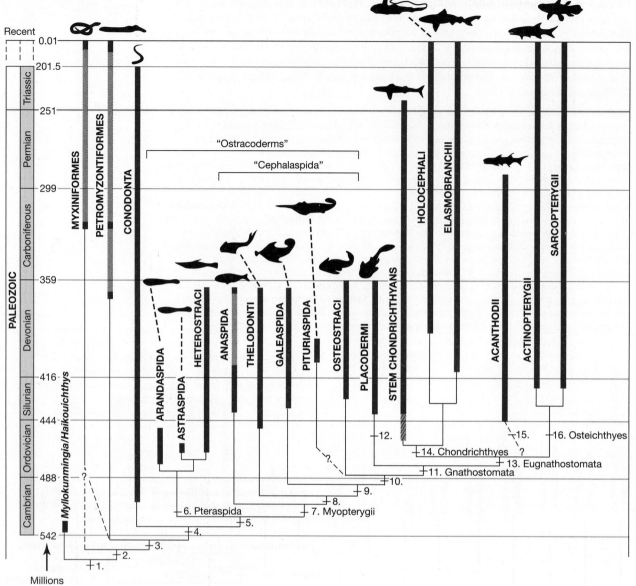

Figure 3–3 Phylogenetic relationships of early vertebrates. This diagram depicts the probable relationships among the major groups of "fishes," including living and extinct jawless vertebrates and the earliest jawed vertebrates. The black lines show relationships only; they do not indicate times of divergence or the unrecorded presence of taxa in the fossil record. Lightly shaded bars indicate ranges of time when the taxon is believed to be present, because it is known from earlier and later times, but is not recorded in the fossil record during this interval. The hatched bar shows probable occurrence based on limited evidence. Only the best-corroborated relationships are shown, and question marks indicate uncertainty about relationships. The numbers at the branch points indicate derived characters that distinguish the lineages.

Legend: 1. Vertebrata (including hagfishes)—distinct head region with cranium incorporating anterior end of the notochord and enclosing the brain and paired sensory organs, neural-crest cells and structures derived from these tissues (including cartilaginous gill skeleton), dorsal fin with fin rays, arcualia (segmental vertebral structures), W-shaped myomeres, gill openings primitively numbering 7 or less, with gill tissue. Characters of living vertebrates (conjectural for Early Cambrian forms): tripartite division of brain with a telencephalon, olfactory receptors, and cranial nerves differentiated from neural tube; paired optic, auditory, and probably olfactory organs; one or more semicircular canals; atrium lost; muscularized hypomere (lateral plate mesoderm), innervated by ventral root spinal nerves; unique type of fibrillar collagen; distinctive endocrine glands; lateral line system; well-developed three-chambered heart; capillaries and blood vessels lined with epithelium; multilayered epidermis; paired kidneys with an archinephric duct. **2. Nonsegmental gonads. 3. Vertebrata (excluding hagfishes)**—presence of arcualia (definitive), extrinsic eye musculature, pineal eye, hypoglossal nerve, Mauthner neurons in the brain stem, sensory-line neuromasts, sensory lines on head and body, capacity for electroreception, two semicircular canals, autonomic innervation of the heart, renal collecting ducts, spleen or splenic tissue, three (versus one) types of granular white blood cells, dilute body fluids, blood comprises more than 10% of body volume, ion transport in gills, pituitary control of pigment cells and gametogenesis. **4. Physiological capacity** to form mineralized tissues in the dermis, horny teeth lost. **5. Dermal-bone** head shield, olfactory tract, cerebellum. **6. Pteraspida**—paired nasal openings; dermal skeleton with characteristic three layers, including spongy bone. **7. Myopterygii**—paired lateral fin folds, dorsal and anal fin; endolymphatic duct in inner ear opens to surface (lost in gnathostomes above level of Chondrichthyes). **8. Stomach** (in some forms—thelodonts are probably polyphyletic). **9. Mineralized braincase (originally of calcified cartilage)**, endolymphatic duct connecting inner ear to exterior. **10. Presence of perichondral bone in braincase**, presence of calcified cartilage in endoskeleton, sclerotic ossicles, cellular dermal bone, three-layered exoskeleton, slitlike (versus pouched) gills, pectoral fins with a narrow concentrated base, large orbits, large head vein (dorsal jugular). **11. Gnathostomata**—jaws formed from mandibular branchial arch (palatoquadrate and Meckel's cartilage);

paired nasal openings; cranium enlarged anteriorly and posteriorly, incorporating one or more occipital neural arches; large, distinct cerebellum in hindbrain; myelinated nerves; mineralized jointed gill skeleton (internal to gill filaments); mineralized vertebral neural arches and haemal arches; pelvic fins; mineralized fin supports (radials); lateral line including head canals. Conjectural for placoderms, or may be present in some ostracoderms: intrinsic eye musculature for focusing the lens, atrium of heart lies posterodorsally (versus laterally) to ventricle, renal portal vein present, spiral valve primitively formed within intestine, pancreas with both endocrine and exocrine functions, stomach and a distinct spleen, male gonads linked by ducts to excretory (archinephric) ducts, female gonads with distinct oviducts, two (versus one) contractile actin proteins (one specific to striated muscle and one to smooth muscle), cartilage based in a proteinaceous matrix of collagen, thicker spinal cord with "horns" of gray matter in cross section, dorsal and ventral spinal nerve roots linked to form compound spinal nerves. **12. Placodermi (note: may be paraphyletic)**—specialized joint between head and trunk dermal shield, a unique arrangement of dermal skeletal plates of the head and shoulder girdle, a unique pattern of lateral line canals on the head, semidentine in the dermal bones. **13. Eugnathostomata**—epihyal element of second visceral arch modified as the hyomandibula, which is a supporting element for the jaw; true teeth rooted to the jaw (Osteichthyes) or in a tooth whorl (some placoderms have teeth on the jaw, and this character may be convergent); fusion of the nasal capsules to the rest of the chondrocranium; axial musculature divided into distinct epaxial and hypaxial components; vertebral centra and ribs (not described in placoderms). **14. Chondrichthyes**—unique perichondral and endochondral mineralization (prismatic calcified cartilage), dermal denticles, unique teeth and tooth replacement mechanisms, distinctive characters of the basal and radial elements of the fins, inner ear labyrinth opens externally via the endolymphatic duct, distinctive features of the endocrine system. **15. Acanthodii (note: may be paraphyletic)**—fin spines on anal and paired fins as well as on dorsal fins. **16. Osteichthyes**—a unique pattern of dermal head bones, including dermal marginal mouth bones with rooted teeth; a unique pattern of ossification of the dermal bones of the shoulder girdle; presence of lepidotrichia (fin rays); differentiation of the muscles of the branchial region; presence of a lung or swim bladder derived from the gut; medial insertion of the mandibular muscle on the lower jaw.

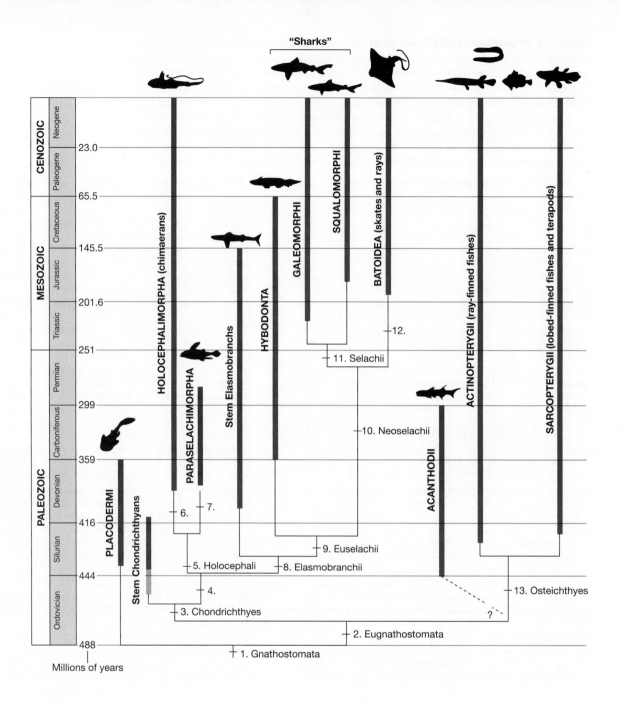

Figure 5-1 Phylogenetic relationships of cartilaginous fishes. This diagram depicts the probable relationships among the major groups of chondrichthyans. The black lines show inter-relationships only; they do not indicate times of divergence or the unrecorded presence of taxa in the fossil record. Only the best-corroborated relationships are shown. The numbers at the branch points indicate derived characters that distinguish the lineages.

Legend: 1. Gnathostomata—jaws formed of bilateral palatoquadrate (upper) and mandibular (lower) cartilages of the mandibular pharyngeal arch, modified hyoid gill arch; branchial arches internal to gill tissue, containing four elements on each side plus one unpaired ventral median element; three semicircular canals; internal supporting girdles associated with pectoral and pelvic fins; many features of the soft anatomy.
2. Eugnathostomata—epihyal element of second visceral arch modified as the hyomandibula, which is a supporting element for the jaw; true teeth rooted to the jaw (Osteichthyes) or in a tooth whorl; fusion of the nasal capsules to the rest of the chondrocranium; axial musculature divided into distinct epaxial and hypaxial components; myelinated nerves; six (reduced from seven) external eye muscles. **3. Chondrichthyes**—unique perichondral and endochondral mineralization of cartilage (prismatic plates of apatite), placoid scales, tooth whorl, distinctive characters of the basal and radial elements of the fins, inner-ear labyrinth opens externally via the endolymphatic duct, distinctive features of the endocrine system, pelvic claspers in males. (Note: Stem chondrichthyans are known mainly from scales and teeth, so some or many of these features may first appear at node 4.) **4. Elasmobranchii plus** Holocephali. **5. Holocephali**—gill cover over four gill openings, complete hyoid arch, hyomandibula not involved in jaw suspensorium, gill arches beneath the braincase. **6. Holocephalimorpha**—holostylic jaw suspension, dentition consists of grinding toothplates.
7. Paraselachimorpha—several unique cranial and dental features. **8. Elasmobranchii**—five to seven separate gill openings on each side. **9. Euselachii**—pectoral fin with three main basal elements, shoulder joint narrowed. **10. Neoselachii**—pectoral fin with three basal elements, the anteriormost of which is supported by the shoulder girdle; mouth underslung, not terminal; characteristics of the nervous system, cranium, and gill arches. **11. Selachii**—united by molecular characters. **12. Batoidea**—body strongly dorsoventrally flattened, gill openings ventral, anterior edge of enlarged pectoral fin attached to the side of the head anterior to the gill openings, eyes and spiracle on dorsal surface, pavement dentition. **13. Osteichthyes**—a unique pattern of dermal head bones, including dermal marginal mouth bones with rooted teeth, an extension of the dermal skull roof into the postcleithral series, which link up with the endochondral shoulder girdle, dermal bones (branchiostegal rays) on the floor of the gill chamber, dermal bones forming operculum over the gills, presence of lepidotrichia (fin rays), differentiation of the muscles of the branchial region, presence of a lung or swim bladder derived from the gut, medial insertion of the mandibular muscle on the lower jaw.

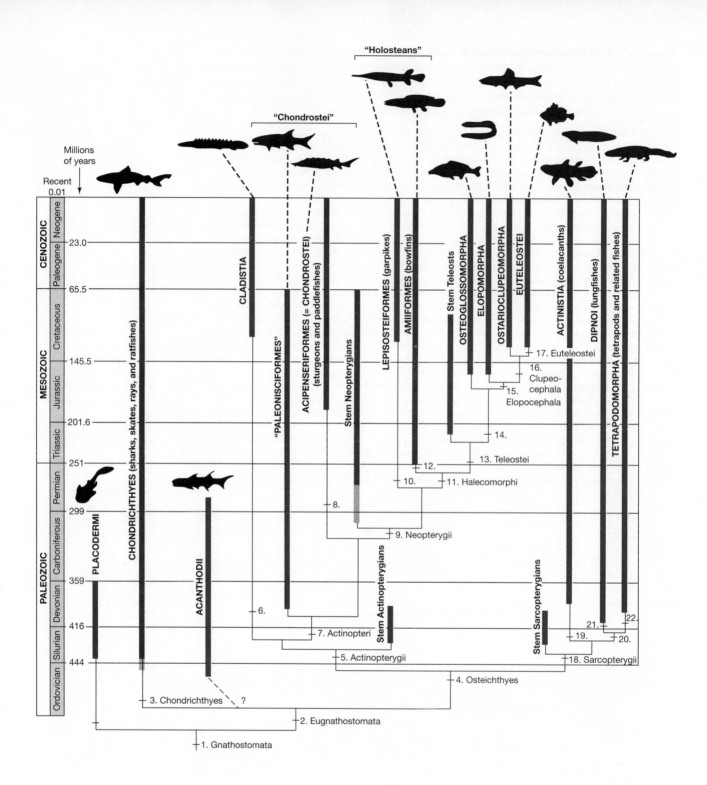

Figure 6–1 Phylogenetic relationships of bony fishes.

This diagram depicts the probable relationships among the major groups of basal gnathostomes. The black lines show interrelationships only; they do not indicate times of divergence or the unrecorded presence of taxa in the fossil record. Lightly shaded bars indicate ranges of time when the taxon is believed to be present, because it is known from earlier times, but is not recorded in the fossil record during this interval. Only the best-corroborated relationships are shown. The numbers at the branch points indicate derived characters that distinguish the lineages.

Legend: 1. Gnathostomata—jaws formed from mandibular branchial arch (palatoquadrate and Meckel's cartilage); pelvic fins; paired nasal openings; cranium enlarged anteriorly and posteriorly, incorporating one or more occipital neural arches; large, distinct cerebellum in hindbrain; myelinated nerves; mineralized jointed gill skeleton (internal to gill filaments); mineralized vertebral neural arches and hemal arches; pelvic fins; mineralized fin supports (radials); lateral line including head canals. Conjectural for placoderms, or may be present in some ostracoderms—intrinsic eye musculature for focusing the lens, atrium of heart lies posterodorsally (versus laterally) to ventricle, renal portal vein present, spiral valve primitively formed within intestine, pancreas with both endocrine and exocrine functions, stomach and a distinct spleen, male gonads linked by ducts to excretory (archinephric) ducts, female gonads with distinct oviducts, two (versus one) contractile actin proteins (one specific to striated muscle and one to smooth muscle), cartilage based in a proteinaceous matrix of collagen, thicker spinal cord with "horns" of gray matter in cross section, dorsal and ventral spinal nerve roots linked to form compound spinal nerves. **2. Eugnathostomata**—epihyal element of second visceral arch modified as the hyomandibula, which is a supporting element for the jaw; true teeth rooted to the jaw (Osteichthyes) or in a tooth whorl; fusion of the nasal capsules to the rest of the chondrocranium; axial musculature divided into distinct epaxial and hypaxial components; vertebral centraa and ribs (not described in placoderms). **3. Chondrichthyes**—unique perichondral and endochondral mineralization (prismatic plates of apatite), dermal denticles, unique teeth and tooth replacement mechanisms, distinctive characters of the basal and radial elements of the fins, presence of ceratotrichia (fin rays), inner ear labyrinth opens externally via the endolymphatic duct, distinctive features of the endocrine system. **4. Osteichthyes**—a unique pattern of dermal head bones, including dermal marginal mouth bones with rooted teeth, an extension of the dermal skull roof into the postcleithral series, which link up with the endochondral shoulder girdle, dermal bones (branchiostegal rays) on the floor of the gill chamber, dermal bones forming operculum over the gills, presence of lepidotrichia (fin rays), differentiation of the muscles of the branchial region, presence of a lung or swim bladder derived from the gut, medial insertion of the mandibular muscle on the lower jaw. **5. Actinopterygii**—basal elements of pectoral fin reduced; median fin rays attached to skeletal elements that do not extend into fin; single dorsal fin; scales with unique arrangement, shape, interlocking mechanism, and histology (outer layer of complexly layered enameloid called ganoine); brain develops with outfolding of cerebral hemispheres; details of posterior braincase structure; specific basal elements of the pelvic fin are fused; and numerous features of the soft anatomy of extant forms that cannot be verified for fossils. **6. Cladistia (Polypteriformes)**—unique dorsal-fin spines, only four gill arches, branchiostegal rays lost, and unique pectoral fin skeleton. **7. Actinopteri**—derived characters of the dermal elements of the skull and pectoral girdle and fins, a spiracular canal formed by a diverticulum of the spiracle penetrating the postorbital process of the skull, other details of skull structure, three cartilages or ossifications in the hyoid. Lung or swim bladder connects dorsally to the foregut, fins edged by specialized scales (fulcra). **8. Acipenseriformes (Chondrostei)**— fusion of premaxillae, maxillae, and dermopalatine bones of the snout; unique anterior palatoquadrate symphysis; endochondral bone lost, dermal bones formed from cartilage. **9. Neopterygii**—rays of dorsal and anal fins reduced to equal the number of endoskeletal supports, basals lost in paired fins, upper lobe of caudal fin containing axial skeleton reduced in size to produce a nearly symmetrical caudal fin, premaxilla with internal process lining the anterior part of the nasal pit, symplectic bone developed as an outgrowth of the hyomandibular cartilage. **10. Lepisosteiformes (Ginglymodi)**—vertebrae with convex anterior faces and concave posterior faces (opisthocoelous), toothed infraorbital bones contribute to elongate jaws, three branchiostegal rays, interoperculum bone absent. **11. Halecomorphi**—modifications of the cheek, jaw articulation, and opercular bones, including a mobile maxilla. Relationships of the Lepisosteiformes, Amiiformes, their fossil relatives, and the Teleostei are currently subject to many differing opinions with no clear resolution based on unique shared derived characters. More conservative phylogenies than those here would represent their relationships as unresolved. Others would unite the lepisosteiformes and the amiiformes as the "Holostei." **12. Amiiformes**—large median gular plate and 10 to 13 branchiostegal rays. **13. Teleostei**—elongate posterior neural arches (uroneurals) contributing to the stiffening of the upper lobe of the internally asymmetrical caudal fin (the caudal is externally symmetrical (= homocercal) at least primitively in teleosts); unpaired ventral pharyngeal tooth plates on basibranchial elements; premaxillae mobile; details of skull foramina, jaw muscles, and axial and pectoral skeleton. **14. Crown-group teleosts**—presence of an endoskeletal basihyal, four pharyngobranchials and three hypobranchials, median tooth plates overlying basibranchials and basihyals. **15. Elopocephala**—two uroneural bones extend anteriorly to the second ural (tail) vertebral centrum; abdominal and anterior caudal epipleural intermuscular bones present. **16. Clupeocephala**—pharyngeal tooth plates fused with endoskeletal gill-arch elements, neural arch of first caudal centrum reduced or absent, distinctive patterns of ossification and articulation of the jaw joint. **17. Euteleostei**—this numerically dominant group of vertebrates is poorly characterized, with no known synapomorphy (shared derived character) unique to this group that is present in all or perhaps even in most forms. However, the following have been used in establishing monophyly: presence of an adipose fin posteriorly on the mid-dorsal line, presence of nuptial tubercles on the head and body, and paired anterior membranous outgrowths of the first uroneural bones of the caudal fin. (These characters are usually lost in the most derived euteleosts.) The nature of these characters leads to a lack of consensus on the interrelationships of the basal clupeocephala, although the group's monophyly is still generally accepted. **18. Sarcopterygii**—fleshy pectoral and pelvic fins have a single basal skeletal element (= monobasal condition), muscular lobes at the bases of those fins, enamel (versus enameloid) on surfaces of teeth, cosmine (unique type of dentine) in body scales, posterior vena cava vein, unique characters of jaws, articulation of jaw supports, gill arches, and shoulder girdles. **19. Actinistia (coelacanths)**—first dorsal fin supported by plate of bone but lacks internal lobe, symmetrical three-lobed tail with central fleshy lobe that ends in a fringe of rays, maxilla and branchiostegal rays lost, unique features of skull and limb skeleton. **20. Dipnoi plus tetrapodomorphs**—heart with separated pulmonary and systemic circulations (although this may be true of early coelacanths). **21. Dipnoi (lungfishes)**—maxilla and premaxilla lost, palatoquadrate fused to chondrocranium (with loss of hyomandibula), tooth plates on palate, alternating (rather than parallel) series of dermal bones on skull roof, unique features skull and of fin skeleton. **22. Tetrapodomorpha**—true choana (internal nostril), labyrinthine folding of tooth enamel, details of limb skeleton.

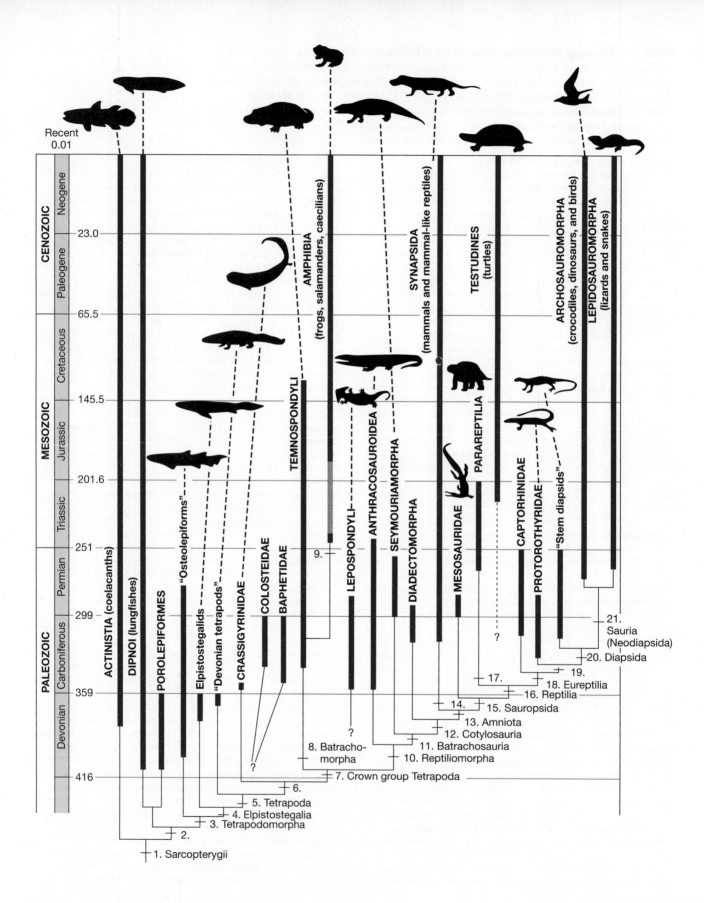

Recent
0.01

CENOZOIC	Neogene
	Paleogene
MESOZOIC	Cretaceous
	Jurassic
	Triassic
PALEOZOIC	Permian
	Carboniferous
	Devonian

23.0
65.5
145.5
201.6
251
299
359
416

ACTINISTIA (coelacanths)
DIPNOI (lungfishes)
POROLEPIFORMES
"Osteolepiforms"
Elpistostegalids
"Devonian tetrapods"
CRASSIGYRINIDAE
COLOSTEIDAE
BAPHETIDAE
TEMNOSPONDYLI
AMPHIBIA (frogs, salamanders, caecilians)
LEPOSPONDYLI
ANTHRACOSAUROIDEA
SEYMOURIAMORPHA
DIADECTOMORPHA
SYNAPSIDA (mammals and mammal-like reptiles)
MESOSAURIDAE
PARAREPTILIA
TESTUDINES (turtles)
CAPTORHINIDAE
PROTOROTHYRIDAE
"Stem diapsids"
ARCHOSAUROMORPHA (crocodiles, dinosaurs, and birds)
LEPIDOSAUROMORPHA (lizards and snakes)

9.

21. Sauria (Neodiapsida)
20. Diapsida
19.
18. Eureptilia
17.
16. Reptilia
15. Sauropsida
14.
13. Amniota
12. Cotylosauria
11. Batrachosauria
10. Reptiliomorpha
8. Batracho-morpha
7. Crown group Tetrapoda
6.
5. Tetrapoda
4. Elpistostegalia
3. Tetrapodomorpha
2.
1. Sarcopterygii

Figure 9-1 Phylogenetic relationships of sarcopterygian fishes and early tetrapods. This diagram depicts the probable relationships among the major groups of sarcopterygian fishes and early tetrapods. The black lines show interrelationships only; they do not indicate times of divergence or the unrecorded presence of taxa in the fossil record. Lightly shaded bars indicate ranges of time when the taxon is believed to be present, because it is known from earlier times, but is not recorded in the fossil record during this interval. Only the best-corroborated relationships are shown. The numbers at the branch points indicate derived characters that distinguish the lineages (An alternative view of lissamphibian relationships is discussed in the text.) Note that some authors prefer to restrict the term *tetrapod* to the crown group, which encompasses only extant taxa and extinct taxa that fall within the range of characters seen in the extant taxa. Thus, under this scheme, the taxonomic term *Tetrapoda* would be shifted to node 7.

Legend: 1. Sarcopterygii—fleshy pectoral and pelvic fins with a single basal element (= monobasal condition), muscular lobes at the base of those fins, true enamel on teeth, posterior vena cava vein, plus features of jaws and limb girdles. **2. Heart** with separated pulmonary and systemic circulations (although this may be true of early coelacanths). **3. Tetrapodomorpha**—true choana (internal nostril), labyrinthine folding of tooth enamel, and details of limb skeleton (one bone/two bone pattern). "Osteolepiforms" is a paraphyletic assemblage including the Osteolepididae, the Rhizodontidae, and the Tristichopteridae. **4. Elpistostegalia**—flattened head with elongate snout, external nares situated on the margin of the mouth, orbits have eye ridges and are on top of the skull, body flattened, humerus with enlarged (deltoid) ridge, absence of dorsal and anal fins, enlarged ribs. The elpistostegalid fishes include *Panderichthys, Livoniana, Elpistostege,* and *Tiktaalik.* **5. Tetrapoda**—limbs with carpals, tarsals, and digits, vertebrae with zygapophyses, large ornamented interclavicle, iliac blade of pelvis attached to vertebral column, bony scale cover restricted to belly region. ("Devonian tetrapods" is a paraphyletic assemblage of Late Devonian genera, including [in order of ancestral to derived] *Acanthostega, Ichthyostega,* and *Tulerpeton,* plus others known from fragmentary material. This node is sometimes called "Stegocephalia.") **6. Absence of anocleithrum (a dermal bone in the shoulder girdle),** five or fewer digits, scales of dermal bone (rather than elasmoid scales). **7. Crown group Tetrapoda**—presence of occipital condyles (projections on the skull for articulation with the vertebral column), notochord excluded from braincase in adults. In living forms and maybe also some stem tetrapods: muscular tongue, salivary glands, lacrimal glands, nasolacrimal duct, Harderian gland (a gland anterior to eye), vomeronasal (Jacobson's) organ, parathyroid gland controlling blood calcium levels, proprioceptive feedback from muscles and tendons, urinary bladder, nitrogen excretion as urea and via the kidney, differentiated hypaxial musculature (including a transverses abdominus for exhalation). **8. Batrachomorpha**—skull roof attached to braincase via the exoccipital bones at back of skull, loss of ancestral mobility (kinesis) within the skull, only four fingers in hand. **9. Lissamphibia**—pedicellate teeth, dermal scales reduced or lost, thin skin allowing for cutaneous gas exchange: (Lissamphibia includes Anura [frogs], Caudata [salamanders], and Gymnophiona [caecilians].) **10. Reptiliomorpha**—several skull characters, plus vertebrae with the pleurocentrum (the posterior central element) as the predominant element. **11. Batrachosauria**—intercentrum (the anterior central

vertebral element) reduced in size, enlarged caninelike tooth in maxilla (upper jaw), dermal scales reduced or lost. (Amniotelike keratinized skin may have evolved at this point, and there is evidence from rib shape of at least some costal ventilation.) **12. Cotylosauria**—sacrum with more than one vertebra, robust claws on feet, distinct astragalus bone in ankle, loss of labyrinthodont teeth, more derived atlas-axis complex, plus other skull characters. **13. Amniota**—hemispherical and well-ossified occipital condyles, frontal bone contacts orbit in skull, transverse pterygoid flange present (reflects differentiation of pterygoideus muscle), three ossifications in scapulocoracoid (shoulder girdle). In living forms and maybe also some stem amniotes: amniotic egg, complete loss of larval phase and lateral line system, skin relatively impervious to water, horny nails, metanephric kidney drained by ureter, uric acid excretion, third eyelid (nictitans), trachea, enlarged lungs with costal ventilation, differentiated epaxial musculature (now important in postural support). **14. Synapsida**—presence of lower temporal fenestra. **15. Sauropsida**—single centrale bone in ankle, maxilla separated from quadratojugal in skull, single coronoid bone in lower jaw. Beta-keratin in skin of living forms. **16. Reptilia**—suborbital foramen in palate, tabular bone in skull small or absent, large post-temporal fenestra in skull. **17. Parareptilia**—loss of caniniform maxillary teeth, posterior emargination of skull, quadratojugal bone in skull expanded dorsally, expanded iliac blade in pelvis: includes the Late Permian Millerettidae and Pareiasauridae and the Late Permian and Triassic Procolophonidae. Opinions vary as to the systematic position of Testudines (turtles). **18. Eureptilia**—supratemporal bone in skull is small, parietal and squamosal bones in skull broadly in contact, tabular bone in skull not in contact with opisthotic, ventral margin of postorbital portion of skull is horizontal, ontogenetic fusion of pleurocentrum of atlas and intercentrum of axis. **19. Postorbital** region of skull short, anterior pleurocentra keeled ventrally, limbs long and slender, hands and feet long and slender, metapodials overlap proximally. **20. Diapsida**—upper and lower temporal fenestrae present, exoccipitals not in contact on occipital condyle, ridge-and-groove tibia-astralagal joint. ("Stem diapsids" is a paraphyletic assemblage of Late Carboniferous and Permian diapsids including [in sequence from ancestral to derived] Araeoscelidia, Coelurosauravidae, and Younginiformes.) **21. Sauria (Neodiapsida)**—dorsal origin of temporal musculature, quadrate exposed laterally, tabular bone lost, unossified dorsal process of stapes, loss of caniniform region in maxillary tooth row, sacral ribs oriented laterally, ontogenetic fusion of caudal ribs, modified ilium, short and stout fifth metatarsal, small proximal carpals and tarsals.

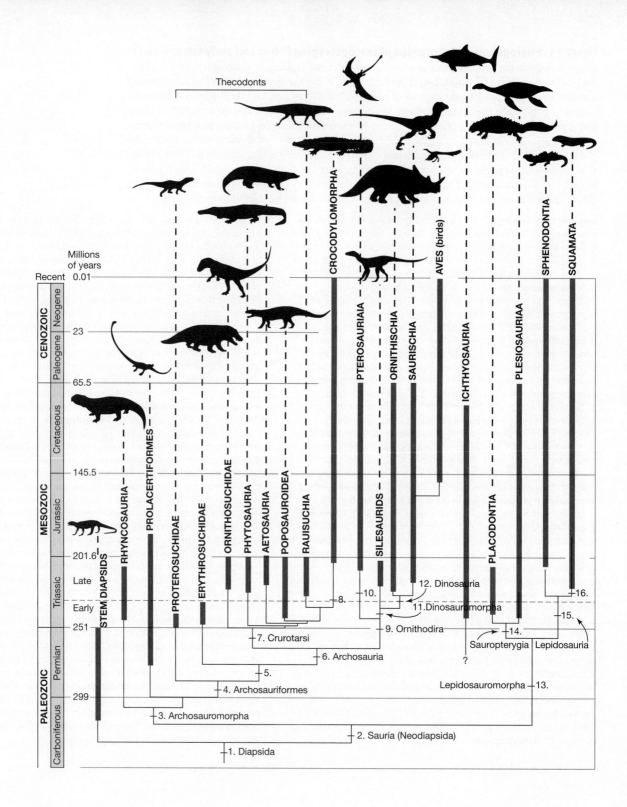

Figure 16–1 Phylogenetic relationships of the Diapsida. This diagram depicts the probable relationships among the major groups of diapsids. (Turtles, which may be diapsids, are not included.) The black lines show interrelationships only; they do not indicate times of divergence or the unrecorded presence of taxa in the fossil record. Only the best-corroborated relationships are shown. The numbers at the branch points indicate derived characters that distinguish the lineages.

Legend: 1. Diapsida—skull with upper and lower temporal fenestrae, upper temporal arch formed by triradiate postorbital and triradiate squamosal, suborbital fenestra, ossified sternum, complex ankle joint between tibia and astragalus, first metatarsal less than half the length of the fourth metatarsal. **2. Sauria (Neodiapsida)**—anterior process of squamosal narrow, squamosal mainly restricted to top of skull, tabular absent, stapes slender, cleithrum absent, fifth metatarsal hooked, trunk ribs mostly single headed. **3. Archosauromorpha**—cervical ribs with two heads, various features of limbs including concave-convex articulation between astragalus and calcaneum. **4. Archosauriformes**—presence of an antorbital fenestra and mandibular fenestra, orbit shaped like an inverted triangle, teeth laterally compressed with serrations. **5. Thecodont tooth implantation.** Pubis and ilium elongated, fourth trochanter on femur. **6. Archosauria**—parietal foramen absent; no palatal teeth on pterygoid, palatine, or vomer. **7. Crurotarsi**—ankle (tarsus) in which the astragalus forms a distinct peg that fits into a deep socket on the calcaneum, characters of the cervical ribs and the humerus. **8. Crocodylomorpha**—secondary palate present and includes at least the maxillae. **9. Ornithodira (= Avemetatarsalia)**—anterior cervical vertebrae longer than mid-dorsals, neck with S-shaped curve, second phalanx of digit 2 in the hand longer than the first, tibia longer than femur, calcaneal tuber rudimentary or absent, metatarsals bunched together and 2–4 elongated, osteoderms absent, gastralia well separated rather than forming interlocking basket. **10. Pterosauria**—hand with three short fingers and elongate fourth finger supporting wing membrane, pteroid bone in wrist, short trunk, short pelvis with prepubic bones. **11. Dinosauromorpha**—details of head/neck articulation and hind foot, including hingelike ankle joint, elongated pubes and ischia. **12. Dinosauria**—long, subrectangular deltopectoral crest on humerus; inturned femoral head; astragalus with acute anteromedial corner, broad ascending process, and reduced fibular articulation; epipophyses on cranial cervical vertebrae; perforated acetabulum in pelvis; arched dorsal iliac margin; additional features (absent in some most basal forms): hand digit 4 reduced, metatarsals II and IV subequal in length, more than two sacral vertebrae. **13. Lepidosauromorpha**—postfrontal enters border of upper temporal fenestra, supratemporal absent, teeth absent on lateral pterygoid flanges, characteristics of the vertebrae, ribs, and sternal plates. **14. Sauropterygia**—elongation of postorbital region of skull, enlargement of upper temporal fenestra, elongate and robust mandibular symphysis, curved humerus, equal length of radius and ulna. **15. Lepidosauria**—determinant growth with epiphyses on the articulating surfaces of the long bones, postparietal and tabular absent, fused astragalus and calcaneum, other characteristics of the skull, pelvis, and feet. **16. Squamata**—loss of lower temporal bar (including loss of quadratojugal), highly kinetic skull with reduction or loss of squamosal, nasals reduced, plus other characteristics of the palate and skull roof, vertebrae, ribs, pectoral girdle, and humerus.

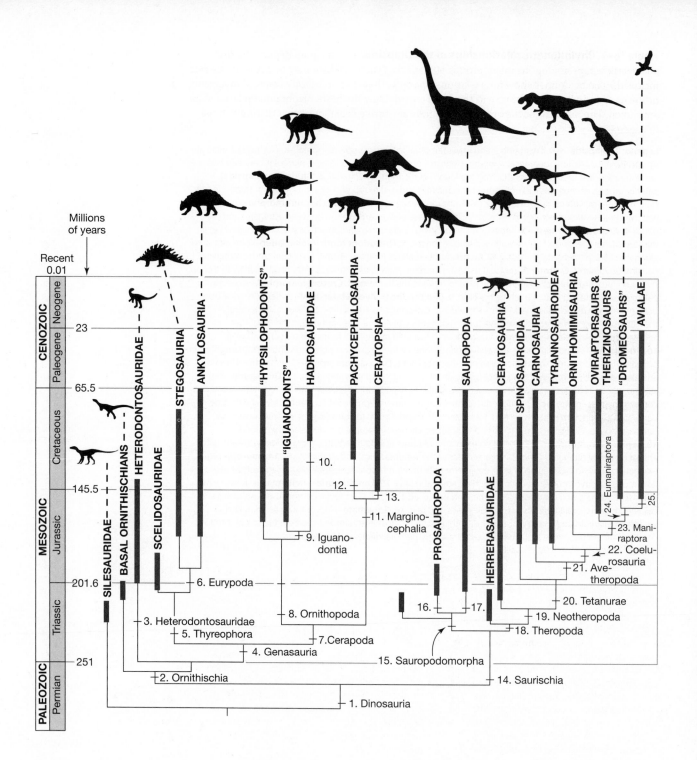

Figure 16–14 Phylogenetic relationships of the Dinosauria. This diagram depicts the probable relationships among the major groups of dinosaurs, including birds. The black lines show interrelationships only; they do not indicate times of divergence or the unrecorded presence of taxa in the fossil record. Only the best-corroborated lineages are shown. The numbers at the branch points indicate derived characters that distinguish the lineages. Quotation marks indicate paraphyletic groups.

Legend: 1. Dinosauria—Long, subrectangular deltopectoral crest on humerus; inturned femoral head; astragalus with acute anteromedial corner, broad ascending process, and reduced fibular articulation; epipophyses on cranial cervical vertebrae; perforated acetabulum in pelvis; arched dorsal iliac margin. Additional features (absent in some most basal forms): hand digit 4 reduced; metatarsals II and IV subequal in length; more than two sacral vertebrae. **2. Ornithischia**—cheek teeth with low subtriangular crowns, reduced antorbital opening, predentary bone, toothless and roughened tip of snout, jaw joint set below level of upper tooth row, at least five sacral vertebrae, ossified tendons above sacral region, pelvis with pubis directed backward, small prepubic process on pubis. (Not all features present in basal Triassic forms.) **3. Heterodontosauridae**—ventral process of predentary reduced or absent, caniniform tooth in both premaxilla and dentary. **4. Genasauria**—buccal emargination on maxilla (indicating muscular cheeks), dentary symphasis spoutlike, reduction in size of mandibular foramen. **5. Thyreophora**—transversely broad postorbital process of jugal, strong ridge on lateral surface of surangular in jaw, parallel rows of keeled scutes on the dorsal body surface. **6. Eurypoda**—fusion between intercentrum of atlas and neural arches, postacetabular process of ilium shortened, short and stocky metapodials, straight (versus bowed) femur with reduction of fourth trochanter. **7. Cerapoda**—five or fewer premaxillary teeth, a diastema between premaxillary and maxillary teeth, loss of squamosal-jugal contact, elongated rodlike prepubic process, robust first metatarsal. **8. Ornithopoda**—premaxillary tooth row offset ventrally compared to maxillary tooth row, narrow and elongated frontal bones, jaw joint set well below level of tooth rows by ventral extension of quadrate; absence of a bony prominence in cheek region, elongation of lateral process of premaxilla to contact lacrimal and/or prefrontal, elongate and narrow prepubic process. **9. Iguanodontia**—premaxillary teeth absent, antorbital opening reduced, denticulate margin of predentary, deep dentary ramus, loss of sternal rib ossification, blade-shaped prepubic process. **10. Hadrosauridae**—three or more replacement teeth per tooth position, posterior extension of dentary toothrow to behind apex of coronoid process, long coracoid process, dorsoventrally narrow proximal scapula. **11. Marginocephalia**—a shelf formed by the parietals and squamosals extends over the occiput, reduced contribution of premaxilla to palate, number of premaxillary teeth reduced to 3, relatively short pubis. **12. Pachycephalosauria**—thickened skull roof (frontals and parietals); short, laterally bowed humerus; other characters of the skull, dorsal vertebrae, and pelvis. **13. Ceratopsia**—rostral bone anterior to premaxilla forming a beak, enlarged premaxilla. jugals flare beyond the skull roof; deep, deeply arched palate beneath beak, frill composed principally of paired parietal bones. **14. Saurischia**—construction of the snout including subnarial foramen, concave facet on axial intercentrum for axis, specialized articulations on dorsal vertebrae, twisting of first phalanx of manual digit I, well-developed supracetabular crest on ilium. **15. Sauropodomorpha**—relatively small skull, anterior end of premaxilla deflected, teeth with serrated crowns, at least 10 cervical vertebrae forming an elongated neck, dorsal and caudal vertebrae added to sacrum, enormous thumb with enlarged claw, enlarged obturator foramen in pelvis, elongated femur. **16. Prosauropoda**—lateral lamina on maxilla, ridge on lateral surface of dentary, phalanx on manual digit 1 with proximal heel, twisting of the large thumb claw. **17. Sauropoda**—four or more sacral vertebrae, modifications of cervical vertebrae, increase in forelimb length, straight femur with lesser trochanter reduced or absent, feet with adaptations for weight-bearing. **18. Theropoda**—bladelike serrated teeth, intramandibular joint, promaxillary fenestra **19. Neotheropoda**—furcula formed from fused clavicles, manual digit V lost, 5 or more sacral vertebrae, pedal digits 1 and 4 reduced. **20. Tetanurae**—low ridge demarcating the maxillary antorbital fossa, prominent acromion on scapula, loss of manual digit IV phalanges, back half of tail stiffened by interlocking zygapophyses, modifications of femur and knee joint. **21. Avetheropoda**—maxillary fenestra, modifications of back of skull and palate, forward shift of zone of stiffening in tail skeleton, other features of pelvis and hind limb. **22. Coelurosauria**—enlargement of antorbital fossa, metcarpal III shorter than metacarpal II, flat articulations between centra of cervical vertebrae. **23. Maniraptora**—enlarged forelimb, well-developed semilunate bone in wrist, large bony sternum, features of ankle joint. **24. Eumaniraptora**— pubis points backwards, stiffened tail vertebrae. **25. Avialae (= extant birds [Aves] plus stem group relatives)**—progressive loss of teeth on maxilla and dentary, well-developed bill, characteristics of skull, jaws, vertebrae, and axial and appendicular skeleton.

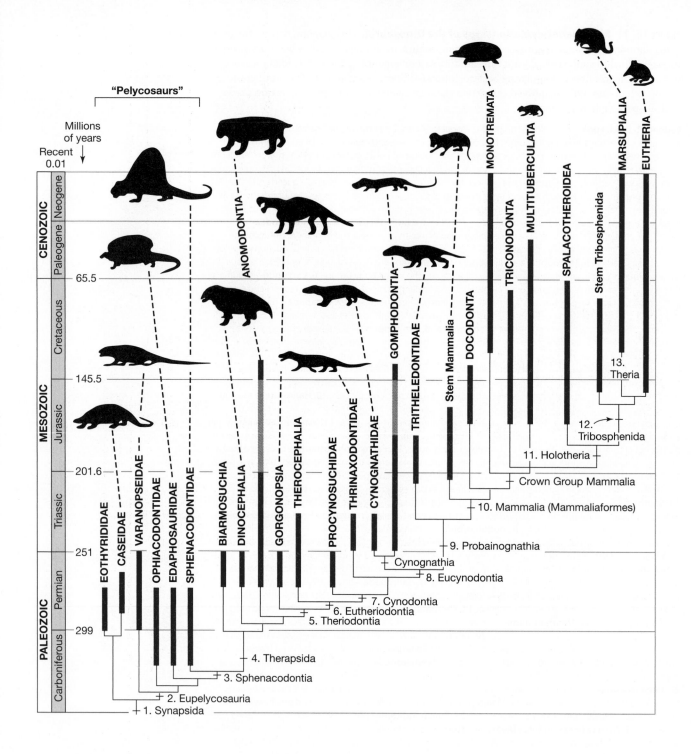

FIGURE 18-1 Phylogenetic relationships of the Synapsida. This diagram depicts the probable relationships among the major groups of synapsids. The black lines show relationships only; they do not indicate times of divergence or the unrecorded presence of taxa in the fossil record. Lightly shaded bars indicate ranges of time when the taxon is believed to be present, because it is known from earlier and later times, but is not recorded in the fossil record during this interval. Only the best-corroborated relationships are shown. Note that some researchers favor classifying the tritylodontids (here included with the Gomphodontia) as the sister taxon to Mammalia. The numbers at the branch points indicate derived characters that distinguish the lineages.

Legend: **1. Synapsida**—lower temporal fenestra present. **2. Eupelycosauria**—snout deeper than it is wide, frontal bone forming a large portion of the margin of the orbit. **3. Sphenacodontia**—a reflected lamina on the angular bone, retroarticular process of the articular bone turned downward, high coronoid eminence on dentary, narrowing of scapular blade. **4. Therapsida**—temporal fenestra enlarged, upper canine plus the bone containing it (the maxilla) enlarged, mobile scapula with loss of pelycosaur screw-shaped glenoid, ossified sternum, limb bones more slender, limbs held more underneath body (indicated by inturned heads of femur and humerus), femur with greater trochanter (for insertion of gluteal muscles), shorter feet. **5. Theriodontia**—coronoid process on dentary, flatter skull with wider snout. **6. Eutheriodontia**—temporal fossa completely open dorsally, differentiation of the vertebral column into distinct lumbar and thoracic regions. **7. Cynodontia**—postcanine teeth with anterior and posterior accessory cusps and small cusps on the inner side, partial bony secondary palate, masseteric fossa on dentary and bowing out of zygomatic arch (evidence for the presence of a masseter muscle), large sagittal crest on top of skull, double occipital condyle, lumbar ribs reduced or lost, coracoid reduced, ilium expanded forward, pubis reduced, femur with inturned head, distinct calcaneal heel. **8. Eucynodontia**—dentary greatly enlarged, postdentary bones reduced, phalangeal formula of 2-3-3-3-3. **9. Probainognathia (also included but not shown here are the families Probainognathidae and Chiniquodontidae)**—pineal foramen absent, posteriorly enlongated secondary palate, at least incipient contact between dentary and squamosal bones; Tritheledontidae and Mammalia share the derived features of prismatic enamel and unilateral action of the lower jaw. **10. Mammalia**—dentary-squamosal jaw articulation with ball-and-socket joint, double-rooted postcanine teeth, specializations of the portion of the skull housing the inner ear. ("Stem Mammalia" is a paraphyletic assemblage of Late Triassic and Early Jurassic genera, including [listed from basal to more derived] *Adelobasileus, Sinoconodon, Megazostrodon*, and *Morganucodon*.) **11. Holotheria**—reversed triangles molar pattern. **12. Tribosphenida**—tribosphenic molars. ("Stem Tribosphenida" is a paraphyletic assemblage of Early Cretaceous genera, including [listed from basal to more derived] *Aegialodon*, *Pappotherium*, and *Holoclemensia*, plus others. The somewhat less derived *Vincelestes* might also be included in this grouping.) **13. Theria**—details of braincase structure, many features of the soft anatomy.

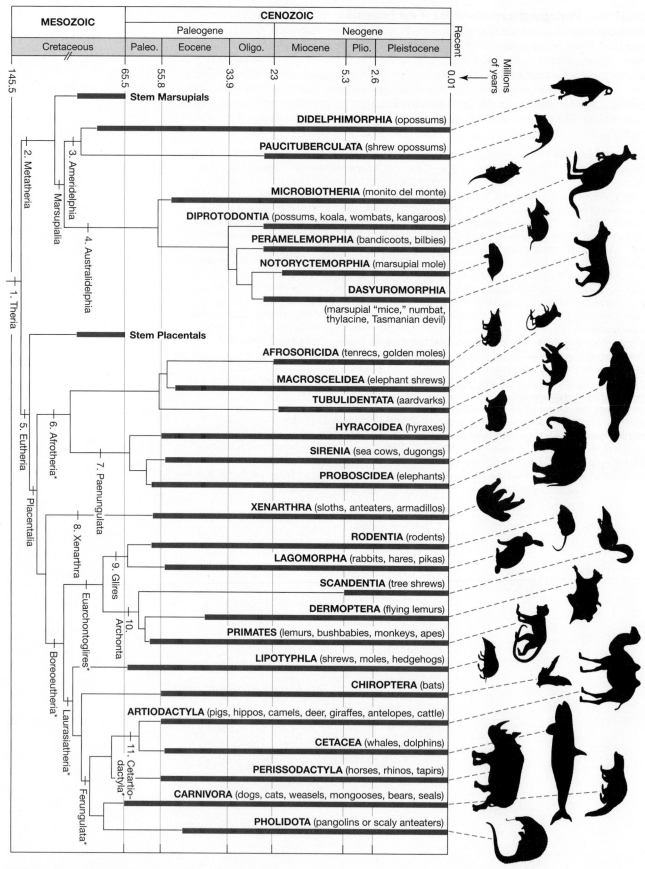

MESOZOIC	CENOZOIC							
	Paleogene			Neogene				
Cretaceous	Paleo.	Eocene	Oligo.	Miocene	Plio.	Pleistocene		Recent

145.5   65.5   55.8   33.9   23   5.3   2.6   0.01

← Millions of years

1. Theria
2. Metatheria
3. Ameridelphia
Marsupialia
4. Australidelphia
5. Eutheria
6. Afrotheria*
7. Paenungulata
Placentalia
8. Xenarthra
9. Glires
10. Archonta
Euarchontoglires*
Boreoeutheria*
11. Cetartio-dactyla*
Laurasiatheria*
Ferungulata*

Stem Marsupials

DIDELPHIMORPHIA (opossums)

PAUCITUBERCULATA (shrew opossums)

MICROBIOTHERIA (monito del monte)

DIPROTODONTIA (possums, koala, wombats, kangaroos)

PERAMELEMORPHIA (bandicoots, bilbies)

NOTORYCTEMORPHIA (marsupial mole)

DASYUROMORPHIA

(marsupial "mice," numbat, thylacine, Tasmanian devil)

Stem Placentals

AFROSORICIDA (tenrecs, golden moles)

MACROSCELIDEA (elephant shrews)

TUBULIDENTATA (aardvarks)

HYRACOIDEA (hyraxes)

SIRENIA (sea cows, dugongs)

PROBOSCIDEA (elephants)

XENARTHRA (sloths, anteaters, armadillos)

RODENTIA (rodents)

LAGOMORPHA (rabbits, hares, pikas)

SCANDENTIA (tree shrews)

DERMOPTERA (flying lemurs)

PRIMATES (lemurs, bushbabies, monkeys, apes)

LIPOTYPHLA (shrews, moles, hedgehogs)

CHIROPTERA (bats)

ARTIODACTYLA (pigs, hippos, camels, deer, giraffes, antelopes, cattle)

CETACEA (whales, dolphins)

PERISSODACTYLA (horses, rhinos, tapirs)

CARNIVORA (dogs, cats, weasels, mongooses, bears, seals)

PHOLIDOTA (pangolins or scaly anteaters)

FIGURE 20–2 Phylogenetic relationships of extant mammalian orders, excluding Monotremata. This diagram depicts the probable relationships among the major groups of living therian mammals. The black lines show interrelationships only; they do not indicate times of divergence or the unrecorded presence of taxa in the fossil record. Lightly shaded bars indicate ranges of time when the taxon is believed to be present, because it is known from earlier times, but is not recorded in the fossil record during this interval. Only the best-corroborated relationships are shown. The numbers at the branch points indicate derived characters that distinguish the lineages. This cladogram is based on a combination of morphological and molecular characters. Higher taxa for which the primary (or only) evidence is molecular are indicated by an asterisk (*). Afrotheria is placed here as the basal placental clade: note, however, that other researchers promote Afrotheria and Xenarthra as sister groups or one or both groups in a more nested position within the other placentals.

Legend: 1. Theria—mammary glands with nipples, viviparity with loss of eggshell, digastric muscle used in jaw opening, anal and urogenital openings separate in adults, spiraled cochlea, scapula with supraspinous fossa, numerous features of skull and dentition. **2. Metatheria**—dentition essentially **monophyodont** (P3 is the only tooth replaced); development of chorioallantoic membrane suppressed; ureters pass medial to Müllerian ducts to enter the bladder; pseudovaginal canal present at parturition; various detailed features of skull, dentition (including upper molars with wide stylar shelves), and ankle joint. **3. Ameridelphia**—sperm paired in epididymis. **4. Australidelphia**—details of dentition and ankle joint. **5. Eutheria**—eggshell membrane lost, intrauterine gestation prolonged with suppression of estrous cycle, corpus callosum connects cerebral hemispheres, ureters pass lateral to Müllerian ducts to enter the bladder, fusion of Müllerian ducts into a median vagina, penis simple (not bifid at tip), details of dentition (including upper molars with narrow stylar shelves). **6. Afrotheria***—abdominal testes, more than 19 thoracolumbar vertebrae, details of fetal membranes, tooth replacement. **7. Paenungulata**—styloglossus tongue muscle bifurcate; details of structure of skull, wrist bones, and placenta. **8. Xenarthra**—details of skull anatomy, sacrum strongly fused to pelvis, tooth development suppressed with loss of anterior teeth and enamel poorly developed or absent. **9. Glires**—enlarged pair of ever-growing upper and lower incisors, which represent the deciduous second incisors of other mammals, details of skull anatomy. **10. Archonta (or Euarchonta)**—pendulous penis, details of ankle structure. **11. Cetartiodactyla***—double-trochleated ("double-pulley") astragalus.

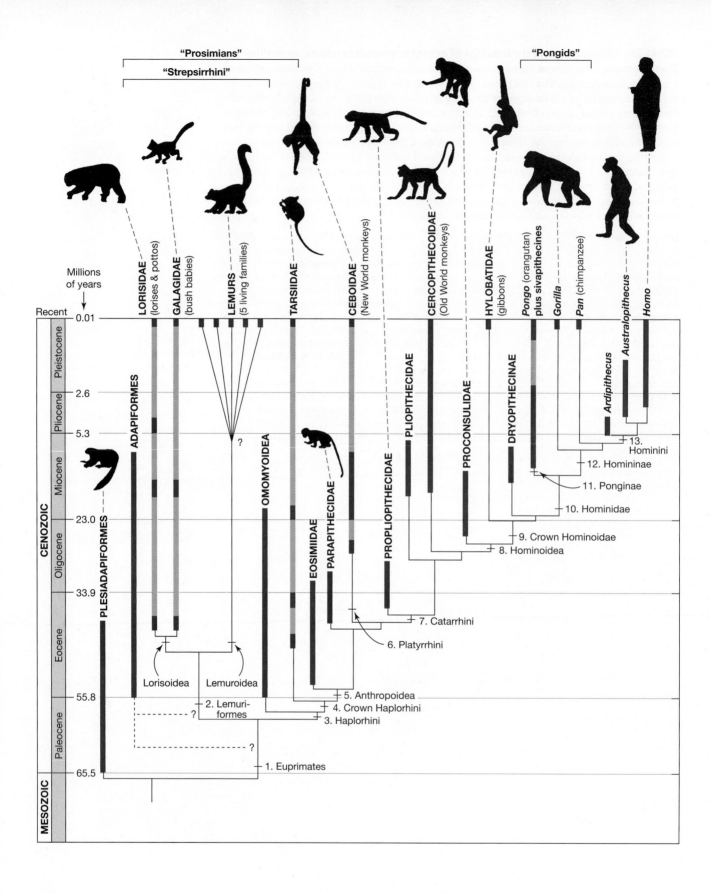

Figure 24–2 Phylogenetic relationships of the primates. This diagram depicts the probable relationships among the major groups of primates. The black lines show interrelationships only; they do not indicate the times of divergence or the unrecorded presence of taxa in the fossil record. Lightly shaded bars indicate ranges of time when the taxon is believed to be present, because it is known from earlier times, but is not recorded in the fossil record during this interval. The numbers at the branch points indicate derived characters that distinguish the lineages. Question marks indicate uncertainties about relationships; quotation marks indicate paraphyletic groupings.

Legend: 1. Euprimates—cheek teeth bunodont; a nail (instead of a claw) always present in extant forms, at least on the pollex (thumb); postorbital bar present. **2. Lemuriformes**—grooming claw present on second toe, lower front teeth modified into a tooth comb. **3. Haplorhini**—cranium short, orbit and temporal fossa separated ventrally by a postorbital wall, dry nose and free (rather than tethered) upper lip. **4. Crown Haplorhini**—eyes greatly enlarged. **5. Anthropoidea (monkeys and apes)**—fused frontal bones; fused mandibular symphysis; lower molars increase in size posteriorly, the third only slightly larger than the second, all with five cusps, the hypoconulid (most posterior cusp) small. **6. Platyrrhini (New World monkeys)**—widely spaced and rounded nostrils, contact between jugal and parietal bones on lateral wall of skull behind orbit, first two lower molars lack hypoconulids.
7. Catarrhini (Old World monkeys and apes)—narrowly spaced nostrils, number of premolars reduced to two, contact between frontal and sphenoid bones in lateral wall of skull, tympanic bone extends laterally to form a tubular auditory meatus (ear tube). **8. Hominoidea (apes and humans)**—lower molars with expanded talonid basin surrounded by five main cusps; broad palate and nasal regions; enlarged brain; broad thorax with dorsally positioned scapula and elongated clavicle; humeral head rounded, with highly mobile wrist joint; short and broad astragalus and calcaneum; reduced lumbar region (five vertebrae), with expanded sacrum (four to five vertebrae) and the absence of a tail (six caudal vertebrae form a coccyx). **9. (Crown Hominoidae)**—features relating to mobility in forearm and wrist joint. **10. Hominidae (great apes and humans)**—shortened face with frontal processes of the maxillae, nasals, and the orbits in the same plane; deepened mandible, spatulate lateral incisors, molars with thick enamel; elbow joint with increased stability; expanded ilium; hind limb reduced in length. **11. Ponginae**—skull with expanded and flattened zygomatic region, giving the face a concave aspect; upper lateral incisors very small relative to the central incisors; deeply wrinkled enamel on molars. **12. Homininae**—skull with true frontal sinus, brow ridges form continuous bar, os centrale and scaphoid fused in wrist. **13. Hominini (previously Hominidae; humans and fossil relatives)**—skeletal adaptations to bipedality (short, broad ilium, long legs in comparison with arms, big toe not opposable); relatively large molars and relatively small canines.

Glossary

abduction   Movement away from the mid-ventral axis of the body. *See also* **adduction**.

acetabulum   A socket in the pelvis that receives the head of the femur.

acid precipitation   Rain and snow acidified by sulfur- and nitrogen-containing gases produced by burning fossil fuels.

acrodont   Teeth loosely attached to the crest or inner edge of the jaw. *See also* **pleurodont**, **thecodont**.

activity temperature range   The range of body temperatures that an ectothermal animal maintains when it is thermoregulating.

adduction   Movement toward the mid-ventral axis of the body. *See also* **abduction**.

adductor mandibularis   The major muscle that closes the jaws.

adipocytes   Fat storage cells.

advertisement call   The vocalization of a male anuran used in courtship and territorial behavior.

aerobic metabolism   The metabolic breakdown of carbohydrates in the presence of oxygen, yielding carbon dioxide and water as the end products.

agnathans   Jawless vertebrates. *See also* **gnathostomes**.

air capillaries   Tiny air passages that intertwine with blood capillaries in the avian lung. Gas exchange takes place in cuplike depressions called faveoli in the walls of the air capillaries.

allantois   The extraembryonic membrane of amniotes that develops as an outgrowth of the hindgut.

allochthonous   Originating somewhere other than the region where found.

allometry   A situation in which the proportions of an object change when its size changes (contrast with **isometry**).

alloparental behavior   Parental care provided by an individual that is not a parent of the young receiving the care.

allopatry   A situation in which two or more populations or species occupy mutually exclusive, but often adjacent, geographic ranges.

Allotheria   The multituberculate mammals (extinct).

altricial   Helpless at birth or hatching, like pigeons and cats. *See also* **precocial**.

alula   The tuft of feathers on the first digit of a bird's wing that reduces turbulence in airflow over the wing.

alveolar lung   A lung in which airflow is tidal (in and out) and gas exchange occurs in closed-ended chambers called alveoli. Mammals have alveolar lungs.

alveoli   Small saclike structures that are sites of gas exchange in the lungs.

amble   A gait of tetrapods, a speeded-up walk in which the animal has at least one foot on the ground and two or three feet off the ground at any one time.

ammocetes larva   The larval form of lampreys.

ammonotelism   The excretion of nitrogenous wastes primarily as ammonia.

amnion   The innermost extraembryonic membrane of amniotes that surrounds the embryo.

amniotes   Vertebrates whose embryos have an amnion, chorion, and allantois (i.e., turtles, lepidosaurs, crocodilians, birds, and mammals) in addition to the yolk sac that is present in all vertebrates.

amniotic egg   An egg that has three extraembryonic membranes (the **amnion**, **chorion**, and **allantois**).

amphicoelous (amphicelous)   The condition in which the vertebral centrum is concave on both the anterior and posterior surfaces.

amphioxus   The lancelet, *Branchiostoma lanceolatum*.

amphistylic   A type of jaw suspension in fishes in which the upper jaw is attached to the chondrocranium and the hyoid arch (contrast with **hyostylic**).

amplexus   Clasping of a female anuran by a male during mating. Axillary amplexus refers to the male clasping the female in the pectoral region; inguinal amplexus is clasping in the pelvic region.

ampullae   (singular *ampulla*) Sense organs within the semicircular canals used for detection of orientation and angular acceleration.

ampullae of Lorenzini   Electroreceptors found in the skin of the snout of chondrichthyans.

anadromous   Migrating from seawater to freshwater to reproduce. *See also* **catadromous**.

anaerobic metabolism   The metabolic breakdown of carbohydrates in the absence of oxygen, usually yielding lactic acid as an end product.

anapsid   A skull that lacks temporal fenestrae, or an animal that has an anapsid skull. *See also* **diapsid**, **synapsid**.

anastomoses   (singular *anastomosis*) Communication among blood vessels in a network of branching and reconnecting streams.

angiosperm   The most recently evolved of the vascular plants, characterized by the production of seeds enclosed in tissues derived from the ovary. The ovary and/or seed are eaten by many vertebrates, and the success of the angiosperms has had important consequences for the evolution of terrestrial vertebrates.

angle of attack   The vertical angle between a chord line of an airfoil and the direction of motion of the fluid through which the airfoil is moving.

anisodactyl   The arrangement of toes seen in perching birds, with three in front opposed to one behind.

anisognathy   The situation in which the tooth rows in the upper and lower jaws are not the same distance apart. *See also* **isognathy**.

annulus   (plural *annuli*) Rings extending around a structure.

Anthropogene   "Age of Man," another name for the Holocene (the past 11,700 years) that is used in popular rather than in scientific writing.

antidiuretic hormone, ADH   (also known as vasopressin) A hypothalamic hormone that causes the kidney to conserve water by increasing the permeability of the collecting tubules.

apatite   The mineral form of calcium phosphate found in bone.

aphotic   "Without light" (e.g., in deep-sea habitats or caves).

apnea   "Without breath" (i.e., holding the breath, as during diving).

apocrine gland   A type of gland in which the apical part of the cell from which the secretion is released breaks down in the process of secretion. *See also* **holocrine gland**.

apomorphy   (also known as *derived character*) A character that has changed from its ancestral condition. *See also* **plesiomorphy, synapomorphy, symplesiomorphy**.

aposematic   Having a character, such as color, sound, or behavior, that is used to advertise an organism's noxious qualities.

appendicular   Within the limbs (e.g., leg bones or muscles).

apteria   Regions of skin without feathers.

arcade   A curve or arch in a structure, such as the tooth row of humans.

archaic   Of a form typical of an earlier evolutionary time.

archinephric duct   The ancestral kidney drainage duct.

archipterygium   A fin skeleton, as in a lungfish, consisting of symmetrically arranged rays that extend from a central skeletal axis.

articular   Pertaining to a joint, or to a bone at the posterior end of the lower jaw.

artiodactyls   Ungulate mammals with an even number of toes, such as cows and sheep (contrast with **perissodactyls**).

aspect ratio   The ratio of the length of a wing to its width.

assortative mating   Mating with like individuals because mates are selected nonrandomly on the basis of a trait that both individuals express.

atrium   (plural *atria*) 1. A chamber of the heart of vertebrates. The atria of the heart receive blood from the sinus venosus or the veins. 2. A chamber surrounding the gill slits of amphioxus and urochordates. The atrium exits to the exterior via a pore. 3. In a general sense, an empty space within a structure.

auditory bulla   An elaboration of bone around the ear region in mammals that may increase their auditory acuity.

aural   Referring to the external or internal ear or sense of hearing.

auricle   (also known as *pinna*) The external flap of the mammalian ear.

australopithecines   Extinct Pliocene/Pleistocene hominids, the sister group of *Homo*.

autapomorphy   An attribute unique to one evolutionary lineage of organisms. *See also* **apomorphy**.

autochthonous   Originating in the region where found.

autonomic nervous system   The part of the peripheral nervous system that controls the glands, smooth muscles, and internal organs and produces largely involuntary responses, including the sympathetic and parasympathetic portions in mammals.

autostylic   A type of jaw suspension in fishes in which the upper jaw is attached to the skull by processes (contrast with **hyostylic**).

autotomy   The voluntary release of a portion of the body to escape a predator, as when a lizard loses its tail. Autotomized structures are subsequently regrown.

axial   Within the trunk region.

baculum   (also known as *os penis*) A bone in the penis of some eutherian mammals.

baleen   (also known as *whalebone*) Sheets of fibrous, hornlike epidermal tissue that extend downward from the upper jaw; used for filter feeding by the baleen whales (mysticetes).

barbs   The primary branches from the rachis of a feather that form the vane.

barbules   Extensions from the **barbs** of a feather. Tiny hooks on the barbules hold adjacent barbs together to form the **vane**.

benthic   Living at the soil/water interface at the bottom of a body of water.

bilateral symmetry (bisymmetry)   Characteristic of a body that can be divided into mirror-image halves.

binomial nomenclature   The Linnean system that assigns generic and species names to organisms (e.g., *Homo sapiens* for modern humans).

biomass   The living organic material in a habitat (available as food for other species).

biome   A biogeographic region defined by a series of spatially interrelated and characteristic life forms (e.g., tundra, mesopelagic zone, tropical rain forest, coral reef).

bipedality   Locomotion on two legs, as in humans.

birdsong   The longest and most complex vocalization produced by a bird. The birdsong identifies the species and in many species is produced only by mature males and only during the breeding season.

blubber   An insulating layer of fat beneath the skin, typical of marine mammals.

bone   A mineralized tissue that forms the skeleton of vertebrates. Bone is about 50 percent mineralized. Not to be confused with the anatomical structure called "a bone," which may be composed of bone tissue, cartilage tissue, or both.

bound   A gait of tetrapods in which the animal jumps off the hind legs and lands on the front legs.

brachial Pertaining to the forelimb.

brachiation Locomotion by swinging from the underside of one branch to another.

brachydont Molar teeth with low crowns. *See also* **hypsodont**.

branchial Pertaining to the gills.

branchiomeric Referring to segmentation of structures associated with, or derived from, the ancestral pharyngeal arches. *See also* **metameric**.

branchiomeric muscles Muscles that power the pharyngeal arches.

branchiostegal rays A series of long, curved bones that support the branchiostegal (gill) membrane.

bristles Specialized feathers with a stiff rachis and without barbs or with barbs on only the proximal portion.

buccal pumping Drawing air or water into and out of the mouth region by raising and lowering the floor of the mouth.

bunodont Molar teeth with low, rounded cusps. *See also* **lophodont, selenodont**.

calamus The tubular base of a feather that remains in the follicle.

calcaneal heel The heel of mammals, formed by the large tarsal bone, the calcaneus.

calcaneus (or calcaneum) The large metatarsal bone that forms the heel of mammals.

cambered airfoil An airfoil that has a curved upper surface and a flat lower surface.

carapace A shell, as of a turtle; specifically, the dorsal part of the shell. *See also* **plastron**.

cardiac Referring to the heart.

cardinal vein A vein found in embryonic vertebrates and adult non-amniotes.

carnassials The teeth of eutherian mammals in the order Carnivora that are specialized as shearing blades.

carotid arteries The large arteries in the neck.

cartilage A firm and flexible skeletal material.

catadromous Migrating from freshwater to seawater to reproduce. *See also* **anadromous**.

catastrophism A hypothesis that major evolutionary change is a result of unique catastrophic events of broad geographic and thus ecological effect.

cavum arteriosum, cavum pulmonale, cavum venosum The chambers formed during ventricular contraction in the hearts of turtles and lepidosaurs.

cellulase An enzyme that can break the bonds between the glucose molecules in the polysaccharide cellulose.

cellulose A polysaccharide (complex carbohydrate) consisting of a polymer of glucose molecules; the main structural component of plant cell walls.

cementum A bonelike substance that fastens the teeth in their sockets.

centrum (plural *centra*) The bony portion of a vertebra that surrounds the notochord.

cephalic Pertaining to the head.

ceratotrichia Keratin fibers that support the web of the fins of Chondrichthyes.

cerebellum The dorsal part of the **metencephalon** of the brain.

cerebrum The two hemispheres of the brain that form most of the **telencephalon**.

character/character state Any identifiable characteristic of an organism. Characters can be anatomical, behavioral, ecological, or physiological.

chloride cells Cells in the gills of fishes and the skin of amphibians that are specialized to transport sodium and chloride ions.

choana (plural *choanae*) Internal nares.

chondrification The formation of cartilage.

chondrocranium (also known as *neurocranium*) A structure that surrounds the brain. Initially formed of cartilage, it is replaced by endochondral bone in most bony fishes and tetrapods. The chondrocranium, dermatocranium, and splanchnocranium are the three basic components of the vertebrate skull.

Chordata The phylum of animals that is characterized by having a notochord at some stage of life.

chorioallantoic placenta A placenta developed from the chorionic and allantoic extraembryonic membranes that replaces the choriovitelline placenta during the embryonic development of all eutherian mammals and some marsupials. *See also* **placenta**.

chorion The outermost extraembryonic membrane of amniotes.

choriovitelline placenta A placenta developed from the yolk sac; characteristic of all therian mammals during early development. *See also* **placenta**.

circadian "About a day." Circadian rhythms are cycles that have a period of approximately 24 hours.

clade A phylogenetic lineage originating from a common ancestral taxon and including all descendants. *See also* **grade**.

cladistics *See* **phylogenetic systematics**.

cladogram A branching diagram showing the hypothesized relationships among taxa.

cleidoic egg An egg that is independent of the environment except for heat and gas (carbon dioxide, oxygen, water vapor) exchange; a characteristic of amniotes.

cline Change in a biological character along a geographic gradient.

cloaca A common opening of the reproductive and excretory tracts.

cochlea (also known as *lagena* in nonmammalian tetrapods) The portion of the inner ear that houses the hair cells.

coelom A body cavity lined with tissue of mesodermal origin.

coevolution The complex biotic interaction through evolutionary time that results in the adaptation of interacting species to unique features of the life histories of the other species in the system.

collagen A fibrous protein that is a component of many structures.

columella The single auditory ossicle of the middle ear of nonmammalian tetrapods; the stapes of mammals. Homologous to the hyomandibula of fishes.

concealed estrus   Estrus that is not revealed by external signals, such as swelling of the genitalia. In the anthropoid lineage concealed estrus is a derived character of humans.

condyle   A rounded articular surface of a bone.

cones   Photoreceptor cells in the vertebrate retina that are differentially sensitive to light of different wavelengths and thus perceive color. *See also* **double cone**.

conodont   Small spinelike or comblike structures formed of apatite and found in marine sediments from the Late Cambrian to the Late Triassic; considered to be toothlike elements of an early vertebrate, the conodont animal.

conspecific   Belonging to the same species as that under discussion. *See also* **heterospecific**.

continental drift   The movement of continental blocks on the mantle of Earth. *See also* **plate tectonics**.

conus arteriosus   An elastic chamber in front of the ventricle of some gnathostomes.

convergent evolution   The appearance of similar characters in widely separated evolutionary lineages (e.g., wings in bats and birds). *See also* **parallel evolution**.

coprolite   Fossilized dung.

coprophagy   Eating the first set of feces that is produced and thereby recycling nutrients that would otherwise be lost.

coronary arteries   The arteries that carry oxygenated blood to the muscles of the heart.

coronoid process   A vertical flange near the rear of the dentary bone that increases the area for attachment of the temporalis muscle.

corpus luteum   (plural *corpora lutea*) A hormone-secreting structure formed in the ovary from the follicular cells that remain after an egg is released.

cosmine   A form of dentine that contains branching canals characteristic of the cosmoid scales of sarcopterygian fishes.

countercurrent exchange   Fluid streams flowing in opposite directions in adjacent vessels to promote the exchange of heat or dissolved substances.

countershaded   Referring to a color pattern in which the aspect of the body that is more brightly lighted (normally the dorsal surface) is darker colored than the less brightly illuminated surface. Countershading makes an animal hard to distinguish from its background.

cranial   Pertaining to the cranium or skull, a unique and unifying characteristic of all vertebrates.

cranial kinesis   Movement within the skull or of the upper jaws independent of the skull.

cranial nerves   The nerves that emerge directly from the brain: 10 pairs in the primitive vertebrate condition and 12 pairs in amniotes.

Craniata   Animals that have a cranium. Hagfishes have a cranium but lack vertebrae and are sometimes classified in the Craniata but not in the Vertebrata.

cranium   A bony, cartilaginous, or fibrous structure surrounding the brain.

crepuscular   Animals that are active at dawn and dusk.

critical period   The restricted time during the ontogeny of an individual when imprinting occurs.

crop   An enlarged portion of the esophagus that is specialized for the temporary storage of food.

crosscurrent exchange   A pattern of flow in an exchange structure in which the fluids flow at an angle to each other rather than in opposite directions, as they do in countercurrent exchange. Blood and air exhibit crosscurrent flow in the avian lung.

cryptodires   Turtles that bend the neck in a vertical plane to retract the head into the shell.

cupula   A cup-shaped, gelatinous secretion of a neuromast organ in which the kinocilium and microvilli are embedded.

cursorial   Specialized for running.

dear enemy recognition   A situation in which a territorial animal responds more strongly to strangers than to its neighbors from adjacent territories.

demersal   More dense than water and therefore sinking—like the eggs of many fishes and amphibians.

dentary   The anteriormost tooth-bearing bone of the lower jaw of vertebrates.

denticles   Small toothlike structures in the skin, as in shark skin.

dentine   A mineralized tissue found in the teeth of extant vertebrates and the dermal armor of some primitive fishes. Dentine is about 90 percent mineralized.

derived character   (also known as *apomorphy*) A character that has changed from its ancestral condition. *See also* **shared derived characters**.

dermal bone   A type of bone that forms in the skin.

dermatocranium   Dermal bones that cover a portion of the skull. The **dermatocranium**, **chondrocranium**, and **splanchnocranium** are the three basic components of the vertebrate skull.

dermis   The deeper cell layer of vertebrate skin of mesodermal and neural crest origin. *See also* **epidermis**.

detritus   Particulate organic matter that sinks to the bottom of a body of water.

deuterostomes   Chordates, echinoderms, and two other phyla (hemichordates and xenoturbellids) linked by unique embryonic features.

diapsid   A skull with two temporal fenestrae, or an animal that has a diapsid skull. *See also* **anapsid, synapsid**.

diastema   A gap between the cheek teeth and the incisors.

diencephalon   The anterior region of the forebrain; contains structures that act in neural-hormonal control and as relay stations for information among sensory areas and higher brain centers.

digitigrade   Standing with the heel off the ground and the toes flat on the ground—like dogs. *See also* **plantigrade, unguligrade**.

diphyodonty   One replacement of the dentition during an animal's lifetime, as in most mammals.

dispersal   The separation of animal populations via movement of the animals themselves. *See also* **vicariance**.

distal   Away from the body. *See also* **proximal**.

distal convoluted tubule   The portion of a mammalian kidney nephron responsible for changing the concentration of the ultrafiltrate by actively transporting salt.

diurnal   Being active during the day.

double cone   A type of retinal photoreceptor in which two cones share a single axon. *See also* **cones**.

down feathers   Entirely plumulaceous feathers in which the rachis is either shorter than the longest barb or entirely absent.

drag   Backward force opposed to forward motion.

durophagous   Feeding on hard materials.

eccrine gland   A gland in the mammalian skin that produces a watery secretion with little organic content; forms the sweat glands of humans.

echolocation   A method of determining location in three-dimensional space by sending out a pulse of sound and listening to the echoes that return from objects in the environment.

ecological speciation   The evolution of barriers to gene flow resulting from divergent selection in different habitats in the absence of a geographic barrier.

ecosystem   A community of organisms and their entire physical environment.

ectoderm   One of the embryonic germ layers, the outer layer of the embryo.

ectotherm   An organism that relies on external sources of heat to raise its body temperature.

edentulous   Lacking teeth.

elastin   A fibrous protein that can stretch and recoil.

electrocytes   Muscle cells modified to produce an electrical discharge.

embryonic diapause   A reproductive strategy in which the embryo is maintained in a stage of arrested development before it implants on the wall of the uterus.

enamel   A mineralized tissue found in the teeth of extant vertebrates and the dermal armor of some primitive fishes. Enamel is about 99 percent mineralized.

encephalization quotient   The ratio of the actual brain size of a species to the brain size expected based on its body size.

endemism   The property of being endemic (i.e., found in only a particular region).

endocasts   Fossil impressions of the insides of body cavities.

endochondral bone   A type of bone that forms in cartilage.

endocrine disruptor   A natural or synthetic chemical that interferes with normal development by duplicating the physiological effect of a hormone.

endocrine glands   Glands that discharge hormones into the blood.

endoderm   One of the embryonic germ layers; the innermost layer of late embryos.

endometrium   The glandular uterine epithelium of mammals that secretes materials that nourish the embryo.

endotherm   An organism that relies on internal (metabolic) heat to raise its body temperature.

epaxial   Referring to muscles on the dorsal portion of the trunk. *See also* **hypaxial**.

epicercal   (also known as *heterocercal*) A tail-fin arrangement in which the upper lobe is larger than the lower lobe. *See also* **hypocercal**.

epicontinental sea (epeiric sea)   A sea extending within the margin of a continent.

epidermis   The superficial cell layer of vertebrate skin of ectodermal origin. *See also* **dermis**.

epigenetic   Pertaining to an interaction of tissues during embryonic development that results in the formation of specific structures.

epiphysis   1. Pineal organ; an outgrowth of the roof of the diencephalon. 2. (plural *epiphyses*) An accessory center of ossification at the ends of the long bones of mammals, birds, and some squamates. In mammals the epiphyses are the actual articulating ends of the long bones themselves, with the cartilaginous zone of growth between the epiphysis and diaphysis. When the ossifications of the shaft (diaphysis) and epiphysis meet, lengthwise growth of the shaft ceases. This process produces a determinate growth pattern.

epiphyte   A plant that grows nonparasitically on another plant.

epipubic bones   Bones in noneutherian mammals that project anteriorly from the pubis.

estivation (aestivation)   A form of torpor; usually a response to high temperatures or scarcity of water.

estrous cycle   The normal reproductive cycle of growth, maturation, and release of an egg.

estrus (oestrus)   The periodic state of sexual excitement in the females of most mammals (but not humans) that immediately precedes ovulation and during which a female is most receptive to mating; also known as *heat*.

estuarine   Pertaining to, or formed in, a region where the fresh water of rivers mixes with the seawater of a coast.

euryhaline   Capable of living in a wide range of salinities. *See also* **stenohaline**.

euryphagous   Eating a wide range of food items; a food generalist. *See also* **stenophagous**.

eurythermal   Capable of tolerating a wide range of temperatures. *See also* **stenothermal**.

eurytopic   Capable of living in a wide variety of habitats.

eusocial   Applied to a species or group of animals that display all of these characters: cooperation in caring for the young, reproductive division of labor, more or less sterile individuals aiding individuals engaged in reproduction, and overlap of two or more generations of life stages capable of contributing to colony labor.

eustachian tube   The passage that connects the middle ear to the pharynx.

Eutheria   The placental mammals and their extinct stem-group relatives.

exocrine glands   Glands that discharge through a duct into a cavity or onto the body surface.

explosive breeding   A very short breeding season.

extant   Currently living; i.e., an extant species has living individuals and an extant lineage has living species.

extra-pair copulation   Mating with an individual other than the partner in a monogamous breeding system.

extraperitoneal   Positioned in the body wall beneath the lining of the coelom (the peritoneum) in contrast to being suspended in the coelom by mesenteries.

fallopian tube   The anterior portion of the oviduct where eggs are fertilized.

faveolar lung   A lung in which air flows in one direction and gas exchange occurs in cuplike depressions (called faveoli) in the walls of the air capillaries. Birds have faveolar lungs.

fenestra   A large opening in a bone (e.g., the temporal fenestra).

ferment   To break down food in the absence of oxygen, as in the stomach of a ruminant.

fever   An increased body temperature in response to infection.

filial imprinting   The process by which a young animal learns to recognize its parents.

filoplumes   Fine, hairlike feathers with a few short barbs or barbules at the tip.

flow-through ventilation   Flow of respiratory fluid (air or water) in one direction, as across the gills of a fish.

foramen   (plural *foramina*) An opening in a bone (e.g., for the passage of nerves or blood vessels).

foregut fermenter   A mammal in which the fermentation of foodstuffs is carried out in a modified stomach (e.g., a cow).

fossa   A groove or depression in a bone or organ.

fossorial   Specialized for burrowing.

fovea centralis   An area of the vertebrate retina that contains only cone cells, where the most acute vision is achieved at high light intensities.

free nerve ending   A sensory nerve ending in the skin that is believed to sense pain.

furcula   The avian wishbone formed by the fusion of the two clavicles at their central ends.

fusiform   Torpedo shaped.

gallop   A gait of mammals that is a modified bound.

gametes   Sex cells—that is, eggs (ova) and sperm.

gastralia   Bones in the ventral abdominal wall of some reptiles.

gastrolith   A stone swallowed to aid digestion by grinding food in the gizzard.

genetic fitness   The contribution of one genotype to the next generation relative to the contributions of other genotypes.

genetic monogamy   A social system in which a male and female share parental responsibilities and do not mate with individuals outside the pair. *See also* **social monogamy**.

genus   A group of related species.

geographic (allopatric) speciation   The formation of new species because a geographic barrier separates populations within the geographic range of a species, allowing the isolated populations to undergo genetic divergence.

geosyncline   A portion of Earth's crust that has been subjected to downward warping. Sediments frequently accumulate in geosynclines.

gestation   The period during which an embryo is developing in the reproductive tract of the mother.

gigantothermy   The ability of an extremely large animal to maintain a constant and relatively high body temperature due to its low surface/volume ratio.

gill arch   The assemblage of tissues associated with a gill. The term may refer to the skeletal structure only or to the entire epithelial muscular and connective tissue complex.

gizzard   The muscular stomach of birds and other archosaurs.

glomerulus   A capillary tuft associated with a kidney nephron that produces an ultrafiltrate of the blood.

gnathostomes   Jawed vertebrates. *See also* **agnathans**.

gonads   The organs that produce gametes—that is, ovaries in females and testes in males.

Gondwana   A supercontinent that existed either independently or in close contact with all other major continental landmasses throughout vertebrate evolution until the middle of the Mesozoic; composed of all the modern Southern Hemisphere continents plus the subcontinent of India.

grade   A level of morphological organization achieved independently by different evolutionary lineages. *See also* **clade**.

Great American Interchange (GAI)   Faunal interchange between North and South America when the Central American land bridge (the Isthmus of Panama) was formed about 2.5 million years ago.

gymnosperms   The group of plants in which the seed is not contained in an ovary—conifers, cycads, and gingkos.

hallux   The big toe.

Harderian gland   A gland associated with the eyes of extant mammals that produces a secretion used to preen and waterproof the fur.

head-starting   Rearing neonatal animals in captivity for a period of time before they are released in the wild.

hemal arch   The structure formed by paired projections ventral to the vertebral centrum and enclosing caudal blood vessels.

hermaphroditic   Having both male and female gonads.

heterocercal   (also known as *epicercal*) A tail-fin arrangement in which the upper lobe is larger than the lower lobe.

heterocoelus   Having the articular surfaces of the vertebral centra saddle-shaped, as in modern birds.

heterodont   A dentition with teeth of different sizes, shapes, and functions in different regions of the jaw. *See also* **homodont**.

heterospecific   Belonging to a different species from that under discussion. *See also* **conspecific**.

heterosporous plants   Plants that have large and small spores; the smaller give rise to male gametophytes and the larger to female gametophytes (equivalent to proto-gymnosperms).

hibernation A period of torpor in the winter when food is scarce.

hindgut fermenter A mammal in which the fermentation of foodstuffs is carried out in the intestine (e.g., a horse).

holocrine gland A type of gland in which the entire cell is destroyed with the discharge of its contents. *See also* **apocrine gland**.

home range The area in which an animal spends most of its time and finds the food and shelter it needs. An animal does not defend its home range. *See also* **territory**.

hominin Modern humans and all of the extinct forms that are more closely related to modern humans than to chimpanzees.

homodont A dentition in which teeth do not vary in size or shape along the jaw. *See also* **heterodont**.

homologous Inherited via common ancestry.

homology The fundamental similarity of individual structures that belong to different species within a monophyletic group.

homoplasy Similarities that do not indicate common ancestry—for example, structures resulting from parallel and convergent evolution and evolutionary reversal.

hormone A chemical messenger molecule carried in the blood from its site of release to its site of action.

hoxgene complex A sequence of DNA that regulates the expression of genes that control the development of body structures.

hydrofoil A water-planing surface, such as the pectoral fins of sharks.

hydrosphere Free liquid water on Earth—oceans, lakes, rivers, and so on.

hyoid arch The second gill arch.

hyostylic A type of jaw suspension in fishes in which the upper jaw is attached to the skull by the hyomandibula. *See also* **autostylic**.

hyostylic jaw articulation A form of jaw attachment seen in sharks that allows great flexibility.

hypapophyses Sharp processes, on the ventral surface of the neck vertebrae of the egg-eating snake (*Dasypeltis*), that slice through the shell of an egg.

hypaxial Referring to muscles on the ventral portion of the trunk. *See also* **epaxial**.

hyperdactyly A condition in which the number of digits is increased above the usual tetrapod complement of five.

hyperosmolal Of lower water potential (higher solute concentration).

hyperphalangy An increase in the number of bones (phalanges) in the digits.

hyperthermia A body temperature higher than normal.

hypertrophy An increase in the size of a structure.

hypocercal A tail-fin arrangement in which the lower lobe is larger than the upper lobe. *See also* **epicercal**.

hypodermis A layer of tissue in the skin beneath the dermis; not technically part of the skin. The hypodermis contains nerves and blood vessels and may serve as a place for fat storage. *See also* **dermis**.

hypophysis The pituitary gland.

hyposmolal Of higher water potential (lower solute concentration).

hypothalamus A structure in the floor of the diencephalon that is involved in neural-hormonal coordination and integration.

hypothermia A body temperature lower than normal.

hypotremate Having the main gill openings on the ventral surface and beneath the pectoral fins, as in skates and rays. *See also* **pleurotremate**.

hypselodont Molar teeth with ever-growing crowns.

hypsodont Molar teeth with high crowns. *See also* **brachydont**.

imprinting A special kind of learning that occurs only during a restricted time (called the critical period) in the ontogeny of an individual.

inclusive fitness The sum of individual fitness plus the effect of kin selection.

index of refraction The amount of deflection of a ray of light as it passes from one medium into another.

individual fitness *See* **genetic fitness**.

infraorbital foramen A hole beneath the eye through which nerves and blood vessels pass to the muzzle.

infrared The portion of the electromagnetic spectrum with wavelengths from 750 nanometers (just beyond visible light) to 1 millimeter (just before microwave radiation); often called *thermal radiation* or *heat*.

infrasound Sound frequencies below the range of human hearing, approximately 20 hertz.

ingroup The group of organisms being considered. *See also* **outgroup**.

insolation Solar radiation that reaches Earth's surface.

intercalary cartilage A cartilage lying between the last two bones in the toes of some tree frogs.

intercalary plates Extra elements in the vertebral column of elasmobranchs that protect the spinal cord and major blood vessels.

interspecific Pertaining to phenomena that occur between members of different species.

intraspecific Pertaining to phenomena that occur between members of the same species.

isognathy The situation in which the tooth rows in the upper and lower jaws are the same distance apart. *See also* **anisognathy**.

isohaline Of the same salt concentration.

isometry The situation in which the proportions of an object remain the same when its size changes. *See also* **allometry**.

isosmolal Of equal water potential (equal solute concentration).

isostasy Condition of gravitational balance between segments of Earth's crust or of return to balance after a disturbance.

isostatic movement Vertical displacement of the lithosphere due to changes in the mass over a point or region of Earth.

isotherm   A line on a map that connects points of equal temperature.

iteropary   Producing several individual babies or litters of young during the lifetime of a female. *See also* **semelpary**.

Jacobson's organ   (also known as *vomeronasal organ*) An olfactory organ in the roof of the mouth of tetrapods.

keratin   A fibrous protein found only in vertebrates that forms epidermal structures such as hair, scales, feathers, and claws.

kin selection   Favoring the perpetuation of one's own genes by helping relatives to reproduce.

kinocilium   A sensory cell located in neuromast organs.

lactation   The production of milk from mammary glands to nourish young; characteristic of mammals.

lagena   (also known as *cochlea* in mammals) The portion of the inner ear that houses the hair cells.

lateral line system   The sensory system on the body surface of aquatic vertebrates that detects water movement.

lateral plate mesoderm   The ventral part of the mesoderm, surrounding the gut.

Laurasia   A northern supercontinent composed of North America, Europe, and Asia.

Laurentia   A Paleozoic continent that included most of present-day North America, Greenland, Scotland, and part of northwestern Asia.

lecithotrophy   Embryonic development nourished by the yolk when eggs are retained within the reproductive tract until they hatch. *See also* **ovoviviparity**.

leptocephalus larva   Specialized, transparent, ribbon-shaped larva of tarpons, true eels, and their relatives.

lift   Vertical force opposed to gravity.

Linnaean system   A system of naming living organisms developed by the Swedish naturalist Carl von Linné (Carolus Linnaeus) in the eighteenth century.

lithosphere   Earth's crust.

littoral   Pertaining to the shallow portion of a lake, sea, or ocean where rooted plants are capable of growing.

loop of Henle   The portion of the renal tubule of mammals that extends into the medulla; essential for establishing the concentration gradient that produces a small volume of highly concentrated urine.

lophodont   Molar teeth with ridges (lophs) that run in a predominantly internal-external direction across the tooth. *See also* **bunodont, selenodont**.

lophophorate   Pertaining to several kinds of marine animals that possess ciliated tentacles (lophophores) used to collect food (e.g., pterobranchs).

lophs   Crests or ridges on mammalian cheek teeth, formed from uniting the individual cusps in a variety of ways.

lower critical temperature   The environmental temperature below which an endotherm must increase its metabolic heat production to maintain a stable body temperature.

lower lethal temperature   The environmental temperature below which even maximum metabolic heat production is inadequate to maintain a stable body temperature.

maculae   (singular *macula*) Sense organs within the sacculus and utriculus used for detection of orientation and linear acceleration.

mammary gland   A gland, found in mammals, that secretes milk. Mammary glands have characters of both **apocrine** and **eccrine** glands.

mammilary bodies   Organic granules attached to the egg membrane that are the sites of the first formation of calcite crystals making up the egg shell.

mandibular arch   The most anterior of the gill arches, forming the jaws of gnathostomes.

marsupium   An external pouch in which the young of marsupial mammals develop.

masseter muscle   A jaw muscle of mammals originating from the zygomatic arch and inserting on the lower jaw.

masticate   To chew thoroughly.

matrilineage   Related females in a group that support one another in social interactions.

matrotrophy   Embryonic development nourished by materials transferred from the maternal circulation. Placentrophic matrotrophy is the situation in which a placenta is the site of transfer of nutrients and wastes between the embryo and maternal circulation.

maxilla   (plural *maxillae*) The left and right maxillae are the tooth-bearing bones in the upper jaw of a vertebrate.

megafauna   Species of large terrestrial animals (mammals, birds, and reptiles) that became extinct when human populations expanded, mostly between 50,000 and 10,000 years ago.

Meissner's corpuscle   A sensory nerve ending in the skin that is believed to sense touch.

melanocyte   A pigment cell that contains melanin.

meninges   Sheets of tissue enclosing the central nervous system. In mammals these are the dura mater, arachnoid, and pia mater.

menstrual cycle   The periodic shedding of the endometrial lining of the uterus; characteristic of humans and some other anthropoid primates.

mesenteries   Membranous sheets derived from the mesoderm that envelop and suspend the viscera from the body wall within the coelom.

mesoblast   Mesodermal cell.

mesoderm   The central of three germ layers of late embryos.

metameric   Referring to ancestral segmentation with serially repeated units along the body axis. *See also* **branchiomeric**.

metamorphic climax   The period in the life of a tadpole that begins with appearance of the forelimbs and ends with disappearance of the tail.

metamorphosis   The developmental transition from larval to adult body form.

metanephric kidney   The adult kidney of amniotes (contrast with **opisthonephric kidney**).

Metatheria   The marsupial mammals and their extinct stem-group relatives.

metencephalon The anterior region of the hindbrain; includes the cerebellum in gnathostomes.

microvilli Sensory cells located in neuromast organs.

mimicry A tripartite system in which one organism (the mimic) counterfeits the signal of a second organism (the model), thereby deceiving a third organism (the dupe). The signal can be any characteristic of the model that the dupe can perceive—for example, color, pattern, scent, and so on.

molting Replacing old hairs or feathers with new ones.

monogamy A mating system based on a pair bond between a single male and female. *See also* **genetic monogamy, social monogamy, polygamy**.

monophyletic Having a single evolutionary origin. *See also* **paraphyletic, polyphyletic**.

monophyletic lineage A taxon composed of a common ancestor and all its descendants.

monophyly Relationship of two or more taxa having a common ancestor.

monophyodonty No replacement of the dentition during an animal's lifetime (compare to **diphyodonty** and **polyphyodonty**).

monotremes Basal living mammals that retain the generalized amniote reproductive feature of laying eggs.

morph A genetically determined variant in a population.

morphotypic Referring to a type of classification based entirely on physical form.

musth An annual period of elevated testosterone levels in the blood of male elephants.

myelencephalon The posterior region of the hindbrain, including the regulatory medulla oblongata.

myomeres Blocks of striated muscle fiber arranged along both sides of the body; most obvious in fishes.

myrmecophagy Eating ants and termites.

naris (plural *nares*) The external opening of the nostril.

nasolabial groove A channel from the external naris to the lip found in plethodontid salamanders.

neocortex The portion of the forebrain involved in sensory integration and nervous control; also involved in thought in humans.

neonates Newborn individuals.

neopallium (also known as *neocortex*) The derived expanded portion of the mammalian cerebral cortex.

neoteny Retention of larval or embryonic characteristics past the time of reproductive maturity. *See also* **paedomorphosis, progenesis**.

nephron The basic functional unit of the kidney.

neural arch A dorsal projection from the vertebral centrum that, at its base, encloses the spinal cord.

neural crest A type of embryonic tissue unique to vertebrates that forms many structures, especially in the head region.

neurocranium (also known as *chondrocranium*) The portion of the head skeleton that encases the brain.

neuromast organs Clusters of sensory hair cells and associated structures on the surface of the head and body of aquatic vertebrates; usually enclosed within the lateral line system.

neuron The basic functional unit of the nervous system.

New Synthesis (also known as *Neo-Darwinism*) The combination of genetics and evolutionary biology developed in the early twentieth century.

niche The functional role of a species or other taxon in its environment; the ways in which it interacts with both the living and nonliving elements.

nocturnal Being active at night.

notochord A dorsal stiffening rod that gives the phylum Chordata its name.

occipital Pertaining to the posterior part of the skull.

odontodes Small toothlike elements in the skin; the original toothlike components of primitive vertebrate dermal armor. The denticles of sharkskin are odontodes.

ontogenetic Pertaining to the development of an individual. *See also* **phylogenetic**.

ontogeny The development of an individual. *See also* **phylogeny**.

operculum (plural *opercula*) A flap or plate of tissue covering the gills.

opisthoglyphs Venomous snakes with enlarged teeth in the rear of the jaw; rear-fanged snakes.

opisthonephric kidney The adult kidney of most non-amniotes (contrast with **metanephric kidney**).

orobranchial chamber The mouth and gill region of a vertebrate.

orogeny The process of crustal uplift or mountain building.

osmosis Movement of water across a membrane from a region of high water potential (low solute concentration) to a region of low water potential (high solute concentration).

osseous Bony.

osteoderm A bone embedded in the skin; characteristic of crocodilians.

ostracoderm Armored jawless aquatic vertebrates known from the Ordovician to the Devonian.

otolith A mineralized structure in the inner ear of teleost fishes.

outgroup A group of organisms that is related to but removed from the group under study. One or more outgroups are examined to determine which character states are evolutionary novelties (apomorphies). *See also* **ingroup**.

ovary The female gonad.

oviduct The tract for the passage of eggs leading from the ovary to the uterus or to the cloaca.

oviparity The form of reproduction in which a mother deposits eggs that develop outside her body.

ovoviviparity Embryonic development nourished by the yolk when eggs are retained within the reproductive tract until they hatch. (*Lecithotrophy* is the preferred term for this type of embryonic nourishment.)

pachyostosis Increased density of bone; characteristic of diving animals.

Pacinian corpuscle A sensory nerve ending in the skin that is believed to sense pressure.

paedomorphosis   The retention of juvenile characters in an adult.

palatoquadrate   The upper jaw element of primitive fishes and Chondrichthyes, portions of which contribute to the palate, jaw articulation, and middle ear of other vertebrates.

pancreas   A glandular outgrowth of the intestine that secretes digestive enzymes.

pancreatic islets   (also known as *islets of Langerhans*) Clusters of endocrine cells in the pancreas that secrete insulin and glucagon.

Pangaea (Pangea)   A single supercontinent that existed during the mid-Paleozoic; composed of all the modern continents apparently in direct physical contact, with a minimum of isolating physical barriers.

parabronchi   The third level of branching of air passages in the avian lung. Air flows from the parabronchi into the air capillaries, where gas exchange takes place in cuplike depressions called faveoli.

parallel evolution   The appearance of similar characters in lineages that have separated recently (e.g., long hind legs in hopping rodents from the North American and African deserts). *See also* **convergent evolution**.

paraphyletic   Referring to a taxon that includes the common ancestor and some, but not all, of its descendants. *See also* **monophyletic, polyphyletic**.

parasympathetic nervous system   The division of the autonomic nervous system that maintains normal body functions, such as digestion.

parsimonious   In evolutionary biology, requiring the fewest changes from ancestral to derived character states.

parthenogenesis   Reproduction by females without fertilization by males.

parturition   The action of giving birth.

pectoralis major   The large breast muscle that powers the downstroke of the wings in a bird.

pelage   The hairy covering of a mammal.

pelagic   Living in the open ocean.

pelvic patch   A vascularized area in the pelvic region of anurans that is responsible for the uptake of water.

pericardial cavity   The portion of the coelom that surrounds the heart.

pericardium   Thin sheets of lateral-plate mesoderm that line the pericardial cavity.

perissodactyls   Ungulate mammals with an odd number of toes, such as horses (contrast with **artiodactyls**).

peritoneal cavity   The portion of the pleuroperitoneal cavity surrounding the viscera.

peritoneum   Thin sheets of lateral-plate mesoderm that line the pleuroperitoneal cavity.

Phanerozoic eon   The period of time since the start of the Cambrian period.

pharyngeal arches   (also known as *visceral skeleton*) The gill supports between the pharyngeal gill slits.

pharyngeal slits   Openings in the pharynx that were originally used to filter food particles from the water.

pharyngotremy   The condition in which the pharyngeal walls are perforated by slitlike openings; found in chordates and hemichordates.

pharynx   The throat region.

pheromone   A chemical signal released by one individual that affects the behavior of other individuals of the species.

photophore   A light-emitting organ.

phylogenetic   Pertaining to the development of an evolutionary lineage. *See also* **ontogenetic**.

phylogenetic systematics   (also known as *cladistics*) A classification system that is based on the branching sequences of evolution.

phylogeny   The evolutionary development of a group. *See also* **ontogeny**.

physoclistous   Lacking a connection between the gut and the swim bladder in adults (of fishes).

physostomous   Having a connection between the swim bladder and the gut in adults (of fishes).

piloerection   Contraction of muscles attached to hair follicles, resulting in the erection of the hair shafts.

pineal organ   A light-receptive organ and/or endocrine gland in the brain; grows out of the dorsal part of the diencephalon.

pinna   (plural *pinnae*) The external ear of mammals.

piscivorous   Having a diet composed of fish.

pitch   Tilt up or down parallel to the long axis of the body. *See also* **roll, yaw**.

pituitary gland   An endocrine organ formed in part from ventral outgrowth of the diencephalon; involved in neural-hormonal coordination and integration.

placenta   Extraembryonic tissue that transfers nutrients from the mother to the embryo and removes waste products. *See also* **choriovitelline placenta, chorioallantoic placenta**.

placentotrophic matrotrophy   Embryonic development nourished by materials transferred from the maternal circulation via a placenta.

placoid scales   A primitive type of scale found in elasmobranchs and homologous with vertebrate teeth.

plantigrade   Standing with the foot flat on the ground, as in humans. *See also* **digitigrade, unguligrade**.

plastron   A shell, as of a turtle; specifically, the ventral part of the shell. *See also* **carapace**.

plate tectonics   A theory of Earth history in which the lithosphere is continuously being generated from the underlying core at specific areas and reabsorbed into the core at other areas, resulting in a series of conveyor-like plates that carry the continents across the face of the Earth.

plesiomorphic   Pertaining to the ancestral character from which an apomorphy is derived.

plesiomorphy   An ancestral character (i.e., one that has not changed from its ancestral condition). *See also* **apomorphy, symplesiomorphy, synapomorphy**.

pleural cavities   Paired portions of the peritoneal cavity surrounding the lungs.

pleurodires   Turtles that bend the neck in a horizontal plane to retract the head into the shell.

pleurodont   Teeth fused to the inner surface of the jaw bones. *See also* **acrodont, thecodont**.

pleuroperitoneal cavity   The portion of the coelom that surrounds the viscera.

pleurotreme   Having the main gill openings on the sides of the body anterior to the pectoral fins, as in sharks. *See also* **hypotreme**.

polarity   The direction of evolutionary change in a character.

polyandry   A mating system in which a female mates with more than one male.

polygamy   A mating system in which an individual has more than one mate in a breeding season. *See also* **monogamy**.

polygyny   A mating system in which a male mates with more than one female.

polymorphism   The simultaneous occurrence of two or more distinct phenotypes in a population.

polyphyletic   Referring to a taxon that does not contain the most recent common ancestor of all the subordinate taxa of the taxon (i.e., not a true taxonomic unit but an assemblage of similar taxa, such as "marine mammals"). *See also* **monophyletic, paraphyletic**.

polyphyodonty   Having more than one replacement set of teeth in a lifetime (contrast with **diphyodonty**).

portal vessels   Blood vessels that lie between the two capillary beds of a portal system.

postzygapophysis   The articulating surface on the posterior face of a vertebral neural arch. *See also* **prezygapophysis**.

precocial   Well developed and capable of locomotion soon after birth or hatching—like chickens and cows. *See also* **altricial**.

premaxilla   (plural *premaxillae*) The left and right premaxillae are a pair of bones at the front of the upper jaw of a vertebrate. The premaxillae of some vertebrates bear teeth.

prezygapophysis   The articulating surface on the anterior face of a vertebral neural arch. *See also* **postzygapophysis**.

progenesis   Accelerated development of reproductive organs relative to somatic tissue, leading to paedomorphosis.

prolonged breeding   A long breeding season.

promiscuity   A breeding system in which both males and females have more than one mate in a breeding season.

proprioception   The neural mechanism that senses the positions of the limbs in space; a derived character of tetrapods.

proteroglyphs   Venomous snakes with permanently erect fangs at the front of the jaw (i.e., cobras and their relatives).

Proterozoic   The later part of the Precambrian, from about 1.5 billion years ago until the beginning of the Cambrian 540 million years ago. *See also* **Phanerozoic**.

protostomy   The condition in which the embryonic blastopore forms the mouth of the adult animal.

Prototheria   The monotreme mammals.

protraction   Movement away from the center of the body, usually in a forward direction. *See also* **retraction**.

protrusible   Capable of being moved away (protruded) from the body.

proventriculus   The glandular stomach of birds.

proximal   Close to the body. *See also* **distal**.

proximal convoluted tubule   The portion of a kidney nephron responsible for changing the concentration of the ultrafiltrate by actively transporting sodium.

pseudovaginal canal   A midline structure in marsupials through which the young are born.

pterylae   Tracts of follicles from which feathers grow.

pygostyle   The fused caudal vertebrae of a bird that support the tail feathers.

rachis   The central shaft of a feather, from which **barbs** extend to form the **vanes**.

ram ventilation   A respiratory current across the gills; created when an animal swims with its mouth open.

rectrices   (singular *rectrix*) Tail feathers.

refugium   An isolated area of habitat fragmented from a formerly more extensive biome.

regional heterothermy   Different temperatures in different parts of the body.

remiges   (singular *remex*) Wing feathers.

resource dispersion hypothesis   The proposal that the size of an animal's home range is determined by its needs for resources, such as food, and by the distribution of resources in the environment.

rete mirabile   "Wonderful net"—a complex mass of intertwined capillaries specialized for the exchange of heat and/or dissolved substances between countercurrent flows.

retina   Light-receptive cells that form a sheet at the back of the eyes; used for vision.

retraction   Movement toward the center of the body or in a backward direction. *See also* **protraction**.

reversal   Return to an ancestral feature (e.g., the streamlined body form of whales and porpoises).

ricochet   A bipedal hopping gait, as in kangaroos and many rodents.

rods   Photoreceptor cells in the vertebrate retina that are specialized to function effectively under conditions of dim light.

roll   Rotate around the long axis of the body. *See also* **pitch, yaw**.

rostrum   Snout, especially an extension anterior to the mouth.

ruminant   A herbivorous mammal with a specialized stomach in which microorganisms ferment plant material.

sacculus   The portion of the vestibular apparatus that contains maculae.

scapulocoracoid cartilage   In elasmobranchs and certain primitive gnathostomes, the single solid element of the pectoral girdle.

sclerophyll   A type of vegetation found in hot, dry regions. Sclerophyll vegetation is characterized by plants with leaves that are resistant to moisture loss.

scutes   Scales, especially broad or inflexible ones.

sebaceous gland   A gland in mammal skin that secretes oily or waxy materials.

sebum   An oily secretion produced by sebaceous glands.

secondary lamellae   Microscopic projections from the gill filaments where gas exchange occurs.

selenodont   Molar teeth with crescentic ridges or lophs rather than cusps that run in a predominantly anterior-posterior direction across the tooth. *See also* **bunodont, lophodont**.

semelpary   Reproducing only once during the lifetime of a female. *See also* **iteropary**.

semicircular canals   The portion of the vestibular apparatus that contains ampullae.

semiplumes   Feathers intermediate in structure between contour feathers and down feathers.

serial   Repeated, as in the body segments of vertebrates.

sexual imprinting   The process by which a young animal learns to recognize a mate of its own species.

shared derived characters   Derived characters shared by two or more taxa. *See also* **synapomorphy**.

shivering   The generation of heat by asynchronous contraction of muscle fibers.

sinus   An open space in a duct or tubular system.

sinus venosus   The posteriormost chamber of the heart of non-amniotes and some reptiles that receives blood from the systemic veins.

sister group   The monophyletic lineage most closely related to the monophyletic lineage being discussed.

sivapithecids   Later Cenozoic Eurasian hominoids related to the extant orangutan.

sociality   The state of living in structured groups.

social monogamy   A mating system in which a male and female share parental responsibility but mate with individuals outside the pair (contrast with **genetic monogamy**).

solenoglyphs   Venomous snakes with long fangs in the front of the jaw that are rotated when the mouth is open; vipers.

solute   A substance dissolved in a liquid.

somatic nervous system   The part of the peripheral nervous system that innervates structures derived from the somatic mesoderm controlling voluntary movements of skeletal muscles and returning sensations from the periphery.

somite   A member of a series of paired segments of the embryonic dorsal mesoderm of vertebrates.

spawning   The process by which fishes deposit and fertilize eggs.

species   In biological time, groups of organisms that are reproductively separated from other groups. In evolutionary time, a lineage that follows its own evolutionary trajectory.

specific dynamic action   Increased heat production associated with digesting food.

speciose   Referring to a taxon that contains a large number of species.

spermatophore   A packet of sperm transferred from male to female during mating by most species of salamanders.

splanchnocranium   The visceral or pharyngeal skeleton associated with the gills. The splanchnocranium, chondrocranium, and dermatocranium are the three basic components of the vertebrate skull.

spleen   The organ in which blood cells are produced, stored, and broken down.

squamation   Scaly covering of the body.

standard metabolic rate   The rate of metabolism that sustains vital functions (respiration, blood flow, etc.) in an animal at rest.

stapes   (called *columella* in nonmammalian tetrapods) The single auditory ossicles of the middle ear of tetrapods other than mammals; part of the ossicular chain of mammals. Stapes are homologous to the hyomandibula of fishes.

stenohaline   Capable of living within only a narrow range of salinity of surrounding water; not capable of surviving a great change in salinity. *See also* **euryhaline**.

stenophagous   Eating a narrow range of food items; a food specialist. *See also* **euryphagous**.

stenothermal   Capable of living or being active in only a narrow range of temperatures. *See also* **eurythermal**.

stratigraphy   Classification, correlation, and interpretation of stratified rocks.

stratum   (plural *strata*) A layer of material.

suckling   Forming fleshy seals against the bony hard palate with the tongue and with the epiglottis, which isolates breathing and swallowing during nursing.

supercooling   Lowering the temperature of a fluid below its freezing point without initiating crystallization.

surface/volume ratio   The ratio of body surface area to body volume.

swim bladder   (also known as *gas bladder*) A buoyancy structure of bony fishes; usually filled with gas, but in coelacanths it is filled with fat.

symbiont   An organism that lives with (usually inside or attached to) another organism, to their mutual benefit.

sympathetic nervous system   The division of the autonomic nervous system that produces largely involuntary responses that prepare the body for stressful or highly energetic situations.

sympatry   The occurrence of two or more species in the same area.

symphysis   A joint between bones formed by a pad or disk of fibrocartilage that allows a small degree of movement.

symplesiomorphy   A character shared by a group of organisms that is found in their common ancestor (i.e., a primitive character). *See also* **plesiomorphy**.

synapomorphy   Derived characters (apomorphies) shared by two or more taxa. *See also* **apomorphy**.

synapsid   A skull with a single temporal fenestra or an animal with a synapsid skull. *See also* **anapsid, diapsid**.

synsacrum   Fused vertebrae and ribs of birds that are fused with the pelvis.

syrinx The vocal organ of birds, lying at the base of the trachea.

tadpole The larval form of anurans.

talonid A basinlike heel on a lower molar tooth; found in therian mammals.

tapetum lucidum A reflective layer of cells behind the retina that increases sensitivity in low light by directing light back through the retina.

tarsometatarsus A bone formed by fusion of the distal tarsal elements with the metatarsals in birds and some dinosaurs. *See also* **tibiotarsus**.

taxon (plural *taxa*) Any scientifically recognized group of organisms united by common ancestry.

telencephalon The front part of the vertebrate brain that contains the cerebral cortex.

temperature-dependent sex determination A situation in which the sex of an individual is determined by the temperature it experiences during embryonic development; universal among crocodilians, widespread among turtles, and occasional among squamates.

temporal fenestra An opening in the bone of the temporal region of the skull that allows for the passage of jaw muscles from the skull to the lower jaw.

tentacle A sensory organ of caecilians that allows chemical substances to be transported from the surroundings to the vomeronasal organ.

territory An area that is defended against incursion by other individuals of the species. *See also* **home range**.

testis (plural *testes*) The male gonad.

tetrapods Terrestrial vertebrates descended from a four-legged ancestor.

thecodont Teeth set in sockets in the jaw bones; also refers to a paraphyletic assemblage of basal, extinct archosaurian reptiles. *See also* **acrodont**, **pleurodont**.

Theria Marsupial and eutherian mammals.

thermoneutral zone The range of ambient temperatures within which an endotherm can maintain a stable body temperature by changing its rate of heat loss to the environment; also called the *zone of physical thermoregulation*.

thermophilic Favoring high temperatures.

thermoregulation Control of body temperature.

tibiotarsus A bone formed by the fusion of the tibia and proximal tarsal elements in birds and some dinosaurs. *See also* **tarsometatarsus**.

tidal ventilation In-and-out flow of respiratory fluid, as in the lungs of a tetrapod.

torpor A period of inactivity accompanied by a fall in the regulated body temperature.

tribosphenic molars A form of molar unique to therian mammals.

troglodyte An organism that lives in caves.

trophic Pertaining to feeding and nutrition.

trophoblast Embryonic tissue of mammals specialized for implanting the embryo on the wall of the uterus, obtaining nutrients from the mother, and secreting hormones to signal the state of pregnancy to the mother.

trot A gait of tetrapods in which an animal moves diagonal pairs of limbs together with a period of suspension between each pair of limb movements when all four feet are off the ground.

tunicate The common name for a member of the subphylum Urochordata (also known as sea squirts or ascideans).

turbinates Scroll-like bones in the nasal passages covered by moist tissues that warm and humidify inspired air.

tympanic membrane, tympanum The eardrum.

ultrafiltrate A fluid produced in the glomerulus of a nephron; composed of blood with the cells and large molecules removed by filtration.

ultrasound Sound frequencies higher than the range of human hearing, approximately 20 kilohertz.

unguligrade Standing with only the tips of the toes on the ground, as in horses. *See also* **digitigrade**, **plantigrade**.

upper critical temperature The environmental temperature at which an endotherm must initiate evaporative cooling to maintain a stable body temperature.

upper lethal temperature The environmental temperature above which cooling mechanisms are insufficient to prevent an explosive rise in body temperature that leads to death.

urea cycle The enzymatic pathway by which urea is synthesized from ammonia.

ureotelism The excretion of nitrogenous wastes as primarily urea.

ureter The duct in amniotes that carries urine from the kidney to the urinary bladder or to the cloaca.

urethra The duct in amniotes that carries urine from the bladder to the outside. In male therian mammals, part of the urethra also carries sperm.

uricotelism The excretion of nitrogenous wastes as primarily uric acid and its salts.

urogenital Pertaining to the organs, ducts, and structures of the excretory and reproductive systems.

urogenital sinus The combined opening of the urethra and vagina in most female mammals. (Primates and some rodents have separate openings.)

urostyle A solid rod formed by fused posterior vertebrae; found in anurans.

utriculus The portion of the vestibular apparatus that contains the maculae.

vacuoles Membrane-bound spaces within cells containing secretions, storage products, and so on.

vanes The flat surfaces of a feather formed by the **barbs** that extend from the **rachis**.

vasa deferentia (singular *vas deferens*) The male reproductive tracts of mammals.

vasa recta The blood vessels surrounding the loop of Henle.

vascular Relating to blood and blood vessels.

vascular plexus Intertwined blood vessels in the skin that are the basis for countercurrent blood flow.

vasodilation Expansion of blood vessels to increase the blood flow to a region.

ventricle A chamber. The ventricle of the heart is the portion that applies force to eject blood from the heart.

Vertebrata Animals that have vertebrae.

vestibular apparatus The combination of sensory structures within the inner ear (sacculus, utriculus, and semicircular canals) that monitor and control balance and orientation.

vibrissae The sensory whiskers of mammals.

vicariance The separation of ancestral animals and plants by being carried passively on moving landmasses. *See also* **dispersal**.

viscera Internal organs suspended within the coelom.

visceral arches Gills and jaws.

visceral nervous system The part of the peripheral nervous system that innervates portions of the body derived from the lateral plate mesoderm; includes the autonomic nervous system and sensory nerves that relay information from the viscera and blood vessels. The system may also include the special branchial motor system of the cranial nerves, although this is now in dispute.

visceral skeleton Skeleton primitively associated with the pharyngeal arches; uniquely derived from the neural-crest cells and forming in mesoderm immediately adjacent to the endoderm lining the gut.

viviparity The form of reproduction in which the mother gives birth to a fully developed baby, as opposed to laying eggs.

vomeronasal organ (also known as *Jacobson's organ*) An olfactory organ in the roof of the mouth of tetrapods.

vulnerable A species is considered vulnerable when it is facing a high risk of extinction in the wild in the medium-term future.

Weberian apparatus A chain of small bones that conducts vibrations from the swim bladder to the inner ear of some bony fishes.

yaw Swing from side to side relative to the long axis of the body. *See also* **pitch**, **roll**.

yolk sac A structure attached to the ventral surface of an embryo that contains the nutritive yolk.

zone of chemical thermogenesis The range of temperatures within which an endotherm can maintain a stable body temperature through metabolic heat production.

zone of evaporative cooling The environmental temperature range between the upper critical temperature and the upper lethal temperature; in this range animals pant, sweat, and employ gular fluttering for cooling.

zone of tolerance The range of environmental temperatures over which an endotherm can maintain a stable body temperature.

Zugdisposition Preparation for migration by accumulating fat.

Zugstimmung The condition in which a bird makes migratory flights.

Zugunruhe Restlessness of caged birds that are prevented from migrating.

zygapophysis Articular process of the neural arch of a vertebra. *See also* **postzygapophysis**, **prezygapophysis**.

zygodactylous A type of foot in which the toes are arranged in two opposable groups.

zygomatic arch A temporal bar that is bowed outward to accommodate a large masseter muscle in mammals.

Credits

Photo Credits

Chapter 1 *1–2a:* Tim Davenport/WCS. *1–2b:* From: A new species of capuchin monkey, genus *Cebus Erxleben* (Cebidae, Primates): found at the very brink of extinction in the Pernambuco Endemism Centre. A.R.M. Pontes, A. Malta & P. H. Asfora. *Zootaxa* 1200: 1–12, 11 May 2006. *1–2c:* Gabriele Gentile. *1–6c:* Nancy Vandermey/wildcatzoo.org. *1–6d:* WWF, Alain Compost, HO/AP Photo.

Chapter 5 *5–9a:* Susan Dabritz/SeaPics.com. *5–9b:* Masa Ushioda/SeaPics.com.

Chapter 10 *10–2:* Gerhard Roth, University of Bremen, Bremen, Germany. *10–8, left and right:* Sharon B. Emerson, University of Utah. *10–14, left and right:* Theodore L. Taigen, University of Connecticut. *10–18e, top and bottom:* Richard J. Wassersug, Dalhousie Medical School. *10–19:* David Pfennig, University of North Carolina. *10–23a–c:* F. Harvey Pough.

Chapter 12 *12–1:* F. Harvey Pough.

Chapter 13 *13–1:* Marcus J. Simons, University of Otago, New Zealand. *13–11a–c:* Benjamin E. Dial, Chapman University. *13–15a–d:* C. J. Cole, American Museum of Natural History. *13–15e:* Charles M. Bogert, American Museum of Natural History. *13–19a:* Daniel H. Janzen, University of Pennsylvania. *13–19b,c:* F. Harvey Pough.

Chapter 14 *14–1a,b:* R. Bruce Bury/U.S. Fish and Wildlife Service. *14–3:* F. Harvey Pough. *14–6:* © Rick & Nora Bowers/Alamy. *14–8a,b:* J. and K. Storey, Carleton University, Ottawa, Ontario.

Chapter 16 *16–11a–c:* Jeffrey W. Lang, University of North Dakota. *16–23:* Barbara Moore, Peabody Museum of Natural History, Yale University. *16–27a:* Siepmann/AGE Fotostock. *16–27b:* Lawrence Witmer.

Chapter 17 *17–1b:* Department of Geology & Paleontology/Natural History Museum Vienna (NHMW). *17–5c:* Alan S. Pooley. *17–6:* Steve Byland/AGE Fotostock. *17–8:* Harry Taylor © Natural History Museum, London. *17–24a:* Steven Katovich, USDA Forest Service/Bugwood.org. *17–25:* Mary Tremaine/Cornell Laboratory of Ornithology. *17–29:* Morales/AG Fotostock. *17–30a:* Dennis Donahue/Shutterstock. *17–30b:* Kane513/Shutterstock. *17–30c:* www.eyethinkphoto.com/Alamy. *17–30d:* EcoView/Fotolia. *17–31a,b:* David W. Winkler, Cornell University.

Chapter 22 *22–4a:* Raymond S. Matlack. *22–4b:* Phil Myers, University of Michigan, http://animaldiversity.org. *22–4c:* Wayne Lynch/All Canada Photos/AGE Fotostock. *22–4d:* Nicholas Bergkessel, Jr./Photo Researchers, Inc. *22–6:* Raven Regan/Design Pics/Alamy. *22–7:* Gail R. Michener, University of Lethbridge, Alberta. *22–11:* Francois Gohier/Photo Researchers, Inc. *22–13:* Michael Fogden/Photoshot. *22–14:* George A. Bartholomew and Mark A. Chappell, University of California

at Los Angeles. *22–16:* Bartosz Hadyniak/iStockphoto. *22–19:* Phil A. Dotson/Photo Researchers, Inc. *22–20:* Jugal Tiwari.

Chapter 23 *23–3a–d:* Sara Cairns. *23–5a:* Jack A. Cranford. *23–5b,d:* Sara Cairns. *23–5c:* EcoPrint/Shutterstock. *23–9a:* Carol A. Saunders. *23–9b:* Carol A. Saunders.

Chapter 24 *24–16a:* Don Hitchcock, donsmaps.com. *24–16b,c:* Philippe Plailly/Photo Researchers, Inc.

Chapter 25 *25–4:* Craig Quirolo, Reef Relief. *25–5:* Kevin Ebi/Alamy. *25–6a:* Steve and Dave Maslowski/Photo Researchers, Inc. *25–6b:* Galina Bonch-Osmolovskaya. *25–6c:* Jeff Foott/Photoshot. *25–7:* Courtesy of Smithsonian Institution. *25–8a:* M. Hosking/FLPA. *25–8b:* Daryl Balfour/Gallo Images ROOTS/Getty Images. *25–9a:* KEN BOHN/UPI/Newscom. *25–9b:* Roger Tidman/FLPA.

Color Insert *CI–1, upper left:* Tom McHugh/Photo Researchers, Inc. *CI–1, upper right:* Bucky Reeves/Photo Researchers, Inc. *CI–1, bottom left:* Solvin Zankl/Naturepl.com. *CI–1, bottom right:* Image Quest Marine. *CI–2, top left:* Michael Lustbader/Photo Researchers, Inc. *CI–2, top right:* Cosmos Blank/National Audubon Society/Photo Researchers, Inc. *CI–2, bottom left, center left, center right:* F. Harvey Pough. *CI–3, top left:* Fred McConnaughey/Photo Researchers, Inc. *CI–3, top right:* Cosmos Blank/National Audubon Society/Photo Researchers, Inc. *CI–3, bottom left:* J. H. Robinson/Photo Researchers, Inc. *CI–3, bottom right:* Jany Sauvanet/Photo Researchers, Inc.

Cover: Rod Williams/Nature Picture Library

Illustration and Text Credits

Chapter 1 *1–6:* "Figure 4" from "Clouded Leopard Phylogeny Revisited: Support for Species Recognitions and Population Division between Borneo and Sumatra" by A. Wilting, et al. in *Frontiers in Zoology* 4:15, 2007.

Chapter 2 *2–3:* "Figure 3" from "Origin and Early Revolution of the Vertebrates: New Insights from Advances in Molecular Biology, Anatomy, and Paleontology" by N. D. Holland and J. Chen from *Bioessays* 23(2):142–151, February 2001. Copyright © 2001 by John Wiley & Sons, Inc. Reprinted by permission of John Wiley & Sons Ltd. *2–5:* From *Studies on the Structure and Development of Vertebrates* by E. S. Goodrich, 1930. *2–6b:* From "Figure 13.8" in *Vertebrates: Comparative Anatomy, Function, Evolution,* 2E by K. V. Kardong. Copyright © 1998 by The McGraw-Hill Companies, Inc. *1–6c:* Nancy Vandermey/wildcatzoo.org. *2–7:* "Figure 7" from *Discovering Fossil Fishes* by John G. Maisey. Copyright © Patricia Wynne. Reprinted by permission of Nevraumont Publishing Company. *2–9a,b:* From "Figure 8.1 and Figure 8.20" from *Comparative Anatomy of the Vertebrates,* 4E by G. C. Kent, 1978. *2–9c:* "Figure 7.3a" from *Functional Anatomy of the Vertebrates*, 2E by W. F. Walker and

Biologists Ltd. Reprinted by permission of The Company of Biologists Ltd.

Chapter 7 7–1: From *Outline of Phanerozoic Biography* by A. Hallam, 1994. 7–4: From *Vertebrate Paleontology*, 2E by M. J. Benton, 1997 and from "Figure 2.14a" from *Basic Paleontology* by M. J. Benton, et al., 1997.

Chapter 8 8–2: "Figure 9.3, part b" from *The Evolution of Vertebrate Design* by L. B. Radinsky. Copyright © 1987 by The University of Chicago Press. Reprinted by permission of The University of Chicago Press via Copyright Clearance Center. 8–3: From "Figure 8.36" in *Vertebrates: Comparative Anatomy, Function, Evolution*, 2E by K. V. Kardong. Copyright © 1998 by The McGraw-Hill Companies, Inc. 8–5: From "Figure 6.4" from *Gaining Ground* by Jennifer A. Clack, 2002. Reprinted by permission of Indiana University Press. 8–7: "Figure 3.3" from "Chapter 3 - Size and Shape" from *Life's Devices: The Physical World of Animals and Plants* by S. Vogel. Copyright © 1989 by Princeton University Press. Reprinted by permission of Princeton University Press. 8–13a: From "Figure 17.34" from *Vertebrates: Comparative Anatomy, Function, Evolution*, 2E by K. Kardong. Copyright © 1998 by The McGraw-Hill Companies, Inc.. Reprinted with permission of The McGraw-Hill Companies, Inc. 8–13b: From "Figure 17.41" from *Vertebrates: Comparative Anatomy, Function, Evolution*, 2E by K. Kardong. Copyright © 1998 by The McGraw-Hill Companies, Inc. Reprinted with permission of The McGraw-Hill Companies, Inc.

Chapter 9 9–3: From "Figure 2" from "The Origin and Early Diversification of Tetrapods" by P. E. Ahlberg and A. R. Milner from *Nature* 368(6471):507–514, April 7, 1994. Copyright © 1994 by Nature Publishing Group. Reprinted by permission of Macmillan Publishers Ltd. 9–4b: From "A Devonian Tetrapod-like Fish and the Evolution of the Tetrapod Body Plan" by E. B. Daeschler et al. from *Nature* 440(7085):757–763, April 6, 2006. Copyright © 2006 by Nature Publishing Group. Reprinted by permission of Macmillan Publishers Ltd. 9–5: From page 238 in *The Rise of Fishes: 500 Million Years of Evolution*, 2E, by John A. Long. Copyright © 2011 by The Johns Hopkins University Press. Reprinted by permission of The Johns Hopkins University Press. 9–6a: From "Figure 5" from "The Origin and Early Diversification of Tetrapods" by P. E. Ahlberg and A. R. Milner in *Nature* 368(6471):507–514, April 7, 1994. Copyright © 1994 by Nature Publishing Group. Reprinted by permission of Macmillan Publishers Ltd. 9–6b: From "Figure 1" from "Polydactyly in the Earliest Known Tetrapod Limbs" by J. A. Clack and M. I. Coates in *Nature* 347:66–69, September 6, 1990. Copyright © 1990 by Nature Publishing Group. Reprinted by permission of Macmillan Publishers Ltd. 9–7top: "Figure 31" from "The Devonian Tetrapod Acanthostega gunnari Jarvik: Postcranial Anatomy, Basal Tetrapod Interrelationships and Patterns of Skeletal Evolution" by M. I. Coates in *Transactions of the Royal Society of Edinburgh: Earth Sciences* 87:363–421. Copyright © 1996 by Cambridge University Press. Reprinted by permission of Cambridge University Press. 9–7 *bottom*: "Figure 1" from "The Axial Skeleton of the Devonian Tetrapod Ichthyostega" by P. E. Ahlberg, J. A. Clack, and H. Blom in *Nature* 437(7055):137–140, September 2005. Copyright © 2005 by Nature Publishing Group. Reprinted by permission of Macmillan Publishers Ltd. 9–8a,b: "Acanthostega" and "Ichthyostega"

from *Atlas of the Prehistoric World* by D. Palmer. Copyright © 1999 by Marshall Editions Developments Limited. Text Copyright © 1999 by Douglas Palmer. Reprinted by permission of Marshall Editions Limited. 9–8c: From "The Pelvic Girdle and Hind Limb of Crassigyrinus scoticus (Lydekker) from the Scottish Carboniferous and the Origin of the Tetrapod Pelvic Skeleton" by A. L. Panchen and T. R. Smithson from *Transactions of the Royal Society of Edinburgh: Earth Sciences* 81(1):31–44. Copyright © 1990 by the Royal Society of Edinburgh. Reprinted by permission of Cambridge University Press. 9–8d,e: From "Chapter 16" in *The Variety of Life* by C. Tudge. Copyright © 2000 by Oxford University Press. Reprinted by permission of Gordon Howes, illustrator. 9–8f: "Cyclotosaurus" from *The Simon and Schuster Encyclopedia of Dinosaurs and Prehistoric Creatures* by Barry Cox et al. Copyright © 1999 by Marshall Editions Developments Limited. Text Copyright © 1999 by Douglas Palmer. Reprinted by permission of Marshall Editions Limited. 9–9a: From "The Tetrapod Assemblage of Nyrany, Czechoslovakia" by Andrew R. Milner in *The Terrestrial Environment and Origin of Land Vertebrates* edited by A. Panchen, 1980. Reprinted by permission of Andrew R. Milner. 9–9b: From *Handbuch der Palaeoherpetologie—Encyclopedia of Palaeoherpetology: Part 5B— Batrachosauria (Anthracosauria)* by R. L. Carroll et al., 1972. 9–9d: From "Chapter 16" in *The Variety of Life* by C. Tudge. Copyright © 2000 by Oxford University Press. Reprinted by permission of illustrator, Gordon Howes. 9–9e: "Westlothiana lizziae" from *Atlas of the Prehistoric World* by D. Palmer. Copyright © 1999 by Marshall Editions Developments Limited. Text Copyright © 1999 by Douglas Palmer. Reprinted by permission of Marshall Editions Limited. 9–10a: From "Figure 1, part D" from "The Order Microsauria" by R. L. Carroll and P. Gaskill in *Memoirs of the American Philosophical Society* 126:1–211, 1978. Reprinted by permission of the American Philosophical Society. 9–10b: From "Chapter 16" in *The Variety of Life* by C. Tudge. Copyright © 2000 by Oxford University Press. Reprinted by permission of Gordon Howes, illustrator. 9–10c: From "The Tetrapod Assemblage of Nyrany, Czechoslovakia" by Andrew R. Milner in *The Terrestrial Environment and Origin of Land Vertebrates* edited by A. Panchen, 1980. Reprinted by permission of Andrew R. Milner. 9–11f: "Procolophon" from *Dawn of the Dinosaurs: The Triassic in Petrified Forest* by Robert A. Long and Rose Houk. Copyright © 2012. Reprinted by permission of the Petrified Forest Museum Association. 9–12: From *Morphogenesis of the Vertebrates* by T. W. Torrey, 1962. Reprinted by permission of John Wiley & Sons Ltd. 9–15: From *Mammal-Like Reptiles and the Origin of Mammals* by T. S. Kemp. Copyright © 1982 by Elsevier. Reprinted by permission of Elsevier.

Chapter 10 10–4: "Figure 4" from "Dear Enemy Recognition and the Costs of Aggression between Salamanders" by R. G. Jaeger in *The American Naturalist* 117(6):962–974, June 1981. Copyright © 1981 by The University of Chicago Press. Reprinted by permission of the University of Chicago Press via Copyright Clearance Center. 10–6: "Figure 14.1" from "Behavioral Energetics" by F. H. Pough in *Environmental Physiology of the Amphibians* edited by M. E. Feder and W. W. Burggren. Copyright © 1992. Reprinted by permission of the University of Chicago Press via Copyright Clearance Center. 10–10b: From *Amphibia and Reptiles* by H. Gadow, 1909. 10–10c,d: From *The Caecilians of the World* by E. H. Taylor, 1968. Reprinted by permission of the University Press of Kansas. *10–11:* From *The Biology of the Amphibia* by G. K. Noble, 1931. *10–15:* From *Tungara Frog* by M. J. Ryan, 1992. Reprinted by permission of The University of Chicago Press via Copyright Clearance Center. *10–16c:* From *The Biology of the Amphibia* by G. K. Noble, 1931. *10–16e:* From *The Hylid Frogs of Middle America* by W. E. Duellman, 1970. Reprinted by permission of the University Press of Kansas. *10–16f:* From *La Terre et la vie* 31:225–311 by M. Lamotte and J. Lescure, 1977. *10–20:* From "The Relationship of Locomotion to Differential Predation on *Pseudacris Triseriata* (Anura: Hylidae)" by R. J. Wassersug and D. G. Sperry in *Ecology* 58(4):830–839, July 1977. *10–24:* From *The Biology of the Amphibia* by G. K. Noble, 1931. *10–25:* From *Amphibia and Reptiles* by H. Gadow, 1909. *10–26:* From "Figure 2 and 3" from "Riding the Wave: Reconciling the Roles of Disease and Climate Change in Amphibian Declines" by Karen R. Lips et al. in *PLOS Biology* 6(3):441–454, March 2008 and personal communication from Karen R. Lips and Douglas C. Woodhams; From "Dramatic Declines in Neotropical Salamander Populations Are an Important Part of the Global Amphibian Crisis" by Sean M. Rovito et al. in *PNAS* 106(9):3231–3236, March 3, 2009 and personal communication from David B. Wake; and From "New Records of Batrachochytrium dendrobatidis in Chilean Frogs" by J. Bourke et al. in *Diseases of Aquatic Organisms* 95:259–261, July 12, 2011. *Table 10–2:* "Phylogenetic Relationships and Numbers of Species" in *Herpetology*, 2E by F. H. Pough et al., 2001. *Table 10–3:* "Phylogenetic Relationships and Numbers of Species" in *Herpetology*, 2E by F. H. Pough et al., 2001. *Table 10–4:* "Phylogenetic Relationships and Numbers of Species" in *Herpetology*, 2E by F. H. Pough et al., 2001.

Chapter 11 *11–1:* "Figure 3" in "The Evolution of Locomotor Stamina in Tetrapods: Circumventing a Mechanical Constraint" by D. R. Carrier from *Paleobiology* 13(3): 326–341, Summer 1987. Copyright © 1987 by The Paleontological Society. Reprinted by permission of The Paleontological Society. *11–2:* "Figure 37" and "Figure 75" from *Mammal-Like Reptiles and the Origin of Mammals* by T. S. Kemp. Copyright © 1982 by Elsevier. Reprinted by permission of Elsevier. *11–3:* "Figure 6" from "Pelvic Aspiration in the American Alligator (Alligator Mississippiensis)" by C. G. Farmer and D. R. Carrier in *The Journal of Experimental Biology* 203:1679–1687, June 1, 2000. Copyright © 2000 by The Company of Biologists Ltd. Reprinted by permission of The Company of Biologists Ltd. *11–4:* "Figure 5" from "The Evolution of Pelvic Aspiration in Archosaurs" by D. R. Carrier and C. G. Farmer in *Paleobiology* 26(2): 271–293, Spring 2000. Reprinted by permission of The Paleontological Society. *11–6a:* From "A Frog's Lung" in *Applied Physiology* by F. Overton, 1910. *11–6b:* From "Mammal Lung" from http://www.probertencyclopaedia.com/cgi-bin/res.pl?keyword=Abdominal+Cavity&offset=0. Copyright © by The Probert Encyclopaedia. Dated used under license. *11–7:* From "Chapter 8 - Respiration and Control of Breathing" by P. Scheid in *Avian Biology*, Volume 6 edited by J. R. King and K. C. Parkes, 1982. Reprinted by permission of P. Scheid. *11–8:* From P. Scheid in *Avian Biology*, Volume 6 edited by J. R. King and K. C. Parkes, 1982. Reprinted by permission of P. Scheid. *11–9:* From P. Buhler in *Form and Function in Birds*, Volume 2 edited by A. S. King and J. McLelland. Copyright © 1981 by Elsevier. Reprinted by permission of Elsevier. *11–11:* "Figure 1" in "The

Evolution of Nasal Turbinates and Mammalian Endothermy" by W. J. Hillenius from *Paleobiology* 18(1):17–29, Winter 1992. Copyright © 1992 by The Paleontological Society. Reprinted by permission of The Paleontological Society. *11–12a,b:* From *The CIBA Collection of Medical Illustrations*, Volume 6 Kidneys, Ureters, and Urinary Bladder by F. H. Netter, 1973. *11–15:* From "Anatomy and Ultrastructure of the Excretory System of the Lizard, Sceloporus cyanogenys" by L. E. Davis et al. in *Journal of Morphology* 149(3):279–326, July 1976. Copyright © 1976 by Wiley-Liss, Inc. Reprinted by permission of John Wiley & Sons Ltd. *11–16:* From "Figure 2.6" in *Evolution of the Brain and Intelligence* by H. J. Jerison, 1973. Reprinted by permission of H. J. Jerison. *11–18:* From "Figure 14.7" in *Evolution of the Brain and Intelligence* by H. J. Jerison, 1973. Reprinted by permission of H. J. Jerison.

Chapter 12 *12–4:* From "A Phylogeny and Classification of the Higher Categories of Turtles" by E. S. Gaffney in *Bulletin of the American Museum of Natural History* 155(5):389–436, 1975. Reprinted by permission of the American Museum of Natural History. *12–7:* From "Figure 9" from "The Mechanism of Lung Ventilation in the Tortoise Testudo Graeca Linne" by C. Gans and G. M. Hughes in *The Journal of Experimental Biology* 47:1–20, August 1967. Copyright © 1967 by The Company of Biologists Ltd. Reprinted by permission of The Company of Biologists Ltd. *12–8:* From *Animal Physiology: Principles and Adaptations*, 4E by M. S. Gordon, et al., 1982. *12–9b:* From "Agonistic Behavior of the Galapagos Tortoise, Geochelone Elephantopus, with Emphasis on Its Relationship to Saddlebacked Shell Shape" by S. F. Schafer and C. O'Neil Krekorian in *Herpetologica* 39(4):448-456, December 1983. Copyright © 1983 by The Herpetologists' League, Inc. Reprinted by permission of Herpetologica, Allen Press Publishing Services and C. O'Neil Krekorian. *12–10a:* From "Sex Determination in Reptiles" by J. J. Bull in *Quarterly Review of Biology* 55:13–21, March 1, 1980. Copyright © 1980 by The University of Chicago Press. Reprinted by permission of The University of Chicago Press via Copyright Clearance Center. *12–10b:* From "Table 1" from "Temperature-Dependent Sex Determination in Turtles" by J. J. Bull and R. C. Vogt from *Science* 206(4423):1186–1188, December 1979. *12–10c:* "Figure 1" from "Temperature-Dependent Sex Determination in Turtles" by J. J. Bull and R. C. Vogt from *Science* 206(4423):1186–1188. Copyright © 1979 by AAAS. Reprinted by permission of AAAS. *12–12:* "Figure 1" from "The Navigational Feats of Green Sea Turtles Migrating from Ascension Island Investigated by Satellite Telemetry" by P. Luschi et al. in *Proceedings of the Royal Society of London*, Series B, 265(1412):2279–2284, page 2280, December 7, 1998. Copyright © 1998 by The Royal Society. Reprinted by permission of The Royal Society. *12–13:* (1) "Figure 1" from "Regional Magnetic Fields as Navigational Markers for Sea Turtles" by K. J. Lohmann et al. in *Science* 294(5541):364–366, October 12, 2001. Reprinted by permission of AAAS; (2) From "Figure 2" from "Orientation of hatchling loggerhead sea turtles to regional magnetic fields along a transoceanic migratory pathway" by M. Fuxjager, B. S. Eastwood, and K. J. Lohmann from *The Journal of Experimental Biology* 214:2504–2508, August 1, 2011. Reprinted by permission of The Company of Biologists, Ltd. *Table 12–1:* "Phylogenetic Relationships and Numbers of Species" in *Herpetology*, 2E by F. H. Pough et al., 2001.

Chapter 13 *13–3:* From *Biomechanics: An Approach to Vertebrate Biology* by C. Gans. Copyright © 1974 by Lippincott Williams & Wilkins. Reprinted by permission of Lippincott Williams & Wilkins. *13–6:* "Figure 2 – Fossil Snake Preserved within a Sauropod Dinosaur Nesting Ground" from *Predation Upon Hatchling Dinosaurs by a New Snake from the Late Cretaceous of India* by J. A. Wilson et al., 2010. *13–7a:* From "Petrolacosaurus, the Oldest Known Diapsid Reptiles" by R. Reisz in *Science* 196(4294):1091–1093, June 3, 1977. Copyright © 1977 by The American Association for the Advancement of Science. Reprinted by permission of AAAS. *13–7b-e:* From *Vertebrate Paleontology*, 3E by A. S. Romer. *13–7f,g:* From *Biomechanics: An Approach to Vertebrate Biology* by C. Gans, 1974. *13–8:* From "Figure 5" in "The Feeding Mechanism of Snakes and Its Possible Evolution" by Carl Gans in *American Zoologist* 1(2):217–227. Copyright © 1961 by Oxford University Press. Reprinted by permission of Oxford University Press. *13–9:* From "Studies on the Morphology and Function of the Skull in the Boidae (Serpentes). Part II. Morphology and Function of the Jaw Apparatus in Python Sebae and Python Molurus" by T. H. Frazzetta in *Journal of Morphology* 118(2):217–296, February 1966. Copyright ©1966 by Wiley-Liss, Inc. Reprinted by permission of John Wiley & Sons Ltd. *13–12:* From "A Comparative Analysis of Aggressive Display in Nine Species of Costa Rican Anolis" by Anthony A. Echelle et al. in *Herpetologica* 27(3):271–288, September 1971. Copyright © 1971 by The Herpetologists' League, Inc. Reprinted by permission of Herpetologica, Allen Press Publishing Services and A. A. Echelle. *13–13:* From "Integration of Internal and External Stimuli in the Regulation of Lizard Reproduction" by D. Crews in *Behavior and Neurology of Lizards* edited by N. Greenberg and P.D. MacLean. Rockville: U.S. Department of Health, Education and Welfare, National Institutes of Health, 1978. *13–14:* From "Inter-and Intraindividual Variation in Display Patterns in the Lizard, Anolis Carolinensis" by D. Crews in *Herpetologica* 31(1):37-47, March 1975. Copyright © 1975 by The Herpetologists' League, Inc. Reprinted by permission of Herpetologica, Allen Press Publishing Services and David Crews. *13–16:* From "Temperature and the Galapagos Marine Iguana—Insights into Reptilian Thermoregulation" by F. N. White in *Comparative Biochemistry and Physiology Part A: Physiology* 45(2):503–513, June 1, 1973. Copyright © 1973 by Elsevier. Reprinted by permission of Elsevier. *13–17:* From "Figure 7A" from "The Thermal Dependence of Locomotion, Tongue Flicking, Digestion, and Oxygen Consumption in the Wandering Garter Snake" by R. D. Stevenson et al. in *Physiological Zoology* 58(1):46–57, January/February 1985. Copyright © 1985 by The University of Chicago Press. Reprinted by permission of The University of Chicago Press via Copyright Clearance Center and R. D. Stevenson. *13–18:* From "Thermal Relations of Five Species of Tropical Lizards" by R. Ruibal in *Evolution* 15(1):98–111, March 1961. Copyright © 1961 by The Society for the Study of Evolution. Reprinted by permission of John Wiley & Sons Ltd. *13–20:* From "Habitat Specificity in Three Sympatric Species of Ameiva (Reptilia: Teiidae)" by Peter E. Hillman from *Ecology* 50(3):476-481, May 1969. Copyright © 1969 by the Ecological Society of America. Reprinted by permission of Ecological Society of America in the format Textbook via Copyright Clearance Center. *13–21:* From "Habitat Specificity in Three Sympatric Species of Ameiva (Reptilia: Teiidae)" by Peter E. Hillman from *Ecology* 50(3):476–481, May 1969. Copyright © 1969 by the Ecological Society of America. Reprinted by permission of Ecological Society of America in the format Textbook via Copyright Clearance Center. *Table 13–2:* "Numbers of Species" in *Herpetology*, 2E by F. H. Pough et al., 2001. *Table 13–4:* From "Body Shape, Reproductive Effort, and Relative Clutch Mass in Lizards: Resolution of a Paradox" by L. J. Vitt and J. D. Congdon in *American Naturalist* 112(985):595–608, May/June 1978; From "Ecological Consequences of Foraging Mode" by R. B. Huey and E. R. Pianka in *Ecology* 62(4):991–999, August 1981; From "The Correlates of Foraging Mode in a Community of Brazilian Lizards" by W. E. Magnusson et al. in *Herpetologica* 41(3):324–332, September 1985; and From "A Comparative Approach to Field and Laboratory Studies in Evolutionary Ecology" by A. F. Bennett and R. B. Huey in *Predator-Prey Relationships: Perspectives and Approaches from the Study of Lower Vertebrates* by M. E. Feder and G. V. Lauder, 1986.

Chapter 14 *14–2:* From "Physiological Ecology of Desert Tortoises in Southern Nevada" by K. A. Nagy and P. A. Medica from *Herpetologica* 42(1):73–92, March 1986. Reprinted by permission of Herpetologica, Allen Press Publishing Services and K. A. Nagy. *14–4:* From "Behavior, Diet and Reproduction in a Desert Lizard, Sauromalus obesus" by K. A. Nagy from *Copeia* 1973(1):93–102, March 5, 1973. Copyright © 1973 by American Society of Ichthyologists and Herpetologists. Reprinted by permission of the American Society of Ichthyologists and Herpetologists. *14–5:* From "Figure 2a and b" from "Water and Electrolyte Budgets of a Free-living Desert Lizard, Sauromalus obesus" by K. A. Nagy in *Journal of Comparative Physiology* 79:39–62, March 1972. Copyright © 1972 by Springer Berlin/Heidelberg. Reprinted with permission. *14–7:* "Figure 4" from "Supercooling and Osmoregulation in Arctic Fish" by P. F. Scholander et al. in *Journal of Cellular Physiology* 49(1):5–24, February 1957. Copyright © 1957 by The Wistar Institute of Anatomy and Biology. Reprinted by permission of John Wiley & Sons Ltd. *14–9:* "Figure 2" from "The Advantages of Ectothermy for Tetrapods" by F. H. Pough in *The American Naturalist* 115(1):92–112, January 1980. Copyright © 1980 by The University of Chicago Press. Reprinted by permission of The University of Chicago Press via Copyright Clearance Center. *14–10:* From F. H. Pough in *Behavioral Energetics: The Cost of Survival in Vertebrates* edited by W. P. Aspey and S. I. Lustick, 1983 and From "How Many Species Are There on Earth?" by Robert M. May in *Science* 241(4872):1441–1449, September 1988. *Table 14–3:* From "Table 3" from "The Advantages of Ectothermy for Tetrapods" by F. H. Pough in *The American Naturalist* 115(1):92–112, January 1980. Copyright © 1980 by The University of Chicago Press. Reprinted by permission of The University of Chicago Press via Copyright Clearance Center.

Chapter 16 *16–1:* From "Figure 58a and b" from "The Early Evolution of Archosaurs: Relationships and the Origin of Major Clades" by S. J. Nesbitt in *Bulletin of the American Museum of Natural History* 352:1–292, April 2011. *16–3a,b:* From "Figures 208 and 235" from *Vertebrate Paleontology*, 3E by A. S. Romer, 1966. Reprinted by permission of The University of Chicago Press via Copyright Clearance Center. *16–5a,b:* From *Paleontology and Modern Biology* by D. M. S. Watson, 1951. *16–5a,b:* From *Traite de Paleontologie*, Volume 5 by J. Piveteau, 1955. *16–7:*

"Mosaur" by Dmitry Bogdanov from https://en.wikipedia.org/wiki/File:Platecarpus2010.jpg. Reprinted by permission of Dmitry Bogdanov. *16–8:* "Geosaurus" by Dmitry Bogdanov from http://news.discovery.com/animals/hypercarnivore-predators-dinosaurs.html. Reprinted by permission of Dmitry Bogdanov. *16–9:* From "Abb. 1a," "Abb. 247," "Abb. 254," and "Abb. 270" from *Schildkroten, Krokodile, Brucken-eschsen* by H. Wermuth and R. Mertens, 1961. *16–12a–e:* From *The Illustrated Encyclopedia of Dinosaurs* by D. Norman, 1985. *16–13:* From http://archosaurmusings.wordpress.com/2010/10/01/interview-with-jeff-martz/. Reprinted by permission of Jeffrey Martz. *16–18g:* From http://australianmuseum.net.au/image/Hadrosaur-skull-cross-section/. *16–25:* "Figure 10" from "Feathered Dinosaurs" by Mark A. Norell and Xing Xu in *Annual Review of Earth and Planetary Sciences* 33:277–299, 2005 by Annual Reviews, Inc. Reprinted by permission of Annual Reviews Inc. via Copyright Clearance Center.

Chapter 17 *17–1a:* From "Early Evolution of Avian Flight and Perching: New Evidence from the Lower Cretaceous of China" by P. C. Sereno and R. Chenggang in *Science* 255(5046):845-848, February 14, 1992. Reprinted by permission of AAAS. *17–2:* From "Figure 1" from "Big Bang' for Tertiary Birds?" by Alan Feduccia in *Ecology and Evolution* 18(4):172–176, April 1, 2003. Copyright © 2003. Reprinted by permission of Elsevier via Copyright Clearance Center. *17–3a:* From *The Age of Birds* by A. Feduccia, 1980. *17–3b:* From "Chapter 9— The Origin and Early Radiation of Birds" by L. D. Martin from *Perspectives in Ornithology: Essays Presented for the Centennial of the American Ornithologists' Union* edited by A. H. Brush and G. A. Clark, Jr. Copyright © 1983 by Cambridge University Press. Reprinted by permission of Cambridge University Press. *17–9:* From *The Life of Birds* by J. Dorst, 1974. Reprinted by permission of Orion Publishing Group, Ltd. *17–10:* From *A New Dictionary of Birds* by Sir Arthur L. Thomson, 1964. *17–17:* From *The Birds*, 2E by R. T. Peterson, 1978. *17–18:* From "Adaptive Radiation of Birds" by R. W. Storer in *Avian Biology*, Volume 1 edited by D. S. Farmer, J. R. King, and K. C. Parkes, 1971. Reprinted by permission of Robert H. Storer. *17–20:* From P. Buhler in *Form and Function in Birds*, Volume 2 edited A. S. King and J. McLelland. Copyright © 1981 by Elsevier. Reprinted by permission of Elsevier. *17–22a–e:* From J. McLelland in *Form and Function in Birds*, Volume 1 edited by A. S. King and J. McLelland. Copyright © 1979 by Elsevier. Reprinted by permission of Elsevier. *17–23:* From "How Do Birds Accommodate?" by Graham R. Martin from *Nature* 328:383–383, July 30, 1987. *17–32:* From "Orientation by Pigeons: Is the Sun Necessary?" by W. T. Keeton in *Science* 165(3896):922–928, August 1969. Reprinted by permission of AAAS and William Scott Keeton. *17–33:* From "The Breeding Cycle of the Short-Tailed Shearwater, Puffinus Tenuirostris (Temminck), in Relation to Trans-Equatorial Migration and Its Environment" by A. J. Marshall and D. L. Serventy from *Proceedings of the Zoological Society of London* 127(4):489–510, December 1956. Copyright © 1956 by The Zoological Society of London. Reprinted by permission of John Wiley & Sons Ltd. *17–34:* "Circular Test Cage" and "Difference in Orientation" illustrations by Adolph E. Brotman from "The Stellar-Orientation System of a Migratory Bird" by Stephen T. Emlen in *Scientific*

American 233(2):102–111, August 1975. Reprinted by permission of Doris Brotman.

Chapter 18 *18–4:* From *Vertebrate Paleontology*, 3E by A. S. Romer, 1966. *18–5:* "Lystrosaurus" by Gregory Paul in *The Dicynodonts: A Study in Paleobiology* by Gillian King, 1990. Reprinted by permission of Gregory S. Paul. *18–6a:* From "A New Haptodontine Sphenacodont (Reptilia: Pelycosauria) from the Upper Pennsylvanian of North America" by Philip J. Currie in *Journal of Paleontology* 51(5): 927–942, September 1977. Copyright © 1977 by The Society of Economic Paleontologists and Mineralogists. Reprinted by permission of The Paleontological Society. *18–6b:* "Lycaenops" by E. H. Colbert in *Bulletin of the American Museum of Natural History* 89:353–404, 1948. Reprinted by permission of the American Museum of Natural History. *18–6c:* "Figure 2" from "Cynodont Postcranial Anatomy and the "Prototherian" Level of Mammalian Organization" by Farish A. Jenkins, Jr. in *Evolution* 24(1):230–252, March 1970. Copyright © 1970 by The Society for the Study of Evolution. Reprinted by permission of John Wiley & Sons Ltd. *18–6d:* From "The Postcranial Skeletons of The Triassic Mammals Eozostrodon, Megazostrodon and Erythrotherium" by F. A. Jenkins, Jr. and F. R. Parrington in *Philosophical Transactions of the Royal Society of London. Series B, Biological Sciences* 273(926):387–431, February 26, 1976. Copyright © 1976 by The Royal Society. Reprinted by permission of The Royal Society. *18–7:* From "Figure 90" in *Mammal-Like Reptiles and the Origin of Mammals* by T. S. Kemp. Copyright © 1982 by Elsevier. *18–8:* From "Origin of Mammals" by A. W. Crompton and F. A. Jenkins, Jr. in *Mesozoic Mammals: The First Two-Thirds of Mammalian History* edited by J. A. Lillegraven, Z. Kielan-Jaworowska and W. A. Clemens, 1979. Reprinted by permission of A. W. Crompton. *18–9:* "Megazostrodon" by Marianne Collins, illustrator, from "Chapter Five: Victors by Default" by C. M. Janis in *The Book of Life: An Illustrated History of the Evolution of Life on Earth* by S. J. Gould, 1993. *18–10e,f:* From J. A. Hopson in *Major Features of Vertebrate Evolution*. Short Courses in Paleontological, Number 7 by D. R. Prothero and R. M. Schoch, 1994. Reprinted by permis-sion of The Paleontological Society. *18–12:* From "Figure 10.40" in *Vertebrates: Comparative Anatomy, Function, Evolution*, 2E by K. V. Kardong. Copyright © 1998 by The McGraw-Hill Companies, Inc. Reprinted with permission of The McGraw-Hill Companies, Inc.

Chapter 19 *19–3:* "Figure 2" from "Climate Response to Orbital Forcing Across the Oligocene-Miocene Boundary" by J. C. Zachos et al. in *Science* 292(5515):274–278, April 2001. Reprinted by permission of AAAS.

Chapter 20 *20–1:* "Ptilodus" by L. L. Sadler from *Science* 220(4598):712–715, May 1983. *20–3:* "Figures 18.14 and 18.15" from *The Life of Vertebrates*, 3E by J. Z. Young, 1981. Reprinted by permission of Oxford University Press. *20–5b–d:* From *The Encyclopedia of Mammals* by David W. Macdonald. Copyright © 1984 by Andromeda Oxford Limited. Color and Line Artwork Panels Copyright © 1984 by Priscilla Barrett. *20–6a:* From *Handbook to the Orders and Families of Living Mammals* by T. E. Lawlor, 1979. Reprinted by permission of Barbara Law-lor. *20–6b,c:* From "Figures 215 and 216" from *The Vertebrate Body*, 6E by A. S. Romer and T. S. Parsons. Copyright © 1986 by Brooks/Cole, a part of Cengage Learning, Inc. Reprinted by permission of Cengage Learning, Inc. www.cengage.com/permissions. *20–9:* From *Histology*, 8E by A. W. Ham and D. W. Cormack, 1979 and From *Man* 2E by R. J. Harrison and E. W. Montagna, 1973. *20–10:* From *Analysis of Vertebrate Structures*, 5E by M. Hildebrand and G. Goslow, 1998. Reprinted by permission of John Wiley & Sons Ltd. *20–11:* From *Analysis of Vertebrate Structures*, 5E by M. Hildebrand and G. Goslow, 1998. Reprinted by permission of John Wiley & Sons Ltd. *20–12:* From *Analysis of Vertebrate Structures*, 5E by M. Hildebrand and G. Goslow, 1998. Reprinted by permission of John Wiley & Sons Ltd. *20–13:* From "Figure 17.36" in *Vertebrates: Comparative Anatomy, Function, Evolution*, 2E by K. V. Kardong. Copyright © 1998 by The McGraw-Hill Companies, Inc. Reprinted by permission of The McGraw-Hill Companies, Inc. *20–14a,b:* From *Looking at Vertebrates* by E. Rogers, 1986. Reprinted by permission of Pearson Education Limited. *20–14c–e:* From *Functional Anatomy of the Vertebrates: An Evolutionary Perspective*, 2E by W. F. Walker, Jr. and K. F. Liem. Copyright © 1994 by Brooks/Cole, a part of Cengage Learning, Inc. *20–15:* From "Figure 10.35" and "Figure 10.28" in *Vertebrates: Comparative Anatomy, Function, Evolution*, 2E by K. V. Kardong. Copyright © 1998 by The McGraw-Hill Companies, Inc. Reprinted by permission of The McGraw-Hill Companies, Inc. *20–16:* "Figure 13" and "Figure 105" from *Handbook to the Orders and Families of Living Mammals* by T. E. Lawlor, 1979. Reprinted by permission of Barbara Lawlor. *20–18:* From "Figure 15.4" in *Life: An Introduction to Biology*, Shorter Edition by G. G. Simpson and W. S. Beck, 1957. *20–19:* From "Mammalian Evolution and the Great American Interchange" by L. G. Marshall et al. from *Science* 215(4538):1351–1357, March 1982 and From "Land Mammals and the Great American Interchange" by L. G. Marshall from *American Scientist* 76(4):380–388, July/August 1988. *Table 20–1:* From "Phylogenetic Relationships of the Families of Marsupials" by L. G. Marshall, J. A. Case, and M. O. Woodburne from *Current Mammalogy*, Volume 2, pp. 433–505, 1990; From *Mammalogy*, 3E by T. A. Vaughan, 1986; and From *Walker's Mammals of the World*, 5E by R. M. Nowak and J. L. Paradiso, 1991. *Table 20–2:* From "The Classification of Nontherian Mammals" by J. A. Hopson from *Journal of Mammalogy* 51:1–9, February 20, 1970; From "Toward a Phylogenetic Classification of the Mammalia" by M. C. McKenna from *Phylogeny of the Primates: A Multidisciplinary Approach* edited by W. P. Luckett and F. Szalay, 1975; From *Mammalogy*, 4E by T. A. Vaughan, 2000; From *The Mammalian Radiations: An Analysis of Trends in Evolution, Adaptation, and Behavior* by John F. Eisenberg, 1981; From *Orders and Families of Recent Mammals of the World* by S. Anderson and J. K. Jones, Jr., 1984; and From "Higher-Level Relationships of the Recent Eutherian Orders: Morphological Evidence" by M. J. Novacek and A. R. Wyss from *Cladistics* 2(4):257–287, September 1986.

Chapter 21 *21–1a:* From "Drawing 47" in "Embryology of the Monotremes. Proceedings of the Medical and Natural Sciences Society" in *Jena* 5:61–74 by R. Semon, 1894. *21–2:* From "Evolution of Viviparity in Mammals" by G. B. Sharman in "Chapter 6: The Evolution of Reproduction" in *Reproduction in Mammals* edited by C. R. Austin and R. V. Short, 1976. *21–3:* From M. B. Renfree in *Mammal Phylogeny* by F. S. Szalay, M. J. Novacek, and M. C. McKenna, 1993. *21–4:* From "Figure 2" from "Constraints

on the Morphological Evolution of Marsupial Shoulder Girdles" by K. E. Sears in *Evolution* 58(10):2353-2370, October 2004. Copyright © 2004 by The Society for the Study of Evolution. Reprinted by permission of John Wiley & Sons Ltd. *21–5a,b:* From "Gestation, naissaince et porgression du jeune vers la poche marsupiale dans la famille des kangourous'" by G. B. Sharman in *Zoo: Societe Royale de Zoologie d'Anvers* 34:124-132, 1969. *21–10a,b:* From *Primate Locomotion* by F. A. Jenkins, Jr. Copyright © 1974 by Elsevier. Reprinted by permission of Elsevier. *21–10c:* From "Figure 4" from "Mammalian Terrestrial Locomotion and Size" by A. A. Biewener in *Bioscience* 39(11):776-783, December 1989. Copyright © 1989 by the University of California Press and the American Institute of Biological Sciences. Reprinted by permission of the University of California Press via Copyright Clearance Center and A. A. Biewener. *21–11:* From *Analysis of Vertebrate Structures*, 5E by M. Hildebrand and G. Goslow, 1998. Reprinted by permission of John Wiley & Sons Ltd. *21–12:* "Figure 29.2" from *Analysis of Vertebrate Structures*, 3E by M. Hildebrand and G. Goslow. Copyright © 1988 by John Wiley & Sons, Inc. Reprinted by permission of John Wiley & Sons Ltd. *21–13:* From *The Encyclopedia of Mammals* by David W. Macdonald. Copyright © 1984 by Andromeda Oxford Limited. Color and Line Artwork Panels Copyright © 1984 by Priscilla Barrett. *21–14:* From *The Encyclopedia of Mammals* by David W. Macdonald. Copyright © 1984 by Andromeda Oxford Limited. Color and Line Artwork Panels Copyright © 1984 by Priscilla Barrett. *21–15:* From *At the Water's Edge: Macroevolution and the Transformation of Life* by C. Zimmer, 1996. Reprinted by permission of Simon & Schuster, Inc.

Chapter 22 *22–8:* "Figure 2" from *Strategies in Cold: Natural Torpidity and Thermogenesis* by L. C. H. Wang and J. W. Hudson, 1978. Reprinted by permission of L. C. H. Wang. *22–9:* From "Figure 3" from *Strategies in Cold: Natural Torpidity and Thermogenesis* by L. C. H. Wang and J. W. Hudson, 1978. Reprinted by permission of L. C. H. Wang. *22–12:* From "Body Size, Medullary Thickness, and Urine Concentrating Ability in Mammals" by C. A. Beuchat in *American Journal of Physiology—Regulatory, Integrative, Comparative, Physiology* 258(2):R298–R308, February 1990. *22–15:* From "Standard Operative Temperature and Thermal Energetics of the Antelope Ground Squirrel Ammospermophilus leucurus" by Mark A. Chappell and George A. Bartholomew in *Physical Zoology* 54(1):81–93, February 1, 1981. Copyright © 1981 by The University of Chicago Press. Reprinted by permission of The University of Chicago Press via Copyright Clearance Center. *22–17:* From "Body Temperature of the Camel and Its Relation to Water Economy" by K. Schmidt-Nielsen et al. in *American Journal of Physiology* 188(1):103–112, December 1956. *22–18:* From C. R. Taylor in *Comparative Physiology of Desert Animals* edited by G. M. O. Maloiy, 1972. *Table 22–1:* From "Table 7" from *Strategies in Cold: Natural Torpidity and Thermogenesis* by L. C. H. Wang and J. W. Hudson, 1978. Reprinted by permission of L. C. H. Wang. *Table 22–2:* From "Table V" in *Desert Animals: Physiological Problems of Heat and Water* by Knut Schmidt-Nielsen.

Chapter 23 *23–1:* From "Figure 6" from "Behavioral Consequences of Adaptation to Different Food Resources" by B. K. McNab in *Advances in the Study of Mammalian Behavior*, Special Publication No. 7 edited by J. F. Eisenberg and D. G. Kleiman,

Index

Latin and Greek Lexicon

Many biological names and terms are derived from Latin (L) and Greek (G). Learning even a few dozen of these roots is a great aid to a biologist. The following terms are often encountered in a vertebrate biology. The words are presented in the spelling and form in which they are most often encountered; this is not necessarily the original form of the word in its etymologically pure state.

An example of how a root is used in vertebrate biology can often be found by referring to the subject index. Remember, however, that some of these roots may be used as suffixes or otherwise embedded in technical words and will require further searching to discover an example. Additional information can be found in a reference such as the *Dictionary of Word Roots and Combining Forms,* by Donald J. Borror (Palo Alto, Calif.: Mayfield Publishing Co.)

a, ab (L) away from
a, an (G) not, without
acanth (G) thorn
actin (G) a ray
ad (L) toward, at, near
aeros (G) the air
aga (G) very much, too much
aistos (G) unseen
al, alula (L) a wing
allant (G) a sausage
alveol (L) a pit
ambl (G) blunt
ammos (G) sand
amnion (G) a fetal membrane
amphi, ampho (G) both, double
amplexus (L) an embracing
ampulla (L) a jug or flask
ana (G) up, upon, through
anat (L) a duck
angio (G) a reservoir, vessel
ankylos (G) crooked, bent
anomos (G) lawless
ant, anti (G) against
ante (L) before
anthrac (G) coal
apat (G), illusion, error
aphanes (G) invisible, unknown
apo, ap (G) away from, separate

apsid (G) an arch, loop
aqu (L) water
arachne (G) a spider
arch (G) beginning, first in time
argenteus (L) silvery
arthr (G) a joint
ascidion (G) a little bag or bladder
aspid (G) a shield
asteros (G) a star
atri, atrium (L) an entrance-room
audi (L) to hear
austri, australis (L) southern
avis (L) a bird
baen (G) to walk or step
bas (G) base, bottom
batrachos (G) a frog
benthos (G) the sea depths
bi, bio (G) life
bi, bis (L) two
blast (G) bud, sprout
brachi (G) arm
brachy (G) short
branchi (G) a gill or fin
buce (L) the check
cal (G) beautiful
calie (L) a cup
capit (L) head
carn (L) flesh
caud (L) tail
cene, ceno (G) new, recent
cephal (G) head
cer, cerae (G) a horn

cerc (G) tail
chir, cheir (G) hand
choan (G) funnel, tube
chondr (G) grit, gristle
chord (G) guts, a string
chorio (G) skin, membrane
chrom (G) color
clist (G) closed
cloac (L) a sewer
coel (G) hollow
cornu (L) a horn
cortic, cortex (L) bark, rind
costa (L) a rib
cran (G) the skull
creta (L) chalk
cretio (L) separate
crini (L) the hair
cten (G) a comb
cut, cutis (L) the skin
cyn (G) a dog
cytos (G) a cell
dactyl (G) a finger
de (L) down, away from
dectes (G) a biter
dendro (G) a tree
dent, dont (L) a tooth
derm (G) skin
desmos (G) a chain, tie, or band
deuteros (G) secondary
di, dia (G) through, across
di, diplo (G) two, double
din, dein (G) terrible, powerful

dir (G) the neck
disc (G) a disk
dory (G) a spear
draco (L) a dragon
drepan (G) a sickle
dromo (G) running
duc (L) to lead
dur (L) hard
e, ex (L) out of, from, without
echinos (G) a prickly being
eco, oikos (G) a house
ect (G) outside
edaphos (G) the soil or bottom
eid (G) form, appearance
elasma (G) a thin plate
eleuthero (G) free, not bound
elopos (G) a kind of sea fish
embolo (G) like a peg or stopper
embryon (G) a fetus
emys (G) a freshwater turtle
end (G) within
enter (G) bowel, intestine
eos (G) the dawn or beginning
ep (G) on, upon
equi (L) a horse
ery (G) to drag or draw
erythr (G) red
eu, ev (G) good, true